IRRIGATION OF AGRICULTURAL CROPS

AGRONOMY

A Series of Monographs

The American Society of Agronomy and Academic Press published the first six books in this series. The General Editor of Monographs 1 to 6 was A. G. Norman. They are available through Academic Press, Inc., 111 Fifth Avenue, New York, NY 10003.

1. C. EDMUND MARSHALL: The Colloid Chemical of the Silicate Minerals. 1949
2. BYRON T. SHAW, *Editor*: Soil Physical Conditions and Plant Growth. 1952
3. K. D. JACOB, *Editor*: Fertilizer Technology and Resources in the United States. 1953
4. W. H. PIERRE and A. G. NORMAN, *Editors*: Soil and Fertilizer Phosphate in Crop Nutrition. 1953
5. GEORGE F. SPRAGUE, *Editor*: Corn and Corn Improvement. 1955
6. J. LEVITT: The Hardiness of Plants. 1956

The Monographs published since 1957 are available from the American Society of Agronomy, 677 S. Segoe Road, Madison, WI 53711.

7. JAMES N. LUTHIN, *Editor*: Drainage of Agricultural Lands. 1957
8. FRANKLIN A. COFFMAN, *Editor*: Oats and Oat Improvement. 1961
9. A. KLUTE, *Editor*: Methods of Soil Analysis. 1986
   Part 1—Physical and Mineralogical Methods. Second Edition.
   A. L. PAGE, R. H. MILLER, and D. R. KEENEY, *Editor*: Methods of Soil Analysis. 1982
   Part 2—Chemical and Microbiological Properties. Second Edition.
10. W. V. BARTHOLOMEW and F. E. CLARK, *Editors*: Soil Nitrogen. 1965
    (Out of print; replaced by no. 22)
11. R. M. HAGAN, H. R. HAISE, and T. W. EDMINSTER, *Editors*: Irrigation of Agricultural Lands. 1967
12. FRED ADAMS, *Editor*: Soil Acidity and Liming. Second Edition. 1984
13. E. G. HEYNE, *Editor*: Wheat and Wheat Improvement. Second Edition. 1987
14. A. A. HANSON and F. V. JUSKA, *Editors*: Turfgrass Science. 1969
15. CLARENCE H. HANSON, *Editor*: Alfalfa Science and Technology. 1972
16. J. R. WILCOX, *Editor*: Soybeans: Improvement, Production, and Uses. Second Edition. 1987
17. JAN VAN SCHILFGAARDE, *Editor*: Drainage for Agriculture. 1974
18. G. F. SPRAGUE and J. W. DUDLEY, *Editors*: Corn and Corn Improvement, Third Edition. 1988
19. JACK F. CARTER, *Editor*: Sunflower Science and Technology. 1978
20. ROBERT C. BUCKNER and L. P. BUSH, *Editors*: Tall Fescue. 1979
21. M. T. BEATTY, G. W. PETERSEN, and L. D. SWINDALE, *Editors*: Planning the Uses and Management of Land. 1979
22. F. J. STEVENSON, *Editor*: Nitrogen in Agricultural Soils. 1982
23. H. E. DREGNE and W. O. WILLIS, *Editors*: Dryland Agriculture. 1983
24. R. J. KOHEL and C. F. LEWIS, *Editors*: Cotton. 1984
25. N. L. TAYLOR, *Editor*: Clover Science and Technology. 1985
26. D. C. RASMUSSON, *Editor*: Barley. 1985
27. M. A. TABATABAI, *Editor*: Sulfur in Agriculture. 1986
28. R. A. OLSON and K. J. FREY, *Editors*: Nutritional Quality of Cereal Grains: Genetic and Agronomic Improvement. 1987
29. A. A. HANSON, D. K. BARNES, and R. R. HILL, JR., *Editors*: Alfalfa and Alfalfa Improvement. 1988
30. B. A. STEWART and D. R. NIELSEN, *Editors*: Irrigation of Agricultural Crops. 1990

IRRIGATION OF AGRICULTURAL CROPS

B. A. Stewart and D. R. Nielsen, co-editors

Editorial Committee

R. R. Bruce   M. H. Niehaus

E. T. Kanemasu   J. R. Gilley

Managing Editor: S. H. Mickelson

Editor-in-Chief ASA Publications: G. H. Heichel

Editor-in-Chief CSSA Publications: C.W. Stuber

Editor-in-Chief SSSA Publications: D. E. Kissel

Number 30 in the series

AGRONOMY

American Society of Agronomy, Inc.
Crop Science Society of America, Inc.
Soil Science Society of America, Inc.
Publishers
Madison, Wisconsin USA
1990

American Society of Agronomy, Inc.
Crop Science Society of America, Inc.
Soil Science Society of America, Inc.
677 South Segoe Road, Madison, WI 53711 USA

Library of Congress Cataloging-in-Publication Data

Irrigation of agricultural crops / B.A. Stewart and D.R. Nielsen, co-
editors.
  p.   cm. — (Agronomy; no. 30)
Includes bibliographical references.
ISBN 0-89118-102-4
  1. Irrigation farming. 2. Irrigation. 3. Crops and water.
I. Stewart, B.A. (Bobby Alton), 1932– . II. Nielsen, Donald R.,
1931– . III. Series.
S613.I724 1990
631.5'87—dc20

Printed in the United States of America

CONTENTS

v

SECTION V. SOIL-PLANT-ATMOSPHERE RELATIONS

SECTION VII. IRRIGATION OF SELECTED CROPS

18 Alfalfa

J. C. Guitjens

19 Corn

F. M. Rhoads and J. M. Bennett

20 Wheat

J. T. Musick and K. B. Porter

FOREWORD

Currently, considerable emphasis, both nationally and internationally, is being placed on water use, conservation, and quality. Efficient irrigation of crop plants for increased productivity while protecting the environment is also being emphasized.

This monograph, *Irrigation of Agricultural Crops*, provides the most current research results and information on many aspects of irrigation. The discussions of the relationships of soil, water, plant, and atmosphere will significantly benefit researchers, practicing professionals, farmers, and decision makers.

On behalf of the tri-Societies, we wish to express our appreciation to the many people who planned, authored, and edited this publication.

A.A. Baltensperger, *president*
American Society of Agronomy

S.A. Eberhart, *president*
Crop Science Society of America

W.R. Gardner, *president*
Soil Science Society of America

PREFACE

Irrigation has expanded dramatically in this century. From 1950 to 1980, it increased worldwide from about 95 million ha to about 250 million ha. Agronomy monograph 11 *Irrigation of Agricultural Lands*, published in 1967 and edited by R.M. Hagan, H.R. Haise, and T.W. Edminster, served as a valuable resource for scientists, engineers, planners, and policymakers during this rapid expansion of irrigation. In the last 10 yr, the focus of irrigation has shifted from expansion to water and energy conservation and its influence on the environment. This monograph addresses these new concerns in both theoretical and practical terms.

Irrigation of Agricultural Crops is intended to update and complement the earlier monograph. Initial planning began in November 1980, when ASA President Roger Mitchell named G.S. Campbell, chair; R.R. Bruce, T.C. Hsaio, J.L. Wright, P.J. Wierenga, and J.W. Van Schilfgaarde as the Feasibility Committee to evaluate the subject of water management and make recommendations regarding future monographs. The Feasibility Committee submitted their report to H.J. Gorz, chair of Monographs Committee, in November 1981, and recommended the planning and publishing of a new monograph relating to irrigation. ASA President C.O. Gardner, in 1982, appointed B.A. Stewart and D.R. Nielsen as co-editors, and R.R. Bruce, E.T. Kanemasu, M.H. Niehaus, and J.R. Gilley as members of the Editorial Committee. A comprehensive planning process allowed chapter topics and authors to be identified in early 1986. Therefore, the publishing of this monograph culminates an 8-yr synthesis.

We are grateful for having had the opportunity of working with the many dedicated and enthusiastic people who have given so much of their time and talents. There were countless people involved in organizing the contents, selecting the authors, reviewing the manuscripts, and producing the monograph. We are especially grateful to the authors, Editorial Committee members, and the ASA Headquarters Staff for their never-ending support.

B.A. Stewart, *co-editor*
USDA-ARS
Conservation and Production Research Laboratory
Bushland, Texas

D.R. Nielsen, *co-editor*
Department of Land, Air, and Water Research
University of California
Davis, California

CONTRIBUTORS

L. R. Ahuja
Soil Physicist, USDA-ARS, Water Quality and Watershed Research Laboratory, Durant, OK 74702

J. M. Bennett
Associate Professor, Agronomy Department, Agronomy-Physiology Laboratory, University of Florida, Gainesville, FL 32611

K. J. Boote
Professor of Agronomy, Agronomy Department, University of Florida, Gainesville, FL 32611

Dale A. Bucks
National Program Leader, Water Management, USDA-ARS, BARC-West, Beltsville, MD 20705

G. S. Campbell
Professor of Soils, Department of Agronomy and Soils, Washington State University, Pullman, WA 99164

Robert N. Carrow
Associate Professor of Agronomy, Agronomy Department, University of Georgia, Georgia Station, Griffin, GA 30223-1797

D. L. Carter
Supervisory Soil Scientist, USDA-ARS, Kimberly, ID 83341

Brent E. Clothier
Research Scientist, Department of Scientific and Industrial Research, Plant Physiology Division, Palmerston North, New Zealand

P. J. Dieleman
Land and Water Development Division, Food and Agriculture Organization of the United Nations, 00128 Rome, Italy

David E. Elrick
Professor of Soil Physics, Department of Land Resource Science, University of Guelph, Guelph, Ontario, Canada N1G 2W1

K. M. El-Zik
Professor of Agronomy and Genetics, Department of Soil and Crop Sciences, Texas A&M University, College Station, TX 77843-2474

Elias Fereres
Professor of Agronomy, University of Cordoba, 14071 Cordoba, Spain

Edwin L. Fiscus
Plant Physiologist, Air Quality Research, 1509 Varsity Drive, Raleigh, NC 27606 (formerly USDA-ARS, Great Plains Systems Research, Fort Collins, CO 80526)

G. J. Gascho
Professor of Agronomy, Coastal Plain Experiment Station, University of Georgia, Tifton, GA 31793

David A. Goldhamer
Extension Irrigation Specialist, Department of Land, Air, and Water Resources, Kearney Agricultural Center, University of California, Parlier, CA 93648

D. W. Grimes
Water Scientist, Kearney Agricultural Center, University of California-Davis, Parlier, CA 93648

J. C. Guitjens
Professor of Irrigation Engineering, 1000 Valley Road, Reno, NV 89512

B. R. Hanson
Irrigation and Drainage Specialist, Department of Land, Air, and Water Resources, University of California-Davis, Davis, CA 95616

J. L. Hatfield
Laboratory Director, USDA-ARS, National Soil Tilth Laboratory, Ames, IA 50011

L. G. Heatherly
Research Agronomist, USDA-ARS, Soybean Production Research Unit, Stoneville, MS 38776

D. F. Heermann
Agricultural Engineer, USDA-ARS, AERC, Colorado State University, Fort Collins, CO 80523

D. W. Henderson
Professor of Water Science Emeritus, Department of Land, Air, and Water Resources, University of California-Davis, Davis, CA 95616

Daniel Hillel	Professor of Plant and Soil Sciences, Department of Plant and Soil Sciences, University of Massachusetts, Amherst, MA 01003
F. J. Hills	Extension Agronomist, Department of Agronomy and Range Science, University of California-Davis, Davis, CA 95616
Arthur G. Hornsby	Professor, Soil Science Department, University of Florida, Gainesville, FL 32611-0151
Terry A. Howell	Agricultural Engineer, USDA-ARS, Bushland, TX 79012
Theodore C. Hsiao	Professor of Water Science and Plant Physiologist, Department of Land, Air, and Water Resources, University of California-Davis, Davis, CA 95616
Ray D. Jackson	Research Physicist, USDA-ARS, U.S. Water Conservation Laboratory, Phoenix, AZ 85040
M. E. Jensen	Director, Colorado Institute for Irrigation Management, Colorado State University, Fort Collins, CO 80523 (formerly National Program Leader, USDA-ARS, Water Management Research)
B. S. Johnson	Postdoctoral Fellow, Department of Crop and Soil Sciences, Michigan State University, East Lansing, MI 48824-1325
C. A. Jones	Professor and Resident Director of Research, Texas Agricultural Experiment Station, Blackland Research Center, Temple, TX 76502
Merrill R. Kaufmann	Principal Plant Physiologist, USDA-Forest Service, Rocky Mountain Forest and Range Experiment Station, Fort Collins, CO 80526
D. L. Ketring	Plant Physiologist, USDA-ARS, Plant Science Research Laboratory, Stillwater, OK 74075
Mark J. King	Research Plant Physiologist, Virginia Polytechnic Institute and State University, Southern Piedmont Center, Blackstone, VA 23824
G. Kingston	Senior Research Officer, Bureau of Sugar Experiment Stations, Bundaberg, Queensland 4670, Australia
M. B. Kirkham	Professor, Evapotranspiration Laboratory, Kansas State University, Manhattan, KS 66506
Betty Klepper	Research Leader, USDA-ARS, Pendleton, OR 97801
Daniel R. Krieg	Professor of Crop Physiology, Agronomy Department, Texas Tech University, Lubbock, TX 79409
E. Gordon Kruse	Agricultural Engineer, USDA-ARS, AERC, Colorado State University, Fort Collins, CO 80523
Robert J. Lascano	Assistant Professor of Soil Physics, Texas Agricultural Experiment Station, Lubbock, TX 79401
Yoseph Levy	Principal Scientist, Agricultural Research Organization, Department of Citriculture Research, Gilat Experiment Station, Mobile Post Negev 2, 85-280, Israel
J. Loveday	Honorary Research Fellow, Division of Soils, CSIRO, Canberra A.C.T., 2601 Australia
D. L. Martin	Associate Professor, Agricultural Engineering Department, University of Nebraska, Lincoln, NE 68583-0726
M. A. Matthews	Associate Professor of Plant Physiology, Department of Viticulture and Enology, University of California-Davis, Davis, CA 95616
D. N. Maynard	Professor of Vegetable Crops, Gulf Coast Research and Education Center, University of Florida, Bradenton, FL 34203

D. J. Mulla	Associate Professor of Soils, Department of Agronomy and Soils, Washington State University, Pullman, WA 99164-6420
J. T. Musick	Agricultural Engineer, USDA-ARS, Conservation and Production Research Laboratory, Bushland, TX 79012
D. R. Nielsen	Professor of Soil and Water Science and Agronomy and Range Science, Departments of Agronomy and Range Science and Land, Air, and Water Resources, University of California-Davis, Davis, CA 95616
K. B. Porter	Professor of Wheat Breeding, Texas Agricultural Experiment Station, Texas A&M Research and Extension Center, Amarillo, TX 79106
W. R. Rangely	Consulting Engineer, Kier Park House, Kier Park, Ascot SL5 7DS, England
D. C. Reicosky	Soil Scientist, USDA-ARS-MWA, North Central Soil Conservation Research Laboratory, Morris, MN 56267
F. M. Rhoads	Professor, Soil Science Department, North Florida Research and Education Center, University of Florida, Quincy, FL 32351
J. D. Rhoades	Director, U.S. Salinity Laboratory, Riverside, CA 92501
J. T. Ritchie	Professor and Homer Nowlin Chair, Crop and Soil Sciences Department, Michigan State University, East Lansing, MI 48824-1325
L. T. Santo	Assistant Agronomist, Hawaiian Sugar Planters' Association, Aiea, HI 96701
Joseph Shalhevet	Director of the Agricultural Research Organization, The Israel Agricultural Research Organization, Volcani Center, Bet-Dagan 50-250, Israel
Robert C. Shearman	Professor of Horticulture and Turfgrass, formerly with Department of Horticulture, University of Nebraska, Lincoln, NE 68583; currently Head, Department of Agronomy, University of Nebraska, Lincoln, NE 68583
Thomas R. Sinclair	Plant Physiologist, USDA-ARS, Agronomy Department, University of Florida, Gainesville, FL 32611
C. D. Stanley	Associate Professor of Soil Science, Gulf Coast Research and Education Center, University of Florida, Institute of Food and Agricultural Sciences, Bradenton, FL 34203
Jeffrey C. Stark	Research Agronomist, Research and Extension Center, University of Idaho, Aberdeen, ID 83210
E. C. Stegman	Professor, Department of Agricultural Engineering, North Dakota State University, Fargo, ND 58105
B. A. Stewart	Laboratory Director, USDA-ARS, Conservation and Production Research Laboratory, Bushland, TX 79012
K. K. Tanji	Professor of Water Science and Assistant Director, Agricultural Experiment Station, Department of Land, Air, and Water Resources, University of California-Davis, Davis, CA 95616
Paul W. Unger	Supervisory Soil Scientist, USDA-ARS, Bushland, TX 79012
R. D. von Bernuth	Professor, Agricultural Engineering Department, University of Tennessee, Knoxville, TN 37901-1071
A. W. Warrick	Professor of Soil Physics, Soil and Water Science Department, University of Arizona, Tucson, AZ 85721

James R. Watson   Vice President, The Toro Company, Minneapolis, MN 55420

L. E. Williams   Associate Professor, Department of Viticulture and Enology, Kearney Agricultural Center, University of California-Davis, Parlier, CA 93648

S. R. Winter   Professor, Texas Agricultural Experiment Station, Bushland, TX 79012

James L. Wright   Research Soil Scientist, USDA-ARS, Soil and Water Management Research, Kimberly, ID 83341

Conversion Factors for SI and non-SI Units

Conversion Factors for SI and non-SI Units

To convert Column 1 into Column 2, multiply by	Column 1 SI Unit	Column 2 non-SI Unit	To convert Column 2 into Column 1, multiply by
Length			
0.621	kilometer, km (10^3 m)	mile, mi	1.609
1.094	meter, m	yard, yd	0.914
3.28	meter, m	foot, ft	0.304
1.0	micrometer, μm (10^{-6} m)	micron, μ	1.0
3.94×10^{-2}	millimeter, mm (10^{-3} m)	inch, in	25.4
10	nanometer, nm (10^{-9} m)	Angstrom, Å	0.1
Area			
2.47	hectare, ha	acre	0.405
247	square kilometer, km^2 (10^3 m)^2	acre	4.05×10^{-3}
0.386	square kilometer, km^2 (10^3 m)^2	square mile, mi^2	2.590
2.47×10^{-4}	square meter, m^2	acre	4.05×10^3
10.76	square meter, m^2	square foot, ft^2	9.29×10^{-2}
1.55×10^{-3}	square millimeter, mm^2 (10^{-3} m)^2	square inch, in^2	645
Volume			
9.73×10^{-3}	cubic meter, m^3	acre-inch	102.8
35.3	cubic meter, m^3	cubic foot, ft^3	2.83×10^{-2}
6.10×10^4	cubic meter, m^3	cubic inch, in^3	1.64×10^{-5}
2.84×10^{-2}	liter, L (10^{-3} m^3)	bushel, bu	35.24
1.057	liter, L (10^{-3} m^3)	quart (liquid), qt	0.946
3.53×10^{-2}	liter, L (10^{-3} m^3)	cubic foot, ft^3	28.3
0.265	liter, L (10^{-3} m^3)	gallon	3.78
33.78	liter, L (10^{-3} m^3)	ounce (fluid), oz	2.96×10^{-2}
2.11	liter, L (10^{-3} m^3)	pint (fluid), pt	0.473

Mass

To convert Column 2 into Column 1, multiply by	Column 1 SI Unit	Column 2 non-SI Unit	To convert Column 1 into Column 2, multiply by
2.20×10^{-3}	gram, g (10^{-3} kg)	pound, lb	454
3.52×10^{-2}	gram, g (10^{-3} kg)	ounce (avdp), oz	28.4
2.205	kilogram, kg	pound, lb	0.454
0.01	kilogram, kg	quintal (metric), q	100
1.10×10^{-3}	kilogram, kg	ton (2000 lb), ton	907
1.102	megagram, Mg (tonne)	ton (U.S.), ton	0.907
1.102	tonne, t	ton (U.S.), ton	0.907

Yield and Rate

To convert Column 2 into Column 1, multiply by	Column 1 SI Unit	Column 2 non-SI Unit	To convert Column 1 into Column 2, multiply by
0.893	kilogram per hectare, kg ha^{-1}	pound per acre, lb acre^{-1}	1.12
7.77×10^{-2}	kilogram per cubic meter, kg m^{-3}	pound per bushel, bu^{-1}	12.87
1.49×10^{-2}	kilogram per hectare, kg ha^{-1}	bushel per acre, 60 lb	67.19
1.59×10^{-2}	kilogram per hectare, kg ha^{-1}	bushel per acre, 56 lb	62.71
1.86×10^{-2}	kilogram per hectare, kg ha^{-1}	bushel per acre, 48 lb	53.75
0.107	liter per hectare, L ha^{-1}	gallon per acre	9.35
893	tonnes per hectare, t ha^{-1}	pound per acre, lb acre^{-1}	1.12×10^{-3}
893	megagram per hectare, Mg ha^{-1}	pound per acre, lb acre^{-1}	1.12×10^{-3}
0.446	megagram per hectare, Mg ha^{-1}	ton (2000 lb) per acre, ton acre^{-1}	2.24
2.24	meter per second, m s^{-1}	mile per hour	0.447

Specific Surface

To convert Column 2 into Column 1, multiply by	Column 1 SI Unit	Column 2 non-SI Unit	To convert Column 1 into Column 2, multiply by
10	square meter per kilogram, m^2 kg^{-1}	square centimeter per gram, cm^2 g^{-1}	0.1
1000	square meter per kilogram, m^2 kg^{-1}	square millimeter per gram, mm^2 g^{-1}	0.001

Pressure

To convert Column 2 into Column 1, multiply by	Column 1 SI Unit	Column 2 non-SI Unit	To convert Column 1 into Column 2, multiply by
9.90	megapascal, MPa (10^6 Pa)	atmosphere	0.101
10	megapascal, MPa (10^6 Pa)	bar	0.1
1.00	megagram per cubic meter, Mg m^{-3}	gram per cubic centimeter, g cm^{-3}	1.00
2.09×10^{-2}	pascal, Pa	pound per square foot, lb ft^{-2}	47.9
1.45×10^{-4}	pascal, Pa	pound per square inch, lb in^{-2}	6.90×10^3

(continued on next page)

Conversion Factors for SI and non-SI Units

To convert Column 1 into Column 2, multiply by	Column 1 SI Unit	Column 2 non-SI Unit	To convert Column 2 into Column 1, multiply by
Temperature			
1.00 (K $-$ 273)	Kelvin, K	Celsius, °C	1.00 (°C $+$ 273)
(9/5 °C) $+$ 32	Celsius, °C	Fahrenheit, °F	5/9 (°F $-$ 32)
Energy, Work, Quantity of Heat			
9.52×10^{-4}	joule, J	British thermal unit, Btu	1.05×10^3
0.239	joule, J	calorie, cal	4.19
10^7	joule, J	erg	10^{-7}
0.735	joule, J	foot-pound	1.36
2.387×10^{-5}	joule per square meter, $J\ m^{-2}$	calorie per square centimeter (langley)	4.19×10^4
10^5	newton, N	dyne	10^{-5}
1.43×10^{-3}	watt per square meter, $W\ m^{-2}$	calorie per square centimeter minute (irradiance), $cal\ cm^{-2}\ min^{-1}$	698
Transpiration and Photosynthesis			
3.60×10^{-2}	milligram per square meter second, $mg\ m^{-2}\ s^{-1}$	gram per square decimeter hour, $g\ dm^{-2}\ h^{-1}$	27.8
5.56×10^{-3}	milligram (H_2O) per square meter second, $mg\ m^{-2}\ s^{-1}$	micromole (H_2O) per square centimeter second, $\mu mol\ cm^{-2}\ s^{-1}$	180
10^{-4}	milligram per square meter second, $mg\ m^{-2}\ s^{-1}$	milligram per square centimeter second, $mg\ cm^{-2}\ s^{-1}$	10^4
35.97	milligram per square meter second, $mg\ m^{-2}\ s^{-1}$	milligram per square decimeter hour, $mg\ dm^{-2}\ h^{-1}$	2.78×10^{-2}
Plane Angle			
57.3	radian, rad	degrees (angle), °	1.75×10^{-2}

Electrical Conductivity, Electricity, and Magnetism

	Column 1 (SI Unit)	Column 2 (non-SI Unit)	
10	siemen per meter, S m^{-1}	millimho per centimeter, mmho cm^{-1}	0.1
10^4	tesla, T	gauss, G	10^{-4}

Water Measurement

	Column 1 (SI Unit)	Column 2 (non-SI Unit)	
9.73×10^{-3}	cubic meter, m^3	acre-inches, acre-in	102.8
9.81×10^{-3}	cubic meter per hour, m^3 h^{-1}	cubic feet per second, ft^3 s^{-1}	101.9
4.40	cubic meter per hour, m^3 h^{-1}	U.S. gallons per minute, gal min^{-1}	0.227
8.11	hectare-meters, ha-m	acre-feet, acre-ft	0.123
97.28	hectare-meters, ha-m	acre-inches, acre-in	1.03×10^{-2}
8.1×10^{-2}	hectare-centimeters, ha-cm	acre-feet, acre-ft	12.33

Concentrations

	Column 1 (SI Unit)	Column 2 (non-SI Unit)	
1	centimole per kilogram, cmol kg^{-1} (ion exchange capacity)	milliequivalents per 100 grams, meq 100 g^{-1}	1
0.1	gram per kilogram, g kg^{-1}	percent, %	10
1	milligram per kilogram, mg kg^{-1}	parts per million, ppm	1

Radioactivity

	Column 1 (SI Unit)	Column 2 (non-SI Unit)	
2.7×10^{-11}	bequerel, Bq	curie, Ci	3.7×10^{10}
2.7×10^{-2}	bequerel per kilogram, Bq kg^{-1}	picocurie per gram, pCi g^{-1}	37
100	gray, Gy (absorbed dose)	rad, rd	0.01
100	sievert, Sv (equivalent dose)	rem (roentgen equivalent man)	0.01

Plant Nutrient Conversion

	Elemental	Oxide	
2.29	P	P$_2$O$_5$	0.437
1.20	K	K$_2$O	0.830
1.39	Ca	CaO	0.715
1.66	Mg	MgO	0.602

1 Scope and Objective of Monograph

B. A. STEWART

USDA-ARS
Bushland, Texas

D. R. NIELSEN

University of California
Davis, California

Irrigation has expanded dramatically in this century and particularly so from 1950 to 1980, during which it increased worldwide from about 95 million ha to about 250 million ha. Irrigated land represents about 18% of the cultivated land, but accounts for one-third of the world's food. The growth in the amount of irrigated land—more than 3% annually during the period mentioned—has now slowed to <1% yr^{-1} (Higgins et al., 1987).

It is interesting to reflect on some of the factors that have influenced the sharp increase in irrigation development and the more recent rapid decrease in rate of expansion. With irrigation, came greater fertilizer usage as well as greater use of herbicides, insecticides, and other inputs. The total energy input other than that from the sun, most of which was derived from petroleum, increased significantly. As energy inputs increased, however, the harvested energy outputs increased severalfold. The relationship between input energy and harvested output energy for the U.S. corn (*Zea mays* L.) crop is shown in Fig. 1-1. This relationship shows that for every unit of energy input, there was a threefold increase in energy output. There is also no indication of an upper limit of the linear relationship. While the U.S. corn crop is largely not grown under irrigation, the area of irrigated corn increased sharply during the period of the analysis shown in Fig. 1-1 because much of the increased irrigated land in the Great Plains was cropped with corn. Depending on the amount of energy required for pumping irrigation water, the slope of the line would change, but the linear relationship between input energy and output energy would likely hold because yields generally increase linearly with water transpired.

The relationships between the cost of oil and the values of wheat (*Triticum aestivum* L. and corn are shown in Fig. 1-2. From 1950 to the early 1970s, the ratio between the price of oil and that of grain was almost constant with 1 kg of wheat worth about 7 L of oil (1 bu of wheat would buy 1 bbl of oil). With each unit of energy input returning three units of energy output as shown in Fig. 1-1, and the favorable ratio between price of oil

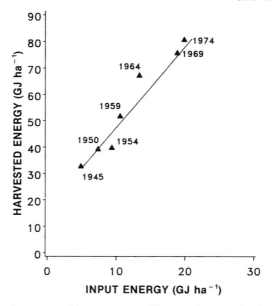

Fig. 1-1. Relation between total input energy and harvested energy for the U.S. corn crop at 4 to 5 yr intervals. Adapted from Smil et al. (1983).

and the value of grain, the expansion of irrigation was propelled at dramatic rates. However, owing to the ratio between energy cost and the value of grains changing unfavorably during the 1970s as shown in Fig. 1-2, the expansion

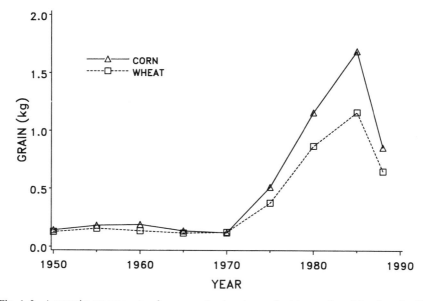

Fig. 1-2. Approximate amounts of corn or wheat grain required to purchase 1 L of crude oil.

of irrigation and even the continuation of some existing irrigation systems have declined. There has been some improvement in the relationship between energy cost and the value of grains in recent years, which may again spur some irrigation expansion, but long-range projections for continued lower oil prices are not favorable.

Although the expansion of irrigated land has slowed to $< 1\%$ yr^{-1}, the world population is growing at about 1.8% annually. In developing countries, however, the annual growth rate is 2.0% and exceeds 3% in parts of Africa (World Bank, 1987). Consequently, the increased demand for food and fiber by an expanding population will have to be met by increased, and more efficient production on both nonirrigated and irrigated land. Certainly, both scenarios will be required. The objective of this monograph is to contribute toward improving the efficient use of irrigation water and to do so in such a manner that production efficiencies will be increased without adverse effects on the environment. To meet this objective, authors from several countries were invited to address important issues.

Section II, "Role of Irrigation in Agricultural Systems," discusses philosophical and technical questions that must be answered before a decision is reached to irrigate. The rationale for reaching a decision to irrigate may be vastly different for various individuals, water districts, areas, and countries. Also, the constraints due to variable costs and competing uses for water must be analyzed and considered.

Sections III, IV, and V consider the basic principles of the soil-water, plant-water, and soil-plant-atmosphere relations. The chapters in these sections include listings of references to aid the reader in obtaining additional information about specific subjects. These chapters were designed to translate knowledge and principles of soil and crop science into useful irrigation practices.

Section VI emphasizes the application of principles to the field level. Focus is on the capabilities of systems to apply water in space over time. Guidance for selecting one type of system over another for specific soil, crop, and water source situations is provided. The effects of climates, soils, and water sources for various irrigation practices are presented and discussed. The consequences of soil and climate differences, particularly precipitation during the irrigation season, have been frequently ignored in irrigation scheduling. Also, energy and water resources have not been sufficiently conserved and must be more carefully managed if many of the present irrigation systems are to remain viable and new systems are to be justified.

Section VII is perhaps the most critical section of the entire monograph. Specific crops respond differently to irrigation and provide unique opportunities for water management. Authors in this section have addressed the irrigation requirements and responses of selected agricultural crops, as well as the cultural practices, that result in the efficient use of water resources.

Section VIII, although presented last, is also an important section because it deals with effects on the environment. Keen (1988) recently reported that of the lands currently irrigated, about half of them need improvements or modernization because of degradation of soil and water quality, inefficient use of water by plants, siltation of reservoirs, and decay of the distribution channels. These adverse effects within irrigated lands need not exist. Properly managed, highly efficient irrigation systems should provide sustainable agricultural production that improves, not deteriorates, land and water resources on a regional basis. With water quality among the highest of public concerns, irrigation systems must continually be redesigned and managed to achieve desirable water quality standards.

The Directors of the American Society of Agronomy charged the editors to develop a monograph that would supercede Agronomy Monograph 11, *Irrigation of Agricultural Lands* published in 1966. This was indeed a challenge because Monograph 11 continues to be a widely used reference. The editors, with the help of an able editorial committee consisting of E.T. Kanemasu, R.R. Bruce, J.R. Gilley, and M.H. Niehaus, have endeavored to develop a monograph that will complement and add to the knowledge base that was covered so well in Monograph 11.

REFERENCES

Higgins, G.M., P.J. Dielman, and C.L. Abernethy. 1987. Trends in irrigation development, their implications for hydrologists and water resource engineers. J. Hydraul. Res. 25:393–406.

Keen, M. 1988. Clearer thoughts flow on irrigation. Ceres (The FAO Review) 21(3):31–35.

Smil, V., P. Nachman, and T.V. Long. 1983. Energy analysis and agriculture: an application to U.S. corn production. Westview Press, Boulder, CO.

World Bank. 1987. World development report. Oxford Univ. Press, New York.

2 Role of Irrigation in Agricultural Systems

DANIEL HILLEL

University of Massachusetts
Amherst, Massachusetts

Irrigation is playing an increasingly important role in the agricultural economy of drought-prone regions. As practiced in many places, however, it is still based largely on traditional methods of distribution and application which fail to measure and optimize the supply of water needed to satisfy the variable demands of different crops. Unmeasured irrigation tends to waste water, nutrients, and energy and may cause soil degradation by water-logging and salinization, particularly where the necessity for drainage is neglected.

The vital task of assuring the adequate and stable production of food and fiber throughout the world must include a concerted effort to modernize existing tradition-bound irrigation schemes so as to achieve higher levels of profitable and sustainable production. New schemes being planned should likewise be based on sound principles and techniques for attaining greater control over the soil-crop-water regime and for optimizing irrigation in relation to all other essential agricultural inputs and operations. Water and soil must be recognized as vital, precious, and vulnerable resources and managed accordingly.

In recent years, revolutionary developments have taken place in the science and art of irrigation. A more comprehensive understanding has evolved of the interactive relationships among the soil, plant, and climate regarding the disposition and utilization of water. These scientific developments have been paralleled by a series of technical innovations in the methodology of water control which have made it possible to establish and maintain nearly optimal soil moisture conditions practically continuously. Among these innovations are techniques for high-frequency, low-volume application of water and nutrients to the root zone in precise and timely response to changing crop needs. The introduction of relatively inexpensive and durable permanent-set distribution and application systems and the advent of self-controlling ancillary devices have made these techniques practical for many crops.

Properly applied, the new irrigation methods can raise yields while avoiding waste, reducing drainage requirements, and promoting the integration of irrigation with essential concurrent operations. The use of bracksih water has become more feasible, as has the irrigation of coarse-textured soils and of steep and stony lands previously considered totally unproductive. Such advances and their consequences could hardly have been foreseen in the literature of prior decades.

I. THE TASK OF IRRIGATION DEVELOPMENT

The needs of growing populations and developing national economies demand intensification of land and water use for the purpose of increasing and stabilizing agricultural production. These needs are most acute in the arid and semiarid regions where, by a cruel stroke of nature, the water requirements of crops are greatest while the supplies by rainfall are least. Under such conditions, even a slight improvement of the water economy may spell the difference between marginal subsistence and profitable production.

Water constitutes a major constraint to increasing crop production in our hungry world. To grow successfully, each crop must achieve a water economy such that the demand made upon it by the climate is balanced by the supply available to it. The problem is that the evaporative demand of the atmosphere is practically continuous, whereas the supply of water by natural precipitation is sporadic. To survive during dry spells, the crop must rely on the limited reserves of extractable moisture temporarily present in the pores of the soil. So tenuous and delicate is the water economy of most crops that even short-term deprivation can cause sufficient stress to impair normal physiological functions and potential yields.

Irrigation is the supply of water to agricultural crops by artificial means, designed to permit farming in arid regions and to offset drought in semiarid or semihumid regions. As such, it plays a key role in feeding an expanding population and seems destined to play an even greater role in the foreseeable future. Even in areas where total seasonal rainfall may seem ample, it is often unevenly distributed during the year so that traditional dryland farming is a high-risk enterprise and only irrigation can provide a stable system of production. Irrigation can prolong the effective growing period in areas with dry seasons, thus permitting multiple cropping (two or three—and sometimes four—crops per year). With the security of cropping under irrigation, additional inputs needed to increase production (fertilizers, improved cultivars, pest control, and better tillage) become economically feasible. Irrigation reduces the risk of these expensive inputs being wasted by drought.

The process of irrigation consists of introducing water into the part of the soil profile that serves as the root zone, for the subsequent use of the crop. A well-managed irrigation system is one that optimizes the spatial and temporal distribution of water so as to promote crop growth and yield, and to enhance the economic efficiency of crop production. The aim is not necessarily to obtain the highest yields per unit area of land, or even per unit volume of water, but to maximize the net returns—not just for a given season but in the long run. Since the physical circumstances and the socioeconomic conditions are site specific (and often season specific) in each case, there can be no single solution to the problem of how best to develop and manage an irrigation project. In different combinations, however, the factors and principles involved are universal.

From its early and primitive antecedents in the great river valleys of the Middle East some seven millennia ago, the practice of irrigation has evolved gradually in the direction of increasing the farmer's control over crop, soil, and even weather variables. Although the degree of control possible today is only partial, as the open field remains ever subject to unpredictable vagaries, modern irrigation is a highly sophisticated operation, involving the simultaneous monitoring and manipulation of numerous factors of production. And yet progress continues. Along with the rising costs of energy, scarcity of good land and water, and increasing demand for agricultural products, the search for new knowledge on how to improve the efficiency of irrigation, and the imperative to disseminate and apply the knowledge gained to date, have become more urgent than ever.

II. ENVIRONMENTAL PROCESSES

A. The Field Water Cycle and Water Balance

Any attempt to control the supply of water to crops must be based on a thorough understanding of soil-water dynamics. The cycle of water in the field consists of a sequence of dynamic processes beginning with the entry of water into the soil, called *infiltration*. The rate of infiltration can be governed by the rate at which water is applied to the surface, as long as the application rate does not exceed the maximum rate that the soil can absorb through its surface. That limiting rate, called the soil's *infiltrability*, is highest for initially dry sandy soils and lowest for wet clayey soils, especially if the soil surface has been compacted by traffic or by the beating action of raindrops. An important design criterion for a sprinkle or drip irrigation system is to deliver water only at the rate that the soil surface can absorb, since an excessive rate of water application can induce ponding, restriction of aeration, runoff, erosion, and inter-row weed infestation.

The water that has entered the soil during infiltration does not remain immobile after the infiltration event has ended. Because of gravity and tension gradients in the soil, this water generally continues to move downward, albeit at a diminishing rate, in a process called *redistribution*. In the course of this process, the relatively dry deeper zone of the soil profile absorbs water draining from the infiltration-wetted upper part. Within a few days, however, the rate of flow can become so low as to be considered negligible. At this time, the remaining water content in the initially wetted zone is termed the *field capacity* and is often taken to represent the upper limit of the soil's capacity to store water. The redistribution process depends on the antecedent (pre-infiltration) soil moisture content, the amount of water infiltrated, and—primarily—the composition and structure of the soil profile. Field capacity tends to be higher in clayey than in sandy soils. Moreover, it is generally greater in layered than in uniform soil profiles of similar texture, as layering inhibits the internal drainage of water.

The amount of water draining out of the root zone during infiltration or redistribution is generally considered to be a loss from the standpoint of immediate crop water use. It is not necessarily a final loss, however. If the area is underlain by an exploitable aquifer, the water drained from the root zone may eventually recharge the groundwater and can be recovered by pumping. Where the water table is close to the surface, some water may enter the root zone by capillary rise and may supply a portion of the water requirements of a crop. This process of *subirrigation* is not an unmixed blessing, however, as it may infuse the root zone with salts.

The pattern and rate of evaporation from bare soil surfaces depend on the external climate as well as on the internal movement of soil moisture and heat. Soon after an infiltration event, while the soil surface is still wet, it is primarily the climate that dictates the rate of evaporation. But as the surface zone desiccates (generally within a few days after the onset of evaporation), the evaporation rate necessarily diminishes to become very slow. Soils that crack as they desiccate may, however, continue to lose water at an appreciable rate for many days. Soils with a high water table can sustain a high evaporation rate still longer. Such soils become sterile as the evaporating groundwater deposits its salts at the soil surface.

Transpiration from plant canopies, rather than direct evaporation of soil moisture, becomes predominant when a crop shades the greater part of the surface. In an arid environment, situations may develop in which the plants cannot draw water fast enough to satisfy the climatically imposed demand. Under such conditions, plants experience stress and must limit transpiration if they are to avoid dehydration. They can do this, to a limited degree and for a limited time, by closing their stomates (Kramer, 1983). The inevitable price of this limitation of transpiration is a reduction of growth, as the same stomatal openings which transpire water also serve for the uptake of CO_2 needed in photosynthesis. While the relative effects of stomatal closure on transpiration and on photosynthesis for different types of crops are still topics for research (Hanks & Hill, 1980), it is clear that conditions of stress limit yield in any case and should be avoided, to the extent possible, in irrigation management (Rawlins & Raats, 1975).

The *field water balance* is an account of all quantities of water added to, subtracted from, and stored within the root zone during a given period of time. The difference between the total amount added and that withdrawn must equal the change in storage. When gains exceed losses, storage increases; conversely, when losses exceed gains, storage decreases. Thus:

$$(\text{Accretion}) = (\text{Gains}) - (\text{Losses}) \tag{1}$$

This general statement can be amplified as follow:

$$(\Delta S + \Delta V) = (P + I + U) - (R + D + E + T) \tag{2}$$

wherein ΔS is accretion of water in the root zone, ΔV the increment of water incorporated in the vegetation, P the precipitation, I irrigation, U the upward capillary flow into the root zone from below, R runoff, D downward drainage out of the root zone, E direct evaporation from the soil surface, and T transpiration by plants. The last two variables are difficult to separate and are therefore generally lumped together and termed *evapotranspiration.* All quantities included in the field water balance are expressed in terms of volume of water per unit area (equivalent depth units) during the period considered. It is convenient to report water quantities in millimeters, i.e., as cubic meters of water per 1000 square meters of land area.

Simple and readily understandable though the field water balance may seem in principle, it is still rather difficult to measure in practice. Often the largest component on the "losses" side of the ledger, and the one most difficult to measure directly, is the evapotranspiration $(E + T)$, simply designated ET. To obtain ET from the water balance we must have accurate measurements of all other terms of the equation. It might seem relatively easy to measure the amount of water added to the field by rain and irrigation $(P + I)$, but this is seldom done on a field-by-field basis, either because of a lack of equipment or trained personnel, or simply through inattention. Even where the input is measured, there remains the problem of how to account for nonuniformities in areal distribution. The amount of runoff generally is (or at least should be) small in agricultural fields, particularly in irrigated fields, so that, justifiably or not, it is most often ignored. So is the change in water content of the vegetation.

For a long period, such as an entire growing season, the change in water content of the root zone is likely to be small in relation to the total water balance. In this case, $(P + I)$ is approximately equal to the sum of ET and deep percolation D. For shorter periods, the change in soil-water storage S can be relatively large and must be measured by periodic sampling or the use of specialized instruments. During dry spells, without rain or irrigation, the reduction in root-zone water storage must equal the sum of ET and *net drainage.* The latter, being the excess of downward percolation from the root zone over the upward capillary rise into the root zone, is seldom measured adequately.

B. Radiation Exchange and Energy Balance

The principal process involving the fate of water in the field, and hence the water requirements of crops, is evapotranspiration, a process that is driven by a constant inflow of energy. In fact, the entire field water balance (see section 2–II.A.) is intimately and reciprocally related to the cycle and balance of energy, since the state and content of water in the soil and its vegetative cover is affected by, and in turn affects, the way the energy flux reaching the field is partitioned and utilized. Control of the soil-plant-atmosphere system must therefore be based on simultaneous consideration of both the water and the energy balances.

A significant fraction of the solar short-wave radiation flux reaching the ground is reflected upward by the surface. That fraction, called *albedo,* varies according to the color, roughness, and inclination of the surface. Albedo is of the order of 10 to 30% for vegetation and 15 to 40% for bare soil (being lower for wet dark clays and higher for dry light-colored sands). In addition to reflecting shortwave radiation, the field emits its own radiation, but at a much longer wavelength than solar radiation (i.e., in the wavelength range of 3 to 50 μm, known as the range of infrared, termed *longwave,* radiation). Between the two radiation spectra, the sun's (shortwave) and the earth's (longwave), there is little overlap. The overall difference between the total incoming and outgoing radiation fluxes (including both the shortwave and longwave components) is termed *net radiation* (J_n) and it represents the rate of radiant energy absorption by the field (Tanner, 1960):

$$J_n = (J_{si} - J_{so}) + (J_{li} - J_{lo}) \qquad [3]$$

where J_{si} is the incoming flux of shortwave radiation from sun and sky, J_{so} the outgoing shortwave radiation (reflected by the surface), J_{li} the incoming longwave radiation from the sky, and J_{lo} the outgoing longwave radiation emitted by the surface. During daytime, the incoming shortwave flux generally dominates the radiation balance and the net radiation is positive but during the night, in the absence of direct solar radiation, the longwave radiation emitted by the surface naturally exceeds that received from the sky and the net radiation flux is negative.

Next, consider the disposition of the radiant energy absorbed by the field. Part of it is transformed into heat, which warms the soil, plants, and atmosphere. Another part is taken up by the plants in their metabolic processes (e.g., photosynthesis). Finally, a major part is generally absorbed as latent heat in the twin processes of evaporation and transpiration. Thus:

$$J_n = S + A + LE + M \qquad [4]$$

where S is the rate at which heat is stored in the soil, water, and vegetation; A is the energy flux that goes into heating the air; LE is the rate of energy utilization in evapotranspiration (a product of the rate of evaporation E and the latent heat of vaporization L); and M represents the metabolic energy uptake (photosynthesis). The energy uptake of photosynthesis generally does not exceed 3% of the daily net radiation. Typically, the major share of the net radiation goes into the latent heat of evaporation and the sensible heating of the air. The proportionate allocation between these terms depends on the availability of water for evaporation.

The relationships we have described for the energy balance apply to extensive uniform areas in which all the fluxes are vertical or nearly so. On the other hand, any small field differing from its surrounding area is subject to lateral effects. Specifically, winds sweeping over a small field can transport energy into or out of it. This phenomenon, called *advection,* can be especially important in arid regions, where small irrigated fields are often surrounded by an expanse of dry land. Under such conditions, the warm and dry incoming air can transfer sensible heat to the crop and thus increase the rate of evapotranspiration relative to that from a large field subject to the same radiation regime. A common sight in arid regions is the poor growth of the plants near the windward edge of an irrigated field, where penetration of warm dry wind enhances evaporation. When advective heat inflow is large, the energy absorbed in evapotranspiration from a limited and sparse stand of vegetation (e.g., a small plot of a widely spaced row crop or a small grove of trees) can greatly exceed that from an extensive stand of smooth and dense vegetation (e.g., mowed grass) and may even exceed the energy input from net radiation. Hence, measurements of water consumption and of apparent irrigation requirements obtained from small experimental plots may not be representative of large fields, unless these plots are buffered, or "guarded," in the upwind direction by an expanse, or "fetch," of vegetation of similar characteristics and water regime.

C. Potential Evapotranspiration and Pan Evaporation

The process of evapotranspiration obviously depends on both the external meteorological regime (radiation, atmospheric humidity, temperature, wind, etc.) and the internal state of the field itself, particularly its degree of wetness and surface properties. Conceptually, therefore, one might suppose that there ought to be a definable evapotranspiration rate for the special case in which the field is maintained perpetually wet, and that this rate should depend only on the meteorological regime. The concept of *potential evapotranspiration* (E_o or PET) is an attempt to characterize the climatic environment in terms of its evaporative power, i.e., the maximal evaporation rate that the atmosphere is capable of extracting from a well-watered field under given conditions. The PET is thus said to represent the climatically imposed "evaporative demand."

Penman (1953, 1956) defined potential evapotranspiration more restrictively as "the amount of water transpired in unit time by a short green crop, completely shading the ground, of uniform height and never short of water." As such, PET is a useful standard of reference for the comparison of different climatic regions (or seasons) and of different crops within a given climatic region.

The PET is conditioned, first of all, by the flux of energy reaching the surface via solar radiation. This, in turn, depends on permanent attributes of the field such as latitude, slope, and aspect—as well as on season of the year and other variables including atmospheric conditions. Despite the field-specific nature of several of the variables affecting the energy balance, PET is often assumed to depend predominantly on the climatic inputs and to be practically independent of field properties.

Various empirical approaches have been proposed for the estimation of potential evapotranspiration. The simplest methods are based on air temperature, since this climatic variable is readily available and, in both moist and arid regions, is practically independent of the wetness of the surface (Sellers, 1965). The formulation of Blaney and Criddle (1950) is still widely used to estimate crop water requirements as related to temperature and day length. The uncertainty involved in any evapotranspiration prediction using

only one or two weather factors is high. Evapotranspiration may vary widely between climates having similar air temperatures if there are differences
in atmospheric humidity, windiness, or sunshine. Thus, the effect of climate
on crop water use is not fully defined by temperature and day length alone.

The method proposed by Penman (1948) is physically based, hence inherently more meaningful than the strictly empirical methods. His equation,
derived from a combination of the energy balance and aerodynamic transport considerations, is as follows:

$$E_o = [(\delta/C_\psi)J_n + 0.35 (P_{vs} - P_{va})(0.5 + U_2/100)]/(\delta/C_\psi + 1)] \quad [5]$$

where δ is the slope of the saturated vapor pressure vs. temperature curve,
c_ψ is the psychrometric constant, J_n is the net radiation, P_{vs} is the saturated
vapor pressure at mean air temperature, P_{va} is the mean vapor pressure in
the air, and U_2 is the mean wind speed at 2 m above the ground. As can
be seen from Penman's equation, the major factors governing E_o are net
radiation, temperature (affecting both δ/C_w and P_{vs}), vapor pressure deficit
of the air, and wind speed. These variables can be obtained from standard
meteorological measurements taken at one level.

The Penman formulation avoids the necessity of determining the value
of T_s, the surface temperature, just as it disregards the possible fluctuations
in the direction and magnitude of the soil heat flux term. Moreover, it makes
no provision for surface roughness, air instability, or canopy resistance
effects. Finally, the Penman theory takes no explicit account of advection.

In recent decades, various modifications have been suggested for the
Penman formulation (e.g., van Bavel, 1966; Szeicz et al., 1973; Monteith,
1980) to allow for such effects as short-term variations in soil heat flux, surface roughness, air instability, and canopy resistance. Such modifications
attempt to account for the fact that the rate of potential evapotranspiration
from an orchard can be expected to differ from that of a corn field, which
in turn might differ from that of a lawn or a smooth bare soil. The advent
of remote-sensing infrared thermometry has made possible continuous
monitoring of surface temperature, and hence also allows a better estimation of the vapor pressure at the surface.

It should be emphasized again that the representation of potential
evapotranspiration purely as an externally imposed "forcing function" is
only an approximation (van Bavel and Hillel, 1976). In actual fact, each field
interacts with the meteorological regime in determining its evapotranspiration rate even when the field is well endowed with water. That interaction
is influenced by each field's own (and often variable) values of surface reflectivity, aerodynamic roughness, thermal capacity and conductivity, crop
phenology, and canopy resistance. The oft-repeated principle that under the
same climate all well-watered fields, regardless of specific characteristics,
exhibit the same rate of evapotranspiration is only "more or less" true.

In view of the complexity of the physically based estimations of potential evapotranspiration (i.e., the Penman method and its derivates), it is not
surprising that many practitioners continue to prefer the simplified empir-

ical methods which depend on correlation with past records rather than on explicit formulation of ongoing processes or mechanisms. Various evaporation-measuring devices, called evaporimeters, have been proposed and tried for the purpose of obtaining an estimate of the climatically driven potential evapotranspiration. Of these, the most frequently used are evaporation pans. They provide an indication of the integrated effect of radiation, wind, temperature, and humidity on evaporation from an open water surface. The most widely adopted type is the "Class A" pan standardized by the U.S. Weather Bureau. Pans are relatively inexpensive and are easy to install, maintain, and monitor. They do, however, have several important shortcomings.

Although a vegetated field responds to the same climatic variables as does a pan, it does not necessarily respond in the same way. A field generally differs from a free water surface in reflectivity, thermal properties, canopy resistance, and aerodynamic characteristics. Hence, the process of evaporation from a water-filled pan is not a true portrayal of evapotranspiration from plants and soil. The daytime storage of heat within the pan can cause considerable evaporation at night (10 to 40% of the diurnal total), while nighttime transpiration from crops is generally below 5% of the diurnal total because of the much smaller heat storage and particularly because of the increase in canopy resistance due to the closure of stomates in the dark. Also, the color of the pan, heat transfer through the sides, turbidity of the water, and possible shading from screens or nearby plants affect the measurement.

Pan evaporation depends greatly on the exact placement of the pan relative to wind exposure and advection from outside the field. Pans surrounded by a tall crop may evaporate 20 to 30% less than pans placed in a fallow area, especially if the climate is dry and windy. To avoid water loss to drinking animals, especially birds, pans are often covered by screens. This reduces the evaporation rate (generally by about 10 to 20%), thus requiring the use of a correction factor, and also interferes with the measurements and the servicing of the pan.

All of these shortcomings notwithstanding, pan evaporimeters, if properly sited and maintained, can indeed be used to assess potential evapotranspiration. A correction factor is needed, however, to account for environmental conditions. For pans placed in the open, this factor generally varies from 0.85 for conditions of high humidity and light winds (below 175 km/d) to about 0.5 for low humidity and strong winds (above 700 km/d). The average is about 0.7.

III. CROP-WATER RELATIONS

A. Water in the Physiology of Crops

Green plants are nature's only autotrophs, able to create living matter from inorganic raw materials. Land plants do this by combining atmospheric CO_2 with soil-derived water while converting solar radiation into chemical energy in the process of photosynthesis. This process not only produces food for all animals, including humans, but also releases into the atmosphere the elemental O_2 needed for respiration by animals and plants alike. Water plays a central role in the metabolism of plants, as a source of H atoms for the reduction of CO_2 in photosynthesis and as a product of respiration. Moreover, water is the solvent, and hence conveyor, of transportable ions and compounds into, within, and out of all living plants. It is also a major structural component, often constituting more than 90% of the vegetative biomass.

Only a small fraction of the water absorbed by plants is used in photosynthesis, while most (often as much as 99%) escapes as vapor in the process of transpiration from plant canopies. Transpiration is made inevitable by the exposure to the dry atmosphere of a large area of moist cell surfaces, necessary to facilitate absorption of CO_2. Hence, transpiration has been described, from the point of view of crop growers in arid regions, as a "necessary evil." Most crop plants are extremely sensitive to lack of sufficient water to replace the amount they must transpire.

Plants live simultaneously in two different realms, the atmosphere and the soil, in each of which conditions vary constantly and not necessarily in coordination. Mesophytes regulate their water economy by developing extensive root systems, thus enhancing water supply from the soil, as well as by reducing the aperture of their stomates, thus limiting vapor loss from the leaves, albeit at the expense of decreased absorption of CO_2. Closure of stomates cannot, however, prevent transpiration entirely, as the leaves continue to lose some water through their cuticular surfaces. Sustained transpiration without sufficient replenishment leads to progressive dehydration, of which the loss of foliar turgidity (i.e., wilting) is the most visible sign.

Many physiological processes are affected by water stress considerably before the plant actually wilts (Hsiao, 1973). By the time the plant approaches the permanent wilting point, cell expansion has long since ceased and the tightly closed stomates severly restrict CO_2 entry as well as water loss. Since transpiration rate may not decrease appreciably until the plant wilts, some researchers in the early 1930s concluded that crop production remains unaffected by moisture stress until the water content of the soil falls to a value near the permanent wilting point. The subsequent observation that growth decreases earlier than does transpiration is at variance with that conclusion. It is now widely accepted that in most crops, growth can proceed unimpaired and crop yield can be maximized only when the soil moisture potential remains high (and water remains readily available) continuously throughout the growing season.

When the plants are stressed and transpiration is curtailed even temporarily (e.g. as a result of stomatal closure at midday), the crop canopy temperature tends to rise appreciably because of the reduction of evaporative cooling. This is the basis for the modern technique of sensing crop water stress by monitoring canopy temperature with an infrared radiation thermometer (Jackson, 1982). The rise in temperature causes an exponential increase in the vapor pressure gradient from the leaf to the atmosphere, partially offsetting the increased stomatal resistance to transpiration. A rise in temperature also inevitably entails an increase in the rate of tissue respiration, so that while the stressed plant produces less owing to curtailed photosynthesis, it actually consumes more of its own reserves. Hence, short periods of stress may tend to reduce net production more than they reduce total transpiration (Huck & Hillel, 1983).

Granted that, for most crops, keeping plant water potential high (by maintaining a moist irrigation regime) results in maximum production per unit area, a question remains as to whether it also results in maximum production per unit amount of water consumed. The answer appears to be that stressed plants do not use water more efficiently than well-watered plants (Rawlins & Raats, 1975). From the standpoint of dry matter production, either per unit of water used or per unit of land occupied, there seems to be no advantage in subjecting plants to water stress. However, except for some forage crops, dry matter production is seldom equivalent to marketable yield. For crops that require water stress to initiate differentiation or maturation of the harvested portion of the plant, (e.g., cotton *Gossypium hirsutum* L.), programming a period of water stress into the growing season may be beneficial. Results from experiments on water stress effects on plant growth (Hsiao, 1973; Hsiao & Acevedo, 1974) suggest that periods of decreased leaf water potential do not necessarily decrease net growth if they are not too long, as assimilates can be stored for a few hours before photosynthesis is decreased and can then be used in accelerated growth when stress is relieved. Hence, short daily stress periods are less damaging overall than an infrequent stress lasting several days.

B. Crop Evapotranspiration and Water Requirements

The daily rate of actual evapotranspiration from a crop will seldom equal the potential rate (E_o or PET). Canopy characteristics, stand density, stage of growth and degree of surface cover, and especially the moisture regime, all affect actual evapotranspiration. In the case of a typical annual crop, the seasonal total evapotranspiration will generally not equal the total PET even if the moisture regime is a wet one. Early in the season, during the germination and seedling-establishment phase, the rate of transpiration is generally small and, as the bare soil surface dries between irrigations, the direct evaporation rate is also lower than PET. Later, as the canopy grows and covers more of the ground, transpiration increases until, at full cover, its rate approximates (and may even exceed) the rate of PET. Finally, as the crop matures, senesces, and dries, its actual evapotranspiration again falls below the PET.

The maximal seasonal evapotranspiration from a well-watered crop stand of optimal density, designated E_{max} or MET, is likely to range between 0.6 and 0.9 of total seasonal PET. To obtain the highest possible yields of many agricultural crops, irrigation should generally be provided in an amount sufficient to prevent water from becoming a limiting factor. Knowledge of the MET for the major crops in a given region can therefore serve as a basis for planning the irrigation regime. In fact, MET is often taken to represent the crop's water requirement. Because MET is affected by both the climate and the characteristics of the crop, it should be measured in the field for each region and major crop.

Using PET (whether calculated as E_o from a Penman-type formula or measured in the field by using a reference crop such as a well-watered, dense, and extensive stand of grass or alfalfa), it is possible to account for the effect of specific crop characteristics on crop water requirements (W_{rc}) using an empirical crop coefficient, K_c (Doorenbos & Pruitt, 1977)

$$W_{rc} = K_c E_o. \qquad [6]$$

In general, K_c is higher in hot, windy, and dry climates than in cool, calm, and humid climates. K_c values vary widely among crops, because of differences in crop height, "roughness" of the stand, reflectivity, degree of ground cover, and canopy resistance to transpiration. In assigning an appropriate K_c value to a crop, one should consider the different stages of growth, particularly as they affect wind and turbulence within and above the crop. For annual crops, K_c typically increases from a low value at seedling emergence to a maximum value when the crop reaches full ground cover, continues at that value during the stage of full activity, then declines as the crop matures and senesces. For most crops, the K_c value for the total growing season lies between 0.6 and 0.9. Higher values are indicated for such crops as bananas, rice, coffee, and cocoa; lower values are indicated for citrus, grapes, sisal, and pineapple. In general, crops and varieties with a high yield potential also tend to be highly sensitive to moisture stress (as well as to other nonoptimal conditions involving fertility and soil aeration). On the other hand, hardy or tolerant crops and varieties are less vulnerable and may be preferable where water is lacking.

Actual evapotranspiration from a crop, designated E_a, is generally lower than the maximal for that crop (E_{max}), as it is constrained by the availability of soil moisture and the degree of canopy cover. The drier the soil moisture regime and the thinner the crop stand, the lower will be the actual evapotranspiration and the smaller the fraction of E_a/E_{max}. The seasonal total of E_a can vary between 0.5 and 1.0 of E_{max} (depending on the amount of irrigation), and the relative yield generally varies correspondingly.

C. Water Use-Yield Relationships

As stated, when the supply of water is plentiful throughout the growing season, a cropped field can be expected to "evapotranspire" at a maximal rate E_{max} and to attain its full potential yield—provided, of course that no additional constraining factors such as pest infestations or nutrient deficiencies interfere. When water is limiting, crop water use falls below E_{max}. Consequently, crop yield is related functionally to crop water use (dictated by the water supply), but this relation may not be a simple one. Interest in the dependence of crop yields on water supply and use has grown in recent years because of the increasing scarcity and cost of water for irrigation.

The first comprehensive analysis of the relation between transpiration and yield was offered by de Wit (1958). He found that in climates with a large percentage of bright sunshine duration (i.e., arid regions) a relation such as

$$Y = m \ (T_a/E_o) \qquad [7]$$

exists between total dry matter yield Y and the ratio of actual transpiration T_a and potential (i.e., free-water) evaporation E_o, with m being the proportionality coefficient. In climates with a small percentage of bright sunshine duration (temperate regions) the relation

$$Y = n(T_a) \qquad [8]$$

was found, i.e., dry matter production is proportional to transpiration and hence the ratio of yield increment to water increment is constant. m and n were reported to be characteristic for each crop. The assumption of proportionality between yield and transpiration has served as the basis for recent efforts toward quantitative modeling of crop growth as influenced by soil water management (e.g., Hanks, 1974).

The relation of yield to evapotranspiration is more complicated than that of yield to transpiration, owing to the variable component of evaporation from the soil surface. An empirically based equation to predict yield from known values of evapotranspiration was given by Stewart et al. (1977) for dry matter production:

$$Y/Y_{max} = 1 - bE_d$$

$$= (1 - b) + b(E_a/E_{max}) \qquad [9]$$

wherein Y is dry matter yield, Y_{max} is maximum attainable yield (water not limiting), E_a actual evapotranspiration, E_{max} maximum evapotranspiration, and b the slope of the relative yield (Y/Y_{max}) vs. the "evapotranspirational deficit" ($E_d = 1 - E_a/E_{max}$). To predict yield from this equation, one must know E_a, E_{max}, Y_{max} and b. As pointed out by Hanks (1980), the ratio E_a/E_{max} where Y/Y_{max} is zero approximates the portion of E_a that is due to direct evaporation from the soil surface.

In a field study reported by Hillel and Guron (1973), total dry-matter yield of corn (*Zea mays* L.) per unit quantity of evapotranspiration increased twofold as the latter was increased 30% in the wetter irrigation regime (in which E_a was nearly equal to E_{max}). The wettest treatment yielded 2.3 times as much grain as the driest treatment, while consuming only 1.3 times as much water. However, when the fraction of E_a due to evaporation from the soil surface was taken into account, the yield appeared to be proportional to transpiration. In this study, grain yield was found to be proportional to dry matter yield, but this cannot be taken for granted in all cases.

Determination of the fraction of evapotranspiration due to evaporation from the soil (as apart from transpiration by the plants) can be important, since evaporation does not relate to plant growth and can therefore be considered a loss. Irrigation applied as spray (as well as rain) is intercepted in part by the foliage and then evaporates rapidly without entering the transpiration stream, so the direct evaporation of intercepted water—as well of soil moisture—can be considered a loss. These facts have a bearing on irrigation method and frequency. For irrigation systems that wet the entire surface, a high irrigation frequency causes greater evaporation from the repeatedly rewetted soil surface and from the canopy. On the other hand, too low an irrigation frequency may cause soil desiccation and plant water stress. Therefore, the most efficient utilization of water may result from some intermediate (optimal) irrigation frequency, even if it produces less than maximum yields. However, irrigation systems that wet only a small fraction of the soil surface, and moreover avoid wetting the foliage (e.g. drip), can result in a considerable reduction of direct evaporation losses even when applied at high frequency.

Doorenbos and Kassam (1979) proposed a method for evaluating the yield response of crops to applied water in terms of the following relationship:

$$1 - (Y/Y_{max}) = f[1 - (E_a/E_o)] \qquad [10]$$

where again Y is actual yield, Y_{max} is maximum attainable yield when full water requirements are met, f is a yield response factor, E_a is actual evapotranspiration, and E_o is potential evapotranspiration. Empirically obtained values of f have been tabulated for different crops and climatic regions.

In many cases, reported relationships between yield and water use pertain to aboveground dry-matter yield. If the yield of interest is, say, fruit or fiber, its relation to water use by the crop can be quite different. Moreover, the linear relationships found between yield and water use under limited water supply may not hold as PET is attained and water ceases to be a limiting factor in crop growth. Beyond the point where transpiration reaches its climatic limit, the promise of increasing production seems to lie in identifying and obviating any other possible environmental constraints, such as soil aeration, nutrient supply, or pest infestation. Finally, we come up against genetic constraints which cannot be obviated by any environmental factors. This is where crop breeders are called upon to work their wonders in developing new varieties with ever greater yield potential. The point to remember is that improved varieties may not succeed unless environmental conditions (chief among them the water regime) are optimized.

The question of what constitutes a "desirable" level of water use is still a matter of controversy. Three differing approaches can be defined (Vaux & Pruitt, 1983): (i) Agronomists are frequently interested in attaining maximum yields per unit area of land by ensuring that water does not become limiting. (ii) An alternative goal is to achieve the maximum degree of "water use efficiency," i.e., to maximize the yield response per unit amount of water applied. (iii) Yet another goal is advanced by economists who argue that water, to be used effectively, should only be applied up to the point where the benefit (revenue) derived from the last increment of water added still exceeds the price of that incremental application.

This issue is most relevant when the amount of irrigable land in a region or project exceeds the area that can be irrigated with the limited amount of water available. The problem then becomes how to spread the water over the land so as to achieve the highest returns. Concentrating the water over a limited area of intensive irrigation (where crop water requirements are fully met) is an approach likely to maximize yields per hectare and to minimize the investment of equipment, energy, and labor, which are often proportional to the area of land under production. The opposite strategy of spreading the water so as to "green-up" more land under a limited per-hectare water supply may produce a greater total yield for the project as a whole but involves additional outlays for water conveyance, tillage, fertilizer, seeds, etc. The problem of determining an optimal allocation of water can have no universal solution, since the economic considerations involved (i.e., the relative costs of the inputs of water, energy, machinery, labor, etc., vs. the income derivable from the yield) are specific in each case.

The problem is further complicated where rainfall may augment the irrigated water supply. Since the timing and quantity of rainfall are generally variable and uncertain, the conjunctive (supplementary) use of rainfall and irrigation becomes an exercise in statistical probability, and the entire management scheme, including the timing and quantities of water applications, must be flexible enough to adjust to changing conditions (Stewart & Musick, 1982). In principle, applying the water in small increments so as to wet the soil partially (rather than saturate the entire root zone) is a preferable tactic, because it allows some potential storage for possible rainfall during the season. Otherwise, even small rainstorms are likely to cause considerable runoff, erosion, and water-table rise. A flexible irrigation system also permits the irrigator to withold water in the days immediately following a drenching rain, so as to save on irrigation costs.

IV. IRRIGATION MANAGEMENT

A. Historical Concepts of Irrigation Management

The concept of soil moisture availability has long served as a criterion for irrigation management, but its limitations should be understood. The classical view, first conceived in the 1920s, was that soil moisture is equally available to most crops in their respective rooting zones throughout a definable range of soil wetness, from an upper limit called field capacity to a lower limit called permanent wilting point. Both of these values were considered to be characteristic and constant for each soil, and independent of crop or climate. The hypotehsis of equal availability rested on the postulate that plant functions remain practically unaffected by any decrease in soil wetness until the point of permanent wilting is reached, at which plant growth is abruptly curtailed (Veihmeyer & Hendrickson, 1927). Later investigators

(Richards & Wadleigh, 1952) produced evidence indicating that soil moisture availability to plants decreases progressively with decreasing soil wetness, and that a plant may suffer water stress and reduction of growth considerably before the permanent wilting point is reached. However, the older concept, being simpler to understand and apply, remained the accepted truism for many years.

In practice, the hypothesis of equal availability resulted in a regimen of infrequent irrigations designed to fill the soil to its field capacity, followed by intervals during which the irrigator awaited depletion to nearly the permanent wilting point before replenishing soil moisture by making up the deficit to field capacity. The traditional irrigation cycle thus consisted of a brief period of infiltration, followed by an extended period of soil moisture extraction by the crop. The soil's surface zone was saturated periodically, with a resulting inhibition of soil aeration, then allowed to desiccate excessively, again to the detriment of the roots in the upper layer. Practical limitations on the frequency of irrigation by the conventional irrigation methods made it difficult to test alternative regimens based on the continuous maintenance of a more nearly optimal level of soil moisture in the root zone.

The traditional mode of irrigation seemed to make good economic sense because many furrow, flood, and sprinkler systems have fixed costs associated with each application of water. With such systems, it is desirable to minimize the number of irrigations per season by increasing the interval of time between successive irrigations. For example, with a portable sprinkler system, the cost of tubing can be a primary consideration. In this case, it obviously pays to make maximal use of the equipment by minimizing the amount of tubes required per unit area irrigated, i.e., by rotating the available tubing from section to section so as to cover the greatest overall area possible before having to return to the same section for re-irrigation. In many cases, traditional irrigation schemes provide water on a fixed schedule and do not permit any change in frequency. In effect, the principle of minimizing irrigation frequency has meant making the greatest use of soil moisture storage by applying heavy irrigations and then maximizing the utilization of the stored water by waiting as long as possible (so as to deplete the soil "reservoir") prior to the re-irrigation.

The classical questions involved in irrigation management are when to irrigate and how much water to apply at each irrigation. To the first question, the traditional irrigator would reply: only when the available soil moisture is practically depleted. To the second question, the traditional answer would be: apply sufficient water to refill the soil root zone to field capacity. These simple (and simplistic) criteria are still followed in many places.

B. Modern Concepts of Irrigation Management

In the last decades, newer concepts of irrigation management have evolved. A fundamental change has taken place in our conception of soil-plant-water relations, leading to a more dynamic and holistic approach. The field is now perceived to be a unified system in which all processes are interdependent (Philip, 1966). In this unified system, called the Soil-Plant-Atmosphere Continuum (SPAC), the availability of soil moisture is not a property of the soil alone but, indeed, a function of the plant, soil, and climate in conjunction.

In principle, the rate of water uptake depends on the ability of the roots to absorb water from the soil with which they are in contact, as well as on the ability of the soil to supply and transmit water toward the roots at a rate sufficient to meet transpiration and growth requirements (Gardner, 1960; Huck & Hillel, 1983). These, in turn, depend on: (i) characteristics of the plant (rooting density, rooting depth, rate of root extension, and the physiological ability of the plant to maintain its vital functions for some time even while its own water potential decreases); (ii) properties of the soil (water retention and conductivity); and (iii) weather conditions, which dictate the rate at which the crop is required to transpire.

When all the controllable variables are optimized so as to avoid any occurrence of moisture stress during the growing season, crops may show a pronounced increase in yield potential. Under continuously favorable moisture conditions, improved varieties can attain their higher yield potential and respond to greater amounts of fertilizer and more intensive management practices. The desired effect can be produced by optimizing the quantity and frequency of irrigation, so as to avoid excessive watering which might impede aeration, leach out nutrients, waste water, or raise the water table. This optimization is difficult to achieve by the traditional surface irrigation methods still dominant in many places, and as a result, the new approach to irrigation management has not yet been applied widely, particularly in developing countries.

The advent of newer irrigation systems (permanent installations of low-intensity sprinklers, drippers or tricklers, microsprayers, bubblers, porous-tube subirrigators, etc.) has made it possible to establish and maintain soil moisture conditions at a more nearly optimal level than heretofore. Since these new systems are capable of delivering water to the soil in controllable, small quantities as often and as long as needed with no significant additional cost for the extra number of irrigations, some of the economic constraints to high-frequency irrigation have been lifted (Rawlins & Raats, 1975). With pressurized pipe irrigation systems that deliver water on demand to all parts of the field, the capital costs depend largely on pipe size, which in turn is governed by the maximum required delivery rate. That rate can be reduced by increasing the duration of each irrigation. Since it costs little more to use a system once it is permanently installed, the best use is almost continuous irrigation during periods of peak water demand.

As the frequency of irrigation increases, the infiltration period becomes a more important part of the irrigation cycle. Changing the irrigation cycle from an extraction-dominant to an infiltration-dominant process brings into play a different set of relationships governing soil moisture. For example, with small daily (rather than massive weekly or monthly) applications of water, the pulses of added water are damped down within a few centimeters or decimeters of the surface, so that flow below this point is essentially steady. By adjusting the rate and quantity of irrigation in accordance with E_o or E_{max}, soil hydraulic conductivity, and the soil solution's salt concentration, it is possible to control the moisture tension prevailing within the root zone, as well as also the amount of through-flow and leaching. Control of the crop's soil environment thus passes into the hands of the irrigator more completely than ever before. Properly managed, the new system can indeed save water while improving growth and increasing yield (Rawlins & Raats, 1975; Bucks et al., 1982).

Since a high-frequency irrigation system can be adjusted to supply water at nearly the exact rate required by the crop, one no longer need depend on the soil's ability to store water during long intervals between irrigations. The consequences of this fact are far-reaching. Soil physical properties such as water-holding capacity or field capacity, formerly considered decisive, are no longer major criteria for determining which soils are irrigable. New lands, until recently considered totally unsuited for irrigation, can now be brought into production. One outstanding example is the case of coarse sands and gravels, where moisture storage capacity is minimal and where the conveyance and spreading of water by surface flooding would involve inordinate losses because of excess and nonuniform seepage. Such soils can now be irrigated quite readily, even on sloping ground without expensive leveling, by means of drip or microsprayer systems.

With high-frequency irrigation, the farmer need no longer worry about when available soil moisture is depleted or when plants begin to suffer stress. Such situations can be avoided entirely. "Field capacity" and "permanent wilting point" lose their relevance. To the old question of when to irrigate, the modern irrigator answers: as frequently as possible, even daily. To the question of how much water to apply, he or she replies: "enough to meet current evaporative demand and to prevent salinization of the root zone." Evaporative demand can be determined by monitoring the weather (as well as the crop) and salinity can be monitored by sampling the soil solution.

By supplying water at a controlled rate that exceeds evapotranspiration by a measured amount, the irrigator can maintain a more-or-less steady rate of drainage out of the root zone. This is in marked contrast with the extreme variability of internal drainage under low-frequency, high-volume irrigation management, where its rate is large right after an irrigation but later declines to small values. With low-frequency irrigation, one attempts to compensate for the periodically high drainage rate by causing the crop to dry the soil profile between irrigations. Only with high-frequency, low-volume irrigation does it become possible to maintain a wet root zone while keeping the drainage rate low.

High-frequency systems have their own shortcomings, however. With less soil moisture storage, the crop depends almost entirely on the continuous operation on the system. Any interruption in that operation, caused by mechanical failure or temporary water shortage, results in immediate deprivation and might cause crop failure. The stringent requirement of continuously perfect operation is especially difficult to achieve where equipment and expertise are lacking. Hence, there is a need to simplify the modern systems so as to adapt them to the needs and circumstances of developing countries.

C. Irrigation Scheduling

Irrigation scheduling is the term commonly used to describe the procedure by which an irrigator determines the timing and quantity of water application. In principle, it is possible to schedule an irrigation program on the basis of monitoring the soil, the plant, and/or the microclimate. We shall review each of these alternative or complementary approaches in turn.

1. Monitoring the Soil

This is the traditional method of determining when and how much to irrigate. The idea is to observe the moisture reserve of the root zone as it gradually diminishes following each irrigation, so as to know when that reserve has been depleted to some level predetermined to serve as the minimum allowable level (i.e., the "time to refill" criterion). At that point, the irrigator is to apply the volume of water calculated to replenish the soil reservoir of the root zone to its "full" level (Campbell & Campbell, 1982).

A precondition to effective management of root zone soil moisture is to establish the rooting depth of the specific crop as it varies during the growing season. Typically, the seedling of an annual crop will begin to develop and extend its roots both vertically and laterally, until its lateral extent impinges upon a region that is already occupied and tapped by the roots of neighboring plants. Thereafter, the major direction of root extension tends to be downward. As it deepens, the root system also proliferates within the moist layers it has invaded and continues thus to ramify and utilize a growing volume of soil until it is constrained from extending any further by environmental, physiological, or genetic factors.

The volume of soil included within the rooting zone of a plant is the first determinant of the size of the soil moisture reservoir potentially available to it. That volume can be assessed by considering the areal density of the crop stand and the depth of root penetration. The latter can be observed by sampling the soil with an auger and examining the extracted samples for the presence of live roots. The phenology of root system development (i.e., the depth of root penetration over time during the growing season) should be determined for the major crops grown in a given region. This determination should be repeated several times and in several fields, as necessary, to assess the range of variation and to gain confidence in the data. Once obtained, the data on root development can be useful not only for the season in which the data were measured but for future seasons as well, provided that planting dates and environmental conditions remain more-or-less the same from season to season.

Once the rooting volume is known, the potential and actual water contents of the root zone can be determined, with a view to establishing the root zone's water deficit at any time. Concurrent measurements can be made of the soil moisture tension, in order to assess and forestall the hazard of subjecting the plants to excessive moisture stress. In addition, the salt concentration of the soil solution can be monitored to determine the adequacy, or inadequacy, of internal drainage and the possible need for leaching.

The potential (full reservoir) water content of the root zone is generally taken as equal to the field capacity, defined in practice as the water content of the specified volume of soil measurable 2 d after a thorough irrigation. The limitations of the field capacity concept are well known. Nevertheless, many irrigators still consider it a useful criterion, or index, of the limit of soil moisture storable for subsequent crop use in a specific depth of soil. Attempts to estimate the field capacity from indirect measurements in the laboratory (e.g., the soil wetness at a tension of 1/3 or 1/10 bar) can be misleading, as the process of internal drainage and the dynamics of soil moisture retention in the field are influenced by the composition of the entire profile rather than of any specific layer in the profile. Hence, the field capacity should be measured in the field itself.

The lower limit of soil moisture presumably available to crops was originally taken to be the permanent wilting point obtainable by growing indicator plants in containers filled with samples of the relevant soil and then observing the extraction of soil moisture by the plants until they wilt permanently (i.e., without being able to recover if placed overnight in a humid atmosphere). The now-prevalent practice in the field is never to allow the crop to deplete all the "available" water in the root zone, so as not to risk subjecting the plants to a level of stress that might cause an appreciable reduction in yield. Hence, the allowable depletion normally recommended is about 50% of the range between field capacity and the permanent wilting point.

Soil moisture content can be determined by using an auger to extract representative samples and then determining their weight loss upon drying in an oven. That is a most laborious, time consuming, and destructive method, because it requires much trampling and augering. Moreover, it yields values of soil wetness by weight rather than by volume, so the results must be multiplied by the soil's bulk density (mass of dry soil per unit bulk volume). This introduces yet another complication since bulk density, which must be measured separately, varies from layer to layer and from location to location in the field.

Much preferable to gravimetric sampling is the use of a neutron moisture meter, which permits the nondestructive sensing of soil wetness in a representative volume of soil. The measurements can be made repeatedly in the same locations and the results are available immediately in volumetric terms. The difficulty here is the relatively high cost of the instrument and of the acess tubing to be installed in the field, and the possible health hazard which might result from inappropriate use of the radioactive probe. A new method of monitoring soil moisture, called time-domain reflectometry may eventually supplant the neutron meter, but at this stage it is still experimental (Topp & Davis, 1985). Numerous other methods have been proposed over the years, but many of these require calibration, are difficult to use, or fail to provide accurate and reliable data over time.

Of fundamental interest is the measurement of soil moisture tension, in addition to moisture content. This can be done in the field by means of tensiometers. After an irrigation, as soil moisture is depleted by evaporation and/or root extraction, the tensiometers register an increase in tension and, if properly interpreted, can help to forecast when the plants might begin to suffer stress. Commercial tensiometers are available in various lengths, allowing the monitoring of soil moisture tension at various depths so as to characterize the root zone as a whole. However, tensiometers are far from being trouble-free instruments. They must be supervised constantly and serviced periodically. Using a sufficient number to characterize the field soil moisture regime properly can be fairly expensive, in terms of the initial investment, the subsequent cost of trained labor needed for operation and maintenance, and the possible obstruction of normal field operations.

2. Monitoring the Crop

In addition to or as an alternative to monitoring the soil, it is possible to monitor the water status of the plants. This can be done visually, as well as instrumentally, to detect early signs of thirst (incipient stress) in time to irrigate and thus prevent any significant reduction of yield. As in the case of soil moisture, numerous methods have been proposed over the years to monitor the state of water in the plant. Included among these are techniques to estimate transpiration using excised leaves, determinations of leaf tissue hydration with punched disks or intact leaves, observations of stomatal aperture, monitoring stem diameter, pressure-cell and psychrometric measurements of leaf water potential, and more. Perhaps the most comprehensive are measurements of total plant transpiration and photosynthesis, using portable tents with transparent plastic walls. Most of these techniques require specialized instrumentation and trained personnel and are difficult to carry out routinely.

One method that may become practical is the monitoring of crop canopy temperature by remote sensing with an infrared radiation thermometer (Jackson, 1982). The device can be hand-held or mounted on a stationary or moving mast, and it can be used to scan the temperature of the canopy over an area large enough to represent the field. This method is based on the fact that the transpiration process has a cooling effect, so the foliage of well-watered plants, transpiring freely at the maximal (potential) rate, are generally cooler than the ambient air. However, as soom as plants sense a deficit of water and begin to close their stomates, the leaves begin to warm up and may rise above the temperature of the surrounding air. The temperature difference between the leaves and the adjacent air increases as the state of stress intensifies, particularly in arid regions.

Still the most common way to monitor the crop is by the tried-and-true method of direct visual inspection. An experienced agronomist or farmer who knows his or her crop can detect early signs of thirst by the appearance of the foliage, especially during the period of peak transpirational demand (usually at midday). Young leaves are most sensitive—they begin to curl or become flaccid. When that happens, an irrigation is indeed overdue, so it is good to be able to discern plant thirst before such signs are too evident. Even at best, however, such detection only indicates when to irrigate, but gives no information on how much water to apply.

3. Monitoring the Weather

The idea here is to follow the meteorologically imposed evapotranspirational demand as it varies over time and to set the quantity of irrigation accordingly. The timing of irrigation can then be determined in reference to the soil's effective storage capacity or its moisture tension (or residual wetness), or in reference to the status of the crop.

The estimation of potential evapotranspiration, as defined and calculated by the Penman formula (see section 2–II.C) or its various derivatives is of great fundamental interest. It can also serve as an irrigation scheduling criterion in some situations, i.e., with a well-equipped and well-staffed central advisory station. However, this approach may be too difficult to apply in the context of small-scale farming in arid regions, particularly in the developing countries. Other agro-climatological methods that seem equally difficult to apply routinely to guide irrigated farming are radiation balance and lysimetry. It appears that the more practical of the microclimatological methods is the use of a pan evaporimeter to characterize the evaporative demand. Experience shows that, despite its shortcomings, this method (described in Section 2–II.C) can provide a basis for assessing crop water requirements if it is used with locally calibrated crop coefficients.

4. General Comments

Flexibility in irrigation scheduling is essential for efficient irrigation management. In some places, water is available most of the time but may become unavailable at unpredictable times. In such cases, a safety factor or buffer of several days is needed to protect the crops. This can be done by lowering the "refill" criterion, i.e., increasing the range of allowable depletion.

The whole issue of irrigation scheduling becomes moot, however, if water is never available on demand or it if is provided on an arbitrary, rigid schedule which is not subject to control and to adjustment in accordance with varying crop needs. Such is the case in some (perhaps too many) of the older irrigation districts where water is diverted from a canal to each farmer for a limited number of hours on specified dates during the growing season. In such a regime, the only choice left to the irrigator is to "take it or leave it," so most irrigators obviously "take it," and as an insuracne against possible future disruptions of delivery they tend to take as much as they can, even much beyond reasonable needs.

Often, an irrigator needs to manage not just one crop but a mix or array of crops in a centrally managed farm. The irrigator's concern is then to develop an overall irrigation schedule that optimizes the allocation of water throughout the season and ensures an adequate supply during the peak water use periods. This may mean that one or another of the crops will not receive water on its own optimal schedule. In view of the fact that some crops require more water than others at certain critical periods and less at other periods, it is necessary to plan the crop mix on the farm so as to balance the demand and the supply. Such a plan involves not only the selection of complementary crops but also the proper allocation of the fractional land areas among them.

D. Efficiency of Water Use and Water Conservation

The term "efficiency" is generally understood to be a measure of the output obtainable from a given input. Efficiency in irrigation can be defined in various ways, depending on the nature of the inputs and outputs considered (Hillel & Rawitz, 1972). For example, one may attempt to define as an economic criterion of efficiency the financial return in relation to the investment in water supply. One problem is that costs and prices vary widely from year to year and from place to place. Another problem is that some of the costs and benefits of irrigation cannot easily be quantified in tangible economic or financial terms, especially in places where a market economy is not yet fully developed. Often, only the short-term costs and benefits are discernible, whereas the long-term advantages or disadvantages are unknown a priori. How can we assign monetary value, for instance, to the possibility that an irrigation project might save the population of a region from the dire effects of a drought if the frequency or probability of droughts of varying degrees of severity cannot be determined?

Quite different from the strictly economic criterion of efficiency is the physiological one, i.e., the plant water-use efficiency. The criterion here is the amount of dry matter produced per unit volume of water taken up by the plant from the soil. As most of the water taken up by plants in the field is transpired (in arid regions, 99% or more) while generally only a small fraction is retained, the plant water-use efficiency is in effect the reciprocal of what has long been known as the *transpiration ratio,* defined as the ratio of the amount of water transpired to the amount of dry matter produced (t/t). That ratio can run as high as 500 or even more in regions and seasons of high evaporativity.

What we shall refer to as the technical efficiency is what irrigation engineers call *irrigation efficiency.* It is generally defined as the net amount of water added to the root zone divided by the amount of water taken from some source. As such, this criterion of efficiency can be applied to complex regional projects, individual farms, or specific fields. In each case, the difference between the net amount of water added to the root zone and the amount withdrawn from the source represents the evaporative losses incurred in conveyance and the losses due to seepage, deep percolation below the root zone within the field, and to runoff from the field.

From the point of view of water use, some large-scale irrigation projects operate in an inherently inefficient way. In many of the surface irrigation schemes, one or a few farms may be allocated large flows representing the entire discharge of a lateral canal for a specified period of time. Where water is delivered to the consumer only at fixed times and charges are imposed per delivery regardless of the actual amount used, customers tend to take as much water as they can while they can. This often results in overirrigation, which not only wastes water but also causes project-wide problems connected with the disposal of return flow, waterlogging of soils, leaching of nutrients, and elevation of the water table requiring expensive drainage. Although it is difficult to arrive at reliable statistics, it has been estimated that the average irrigation efficiency in such schemes is probably well below 50% (and may be as low as 30%). Since it is a proven fact that, with proper management, it is possible to achieve irrigation efficiencies as high as 90%, there is obviously much room for improvement.

Particularly difficult to change are management practices that lead to deliberate waste, not necessarily because of insurmountable technical problems or lack of knowledge, but simply because it appears more convenient, or even more economical in the short run, to waste water rather than to apply proper management practices of strict water conservation. Such situations typically occur when the price of irrigation water is lower than the cost of labor or of the equipment needed to avoid overirrigation. Often the price of water does not reflect its true cost but is kept deliberately low by direct or indirect government subsidy, which can be self-defeating. (Incidently, the cost of water may be distorted even in the absence of government subsidy. For example, in the case of an operator drawing water from an aquifer in excess of the rate of natural recharge, the cost of pumping may be only a small fraction of the cost of replenishing the aquifer after it is depleted).

Where open and unlined distribution ditches are used, uncontrolled seepage and evaporation, as well as transpiration by riparian *phreatophytes,* can cause major losses of water. Even pipeline distribution systems do not always prevent loss. Leaky joints resulting from poor workmanship, corrosion, ill-maintained valves, or mechanical damage by farm machinery may cause large losses. Sometimes the damage is not immediately apparent, as when a buried pipe under pressure fails at night, with no one in attendance.

Surface runoff resulting from the excessive application of water ideally should not occur. Sprinkler irrigation systems should be designed to apply water at rates that never exceed soil infiltrability. In the case of gravity irrigation systems, however, it is often virtually impossible to achieve uniform water distribution over the field without incurring some runoff ("tail water"). Only when provision is made to collect irrigation and rainwater surpluses at the lower end of the field and guide them as controlled return flow can this water be considered anything but a loss.

Evaporative losses associated with water application include any evaporation from open water surfaces or border checks or furrows, evaporation of water droplets during their flight from sprinkler to ground surface, wind drift of droplets away from the target area, and evaporation from wetted crop canopies or from the wet soil immediately after irrigation. While some of these losses cannot be totally eliminated, most can be greatly reduced. Transpiration by weeds is also a largely preventable loss.

In the open field, little can be done to decrease transpiration if the conditions required for high yields are to be maintained. Attempts to use chemical sprays known as "antitranspirants" have failed, and the use of windbreaks to control wind movement above and through a crop stand does not always produce the desired effect economically.

It appears in many cases that the greatest promise for increasing water use efficiency lies in allowing the crop to transpire freely at the climatic limit. This can be achieved by alleviating any water shortages, while at the same time controlling all other processes of water loss and obviating the other environmental constraints to attainment of the full productive potential of the crop. Some of the new and superior varieties, in particular, can provide high yields only if water stress is eliminated and such other factors as soil fertility, aeration, salinity, and tilth are optimized. Plant diseases and pests may depress yields without a proportionate decrease in transpiration and water use. All management practices can thus influence the efficiency of water use in irrigation. The practice of irrigation, therefore, should not be regarded merely as the provision of water to thirsty crops, but as an integrated production system designed to maximize the efficiency of land, water, labor, machinery, and energy utilization.

REFERENCES

Blaney, H.F., and W.D. Criddle. 1950. Determining water requirements from climatological and irrigation data. U.S. Soil Conserv. Serv. Publ. 96. U.S. Gov. Print. Office, Washington, DC.

Bucks, D.A., F.S. Nakayama, and A.W. Warrick. 1982. Principles, practices, and potentialities of trickle (drip) irrigation. Adv. Irrig. 1:220–298.

Campbell, G.S., and M.D. Campbell. 1982. Irrigation scheduling using soil moisture measurements: theory and practice. Adv. Irrig. 1:25–85.

Doorenbos, J., and A.H. Kassam. 1979. Yield response to water. Irrig. Drain. Pap. 33. FAO, Rome.

Doorenbos, J., and W.O. Pruitt. 1977. Crop water requirements. Irrig. Drain. Pap. 24. FAO, Rome.

de Wit, C.T. 1958. Transpiration and crop yields. Versl. Landbouk. Onderz. 64. Wageningen, Netherlands.

Gardner, W.R. 1960. Dynamic aspects of water availability to plants. Soil Sci. 89:63–73.

Hanks, R.J. 1974. Model for predicting plant yield as influenced by water use. Agron. J. 66:660–665.

Hanks, R.J., and R.W. Hill. 1980. Modeling crop response to irrigation. Int. Irrig. Inf. Ctr. no. 6. Bet Dagan, Israel.

Hillel, D., and Y. Guron. 1973. Relation between evapotranspiration and maize yields. Water Resour. Res. 9:743–748.

Hillel, D., and E. Rawitz. 1972. Soil moisture conservation. p. 307–338. *In* Water deficits and plant growth. Academic Press, New York.

Hsiao, T.C. 1973. Plant responses to water stress. Annu. Rev. Plant Physiol. 24:519–570.

Hsiao, T.C., and E. Acevedo. 1974. Plant responses to water deficits, water use efficiency, and drought resistance. Agric. Meteorol. 14:59–84.

Huck, M.G., and D. Hillel. 1983. A model of root growth and water uptake accounting for photosynthesis, respiration, transpiration, and soil hydraulics. Adv. Irrig. 2:273–335.

Jackson, R.D. 1982. Canopy temperature and crop water stress. Adv. Irrig. 1:43–86.

Kramer, P.J. 1983. Water relations of plants. Academic Press, New York.

Monteith, J.L. 1980. The development and extension of Penman's evaporation formula. p. 247–253. *In* D. Hillel. Applications of soil physics. Academic Press, New York.

Penman, H.L. 1948. Natural evaporation from open water, bare soil, and grass. Proc. R. Soc. London Ser. A 193:120–146.

Penman, H.L. 1956. Evaporation: An introductory survey. Neth. J. Agric. Sci. 4:9–29.

Philip, J.R. 1966. Plant water relations: Some physical aspects. Annu. Rev. Plant Physiol. 17:245–268.

Rawlins, S.L., and P.A.C. Raats. 1975. Prospects for high frequency irrigation. Science 188:604–610.

Richards, L.A., and C.H. Wadleigh. 1952. Soil water and plant growth. *In* Soil physical conditions and plant growth. Agronomy 2:73–251.

Sellers, W.D. 1965. Physical climatology. Univ. Chicago Press, Chicago.

Stewart, B.A., and J.T. Musick. 1982. Conjunctive use of rainfall and irrigation in semiarid regions. Adv. Irrig. 1:1–24.

Stewart, J.I. (ed.). 1977. Optimizing crop production through control of water and salinity levels in the soil. Utah Water Res. Lab., Logan.

Szeicz, G., C.H.M. van Bavel, and S. Takami. 1973. Stomatal factor in water use and dry matter production by sorghum. Agric. Meteorol. 12:361–389.

Tanner, C.B. 1960. Energy balance approach to evapotranspiration from crops. Soil Sci. Soc. Am. Proc. 24:1–9.

Topp, G.C., and J.L. Davis. 1985. Time-domain reflectometry and its application to irrigation scheduling. Adv. Irrig. 3:107–129.

van Bavel, C.H.M. 1966. Potential evapotranspiration: The combination concept and its experimental verification. Water Resour. Res. 2:455–467.

van Bavel, C.H.M., and D. Hillel. 1976. Calculating potential and actual evaporation from a bare soil by simulation of concurrent flow of water and heat. Agric. Meteorol. 17:453–476.

Vaux, H.J., and W.O. Pruitt. 1983. Crop-water production functions. Adv. Irrig. 2:61–69.

Veihmeyer, F.J., and A.J. Hendrickson. 1927. Soil moisture conditions in relation to plant growth. Plant Physiol. 2:71–78.

3 Irrigation Trends in World Agriculture

M. E. JENSEN

USDA-ARS
Fort Collins, Colorado

W. R. RANGELEY

Rangeley Associates
Berkshire, Great Britain

P. J. DIELEMAN

FAO
Rome, Italy

Irrigation enables farmers to eliminate or minimize the effects of insufficient soil water on plant growth. Regional or national average crop yields on irrigated lands typically are two to three times higher than average yields on rain-fed lands. Irrigation enables control of the quality of the marketable component of high-value crops as needed. Irrigation development over the past 30 yr has played a major role in increasing food production in developing countries whose populations have been increasing at annual rates of 2 to 4%. Irrigation is a key factor in stabilizing national and regional food and fiber production.

Irrigation reduces the risks that farmers take in their seasonal investment in land preparation, seed, fertilizer, and weed control. Irrigation enables timely planting in warm arid areas so that crops will mature and be ready to market when prices are highest and/or the market demands are at their peak for specialized crops. Irrigation is essential for many specialized crops that require both arid climates and water control throughout their growing cycles.

Irrigation projects in most developing countries are government initiated and sponsored. They are planned and facilities are constructed for several main purposes. First, irrigation projects usually provide food security (World Bank, 1986b). Second, in many developing countries, agriculture makes a major contribution to exports that generate foreign currency needed to pay for other essential imports. Thus, national views on the justification of irrigation development are often different from private views in which the driving force is competition in a free market. Private irrigation development in a free economy must compete economically with rain-fed agriculture.

I. COMMON PROBLEMS

Irrigation has been practiced for over two millenia. History has documented long-term successes, but more often, failures have been recorded and publicized. Irrigated agriculture has been successful for centuries in many areas (ICID, 1981, 1984). Some of the Iranian ganats, a 3000-yr-old technique for developing groundwater, are still being used today (Kuros, 1984). Earth dams were constructed in the second and third centuries in Japan for irrigation of rice (Takase, 1967). Some of these are still being used (Takase, 1987, personal communication). Similarly some "tanks" constructed in Sri Lanka centuries ago are still used. Irrigated agriculture can be permanent if the basic principles of water management and salinity control are recognized and required management principles are implemented and continued on a timely basis. The most common irrigation problems summarized in this section can be prevented or controlled by implementing timely corrective measures.

A. Waterlogging and Soil Salination

Probably the greatest technical cause of declining agricultural production on irrigated land, or irrigation failure, is waterlogging and salination of the soil in arid and semiarid regions. Waterlogging and salination are not inevitable. The primary cause of waterlogging is excessive water input into a system that has finite storage and limited natural drainage capacities. Excess water recharges the groundwater underlying irrigated land, causing the water table to rise from its previous natural or equilibrium state. Typically, the water table rises in proportion to the recharge rate until plants extract water from the capillary fringe, water reaches the surface by capillary action which increases annual evaporation, and seepage rates from canals decrease because of smaller hydraulic gradients.

A classic example of the rise in the water table following irrigation expansion is illustrated in the world's largest contiguous irrigated area in Pakistan (Fig. 3–1). The depth to the water table between the rivers (doabs) from pre-irrigated time, about 1860, to the early 1960s decreased at a relatively constant rate until the water table approached the land surface (Greenman et al., 1967). The total rise ranged from 20 to 30 m over this 80 to 100-yr period. In one region, the water table rose nearly linearly from 1920 to 1950 with only a slight decrease in the rate, demonstrating that seepage from the irrigation distribution system was the primary cause (Awan, 1984). About 50% of the water diverted into the canals goes to the groundwater by seepage and deep percolation (Ahmad, 1986). The warnings of impending waterlogging and salinity problems were evident for as long as 50 yr but little was done to implement corrective actions, particularly reducing seepage, until the problem reached critical stages. Salinity control and reclamation projects were initiated in Pakistan in about 1960. In 1986, Aziz (1986) estimated that about 10 million ha of cultivatable land were waterlogged.

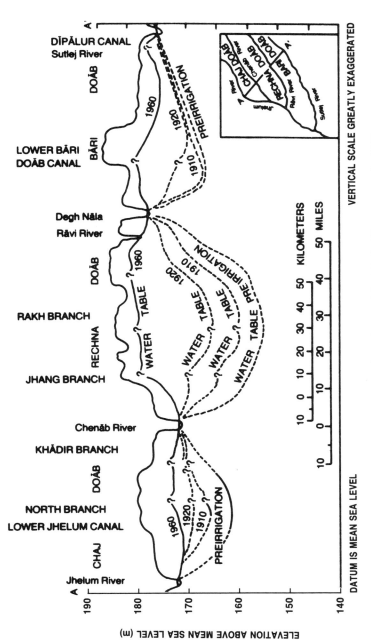

Fig. 3–1. Water table profiles, Chaj, Rechna, and Bari doabs (Greenman et al., 1967).

Waterlogging and salination problems are not unique to developing countries. Drainage problems developed on the west side of the Central Valley of California soon after irrigation began in the 1870s (Nelson & Johnston, 1984). By 1900, extensive areas were abandoned because of alkalinity and salinity problems. Irrigation continued to expand with some water table control provided by deep wells and deep open drains. Use of subsurface drains expanded in the 1950s. A detailed report of the drainage problems was prepared by Letey et al. (1986). Parts of that report are abstracted in the following paragraphs.

The Central Valley Project—involving dams, canals, drains, and other structures—was proposed in the 1930s. Initially, water was derived mainly from the underlying aquifer which began to decline from 1 to 9 m/yr. By 1950, basin-wide salt balance problems became apparent and the 1956 San Luis Feasibility Study acknowledged that drainage systems would be needed. Farmers began installing subsurface drainage systems. A main drain was planned by the California Department of Water Resources (DWR). Investigations were initiated by DWR on the extent of the drainage problems from 1957 to 1965. The DWR director announced that California would participate in a Federal-State single 470-km drain (San Luis Drain) from Bakersfield to the San Francisco Bay area. The San Luis Drain construction was started in 1968 and was completed to a regulating reservoir known as the Kesterson Reservoir. This reservoir involved a wildlife habitat that was secondary to drain water regulations. The drain was never completed to the Bay area.

Kesterson began receiving irrigation runoff water in 1973 and subsurface drainage water in 1978. Tile drain water became the sole source of water in 1981. In 1983, the high incidence of bird deformities resulted from the selenium that was present in the drainage water. In 1985, a court order was issued to prevent agricultural effluent from entering the Kesterson Reservoir. In 1986, all subsurface drains were plugged, terminating all drainage into the reservoir.

The Westlands Water District, which was organized in 1952, lacked adequate drainage water disposal capacity. In 1983, more than 10 000 ha had a perched saline water table within 1.5 m from the ground surface. An additional 47 500 ha had a perched water table between 1.5 and 3 m below the surface and 36 800 ha had a water table between 3 and 6 m from the surface. The 1983 survey showed that additional drainage was needed and completion of the San Luis Drain to the Bay area was expected. However, because of the closure of the drain to the Kesterson Reservoir, the Westlands farmers now are required to greatly improve irrigation management to minimize the quantity of drainage effluent that is unsuitable for reuse. The alternative disposal options are expensive. In this case, canals are lined and seepage is minimal. Nonuniform and excessive irrigation applications are the main sources of drainage water. Investments in higher-cost irrigation systems and improved irrigation scheduling techniques that enable more uniform and accurate applications of irrigations now must be weighed against the costs of drainage water disposal.

The Pakistan and California examples illustrate that long-range planning and implementing timely water table control actions are essential for a sustained, successful irrigation project. Also, earlier improvements in canal and on-farm systems and management may not only have increased and sustained the productivity of some projects, but may have reduced the later needs for an extensive and expensive drainage system.

Waterlogging and salinity can be prevented or reduced by either or a combination of two basic techniques: (i) reducing the application of water in excess of crop needs and reducing seepage from canals and (ii) supplementing the natural drainage capacity by constructing tubewells or subsurface drains. Once waterlogging has occurred, enhanced drainage by vertical and/or horizontal methods is essential to regain the productive capacity that has been lost.

B. Disposal of Drainage Effluent

On large irrigation projects with small topographic gradients, disposal of saline drainage water is often a problem. In the past, saline effluent was discharged directly into the rivers. Today, because of increasing competition for good quality water, saline drain waters are being diverted to evaporation ponds (Collett & Earl, 1984) and where feasible, separate drainage channels to the ocean or salt sinks have been constructed (Effertz et al., 1984; McNicoll & Abernethy, 1985). For example, an outlet drain was constructed to enable Mexico to bypass all or a portion of the Wellton-Mohawk drainage around Morelos Dam on the Colorado River. The Left Bank Outfall Drain is being constructed in Pakistan and is expected to benefit some 525 000 ha. In some cases, the saline effluent is mixed with the supply water for reuse or released to the river only during high river flows as in the Murray River basin in Australia (Collett & Earl, 1984). The occurrence of materials like selenium in drainage waters further complicates this problem.

C. Low Crop Yields

In many developing countries, crop yields per unit of irrigated land are low (Table 3–1). The causes of low yields usually are complex and often are the results of both technical and nontechnical factors.

A major irrigation factor that adversely affects crop yields is the untimely delivery of irrigation water. Yields are reduced when the amount of water needed by a crop between irrigations is greater than that which can be extracted from the soil because of limited root systems or soil water-holding capacities. Substantial reductions in yield due to plant water stress at critical growth stages may occur even though the total amount of water delivered during the cropping season may be adequate. Basically, water must be made available to farms in proportion to the average rate of evapotranspiration (ET) expected for well-watered crops that is not supplied by rainfall. An example of crop growth and the average rate of reference crop ET and average crop ET on a large-scale irrigated area in southern Idaho is illustrated in Fig.

Table 3-1. Average crop yields† in People's Republic of China (PRC), India, Pakistan, Egypt, Australia, Mexico, and the USA. Source: Agric. Statistics, USDA, 1985.

Crop	1984 to 1985 Data						
	PRC	India	Pakistan	Egypt	Australia	Mexico	USA
	t/ha						
Wheat (*Triticum aestivum* L.)	2.99	1.85	1.49	3.70	1.52‡	4.42	2.61‡
Rice (*Oryza sativa* L.)	5.51	2.17	2.57	5.66	6.31	2.89	5.52
Cotton (*Gossypium hirsutum* L.)	0.893	0.193	0.429	0.931	--	0.850	0.673
Sugarcane (*Saccharum officinarum* L.)	59.5	55.0	38.9	78.5	79.5	79.2	65.1

† Some of the yields differ from those reported by FAO (1986a).
‡ Much of the wheat is grown under rain-fed conditions.

3-2. Delivery of water on a fixed schedule and rate throughout the season for this project will produce excesses and/or deficits during the growing season (Brockway et al., 1985).

Other technical factors contributing to low crop yields are inadequate plant nutrition, inadequate weed control, and damaging plant pests and diseases. Planning and implementing measures to correct low crop yields first require identifying the principal causes or constraints. Principal causes usually are not apparent when reported crop yields and irrigation inputs are integrated over large areas. The real causes of low yields either go undetected, or are

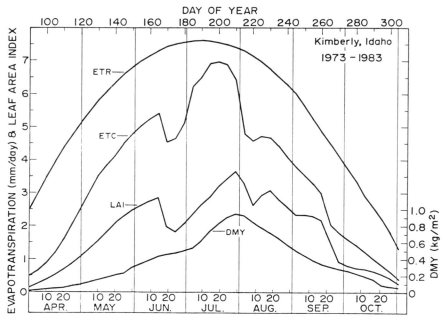

Fig. 3-2. Mean curves of daily reference evapotranspiration (ETR), crop evapotranspiration (ETC), leaf-area-index (LAI), and dry-matter yield (DMY).

erroneously attributed to other factors. Corrective actions implemented under these conditions may have limited impact on improving yields.

D. Erosion and Sedimentation

Erosion and sedimentation of reservoirs and channels are believed to have caused failures of some ancient irrigation systems (Fukuda, 1976). Erosion in water source areas and the resulting sedimentation have become major problems affecting water storage and distribution systems. Increased deforestation and heavy grazing have accelerated the rate of erosion in watersheds. Accelerated erosion in water source areas has permanently damaged many water storage reservoirs. The permanence of some irrigation projects will require greatly reduced erosion in water source areas.

Erosion on undulating irrigated lands also reduces productivity of soils. Carter et al. (1985) and Carter (1986), in a study in southern Idaho, showed that 75% of the fields now have whitish subsoils exposed on their upper ends caused by erosion after 80 seasons of furrow irrigation. The original topsoil thickness was about 38 cm. Some fields had lost all of the topsoil and even some of the subsoil near the upper end. Most fields had lost about 20 cm of topsoil. Topsoil thickness on downslope parts had increased to 60 to 150 cm. With about one-third of the fields showing whitish subsoils, crop yields today are estimated to be only 75% of what they could have been without erosion.

E. Socio-economic and Institutional Issues

Good irrigation designs and facilities do not guarantee optimum crop production. To make irrigation development successful, conditions must be met that permit the adequate use and management of the schemes.

In many developing countries, these conditions are not available or are insufficient. An inadequate number of trained staff is one of the greatest limitations to irrigation development and management in sub-Saharan Africa. Poor management has increasingly become the focus of attention in recent years. Inefficient water use, unequal distribution of water, and low production levels are directly related to inadequate management.

The planning, decision making, and implementation procedures of most irrigation agencies are centralized. The agency structures, are often "top-down" and, as a result, policies and procedures governing water supplies are not necessarily linked to the needs of crops and farmers. Operational adjustments to meet field level needs are not easily introduced. There is an urgent need to adapt existing administrative procedures at all levels, and to encourage farmers to initiate planning and action.

While many irrigation agencies have little responsibility downstream of the "watercourse" inlet, the farmers are usually poorly organized and equipped to operate and maintain the distribution channels they have in common. Recent experience in the Philippines indicates that farmers' involvement in the planning, construction, and management of the system may be positive and even a crucial factor in production (Food and Agriculture Organization of the United Nations, FAO, 1985a).

Irrigation systems in developing countries often suffer from lack of maintenance because the responsible agencies' maintenance capacities are limited and because available funding is limited. While some agencies depend entirely on the government for funding, others obtain all, or part of it, through the collection of water charges—an activity often beset with difficulties.

A management facet requiring particular attention is the provision of extension, credit, marketing, and related support services to the farmers. Experience in Africa shows that deficiencies in credit, supply, marketing and transport networks, and local manufacturing and servicing enterprises are important obstacles to the success of irrigation schemes (FAO, 1987). Farmers can be responsive to incentives as is evidenced by the recent increased agricultural production in the People's Republic of China (PRC). Typically, in countries with low yields, farmers receive low prices for their products. Examples in many countries show that, in the absence of adequate incentives, farmers are unlikely to give irrigated plots the attention needed to produce economically.

F. Human Health

In addition to water and soil, irrigation development may also have adverse impacts on human health. There are various diseases that are likely to increase in prevalence due to changes in the environment. They include such water-related diseases as malaria, onchocerciasis (river blindness), and schistosomiasis (bilharzia) in which the transmission chain includes an insect or, as in schistosomiasis, a snail. While occurring on all continents, these vector-borne diseases are particularly endemic in Africa (FAO, 1987).

Malaria is probably the most important of all water-related diseases. Of a world total of more than 1 million deaths caused annually, most are in tropical Africa. Schistosomiasis is predominantly associated with irrigated areas and associated water storages, canals, and drains, but natural waters, particularly marshes and swamps, also provide the snail host with suitable habitats and serve as transmission sites.

Since the early 1970s, most water development projects have been subject to a much closer environmental scrutiny and the requirements for environmental analyses are becoming more stringent. While not all answers to the problem are known, a major approach to disease control includes improved water management. This implies, specifically, preventing the occurrence of unnecessary water bodies, such as depression pools or standing water caused by seepage below storages or canals. This may require local drainage or land-fill—together with land shaping—to facilitate surface water runoff. To this should be added measures such as irrigation canal lining or raised flow velocities, where practicable, to discourage the attachment of snails to surfaces of watercourses or to aquatic weeds.

G. Reuse of Water

As consumptive water use increases, degradation of surface and ground-water quality is inevitable. Use of return flows and sewage effluents will become increasingly common. These issues also are receiving increased attention and must be considered in planning new projects and in rehabilitating old irrigation projects.

II. IRRIGATION DEVELOPMENT

A. Total Irrigated Land Area and Trends

The total gross area irrigated in the world expanded rapidly from 1950 to 1970, increasing from 94 million ha in 1950 to an estimated 198 million ha in 1970. Most of this expansion occurred in developing countries (Rangeley, 1986b, 1987). Today, there are about 220 million ha of irrigated land in the world. More than two-thirds of this total is in developing countries and over 60% is in Asia (Higgins et al., 1987).[1]

The World Bank has loaned more than $12 billion for some 320 projects and 90% of this has been within the past 14 yr (Le Moigne, 1986). During Fiscal Year 1985, World Bank lending for irrigation was about $1.1 billion, of which about 85% was for irrigation development in four Asian countries.

Expansion of irrigated land has occurred at about the same rate in most continents, but in absolute terms the increase from 1950 to 1970 has been most dramatic in Asia (Table 3–2). The irrigated area in Asia increased at an annual compounded rate of 3.5% from 1950 to 1970. Rangeley (1986a) estimated that the current rate of expansion in total irrigated area has decreased to about 1.5%/yr. The rate of expansion from 1980 to 1984, based on FAO statistics, is 1.0%/yr. From 1982 to 1984, it has decreased to 0.8%/yr (Higgins et al., 1987).

Rates of expansion during periods of most rapid development in three regions of the USA are summarized in Table 3–3. The expansion rate in the Pacific Northwest was only 1.67%/yr. This area is characterized by a steady increase in irrigated area since 1944. Most of the water is pumped from rivers with high lifts and low-cost electrical energy has been available until recently. Sprinkler irrigation was the main irrigation method. Expansion in the Southern Plains was 4.36%/yr. Here, irrigation using groundwater began expanding rapidly during the drought years of the 1950s. Furrow irrigation could be used without land leveling because of ideal topography and energy costs were low for pumping. Expansion in the Central Plains has been the most rapid—5.77% per yr. In the Central Plains, groundwater was the primary source of water and the advanced center pivot sprinkler system was the main irrigation method used. The center pivot system made it possible

---

[1] Lower figures published by FAO (1986a) compared to those by Rangeley (1986b) in Table 3–2 appear to be due to the treatment of gross irrigated area vs. land area irrigated as reported by ministries of agriculture.

Table 3-2. Land area irrigated. From Rangeley (1986b) and FAO (1986a).

Continent/country	From Rangeley (1986b)†			From FAO (1986a)‡		
	1950	1960	1970	1972	1980	1984
	million ha					
Europe (including part of USSR)	8	12	20			
Europe				11.4	14.7	15.6
USSR				12.0	17.5	19.5
Asia (including part of USSR)	66	100	132			
Asia				116.6	132.6	137.0
Africa§	4	5	9	7.3	9.3	10.4
North America	12	17	29			
North and Central America				21.8	27.9	27.4
South America	3	5	6	5.9	7.4	8.0
Australia and Pacific	1	1	2	1.6	1.7	1.9
Total	94	140¶	198	176.6	211.1	219.8
	%					
Rate of increase#		4.1	3.5		2.3	1.0

† Gross irrigated area.
‡ Land area provided with water whether irrigated once or several times during the year.
§ Excluding the Republic of South Africa.
¶ FAO reported 137.2 million ha in 1961 (Higgins et al., 1987).
# Percent compounded annually.

to irrigate land that otherwise was topographically unsuitable for irrigation with surface methods (Jensen, 1987).

Major reasons for the decreasing rate of expansion appear to be high cost of development and low prices for farm products and, in some areas, limited land and water resources. Lowering the cost of new irrigation development and increasing the productivity per unit area have become major issues. Rehabilitation of existing projects is favored over expansion into new areas because it builds on existing infrastructure, tradition, and expertise. Also, there is an increasing tendency to favor small-scale irrigation developments, for which operation and maintenance costs are less of a burden on government authorities and resources.

Table 3-3. Rates of expansion of irrigated areas in several regions of the USA during periods of most rapid development. From Jensen (1987).

Region	Period	Years	End of period area	Expansion rate†
			million ha	%
Pacific Northwest	1944–82	38	1.9	1.67
Southern Plains	1949–69	20	2.3	4.36
Central Plains	1964–82	18	3.5	5.77

† Expansion rate compounded annually.

B. Irrigated Land Relative to Total Agricultural Lands and Trends

Arable land as a percent of total land area varies greatly from 2% in Egypt to nearly 60% in India (Table 3-4). Essentially all of the arable land in Egypt is irrigated. A high percentage of the arable land in Asia is irrigated because agricultural development has been underway for a long time and rice is a major crop. Expansion of irrigated land and intensification are expected to continue for several more decades in countries in which a high proportion of food production is from irrigated land (Table 3-5). The highest relative rates will be in countries like Brazil where there is recent major emphasis on irrigation development. In Brazil, the total area cultivated in 1960 was 70 million ha. In 1962, only 0.546 million ha, or 0.8%, were irrigated (Framji et al., 1981). In 1984, 2.2 million ha were irrigated, or 2.9% of the current cultivated area (FAO, 1986a).

Table 3-4. Irrigated land relative to arable land in selected countries. From FAO (1985b).

Continent/Country	Arable land relative to total land, 1982	Irrigated land relative to arable land, 1982
	——————————— % ———————————	
Africa		
Egypt	2	100
Gambia	16	21
Morocco	58	15
South Africa	11	8
Sudan	5	15
North America		
Cuba	28	31
Dominican Republic	30	12
Mexico	12	22
Panama	8	5
USA	21	11
South America		
Argentina	13	5
Brazil	9	3
Guyana	3	25
Peru	3	34
Venezuela	4	8
Asia		
People's Republic of China	11	44
India	57	24
Japan	13	67
Pakistan	26	72
Thailand	37	18
Europe		
Bulgaria	38	29
France	34	6
Romania	46	23
Spain	41	15
United Kingdom	29	3
Australia and New Zealand		
Australia	6	4
New Zealand	2	36
USSR	10	8

Table 3-5. Estimated contribution of irrigated land to food supply in selected countries. From Rangeley (1986b).

Country	Food production on irrigated land relative to total production
	%
India	55
Pakistan	80
PRC	70
Indonesia	50
Chile	55
Peru	55

The continued rate of expansion of irrigated land in individual countries will be strongly dependent on the need to increase production to keep up with food demands caused by population increases and the limits of resources available to the country. Recently, Puli (1986) summarized the expansion of irrigated land and increased grain production in the PRC relative to population growth from 1950 to 1983 (Table 3-6). The PRC, which has one-fourth of the world's population, had a population growth rate during this period of 1.89% compounded annually. The increase in irrigated area during this time period was 3.25%/yr compared to the world increase, as calculated by Puli, of 2.44%. Grain production increased at a rate of 3.31%/yr. The PRC has recently imposed a stringent population control program because it has recognized the limitations of its production resources (0.07 ha per capita) and desires to maintain food self-sufficiency.

In contrast to the situation in the PRC, the population of Pakistan expanded from 33.8 million in 1951 to 92 million in 1984, an annual growth rate of 3.1%/yr. The total cultivated area, irrigated and rain fed, during this

Table 3-6. Expansion of irrigated land and increased grain production relative to population growth. From Puli (1986).

Parameters	Year		Ratio (1983/1950)	Rate of increase†
	1950	1983		
				%
Population, billions				
PRC	0.55	1.02	1.85	1.89‡
World	2.51	4.66	1.86	1.89‡
Irrigated area, million ha				
PRC	16	46	2.88	3.25
World	96	213	2.22	2.44
Grain production, million t				
PRC	132	387	2.93	3.31
World	623	1447	2.32	2.59
Grain/capita, kg				
PRC	239	379	1.59	1.41
World	248	310	1.25	0.68

† Compounded annually.
‡ World Bank (1986c) and Holden (1986) estimate the population growth rate to be about 1.2%/yr from 1980 to 2000.

Table 3–7. Expansion of irrigated land relative to population growth in Pakistan. From Ahmad (1986) and World Bank (1986c).

Parameters	Year 1951	1984	Ratio (1984/1951)	Rate of increase†
				%
Population, millions	33.8	92	2.72	3.08
Cultivated area, million ha	12.86	19.2‡	1.49‡	1.34‡
Cultivated area per capita, ha	0.38	0.21	0.55	−1.96
Irrigated area, million ha	12.9§	14.35	1.11¶	1.07
	1990	2000	Ratio (2000/1990)	Rate of increase
				%
Projected population, millions	108	138	1.28	2.5 (est.)
Cultivated area per capita, ha	0.19#	0.14#	0.78	−2.4 (est.)

† Compounded annually.
‡ 1981 data from Ahmad (1986).
§ Data from Pakistan National Committee (1983) and Framji et al. (1982).
¶ 1981/1971.
# Assuming 20 million ha cultivated area.

period increased from 12.86 to 19.20 million ha (Ahmad, 1986). The cultivated area per capita during these 33 yr decreased from 0.38 to 0.21 ha. The population in Pakistan is projected to be 108 million in 1990 and 138 million in the year 2000 (World Bank, 1986c). If it is assumed that the cultivated area cannot be changed significantly and remains at 20 million ha, the cultivated area per capita will decrease to 0.14 in the year 2000 (Table 3–7).

Countries like Pakistan are facing serious future problems. By significantly improving the irrigation system and its on-farm management, relief in the short term is possible because of low current yields, but this alone is not a long-term solution.

In Africa, exclusive of the Republic of South Africa, there are about 10 million ha of irrigated land which represent only 5% of the area under temporary and permanent cash crops. Over 6 million ha are in four countries: Egypt, Sudan, Madagascar, and Nigeria. The remaining 4 million ha are distributed over 47 countries (Dieleman, 1986). One reason for the small amount of irrigated land in Africa is the relatively low population density—0.17 person/ha in sub-Saharan Africa which is about one-tenth that of monsoon Asia and one-fifth that of India. Water resources are unevenly distributed. The Zaire basin, which occupies 16% of the surface area of sub-Saharan Africa, has 55% of the main discharge. Water-bearing formations underlying more than half of the continent contain only small, discontinuous aquifers. Groundwater supplies in extended geological formations are deep. The potential available resources are limited, but they could permit expansion of irrigation to 30 to 150 million ha of land. At present, the small percentage of the total cultivated area under irrigation produces 20% of the total production value, or three times that of a rain-fed hectare.

Production of staple food in sub-Saharan Africa, largely under rain-fed conditions in the decade 1970 to 1980 increased 1.6%/yr (World Bank, 1986b), which is about half the 1980 to 2000 population growth rate (World Bank, 1986a). The population in 1982 was estimated at 385 million (World Bank, 1986a). The estimated growth rates for the 1980 to 2000 period vary from 2 to over 4%/yr for a total population of 690 million in 2000 (Dieleman, 1986; World Bank, 1986a).

A technical calculation model, discussed at a Food and Agriculture Organization (FAO) Consultation on Irrigation in Africa in April 1986, suggests that Africa as a whole, and the sub-Saharan region, has sufficient rain-fed land resources to produce food for the estimated peak populations in the future, provided the level of input used is increased. The potential for food self-sufficiency can only be translated into practical realization through substantial rural and nonrural development. The estimate, however, includes some major assumptions, namely that all suitable land is cleared and cultivated to food crops and there is unrestricted movement of surplus food and labor. A more realistic picture is obtained through assessment of the potentials in individual countries. This shows that several countries in all regions face serious problems at the intermediate and high levels of input, particularly Mediterranean Africa which appears unable to support its population from its own resources after about 1990. In 1982, the five nations of this region imported enough cereals and other food to feed nearly half their present populations.

The role of irrigation in Africa is considered essential in achieving stated objectives of food self-sufficiency. However, irrigation in sub-Saharan Africa is relatively new and experience is limited, the data-base is inadequate, the infrastructure facilities are poorly developed, there is a shortage of local manufacturing and servicing enterprises, and the cost of development is high. These factors were not included in the model (FAO, 1987).

Sub-Saharan Africa is entering a new era in which its production can be based on the complementation of irrigated and rain-fed agriculture. Lessons learned from irrigation experience gained in various areas can be used to enhance irrigation development. Irrigation must not be developed in isolation, but should be integrated in the overall development plans. If existing farmers are to change successfully from nonirrigated to irrigated farming, they must be involved in all stages of irrigation development and management (Dieleman, 1986).

C. Associated Problems

Technically, as previously indicated, waterlogging and soil salination are the main problems associated with irrigation projects. The World Bank considers poor mangement, operation, and maintenance as its next major problem although both groups of problems usually are closely associated (Le Moigne, 1986). An estimated 50% of the Bank's projects are in need of rehabilitation, partly because maintenance is often viewed as low priority and postponable. Another reason is that the best engineering staffs usually are assigned to design and construction.

During the past 30 yr, there have been significant technical advances in irrigation and drainage in developed countries. However, a major issue is why these advances in technology have not had a greater impact in developing countries (Rangeley, 1986b). Why are new projects being built with about the same concepts and designs as those developed at the turn of the century? One explanation is that technological advances of recent years have not been appropriate in developing countries. Specific examples are the expansion of sprinkler and trickle or drip irrigation. They are either unsuited to local conditions or are not economically viable where labor costs, crop yields, and artificially set commodity prices and gross revenues per unit area are low (Rangeley, 1986b). Surface systems in most developing countries can be modernized and on-farm management using these systems can be greatly improved to achieve better performance.

Because of rapidly expanding populations in developing countries, the traditional rationale that modern technology is unsuitable for developing countries is no longer acceptable. Developing countries must adopt improved methods and practices if they are to achieve and retain food self-sufficiency. Training must be provided at all levels and supporting infrastructure must be established. In most cases, modern technologies need to be adapted to local conditions. This will require much more research than is at present being undertaken by developing countries. Adapting modern technology includes improving surface systems to enable operating them more efficiently.

D. Institutional and Social Processes

Irrigation development will be essential in some regions if food production is to keep up with population growth. Many difficulties are encountered when changing from subsistence to intensive agriculture. An example of the many problems that can be encountered is found in the massive program to develop the Senegal River Valley and the stakes riding on that venture (Walsh, 1986). Three countries are involved—Mali, Mauritania, and Senegal. Farmers historically practiced flood recession agriculture to supplement rain-fed production, although with low crop yields. Better agricultural technology is needed. Crop storage, marketing, and transportation services must be made available to cope with surplus production in parts of the region. Credit systems must be established. A system for transferring technology to farmers will be needed. The farmers will need to produce good crop yields twice a year. Social impacts of these improvements will be large. For example, women used to cultivate plots for their families. This role will change as men will make the decisions for the larger land areas.

Ministries of irrigation or agriculture are commonly involved in developing and controlling irrigation in countries where the irrigated subsector is large such as in Egypt and Sudan. The optimal governmental or regional authority, community or private organization, or combinations of these, for irrigation development in an individual country is not always a clear choice.

Rights to land and water resources and community relationships are all involved. Infrastructure for creating the necessary technical base, improving farmers' skills, providing extension services, developing professional, technical and support staffs, and transferring experience from successful schemes and pilot projects are essential to successful expansion of irrigated agriculture.

E. Financial Factors

The World Bank considers delays in execution and subsequent cost overruns to be a major financial problem associated with irrigation development (Le Moigne, 1986). One of the difficulties associated with recovering the costs of irrigation development is the low prices typically paid to farmers for their products in developing countries. Farmers need adequate prices for their products to enable them to amortize the investments made in land preparation, machinery, seed, fertilizer, etc. (Walsh, 1986). Basically, policy changes will be needed to reduce the subsidy to urban consumers. This will be difficult in many countries because of a long history of such practices and the low purchasing power of large parts of the urban population, coupled with the present trend toward low international prices.

Financial issues that must be considered at the project level include: (i) management options; (ii) cost reduction or transfer, and cost recovery; (iii) improving returns from the high fixed costs; (iv) selecting and changing technology; and (v) farmer involvement and government support (FAO, 1987).

Most developing countries have limited resources for operation and maintenance. Politically supportable water-charge policies must be established for investment cost recovery and system management and maintenance. Policy changes should lead to operating budgets that vary directly with the revenues generated by the operations of the agency. All farmers who benefit should contribute to the cost of providing water for irrigation. Least-cost operation and maintenance in relation to social, environmental, and production goals need to be developed. Generally, decentralization of operational and financial responsibilities is encouraged, as is the involvement of water users in establishing water rates. Clear linkages between financial responsibilities of water users for irrigation costs, and accountability to water users of those responsible for operation and maintenance are needed (FAO, 1986b).

F. Decreasing Water Quantities and Quality

Total, long-term annual national water supplies are essentially fixed. Water supplies that can be managed for irrigated agriculture include surface flow that coincides with crop water requirements, or flow that can be stored in surface or groundwater reservoirs and released or pumped as needed. Conjunctive surface and groundwater use will be essential. As populations continue to increase, water supplies for agriculture will decrease as competitive domestic and industry demands increase. Also, as the proportion of available water supply consumption increases, i.e., used in evaporation and transpiration, the salinity of the remaining surface water supply will increase.

As part of the groundwater supply is recycled, its salinity will also increase. The future quality of water available for agriculture will inevitably decrease in most large irrigated areas.

Groundwater must be managed to create aquifer space to store annual flows that now are being wasted. However, groundwater use in excess of annual recharge, or groundwater mining, cannot be considered as a long-term continuous supply. Some groundwater supplies are vast and will provide water for decades. But even extensive groundwater reservoirs that are being mined for large-scale agriculture can be depleted beyond economic pumping levels for agriculture in a few decades (Jensen, 1987).

III. STATE OF THE ART

The two major physical components of an irrigation project are: (i) the canal and lateral water distribution system and (ii) water courses and on-farm irrigation systems. Both components affect the productivity of an irrigation project and contribute to the major irrigation problems of waterlogging and soil salinity.

A. Physical Evaluation Parameters

Two basic parameters often are used to evaluate the effectiveness at which the two main project components are operated: (i) canal or water distribution efficiency and (ii) on-farm irrigation efficiency. On-farm irrigation efficiency may in turn be subdivided into several components.

1. Irrigation and Canal Efficiencies

Irrigation efficiency is a concept that is often misunderstood and misused. Misuse can lead to ineffective policies and decisions concerning irrigation systems and practices.

Irrigation efficiency (E_i) is the ratio of the volume of irrigation water beneficially used by a crop (V_{et}) as evapotranspiration (ET) in a specified area to the volume of irrigation water delivered to this area (V_d).

$$E_i = V_{et}/V_d. \qquad [1]$$

Irrigation efficiency as defined in Eq. [1] has been used for designing and evaluating irrigation systems for nearly 100 yr. The term, E_i, by itself is not adequate to estimate or predict available water supplies because potentially usable return flows are not included. Therefore, E_i alone cannot be used to estimate either true losses or savings of water, nor can E_i alone be used to estimate the amount of additional land that can be irrigated as a result of a change in E_i, except on individual land holdings.

The ET for a well-watered, closely spaced crop with a dense canopy will be about the same with any irrigation method. The ET for crops with incomplete canopies such as widely spaced young trees and vines will be less if they can be irrigated with drip and microsprinkler systems that do not wet the soil between the trees and vines. Although there is not a set irrigation efficiency associated with a given irrigation method, there is a potential, or an attainable efficiency, that can be achieved with each of the various irrigation methods. The actual irrigation efficiency achieved depends on both the on-farm system used and how it is managed and operated. Typically, E_i tends to be higher as more control is built into the on-farm irrigation system. However, the initial cost of a system usually increases as its intrinsic efficiency increases, and operating costs may increase depending on operating pressures and pumping lifts. Higher capital costs for more efficient systems may be offset by lower drainage costs or greatly delayed need for drainage systems.

2. Water Use Efficiency

The production of the marketable unit of a crop per unit of water consumed in ET is called *water use efficiency* (WUE) (also called water utilization efficiency). The magnitude of WUE for an individual crop indicates the overall effectiveness at which water is used for agricultural production in an irrigation project. It may be the best single index of how well a project with limited water supplies is managed in a given climate regime. It is affected mainly by crop yields and thus is also a measure of productivity.

The most common WUE expression relates the marketable crop yield per unit area to the water consumed in ET from planting to harvest. This term has been defined as net WUE (Jensen, 1984, 1987).

$$\text{WUEnet} = (\text{Marketable Yield in kg})/(\text{ET in cubic meters}) \quad [2]$$

When evaluating the effect of irrigation water alone, the increase in irrigated crop yield over rain-fed yield of the same crop in the same general area divided by the increase in ET is a measure of the WUE of irrigation water (Bos, 1980).

The ET from planting to harvest for a well-watered crop is not influenced much by the amount of fertilizer used or by different cultivars. New cultivars generally will use less water only if they have a shorter growing season than old cultivars. During the past 100 yr, the *harvest index* (proportion of dry matter produced that becomes the marketable product) of new cultivars has been increased through plant breeding. Essentially the same total dry matter is produced by new cultivars and, if the growing season has not been shortened, ET remains about the same. The resulting increases in WUE with new cultivars have been mainly from increases in the harvest index. The WUE assessments should be an integral part of overall project evaluations (Jensen, 1987).

B. Projects and Project Management

One of the primary factors affecting production levels on existing irrigated land is project management, particularly poor management. Major national and international efforts must address this problem if the full potential of the contribution of irrigation to agricultural production is to be achieved. Emphasis is now being placed on improving project management, as is indicated by numerous recent specialized training courses and publications. Increased resources have been directed to irrigation management by loan and aid organizations such as The World Bank, U.S. Agency for International Development (USAID) and the Inter-American Development Bank (Le Moigne, 1986; Jordan, 1987; FAO, 1985a, 1987; Nobe & Sampath, 1986; Coward, 1980).

In many developing countries, increasing management problems have adversely affected various components of the economy. Management standards have stagnated—or at times declined—because of a decline in management discipline. In irrigation, the primary factor appears to be associated with the long-established bureaucratic nature of irrigation agencies in many countries with inadequate integration of all disciplines needed to make crop production economical. These disciplines involve not only agriculture and engineering, but bankers, economists, accountants, and sociologists. Also, the irrigation system needs to be treated as a whole. Piecemeal strategies can cause large unbalances in economies (Rawlins, 1984).

Increased emphasis on user participation through cooperatives and associations is expected to enhance progress in irrigation management (Rangeley, 1986b). In 1984, the International Irrigation Management Institute (IIMI) was established with headquarters in Sri Lanka. The objective of IIMI is to find ways of improving the performance of irrigation systems and projects through management innovations. A branch of IIMI was established in Pakistan in 1986. Also in 1986, Colorado State University established the Colorado Institute for Irrigation Management. In Australia, the Centre for International Irrigation Training and Research was established at the University of Melbourne in October 1987.

Research in irrigation management will stress two approaches: (i) identify the constaints that currently control production and determine how they can be removed or relaxed to increase production without major modernization of physical works or equipment and (ii) identify new forms of modernization required to increase or maximize economic production. The second stage involves not only the hydraulic works, but all of the associated infrastructure including processing, transportation, and marketing. It is estimated that irrigation systems on some 150 million ha of irrigated land in the world need some form of rehabilitation and modernization (Rangeley, 1986b). Training at all levels within the project organization is a must. Training also must be provided on a continuing basis because of turnover in personnel and the need to retrain current personnel.

C. Water Storage, Distribution, and Conveyance System Management

Surface water storage management is not solely within the jurisdiction of irrigation project managers because many surface storage facilities have multipurpose objectives such as flood control, power generation, and recreation. Water released for irrigation is managed by conveyance system or project managers until that water is delivered to the farmer or to a watercourse.

Conveyance systems in many developing countries are designed to deliver water on a fixed schedule, such as the warabandi system used in India (Malhotra, 1982). The warabandi system evolved in the Indian continent as irrigation began expanding in the late 1880s. The government delivers water on a rotation basis to the heads of water courses from distributary canals. The rotation is usually on a 7-d basis. The farmers assume responsibility for the distribution of water from the water courses. The time that a land holder receives water usually is proportional to his land holding. This does not assure equitable delivery of water because the flow rate is not constant throughout the water course. These systems have little flexibility to enable changes in cropping sequences. Generally, they deliver water with little regard to seasonal variations in crop water requirements. Unfortunately, it is often the farmers who are accused of being inefficient under this mode of operation. In subhumid areas, these systems continue to operate and deliver water whether or not rains have occurred because surface storage or surface release channels do not exist. In a sense, these systems "dispose" of water and its silt load into watercourses and onto farmland. The rigidity of the system is partly justified because the canal system, which draws water from silt-laden rivers, is designed to operate in a stable regime—nonsilting and nonscouring. When water demands are low, the deliveries may be rotated independently of crop water needs. Most of the silt problems are transferred to farmers who must regularly clean water courses and relevel their small fields.

The warabandi system functions reasonably well when the cropping intensities are low because the farmers have some flexibility to adjust their cropping pattern and crop water needs to the constant flows. As cropping intensities rise, there is a greater need to provide for adjustments in the rates of water deliveries. Rehabilitation of watercourses without fundamental changes in system design and the organization of irrigation-related tasks presents long-term social problems (Merrey, 1986).

Increased cropping intensities and limited water resources also require systems to be more flexible. Automation will be playing an increasing role in achieving greater flexibility and efficiencies in future conveyance systems. Private tubewells provide some of the flexibility needed in water supplies that are not provided by a canal distribution system. Eventually, most continuous flow systems must be modified to perform more like demand systems if water is to be delivered to avoid plant water stress and consequence yield reductions. To avoid losses in crop production caused by water stress, water must be delivered in proportion to integrated water needs of crops being grown that are not supplied by rainfall as illustrated in Fig. 3–2 (Brockway et al., 1985). In this example, the decreases in mean crop evapotranspiration in June and August are due to the first and second cuttings of alfalfa. When sudden heavy rains occur, provisions for spilling water to drains must be provided. Continued water delivery to farms after rains and unnecessary irrigation leaches limited plant nutrients, and excessively wet soil may cause plant water stress due to limited soil oxygen.

D. Operation and Maintenance

An estimated 50% of the World Bank's irrigation projects are in need of rehabilitation. Maintenance often is viewed as low priority and easily postponable. The Bank believes that farmers should contribute to the costs of supplying them with water (Le Moigne, 1986). Prices received for agricultural crops must compensate for these production costs.

Modern, more sophisticated irrigation systems that enable more timely and uniform distribution of water over irrigated fields, as compared with traditional surface irrigation methods, will require higher standards of maintenance. Also, if new irrigation technology is to be adopted, training at all levels must be provided. Future improvements in operation and maintenance will require significant changes in priorities and resources assigned to this activity. Irrigation distribution systems must have more flexibility than at present. Target objectives should be to move toward "demand" systems in which farmers can obtain adequate water at proper times to minimize plant water stress and consequent decreased crop yields. This is the main purpose of irrigation and it should not be given second priority. However, in most irrigated areas of Asia, conversion to full-demand systems will not occur in the near term.

Computer software is available to facilitate calculating irrigation water demands on a real-time basis for both individual farms and groups of farms on lateral systems or water courses using real-time climatic data and crop conditions. Low-cost microcomputers also are available. This technology is simple to apply, but major constraints in implementation lie in the lack of flexibility in canal delivery systems and in training farmers to schedule irrigations according to crop needs.

Remodeling or rehabilitation of water distribution and control facilities will receive increasing emphasis in old irrigation systems. The alternative is continual deterioration of the system and decreased agricultural productivity. However, rehabilitation of existing physical facilities is unlikely to produce long-term improvements unless accompanied by equally important changes in the management system.

E. Land Drainage and Salinity Control

Where adequate natural drainage or a system of subsurface drainage has been installed, a net downward flux of water removes salts that accumulate as plants extract water from the soil (i.e., leaching). The amount of water required for leaching depends on the salinity of the irrigation water and the salinity tolerance of the crops to be grown. Persons or agencies responsible for constructing an irrigation project in today's world must take into account the immediate and eventual required drainage capacity to remove excess water. Plans must be developed to either control both canal and lateral seepage and on-farm deep percolation or to provide essential additional drainage capacity. Neglecting the consideration of drainage requirements is in some cases a result of resources control being removed from the responsible person or organization.

Drainage requirements can be reduced by using more efficient irrigation systems and by lining canals. The natural drainage capacity may be adequate, or the installation of expensive drains can be delayed for years, if canal seepage is reduced and deep percolation is minimized. Conjunctive surface and groundwater should be part of the system planning where canal losses and deep percolation are high. Conjunctive use also can provide more flexible water supplies and reduce drainage requirements.

Today, computer programs are available to facilitate the prediction of water-table levels between subsurface drains and crop response as a function of drain spacing, soil parameters, and normal climate including irrigations (Skaggs, 1980, 1987; Skaggs & Nasselzadeh-Tabrizi, 1983). Similarly, groundwater models are available to simulate changes in groundwater levels as a function of inputs and aquifer parameters and conjunctive use of surface and groundwater in a river basin (Labadie et al., 1986). Principles and guidelines for reclaiming salt-affected soils also are available (Ayers & Westcot, 1985; Hoffman, 1986; Rhoades, 1987).

F. Water Supply Systems and Transfers

Water supply and transfer systems have been used for centuries. Basically, they transfer water from areas of surplus to areas of deficit for domestic, municipal, and irrigation use. Water transfer systems have been controversial in the past and are expected to become more controversial in the future, especially with increased emphasis being placed on environmental issues.

Some of the more recent major systems and their primary purposes include the Central Valley Project of California which transfers surplus water from Northern California to the San Joaquin Valley and southern California, the Central Arizona Project on the Colorado River which lifts water from the Colorado River for transfer to the area around Phoenix and Tucson, the Tajo-Seguro Project in Spain which transfers water from areas of surplus in central Spain to the arid southeastern area, and the Link Canals in Pakistan which transfer water from western rivers to eastern rivers that are now used by India as a result of the 1960 treaty.

Major water supply and transfer systems were discussed in 1978 at a special session organized by the International Commission on Irrigation and Drainage. Papers describing water transfer projects in Australia, California, Canada, Federal Republic of Germany, France, India, Iraq, Morocco, Sudan, and the USSR were prepared and discussed (ICID, 1978; Teerink,

1978). Generally, these papers described a wide range of solutions to water supply distribution problems. Most authors recognized the criticism of these projects because of their impacts on human and natural environments (Teerink, 1978).

One of the major problems associated with water supply systems is the sediment load in the water. Some of the heaviest sediment loads are encountered in the Yellow River in the PRC. The Yellow River carries some 1.6 billion t of sediment annually. The river bed is 3 to 5 m above the alluvial plains which are subject to waterlogging and salinity. The sediment in the water is used for improving the land by a process called *warping*. Water is diverted through a sedimentation basin where about half of the sediment is deposited, raising the ground levels 2 to 3 m (Tong, 1986). Sedimentation in canals has been a major problem in Pakistan. A 2-yr research and data collection project was conducted as part of the Irrigation Systems Management Project (Bakker et al., 1986). The Tarbela Dam is estimated to have a half life of 60 yr because of sedimentation (Rangeley, 1986a).

G. Constraints to Improving Irrigation Water Management

Lee et al. (1985) recently evaluated the economics of improving irrigation efficiency where an exhaustible groundwater source is involved. They concluded that, under existing low prices, it is more economical to improve the application efficiency of existing systems (furrow or sprinkler) than to convert from one system to another. The net benefits of improving existing systems are large where pumping is the main source of water. However, many authors still suggest changing from gravity or surface systems to drip or sprinkler systems without considering the economic benefits from such a change or the increased training requirements for operating and managing a more complex system. With low prices paid for agricultural field crops, the benefits to the farmer from changing to drip or sprinkler systems often do not warrant the capital investment and increased operating and maintenance costs associated with a pressurized system.

Lack of sufficient numbers of adequately trained personnel, particularly in Africa, is a major constraint to improving irrigation management (FAO, 1986b). Attempts to introduce new technology without providing adequate training and maintenance requirements often result in failures.

H. Technology Transfer and Future Trends

During the past 10 yr, major efforts have been initiated by national and international organizations to transfer improved irrigation system and management technology to developing countries. For example, following successful water management programs in several countries in the 1970s, the FAO sponsored an Expert Consultation on Farm Water Management in May 1980 (FAO, 1980). In 1984, an Expert Consultation on Irrigation Water Management sponsored by the Government of Indonesia, FAO, and USAID was held in Indonesia (FAO, 1985a).

Power (1986), in summarizing his experiences as president of a large international firm specializing in irrigation equipment, found that in some countries there has been almost no input from farmers in planning some projects. The organizations responsible for developing water resources tend to be well funded and have considerable prestige and influence. Organizations responsible for on-farm activities are under funded and have little prestige and influence. He stated that the challenge is to improve communications between engineers and agriculturalists in implementing better irrigation systems and practices. Technology transfer is a slow process, but modern communications and transportation systems are improving this process.

Garrett (1986) reported that changes affecting people are not always greeted with universal enthusiasm and that engineers developing new agricultural technologies must consider the social implications of their work. Walsh (1984) feels that insufficient attention to local circumstances has led to project failures and that a better understanding of local conditions, particularly human conditions, may be needed before accepting project designs.

Tolba (1986), concerned that so little progress has been made in increasing agricultural production in Africa, suggested that if African governments agree on priority activities, they must commit themselves and their financial and human resources to the implementation of these activities. If they are to succeed, in addition to seeking the assistance of the international community, they must first enlist the understanding and then the active cooperation and assistance of the African people who are most directly and personally affected by the food production problems.

Why has the transfer and adoption of new irrigation technology been successful in some areas? Basically, there must be motivation to facilitate transfer and adoption of new technology. In the USA, the main purpose for adopting new technology has been to reduce the labor required for irrigation. The second purpose has been to reduce energy costs for pumping water. Escalating energy costs in the 1970s caused irrigators with large pumping lifts to change to more efficient systems that enable applying the targeted amount of water more uniformly and accurately. Other factors that have influenced the adoption of new technology are the type of crops grown, their market value, and the need to maintain better water control to assure high-quality produce that brings the highest prices. Changes in laws also have resulted in changes in irrigation systems.

Changes in agricultural technology occur at relatively slow rates except where simple changes are needed to adopt a new technology. There must be strong incentives and desires by irrigators and managers of irrigation projects to make improvements. Then, policies and institutions must help facilitate the adoption of new technology. Without the latter, little change in the status quo can be expected during the next few decades.

IV. POTENTIAL FOR CONTINUED IRRIGATION DEVELOPMENT

Irrigated land will continue to expand where climate now limits rainfed agriculture if there is a market for the agricultural products. In some countries, irrigation will expand if there is a need to attain food self-sufficiency, and adequate soil and water resources exist to permit agricultural expansion. In some developing countries where food production per capita is a critical issue and where existing resources are approaching full development, the only viable remaining alternative appears to be population control.

A. Developed Countries

Potential for continued irrigation development in the USA is typical of many developed countries. Many of the examples given can be found in most developed countries. In 1982, only 2.4% of the population lived on farms compared with 10% in 1959 and 5% in 1969 (USDC, 1986a). Increasing population in the USA is no longer a major driving force for increasing food production. Fertility rates have decreased from a post-World War II high of 3.8 births/woman to 1.8 where it has remained for the past 10 yr (Westoff, 1986). The U.S. population in the year 2025, assuming a fertility rate of 1.9 and a moderate net annual immigration of 450 000, is expected to increase from 231 million in 1982 and stabilize at about 310 million (Westoff, 1986). With the current surplus food production capacity, there will be little need to expand irrigation to assure food security, provided there are sufficient food storage and transportation capacities to provide a food supply during extended drought periods. Irrigation may be an alternative to increasing food storage capacity because it can provide food production during droughts, especially where groundwater storage is available. Irrigation costs also must be weighed against investments in transport infrastructure or international trade. Large buffer food stocks may be less cost effective than increased trade policy (World Bank, 1986b).

Irrigation will continue to play an important role in maintaining a reliable source of good-quality seed for those crops that require arid environment for disease control and in providing a continuous supply of specialized crops. Continued irrigation development will depend on availability of water resources, irrigation costs, and the domestic and export markets for the irrigated agricultural production. The future role of the U.S. Bureau of Reclamation will change from that of primarily a construction organization to that of a more efficient water resource management agency (Klostermeyer, 1986).

In the USA, irrigation development in the arid and mountain areas occurred rapidly at the beginning of this century and has remained relatively stable for the past few decades (Fig. 3–3). Expansion of irrigation started later in semiarid and humid regions of the USA and expansion in those areas is expected to continue for several more decades. Because of reduced exports during the past few years and low prices paid for agricultural crops, total irrigation expansion in the USA nearly ceased in the 1980s. The expansion during the past 20 yr has been almost exclusively by the private sector except for completion of some large Federal projects that were started many years ago. Some Federal projects, like the Garrison project in North Dakota, in which the initial stage was authorized in 1965 for 101 000 ha of irrigated land, have been scaled back in size because of environmental concerns and project costs (Krull, 1986).

Future water supplies for irrigation in the western USA will be declining because of increasing competition for water and depletion of groundwater in some areas. For example, the land area irrigated in the Texas High Plains, the main area of groundwater mining, has decreased during the past few years (Texas Water Development Board, 1986). Irrigation will continue to expand in semiarid and humid areas where groundwater supplies are readily available as farmers try to reduce their risks of losses due to droughts and increase net returns by producing higher-value and better-quality crops (Jensen, 1984). With a highly developed transportation system, high-value, specialized crops will continue to be grown in warm arid areas to satisfy market demands year round. Foreign imports from tropical areas and the Southern Hemisphere are beginning to satisfy some of the market demands for fresh fruit and vegetables during the winter season.

The quality of water available for irrigation will decline in some river basins such as the Colorado River as consumptive use of water continues

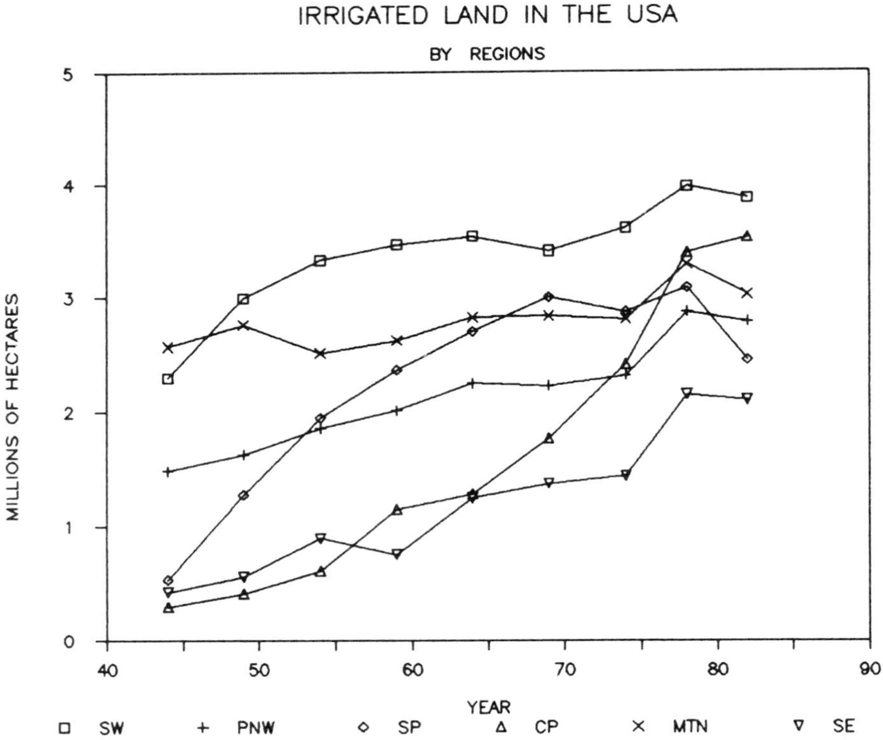

Fig. 3-3. Development of irrigated land by regions in the USA. States in each region are: SW, Arizona and California; PNW, Idaho, Oregon and Washington; SP, Oklahoma and Texas; CP, Kansas and Nebraska; MTN, Colorado, New Mexico, Nevada, Utah, and Wyoming; and SE, Alabama, Arkansas, Florida, Georgia, Mississippi, North Carolina and South Carolina. (From USDC, 1983.)

to increase. Increasing salinity causes damage to all water users, but the largest share of damage is to municipal water users (Kleinman & Brown, 1980). Similar water quality problems have been encountered in Australia on the Murray River (O'Brian, 1984; Collett & Earl, 1984). With the bulk of the population located at the lower end of the Murray River, restrictions have been imposed on Australian irrigators returning saline drainage water to the river. Evaporation ponds with a life expectancy of about 100 yr are being considered as an alternative for disposal of some saline drainage effluents.

In the Central Valley of California, even if the selenium problem is solved, drainage problems still exist. Today, about 0.4 million ha of irrigated land are threatened (Squires & Johnston, 1986). The 250 000-ha Westlands Water District currently does not have adequate drainage outlets. The san Luis Drain has not been extended beyond the Kesterson Reservoir to the Sacramento-San Joaquin Delta estuary, which is the natural drainage path for the area.

In Canada, there has been opposition to conversion of rain-fed farming to irrigated farming on the grounds that it is ecologically damaging, is unnecessary or excessively costly, and because it discriminates among farmers (Canada, 1985). These are examples of human and environmental concerns that will limit further irrigation expansion in many developed countries.

Although 10 yr ago some scientists predicted large decreases in irrigated land in the USA during the ensuing 20 or 30 yr, this has not occurred (NAS, 1976). A more recent study indicates that the decrease in irrigated land in the southern High Plains will be more than offset by increases in Nebraska in the Central Plains (High Plains Associates, 1982). When combined with the expansion of irrigation in the eastern states, there probably will continue to be an overall increase in irrigated land in the USA through 2030 even though there will be some decreases in arid areas and the southern High Plains (Jensen, 1984). Future expansion will involve mainly sprinkler and some drip irrigation systems.

In Canada, 0.65 million ha of land is irrigated today, compared with 0.26 million ha in 1965. Most of the irrigated land is in the Province of Alberta (0.53 million ha) (Thiessen, 1986). This expansion was made possible by large capital investments by the Alberta government and the irrigation districts under a cost-share program. Farmers also made large investments in sprinkler equipment. In 1984, Pohjakas (1984) reported there were 900 center pivot machines in Alberta since they were first introduced in 1964. In 1980, a new water resource program was established which included the development or rehabilitation and expansion of diversion works, main canals, and storage projects. Expenditures by the Province of Alberta over the past 5 yr have totalled about $250 million (Thiessen, 1986).

In Australia, the border check method of irrigation has been the traditional system. It is used for pastures or close-growing crops on gentle slopes. The contour bay system is the most economical system for nearly level land. It is used mainly for producing rice and for some pastures and winter cereals. About 80% of the irrigated land in Australia is irrigated by surface methods (Barrett & Purcell, 1987). Sprinkler irrigation is used on about 60% of the irrigated land in South Australia and in Queensland. Land forming using laser-control equipment has increased rapidly, resulting in improved layouts and better use of land, water, and labor. It has also been a factor in reducing waterlogging and soil salination (Framji et al., 1981).

Significant improvements in irrigation technology have been made during the past 30 yr and more improvements can be expected. First, regardless of the on-farm irrigation method used, better information on a real-time basis will be available to the irrigators to aid in decision making. Software programs for developing preseason normal irrigation schedules for the crops to be grown have been available since about 1970 (Jensen, 1972). These programs were designed for use by irrigation specialists. Recently, software programs for use on microcomputers have been developed for use by irrigators or farm managers. California has developed a real-time weather and evapotranspiration data network for use by irrigators or consultants, known as the California Irrigation Management Information System (CIMIS) (Snyder et al., 1985). About half of the growers in California have their own microcomputers that can be used to develop normal irrigation schedules or to access the CIMIS network for real-time weather and evapotranspiration data. Other states already have similar computer-generated information systems (Stegman and Coe, 1984; Thompson, 1979). To date, less than 10% of the farms use commercial scheduling services or media reports (USDC, 1986b). About half of the farms that use commercial scheduling services are in Kansas and Nebraska. About half of the farms that use media reports are in California, Idaho, and Nebraska. This is relatively new technology and its use is expected to expand.

On-farm irrigation systems will continue to be improved. Surface irrigation methods will continue to be used where the land has been leveled, water supplies are not limited, water costs are low, and waterlogging is not a problem. This is especially true where relatively low-value crops and forages are grown. Today, except on very large farms in the arid Southwest, most unlined farm ditches have been replaced with concrete-lined channels or pipelines. With recent developments in electronics and microcomputers, significant changes are being made in automating the control of flow to furrows, border strips, and level basins. For example, surge flow to borders and furrows has become a common practice in some areas of the USA and Australia. Future controllers will automatically adjust flow rates based on feedback sensors to compensate for the large changes in infiltration which occur from one irrigation to the next and the retardance caused by increasing vegetative growth during the season. More flow rate measurement, or volume integration, will be used as water costs increase.

Sprinklers will continue to be the primary irrigation method used in semiarid and humid regions. In 1984, 37% of the irrigated land in the USA was irrigated with sprinkler systems and 55% of sprinkler-irrigated land was irrigated with center pivot systems (USDC, 1986b). Gravity flow, or surface systems, are still used on 60% of the irrigated land. Drip or trickle systems are used on about 2% of the irrigated land. About half of this area is in California and Florida (Jensen, 1985).

Further improvements in irrigation uniformity will be made and automatic controls will be linked directly to real-time weather data and rainfall received on the area irrigated by each system. Use of drip or trickle irrigation systems on high-value crops will continue to expand where water costs are high and on sandy soils where limited irrigation is needed to produce high-quality crops. Increased use of buried drip systems is expected in semiarid and subhumid areas to supplement natural precipitation. Buried drip lines currently are used to irrigate sugarcane (*Saccharum officinarum* L.) in Hawaii and, to a limited extent cotton (*Gossypium hirsutum* L.) in Arizona and California.

Where physiographic conditions permit and where drainage is essential for part of the growing season, controlled subsurface drainage systems with automatic controls linked to weather forecasts will become more common because these systems usually cost less than sprinkler or drip systems (Fouss et al., 1986). Where water supplies are available, the same system is used for both drainage and subirrigation.

Application of fertilizers and other chemicals through the irrigation systems will become more common as improved techniques to achieve greater irrigation uniformity and application of precise amounts are implemented. This practice, coupled with plant tissue analyses, can reduce the loss of soluble fertilizers like N and reduce the amounts of pesticides that must be applied for pest control (Lyle & Bordovsky, 1986). Remote sensing for both scheduling irrigation and detecting nutrient deficiencies will become a common practice (Hatfield, 1983).

B. Developing Countries

In countries such as Egypt, almost all of the food production is from irrigated land. With a population growth rate of 2.2%/yr, the population in Egypt will increase from its present 48 million to 65 million in the year 2000 (World Bank, 1986a). Currently, Egypt imports about one-half of its food supplies. It is claimed that Egypt has the potential for increasing its agricultural production to eliminate this deficit in the next 10 to 20 yr (USAID/CAIRO, 1985).

Irrigation will continue to expand in many developing countries during the next 10 to 20 yr because it is a major way of increasing food production to meet demands, and the only way to achieve stability of production in areas having unreliable rainfall. However, with some exceptions in sub-Saharan Africa and in such countries as Brazil, possibilities to expand into new areas are fairly limited. Therefore, expansion will importantly imply the conversion of rain-fed land.

The high cost of irrigation development will contribute to further emphasis on achieving higher production per unit of land and water. The implications are likely to include a trend to emphasize small-scale irrigation and the development of groundwater resources. In countries that have already developed much of their water resources, like Pakistan and Egypt, production will need to be greatly increased on the existing irrigated lands. Rehabilitation of systems performing well below their potential will be a major tool to achieve this. Since poor management is often a major cause of the deterioration of the physical system, substantial institutional and organizational improvements will need to be undertaken along with physical rehabilitation.

At the same time, future development of irrigation projects will require more attention to the environmental and indigenous impacts (Norman, 1986). Organizational changes are expected to handle environmental and natural resources issues more effectively (Norman, 1986). The net effect of environmental and indigenous impacts on continued future irrigation development and agricultural productivity can only be a guess at this time. Requirements of greater consideration to these issues will probably raise the cost of irrigation projects and may cause greater delays in irrigation development.

In countries that do not have the land and water resources to expand irrigation to meet food demands, some control may need to be placed on increasing demand (population). Yet, population control is highly controversial. The PRC plans to bring down its present growth rate of 1.17%/yr to near 1% (Holden, 1986). Even with this population control program, it is estimated that the PRC's 1 billion population will reach 1.2 billion by the year 2000 (Keyfitz, 1984). Future agricultural planning must involve both production capacities and food demands. A majority of developing countries have been spending an increasing amount of their export earnings on food imports since the early 1970s (FAO, 1985b). In many countries, the gap between resources available and resources demanded is widening (Holden, 1986). A summary of the annual rate of change of food production relative to population growth for 105 developing countries for the period 1974 to 1984 is presented by FAO (1985b). Nearly half of the 105 developing countries had population growth rates that were greater than food production growth rates. Many of these countries are very small, however, compared to the PRC, India, and Pakistan where increases in production have been impressive.

V. IRRIGATION RESEARCH AND DEVELOPMENT

A. Current Status in Developing Countries

During the last 50 yr there has been little technological progress in irrigation in many developing countries. Large expansion in irrigated areas in developing countries has been based on planning and design concepts that were evolved at the beginning of the century. Evidence indicates that there has been inadequate investment in irrigation research and development within these countries or irrigation research has not been well planned and implemented. In contrast, the investment in the "green revolution" package and coordinated planning of associated research has advanced agronomic practices in some areas far ahead of the ability of the irrigation systems to provide the crop water requirements. In many countries, the irrigation systems represent a major basic component of the farm infrastructure.

Countries that have large irrigated areas have many institutions that deal with irrigation research, but these are strongly biased toward hydraulic engineering aspects of channel design, canal lining, diversion weirs on permeable foundations and tubewells, and toward hydrology. Considerable effort has gone into soil and land classification methods and reclamation of saline/alkaline soils, but less effort has been devoted to nontechnical issues associated with the prevention of soil salination. More recently, increased effort has been directed toward improving on-farm irrigation systems. Where commercial products are involved, the private sector in developed countries has been active in developing improved irrigation equipment such as the automatic water level control gates in France, automated volume flow controls in Israel, drip irrigation in Italy, and the center pivot sprinkler systems in the USA.

As a whole, world-wide research and development have had limited impact on the design of new irrigation projects involving traditional irrigation. Irrigation research tends to be site specific and thus the major impacts have been in developed countries. Exceptions are found in Brazil, where major progress has been made in adapting European and North American technology to suit local conditions.

B. Irrigation Research Needs

Regarding physical systems, research is needed to achieve: (i) improved water control to enable greater diversification in cropping patterns, especially in traditional rice-producing area and (ii) improved on-farm systems to enable applying lighter and more timely irrigations in order to minimize crop water stress and excess water applications.

A significant component of irrigation research should be focussed on project management, which presents the biggest constraint in increasing production levels. Only recently has irrigation management been treated as a research subject. Previously, it was generally believed that rigid application of conventional rules or practices that were effective in some projects could correct management deficiencies. Research into causes and effects and evaluation of new concepts has been initiated on a limited scale. This research has focussed decision makers' attention on issues involved and, as a result, some improvements are being made. The International Irrigation Management Institute is one of the more recently established institutions with the objective of developing ways to improve the performance of irrigation systems through management innovations.

An essential component in implementation of improved management is the development and dissemination of information on irrigation management. This will require training and retraining programs for new and existing personnel. Efforts on the development and dissemination of irrigation management information have been fairly limited compared to other facets of irrigated agriculture.

C. Irrigation Research Planning

The main purpose of irrigation and irrigation projects (schemes) is to increase and stabilize agricultural production. The production system involves plants, soils, and the crop and human environments. The primary resource inputs involve seeds, water, plant nutrients, and labor. The production system is operated mainly by farmers until the products are sold in a free market or to government-operated storage and distribution systems. The markets for the products are the processors, local people, and distant urban people who are the consumers. A supply and transportation system is required for the inputs and a storage transportation and marketing system is required for the products.

Planning and implementing an effective research program to improve production of irrigation projects will require inputs from all segments of the production and marketing system. Multidisciplinary and integrated research will be required. Research priorities must be developed. A research strategy for allocating available resources and an implementation plan must be prepared and implemented.

An example of irrigation research priorities recently developed in Australia illustrate the typical scope of irrigation-related research needs (von Mengersen et al., 1987). The four primary categories of irrigation-related research that were identified with specific topics in each are:

1. Physical processes (climatic factors, crop water uses, physical processes associated with salt and water movement, salinity, and waterlogging)
2. Crop agronomy (soil management, crop selection, input scheduling, water and energy use efficiency)
3. Irrigation design and equipment (soil characteristics, economic design, and comparison of irrigation methods and standards)
4. Community impacts of irrigation (history of irrigation, institutional structures, conjunctive use of water resources, simulation modeling, and technology transfer)

The above example clearly illustrates that needed irrigation-related research generally is beyond the mission and scope of most irrigation agencies. Successful and effective irrigation-related research should be an essential component of a national irrigation development or improvement program.

VI. CONCLUSION

The world area of irrigated land has expanded from 94 million ha in 1950 to about 220 million ha in 1984. The annual rate of expansion was 4.1, 3.5, and 2.3% for the periods 1950 to 1960, 1960 to 1970, and 1974 to 1984, respectively. The current rate of expansion is about 1% annually. Most of the expansion during the past 30 yr has been in developing countries as they try to increase food production to meet exponentially increasing food demands caused by exponentially increasing populations.

Irrigation is one of the oldest agricultural technologies that, when practiced according to basic and well-established principles, can enable continued, high agricultural production. When basic principles are ignored, the productivity of irrigated lands often declines. The principal technical causes of declining productivity are waterlogging and soil salinity. Water-logged and salinized soils can be reclaimed, but the costs in terms of lost production and physical systems usually are high. Better and more sophisticated water control and irrigation systems, along with greatly improved management, could delay, minimize, or even—in some cases—prevent these common problems.

Improved irrigation management will be indispensable for sustained agricultural production and to meet increasing food demands in developing countries. New programs are needed to develop improved technology and management practices. Increased and coordinated research, coupled with information dissemination and training programs, are needed to transfer new technologies from developed to developing countries and between developing countries. These steps are required to assure that new and rehabilitated projects will regain and sustain increased productivity or continue to improve productivity as better agricultural practices are implemented. Food storage and transport facilities must be developed and early warning systems relative to weather, crop, and range conditions must be improved.

In some developed countries, greater economies of agricultural production on rain-fed lands and improved transportation systems will enable more of the common field crops grown on these lands to be distributed to demand centers. Thus, as competition for available water resources for municipal, recreational, and other uses increases, irrigated lands will be used primarily to grow specialized crops. In developing countries, despite current indications of overproduction in Asia, irrigation development will continue in the long term in response to ever-increasing food demands.

REFERENCES

Ahmad, N. 1986. Planning for future water resources of Pakistan. p. 279–294. *In* Proc. Darves Bornoz. Spec. Sess., Natl. Cmte. of Pakistan, ICID, Vol. 2, Lahore, Pakistan. ICID, New Delhi.

Awan, N.M. 1984. Some technical and social aspects of water management in salinity control & reclamation project no. 1. Pakistan. Int. Comm. Irrig. Drain. Bull. 33(1):22–33.

Ayers, R.S., and D.W. Westcot. 1985. Water quality for agriculture. Irrig. Drain. Pap. 29, Rev. 1. FAO, Rome.

Aziz, S. 1986. Keynote address. Proc. Darves Bornoz Spec. Sess., Natl. Cmte. of Pakistan, ICID, Lahore, Pakistan. 4 Oct. ICID, New Delhi.

Bakker, B., H. Vermaas, and A.M. Choudri. 1986. Regime theories updated or outmoded. p. 147–175. *In* Proc. Darves Bornoz Spec. Sess., Natl. Cmte. of Pakistan, Lahore, Pakistan. 4 Oct. ICID, New Delhi.

Barrett, J.W.H., and J.D. Purcell. 1987. Design of surface irrigation systems. Irrig. Assoc. Austr. J. 2(4):17–21.

Bos, M.G. 1980. Irrigation efficiencies at crop production level. Int. Comm. Irrig. Drain. Bull. 29(2):18–25, 60.

Brockway, C.E., G.S. Johnson, J.L. Wright, and A.L. Comer. 1985. Remote sensing for irrigated crop water use, Phase 1. Tech. Complet. Rep., Univ. Idaho Water Resour. Inst.

Canada. 1985. Hearings about water, a synthesis of public hearings of the Inquiry on Federal Water Policy. Ottawa, Canada.

Carter, D.L. 1986. Effects of erosion on soil productivity. p. 1131–1138. *In* Proc. Water Forum '86. Long Beach, CA. 4–6 Aug. Vol. 2. Am. Soc. Civil Eng., New York.

Carter, D.L., R.D. Berg, and B.J. Sanders. 1985. The effect of furrow irrigation erosion on crop productivity. Soil Sci. Soc. Am. J. 49:207–211.

Collett, K.O., and G.C. Earl. 1984. Salinity control and drainage in northern Victoria (Australia). p. 315–323. *In* R.H. French (ed.) Salinity in watercourses and reservoirs. Butterworth Publishers, Boston.

Coward, E.W. 1980. Irrigation and agricultural development in Asia. Cornell Univ. Press, Ithaca.

Dieleman, P.J. 1986. Shared government-farmer management of irrigation resources: The potential for farmer participation. p. 347–369. *In* Proc. Darves Bornoz Spec. Session, Vol. 2, Lahore, Pakistan. ICID, New Delhi.

Effertz, R.J., W.K. Sidebottom, and M.D. Turley. 1984. Measures for reducing return flows from the Wellton-Mohawk irrigation and drainage district. p. 305–314. *In* R.H. French (ed.) Salinity in watercourses and reservoirs. Butterworth Publishers, Boston.

Food and Agricultural Organizaiton. 1980. Farm water management. Rep. on expert consultation, Beltsville, MD. 13–15 May. FAO, Rome.

Food and Agricultural Organization. 1985a. Participatory experiences in irrigation water management. Proc. expert consultation on irrigation water management, Yogyakarta and Bali, Indonesia. 16–22 July 1984.

Food and Agricultural Organization. 1985b. The state of food and agriculture 1984. FAO Agric. Ser. 18. FAO, Rome.

Food and Agricultural Organization. 1986a. FAO production yearbook. 1985, Vol. 39, FAO Stat. Ser. 70, and 1978, Vol. 32, FAO Stat. Ser. 22.

Food and Agricultural Organization. 1986b. Expert consultation on water charges. 22–26 Sept. FAO, Rome.

Food and Agricultural Organization. 1987. Consultation on irrigation in Africa. Proc., Irrig. Drain. Pap. 42. Lome, Togo.

Fouss, J.L., C.E. Carter, and J.S. Rogers. 1986. Simulated water table management by controlled-drainage based on forecasts. Trans. Am. Soc. Agric. Eng. 29:988–994.

Framji, K.K., B.C. Garg, and S.D.L. Luthra. 1981. p. i–cxxiii, 1–491. *In* Irrigation and drainage in the world. Vol. 1.

Framji, K.K., B.C. Garg, and S.D.L. Luthra. 1982. p. 493–1159. *In* Irrigation and drainage in the world. Vol. 2. ICID, New Delhi.

Fukuda, H. 1976. Irrigation in the world. Univ. Tokyo Press.

Garrett, R.E. 1986. Social impacts: assessing the "people factors" in agricultural technology. Agric. Eng. 67(7):15–18.

Greenman, D.W., V.W. Swarzenski, and G.D. Bennett. 1967. Groundwater hydrology of the Punjab, West Pakistan, with emphasis on problems caused by canal irrigation. U.S. Geol. Surv. Water Supply Pap. 1608-H. Washington, DC.

Hatfield, J.L. 1983. Evapotranspiration obtained from remote sensing methods. p. 345–429. *In* D. Hillel (ed.) Advances in irrigation. Vol. 2. Academic Press, New York.

Higgins, G.M., P.J. Dieleman, and C.L. Abernethy. 1987. Trends in irrigation development, and their implications for hydrologists and water resources engineers. J. Hydraul. Res. 25(3):393–406.

High Plains Associates. 1982. Six-state high plains Ogallala Aquifer regional resources study, summary. Rep. to U.S. Dep. Commerce. Camp Dresser & McKee Inc., Black & Veatch, and Arthur D. Little.

Hoffman, G.J. 1986. Guidelines for reclamation of salt-affected soils. Appl. Agric. Res. 1:65–72.

Holden, C. 1986. A revisionist look at population and growth. Science 231:1493–1494.

International Commission on Irrigation and Drainage. 1978. Mass transfer of water over long distances for regional development and its effects on human environment. Proc. Spec. Sess., 10th Congr., Int. Comm. Irrig. Drain., Athens, Greece. 24 May–4 June. ICID, New Delhi.

International Commission on Irrigation and Drainage. 1981. History of irrigation. Proc. Spec. Sess., 11th Congr., Int. Comm. Irrig. Drain., Grenoble, France. 31 Aug.–6 Sept. ICID, New Delhi.

International Commission on Irrigation and Drainage. 1984. History of irrigation. Proc. Spec. Sess., 12th Congr., Int. Comm. Irrig. Drain., Fort Collins, CO. ICID, New Delhi.

Jensen, M.E. 1972. Programming irrigation for greater efficiency. p. 133–162. *In* D. Hillel (ed.) Optimizing the soil physical environment toward greater crop yields. Academic Press, New York.

Jensen, M.E. 1984. Water resources technology and management. p. 142–146. *In* B.C. English et al. (ed.) Future agricultural technology and resource conservation. The Iowa State Univ. Press, Ames.

Jensen, M.E. 1985. Design and performance of irrigation and drainage systems—The window of success. Int. Comm. Irrig. Drain. Bull. 34(2):1–10.

Jensen, M.E. 1987. New technology related to water policy—engineering. p. 43–49. *In* W.R. Jordan (ed.) Proc. Water and Water Policy in World Food Supplies Conf., College Station, TX. 26–30 May. Texas A&M Univ. Press, College Station.

Jordan, W.R. (ed.). 1987. Water and water policy in world food supplies. Conf. Proc. Texas A&M Univ. Press, College Station.

Keyfitz, N. 1984. The population of China. Sci. Am. 250(2):38–47.

Kleinman, A.P., and F.B. Brown. 1980. Colorado River salinity: Economic impacts on agriculture, municipal, and industrial users. Water Power Resour. Serv. (U.S. Bur. of Reclamation) Denver, CO.

Klostermeyer, W.C. 1986. The Bureau of Reclamation—1987 update. Speech to Am. Soc. Civil Eng., 9 Dec. 1986, Washington, DC.

Krull, D.L. 1986. The changing objectives of Garrison diversion unit. p. 2151–2157. *In* Proc. Water Forum '86. Long Beach, CA. 4–6 Aug. Vol. 2. Am. Soc. Civil Eng., New York.

Kuros, G.R. 1984. Qanats—A 3000 year-old invention for development of groundwater supplies. 12th Congr. p. 495–515. *In* Trans. Int. Comm. Irrig. Drain. Spec. Sess., Fort Collins, CO. 28 May–2 June. ICID, New Delhi.

Labadie, J.W., S. Patamatamkul, and R.C. Lazaro. 1986. River basin network model for conjunctive use of surface and groundwater. p. 310–319. *In* Proc. Water Forum '86. Long Beach, CA. 4–6 Aug. Vol. 1. Am. Soc. Civil Eng., New York

Lee, J.C., J.R. Ellis, and R.D. Lacewell. 1985. Valuation of improved irrigation efficiency from an exhaustible groundwater source. Am. Water Resour. Assoc. Water Resour. Bull. 21:441–447.

Le Moigne, G.J.M. 1986. World bank involvement in irrigation and drainage. Int. Comm. Irrig. Drain. Bull. 35:1–5, 24.

Letey, J., Jr., C. Roberts, M. Penberth, and C. Vasek. 1986. An agricultural dilemma: Drainage water and toxics disposal in the San Joaquin Valley. Univ. Calif. Agric. Exp. Stn. Spec. Publ. 3319.

Lyle, W.M., and J.P. Bordovsky. 1986. Multifunction irrigation system development. Trans. Am. Soc. Agric. Eng. 29:512–516.

Malhotra, S.P. 1982. The warabandi and its infrastructure. Publ. 157. Cent. Board Irrig. Power. New Delhi.

Malik, B.A. 1986. Pakistan's limiting water resources and growing demand. p. 227–248. *In* Proc. Darves Bornoz Spec. Sess., Natl. Cmte. of Pakistan, ICID, Vol. 2. Lahore, Pakistan. ICID, New Delhi.

McNicholl, I.H., and C.L. Abernethy. 1985. Distribution of indirect benefits from the left bank outfall drain project in the provincial economy of Pakistan. Rep. 9. Proc. Spec. Sess. Int. Comm. Irrig. Drain., Vina del Mar, Chile. 18 Oct. ICID, New Delhi.

Merrey, D.J. 1986. The sociology of warabandi: A case study from Pakistan. p. 44–61. *In* Res. Pap. No. 4. Int. Irrig. Manage. Inst., Sri Lanka.

National Academy of Sciences. 1976. Climate and food, climate fluctuations and U.S. agricultural production. NAS, Washington, DC.

Nelson, D.G., and W.R. Johnston. 1984. San Joaquin drainage—development and impact. p. 424–432. *In* Proc. Specialty Conf. Flagstaff, AZ. 24–26 July. Am. Soc. Civil Eng., New York.

Nobe, K.C., and R.K. Sampath (ed.). 1986. Irrigation management in developing countries. Westview Press, Boulder, CO.

Norman, C. 1986. World Bank pressed on environmental reforms. Science 234:813–815.

O'Brian, T.A. 1984. The problem of salinity and its control, River Murray, Australia. p. 33–42. *In* R.H. French (ed.) Salinity in watercourses and reservoirs. Butterworth Publishers, Boston.

Pakistan National Committee. 1983. p. 1034–1069. *In* K.K. Framji et al. (ed.). Irrigation and drainage in the world. Vol. 2. Int. Comm. Irrig. Drain., New Delhi.

Pohjakas, K. 1984. Evaluation of center pivot irrigation in Alberta. R18, Q38. p. 313–324. *In* Trans. Int. Comm. Irrig. Drain. Vol. 1.

Power, J.W. 1986. Sharing irrigation know-how with developing countries. Agric. Eng. 67(6):15–18.

Puli, L. 1986. Water conservancy development for food self-sufficiency in China. p. 17–32. *In* Proc. Darves Bornoz Spec. Sess., Natl. Cmte. of Pakistan, ICID, Vol. 2. Lahore, Pakistan, 4 Oct. ICID, New Delhi.

Rangeley, W.R. 1986a. Global water issues. Civil Eng. 96(12):60–62.

Rangeley, W.R. 1986b. Scientific advances most needed for progress in irrigation. Philos. Trans. R. Soc. London. A316:355–368.

Rangeley, W.R. 1987. Irrigation and drainage in the world. p. 29–35. *In* W.R. Jordan (ed.) Proc. Water and Water Policy in World Food Supplies Conf., College Station, TX. 26–30 May. 1985. Texas A&M Univ. Press, College Station.

Rawlins, S.L. 1984. Strategies to cope with the impact of the energy crisis on irrigation. p. 509–521. *In* Trans. 12th Congr. Int. Comm. Irrig. Drain., Spec. Sess. Vol. 1.

Rhoades, J.D. 1987. Principles and practices of salinity control in food production in North America. p. 141–151. *In* Jordan (ed.) Proc. Water and Water Policy in World Food Supplies Conf., College Station, TX. 26–30 May 1985. Texas A&M Univ. Press, College Station.

Skaggs, R.W. 1980. A water management model for artificially drained soils. Tech. Bull. 267, North Carolina Agric. Res. Serv., North Carolina State Univ.

Skaggs, R.W. 1987. Drainage. p. 319–324. *In* W.R. Jordan (ed.) Proc. Water and Water Policy in World Food Supplies Conf., College Station, TX. 26–30 May. Texas A&M Univ. Press, College Station.

Skaggs, R.W., and A. Nassehzadeh-Tabrizi. 1983. Optimum drainage for corn production. North Carolina Agric. Res. Serv. Tech. Bull. 274.

Snyder, R.L., D.W. Henderson, W.O. Pruitt, and A. Dong. 1985. Final Rep. Calif. Irrig. Manage. Inf. Syst., Exec. Sum. Univ. of California, Davis.

Squires, R.C., and W.R. Johnston. 1986. Agricultural drainage water treatment—are toxic elements useful? p. 358–365. *In* Proc. Water Forum '86. Long Beach, CA. 4–6 Aug. Vol. 1. Am. Soc. Civil Eng.

Stegman, E.C., and D.A. Coe. 1984. Water balance irrigation scheduling based on Jensen-Haise equation: software for Apple II, II + and IIE computers. North Dakota Agric. Exp. Stn. Res. Rep. 100.

Takase, K. 1967. Statistic study on failure, damage and deterioration of earth dams in Japan. p. 1–19. *In* Trans. Int. Comm. Large Dams.

Teerink, J.R. 1978. Mass transfer of water over long distances for regional development and its effects on human environment. p. GS1–GS11. *In* Gen. Rep. Spec. Sess., Int. Comm. Irrig. Drain., Athens, Greece.

Texas Water Development Board. 1986. Surveys of irrigation in Texas 1958, 1964, 1969, 1974, 1979 and 1984. Rep. 294. Texas Water Dev. Bd., Austin.

Thiessen, J.W. 1986. Irrigation development in Alberta, Canada. p. 1395–1402. *In* Proc. Water Forum '86. Long Beach, CA. 4–6 Aug. Vol. 2. Am. Soc. Civil Eng., New York.

Thompson, T.L. 1979. AGNET—Weather related uses. Am. Soc. Agric. Eng. Pap. 79-4542.

Tolba, M.K. 1986. The African environmental situation. Int. Comm. Irrig. Drain. Bull. 35(2):13–18, 32.

Tong, E. 1986. Diversion for irrigation and agricultural production in lower yellow river. p. 33–51. *In* Proc. Darves Bornoz Spec. Sess., Natl. Cmte. of Pakistan, ICID, Vol. 2, Lahore, Pakistan. ICID, New Delhi.

U.S. Agency for International Development/Cairo. 1985. Irrigation briefing paper.

USDA. 1985. Agricultural statistics 1985. U.S. Dep. Agric., Washington, DC.

U.S. Department of Commerce. 1983. 1982 census of agriculture. Vol. 1, Part 51, United States, summary and state data. Superintendent of Doc., U.S. Gov. Print. Office, Washington, DC.

U.S. Department of Commerce. 1986a. 1984 Farm and ranch irrigation survey. AG84-SR-1, Spec. Rep. Ser., Bur. Census. Superintendent of Doc., U.S. Gov. Print. Office, Washington, DC.

U.S. Department of Commerce. 1986b. America's agriculture, a portrait of the past and present. AG-86-PP-1. Agric. Div., Bur. of Census, U.S. Dep. Commerce, Washington, DC.

von Mengersen, A.G., B.J. Button, P.J. Watts, M.E. McKay, and Associates. 1987. Research priorities for irrigation in Australia. Irrig. Assoc. Austr. J. 3(1):17–19.

Walsh, J. 1984. World Bank puts priority on Africa program. Science 226:148–149, 152.

Walsh, J. 1986. A project born of hope, desperation. Science 232:1081–1083.

Westoff, C.F. 1986. Fertility in the United States. Science 234:554–559.

World Bank. 1986a. Population growth and policies in sub-Saharan Africa. World Bank, Washington, DC.

World Bank. 1986b. Poverty and hunger: Issues and options for food security in developing countries. World Bank, Washington, DC.

World Bank. 1986c. World development report 1986. World Bank, Washington, DC.

4

Nature and Dynamics of Soil Water

A. W. WARRICK

University of Arizona
Tucson, Arizona

The soil is porous material composed of a skeleton of solids, with air and water filling the interspaces. In this section, we are concerned primarily with the pore space because that is where most of the water resides and transport occurs. The entire system, however, is interactive and the solid cannot be taken as inert. A permeable system may be transformed into an impermeable or problem soil when conditions are slightly changed, perhaps by dispersion, compaction, or wettability changes. However, here the chemical and biological reactivity of the soil system will be downplayed and changes in the soil matrix itself largely ignored.

I. SOLID–PORE SYSTEM IN SOILS

To describe water flow in soils, it is necessary to characterize the pore space, which in turn is related to the geometry of the solid particles. Use of a classic porous media approach requires an idealization of the solid-pore system as an equivalent, uniform material at a scale large relative to individual grains and pores, but at the same time small enough to have a locally defined value. This is commonly referred to as a *representative elementary volume* (REV).

As an example, consider the plot of apparent density as a function of sample volume in Fig. 4–1. For a small volume in a porous medium, the apparent density (mass/volume considered) will be essentially zero if located in a pore, or equal to the solid particle density if located within a particle. As the sample volume increases, an ambiguity remains until eventually a somewhat constant value is found at the REV. The value may eventually change when still larger volumes are taken due to changes in the basic parent material or overall structure, but a reasonably repeatable value would be found for REV's in nearby locations. Measurement methodology is closely related to REV in that the volume sensed by an instrument influences repeatability and scatter, perhaps even the expected value.

In a porous system, two densities are of primary interest. The first is the apparent or bulk density ρ_b defined by

Irrigation of Agricultural Crops—Agronomy Monograph no. 30.

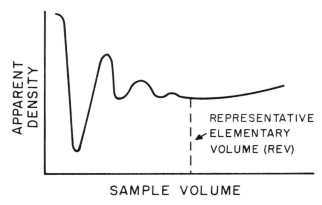

SAMPLE VOLUME

Fig. 4–1. Apparent density as a function of sample volume.

$$\rho_b = \text{(Mass of solids)/(Volume of pores and solids).} \qquad [1]$$

The second is the particle density ρ_p given by

$$\rho_p = \text{(Mass of solids)/(Volume of solids alone).} \qquad [2]$$

Values for ρ_b are mostly between 1 and 1.8 Mg/m^3. Finer soils tend to have more pore space and hence smaller ρ_b values. Also, large amounts of organic matter can lead to small values. Coarse-textured soils tend to have less total pore space and consequently larger bulk densities.

Minerals typically have particle densities of 2.5 to 2.7 regardless of individual grain size; in fact, unweathered rocks have similar densities. Mineralogical composition is important; for example, hematite and iron-rich materials generally have high values considerably outside the above range.

Based on the above definitions, the total porosity f may be derived

$$f = 1 - \rho_b/\rho_p. \qquad [3]$$

Not only the total pore space, but also the distribution of the pore size is of importance. Generally, coarse-textured soils have a preponderance of large pores even though, as noted above, f tends to be small. In a later section, the pore size distribution will be described in greater detail.

II. SOIL WATER CONTENT

Soil water content describes the amount of water present in the system. There are several commonly used definitions. The choice of definition is usually made on the basis of convenience. Some of these are

$\theta_m = \text{(Mass of water in sample)/(Dry mass of soil in sample)}$
$\theta_v = \text{(Volume of water in sample)/(Apparent volume of sample)}$

W = (Volume of water in sample − residual water content)/(Total pore space − residual water content)

$d = \theta_v D$ (D = Depth increment of the profile).

The first (θ_m) is the most commonly used and is referred to as water content on a mass basis or as *gravimetric water content*. Normally, if no other qualifications are given or the context does not infer otherwise, "water content" refers to θ_m. In some engineering applications, the same name is used when a wet mass of soil is used in place of the dry mass in the definition, but this is somewhat rare and causes little confusion if the context is clear. Numerical values of θ_m are always positive and normally <0.5, although in the case of organic soils the values can be 1 and larger. With little added confusion, θ_m (as well as θ_v and W) can be expressed as a percentage as well as a fraction.

The second most common definition is water content on a volume basis θ_v. This is most often called *volumetric water content*. When shortened to just "water content," θ_v is distinguished from the mass basis water content by usage. The θ_v is particularly convenient when inputs or capacities are expressed on a volume basis or storage in a given volume of soil. As irrigation amounts are normally expressed on a volume basis and the water stored often considered on a root-zone basis, volumetric water content is often the preferred term. Also, θ_v is more closely related to water content determined by neutron thermalization than is θ_m. The definition is unaffected, but some complications obviously arise if the soil matrix changes its volume with wetting or drying, as θ_v will not change directly in accordance with the change in the volume of water.

Clearly, θ_v and θ_m are related by

$$\theta_v = (\rho_b/\rho_w)\, \theta_m \qquad [4]$$

where ρ_w is the density of water. Both of the water contents are dimensionless. Often they are followed by kg kg^{-1} or m^3 m^{-3} for specificity. Also, note by [Eq. 4] than if the bulk density is greater than the density of water, then θ_v will be greater than θ_m; this is nearly always true when dealing with mineral soils. The volumetric water content will always be positive (or zero) and less than the porosity.

The degree of saturation W is defined so as to result in a normalized value between 0 and 1:

$$W = (\theta_v - \theta_r)/(\theta_{sat} - \theta_r). \qquad [5]$$

The residual (reduced) water content θ_r is most often taken to be zero, but may be chosen as a value greater than zero to denote a limit below which the water content does not reduce appreciably for decreasing matric potentials. Such normalized values are sometimes convenient to use, particularly in generalized forms for soil water characteristics or unsaturated hydraulic conductivities as we will demonstrate shortly. Of course, W is dimensionless also.

Finally, d as defined above is a depth of water. It is defined with respect to specific profile depth D and is the amount stored expressed as a depth. This is convenient with respect to irrigation water delivered, which is often given as a depth of delivery (i.e., volume of water per unit area of land). The relevant D would normally be taken as an effective rooting zone and thus d can be used to calculate water needed, water stored, etc. Obviously, this also allows direct comparisons to precipitation or evapotranspiration amounts, which are normally expressed as depth for a given time period.

III. SOIL WATER POTENTIAL

Soil water potential is an expression of the energy level of water in the soil system. This is contrasted to the amount of water present in the system for which water content is the fundamental parameter. The energy level of water is particularly useful for describing the dynamics of flow. We expect movement to occur from regions of higher to lower potential. This is true not only with the soil in isolation, but also true with respect to the rhizosphere and the atmosphere. The entire soil-plant-atmosphere system can be formulated on the basis of energy levels, with water flow from the soil to the root to the upper plant and into the atmosphere in directions of decreasing potentials.

The *soil water potential* is formally defined as the work necessary to transfer a unit quantity of water from a reference state to the situation of interest. The reference state may conveniently be taken as pure water under a standard atmosphere of air pressure and at a given elevation. Factors which affect work necessary to make the appropriate (virtual) transfer are elevation, soil matric effects, liquid and air pressures, solutes present, temperature, and the overburden. The usual approach is to consider each factor separately to define individual components and simplify the analysis.

With this in mind, define the total potential ϕ_T as

$$\phi_T = \phi_g + \phi_p + \phi_m + \phi_\pi + \Sigma \, \phi_i \qquad [6]$$

where the components on the right-hand side refer to gravity, pressure, matric, osmotic, and the sum of whatever other factors are to be included. The units of soil water potential are depenent upon the unit quantity in the basic (energy/unit quantity) definition. Logical and commonly used unit quantities are mass, volume, or weight resulting in ϕ_T and individual components as joules per kilogram, pascals, or meters, respectively, in meter-kilogram-second system units. Choosing energy/unit weight and grouping the pressure and matric effects together, we simplify Eq. [6] to

$$\phi_T = z + h + \pi \qquad [7]$$

where z is the elevation above a specified reference, h is the pressure head, which includes pressure and matric components, and π is the osmotic head.

The user must specify the datum for the gravitational component and the resulting z can be positive or negative depending on whether the point of interest is above or below that datum. For most field applications h will be positive and the water pressure will be positive for saturated conditions; conversely, h will be negative for unsaturated conditions. For the above generalizations with respect to h, the assumption is that the soil air pressure (or pressure acting on the water for the saturated case) is the same as the reference air pressure taken as the local atmosphere, i.e., the gage pressure is zero. Cases where the gage pressure is not zero may typically be within pressure chambers in the laboratory or for entrapped air during infiltration.

Figure 4-2 demonstrates h for saturated and unsaturated (but fairly wet) conditions. For saturated conditions (Fig. 4-2A), water will stand in a piezometer to depth h above the test point. If static conditions exist, the water level will correspond to a water-table level. The water in the tube is under the local atmospheric pressure. For unsaturated conditions, water will not rise in a piezometer tube; in fact, it will equilibrate with water at less than atmospheric pressure through a porous plate having sufficiently small water-filled pores which air cannot go through when wet. Such a condition may be measured with a hanging water column illustrated to the left of the test point where $h = -|h|$ (Fig. 4-2B). In this case the air pressure in the soil and that acting on the water column is the same. For these conditions h is the matric potential. Also shown is a tensiometer for which the pressure in the water columns is measured by a vacuum gage. The gage expresses a deficit pressure relative to the atmosphere which may be given as h which will also be $-|h|$. The deficit pressure is commonly measured with a mechanical vacuum gage, a mercury-water column, or an electronic transducer.

The hanging water concept as in Fig. 4-2B breaks down for drier conditions. Nevertheless, the basic definitions of soil water potential and the individual components remain valid for drier conditions. Pressure deficits beyond an atm. may be achieved in the laboratory by using higher air pressures in the system, as is commonly done with a pressure-plate apparatus, for example. For such a system, the water pressure deficit with respect to the soil atmosphere can be made large, but the absolute water pressure remains at the local atmospheric pressure.

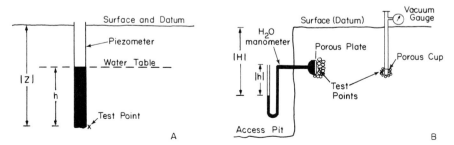

Fig. 4-2. (A) Piezometer useful for measuring pressure head h for saturated conditions with $h \geq 0$. The depth is z. (B) Tensiometers useful for measuring pressure head h for unsaturated conditions for $h \leq 0$ (but not overly dry). The hydraulic head is H.

The osmotic component may be conceptualized with a semipermeable membrane separating the soil solution and a specified reference solution which is most logically taken to be pure water. The π is defined as the excess pressure head (i.e., pressure expressed in terms of a height of water) within the solution which will be in equilibrium across the membrane. Basic fundamental differences in the definition and application for osmotic effects were discussed by Corey and Klute (1985).

Terminology regarding soil water potential can be ambiguous and confusing. The primary source of confusion is the multiplicity of dimensions and units. Additionally, gage and absolute pressures are mixed, as well as alternative terms. Alternative dimensions have been discussed and arise from whether a unit mass, volume, or weight is used in the basic definition of potential. Equivalences in terms of water status that are useful to remember are that 10^5 Pa is 1 bar, which is about 1 atm., which is approximately equivalent to 10 m of water (or 100 J kg^{-1}). The terms *tension* and *suction* are used intuitively and are normally the negative of the matric potential expressed in various units of length or pressure. Suction is used as the negative of the osmotic component as well, although less frequently. Absolute pressure is with respect to a perfect vacuum and can never be less than zero. Gage pressure is always evaluated relative to another pressure, the most obvious being the local atmosphere. Thus, gage pressure can be negative or positive, but will always be less than the absolute pressure.

IV. SOIL WATER CHARACTERISTICS AND PORE SIZE DISTRIBUTION

Two primary ways of describing soil water have just been discussed: the water content and the soil water potential. The relationship between soil water content and the matric potential is of special interest and is referred to as the *soil water characteristic*. Other names for the relationship include water retention curves and h-θ or ψ-θ curves. By defining an effective radius with

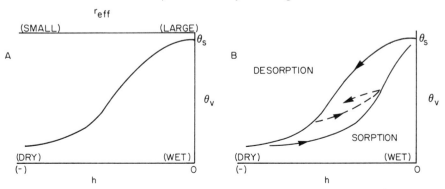

Fig. 4–3. (A) Soil water characteristic curve. The relationship of pressure h to effective radius, r_{eff}, is by Eq. [8]. (B) Sorption and desorption curves. Primary curves form the outside envelope.

the capillary-rise equation, the matric potential corresponds to a given radius and the soil water characteristic curve is equivalent to the pore-size distribution.

A typical soil water characteristic curve is given as Fig. 4–3A. The water content θ_v is a maximum of θ_{sat} at $h = 0$ and decreases as h decreases. For a sand or coarse-textured soil the maximum value of θ_v is expected to be somewhat less and the curve also expected to drop off somewhat more sharply than for the finer-textured soil. The effective radius r_{eff} for a given matric potential h is defined by the capillary rise equation as

$$r_{eff} = 2\sigma \cos(\gamma)/(\rho_w g\, |h|) \qquad [8]$$

where σ is the surface tension of water (kg s^{-2}) and γ the contact angle of water formed with air and the soil material. For lack of a better value, γ is often taken as zero leading to $\cos(\gamma) = 1$. Use of Eq. [8] results in values of effective radius as shown by the upper abscissa in Fig. 4–3A. The larger h (close to zero) corresponds to a larger r_{eff}, and a smaller h (to the left) corresponds to a smaller r_{eff}. The fact that a coarser-textured soil drops in water content abruptly under a small tension, corresponds to a preponderance of larger pores. Conversely, the more gradual decrease in water content for finer-textured soils corresponds to a wider distribution of pore sizes and a greater fraction of smaller pores.

Functional forms for the soil water characteristic curves are often convenient for generalizations and for modeling. These are necessarily empirical in form. An example of one such form is by van Genuchten (1980) and relates h to the reduced water content W by

$$W = (\theta_v - \theta_r)/(\theta_{sat} - \theta_r) = [1 + |\alpha h|^n]^{-m}. \qquad [9]$$

Conversion of water content to matric potential or vice versa is complicated by the fact that the relationship often depends on the previous history of the system. This gives rise to hysteresis or a family of relationships rather than a 1:1 correspondence. In Fig. 4–3B, a characteristic curve is repeated but an arrow added to denote a drying (desorption) curve. Also shown is the other part of a closed hysteresis envelope which results from a wetting or sorption curve. Within the large loop are secondary curves showing what could result if wetting or drying was started from intermediate positions. Several mechanisms exist to explain the hysteretic phenomena, including an effective radius which is different whether filling or emptying, possible changes in contact angles on wetting and drying, and entrapped air.

V. SOIL WATER DYNAMICS—BASIC FORMULATIONS

Hydraulic head H may be defined as

$$H = h + z. \qquad [10]$$

In the previous section, we used pressure head h to be positive for saturated cases, for which the water pressure is likewise positive, and to be negative for unsaturated conditions. Assumptions are that the air pressure acting on the soil water is atmospheric. The hydraulic head will be the total soil water potential if the further assumptions are made that the osmotic, thermal, and other potentials are negligible.

Darcy's Law relates flux density q to the hydraulic head by

$$\vec{q} = -K(\theta_v)\nabla H \qquad [11]$$

where $\vec{q}$ is the flux density (dimensions LT^{-1}), $K(\theta_v)$ is the hydraulic conductivity (LT^{-1}), and ∇ the vector gradient operator (L^{-1}). The flux density is commonly called flow velocity or, more specifically, it can be referred to as the Darcian velocity since other types of velocity are sometimes of interest. The velocity is a vector quantity in the direction of flow with magnitude of volumetric water flux per unit area perpendicular to the maximum gradient of H. (Isotropic conditions are also assumed.)

The conductivity K is a constant for saturated flow; otherwise it is strongly dependent on water content. Saturated conditions when Darcy's Law is known to be invalid include extreme velocities—either very high or very low. Darcy's Law is borrowed from the saturated for the unsaturated case.

The dependence on water content is well recognized and K may vary several orders of magnitude over a field cycle. In Fig. 4-4, K is given as a

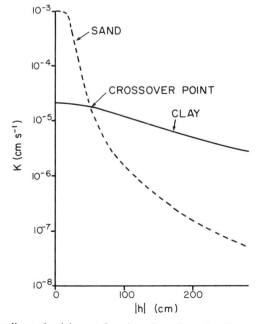

Fig. 4-4. Hydraulic conductivity as a function of matric suction for a sand and clay soil. After Hillel (1982).

function of matric suction ($|h|$). Typical coarse-textured soils have larger conductivities near saturation than fine-textured soils as shown. For larger suctions, typically a cross-over point exists, corresponding to more fine pores in the finer soils. Tests have shown Darcy's Law can be invalid for some unsaturated regimes. Its widespread use is in part due to convenience and lack of any viable alternative.

Darcy's Law may be coupled with conservation of mass principles to derive a continuity equation. With this in mind, consider the closed volume V in Fig. 4-5. The Darcian velocity $\vec{q}$ is illustrated at an arbitrary point on the surface along with an elemental surface vector of magnitude dS and directed on an outward normal. If mass is conserved and no sources or sinks exist within, the total normal component of the outward flux is equal to the rate of depletion within the volume element:

$$\int_s \rho_w\, \vec{q} \cdot d\vec{S} = -\frac{\partial}{\partial t} \int_v \rho_w \theta_v\, dV \qquad [12]$$

where the vector "dot" product gives the normal component of the outward flux density. By Gauss' Theorem (Wylie, 1966), the surface integral may be replaced by a volume integral of the divergence of $\vec{q}$ resulting in

$$\int_v \nabla \cdot (\rho_w \vec{q})\, dV = -\frac{\partial}{\partial t} \int_v \rho_w \theta_v\, dV. \qquad [13]$$

As the above equation holds for all closed surfaces, the time derivative can be moved inside and the integrands equated, resulting in a continuity equation as

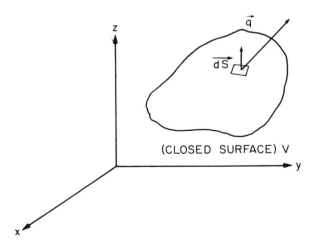

Fig. 4-5. Control volume considered for continuity relationships. The q is the outward flux density vector and dS is a normal surface element.

$$\frac{\partial}{\partial t} (\rho_w \theta_v) = - \nabla \cdot (\rho_w \vec{q}). \qquad [14]$$

Substitution of Darcy's Law (Eq. [11]) and assuming ρ_w a constant gives

$$\frac{\partial \theta_v}{\partial t} = \nabla \cdot (K \nabla H). \qquad [15]$$

Richards (1931) derived a relationship equivalent to Eq. [15]. Several different equivalent forms may be derived and we can use Richards' Equation in a generic sense for all of them. The solution of Eq. [15] is difficult in general as it is nonlinear and has two dependent variables, θ_v and h. For the saturated case (and when compressibility of water may be neglected), it is much simplified because θ_v is constant, resulting in Laplace's Equation. Once Laplace's Equation is obtained, abundant techniques and available solutions make numerical treatment relatively easy—especially for common geometries.

The number of dependent variables in Eq. [15] may be reduced from 2 to 1 provided a soil water characteristic relationship exists, either as $h = h(\theta_v)$ or $\theta_v = \theta_v(h)$. With this assumption, we may eliminate h and have a "θ-based" equation or eliminate θ_v and have an "h-based" equation. Specific water capacity C (dimensions L^{-1}) and soil water diffusivity D (dimensions $L^2 T^{-1}$) can be defined for convenience

$$C = d\theta_v / dh \qquad [16]$$

$$D = K/C. \qquad [17]$$

These terms have analogs in other areas of physics, for example, in a form identical to Eq. [17], thermal diffusivity is the ratio of thermal conductivity to the volumetric specific heat. Both C and D are highly dependent on water status. Table 4–1 gives alternative forms of Richards' Equation including general relationships, θ-based, h-based, and matric flux-based.

The specification of boundary and initial conditions is critical for describing soil water dynamics. Boundary conditions are generally

1. Constant head (or water content). The value of h or θ_v is specified on the boundary. (This is called a Dirichlet condition.)
2. Gradient specified. The value of $\vec{q}$ is specified, perhaps as zero (no flow) or as a function of time. (This is called a Neumann condition.)
3. Mixed. Part of the boundary is head specified and part is gradient specified.

A water-table boundary would be the first type. Flux conditions would generally be of the second type. A unit hydraulic gradient useful for a lower condition above a deep water table would be of the second type.

The above framework is easily modified to include plant root uptake. If the volume of uptake per bulk volume of soil per unit time is S then a

Table 4–1. Some alternative forms of Richards' Equation (Eq. [15]).

Equation	Basis
1. $\partial\theta_v/\partial t = \dfrac{\partial}{\partial z}\left(D\dfrac{\partial\theta_v}{\partial z}\right) - \dfrac{\partial K}{\partial z}$	θ-based, 1-D
2. $C(\partial h/\partial t) = \dfrac{\partial}{\partial z}\left(K\dfrac{\partial h}{\partial z}\right) - \dfrac{\partial K}{\partial z}$	h-based, 1-D
3. $(d\theta/d\phi)(\partial\phi/\partial t) = \dfrac{\partial^2\phi}{\partial z^2} - \dfrac{\partial K}{\partial z}$ $\phi = \int_{-\infty}^{h} K\,dh = \int_{o}^{\theta} D\,d\theta$	Matric flux-based, 1-D
4. $\nabla^2\phi - \alpha\,\partial\phi/\partial z = 0$ $K = K_{wet}\exp(\alpha h)$	Matric flux-based, steady state, K exponential with h

term "-S" is added to the right side of Eq. [15]. The simplest condition is for S explicitly defined in space and time, but for the most part, it will be interactive with the atmospheric conditions, soil water status, and the plant system itself. Thus, although S is easily put in the differential equation, the adequate formulation of the sink term is generally complex.

VI. INFILTRATION

A. One Dimensional

Infiltration is the entry of water from above the soil into the subsurface. The analysis, for the most part, is a special example of unsaturated flow. As the primary mechanism of water input to the rhizosphere and the root zone, the process is intensely studied by soil scientists, watershed hydrologists, and irrigation engineers.

Consider the idealized and simplified situation depicted in Fig. 4–6A. Water is ponded at the surface to a depth h_o and infiltrates into the soil. At time t the given profile exists, with a wetting depth at depth z_f. The moisture profile is given as Fig. 4–6B, and shows a water content $\theta_v = \theta_{wet}$ behind the front and θ_{dry} ahead of the front, corresponding to the antecedent conditions.

Let us assume a one-dimensional analysis and take the reference level for the gravitational potential at the soil surface. Furthermore, take uniform hydraulic conductivities K_{wet} and K_{dry} corresponding to the wet and dry portions of the profile, respectively. Assume the matric potential at the front can be replaced simply by a constant value h_f. With these simplifications, Darcy's Law (Eq. [11]) applied from Points 1 to 2 for one-dimensional flow results in (Point 1 at the front and Point 2 at the surface)

$$q = -K_{wet}\ (\text{Potential at 2} - \text{Potential at 1})/z_f.$$

We define the surface Darcian velocity as the infiltration rate i and reduce the above to

$$i = -K_{wet} (h_f - z_f - h_o)/z_f$$
$$= K_{wet} - K_{wet} (h_f - h_o)/z_f. \qquad [18]$$

Observe that for these idealized conditions i will take the form of Fig. 4–6C. As K_{wet}, $-(h_f - h_o)$ and z_f are all positive constants, i will be large for small times when z_f is small. As t increases, z_f becomes larger and i becomes smaller, with i eventually approaching asymptotically a constant value, K_{wet}.

The cumulative infiltration I (units of length) is defined as the integral of i over time, i.e.,

$$I = \int_0^t i \, dt \qquad [19]$$

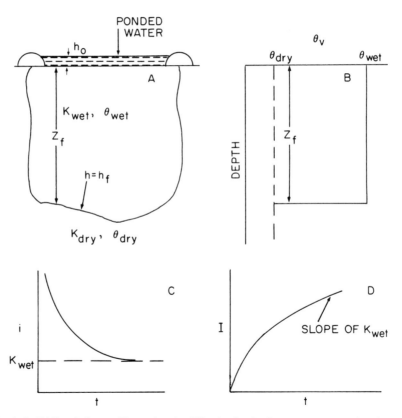

Fig. 4–6. (A) Ponded water illustrating simplification by the Green-Ampt approximations. The subscript "f" is for the wetting front, "wet" is for the wetted area, and "dry" for the initial conditions. (B) Water content profile corresponding to A. (C) Typical relationship of infiltration rate to time for surface ponding and a homogeneous soils profile. (D) Cumulative infiltration I corresponding to C.

which is equivalent to

$$I = K_{wet} t + |h_f - h_o| K_{wet} \int_o^t (1/z_f)dt. \qquad [20]$$

By continuity, the amount of water entering the profile must be equal to the change in storage minus any amount leaving the storage area. The latter includes deep seepage corresponding to the antecedent conditions. As we took the initial value to be a constant θ_{dry}, we write the corresponding conductivity value as K_{dry} which corresponds to the Darcian velocity ahead of the wetting front. Thus, the surface input, change in storage, and water moving ahead of the front z_f for the time period 0 to t results in a second relationship for I

$$I = K_{dry}t + (\theta_{wet} - \theta_{dry})z_f. \qquad [21]$$

By solving Eq. [21] for z_f and substituting back into Eq. [20], we find an implicit relationship for I of

$$I = K_{wet} t + b \int_o^t dt'/(I - K_{dry}t') \qquad [22]$$

with b a positive constant defined by

$$b = |h_f - h_o|(\theta_{wet} - \theta_{dry})K_{wet}. \qquad [23]$$

A reasonable assumption under most conditions is that $I >> K_{dry}t$. By making this assumption, we can differentiate Eq. [22] with the result

$$i = dI/dt = K_{wet} + b/I. \qquad [24]$$

The above equation was introduced by Green and Ampt (1911) and the various forms are referred to as the Green-Ampt infiltration formulas. The above nonlinear differential equation may be integrated to give an implicit relationship of I and t. To do this, use the middle and final expressions to solve for dt and integrate with the result:

$$t = (b/K_{wet}^2)\{(K_{wet}I/b) - \ln[1 + (K_{wet}I/b)]\}. \qquad [25]$$

The general relationship of I to t is given as Fig. 4-6D.

Although greatly simplified, the Green-Ampt relationships are easy to apply and preserve many of the gross properties of infiltration. The infiltration rate and cumulative infiltration generally are similar to Fig. 4-6C and D for uniform soil conditions. The mass conservation equation (Eq. [21]) is useful for estimating wetting depth as a function of water added I, especially when $K_{dry}t$ can be taken as negligible. For most applications, the physical interpretation of h_f would not be sought, but its integrated effect on b would be used empirically.

A more comprehensive analysis of infiltration follows Richards' Equation (Eq. [15]). The best known is after Philip (1968). The solution was based on a perturbation analysis consisting of the solution without gravity followed by a series of corrections to account for gravity. The complete solution is of the form:

$$I = St^{0.5} + (A_o + K_{dry})t + \sum_{n=1}^{N} A_n t^{(0.5n+1)}. \qquad [26]$$

The S is the "sorptivity" $(LT^{-0.5})$ which is dependent on the soil hydraulic properties as well as the initial and boundary water content. More specifically, it is given by

$$S = \int_{\theta_{dry}}^{\theta_{wet}} \lambda d\theta \qquad [27a]$$

where the Boltzmann transformation $\lambda = xt^{-0.5}$ is for the horizontal absorption and determined from the solution of the integral equation

$$\int_{\theta_{dry}}^{\theta_v} \lambda d\theta = -2D \, d\theta_v/d\lambda. \qquad [27b]$$

The coefficients A_o, A_1, A_2, . . . A_N also are dependent on the hydraulic properties and the initial and boundary values.

The results of the comprehensive analysis is schematically shown as Fig. 4–7A. The moisture profile is similar to Fig. 4–6B, but the water content monotonically decreases with depth. The sharpness of the wetting front is a function of the nonlinearity of the system, with Fig. 4–6B representing the limiting condition for which K abruptly increases as θ_v approaches θ_{wet}.

The application of Eq. [26], generally takes two different forms. The simpler is to assume the form is correct, that the summation term is negligible, and apply as an empirical relationship similar to the Green-Ampt Equation. In that case, the sorptivity S and the coefficient $(K_{dry} + A_o)$ would

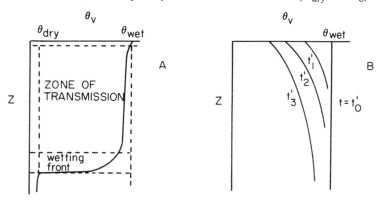

Fig. 4–7. (A) Wetting front profile for one-dimensional infiltration under ponded conditions. (B) Time sequence of water profiles after the profile is initially at θ_{wet} to depth of interest.

be found by fitting measured infiltration relationships similar to Fig. 4–6C or D. In the second approach S, K_{dry}, A_o, A_1, A_2 . . ., A_N would be found from the hydraulic properties of the soil, namely two of the three functions K, D, and C, coupled with the initial and boundary conditions. The complication of the divergence of the series Eq. [26] as $t \rightarrow \infty$ is overcome by limiting application only until some cutoff time beyond which the wetting front assumes a constant shape and moves at a velocity of $(K_{wet} - K_{dry})/(\theta_{wet} - \theta_{dry})$ through the profile.

Numerous other infiltration models exist. A useful empirically based relationship often used for irrigation engineering is the Kostiakov-Milne Equation:

$$i = At^{-n} \qquad [28]$$

with A and n empirical constants dependent upon the soil and initial conditions.

Numerical techniques may be applied to Richards' Equation directly. These allow the inclusion of complex boundary, initial, and profile conditions—for example, surface crusting, layering, and heterogeneous antecedent water. The general availability of computers adds to their practicality. Results are generally in the form of tables of depth and time, and water content or matric pressure, which of course can be interfaced with graphical output. Generally, the solution of Richards' Equation is input intensive, especially for complex profiles.

The above relationships are for a boundary maintained at a given water content or potential. Results similar to Eq. [26] have also been developed for a flux-driven boundary condition. Likewise, numerical solutions can be used for such conditions.

B. Point Source Example

The classical infiltration relationships are in connection with one-dimensional flow. For very large times, the moisture profile for uniform soils is trivial; it is simply a constant water content over measurable distances. Associated with large times, is a front deep within the soil, moving ahead at the rate $(K_{wet} - K_{dry})/(\theta_{wet} - \theta_{dry})$. For a point (or line) application, a "wet" region develops near the source, but for large times the profile is no longer trivial. Steady water conditions develop with drier conditions prevailing away from the source. Water enters the system and moves through the system by capillarity and gravity as before, but the profile distribution is space dependent. The concept of "wetting front" is meaningful for only relatively small times and is more difficult to identify either physically or quantitatively at large times.

A solution to the steady-state, matric flux form of Richards' Equation (Line 4 of Table 1) for a point source of strength q (units $L^3 T^{-1}$) at the origin is

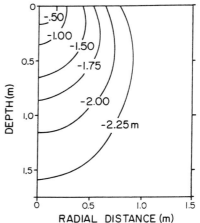

Fig. 4-8. Steady-state distribution of pressure head for point source.

$$\phi = \left(\frac{\alpha q}{4\pi}\right) \{\exp(Z - \rho)/\rho - \exp(2Z)E_1(Z + \rho)\} \qquad [29]$$

with E_1 an exponential integral as defined as

$$E_1(x) = \int_x^\infty \frac{\exp(-x)}{x} \, dx. \qquad [30]$$

The dimensionless depth Z and total radius ρ are by

$$Z = \alpha z/2 \qquad \rho^2 = (\alpha^2/4)(r^2 + z^2). \qquad [31]$$

The solution is valid only for a special conductivity function $K = K_o\exp(\alpha h)$ for which the steady-state Richards' Equation becomes linear.

Numerical contours for h are given in Fig. 4-8. The results are for an input of $q = (5.6)(10)^{-7}$ m^3 s^{-1} (2 L/h) and $K = 2.75(10)^{-5} \exp(1.4\ h)$ m s^{-1}. Largest (algebraic) values of h are found near the origin with drier conditions prevailing at larger distances. Similar results can be developed for line sources (a two-dimensional problem) as well as for arrays of points or lines, or for more complex surface conditions.

Generally, time-dependent conditions, heterogeneous profiles, and other hydraulic conductivity forms can only be addressed using numerical techniques. Semi-analytical approaches can be used when gravity is ignored.

VII. FIELD WATER REGIME, FIELD CAPACITY, AND REDISTRIBUTION

At the cessation of infiltration, movement of water in the soil profile continues. The near-surface regions start to become depleted of water as drainage occurs. The time sequence of water redistributions for one-dimensional conditions follows a pattern similar to Fig. 4–7B. In this particular sequence, infiltration stops at time t_o'. Drainage occurs to times t_1', t_2', . . ., with a gradual decrease in the profile water. The decrease in the profile normally occurs by drainage to deeper depths, surface evaporation, and plant water uptake. The rate of change becomes much slower as the profile becomes drier; for example, the amount of change the first hour may be equivalent to that for the next day, which may be equivalent to that for the next week.

The analysis of drainage (and redistribution) can be analyzed by a numerical solution of Richards' Equation. Fewer empirical relationships are relevant than for infiltration. One such simplification is based on the *unit gradient assumption*. In this case evaporation is assumed negligible, and the magnitude of the hydraulic gradient is taken to be unity to the depth of interest L. Thus, the rate of drainage q_L past the depth L is given by $q_L = K(\theta_v)$ which by conservation of mass is equal to $-L d\bar{\theta}_v/dt$ with $\bar{\theta}_v$ the average water content from the surface to depth L. (Here, for convenience, we choose q_L to be positive, which is equivalent to replacing z by $-z$ in Eq. [7].) Thus, we have

$$q_L = K = -L \, d\bar{\theta}_v/dt. \qquad [32]$$

Integration results in

$$t = L \int_{\theta_v}^{\theta_{wet}} d\theta_v/K. \qquad [33]$$

If the form of K is known, then t may be found from the integral as a function of θ_v (or h), from which q_L also follows. For example, if K is given as $K = K_{wet} \exp[\alpha(\theta_v - \theta_{wet})]$ the solution of Eq. [32] leads to

$$q_L = K_{wet}/[1 + (\alpha K_{wet} t/L)] \qquad \text{(unit gradient)}. \qquad [34]$$

The average water content in the profile is determined by the integration of Eq. [33] and solving for θ_v with a result of

$$\theta_v = \theta_{wet} - (1/\alpha) \, ln[1 + (\alpha K_{wet} t/L)] \qquad \text{(unit gradient)}. \qquad [35]$$

Similar, purely empirical relationships have been used of the form $\theta_{avg} \approx a/(b + t)^c$.

If initially the profile is thoroughly wetted, drainage will continue indefinitely, albeit at ever slower rates. "Field capacity" was defined for the purpose of calculating "the amount of water held in soil after excess water has drained away and the rate of downward movement has materially decreased. . ." (Veihmeyer & Hendrickson, 1949). A problem arises in that a definitive "field capacity" is often needed, but we know that the water stored indefinitely changes with time. An example is given in Table 4–2, demonstrating water content observed over time at the 60 to 90 cm depth in a draining loessial silt loam. The field capacity is dependent on the appli-

Table 4–2. Moisture at the 60 to 90 cm depth as a function of time for an initially wet loessial soil. After Hillel (1982).

Time, d	θ_m
0	0.290
1	0.202
2	0.187
7	0.175
30	0.159
60	0.147
156	0.136

cation; for irrigated agriculture an appropriate figure may be 0.18 or 0.19 (for θ_m), while for dryland farming the relevant time involved would be much longer and perhaps 0.14 would be appropriate. The general rule with respect to field capacity, is that it should be measured in situ and under the conditions of interest. In addition to the particular application, the value is influenced by texture, types of clay, organic matter content, and the soil profile development.

VIII. VAPOR FLOW

The previous sections imply liquid transport, but vapor transport is also possible—notably for dry conditions near the soil surface. Under arid conditions, output by evaporation from the soil surface may fully balance input by precipitation over reasonably long periods of time.

Although vapor can be transported by advective flow resulting from air pressure gradients, the dominant mechanism is normally believed to be diffusion. This is described by Fick's Law, which we simplify for one-dimensional flow in the z-direction as

$$q_v = -D_v \, d\rho_v/dz \qquad [36]$$

where q_v is mass of vapor per unit area per unit time, D_v is the vapor diffusion coefficient ($L^2 T^{-1}$), and ρ is the soil vapor concentration (ML^{-3}) in the gaseous phase. The diffusion coefficient is generally less than for water vapor in open air because only the pore space is available in which diffusion can occur. An empirical relationship credited to Penman (1940) is

$$D_v = 0.66 \, (\theta_{sat} - \theta_v)D_o \qquad [37]$$

with D_o the diffusion coefficient of water vapor in air.

Equation [36] may be reformulated in terms of a matric potential gradient by use of the relationship between the relative humidity r and matric potential (in the absence of salts):

$$r = \rho_v/\rho_o = \exp(M_w gh/RT) \qquad [38]$$

with ρ_0 the saturated vapor pressure, M_w the mass of a mole of water, R the universal gas constant [8.31 J (mol K)$^{-1}$], and T the absolute temperature. By Eq. [36] the vapor flux density is

$$q_v = -K_v \, dh/dz \qquad [39]$$

with the effective vapor conductivity defined as

$$K_v = D_v \, \rho_v M_w g/(\rho_w R T). \qquad [40]$$

The total mass water flux density is given by $q_v + q_l$ where q_l is the liquid mass water flux found by Eq. [11] to be

$$q_l = -\rho_w K \, dH/dz. \qquad [41]$$

Together, the liquid and vapor transport for isothermal conditions may be combined (Milly, 1982) as

$$q_m = q_l + q_v = -\rho_w (K + K_v) \, dh/dz - \rho_w K. \qquad [42]$$

These may alternatively be expressed as effective diffusivities.

Figure 4–9 compares liquid and vapor conductivities for typical sand, silt loam, and clay soils (Campbell, 1985). For wet conditions (to the left of the figure), the vapor conductivities given by the solid lines are much less than the liquid conductivities, which are large and above the scale plotted. Under drier conditions, the vapor conductivity increases and eventually overtakes the liquid component, becomes dominant, and then decreases as the soil becomes so dry that essentially the vapor conductivity drops to zero as well. It is debatable whether soils ever become dry enough that vapor trans-

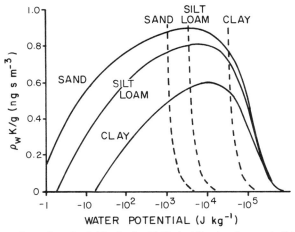

Fig. 4–9. Comparison of conductivities for liquid (dashed lines) and vapor (solid lines) for typical sand, silt loam, and clay soils. (1 J kg^{-1} is equivalent to $h = 0.103$ m). After Campbell (1985).

port is negligible under natural conditions. An additional point is that the liquid and vapor transport are closely interrelated, and it is often preferable from a practical standpoint to lump them together.

IX. NONISOTHERMAL FLOW

For nonisothermal conditions, it is convenient to use

$$d\rho_v/dz = d(r\rho_o)/dz = r\, d\rho_o/dz + \rho_o\, dr/dz \qquad [43]$$

which leads to

$$q_v = q_{v\theta} + q_{vT} \qquad [44]$$

where $q_{v\theta}$ is the isothermal part given by the right side of Eq. [39] and the q_{vT} is due to the nonisothermal part given by

$$q_{vT} = -D_v r(d\rho_o/dT)(dT/dz). \qquad [45]$$

Experimental results generally show q_{vT} as given by Eq. [45] to underestimate actual temperature induced diffusion; consequently, a multiplicative "enhancement" factor is usually added. Similarly, the liquid phase flow may be taken as

$$q_l = q_{l\theta} + q_{lT} \qquad [46]$$

with $q_{l\theta}$ given by Eq. [41] and q_{lT} by

$$q_{lT} = -\rho_w K\, (\partial h/\partial T)\, dT/dz. \qquad [47]$$

A. Nonisothermal Flow Near the Soil Surface—the Phoenix Example

In March 1971, an experiment was conducted near Phoenix, AZ to carefully examine diurnal soil water content changes during a general drying of the soil profile over 2-wk (Jackson, 1973). A bare soil profile was thoroughly wetted and water contents were measured gravimetrically. Samples were taken for the 0 to 0.5 cm, as well as 1-cm increments to 5 cm, and 2-cm increments from 5 to 9 cm. Samples were taken at 0.5-h increments from 2300 to 0130 h on 4 to 10 March, 24 to 26 March, and 7 to 9 April. Six sites were sampled each time and composited. Generally, skies were clear and winds were light. Maximum and minimum air temperatures ranged from 17 to 24 and -2 to 5 °C, respectively. In addition to the water content values, intensive soil temperature data were collected.

In Fig. 4–10 are the results for the surface 0 to 0.5 cm for the 3 d commencing 7 March. There is an overall drying trend for the 3 d shown from left to right on the figure. However, there is a dramatic fluctuation diurnal-

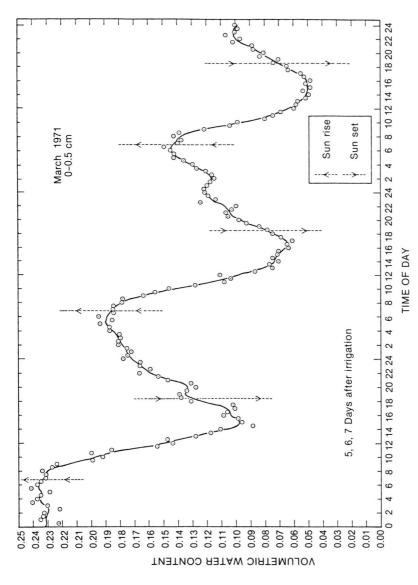

Fig. 4–10. Volumetric water content in the 0- to 0.5-cm increment *vs.* time for 3 d during March 1971. The solid line represents smoothed data and the symbols represent the measured values. After Jackson (1973).

ly. A daily maximum occurs at approximately sunrise, followed by a rapid decrease until early to mid-afternoon and then a recovery through the night towards another diurnal maximum. The overall behavior can be explained as nonisothermal effects. The diurnal drying occurs during the warmer part of the day when the soil surface is warm and rapid evaporation occurs. Conversely, when the soil surface cools and the evaporation rate tends to decrease, the water content increases. Thus, the drying is due to the simultaneous evaporation to the atmosphere and the downward movement of water from hot to cooler regions in the profile. The increase in water content occurs from water moving back towards a cool surface from within the soil profile after the sun retreats.

The above explanation is somewhat simplistic. More complete results, explanations, and analyses are given in the original references. Obviously, simultaneous liquid and vapor transport occur. Temperature differences play a role; otherwise, the water decrease would be monotonic. Also, the evaporative demand is of importance. The osmotic effects should perhaps be considered also, although nonsaline conditions prevailed.

X. OSMOTIC AND SALT EFFECTS

Thus far, we have neglected the effects of salt concentration gradients on water flow. Of particular interest, are osmotic potential gradients which can induce flow across semipermeable membranes. More general transport of water and solutes is discussed in chapter 5 of this book.

When a solute is restricted, water can flow in response to a gradient of the osmotic potential. For these conditions, the water flux density (simplified for one-dimensional flow) is

$$q = -K[(dH/dz) + F_o (d\pi/dz)] \qquad [48]$$

where H is the hydraulic head and π is the osmotic potential expressed as in Eq. [7]. The osmotic efficiency factor F_o ($0 \leq F_o \leq 1$) is unity when solute transport is totally restricted and 0 for the totally nonrestricted case.

The ability of the soil matrix to restrict flow and serve as a partially semipermeable membrane becomes the crucial factor as to whether an osmotic gradient can be sustained. The expert consensus is that, except for biological membranes (such as plant roots), hydraulic forces are generally dominant. The next most likely scenario of significance of the effect is for clayey soils and dry conditions. Further discussion, including experimental F_o values are by Letey (1968), Nielsen et al. (1972), and Hillel (1980).

XI. SUMMARY AND OBSERVATIONS

We have presented the nature and dynamics of soil water in a conventional and reasonably uncomplicated form. Complex models of water balance, plant growth, modeling, and drainage can be based on the same principles. Additionally, the analysis of soil water dynamics serves as the first step in determining directions and maximum travel velocities for a host of water-carried chemicals and their effects on groundwater quality. Solute movement is the next topic to be discussed in this monograph.

Modern technology offers continuing challenges for practical applications of these principles. Measurement, monitoring, and data-handling techniques will be discussed elsewhere. Computationally, the evaluation of complex equations is greatly enhanced by the common usage and availability of computers. Not only are the capabilities expanding rapidly, but costs for given computational levels are generally decreasing. Furthermore, user-friendliness and automatic solutions are closer to reality than ever before.

A long list of complications remain to be dealt with. The quantification of field variability and its effects on problems of interest have been vigorous areas of research activity over the past decade. Recent studies include an analytical approach to a layered profile (Sisson, 1987) and formulation of directional hysteresis induced by variability itself (Mantoglou & Gelhar, 1987). Another keen area of interest in which a great deal of progress has been made is flow by preferred paths, such as through cracks or macropores (Germann & Beven, 1985). Formulations of infiltration which account for the simultaneous displacement of air have been well documented for some time (Morel-Seytoux & Vauclin, 1983) and water transport in connection with various organic liquids has been quantified in considerable detail (Parker et al., 1987). Difficult areas which must continue to receive attention include surface crusting, plugging, soil matrix deformation, and interactions of solutes with the hydraulic properties.

REFERENCES

Campbell, G.S. 1985. Soil physics with BASIC. Elsevier Sci. Publ. CO., New York.

Corey, A.T., and A. Klute. 1985. Application of the potential concept to soil water equilibrium and transport. Soil Sci. Soc. Am. J. 49:3–11.

Germann, P.F., and K. Beven. 1985. Kinematic wave approximation to infiltration into soils with sorbing macropores. Water Resour. Res. 21:990–996.

Green, W.H., and G.A. Ampt. 1911. Studies on soil physics. I. Flow of air and water through soils. J. Agric. Sci. 4:1–24.

Hillel, D. 1980. Fundamentals of soil physics. Academic Press, New York.

Hillel, D. 1982. Introduction to soil physics. Academic Press, New York.

Jackson, R.D. 1973. Diurnal changes in soil water content during drying. p. 37–55. In R.R. Bruce et al. (ed.) Field soil water regime. SSSA Spec. Publ. 5. SSSA, Madison, WI.

Letey, J. 1968. Movement of water through soil as influenced by osmotic pressure and temperature gradients. Hilgardia 39:405–418.

Mantoglou, A., and L.W. Gelhar. 1987. Effective hydraulic conductivities of transient unsaturated flow in stratified soils. Water Resour. Res. 23:57–67.

Milly, P.C.D. 1982. Moisture and heat transport in hysteretic, inhomogenous porous media: A matric head-based formulation and a numerical model. Water Resour. Res. 18:489–498.

Morel-Seytoux, H.J., and M. Vauclin. 1983. Superiority of two-phase formulation for infiltration. p. 34–47. In Advances in infiltration. ASAE, St. Joseph, MI.

Nielsen, D.R., R.D. Jackson, J.W. Cary, and D.D. Evans (ed.). 1972. Soil water. ASA, Madison, WI.

Parker, J.C., R.J. Lenhard, and T. Kuppusamy. 1987. A parametric model for constitutive properties governing multiphase flow in porous media. Water Resour. Res. 23:618–620.

Penman, H.L. 1940. Gas and vapor movements in the soil. I. The diffusion of vapors through porous solids. J. Agric. Sci. 30:437–461.

Philip, J.R. 1968. Theory of infiltration. Adv. Hydrosci. 5:150–191.

Raats, P.A.C. 1970. Steady infiltration from line sources and furrows. Soil Sci. Soc. Am. Proc. 34:709–714.

Richards, L.A. 1931. Capillary conduction of liquids through porous media. Physics 1:318–333.

Sisson, J.B. 1987. Drainage from layered field soils: Fixed gradient models. Water Resour. Res. 23:2071–2075.

van Genuchten, M.Th. 1980. A closed-form equation for predicting the hydraulic conductivity of unsaturated soils. Soil Sci. Soc. Am. J. 44:892–898.

Veihmeyer, F.J., and A.H. Hendrickson. 1949. Methods of measuring field capacity and wilting percentages of soils. Soil Sci. 68:75–94.

Wylie, C.R. 1966. Advanced engineering mathematics. 3rd ed. McGraw-Hill, New York.

5 Solute Transport and Leaching

DAVID E. ELRICK

University of Guelph
Guelph, Canada

BRENT E. CLOTHIER

DSIR
Palmerston North, New Zealand

Water is vital for life in the soil. Water links the crop to the soil. Not only is water a catalyst for soil biological and chemical reactions, but it is also a solvent of, and vehicle for, agricultural amendments such as nutrients, pesticides, and nematicides. Thus the physical well-being and nutrition of plants are vitally dependent upon the small-scale physical and chemical processes engendered by water and its movement in the surface soil of the root zone.

The soil of the root zone of agricultural crops acts both as a transmission medium and as a storage reservoir for any water that falls on the soil surface. When natural rainfall is insufficient, irrigation can be used to maintain a hospitable water status in the root zone. It is imperative that water-driven solute transport processes be understood so that through good irrigation management, crop nutrition can be optimized without groundwater quality being compromised. Salinity control and reclamation are specific examples of solute transport processes that are vitally important in many irrigated regions of the world.

The description, if not the discovery, by Slichter in 1905 that a quantity of salt solution, when added to a soil, will consequently spread out away from the source, has led to numerous studies of the mixing (or dispersion) of solutes in soil and other natural porous media. In particular, the recent concern over soil and groundwater pollution has spurred numerous experimental studies, both in the laboratory and, more importantly, in the field.

Analytically, the conceptual aspects of various deterministic mathematical approaches have received much attention during the past few decades and progress on transport processes in the unsaturated zone has been reviewed recently by Nielsen et al. (1986). More recent statistical approaches, initiated about 10 yr ago, are receiving much attention. The statistical approaches have been motivated by our inability to extrapolate laboratory results to field conditions and the observation that soils and porous geological formations can be very heterogeneous in nature, i.e., they can vary in an irregular manner in space. The soil factors affecting transport can also vary with time, generally in a complex manner as well. Discussions of the statistical approach are given in recent reviews by Dagan (1986), Gelhar (1986), and Sposito et al. (1986). It is convenient to describe soil transport parameters (such as hydraulic conductivity) statistically in terms of a mean and variation about that mean. Our "deterministic" equations are simply approximations for these mean values. They became exact in the limit of zero variation about the mean.

In this chapter we review solute transport and leaching processes from the microscale of the pore-matrix complex, through to the scale of the irrigated field. En route a dilution occurs in the description of the specific transport mechanisms, for spatial and temporal variability in these processes assumes greater importance as the scale approaches that found in the field.

I. LENGTH SCALES

Several length scales need to be considered when examining soils. The first is concerned with some medium characterization scale of the property to be examined, the second is concerned with the size over which the property must be averaged in order to give a proper continuum representation, and the third can be thought of as a length that is representative of the extent over which the measurements are taken. For example, if the property that we are interested in measuring is the porosity, then some average pore (or particle) diameter would be representative of the "medium characterization scale." Clearly, the diameter of some averaging volume must be considerably larger than the average pore (or particle) diameter and the third length, over which the porosity is strongly correlated could, for example, be the length of a packed column of soil.

In what follows we will consider liquid water in soil at the microscopic (pore) scale to be homogeneous and assume that uniformly packed, sieved, oven-dried soil in the laboratory is also homogeneous (within experimental limitations). Thus the properties of liquid water, such as density and viscosity, will be assumed to be independent of position. In the macroscopic (laboratory) domain we will also assume that soil properties such as bulk density and hydraulic conductivity are indepent of position (assuming uniformly packed columns). In both the pedon and field domains, however, soils are heterogeneous in nature and soil properties such as bulk density and hydraulic conductivity will be assumed to be space dependent. Time dependence may also need to be considered.

The length scales of interest are summarized in Table 5–1 (adapted from Dagan, 1986). We have identified four levels of interest, i.e., microscopic, macroscopic, pedon, and field, and approximate length scales as indicated. We begin our discussion at the microscopic domain where the fundamental principles of solute transport can be more easily explained and then move progressively to the larger scales where spatial (and perhaps temporal) variability plays an increasingly important role.

Table 5-1. Continuum approach—a matter of scale.

Domain	Medium characteriza-tion scale	Minimum size for averaging	Usual ex-tent of flow domain	Homogeneous or heterogeneous
		m		
Microscopic	10^{-9}	10^{-7}	10^{-6}	Liquid water in soil assumed to be homogeneous
Macroscopic	10^{-4}–10^{-3}	10^{-2}– 5×10^{-2}	10^{-1}–10^{0}	Uniformly packed columns assumed to be homogeneous
Pedon $(1 \times 1 \times 1$ m) (in the field)	10^{0}	NA	NA	Homogeneous horizontally, layered vertically, macropores present, flow not one-dimensional vertically
Field	10^{1}–10^{3}	NA	NA	All the above for pedon domain but now spatially variable horizontally, surficial effects; e.g., slopes

Although we have limited our discussion to the four length scales noted above, the next larger scale could be at the watershed or irrigation project level and would generally include several soil types. At an even larger landscape level, such as a region of a country, spatial variability studies of soil properties have been shown to yield useful information (Trangmar et al., 1986).

II. PRINCIPLES OF TRANSPORT—THE CAPILLARY TUBE (MICROSCOPIC) ANALOGY

We begin by examining the mixing (i.e., diffusion/dispersion) of solute during flow in a long, straight, narrow cylindrical tube (i.e., a capillary tube)—not a very good approximation of soil. The complexities of a pore network are vastly oversimplified. However, examining the flow behavior of the microscopic system whose geometry and flow velocity can be completely specified gives us an understanding of the basic processes that take place during mixing in larger-scale soil systems where the geometry of the pore space and spatial variability at an even larger scale are extremely complex. We will discover later that many of the equations that we develop to examine dispersion in capillary tubes carry over with only minor modifications to similar situations involving laboratory packed columns. More ingenuity is required when extrapolating to the local and regional scales, however.

A. Solute Diffusion

The process called diffusion can take place in either of the solid, liquid, or gaseous phases as a result of the random thermal motion (called Brownian motion) of molecules. In this section we are concerned only with diffusion in a tube filled with water (solution) in which the water is stationary. If concentration gradients exist in the solution, then solutes will diffuse (or move on average) from a region of high concentration to one of a lower concentration. In stationary bulk water, the solute flux density in the s direction, J_{ss} (mol m^{-2} s^{-1}) is proportional to the concentration gradient and is called Fick's first law:

$$J_{ss} = -D_o \partial c / \partial s \qquad [1]$$

where D_o (m^2 s^{-1}) is the diffusion coefficient for the solute in bulk water, c (mol m^{-3}) is the concentration in the solution, and s is a space coordinate such as x, z, or r.

Combining the above equation with the equation of continuity (conservation of mass)

$$\frac{\partial c}{\partial t} = - \frac{\partial J_{ss}}{\partial s} \qquad [2]$$

gives

$$\frac{\partial c}{\partial t} = D_o \frac{\partial^2 c}{\partial s^2} \qquad [3]$$

or

$$\frac{\partial c}{\partial t} = D_o \nabla^2 c \qquad [4]$$

in general coordinate form where D_o has been taken, as is customary, to be constant.

B. Solute Dispersion

The dispersal of material in a moving fluid (liquid or gas) is a complicated process involving the interacting effects of both convective flow and molecular diffusion. Here we limit the analysis to steady laminar flow in a straight, narrow, circular tube that is filled with water (a solution). With reference to Fig. 5–1, the x component of the velocity at a distance r from the center of the tube is given by Poiseuille's relation (Hillel, 1980 p. 166–168):

$$u(r) = u_o (1 - r^2/a^2) \qquad [5]$$

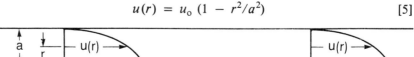
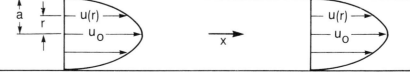

Fig. 5-1. Poiseuille flow in a capillary tube.

where u (m s^{-1}), the x component of the velocity, is a function of the radius r only, u_o is the maximum velocity (at $r = 0$), and r is the radial coordinate.

In the capillary tube, the solute flux densities in the x (longitudinal) and r (radial) directions are given by

$$J_{sx} = -D_o \frac{\partial c}{\partial x} + u(r)c \qquad [6]$$

$$J_{sr} = -D_o \frac{\partial c}{\partial r}. \qquad [7]$$

Note that both diffusive and convective components contribute to J_{sx}. Combining Eq. [6] and [7] with the equation of continuity (Eq. [2]) gives

$$\frac{\partial c}{\partial t} + u(r) \frac{\partial c}{\partial x} = D_o \left[\frac{1}{r} \frac{\partial}{\partial r} \left(r \frac{\partial c}{\partial r} \right) + \frac{\partial^2 c}{\partial x^2} \right] \qquad [8]$$

where axisymmetric cylindrical coordinates have been used to take advantage of the fact that the velocity is constant on cylindrical surfaces which are concentric about the x axis.

Subject to the usual boundary conditions of step or plug solute inputs, Eq. [8] has proven difficult to solve analytically. The approach has been twofold: the first makes use of simplifying assumptions in order to obtain an approximate differential equation which can be solved analytically (Taylor, 1953); the second uses numerical techniques to solve Eq. [8] directly (Ananthakrishnan et al., 1965; Nunge & Gill, 1970). In our brief summary we will concentrate on the approximate analytical techniques because they are more relevant to the problems encountered at the macroscopic and field levels.

At this point we introduce the dimensionless ratio known as the Péclet number, P_e, for solute diffusion (the analogue of the Péclet number for heat transfer):

$$P_e = au_m/D_o, \qquad [9]$$

where a is the tube radius, u_m is the mean velocity, and D_o is the molecular diffusion coefficient. Thus, P_e is the ratio of the convective dispersal processes au_m, to the diffusive dispersal processes D_o. This number is useful in depicting the ranges over which certain processes may dominate.

C. Convective Dominant Flow

By far the simplest analytical approximation is to ignore diffusion completely and assume that all the transport is by convective flow. This holds under conditions of large P_e and/or small t.

Neglecting diffusion in Eq. [8] gives

$$\frac{\partial c}{\partial t} + u(r)\frac{\partial c}{\partial x} = 0. \qquad [10]$$

1. Piston Flow

The simplest solution of Eq. [10] is for piston flow where u is a constant (independent of r) and the initial concentration distribution moves unchanged through the tube at a constant speed in the direction of flow. The solution is given by

$$c = f(x - ut) \qquad [11]$$

where f is a function determined by the initial condition.

Coupled with the initial condition

$$t = 0, c = \begin{cases} c_0, & x < 0 \\ 0, & x > 0 \end{cases} \qquad [12]$$

the solution is given by

$$c = \begin{cases} c_0, & x - ut < 0 \\ 0, & x - ut > 0 \end{cases} \qquad [13]$$

2. Capillary Flow

For capillary flow, with the water velocity given by Eq. [5], the solution of Eq. [10] subject to Eq. [12] is given by

$$c = \begin{cases} c_0, & x - u_0(1 - r^2/a^2)t < 0 \\ 0, & x - u_0 (1 - r^2/a^2)t > 0 \end{cases} \qquad [14]$$

In what follows, the volume-averaged concentration represents the mean concentration at some position in the tube whereas the flux-averaged concentration represents the mean concentration flowing across a given cross-sectional area. Parker and van Genuchten (1984) discussed the interpretation of these concentrations in soil systems.

D. Dispersive Flow—Approximate Analytical Approach

Taylor's (1953) approach was to apply certain limiting conditions which would then simplify Eq. [8] to a form that could be solved analytically. His approach was based on earlier work which showed that a dye tracer injected into a stream of water undergoing laminar flow appeared to spread out in a symmetrical manner about the plane moving at the average flow velocity. The interaction between convective flow and molecular diffusion under certain (but not all) circumstances produces a concentration distribution that is symmetric about the plane moving at the average velocity. The final approximate form of Eq. [8], called the Taylor-Aris equation (Aris, 1959), is given by

$$\frac{\partial c_m}{\partial t} + u_m \frac{\partial c_m}{\partial x} = D_2 \frac{\partial^2 c_m}{\partial x^2} \qquad [15]$$

where

$$D_2 = D_o + \frac{a^2 u_o^2}{192 D_o} = D_o + \frac{a^2 u_m^2}{48 D_o} \qquad [16]$$

and c_m is the mean concentration.

What is intriguing about Eq. [15] is that its form is similar to the equation that would apply to the distribution about the plane moving at the mean velocity, u_m, with the mixing brought about by the dispersion coefficient D_2. Note that $D_2 - D_o$ is proportional to u_m^2 and that the molecular diffusion coefficient, D_o, appears in the denominator in the second term on the right hand side of Eq. [16]. At low values of u_m, D_2 reduces to D_o, which is what one would expect.

E. Analytical Solutions for Steady Flow

Several important analytical solutions of equations similar in form to Eq. [15] for infinite, semi-infinite, and finite systems were summarized by van Genuchten and Wierenga (1986).

III. LABORATORY (MACROSCOPIC) SCALE LEVEL—HOMOGENEOUS SOIL

In the previous section we were able to specify completely the flow velocity everywhere within the capillary tube. In a soil, or even in a regular packing of spheres, it is not possible to do so and we avoid this complexity by developing a macroscopic continuum on a scale considerably larger than the average pore size of the soil (see section I in this chapter).

Most approaches use probability theory applied to a large number of steps or systems and may be classified as "statistical" theories. Dullien (1979) classified the statistical approaches into three broad categories:

1. Random phenomena-type (stochastic) models.
2. Network models of capillary tubes with randomly distributed diameters.
3. Statistical models based on spatial (or volume) averaging.

In (1) to (3) above, the problem of solute transport was treated without including any microscopic pore structure parameters in the model. However, the assumptions made were in most cases based on the quantitative knowledge obtained by examining dispersion in capillary tubes.

For example, one of the earliest statistical approaches is due to Scheidegger (1960) who considered the passage of each fluid element through the porous system as a random walk consisting of a succession of statistically independent straight steps in equal small intervals of time. He also assumed that Darcy's Law was satisfied for this macroscopic flow and that the medium was homogeneous and isotropic. Scheidegger (1960) invoked the central limit theorem for the law of large numbers, which leads to a probability distribution function that is Gaussian (normal).

A. Solute Transport During Steady One-Dimensional Flow

The result of Scheidegger's (1960) analysis, as well as that of the many others who used various statistical approaches (Dullien, 1979), is the following differential equation which has been generally used to describe the mixing of solute during macroscopic, one-dimensional, steady flow in homogeneous soil:

$$\frac{\partial c}{\partial t_1} = D_s \frac{\partial^2 c}{\partial s_1^2} \qquad [17]$$

where s_1 and t_1 are coordinates in a system moving at the mean pore water velocity v and where

$$s_1 = s - vt \text{ and } t_1 = t. \qquad [18]$$

In a fixed coordinate system Eq. [17] is written:

$$\frac{\partial c}{\partial t} + v \frac{\partial c}{\partial s} = D_s \frac{\partial^2 c}{\partial s^2} \qquad [19]$$

where D_s is the longitudinal dispersion coefficient ($m^2 s^{-1}$), v is the constant pore water velocity ($m s^{-1}$), and c is the liquid concentration ($mol m^{-3}$). The above equation bears a strong resemblance to Eq. [15], the latter developed by Aris (1959) as an approximate equation to describe dispersion in a single capillary tube.

Considerable effort has recently been devoted to the derivation of Eq. [19] in saturated porous media from molecular or fluid mechanical principles applied at the microscopic level. The physical and mathematical issues raised in the derivations of Eq. [19] were discussed critically by Sposito et al. (1979). More recently Chu and Sposito (1980) and Bhattacharya and Gupta (1983) made progress in deriving Eq. [19] for homogeneous, saturated porous media at the macroscopic (laboratory) scale. The lazy person's approach is to invoke heuristic arguments and simply assume that Eq. [19] describes dispersion at the macroscopic level in homogeneous soil columns (Aronofsky

& Heller, 1957) and then let the errors in the approximation collect in the dispersion coefficient, D_s. Regardless of the approach, the convection-dispersion equation for nonreactive soil conditions (Eq. [19]) forms the mechanistic basis for describing solute movement in soils.

Thus, the parameter that describes solute mixing in soils is the macroscopic dispersion coefficient, D_s. Theoretically, if we know D_s and its dependence, if any, on θ and J_w, we are then in a position to predict the distribution of solute in a soil column as a function of distance and time.

In the capillary tube (microscopic level), Eq. [16] predicts a quadratic relationship between D_2 and u_m. One would then perhaps expect a similar relationship between D_s and v in soils. By analogy to the capillary tube model it appears reasonable to write:

$$D_s = D_o\tau + \lambda v^\eta \qquad [20]$$

where D_o is the diffusion coefficient (m^2 s^{-1}), τ is the tortuosity factor (which depends on the water content θ), and λ and η are simply empirical constants to be determined experimentally. Thus the complexity of the flow in soil is rendered treatable by empiricism. For saturated homogeneous systems, the exponent η has been shown to be closer to 1.0 than 2.0 and the above equation is often written as

$$D_s = D_o\tau + \lambda' v \qquad [21]$$

where λ' is called the dispersivity (m). The value of λ' is reported to be about 0.005 m or less for laboratory scale experiments with packed soil columns (Nielsen et al., 1986). Theoretically, Bhattacharya and Gupta (1983) have shown that D_s is expected to be linearly related to v for large values of v in saturated porous media. The tortuosity, τ, for a glass bead mixture was reported to be 0.84 (De Smedt & Wierenga, 1984). In soils the tortuosity is generally smaller and decreases as the aggregate and/or particle size becomes smaller and the shapes less spherical and more plate-like. The tortuosity for most soils ranges from 0.7 for sandy soils to about 0.3 or less for unstructured clayey soils.

Procedures for the calculation of D_s from experimental data are given in a recent review by van Genuchten and Wierenga (1986).

1. Immobile Water—Nonequilibrium, Diffusion-Controlled Transport

The simplest inert porous medium consists of a pack of glass beads or beach sand, and even here the analysis can become extremely complex. In the straight capillary tube the water velocity can be completely defined everywhere in the tube; this is not true in soils and it is probable that even in the simplest system of a pack of glass beads not all of the water is completely mobile as is assumed in the capillary tube. De Smedt and Wierenga (1984) assumed that the water in a porous medium, such as glass beads, could be divided into a mobile, θ_m, and immobile θ_{im}, fraction where the total water content, θ, is given by

$$\theta = \theta_m + \theta_{im}. \qquad [22]$$

The convective-dispersion equation (Eq. [19]) is then replaced by

$$\frac{\partial c_m}{\partial t} = D_m \frac{\partial^2 c_m}{\partial x^2} - v_m \frac{\partial c_m}{\partial x} - \theta_{im} \frac{\partial c_{im}}{\partial t} \qquad [23]$$

where

$$\theta_{im} \frac{\partial c_{im}}{\partial t} = \alpha \theta_m (c_m - c_{im}) \qquad [24]$$

and

$$v_m = J_w / \theta_m. \qquad [25]$$

Equation [24] describes the transfer of solute between the mobile and immobile fractions where α is a mass transfer coefficient (s^{-1}). In this model, nonequilibrium conditions were assumed by including diffusion controlled movement into the postulated stagnant water zones. This immobile water, although formulated on a macroscopic scale, has its origins at the microscopic level and has been visualized as dead-end pores (Coats & Smith, 1964), thin, liquid films around soil particles, stagnant intra-aggregate water (Passioura, 1971), or isolated portions of water found during unsaturated flow (Nielsen & Biggar, 1961; Rao et al., 1980). At the macroscopic scale, the process can be visualized as one-dimensional flow, v_m, in the x direction, with dispersion as predicted by the convection-dispersion equation complicated by sorption/desorption into immobile water zones distributed uniformly throughout the porous material (Nkedi-Kizza et al., 1984).

The observed and calculated breakthrough curves for 100-μm glass beads, obtained by Krupp and Elrick (1968) and analyzed by De Smedt and Wierenga (1984) using Eq. [23], [24], and [25], are shown in Fig. 5-2. An excellent fit to the experimental data is obtained by assuming that the volumetric water content of 0.24 contains 87% mobile water and 13% immobile water. The transfer coefficient, α, was estimated at 20.9 d^{-1} and the dispersion coefficient for the mobile water fraction was calculated to be 8.06 $cm^2\ d^{-1}$.

A linear fit to the D_s vs. v data gave

$$D_s = 1.2 + 0.21\ v \qquad [26]$$

where D_s is in $cm^2\ d^{-1}$ and v in $cm\ d^{-1}$. If the classical equation was used [Eq. 19], the calculated values of D_s in unsaturated soil were about 20 times larger than those calculated for saturated columns leached at comparable pore water velocities. However, when Eq. [23] and [24] were used, the dis-

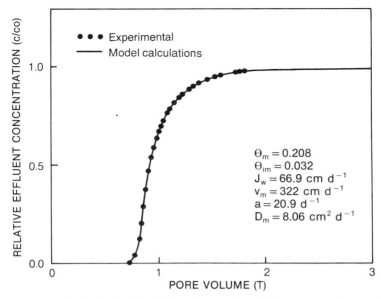

Fig. 5-2. Observed and calculated breakthrough curves for a displacement experiment of Krupp and Elrick (1968) (redrawn from De Smedt & Wierenga, 1984).

persion coefficients were found to be consistent with those obtained from comparable saturated flow experiments.

Thus far we have found that the macroscopic porous medium differs from the microscopic capillary tube in that immobile water should be included in the description of even the simplest system such as inert, homogeneous, isotropic, and rigid glass bead media. We now move on to further complications.

2. Reactive Porous Media

During steady, one-dimensional flow in soils, in which some of the solute is sorbed onto the particle surface, the classical convective-dispersion equation (Eq. [19]) can be modified by adding a term to account for the sorption:

$$\frac{\partial c}{\partial t} + \frac{\rho}{\theta} \frac{\partial c}{\partial t} + v \frac{\partial s}{\partial x} = D \frac{\partial^2 c}{\partial x^2} \qquad [27]$$

where ρ is the soil bulk density (kg m^{-3}), θ is the volumetric water content (m^3 m^{-3}), and s is the solute concentration associated with the solid phase of the soil (mol kg^{-1}). Equation [27] could also be expanded to include mobile-immobile water (Eq. [23]) but for simplicity Eq. [19] is used.

A further simplification is to assume: (i) that the reaction rate is sufficiently fast so that instantaneous equilibrium between the concentrations in the liquid and solid phases can be assumed or (ii) that the rate of reaction is either diffusion or chemically controlled. A more detailed review than the following is given in Nielsen et al. (1986).

3. Equilibrium Transport with a Linear Isotherm

The most common approach has been to assume a linear relationship between s and c:

$$s(c) = K_d c \qquad [28]$$

where K_d is the slope of the isotherm and is referred to as the distribution coefficient. As shown below, the simple linear isotherm above has mathematical conveniences but the linear relationship rarely holds over a large range of c values. However, linearization of the nonlinear isotherm at the appropriate concentration can often achieve usable results. For a more detailed analysis, isotherms of the Freundlich or Langmuir forms, or whatever form is appropriate, may be used to describe the $s(c)$ equilibrium relationship. It should also be noted that the process is often not reversible and that different coefficients may apply to desorption as compared to adsorption, particularly for organic solutes (van Genuchten et al., 1974).

Substituting Eq. [28] into [27], gives

$$R\frac{\partial c}{\partial t} + v\frac{\partial c}{\partial x} = D\frac{\partial^2 c}{\partial x^2} \qquad [29]$$

where R, the retardation factor (dimensionless) is given by

$$R = 1 + \frac{\rho K_d}{\theta}. \qquad [30]$$

Wagenet et al. (1977) found that whereas urea was only weakly adsorbed by Tyndall silty clay loam (coarse-loamy, mixed [calcareous], thermic Aeric Haplaquepts) ($K_d = 0.08$ L kg^{-1}), the cation NH_4^+ was more strongly sorbed ($K_d = 3.3$ L kg^{-1}). In a leaching experiment they injected a solution of urea (1000 mg L^{-1}) and NO_3^- (500 mg L^{-1}) at a pore water velocity of 2.6 mm h^{-1} into a 150-mm column of soil. The water content increased to $\theta = 0.4$ with $\rho = 1500$ kg m^{-3}. In the resulting effluent breakthrough curves (Fig. 5–3) the inert NO_3^- ($R = 1$) preceded the slightly retarded urea ($R = 1.3$). The NH_4^+ rapidly generated from the urea lagged even further ($R = 13.4$). The lines on Fig. 5–3 are obtained from solving equations similar in form to Eq. [36] and [37], which are discussed in section III.A.5. of this chapter.)

Noting that R is constant with respect to c for the case of a linear (or linearized) isotherm, it is obvious that all our previous solutions can be sim-

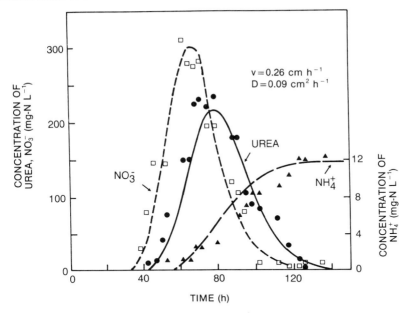

Fig. 5-3. Breakthrough curves of NO_3^-, urea, and NH_4^+ in a 150-mm column of Tyndall silty clay loam (redrawn from Wagenet et al., 1977).

ply modified by including the retardation coefficient R which merely delays the appearance of the solute. In fact, it is Eq. [29] that is used as the basic equation by van Genuchten and Wierenga (1986) in the calculation of D_s from experimental solute effluent data. For nonreactive solutes in soil R is unity, and the solutions are applicable to inert media. Based on the concept of a retardation constant, Baes and Sharp (1983) published a proposal for estimating soil leaching and leaching constants for solutes in agricultural soils.

4. Nonequilibrium Transport

The mobile-immobile water concept is an example of diffusion-controlled sorption in which Eq. [24] gives the rate of transport of the solute from the one fraction of the soil water to the other for reactive solutes. Equilibrium adsorption-desorption processes could easily be considered in Eq. [23] and [24] by including R where appropriate (Nielsen et al., 1986).

The simplest formulation of nonequilibrium transport occurs when first order, linear kinetics are assumed, i.e.,

$$\frac{\partial s(c)}{\partial t} = \alpha \, (kc - s) \qquad [31]$$

where α is a first order rate coefficient (s^{-1}). The use of Eq. [31] in Eq. [27] has resulted in only modest improvements in the fitting of laboratory effluent data, and success has generally been limited to experiments carried out at relatively low values of v (Hornsby & Davidson, 1973; van Genuchten et al., 1974). The similar form of the nonequilibrium representation (Eq. [31]) and the mobile-immobile water model (Eq. [24]) means that it is difficult to distinguish between these two processes.

Nielsen et al. (1986) reported that the two-site model, in which the adsorption term is thought to consist of two components, one governed by equilibrium adsorption and one by first-order kinetics, has led to improved predictions of the effluent curve (Nkedi-Kizza et al., 1984). The model assumes that the sorption/exchange sites can be divided into two fractions; adsorption on one fraction (type-1 site) is assumed to be instantaneous, whereas adsorption on the other fraction (type-2 sites) is assumed to be time dependent. The two-site model has been used with relatively good success to describe the transport of various organic and inorganic chemicals (Rao et al., 1979; Hoffman & Rolston, 1980; Nkedi-Kizza et al., 1983). Nielsen et al. (1986) concluded, however, that the nonequilibrium transport problem, as approached at present, remains largely unsolved.

Much of the previous work has been summarized in a recent series of papers by Crittenden et al. (1986) and Hutzler et al. (1986) in which a family of numerical models is developed for computing solute transport in quasi-homogeneous saturated porous media consisting of reactive spherical porous aggregates packed into long columns. The model includes the transport and retardation mechanisms of convective flow, longitudinal dispersion, liquid-phase mass transfer, diffusion into immobile liquid and local adsorption equilibrium (in essence all of the effects that have been previously described and more). The model does not include the effects of degradation, which will be discussed in section III.A.5. of this chapter.

The complete model developed by Crittenden et al. (1986) carries the acronym DFPSDM (Dispersed flow, Film transfer, Pore and Surface Diffusion Model). The DFPSDM includes the following mechanisms: (i) constant convective flow; (ii) axial dispersion and diffusion; (iii) film transfer resistance to mass transport from the mobile to the stationary phase; (iv) local adsorption equilibrium between the solute adsorbed onto the soil matrix surface and the solute in the intra-aggregate stagnant fluid; and (v) both surface and pore diffusion as intrastationary phase mass transport mechanisms, all in saturated porous media. The set of simultaneous, nonlinear partial differential equations is solved by orthogonal collation.

Other models such as Dispersed Flow, Local Equilibrium Model (DFLEM), which is comparable to solving Eq. [29], Segregated Flow Model (SFM), and Local Equilibrium Model (LEM) (Crittenden et al., 1986) are based on simplifications of the DFPSDM model. Hutzler et al. (1986) measured breakthrough curves of Cl^- and organic compounds from packed columns of a sandy loam soil. Both the LEM and the SFM models approximated the retardation of the organic chemicals but did not account for the amount of spreading in the breakthrough data. The DFLEM simulated the breakthrough of the organic chemicals and Cl^- but only if D_s was adjusted to match the breakthrough data. The DFLEM model with independently predicted D_s values greatly underestimated the dispersion. The DFPSDM also simulated well the breakthrough data if the aggregate radius was adjusted to fit the data. Neither the adjusted radius nor D_s could be related to the hydraulic characteristics of the column or to the soil properties.

Recently, Roberts et al. (1987) applied the DFPSDM transport model to interpret the results of a set of miscible displacement experiments con-

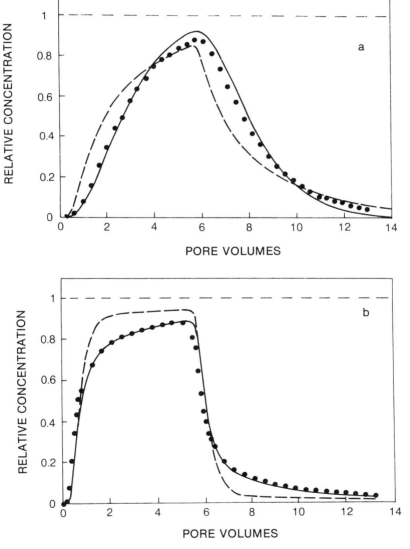

Fig. 5-4. Predicted and observed effluent concentration histories for Ca. (a) $v = 3.2$ cm h^{-1}; (b) $v = 127.5$ cm h^{-1}. Curves represent DFPSDM predictions for high (solid) and low (dashed) end of D_p range; solid circles are the data of Nkedi-Kizza et al., (1982) (redrawn from Roberts et al., 1987).

ducted with an aggregated soil (the data of Nkedi-Kizza et al., 1982). Roberts et al. (1987) used, wherever possible, parameter values obtained independently of the column data. The contribution of surface diffusion was neglected on the grounds that pore diffusion was expected to dominate internal transport where the sorptive interactions were relatively weak, as in the aggregated soil of Nkedi-Kizza et al. (1982). The incorporation of these several effects results in a relatively good fit of the predicted and experimental data as shown in Fig. 5-4, in which the effluent curves for Ca are shown. For each experimental curve, two predictions are shown corresponding to the high (solid curve) and low (broken curve) range of feasible values of the effective pore diffusion coefficient, D_p. In all cases (i.e., for tritiated water and Cl$^-$ as well as for the Ca results shown in Fig. 5-4) the experimental data are fairly well bracketed by the two D_p values. One of the problems, however, with this approach is that any inaccuracy in estimating one of the parameters can interact to confound the estimates of the other parameters. One of the conclusions from the analysis of Roberts et al. (1987) is that hydrodynamic disperson dominated the front-spreading behavior at low velocities (approximately 3 cm h^{-1}) and internal diffusion dominated at high velocities (>120 cm h^{-1}).

The above studies indicate some of the complexities that need to be employed when attempting to model solute transport in relatively simple soil systems; i.e., homogeneous, aggregated, saturated soil.

We have one important factor yet to consider in soil columns; the source term which is required in order to include the effects of chemical precipitation and dissolution, adsorption of solutes by plant roots, and the transformation and utilization of solutes by soil microorganisms.

5. Sources and Sinks

We now include the effects of the above processes by adding the source term ϕ_i [mol m^{-3} s^{-1}] to Eq. [29]:

$$R \frac{\partial c}{\partial t} = D_s \frac{\partial^2 c}{\partial x^2} - v \frac{\partial c}{\partial x} + \Sigma\phi_i (c, s, \ldots) \qquad [32]$$

where ϕ_i accounts for sources and/or sinks in the soil.

A relatively simple form for ϕ_i is given by (Nielsen et al., 1986)

$$\phi_i = -\mu_l \theta c - \mu_s \rho s + \gamma_l \theta + \gamma_s \rho = \mu c + \gamma \qquad [33]$$

where μ_l and u_s are rate coefficients for first-order decay, and γ_l and γ_s are zero-order production terms for the liquid (l) and solid (adsorbed) (s) phases, respectively. The source term has been simplified by setting

$$\mu = \mu_l + \mu_s \rho K_d / \theta \qquad [34]$$

$$\gamma = \gamma_l + \gamma_s \rho / \theta \tag{35}$$

where use has been made of the linear adsorption isotherm, $s = K_d c$. For either chemical or microbial degradation, each of the above coefficients is likely to have a different value, which leads to a complex analysis. It is also likely that the coefficients would be time dependent as the microbial population changes (McLaren, 1970; Blanch, 1981).

If the consecutive decay chains of organic or inorganic compounds are of interest, Eq. [32] can be replaced by the following system of partial differential equations (Nielsen et al., 1986):

$$R_1 \frac{\partial c_1}{\partial t} = D_s \frac{\partial^2 c_1}{\partial x^2} - v \frac{\partial c_1}{\partial x} - \mu_1 c_1 \tag{36}$$

$$R_i \frac{\partial c_i}{\partial t} = D_s \frac{\partial^2 c_i}{\partial x^2} - v \frac{\partial c_i}{\partial x} + \mu_{i-1} c_{i-1} - \mu_i c_i, \quad i = 2, 3, 4, \ldots n \tag{37}$$

where n is the number of chemical species in the decay chain. As an example, the nitrification process can be modeled in soils by setting c_1, c_2, and $c_n = c_3$ as the concentration in the liquid phase of NH_4^+, NO_2^-, and NO_3^- nitrogen, respectively (McLaren, 1970; Cho, 1971). Selim and Iskander (1981) developed a model for predicting N transport and transformations during infiltration, or transient flow, which leads us into the next section in which the steady flow of this section is replaced by nonsteady or transient flow. In nature, steady flow conditions seldom prevail and the transient processes of infiltration, evaporation, redistribution, and drainage dominate, but generally only in the short term. Long-term effects are often evaluated using steady-flow conditions as a first approximation.

B. Solute Transport During Transient One-Dimensional Flow

The one-dimensional, transient transport of solute in laboratory columns of uniformly packed soil can be approximately described by a convection-dispersion equation, equivalent to Eq. [32] for steady flow, but now written in the more general form where J_w and θ are functions of x and t (Laryea et al., 1982):

$$\frac{\partial(\theta c)}{\partial t} + \rho \frac{\partial s}{\partial t} = \frac{\partial}{\partial x} \left[\theta D_s \frac{\partial c}{\partial x} \right] - \frac{\partial}{\partial x} (J_w c) + \Sigma \phi_i. \tag{38}$$

Note that the mass balance on the above equation is based on moles of solute per unit bulk volume of soil.

A critical concern in deriving the analytical solutions of Eq. [38] is the dependence of D_s on water content and velocity. Equation [21] has been generally accepted as a strictly empirical law to describe solute transport through water-saturated porous media in laboratory packed columns. In unsaturated soils, however, the use of Eq. [21] is controversial (Smiles et al., 1978; Elrick et al., 1983; Sposito et al., 1986). Beese and Wierenga (1983) reviewed a number of studies which support the accuracy of Eq. [21] when $v > 0.01$ m d^{-1}. Smiles and Philip (1978), Elrick et al. (1979), Smiles et al. (1981), and Watson and Jones (1984) concluded, on the basis of the observed scaling of unsaturated solute profile data from packed soil columns, that D_s is independent of v at least for values of $v < 1.0$ m d^{-1}. Therefore, D_s can be considered a function of θ only. It is probable that some velocity dependence of D_s exists during the initial infiltration stages when v is largest but the overall effect is so small that it is masked by the scatter in the experimental data.

For free-water adsorption, use of the Boltzmann transformation

$$\lambda = xt^{-1/2} \qquad\qquad [39]$$

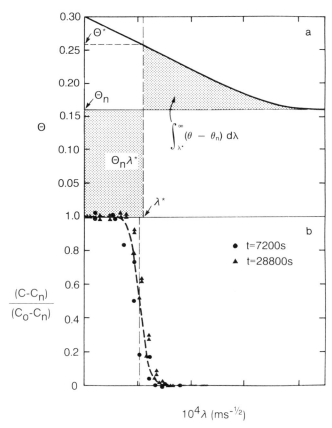

Fig. 5-5. (a) Absorption water content profile and the plane $g = 0$ ($\lambda = \lambda^*$) that separates the invading solution from the water initially present in the soil. (b) Dispersion of KCl about λ^* with $\theta D_s = 5 \times 10^{-10}$ m^2 s^{-1}. The broken curve is the theoretical solution (redrawn from Smiles & Philip, 1978).

allows both x and t to be removed explicitly from Eq. [38] and be replaced by λ. With $\phi_i = 0$ this gives

$$\frac{d}{d\lambda}\left[\theta D_s(\theta)\frac{dc}{d\lambda}\right] + \frac{1}{2}\left[g(\theta) + \rho\lambda\frac{ds}{d\lambda}\right]\frac{dc}{d\lambda} = 0 \qquad [40]$$

Smiles et al. (1978) defined

$$g[\theta(\lambda)] = \theta\lambda + 2D(\theta)d\theta/d\lambda = \theta\lambda - \int_{\theta_n}^{\theta_s}\lambda d\theta \qquad [41]$$

where $D(\theta)$ is the soil water diffusivity and θ_s and θ_n are the saturated and initial water contents, respectively. The function g of Eq. [41] is a material coordinate based on the distribution of the invading water. This reduces the problem to one of describing solute movement relative to the invading water. Smiles and Philip (1978) recognized that salt movement relative to the absorbing water would be approximately centered about $g = 0$. When $g = 0$ let $\lambda = \lambda^*$ and $\theta = \theta^*$. These values can be found directly from the wetting profile $\lambda(\theta)$ using Eq. [41], hence

$$\theta^*\lambda^* = \int_{\theta_n}^{\theta^*}\lambda d\theta, \text{ or } \theta_n\lambda^* = \int_{\lambda^*}^{\infty}(\theta - \theta_n)\,d\lambda. \qquad [42]$$

The λ^* represents a plane separating the antecedent water (which appears to be completely shunted ahead of the invading solution) from the invading solution. Figure 5–5 shows this arrangement and the dispersion of KCl during wetting of a repacked soil (Smiles & Philip, 1978).

This approach identifies clearly the role of the antecedent water on the penetration of solute, or on the leaching of the initial water by the invading irrigation water. The drier the soil the closer the solute plane-of-separation is to the wetting front (Raats, 1984). This variation in velocity between the solute and wetting fronts arises because solute movement represents displacement of fluid through the entire wetted pore space, whereas the wetting profile is just a dynamic propagation of water in response to gravity and capillarity (Wilson & Gelhar, 1981).

Independently, De Smedt and Wierenga (1978) developed a plane-of-separation approach to solute flow and applied it to infiltration and redistribution. Various forms of these displacement models have been applied to one-dimensional flux absorption (Smiles et al., 1981; Elrick et al., 1987); to cases of mobile-immobile water (Smiles & Gardiner, 1982; Bond et al., 1982); and to tracer movement during axisymmetric three-dimensional flux absorption (Clothier & Elrick, 1985).

It is possible to simplify the one-dimensional absorption analysis of Eq. [40]. Taking x^* as the plane-of-separation found from the general wetting profile, Watson and Jones (1984) derived the simple expression for the normalized concentration C,

$$C = (c - c_i)/(c_o - c_i) = (1/2)\ erfc\ [(x - x^*)/(2\sqrt{D^*t}\,)] \qquad [43]$$

giving solute dispersion about x^* where $D^* = D_s\ (\theta^*)$. The solution is for a step function input of solution of concentration c_o into an infinite column having an initial concentration of c_i. To find a simple approximation of x^*, i.e., x_a^*, let

$$x_a^* = I/\theta^*, \qquad [44]$$

where I is the cumulative infiltration (mm), and θ^* is the average water content of the wetted soil. This assumes that the invading solution has occupied

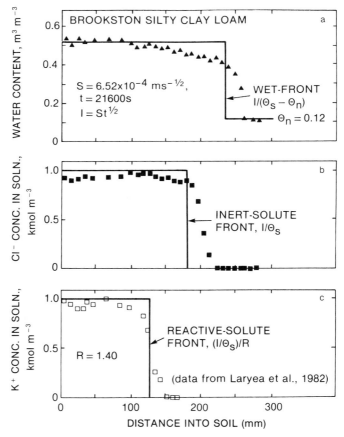

Fig. 5–6. Absorption of water, Cl^-, and K^+ in Brookston silty clay loam (from Laryea et al., 1982). The Green-Ampt approximation of soil wetting is given in (a), and in (b), Eq. [44] locates the Cl^- front. The retarded penetration of the K^+ front is predicted in (c) (redrawn from Elrick et al., 1987).

the entire wetted space $\theta^* x_a^*$, which is assumed constant over the zone of solute dispersion. For absorption into packed laboratory columns we assume that $\theta^* = \theta_s$, as a Green-Ampt rectangular profile has been assumed. However, in the following sections where natural field soils are examined, the average water content of the soil may be considerably less than saturation and quite variable in time and space as well.

As we proceed towards the field scale, even further simplifications are justified when balancing the need for simplicity with the complexity of the field. The second order facet of solute dispersion can be dropped to concentrate primarily on the more important factor of the depth of solute penetration by convection. Often it is reasonable to approximate the wetting profile by the Green-Ampt rectangle: $I = (\theta^* - \theta_n)x_f$, where x_f is the position of the wetting front. Figure 5-6 shows how well this fits the wetting of the Brookston silty clay loam (fine-loamy, mixed, mesic Typic Argiaquolls) found by Laryea et al. (1982). Also x_a^* from Eq. [44] reasonably approximates the depth of penetration of Cl^-. This simple analysis can be extended to reactive chemical transport. Laryea et al. (1982) independently determined that K^+ was adsorbed by this soil with $K_d = 0.2$ L kg^{-1} so that here $R = 1.43$ for K^+. This places the K front at x_a^*/R, which is seen to be reasonable (Fig. 5-6c). Estimation of θ^*, the water content of the transport zone, is critical in such calculations of solute movement. Note that in the absorption profile of Fig. 5-6 that θ^* was set equal to θ_s. Determination of this volume fraction of the transport phase is critical in the field and it is probable that θ^* would be lower than θ_s as well as temporally and spatially variable.

Rose et al. (1982a) extended this approach to predict the displacement of the peak concentration of solute simply from a knowledge of water inputs and drainage and evapotranspiration losses over time, Δt. The shift in the location of the solute peak, Δz, they considered to be

$$\Delta z = \Delta W / \theta^* \qquad [45]$$

where ΔW is the equivalent depth of ponded water that passed beyond the previous depth of the peak over interval Δt. Determination of the transporting fraction of water, θ^*, is again critical. Rose et al. (1982b) further developed the model to include dispersive effects during displacement of the peak. Cameron and Wild (1982) found this approach described Cl movement and dispersion through field soil. Barry et al. (1985) carried the analysis further to interpret the experiments of Saffigna et al. (1977) of leaching under multiple fertilizer applications.

IV. NATURAL SOIL—PEDON SCALE

At the pedon scale we consider a volume of soil with surface area of approximately 1 m^2, and reasonably uniform in horizontal extent. The soil may have natural pedogenic profiling, layering, and contain both macropores and interstitial pores characteristic of the particle size distribution.

Macropores we class as the large, often interconnected, voids that result from soil floral and faunal activity, or pedogenic processes such as cracking.

The quantitative physical understanding of the previous sections provide a sound physical basis upon which to analyze solute transport and leaching in soil at the pedon scale. Consider the movement of soluble N-P-K fertilizer away from a surface drip irrigation emitter. Guennelon et al. (1979) measured the N, P, and K^+ concentrations in soil around a drip emitter after one summer's application of soluble fertilizer in the irrigation water applied to apples growing in clay near Avignon, France (Fig. 5-7). Many of our prejudices are confirmed. The mobile (inert) nitrate travelled extensively away from the emitter. The K^+, and to a greater extent the P, experienced retarded movement. Despite this broad-scale accord, there are significant deviations from our theoretical expectations.

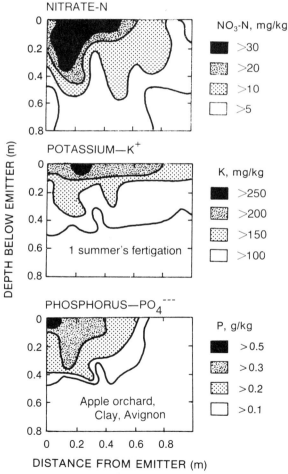

Fig. 5-7. Measured pattern of N, P, and K^+ in the soil after one summer's application of soluble fertilizer via a drip emitter (redrawn from Guennelon et al., 1979).

Nonuniformity in soil wetting by the irrigation emitter is considered to have led to the preferential movement of fertilizer along twin lobes away from the emitter. Rapid macropore flow is a suspected mechanism, as could be fingering resulting from frontal instability. These could account for the greater-than-expected mobility of P, normally a strongly reactive chemical.

Such nonuniformity in soil water flow results in local multidimensional flow, rendering inappropriate direct translation of our understanding gained from studying one-dimensional flow in packed soil columns in the laboratory. Macropores, both pedogenic and biogenic, can be considered the sine qua non of irrigated field soil. They can provide for rapid transmission of air, water, and nutrients through the soil to plant roots. This so-called "short-circuiting" may well, in some soils, be vital for the maintenance of plant vigor and crop growth. It provides for rapid transmission of applied water and nutrients to the roots of the plant.

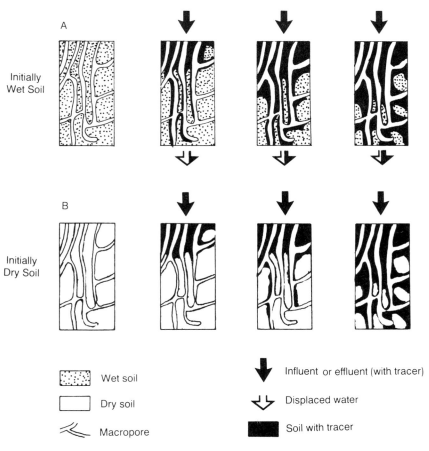

Fig. 5–8. Effect of macropores on water flow and breakthrough in two unsaturated soils (moist and dry) following ponding at the soil surface (redrawn from Bouma, 1981).

Surface-vented macropores play an important role in the entry and transmission of ponded water containing solutes (Dixon, 1972; Thomas & Phillips, 1979; Beven & Germann, 1982; White, 1985). Bouma (1981) qualitatively analyzed the entry of ponded water into both a wet and dry soil containing macropores. He suggests that the problem of field-scale variability may well be reduced when the effects of macropores are isolated. Furthermore, macropores substantially alter the fraction of the soil volume that effectively participates in flow (Fig. 5–8). This markedly alters the breakthrough of water and solute. The effect of antecedent water highlighted in this diagram follows qualitatively from that already outlined theoretically in section III.B. of this chapter. The wetter the soil initially, the greater the lag of the solute front behind the wet front.

Recently, quantitative analyses have been applied to discern the effect of surface-vented macropores on water entry into soil. The geometric complexity of macropore systems as that depicted in Fig. 5–8 presently defies quantitative description and analysis. But the role of simple vertical slots and holes on infiltration and leaching has been treated analytically (Germann, 1983; Davidson, 1984; Smettem, 1986) and numerically (Scotter, 1978; Edwards et al., 1979). Fig. 5–9a, from Edwards et al. (1979), shows the simulated pressure potential distribution in uniform soil with no macropores 30 min after application of free water at the surface. The wet front of irrigated water penetrated to only about 200 mm. Figure 5–9b shows the dramatic effect of a single 600-mm-long, 10-mm-diam. "macropore" on the entry of the surface ponded irrigation water. Thus a few surface-vented macropores can markedly alter the penetration of ponded water and surface-applied solute and lead to substantial spatial variability in soil wetting and solute concen-

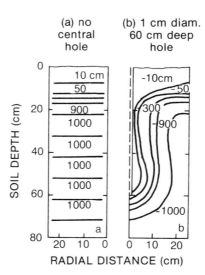

Fig. 5–9. Simulated pressure head in 0.5 m soil columns (*a*) with no central hole and (*b*) a "macropore," after 30 min (redrawn from Edwards et al., 1979).

trations (Kanchanasut & Scotter, 1982). The fraction of the soil "seen" by the invading water can be quite low in these cases. This can lead to substantial mispredictions of solute transport and leaching if uniform flow through the entire pore system is assumed. Watson and Luxmoore (1986) found that large macropores (>0.5 mm) comprised only 0.04% of the soil volume, yet they conducted 73% of the water flux. Such a phenomenally low θ^* (Eq. [44]) can account for the rapid and far-reaching transport of solutes under irrigation.

The contribution of large macropores to this preferential flow is critically dependent upon the local presence of free water at the entrance of the macropore. Locally, if the irrigation rate (or rainfall rate) is less than the saturated matrix hydraulic conductivity, then surface ponding will not occur. The role of large macropores will then be diminished. Scotter and Kanchanasut (1981) used two dyes to mimic anion movement in soil cores obtained under pasture. Methylene blue in ponded water identified those macropores contributing to saturated flow. Figure 5–10 shows this small volumetric fraction. When water containing rhodamine B dye was applied to the same core at the slight unsaturated potential of just -2 J kg^{-1}, quite a different pattern of flow was detected and is shown in Fig. 5–10. White et al. (1986) examined Cl$^-$ transport through undisturbed Evansham clay (fine, montmorillonitic, mesic Typic Pelluderts), a soil with $\theta_s = 0.49$. At a high rate of irrigation they found that only $\theta = 0.13$ contributed to Cl$^-$ transport, whereas at a much lower rate the flow was through $\theta = 0.25$; both of these transport volumes are well below θ_s.

Thus, the rate of irrigation can markedly affect the pattern of soil wetting. When the rate is high relative to the soil's hydraulic conductivity, substantial surface ponding can occur, allowing full expression of macropores and leading to preferential flow of water and solute through a small fraction of the soil. A new kind of permeameter has been developed (Perroux & White,

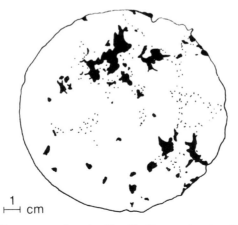

Fig. 5–10. Contrasting pattern of wetting identified by two dyes in a field soil core. The dark area was stained by saturated flow, and the speckled area by flow at a slightly unsaturated potential (redrawn from Scotter & Kanchanasut, 1981).

1988) which gives appropriate measurement and better description of the surface soil macropore-matrix flow regime. This approach highlights the dominant role of surface-vented macropores in transmitting surface free water through the soil (Watson & Luxmoore, 1986). Clothier and Heiler (1983) traced the wet-front penetration in field soil after sprinkler irrigation of about 15 mm of water; at one site the rate of irrigation was 102.5 mm h^{-1}, and at the other it was much lower at 4.1 mm h^{-1}. The contrasting pattern of wetting due to the differing role of surface-vented macropores (Fig. 5–11) would have resulted in quite different patterns of solute transport and leaching. The fraction of the soil involved in convective transport varied greatly between the high and low rates.

A key consequence of this matrix-macropore dichotomy is that the character of the immediate soil surface is crucial for water entry into the rapid preferential pathways of the macropore system. Critically, the physical condition of the soil surface is intimately dependent upon the soil and irrigation management practices. Macroporosity can obliterate the effects of soil textural control on the infiltration rate. Surface-vented macropores can cause a clay to behave hydraulically like a "sand"; conversely, a slaked silt cap can transform a sand into behaving like a "clay."

Nonuniformity in soil wetting during irrigation can also result from other factors, such as wet-front fingering engendered by frontal instability (Raats, 1973; Philip, 1975). Frontal instability leading to fingering can result, inter alia, from air compression ahead of the front (White et al., 1977) or from a depth-wise increase in conductivity such as in layering (Hill & Parlange, 1977). In the laboratory, Starr et al. (1986) studied the pattern of ponded water and solute flow in finer-textured soil overlying coarser-textured material. The onset of frontal instability at the interface produced fingers upwards

WET-FRONT TRACINGS

Chertsey Silt Loam: UDIC USTOCHREPT

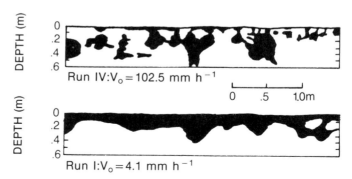

Fig. 5–11. Tracings of the soil wetting in field soil sprinkler-irrigated at two different rates (redrawn from Clothier & Heiler, 1983).

of 50 mm diam. From analysis of field fingering experiments, Starr et al. (1986) cautioned against assuming a uniform wetting of the entire soil as this could lead to serious mass balance calculation errors.

All these examples serve to highlight the care with which the fraction of soil water involved in rapid and far-reaching solute transport needs to be identified. Correct estimation of this flow fraction is critical for the success of simple models such as those of Nofziger and Hornsby (1986) and Barry et al. (1985) which are touted as management tools.

Nonuniformity in solute concentration in the root zone can also arise from factors other than nonuniformity in the transport mechanism of wetting. Chemical reactions in the soil will result in temporal and spatial variations in the solute concentration of any given solute species (Skopp, 1986). For example, McLaren (1970) analytically treated the temporal changes in N resulting from enzymatic hydrolysis of urea to NH_4^+ and the subsequent microbial oxidation to NO_2^- and NO_3^-. Darrah et al. (1985) derived a mathematical model for simultaneous nitrification and diffusion that included nitrifier growth and activity, NH_4^+ adsorption and the influence of pH, and the role of other ions upon diffusion. A challenge for the future will be to incorporate biological facets into physical models of chemical transport in soil (Starr & Parlange, 1976).

V. NATURAL SOIL—FIELD SCALE

Translation of uniform, one-dimensional transport theory to the pedon scale has required significant reconsideration of the significant processes involved in solute transport during irrigation. In moving from the pedon scale to that of the larger field scale we also have to face substantial spatial variation in the horizontal plane. There is also temporal variability. Philip (1975), in pondering this jump in spatial scale, noted that the beautiful economy of analytical scientific methods is soon lost in the magnitude, complexity, and imprecision of the task of synthesis and integration at the larger scale.

Pioneers in field-scale water and solute movement were Nielsen et al. (1973), and Biggar and Nielsen (1976), respectively. Their experiments were carried out on a 150-ha field containing 20 separate 6.5 m^2 plots. Steady-state ponded infiltration was achieved in these plots and the flow ranged from 5.4×10^{-3} to 4.6×10^{-1} m d^{-1}. A 75-mm pulse of water containing elevated Cl^- and NO_3^- levels was then steadily infiltrated and followed again by ponded infiltration with the original solution. Samplers at 359 locations were used to observe the passage of the solute pulse. By fitting the solution of the convective-dispersion equation (Eq. [19]) to these inert tracers, Biggar and Nielsen (1976) obtained 359 values of the convective pore water velocity, v, and the apparent dispersion coefficient D. They found solute transport properties to be log-normally distributed (see Fig. 5–12). The available data at this field scale showed D to be related to v by $D = 0.6 + 2.83$ $v^{1.11}$. Biggar and Nielsen (1976) cautioned that recognition of the form of the frequency distribution of solute transport properties is critical. Substan-

tial errors would result from calculations of solute fluxes simply from multiplication of the mean concentration by the mean water flux.

A different statistical approach was employed by Amoozegar-Fard et
al. (1982) in which a Monte Carlo simulation was carried out in order to
obtain solute concentration and solute transport properties as affected by
the spatial variability of the pore water velocity v and the dispersion coefficient D_s of Eq. [19]. They assumed that the soil water properties were random variables characterized by their relative frequency distribution, i.e.,
normal distributions of ln v, ln D, and θ were used to calculate 2000 values
of C using Eq. [19]. The results showed large differences between the solute
profiles when deterministic values of v were used as compared to average
salt profiles for the 2000 random values of v. The authors conclude that with
a sufficient number of observations of the soil water parameters, the leaching of an entire field may be estimated but not necessarily that of a specific
site.

Sposito and Jury (1986) showed that if solute transport is considered
an ensemble of local convective-dispersion processes, then log-normality in
the distribution of v and D should be expected. Reviews of spatial variability

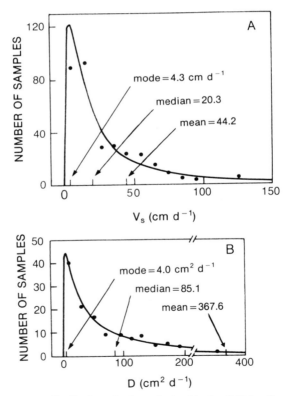

Fig. 5–12. (*A*) Frequency distribution of values obtained in the field for the convective pore
water velocity *v*; and (*B*) for the apparent dispersion coefficient, *D*. (redrawn from Biggar
& Nielsen, 1976).

and the field sampling problem have been presented by Warrick and Nielsen (1980), McBratney and Webster (1983), Peck (1983), Nielsen et al. (1983), and Nielsen and Bouma (1985).

Such extreme spatial variability in field solute transport has led recently to its characterization using mathematical techniques for analyzing random processes. A volte face was heralded by Jury (1982), who developed a stochastic transfer function model to describe solute moving in field soil under steady convective water flow. This model relies purely on elucidation of the probability density function (pdf) of the travel time of solute moving through a volume of soil, and represents a stochastic extension of the deterministic solute travel time work of Jury (1975) and Raats (1978). In the stochastic transfer function model of Jury (1982), Jury et al. (1982) and Jury et al. (1986) no reference to physico-chemical mechanisms is required. Only two time parameters are defined: t', the time when a solute parcel first enters the soil volume, and the time $t + t'$, when it exits. By measuring the passage of solute at one depth, Jury et al. (1982) deduced $g(\tau:t')$, the pdf of the lifetime (τ) of solutes that entered the soil at t'.

A conceptualization of this is given in Fig. 5–13. From measurement at one depth, $z = L$, Jury et al. (1982) used this lifetime function approach to transfer solute profile leaching characteristics to depths both $z < L$ and $z > L$. No consideration can be given to the intervening transport mechanisms. The use of such nonmechanistic models has been controversial.

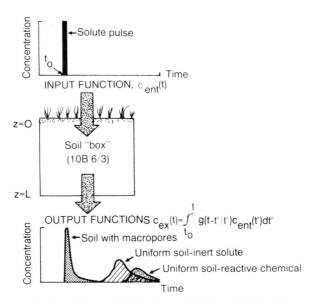

Fig. 5–13. Conceptualization of the transfer function model (Jury, 1982) for characterizing solute transport through a "black-box" (Munsell color 10B6/3) of field soil. The input solute pulse, $c_{ent}(t)$ is transferred to an output pulse, $c_{ex}(t)$, at depth $z = L$. Idealized transfers are given for three common field circumstances of solute leaching.

However, recently there has been a narrowing of the gap between the Convective Dispersion Equation and the Transfer Function Model analyses (Raats, 1985). Sposito et al. (1986) showed that convective-dispersive transport has its own set of characteristic transfer functions. The mechanistic effects of convection, dispersion, adsorption, and mobile-immobile water can be represented in the travel time pdf. Sposito et al. (1986) deduced that the fractional volume of water taking part in solute transport was critical in determining the shape of the travel time pdf (see Fig. 5–13). This emphasis on the fractional flow volume was highlighted in section IV in this chapter, for rarely is the entire volumetric water content involved in field solute transport and leaching during and after irrigation. Use of the transfer function model in this way to elicit the physical processes involved in solute transport is encouraging and should lead to a better understanding of the spatial and temporal variability of observations of wetting and solute concentration in irrigated fields. Concurrently, a more incisive description of field processes is being achieved. This is in part deriving from an improved ability to measure field-soil properties (Topp & Davis, 1981; Dalton et al., 1984), even in structured, or shrinking/swelling soils (Bouma & Raats, 1984).

Better descriptions of unsaturated, solute (e.g., fertilizer) transport through spatially variable soils have been developed (Dagan and Bresler, 1979) and these have been applied successfully to heterogeneous fields (van der Zee and Riemsdijk, 1986). Similarly, the impact of irrigation management on crop yield, for plants growing on heterogeneous soils, can also be analyzed stochastically (Russo, 1986).

VI. CONCLUDING REMARKS

Under both irrigated and natural rainfall conditions, the movement of soil solutes is driven principally by the movement of soil water. Until the 1970s, research studies on solute transport focused generally on the principles of chemical diffusion and dispersion in porous materials. However, the mechanistic principles that worked so well in laboratory column studies have been found wanting when transferred directly to the field. Large macropores such as root channels and cracks short-circuit the flow under saturated conditions, and spatial variability of a generally unpredictable nature introduces unknown factors into the leaching process. In this review, a length-scale approach from the soil pore to the field level has been used, which conveniently categorizes the analyses of solute transport. This approach provides a basis for analyzing the complexity of solute transport and leaching under irrigated conditions.

As concern mounts to minimize leaching for reasons of both efficient use of fertilizer and the maintenance of groundwater quality, greater effort will be directed towards understanding field-scale solute transport. There should in the future be a fruitful combination of stochastic and deterministic, mechanistic, and descriptive models of the biological, chemical, and physical factors involved in solute transport and leaching under irrigated crops.

REFERENCES

Amoozegar-Fard, A., D.R. Nielsen, and A.W. Warrick. 1982. Soil solute concentration distributions for spatially varying pore water velocities and apparent diffusion coefficients. Soil Sci. Am. J. 46:3-9.

Ananthakrishnan, V., W.N. Gill, and A.J. Barduhn. 1965. Laminar dispersion in capillaries. Am. Inst. Chem. Eng. J. 11:1063-1072.

Aronofsky, J.S., and J.P. Heller. 1957. A diffusion model to explain mixing of flowing miscible fluids in porous media. Pet. Trans. AIME. 210:345:349.

Aris, R. 1959. The longitudinal diffusion coefficient in flow through a tube with stagnant pockets. Chem. Eng. Sci. 11:194-198.

Baes, C.F., and R.D. Sharp. 1983. A proposal for estimation of soil leaching and leaching constants for use in assessment models. J. Env. Qual. 12:17-28.

Barry, D.A., C.W. Rose, P.G. Saffigna, and J.-Y. Parlange. 1985. Interpretation of leaching under multiple fertilizer applications. J. Soil Sci. 36:9-20.

Beese, F., and P.J. Wierenga. 1983. The variability of the apparent diffusion coefficient in undisturbed soil columns. Z. Pflanzenernaehr. Bodenkd. 146:302-315.

Beven, K., and P. Germann. 1982. Macropores and water flow in soils. Water Resour. Res. 18:1311-1325.

Bhattacharya, R.N., and Gupta, V.K. 1983. A theoretical explanation of solute dispersion in saturated porous media at the Darcy scale. Water Resour. Res. 19:938-944.

Biggar, J.W., and D.R. Nielsen. 1976. Spatial variability of the leaching characteristics of a field soil. Water Resour. Res. 12:78-84.

Blanch, H.W. 1981. Microbial growth kinetics. Chem. Eng. Commun. 8:181-211.

Bond, W.J., B.N. Gardiner, and D.E. Smiles. 1982. Constant-flux absorption of a tritiated calcium chloride solution by a clay soil with anion exclusion. Soil Sci. Soc. Am. J. 46:1133-1137.

Bouma, J. 1981. Soil morphology and preferential flow along macropores. Agric. Water Manage. 3:235-250.

Bouma, J., and P.A.C. Raats (ed.). 1984. Proceedings of the ISSS Symposium on water and solute movement in heavy clay soil. ILRI Publ. 37. Wageningen, Netherlands.

Cameron, K.C., and A. Wild. 1982. Prediction of solute leaching under field conditions: An appraisal of three methods. J. Soil Sci. 33:659-669.

Cho, C.M. 1971. Convective transport of ammonium with nitrification in soil. Can. J. Soil Sci. 51:339-350.

Chu, S.-Y., and G. Sposito. 1980. A derivation of the macroscopic solute transport equation for homogeneous saturated porous media. Water Resour. Res. 16:542-546.

Clothier, B.E., and D.E. Elrick. 1985. Solute dispersion during axisymmetric three-dimensional unsaturated water flow. Soil Sci. Soc. Am. J. 49:552-556.

Clothier, B.E., and T.D. Heiler. 1983. Infiltration during sprinkler irrigation: Theory and field results. p. 275-283. In Advances in infiltration. ASAE, St. Joseph, MI.

Coats, K.H., and B.D. Smith. 1964. Dead-end pore volume and dispersion in porous media. Soc. Pet. Eng. J. 4:73-84.

Crittenden, J.C., N.J. Hutzler, D.G. Geyer, J.L. Oravitz, and G. Friedman. 1986. Transport of organic compounds with saturated groundwater flow: Model development and parameter sensitivity. Water Resour. Res. 22:271-284.

Dagan, G. 1986. Statistical theory of groundwater flow: Pore to laboratory, laboratory to formation, formation to regional scale. Water Resour. Res. 22:120s-134s.

Dagan, G., and E. Bresler. 1979. Solute dispersion in unsaturated heterogeneous soil at field scale. I. Theory. Soil Sci. Soc. Am. J. 43:461-467.

Dalton, F.N., W.N. Herkelrath, D.S. Rawlins, and J.D. Rhoades. 1984. Time domain reflectometry: Simultaneous assessment of the soil water content and electrical conductivity with a single probe. Science 224:989-990.

Darrah, P.R., R.E. White and P.H. Nye. 1985. Simultaneous nitrification and diffusion in soil. I. The effects of the addition of a low level of ammonium chloride. J. Soil Sci. 36:281-292.

Davidson, M.R. 1984. A Green-Ampt model of infiltration in a cracked soil. Water Resour. Res. 20:1685-1690.

De Smedt, F., and P.J. Wierenga. 1978. Approximate analytical solution for solute flow during infiltration and redistribution. Soil Sci. Soc. Am. J. 42:407-412.

De Smedt, F., and P.J. Wierenga. 1984. Solute transfer through columns of glass beads. Water Resour. Res. 20:225–232.

Dixon, R.M. 1972. Controlling water infiltration in bimodal porous soils: Air-earth interface concept. p. 102–17. *In* Proc. Symp. Fundamentals of Transport Phenomena in Porous Media, Univ. of Guelph, Canada. Aug. Int. Assoc. of Hydrol. Sci. and ISSSA,

Dullien, F.A.L. 1979. Porous media: Fluid transport and pore structure. Academic Press, Orlando, FL.

Edwards, W.M., R.R. van der Pleog, and W. Ehlers. 1979. A numerical study of the effects on noncapillary-sized pores upon infiltration. Soil Sci. Soc. Am. J. 43:851–856.

Elrick, D.E., B.E. Clothier, and J.E. Smith. 1987. Solute transport during absorption and infiltration: A comparison of analytical approximations. Soil Sci. Soc. Am. J. 51:282–287.

Elrick, D.E., K.B. Laryea, and P.H. Groenevelt. 1979. Hydrodynamic dispersion during infiltration of water into soil. Soil Sci. Soc. Am. J. 43:856–865.

Elrick, D.E., M.J.L. Robin, and K.B. Laryea. 1983. Hydrodynamic dispersion during absorption of water by soil: 1. Model moisture profiles. J. Hydrol. 65:313–331.

Gelhar, L.W. 1986. Stochastic subsurface hydrology from theory to applications. Water Resour. Res. 22:1355–1455.

Germann, P.F. 1983. Slug approach to infiltration into soils with macropores. p. 171–177. *In* Advances in infiltration. ASAE, St. Joseph, MI.

Guennelon, R., R. Habib, and A.M. Cockborn. 1979. Aspects particuliers concernant la disponibilitie de N, P et K en irrigation localisee fertisante sur arbres fruitiers. p. 21–34. *In* Seminaires sur l'Irrigation Localisee I: par L'Institut d'Agronomie de l'Universitie de Bologne (Italie).

Hill, D.E., and J-Y. Parlange. 1977. Wetting front instability in layered soil. Soil Sci. Soc. Am. Proc. 36:697–702.

Hillel, D. 1980. Fundamentals of soil physics. Academic Press, Orlando, FL.

Hoffman, D.L., and D.E. Rolston. 1980. Transport of organic phosphate in soil as affected by soil type. Soil Sci. Soc. Am. J. 44:46–52.

Hornsby, A.G, and J.M. Davidson. 1973. Solution and adsorbed fluometuron concentration distribution in water-saturated soil: Experimental and predicted evaluation. Soil Sci. Soc. Am. Proc. 37:823–828.

Hutzler, N.J., J.C. Crittenden, and J.S. Gierke. 1986. Transport of organic compounds with saturated groundwater flow: Experimental results. Water Resour. Res. 22:285–295.

Jury, W.A. 1975. Solute travel-time estimates for tile-drained fields. II. Application to experimental studies. Soil Sci. Soc. Am. Proc. 39:1020–1024.

Jury, W.A. 1982. Simulation of solute transport using a transfer function model. Water Resour. Res. 18:363–368.

Jury, W.A., G. Sposito, and R.E. White. 1986. A transfer function model of solute transport through soil 1: Fundamental concepts. Water Resour. Res. 22:243–247.

Jury, W.A., L.H. Stolzy, and P. Shouse. 1982. A field test of the transfer function model for predicting solute transport. Water Resour. Res. 18:369–374.

Kanchanasut, P., and D.R. Scotter. 1982. Leaching patterns in soil under pasture and crop. Aust. J. Soil Res. 20:193–202.

Krupp, H.K., and D.E. Elrick. 1968. Miscible displacement in an unsaturated glass bead medium. Water Resour. Res. 4:809–815.

Laryea, K.B., D.E. Elrick, and M.J.L. Robin. 1982. Hydrodynamic dispersion involving cationic adsorption during unsaturated, transient water flow in soil. Soil Sci. Soc. Am. J. 46:667–671.

McBratney, A.B., and R. Webster. 1981. Spatial dependence and classification of the soil along a transect in northeast Scotland. Geoderma 26:63–82.

McLaren, A.D. 1970. Temporal and vectorial reactions of nitrogen in soil: A review. Can. J. Soil Sci. 50:97–109.

Nielsen, D.R. and J.W. Biggar. 1961. Miscible displacement in soils 1: Experimental information. Soil Sci. Soc. Am. Proc. 25:1–5.

Nielsen, D.R., P.J. Wierenga, and J.W. Biggar. 1983. Spatial soil variability and mass transfers from agricultural soils. p. 65–78. *In* D.W. Nelson et al. (ed.) Chemical mobility and reactivity in soil systems. SSSA Spec. Publ. 11. ASA and SSSA, Madison, WI.

Nielsen, D.P., J.W. Biggar, and K.T. Erh. 1973. Spatial variability of field-measured soil-water properties. Hilgardia 42:215–259.

Nielsen, D.R., and J. Bouma (ed.). 1985. Soil spatial variability. Proc. Workshop Int. Soil Sci. Soc. and Soil Sci. Soc. Am., Las Vegas, NV. 30 Nov.–1 Dec. 1984. PUDOC, Wageningen, Netherlands.

Nielsen, D.R., M.Th. van Genuchten, and J.W. Biggar. 1986. Water flow and solute transport processes in the unsaturated zone. Water Resour. Res. 22:89S–108S.

Nkedi-Kizza, P., J.W. Biggar, M.Th. van Genuchten, P.J. Wierenga, H.M. Selim, J.M. Davidson, and D.R. Nielsen. 1983. Modeling tritium and chloride 36 transport through an aggregated oxisol. Water Resour. Res. 19:691–700.

Nkedi-Kizza, P., J.W. Biggar, H.M. Selim, M.Th. van Genuchten, P.J. Wierenga, J.M. Davidson, and D.R. Nielsen. 1984. On the equivalence of two conceptual models for describing ion exchange during transport through an aggregated oxisol. Water Resour. Res. 20:1123–1130.

Nkedi-Kizza, P., P.S.C. Rao, R.E. Jessup, and J.M. Davidson. 1982. Ion-exchange and diffusive mass transfer during miscible displacement through an aggregated oxisol. Soil Sci. Soc. Am. J. 46:471–476.

Nofziger, D.L., and A.G. Hornsby. 1986. A micro-computer-based management tool for chemical movement in soil. Appl. Agric. Res. 1:50–56.

Nunge, R.J., and W.N. Gill. 1970. Mechanisms affecting dispersion and miscible displacement. p. 179–195. In Flow through porous media. Am. Chem. Soc., Washington, DC.

Parker, J.C., and M.Th. van Genuchten. 1984. Flux-averaged and volume-averaged concentrations in continuum approaches to solute transport. Water Resour. Res. 20:866–872.

Passioura, J.B. 1971. Hydrodynamic dispersion in aggregated media. Soil Sci. 11:339–344.

Peck, A.J. 1983. Field variability of soil physical properties. Adv. Irrig. 2:189–221.

Perroux, K.M., and I. White. 1988. Designs for disc permeameters. Soil Sci. Soc. Am. J. 52:1205–1215.

Philip, J.R. 1975. Some remarks in science and catchment prediction. p. 23–30 In T.G. Chapman and F.X. Dunin (ed.) Prediction in catchment hydrology. Aust. Acad. of Sci., Canberra.

Raats, P.A.C. 1973. Unstable wetting fronts in uniform and non-uniform soils. Soil Sci. Soc. Am. Proc. 37:681–685.

Raats, P.A.C. 1978. Convective transport of solutes by steady flows. I. General theory. Agric. Water Manage. 1:201–218.

Raats, P.A.C. 1984. Tracing parcels of water and solutes in unsaturated zones. p. 4–16. In B. Yaron et al. (ed.) Pollutants in porous media: The unsaturated zone between soil surface and groundwater. Springer-Verlag, Berlin.

Raats, P.A.C. 1985. Response of hydrological systems to changes in water quantity and quality. p. 3–12. In Proc. Hamburg Symp. on Relation of Groundwater Quantity and Quality. IAHS Publ. 146.

Rao, P.S.C., J.M. Davidson, and H.M. Selim. 1979. Evaluation of conceptual models for describing nonequilibrium adsorption-desorption of pesticides during steady flow in soils. Soil Sci. Soc. Am. J. 43:22–28.

Rao, P.S.C., D.E. Rolston, R.E. Jessup, and J.M. Davidson. 1980. Solute transport in aggregated porous media: Theoretical and experimental evaluation. Soil Sci. Soc. Am. J. 44:1139–1146.

Roberts, P.V., M.N. Goltz, R.S. Summers, J.C. Crittenden, and P. Nkedi-Kizza. 1987. The influence of mass transfer on solute transport in column experiments with an aggregated soil. J. Contam. Hydrol. 1:375–393.

Rose, C.W., F.W. Chicester, J.R. Williams, and J.T. Ritchie. 1982a. A contribution to simplified models of field solute tranport. J. Environ. Qual. 11:146–150.

Rose, C.W., F.W. Chichester, J.R. Williams, and J.T. Ritchie. 1982b. Application of an approximate analytic method of computing solute profiles with dispersion in soils. J. Environ. Qual. 11:151–155.

Russo, D. 1986. A stochastic approach to the crop yield-irrigation relationships in heterogeneous soils: I. Analysis of the field spatial variability. Soil Sci. Soc. Am. J. 50:736–745.

Saffigna, P.G., D.R. Keeney, and C.B. Tanner. 1977. Lysimeter and field measurements of chloride and bromide leaching in an uncultivated loamy sand. Soil Sci. Soc. Am. J. 41:478–482.

Scheidegger, A.E. 1960. Physics of flow through porous media. Univ. Toronto Press.

Scotter, D.R. 1978. Preferential solute movement through larger soil voids. I. Some computations using simple theory. Aust. J. Soil Res. 16:257–269.

Scotter, D.R., and P. Kanchanasut. 1981. Anion movement in a soil under pasture. Aust. J. Soil Res. 19:299–307.

Selim, H.M., and I.K. Iskander. 1981. A model for predicting nitrogen behaviour in slow and rapid infiltration systems. p. 478–507. *In* I.K. Iskander (ed.) Modeling wastewater renovation. Wiley, Interscience, New York.

Skopp, J. 1986. Analysis of time-dependent chemical processes in soils. J. Environ. Qual. 15:205–213.

Slichter, C.S. 1905. The rate of movement of underground waters. U.S. Geol. Surv., Washington, DC.

Smettem, K.R.J. 1986. Analysis of water flow from cylindrical macropores. Soil Sci. Soc. Am. J. 50:1139–1142.

Smiles, D.E., and B.N. Gardiner. 1982. Hydrodynamic dispersion during unsteady, unsaturated water flow in a clay soil. Soil Sci. Soc. Am. J. 46:9–14.

Smiles, D.E., K.M. Perroux, S.J. Zegelin, and P.A.C. Raats. 1981. Hydrodynamic dispersion during constant rate absorption of water by soil. Soil Sci. Soc. Am. J. 45:453–458.

Smiles, D.E., and J.R. Philip. 1978. Solute transport during absorption of water by soil: Laboratory studies and their practical implications. Soil Sci. Soc. Am. J. 42:537–544.

Smiles, D.E., J.R. Philip, J.H. Knight, and D.E. Elrick. 1978. Hydrodynamic dispersion during absorption of water by soil. Soil Sci. Soc. Am. J. 42:229–234.

Sposito, G., and W.A. Jury. 1986. Group invariance and field-scale solute transport. Water Resour. Res. 22:1743–1748.

Sposito, G., V.K. Gupta, and R.N. Bhattachaya. 1979. Foundational theories of solute transport in porous media: A critical review. Adv. Water Resour. 2:59–68.

Sposito, G., R.E. White, P.R. Darrah, and W.A. Jury. 1986. A transfer function model of solute transport through soil, 3. The convection-disperson equation. Water Resour. Res. 22:255–267.

Starr, J.L., and J.-Y. Parlange. 1976. Relation between the kinetics of nitrogen transformation and biomass distribution in a soil column during continuous leaching. Soil Sci. Soc. Am. J. 40:458–460.

Starr, J.L., J.-Y. Parlange, and C.R. Frink. 1986. Water and chloride movement through a layered field soil. Soil Sci. Soc. Am. J. 50:1384–1390.

Taylor, G.I. 1953. Dispersion of soluble matter in solvent flowing slowly through a tube. Proc. R. Soc. (London) 229A:186–203.

Thomas, G.W., and R.E. Phillips. 1979. Consequences of water movement in macropores. J. Environ. Qual. 8:149–152.

Topp, G.C., and J.L. Davis. 1981. Detecting infiltration of water through soil cracks by time domain reflectometry. Geoderma 26:13–23.

Trangmar, B.B., R.S. Yost, and G. Uehara. 1986. Spatial dependence and interpolation of soil properties in West Sumatra, Indonesia: I. Anisotopic variation. Soil Sci. Soc. Am. J. 50:1391–1395.

van Genuchten, M.Th., J.M. Davidson, and P.J. Wierenga. 1974. An evaluation of kinetic and equilibrium equations for the prediction of pesticide movement in porous media. Soil Sci. Soc. Am. Proc. 38:29–35.

van Genuchten, M.Th., and P.J. Wierenga. 1986. Solute dispersion coefficients and retardation factors. *In* A. Klute (ed.) Methods of soil analysis. Part 1. 2nd ed. Agronomy 9:1025–1054.

van der Zee, S.E.A.T.M., and W.H. Riemsdijk. 1986. Transport of phosphate in a heterogeneous field. Transp. Porous Media 1:339–359.

Wagenet, R.J., J.W. Biggar, and D.R. Nielsen. 1977. Tracing the transformations of urea fertilizer during leaching. Soil Sci. Soc. Am. J. 41:896–902.

Warrick, A.W., and D.R. Nielsen. 1980. Spatial variability of soil physical properties in the field. p. 319–244. *In* D. Hillel (ed.) Applications of soil physics. Academic Press, New York.

Watson, K.K., and M.J. Jones. 1984. Algebraic equations for solute movement during absorption. Water Resour. Res. 20:1131–1136.

Watson, K.W., and R.J. Luxmoore. 1986. Estimating macroporosity in a forest watershed by use of a tension infiltrometer. Soil Sci. Soc. Am. J. 50:578–582.

White, I., P.M. Colombera, and J.R. Philip. 1977. Experimental studies of wetting front instability induced by gradual change of pressure gradient and by heterogeneous porous media. Soil Sci. Soc. Am. J. 41:483–489.

White, R.E. 1985. the influence of macropores on the transport of dissolved suspended matter through soil. Adv. Soil Sci. 3:95–120.

White, R.E., J.W. Dyson, Z. Gerste, and B. Yaron. 1986. Leaching of herbicides through undisturbed cores of a structured clay soil. Soil Sci. Soc. Am. J. 50:277–283.

Wilson, W.H., and L.W. Gelhar. 1981. Analysis of longitudinal dispersion in unsaturated flow. 1: The analytical method. Water Resour. Res. 17:122–130.

6 Measurement of Soil Water Content and Potential

G. S. CAMPBELL and D. J. MULLA

Washington State University
Pullman, Washington

At least three types of measurements relating to soil water are important in planning and management of irrigation. These are the measurements of soil water content, soil water potential, and soil water diffusivity or conductivity. The first two relate to the state of water in soil, and the last to its movement, as described in chapter 4 in this book.

Water content is generally described in terms of the mass of water in unit mass of soil, or as the volume of water in unit volume of soil. This measurement is needed to describe the amount of water stored by the soil. Direct measurement of water content is possible by sampling the soil and weighing, drying, and reweighing the samples. Indirect methods of estimating water content include neutron scattering, γ attenuation, and electromagnetic interactions. In addition, water content can be inferred from water potential measurements using a soil water characteristic (chapter 4).

Soil water potential is the potential energy per unit quantity of water, and is useful for describing the availability of water to plants and the driving forces which cause water to move in soil (chapter 4 in this book). Several components of the water potential are typically identified (chapter 4). Of those, only methods for measuring matric potential and the sum of osmotic and matric potential are considered in this chapter. The matric potential is the main driving force for water movement in soil, and is therefore important in determining direction and magnitude of water flow. The sum of osmotic and matric potential is important in determining the availability of water to plants. Methods for measuring these potentials that have application in irrigation are tensiometers, electrical resistance sensors, thermal conductivity sensors, thermocouple psychrometers, and filter paper equilibration methods. In addition, it is possible to infer water potential from a water content measurement and a soil water characteristic (chapter 4 in this book).

The hydraulic conductivity of the soil is important for determining maximum infiltration rates, field capacity, resistance to flow to plant roots, and drainage of saturated soil. The first and last are determined by the saturated conductivity of the soil; the middle two are determined by unsaturated conductivity.

Since there are no sensors, as such, for measuring either saturated or unsaturated conductivity, and the procedures for making the measurement are somewhat involved and have been recently reviewed in considerable detail (Klute, 1986a), we will not attempt to deal further with these measurements in this chapter.

I. METHODS FOR MEASURING SOIL WATER CONTENT

Methods for measuring water content in soils have been reviewed in several texts and papers (Holmes et al., 1967; Marshall & Holmes, 1979; Hanks & Ashcroft, 1980; Schmugge et al., 1980; Hillel, 1982; Gardner, 1986). In this paper, a comparison and contrast of selected gravimetric, nuclear, and microwave techniques for measuring water content will be presented. The discussion will emphasize principles of operation, expected precision, and advantages or disadvantages of each technique for regulating soil water by irrigation.

A. Gravimetric Methods

The gravimetric method involving oven-drying soil samples to a soil constant mass (for 12–24 h) at 105 °C is the standard method for measuring soil water content. As noted by Gardner (1986), attaining constant mass depends upon many factors including uniformity of oven temperature, particle size and mineralogy of the sample, oxidation of organic matter, and initial mass of the sample. Even after constant mass is obtained, oven drying will not completely remove the last layer of adsorbed water or structural water in the sample.

The errors in gravimetric measurement increase as dry mass of the sample increases, and decrease as the precision of the analytical balance increases (Gardner, 1986). Typically, these errors are less than 0.4% for samples having a dry mass of 20 g or more. Since the coefficient of variation (CV) for water content of field samples may exceed 35% (Wilding, 1985), a far greater source of error in water content measurements is that which results from collecting an insufficient number of samples to adequately characterize the soil volume in question.

Two types of gravimetric measurements may be obtained, depending upon the method used to collect soil from the field. For disturbed samples, water content is usually determined on a mass basis (kg/kg). When undisturbed soil samples are collected using a core sampler of known volume, the water content can be expressed as a volumetric water content in m^3 of water per m^3 of soil. Alternatively, the volumetric water content (θ) can be calculated if the mass wetness (w), density of water (ρ_w) and soil bulk density (ρ_b) are known, using the relation:

$$\theta = (\rho_b/\rho_w)w. \qquad [1]$$

Gravimetric method	
Works best for:	Infrequent sampling in uniform soils.
Not suited for:	Frequent sampling or work in rocky, gravelly soils.
Advantages:	Accurate; low equipment cost.
Disadvantages:	Destructive sampling required.
	Care must be taken in heterogeneous profiles to take representative samples from each soil layer.
	Time consuming to adequately characterize field variability in water content.

B. Nuclear Methods

Two useful nuclear methods for measuring volumetric water content involve neutron scattering and γ-ray attenuation. Both methods involve principles and measurements which indirectly relate to the amount of water in the soil.

1. Neutron Scattering

High energy neutrons emitted into the soil from an ^{241}Am/Be source rapidly lose kinetic energy when they experience inelastic collisions with low atomic mass substances such as hydrogen. After approximately 18 collisions with hydrogen atoms, the kinetic energy of the neutrons will be reduced to values representative of the thermal energy of atoms at room temperature. These slow neutrons can be counted using a detector for thermalized neutrons. Since protons on water molecules are mainly responsible for thermalizing fast neutrons emitted into soil, the count rate provides an indirect measure of volumetric soil water content.

The counting rate (I) and volumetric water content are related by a calibration curve having the form (Gardner, 1986):

$$\theta = a + b(I/I_{std}) \qquad [2]$$

where I_{std} is the counting rate for an access tube placed in soil of known water content, a is a constant that depends upon soil bulk density, and b is a parameter that depends upon substances in the soil (other than protons) that are capable of thermalizing or capturing neutrons. The most important of these substances include Fe, B, Mo, and Cd. Manufacturers supply calibration curves with their instruments, but researchers should establish new curves if significant amounts of Fe, B, or Cd are present. Special precautions are required for constructing calibration curves in swelling soils (Jaywardane et al., 1983).

The spatial resolution of the neutron probe measurement varies with water content, since neutrons travel farther before thermalization in dry soils than in wet soils. Typically, the diameter of the sphere of influence ranges from 16 cm at saturation to 70 cm at near zero water content (Gardner, 1986). Thus, even in wet soils, the spatial resolution of the neutron probe is such that locating the precise depth of the wetting front is difficult. On the other hand, this averaging can be useful when sampling spatially variable soils in the field because it may reduce the number of measurements needed to determine average water content with some specified precision. Neutron probes lowered through an access tube are not desirable for measurements near the soil surface (15 cm) unless calibrations for shallow depths have been developed, since a significant fraction of the neutrons escape to the air.

Errors in neutron probe measurements of water content can be minimized by accurately determining values for the parameters a and b in Eq. [2] and by using long counts to determine I_{std}. As noted by Gardner (1986), it is possible to obtain a precision of better than 1% when measuring water content if the parameters a and b do not vary spatially. Generally, neutron measurements are precise, even when they are not accurate, so they work well for measuring changes in water content at a given depth. Since accurate measurements of change in water content are important for irrigation scheduling and water budget studies, the neutron probe is an important tool for irrigation applications.

Neutron probe method	
Works best for:	Uniform, coarse, or medium-textured soils.
	Long-term experiments at fixed locations and depths.
	Measurements of average water content or changes in average water content.
Not suited for:	Measurements near the soil surface or in shallow soils without specialized equipment.
	Rocky or gravelly soils.
	Soils containing appreciable B of Fe.
	Highly stratified or layered soils.
Advantages:	Frequent sampling at one depth or position possible.
	Measurements are made in situ and are nondestructive.
	Averages out short-range spatial variation.
Disadvantages:	Radiation hazard to operator, expensive, unable to measure soil water separately from water held in dense root systems or tubers.

2. Gamma-Ray Attenuation

Gamma rays emitted into a soil column are attenuated in intensity to an extent which depends mainly upon the thickness of the soil, the thickness of the wall of its container, and the bulk density and volumetric water content of the soil (Gardner, 1986; Schmugge et al., 1980). In nonswelling, homogeneous soil a measurement of the intensity of γ rays from a ^{137}Cs source passing through the column can be used to compute volumetric water content of the sample using the expression:

$$\theta = \ln(I_d/I_m)/(\mu_w S).$$ [3]

In this expression I_m and I_d refer to the intensities of γ ray radiation passing through the wet and dry soils, respectively, μ_w is the mass attenuation coefficient for the water, and S is the soil thickness. The intensities in Eq. [3] are corrected for instrument dead time and background noise.

In swelling or layered soil columns it is advisable to use dual beam γ ray techniques in which simultaneous measurements of bulk density and water content are possible (Gardner, 1986). This technique involves using γ rays emitted from ^{241}Am and ^{137}Cs sources, and solving simultaneous equations involving measured intensities at both energy levels in terms of water content and bulk density. Corrections for down-scattering from the Cs source are necessary (Nofziger & Swartzendruber, 1974).

Errors in determining water content by γ-ray attenuation can be minimized by selecting the optimum thickness of the soil column, accurately measuring the thickness of the column, and increasing the counting time or intensity of the attenuated γ-ray beam. According to Gardner (1986), a column thickness of from 5 to 10 cm is satisfactory for most purposes. Furthermore, the errors are potentially larger in using cylindrical soil columns than rectangular columns, since the uncertainties in measuring column thickness are greater for the cylindrical column. Generally, the errors in measuring water content with γ-ray attenuation are <1% (Schmugge et al., 1980).

While both field and laboratory applications of γ attenuation have been described, the laboratory applications have been by far the most successful. The technique, at present, is mainly useful for specialized laboratory studies, and is not suitable for most field applications.

Gamma-ray attenuation	
Works best for:	Laboratory soil columns or soil cores.
Not suited for:	Rocky or gravelly soil. Highly stratified or layered soil. Most field applications.
Advantages:	Quick, simultaneous measurement of water content and bulk density with depth possible, excellent spatial resolution, measurements are nondestructive.
Disadvantages:	Possible radiation hazard to operator, expensive.

C. Microwave Methods

Water is a strong absorber of electromagnetic radiation in the microwave portion of the spectrum. The use of microwave radiation for nondestructive determination of water content in soils is a new and promising approach. Remote-sensing methods involving active or passive microwave techniques have been reviewed by Schmugge et al. (1980). Time-domain reflectometers for in situ determination of water content using microwave radiation will be discussed here.

1. Time-Domain Reflectometry

When a nonconductive dielectric medium is placed between the plates of a capacitor and an electric field is applied, the dielectric will become polarized by induction and/or orientation effects (Moore, 1972). As compared to the properties of the same capacitor in a vacuum (denoted by subscripts o), the new capacitance increases to $C = \kappa^* C_o$, the electric field normal to the plates decreases to $E = E_o/\kappa^*$, and the characteristic impedance decreases to $Z = Z_o/\sqrt{\kappa^*}$. The factor κ^* is known as the complex dielectric response function or permittivity of the dielectric medium. For a detailed discussion of electromagnetic theory the reader is referred to Kraus and Carver (1973).

Water is a conductive, polar molecule which has a relatively large dielectric response, $\kappa^*(\omega) = \kappa'(\omega) + i\kappa''(\omega)$, in the liquid state. The real component, κ', of the dielectric response is a measure of the capacity of the medium to absorb electromagnetic energy at a given frequency (ω), while the ingainary component, κ'', is a measure of the energy dissipation rate or dielectric loss (Schmugge et al., 1980). The real part of the dielectric response for liquid water has a nearly constant value of 78 at applied angular frequencies (ω) below 18 GHz. In view of this behavior, κ' is often referred to as the dielectric constant. Thin films of water adsorbed on clay particles have a dielectric constant of about 10 (Calvet, 1975), and this value is nearly constant at frequencies below 1 GHz. For comparison, the dielectric constants of dry soil and ice are both about 3 at microwave frequencies. Clearly, the dielectric constant of mixtures of liquid water and dry soil or ice will be extremely sensitive to the amount of liquid water present.

Time-domain reflectrometry (TDR) is a method for measuring the complex dielectric response function using a broadband pulse of input microwave energy that typically includes frequencies ranging from 1 MHz to 1 GHz. A TDR unit consists of a pulse generator, a sampling head, and an oscilloscope to record voltage amplitudes and transit times as the input energy is transmitted into the soil dielectric medium (Dalton et al., 1984; Rhoades & Oster, 1986). The measured dielectric response is calibrated to obtain a measurement of volumetric water content as described below.

Two methods are available for transmitting microwave energy into the soil medium. The first involves coaxial transmission lines (Topp et al., 1980) which must be packed with soil using a fairly tedious procedure. The second,

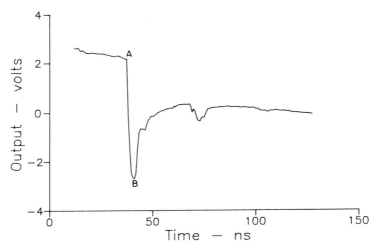

Fig. 6–1. Typical TDR signal waveform for parallel transmission probes inserted into soil. Point A is the time at which the pulse enters the soil medium. Point B indicates the reflection at the open end of the parallel probes.

a more useful, method involves portable probes consisting of 0.5-cm diam. parallel rods of brass or steel which can be quickly inserted into the soil (Topp et al., 1984; Dalton et al., 1984). The length, L, of these parallel rods typically ranges from 10 to 100 cm, and they are typically separated by a distance, S, of 5 cm. The measured volumetric water content is representative of the average value for the volume of soil enclosed between the parallel rods. Probes may be inserted either vertically or horizontally (Topp and Davis, 1982), depending on the depth resolution desired, but they must always be in good contact with the soil. The remainder of the discussion will deal only with pulses transmitted along parallel probes.

A typical TDR signal display from an oscilloscope is shown in Fig. 6-1. Changes in voltage occur when the input microwave pulse is reflected back to the TDR unit due to changes in impedance along the path of signal propagation (Fellner-Feldegg, 1969). As a result, it is important that the TDR unit, the transmission line, and the parallel probe be balanced in impedance using a balun (Topp et al., 1984) or impedance-matching transformer (Dalton et al., 1984). For a properly balanced system, the transmission line which carries the input pulse has a uniform characteristic impedance, Z_o, while the soil medium has a lower impedance, Z, given by

$$Z = Z_o / \sqrt{(\kappa^* - i4\pi\sigma/\omega)}. \qquad [4]$$

The term σ in Eq. [4] represents the low frequency conductivity of the soil medium. The voltage at point A in Fig. 6-1 decreases because the pulse experiences a discontinuity in impedance when the parallel transmission line enters the soil medium. The voltage increases at point B due to reflection of energy from the open termination at the end of the parallel probes.

The time on the abscissa between the first reflection (point A) and the second reflection (point B) represents the time required for the pulse to travel twice the probe length ($2L$). As the liquid water content, and thus the dielectric constant of the soil medium, increases, this transmission time (t) increases. For soils, the dielectric constant, κ', is much greater than the terms $\kappa'' + 4\pi\sigma/\omega$. Soils with this behavior are known as low-loss soils. The velocity, v, of wave propagation in low-loss soils is given by the approximation $v \simeq c/\sqrt{K}$, where K ($K \simeq \kappa'$) is the apparent dielectric constant of the soil medium and c is the speed of light (Topp et al., 1980). Using these approximations, the apparent dielectric constant (K) can be computed from the transmission time (t) and the probe length (L) using the relation:

$$K = [ct/(2L)]^2. \qquad [5]$$

A calibration curve is constructed relating volumetric water content and the apparent dielectric constant. The form of this curve for a wide variety of soils (Topp et al., 1980) and for unfrozen water content in partially frozen soil (Patterson & Smith, 1981) is given by

$$\theta = -5.3 \times 10^{-2} + 2.92 \times 10^{-2}K$$
$$- 5.5 \times 10^{-4}K^2 + 4.3 \times 10^{-6}K^3. \quad\quad [6]$$

A distinct advantage of the TDR method is that this calibration curve is unaffected by factors such as soil texture, salinity, bulk density, temperature, or organic matter content. In practice, Eq. [6] provides a quick method for estimating soil volumetric water content by using TDR. Topp et al. (1984) showed that TDR measurements of water content in the field are as accurate and precise as those from gravimetric determinations.

In very wet or saline soils, or when the parallel probe length is too long, the amount of TDR signal attenuation may become excessive. In such situations, the amplitude of signal reflections will be too small to detect and TDR measurements will fail. Recent investigations by Dalton et al. (1984), Dasberg and Dalton (1985), and Dalton & van Genuchten (1986) have used TDR measurements of signal attenuation to estimate soil electrical conductivity. Although it appears that TDR can be used to simultaneously estimate volumetric water content and electrical conductivity, further research is needed to understand and remove uncertainties caused by multiple signal reflections (Fellner-Feldegg, 1972) in layered soil media. Extensive research by Topp et al. (1982a, b) to account for multiple reflections in layered soils has not yet led to practical improvements in TDR technology.

Time domain reflectometry	
Works best for:	Measurements in uniform soils.
	Rapid soil survey measurements.
Not suited for:	Highly stratified soils.
	Rocky, gravelly, or saline soils.
Advantages:	One calibration curve applies to all soils.
	Rapid, easy technique.
	Some spatial averaging.
	Nondestructive, in-situ measurements.
	Can be used to rapidly assess spatial variability.
Disadvantages:	Probes inserted vertically have poor depth resolution.
	Expensive.
	Signal attenuation may occur in very wet or saline soil.

II. METHODS FOR MEASUREMENT OF WATER POTENTIAL

At the present time, no method is available for direct measurement of the energy status of water in soil. All of the methods discussed in this section require that a reference phase be equilibrated with the soil until both reach the same water potential, at which time the potential of the water in the reference phase is determined to find the potential of the soil water. In the psychrometer, the reference phase is gas; in the tensiometer it is liquid water. In the electrical resistance and thermal conductivity sensors, it is a standard porous material.

A. Tensiometers

Of all the methods available for monitoring water potential for irrigation, the tensiometer is perhaps the most widely used. Principles and applications of tensiometry have recently been reviewed in detail (Cassell & Klute, 1986).

The tensiometer consists of a sealed, water-filled tube with a porous cup on one end, and some means of measuring pressure (a gauge, manometer, or electronic pressure transducer) on the other. The porous cup is permeable to water and to solutes in the soil solution, but not to the soil matrix or to gases. Water moves through the cup until the water pressure inside the tensiometer is equal to the matric potential of the soil water (gas-phase and overburden pressure can also affect tensiometer pressure, but generally are not factors to consider in irrigation). At equilibrium, the water pressure (suction) in the tensiometer is equal in magnitude to the soil matric potential.

Carefully prepared water columns are capable of withstanding tensions in excess of 25 MPa (Briggs, 1950). The water columns in the xylem of plants also demonstrate the ability to withstand large tensions (2–5 MPa) without cavitation. The water in tensiometers, however, cavitates at tensions around 85 kPa. This limits the useful range of measurement to matric potentials between 0 and −85 J/kg. Apparently there is scope for considerable improvement in the tensiometer, if conditions that prevent cavitation in plants were better understood.

Even though the range of the tensiometer is only a small fraction of the total range of water potentials over which plants can extract water from soil, it is an important range for irrigation management, since water uptake by plants often begins to drop below potential uptake rate before soil water potential in the root zone gets out of the tensiometer range. Within the tensiometer range, the tensiometer is the most precise of any of the devices available for making this measurement.

If cavitation occurs, and gas pockets form in the tensiometer, the approach to equilibrium with the soil water is slowed markedly. It is therefore important to fill tensiometers with deaerated water, purge then with a hand-operated vacuum pump when water is added, and to check and add water a few times a week.

Commercial sources of tensiometers are listed in the Appendix. Construction details for home-built units are given by Cassell and Klute (1986). Two commercial types are in common use, one which has a pressure gauge in each unit, and another which has only the water-filled tube and porous cup, the top of the tube being sealed by a septum. An electronic pressure transducer, with digital readout, is connected to a hypodermic needle. The pressure of the tensiometer is registered when the needle is inserted through the septum. Thus, a large number of inexpensive tensiometers can be read with a single gauge. New versions allow the direct electronic recording of the data in computer-readable form. One variant of the gauge tensiometer is called the "Quick-draw," and is designed for portable, survey-type measurements. The instrument is inserted in the soil and read within about 2 min.

As pointed out in chapter 7 of this book, matric potential is spatially variable in the field. The tensiometer is essentially a point measurement (unlike the neutron and TDR methods discussed in the previous section). Several measurements are therefore necessary to characterize the matric potential of a location or field.

Tensiometer

Works best for:	In-situ measurement of matric potential above -85 J/kg, irrigation scheduling, water flow studies.
Not suited for:	Measurement of potentials lower than -85 J/kg.
Advantages:	Very accurate, easily installed, relatively inexpensive.
Disadvantages:	Limited water potential range, frequent servicing required, no spatial averaging.

B. Pressure Plate

The pressure plate is not a device for measuring soil water potential; it allows equilibration of the matric potential of soil samples to some specified water potential. The water content of the samples can then be determined in order to establish a soil water characteristic or water release (desorption) function. The water characteristic can be used, along with water content measurements, to infer water potential.

It is generally recognized that field capacity, permanent wilting point, and available water are dynamic properties of the soil profile (chapter 7 in this book), and are not simply related to water contents at specified matric potentials. Still there is a strong tendency to rely, in practice, on correlations of these variables with water contents at specified water potentials. Thus, the pressure plate is often used to infer field capacity (water content at -10 to -30 J/kg) and permanent wilting point (water content at -1500 J/kg) water contents.

A detailed description of pressure plate methods is available (Klute, 1986b), so again, those details will not be repeated here. Briefly, the *pressure plate* consists of a chamber in which soil samples are placed. The samples are placed on a porous plate which is permeable to solutes and water, but not the soil matrix or gases. The soil samples are initially wet. Pneumatic pressure is applied to the samples, which raises their water potential. Water flows through the porous plate until the sum of the pneumatic pressure potential and the soil matric potential is equal to zero (The potential of the free water on the atmospheric pressure side of the porous plate). The matric potential of the samples is then equal to the negative of the applied pressure.

We will ignore, for the moment, errors which are likely if one infers field capacity and permanent wilting water contents from pressure plate data (chapter 7 in this book) and concentrate only on possible errors in determining soil water characteristics with the pressure plate. Errors result mainly from failure of samples to attain equilibrium with the specific potential, and changes in the matrix which may occur as a result of sampling. The former are most likely on samples at low potential and the latter on samples at high

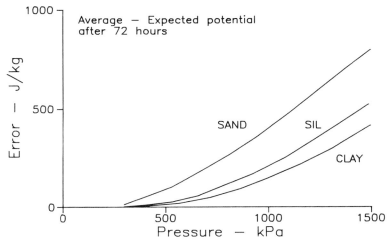

Fig. 6–2. Simulated pressure plate error after 3 d-equilibration as a function of sample final water potential and texture. The simulation used a program from Campbell (1985) with boundary conditions modified to match those of a pressure plate apparatus. Hydraulic properties were those given by Campbell (1985) for the textures shown.

potential. Effects of changing the matrix on the moisture characteristic can be minimized by taking undisturbed soil cores.

Some idea of the time required for equilibration of samples on a pressure plate can be had by simulation. Figure 6–2 shows the simulated difference between average sample water potential and applied pressure of 3.5-cm thick soil samples after 3 d of equilibration. While this simulation can only be used to indicate possible magnitudes of errors, the results are consistent with measurements by Madsen et al. (1986) and indicate that actual potentials of samples taken from pressure plates may be quite different from expected values when potentials are low. Differences will be largest when the soil is sandy or loosely packed. We conclude that the pressure plate is best suited for determining water release at high potentials, and that water contents at −1500 J/kg may be somewhat higher than the equilibrium values because of lack of equilibration. Since the change of water potential with water content is large in the dry range, the error in estimation of available water may be fairly small, except for sandy soils.

Once data for characteristic curves have been obtained, it is often useful to fit them with some mathematical function. A power function (log water potential vs. log water content are linearly related) is sufficiently accurate for many purposes, but Bruce and Luxmoore (1986) gave several other useful functions.

Pressure plate	
Works best for:	Determination of soil water characteristics in the range 0 to −500 J/kg.
Not suited for:	Measurement of low water potentials, in-situ measurement.
Advantages:	Accurate, simultaneous measurement of multiple samples.
Disadvantages:	Disturbed samples, permits only indirect determination of potential, slow equilibration on dry samples.

C. Thermal Conductivity

Matric potential sensors that use heat dissipation are generally of the type described by Phene et al. (1971).The thermal conductivity of a reference matrix is measured by the rate of heat dissipation in the matrix. If the reference matrix is in equilibrium with the soil, then the thermal conductivity of the matrix can be used to infer the soil water potential. The porosity of the matrix is selected to give desorption over the range of potentials of interest, but no matrix has been found which gives good response at all potentials. Materials have been found, however, which give good results over the range of potentials of interest for irrigation scheduling. Details on construction, calibration, and use of the sensors can be found in Phene et al. (1971).

Thermal conductivity methods	
Works best for:	In-situ measurement of matric potential, irrigation scheduling.
Not suited for:	Measurement in the presence of rapidly varying temperature, high water potentials in coarse soil.
Advantages:	Fairly precise, easily installed, unaffected by salts.
Disadvantages:	Limited water potential range, expensive, and somewhat complex electronics required, no spatial averaging.

D. Electrical Resistance Methods

The most popular method for electronic monitoring of soil water potential uses *electrical resistance sensors* (Bouyoucos & Mick, 1940). These are typically made from gypsum, though fiberglass and nylon have also been used. The matrix equilibrates with the soil, and then the electrical resistance of the matrix is measured. The resistance is related to the water content of the matrix, which, in turn is determined by the water potential of the surrounding soil.

Since it is the ions in the soil solution which determine its electrical conductivity, the resistance can be strongly affected by the composition of the soil solution. Gypsum provides some buffering by maintaining an electrical conductivity around 2 dS/m, but when salts are more concentrated than this, even the gypsum block calibration is affected.

It is important to measure the resistance of the units with alternating current at a frequency at or above 1000 Hz. Measurements with direct current will polarize the electrodes, and can eventually cause block failure through electrolysis. Temperature should also be measured and corrections

Electrical resistance methods	
Works best for:	In-situ measurement of changes in matric potential, irrigation scheduling.
Not suited for:	Measurement in the presence of high or varying soil solute concentrations, high water potentials in coarse soil.
Advantages:	Inexpensive, easily installed, automatic recording.
Disadvantages:	Sensitive to changes in soil solute concentrations. Calibration needed for precision. No spatial averaging.

applied for temperature variation. Methods for calibrating, installing, and reading the sensors, and for temperature corrections to the reading are given by Taylor et al. (1961), Campbell and Gee (1986), and Willoughby and Cuming (1985).

E. Filter Paper Methods

The filter paper method provides a convenient and simple method for determining the matric potential of soil samples. The method was first described by Gardner (1937). Details of calibration and use are given by Hamblin (1981). Matric potential is measured by equilibrating water in filter paper with the water in a soil sample. The water content of the filter paper is then determined by drying. Single disks of Whatman No. 42 filter paper are equilibrated, either in situ, or with samples taken from the soil profile, for 1 to 2 d. At the end of the equilibration time, the filter paper is quickly removed from the soil, loose soil removed by tapping, and the filter paper is sealed in a container for weighing and drying. If soil samples are used, it is important to place them in an insulated container for equilibration, since temperature gradients can cause water movement in the sample. Obviously, care must be taken to minimize water loss during transfer of papers. The small amount of soil that adheres to the paper after tapping generally does not influence the measurement appreciably. The paper water content, θ (g/g) is converted to water potential using the relation:

$$\psi_m = 11 \, \theta^{-3.68}.$$

The filter paper method certainly has many applications in irrigation research, but would generally not be suitable for irrigation scheduling because of the time required to obtain a measurement.

Filter paper methods	
Works best for:	Measurement of matric potential on samples or in situ.
Not suited for:	Continuous or electronic monitoring of soil moisture, rapid results.
Advantages:	Inexpensive, simple, accurate.
Disadvantages:	No electronic recording, 2–3 d required for results, no spatial averaging.

F. Thermocouple Psychrometer

Since the first descriptions of thermocouple psychrometers by Spanner (1951) and Richards and Ogata (1958), considerable effort has gone into the development of this technique. The principle involved is simple. Air, in vapor and temperature equilibrium with soil, has a relative humidity which is determined by the water potential of the soil. The relationship is

$$\psi = (RT/M_w) \ln h,$$

where h is the relative humidity, R is the gas constant, T is kelvin temperature, and M_w is the molecular mass of water. Extreme precision, however, is required in the humidity measurement, since the full range of soil water potential over which plants can extract water corresponds to the humidity range 0.99 to 1.0. This requirement limits the precision generally achievable with psychrometers to 50 to 100 J/kg. Such low precision makes the psychrometer relatively useless for irrigation scheduling on most crops, where one generally wants to keep soil water potentials within the tensiometer range. However, the psychrometer is useful in determining the lower limit of available water for crops. Since it starts working well at about the water potential where the pressure plate starts to fail, the psychrometer is useful for determining soil water characteristics. It is important to remember, though, that the psychrometer measures the sum of osmotic and matric potential, while the other methods discussed (including the pressure plate) respond only to matric potential. If solutes are present in the soil in appreciable concentrations, these will influence the measurement.

Units are now commercially available for both in situ and sample measurements. Names of manufacturers are given in the Appendix. Details of theory, construction, calibration, installation, and errors are given by Rawlins and Campbell (1986).

Thermocouple psychrometer	
Works best for:	Measurement of matric and osmotic potential, on samples or in situ, at potentials below -100 J/kg.
Not suited for:	Measurements at high water potential, when high precision is required, or when temperatures are changing rapidly.
Advantages:	Covers a wide range of water potentail, rapid measurement, electronic recording possible, measures osmotic and matric potentials.
Disadvantages:	Limited precision in the wet range, sophisticated electronics required, no spatial averaging.

APPENDIX

Commercial sources of equipment for measuring soil water content and potential.

Agwatronics
P.O. Box 2807
Merced, CA 95344
Thermal conductivity

Cambell Pacific Nuclear
130 South Buchanan Circle
Pacheco, CA 94553
Neutron probe

Decagon Devices, Inc.
P.O. Box 835
Pullman, WA 99163
Thermocouple psychrometer

Irrometer Co.
P.O. Box 2424
Riverside, CA 92516
Tensiometer

G.F. Larson Co., Inc.
P.O. Box 4453
Santa Barbara, CA 93103
Electrical resistance

Soil Measurement Systems
7344 N. Oracle Rd., Suite 170
Tucson, AZ 85704
Tensiometer

Soil Moisture Equipment Corp.
P.O. Box 30025
Santa Barbara, CA 93105
Tensiometer, pressure plate,
  TDR, electrical resistance

Soiltest, Inc.
2205 Lee St.
Evanston, IL 60602
Tensiometer, electrical resistance,
  pressure plate

Troxler
P.O. Box 12057
Research Triangle Park, NC 27709
Neutron probe

Wescor, Inc.
P.O. Box 459
Logan, UT 84321
Thermocouple psychrometer

REFERENCES

Bouyoucos, G.J., and A.H. Mick. 1940. An electrical resistance method for the continuous measurement of soil moisture under field conditions. Michigan Agric. Exp. Stn. Tech. Bull. 172.

Briggs, L.J. 1950. Limiting negative pressure of water. J. Appl. Phys. 21:721–722.

Bruce, R.R., and R.J. Luxmoore. 1986. Water retention: Field methods. In A. Klute (ed.) Methods of soil analysis. Part 1. 2nd ed. Agronomy 9:663–686.

Calvet, R. 1975. Dielectric properties of montmorillonites saturated by bivalent cations. Clays Clay Miner. 23:257–265.

Campbell, G.S. 1985. Soil physics with BASIC: Transport models for soil-plant systems. Elsevier, Amsterdam.

Campbell, G.S. and G.W. Gee. 1986. Water potential: Miscellaneous methods. In A. Klute (ed.) Methods of soil analysis. Part 1. 2nd ed. Agronomy 9:619–633.

Cassell, D.K., and A. Klute. 1986. Water potential: Tensiometry. In A. Klute (ed.) Methods of soil analysis. Part 1. 2nd ed. Agronomy 9:563–596

Dalton, F.N., W.N. Herkelrath, S.D. Rawlins, and J.D. Rhoades. 1984. Time-domain reflectometry: Simultaneous measurement of soil water content and electrical conductivity with a single probe. Science 224: 989–990.

Dalton, F.N., and M. Th. van Genuchten. 1986. The time-domain reflectometry method for measuring soil water content and salinity. Geoderma 38:237–250.

Dasberg, S., and F.N. Dalton. 1985. Field measurement of soil water content and bulk electrical conductivity with time-domain reflectometry. Soil Sci. Soc. Am. J. 49:293–297.

Fellner-Feldegg, H. 1969. The measurement of dielectrics in the time domain. J. Phys. Chem. 73(3):616–623.

Fellner-Feldegg, H. 1972. A thin-sample method for the measurement of permeability, permittivity, and conductivity in the frequency and time domain. J. Phys. Chem. 76(15): 2116–2122.

Gardner, R. 1937. A method of measuring capillary tension of soil moisture over a wide moisture range. Soil Sci. 43:238–277.

Gardner, W.H. 1986. Water content. In A. Klute (ed.) Methods of soil analysis. Part 1. 2nd ed. Agronomy 9:493–544.

Hamblin, A.P. 1981. A filter paper method for routine measurement of field water potential. J. Hydrol. 53:355–360.

Hanks, R.J., and G.L. Ashcroft. 1980. Applied soil physics. Springer-Verlag, Berlin.

Hillel, D. 1982. Soil water: content and potential p. 57–89. In Introduction to soil physics. Academic Press, New York.

Holmes, J.W., S.A. Taylor, and S.J. Richards. 1967. Measurement of soil water. In R.M. Hagan et al. (ed.) Irrigation of agricultural lands. Agronomy 11:275–303.

Jaywardane, N.S., W.S. Meyer, and H.D. Barrs. 1983. Moisture measurement in a swelling clay soil using neutron moisture meters. Aust. J. Soil Res. 22:109–117.

Klute, A. (ed.) 1986a. Methods of soil analysis. Part 1. 2nd ed. Agronomy 9.

Klute, A. 1986b. Water rentention: laboratory methods. In A. Klute (ed.) Methods of soil analysis. Part 1. 2nd ed. Agronomy 9:635–662.

Kraus, J.D., and K.R. Carver. 1973. Electromagnetics. 2nd ed. McGraw-Hill, New York.

Madsen, H.B., C.R. Jensen, and R. Boysen. 1986. A comparison of the thermocouple psychrometer and the pressure plate methods for determination of soil water characteristic curves. J. Soil Sci. 37:357–362.

Marshall, T.J., and J.W. Holmes. 1979. Soil physics. Cambridge Univ. Press, Cambridge, Great Britain.

Moore, W.J. 1972. Physical chemistry. 4th ed. Prentice-Hall, Eaglewood Cliffs, NJ.

Nofziger, D.L., and D. Swartzendruber. 1974. Material content of binary physical mixtures as measured with a dual energy beam of γ rays. J. Appl. Phys. 45:5443–5449.

Patterson, D.E., and M.S. Smith. 1981. The measurement of unfrozen water content by time domain reflectometry: Results from laboratory tests. Can. Geotech. J. 18:131–144.

Phene, C.J., G.J. Hoffman, and S.L. Rawlins. 1971. Measuring soil matric potential in situ by sensing heat dissipation within a porous body: I. Theory and sensor construction. Soil Sci. Soc. Am. Proc. 35:27–32.

Rawlinds, S.L., and G.S. Campbell. 1986. Water potential: Thermocouple psychrometry. *In* A Klute (ed.) Methods of Soil Analysis. Part 1. 2nd ed. Agronomy 9:597–618.

Rhoades, J.D., and J.D. Oster. 1986. Solute content. *In* A. Klute (ed.) Methods of soil analysis. Part 1. 2nd ed. Agronomy 9:985–1006.

Richards, L.A., and G. Ogata. 1958. Thermocouple for vapor-pressure measurement in biological and soil systems at high humidity. Science 128:1089–1090.

Schmugge, T.J., T.J. Jackson, and H.L. McKim. 1980. Survey of methods for soil moisture determination. Water Resour. Res. 16:961–979.

Spanner, D.C. 1951. The Peltier effect and its use in measurement of suction pressure. J. Exp. Bot. 11:145–168.

Taylor, S.A., D.D. Evans, and W.D. Kemper. 1961. Evaluating soil water. Utah Agric. Exp. Stn. Bull. 426.

Topp, G.C., and J.L. Davis. 1982. Measurement of soil water content using time domain reflectometry. p. 269–287. *In* Proc. Canadian Hydrology Symp., Fredericton, New Brunswick. 14–15 June. Nat. Res. Council of Canada, Ottawa.

Topp, G.C., J.L. Davis, and A.P. Annan. 1980. Electromagnetic determination of soil water content: Measurement in coaxial transmission lines. Water Resour. Res. 16:574–582.

Topp, G.C., J.L. Davis, and A.P. Annan. 1982a. Electromagnetic determination of soil water content using TDR: I. Applications to wetting fronts and steep gradients. Soil Sci. Soc. Am. J. 46:672–678.

Topp, G.C., J.L. Davis, and A.P. Annan. 1982b. Electromagnetic determination of soil water content using TDR: II. Evaluation of installation and configuration of parallel transmission lines. Soil Sci. Soc. Am. J. 46:678–684.

Topp, G.C., J.L. Davis, W.G. Bailey, and W.D. Zebchuk. 1984. The measurement of soil water content using a portable TDR hand probe. Can. J. Soil Sci. 64:313–321.

Wilding, L.P. 1985. Spatial variability: Its documentation accommodation and implication to soil survey. p. 166–194. *In* D.R. Nielsen and J. Bouma (ed.) Soil spatial variability. PUDOC, Wageningen.

Willoughby, P., and K.J. Cuming. 1985. Automatic irrigation in a peach orchard. HortScience 20:445–446.

7   Field Soil-Water Relations

L. R. AHUJA

USDA-ARS
Durant, Oklahoma

D. R. NIELSEN

University of California
Davis, California

The theories of soil water flow and chemical transport presented in the preceding chapters have vital field applications for irrigation management and conservation of resources. The increase in our basic understanding, during the last 25 yr, of infiltration, redistribution, retention, and uptake of water by plant roots has revolutionized the concepts of field capacity and available water so ingrained in irrigation practice. The applications of the new concepts for irrigation management are enormous, as will be elucidated in this chapter. The water flow theory is also being used in the design of irrigation methods for improving uniformity of water application, whether border (e.g., Bassett et al., 1980; Jaynes, 1986), furrow (e.g., Elliott & Walker, 1982), sprinkler (e.g., Slack, 1980), or drip (e.g., Amoozegar-Fard et al., 1984). Furthermore, theories are being employed to develop models that incorporate the transport of water, mass (salts, chemicals), and energy through the entire soil-plant-atmosphere continuum in relation to plant growth. Models of this nature are useful as research and management tools for efficient use of water, soil, and nutrient resources, while maintaining surface and groundwater quality (e.g., Watts & Hanks, 1978; Shaffer & Larson, 1987; Lemmon, 1986). The technological advancement and efficiency of today's digital computers have helped make this possible.

It is important that field-scale applications of flow and transport theories take into account the heterogeneity and spatial variability of field soils (Nielsen et al., 1973; Warrick & Nielsen, 1980). The variability generally consists of some directional trends in the values of soil water properties, as well as a degree of spatially correlated random fluctuations in these values. Considerable research effort has been made in the last 15 yr to measure and characterize spatial variability. Also important, but studied to a lesser extent, is the temporal variability of soil water properties induced by tillage, cropping, and other management practices (Klute, 1982; Rawls et al., 1983; Brakensiek & Rawls, 1983; Hines, 1986). Surface sealing and crusting of tilled soils is a predominant phenomenon that affects water flow (Ahuja & Ross,

1983). Related to the degree of tillage are the effects of large macropores consisting of interpedal voids, drying cracks, root channels, and worm holes (Beven & Germann, 1982; White, 1985).

The measurement of field soil water properties is time consuming and expensive, especially when characterizing their spatial and temporal variability which requires a large number of measurements. This characterization is probably the greatest hurdle for general and large-scale application of soil water theory. Research, however, is being conducted to devise simpler methods of measurement or estimation for this purpose (Libardi et al., 1980; Ahuja et al., 1980, 1984a, 1985; Naney et al., 1983; Rawls et al., 1982).

In this chapter we summarize present knowledge of the important features of the field soil water relations and applications for irrigated agriculture. The gaps in knowledge and the limitations to be overcome by future research, to the best of our understanding, are also enumerated.

I. THE OLD AND THE NEW DYNAMIC CONCEPT OF FIELD CAPACITY

Hillel (1980) provided a good discussion of these concepts. Early experimental results showed that, following wetting, changes in soil water content decrease rapidly with time (Alway & McDole, 1917; Israelson & West, 1922; Veihmeyer & Hendrickson, 1927, 1931). These results were interpreted to mean that the internal drainage rate generally becomes negligibly small or even zero in just a few days. The water content at which the internal drainage supposedly stopped was termed the *field capacity*. For example, Veihmeyer & Hendrickson (1949) defined field capacity as "the amount of water held in soil after excess water has drained away and the rate of downward movement has materially decreased, which takes place within 2–3 days after a rain or irrigation in pervious soils of uniform structure and texture." As a result, field capacity has been widely accepted as a physical characteristic of soil. Some workers believed it to result from a static equilibrium or a discontinuity in capillary water. A widely used laboratory method for the estimation of field capacity has been the soil water content in equilibrium with a matric suction of 33 or 10 kPa. More recent field measurements with tensiometers indicate that in most soils the soil water suction attained after 2 to 3 d of drainage is in the range of 5 to 10 kPa (e.g., Bruce et al., 1983; Nofziger et al., 1983). For irrigations, it has been commonly accepted that the application of a certain quantity of water to the soil will only fill the deficit to field capacity, to a certain definable depth. The amount of irrigation to be applied was determined at any time on the basis of deficit from the field capacity of the root zone to be wetted.

With the development and predictions of theory and more precise experimental measurements (Richards et al., 1956; Ogata & Richards, 1957), field capacity is no longer considered a constant or an intrinsic soil property, but rather an arbitrary value. It is recognized that internal drainage (or redistribution) is indeed continual and shows no sharp changes or constant levels.

In the absence of a water table, the process continues (albeit at a decreasing rate) indefinitely. In most soils, except for deep coarse-textured soils, internal drainage can persist at an appreciable rate for many days. The rate at which drainage becomes negligible depends upon the process being studied. The time of its occurrence is not constant and depends upon the interplay of several factors in the equation for unsaturated flow (see chapter 4 in this book). For instance, the rate of drainage is influenced by the configuration of impeding or coarse-textured layers in the soil profile (Eagleman & Jamison, 1962; Miller, 1969, 1973; Clothier et al., 1977), preinfiltration wetness of soil and depth of wetting (Richards & Moore, 1952; Biswas et al., 1966; Carbon, 1975), and depth of a permanent or temporary groundwater table. It is also affected by evaporation from the soil surface and the uptake of water by plants. An example of the dynamic nature of the internal drainage and water storage process is presented in Fig. 7-1. The solid lines are calculated using

$$W = W_o \, t^{-b} \qquad\qquad [1]$$

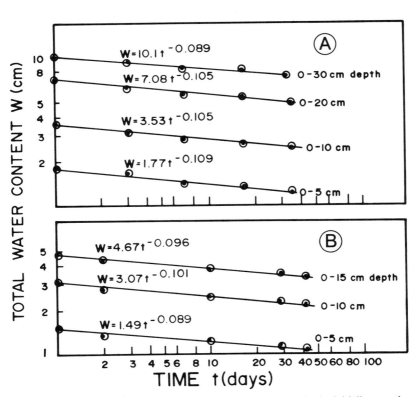

Fig. 7-1. Amount of water retained in various depth intervals within the initially wetted zone of a fine sandy loam during redistribution following irrigations of (*A*) 10 and (*B*) 5 cm of water. After Gardner et al. Water Resour. Res. 6:851–861, 1970. Copyright by the American Geophysical Union.

earlier suggested by the work of Richards et al. (1956) and Ogata & Richards (1957), where W is the depth of water (cm) stored within the soil profile depth, t is the redistribution time following irrigation, and W_o the value of W at $t = 0$. Gardner et al. (1970) showed that a soil profile irrigated with 5 cm of water to an initial wetting depth of 16 cm retained only 56% of its water after 30 d whereas the same profile irrigated with 10 cm of water retained in its initially wetted zone 69% of its water after the same redistribution period.

Field capacity has commonly been assumed to be the upper limit of soil water available to plants. Ratliff et al. (1983) defined the upper limit of availability in the field as the soil water content where the daily rate of drainage, following a thorough wetting of the soil, was reduced to between 0.1 and 0.2% of the stored water. Using field data for 61 soil profiles, obtained from 15 states in the USA, they reported that it typically took 2 to 12 d for soils to reach this criterion. Some fine soils with restrictive layers required up to 20 d. These findings make the definition of the upper limit of water availability invalid and support the reservations expressed above concerning the old concept of field capacity. The drainage of soil has to be treated as a dynamic process, computed from the soil's hydraulic properties and boundary conditions.

II. OLD AND NEW CONCEPTS OF SOIL WATER AVAILABILITY TO PLANTS

Veihmeyer and Hendrickson (1927, 1949, 1950, 1955) postulated that soil water is equally available to plants in the range of soil water contents from field capacity to a permanent wilting point, both of which are constant values for a given soil. Based on the work of Briggs and Shantz (1912) and Furr and Reeve (1945), the permanent wilting point was defined as the soil water content at which plants wilt and do not recover turgidity even when placed in a 100% relative humidity atmosphere for 12 h. This concept was widely accepted for many years. Later on, it was recognized that the energy status of soil water, rather than the soil water content, may be the better and more universal criterion of availability to plants. The upper and lower limits of water availability were, therefore, defined in terms of soil water suction, as 10 or 33 kPa and 1500 kPa, respectively (Richards & Weaver, 1944; Slater & Williams, 1965).

Richards and Wadleigh (1952) and other investigators showed that soil water availability to plants decreases as the soil water content decreases from field capacity to permanent wilting point, i.e., the plant growth may be reduced before the wilting point is reached. This led some workers to propose that between the upper limit and a certain point within the so-called *available range* the water is equally available, and from that point up to the lower limit the availability decreases (Thornthwaite & Mather, 1955; Pierce, 1958; Ritchie et al., 1972). Ritchie et al. showed experimentally that this concept held approximately for cotton (*Gossypium hirsutum* L.) and grain sorghum (*Sorghum bicolor* L.) crops. Soil water in the upper three-fourths of the available range was readily available, after which the availability decreased approximately linearly. The break-point value of available water may, however, vary with crop, soil, and atmospheric conditions.

Advances in experimental measurements and theoretical understanding of the flow of water in soil, plant, and atmosphere led to dynamic concepts of soil water availability to plants (Philip, 1957; Gardner, 1960; Penman, 1949). It was postulated that the rate of water uptake that would sustain normal plant growth at any given time depends not only upon soil water status but also upon the atmospheric conditions and properties of the plants. Atmospheric conditions dictate the evapotranspiration (ET) demand, or the rate at which the plant is required to transpire and absorb water from the soil to maintain its turgidity. Rooting depth, density, proliferation and extension, physiological adjustment of plant to water stress, and soil's water conducting properties at varying moisture contents determine the rate of actual uptake and transpiration in response to the imposed demand.

The classical experiments of Denmead and Shaw (1962) presented evidence of the above dynamic concept. They measured transpiration rates of corn plants grown in containers and placed in the field under different conditions of water application and ET demand. Their results are shown in Fig. 7-2a. Under an ET demand of 3 to 4 mm d^{-1}, the measured transpiration rate began to fall below the demand rate at a soil-water suction of about 100 kPa (1.0 bar). Under a demand of 6 to 7 mm d^{-1}, the fall in the measured rate began at a suction of only 20 kPa. When the ET demand was very low (1.4 mm d^{-1}) actual transpiration was maintained at the demand rate up to a soil water suction of 500 kPa or higher. The soil water content at which the actual transpiration began to fall below the potential demand rate varied between 23 and 34% (Fig. 7-2b). Figure 7-2a also indicates that a soil water suction at which transpiration approaches zero (wilting point) is also lower for higher ET demand conditions.

It is now recognized that the soil, plant, and atmosphere form an integrated physical system or continuum (Philip, 1966). Water flow occurs in the system from any location of higher potential energy to another of lower potential energy. The energy level of water is the effective potential of water, since not all components of the total potential of water are effective in different parts of the system. The rate of water flow from soil to plant roots or from plant leaves to the atmosphere, as well as from one location to another within the soil and plant, is directly proportional to the effective water potential gradient between them and inversely to the resistances to flow between them. The flow can be in a liquid or vapor form. Two alternative approaches have been taken to model uptake of soil water by plants (Hillel, 1980; Molz, 1981; Feddes, 1981)—a microscopic single-root approach and a macroscopic root-system approach. In the first approach, the absorbing roots having a certain average diameter are all parallel to each other and uniformly distributed within each soil layer. It is further assumed that water flows radially through the surrounding soil to each root. Solution of the flow problem can be obtained analytically if certain linearizing assumptions are made (Gardner, 1960; Cowan, 1965; Philip, 1966), or numerically using a computer if fewer assumptions are made (Molz & Remson, 1970; Hillel et al., 1975). In the macroscopic approach, the root system as a whole is considered a distributed sink, whose strength varies with depth (Ogata et al., 1960; Gardner, 1964; Whisler et al., 1968; Molz & Remson, 1970, 1971; Nimah & Hanks, 1973; Feddes et al., 1974; Hillel et al., 1976). To illustrate the dynamic con-

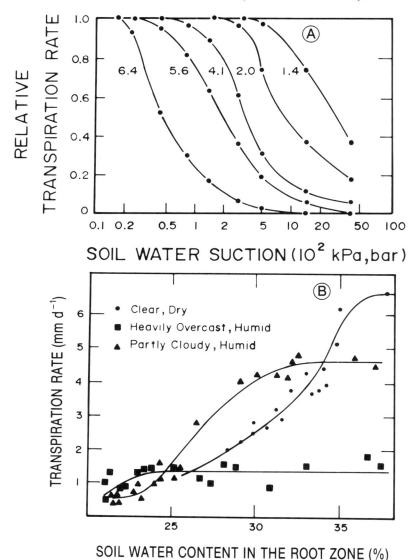

Fig. 7-2. (A) Relative transpiration rate as a function of average soil water suction, under different meteorological conditions. The numbers on the curves represent different rates of potential evapotranspiration in millimeter per day. (B) Actual transpiration rate as a function of soil-water content, under different meteorological conditions. After Denmead and Shaw (1962).

cept of soil water availability, we will briefly summarize only the macroscopic approach. Hillel (1980) gave a good description of both approaches.

The unsaturated soil water flow in any given root horizon is described by the equation:

$$\frac{\partial \theta}{\partial t} = \frac{\partial}{\partial z} \left[K(\theta) \frac{\partial \phi_m}{\partial z} - K(\theta) \right] - S_w \qquad [2]$$

where θ is the soil water content, t the time, z the depth, $K(\theta)$ the hydraulic conductivity as a function of θ, $\phi_m(\theta)$ the matric potential head, and S_w the rate of water uptake by roots per unit volume of soil. The rate of uptake is governed by the relation:

$$S_w = (\phi_{soil} - \phi_{stem})/(R_{soil} + R_{roots}) \qquad [3]$$

where ϕ_{soil} is the total potential head of soil water (i.e., sum of matric, gravitational, and osmotic components), ϕ_{stem} is the total potential head of water in the plant at the base of the stem, R_{soil} is the hydraulic resistance of water flow in soil, and R_{roots} is the hydraulic resistance of roots. R_{soil} may be defined as (Gardner, 1964)

$$R_{soil} = [BLK(\theta)]^{-1} \qquad [4]$$

where L is the total length of active roots per unit volume of soil and B is an empirical constant.

The total rate of flow from all root zone depths to the stem (say Q) is obtained by summing S_w for all depths. An equation similar to [3] can be written for the flow of water from the base of the stem to the leaves, and from the leaves to the atmosphere:

$$Q = (\phi_{stem} - \phi_{leaves})/R_{plant}$$

$$= (\phi_{leaves} - \phi_{atm})/(R_{leaves} + R_{atm}). \qquad [5]$$

Knowing the soil, plant, and atmospheric parameters, Eq. [2], [3], and [5] can be solved iteratively for ϕ_{leaves}, ϕ_{stem}, $\phi_{roots}(z)$, and $\phi_m(z)$ for a given Q (potential ET demand) until ϕ_{leaves} is above a certain limiting value (wilting point), and also for a variable Q after the limiting ϕ_{leaves} is reached.

Hillel et al. (1975) used the above model with hypothetical soil, plant, and atmospheric conditions and showed that the results qualitatively resembled the experimental finding of Denmead and Shaw (1962). Attempts made for experimental validation of the model by Belmans et al. (1979) and Feyen et al. (1980) indicated our inadequate understanding of the hydraulic properties of roots and the root growth. With independently measured hydraulic resistance of roots, Lascano and van Bavel (1984) found that the hypothesis of Eq. [3] for fractional water uptake for a portion of the root system adequately matched the experimental uptake by a cotton split-root system. The model also seemed to work well in predicting ET and soil water balance in a sorghum field, wherein it was also assumed that the water can be redistributed in the soil from one point to another by flowing through the root system (van Bavel et al., 1984).

The new dynamic concept of soil water availability has tremendous implications for irrigation management. The old concept of equal availability of water to plants between field capacity and permanent wilting point resulted in large, infrequent irrigations. Water was applied when the so-called *available water* was nearly depleted. The amount applied equalled that needed to fill the current root zone to field capacity, plus a leaching fraction. Although this method of irrigation management made good economic and operational sense in conjunction with the prevailing methods of irrigation and water delivery, it was possible that the crop could be subjected to extreme conditions of poor aeration and excessive desiccation (Rawlins & Raats, 1975; Hillel, 1980). The dynamic concept of soil water availability requires that irrigation application be more flexible, based on meteorological conditions, plant growth stage, root system, and soil water flow rates. With this concept, modeling of the soil-plant-atmosphere system on a small computer could help in deciding the time and amount of irrigation (Lemmon, 1986). Alternatively, such calculations provide guidelines for frequent, small irrigations that maintain nonstress moisture levels in the root zone, good aeration, and necessary deep drainage. However, these calculations should be supported by field measurements of soil moisture tension with modern tensiometers (Marthaler et al., 1983), soil water content with a neutron probe or other means, and/or plant water status with infrared thermometry or other means (Jackson, 1982). Details about these methods are given in chapter 17 of this monograph.

Newer methods of irrigation, including solid-set, low-intensity sprinklers and surface or subsurface drip irrigation, have made either of the above alternatives possible. With these approaches, we can optimize the use of soil moisture as well as fertilizers, and control salinity and drainage water quality (Rawlins & Raats, 1975). This can be done even on coarse-textured, gravelly, or sloping soils that were considered unsuitable for irrigation by traditional methods, as well as under conditions of spatially variable soils and topography (Hillel, 1980). There is growing experimental evidence of increased yields as a result of frequent, small irrigations (e.g., Viets, 1966; Rawitz, 1969; Acevedo et al., 1971; Hillel & Guron, 1973).

III. SPATIAL VARIABILITY OF SOIL WATER PROPERTIES AND THEIR STATISTICAL REPRESENTATION

Natural field soils encompass considerable inherent variability in their texture, structure, and physical and chemical properties, as a result of variability in parent material and other soil-forming factors. This variability may be further enhanced by tillage, cultivation, cropping, and related management practices. The variability greatly affects irrigation efficiency and, thus, needs to be well described for good management.

The recognition and study of soil variability is not new (e.g., Montgomery, 1913; Pendleton, 1919). However, since the early 1970s there has been a large number of studies conducted. The work of Nielsen et al. (1973) provided a great stimulus. The new methods being used for analysis and representation of spatial variability are based on time-series approaches developed for atmospheric and hydrologic variability (Lumley & Panofsky, 1964; Yevjevich, 1972), and the geostatistical approach, or the regionalized variable theory, developed in geology and mining (Matheron, 1971; Davis, 1973; Journel & Huijbregts, 1978). A simpler method of similar-media scaling (Miller & Miller, 1956; Miller, 1980) has also shown a great deal of promise in compressing variability of soil water properties to a narrow range (Peck et al., 1977; Warrick et al., 1977a; Simmons et al., 1979). The various methods of analysis and their application to field measurements have been presented in several recent reviews (Warrick & Nielsen, 1980; Webster, 1985; Trangmar et al., 1985; Warrick et al., 1986). Here, we present a summary of the methodology and some examples of application.

A. Frequency Distribution

Conventionally, the most common way of expressing variability of a property is to obtain its sample mean and standard deviation. Implicit in this expression is the assumption that the property has a normal or Gaussian distribution, for which the mean and standard deviation completely describe the variability. The equation for this frequency distribution is

$$f(x) = (\sigma\sqrt{2\pi})^{-1} \exp [-(x-\mu)^2/2\sigma^2] \qquad [6]$$

where $f(x)$ is the frequency of occurrence of a value of the property equal to x, μ is the mean, and σ the standard deviation of the population. This expression is also called the *normal probability density function*. The soil properties of bulk density, porosity, water content at a certain suction or time, and textural components have been shown to obey Eq. [6] approximately (Nielsen et al., 1973; Coelho, 1974). An example is shown in Fig. 7–3. The distribution curve is bell shaped and symmetrical about the mean. The mode and median are the same as the mean. The above-noted soil properties that follow normal distribution also seem to have low to medium variability, with the coefficient of variation (σ/μ) ranging from about 0.05 to 0.60 (Warrick & Nielsen, 1980).

Recent studies have indicated that several soil properties do not follow a normal frequency distribution. A typical case is that of steady infiltration rates or pore water velocities (Fig. 7–4a). The distribution is highly skewed towards smaller values. The mean value is much greater than the mode (the peak of the distribution curve). The median falls in between the mean and mode. There is now extensive evidence in the literature to show that the dynamic transport rate parameters of soil, such as the saturated or unsaturated

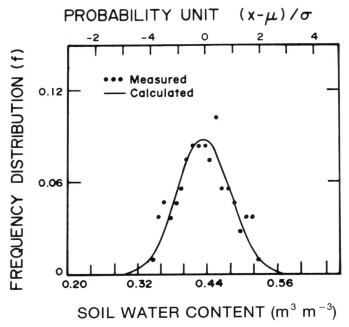

Fig. 7–3. Measured and calculated frequency distribution for soil-water contents manifested at all soil depths during steady-state infiltration. The curve is described by Eq. [6] for values of μ and σ equal to 0.433 and 0.0455 m^3 m^{-3}, respectively. From Nielsen et al. (1973).

hydraulic conductivity, pore water velocity, and apparent diffusion coefficient, have skewed frequency distributions (Nielsen et al., 1973; Biggar & Nielsen, 1976; Sharma et al., 1980). In most cases, the skewed distribution could be approximated as a log-normal distribution, which means that the distribution of logarithms of the values obeys Eq. [6]. However, other transformations of the values, such as square roots, other power roots, or a γ function, could be more suitable in some cases (Hald, 1952). Soil variables that have such skewed distributions generally also have high variability, with the coefficient of variation ranging from 10^2 to 10^6% (Warrick & Nielsen, 1980).

A simple way to find out whether the observed data follow a normal distribution is to graph the data as a fractile diagram (Hald, 1952; Warrick & Nielsen, 1980). The observed values, x, of the soil property are arranged in an increasing order i, $i = 1, 2, \ldots n$. For each value the cumulative probability is approximated by $(i-0.5)/n$. Corresponding to this cumulative probability, values of the reduced variable $u = [(x-\mu)/\sigma]$ are obtained from the cumulative probability tables for normal distribution. A plot of u vs. x is a fractile diagram (Fig. 7–4b). If the plot can be approximated as a straight line, the observed data are assumed to obey a normal distribution. If not, the above procedure is repeated after the observed values are transformed to logarithmic or another functional form, to see if the distribution is log-normal or other-transform normal. Goodness-of-fit tests of normality for

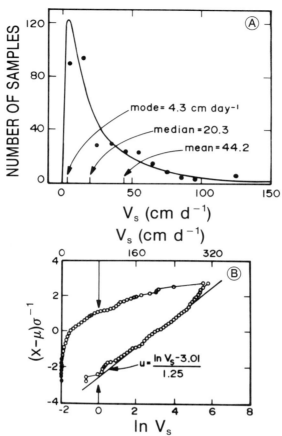

Fig. 7-4. (A) Frequency distribution of values of the pore water velocity V_s for a class length of 10 cm d^{-1}. (B) Fractile diagram stemming from Fig. 7-4A. Probability units [$(x - \mu)\sigma^{-1}$] are plotted vs. V_s and $ln\ V_s$ for $x = V_s$ and $x = ln\ V_s$, respectively. From Biggar and Nielsen. Water Resour. Res. 12:78-84, 1976. Copyright by the American Geophysical Union.

the original or transformed distribution can be made by computing Chi-square, Kolmogorov and/or other statistics (Rao et al., 1979).

The knowledge of the frequency distribution of a soil property is very useful for statistical interpretations and as input to modeling if the variability at the scale of interest is random and independent, i.e., there are no spatial trends in the values and the neighboring values are not correlated with each other. For two or more random variables that are mutually correlated, a joint frequency distribution can be used. An example of this is the joint normal distribution of soil water content and log hydraulic conductivity at saturation (Nielsen et al., 1973).

B. Spatial Trends or Drift

In many cases, a soil property shows a consistent trend of increase or decrease in value in a certain direction. The data and spatial trends for soil texture components in a 1.6-ha field along the main slope (3–4%) are shown in Fig. 7–5 (Williams et al., 1985). In such cases, the trends can be represented by some functional form, such as a polynomial, and the random fluctuations around the trends represented by a frequency distribution. When the random fluctuations around the trends are not independent, the above methods of defining trends are not strictly correct. The trends and spatial correlation structure have to be defined simultaneously by an iterative procedure (Webster & Burgess, 1980).

C. Spatial Correlation Structure

1. Autocovariance and Autocorrelation

The representation of variability of a soil property by a frequency distribution does not assume that the values of the property are random and independent, although this assumption greatly increases its usefulness by tapping many powerful methods of analysis developed in conventional statis-

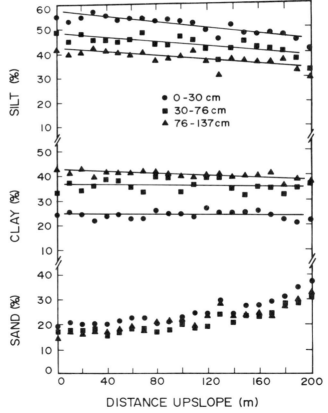

Fig. 7–5. Sand, silt, and clay contents in 0 to 30, 30 to 76, and 76 to 137 cm depth intervals along the slope in a 1.6-ha watershed. From Williams et al. (1985).

tics. Physically, we will expect the values of the property at two neighboring points to be closer to each other than those at two distant points. Limiting our consideration of the above spatial relationships only to the second order (i.e., second moments), the relationship is expressed by the so-called *auto-covariance*, as a function of separation distance, $C(h)$:

$$C(h) = E \{[z(x) - \mu][z(x + h) - \mu]\}$$

$$= E [z(x) . z(x + h)] - \mu^2 \qquad [7]$$

where E stands for the expected or mean value of the argument, $z(x)$ is the value of the soil property at point x, $z(x + h)$ at point $x + h$, h is the distance separating the points, commonly called lag, and μ is the population mean of the variable. In practice, $C(h)$ is obtained from a series of values $z(x_i)$, $i = 1, 2, \ldots n,$, taken at a regular interval as

$$C(h) = \frac{1}{(n - h)} \sum_{i=1}^{n-h} [z(x_i) - m][z(x_i + h) - m] \qquad [8]$$

where h is any integral multiple of the sampling interval, n-h is the number of pairs of points, and m is the mean of all values.

From Eq. [7] and [8], it is clear that $C(O)$ is the variance, σ^2 or its sample estimate s^2, of the variable. The ratio $C(h)/C(O)$ is the autocorrelation, $\rho(h)$, having values between $+1$ and -1. It is also apparent that for $C(h)$ and $\rho(h)$ to exist, the mean and variance of the population must be finite and constant in the area of consideration. This requirement is called the condition of full second-order homogeneity or stationarity. In relation to our earlier discussion, any trends in the field for either mean or variance must be removed before this approach can be applied.

Sisson and Wierenga (1981) applied the above approach to a series of field-measured steady infiltration rates at 5-cm intervals. A plot of $\rho(h)$ vs. h, commonly called a *correlogram*, of their data is shown in Fig. 7-6a. The plot shows that the infiltration rates are autocorrelated up to a separation distance of about 1.0 m (20 lags). This distance is called the correlation length or range. The solid curve shows an idealized shape of the function. Vieira et al. (1981) measured a series of 160 infiltration rates at 1-m intervals along eight different transects (columns). Their one-dimensional correlograms, presented in Fig. 7-6b, show a range between 10 and 30 m.

2. Semivariograms

The Regionalized Variable Theory (Matheron, 1971) concentrates on differences $[z(x) - z(x + h)]$ and their variances. A semivariance, $\gamma(h)$, is defined as

$$\gamma(h) = \frac{1}{2} E \{[z(x) - z(x + h)]^2\}$$

$$= \frac{1}{2} \text{Var} [z(x) - z(x + h)]. \qquad [9]$$

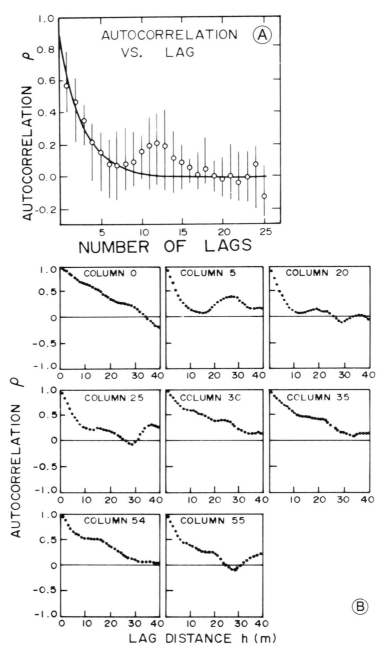

Fig. 7-6. (A) Autocorrelation function for the 5-cm infiltration rates obtained by averaging the five autocovariance functions for the five transects. The solid curve is the estimated first-order autocorrelation function. Vertical bars show range of autocorrelations obtained. From Sisson and Wierenga (1981). (B) Autocorrelations for the eight columns of infiltration rate data. Calculations were based on 160 observations for each column. From Vieira et al. (1981).

A plot of $\gamma(h)$ vs. h is called a *semivariogram*. The semivariance is related to $C(h)$ defined earlier as

$$\gamma(h) = C(O) - C(h). \tag{10}$$

The use of $\gamma(h)$ requires the assumption of a constant population mean μ, but not of a constant and finite population variance σ^2. It only requires that for any h, the difference $[z(x) - z(x + h)]$ have a finite variance, independent of x (intrinsic hypothesis). It frequently happens in earth sciences that a variable may have no finite variance or covariance, but still have a finite semivariance (Webster, 1985). The semivariogram is, therefore, more commonly used.

In the last 10 yr, semivariograms have been obtained for several different soil properties and many soil types. Presented in Fig. 7-7 are an idealized semivariogram and a semivariogram form commonly observed for soil properties. Nugget variance is the variance at a scale smaller than the scale of measurement. In Fig. 7-8, the semivariogram of 1280 values of infiltration rate measured by Vieira et al. (1981), pooled over all directions, is pre-

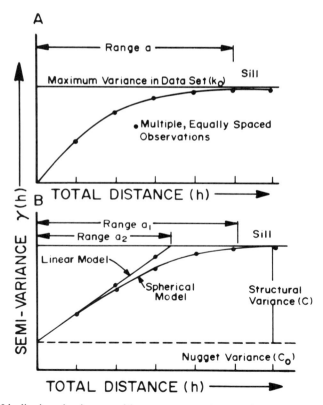

Fig. 7-7. (*A*) Idealized semivariogram with zero nugget variance and (*B*) observed semivariograms for soil properties with nugget variance. From Wilding and Drees (1983).

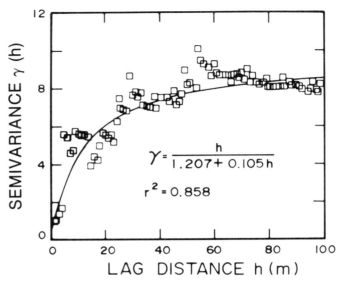

Fig. 7–8. Variogram for the 1280 measured values of the infiltration rate V (mm h^{-1}) for both x and y directions. The value of the variance is 7.77 (mm h^{-1})^2. From Vieira et al. (1981).

sented. It shows a range of 50 m. For some soil properties, directional differences or anisotropy in semivariograms is observed (Webster, 1985).

For details of calculating semivariograms and many models or equations that can be used to describe them, as well as some practical problems involved, consult the above citations or Webster (1985) and Warrick et al. (1986).

3. Cross Correlation

Two different soil properties may often be correlated, so that the values of one at a given place depend upon values of the other at other places nearby. This spatial interdependance may be expressed by cross covariance:

$$C_{uv}(h) = E\{[z_u(x) - \mu_u][z_v(x + h) - \mu_v]\} \qquad [11]$$

where u and v are the two properties being considered. The cross semivariance is then given by:

$$\gamma_{uv}(h) = \tfrac{1}{2}E\{[z_u(x) - z_u(x + h)][z_v(x) - z_v(x + h)]\}. \qquad [12]$$

The method of calculation of γ_{uv} is similar to that of the semivariance. Vauclin et al. (1983) obtained cross-covariance of available water and 33-kPa soil water content with sand content.

D. Similar-Media Scaling

Scaling is a simple but powerful tool for approximately describing field spatial variability of suction-water content and hydraulic conductivity-water content relationships, as well as the characteristics derived from these relations, such as the available water, infiltration, and drainage. The measured curves for these properties, from many experimental sites, are combined into representative mean curves by choosing a single factor for each site. The frequency distribution and spatial-correlation structure of these factors describe the variability in the field, thus resulting in considerable simplicity. Ideally, if the soil media at different sites are "similar," meaning that they differ only in the scale of their internal microscopic geometries (Miller & Miller, 1956), the scale factors for all above properties are the same. In reality, these factors may be different for different properties.

According to the extended scaling concepts (Warrick et al., 1977a; Simmons et al., 1979), soil water pressure head at any given degree of saturation (volumetric soil water content/volumetric soil water content at saturation), S, at a given site i, $h(S)_i$, is related to a scaled mean value for the field at that S, $h(S)_m$, as

$$h(S)_i = h(S)_m/\alpha_i \qquad [13]$$

where α_i is a scaling factor for site i. Setting the average of all α_i values equal to 1.0, $h(S)_m$ can be obtained from a known set of N $h(S)_i$ as

$$h(S)_m = N/\left\{ \sum_{i=1}^{N} [h(S)_i]^{-1} \right\} . \qquad [14]$$

The scaling factor α_i can then be obtained from Eq. [13]. However, this α_i is for h at one value of S. A least-squares fitting approach is used to obtain one value of α_i that brings the complete $h(S)$ curve for site i, closest to $h(S)_m$ (Russo & Bresler, 1980).

The hydraulic conductivity at any given S and site i, $K(S)_i$, is related to the scaled mean $K(S)_m$ as

$$K(S)_i = K(S)_m \alpha_i^2. \qquad [15]$$

With the condition of sum of α_i equal to N, the $K(S)_m$ is derived to be

$$K(S)_m = \left\{ \left[\sum_{i=1}^{N} (K(S)_i)^{1/2} \right] /N \right\}^2 . \qquad [16]$$

Thus, the α_i can also be obtained from $K(S)$ data using Eq. [15] and the least-squares technique. For "similar" soils, these α_i values should be the same as those obtained from $h(S)$ data for the same sites. For such soils the

$K(S)$ function need not be measured at all the sites. From α_i obtained from $h(S)$ data and the measurement of $K(S)$ at only one site, $K(S)$ at all other sites can be estimated. The reverse is also true. For appreciably dissimilar soils, this is not possible and a different set of scale factors for $h(S)$ and $K(S)$ need to be obtained (Rao et al., 1983). The scaling is still worthwhile to obtain one factor to represent the entire $h(S)$ or $K(S)$ function. Also, for infiltration and other flow processes the $K(S)$ function would be more important, so that one could obtain scaling factors from $K(S)$ data for different locations and use the same factors to generate $h(S)$ for use in the predictions without too much error (Ahuja et al., 1984b). An example of scaling $h(S)$ functions is presented in Fig. 7–9. Further examples and details are provided by Warrick et al. (1977a), Simmons et al. (1979), Russo and Bresler (1980), and Rao et al. (1983).

Similar-media assumption also enables transformation of the differential equation of soil-water or chemical transport in uniform soils into a generalized equation containing scaled mean soil properties and scaled time and position variables. For example, the Richards equation for unsaturated flow (Eq. [2] with S_w term $= 0$) is transformed to

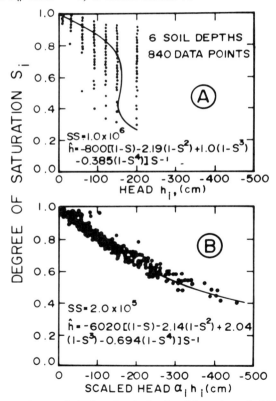

Fig. 7–9. Soil water characteristic data for six depths of Panoche soil: (A) unscaled and (B) scaled. From Warrick et al. Water Resour Res. 13:355–362, 1977. Copyright by the American Geophysical Union.

$$\frac{\partial S}{\partial T} = \frac{\partial}{\partial Z}\left[K(S)_m \frac{\partial h(S)_m}{\partial Z} - K(S)_m\right] \qquad [17]$$

where $T = (\alpha_i^3/\theta_{si})t$ is the scaled time, $Z = \alpha_i z$ is the scaled depth variable, and θ_{si} is the saturated soil water content for site i. A solution of this generalized equation for a particular problem, such as infiltration, obtained only once, can provide solutions for all sites in the field by back transforming T and Z to t and z for each site i using appropriate α_i and θ_{si} (Warrick & Amoozegar-Fard, 1979).

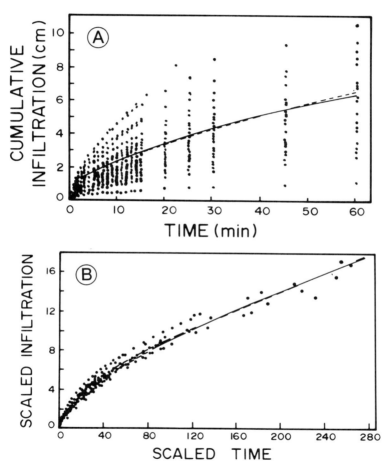

Fig. 7-10. (A) Field-measured cumulative infiltration vs. time plots of 26 sites on a watershed. The solid line is an "average" curve computed from average infiltration parameters, S and A, of the Philip equation, $I = St^{1/2} + At$, while the broken line is the least-squares fit through the data. Total data points were 618, but only 616 appear on the graph; the remaining points fall beyond the selected graph scale. (B) Infiltration data from 26 sites scaled by Eq. [18]. Number of data points appearing on the graph is 611. The solid line is computed from S and A, while the broken line is the least-squares fit through the scaled data. From Sharma et al. (1980).

Based on Eq. [17], the scaled infiltration rate into the soil, q_m, at time T is given by

$$q_m = -K(S)_m \frac{\partial h(S)_m}{\partial Z} + K(S)_m, \quad Z = 0$$

$$= q_i/\alpha_i^2, \quad z = 0 \tag{18}$$

Thus, the infiltration rates at all sites can be scaled by transforming q_i to q_m and t_i to T for each time. The field-measured infiltration or drainage rates can be scaled in this fashion even when the soil media at different sites are not strictly "similar," i.e., when the scaling factors for $K(S)$ and $h(S)$ are not the same and Eq. [17] is not valid. The results of Sharma et al. (1980) on the scaling of cumulative infiltration vs. time are presented in Fig. 7–10. Simmons et al. (1979) scaled the soil water content vs. time during drainage in a similar way. The scale factors obtained this way agreed with those obtained from unsaturated hydraulic conductivity functions. Scaling can also be done for parameters of the simple infiltration equations, such as the Green-Ampt equation (Ahuja et al., 1984b) or the Philip's two-term equation (Sharma et al., 1980).

E. Applications of the Spatial Variability Analyses

The correlograms and variograms, along with scaling, trend analysis, and frequency distributions, provide a logical quantitative description of soil attributes for both research and management purposes. They serve as realistic input into simulation models of an irrigated field, which can be used to get an insight into the effect of natural and management factors on its performance and productivity. Bresler et al. (1979) and Wagenet & Rao (1983) used scaling approach in modeling water and solute transport in a heterogeneous irrigated cropped soil. Dagan & Bresler (1983) and Bresler & Dagan (1983a, b) have presented models of transport in a bare field that use only a frequency distribution of hydraulic conductivity. Other efforts to incorporate spatial variability in water flow models are reviewed in section VI of this chapter. An important application of semivariograms also is in the optimal estimation of values of a variable at unknown points by the procedure called *kriging* (Webster, 1985; Warrick et al., 1986). Maps of a particular soil property can thus be obtained within prescribed limits of accuracy (Fig. 7–11) for use in irrigation management decisions, such as irrigation design (Russo, 1983, 1986b), and optimum locations for soil sampling and tensiometers (Saddiq et al., 1985). Variances can also be estimated for different-size blocks of land, which could be the basis for forming optimum management units (Gurovich et al., 1983; Hendrickx et al., 1986). Applications of the spatial variability analyses in improving soil and water management are largely a challenge for the future.

In all of the four approaches as well as potential analyses of spatial variability not mentioned above, the greatest challenge is to ascertain the statistical parameters as economically and effectively as possible. Persons familiar only with the first approach (frequency or probability density functions) may be skeptical about the utility of the others owing to their initial preceptions that excessively large numbers of measurements are required. They may even believe that the numbers are prohibitively large. Indeed, the number of observations must be adequate to estimate the spatial variance structure over distances and directions applicable to the physical and biological processes of concern to the investigator. However, that number may be smaller or larger than that required to reliably estimate the mean or the probability density function of a soil attribute for a given irrigation management unit. To date, research results are inadequate to provide reliable guidelines for the number of observations needed in an irrigated field. That topic is a foremost research need and is one most likely to succeed when consideration is given to spatial and temporal variations expected within soil domains identified across similar soil mapping units.

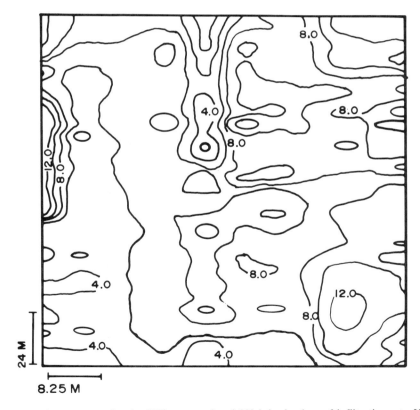

Fig. 7–11. Contour map for the 1280 measured and 800 kriged values of infiltration rate V (mm h^{-1}). Contour interval is 2. From Vieira et al. (1981).

IV. CHANGES IN SOIL WATER PROPERTIES RESULTING FROM TILLAGE AND SUBSEQUENT CONSOLIDATION

Tillage and related management practices cause a temporal variability in soil-water properties and contribute to spatial variability on a local scale under both irrigated and nonirrigated conditions. Klute (1982) and Hines (1986) have presented good reviews of the effects of tillage on soil structure and soil water properties. Unfortunately, the experimental data concerning the effects on soil-water properties are limited, generally qualitative, and often confusing or contradictory. Reasons for the contradictions are that the effects depend upon several factors—soil texture, initial soil structure and moisture conditions, type of tillage implement, history of tillage and cropping, and climatic conditions.

A. Changes Below the Soil Surface

In soils continuously cultivated over long periods of time, tillage generally increases total porosity of the plow layer by loosening the soil mass and promoting formation of large aggregates. For example, Unger (1984) found mean weight diameter of aggregates in the tilled layer of a silty loam in TX to increase with degree of tillage, from no-tillage to sweep, rotary, disk, and moldboard-plow tillage. Hamblin and Tennant (1981) measured the highest total porosity of a sandy loam soil under conventional tillage, followed by reduced tillage and no tillage. Lindstrom and Onstad (1984) had similar results on a loam soil—the total porosity being 61, 55, and 49%, respectively. However, the increase in porosity due to a primary tillage operation can be reduced by a secondary tillage operation (Ehlers, 1976; Allmaras et al., 1966, 1967; Boone et al., 1976; Millette et al., 1980). For instance, Allmaras et al. (1966) observed an increase in porosity from 53 to 84% by moldboard plowing, but only to 73% by plowing followed by disking and harrowing. The increase in porosity brought about by tillage also degrades with time during the course of a season, as a result of aggregate breakdown by rainfall, irrigation, clogging of large pores, and consolidation by water flow through soil (e.g., Moldenhauer & Kemper, 1969; Gumbs & Warkenton, 1973; Cassel, 1983). For modeling purposes, Rawls et al. (1983) summarized available information on porosity changes due to moldboard plowing in the form of a nomograph built on the textural triangle (Fig. 7–12). The initial increase in porosity and the subsequent decrease varied from 15 to 2%, respectively, for a clayey soil, to 30 and 10%, respectively, for a sandy loam. Increases due to a plow-disk-harrow operation were noted to approximate about 97% of the moldboard plow alone, for rotary 95%, for plow and pack 82%, and for chisel 102% (Rawls et al., 1983).

The rate at which the initial increase in porosity due to tillage degrades with time is not well defined. On a tilled loamy sand cropped to soybean (*Glycine max* L.), Cassel (1983) measured major changes in soil bulk density (and hence, porosity) during the first month after tillage when two rainfalls totaling 6.5 cm were received. Onstad et al. (1984) subjected replicated plots of four freshly tilled soils to four applications of simulated rainfall totaling 15.2 cm. Most of the change in soil bulk density occurred with the first 5 cm of water application. Rousseva et al. (1986) used a surface γ-neutron gauge to monitor bulk densities of a tilled soil with time at a large number of fixed locations. The average bulk density of the 0 to 10 cm layer gradually increased with water application up to 140 cm.

Lindstrom and Onstad (1984) observed that the increase in porosity by conventional tillage was mostly in the range of large pores, corresponding to less than 6 kPa suction. Results of Hamblin and Tennant (1981) were similar, but there was also a decrease in smaller size pores. Results of Mapa et al. (1986) indicated that the changes in soil water characteristics (decrease in soil water contents) after tillage were mainly at soil water suctions less than 30 kPa.

Lindstrom and Onstad (1984) obtained an increase in saturated hydraulic conductivity from 0.3 to 18.0 cm h^{-1} by conventional tillage. Negi et al. (1981) found unsaturated hydraulic conductivities of a tilled clay soil to be an order of magnitude higher than those of the untilled soil in the entire practical range of suctions. Hamblin and Tennant (1981) obtained similar results for a sandy soil, but the magnitude of difference was smaller. On the other hand, Ehlers (1976) reported that hydraulic conductivity of a tilled soil was lower than that of an untilled soil at suctions ranging from 0 to 100 kPa, after which the reverse occurred. Mapa et al. (1986) observed that satu-

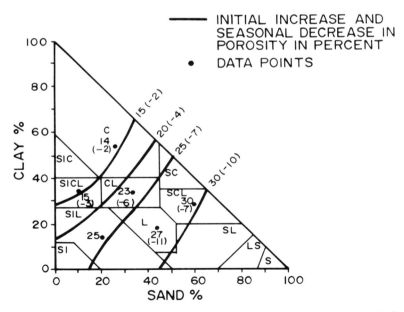

POROSITY CHANGES CAUSED BY PLOWING

Fig. 7-12. Percent porosity increases due to a moldboard plow (and percent decreases during the growing season) as a function of soil texture. From Rawls et al. (1983).

rated hydraulic conductivity of a tilled soil decreased 100-fold with time subsequent to tillage, by wetting and drying. At higher suctions, the decrease in conductivity was progressively less.

B. Changes at the Soil Surface

Tillage destroys the soil surface cover. Rainfall or sprinkler irrigation on a freshly tilled soil results in formation of a dense surface seal by aggregate breakdown, washing of fine particles into the soil matrix, and raindrop compaction (Duley, 1939; McIntyre, 1958a, b; Tackett & Pearson, 1965). On drying, this surface seal becomes a crust. McIntyre (1958a) measured the compacted skin to be only 0.1 mm-thick and the wash-in layer 1.5- to 2.5-mm thick. However, the hydraulic conductivity of these layers was as much as 2000-fold smaller than that of the original soil. Other investigators have reported 5-fold or more reduction in hydraulic conductivity (Hillel, 1960; Tackett & Pearson, 1965; Falayi & Bouma, 1975; Sharma et al., 1981). Whatever the magnitude, infiltration of water into a sealed-crusted soil can be markedly reduced (Duley, 1939; Mannering & Wiersma, 1970; Morin & Benyamini, 1977). Sealing and crusting also occur under surface irrigation (Eisenhauer et al., 1984).

A thin, dense crust will be quickly saturated at the start of rainfall or irrigation. Therefore, its unsaturated hydraulic conductivity and water storage can be neglected, and only saturated hydraulic conductivity or hydraulic resistance need to be characterized. Edwards and Larson (1969) measured the change in saturated conductivity of a seal-crust formed under simulated rainfall in bare Ida silt loam [fine-silty, mixed (calcareous), mesic Typic Udorthents] soil. The conductivity decreased approximately as an exponential function of rainfall time. The conductivity value was not quite steady after 90 min of rainfall, with an intensity of 6 cm h^{-1}. Van Doren and Allmaras (1978) related the transient seal-crust hydraulic conductivity (K_{sc}) to rainfall kinetic energy instrumental in crusting (E_{sc}) as

$$K_{sc} = K_f + (K_o - K_f) \exp(-CE_{sc}) \qquad [19]$$

where K_f is the final steady value of K_{sc}, K_o is the initial value of K_{sc}, and C is a constant for the soil. For loam to clay loam soils having K_o value between 1.4 and 30 cm h^{-1}, the value of K_f was between 0.5 and 3 mm h^{-1}, and value of C between 40 and 70 cm^2 J^{-1} (Moore, 1981). Linden (1979) expressed the rainfall crusting energy E_{sc} as

$$E_{sc} = E_o B \cos R/A \qquad [20]$$

where E_o is the rainfall energy, B is the fraction of soil surface exposed, R is the average angle of inclination of the rough soil surface, and A is the surface area of soil exposed to rainfall per unit horizontal area. The cos R/A is a factor to account for surface roughness of a tilled soil. Work of Amemiya (1968) and Burwell et al. (1968) had indicated that soil surface roughness deters sealing-crusting. Using the simulation model and field data, Linden (1979) related cos R/A to random roughness, RR (Allmaras et al., 1966), as

$$\cos R/A = 1.11 - 0.247\ RR. \qquad [21]$$

Equation [21] indicates that no seal-crust is formed when RR is nearly 4.5 cm. For an average RR value of 1.0 cm, cos $R/A = 0.86$. The random roughness generally also changes with time after tillage, and needs to be defined empirically.

C. Soil Macropores

The term *macropores* is used here to refer to relatively large noncapillary pores or channels, such as interaggregate pores, interpedal voids, drying cracks in clay soils, worm holes, animal burrows, and channels created by plant roots. Most natural soils contain some macropores. Recent reviews of the experimental work have indicated that water flow and transport in soils containing a network of macropores that are open at the soil surface can deviate substantially from predictions of the current theories for these processes, which hold adequately in soils without a large number of macropores (Thomas & Phillips, 1979; Beven & Germann, 1982; White, 1985). The macropores allow rapid gravitational flow of the free water available at the soil surface during prolonged high-intensity rainfalls or above an impeding soil horizon, which bypasses the soil matrix. Obviously, a less permeable soil matrix will permit more macropore flow. This phenomenon is receiving attention, especially because of the possibilities of rapid transport of a portion of the chemicals applied on the soil surface to the groundwater.

Conventional tillage of soil, consisting of plow-disk-harrow, under most conditions destroys the continuity of macropores in the topsoil. As a result, the flow of free water from the soil surfce to the macropore channels in the subsoil is greatly reduced (Bouma et al., 1982). The initial plowing, however, may create some new large voids in the plow layer. Methods of characterizing the distribution of effective macropores and the transport through them are still in early stages of development.

The above review of literature indicates that the tillage and subsequent consolidation induce appreciable temporal changes in soil water properties, which will in turn greatly influence irrigation efficiency and management. However, the information available on the effects of tillage is scanty. Further studies and quantification of temporally variable changes in soil water properties from tillage, as well as from perturbations in the climate and weather, remain a goal for future research. The concept of an autocorrelation length in terms of time rather than spatial distance has yet to be analyzed for any soil location. In other words, a method to obtain the optimal frequency of making observations of the soil pore size distribution in order that its changes can be related to a particular tillage system has not been examined. And indeed, coupling the spatial variance structure with the temporal variance structure of the pore size distribution is an immediate need but will probably await a great deal of future research effort.

V. SIMPLIFIED MEASUREMENT OR ESTIMATION OF SOIL WATER PROPERTIES AND THEIR DISTRIBUTION FOR FIELD APPLICATIONS

Rigorous application of the soil water flow theory to describe water flow and uptake processes in the field requires the determination of soil hydraulic conductivity as a function of soil water content or matric potential, the matric potential-water content relationships, and their spatial variability. Standard methods of measuring these basic soil hydraulic properties in the laboratory and field are detailed in the new Agronomy Monograph *Methods of Soil Analysis, Part 1* (Klute, 1986) and are also covered in chapter 6 in this book. The most important range of determinations is the wet region at matric potentials greater than -80 kPa. Perhaps the most reliable method for determining hydraulic conductivities in this region, for field conditions, is the Darcian analysis of in-situ tensiometric measurements during steady state infiltration and the subsequent drainage, using the matric potential-water content relationships (Richards et al., 1956; Ogata & Richards, 1957; Nielsen et al., 1964; Rose et al., 1965; Watson, 1966; van Bavel et al., 1968; Fluhler et al., 1976). The matric potential-water content relationships, required for the above analysis, can be obtained by periodic measurement of soil water content during the drainage phase by gravimetric, neutron meter, or γ-ray attenuation techniques. More commonly, however, this relationship is measured in the laboratory on carefully excavated, undisturbed soil cores. These methods for determining soil hydraulic and storage properties are time consuming, tedious, and expensive, especially since a large number of measurements is required to characterize the spatial variability in these properties within a field. Innovative approaches that require less time and effort would increase applications of the current theory. A summary of some methods for determining soil hydraulic parameters from simpler and limited data is given below.

A. Simplified Methods of Field Measurement of $\theta(h)$ and $K(h)$

Several investigators (e.g., Black et al., 1969; Davidson et al., 1969) have shown that the hydraulic gradient during the drainage process in a uniform soil profile, without a shallow water table, is nearly equal to unity. Assuming unit gradients, Nielsen et al. (1973) tested a simple field method to determine the unsaturated hydraulic conductivities from analysis of only the soil water content measurements during drainage and found it to work well for Panoche clay loam [fine-loamy, mixed (calcareous), thermic Typic Torriorthents], a rather uniform soil. Since then, this unit-gradient method has been developed further, using certain functional forms for hydraulic conductivity as a function of soil water content, $K(\theta)$, and applied to even layered soils (Libardi et al., 1980; Sisson et al., 1980; Chong et al., 1981). However, some recent work indicates that the unit-gradient methods may not be adequate for complex stratified soils (Schuh et al., 1984; Jones & Wagenet, 1984).

With certain reasonable assumptions about the functional forms for hydraulic conductivity as a function of soil water pressure head, $K(h)$, soil water characteristic, $\theta(h)$, and the drainage process, the $K(h)$ and the diffusivity, $D(h)$, can be determined from analysis of only field tensiometric data taken during infiltration and subsequent drainage (Ahuja et al., 1980; Schuh et al., 1984). With measurement of soil water profile at one time, this analysis can also provide estimates of soil water storage with time as a function of average pressure head, i.e., average $\theta(h)$, in each soil horizon.

Progress is also being made to identify multiple unknown parameters simultaneously of some suitable functional forms of $K(\theta)$ or $K(h)$ and $\theta(h)$ by repeated numerical solutions of the flow equations from some simple field data. Zachman et al. (1981, 1982) used the cumulative discharge at the bottom of a soil column during gravity drainage to identify parameters of $K(h)$ and $\theta(h)$. Dane and Hruska (1983) and Wall and Miller (1983) used measured soil water content profiles during drainage for this purpose. These methods may be better than the currently popular method of measuring $\theta(h)$ of each soil horizon on undisturbed cores in the laboratory, and then determining $K(h)$ from measured $\theta(h)$ using the methods described by Mualem (1986). The latter methods also require measurements of in-situ saturated hydraulic conductivities as matching factors to determine $K(h)$.

B. Estimating $\theta(h)$ and Its Distribution in a Watershed from Limited Data

The method of extended similar-media scaling can be used to approximate the complete $\theta(h)$ function at any specific site and depth within a watershed. The approximation depends upon the soil bulk density and one value of $\theta(h)$, e.g., at $h = -33$ kPa, being measured or available from soil survey at each location and $\theta(h)$ being measured or available at one representative location. This method seemed to work well for two watersheds containing several related soil series each having strongly layered profiles, and horizons of widely varying texture and bulk density (Ahuja et al., 1985). The results are presented in Tables 7-1 and 7-2. An additional advantage of the scaling method is that the spatial variability of the soil water storage characteristics is already described, in terms of the scaling factors, for use in modeling or other applications (see Sharma & Luxmoore, 1979).

C. Estimationg $\theta(h)$ from Soil Composition and Bulk Density

Several efforts have been made to estimate the soil water characteristic curves or selected points on them from easily measured soil textural and structural properties. Rawls and Brakensiek (1982) presented a good summary of these past efforts. Noteworthy among them is the work of Gupta and Larson (1979) who developed regression equations based on percent sand, silt, clay, organic matter, and bulk density for predicting soil water retention at 12 matric potentials ranging from -4 to -1500 kPa. Rawls et al. (1982) and Rawls and Brakensiek (1982) collected a national data base of 2543

Table 7-1. Mean errors, mean relative errors, and the standard deviation of these errors in soil water contents at fixed matric potentials calculated by the method of scaling applied to data for different soil depth intervals. El Reno Watershed data from Ahuja et al. (1985).

S. No.	Depth interval	Matric potential, kPa							
		−10		−60		−100		−1500	
		Mean error	Mean relative error	Mean error	Mean relative error	Mean error	Mean relative error	Mean error	Mean relative error
	m	$m^3\ m^{-3}$							
1	0–1.35 (all depths combined)	−0.0027 (0.0305)†	−0.0096 (0.0801)†	0.0136 (0.0207)	0.0539 (0.0840)	0.0135 (0.0208)	0.0575 (0.0919)	0.0052 (0.0444)	0.0910 (0.3033)
2	0–0.3	−0.0179 (0.0266)	−0.0485 (0.0712)	0.0149 (0.0217)	0.0716 (0.1056)	0.0112 (0.0253)	0.0598 (0.1192)	−0.0027 (0.0412)	0.0628 (0.3452)
3	0.3–0.75	−0.0059 (0.0217)	−0.0164 (0.0566)	0.0160 (0.0198)	0.0592 (0.0718)	0.0175 (0.0193)	0.0686 (0.0759)	0.0126 (0.0414)	0.0978 (0.2175)
4	0.75–1.35	0.0099 (0.0134)	0.0237 (0.0328)	0.0105 (0.0178)	0.0322 (0.0533)	0.0107 (0.0137)	0.0338 (0.0425)	0.0032 (0.0378)	0.0364 (0.1696)

† Standard deviation of errors in parenthesis beneath the associated mean for all cases.

Table 7-2. Mean errors, mean relative errors, and the standard deviation of these errors in soil water contents at fixed matric potentials calculated by the method of scaling applied to data for different soil depth intervals. Chickasha Watershed data from Ahuja et al. (1985).

		Matric potential, kPa										
		−3		−6		−15		−33		−100		−1500
S. no.	Depth interval	Mean error	Mean relative error	Mean error	Mean relative error	Mean error	Mean relative error	Mean error	Mean relative error	Mean error	Mean relative error	Mean error	Mean relative error
	m	$m^3\ m^{-3}$											
1	0–0.6 (all depths combined)	0.0035 (0.0388)†	0.0102 (0.0948)†	−0.0036 (0.0264)	−0.0087 (0.0660)	−0.0051 (0.0173)	−0.0144 (0.0482)	--	--	0.0010 (0.0108)	0.0073 (0.0490)	0.0031 (0.0149)	0.0287 (0.1216)
2	0–0.3	0.0379 (0.0480)	0.0909 (0.1154)	0.0082 (0.0237)	0.0210 (0.0591)	−0.0032 (0.0134)	−0.0090 (0.0374)	--	--	−0.0014 (0.0100)	−0.0039 (0.0421)	−0.0027 (0.0104)	−0.0154 (0.0632)
3	0.3–0.6	0.0232 (0.0412)	0.0590 (0.1016)	0.0077 (0.0220)	0.0197 (0.0567)	0.0001 (0.0114)	−0.0001 (0.0315)	--	--	−0.0021 (0.0085)	−0.0056 (0.0292)	−0.0010 (0.0108)	0.0011 (0.0878)

† Standard deviation of errors in parenthesis beneath the associated mean for all cases.

Table 7–3. Coefficients for independent variables of the multiple linear regression equations for prediction of soil water content at specific matric potentials. From Rawls et al. (1982).

Matric potential kPa	Model no.	Intercept	Sand, %	Silt, %	Clay, %	Organic matter, %	Bulk density, Mg m^{-3}	−33 kPa Water retention, m^3 m^{-3}	−1500 kPa Water retention, m^3 m^{-3}	Correlation coefficient, R
						Coefficients				
−4	1	0.7899	−0.0037			0.0100	−0.1315			0.58
	2	0.6275	−0.0041			0.0239	−0.0376	1.89	−0.08	0.57
	3	0.1829				−0.0246			−1.38	0.77
−7	1	0.7135	−0.0030		0.0017		−0.1693			0.74
	2	0.4829	−0.0035			0.0263		1.53	0.25	0.74
	3	0.8888	−0.0003			−0.0107			−0.81	0.91
−10	1	0.4118	−0.0030		0.0023	0.0317				0.81
	2	0.4103	0.0031			0.0260		1.34	0.41	0.81
	3	0.0619	−0.0002			−0.0067			−0.51	0.95
−20	1	0.3121	−0.0024		0.0032	0.0314				0.86
	2	0.3000	−0.0024			0.0235		1.01	0.61	0.89
	3	0.0319	−0.0002						−0.06	0.99
−33	1	0.2576	−0.0020		0.0036	0.0299				0.87
	2	0.2391	−0.0019			0.0210			0.72	0.92
−60	1	0.2065	−0.0016		0.0040	0.0275				0.87
	2	0.1814	−0.0015			0.0178			0.80	0.94
	3	0.0136					−0.0091	0.66	0.39	0.99
−100	1	0.0349			0.0055	0.0251				0.87
	2	0.1417	−0.0012	0.0014		0.0151			0.85	0.96
	3	−0.0034				0.0022		0.52	0.54	0.99
−200	1	0.0281			0.0054	0.0200				0.86
	2	0.0986	0.0009	0.0011		0.0116			0.90	0.97
	3	−0.0043				0.0026		0.36	0.69	0.99
−400	1	0.0238			0.0052	0.0190				0.84
	2	0.0649	−0.0006	0.0008		0.0085			0.93	0.98
	3	−0.0038				0.0026		0.24	0.79	0.99
−700	1	0.0216			0.0050	0.0167				0.81
	2	0.0429	−0.0004	0.0006		0.0062			0.94	0.98
	3	−0.0027				0.0024		0.16	0.86	0.99
−1000	1	0.0205			0.0049	0.0154				0.81
	2	0.0309	−0.0003	0.0005		0.0049			0.95	0.99
	3	−0.0019				0.0022		0.11	0.89	0.99
−1500	1	0.0260			0.0050	0.0158				0.80

Table 7-4. Mean errors, mean relative errors, and the standard deviation of these errors in soil water contents at fixed matric potentials calculated by Models of Rawls et al. (1982) and Ahuja et al. (1985). El Reno Watershed Data from Ahuja et al. (1985).

	Matric potential, kPa									
	−10		−33		−60		−100		−1500	
Model no.	Mean error	Mean relative error	Mean error	Mean relative error	Mean error	Mean relative error	Mean error	Mean relative error	Mean error	Mean relative error
	$m^3\ m^{-3}$									
1	0.1042 (0.0524)†	0.2924 (0.1758)†	0.0598 (0.0554)	0.2079 (0.2052)	0.0567 (0.0526)	0.2211 (0.2189)	0.0478 (0.0497)	0.2057 (0.2221)	0.0073 (0.0532)	0.1177 (0.3599)
2	0.2391 (0.0457)	0.6502 (0.1916)	0.0616 (0.0432)	0.2021 (0.1561)	0.0542 (0.0347)	0.1949 (0.1425)	0.0466 (0.0329)	0.1798 (0.1438)	-- --	-- --
3	−0.0002 (0.0298)	−0.0008 (0.0790)	-- --	-- --	0.0007 (0.0127)	0.0030 (0.0464)	0.0093 (0.0180)	0.0345 (0.0726)	-- --	-- --

† Standard deviation of errors in parenthesis beneath the associated mean for all cases.

horizons of 500 soils from 18 states, and used them to develop regression equations along the lines of Gupta and Larson (1979). Rawls et al. (1982) presented three different models (or equations). In Model 1, all of the above five variables were used in the regression, but the variables whose contributions to the regression sum of squares were statistically nonsignificant were deleted from the final equation. Model 2 stemmed from a similar regression analyses based on the above five variables and the measured soil water content at -1500 kPa potential as an additional variable. Model 3 resulted from regression which incorporated two measured water contents, at -33 and -1500 kPa potentials, as additional independent variables. The coefficients of the above regression models (linear multiple regression equations) for selected matric potentials are reproduced from Rawls et al. (1982) in Table 7-3.

Results of $\theta(h)$ estimations by the equations of Rawls et al. (1982) and Rawls and Brakensiek (1982) in comparison with experimental data for one watershed are summarized in Table 7-4. With the equations based on soil texture, bulk density, and organic matter content, the soil water contents calculated at different matric potentials were generally larger than the measured values. The mean relative error ranged from 8 to 29%, with the standard deviation of errors ranging from 17 to 36%. The model that incorporated one measured soil water content (at -1500 kPa potential), as an additional independent variable, did not greatly improve the results. The model that incorporated two measured soil water contents (at -33 and -1500 kPa potentials) as additional variables reduced the errors in calculated values considerably.

D. Estimating Distribution of K_s from Effective Porosity

Cumulative frequency distribution of scaling factors for saturated hydraulic conductivity (K_s) in a soil may be adequately estimated from measurement of its effective porosity (θ_{se}) distribution (Ahuja et al., 1984a). The θ_{se} is defined as total porosity minus soil water content at -33 kPa pressure head. The K_s is related to θ_{se} by a generalized Kozeny-Carman equation:

$$K_s = B \, \theta_{se}^n.$$ [22]

The exponent of this relationship is assumed to vary within a narrow range (value of 4 or 5). The equation is combined with scaling theory to derive frequency distribution of K_s scaling factors from the θ_{se} distribution. One or a few measurements of K_s corresponding to known values of scaling factors will then yield the complete distribution of K_s.

Figure 7-13 contains a graph of log K_s vs. log θ_{se} for a small watershed in Oklahoma (Ahuja et al., 1984a). In Fig. 7-14, a fractile or a cumulative frequency distribution of scaling factors obtained from experimental K_s data is compared with the distribution obtained from θ_{se} values. The results show that the generalized Kozeny-Carman Eq. [22], with the exponent n either 4 or 5, was successful in characterizing spatial distribution of saturated hydraulic conductivity from effective porosity measurements. Recent unpublished work by the authors indicates a general applicability of Eq. [22] in eight soil series. Experimental data from eight different soils could also be combined into one relationship.

Otto Baumer of the USDA National Soil Survey Laboratory (1986, personal communication) has fitted a form of Eq. [22] to available experimental data on K_s. He assumed $n = 3$, as in the original Kozeny equation, and defined θ_{se} as total porosity minus a residual water content θ_r. The value of θ_r is much smaller than the θ at 33 kPa suction. The soils were grouped based on clay content and clay activity for the purpose of fitting, so that a different value of intercept B is obtained for each group.

Although each of the approaches relying on simplified measurements to quantify soil water properties outlined above requires additional research and development, a major unmet challenge is common with all of them including the classical approaches. That challenge is the identification of the sample size or the scale of observation best used to ascertain the soil water attribute or property. Concepts of representative elementary volume and autocorrelation length need to be reconciled for each term used in any of the mathematical formulations. Does a measurement of soil water content apply to the same soil volume as a measurement of soil water potential? How

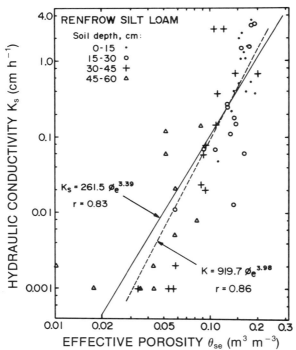

Fig. 7–13. Saturated hydraulic conductivity as a function of effective porosity for soil cores of Renfrow silt loam (fine, mixed, thermic Udertic Paleustolls). Solid line is the least-squares fitted line for the entire data set, while the dashed line is fitted line when one point, corresponding to $\theta_{se} = 0.01$, was excluded. From Ahuja et al. (1984a).

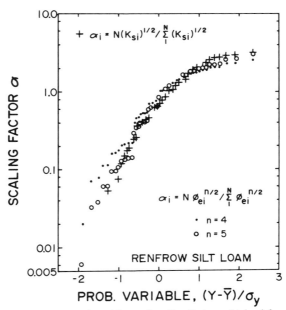

Fig. 7-14. Fractile diagram of the logarithms of scaling factors obtained from K_s data compared with the fractile diagram for K_s factors obtained from θ_{se} data. The symbol Y in the theoretical probability variable stands for the transformed values of α (e.g., log α or others) that would have a normal distribution. The $\overline{Y}$ and σ_y stand for mean and standard deviation of Y. From Ahuja et al. (1984a).

far away should a neutron moisture meter access tube be placed from a tensiometer (Greminger et al., 1985)? Considering local scales of variation, what size sample to identify the pore size distribution must be used to relate to the hydraulic conductivity, which is sensitive to a few macropores distributed uniquely but not uniformly through the soil? Until that challenge is met through field research, the management of field soil-water relations will remain incomplete.

VI. FIELD MODELING AS A MANAGEMENT TOOL

Modeling has been used extensively in research as an aid to understanding of physical phenomena, and in developing and testing new ideas or methods. Use of modeling for analysis of a complex system and making management decisions is a relatively new idea, made possible by the availability of large computers. In the soil water area, modeling may be used in a variety of ways from designing an irrigation system for uniformity of application, and comparison of water and soil conservation practices, to simulation of the entire soil-plant-atmosphere continuum on a day-to-day basis for making decisions on water, fertilizer, or pesticide applications. Although there are many gaps in our knowledge at present, the theory is developed well enough to enable us to achieve optimum use of resources while maintaining or improving environmental quality.

The soil water and chemical transport theories and their simplications described in chapters 4 and 5, together with sections II and IV of this chapter, form the basis over which a field-scale model is built. Special features of such a model, including the boundary conditions at the soil surface, dynamic interactions between several processes, subsurface lateral flow, and some methods of handling of spatial variability, are summarized below. Following these, some currently available field-scale models and examples of their useful results are briefly discussed.

A. Boundary Conditions at the Soil Surface and Interaction of Soil Water Flow with Other Processes

The upper boundary condition for soil water movement under field conditions is usually not that of a constant soil water pressure or flux, except under certain solid-set sprinkler systems. Natural rain falls with an intensity that may vary highly with time. Under a center pivot irrigation system, the sprinkling at any point consists of interrupted events of time-varying intensity (Heerman & Duke, 1983). In such cases, the boundary condition at the soil surface is that of a time-varying flux, equal to rainfall or sprinkling rate, when and as long as the soil surface water content is less than the field saturated value, and of a constant or varying pressure head equal to or greater than zero otherwise. Between periods of rainfall or sprinkling, evaporation occurs at the soil surface and redistribution of infiltrated water occurs in the soil profile. Thus, the initial condition of the soil changes before the next spurt of water is received. Hysteresis of soil water relations also comes into play but is often neglected.

When rainfall or sprinkling occurs on a freshly tilled soil, a seal-crust develops on the surface with time (see section IV of this chapter). Boundary condition at the subcrust soil surface is then a flux which is the smaller of the rainfall (or sprinkling) rate and the value q given by

$$q = (-h_c + d)/r \qquad [23]$$

where $-h_c$ is the soil water suction just below the seal, d is the thickness of the seal, and r is its hydraulic resistance. Initially the r will be a function of cumulative rainfall energy (and hence, time) as well as surface roughness and residue cover by Eq. [19]. Tillage of soil between crop rows is similarly subject to sealing and crusting, except that the part of soil under the canopy cover is partially protected. The temporal changes in soil hydraulic properties of the tilled zone also need to be taken into account (see section IV of this chapter).

In flooding surface irrigation methods (border or basin), the surface boundary condition is a nearly constant pressure condition in the ponded area. However, hydrodynamic equations of surface water advance and recession have to be solved in combination with infiltration into the soil (Heerman & Duke, 1983; Shafique & Skogerboe, 1983). Similar flow conditions exist in furrow irrigation except that the flow is two-dimensional. In a drip or trickle irrigation system, the boundary condition is a constant flux at the source point, which, for high enough flux, changes with time to a constant pressure head (equal to zero) over an expanding circle around the source (Brandt et al., 1971). The water flow from an isolated emitter is three-dimensional. However, when the emitters are close together in a line, the flow may be approximated as two-dimensional occurring from a line or a strip source (Warrick & Lomen, 1983).

Related to boundary condition at the soil surface is the process of overland flow of runoff downslope under natural rainfall (and sometimes under sprinkler irrigation). The overland flow rate depends upon soil slope, surface roughness, and crop structure and density. Between short periods of rainfall, continued overland flow of water from higher parts of a field will be subject to infiltration in lower parts. Since the volume and depth of overland flow depend upon infiltration rates into the soil, the equations for infiltration and overland flow have to be solved interactively (Rovey et al., 1977).

The above discussion has indicated the existence of important dynamic interactions between soil water flow and other processes, such as tillage, crop growth, and residue management. Crop growth, of course, interacts with the water flow not only at the soil surface but also within the profile, by way of differential plant uptake. Another related process that is effective both at the surface and below is heat absorption and transport.

B. Subsurface Lateral Flow

In fields having appreciable slope or impeding layers in the soil profile, there can be significant subsurface lateral flow. In general, the topographic variability is such that one has to solve for three-dimensional surface and subsurface flow. However, in many cases the flow may be approximated as two-dimensional along the main slope. This simplification is carried further in some existing field-scale models (e.g., Huff et al., 1977) wherein the flow is treated as one-dimensional vertical, but a separate, approximate accounting is made of subsurface lateral flow.

C. Incorporation of Spatial Variability of Soil

The most realistic treatment of spatial variability of soil water properties in a field-scale model will require their representation by a measured three-dimensional correlated structure, with a known multivariate frequency distribution superimposed on prevailing trends. Input data for such a representation would be extremely expensive to obtain and the resulting model would be complex and cumbersome. To be practical, therefore, simplifications have to be made which retain and incorporate the major effects of variability.

Essentially all the work done so far has been with research-type models with limited objectives. Early investigations were those of Freeze (1975), Gelhar (1976), Warrick et al. (1977b), Bakr et al. (1978), Smith and Hubbert (1979), and Smith and Freeze (1979a, b). Freeze (1980) presented a stochastic-conceptual model of rainfall-runoff processes that incorporates autocorrelated spatial variability of soil properties as well as rainfall. His results indicated that the frequency distribution parameters for hydraulic conductivity were much more important than the autocorrelation. The calculations are made for soil blocks in which the flow is assumed one-dimensional vertical. This model could be enhanced by adopting conditional simulation (Delhomme, 1979) in which the statistical structure of soil properties that is generated is forced to reproduce the measured values at given points. Russo (1986a) used variograms and conditional simulation in conjunction with a simplified crop-response model to study crop yield-irrigation relationships in a heterogenous field. Dagan and Bresler (1983) and Bresler and Dagan (1983a, b) have presented simplified field models for infiltration, redistribution, and solute transport, which assume that the spatial variability of soil exists only in the horizontal plane, and in which the variability is considered only for K_s. The water flow is assumed vertical everywhere. This assumption, thus, does away with the need to know the autocorrelation structure of soil if we are interested only in the field mean and variance of a result. Only a frequency distribution of K_s is required. The field is assumed to have no trends in K_s or other soil properties.

Several other modelers have resorted to similar-media scaling to describe spatial variability of both $h(\theta)$ and $K(\theta)$, and used frequency distribution of the resulting scaling factors for calculations (Peck et al., 1977; Sharma & Luxmoore, 1979; Luxmoore & Sharma, 1980; Warrick & Amoozegar-Fard, 1979; Ahuja et al., 1984b). Autocorrelation of scaling factors can also be considered, but has not been done. All studies so far have included spatial variability only in the horizontal plane, in that one scaling factor is assigned to a site even in a layered soil. In field test of flow models, Bresler et al. (1979) and Wagenet and Rao (1983) used just the mean and two extreme values of scaling factors to obtain a range in predicted results.

At this stage, the use of scaling along with frequency distribution and trend analysis seems to be a practical way to incorporate soil spatial variability in models.

D. Examples of Models for Management Applications

The models noted below are physically based deterministic models of soil-water-plant relationships of a cropped field. However, any of these models can be used with frequency distributions and trends of parameters to account for spatial variability.

1. Model of Watts and Hanks (1978)

This is a soil-water-N-model for irrigated corn. Water flow and uptake, and N transformations, uptake, and transport are treated in relation to crop growth under given weather conditions. The transformations include mineralization of organic matter, hydrolysis of urea, and nitrification. Soil temperature is calculated with depth. A potential N uptake function—fraction of total uptake vs. fraction of growing season—developed from field data is used to determine maximum uptake for nonlimiting soil water and N conditions. Actual uptake is calculated as less than potential when soil water content, and/or mineral N concentration and distribution with depth limit convective and diffusive movement of NO_3 to the root system. The model gave good results for plant uptake, distribution in soil, and losses below the root zone in field tests. The model can be used as a planning tool, as well as run on a continuing basis during a crop season to decide on applications of water and/or N. Salt transport in the soil profile can be easily accommodated.

The model of Tillotson and Wagenet (1982) is similar to the above model in most respects. Wagenet and Hutson (1986) have extended the model to include pesticides and some other solutes. Model of Bresler et al. (1979) deals with only water and salt transport in a cropped field.

3. Comax—An Expert System for Cotton Crop Management (Lemmon, 1986)

This is a computer program developed by the USDA-ARS, that advises cotton growers on crop management at the farm level on a daily basis. Heart of the Comax (Fig. 7-15) is a cotton growth simulation model, called Gossym. The Gossym was developed by a team of scientists in Mississippi and South Carolina (Baker et al., 1983). It simulates the growth and development of cotton on organ-by-organ basis (roots, stems, leaves, blooms, squares, and bolls), as well as the transport and uptake of water and nutrients through the soil profile. The model is driven by daily weather variables, such as solar radiation, rainfall, and maximum-minimum air temperatures, and requires input data on soil properties and soil fertility. Gossym is used to determine irrigation schedules, N requirements, and crop maturity dates.

The knowledge base of Comas (Fig. 7-15) is a set of rules and facts. The inference engine (Fig. 7-15) examines the rules and facts to determine

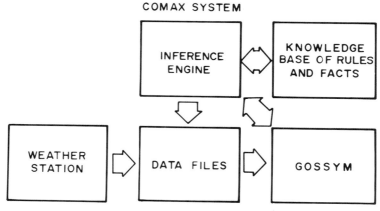

Fig. 7-15. The Comax components. The four components to the right "reside" in a microcomputer located at the grower's farm. From Lemmon (1986).

what is to be done. Each day it hypothesizes the weather scenario for the near future, and then calls Gossym, which simulates the expected growth of cotton plant from the given current conditions. The results of simulation, such as the day the simulated crop goes into water or N stress, are used to compute expected irrigation, fertilization, and crop maturity dates.

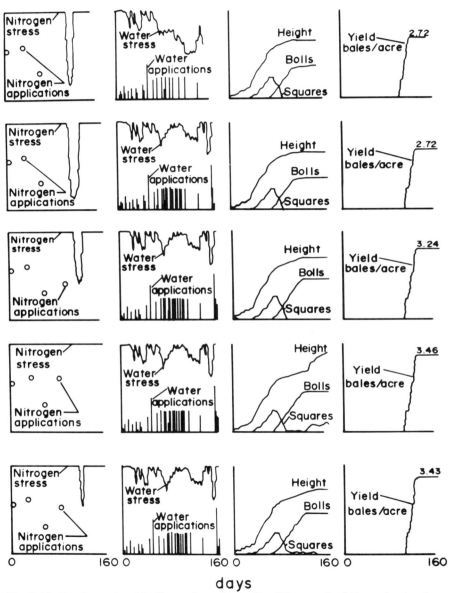

days

Fig. 7-16. Graphs produced by Comax from the results of Gossym simulations, showing the process whereby Comax reduces the water stress and then the N stress, as described in text. From Lemmon (1986).

An example of how Comax selects irrigation and N schedules from the results of Gossym simulations are shown in Fig. 7–16 (after Lemmon, 1986). The N or water stress values are ratios of the amount of N or water used to the amount needed for full growth at different times. The first row of graphs was produced by Comax just after the third application of N. The second row of graphs is the last of a series of Gossym runs in which Comax has relieved the water stress problem by hypothesizing an increased irrigation schedule. Increased irrigation, however, results in increased N stress. In the third row, Comax has hypothesized a small application of N, which reduces the N stress and increases the yield. In the fourth row, Comax has increased the amount of the fourth N application, which eliminates the N stress and increases yield. However, this increased application also resulted in some new vegetative growth and squares near the harvest time. This undesirable hypothesis is rejected and the Comax, in the fifth row, reduces the amount of N applied to a moderate level.

The Comax was tested on a 2430-ha farm in Mississippi in 1985. Based on Comax prediction, the grower applied an additional dose of N that he had not planned to apply. The result was an increased yield. The prediction of the date of maturity is also a substantial benefit in achieving timely har-

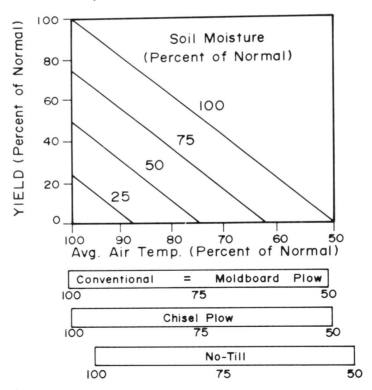

Fig. 7–17. Corn yield response to climate and tillage, as computed by NTRM model. From Shaffer and Larson (1987).

vest of cotton (Lemmon, 1986). Programs similar to Comax are also being developed for soybean and other crops (Swaney et al., 1983).

3. Nitrogen, Tillage, and Residue Management Model (NTRM) (Shaffer & Larson, 1987)

This is another comprehensive model of soil-water-crop-atmosphere continuum, developed by a team of USDA-ARS scientists. The emphasis is on management of N sources and crop residues on the soil surface in conventional and reduced tillage systems. The model presently simulates the growth (both below and above ground) of corn (*Zea mays* L.). Carbon and N transformations, including the decomposition of residues, transport of water and nutrients in soil, ET, soil temperature-crop residue interactions, and the effects of surface roughness and residue cover on infiltration and evaporation are the processes treated in detail. The model contains simplified versions for certain purposes. Application of this model is illustrated by a nomograph on corn yield response to climate and tillage (Fig. 7–17).

VII. PROSPECTIVE

During the last 20 yr we have made progress in the study of field soil water relations, but more remains to be done. Further research under field conditions must continue for each of the topics reviewed above, targetted at the gaps in our understanding of the processes and complex interactions. A major emphasis must be on research and development of economical and effective methods of determining soil properties and their statistical spatial structure. Quantification of the effects of tillage, cropping, and related temporal changes also needs immediate attention. Improved sampling procedures and improved field-scale modeling should result from this effort. An interactive use of experiments and theoretical modeling would enhance the achievement of the above objectives, and improve the models designed for management applications.

REFERENCES

Acevedo, E., T.C. Hsiao, and D.W. Henderson. 1971. Immediate and subsequent growth response of maize leaves to changes in water status. Plant. Physiol. 48:631–636.

Ahuja, L.R., R.E. Green, S.K. Chong, and D.R. Nielsen. 1980. A simplified functions approach for determining soil hydraulic conductivities and water characteristics *in situ*. Water Resour. Res. 16:947–953.

Ahuja, L.R., J.W. Naney, R.E. Green, and D.R. Nielsen. 1984a. Macroporosity to characterize spatial variability of hydraulic conductivity and effects of land management. Soil Sci. Soc. Am. J. 48:699–702.

Ahuja, L.R., J.W. Naney, and D.R. Nielsen. 1984b. Scaling soil water properties and infiltration modeling. Soil Sci. Soc. Am. J. 48:970–973.

Ahuja, L.R., J.W. Naney, and R.D. Williams. 1985. Estimating soil water characteristics from simpler properties or limited data. Soil Sci. Soc. Am. J. 49:1100–1105.

Ahuja, L.R., and J.D. Ross. 1983. A new Green-Ampt type model for infiltration through a surface seal permitting transient parameters below the seal. p. 147–162. *In* Advances in infiltration. ASAE Publ. 11-83. ASAE, St. Joseph, MI.

Allmaras, R.R., R.E. Burwell, and R.F. Holt. 1967. Plow-layer porosity and surface roughness from tillage as affected by initial porosity and soil moisture at tillage time. Soil Sci. Soc. Am. Proc. 31:550–556.

Allmaras, R.R., R.E. Burwell, W.E. Larson, and R.F.Holt. 1966. Total porosity and random roughness of the interrow zone as influenced by tillage. Conserv. Res. Rep. 7. USDA-ARS, Washington, DC.

Alway,F.J., and G.R. McDole. 1917. Relation of water-retaining capacity of soil to its hygroscopic coefficient. J. Agric. Res. 9:27–71.

Amemiya, M. 1968. Tillage-soil water relations of corn as influenced by weather. Agron. J. 60:534–537.

Amoozegar-Fard, A., A.W. Warrick, and D.O. Lomen. 1984. Design and operation nomographs for trickle irrigation systems. J. Irrig. Drain. Eng. 110:107–120.

Baker, D.N., J.R. Lambert, and J.M. McKinion. 1983. A simulator of cotton crop growth and yield. South Carolina Agric. Exp. Stn. Tech. Bull. 1089.

Bakr, A.A., L.W. Gelhar, A.L. Gutjahr, and J.R. McMillan. 1978. Stochastic analysis of spatial variability in subsurface flows. I. Comparison of one and three dimensional flows. Water Resour. Res. 14:263–271.

Bassett, D.L., D.D. Fangmeier, and T. Strelkoff. 1980. Hydraulics of surface irrigation. p. 447–498. *In* M.E. Jensen (ed.) Design and operation of farm irrigation systems. ASAE Monogr. 3. ASAE, St. Joseph, MI.

Belmans, C., J. Feyen, and D. Hillel. 1979. An attempt at experimental validation of macroscopic-scale models of soil moisture extraction by roots. Soil Sci. 127:174–186.

Beven, K.J., and P. Germann. 1982. Macropores and water flow in soils. Water Resour. Res. 18:1311–1325.

Biggar, J.W., and D.R. Nielsen. 1976. Spatial variability of leaching characteristics of a field soil. Water Resour. Res. 12:78–84.

Biswas, T.D., D.R. Nielsen, and J.W. Biggar. 1966. Redistribution of soil water after infiltration. Water Resour. Res. 2:513–514.

Black, T.A., W.R. Gardner, and G.W. Thurtell. 1969. The prediction of evaporation, drainage, and soil water storage for a bare soil. Soil Sci. Soc. Am. Proc. 33:655–660.

Boone, F.R., S. Slager, R. Meidema, and R. Elevald. 1976. Some influences of zero tillage on the structure and stability of a fine textured river levee soil. Neth. J. Agric. Sci. 24:105–119.

Bouma, J., C.F.M. Belmans, and L.W. Dekker. 1982. Water infiltration and redistribution in a silt loam subsoil with vertical worm channels. Soil Sci. Soc. Am. J. 46:917–921.

Brakensiek, D.L., and W.J. Rawls. 1983. Agricultural management effects on soil water processes. Part II: Green and Ampt parameters for crusting soils. Trans. ASAE 26:1753–1757.

Brandt, A., E. Bresler, N. Diner, and I. Ban-Asher. 1971. Infiltration from a trickle source. I. Mathematical models. Soil Sci. Soc. Am. Proc. 35:683–689.

Bresler, E., H. Bielorai, and A. Laufer. 1979. Field test of solution flow models in a heterogenous irrigated cropped soil. Water Resour. Res. 15:645–652.

Bresler, E., and G. Dagan. 1983a. Unsaturated flow in spatially variable fields. II. Application of water flow models to various fields. Water Resour. Res. 19:421–428.

Bresler, E., and G. Dagan. 1983b. Unsaturated flow in spatially variable fields. III. Solute transport models and their applications to two fields. Water Resour. Res. 19:429–435.

Briggs, L.J., and H.L. Shantz. 1912. The wilting coefficient for different plants and its indirect determination. p. 26–33. *In* Bur. Plant Industry Bull. 230. USDA, Washington, DC.

Bruce, R.R., J.H. Dane, V.L. Quisenberry, N.L. Powell, and A.W. Thomas. 1983. Physical characteristics of soils in the southern region: Cecil. Georgia Agric. Exp. Stn. South. Coop. Ser. Bull. 267.

Burwell, R.E., L.L. Sloneker, and W.W. Nelson. 1968. Tillage influences water intake. J. Soil Water Conserv. 23:185–187.

Carbon, B.A. 1975. Redistribution of water following precipitation on previously dry sandy soils. Aust. J. Soil Res. 13:19.

Cassel, D.K. 1983. Spatial and temporal variability of soil physical properties following tillage of Norfolk loamy sand. Soil Sci. Soc. Am. J. 47:196–201.

Chong, S.K., R.E. Green, and L.R. Ahuja. 1981. Simple *in situ* determination of hydraulic conductivity by power-function description of drainage. Water Resour. Res. 17:1109–1114.

Clothier, B.E., D.R. Scooter, and J.P. Ken. 1977. Water retention in soil underlain by a coarse-textured layer: Theory and a field application. Soil Sci. 123:393–399.

Coelho, M.A. 1974. Spatial variability of water related soil physical properties. Ph.D. diss. Univ. of Arizona, Tucson (Diss. Abstr. 35(11B):Order no. 7511061).

Cowan, I.R. 1965. Transport of water in the soil-plant-atmosphere system. J. Appl. Ecol. 2:221–229.

Dagan, G., and E. Bresler. 1983. Unsaturated flow in spatially variable fields. I. Derivation of models of infiltration and redistribution. Water Resour. Res. 19:413–420.

Dane, J.H., and S. Hruska. 1983. *In-situ* determination of soil hydraulic properties during drainage. Soil Soc. Am. J. 47:619–624.

Davidson, J.M., L.R. Stone, D.R. Nielsen, and M.E. LaRue. 1969. Field measurement and use of soil water properties. Water Resour. Res. 5:1312–1321.

Davis, J.C. 1973. Statistics and data analysis in geology. John Wiley and Sons, New York.

Delhomme, J.P. 1979. Spatial variaiblity and uncertainty in groundwater flow parameters: A geostatistical approach. Water Resour. Res. 15:269–280.

Denmead, O.T., and R.H. Shaw. 1962. Availability of soil water to plants as affected by soil moisture content and meteorological conditions. Agron. J. 54:385–390.

Duley, F.L. 1939. Surface factors affecting the rate of intake of water by soils. Soil Sci. Soc. Am. Proc. 9:60–64.

Eagleman, J.R., and V.C. Jamison. 1962. Soil layering and compaction effects on unsaturated moisture movement. Soil Sci. Soc. Am. Proc. 26:519–522.

Edwards, W.R., and W.E. Larson. 1969. Infiltration of water into soils as influenced by surface seal development. Trans. ASAE 12:463–465, 470.

Ehlers, W. 1976. Rapid determination of unsaturated hydraulic conductivity in tilled and un-tilled loess soil. Soil Sci. Soc. Am. J. 40:837–840.

Eisenhauer, D.E., D.F. Heermann, and A. Klute. 1984. Surface sealing effects on infiltration with surface irrigation. ASAE Pap. No. 84-2510. ASAE, St. Joseph, MI.

Elliott, R.L., and W.R. Walker. 1982. Field evaluation of furrow infiltration and advance function. Trans. ASAE 25:396–400.

Falayi, O., and J. Bouma. 1975. Relationships between the hydraulic conductance of surface crusts and soil management in a typic Hapludalf. Soil Sci. Soc. Am. Proc. 39:957–963.

Feddes, R.A. 1981. Water use models for assessing root zone modification. p. 347–390. *In* G.F. Arkin, and H.M. Taylor (ed.) Modifying the plant environment to reduce crop stress. ASAE Monogr. 17. ASAE, St. Joseph, MI.

Feddes, R.A., E. Bresler, and S.P. Newman. 1974. Field test of a modified numerical model for water uptake by root systems. Water Resour. Res. 10:1199–1206.

Feyen, J., C. Belmans, and D. Hillel. 1980. Comparison between measured and simulated plant water potential during soil water extraction by potted ryegrass. Soil Sci. 129:180.

Fluhler, H., M.S. Ardakani, and L.H. Stolzy. 1976. Error propagation in determining hydraulic conductivities from successive water content and pressure head profiles. Soil Sci. Soc. Am. J. 40:830–836.

Freeze, R.A. 1975. A stochastic-conceptual analysis of one-dimensional groundwater flow in nonuniform homogenous media. Water Resour. Res. 11:725–741.

Freeze, R.A. 1980. A stochastic-conceptual analysis of rainfall-runoff processes on a hillslope. Water Resour. Res. 16:391–408.

Furr, J.A., and J.O. Reeve. 1945. The range of soil moisture percentages through which plants undergo permanent wilting in some soils from semi-arid, irrigated areas. J. Agric. Res. 71:149–170.

Gardner, W.R. 1960. Dynamic aspects of water availability to plants. Soil Sci. 89:63–73.

Gardner, W.R. 1964. Relation of root distribution to water uptake and availability. Agron. J. 56:35–41.

Gardner, W.R., D. Hillel, and Y. Benyamini. 1970. Post irrigation movement of soil water: I. Redistribution. Water Resour. Res. 6:851–861.

Gelhar, L.W. 1976. Effects of hydraulic conductivity variations on groundwater flows. *In* Proc., 2nd Int. IAHR Symp. on Stochastic Hydraulics. Int. Assoc. of Hydraulic Research, Lund, Sweden.

Greminger, P.J., Y.K. Sud, and D.R. Nielsen. 1985. Spatial variability of field measured soil-water characteristics. Soil Sci. Soc. Am. J. 49:1075–1082.

Gumbs, F.A., and B.P. Warkenton. 1973. Prediction of infiltration of water into aggregated clay soil samples. Soil Sci. Soc. Am. Proc. 39:255–263.

Gupta, S.C., and W.R. Larson. 1979. Estimating soil water retention characteristics from particle size distribution, organic matter percent, and bulk density. Water Resour. Res. 15:1633–1635.

Gurovich, L., J. Stern, and R. Ramos. 1983. Simulated optimization of crop yield through irrigation system design and operation based on spatial variability of soil hydrodynamic properties. p. 401–415. *In* Isotope and radiation techniques in soil physics and irrigation studies. International Atomic Energy Agency, Vienna.

Hald, A. 1952. Statistical theory with engineering applications. John Wiley and Sons, New York.

Hamblin, A.P., and D. Tennant. 1981. The influence of tillage on soil water behavior. Soil Sci. 132:233–239.

Heerman, D.F., and H.R. Duke. 1983. Applications in irrigated and dryland agriculture. p. 254–265. *In* Advances in infiltration. ASAE Publ. 11-83. ASAE, St. Joseph, MI.

Hendrickx, J.M.H., P.J. Wierenga, M.S. Nash, and D.R. Nielsen. 1986. Boundary location from texture, soil moisture, and infiltration data. Soil Sci. Soc. Am. J. 50:1515–1520.

Hillel, D. 1960. Crust formation in loessial soils. Trans. Int. Congr. Soil Sci. 7th, 1:330–339.

Hillel, D. 1980. Applications of soil physics. Academic Press, New York.

Hillel, D., and Y. Guron. 1973. Relation between evapotranspiration rate and maize yield. Water Resour. Res. 9:743–748.

Hillel, D., H. Talpaz, and H. van Keulen. 1976. A macroscopic-scale model of water uptake by a nonuniform root system and of water and salt movement in the soil profile. Soil Sci. 121:242–255.

Hillel, D., C. van Beek, and H. Talpaz. 1975. A microscopic-scale model of water uptake and salt movement to plant roots. Soil Sci. 120:385–399.

Hines, J.W. 1986. Measurement and modeling of soil hydraulic conductivity under different tillage systems. M.S. thesis. Univ. of Minnesota, St. Paul.

Huff, D.D., R.J. Luxmoore, J.B. Mankin, and C.L. Begovich. 1977. TEHM: A terrestrial ecosystem hydrology model. ORNL/NSF/EATC-27. Oak Ridge Natl. Lab., Oak Ridge, TN.

Israelson, O.W., and F.L. West. 1922. Water holding capacity of irrigated soils. Utah State Univ. Agric. Exp. Stn. Bull. 183.

Jackson, R.D. 1982. Canopy temperatures and crop water stress. Adv. Irrig. 1:43–85.

Jaynes, D.B. 1986. Simple model of border irrigation. J. Irrig. Drain. Eng. 112:172–184.

Jones, A.J., and R.J. Wagenet. 1984. *In situ* estimation of hydraulic conductivity using simplified methods. Water Resour. Res. 20:1620–1626.

Journel, A.G., and C.J. Huijbregts. 1978. Mining geostatistics. Academic Press, New York.

Klute, A. 1982. Tillage effects on the hydraulic properties of soil: A review. p. 29–43. *In* Predicting tillage effects on soil physical properties and processes. ASA Spec. Publ. 44. ASA and SSSA, Madison, WI.

Klute, A. (ed.). 1986. Methods of soil analysis. Part 1. 2nd ed. Agronomy 9.

Lascano, R.J., and C.H.M. van Bavel. 1984. Root water uptake and soil water distribution: Test of an availability concept. Soil Sci. Soc. Am. J. 48:233–237.

Lemmon, H. 1986. Comax: An expert system for cotton crop management. Science 233:29–33.

Libardi, P.L., K. Reichardt, D.R. Nielsen, and J.W. Biggar. 1980. Simple field methods of estimating soil hydraulic conductivity. Soil Sci. Soc. Am. J. 44:3–7.

Linden, D.R. 1979. A model to predict soil water storage as affected by tillage practices. Ph.D. thesis. Univ. of Minnesota, St. Paul (Diss. Abstr. 40(11B):Order no. 8011844).

Lindstrom, M.J., and C.A. Onstad. 1984. Influence of tillage systems on soil physical parameters and infiltration after planting. J. Soil Water Conserv. 39:149–152.

Lumley, J.L., and A. Panofsky. 1964. The structure of atmospheric turbulence. John Wiley and Sons, New York.

Luxmoore, R.J., and M.L. Sharma. 1980. Runoff responses to soil heterogeneity: Experimental and simulation comparisons for two contrasting watersheds. Water Resour. Res. 16:675–684.

Mannering, J.V., and D. Wiersma. 1970. The effect of rainfall energy on water infiltration into soils. Proc. Indiana Acad. Sci. 79:407–412.

Mapa, R.B., R.E. Green, and L. Santo. 1986. Temporal variability of soil hydraulic properties with wetting and drying subsequent to tillage. Soil Sci. Soc. Am. J. 50:1133–1138.

Marthaler, H.P., W. Vogelsanger, F. Richard, and P.J. Wierenga. 1983. A pressure transducer for field tensiometers. Soil Sci. Soc. Am. J. 47:624–627.

Matheron, G. 1971. The theory of regionalized variables and its applications. Les Cahiers du Centre de Morphologic Mathematique, Fas. 5, C.G. Fontainebleau, France.

McIntyre, D.S. 1958a. Permeability measurements of soil crusts formed by raindrop impact. Soil Sci. 85:185–189.

McIntyre, D.S. 1958b. Soil splash and the formation of surface crusts by raindrop impact. Soil Sci. 85:261–266.

Miller, D.E. 1969. Flow and retention of water in layered soils. Conserv. Res. Rep. 13. USDA-ARS, Washington, DC.

Miller, D.E. 1973. Water retention and flow in layered soil profiles. p. 107–117. *In* R.R. Bruce et al. (ed.) Field soil water regime. SSSA Spec. Publ. 5. SSSA, Madison, WI.

Miller, E.E. 1980. Similitude and scaling of soil-water phenomena. p. 300–318. *In* D. Hillel. Applications of soil physics. Academic Press, New York.

Miller, E.E., and R.D. Miller. 1956. Physical theory for capillary flow phenomena. J. Appl. Phys. 27:324–332.

Millete, G.J.F., R.G. Leger, and R.E. Bailey. 1980. A comparative study of physical and chemical properties of two cultivated and forested orthic podzols. Can. J. Soil Sci. 60:707–719.

Moldenhauer, W.C., and W.D. Kemper. 1969. Interdependence of water drop energy and clod size on infiltration and clod stability. Soil Sci. Soc. Am. Proc. 33:297–301.

Molz, F.J. 1981. Models of water transport in soil-plant system: A review. Water Resour. Res. 17:1245–1260.

Molz, F.J., and I. Remson. 1970. Extraction term models of soil moisture use by transpiring plants. Water Resour. Res. 6:1346–1356.

Molz, F.J., and I. Remson. 1971. Application of an extraction term model to the study of moisture flow to plant roots. Agron. J. 63:72–77.

Montgomery, E.G. 1913. Experiments in wheat breeding: Experimental error in the nursery and variation in nitrogen and yield. USDA Bur. Plant Indust. Bull. 269. USDA, Washington, DC.

Moore, I.D. 1981. Effect of surface sealing on infiltration. Trans. ASAE 24:1546–1553.

Morin, J., and Y. Benyamini. 1977. Rainfall infiltration into bare soils. Water Resour. Res. 13:813–817.

Mualem, Y. 1986. Hydraulic conductivity of unsaturated soils: Prediction and formulas. *In* A. Klute (ed.) Methods of soil analysis. Part 1. 2nd ed. Agronomy 9:799–823.

Naney, J.W., L.R. Ahuja, and B.B. Barnes. 1983. Variability and interrelation of soil-water and some related soil properties in a small watershed. p. 92–101. *In* Advances in infiltration. ASAE Publ. 11-83. ASAE, St. Joseph, MI.

Negi, S.C., G.S.V. Raghovan, and F. Taylor. 1981. Hydraulic characteristics of conventionally and zero-tilled field plots. Soil Tillage Res. 2:281–292.

Nielsen, D.R., J.W. Biggar, and K.T. Erh. 1973. Spatial variability of field measured soil-water properties. Hilgardia 42:215–259.

Nielsen, D.R., J.M. Davidson, J.W. Biggar, and R.J. Miller. 1964. Water movement through Panoche clay loam soil. Hilgardia 35:491–506.

Nimah, M.N., and R.J. Hanks. 1973. Model for estimating soil, water, plant and atmosphere interactions. 1. Description and sensitivity. Soil Sci. Soc. Am. Proc. 37:522–527.

Nofziger, D.L., J.R. Williams, A.G. Hornsby, and A.L. Wood. 1983. Physical characteristics of soils of the southern region: Bethany, Konawa, and Tipton series. Oklahoma State Univ. Agric. Exp. Stn. South. Coop. Ser. Bull. 265.

Ogata, G., and L.A. Richards. 1957. Water content change following irrigation of bare field soil that is protected from evaporation. Soil Sci. Soc. Am. Proc. 21:355–356.

Ogata, G., L.A. Richards, and W.R. Gardner. 1960. Transpiration of alfalfa determined from soil water content changes. Soil Sci. 89:179–182.

Onstad, C.A., M.L. Wolfe, C.L. Larson, and D.C. Slack. 1984. Tilled soil subsidence during repeated wetting. Trans. ASAE 27:733–736.

Peck, A.J., R.J. Luxmoore, and J.L. Stolzy. 1977. Effects of spatial variaiblity of soil hydraulic properties in water budget modeling. Water Resour. Res. 13:348–453.

Pendleton, R.L. 1919. Are soils mapped under a given name by Bureau of Soils Method closely similar to one another? Agric. Sci. 3:369–498.

Penman, H.L. 1949. The dependence of transpiration on weather and soil conditions. J. Soil Sci. 1:74–89.

Philip, J.R. 1957. The physical principles of soil water movement during the irrigation cycle. 3rd Congr. Int. Comm. Irrig. Drain. Question 8:8.125–8.154.

Philip, J.R. 1966. Plant water relations: Some physical aspects. Annu. Rev. Plant Physiol. 17:245–268.

Pierce, L.T. 1958. Estimating seasonal and short-term fluctuations in evapotranspiration in meadow crops. Bull. Am. Meterol. Soc. 39:73–78.

Rao, P.S.C., R.E. Jessup, A.C. Hornsby, D.K. Cassell, and W.A. Pollans. 1983. Scaling soil microhydrologic properties of Lakeland and Konawa soils using similar media concepts. Agric. Water Manage. 6:277–290.

Rao, P.V., P.S.C. Rao, J.M. Davidson, and L.C. Hammond. 1979. Use of goodness-of-fit tests for characterizing the spatial variability of soil properties. Soil Sci. Soc. Am. J. 43:274–278.

Ratliff, L.F., J.T. Ritchie, and D.K. Cassel. 1983. Field-measured limits of soil water availability as related to laboratory measured properties. Soil Sci. Soc. Am. J. 47:770–775.

Rawitz, E. 1969. The dependence of growth rate and transpiration on plant and soil physical parameters under controlled conditions. Soil Sci. 110:172–182.

Rawlins, S.L., and P.A.C. Raats. 1975. Prospects for high frequency irrigation. Science 188:604–610.

Rawls, W.J., and D.L. Brakensiek. 1982. Estimating soil water retention from soil properties. J. Irrig. Drain. Eng. 108:166–171.

Rawls, W.J., D.L. Brakensiek, and K.E. Saxton. 1982. Estimation of soil water properties. Trans. ASAE 25:1316–1320, 1328.

Rawls, W.J., D.L. Brakensiek, and B. Soni. 1983. Agricultural management effects on soil water processes. Part I: Soil water retention and green and ampt infiltration parameters. Trans. ASAE 26:1747–1752.

Richards, L.A., and D.C. Moore. 1952. Influence of capillary conductivity and depth of wetting on moisture retention in soil. Trans. Am. Geophys. Union 33:4.

Richards, L.A., W.R. Gardner, and G. Ogata. 1956. Physical processes determining water loss from soils. Soil Sci. Soc. Am. Proc. 20:310–314.

Richards, L.A., and C.H. Wadleigh. 1952. Soil water in plant growth. *In* Soil physical conditions and plant growth. Agronomy 2:13.

Richards, L.A., and L.R. Weaver. 1944. Fifteen atmosphere percentage as related to the permanent wilting percentage. Soil Sci. 56:331–339.

Ritchie, J.T., E. Burnett, and R.C. Henderson. 1972. Dryland evaporative flux in a subhumid climate: III. Soil water influence. Agron. J. 64:168–173.

Rose, C.W., W.R. Stern, and J.E. Drummond. 1965. Determination of hydraulic conductivity as a function of depth and water content for soil *in situ*. Aust. J. Soil Res. 3:1–9.

Rousseva, S.S., L.R. Ahuja, and G.C. Heathman. 1986. Temporal changes in bulk density and water content of a tilled soil as measured with a surface gamma-neutron gauge. Res. Rep., USDA-ARS, Durant, OK.

Rovey, E.W., D.A. Woolhiser, and R.E. Smith. 1977. A distributed kinematic model of upland watersheds. Hydrol. Pap. 93. Colorado State Univ., Fort Collins.

Russo, D. 1983. A geostatistical approach to the trickle irrigation design in heterogenous soil. 1. Theory. Water Resour. Res. 19:632–642.

Russo, D. 1986a. A stochastic approach to crop yield-irrigation relationships in heterogenous soils. 1. Analysis of field spatial variability. Soil Sci. Soc. Am. J. 50:736–745.

Russo, D. 1986b. A stochastic approach to crop yield-irrigation relationships in heterogenous soils: II. Application to irrigation management. Soil Sci. Soc. Am. J. 50:745–751.

Russo, D., and E. Bresler. 1980. Scaling soil hydraulic properties of heterogeneous field. Soil Sci. Soc. Am. J. 44:681–684.

Saddiq, M.H., P.J. Wierenga, J.M.H. Hendrickx, and M.Y. Hussain. 1985. Spatial variability of soil water tension in an irrigated soil. Soil Sci. 140:126–132.

Schuh, W.M., J.W. Bauder, and S.C. Gupta. 1984. Evaluation of simplified methods for determining unsaturated hydraulic conductivity of layered soils. Soil Sci. Soc. Am. J. 48:730–736.

Shaffer, M.J., and W.E. Larson (ed.). 1987. NTRM: A soil-crop simulation model for nitrogen, tillage, and crop residue management. Conserv. Res. Rep. 34-1. USDA-ARS, Washington, DC.

Shafique, M.S., and G.V. Skogerboe. 1983. Impact of seasonal infiltration function variation on furrow irrigation performance. p. 292–301. *In* Advances in infiltration. ASAE Publ. 11-83. ASAE, St. Joseph, MI.

Sharma, M.L., G.A. Gander, and C.G. Hunt. 1980. Spatial variability of infiltration in a watershed. J. Hydrol. 45:101–122.

Sharma, P.P., C.J. Gantzer, and G.R. Blake. 1981. Hydraulic gradients across simulated rainformed soil surface seals. Soil Sci. Soc. Am. J. 45:1031–1034.

Sharma, M.L., and R.J. Luxmoore. 1979. Soil spatial variability and its consequences on simulated water balance. Water Resour. Res. 15:1567–1573.

Simmons, C.S., D.R. Nielsen, and J.W. Biggar. 1979. Scaling of field-measured soil water properties: 1. Methodology. II. Hydraulic conductivity and flux. Hilgardia 47:77–173.

Sisson, J.B., A.H. Ferguson, and M.Th. van Genuchten. 1980. Simple method for predicting drainage from field plots. Soil Sci. Soc. Am. J. 44:1147–1152.

Sisson, J.B., and P.J. Wierenga. 1981. Spatial variability of steady-state infiltration rates as a stochastic process. Soil Sci. Soc. Am. J. 45:699–704.

Slack, D.C. 1980. Modeling infiltration under moving sprinkler irrigation systems. Trans. ASAE 23:596–600.

Slater, P.J., and J.B. Williams. 1965. The influence of texture on the moisture characteristics of soils. I. A critical comparison of techniques for determining the available water capacity and moisture characteristic curve of a soil. J. Soil Sci. 16:1–12.

Smith, L., and R.A. Freeze. 1979a. Stochastic analysis of steady state groundwater flow in a bounded domain. I. One-dimensional simulations. Water Resour. Res. 15:521–527.

Smith, L., and R.A. Freeze. 1979b. Stochastic analysis of steady state groundwater flow in a bounded domain. II. Two-dimensional simulations. Water Resour. Res. 15:1543–1559.

Smith, R.E., and R.H.B. Hubbert. 1979. A Monte Carlo analysis of the hydrologic effects of spatial variability of infiltration. Water Resour. Res. 15:419–429.

Swaney, D.P., J.W. Jones, W.G. Boggess, G.G. Wilkerson, and J.W. Mishoe. 1983. Real-time irrigation decision analysis using simulation. Trans. ASAE 26:562–568.

Tackett, J.L., and R.W. Pearson. 1965. Some characteristics of soil crusts formed by simulated rainfall. Soil Sci. 99:407–413.

Thomas, G.W., and R.E. Phillips. 1979. Consequences of water movement in macropores. J. Environ. Qual. 8:149–152.

Thornthwaite, C.W., and J.R. Mather. 1955. The water budget and its use in irrigation. USDA Yearb. Agric. 1955:346–358.

Tillotson, W.R., and R.J. Wagenet. 1982. Simulation of fertilizer nitrogen under cropped situations. Soil Sci. 133:133–143.

Trangamar, B.B., R.S. Yost, and G. Uehara. 1985. Application of geostatistics to spatial studies of soil properties. Adv. Agron. 38:45–94.

Unger, P.W. 1984. Tillage effects on surface soil physical conditions and sorghum emergence. Soil Sci. Soc. Am. J. 48:1423–1432.

van Bavel, C.H.M., R.J. Lascano, and L. Stroosnijder. 1984. Test and analysis of a model of water use by sorghum. Soil Sci. 137:443–456.

van Bavel, C.H.M., G.B. Stirk, and K.J. Brust. 1968. Hydraulic properties of a clay loam soil and field measurement of water uptake by roots: 1. Interpretation of water content and pressure profiles. Soil Sci. Soc. Am. Proc. 32:310–317.

van Doren, D.M., and R.R. Allmaras. 1978. Effect of residue management practices on the soil physical environment, micro-climate, and plant growth. p. 49–84. *In* Crop residue management systems. ASA Spec. Publ. 31. ASA, CSSA, and SSSA, Madison, WI.

Vauclin, M., S.R. Vieira, G. Vachaud, and D.R. Nielsen. 1983. The use of cokriging with limited field soil observations. Soil Sci. Soc. Am. J. 47:175–184.

Veihmeyer, F.J., and A.H. Hendrickson. 1927. Soil moisture conditions in relation to plant growth. Plant Physiol. 2:71–78.

Veihmeyer, F.J., and A.H. Hendrickson. 1931. The moisture equivalent as a measure of the field capacity of soils. Soil Sci. 32:181–193.

Veihmeyer, F.J., and A.H. Hendrickson. 1949. Methods of measuring field capacity and wilting percentages of soils. Soil Sci. 68:75–94.

Veihmeyer, F.J., and A.H. Hendrickson. 1950. Soil moisture in relation to plant growth. Annu. Rev. Plant Physiol. 1:285–304.

Veihmeyer, F.J., and A.H. Hendrickson. 1955. Does transpiration decrease as the soil moisture decreases? Trans. Am. Geophys. Union 36:425–448.

Vieira, S.R., D.R. Nielsen, and J.W. Biggar. 1981. Spatial variability of field-measured infiltration rate. Soil Sci. Soc. Am. J. 45:1040–1048.

Viets, F.G., Jr. 1966. Increasing water use efficiency by soil management. p. 259–274. *In* W.H. Pierre et al. (ed.) Plant environment and efficient water use. ASA and SSSA, Madison, WI.

Wagenet, R.J., and J.L. Hutson. 1986. Predicting the fate of nonvolatile pesticides in the unsaturated zone. J. Environ. Qual. 15:315–322.

Wagenet, R.J., and B.K. Rao. 1983. Description of nitrogen movement in the presence of spatially variable soil hydraulic properties. Agric. Water Manage. 6:227–242.

Wall, B.H., and A.J. Miller. 1983. Optimization of parameters in a model of soil water drainage. Water Resour. Res. 19:1565–1572.

Warrick, A.W., and A. Amoozegar-Fard. 1979. Infiltration and drainage calculations using spatially scaled hydraulic properties. Water Resour. Res. 15:1116–1120.

Warrick, A.W., and D.O. Lomen. 1983. Linearized moisture flow with water extraction over two-dimensional zones. Soil Sci. Soc. Am. J. 47:869–872.

Warrick, A.W., G.J. Mullen, and D.R. Nielsen. 1977a. Scaling field-measured soil hydraulic properties using a similar-media concept. Water Resour. Res. 13:355–362.

Warrick, A.W., G.J. Mullen, and D.R. Nielsen. 1977b. Prediction of the soil water flux based upon field-measured soil-water properties. Soil Sci. Soc. Am. J. 41:14–19.

Warrick, A.W., D.E. Myers, and D.R. Nielsen. 1986. Geostatistical methods applied to soil science. In A. Klute (ed.) Methods of soil analysis. Part 1. 2nd ed. Agronomy 9:53–82.

Warrick, A.W., and D.R. Nielsen. 1980. Spatial variability of soil physical properties in the field. p. 319–344. In D. Hillel. Applications of soil physics. Academic Press, New York.

Watson, K.K. 1966. An instantaneous profile method for determining the hydraulic conductivity of unsaturated porous materials. Water Resour. Res. 2:709–713.

Watts, D.C., and R.J. Hanks. 1978. A soil-water-nitrogen model for irrigated corn on sandy soils. Soil Sci. Soc. Am. J. 42:492–499.

Webster, R. 1985. Quantitative spatial analysis of soil in the field. Adv. Soil Sci. 3:1–70.

Webster, R., and T.M. Burgess. 1980. Optimal interpolation and isarithmic mapping of soil properties. III. Changing drift and universal kriging. J. Soil Sci. 31:505–524.

Whisler, F.D., A. Klute, and R.J. Millington. 1968. Analysis of steady-state evapotranspiration from a soil column. Soil Sci. Soc. Am. Proc. 32:167–174.

White, R.E. 1985. The influence of macropores on the transport of dissolved and suspended matter through soil. Adv. Soil Sci. 3:95–120.

Wilding, L.P., and L.R. Drees. 1983. Spatial variability and pedology. p. 83–116. In L.P. Wilding et al. (ed.) Pedogenesis and soil taxonomy. 1. Concepts and interactions. Elsevier, Amsterdam.

Williams, R.D., J.W. Naney, and L.R. Ahuja. 1985. Soil properties and productivity changes along a slope. p. 96–106. In Proc. Natl. Symp. Soil Erosion Productivity, New Orleans. 10–11 Dec 1984. ASAE Publ. 8-85. ASAE, St. Joseph, MI.

Yevjevich, V. 1972. Stochastic processes in hydrology. Water Resour. Publ., Fort Collins, CO.

Zachmann, D.W., P.C. DuChateau, and A. Klute. 1981. The calibration of the Richards flow equation for a drainage column by parameter identification. Soil Sci. Soc. Am. J. 45:1012–1015.

Zachmann, D.W., P.C. DuChateau, and A. Klute. 1982. Simultaneous approximation of water capacity and soil hydraulic conductiity by parameter identification. Soil Sci. 134:157–163.

8

The Nature and Movement of Water in Plants

EDWIN L. FISCUS

USDA-ARS
Fort Collins, Colorado

MERRILL R. KAUFMANN

United States Forest Service
Fort Collins, Colorado

Much of the subject matter that is called plant-water relations could more aptly be titled water transport in plants. Usually, we are interested in dynamic systems where water and solutes are being shunted from one place to another to meet the various requirements of growth and atmospheric demand. Most often we are interested in knowing in which direction in space water, or solution, will move, how much moves, and how fast it moves. These three questions form the basis of most studies in plant water transport, and indeed much of science in general. It is also these same three questions that will form the central theme of this chapter. When water transport is compromised the ensuing stresses have far-reaching physiological significance, but these will be discussed elsewhere in this volume. First we will discuss some of the characteristics of water and aqueous solutions that are relevant to these questions and then try to show in general terms how to answer them. This chapter is not meant to be comprehensive in scope, either in terms of the material covered or in the literature cited. It is our intention to present basic information that will allow the reader to understand the major processes which occur. The literature citations will be confined to those which the authors feel either illustrate a particular point or provide additional review or background. Additional preference will be given to texts and reviews.

The water component of an aqueous solution has a thermodynamic property called its *chemical potential* which is a measure of the ability of that water to do chemical work. One form of chemical potential that has proven useful to physiologists is called the *water potential,* which is the chemical potential of the water component of a solution compared to some reference state, usually pure water at the same temperature and elevation. The water potential can be expressed in various units such as energy per unit mass $(J mol^{-1})$ but for much of plant water relations research the units of pressure (Pa) are more convenient. For those who are interested there is a somewhat abbreviated derivation of the water potential in the appendix which may clarify such things as the relationship between the water potential of a solution and the capacity of that solution to perform work. It is sufficient at this point, however, to recognize that it is the difference in water potential between two phases of a system or between two parts of the same phase that determines the direction in which water will diffuse spontaneously. Diffusive fluxes, however, do not account for all cases of water movement in plants and when bulk flow of *solution* is involved, water may move in directions that are not determined by the difference in the total water potential. This will be discussed later, but for now we will concentrate on some of the common properties of water and aqueous solutions which will be useful in our understanding of plant water transport processes.

I. THE CHARACTERISTICS OF WATER

A. Structure and Behavior of Water

Water is an extremely important component of most biological systems, serving not only as the milieu in which organelles exist and through which they conduct such communications as they are capable of, but as a reaction solvent, a reagent, and as a hydraulic fluid. Water as a hydraulic fluid is used to maintain structure (turgor), as a general growth component (50–95% of plant parts), and to perform special functions such as stomatol opening, spore dispersal, and others.

Much of the importance of water as a biological fluid is due to many of its physical properties which are considered anomalous. That is, water acts differently qualitatively and quantitatively from other molecules of similar size and composition such as CH_4, NH_3 and HF. The reason for this is a phenomenon called *hydrogen bonding,* which is an electrostatic attraction between the O atom of one water molecule and the H atom of an adjacent molecule. Such attractions are possible because O is very electronegative and does not share equally the electrons contributed to the molecule by the H atoms. Rather, the O tends to hoard the negatively charged electron set so that the H atoms are left with a residual positive charge and the O a residual negative charge. Because of the residual charges adjacent molecules orient themselves with respect to each other. In ice this orientation reaches a maximum and frozen water may be characterized as a crystalline lattice.

Upon melting, only about 15% of the hydrogen bonds in ice are broken, rising to about 30% at 373 K, so that the molecules of liquid water form a fairly well-ordered system. It is the extensive hydrogen bonding in liquid water which accounts for most of the anomalous properties which are so beneficial to biological systems. Only on vaporization are the remainder of the hydrogen bonds broken with the subsequent exchange of large amounts of energy (approximately 41 kJ mol^{-1}).

B. Effects of Gravity, Pressure, and Solutes

The water potential derived in the appendix was first used to describe the physiology of water relations by Tang and Wang (1941). It can be shown that the water potential can be split into three parts reflecting the influence of dissolved solutes, hydraulic pressures, and gravitational fields. The total water potential then is usually written as

$$\Psi_t = \Psi_s + \Psi_p + \Psi_g \qquad [1]$$

where the subscripts refer to the aforementioned variables. It is the difference in Ψ_t between two phases or regions of a phase that determines the direction of spontaneous diffusion of water.

1. Gravitational Potential

The Ψ_g represents the increment of water potential due to the position of that water in a gravitational field. It is the potential energy of position found in classical physics (mgh) and is calculated in a similar way, $\Psi_g = (mgh)/V_w$. Here, m is the mass per mole of water (0.01802 kg mol^{-1}), g the gravitational acceleration (9.8 m s^{-2} at sea level), and h is the height in meters above ($+$) or below ($-$) the reference level. The division by V_w is necessary to convert to water potential units (J mol^{-1}/m^3 mol^{-1} = Pa). A convenient approximation for each meter of elevation is to take $\Psi_g = 0.01$ MPa.

2. Pressure Potential

The interpretation of Ψ_p is fairly obvious and represents the effect on Ψ_t of hydrostatic pressures exerted on the system. Cellular turgor pressures and hydrostatic pressures ($+$) or tensions ($-$) developed in the xylem or cell wall matrices are the most often considered. Usually turgor pressures are positive while xylem pressures may vary from slightly positive (a few tenths of 1 MPa due to root pressure), to highly negative (several MPa in the xylem due to transpiration and growth demands).

Understanding the development of these negative pressures or tensions in the xylem requires that we digress and discuss some of the things that happen when water or solution interfaces with either solid or gaseous phases.

When a solution encounters a wettable surface, the energy of the water is lowered in the immediate vicinity of that surface. Water sticking to a surface (adhesion) may be held so tightly that energies equal to several tens of MPa must be applied to strip it away. Because of the extensive hydrogen bonding which occurs in the liquid phase, the water molecules try to stick to each other (cohesion) and a structured layer of water may exist near a wettable surface.

Although wettability is important throughout the various plant tissue complexes, it is only in the presence of a gas-liquid interface that adhesion is able to generate the tensions necessary to drive bulk water transport through the xylem. To see how this is possible we first must discuss the concepts of surface tension and capillarity.

a. Surface Tension. A water molecule in the liquid phase has its lowest energy when surrounded by the maximum number of other water molecules.

For the sake of illustration, let's assume that the water molecules are spherical and *close packed*. Being close packed means that they are arranged so that each unit of volume contains the maximum number of molecules. In the case of spheres close packing results in each molecule being in contact with 12 other molecules. As a result each molecule will be bound by a cohesive energy of 12E, where E is the energy of an individual bond.

At a surface, however, close packing results in each molecule being surrounded by only nine neighbors so that it will be bound by a cohesive energy of only 9E. The surface molecule then is bound by only 75% of the energy of the molecules in the liquid interior. The surface molecule therefore has a higher energy. Consequently, to move a molecule from the interior of a liquid to the surface requires an expenditure of energy.

It is a characteristic of physical systems that they try to attain their lowest possible energy level. A wonderful illustration of this principle, which is immediately relevant, is the formation of a liquid into spherical droplets in the absence of gravity. The sphere presents the least surface area possible for a given volume of water and so is the least energetic configuration possible. Any attempt to increase the surface area will require the expenditure of energy. Thus the surface tension may be regarded as the energy per unit surface area and has the units of J m^{-2}. More often though, the alternate units of force per unit length (N m^{-1}) are used. The meaning of these units may be illustrated by the classic film-stretching experiment.

In Fig. 8–1 we show a thin film of liquid stretched between the sides of a wire frame. The left boundary of the frame is movable so that the film may be stretched by exerting a force on the handle. The force required to expand the surface is $f = 2l\gamma$. The proportionality constant γ is the surface tension and may be considered the force exerted by a surface of unit length, not really a great intuitive help but a common definition. Of more use to us is the relationship between the surface tension and the surface energy of the system. An easy way of arriving at such a relationship is to consider the amount of work necessary to change the surface by a unit amount. In the case of the wire frame, the work required to move the left boundary by dx is $W = fdx = 2l\gamma dx$ (above). The change in surface area is $2ldx$ (remember there are two surfaces). Therefore the work per unit surface area increase (W_u) is

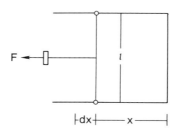

Fig. 8–1. Diagram of a film-stretching experiment. The liquid film is spread over the wire frame with dimensions 1 × x. When force is applied to the handle attached to the moveable part of the frame the film is stretched over an additional area *ldx*.

$$\frac{21\gamma dx}{21 dx} = \gamma \qquad\qquad [2]$$

Since work, like heat, is merely energy in transit, γ also represents the energy required to exapnd the surface area of the film.

b. Capillary Rise. Capillary rise in a small diameter glass tube results from a combination of effects working against gravity: adhesion of the water molecules to the glass surface and cohesion of the water molecules to each other.

A difference in pressure exists across a curved interface. Although this is true for all types of curved interfaces, we will confine our considerations to a liquid (phase 2)—gas (phase 1) interface. The pressure in the two phases is related to the surface tension and the radius of curvature of the liquid surface (R) by $P_2 = P_1 + 2\gamma/R$ (see Castellan, 1971, p. 426–429). The radius of curvature is defined such that if the surface of the liquid is concave (Fig. 8–2), $R < 0$ and $P_2 < P_1$. Conversely, if the surface of the liquid is convex, $R > 0$ and $P_2 > P_1$. It is this difference in pressure which accounts for the phenomena of capillary rise and depression. We may arrive at a quantitative understanding of this statement by examing Fig. 8–2 which shows the effect of immersing a small bore glass tube in a reservoir of water. The water rises in the tube to a height h above the free water surface. By analyzing the pressures in the system at various places we can arrive at a relationship between h and the surface tension of the liquid.

The pressure P_1 is the ambient air pressure at the height of the meniscus, P_2 the pressure in the liquid just below the surface of the meniscus, P_1' the air pressure at elevation z-h, and P_2' the liquid pressure just below the flat surface. Because the surface at z-h is flat $P_2' = P_1'$. At the reference elevation, Z_0, P_z is the pressure in the liquid. At equilibrium, P_z must be the same at all points in the plane Z_0, otherwise water would move horizontally and we might have the essential component for a perpetual motion machine. Now let's examine the components of the pressure at points a and b in the Z_0 plane:

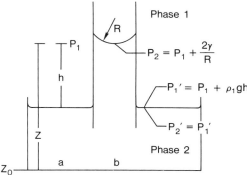

Fig. 8–2. Analysis of the pressure components related to the capillary rise of water in a glass tube. Phase 2 is pure water and phase 1 is air. Water in the capillary tube rises to height h above the free water surface in the reservoir.

$$P_z^a = P_1' + \rho_2 g\,(z - h)$$

$$P_z^b = P_2 + \rho_2 gz \qquad\qquad [3]$$

where ρ is the density of the phase in question and g is the gravitational acceleration. Now, $P_1' = P_1 + \rho_1 gh$ and $P_2 = P_1 + 2\gamma/R$. We can now set $P_z^a = P_z^b$, make the substitutions, and rearrange to obtain the relationship

$$h = \frac{2\gamma}{Rg\,(\rho_1 - \rho_2)} \qquad\qquad [4]$$

which gives the capillary rise, or depression, in terms of the surface tension, the radius of curvature of the interface, and the density of the two phases. The radius of curvature in this case is negative (the liquid surface is concave) and since $\rho_2 \gg \rho_1$, h is positive. If the liquid had been Hg, the surface would

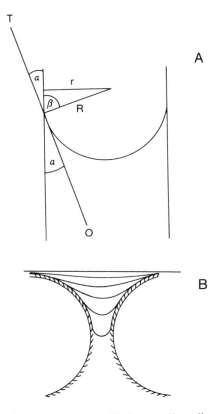

Fig. 8–3. (A) The contact angle α and the relationship between the radius of curvature of the interface (R) and the radius of the tube (r). (B) Tapered pores as they might exist between the microfibrils of a plant cell wall. As water evaporates from the surface, the interface is drawn farther down into the "pore" with a continually decreasing radius. The hatched areas on the microfibrils surfaces represent a layer of adsorbed water which is held tightly, and effectively reduces the radius of the pore.

have been convex (the radius of curvature positive); h would then be negative and a capillary depression would result.

For convenience we would like to have h as a function of the radius of the tube (r) rather than the radius of curvature. Figure 8-3A illustrates the relationship between the contact angle of the liquid, the radius of curvature of the interface, and the radius of the tube. The line 0-T is drawn tangent to the surface of the liquid at the point of contact with the glass tube. The R is perpendicular to 0-T and the objective is to solve the triangle for r. The angle between the side of the tube and the line 0-T is called the contact angle, α. The angle β may be written as $90° - \alpha$ so we now have sin $\beta = r/(-R) = \cos \alpha$, or $R = -r/\cos \alpha$. The negative sign results from the fact that the radius of curvature, R, is negative for a concave surface. Reversing the order of the density difference allows us to eliminate the negative sign and leads to the expression we want from Eq. 4:

$$ h = \frac{2\gamma \cos \alpha}{rg \, (\rho_2 - \rho_1)} . \qquad [5] $$

In the case where the surface is wettable, such as water and glass, the angle α is less than 90% and in the case where the surface is not wettable (Hg and glass) the contact angle is $> 90°$. Also, if the radius of the tube is small, the contact angle may be taken as $0°$ for a wettable surface and $180°$ for a nonwettable surface. Other contact angles are difficult to determine experimentally. Finally, the density of phase 1 for an air-water interface is much less than that for phase 2. Taking all of these conditions into account leads to the condensed form Eq. [5],

$$ h = \frac{2\gamma}{rg\rho_2} \qquad [6] $$

which numerically reduces to $h = 1.49 \times 10^{-5}/r$ (m) for water.

Inspection of the numerical form of Eq. [6] shows that a tube with a radius of slightly < 15 μm, about the size of a relatively small xylem element, will result in a capillary rise of only 1 m. Capillary rise in the xylem of plants is therefore unlikely to be of great significance.

The real importance of surface tension and curved interfaces in plants becomes apparent when we consider the dimensions of the interstices of the cell walls in the leaf where evaporation of water into the vapor phase occurs. Here the diameter of the pore spaces are several orders of magnitude less than the xylem elements and have the potential for developing pressure differences of several tens of MPa across the interface. Furthermore, these "pores" tend to be tapered so the radius of curvature decreases as water is lost from the pore (Fig. 8-3B). And, as we have just seen, a decrease in the radius of curvature results in an increased pressure difference across the interface. Therefore, in a tapered pore it is possible to produce increasingly large tensions in the liquid phase with a loss of water from the pore.

3. Solute Potential

The presence of dissolved solutes lowers the vapor pressure of the solvent (water), raises the boiling point, and lowers the freezing point and water potential of aqueous solutions. These properties of solutions, called *colligative properties,* are all related to the concentration of the dissolved substances.

The water potential is expressed in the appendix in terms of the relative vapor pressure, and the vapor pressure is linked to dissolved substances, as for example by Raoult's law for ideal dilute solutions. The boiling point elevation and the freezing point depression are related to the vapor pressure as may be seen in Fig. 8–4.

The major significance of the changes in vapor pressure, boiling point and freezing point lies in the fact that these are easily measured properties and thus provide a convenient means for assessing the water potential. Colligative properties are thoroughly discussed in most physical chemistry textbooks but because of the importance of the osmotic potential and the osmotic pressure to cell water behavior we will spend more time discussing those concepts.

a. Osmotic Potential and Osmotic Pressure. The component of the water potential reduction caused by the presence of dissolved substances is referred to as the *solute potential* or *osmotic potential* and is indicated by Ψ_s or Ψ_π. The degree of reduction is related to the mole fraction of water in the solution, which will be decreased in proportion to the concentration or activity of the dissolved substances (see appendix). Therefore solutes act by decreasing the ability of the water in the solution to perform work.

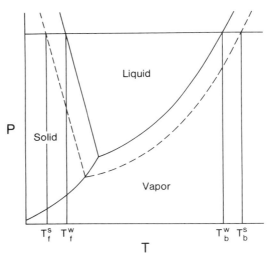

Fig. 8–4. Phase diagram. The triple point for water is the temperature and pressure where the solid lines separating the three phases converge ($T = 0.0098C$, $P = 0.603$ kPa). When solutes are added (dashed lines) the vapor pressure is lowered and the area of the liquid phase in the diagram increases. The dashed lines run approximately parallel with their solid counterparts and it is easy to see that the freezing point of the solution (T_f^s) is lowered (-1.86 K mol^{-1}) and the boiling point (T_b^s) elevated (0.51 K mol^{-1}) relative to pure water.

Closely linked to the osmotic potential is the osmotic pressure, which is illustrated in Fig. 8–5A. In this illustration a solution is enclosed in a container with rigid walls in contact with a reservoir of pure water. The contacting surface of the container is composed of a rigid semipermeable membrane. Such a membrane allows free passage of water but restricts the passage of the solute. The degree to which solutes are restricted is a characteristic of the particular membrane-solute combination and is indicated by σ, the "reflection coefficient" (Stavermann, 1951). For different membrane-solute combinations σ may have any value from 0 to 1. The former case is nonrestrictive where the solutes may move freely with the solution, while in the latter, solutes are totally screened out by the membrane and the membrane is referred to as ideally semipermeable.

Assuming that the membrane in the illustration is ideally semipermeable, we can expect water to diffuse from the pure water, across the membrane into the solution phase until the water potentials on both sides of the membrane are equal. As the water enters the solution side, the solution will rise and continue to rise in the tube until the pressure created by the height of the solution column just balances the difference in osmotic potential across the membrane. Of course the solution will have to be continuously stirred since the entry of water will dilute it. Thus the final equilibrium will occur

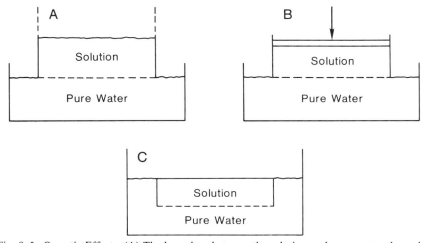

Fig. 8–5. Osmotic Effects: (A) The boundary between the solution and pure water phases is a rigid, ideally semipermeable membrane (dashed line). Water will flow across the membrane into the solution, which will rise to some definite height. At equilibrium, the hydrostatic pressure bearing on the solution side of the membrane exactly balances the osmotic potential difference acting across the membrane in the opposite direction. The balancing hydrostatic pressure is called the osmotic pressure of the solution. (B) In this case the hydrostatic pressure is provided by the piston rather than the solution column. The balancing pressure (osmotic pressure) will be somewhat greater than in (A) because dilution of the solution by entering water is prevented. (C) The solution container is rigid and all walls are impermeable except for the one covered by the membrane. In this case only a small quantity of water will enter the solution phase. The exact quantity will depend on how much the walls can deform under pressure. Again, at equilibrium, the hydrostatic pressure developed on the solution side (osmotic pressure) will balance the osmotic potential difference acting across the membrane. This is analogous to the situation in a plant cell.

at a pressure somewhat less than would have been necessary to balance the original concentration.

An alternative to allowing the hydrostatic pressure of the water column to determine the equilibrium would be to apply the pressure with a piston (Fig. 8-5B). In either case, the *osmotic pressure* of the solution is defined as the hydrostatic pressure necessary to balance the osmotic potential difference across the membrane. If the piston is used to generate the counter pressure, the force applied to the piston may be made just sufficient to prevent the passage of water across the membrane and so should prevent dilution of the solution and give a more accurate measure of the osmotic potential difference. The osmotic pressure, therefore, will have the same numerical value but an opposite sign from the osmotic potential of the solution. Thus the van't Hoff approximation given in the appendix may be written

$$\pi = CRT = -\psi_s \qquad [7]$$

where π is the traditional symbol for osmotic pressure, C is the molar concentration, R is the gas constant and T the temperature in K.

If the solution container were completely enclosed, then the system in Fig. 8-5C would resemble a single cell and the osmotic pressure developed as a result of the osmotic potential difference would be described as the *turgor pressure* of the cell. Therefore, according to Eq. [1], for a single cell in equilibrium with distilled water the osmotic potential will be exactly balanced by the turgor pressure.

Thus far we have been talking about the thermodynamic potential of the water in plants, and while the water potential is extremely useful when dealing with liquid phase diffusive processes, there are instances when the mechanical potential of the solution—and not just the water in the solution—becomes of prime importance. In those instances where convective or mass flow of solution are involved, the "water potential" per se may not even indicate the direction of solution flow. We will say more about this in the section on transport.

II. DISTRIBUTION AND COMPOSITION OF SOLUTIONS IN PLANTS

Except where specifically noted, the data on the distribution and composition of water in plants were gathered from the following sources: Kramer, 1983, p. 28-32; Meidner & Sheriff, 1976, p. 104-114; Milburn, 1979, p. 37-45; and Slatyer, 1967, p. 127-137. These publications should be consulted for further details and additional references.

A. Distribution

Water in plants is partitioned between two compartments, the *symplast* and the *apoplast*. The symplast includes all the water inside the *plasmalemma,* the membrane bounding the protoplast, and may range from 50 to 95% of the total cell water. The liquid phase of the symplast is continuous since the cytoplasm of adjacent cells are usually joined by tubular cytoplasmic connections, called *plasmodesmata,* extending through the cell walls. Plasmodesmata are lined with cell membranes and serve to join directly the cytoplasm of adjacent cells, forming an extensive network of linked cells, collectively termed the symplast.

The apoplastic water includes that occurring outside the plasmalemma, mainly in the cell walls and xylem tissues. The proportion of the total cell water occurring in the walls may range from 5 to 40%, depending on the type of tissue, age, wall thickness, and composition. The apoplast too, is continuous throughout the plant except where it is interrupted by the Casparian bands in the endodermis of young roots. Table 2.1 in Kramer (1983) summarizes some of these data on symplastic-apoplastic water distribution.

Figure 8-6 compares some features of a young meristematic cell and a mature parenchyma cell. In the young cell the wall is generally thin and contains a high percentage of water in its structure. Because it is thin the proportion of the total cell water in the wall is relatively small. Therefore, although a young cell wall may be as much as 50% water, by fresh weight, that water may comprise as little as 5% of the total cell water. The rest of the cell water is found in the symplastic phase, consisting of the cytoplasm

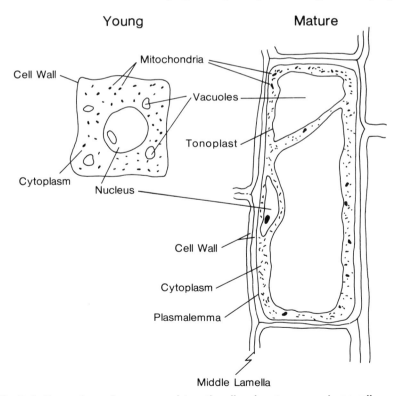

Fig. 8-6. Comparison of a young meristematic cell and mature parenchyma cell.

and vacuole. In a meristematic cell, most of the water occurs in the cytoplasm. Although there may be more than one vacuole in such cells, they are small and contain little of the symplastic water.

As a cell matures it increases in size, the wall normally undergoes various degrees of secondary thickening, the small vacuoles grow and coalesce, and the cytoplasm becomes a thin layer appressed to the periphery of the cell. The cytoplasmic layer may be as little as 10 μm thick, but it still contains all the apparatus necessary for physiological functions. In this condition the cytoplasm may contain only 5 to 10% of the cell water while 50 to 80% is in the vacuole.

Although most water taken up by plants passes through the xylem conduits, the actual amount of water in the xylem tissue usually makes up only a small part of the total plant water in herbaceous species. This proportion, however, may become significant in woody plants where a considerable portion of the plant volume is xylem tissue.

B. Composition

1. Xylem Solution

Although the water in the xylem tissue may be a small percentage of the total plant water, it is responsible for the initial distribution of nutrient ions to the rest of the plant. As one might suspect, then, the major proportion of solutes in the xylem are inorganic ions placed there by active accumulation from the soil. Even though there are also a wide variety of organic solutes in the xylem, including plant growth regulators, amino acids, amides, ureides, organic acids, etc., they are normally present in such low concentrations that they do not need to be considered as osmotically important. Inorganic ions, on the other hand, may occur in concentrations such that they give the xylem sap an osmotic potential as low as -0.4 MPa. Such low osmotic potentials would usually be found only in the root xylem and then only under conditions of very low transpiration.

As the xylem sap moves upward, many of the salts are selectively removed by tissues adjacent to the xylem so that the composition as well as the concentration is changed by the time the sap reaches the evaporating sites of the leaves. The osmotic potential at this point is normally close to zero.

In slowly transpiring plants, root pressure may develop and force more highly concentrated sap to the leaves. Since the solution is under positive pressure, droplets of it may be forced out the ends of veins to the leaf surface. This phenomenon, often confused with dew, is called *guttation* and the droplets accumulate mostly along the margins of the leaf at the sites of specialized cells called *hydathodes*. Residues left along the leaf margins provide ample evidence that under these circumstances there are still significant amounts of salts in the xylem sap. In fact, under the proper circumstances of soil moisture, fertility, and weather, the leaf tips and margins may become noticeably damaged by these residues (unpublished observations from field and greenhouse).

The composition of the xylem sap also varies greatly with the species and the season, especially among perennials. A prominent example is provided by the high carbohydrate concentrations found in the sap of sugar maple (*Acer saccharum*) trees in the winter and spring.

a. Cell Wall Water. The composition of cell wall solution varies with the tissue. In the walls of the xylem elements it will vary with the composition of the xylem sap and may be more concentrated than the sap itself because of a local Donnan equilibrium (see Nobel, 1983, p. 134–136).

Cosgrove and Cleland (1983a, 1983b) found solutes in the apoplast of growing pea (*Pisum sativum* L.) stem tissue with osmotic potentials ranging from −0.18 to −0.29 MPa for basal and apical segments respectively. Of the solutes extracted from the apoplast about 25% were inorganic electrolytes and the remainder were organic nonelectrolytes about the size of glucose molecules. They found similar concentrations in the apoplastic water from etiolated soybean (*Glycine max* L.) and cucumber (*Cucumis sativus* L.) tissue.

In the cell walls of roots the solution can vary greatly and the only possible generalization, however weak, is to say that it resembles the soil solution from which it was drawn but is depleted of many ions because of active accumulation by the root tissues.

Finally, the solution in the cell walls of plant leaves was formerly thought to be almost devoid of solutes, and the ones which were present probably disappeared rapidly into the symplasm by active absorption. However, several studies indicate that there may be appreciable quantities of sugars, amino acids, and mineral salts in the apoplast of photosynthesizing leaves (Kursanov & Brovchenko, 1970; Tukey et al., 1957; Tukey et al., 1965). Later, Bernstein (1971) studied leaves of sunflower (*Helianthus annuus* L.), cabbage (*Brassica oleracea*), and castor bean (*Ricinus communis* L.) under saline and nonsaline conditions and found osmotic potentials of cell wall solutions to range from −0.03 to −0.2 MPa. Under nonsaline conditions, 22% of the osmotic potential was due to the presence of electrolytes in sunflower and 40% in cabbage leaves.

2. Protoplasmic Water

The *cytoplasm* itself bears little resemblance to a true solution. It is rather a suspension, able to move between a sol and a gel state, interwoven with a network of membranes, the *endoplasmic reticulum*. There are large protein molecules suspended in the cytoplasm on which are adsorbed fairly large quantities of water, mainly through hydrogen bonding. Also suspended in this soup are all of the cellular organelles, which are generally separated from the cytoplasmic suspension by their own membranes. The organelles therefore maintain an osmotic equilibrium with the cytoplasm. In addition to the proteins which constitute the majority of the suspended particles, there also may be sugars, salts, lipids, and other compounds either in solution or suspension.

The state of the cytoplasm is strongly dependent on its water content and so is subject to the influence of the vacuole, since the cytoplasm and the vacuole must maintain an osmotic equilibrium. Also, it is the osmotic potential of the vacuole, which may range from -1 to -3 MPa, which controls osmotic water movement and so to a large extent the water relations of a cell.

The water in the vacuole may contain some colloidal material but in general it may be described as a true solution, as opposed to the cytoplasm. This solution may contain a wide range of materials such as sugars, ions, organic acids, pigments, and in some cases crystalline material. The *tonoplast* is a true semipermeable membrane and once solutes are accumulated in the vacuole they are released only with great difficulty. As cells become older the amount of suspended material may increase to the point where the vacuolar contents begin to resemble a gel rather than a solution.

III. NATURE OF THE CONDUCTING SYSTEM

The most prominent part of the water conducting system in plants is the xylem tissue. However, that small fraction of absorbed water which constitutes the plant structure must be distributed to individual cells which may be somewhat distant from the xylem. Therefore, the cell membranes and aggregates of cells through which the water must pass also will be considered as part of the water conducting system.

A. Cellular Membranes

In this section we will deal primarily with the plasmalemma, but bear in mind that many of the statements also apply to the tonoplast. The plasmalemma is composed primarily of phospholipids and proteins. The exact pattern of these components in membranes is a question of great interest to physiologists because the configuration has a great influence on the physical properties that the membrane as a whole exhibits. Early work revealed that the quantity of lipid in the membrane was just sufficient to form a double layer completely enclosing the cell. The hydrophobic ends of the phospholipid molecules are oriented toward the inside of the membrane while the hydrophyllic ends form the outside layers in contact with the surroundings. Originally, it was assumed the protein components were attached to the outside of the membrane and formed a layer across both surfaces. However, the globular nature of the proteins isolated from membranes indicated that they could not be spread uniformly. This latter fact and the results of thermodynamic studies led Singer (1974) to develop the fluid mosaic concept which constitutes the most widely accepted current model for membrane structure. This model still retains the phospholipid bilayer but has the globular proteins embedded in the bilayer. In some cases the proteins may form hydrophilic channels through the membrane or in others they may simply be carrier enzymes which can transfer substances through the membrane. Furthermore the "pores" in a membrane may or may not be stable. They may fluctuate in time and space.

The phospholipid bilayer has important implications in the penetration of different ionic or molecular species through the membrane. Hydrophilic species generally have a low ability to penetrate membranes by diffusion. Passive penetration increases dramatically as a function of the lipid solubility of the substance. For more background and relevant literature on membrane structure and properties, including those just discussed, the reader is referred to the following: Clarkson, 1984; Kramer, 1983, p. 32–34: Nobel, 1983 p. 20–26; Stein, 1967, p. 31–35.

Since the plasmalemma forms the boundary between the cytoplasm and the exterior, the specific ways in which water and solutes penetrate a membrane are of interest and will be discussed in section VI of this chapter.

B. Xylem

1. Vessels

Plant anatomy texts provide detailed discussions of xylem structure, development, and evolution; hence, we will only highlight those features of the xylem tissue that are relevant to water transport phenomena. There are three main types of cells occurring in xylem tissue apart from the parenchyma: vessel elements, tracheids, and fibers. The dimensions and distribution of each vary widely among species. Vessel elements usually have the largest diameters of the cells in a xylem tissue and carry the largest proportion of the water transported, so we will describe them first.

The mature *vessel element* is an elongate nonliving cell having a lignified secondary wall. The diameters may vary over an approximate thousand-fold range, from 1 μm in small grasses to 1 mm or more in certain vines and lianas. A typical value, however, might be about 10 to 100 μm. The length of vessel elements is also highly variable, ranging from a few hundred to a few thousand μm.

Vessel elements are arranged end to end and during the final stages of development most of the end walls may disappear, forming a *perforation plate.* Thus contiguous vessel elements form a long, hollow tube called a *vessel,* which may be as long as several meters. In the case of certain ring-porous trees, vessel length may reach more than 18 m and extend the entire height of the tree (see Milburn, 1979, p. 82; Kramer, 1983, p. 265). More commonly, though, in woody shrubs and diffuse-porous trees the majority of vessels are in the 0 to 10 cm length class (Zimmerman & Jeje, 1981). The perforation plates may be *simple,* consisting of a single large hole nearly the diameter of the vessel element itself. Alternately, the perforation plate may be *scalariform,* consisting of several slit-like openings arranged parallel to each other; or it may be *reticulate,* composed of more than one circular pore. In addition, there may be pits in the longitudinal walls of vessel elements which join with pits in adjacent elements or parenchyma cells, allowing the exchange of water and solutes. The pits are not open pores but areas in the primary cell wall that have not undergone secondary thickening. The remaining primary wall is referred to as the pit membrane. The pits may be bordered, where the secondary thickening overhangs the pit membrane so that the pit aperture is narrower than the pit membrane. The pattern of secondary thickening in the vessel elements can vary greatly from simple rings and helices to a more or less solid surface with retained scalariform or circular pits.

The vessels do not usually extend through the entire length of the plant but are limited in length and joined to each other through nonperforate end walls. Depending on the species, these end walls may be more or less oblique and overlap the succeeding vessel for a considerable length. For example, Zimmermann (1976) observed vessels in *Rhapis* which were 10 cm long and overlapped the vessels on either end by 2 cm.

2. Tracheids

The other type of water-conducting cell found in the xylem is the *tracheid*. Generally, it has a smaller diameter (10–20 μm) and greater length (as much as 5 mm) than a vessel element in the same plant and does not possess a perforated end wall. Tracheids tend to be more spindle shaped with tapered end walls. The end walls, which typically overlap the next tracheid in a series, have a high concentration of pits which facilitates the longitudinal transfer of solution. As with vessel elements, the longitudinal walls also have pits which allow lateral exchange. Gymnosperms have no vessels and depend entirely on tracheids for water conduction. Therefore, the resistance to longitudinal transport of water is greater in Gymnosperms than in Angiosperms.

3. Fibers

Fibers are also elongate spindle shaped cells but, unlike the tracheids, are more specialized for support. They still possess a protoplast in the mature state and the walls frequently have undergone considerable secondary thickening. In addition to support, fibers may also serve as storage cells. Because they do not function as water transport cells we will not discuss them further.

C. Balance of Transport System Components

It has long been recognized that the relative size of plant parts remains approximately the same over time. For example, in rapidly growing plants, the ratio of shoot-to-root dry weight is relatively constant because roots and shoots grow at about the same relative rates. On closer examination, it appears that the relative size of the components of the water transport system in plants also remains fairly constant (Kaufmann & Fiscus, 1984).

The water transport system consists of the absorbing surface of roots, the root and shoot xylem system, and the leaf transpiring surface. During transpiration, nearly equal amounts of water move through each segment of the transport system in a day. The only factor altering this is tissue growth, which even in rapidly growing plants probably utilizes only a few percent of the water absorbed by the root system.

According to the so-called *pipe theory,* a balance is achieved in the relative size of the absorbing, transporting, and transpiring tissues in plants. While there may be short-term imbalances in absorption and transpiration, e.g., absorption lagging behind or exceeding transpiration at different times of the day, each part of the transport system must have a similar capacity for water flux.

Research has confirmed that the amounts of conducting tissues in roots, stems, and leaves are strongly correlated. Much of the research has been on forest trees, which have generally linear relationships within each species between leaf area and the cross-sectional area of sapwood conducting tissue. However, there is no reason to suspect that similar relationships do not exist in fruit trees and other woody horticultural crops. In fact, observations have confirmed that the ratio of root absorbing surface and leaf transpiring surface is nearly constant in herbaceous plants as well. Fiscus (1981) showed that the leaf area of bean (*Phaseolus vulgaris* L.) plants was linearly related to the root surface area.

The balance in size among plant parts is maintained largely through two important feedback processes. One involves the regulatory effects of carbohydrate supply for growth, and the second involves the effects of water stress on growth and the retention of roots and foliage. Under conditions favorable for growth, photosynthetic rates are high on a unit leaf area basis, favoring leaf production as well as increases in stem size and amount of roots, unless reproductive growth is competing for photosynthate.

The balance of growth among the various organs is strongly affected by water relations of the various tissues and by carbohydrate supply. If leaf growth is more rapid than root growth, for example, the transpiring capacity of the shoot system may exceed the absorbing capacity of the roots. The direct consequence of this is increased water stress in the foliage, reduced photosynthesis as a result of stomatal closure, and reduced leaf growth as a result of turgor loss and lower photosynthesis. Root growth is reduced less than shoot growth, thus reestablishing a favorable root/shoot ratio. Alternatively, root growth cannot be excessive compared to shoot growth because it is limited by the availability of carbohydrates for growth and respiration.

Nutrient balance, cell water and solute relations, and hormonal effects are also important in regulating the size of various plant parts, but the C and water balance among plant organs must be considered the primary factors determining shoot/root ratios and plant growth. Because water stress effects on stomatal function and on root and shoot meristematic tissues may last for several days or more following stress, and because growth is an integral result of current photosynthesis and the utilization of transported and stored carbohydrate, the relative growth of plant parts and the balance of their size is the result of the continual integration of environmental effects on internal physiological processes.

D. Interconnection of the Xylem System

The structure of the vascular network in plants may be important in water transport and, consequently, in the response of the shoot of plants to environmental condition. In section III of this chapter anatomical aspects of different cell types in xylem conducting tissues were discussed. Features of the whole conducting system should also be considered. Dye transport experiments in trees indicate that dye, and presumably water, often move upward in narrow portions of the sapwood with little lateral exchange (Kramer & Kozlowski, 1979). Thus, certain branches may receive nearly all of their water from only a portion of the xylem. Experiments on tobacco (*Nicotiana tabacum* L.) (Fiscus et al., 1973) and on cottonwood (*Populus deltoides*) (Larson, 1980) provide clear evidence that individual leaves have specific connections to vascular traces in the stem. Fiscus et al. determined that resistance to vertical movement of water is much lower than resistance to lateral movement between traces. Consequently, water movement between adjacent leaves is much more difficult than between leaves that are phyllotactically related.

Other sections of this chapter discuss resistances to water flow in various parts of the water transport pathway. It is important to note that within a portion of the transport system there may be considerable restriction to water exchange, causing marked differences in water potential. Kaufmann (1976) discussed evidence showing that leaf water potentials of sunlit and shaded leaves of plants regularly differed by as much as 0.4 MPa. Thus flow resistances, including those associated with lateral exchange of water in the xylem, may permit leaves and sections of plant crowns to behave somewhat independently from one another. An understanding of water movement in plants requires an awareness of anatomical effects on transport phenomena.

IV. WATER TRANSPORT IN THE WHOLE PLANT

In the two sections that follow, we discuss specific aspects of the driving forces and resistances for water movement in the water transport system. To introduce that material, it is helpful first to examine whole-plant water transport to provide a perspective within which the details are relevant.

Water movement occurs in proportion to a driving force and in inverse proportion to the resistances that are encountered. In an intact plant adequately supplied with water, stomata open and close diurnally in response to light, humidity, and perhaps other environmental conditions. Transpiration creates the internal conditions under which water is absorbed by the root system and transported to the foliage. As water is lost from the foliage, leaf water potential and the pressure in the xylem are reduced. Except in situations where root pressure occurs, transpiration establishes a hydraulic gradient in the xylem which results in water flow from the roots to leaves. The size of the gradient, and the rate of flow, depend upon the rate of transpiration, the liquid phase flow resistances in the root system and xylem, the soil water availability, and lateral exchange of water with other tissues.

Experiments provide data showing that, for well-watered plants both in natural and artificial environments, increased transpiration rates generally are associated with decreased leaf water potentials (Kaufmann, 1976). The difference between leaf water potential and soil water potential (the latter near zero for moist soil) is an estimate of the driving force for water movement from soil to the foliage. Note that transpiration depends upon those factors regulating water loss from the leaf surface. The driving force through the soil-plant system is determined by the transpirational flux and the resistances encountered in the soil and plant. Therefore, the driving force affects flow of liquid water to the evaporating surfaces, but it does not affect transpiration except when leaf water stress causes stomatal closure.

In a typical diurnal progression, transpiration increases during the morning and decreases in the afternoon while leaf water potential becomes more negative in the morning and recovers in the afternoon and evening. For well-watered plants, a plot of leaf water potential against transpiration generally indicates that the decline and recovery of water potential follow the same curve, i.e., there is no hysteresis in the potential-flux relationship. This suggests that the water transport system in plants frequently remains in a quasisteady state.

Several factors are known to change the relationship between leaf water potential and transpiration rate for a given plant. For example, low soil tem-

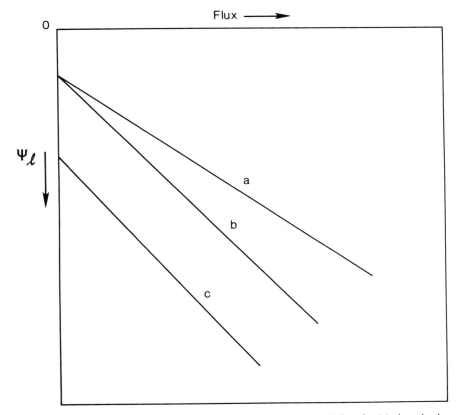

Fig. 8-7. Relationship between leaf water potential and transpirational flux, for (*a*) plants having an adequate supply of soil water and a warm root system, (*b*) plants having roots in cold soil, and (*c*) plants in dry soil.

peratures may decrease root permeability and increase water viscosity, both of which increase resistance to water movement from the soil to leaves. At zero flux, this has no effect on leaf water potential. When there is flux, however, the gradient must be higher than is required at higher soil temperatures to deliver an equal amount of water to the leaves. This occurs in the form of more negative leaf water potentials and a steeper slope of the water potential-flux relationship (Fig. 8–7, line b vs. line a). In contrast, reducing the supply of water in soil (lower soil water potential) reduces the equilibrium leaf water potential at zero flux as well as at finite fluxes (Fig. 8–7, line c). Furthermore, flow resistances for movement of water from soil into roots may increase as the soil and roots dry, also steepening the water potential-flux slope.

Each segment of the water transport system in plants has a volume of water occupying the actual transport pathway and the tissues immediately around the pathway. When the change in the volume of water in these segments is small during the course of the day, changes in transpiration are quickly accompanied by changes in the gradient of driving force throughout the transport system. When the water content of the segments is large relative to other segments, for example in the boles of large trees, there may be a large quantitative change in water content of the segment in response to changes in the water potential.

Changes in water content in various parts of the transport pathway affect water transport characteristics. An increase in transpirational flux reduces the pressure (increases the tension) in the xylem, thereby establishing a gradient for water movement from surrounding tissues into the xylem. The reverse happens when transpiration decreases. Thus in the diurnal course of transpiration, tissues surrounding the xylem may act as a variable source and sink for water, and the gradients and resistances for flow through the system are accompanied by a series of gradients and resistances for lateral water exchange within each segment. In a sense, lateral water exchange partially "uncouples" flux from the soil-to-leaf driving gradient and resistances. For instance, it has been reported that the total water storage for *Larix* and *Picea* trees was 24 and 14%, respectively of the total daily transpiration (Schulze et al., 1985). As a result of this storage, water flow in the xylem below the crown did not start until 2 to 3 h after the onset of transpiration. Further, transpiration decreased in the afternoon long before xylem water flow reached a peak, and ceased at night 2 to 3 h before xylem water flow.

The importance of this uncoupling in different plants depends upon the magnitude of lateral water exchange in relation to the volume flux. If large amounts of water are exchanged laterally, hysteresis may be induced in the water potential-flux relationship. Hysteresis has been observed in large forest trees, but as a general rule it is not observed in species used in horticulture or in annual crops except when soil water is limiting. Presumably the volumetric capacity for lateral exchange is small relative to transpiration volume, and the changes in water content of each transport segment have a negligible effect on the driving force for water movement from soil to leaves. Hysteresis in moderately dry soil may occur as a result of a shift in water availability in the soil from near the roots early in the day to farther from the roots later in the day. It is possible to treat mathematically the relationships between transpiration and water potential, as affected by plant water storage capacity, and the reader is referred to one of the more recent publications of this type (Wronski et al., 1985).

V. EQUILIBRIUM CELL WATER RELATIONS

The vast majority of water absorbed by the plant is simply lost to the atmosphere by transpiration. However, most of the physiological processes essential to growth and development occur within the confines of individual cells. The state of hydration of these cells has important implications about the success and efficiency of these processes. We will, therefore, spend some time discussing the equilibrium water relations of individual cells and later approach the matter of water transport through organized collections of cells (tissues). The following discussion of cell water relations is general in nature but more quantitative treatments of static and dynamic cell water relations are discussed in the review by Molz and Ferrier (1982).

A cell has many of the characteristics of a closed osmotic bag. There are, of course, substantial differences between a purely mechanical osmometer and a cell which continues to execute vital physiological functions. One of the more important characteristics of plant cells, and one which profoundly affects their water relations, is that they are enclosed in a cellulosic cell wall which is capable of both elastic (reversible) and plastic (permanent) deformations. The thickness and stretching characteristics of cell walls vary enormously from tissue to tissue and especially between ecotypes (xeric, mesic, and hydric). Generally, though, it is the cell walls that restrict the size of the cell and determine the structure of the plant.

Returning to the osmometer discussed in section I, we see that most of the principles of equilibrium cell water relations were covered there. To make the system more realistic all we need do is specify that the solution container is completely enclosed in a cellulosic wall and that the semipermeable membrane covers the entire surface of the solution phase. For present purposes we will consider that the cell wall is capable of limited elastic extension. Let us further assume that the cell is spherical with a 50 μm diam. and that the cell contents have an osmotic potential of -1.2 MPa at its maximum size.

If such a cell exists in equilibrium with pure water, then the total water potential outside the cell must equal the total water potential inside. We can then see from Fig. 8–8A that the turgor potential is at the maximum permissible by the cell wall characteristics. That is, the cell is at the limits of its elastic capabilities. Since the potential outside the cell is zero, and we have stated that the osmotic potential of the cell is -1.2 MPa, then the turgor potential must be $+1.2$ MPa. This represents the condition of maximum or "full" turgor.

Placing the same cell in a solution of nonpermeating solutes with an osmotic potential of -0.6 MPa (Fig. 8–8B) will result in water flowing out of the cell down the gradient of total water potential. This flow will continue until the new equilibrium state is reached where the total water potential of the cell is also -0.6 MPa.

As a result of water moving out of the cell there is a decrease in volume of the cell, and two things happen simultaneously; there is a decrease in the turgor of the cell because the wall is no longer stretched to its elastic limits, and there is a decrease in the osmotic potential of the cell contents because all the solutes remain inside but are now dissolved in a smaller quantity of water (i.e., the cell contents are now more concentrated). At the new equilibrium we know only that the total cell water potential must be -0.6 MPa.

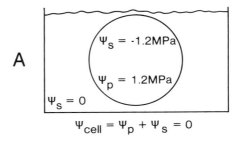

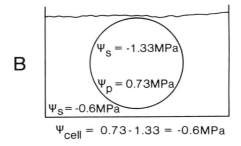

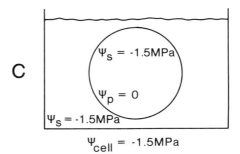

Fig. 8–8. Equilibrium water relations of a hypothetical cell in three solutions of different osmotic potentials. In each case, the water potential of the cell is given as $\psi_{cell} = \psi_P + \psi_S$.

Without additional information we can infer neither the cell's osmotic potential nor its turgor potential.

The possibility exists that we could calculate an approximate value for the new osmotic potential since we know the cell volume at full turgor. The reasoning may be based on the van't Hoff approximation mentioned earlier. We know that $-\Psi_s = CRT$ and that $C = n/V$, where n is the number of moles of solute present and V is the cell volume. Therefore, in the first instance $-\Psi_{s1} = (n/V_1)RT$ and in the second $-\Psi_{s2} = (n/V_2)RT$. Since n, R, and T are all constant we may write $\Psi_{s1} V_1 = \Psi_{s2} V_2$. Also we know Ψ_{s1} (at full turgor), V_1, and V_2. It is therefore a simple matter to calculate $\Psi_{s2} = (\Psi_{s1} V_1)/V_2$. If we measure V_2 as 90% of its value at full turgor, we estimate $\Psi_{s2} = -1.33$ MPa.

If we place the same cell in successively more concentrated solutions, we will eventually observe that the cell membrane begins to pull away from the cell wall. At this point, called *incipient plasmolysis,* the turgor potential of the cell is zero and the cellular osmotic potential is equal to the osmotic potential of the external solution. At incipient plasmolysis we needn't know the cell volume to infer both the osmotic and turgor potentials. As an exercise, we can estimate what the Ψ_s might be at incipient plasmolysis if the cell volume had decreased to 80% of its maximum. Using exactly the same method of calculation as before we estimate $\Psi_{sip} = -1.5$ MPa.

Cells may be taken beyond the point of incipient plasmolysis simply by placing them in more concentrated solutions. When this occurs and the plasmalemma pulls away from the cell wall, the condition is called a *true plasmolysis.* In many instances continued plasmolysis can damage the cell but Palta and Lee-Stadelmann (1983) were able to plasmolyse cells of onion (*Allium cepa* L.) and pea to 15 to 45% of their original volume. During the course of these plasmolyses the cells behaved as ideal osmometers and the membranes retained their semipermeable properties. Further, the cells resumed protoplasmic streaming upon deplasmolysis so that they were able to withstand levels of dehydration that would normally be fatal if extended throughout the plant.

In other situations where the plasmolysing solute cannot penetrate the cell wall, water will still be withdrawn from the cell but the cell wall may collapse because of the development of negative turgor, a condition referred to as *cytorhyssis.*

The above calculations assume a linear relation between cell volume and Ψ_s. Although this is not strictly true for most tissues, it serves to illustrate the basic equilibrium water relations of cells or small pieces of tissue. The actual relationships between Ψ_p, Ψ_s, and Ψ_t are frequently plotted as in Fig. 8–9 which is called a Höfler diagram (after Höfler, 1920). Although the exact values shown in this diagram are synthesized, they do resemble in magnitude many published data. Also see the later treatment by Richter (1978).

The Höfler diagram is also useful for illustrating the differences in the water content-turgor relations of cells whose walls are of different thickness and rigidity. These differences may be characteristic of different tissues within the same plant, differences within a species due to acclimation, or differences among species. Typically, xeric species have thicker, more rigid cell walls than their mesic counterparts, leading to greater proportional losses of turgor per unit water loss from the tissue.

Referring to Fig. 8–9 we can imagine that the loss in relative cell volume will be much greater for thin-walled cells than for thick ones. By way of comparison, suppose that the dashed lines in Fig. 8–9 represent a thick-walled cell. The change in turgor from full to zero is compressed over a much narrower range of relative cell volume. The relative volume may be decreased by as little as 5%. A thin-walled cell, on the other hand, might change by as much as 30% from full to zero turgor.

It is possible to briefly extend the concept of cell water relations to groups of cells. If indeed we placed a cell, not in a solution, but within a group of other cells, water exchange between the cells would occur until they were all in equilibrium at the same total thermodynamic water potential. The direction of water diffusion would depend on the initial imbalance between the water potentials of the various cells. Although the direction of water diffusion will be toward decreasing water potential, we still know nothing about how long it will take to reach the new equilibrium. That subject will resurface in section VI of this chapter.

Now we want to consider a group of cells, arranged in a line for convenience, initially at water potential equilibrium. Assume that the cells are isolated from their surroundings but not from each other. Also assume that something occurs in the cell at the end of the line to reduce its water potential. This could be due to a decrease in either the osmotic or turgor potentials. The osmotic potential could be decreased by an influx or production of additional solutes in the cell and the turgor reduction could be the result

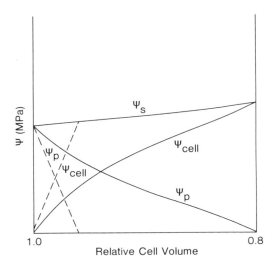

Fig. 8–9. Höfler diagram showing the relationships between the total water potential, osmotic potential, and pressure potential over the range from full turgor to incipient plasmolysis. The dashed lines show the same relationships as they might exist in a cell with a thicker, more rigid cell wall. The signs of ψ_s and ψ_{cell} are negative.

of a weakening or loosening of the cell wall microfibrils. The cause of the reduction is unimportant for present purposes. The result, however, is that a gradient of water potential is established between the cell in question and its neighbor which causes water to diffuse down the gradient into the cell at the end of the line. Obviously the loss of water from the neighbor cell lowers its water potential and a chain reaction is set up in which each cell down the line moves towards a new equilibrium with all its neighbors. The tissue will come to a new equilibrium in which the water potential of the tissue will be somewhat less than the original. The total volume of the system will remain the same since there is no additional water available. If, however, the cells are in contact with a water source, they will move toward equilibrium with the potential of that water. There will be a net influx of water into the tissue, if the potential of the source is sufficiently high, with the result that the cell at the end of the line will increase in size. This example illustrates the fundamental process by which cells grow (see chapter 10 in this book for further discussion of cell and plant growth). Of course there are complicating factors such as more complex tissue geometries, plasmodesmata, cytoplasmic streaming, growth of neighboring cells, and mutual pressures from adjacent cells which must be considered in more exact treatments.

VI. TRANSPORT, DRIVING FORCES, AND RATE COEFFICIENTS

A. Flows and Forces

In general, objects change their state of motion in response to the imposition of a force. Classical mechanics tells us that the force is proportional to the acceleration of the object, the constant of proportionality being the mass ($f = ma$). In the day-to-day world, however, where we commonly encounter frictional forces, the relationship is more accurately expressed as $f = ma + hv$. Because of the presence of friction, the acceleration term is quickly damped out and the continued steady motion of the object requires that

$$f = hv \qquad [8]$$

where v is the velocity and h is some constant of the system related to the frictional forces and frequently referred to as the resistivity. Thus we find a series of empirical physical laws in which the force is proportional to velocity. Ready examples of this type of law are Fick's law of diffusion, Fourier's law of heat conduction, Poiseuille's law of fluid flow, and Ohm's law of electrical conduction. The respective driving forces for these processes are the gradients of concentration, temperature, pressure, and electrical potential along the axis of movement. More commonly, however, these physical laws are expressed with the flows (velocities) as functions of the driving forces so that in general we may write

$$\text{Flux} = \text{coefficient} \times \text{driving force} \qquad [9]$$

where the driving force is the gradient of some potential, the flux is the amount of some physical quantity transported per unit time through a unit of area perpendicular to the direction of transport, and the coefficient is related to the inverse of the frictional forces and in this context is referred to as the conductivity. In the specific examples mentioned, the respective constants are the diffusion coefficient, the thermal conductivity coefficient, a frictional coefficient related to the viscosity of the fluid, and the electrical conductivity. Thus we can write an expression specific to each of these laws:

Diffusion	$J_D = -D \, (\partial C/\partial z)$	[10]
Heat conduction	$J_H = -k_T \, (\partial T/\partial z)$	[11]
Fluid flow	$Q = -C \, (\partial P/\partial z)$	[12]
Electrical current	$J_e = -k \, (\partial E/\partial z)$	[13]

A dimensional analysis of Eq. [12] (Poiseuille's law) shows that Q has the units of $1^3 \, 1^{-2} \, t^{-1}$ and reduces to $1 \, t^{-1}$, which is a velocity with the preferred SI units of m s^{-1}. Similarly, the other flows will reduce to the form of a velocity, although not always in such an obvious manner.

Recently (historically speaking) the formalism of irreversible thermodynamics (IRT) has provided a way to characterize not only the relationships (Eq. [10]–[13]) but many other force-flow relationships that occur in nature. More importantly, perhaps, is that is provides a mechanism for characterizing the interactions between seemingly unrelated forces and flows.

In general, a set of equations, called *Phenomenological equations,* may be written in which each of the flows is described as a linear function of its "conjugate force." If only one flow, driven by its conjugate force, occurs in a system, then the flow may be described as $J = LX,$ where X is the relevant force and L is a generalized conductance, or in the formalism of IRT, a "straight" coefficient (compare to relationships 10–13).

An alternate way of relating the flow and force is by using resistances or generalized frictional coefficients. Thus for the same system we may write $X = RJ$ (note the similarity to Eq. [8]). The R is still called the straight coefficient and the choice of using R or L is strictly a matter of convenience.

Now consider a system where two flows and forces operate, each flow linearly related to its conjugate force. In this case, however, each flow is also linearly related to the nonconjugate force by a "cross coefficient." The system may then be described by the set of two equations:

$$J_1 = L_{11} X_1 + L_{12} X_2$$

$$J_2 = L_{21} X_1 + L_{22} X_2. \qquad [14]$$

Here, L_{11} and L_{22} are the straight coefficients relating J_1 and J_2 to X_1 and X_2, respectively. The interactions between the nonconjugate forces and flows, that is between X_1 and J_2 and X_2 and J_1, are described by the cross coefficients L_{21} and L_{12}, respectively.

The number of equations may be expanded to cover all the forces and flows operating in a system, and the resultant system of equations may be summarized as

$$J_i = \Sigma L_{ik} X_k \tag{15}$$

or as

$$X_i = \Sigma R_{ik} J_k. \tag{16}$$

The choice of whether to use Eq. [15] or [16] is a matter of convenience and it is possible to move between systems of coefficients by the rules of matrix algebra. These rules tell us that

$$R_{ik} = \frac{|L|_{ik}}{|L|} \tag{17}$$

where $|L|$ is the determinant of the matrix of coefficients and $|L|_{ik}$ is the minor of the determinant about L_{ik}.

The exact choice of which flows and forces to use in describing a system can also be arbitrary to a certain extent, with the proviso that the product $J_i X_i$ must have the units of entropy production or decrease in free energy. For a much more extensive treatment see Katchalsky and Curran (1965).

These remarks concerning IRT have been brief. However, the mere recognition of the existence of such formalism allows us to proceed with our discussion of water flow through biological membranes. The formalism of IRT allows the development of an equation relating the total flow of volume (solute + solvent) to both the osmotic and hydrostatic pressure differences across a membrane. To wit

$$J_v = L_p (\Delta P - \sigma \Delta \Pi) \tag{18}$$

Here, J_v is the volume flux in $m^3 \, m^{-2} \, s^{-1}$ [sometimes written as $(1/A)$ (dV/dt)], L_p the hydraulic conductance (the conductance with respect to flow driven by a hydrostatic pressure difference) in $m \, s^{-1} \, MPa^{-1}$, ΔP and $\Delta \Pi$ are the hydrostatic and osmotic pressure differences across the membrane, and σ is the reflection coefficient of Stavermann (1951) which describes the selectivity of the membrane to each solute. It is important to remember that σ is specific to each combination of solute and membrane and that the overall selectivity of any biological membrane represents a summation of the products of the reflection coefficients and the osmotic pressure differences for every solute present ($\Sigma \sigma \Delta \Pi$).

In a sense related to the last comment, we must note that the difference terms in Eq. [18] could just as easily have been written in terms of the osmotic and pressure potentials, noting that the sign of the osmotic potential is opposite to that of the osmotic pressure. Having done so, we might be tempted to conclude that the term inside the parentheses represents the total thermo-dynamic potential difference and that the direction of flow thus should be determined. Unfortunately there is a membrane property included within these parentheses (σ) and so even the direction of water flow is indicated only under special circumstances. In fact, to determine even the direction of flow, as we will see later, several properties of the membrane, as well as the solutes present, must be specified. For present purposes, however, we will assume that the membrane is ideally semipermeable, allowing the passage of no solutes either by convective, diffusive, or any other processes. For quantita-tive purposes we need to be able to place some value on L_p. Fortunately, there are numerous values available in the literature for cell membranes so we will artibrarily pick a feasible value within the range of those available. For many values of published data the reader is referred to: Baker, 1984; Dainty et al., 1974; Milburn, 1979, p. 56; and Steudle et al., 1983. Since most published values fall within the range of 10^{-6} to 10^{-8} m s^{-1} MPa^{-1}, choos-ing a value of 10^{-7} will offend as many people as possible.

Having obtained a value for the hydraulic conductance of biological membranes, it is time to return to the relationships (Eq. [10]-[13]), in par-ticular Poiseuille's law of fluid flow, and compare it to the characteristics of volume flow through a membrane.

Poiseuille's law describes the volumetric flow per unit time through a tube of circular cross section as

$$V = -\frac{\pi r^4}{8\eta}\left(\frac{\partial P}{\partial z}\right),$$ [19]

where r is the radius of the tube (m), η the viscosity of the fluid (Pa s), and ($\partial P/\partial z$) the gradient of pressure through the tube (Pa m^{-1}). We see that in this case the conductance $C = \pi r^4/8\eta$. Compare this to Eq. [12] where the conductivity (conductance per unit cross-sectional area of the tube) $C = r^2/8\eta$. The units of V are m^3 s^{-1}, but those of Q and, incidentally of J_v in equation 18, are m^3 m^{-2} s^{-1}. In both cases (Q and V), however, the coefficient for fluid flow is related to the size of the tube and the viscosity of the fluid.

For use later in the discussion, it is convenient now to calculate a value of C. Using a value of 1.002×10^{-3} Pa s for η and assuming a tube diameter of 30 μm, the size of a typical xylem vessel, C becomes about 1.98×10^{-11} m^4 s^{-1} MPa^{-1}.

Now we are in a better position to discuss the types of water movement that occur in plants and consider the relative magnitudes of each.

B. Convection and Diffusion

There are two fundamental types of water movement of concern in the study of plant water relations—diffusive and convective. The *diffusive flow* of a solution component, including water, may be defined as the movement of that component relative to the average motion of the solution; *convective flow* of that same component is its movement due to the average motion of the solution. It is important to distinguish between these two because the driving forces and flow characteristics can be quite different for each.

The total flux of the water component in an isothermal system is the sum of the diffusive and convective fluxes and may be written

$$F_w = J_w + C_w Q \qquad [20]$$

where C_w is the concentration of water in the solution phase, Q is the volumetric flux of the solution phase, and J_w is the diffusion flux of water (Corey & Klute, 1985).

The diffusive flux of water in Eq. [20] occurs in response to gradients in the water potential, which has osmotic, gravitational, and pressure components. The convective flux, however, occurs because water is a component of the solution phase that moves in response to mechanical forces, i.e., gravitational and pressure gradients.

Distinguishing between these types of flows is important because in the xylem and along cell walls of plants, water moves primarily by convection, whereas diffusive fluxes predominate in cell-to-cell movement across membranes in living tissues. In the transpiration stream both types of flow occur in series.

Earlier we mentioned that water does not always flow down the gradient of total water potential and at this point we will consider the significance and validity of this statement within the context of Eq. [20]. A simple thought experiment will serve to illustrate both the meaning of Eq. [20] and the fact that water does not always flow down the gradient of its own potential.

Imagine that the horizontal tube in Fig. 8–10A is of the dimensions given in the previous section (d = 30 μm) and filled on the right side with a solution of osmotic potential equal to -1.0 MPa and on the left side with pure water. The profile of the osmotic potential is shown above the tube.

If the system were left in such a state there would be a slow net diffusion of water from the left to the right side of the tube, down the gradient of total thermodynamic water potential. If, however, a small pressure gradient, say 0.01 MPa m^{-1}, is applied from the right hand side of the tube, then it is obvious that the solution will move to the left against the gradient of water potential and at a rate, based on the previously determined value of C, of 1.98×10^{-13} m^3 s^{-1}. A moment's consideration will show that the thermodynamic pressure potential which results from the mechanical pressure is insufficient to reduce and reverse the water potential gradient. The important point here is that the mechanical pressure is applied to the entire solution and that the water moves by virtue of being part of the solution. If we establish a frame of reference within the solution, say at the midpoint of the concentration gradient (P), we could still observe a net diffusion of water to the right, down its water potential gradient. However, the entire frame of reference will be moving to the left and the sum of the diffusive and convective fluxes, using the tube as the fixed frame of reference, will

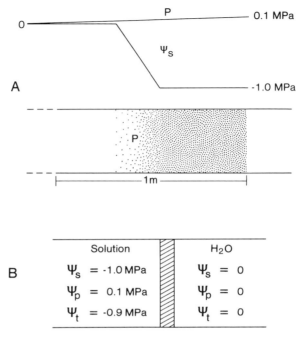

Fig. 8-10. (A) Horizontal tube filled with a solution on the right and interfaced with pure water on the left. Both ends extend to infinity. The osmotic potential (ψ_s) profile and the hydraulic pressure (P) gradient of the solution in the tube are shown above. (B) Simple system of two compartments separated by a porous membrane. The characteristics of the membrane are variable and specified in the text.

show a net movement to the left. Clearly the convective flow completely overwhelms the diffusive flow and from this it is clear that the statement, so often found in text books, that "water always flows down the gradient of its own potential" is true only in a limited context and therefore not true at all.

Are there other circumstances in which water will flow in a direction not indicated by the thermodynamic water potential gradient? To answer this question let us consider another similar experiment shown in Fig. 8-10B. In this case, the diameter of the tube is the same as in Fig. 8-10A; the right compartment is filled with pure water and the solution in the left compartment is separated from the right one by a semipermeable membrane which, for the moment, we will assume is ideal ($\sigma = 1$). In addition, the solution in the left compartment is subjected to a hydrostatic pressure of 0.1 MPa so we may infer that the total thermodynamic water potential difference across the membrane is -1.0 MPa ($\Delta\psi_s$) + 0.1 MPa ($\Delta\Psi_p$) = -0.9 MPa, which indicates that there will be a net movement of water from the right compartment into the left. Indeed, under the present conditions, that is what will happen. We can compare the total flow in this case with that through the open tube by calculating a total conductance for the membrane (L_v). This is done simply by multiplying L_p by the total membrane area, which will be the same as the cross-sectional area of the tube (7.07×10^{-10} m^2), to get

$L_v = 7.07 \times 10^{-17}\,\text{m}^3\,\text{s}^{-1}\,\text{MPa}^{-1}$. With a potential difference of 0.9 MPa across the membrane, the resultant flow will be $6.36 \times 10^{-17}\,\text{m}^3\,\text{s}^{-1}$ at least three orders of magnitude less than the convective flow through a tube of equal diameter and of a similar magnitude to a purely diffusive process. Of course, if the conductance of the tube is reduced sufficiently, by reducing the radius, then the convective flow will approach the order of magnitude of the diffusive flow and the distinction between the two will become blurred.

Now consider the case where the membrane is completely nonselective ($\sigma = 0$). Given that the other initial conditions are the same as in the last case, it is easy to see that the initial osmotic pressure difference can be ignored and the water will be driven by the gradient of pressure against the total water potential gradient. Of course there is an infinity of circumstances for $0 \leq \sigma \leq 1$ for which there may be a net movement of water either with or against the gradient of water potential even though our fixed frame of reference is the membrane and flows are of a diffusive magnitude. There is also one value of σ for which the pressure driven flow in one direction will exactly balance the osmotic pressure-driven flow in the opposite direction. So, even though there is a water potential difference, there is no net movement of water.

It is also easy to demonstrate with this system that equality of water potential in the two compartments, although a necessary condition for equilibrium, is not sufficient to ensure equilibrium. For instance, if the right-hand compartment is filled with a solution of osmotic potential equal to -0.9 MPa, we will have achieved a water potential equality on either side of the membrane. However, for any value of $\sigma < 1$, water will flow from the left compartment to the right down the gradient of pressure. This provides another example in which water does not flow down the gradient of its own potential and where special conditions must be introduced to support the statement that water will not flow if the water potentials are equal.

Reiterating that the reflection coefficient is a property of both the membrane and the solute, we can see that water does not always flow down the gradient of its own potential. "Thus transport may be either with the gradient of thermodynamic potential or against it, depending on the relative magnitudes of the forces causing convection and the thermodynamic gradients, as well as the magnitudes of the pertinent transport coefficients. However, even if the water component is transported by convection in a direction of increasing thermodynamic potential of the water component, the entropy of the system increases, and the free energy of the system . . . decreases." (Corey & Klute, 1985). For a more rigorous discussion of these points and the attendant historical controversy, consult Corey and Klute (1985).

C. Transport Coefficients

In its simplest form a resistance is a coefficient of proportionality relating the force applied to the resultant velocity. Equation [8] is an example, and in this case the inverse of the resistance may be taken as the conductance if one wishes to express the velocity as a function of the force (Eq. [10]–[13]).

If Eq. [8] is graphed it will resemble Fig. 8–11A, a straight line passing through the origin. The resistance h may then be expressed simply as the ratio F/v. The resistance may also be expressed as the first derivative (slope) of the line, dF/dv which will again be equal to h. This rather trivial statement takes on significance later when we touch upon relationships between forces and flows that are not linear or do not pass through the origin.

Throughout the remainder of this discussion we will sometimes talk about conductance and sometimes about resistance. The meaning should be clear from the context and it should also be clear that similar, but inverse, arguments apply to either concept.

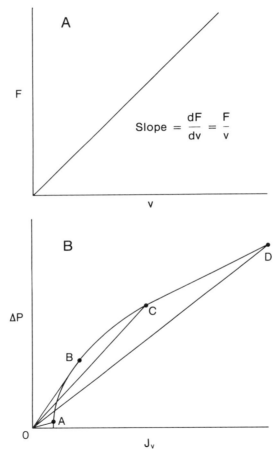

Fig. 8–11. (A) Simple force-velocity function. Either the slope of the line or the ratio F/v at any point may be taken as the resistance. (B) A more complicated force-velocity function showing the difference between defining the resistance as the slope of the curve and the ratio of the applied force to the velocity. The slopes of the lines joining points A-D on the curve with the origin are given by the ratio F/v and are seen to rise and then decline in sequence while the slope of the curve is seen to decline smoothly. Interpretation of the resistance from experimental data could vary greatly depending on which method of calculation is used.

We have already touched on several types of resistances and conductances in this chapter and treated them as constants. Generally speaking, however, they are not constant. For example the conductance of Eq. [19] is a function of the radius of the tube as well as the viscosity of the fluid flowing through it. A more detailed consideration of this conductance will reveal that the viscosity itself is a function of composition and especially temperature. The radius of the tube may vary somewhat with temperature as well as the pressure within the system. The magnitude of these functions will vary with the composition of the tube and the fluid, but for a xylem vessel filled with a dilute aqueous solution, the effect of temperature on the viscosity is the most significant variable. For example, a temperature increase from 20 to 30 °C will decrease the viscosity about 20% with a resultant increase of about 25% in the conductance. Likewise the effect of temperature on L_p in Eq. [18] should be considered since it can be substantial (see section VIII of this chapter on temperature effects). There is also evidence that L_p is a function of the cellular turgor pressure. See the review by Zimmermann (1978) for more discussion on this point.

In the case of Eq. [18] the resistance coefficient R_p may be taken as the inverse of the conductance coefficient L_p. However, other resistance functions may be formed from Eq. [18]. In particular, consider the situation where water flow across a membrane is driven by an imposed pressure difference, such as water exchange between the xylem and adjacent cells. In such cases the osmotic pressure difference becomes a function of J_v as well as being part of the force driving J_v. This is seen easily by examining the three components of solute transport—convective, diffusive, and active. The total solute transport (J_s) may be described (Katchalsky & Curran, 1965, p. 215) as

$$J_s = \overline{C}_s (1 - \sigma)J_v + \omega(\pi_1 - \pi_2) + J_s^* \qquad [21]$$

where σ is the reflection coefficient, $\overline{C}_s$ is the average of the solute concentrations on either side of the membrane, ω the coefficient of solute mobility in the membrane, and J_s^* the component of transport requiring the expenditure of metabolic energy (active transport). The π has its usual meaning and the subscripts refer to the (1) upstream, or high pressure, and (2) downstream compartments which the membrane separates.

If we expand the osmotic term in Eq. [18] and [21] to account for the different ways in which solutes may move across the membrane in a simple two-compartment system, then we obtain a more complex expression describing J_v:

$$J_v^2 (1 + \sigma) + J_v[2\omega RT - L_p\Delta P(1 + \sigma) + 2\sigma^2 L_p\pi_1]$$
$$- 2L_p RT(\omega\Delta P + \sigma J_s^*) = 0. \qquad [22]$$

This function would resemble the curve in Fig. 8–11B. We can then form a resistance function with respect to the pressure difference ($R_{\Delta P}$) by first solving Eq. [22] for ΔP and taking the first derivative with respect to J_v (Fiscus, 1975). Thus

$$R_{\Delta P} = \frac{d\Delta P}{dJ_v} = \frac{1}{L_p} + \frac{2\sigma RT[2\sigma\omega\pi_1 + J_s^*(1 + \sigma)]}{[J_v(1 + \sigma) + 2\omega RT]^2} \,. \qquad [23]$$

Note that the hydraulic conductance coefficient (L_p) forms a part of the expression but there is also a variable term. Note too that the coefficients within the variable term may also be subject to the effects of the physical variables, pressure, and especially temperature (more on temperature effects in section VIII of this chapter).

In the case of the relationship in Fig. 8–11B, however, the ratio $\Delta P/J_v$ is not the same as the "differential resistance" we have just formed. In fact, if we examine the lines O-A, O-B, and O-C, the slopes of which are described as the ratio $\Delta P/J_v$, we might imagine the "resistance" to be low at first, rising to a maximum and then declining asymptotically. The asymptote will approach the first derivative near infinite J_v where the second term in Eq. [23] approaches zero. The point to be made here is that Eq. [23] describing the differential resistance allows insight into how the system operates and offers clues as to how resistance might be affected by physical and physiological factors. Taking the simple force-flux ratio as the resistance may not be very helpful and indeed may be misleading.

We have seen two ways of forming a resistance so far—as a simple ratio and as the slope of the force-flux curve. In the case of linear functions passing through the origin, either one will suffice. However, since taking the slope of the line works for all functions, that would seem to be the preferred method.

There are other terms used for rate coefficients depending on the nature of the system and, in some cases, historical precedence. There are diffusivities, conductivities, resitivities, and permeabilities as well as conductances and resistances. As a general rule, each of these terms has a specific well defined meaning, but sometimes they are used interchangeably without regard to their suitability to the particular circumstances. Students would be well served to pay particular attention to precise usage.

There are other ways in which resistances, or rate coefficients, may be formed and we will touch upon these in other sections as they arise.

VII. TRANSPORT OF WATER

A. Intercellular and Tissue Transport

In section V of this chapter we touched upon the concept of water diffusion through a linear array of cells. It is now time to approach that subject in greater detail.

The first attempts at detailed analysis of water movement through plant tissues were those of Philip (1958a, b). Although it represents a substantial advance in the study of plant water relations, it was incomplete in the sense that it did not allow for significant water flow through the cell wall pathway. Nevertheless, these papers formed the foundation for further development and Molz and Hornberger (1973) extended Philip's theory to include a diffusible solute and later, Molz (1976) further developed the theory to address explicitly the interrelations between symplastic and apoplastic flows. It is this latter analysis which we will discuss for purposes of illustration.

Considering the diagram of several cells in series (Fig. 8–12), it is evident that in moving from the left to the right sides there are several alternate pathways along which water might flow. Specifically, water may flow along the cell walls, through the symplast, or the total flow might be partitioned between the two pathways. The proportions flowing in each pathway would depend on the relative conductances of each part, the distribution of driving forces, and the ease with which water could be exchanged between the two pathways. Water flow in the apoplast was considered by Molz to be driven by a gradient of negative hydrostatic pressure ($\partial \tau / \partial x$) while in the symplast, he considered the difference in the total water potential the driving force. Therefore, the total water flux was divided into a convective term (apoplast) and a diffusive one (symplast).

In his analysis, Molz identified five relevant parameters which control water flow through tissue. These parameters were: the resistance to flow of each pathway separately (R_1 and R_2); the resistance to exchange flow between the pathways (R); and the water storage capacity of each pathway (C_1 and C_2). Therefore, for the propagation of the water potential through the symplasm and the propagation of the hydrostatic pressure through the apoplasm, respectively, he wrote

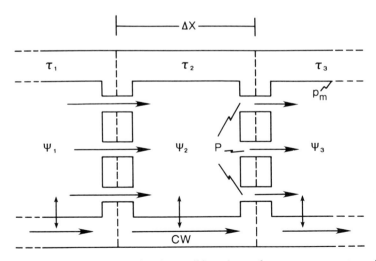

Fig. 8–12. Three cells in series showing the possible pathways for water movement considered in the development of Eq. [24]. From each protoplast the water may move either to the next cell across a plasma membrane (Pm) or through plasmodesmata (P) or exchange with the cell wall (CW) solution across a plasma membrane. In the cell walls the water may exchange with the protoplast or move along the cell walls down a pressure gradient.

$$\frac{\partial \psi}{\partial t} = \frac{(\Delta x)^2}{R_2 C_2} \frac{\partial^2 \psi}{\partial x^2} - \frac{(\psi - \tau)}{R C_2}$$

and

$$\frac{\partial \tau}{\partial t} = \frac{(\Delta x)^2}{R_1 C_1} \frac{\partial^2 \tau}{\partial x^2} + \frac{(\psi - \tau)}{R C_1}. \qquad [24]$$

Each of the resistances is composed of an area term and the inverse of a conductance or permeability. The capacity terms relate the change in potential to a unit change in water content, and in the case of the symplasm the capacity is a function of the elastic modulus of the cell walls.

Using these five parameters, Molz formed a complex coefficient which he called a tissue diffusivity, D (a form of conductance). Thus he could write a single equation describing the transport process through a tissue as

$$\frac{\partial \phi}{\partial t} = D \frac{\partial^2 \phi}{\partial x^2} \qquad [25]$$

where ϕ is the tissue water potential. The utility of such an equation is obvious when we consider the difficulties of measuring all of the parameters necessary to synthesize a value for D (conductances of plasmodesmata and plasmalemmae, cross sectional areas of cells and cell walls, elastic moduli of cells, and the capacity factor of the cell walls). Thus we find another way of forming a rate coefficient and we reveal another important point concerning transport models; that is, the amount of detail necessary in any model is determined by our purpose in formulating the model in the first place. For many purposes an experimental determination of D in Eq. [25] would be adequate. If, however, we are interested in the effects of environmental parameters, growth regulators, or some other external influence on D itself, we would then need to analyze all of the hidden components of D.

This diffusivity is very similar in principle to the resistance function (Eq. [23]) found earlier. In fact, with some adjustment of perspective it is possible to use Eq. [22] and [23] to describe water and solute transport through tissues, as well as across individual membranes. In such a case, we can imagine the tissue as a highly complex macroscopic membrane and the individual transport coefficients no longer represent the properties of individual biological membranes, but the collective effects of numerous parallel and serially arranged membranes and pathways. Detailed interpretation of the various coefficients becomes difficult in terms of individual membrane properties but the general meaning of these "effective" or collective coefficients is consistent with their origins. Thus, the effective reflection coefficient for the tissue describes the ability of that tissue to restrict the passage of solutes flowing with the water; the effective coefficient of solute mobility describes the diffusive flux of solutes through the tissue, analogously to Philip's tissue diffusivity term; the tissue L_p describes the response of the volume flux to hydrostatic pressure gradients; etc.

It is also possible to use kinetic analyses to determine a composite tissue diffusivity by measuring the half-time for water transport after a sudden change in experimental conditions. The general form for the half-time relationship, based on heat transfer theory is

$$t_{1/2} = Sl^2/D \qquad [26]$$

where S is a constant based on the geometry of the tissue and l is the distance the water must move. A similar relationship was used by Boyer (1969) to estimate the resistance to water flow of various parts of the soil-plant water transport system. The review by Boyer (1985) covers both the rationale for Eq. [26] and the effects of tissue capacitances on the transient analysis, and the effects of growth components on water transport. Additional discussion relevant to these particular points and to cell growth in general may be found in the review by Cosgrove (1986).

B. Water Transport Through the Xylem

The largest fraction of water absorbed by the plant root system, however, passes through the xylem to other parts of the plant and to the atmosphere. It is the subject of transport through the xylem that we now wish to address.

Since the xylem of many species is composed more or less of a series of hollow pipes, water flow through it is frequently described by the Poiseuille equation (Eq. [12] and [19]). Because the xylem tissue is composed of numerous elements of varying diameter, Eq. [19] needs to be modified. The appropriate modification is in the conductance term, which becomes $(\pi \Sigma r^4)/8\eta$, where the summation is over all the individual conducting elements present. If there is a wide range of vessel radii, most of the water will be carried by the large vessels. This is easily seen because the r^4 factor in the conductance indicates that a doubling of radius will increase the conductance 16 times.

It has been found experimentally that actual flows, measured under a known pressure, varied from 10 to 100% (Zimmerman & Brown, 1971) of that expected on the basis of Poiseuille assumptions and anatomical measurements of the numbers and diameters of conducting elements.

There are a number of factors that may cause deviations from Poiseuille flow in various plants. Since conifers have no vessels, the xylem solution must pass through a pit field at the end of each trachied. These pits, although very permeable to solution, can offer a relatively high resistance to flow compared to the lumens of the tracheid. Poiseuille expectations are usually based on anatomical measurements of the tracheary lumena and the deviations should not be surprising.

Angiosperms, with their vessel elements stuck end to end to form relatively lengthy vessels, might be expected to approach more closely Poiseuille behavior. This is not necessarily the case since the vessels are of finite length and individual vessels do not normally extend throughout the plant. Where successive vessels join, the xylem solution must pass through pit membranes just as they do in conifers. The degree of overlap of successive vessels and the density of pits in the overlapping walls will affect the conductance between vessels in a direct manner.

In addition, the perforation plates between vessel elements can cause deviations from Poiseuille flow if they are not simple and more-or-less completely open. One might anticipate that even the effects of scalariform or reticulate perforation plates might be negligible compared to the effect of the end walls. This relationship too, is uncertain since the relative resistances of the perforation plates vs. the end walls will depend on the number of vessel elements (the number of perforation plates) per vessel and, of course, the conductance and extent of overlap of the end walls.

With so many possible factors causing reductions in the Poiseuille expectations, it is clear that any predictions of xylem conductance based on counts and diameter measurements alone must represent an upper limit to our expectations. The anatomical information necessary to accurately predict xylem water flow can be quite complex and detailed, and for physiological purposes it would usually be more convenient to make some sort of direct physical measurement.

C. Ohm's Law Analogy

Before leaving this general discussion of water transport, it may be useful to express some views concerning the practice of conceptualizing plant water transport phenomena in terms of the so-called Ohm's law analogy. Although use of the analogy historically has been credited with stimulating plant physiologists to think in terms forces, fluxes, and resistances, we really need to examine some of the assumptions and shortcomings of the analogy with the view toward encouraging a more forthright approach to plant water transport theory, both for teaching and research.

The analogy is generally formulated such that the electrical current is the analog of water movement in a plant system, the electro-motive force (EMF) is taken as the water potential difference between two points in the system, and of course the electrical resistance is taken as analogous to the fluid resistance. At the outset it is useful to examine the reasons for communicating by analogy. First, the analogy should somehow be easier to understand and/or more complete than the proper theory, and next, we must assume that the student or other researcher is more familiar with the properties of the analog than with the proper theory. We have seen that most basic transport phenomena can be described by a limited number of relationships (Eq. [10]–[13]) which are very similar in form and of which both Ohm's law and the fluid transport laws are but specific examples. It is therefore difficult to imagine that any of the relationships is easier or more difficult to understand, or that any one of them contains some mysterious feature that makes all the others easier to understand. In short, we would not expect an electrical engineer to teach simple direct current theory by analogy with Poiseuille's law, although he might mention in passing that there is a similarity in the basic transport relationships. We would expect him or her to use the formulations of transport phenomena appropriate to his or her own subject matter. We should do the same, not simply because we can avoid the necessity of teaching additional subject matter that is not directly related to our concerns, but because most analogies, including the one at hand, are not complete. For instance, if we deal with diffusive processes as being the current analog, and the gradient of water potential the EMF analog, what is the logical and useful analog for convective flow and the force driving that flow? Finally, we see no reason to assume greater knowledge of electrical theory among plant physiology students and researchers than any of the other transport phenomena mentioned.

VIII. TEMPERATURE EFFECTS ON WATER TRANSPORT

A. Characterization of Temperature Responses

There are two common and convenient ways of characterizing the temperature response of a process, by determining the Q_{10} and/or the activation energy, E^a, of the process. Both of these characteristics of a process may be determined by measuring the rate of the process at two different temperatures.

The Q_{10} is simply the ratio of the rates of a process over a 10 K temperature difference. A Q_{10} of 2, for example indicates that the rate of the process doubles when measured at some $(T + 10)$ K as compared to the rate at T. Since it is not always convenient to measure processes at 10 K intervals we may calculate the Q_{10} from measurements at any two temperatures by

$$Q_{10} = \left(\frac{k_2}{k_1}\right)^{\frac{10}{T_2 - T_1}} \qquad [27]$$

where k_1 and k_2 are the rates at the lower (T_1) and higher (T_2) temperatures, respectively.

As a rule of thumb we may say that processes for which $Q_{10} < 1.5$ are limited by purely physical phenomena such as diffusion or Poiseuille flow, while a higher Q_{10} might indicate the involvement of some biological phenomenon (e.g., an enzyme system or living membrane barrier) that is limiting the process.

The activation energy (E^a) of a process is involved in the Arrhenius rate equation:

$$k = B \exp\left(-E^a/RT\right) \qquad [28]$$

where k may represent either a rate constant or a rate and B is a constant sometimes referred to as the frequency factor. Taking the natural log of both sides leads to the result

$$\ln k = \ln B - E^a/RT \qquad [29]$$

which gives ln k as a linear function of $1/T$, with slope equal to $-E^a/R$. The energy of activation of a process may therefore be determined from an experimental plot of the temperature response (ln k vs. $1/T$), which is called an Arrhenius plot. In the case where the system is complex or ill defined and there is doubt as to the limiting process, the term *temperature characteristic* should be used instead of the more precise *activation energy*.

Integrating Eq. [29], assuming B and E^a are constant over the temperature range of interest, yields

$$ln\left(\frac{k_2}{k_1}\right) = \frac{E^a}{R}\left(\frac{1}{T_1} - \frac{1}{T_2}\right), \qquad [30]$$

Also, taking the natural log of Eq. [27] for the Q_{10} and substituting for $ln(k_2/k_1)$ gives the relationship between Q_{10} and E^a as

$$E^a = \frac{RT_1T_2 \ln Q_{10}}{10}. \qquad [31]$$

B. Temperature Responses of Plant-Water Transport

Just as there are two major compartments for water in the plant, the symplast and apoplast, there are two major types of transport which we need to consider in terms of the temperature response, movement in the apoplast and movement between the apoplast and the symplast. Movement in the apoplast occurs in response to pressure gradients while movement between the compartments is membrane limited and can involve osmotic as well as pressure gradients.

Relatively little is known about the effect of temperature on water flow from cell to cell through the symplast except that in certain chilling-sensitive species, cytoplasmic streaming is greatly reduced below a critical temperature. If cytoplasmic streaming significantly influences water movement in the symplast, then we might expect movement to be greatly curtailed below the critical temperature. We will discuss this further in reference to the effects of temperature on membrane-limited water transport processes, but first let us look at the possible temperature responses of apoplastic flow.

In a continuous liquid phase system it is possible to demonstrate the phenomenon of *thermal diffusion*. Thermal diffusion is driven by a temperature difference between two parts of a system and if the temperature gradient can be maintained, at the steady state, a steady concentration gradient may also be detected. Since large temperature differences occur in different parts of a plant, the question arises as to the effectiveness of that temperature difference in driving water flow. Fortunately for the plant, the direct effects of temperature on apoplastic flow appear to be small. The reason for this is that the coefficient for thermal diffusion normally is in the range of 0.01 to 0.001 times the ordinary diffusion coefficient (Katchalsky

& Curran, 1965, p. 185) in the same medium. Therefore, any gradients built up by a massive temperature gradient would be rapidly dissipated by ordinary diffusion and it is unlikely that temperature gradients will have much direct influence on water transport in a continuous liquid phase.

There is also the problem of *thermal osmosis,* which is the same as thermal diffusion except that it takes place in a discontinuous medium (i.e., across a membrane). In this case, it is possible to establish a concentration gradient across a membrane by imposing a temperature gradient across the system. Here again, though, the importance of this effect on water movement in biological systems is likely to be small. Unless the thermal properties of the membrane are considerably different from other solids or liquids, it would be very difficult to maintain significant temperature differences across them, owing primarily to their thickness.

Based on Spanner's estimates of the heat of transfer for plant cell membranes (Spanner, 1954) we can make the following calculations, similar to those of Spanner, of thermal osmosis through a sheet of rectangular parallelepipedal cells. If it is assumed that each cell has a length of 50 μm, the cell membrane has a thickness of 10 nm, and that we are dealing with a linear temperature gradient along the long axes of the cells of 10 K cm^{-1}, we can estimate the pressure difference due to thermal osmosis across the tissue. Given the dimensions and the gradient we can easily calculate a temperature difference across each membrane (1000 K m^{-1}/1 $\times$ 10^{-8} m membrane^{-1}) of 1 $\times$ 10^{-5} K. Taking Spanner's value of 13.2 MPa K^{-1}, we find that thermal osmosis in this case will result in a pressure difference of only 0.00026 MPa across each membrane. Further, to maintain a ΔP across the membrane of 0.1 MPa through thermal osmosis would require a ΔT of 0.008 K or a gradient of 8000 K cm^{-1}.

Indirect effects of temperature in the apoplast will be confined primarily to resultant changes in viscosity. Indeed, this may also be the case for some types of membrane-limited transport at ordinary room temperatures.

Numerous studies have been done on the effects of temperature on water uptake by root systems, and these will form the basis of the remainder of this discussion (see Kramer, 1983, p. 243–244 and p. 254–259 for further details and references). It has been shown that there are substantial differences in the response to low temperatures between chilling-tolerant (cool season) and chilling-intolerant (warm season) species. Generally, the warm season species show a much greater decrease in water uptake at low temperatures than do the cool season species. Frequently, an abrupt change in slope, at around 12 to 15 °C, occurs in an Arrhenius plot (Fig. 8–13) and is characteristic of the warm season types. The cool season species, however, show the characteristic break at much lower temperatures, or frequently, not at all. They do exhibit an Arrhenius slope which is typically intermediate between the high temperature slope and the low temperature slope of the warm season types.

The break point, at the critical temperature (T_c), in the Arrhenius slopes of warm season types may be shifted to lower temperatures, or eliminated entirely, through a process of acclimation which consists simply of growing the plants at reduced temperatures (Kuiper, 1964; Markhart et al., 1979a). In cases where it has been examined, this acclimation is accompanied by increases in the degree of unsaturation of the root tissue fatty acids (Markhart et al., 1980). A similar change in response may also be induced by the external application of abscisic acid (Markhart et al., 1979b).

Interpretation of these experiments usually involves the concept of a phase transition of the phospholipid component of cellular membranes. Similar phase transitions have been studied in purely physical systems (e.g., Träuble & Haynes, 1971) which seem to bear a close resemblance to biological structures. Indeed, these temperature response studies provide some of the strongest evidence that water uptake in the root is membrane limited. However, before inferring too much about biological membranes from water transport experiments, consult the discussions by Lyons (1973) and Lyons

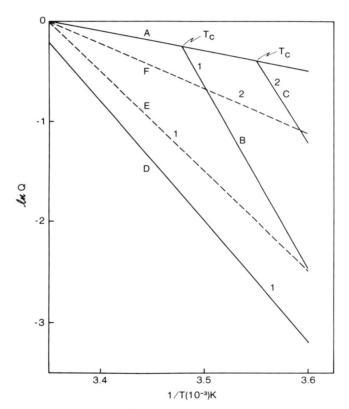

Fig. 8-13. Arrhenius plots of water flow through soybean root systems. The breakpoint in the curves for soybean can be shifted to a lower temperature by lowering the growth temperature of the plants (curve C compared to curve B). Application of abscisic acid results in elimination of the breakpoints and a slope intermediate between the normal high and low temperature regions (curve D). Cool season species exhibit slopes similar to the ABA treatment when grown at higher temperatures (curve E) but are much less sensitive when grown at low temperatures (curve F). The numbers on the curves refer to the day/night growth temperatures; 1 = 28/23, 2 = 17/11. (Figure redrawn after Markhart et al., 1979a, b).

et al. (1979) which cast doubt on the idea that "bulk melts" of membrane lipids occur in the range of temperatures where rate discontinuities are usually observed in root water transport experiments. Indeed, some workers question the validity of using the Arrhenius approach for interpreting many temperature responses (e.g., Wolfe & Bagnall, 1980) but there can be little doubt that the temperature data of Markhart et al. (1979a), just mentioned, and others, consist of straight line segments with distinct break points.

Above T_c, the cell membranes are thought to exist in a more fluid, more porous state which presents a relatively low resistance to water flow. At T_c a phenomenon similar to freezing of the phospholipid component is thought to take place, resulting in a more rigid, less porous membrane below T_c.

An examination of the Arrhenius slopes or of the $Q_{10}s$ for these experiments tend to bear out this interpretation. In most of the cases studied, the Q_{10} above T_c is about 1.2 to 1.4, which coincides well with the changes in viscosity expected over the same temperature ranges. The temperature response above T_c, in those species exhibiting a sharp temperature break, appears to be limited by changes in solution viscosity—a fact that argues for some sort of water filled pores in cell membranes. Below T_c the Q_{10} may range from 4 to 13 or more. This certainly suggests that at lower temperatures the water must traverse a much higher energy barrier, such as would be typical of a more rigid membrane without identifiable porous structures.

There are, however, other factors to consider in the interpretation of the temperature characteristics of water transport in plants, especially in roots. Dalton and Gardner (1978), for example, working with the data of Kramer (1940), concluded that the temperature characteristic curve for live roots could be explained best as a result of the temperature response of the solute uptake system of the roots. Dead roots, on the other hand, showed temperature responses that could be explained simply by viscosity changes, similar to the data just discussed for temperatures above T_c. Reconciling these two points of view may be easier than it first appears if we examine the relative flow rates through the two systems under consideration in terms of the implications of Eq. [22] and [23]. These two equations suggest that at low relative flows, the differential resistance, and therefore the flow rate, is controlled by the solute transport coefficients, consistent with Dalton's interpretation of Kramer's data, which would fit at the low end of the flow scale. However, at high flow rates, such as occurred in the experiments of Markhart et al. (1979a, b), the system flow was limited by L_p and so the temperature characteristic above T_c can be explained in terms of viscosity changes. There is currently not enough data below the critical temperatures of various species to allow a consistent evaluation of the transport coefficients in Eq. [22]. However, it seems reasonable to suppose that any radical change in membrane configuration at T_c would affect both L_p and the solute transport coefficients. Certainly, below T_c, the Q_{10} is much too high to be explained by viscosity changes. The same is generally true for the cool season species (line E and F in Fig. 8–13) which do not exhibit the break in the Arrhenius curves but have slopes too steep to be explained by viscosity changes alone.

IX. EFFECTS OF SYSTEM DISTURBANCES

Thus far we have been discussing the characteristics of solutions and flows, and the systems through which those solutions flow as though the systems themselves were more or less fixed in their dimensions. We would now like to discuss briefly some of the effects of disturbing the transport system.

Because the balance of the parts of the water transport system occurs through feedback mechanisms (see section III.C of this chapter), it is clear that any disturbance that alters the size of plant parts or the resistance to water flow along the water transport pathway will alter the water transport characteristics of the plant, perhaps drastically and fatally. Generally, disturbances from insects, diseases, or physical agents have one of three effects: (i) reduction of leaf area, (ii) reduction of root absorbing surface, or (iii) blockage of the xylem transport system. The effects may be direct, including, for example, defoliation by insects, xylem emboli caused by stress-induced cavitation, or direct attack of fine roots by fungus. Or, the effects may be indirect, such as a phloem disorder which results in root death because carbohydrate cannot reach the root system.

The impact of the disturbance depends on the portion of the water transport system that is altered and on the timing and duration of the disorder. Defoliation, whether by insects, disease, grazing, or even by physical factors such as hail or wind, reduces transpiration roughly in proportion to the amount of leaf area removed. The actual reduction depends upon a number of factors, including the importance of aerodynamic effects on water vapor loss (Kaufmann & Kelliher, 1989), transpiration rate of the remaining foliage, etc. Defoliation does not result in plant water stress, and in fact may decrease it, because the rest of the transport pathway becomes oversized relative to the remaining leaf area.

The long-term effects of defoliation depend on the type of plant. For trees, defoliation may result in reduced photosynthesis and availability of carbohydrate for growth, but stored carbohydrate may be adequate for new leaf production if photoperiod and other factors are favorable. For example, sycamore (*Platanus occidentalis* L.) trees in some regions are attacked by *Gnomonia venata* (Sace. Speg.) Kleb, causing anthracnose, which is characterized by loss of leaves early in the growing season (Boyce, 1961). A second flush of leaves results in fairly normal tree behavior the balance of the season, except that growth is reduced. In annual crops, defoliation-may have far more serious effects on the plant, but the effects occur largely through factors other than water stress, such as nonexistence or destruction of primordia for new leaf development, time limitations, or an inadequate supply of stored food reserves to support new foliage production.

Loss of part of the absorbing root surface or blockage of the vascular system may have rapid effects and far more serious consequences on plant water relations than loss of foliage. A reduction in the root system lowers the root capacity to supply water to the shoot. Similarly, partial blockage of the xylem limits the capacity of the vascular network to transport water to the foliage. If the reduction in roots or xylem transport tissue occurs too rapidly for a timely adjustment of leaf area through normal feedback mechanisms, excessive water stress and even death may occur. Furthermore, if the necessary reduction in foliage size is too large, even if it occurs gradually, it may leave a plant in a hopeless situation in which photosynthesis is too low for recovery. Thus, it is much more difficult for a plant to recover from a significant loss of absorbing or transporting tissue than from loss of foliage.

APPENDIX

Thermodynamic systems may be characterized by system variables. The more intuitively recognizable of these would be the system temperature (T), pressure (P), and volume (V). Less intuitive, perhaps, but equally legitimate system variables are the internal energy of the system (U) and its entropy (S). These state variables may be combined in various ways to form new state variables. The combinations cannot be performed randomly but must conform to a mathematical system called Legendre transforms, which for present purposes will remain mysterious. Our only concern at this time is with some of the results of the process, to wit

$$H = U + PV$$
$$A = U - TS$$
$$G = U + PV - TS.$$

The first of these (H) is called the enthalpy of the system. The second (A), is known as the Helmholtz function, the Helmholtz free energy, or the Arbeit function. Finally, the last (G), and most relevant for our purposes, is the Gibb's function, also called the Gibb's free energy, or sometimes the free enthalpy of the system. All of these functions are for systems of constant composition, a rarity in the biological world where the composition of a system or phase may change with time, or where there are compositional differences between phases. Therefore, if we are to make good use of the Gibb's free energy function, it is necessary to look at it more closely with an eye toward including the necessary compositional variables.

We may take the differential of G as

$$dG = dU + PdV + VdP - TdS - SdT. \qquad [32]$$

The differential of the internal energy (U) may be expanded by considering a statement of the first law of thermodynamics

$$dU = \delta Q - \delta W \qquad [33]$$

which simply states that any change in the internal energy of a system is the sum of the heat transferred to the system (δQ) and the work done by the system (δW) during the transition from one state to another. The different differential symbols used for Q and W indicate that these are *inexact differentials*. Unlike *exact differentials,* which yield a finite difference upon integration, the result of integrating an inexact differential is a finite quantity. The value of the integral for an inexact differential depends not only on the end points of the process but on the path followed between equilibrium states. Thus, different quantities of heat and work may be extracted during a process depending on how that process is carried out.

Just as fundamental as the first law statement is the definition of the entropy, which when rearranged to

$$\delta Q = TdS \tag{34}$$

will be used shortly to replace δQ in Eq. [33]. The work term in Eq. [33] may be expanded to include many possible types of work:

$$\delta W = PdV - fdl - \psi de - \Sigma \mu_i dn_i + \ldots \tag{35}$$

which may be summarized as

$$\delta W = \Sigma Y_i dX_i \tag{36}$$

where the Y_i represents a generalized force, which may also be referred to as a potential or intensity factor. The dX_i then are generalized displacements, extensive, or capacity factors. The important point is that any type of work may be added to the summation as long as the force-displacement pair is properly selected.

Specifically, in Eq. [35], PdV represents the work of expansion where P is the potential and dV the displacement; fdl is mechanical work and the potential might be exerted by a spring, a gravitational field, or even a contractile fiber, and the displacement is movement through space; ψde represents electrical work where a quantity of electricity is moved under the influence of an electrical potential; and the last term represents chemical work. For example, if dn moles of substance i are transported to the surroundings, chemical work equal to $\mu_i dn_i$ is performed and the intensity factor is called the chemical potential of the ith component. Thus μ_i represents the ability to do work.

Substituting Eq. [34] and [35] into the first law statement (Eq. [33]) yields

$$dU = TdS - PdV + fdl + \psi de + \Sigma \mu_i dn_i + \ldots \tag{37}$$

which is known as the Gibb's equation.

From now on we will ignore the mechanical and electrical work terms, because they are not relevant at this time, and confine our considerations to one chemical species. Then, substituting the remaining parts of the Gibb's equation into the differential of the Gibb's free energy function (Eq. [32]) gives the relatively simple relationship

$$dG = VdP - SdT + \mu_i dn_i. \tag{38}$$

The differential of the Gibb's free energy may also be written in terms of partial differential coefficients as

$$dG = \left(\frac{\partial G}{\partial P}\right)_{T, n_i} dP + \left(\frac{\partial G}{\partial T}\right)_{P, n_i} dT + \left(\frac{\partial G}{\partial n_i}\right)_{T, P} dn_i \tag{39}$$

Comparing Eq. [38] and [39] yields the relationships

$$\left(\frac{\partial G}{\partial P}\right)_{T, n_i} = V; \left(\frac{\partial G}{\partial T}\right)_{P, n_i} = -S; \text{ and } \left(\frac{\partial G}{\partial n_i}\right)_{T, P} = \mu_i \tag{40}$$

the last term of which is the chemical potential of species i in terms of the Gibb's free energy function. Other less familiar, but equally valid chemical potentials may be formed from $U = U(S, V, n_i)$ or $A = A(T, V, n_i)$ by similar reasoning. The result would be the other two chemical potentials

$$\left(\frac{\partial U}{\partial n_i}\right)_{S, V} = \mu_i \text{ and } \left(\frac{\partial A}{\partial n_i}\right)_{T, V} = \mu_i.$$

The importance of the Gibb's chemical potential lies in the fact that in systems at constant T and P, it is the difference in chemical potential between two phases (or regions of a single phase) that determines the direction in which a substance will spontaneously diffuse.

To make good use of the concept of chemical potential requires that we know how it is affected by the common physical variables T and P, as well as by compositional variables such as the mole fraction of i in the system. We may start by writing the total differential of μ as a function of T, P, and composition n_i

$$d\mu_i = \left(\frac{\partial \mu_i}{\partial T}\right)_{P, n_i} dT + \left(\frac{\partial \mu_i}{\partial P}\right)_{T, n_i} dP + \left(\frac{\partial \mu_i}{\partial n_i}\right)_{T, P} dn_i \qquad [41]$$

where n_i is the number of moles of i in the system. Since $\mu_i = (\partial G/\partial n_i)_{T, P}$ we might just as well write for the first partial differential coefficient

$$\left(\frac{\partial \mu_i}{\partial T}\right)_{P, n_i} = \frac{\partial}{\partial T}\left(\frac{\partial G}{\partial n_i}\right)_{T, P}$$

and because of a mathematical rule (Cauchy's rule) which states that the order of differentiation doesn't matter, we may also write this as

$$\left(\frac{\partial \mu_i}{\partial T}\right)_{P, n_i} = \frac{\partial}{\partial n_i}\left(\frac{\partial G}{\partial T}\right)_{P, n_i}.$$

From the relations of Eq. [39] we see that $(\partial G/\partial T)_{P, n_i} = -S$, so that

$$\left(\frac{\partial \mu_i}{\partial T}\right)_{P, n_i} = -\left(\frac{\partial S}{\partial n_i}\right)_{P, n_i} = -\bar{S}_i.$$

The special symbol $\bar{S}_i$ is used to indicate what is called the *partial molar entropy*. All partial molar quantities are defined in this way—as the compositional differential of some state function.

By following exactly the same process that gave us $\bar{S}_i$ we can also define a partial molar volume

$$\left(\frac{\partial \mu_i}{\partial P}\right)_{T, n_i} = \left(\frac{\partial V}{\partial n_i}\right)_{T, n_i} = \bar{V}_i$$

and rewrite Eq. [41] as

$$d\mu_i = \bar{V}_i \, dP - \bar{S}_i \, dT + \left(\frac{\partial \mu_i}{\partial n_i}\right)_{T, P} dn_i. \qquad [42]$$

We should also note that according to the relationships of Eq. [40], μ_i is the partial molar Gibb's free energy, usually indicated by $\bar{G}_i$. It is also the intensive factor in the chemical work term of Eq. [35] and, as such, represents the ability of species i to do some type of chemical work.

Sometimes it is more convenient to use the mole fraction of a substance (N_i) instead of the mole number (n_i) defined such that

$$N_i = \frac{n_i}{n_i + \Sigma n_j}$$

where the summation represents all the other chemical species in the system. In this case the last term in Eq. [42] will be written as

$$\left(\frac{\partial \mu_i}{\partial N_i}\right)_{T, P} dN_i.$$

Since our main interest is in characterizing the water component of a solution or biological system, we will change the subscript of the component i to w for the rest of this discussion.

Integration of Eq. [42] will yield a difference between the chemical potential of the water in a system and in some reference system. Generally the reference state is taken as pure free water at the same temperature and elevation as the solution in question. The result of the integration may then be written as $\mu_w - \mu_w^o$ and is a measure of the capacity of the water in the solution to do work against water in the reference state. Because of the importance of this potential in describing the direction of spontaneous diffusive flow, the difference which has the units of J mol^{-1} is divided by $\bar{V}_w$ with the units of m^3 mol^{-1} to put it into the more familiar terms of pressure (Pa). The quotient is then given the special name *water potential* and usually is designated by the Greek letter Ψ.

We now consider a simple isothermal liquid-vapor equilibrium system of constant composition. We recall from the ideal gas law that

$$P_w \bar{V}_w = RT$$

where P_w is the partial pressure of the water vapor in the gas phase. Rearranging and substituting for $\bar{V}_w$ in Eq. [42] we find

$$d\mu_w = (RT/P_w) \, dP_w.$$

Integrating between μ_w^o and μ_w on the left and between P_w on the right leads us to

$$\mu_w - \mu_w^o = RT \ln (P_w/P_w^o)$$

which expresses the very important result that the chemical potential of water, or of any component in the system, can be expressed in terms of its partial gas pressure. Dividing both sides by V_w then defines the water potential in terms of the relative vapor pressure of the water.

In addition, Raoult's law states that the partial pressure of a component is proportional to the mole fraction of that component in the liquid phase. The proportionality constant is the vapor pressure of the pure substance so that $P_w/P_w^o = N_w$, and

$$\mu_w - \mu_w^o = RT \ln N_w.$$

Nonideal solutions require the use of activities rather than concentrations or mole fractions so we now have three expressions relating the water potential to the vapor pressure and dissolved substances in a solution:

$$\Psi_w = (RT/\overline{V}_w) \ln(e/e^o)$$

$$\Psi_w = (RT/\overline{V}_w) \ln N_w$$

$$\Psi_w = (RT/\overline{V}_w) \ln a_w.$$

In this case e/e^o, the relative humidity of the vapor phase, is used. The importance of these last three expressions is that they allow measurement of the water potential in terms of parameters that are more easily measured (humidity, concentration, and activity).

REFERENCES

Baker, D.A. 1984. Water relations. p. 296–318. *In* M.B. Wilkins (ed.) Advanced plant physiology. Pitman Publ., Marshfield, MA.

Bernstein, L. 1971. Method for determining solutes in the cell walls of leaves. Plant Physiol. 48:361–365.

Boyce, J.S. 1961. Forest pathology. McGraw-Hill, New York.

Boyer, J.S. 1969. Free energy transfer in plants. Science 163:1219–1220.

Boyer, J.S. 1985. Water transport. Annu. Rev. Physiol. 36:473–516.

Castellan, G.W. 1971. Physical chemistry (2nd ed.) Addison-Wesley, Reading, MA.

Clarkson, D.T. 1984. Ionic relations. p. 319–327. *In* M.B. Wilkins (ed.) Advanced plant physiology. Pitman Publ., Marshfield, MA.

Corey, A.T., and A. Klute. 1985. Application of the potential concept to soil water equilibrium and transport. Soil Sci. Soc. Am. J. 48:3–11.

Cosgrove, D.J., and R.E. Cleland. 1983a. Solutes in the free space of growing stem tissue. Plant Physiol. 72:326–331.

Cosgrove, D.J., and R.E. Cleland. 1983b. Osmotic properties of pea internodes in relation to growth and auxin action. Plant Physiol. 72:332–338.

Cosgrove. D.J. 1986. Biophysical control of plant cell growth. Annu. Rev. Plant Physiol. 37:337–405.

Dainty, J., H. Vinters, and M.T. Tyree. 1974. A study of transcellular osmosis and the kinetics of swelling and shrinking in cells of *Chara corallina*. p. 59–63. *In* U. Zimmerman and J. Dainty (ed.) Membrane transport in plants. Springer-Verlag New York, New York.

Dalton, F.N., and W.R. Gradner. 1978. Temperature dependence of water uptake by plant roots. Agron. J. 70:404–406.

Fiscus, E.L. 1975. The interaction between osmotic and pressure-induced water flow in plant roots. Plant Physiol. 55:917–922.

Fiscus, E.L. 1977. Determination of hydraulic and osmotic properties of soybean root systems. Plant Physiol. 59:1013–1020.

Fiscus, E.L. 1981. Analysis of the components of area growth of bean root systems. Crop Sci. 21:909–913.

Fiscus, E.L., L.R. Parsons, and R.S. Alberte. 1973. Phyllotaxy and water relations in tobacco. Planta 112:285–292.

Höfler, K. 1920. Ein Scheme für die osmotische Leistung der Pflanzenzellen. Ber. Dt. Bot. Ges. 38:288–298.

Katchalsky, A., and P.F. Curran. 1965. Nonequilibrium thermodyanmics in biophysics. Harvard Univ. Press, Cambridge, MA.

Kaufmann, M.R. 1976. Water transport through plants: Current perspectives. p. 313–327. In I.F. Wardlaw and J.B. Passioura (ed.) Transport and transfer processes in plants. Academic Press, New York.

Kaufmann, M.R., and E.L. Fiscus. 1984. Water transport through plants—internal integration of edaphic and atmospheric effects. Acta Hortic. 171:83–93.

Kaufmann, M.R., and F.M. Kelliher. 1989. Estimating tree transpiration rates in forest stands. In J.P. Lassoie and T.M. Hinckley (ed.) Techniques and approaches in forest tree ecophysiology. CRC Press, Boca Raton, FL. (In press).

Kramer, P.J. 1940. Root resistance as a cause of decreased water absorption in plants at low temperatures. Plant Physiol. 15:63–79.

Kramer, P.J. 1983. Water relations of plants. Academic Press, New York.

Kramer, P.J., and T.T. Kozlowski. 1979. Physiology of woody plants. Academic Press, New York.

Kuiper, P.J.C. 1964. Water uptake of higher plants as affected by root temperature. Meded Landbouwhoge School, Wageningen. 64:1–11.

Kursanov, A.L., and M.I. Brovchenko. 1970. Sugars in the free space of leaf plates: Their origin and possible involvement in transport. Can. J. Bot. 48:1243–1250.

Larson, P.R. 1980. Interrelations between phyllotaxis, leaf development and the primary-secondary vascular transition in Populus deltoides. Ann. Bot (London) 46:757–769.

Lyons, J.M. 1973. Chilling injury in plants. Annu. Rev. Plant Physiol. 24:445–466.

Lyons, J.M., J.K. Raison, and P.L. Steponkus. 1979. The plant membrane in response to low temperature: An overview. p. 1–24. In J.M. Lyons et al. (ed.) Low temperature stress in crop plants: The role of the membrane. Academic Press, New York.

Markhart, A.H. III, E.L. Fiscus, A.W. Naylor, and P.J. Kramer. 1979a. Effect of temperature on water and ion transport in soybean and broccoli systems. Plant Physiol. 64:83–87.

Markhart, A.H. III, E.L. Fiscus, A.W. Naylor, and P.J. Kramer. 1979b. Effect of abscisic acid on root hydraulic conductivity. Plant Physiol. 64:611–614.

Markhart, A.H. III, M.M. Peet, N. Sionit, and P.J. Kramer. 1980. Low temperature acclimation of root fatty acid composition, leaf water potential, gas exchange and growth of soybean seedlings. Plant Cell Environ. 3:435–441.

Meidner, H., and D.W. Sheriff. 1976. Water and Plants. John Wiley & Sons, New York.

Milburn, J.A. 1979. Water flow in plants. Longman Press, London.

Molz, F.J., and G.M. Hornberger. 1973. Water transport through plant tissues in the presence of a diffusible solute. Soil Sci. Am. Proc. 37:833–837.

Molz, F.J. 1976. Water transport through plant tissue: The apoplasm and symplasm pathways. J. Theor. Biol. 59:277–292.

Molz, F.J., and J.M. Ferrier. 1982. Mathematical treatment of water movement in plant cells and tissue: A review. Plant Cell Environ. 5:191–206.

Nobel, P.S. 1983. Biophysical plant physiology and ecology. W.H. Freeman & Co., New York.

Palta, J.P., and O.Y. Lee-Stadelmann. 1983. Vacuolated plant cells as ideal osmometer: Reversibility and limits of plasmolysis, and estimation of protoplasm volume in control and water-stress-tolerant cells. Plant, Cell Environ. 6:601–610.

Philip, J.R. 1958a. The osmotic cell, solute diffusibility, and the plant water economy. Plant Physiol. 33:264–271.

Philip, J.R. 1958b. Propagation of turgor and other properties through cell aggregations. Plant Physiol. 33:271–274.

Richter, H. 1978. A diagram for the description of water relations in plant cells and organs. J. Exp. Bot. 29:1197–1203.

Schulze, E.-D., J. Cermak, R. Matyssek, M. Penka, R. Zimmerman, and F. Vasicek. 1985. Canopy transpiration and water fluxes in the xylem of the trank of Larix and Picea trees—a comparison of xylem flow, porometer and cuvette measurements. Oecologia 66:475–483.

Singer, S.J. 1974. The molecular organization of membranes. Annu. Rev. Biochem. 43:805–833.

Slatyer, R.O. 1967. Plant-water relationships. Academic Press, London.

Spanner, D.C. 1954. The active transport of water under temperature gradients. Symp. Soc. Exp. Biol. 8:76–93.

Stavermann, A.J. 1951. The theory of measurement of osmotic pressure. Rev. Trav. Chim. 70:344–352.

Stein, W.D. 1967. The movement of molecules across cell membranes. Academic Press, New York.

Steudle, E., S.D. Tyreman, and S. Wendler. 1983. Water relations of plant cells. p. 95–109. *In* R. Marcelle et al. (ed.) Effects of Stress on Photosynthesis. Martinus Nijhoff, The Hague.

Tang, P.S., and J.S. Wang. 1941. A thermodynamic formulation of the water relations in an isolated living cell. J. Phys. Chem. 45:443–453.

Träuble, H., and D.H. Haynes. 1971. The volume change in lipid bilayer lamellae at the crystalline-liquid crystalline phase transition. Chem. Phys. Lipids 7:324–335.

Tukey, H. Jr., S. Wittwer, and H. Tukey. 1957. Leaching of carbohydrates from plant foliage as related to light intensity. Science 126:120–121.

Tukey, H. Jr., R. Mecklenburg, and J. Morgan. 1965. A mechanism for the leaching of metabolites from foliage. p. 371–385. *In* Isotopes and radiation in soil-plant nutrition studies. Int. Atomic Energy Agency, Vienna.

Wolfe, J., and D.J. Bagnall. 1980. Arrhenius plots—curves or straight lines? Ann. Bot. (London) 45:485–488.

Wronski, E.B., J.W. Holmes, and N.C. Turner. 1985. Phase and amplitude relations between transpiration, water potential and stem shrinkage. Plant Cell Environ. 8:613–622.

Zimmerman, M.H., and C.L. Brown. 1971. Trees, structure and function. Springer-Verlag New York, New York.

Zimmerman, M.H. 1976. The study of vascular patterns in higher plants. p. 221–235. *In* I.F. Wardlaw and J.B. Passioura (ed.) Transport and transfer processes in plants. Academic Press, New York.

Zimmerman, M.H., and A.A. Jeje. 1981. Vessel length distribution in stems of some American Woody Plants. Can. J. Bot. 59:1882–1892.

Zimmermann, U. 1978. Physics of turgor and osmoregulation. Annu. Rev. Plant Physiol. 29:121–148.

9   Measurements of Plant Water Status

THEODORE C. HSIAO

University of California
Davis, California

The growth, function, productivity, and water use of a plant are intimately related to its water status. Ever since plants became a subject for serious scientific investigation, efforts have been made to measure and interpret their water status. The earlier measurements, however, were either inexact or exact but ambiguous in meaning. Only with the further development of measurement instruments and understanding of water physics and transport over the last several decades has it been possible to follow the plant water status in the field with some confidence. Even then, measurements of plant water status still involve difficulties, and a substantial understanding of the interactions of soil water status, evaporative demand, and physiological factors is necessary to interpret the measurement results with acumen. This chapter begins with a discussion of the various indicators (parameters) of plant water status and their significance. The more common techniques for measuring plant water potential are then considered and the results obtained with these methods compared. That is followed by discussions on the measurements of relative water content and components of water potential. Selected less direct methods of monitoring plant water status, including canopy temperature and expansive growth, are covered last. The final section considers the physiological framework of crop growth and water relations for irrigation scheduling and the promise and problems of using plant water status indicators in scheduling.

Since this chapter is written for use in irrigation research and management, emphasis will be placed on relatively common measurement methods not requiring highly specialized knowledge and instruments. Particular attention will be given to methods suitable for routine field use. The literature on mesurements of plant water status is extensive and has been reviewed frequently. This chapter is not written as a comprehensive review, but rather as a blend of principles and procedural leads which hopefully will serve the practical users and be a basis for the better interpretation of results. Discussions are often directed at the less clear and confusing aspects in the literature. Principles and practices for which there are strong consensus and those that are widely known in the literature receive only cursory attention. For an exhaustive compendium of the measurement methods, the reader is referred to the book by Slavik (1974a). Some older reviews (Barrs, 1968;

Boyer, 1969) are still of value, as is a pamphlet (Wiebe et al., 1971) describing the principles and steps of many of the methods. More recent general reviews are those of Turner (1981) and Spomer (1985). Thermocouple psychrometers have been reviewed in detail by Savage and Cass (1984) with emphasis on the in situ type, and are also addressed extensively in a symposium volume (Brown & Van Haveren, 1972). Pressure chambers as measurement instruments have been thoroughly reviewed by Ritchie and Hinkley (1975). Finally, the proceedings of a recent international meeting on the measurements of water status (Hanks & Brown, 1987) contain much up-to-date information.

I. WATER STATUS PARAMETERS

Various parameters are used as indicators of plant water status; the most common is tissue or organ water potential (Ψ). As a variant of the chemical potential of water, Ψ is well rooted in thermodynamics, and its gradient is all important for water transport. From the view point of physiology and metabolism, however, a direct relationship between Ψ and plant processes is not clear (Hsiao & Bradford, 1983). A recent paper (Sinclair & Ludlow, 1985) addressed this point specifically. Regardless, Ψ is the most commonly measured parameter, is quantitatively linked to other parameters more closely tied to plant functions, and its reduction relative to well-watered controls can be correlated with productivity and yield (Hiler et al., 1972). Another indicator of water status is tissue water content, which can easily be measured with high accuracy, either on a fresh weight or a dry weight basis. Earlier work, however, has shown that the water-holding capacity of different organs or tissue of a plant can vary widely, largely because of variation in the dry matter content of the tissue. Consequently the parameter has been normalized, to *relative water content* (RWC), the water content relative to that when the tissue is saturated with water and Ψ is zero. As implied in its definition, RWC is a measure of tissue volume maintenance and water retention and is therefore closely linked with plant functions.

Components of water potential may also be indicative of plant water status. A reduction in solute potential (ψ_s) indicates a concentration of solutes by dehydration or by osmotic adjustment, either of which can arise from plant water stress (deficit). The other component of water potential, pressure potential or turgor pressure (ψ_p), is an even more meaningful measure of plant water status (Hsiao & Bradford, 1983). Unfortunately, it cannot yet be directly determined with ease. High ψ_p is indicative of good water status except in the case of plants undergoing strong osmotic adjustment in response to water or salinity stress. Values of ψ_p approaching zero indicate substantial water stress.

Water potential in living cells and tissue is the algebraic sum of its component potentials, ψ_s and ψ_p, with the matric component being negligible (Wiebe, 1966; Boyer, 1967b). The behavior of ψ_s follows essentially the Morse equation, being proportional to the molar solute content and inversely proportional to the volume of solvent water (Craft et al., 1949). Pressure potential is a function of water or protoplast volume and of the elastic properties of the cell wall (Dainty, 1976). Only when the cell volume rises above a minimal level is ψ_p positive and detectable. Below the minimal volume, ψ_p is zero in parenchyma cells. The more elastic the cell wall, the smaller is the change in ψ_p for a given change in cell volume. The less elastic the wall, the greater is the change in ψ_p. Since both ψ_s and ψ_p are dependent on water volume and hence RWC, Ψ must be also. The relationship among Ψ, RWC, ψ_s and ψ_p is often depicted in so-called *pressure-volume curves* (a variant of the Höfler diagram) which in reality are curves of water and component potentials vs. water content. These curves will be considered further later.

II. PLANT WATER POTENTIAL

The most commonly measured plant water status parameter in recent times is undoubtedly Ψ. Thermocouple psychrometry or hygrometry is perhaps the most definitive for determining Ψ, when the necessary conditions are met. On the other hand, the only method suitable for rapid and routine field use is the pressure chamber method (Scholander et al., 1965). Discussed here first are the thermocouple-based methods, followed by the old Shardakov dye method and the pressure chamber technique.

A. Thermocouple Psychrometry and Hygrometry

As the names imply, these methods determine Ψ by measuring humidity of the air in equilibrium with the tissue sample. With the psychrometric technique, thermocouples function as wet and dry bulbs to measure the humidity. With the hygrometric technique, on the other hand, humidity is measured by determining the dew point of the air with thermocouples. The distinction between the psychrometric and hygrometric methods is not always that clear, however, and one particular commercial instrument can be used for both modes of measurements.

1. Theoretical Background

One advantage of the water potential concept is that at equilibrium, Ψ of all phases of the water must be equal. This means Ψ of the liquid phase can be measured by determining Ψ of the vapor phase with which the liquid phase is in equilibrium. A thermocouple psychrometer measures the relative humidity (RH) in the vapor phase. RH (e/e_o, in fraction) is fundamentally linked to the chemical activity of water (a) and to Ψ:

$$\Psi = \frac{RT}{\bar{v}} \ln \left(\frac{a}{a_o}\right) = \frac{RT}{\bar{v}} \ln \left(\frac{e}{e_o}\right) \qquad [1]$$

where R is the universal gas constant, T is absolute temperature, and $\bar{v}$ is the partial molal volume of water. The subscript o denotes the reference state which is pure water at atmospheric pressure and the same temperature as the sample. The ratio (a/a_o) and (e/e_o) are, respectively, the activity and vapor pressure of water of the sample relative to that of water at the reference state.

As in all psychrometry, the measurement is based on readings of the dry and wet temperature sensors which, in this case, are thermocouple junctions instead of the classical Hg thermometer bulbs. The wet bulb (junction) depression, together with the dry junction or air temperature, define the RH of the air and therefore Ψ of the tissue. The distinction between psychrometers used for the ordinary determination of RH and thermocouple psychrometers for measuring Ψ lies in the required measurement sensitivity. The logarithmic relationship between Ψ and relative humidity described by Eq. [1] dictates that a difference in Ψ of a few fractions of a megapascal is associated with only slight differences in RH. For example, at 25 °C, the equilibrium RH for a Ψ value of -0.5 MPa is 99.61%, which is very similar to the RH of 99.22% for a Ψ value of -1.0 MPa. According to the theoretical analysis of Rawlins (1966), the corresponding wet junction depression should be approximately 0.04 and 0.08 °C, respectively. This small wet junction depression per unit of change in Ψ dictates that the psychrometer temperature must be precisely controlled and any temperature gradient within the psychrometer be eliminated or accurately compensated for in order to obtain reliable measurements. The task is made more difficult by the fact that the voltage output of the commonly used chromel-constantan thermocouple is only about 60 μV per degree C, corresponding to -5.6 μV MPa^{-1} (Rawlins, 1966). So, in addition to the precise control of temperature, it is also critical to use high quality microvolt or nanovolt meters for the measurement, and to control electrical noises to the nanovolt level.

2. Wet Loop Thermocouple Psychrometer

Two types of thermocouple psychrometers have been developed, the wet loop (Richards & Ogata, 1958) type and the Spanner (1951) or Peltier type. The wet loop type is slightly simpler and will be discussed first. Its wet junction is actually made into a small loop to hold a drop of water and the dry junction is situated in close vicinity. A drop of pure water is placed on the loop and the wet and dry junctions are sealed with the tissue sample into the psychrometer cup. The temperature of the psychrometer and cup is normally controlled in a constant temperature bath. Alternatively, the temperature may be controlled by placing the psychrometer inside a tight-fitting hole drilled in an aluminum (for its high heat capacity and conductivity) block, which in turn is placed in a well-insulated container. To avoid errors caused by heat conduction along the wires, 0.3 m or more of the lead wires of the psychrometer should be coiled inside the insulated container. The sample Ψ is almost always below zero and therefore lower than Ψ of the water drop on the wet loop. Water will evaporate from the drop and be eventually absorbed by the sample. The latent heat of evaporation cools the drop and a wet junction depression is recorded. This depression is nearly a linear function of the Ψ of the sample (Rawlins, 1966). However, the psychrometers should be individually calibrated with standard solutions of known Ψ.

A variation of the wet loop psychrometer is now commercially available, instead of the loop to hold a water drop, the wet bulb is made of absorbing material covering a thermocouple and is dipped in water periodically when in use.

The measured parameter, wet junction depression, is dependent on the rate of evaporation from the junction and hence on geometry and resistances of the water vapor pathway. One aspect of geometry is the shape and the placement of the tissue sample. If the sample placement in the chamber is substantially different from the placement of the standard solution for calibration, significant errors may be introduced. For example, to measure Ψ of leaves, if the calibration was done with the standard solution absorbed on filter paper lining the side and bottom of the psychrometer chamber, then the side and bottom of the chamber must be similarly lined with leaf tissue for the measurement. Leaving the wall unlined would lead to an overestimation of leaf Ψ (less negative than true Ψ value).

The problem of resistances to water vapor transfer is more difficult to deal with. The cutinized epidermis of plant samples offers a substantial resistance to water vapor transfer. With leaf samples, the resistance is variable, depending on the degree of stomatal opening and permeability of the leaf cuticle. When the leaf sample is sealed in the psychrometer chamber, stomata close in response to darkness and to the buildup of respiratory CO_2. Therefore, epidermal resistance to water vapor transfer is high and can cause a substantial error since there is no known way to provide a similar resistance in calibration. The presence of epidermal resistance slows down water vapor absorption by the sample and hence the evaporation from the wet junction, leading to possible overestimations of Ψ.

3. Isopiestic Thermocouple Psychrometry

The equal potential or isopiestic method also uses the wet loop psychrometer, but in a different mode of operation. The objective of this method is to find a solution whose Ψ is the same as Ψ of the sample, and hence, when placed on the wet loop in the psychrometer, it neither gains from nor loses water to the sample; so, the wet junction depression would be zero. In practice, a series of two or more known solutions with Ψ values bracketing the Ψ of the sample are placed sequentially on the wet loop and the steady state output of the thermocouple recorded each time. The isopiestic point of zero output is obtained by interpolation, which is linear, in accordance with theory (Boyer & Knipling, 1965). Of course, the closer to the isopiestic point the chosen solutions fall, the more accurate is the measurement. The major advantage of the isopiestic technique is that the problems associated with thermocouple-sample geometry and resistances to water vapor transfer are minimized since the determination is made at or near the point where there is no water vapor transfer. Another advantage is that no calibration is necessary, thus eliminating additional errors associated with the calibration. For these reasons, the isopiestic technique is thought to be the most definitive of the psychrometric methods (Boyer, 1972). On the other hand, the isopiestic mode of operation requires the use of at least two solutions differing in Ψ on the loop, and hence, is the most time consuming of all the psychrometric techniques. Furthermore, wet loop psychrometers capable of isopiestic operation are not made commercially.

4. Spanner Thermocouple Psychrometer

The Spanner psychrometer operates on the same principle as the wet loop ones to measure wet junction depression. The difference is that instead of placing a drop of water on the wet junction, a Peltier cooling current is passed through the thermocouple to condense a tiny drop of water from the atmosphere of the same cup onto the junction. The temperature depression caused by the evaporation of this drop is measured after turning off the cooling current. In contrast to the wet loop instrument, where the sample gains water through vapor transfer and can increase its Ψ significantly if an abnormally long equilibration time is used, the Spanner psychrometer provides no source of water for absorption. On the other hand, if the sample has a high epidermal resistance, transpiration from the sample will be slow and true water equilibrium between the sample and the chamber air may take more than several hours to achieve. Reading the psychrometer before equilibrium is achieved will underestimate sample Ψ. Another drawback of the Spanner psychrometer is that the required condensation of a water drop, though small, does create suddenly a pocket of low vapor pressure which is not readily replenished by vapor transfer from the sample if the epidermal resistance is high. This may be the cause of the anomalously low values measured on samples of near zero Ψ with Spanner psychrometers (see Table 9–1). Also, the wet junction must be kept immaculately clean; otherwise, the water may not nucleate and condense at the same areas of the junction when the cooling current is turned on, and this change in geometry would give rise to errors.

Other than the differences mentioned here, the comments made regarding the wet loop psychrometer are equally applicable to the Spanner psychrometer.

5. Thermocouple Dewpoint Hygrometer

A thermocouple dewpoint hygrometer with high resolution was designed by Neumann and Thurtell (1972) for the measurement of plant and soil Ψ. The hygrometer measures the dewpoint, which is reduced by the lowering of vapor pressure associated with Ψ values less than zero (Eq.[1]). The thermocouple used has four leads, two for measuring the junction temperature and two for the passing of Peltier cooling currant. A curve of junction temperature vs. cooling current is first obtained with the hygrometer chamber dry, by varying the cooling current. Then another curve is obtained with the sample in the chamber and after a drop of water has been condensed on the junction by Peltier cooling. When the junction temperature is below the dewpoint, there is continuous condensation and the junction is warmer than it would be in the dry chamber at the same cooling current. When the junction temperature is above the dewpoint, there is continuous evaporation and the junction is colder than it would be in the dry chamber at the same cooling current. Only when the junction temperature is the same as the dewpoint does the curve of thermocouple output vs. cooling current for the wet chamber cross the curve for the dry chamber. The output at the crossing point is the measure of the dewpoint depression (Newmann & Thurtell, 1972).

Since the dew point hygrometer is designed to cool the wet junction by a Peltier current and measure its temperature, it can also be used as a Spanner psychrometer. The current commercial dewpoint hygrometer, based on the same principle as the model of Neumann and Thurtell, has the same dual capability. When operating in the hygrometric mode, the cooling current is adjusted electronically to locate the endpoint (Campbell et al., 1973), therefore eliminating the plotting of the dry and wet curves. In my experience, however, the Neumann and Thurtell model is capable of better resolution.

One advantage the dewpoint hygrometer has is that, similar to the isopiestic thermocouple psychrometer, there is no net water transfer at the endpoint. The problems of geometry and resistances to water transfer are therefore minimized. Another advantage is that the temperature dependence of the dewpoint depression (Newmann & Thurtell, 1972) is less than that of the wet junction depression measured psychrometrically (Rawlins, 1966). A third and minor advantage is that the output in the hygrometer mode is higher—7.5 μV MPa^{-1} (Campbell et al., 1973) compared to the 5.6 μV MPa^{-1} (Rawlins, 1966) in the psychrometric mode.

6. In Situ Thermocouple Psychrometry and Hygrometry

The psychrometers and hygrometers can be designed to monitor Ψ of an intact plant organ in situ. Lang and Barrs (1965) appeared to be the first to have measured intact leaf Ψ successfully with a Spanner psychrometer. Boyer (1968) used an isopiestic psychrometer, with its temperature precisely controlled by a circulation water bath, to monitor Ψ of growing leaves. Since then, Spanner psychrometers and dewpoint hygrometers have been designed to measure leaf Ψ, root Ψ, as well as stem Ψ in situ. Since a review of these measurements and instruments is available (Savage & Cass, 1984), it will not be attempted here. It is sufficient to say that these in situ methods are not easily used on a routine basis in the field. Good temperature equilibration and insulation are a must and not always achievable. Even more difficult is the problem, when monitoring leaves, of high epidermal resistance to diffusion when the intact leaf is enclosed in the measurement chamber. The lack of cut edge precludes quick vapor exchange between the leaf and the wet junction, causing the measured Ψ to lag behind the actual leaf Ψ when the leaf water status changes. Gentle rubbing of the leaf with a fine abrasive

(Shackel, 1984) or organic solvent (Newmann & Thurtell, 1972) can overcome this problem; but this can be easily overdone, resulting in desiccation of the rubbed spot in the chamber and abnormally low Ψ readings.

B. Shardakov Dye Method

In contrast to the thermocouple based methods, the Shardakov dye method (Shardakov, 1948; Knipling & Kramer, 1967) requires no special equipment and furthermore, is fairly quick and can be used in the field. The method is also isopiestic, but water exchange occurs in the liquid phase and the direction of water flow between the sample and the solutions of known Ψ is determined by changes in solution density. True water equilibrium is not necessary for the measurement; mere indication of the direction of water movement is sufficient. In my opinion the method should be more widely used. The procedure I found to be most workable and precise is described in detail here.

First a series of known sucrose solutions, graded in Ψ in steps of 0.1 MPa, are made up. A sequence of five to seven solutions, chosen to bracket the expected Ψ of the sample, are used to fill five to seven small test tubes two-thirds full. The leaf sample is excised, cut into small pieces in a humid chamber, divided into five to seven subsamples of approximately equal size, and quickly placed in another set of five to seven small test tubes. A few drops of the known solution, just sufficient to wet the cut edges of the tissue in a subsample and provide about 0.1 ml extra, are added with a Pasteur pipet from the test tubes containing the solution. The amount of solution added is much less than that required to immerse the sample, if the sample is bulky. The Pasteur pipets are returned to the solution test tube and kept there for further use. The sample test tubes are covered with a loose cap lined on the inside with a piece of filter paper moistened with distilled water to humidify the air and prevent excessive evaporation from the small volume of the solution bathing the sample. The sample tubes are shaken and tapped hard with the fingers every 2 or 3 min to facilitate mixing and water transport. After 10 to 15 min, a minute amount of methylene blue, ground to fine crystals, is tapped onto the sample from the tip of a hypodermic needle. After swirling and tapping to disperse the dye, a drop of the colored bathing solution is removed with the original Pasteur pipet and introduced into the middle of the original solution, to see whether it sinks or floats to the top. Sinking indicates that the solution has lost water to the sample and become denser, and therefore has a higher Ψ than the sample. Floating indicates the opposite. When the drop neither sinks nor floats upward, Ψ of the sample is equal to Ψ of the known solution. With known solutions graded in steps of 0.1 MPa, Ψ of the sample can be interpolated to 0.03 to 0.05 MPa, by the relative speed of sinking or floating of the colored drop. It is not necessary to have subsamples of exactly the same amount, but the proportion of bathing solution to sample should be similar. The sample should be cut with a sharp razor blade to sizes small enough for quick water exchange. For leaves this may be pieces 3 to 4 mm wide and 10 mm long. With maize (*Zea mays*

L.) leaves and seedling mesocotyls, duplicate measurements by the Shardakov method usually agreed within 0.1 MPa, provided care was taken to ensure uniformity of the subsamples and to maximize the volume ratio of sample to bathing solution.

Some workers have found the Shardakov dye method to give erratic results and poor reproducibility. Most likely this is caused by the use of too much solution for the size of the sample, leading to changes in solution density too small for easy evaluation. Small changes in density can be obscured by temperature gradients in the solution or the liquid motion induced by the introduction and removal of the pipet. The literature commonly recommends a bathing time in the range of 30 min to 3 h (Slavik, 1974b) to ensure a large or maximum density change. With the minimum bathing solution volume recommended here, detectable density change occurs within a few minutes. The short bathing time is best since the sucrose or other solutes (Slavik, 1974b) commonly used to make up the known solutions are absorbed by the tissue, but only slowly.

Because of its short bathing time and the fact that it works for a wide range of temperature, the Shardakov method is particularly suitable for measuring Ψ of actively growing (expanding) tissue. Because of cell wall relaxation (Cosgrove et al., 1984) after the sample is excised from its water source and the consequent reduction in ψ_p, Ψ of the sample drops with time. The short time required to make the Shardakov measurement minimizes this problem. Furthermore, measurements can be made at a temperature cold enough (e.g., 5 °C) to virtually eliminate cell wall relaxation, which depends on metabolism.

There can be problems with the Shardakov method. Often the measured Ψ became more negative when the bathing time was increased (Slavik, 1974b). This may be caused by the absorption of solute from the bathing solution and would be minimized if the recommended short bathing time is used. A more serious problem is that caused by cutting the sample into small pieces. The sap from the cut cells will differ from the density of the known solutions, and additionally, provides a source of readily absorbable solutes. This potential error should be evaluated by blotting off the cut edge before adding the bathing solution and by comparing the values of Ψ measured on the same tissue cut to different sizes.

C. Pressure Chamber

The method for measuring plant water potential that is convenient for field use and very rapid is the pressure chamber method (Scholander et al., 1965). To measure the Ψ of a leaf or even a small branch, the sample is excised from the plant and quickly sealed in the pressure chamber, but with the cut end of the leaf or stem protruding out of the chamber. Upon excision, the tension in the petiole or stem xylem is immediately and partly released, and the xylem sap retracts into the xylem. Pressure is raised in the chamber until the sap just appears at the cut end of the xylem. This *balancing pressure*, corrected for the ψ_s of the xylem sap, is a measure of Ψ.

1. Underlying Principle

The theoretical basis for the pressure chamber method is not as clear as those for the other methods described above. There is confusion in the literature of what the method actually measures, the Ψ of the sample as a whole, or of the xylem of the sample. The following discussion will make clear it is the volume-averaged Ψ of the whole sample that is measured.

To explain the principle underlying the method, one may visualize the leaf or branch, in the simplest of terms, to consist of a semipermeable sack encased in a porous and slightly elastic container which in turn is connected through a tube to the rest of the plant (Fig. 9-1). The sack, with its content of a concentrated osmotic solution, represents the cell plasmalemma and the protoplasm within. The porous container enveloping the sack represents the cell wall, with its pores filled with apoplastic sap which is dilute in solutes. The tube represents the xylem, also filled with the same apoplastic sap. The narrow partitions in the tube represent restrictions due to the end wall of the vessel, tracheids, or simply the narrow portion of vessel elements. If the model leaf is on a model plant low in water status, the water in the tube would be under tension. If the container (leaf) is covered to prevent evaporation (transpiration) and water movement, Ψ of the container would equal Ψ of the tube and be substantially below zero. Theoretically, one can raise Ψ of the model leaf by enclosing it in a chamber and applying pressure, to

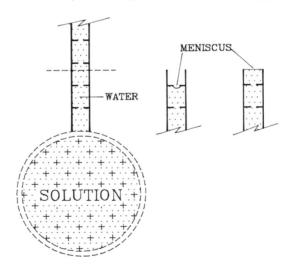

Fig. 9-1. Simplified representation of a leaf or branch for pressure chamber measurements. The sphere represents the collective protoplasm volume. The double dashed lines enclosing the sphere represents porous cell wall, with the inner dashed line also representing the plasmalemma. The crosses inside the sphere represent solutes. Xylem vessels are shown as a tube with perforated cross walls attached to the sphere. The horizontal dashed line indicates the location where the xylem of the stem is excised for the pressure chamber measurement. The figures to the right represent the location of the xylem sap at the cut end of the stem, after excision but before pressurization, and after the pressure is raised to the balancing point, respectively.

some known reference point, and the applied pressure would be a measure of how much Ψ had been depressed below the reference point prior to the pressure application. The question is how to select the reference point whose Ψ is known. Fortunately, when a leaf is cut off from the plant, deduction of ψ_p and measurement of ψ_s can be made for the sap in the xylem at the cut end. The sum of these two components of water potential then gives Ψ of the xylem, which then serves as the reference point.

When a leaf is excised, tension of the water at the cut end of the xylem is released and the consequent increase in its Ψ causes the water to move to the leaf which remains at low Ψ. This movement, however, is stopped at the narrow openings or end walls of the xylem, and a curved water-air meniscus is formed there (Fig. 9-1). The curvature is indicative of a pressure discontinuity across the water-air interface, with the pressure being lower in the water than that in the air. On the other hand, if the xylem water is made to move away from the narrow opening and back exactly to the cut end, the water-air interface would be flat and the pressure in the xylem water would be atmospheric, i.e., equal to that of the air. Consequently, ψ_p of the xylem water would be zero. If ψ_s of the xylem sap is measured, Ψ of the sap would be known and could serve as the reference point. So all one has to do is to apply just enough pressure on the leaf to push the xylem sap back to the cut end (Fig. 9-1), and measure ψ_s of the sap to get the known reference point. If the pressure application is carried out slowly so there is water equilibrium, $\Psi_{xylem} = \Psi_{leaf} = \psi_{s,xylem}$. Hence, the applied pressure is a measure of the depression (below the reference point) of Ψ of the leaf prior to excision.

It is important to note that the pushing back of the xylem sap exactly to the cut end is also necessary to restore the leaf to the original condition before excision. Even if somehow a flat water-air interface can be achieved within the xylem, as long as not enough pressure is applied to push the water to the cut end, the leaf parenchyma would contain more water (from the empty portion of xylem lumen) than it did while attached to the plant. Hence, its solutes would be more dilute and ψ_s more positive, and its volume larger and ψ_p also more positive than they would be had the water been pushed back to the cut end. As the result, the applied pressure would underestimate the depression of the Ψ prior to detachment. If the pressurization moves the water beyond the cut end, the leaf parenchyma would contain less water than it did and the depression of Ψ would be overestimated.

The discussion so far is on a leaf covered to prevent transpiration. Since there is no water transport and leaf Ψ is equal to xylem Ψ, the negative of the balancing pressure plus xylem ψ_s is an estimate of Ψ of the leaf as well as of the xylem prior to detachment from the plant (Boyer, 1967a). Indeed, Ψ of the covered leaf has been used to estimate stem xylem Ψ at the point of leaf attachment (Begg & Turner, 1970). With transpiring leaves, however, before excision, xylem Ψ is higher than that of the leaf cells by specific amounts dictated by the rate of water transport and hydraulic conductances. After excision and pressurization to reach the balancing pressure, the whole sample is in water equilibrium and Ψ of the xylem is now equal that of the rest of the leaf. Since the parenchyma cells dominate in volume over the xylem in a leaf, the measured volume-averaged Ψ must reflect mostly Ψ of the cells before excision. Xylem Ψ before excision would be higher than the measured volume-averaged Ψ. It is clear then that designating Ψ measured with the pressure chamber *xylem* Ψ, as is frequently done in the literature, is erroneous except in the case where water equilibrium was attained before excision.

The theory that the pressure chamber measures the volume-averaged Ψ of the leaf before excision has been supported by findings that Ψ measured with the pressure chamber agree generally with the values measured with thermocouple psychrometers, usually within one to several tenths of a megapascal (Duniway, 1971; Fig. 9–3). In some studies, however, Ψ determined with a pressure chamber were lower than those determined psychrometrically (Kaufman, 1968). Boyer (1967a) pointed out that for the pressure chamber to measure Ψ in the intact plant, water in the leaf or branch must be arranged spatially the same way under the balancing pressure as before excision. In the case of rhododendron (*Rhododendron roseum* Rehd.) leafy shoots, he observed sap infiltration of stem pith tissue under pressurization and concluded that this infiltration diverted sap from the xylem and, hence, extra pressure was needed to reach the balancing point. The measured Ψ would then be too low, by as much as 0.4 MPa, as his results showed. He also found that when more stem with its pith tissue was included in the sample, the determined Ψ was more negative compared to that measured with an isopiestic thermocouple psychrometer. These findings highlight the fact that pressure chamber measurements can have large systematic errors for species with anatomies that allows the infiltration of water into space normally occupied by air when pressurized.

2. Procedure

The procedure I use to make pressure chamber measurements is given here. The sample is excised with a sharp razor blade and immediately covered with several layers of moist—but not excessively wet—cheesecloth, taking care to avoid direct contact between the cut surface of the sample and the moist cloth. This stops transpiration and protects the sample from heating when the chamber is pressurized. Without contact with a hydrophilic surface on the sample, the amount of moisture the sample can absorb from the cheesecloth is insignificant. Alternatively, the sample can be enclosed in a plastic bag. The protected sample is sealed in the chamber, with only the minimum amount of petiole or stem protruding out. If necessary, the protected samples can be kept for some time (e.g., 30 min) without a detectable change in Ψ. With the sample in the chamber, pressure is applied, preferably at a rate of less than 0.1 MPa s^{-1} initially, then slowed to less than 0.02 MPa s^{-1} as the balancing pressure is approached. The end point is when sap just appears at the cut end of most of the xylems, as indicated by light reflection or glitter. The end point should always be observed carefully under a magnifying lens and adequate lighting, and may be hard to see if there is too much bubbling due to the escape of gas through the intercellular space network. At the correct end point, the xylem sap should just wet the cut ends and not collect as exudate, when the balancing pressure is maintained for several seconds. If the exudate forms into a lens or a hemisphere, the sample has been overpressurized and the pressure may be reduced gradually until the excessive sap is just reabsorbed. The pressure at that point is recorded as the balancing pressure. The importance of slowly reaching the end point for this equilibrium method cannot be overemphasized.

I use N_2 as the pressurizing gas. Using air instead of N_2, as was done in some studies, would mean that the partial pressure of O_2 in the chamber rises proportionally with pressure, to several atmospheres or more in the case of samples with very low Ψ. This abnormally high O_2 level could alter the cells through oxidation. For example, the plasmalemma membrane may become leaky.

3. Potential Errors and Problems

A major source of error in some literature data is that the sample was not protected after excision and during the measurement, leading to an underestimation (more negative values) of Ψ. Since this is most likely due to transpirational losses outside and inside the chamber, quick handling and humidification of the chamber and the pressurizing gas (Boyer, 1967a) had been thought to be sufficient to minimize the error. Later, it was realized that during pressurization, the heat of compression, essentially adiabatic, causes a rapid warming of the sample inside the chamber (Puritch & Turner, 1973). The sudden increase in sample temperature can be as much as 20 or 30 °C (Wenkert et al., 1978). The higher is the rate of pressurization, the higher is the temperature rise. If a sample is originally at 25 °C and its transpiration driven by a vapor pressure difference from the leaf to the air of 2 kPa, a rise of the sample temperature to 35 °C would increase the intercellular water vapor concentration from 3.2 to 5.6 kPa, leading to more than a doubling of the vapor pressure difference, therefore giving rise to a burst of transpiration. Negative errors of as much as 0.5 to 0.7 MPa was reported by Turner and Long (1980) when leaves of several species were left uncovered after excision and during measurement. They attributed this to rapid loss of water from the leaves immediately after excision while the leaf was still outside the chamber and ignored the heating problem effected by compression. It is more likely that rise in leaf temperature inside the chamber had also played a major role in causing the negative error. Covering the sample with a plastic bag reduced markedly the temperature rise (Wenkert et al., 1978), but, as shown in Fig. 9–2, covering with moist cheesecloth was even more effective.

Much of the literature data were not corrected for xylem ψ_s. This may be justified in many cases, since xylem solutes concentration is often low and ψ_s is in the range of -0.05 to -0.2 MPa (Scholander et al., 1964; Boyer, 1967a; Duniway, 1971), a magnitude similar to the measurement errors and sample-to-sample variations, especially in the field. On the other hand, xylem ψ_s can be considerably more negative under water stress (Fischer, 1973) and in the growth zone of plant organs (Cosgrove & Cleland, 1983) and should be corrected for, as also is the case when more accuracy is desired. Because of its small volume, xylem ψ_s is hard to determine without the right equipment. The wide availability of a commercial vapor pressure osmometer capable of measuring volumes as small as a few microliters has alleviated this problem. It is also possible to determine ψ_s of such small samples by freezing point depression (Silk et al., 1986).

The pressure chamber is particularly suited for measuring branches and leaves with nearly round petioles. Petioles with deep indentations or pronounced protrusions may present problems because the chamber gasket would not seal tightly around the irregular shape to prevent gas leakage. A slight leakage is acceptable. A substantial leakage would mean there is a pressure gradient and the pressure sensed by the gauge may be different from that experienced by the sample. The application of quick setting silicone rubber or rubber putty around the petiole before tightening the gasket can eliminate the leaks but will lengthen the time required per measurement. For chambers using rubber stoppers as gaskets, leakages will be minimized by cutting and using a thin section (e.g., 2 mm) of the stopper as gasket. That way the applied pressure will translate into greater force per unit of gasket surface around the petiole, ensuring a tighter seal.

Significant gas leakage from the chamber may also cause a change of oxygen pressure in the chamber. If N_2 is used for pressurization, the chamber air will be gradually depleted of O_2 as the leakage continues and N_2 replaces the lost O_2. The sample may be damaged by the resultant anaerobic condition.

Monocotyledonous leaves present special problems for the pressure chamber. Gaskets with slit openings are used for sealing the leaf blades in.

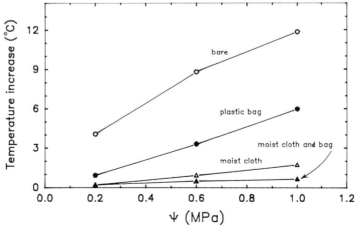

Fig. 9–2. Temperature increase of leaves during pressurization in a pressure chamber. The leaf was placed in the chamber either bare, enclosed in a plastic bag, in moist cheese cloth, or in moist cheesecloth then a plastic bag. Fine-wire thermocouples were taped onto the leaf to monitor leaf temperature. The chamber was pressurized to the indicated pressure with each increment in pressure taking about 30 s.

With the blades of cereals, however, the blades often rip transversely when pressurized, severing the veins. Here too, reducing the depth of the gaskets helps, especially when sections of rubber stoppers are used. Another thing we found helpful is to fold a thin plastic film such as Saran Wrap around the length of the blade and insert the cut end of the blade, with its film wrapping, in the slit of the gasket. When pressurized, the film prevents excessive stretching of the leaf, effected by the distortion of the gasket under pressure, and reduces blade ripping.

The leaf blades of some grass species [e.g., maize and sorghum, *Sorghum bicolor* (L.) Moench] are broad and too wide to fit the standard slit gaskets of pressure chambers. Measurements have been made by trimming the blade from the basal portion of the midrib and pressurizing with only the midrib protruding outside the chamber. Because the vascular bundles of the blade inside the chamber have been cut open, the applied pressure would empty at least many of the vessel elements and therefore give rise to a premature end point. A better procedure is to cut the blade transverse first, then pull a portion of the blade away from the midrib along the length of the vein toward the leaf apex to obtain a strip for measurement. That way the vascular breakage would be limited to a few veins and to the connecting small vascular bundles between veins. Xylems in the connecting bundles should be narrow and may have imperforate walls, thus minimizing air entry and prevent their emptying. The measured Ψ should still be more positive than the true Ψ of the leaf, but the error should be less compared to the sample with the basal portion of the blade trimmed off the midrib.

III. COMPARING Ψ MEASURED BY THE DIFFERENT METHODS

Numerous comparisons have been made of the Ψ measured by different methods. Usually only two methods were compared in a given study. When the results of the two methods disagree, there is first the question of whether the compared samples were truly nearly identical in Ψ. In some studies, different leaves on a branch were sampled for each of the methods, but the branch was not covered and given sufficient time to allow water equilibrium among its parts. If a disagreement truly exists and is not because of a lack of equilibration between the samples, then there is the fundamental dilemma of value measured by which method is the more correct. There is a tendency to use values determined by thermocouple psychrometry or hygrometry as the primary standard. However, as the preceding sections made clear, each technique has its own drawbacks and sources of errors. Even isopiestic thermocouple psychrometry in the hand of an experienced operator cannot be free of errors caused by cut-edge effects for excised samples and nonisothermal conditions due to respiration. The situation is not helped by the lack of plant tissue standards of known Ψ. The only plant standard one can use in checking the various methods is a mature leaf that has been fully equilibrated with pure water and, hence, should be at zero water potential. In a test on rice (*Oryza sativa* L.) using this as the standard (Table 9–1),

Table 9-1. Comparison of mean Ψ and variance (s^2) of fully hydrated rice leaves as measured by different methods. Tillers were excised at the base and recut under water twice, each time removing a basal node and the attached leaf, then allowed to rehydrate with base in water and the top loosely covered in a plastic bag for at least 1 h. The leaves measured were the same as those described for Fig. 9-3. Unpublished data of E. Yambao, T.C. Hsiao, and J.C. O'Toole.

Method	No. of measurements	Mean Ψ (MPa)	s^2
Isopiestic psychrometer	4	−0.13	0.058
Commercial swedgelock psychrometer	14	−0.73	0.820
Commercial dewpoint hygrometer	6	−0.35	0.087
Pressure chamber	13	−0.10	0.012
Shardakov dye method	8	−0.28	0.086

the pressure chamber method and isopiestic thermocouple psychrometry gave results closer to the expected value of zero than the Shardakov method, which in turn was better than the commercially available dewpoint hygrometer or thermocouple psychrometer (spanner type).

A few of the comparisons devoid of obviously avoidable errors are discussed briefly here. Duniway (1971) found almost a perfect correspondence between Ψ measured by a pressure chamber and that measured by a Spanner psychrometer immediately afterward on the same leaflet of tomato (*Lycopersicon esculentum* Mill.), with a maximum deviation of 0.1 MPa. He pointed out that the use of leaflets with petiolules cut short (5 mm long) might have improved the correspondence because little xylem water was present to be absorbed by the mesophyll cells when xylem tension was released by excision; so leaf Ψ was not raised significantly for the psychrometric mea-

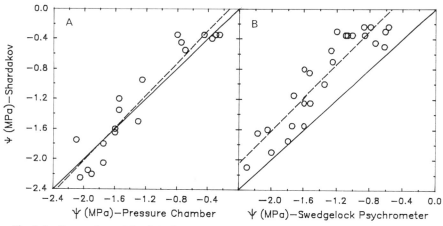

Fig. 9-3. Comparison of Ψ of rice leaves measured with Shardakov dye method vs. that measured with a pressure chamber (*A*) and a swedgelock psychrometer (*B*). Tillers of rice was dried on a laboratory bench to the desired water status, then covered with a plastic bag for some time to ensure water equilibrium among the plant parts. The most recently matured leaf was measured in the pressure chamber. The next older leaf was sampled for measurement by the Shardakov method or thermocouple psychrometry. Dashed lines represent regressions of the data and solid lines represent equal values. Unpublished data of E. Yambao, T.C. Hsiao, and J.C. O'Toole.

Table 9-2. Regression coefficients for comparisons between various methods of estimating water potential of rice leaves. Unpublished data of E. Yambao, T.C. Hsiao, and J.C. O'Toole.

Method	n	r
Isopiestic vs. swedgelock	30	0.81
Isopiestic vs. pressure chamber	25	0.98
Pressure chamber vs. dewpoint hygrometer	52	0.87
Pressure chamber vs. swedgelock	55	0.80
Pressure chamber vs. Shardakov	31	0.94
Swedgelock vs. Shardakov	26	0.92

surement. Boyer (1967a) also found quite good correspondence between the values measured by isopiestic thermocouple psychrometers and by a pressure chamber on sunflower (*Helianthus annuus* L.) and yew (*Taxus cuspidata* Sieb. & Zucc.). Rhododendron was the exception, as mentioned earlier, apparently because infiltration of the space in the stem pith tissue normally occupied by air when pressurized. In work with rice leaves, we (E. Yambao, T.C. Hsiao, & J.C. O'Toole, unpublished data) found pressure chamber measurements to be in good agreement with that measured by isopiestic psychrometers, when account was made for the imperfect seals of the psychrometers used. Agreement of the Shardakov method with the pressure chamber was acceptable (Fig. 9–3) whereas agreement with the commercial swedgelock psychrometer (Spanner type) was poor, with the swedgelock psychrometers giving more negative values (Fig. 9–3). The regression coefficients (Table 9–2) showed that the scatter in the data points was the least in the pressure chamber-isopiestic comparison, followed by the pressure chamber-Shardakov comparison. The worst scatter was found for the pressure chamber-swedgelock and isopiestic-swedgelock comparisons. These results led to the conclusion that for rice leaves, the pressure chamber and isopiestic psychrometry are the best methods for measuring Ψ, followed by the Shardakov dye method.

In reviewing the literature on comparisons between the pressure chamber and psychrometers, Ritchie and Hinkley (1975) found many kinds of deviations from the 45-degree 1:1 line. Again, in many cases it is not clear whether the pressure chamber or the psychrometer should be considered to be more correct. On the other hand, in the more extreme cases (e.g., Kaufmann, 1968) the error was too large to be attributed to the psychrometer.

IV. RELATIVE WATER CONTENT

Relative water content is indicative not only of tissue water content but also of volume maintenance. As mentioned before, unfortunately this parameter suffers from inaccuracy caused by the uncertainties involved in determining the saturation water content of the sample. Commonly the sample consists of leaf discs or pieces. After weighing, the discs are floated on water for a set time considered necessary to obtain water equilibrium. The discs are then quickly blotted to remove external water, weighed for saturated weight, and oven dried for dry weight. The time required to saturate the discs is supposedly the time necessary for the discs to reach a maximum weight floating on water. It turned out this end point is elusive. Upon floating, the weight gain by the discs is initially rapid, then slows as expected; but often the slow gain continues for many hours, with no clear stopping point (Barrs & Weatherley, 1962). One obvious cause in the case of growing leaves is that after excision the leaf will continue to grow and absorb water if water is provided. But the problem is encountered even with mature leaves (Millar, 1966; Catsky, 1974b) and has been attributed to infiltration by water of the intercellular air space along the cut edge, which increases with time. These and other problems have been reviewed in detail by Catsky (1974b), including some fairly elaborate procedures used to minimize them. Even when sampling is confined to fully matured leaves, the edge infiltration problem remains. The literature data (Catsky, 1974b) indicate that in some cases the phase of fast weight gain can take 4 to 6 h and, in most cases, at least 2 h are necessary, even for small leaf discs (8 mm diam). Since the half time of water exchange for intact and whole leaves are only in the order of several minutes (Newmann & Thurtell, 1972), one may argue that the apparent half times of many minutes to hours for leaf discs indicate interrupted hydraulic conduits, probably due to air entry into and embolism of the xylem system during excision. If the half time can be shortened and the floating time reduced, the error in determining the saturated water content would be minimized. Unfortunately, there is yet no simple way to ensure xylem integrity upon sampling without changing the sample water status. One possible improvement is to take a large leaf disc or piece as the sample, determine the fresh weight, then cut a much smaller subsample under water from the center for the determination of saturated weight using only a short floating time. The subsampling should ensure that some of the xylem remain water filled for the quick attainment of water equilibrium. The fresh weight of the subsample can be calculated from the total fresh weight and the proportion of the total dry weight in the subsample.

V. COMPONENT WATER POTENTIALS

Only the solute or osmotic (ψ_s) and the pressure (ψ_p) components are considered here. The matric component can usually be included in ψ_p (Passioura, 1980) and is anyhow negligible on a whole tissue or leaf basis unless there is a severe water deficit (Wiebe, 1966; Boyer, 1967b). Solute potential and ψ_p can be measured independently. More common is the practice of measuring Ψ and ψ_s and calculating ψ_p as the difference.

A. Independent Measurements of ψ_s and ψ_p

Usually ψ_s is measured on sap extracted from tissue after freezing and thawing to rupture the cell membrane and reduce ψ_p to zero. Sap Ψ is determined by vapor pressure osmometry, thermocouple psychrometry, or freezing point depression. Since ψ_p has been eliminated, the measured value is taken as ψ_s. As is the case whenever the membrane is ruptured, there would be an error because of the dilution by the apoplastic solution, if measuring ψ_s of the protoplasm is the objective. It is now possible to suck off a part of the sap from large single cells using the pressure microprobe and determining the sap ψ_s by a freezing depression apparatus capable of measuring samples down to the size of fractions of a nanoliter (Shackel, 1987). As for ψ_p, earlier it was only feasible to measure ψ_p of giant algal cells, for the cells had to be large enough for the insertion of a capillary mercury manometer. With the development of the pressure microprobe (Zimmerman & Steudle, 1978), it is now also possible to determine directly ψ_p of individual cells of higher plants, especially those of large size. However, these micro-techniques require much expertise and extreme care, and are not designed for routine measurements.

B. Estimating ψ_p from Measurements of Ψ and ψ_s

Since its direct determination is not yet routine, ψ_p is still normally calculated as the difference between measured Ψ and ψ_s. To minimize the errors introduced by sample-to-sample variation, it is now common to measure both Ψ and ψ_s on the same sample, by either thermocouple psychrometry (or hygrometry) or by measuring the moisture release curve of the sample with a pressure chamber.

1. By Thermocouple Psychrometry

The Ψ of the sample is measured first in the psychrometer. Then the sample is frozen and thawed while still enclosed in the psychrometer cup, and measured again for Ψ. The second measurement is of ψ_s since ψ_p has been eliminated by freezing and rupturing of the plasmalemma. Because the same sample and the same psychrometer are used for both measurements, errors because of sample-to-sample and instrument-to-instrument variations

Table 9–3. Comparison of solute potential of portions of growing maize leaves measured directly with that measured after the measurement of water potential, by isopiestic thermocouple psychrometry. Two matched samples were taken from each location of the leaf. One sample was measured first for Ψ over a period of approximately 5 h, then frozen and thawed for the measurement of ψ_s. Measurements were at 29 °C. Values are averages of three measurements on consecutive 20 mm segments of the growth and mature zones, and of equally spaced 10 mm segments of the old zone. The unstressed leaf was growing at 40 μm min^{-1} and the water stressed leaf was growing at 2 μm min^{-1}. Unpublished data of J. Jing and T.C. Hsiao.

		Unstressed		Water stressed	
Zone	Location from leaf base (mm)	ψ_s (MPa) measured directly	ψ_s (MPa) measured after measuring Ψ	ψ_s (MPa) measured directly	ψ_s (MPa) measured after measuring Ψ
Expanding	0–60	−1.23	−1.02	−1.45	−1.29
Mature	80–140	−1.19	−1.03	−1.39	−1.29
Old	210–300	−1.20	−1.15	−1.40	−1.35

are avoided. One disadvantage of the technique was realized only recently. Each of the two measurements commonly required 3 h or more for equilibration. That time is sufficient to allow changes in ψ_s of 0.1 to 0.3 MPa as the result of metabolism. The change caused the measured value to be more negative (Grange, 1983; Bennett et al., 1986). Working with growing tissue, Jing and Hsiao (unpublished data) noted a change in ψ_s in the opposite direction (Table 9–3). The rise in ψ_s, apparently during the first measurement of Ψ, was possibly associated with respiration (data not shown), which was higher in the younger and more actively growing tissue. The error, of course, increases as the measurement time and temperature are increased.

2. By Moisture Release Curve Determined with a Pressure Chamber

The moisture release curve (so called pressure-volume curve) represents the relationship between Ψ and water content of the tissue, analogous to the moisture release curve of the soil. The moisture release curve of a leaf or branch is conveniently determined with a pressure chamber. Analysis of the curve can yield values for ψ_s and ψ_p at any given Ψ or water content of the sample.

   a. Theoretical Background. An example of moisture release curve for a leaf is given in Fig. 9–4A for cotton (*Gossypium hirsutum* L.), with the water content normalized to RWC. The portion between RWC of approximately 65 and 90% is not linear but shows a slight downward curvature. As RWC rises above 90%, the curvature turns upward. This change in curvature is the result of a rising ψ_p, which started at the inflection point near 90% RWC. Below the inflection point ψ_p is zero and the only significant

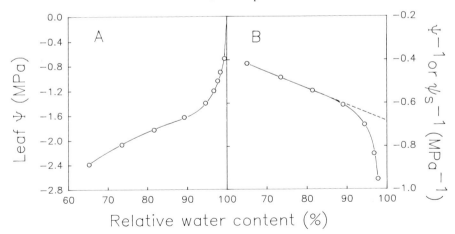

Fig. 9–4. Relation of Ψ to relative water content (*A*) and of the inverse of water potential (Ψ^{-1}) to relative water content (*B*) for a field grown and well-irrigated cotton leaf. The leaf was first saturated with its petiole in water, then wrapped in moist cheesecloth, placed in a pressure chamber, and dehydrated stepwise, along with the measurement of water loss and balancing pressure at each step. In *B*, dashed line is extrapolated from the linear portion of the data.

component of Ψ is ψ_s. Above the inflection point ψ_p rises continuously with increases in RWC, causing the upward curvature in the line for Ψ, although the curvature of the line for ψ_s in that region of RWC would still be slightly downward. The downward curvature in the line for ψ_s as RWC increases is consistent with the theoretical behavior of any osmotic solution when its water volume is raised, in accordance with the Morse equation relating ψ_s to the number of moles of solutes (n_s) and the volume of water (V_w) in a system:

$$\psi_s = -RT \frac{n_s}{V_w} \tag{2}$$

where R is the universal gas constant, T, the absolute temperature, and V_w, the partial molal volume of water (a constant for our purpose). The equation shows that, when n_s is constant, the relationship between ψ_s and water volume or content is inverse and not linear. That explains qualitatively the downward curvature of the line between 65 and 90% of RWC in Fig. 9-4A. On the other hand, the equation indicates that the relationship between the inverse of ψ_s and water content would be linear.

Equation [2] is for a homogeneous solution and needs to be modified for the quantitative application to a plant sample. The sample must be considered to be made up of at least two compartments (Fig. 9-1), the symplast inside the membrane (plasmalemma) system, and the apoplast outside, consisting of the xylem lumens and the porous space within the cell wall. To apply Eq. [2] to the symplast, water conent of the apoplast must be excluded as it is unavailable as solvent for the symplastic solutes. In practice, it is easier to use the inverse form of Eq. [2] to estimate the values of components of Ψ at any given Ψ or water content. The unavailable solvent volume is accounted for in the inverse of Eq. [2] (Hsiao & Bradford, 1983) as follows:

$$\frac{1}{\psi_s} = -\frac{V_{tw}}{RTn_s} - \frac{V_{aw}}{RTn_s} \tag{3}$$

where V_{tw} is the total volume of water in the sample and V_{aw} is the volume in the apoplast. Equation [3] is a linear function if n_s and V_{aw} remain constant as ψ_s changes with changes in sample water content. Since only ψ_s and ψ_p are the significant components of Ψ, the linear portion of the plot of the inverse of Ψ vs. water content is assumed to be identical to the plot of the inverse of ψ_s vs. water content, and the deviation from linearity in the curve is attributed to ψ_p. An example of the inverse of Ψ plot, based on the data of Fig. 9-4A, is shown in Fig. 9-4B, where the sample water content has been normalized to RWC.

To determine the ψ_s and ψ_p of the sample at a given Ψ, the value of the inverse of Ψ is located on the curve of Ψ^{-1} vs. water content. If this falls on the straight line portion of the experimental data, ψ_p is zero and ψ_s

equals the value of Ψ. If ψ_p is above zero, the inverse of Ψ would fall on the curved portion of the experimental data, and the inverse of ψ_s is read off the extrapolated portion of the straight line, at the same water content. ψ_p is then calculated as the difference between Ψ and ψ_s.

A plot such as that in Fig. 9–4B, or with water content as the abscissa, could also be used to estimate the volume of apoplastic water by extrapolation (Boyer, 1969). From Eq. [3], it can be seen that when ψ_s^{-1} (hence, Ψ^{-1}) equals zero, $V_{tw} = V_{aw}$. This extrapolation to $\psi_s^{-1} = 0$, however, is wrought with uncertainty because the linear portion of the data normally does not go below RWC of 0.6 or 0.5. Below this level, the data do not obey Eq. [3], possibly because of a loss of membrane integrity or collapse of xylem vessels under the high pressure applied in the pressure chamber. The long distance extrapolation to V_{aw}, which tends to fall in the range corresponding to RWC of 0.1 to 0.2, amplifies the inaccuracies of the original data.

b. Procedure. The procedure I use to obtain the data for moisture release curve is given here. After excision, the protected leaf or branch is placed in the pressure chamber and its balancing pressure determined. The pressure is then increased by 0.1 to 0.2 MPa after a small tube, sealed at one end and stuffed partly full with absorbing tissue paper, is placed over the protruding cut end of petiole or stem of the sample, with the absorbing paper in good contact with the cut end. The pressure is held steady while the paper absorbs the xylem exudation caused by the over pressure. For a sample at high turgor (ψ_p), a relatively small volume of exudate would be pressed out by this overpressure; for a sample at low turgor, a large volume would be pressed out. When the exudation ceases, water in the system is again at equilibrium, but at a smaller sample volume and hence a lower water content, and under a new and higher balancing pressure, which is recorded. The absorbent tube is removed, its open end covered, and weighted on an analytical balance to measure the water content change of the sample. The pressure is again raised by a similar amount after the same tube with absorbing paper is placed back on the cut end of the sample, and the whole process repeated. This progressive dehydration and measurements of the water potential and content of the sample should be carried out preferably in at least 8 or 10 steps. At the end, the sample is weighed, dried, and weighed again to determine the remaining water content. Because of the small water volume involved, it is important to cover the cut end of the sample and the open end of the absorbent tube to prevent evaporation when weighing the latter. The sample cut end can be covered by an inverted small vial fitted with a disc of moistened filter paper at the bottom to humidify the air in the vial but not touching the cut end.

The dehydration steps are rather slow because of the need to wait until the system reaches water equilibrium. Cutler et al. (1979) speeded up the process at each step by applying first an overpressure greater than the desired equilibrium value, and then reducing it after a few minutes until the xylem sap just stopped to exude from the cut end (the balancing point). The balancing pressure and the weight of the sap collected at that point are recorded.

An alternative to dehydration in the pressure chamber is to dry out the sample in open air at each dehydration step. After its Ψ is measured in the chamber, the sample is removed and allowed to dry to a desired level on the laboratory bench. It is then weighed for water content and placed back in the chamber to measure its new Ψ. The procedure is repeated until the desired range of data is obtained. This approach has the advantage of allowing the determination of the moisture release curve of a number of samples at one time using only one pressure chamber. The limitation is that the method works only if the stem or petiole of the sample is sufficiently strong to withstand repeated insertion and removal from the chamber, and compression and decompression.

The data collected by the procedures given above cover only the span of water status ranging from a maximum at the time of sampling downward. The data would not include the curved portion of the inverse of Ψ plot as exemplified in Fig. 9–4B if the tissue was at zero turgor at the time of sampling. Further, it is not possible to express the water content as RWC since saturation water content is not known. Nonetheless, the data provide values of Ψ and ψ_s at the time of sampling, and at sample water contents lower than that at sampling. The data are also sufficient for the estimation of ψ_p at sampling or at water contents lower than that at sampling.

To obtain moisture release curves up to 100% RWC, the sample must be saturated first before being subjected to step-wise dehydration in the pressure chamber. In my experience, saturation is easily achieved with the following steps, which unfortunately do not allow an accurate estimate of Ψ of the leaf before excision. Instead of the normal procedure, the basal node of the petiole is included when excising the sample from the plant. Afterwards, the node and excessive petiole length is cut off with a sharp razor under distilled water. The leaf is then covered loosely with a plastic bag and left with its petiole in water. Saturation usually takes <30 min, even with wilted leaves. Excision in air when the xylem is under tension allows the air to fill the xylem vessel elements, perhaps for the full length of the vessel which can be the length of the petiole. Tracheary elements in the node are apparently narrow or connected by imperforate walls, preventing air entry and intrusion into the petiole after the initial excision. The vessels remain filled when the second cut is made under water, ensuring fast water uptake and saturation.

An assumption underlying the estimates made from the moisture release curve is that the amount of solutes in the sample remains constant. In view of the change in total solute content with time in some excised plant material (Table 9–3) and because of the long time required for collecting the data to cover a sufficient range of water status, this method is potentially subjected to errors because of changing ψ_s. The problem warrants systematic investigation.

VI. OTHER MEASUREMENT METHODS

Numerous other ways of measuring plant water status exist (Slavik, 1974a); many of them are indirect and measure parameters which can be correlated with plant Ψ or water content. Remarks here are confined to only six of these: the hydraulic press method, and measurements of organ thickness or water content, of stomatal opening, of canopy temperature, of cavitation in the xylem and of growth.

A. Hydraulic Press

This method is related to the pressure chamber method but the theoretical basis is not clear. A small excised leaf or a piece of several cm^2 in area cut from a large leaf is placed between a clear acrylic plastic window and a piece of filter paper atop a rubber diaphragm of the hydraulic press. Pressure is applied by the press to the diaphragm until tissue sap appears at the cut edge and begins to be absorbed by the filter paper. The pressure at this or some other similarly subjective end point is taken as a measure of the leaf water status. It is expected that the lower is the Ψ of the cells, the harder must they be squeezed to obtain an outflow of sap; so, there should be a correlation between the hydraulic press reading and leaf Ψ. What is not clear is how much Ψ of the cells is raised by an increment in the applied pressure. This is because the applied pressure is not uniform as is the case for the pressure chamber, but instead, is transmitted only by the areas of the leaf that are in contact with the filter paper supported by the rubber diaphragm, and by the areas in contact with the acrylic plastic window. The contact areas are likely to be confined to the raised parts of the leaf surface, with the microdepressions not in contact or in poor contact. As the hydraulic pressure is increased, the cells would distort more and more to equalize the pressure within the cells and among the cells. At the end point, equilibrium is approached between the water inside the cells and that at the cut edge flowing onto the filter paper. The distortions of the cells and tissue, however, have changed the volume-pressure relationship. This altered relationship precludes a direct estimate of the original leaf Ψ.

To interpret the values measured with the hydraulic press, calibrations must be made against a physiologically more meaningful measure such as Ψ (Grant et al., 1981; Rajendrudu et al., 1983) or RWC. The calibration is crop specific, at least in many cases, and may even be dependent on growth stages because of likely changes in the mechanical properties of leaves (Yegappan & Mainstone, 1981). On the other hand, there is no question that the hydraulic press method is the most rapid and adapted to field use among all the techniques for excised samples discussed here.

B. Organ Dimensions

Methods have been devised for the nondestructive monitoring of changes in organ water content or volume over periods of days or even weeks.

The changes in water content of a leaf can be detected by β gauging (Mederski, 1961; Nakayama & Ehrler, 1964). A radioactive source of β particles is positioned at a fixed distance of a few millimeters on one side of the leaf and a Gaiger Muller tube is positioned on the other side, also a short fixed distance away. The β particles emitted are partly absorbed by the water and the other leaf constituents, so the counts detected by the detector are related to the leaf water content. With β particles of the right energy range, the logarithm of the detected counts is linearly related to leaf water content, with a negative slope (Nakayama & Ehrler, 1964) and, when calibrated, is a measure of RWC (Catsky, 1974a). Because the nonaqueous material of the leaf also absorbs β particles, the calibration is not only species specific, but is also influenced by leaf age (Mederski & Alles, 1968) or even the location on the leaf.

For stems, the changes in water content are more easily followed by measuring their diameters. The traditional measurement of tree stems with dendrometers can detect the contraction and swelling associated with the diurnal cycling of tree water status (Verner et al., 1962). With the advent of modern electronics, it is now feasible to monitor changes in diameter of small stems with position transducers (Namken et al., 1969; Klepper et al., 1971).

The diurnal cycle and ontogenetic changes, such as growth, confound the interpretation of dimensional and water content measurements. With a continuous record, however, it is possible to follow the long-term trends of selected points in the diurnal cycle, such as the maximal water content or thickness at dawn, as an indicator of the development of water stress. To rule out ontogenetic effects, it may be necessary to have a well-watered control for comparison.

C. Stomatal Opening

It is well known that water stress effects stomatal closure. So the degree of stomatal opening can be indicative of plant water status. More significantly, stomatal conductance is well correlated with the rate of photosynthesis (Wong et al., 1979), which also depends on favorable water status. The standard way of measuring stomatal conductance with a diffusion porometer, however, yields only spot readings. Since stomata are also affected by many other environmental and physiological factors (Hsiao, 1975), continuous monitoring of stomatal opening is necessary to sort out the effect of water stress from those due to other variables. Viscous flow porometers, which measure the ease of air passage through amphistomotous (stomata on both side) leaves, are particularly suited for this purpose (Hsiao & Fischer, 1975) and had been automated earlier (Allaway & Mansfield, 1969). Recently, Fiscus et al. (1984) developed a version of the porometer for field use with data loggers. Their tests suggest that it may be useful for irrigation scheduling.

D. Canopy Temperature

When stomata close partially or fully under water stress, the energy balance of the crop canopy will be altered and, hence, the canopy temperature will also be altered (Gates, 1968). The difference between leaf and air temperature was shown to be relatable to leaf water potential by Ehrler et al. (1978). Jackson et al. (1977) proposed the use of the summation of canopy-air temperature difference over time—termed *stress degree day* (SDD) as an indicator for irrigation scheduling. It is clear, however, that other factors such as vapor pressure, radiation, and wind would also affect the canopy-air temperature difference. Later it was shown that air vapor pressure deficit had major effects and should be corrected for (Idso et al., 1981). The minimum measurements required for the use of the SDD concept, therefore, are canopy temperature, air vapor pressure, and air temperature. Canopy temperature is measured with an infrared thermometer. To avoid including soil in the field of view of the instrument, the sensor is usually held obliquely and aimed at the upper part of the plant canopy.

Because the sensing of the canopy is remote, and the measurement easy to make and readily automated, the SDD concept has received much attention as a potential tool for assessing productivity under water-limiting conditions and for irrigation scheduling. The canopy-air temperature difference, however, is subject to the influence of factors other than plant water status and air vapor pressure deficit, as mentioned before. Published data (e.g., Idso et al., 1981) show a noise level of several degrees C after normalizing for the air vapor pressure deficit effect. Furthermore, the method, similar to the stomatal opening measurement, probably lacks the sensitivity to detect mild water stress, which could have significant impact on vegetative growth and canopy development. This will be discussed further in the last section of this chapter.

E. Xylem Cavitation

When a plant is under water stress, its Ψ is low and tension in the water of its xylem is high. The high tension can cause xylem water columns to break and pull apart suddenly, leaving vapor and air filled cavities in some xylem vessels to block transport. A higher tension in the xylem is associated with more cavitated vessels. Cavitation was first detected by amplifying the minute sound associated with each of the explosive event (Milburn & Johnson, 1966). Environmental noises, however, interfere seriously with the detection of cavitation in the audible range. The use of ultrasound detectors minimize this interference problem (Tyree & Dixon, 1983). Ultrasonic cavitation detectors are now commercially available. The susceptability of the vessels to cavitation appears to be species specific, as is apparently the frequency of cavitation events for a given change in leaf water status.

F. Expansive Growth of Leaves or Stems

Among all the plant processes, expansive growth of leaves and stems is one of the most sensitive to water stress (Hsiao, 1973; Hsiao & Bradford, 1983). In many studies, any reduction in leaf Ψ below the maximum value attained by well-watered plants slowed down expansive growth (Boyer, 1968;

Jordan, 1970; Acevedo et al., 1971; Hsiao & Jing, 1987). It follows that plant water status may be inferred from measurements of leaf or stem growth. Leaves of grass species usually grow several tens of millimeters per day. This amount of increase in length is easily followed with a ruler. For accuracy, it is important to measure the same leaf repeatedly and to measure from a fixed reference point, such as stake driven into the soil at the base of the plant or a mark on the sheath of a mature leaf.

VII. USE OF PLANT WATER STATUS MEASUREMENTS IN IRRIGATION MANAGEMENT

Rational irrigation management must be based on productivity behavior of crops in response to changes in soil and plant water status as modulated by the aerial environment. The plant, situated midway in the pathway of water between the soil and the atmosphere, is the integrator of its hydroenvironment. Knowledge of the plant water status not only provides inference of the soil water conditions, but also of the current performance and health and well being of the plant. On the other hand, the close coupling of the plant with the evaporative demand of the atmosphere makes the plant water status exceedingly dynamic and not a simple static indicator of when soil water becomes limiting and an irrigation is called for. Even more challenging is the fact that crop species, and plant productivity processes, differ in their sensitivity to water stress. The effects of water supply and deficit on plant productivity have been reviewed and discussed by a number of authors (Fischer & Hagan, 1965; Fischer & Turner, 1978; Hsiao et al., 1976). Some of the salient features are summarized here, to provide the necessary framework for the subsequent discussion of how measured plant water status may be used to aid irrigation scheduling.

Crop productivity and yield may be considered in terms of the cumulative carbon (C) gain through photosynthesis, and the partition of the C to the harvested plant organs. Productivity in a given climate is intimately linked to the water use by the crop, a fact attested by the numerous empirical water production functions (Hanks et al., 1983; Stewart et al., 1974). The basis for this linkage between water loss and C gain is more than just the commonality for water vapor and CO_2 in the leaf-air transport pathways. As discussed by Hsiao and Bradford (1983), crop foliage coverage of the ground and the interception of radiation by the foliage canopy also play a major, or even dominant, role because both transpiration and photosynthesis are dependent on the same canopy surface. In addition, radiation absorbed by the canopy powers both processes. This hand-in-hand relationship between biomass production and cumulative transpiration means that any reduction in transpirational water use would reduce yield of forage and silage crops proportionally, when other factors are not limiting. For grain or other crops whose yields constitute only a part of the biomass, reductions in water use effected by water deficits can be more complex, depending also on how the partition of assimilated C to the harvested organ is affected, i.e., how the harvest index is affected.

Plant water status should be managed to: (i) speed up the develoment of foliage canopy or effective photosynthetic surface area (area of the source of C, or simply, source size); (ii) enhance the rate of photosynthesis per unit of source area (source intensity), and (iii) maintain the photosynthetic surface for the full duration of the season (source duration). If the harvestable part of the crop is considerably less than the total biomass, plant water status should also be managed to: (i) increase the number of the sinks for assimilates, such as grains or fruits (sink size) and (ii) enhance the filling of the sinks (sink intensity). The sensitivity of these productivity parameters to water stress, and some of their major interactions are depicted for the life cycle of a generalized annual grain or fruit crop in Fig. 9-5.

In the figure, the width of the arrow shaft indicates the relative sensitivity of the given parameter to water stress. Although attention in the past

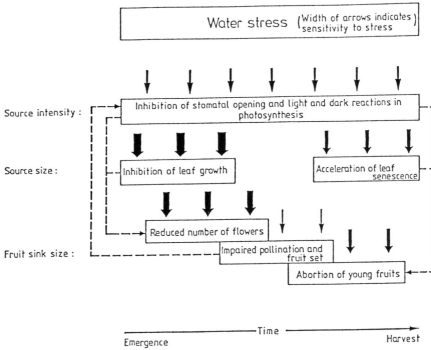

Fig. 9-5. Effects of water stress on parameters underlying source intensity, source size, and sink size for assimilation at various times of ontogeny, generalized for annual crops grown for grain or fruit. The time intervals within the crop ontogeny when water stress can cause physiological and morphological changes are indicated by the locations of the outlining rectangles. Solid short arrows point to the changes. Arrow shaft widths indicate the sensitivity of the parameter to water stress. For example, leaf growth is the most sensitive to stress and flower number is the next most sensitive. Dashed arrows and lines (-----) indicate causal relations among the parameters. For example, inhibition of leaf growth, of stomatal opening, and of photosynthesis results in fewer flowers being differentiated, probably because the number of reproductive axes is determined by plant size and the amount of assimilates available. Another example of causal relations is that impaired fruit setting reduces the number of sinks for assimilates and usually leads to a reduction in stomatal opening and photosynthesis via feedback inhibition. Based on Hsiao (1982).

has been directed at effects of water stress on source intensity, source size is considerably more sensitive to water stress and therefore would be more critical during the period of canopy development. Furthermore, reductions in source size by water stress can be reversed only slowly in indeterminate crops and possibly not at all in determinate crops, whereas source intensity usually recovers in a matter of one to a few days after the crop is irrigated. If canopy size is restricted, there would be fewer branches or tillers on the plant, and the number of potential fruiting points would be reduced, leading to a smaller sink size. It has often been assumed that pollination and fruit setting are especially sensitive to water stress, partly because the number of grains produced decreases with water stress at flowering time. Recent careful studies (Westgate & Boyer, 1986), however, have shown that pollination is much less sensitive to water stress (as depicted in Fig. 9–5) than previously believed. The often-observed reduction in grain number under water stress is probably largely due to the abortion of the younger fruits when assimilate source is limited because of shortened source duration and, possibly, decreased source intensity under water stress. The shortening of source duration is due to stress-induced acceleration of senescence, which often is aggravated by N deficiency (Wolfe et al., 1988).

The differential sensitivity to water stress at different growth stages, qualitatively depicted in Fig. 9–5, indicates that generally, to maximize yields, plant water status should be maintained high through irrigation during the canopy development stage. That is to maintain fast leaf expansive growth and canopy development. Once the canopy is complete, plant water status may be allowed to drop to a lower level, but still must be maintained high enough so that stomatal opening and photosynthesis per unit canopy cover (source intensity) are not inhibited. Because pollination and fruit growth is more resistant to water stress, there is no need to apply extra irrigation at or near anthesis time, as long as photosynthesis is maintained. During the maturation stage, it is important not to cut off irrigation too early, so as not to bring about premature senescence of the leaves. A reduction in assimilate supply due to the shortened source duration can be particularly detrimental since all the current assimilates go to the harvestable organs at that stage, and the reduction in yield is usually more than proportional to the reduction in total biomass because of a reduction in harvest index.

So far, this discussion has dealt with plant water status in qualitative terms, for a good reason—there is no easy way to refer to quantitative values without being crop specific. This problem is less serious with the water status indicators more closely tied to physiological functions, but is quite considerable with the most commonly measured parameter, leaf Ψ. One of the problems is that crop species differ in leaf Ψ under similar evaporative demand, apparently because of differences in hydraulic conductances of their root systems and stems. One of the larger differences I observed is between leaf Ψ of cotton and processing tomato in California. With virtually all the soil of the root zone at field capacity, Ψ of fully exposed leaves of cotton under a full sun at midday was -1.0 to -1.2 MPa, whereas that of tomato was -0.5 to -0.7 MPa. Those tomato leaves wilted visibly and its ψ_p became negligible when its Ψ dropped to -1.0 MPa, whereas ψ_p of cotton leaves did not become negligible until leaf Ψ dropped to -1.6 to 1.8 MPa (under conditions when there was no significant osmotic adjustment). Another problem is that in some crop species, Ψ of leaves on top of the plant growing in a soil at field capacity changes with ontogeny. Exposed top leaves of maize exhibited, at midday, Ψ values of around -0.5 MPa before rapid stem elongation, but values of approximately -1.1 MPa after the plant reached full height, with the soil maintained near field capacity (Fereres et al., 1978). Cotton and tomato, on the other hand, maintain Ψ of their top leaves about constant near noon on sunny days throughout the season when well watered. The differences in Ψ among species and at different ontogenetic stages discussed here are of magnitudes similar to the reductions in Ψ necessary to cause a well-watered plant to reduce its stomatal opening and photosynthesis in the absence of substantial osmotic adjustment (Hsiao, 1973). It is clear that as an indicator of the beginning of plant water stress, a single value—or even a narrow range of values—of midday Ψ is not sufficient. Instead, a family of curves covering different growth stages would have to be developed, each specific for one or several crops. Complicating the matter further is the fact that various productivity parameters have differing sensitivities to water stress, as already discussed.

Instead of the measurements taken at midday, which is strongly influenced by the aerial environment, leaf Ψ measured predawn or at dawn has been used as an indicator of the long-term trend in plant water status and also of the need for irrigation. However, because large variations in midday Ψ are associated with small variations in predawn Ψ, it remains to be tested whether predawn Ψ is a sufficiently sensitive indicator of plant water status to serve as a guide for irrigation scheduling. Also, it is possible that the predawn values are species specific and ontogeny dependent as well.

So far, scheduling irrigation by leaf Ψ measurements has only been advocated for cotton (Grimes & Yamada, 1982) and some farmers are currently trying the method in California. Grimes and Yamada (1982) gave a midday value of -1.9 MPa (leaves measured unprotected in a pressure chamber) as the threshold for applying water. That value probably corresponds to a leaf Ψ of -1.7 MPa (Radulovich, 1984) and may be too low for the fast development of canopy in the vegetative stage (Grimes et al., 1978). Cotton is particularly suited for this approach because it undergoes strong osmotic adjustment under water stress and hence its Ψ can drop substantially without a substanial reduction in photosynthesis while its harvest index is improved due to the restriction of vegetative growth.

Relative water content, as an indicator of the level of water stress that triggers a specific physiological response, appears to be more consistent from species to species and from one ontogenetic stage to another. This is probably due to the fact that RWC is a direct measure of cell hydration and relative volume. To generalize, leaf RWC of well-watered plants fall in the range of 88% or higher at midday. When water deficit is sufficient to reduce ψ_p to near zero, RWC usually is in the 72 to 88% range. At this point, the leaves are visibly wilted or rolled, and stomatal opening and photosynthesis are substantially reduced. When RWC is reduced to 50 to 60% for several hours, cells in the leaf normally die and the damage becomes irreversible (Catsky, 1974b; Hsiao et al., 1984). Note that only a few percentage points may spell the difference between a leaf that is well supplied with water and one that is beginning to experience deficiency, at least enough to inhibit leaf expansive growth and canopy development. Unfortunately, as pointed out before, the determination of RWC can involve errors of this magnitude because of the uncertainty in knowing when the sample is saturated.

The remaining water status indicators are either more directly linked to physiological functions or the actual measurements thereof. On the other hand, like most physiological functions, they may be affected by other factors such as inorganic nutrition and, in addition, each has its own drawbacks. The dimensional measurements are relatively easy to make (except for the leaf thickness measurements) but difficult to interpret unless extensive comparisons have been made among several irrigation regimes for the same crop at the same location. The interpretation of stomatal opening, in contrast, can be made in a general way even without establishing values for a well-irrigated control. The drawback is, however, that stomatal opening is not sensitive enough as an indicator of water stress for the canopy development phase since expensive growth is inhibited markedly by the time stress becomes severe enough to inhibit stomatal opening. Canopy temperature has this same drawback. Additionally, its monitoring is more complex because of the need to also monitor humidity and establish the base line of well-watered control for a range of humidities. In published studies, yields have been well correlated with the cumulative or mean canopy-air temperature difference, whether normalized (Smith et al., 1985) or not (Walker & Hatfield, 1979). The correlation was done for a wide range of yields, ranging from very high to very low. In irrigated agriculture, however, the concern is to manage the irrigation to keep the yield at or near the maximum. For this narrow range (e.g., 80–100%) of yields at the top, the published data, due to scatter, appeared to show very poor correlations between yield and canopy-air temperature difference.

Leaf and stem elongations are undoubtedly the most sensitive for detecting the onset of water stress. Inhibition of elongation by mild water stress is so marked that direct comparison with well-watered control is not needed. On the other hand, as with stomatal opening or canopy temperature, factors other than water can be responsible for the slowing in expansive growth. Furthermore, in many determinate crops, elongation ceases shortly after the onset of anthesis and therefore cannot be used as an indicator thereafter. Even in many indeterminate crops, elongation can cease or become uselessly slow in the later growth stages. Finally, automated monitoring of leaf elongation is complicated by the fact that the instrument must be moved from leaf to leaf at regular intervals as the leaves progressively slow down and stop growth when reaching maturity.

The desired levels of plant water status at different growth stages for high productivity have been described in general and qualitative terms earlier. The preceding discussion makes it clear that the quantitative use of plant water status measurements for irrigation management is more complicated, and no values of the measurement parameters can be given for general use across different crops. Of the different measurements, it is better to use the ones more closely tied to crucial physiological functions. Leaf or stem expansive growth is perhaps the most useful measurement for irrigation management at the vegetative growth stage. The only need is to expand the data base on the normal leaf growth rates of different crops when well watered, for comparison in determining when stress is sufficient to call for irrigation. Xylem cavitation may be a measurement with enough sensitivity for use during the vegetative growth stage as well; but more evaluation is necessary and the number of cavitation event at a given water status will likely be crop specific. The other function-related measurements, such as stomatal opening or canopy temperature, are apparently not sensitive enough for the vegetative growth stage, but should provide good indications for the need for irrigation at later stages, when the canopy has closed.

As for water potential and its components, ψ_p, though intimately linked to some key physiological functions, is too difficult to measure for routine use. Both Ψ and ψ_s offer possibilities, as attested by the scheduling of irrigation of cotton based on Ψ measured in a pressure chamber. Because the threshold values triggering irrigation are crop and growth stage dependent, substantial amount of additional research is necessary before Ψ and ψ_s can be widely used for this purpose.

In spite of the present limitations, plant water status measurements provide a critical part of the total information package necessary for the rational management of irrigation. They complement the measurements taken in the soil and estimates based on weather, and provide the only direct indication of the health and well being of the crop.

VIII. SUMMARY

This chapter has discussed many of the direct and indirect measurements of plant water status and their potential uses in irrigation research or management. Emphasis has been placed on the more common methods and those with good potential for adaptation for field use. When relevant, the focus was on principles. The views expressed are personal, as are some of the procedural details, which are given to aid the nonspecialists.

For the measurements of water potential, thermocouple psychrometry and dewpoint hygrometry were reviewed briefly first. This was done because of the prominence of these methods in the literature, although in my opinion they are not yet developed to a stage such that routine field measurements can be made. Next, the Shardakov dye method was described, much more thoroughly, because it is simple and requires virtually no equipment, yet is capable of a precision comparable to some of the thermocouple psychrometers. The special details needed to achieve this precision were given. The pressure chamber technique also received extra attention, but for quite different reasons. It is presently the most widely used field method, yet confusion exists regarding the principle of the method and what parameter it really measures. Hopefully my discussions alleviated some of the confusions. In addition, some useful but uncommon procedural details were described for the pressure chamber method. In comparing the different methods of measuring leaf or branch Ψ, it was pointed out that when the results of two methods disagree, it is exceedingly difficult to know which of the two is the more reliable, since primary standards are lacking. The common assumption that thermocouple psychrometers or hygrometers are the most accurate appears unwarranted, at least in some studies with some types of these instruments.

The next section was on the measurement of RWC. Highlighted were the difficulties in knowing when full hydration of the sample is achieved for measuring the saturated weight of the sample. Coverage was also given to the techniques of determining the components of water potential, ψ_s and ψ_p. Again some emphasis was placed on using the pressure chamber for this purpose, by determining the water potential-volume curves. Recent findings on changes in ψ_s with time during measurements were also discussed. A number of more indirect indicators of plant water status, though often functionally more significant than tissue Ψ, were covered last. These included changes in stem diameter and leaf thickness, stomatal opening, canopy temperature, and leaf and stem elongation.

After discussing the measurement techniques, the use of plant water status indicators in irrigation management was considered. As background, how yields derive from the accumulation and partition of photosynthetic assimilates and the hand-in-hand relationship between biomass productivity and cumulative transpiration were sketched out. The relative sensitivity of the important productivity parameters to water stress at various times of the crop life cycle was also summarized as a part of the reference frame. The significance of the differential sensitivity to irrigation scheduling was pointed out, with some emphasis on the fact that the development of foliage canopy is particularly sensitive to water stress.

Critical values of Ψ, its components, or RWC for maintaining productivity differ among crop species, as well as between growth stages. Because of this complexity, it is not yet possible to specify for most crops the series of plant water status threshold values at which to apply irrigation. A notable possible exception is cotton, for which there is a fair data base to make recommendations. Measurements based on physiological functions, being more closely tied to productivity of the crop, are probably more valuable for the purpose of irrigation scheduling. Much more research is needed, however, to build the necessary data base. For the present, measurements of plant water status complement the measurements taken in the soil and of the weather as the basis for making scientific irrigation management decisions.

ACKNOWLEDGMENT

I thank Bruce Lampinen for his timely assistance in the preparation of the manuscript and in carrying out the experiment depicted in Fig. 9-2.

REFERENCES

Acevedo, E., T.C. Hsiao, and D.W. Henderson. 1971. Immediate and subsequent growth responses of maize leaves to changes in water status. Plant Physiol. 48:631–636.

Allaway, W.G., and T.A. Mansfield. 1969. Automated system for following stomatal behavior of plants in growth cabinets. Can. J. Bot. 47:1995–1998.

Barrs, H.D. 1968. Determination of water deficits in plant tissues. p. 235–368. *In* Water deficits and plant growth. Vol. 1. Development, control and measurement. Academic Press, New York.

Barrs, H.D., and P.E. Weatherley. 1962. A re-examination of the relative turgity technique for estimating water deficits in leaves. Aust. J. Biol. Sci. 15:413–428.

Begg, J.E., and N.C. Turner. 1970. Water potential gradients in field tobacco. Plant Physiol. 46:343–346.

Bennett, J.M., P.M. Cortes, and G.F. Lorens. 1986. Comparison of water potential components measured with a thermocouple psychrometer and a pressure chamber and the effects of starch hydrolysis. Agron. J. 78:239–244.

Boyer, J.S. 1967a. Leaf water potentials measured with a pressure chamber. Plant Physiol. 42:133–137.

Boyer, J.S. 1967b. Matric potentials of leaves. Plant Physiol. 42:213–217.

Boyer, J.S. 1968. Relationships of water potential to growth of leaves. Plant Physiol. 43:1056–1062.

Boyer, J.S. 1969. Measurement of the water status of plants. Annu. Rev. Plant Physiol. 20:351–364.

Boyer, J.S. 1972. Use of isopiestic technique in thermocouple psychrometry. I. Theory. p. 51–55. *In* R.W. Brown and B.P. Van Haveren (ed.) Psychrometry in water relations research: Proceedings of the symposium on thermocouple psychrometers. Utah State Univ. 17–19 Mar. 1971. Utah Agric. Exp. Stn., Logan

Boyer, J.S., and E.B. Knipling. 1965. Isopiestic technique for measuring leaf water potentials with a thermocouple psychrometer. Proc. Natl. Acad. Sci. USA 54:1044–1051.

Brown, R.W., and B.P. Van Haveren (ed.). 1972. Psychrometry in water relations research: Proceedings of the symposium on thermocouple psychrometers. Utah State Univ. 17–19 Mar. 1971. Utah Agric. Exp. Stn., Logan.

Campbell, E.C., G.S. Campbell, and W.K. Barlow. 1973. A dewpoint hygrometer for water potential measurement. Agric. Meteorol. 12:113–121.

Catsky, J. 1974a. Indirect methods of water content determination. p. 123–131. *In* B. Slavik (ed.) Methods of studying plant water relations. Springer Verlag New York, New York.

Catsky, J. 1974b. Water saturation deficit (relative water content). p. 136–156. *In* B. Slavik (ed.) Methods of studying plant water relations. Springer Verlag New York, New York.

Cosgrove, D.J., and R.E. Cleland. 1983. Osmotic properties of pea internodes in relation to growth and auxin action. Plant Physiol. 72:332–339.

Cosgrove, D.J., E. Van Volkenburgh, and R.E. Cleland. 1984. Stress relaxation of cell walls and the yield threshold for growth: Demonstration and measurement by micro-pressure probe and psychrometer techniques. Planta 162:46–54.

Craft, A.H., H.B. Currier, and C.R. Stocking. 1949. Water in the physiology of plants. The Ronald Press Company, New York.

Cutler, J.M., K.W. Shahan, and P.L. Steponkus. 1979. Characterization of internal water relations in rice by a pressure-volume method. Crop Sci. 19:681–685.

Dainty, M. 1976. Water relations of plant cells. p. 12–35. *In* U. Lüttge and M.G. Pitman (ed.) Encyclopedia of plant physiology. New series, Vol. 2. Transport in Plant II. Part A. Cells. Springer-Verlag New York, New York.

Duniway, J.M. 1971. Comparison of pressure chamber and thermocouple psychrometer determinations of leaf water status in tomato. Plant Physiol. 48:106–107.

Ehrler, W.L., S.B. Idso, R.D. Jackson, and R.J. Reginato. 1978. Wheat canopy temperature: relation to plant water potential. Agron. J. 70:251–256.

Fereres, E., E. Acevedo, D.W. Henderson, and T.C. Hsiao. 1978. Seasonal changes in water potential and turgor maintenance in sorghum and maize under water stress. Physiol. Plant. 44:261–267.

Fischer, R.A. 1973. The effect of water stress at various stages of development on yield processes in wheat. p. 233–241. *In* Plant response to climatic factors. Proc. Symp., Uppsala, Sweden, 1970. UNESCO, Paris.

Fischer, R.A., and R.M. Hagan. 1965. Plant water relations, irrigation management and crop yield. Exp. Agric. 1:161–177.

Fischer, R.A., and N.C. Turner. 1978. Plant productivity in the arid and semiarid zones. Annu. Rev. Plant Physiol. 29:277–317.

Fiscus, E.L., S.D. Wullschleger, and H.R. Duke. 1984. Integrated stomatal opening as an indicator of water stress in Zea mays. p. 278–295. In Agricultural electronics—1983 and beyond. Vol. I. St. Joseph, MI.

Gates, D.M. 1968. Transpiration and leaf temperature. Annu. Rev. Plant Physiol. 19:211–238.

Grange, R.I. 1983. Solute production during the measurement of solute potential on disrupted tissue. J. Exp. Bot. 34:757–764.

Grant, R.F., M.J. Savage, and J.D. Lee. 1981. Comparison of hydraulic press, thermocouple psychrometer, and pressure chamber for the measurement of total and osmotic leaf water potential in soybeans. S. Afr. J. Sci. 77:398–400.

Grimes, D.W., W.L. Dickens, and H. Yamada. 1978. Early-season water management for cotton. Agron. J. 70:1009–1012.

Grimes, D.W., and H. Yamada. 1982. Relation of cotton growth and yield to minimum leaf water potential. Crop Sci. 22:134–139.

Hanks. R.J. 1983. Yield and water-use relationships: An overview. p. 393–411. In H.M. Taylor, W.R. Jordan, and T.R. Sinclair (ed.) Limitations to efficient water use in crop production. ASA, CSSA, and SSSA, Madison, WI.

Hanks, R.J., and R.W. Brown (ed.). 1987. Proceedings of international conference on measurement of soil and plant water status. Vol. 2. Plants. Utah State Univ., Logan.

Hiler, E.A., C.H.M. Van Bavel, M.M. Hossain, and W.R. Jordan. 1972. Sensitivity of southern peas to plant water deficit at three growth stages. Agron. J. 64:60–64.

Hsiao, T.C. 1973. Plant responses to water stress. Annu. Rev. Plant Physiol. 24:519–570.

Hsiao, T.C. 1975. Variables affecting stomatal opening—complicating effects. p. 28–31. In Measurement of stomatal aperture and diffusive resistance. Wash. State Univ., Bull. 809.

Hsiao, T.C., E. Acevedo, E. Fereres, and D.W. Henderson. 1976. Water stress, growth and osmotic adjustment. Phil. Trans. R. Soc. London Ser. B. 273:479–500.

Hsiao, T.C. 1982. The soil-plant-atmosphere continuum in relation to drought and crop production. p. 39–52. In Drought resistance in crops, with emphasis on rice. Int. Rice Res. Inst., Los Baños, Philippines.

Hsiao, T.C., and K.J. Bradford. 1983. Physiological consequences of cellular water deficits. p. 227–265. In H.M. Taylor, W.R. Jordan, and T.R. Sinclair (ed.) Limitations to efficient water use in crop production. ASA, CSSA, and SSSA, Madison, WI.

Hsiao, T.C., and R.A. Fischer. 1975. Mass flow porometers. p. 5–11. In Measurement of stomatal aperture and diffusive resistance. Washington State Univ. Bull. 809.

Hsiao, T.C., and J. Jing. 1987. Leaf and root expansive growth in response to water deficits. p. 180–192. In D.J. Cosgrove and D.P. Knievel (ed.) Physiology of cell expansion during plant growth. Am. Soc. Plant Physiol., Rockville, MD.

Hsiao, T.C., J.C. O'Toole, E.B. Yambao, and N.C. Turner. 1984. Influence of osmotic adjustment on leaf rolling and tissue death in rice (Oryza sativa L.). Plant Physiol. 75:338–341.

Idso, S.B., R.D. Jackson, P.J. Pinter Jr., R.J. Reginato, and J.L. Hatfield. 1981. Normalizing the stress-degree-day parameter for environmental variability. Agric. Meteorol. 24:45–55.

Jackson, R.D., R.J. Reginato, and S.B. Idso. 1977. Wheat canopy temperature: A practical tool for evaluating water requirements. Water Resour. Res. 13:651–656.

Jordan, W.R. 1970. Growth of cotton seedlings in relation to maximum daily plant-water potential. Agron. J. 62:699–701.

Kaufmann, M.R. 1968. Evaluation of the pressure chamber method for measurement of water stress in citrus. Proc. Am. Soc. Hortic. Sci. 93:186–190.

Klepper, B., V.D. Browning, and H.M. Taylor. 1971. Stem diameter in relation to plant water status. Plant Physiol. 48:683–685.

Knipling, E.B., and P.J. Kramer. 1967. Comparison of the dye method with the thermocouple psychrometer for measuring leaf water potentials. Plant Physiol. 42:1315–1320.

Lang, A.R.G., and H.D. Barrs. 1965. An apparatus for measuring water potentials in the xylem of intact plants. Aust. J. Biol. Sci. 18:487–497.

Mederski, H.J. 1961. Determination of internal water status of plants by beta ray gauging. Soil Sci. 92:143–146.

Mederski, H.J., and W. Alles. 1968. Beta gauging leaf water status: Influence of changing leaf characteristics. Plant Physiol. 43:470–472.

Milburn, M.A., and R.P.C. Johnson. 1966. The conduction of sap. II. Detection of vibrations produced by sap cavitation in *Ricinus* xylem. Planta 69:43–52.

Millar, B.D. 1966. Relative turgidity of leaves: Temperature effects in measurement. Science 154:512–513.

Nakayama, F.S., and W.L. Ehrler. 1964. Beta ray gauging technique for measuring leaf water content changes and moisture status of plants. Plant Physiol. 39:95–98.

Namken, L.N., J.F. Bartholic, and J.R. Runkles. 1969. Monitoring cotton plant stem radius as an indication of water stress. Agron. J. 61:891–893.

Neumann, H.H., and G.W. Thurtell. 1972. A peltier cooled thermocouple dewpoint hygrometer for in situ measurement of water potentials. p. 103–112. *In* R.W. Brown and B.P. Van Haveren (ed.) Psychrometry in water relations research: Proceedings of the symposium on thermocouple psychrometers. Utah State Univ. 17–19 Mar. 1971. Utah Agric. Exp. Stn., Logan.

Passioura, J.B. 1980. The meaning of matric potential. J. Exp. Bot. 31:1161–1169.

Puritch, G.S., and J.A. Turner. 1973. Effects of pressure increase and release on temperature within a pressure chamber used to estimate water potential. J. Exp. Bot. 79:342–348.

Radulovich, R.A. 1984. Reproductive behavior and water relations of cotton. Ph.D. diss., Univ. of California, Davis.

Rajendrudu, G., M. Singh, and J.H. Williams. 1983. Hydraulic press measurements of leaf water potential in groundnuts. Exp. Agric. 19:287–291.

Rawlins, S.L. 1966. Theory for thermocouple psychrometers used to measure water potential in soil and plant samples. Agric. Meteorol. 3:281–292.

Richards, L.A., and G. Ogata. 1958. Thermocouple for vapor pressure measurement in biological and soil systems at high humidity. Science 128:1089–1090.

Richtie, G.A., and T.M. Hinckley. 1975. The pressure chamber as an instrument for ecological research. Adv. Ecol. Res. 9:165–254.

Savage, M.J., and A.Cass. 1984. Measurement of water potential using in situ thermocouple hygrometers. Adv. Agron. 37:73–126.

Scholander, P.F., H.T. Hammel, E.A. Hemmingsen, and E.D. Bradstreet. 1964. Hydrostatic pressure and osmotic potential in leaves of mangroves and some other plants. Proc. Natl. Acad. Sci. U.S.A. 52:119–125.

Scholander, P.F., H.T. Hammel, E.D. Bradstreet, and E.A. Hemmingsen. 1965. Sap pressure in vascular plants. Science 148:339–346.

Shackel, K.A. 1984. Theoretical and experimental errors for in situ measurements of plant water potential. Plant Physiol. 75:766–772.

Shackel, K.A. 1987. Direct measurement of turgor and osmotic potential in individual epidermal cells. Plant Physiol. 83:719–722.

Shardakov, V.S. 1948. Novyi polevoi metod opredeleniya sosushchei sily rastenii. [New field method of suction force determination]. Dokl. Akad. Nauk SSSR 60:160–172.

Silk, W.K., T.C. Hsiao, U. Diedenhofen, and C. Matson. 1986. Spatial distributions of potassium, solutes, and their deposition rates in the growth zone of the primary corn root. Plant Physiol. 82:853–858.

Sinclair, T.R., and M.M. Ludlow. 1985. Who taught plants thermodynamics? The unfulfilled potential of plant water potential. Aust. J. Plant Physiol. 12:213–217.

Slavik, B. (ed.). 1974a. Methods of studying plant water relations. Springer Verlag New York, New York.

Slavik, B. 1974b. The determination of water potential of cells and tissues. p. 15–74. *In* B. Slavik (ed.) Methods of studying plant water relations. Springer Verlag New York, New York.

Smith, R.C.G., H.D. Barrs, J.L. Steiner, and M. Stapper. 1985. Relationship between wheat yield and foliage temperature: Theory and its application to infrared measurements. Agric. For. Meteorol. 36:129–143.

Spanner, D.C. 1951. The Peltier effect and its use in the measurement of suction pressure. J. Exp. Bot. 2:145–168.

Spomer, L.A. 1985. Techniques for measuring plant water. HortScience 20:1021–1028.

Stewart, J.I., R.M. Hagan, and W.O. Pruitt. 1974. Function to predict optimal irrigation programs. J. Irrig. Drain. Div. Am. Soc. Civ. Eng. 100:179–199.

Turner, N.C. 1981. Techniques and experimental approaches for the measurement of plant water status. Plant Soil 58:339–366.

Turner, N.C., and M.J. Long. 1980. Errors arising from rapid water loss in the measurement of leaf water potential by the pressure chamber technique. Aust. J. Plant Physiol. 7:527–537.

Tyree, M.T., and M.A. Dixon. 1983. Cavitation events in *Thuja occidentalis* L.? Plant Physiol. 72:1094–1099.

Verner, L., W.J. Kochan, D.O. Ketchie, A. Kamal, R.W. Braun, J.W. Berry Jr., and M.E. Johnson. 1962. Trunk growth as a guide to orchard irrigation. Idaho Agric. Exp. Stn. Res. Bull. 52.

Walker, G.K., and J.L. Hatfield. 1979. Test of the stress-degree-day concept using multiple planting dates of red kidney beans. Agron. J. 71:967–971.

Wenkert, W., E.R. Lemon, and T.R. Sinclair. 1978. Changes in water potential during pressure bomb measurement. Agron. J. 70:353–355.

Westgate, M.E., and J.S. Boyer. 1986. Reproduction at low silk and pollen water potentials in maize. Crop Sci. 26:951–956.

Wiebe, H.H. 1966. Matric potential of several plant tissues and biocolloids. Plant Physiol. 41:1439–1442.

Wiebe, H.H., G.S. Campbell, W.H. Gardner, S.L. Rawlins, J.W. Cary, and R.W. Brown. 1971. Measurement of plant and soil water status. Utah Agric. Exp. Stn. Bull. 484.

Wolfe, D.W., D.W. Henderson, T.C. Hsiao, and A. Alvino. 1988. Interactive water and nitrogen effects on senescence in maize. II. Photosynthetic decline and longevity of individual leaves. Agron. J. 80:865–870.

Wong, S.C., I.R. Cowan, and G.D. Farquhar. 1979. Stomatal conductance correlates with photosynthetic capacity. Nature (London) 282:424–426.

Yegappan, T.M., and B.J. Mainstone. 1981. Comparisons between press and pressure chamber techniques for measuring leaf water potential. Exp. Agric. 17:75–84.

Zimmermann, U., and E. Steudle. 1978. Physical aspects of water relations of plant cells. Adv. Bot. Res. 6:45–117.

10   Root Growth and Water Uptake[1]

BETTY KLEPPER

USDA-ARS
Pendleton, Oregon

Roots are an important component in all crop management systems. Decisions made by growers impact root growth or function directly, especially in the management of soil amendments, fertilizers, pesticides, tillage options, and residues (Fig. 10–1). The root system serves as a bridge between management impacts on soil biological, chemical, and physical properties and changes in shoot function and harvested yield. For example, a decision to apply irrigation water at a particular time will increase the water content of the upper part of the soil profile. This increase in watter content may increase the elongation rate of roots in that part of the profile (root growth). At the same time, the rate of water extraction (root function) will also increase to raise the replacement rate of plant water lost to the atmosphere. This increase in water supply can, in turn, relieve shoot water deficits and increase shoot growth and function to result in improved yields. As researchers improve understanding of root-soil relationships, the important role of roots in crop growth will become more clear.

This chapter concerns growth and water extraction by field crop roots. Root systems explore soils by a combination of two processes: (i) downward penetration of main vertical axes and (ii) proliferation of roots at any given depth by production of branches. Root length density profiles at any time result from the number of vertical axes which have penetrated the various layers, the branching history and elongation rates of roots in each horizontal layer, and the decay or loss history of the roots in each layer. These principles lead to relatively characteristic developmental patterns for a given species, with fairly clear-cut differences between monocots and dicots. In addition to genetically determined factors, there are numerous soil and cultural factors that influence root distributions in field soils.

These aspects of crop root system development will be discussed prior to the topic of water uptake. Discussion of water movement to root surfaces will be relatively brief because water movement through soils has been covered in depth in Section III of this volume. The primary concern of this chapter will be on water movement into and through root systems and on interactions between root growth and water uptake.

---

[1] Contribution from Columbia Plateau Conservation Res. Ctr., USDA-ARS, Pendleton, OR 97801.

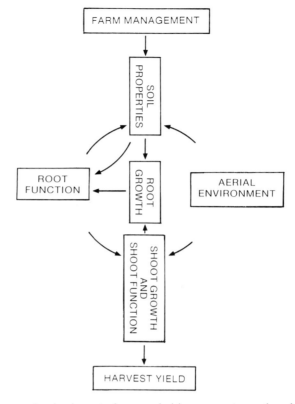

Fig. 10-1. Diagram showing impact of grower decisions on root growth and yield.

I. ROOT ANATOMY

Elongation of both main roots and branches occurs as a result of cell enlargement immediately behind the root tip (Fig. 10–2). The length of the elongation zone varies with species and growing conditions, but is generally short; for example, in maize (*Zea mays* L.) it is about 4 mm long and begins about 2 mm behind the tip (Versel & Mayor, 1985).

If we follow the history of a particular cell destined to become an epidermal cell bearing a root hair, we find that the cell begins as a more-or-less isodiametric daughter cell generated by mitotic division from the outermost cell layer of the apical meristem. The volume of a typical meristematic cell would be less than 2×10^{-5} mm^3 (Pilet & Senn, 1980). It contains dense, evacuolate protoplasm and is surrounded by a relatively thin cell wall. The structure of the microfibrils in this cell wall determines the final shape of the cell. Root cell walls are built with reinforcing rings or spirals which surround the cell on the lateral tangential walls. These reinforcements prevent these cells from expanding to the side as readily as they expand in length;

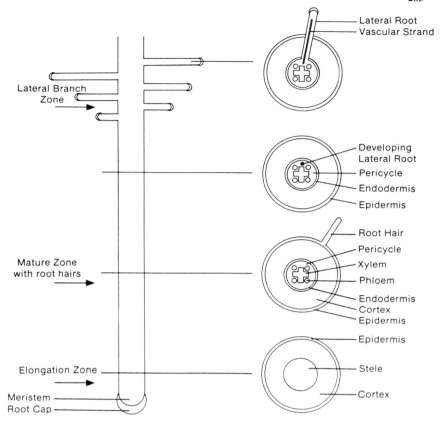

Fig. 10-2. Major anatomical zones of dicot roots showing the tissues present in cross-sections.

i.e., they make the cells elongate rather than enlarge isodiametrically. Cells in the interior of the root are further constrained by the pressures of the surrounding tissues (Veen, 1982).

After the new epidermal cell has been generated, it begins to enlarge both by increasing in dry matter and by increasing in water content. The dry material consists of organic materials such as soluble carbohydrates imported from the shoot by way of the phloem and minerals such as K taken up primarily from the surrounding soil solution. The source of water is from the surrounding soil although water may also be exchanged with adjacent tissues. Most of the increase in cell volume comes from the increase in water content. At 1 cm behind the tip, cell volume may have increased 10-fold to over 2×10^{-4} mm^3 (Pilet & Senn, 1980).

As the cell slows in elongation, it begins to mature. The exterior lateral wall of a root-hair-bearing epidermal cell develops a tubular outgrowth which expands by growth at the tip of the hair (Jaunin & Hofer, 1986). In rice, (*Oryza sativa* L.), this hair originates at the distal end of the epidermal cell (Tanaka, 1979). After the hair elongates, the walls become set so that further expansion does not occur. Meanwhile, the root has continued to add new cells near the tip and these new cells have elongated to push the tip forward so that the hair-bearing epidermal cell is now several centimeters behind the tip. Because of this anatomical arrangement, it is possible to follow the history of the process of elongation and maturation by studying serial lines of cells in longitudinal root sections (Feldman, 1977).

A. Mucilages

One important anatomical feature of growing roots is the lubrication provided to the root surface as it penetrates or moves alongside soil particles. Two processes are involved: (i) the production of slippery mucilaginous materials and (ii) the erosion of sloughing of whole cells from root surfaces, especially root cap surfaces. Mucilage on root surfaces varies from 1 to 10 μm in thickness and is exuded primarily from the root tip. Both the outer root cap cells and the distal epidermal cells contribute (Oades, 1978). Chemically, mucilages are mostly polymers, generally polysaccharides, and they provide part of the substrates that support rhizosphere organisms (Rougier & Chaboud, 1985).

B. Zone of Maturation

Behind the elongation zone is the *zone of maturation,* which is externally seen as a region of root hairs. These structures are usually about 0.01 mm diam. and 1 mm in length (Table 10–1). The abundance and length of these hairs is determined by both genetic and environmental factors. For example, Caradus (1981) described populations of white clover (*Trifolium repens* L.) with long (0.38 mm) and short (0.28 mm) hairs. In experiments in a silt-loam soil with maize, Mackay and Barber (1985) showed that the percentage of total length of roots that bear hairs, the number of the hairs per unit surface area, and their length all increased as volumetric soil moisture fell from 32 to 22%. They also found that the number of root hairs per unit of root surface fell as soil P concentrations rose from 0.81 to 203.3 μM P L^{-1}. Soil P concentrations did not affect root hair length in these experiments with maize nor with wheat (*Triticum aestivum* L.) (Ewens & Leight, 1985). On the other hand, in nutrient solution experiments on rape (*Brassica napus* L.), spinach (*Spinacea oleracea* L.), and tomato (*Lycopersicon esculentum*), Foehse and Jungk (1983) showed that increasing the P concentration from 2 to 100 μM P L^{-1} decreased both root hair length and density for all three species. Using a split-root experiment, they demonstrated that the P content of the plant, and not the P content of the local soil surrounding the growing root, controls the production of root hairs (Foehse & Jungk, 1983). The length of wheat root hairs is decreased by high pH (>7.0), low Ca concentrations and extremely low concentrations of NO_3 (below 0.12 M N m^{-3}) but is unaffected by changes in supply of other elements (Ewens & Leigh, 1985).

Table 10-1. Root hair density (number per millimeter) and length (mm) for several species.

Crop	Density	Length	Reference(s)
	mm^{-1}	mm	
Maize	162 ($\pm$ 6)	0.90 ($\pm$ 0.02)	Reid, 1981
	20-160	0.15-0.49	Mackay & Barber, 1985
	55-65	0.56 ($\pm$ 0.01)	Shierlaw & Alston, 1984
Perennial ryegrass (Lolium perenne L.)	88 ($\pm$ 6)	1.12 ($\pm$ 0.05)	Reid, 1981
Wheat (Triticum aestivum L. em Thell)	30-80	0.2-0.8	Bok, 1973
Annual ryegrass (Lolium multiflorum Lam.)	121 ($\pm$ 4)	0.67 ($\pm$ 0.02)	Shierlaw & Alston, 1984

Also, the internal features of the root change in the region of maturation so that one can discern mature primary tissues: epidermis, cortex (including its innermost layer, the endodermis), and the stele comprised of pericycle, xylem, and phloem (see Fig. 10-2). These tissues mature at different distances from the root tip. For example, in peanut (*Arachis hypogea* L.) as the rate of root growth increased from 2 to 18 mm d^{-1}, phloem was found ranging from 0.75 to 1.0 mm from the tip (Salim & Oryem-Origa, 1981). The early maturity of phloem provides for delivery of food materials close to the growing tip of the root where there are frequent plasmodesmata in walls for symplastic delivery of organic materials to growing tissues (Warmbrodt, 1985).

Water and nutrients that enter the plant must traverse the epidermis, in many species also the hypodermis, the cortex (including its innermost layer the endodermis), and the pericycle prior to entering the xylem. Similarly, organic materials from the phloem must move across these same tissues in order to supply food to peripheral cells.

C. Older Zones

Behind the mature zone is a zone showing lateral branches. Even further back is a zone where primary tissues senesce, for example where the epidermis and cortex slough off in cereal root axes. In those plants where secondary growth occurs (as in taprooted dicot species), the vascular cambium located between the xylem and phloem is activated in this older part of the root to generate new xylem and phloem cells and may eventually produce a root with bark similar to the bark in shoots.

D. Branch Anatomy

After roots have elongated and tissues have differentiated, branch root initials arise in the outermost cell layer (*pericycle*) of the stele about 10 to 20 mm proximal to the root tip. These initials are readily observed in cross sections because they are made up of dense, generally evacuolate meristematic cells. These initials then begin to grow and literally tear their way out of the

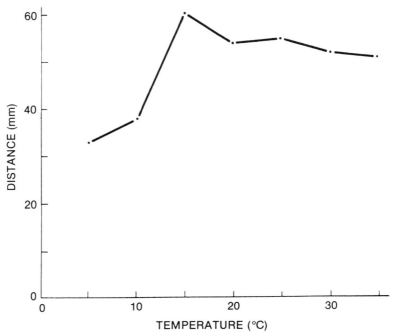

Fig. 10-3. Effect of temperature on distance between root apex and youngest branch of bean roots. Redrawn from Brouwer and Hoogland (1964).

parent root as a result of growth pressures generated. The distance between the tip of the parent root and the point of appearance of young lateral roots depends on species and growing conditions. Figure 10-3 shows the effect of temperature on this distance for roots of bean (*Phaseolus vulgaris* L.) (Brouwer & Hoogland, 1964). For this species, temperature effects are significant only below 15 °C. Table 10-2 shows features of primary roots of four species grown at room temperature. Under the conditions of this experiment, the distance from the tip of the youngest lateral root ranges from 5.5 to 8.2 cm for the three legumes, but is greater (11.4 cm) for maize (MacLeod & Thompson, 1979).

As the distal end of the new branch root presses outward through endodermal, cortical, and epidermal layers, the proximal end differentiates to produce xylem cells which abut the already-matured xylem of the parent root (Esau, 1965). This junction of new to old xylem is a potential site for resistances to water flow in the soil-root-shoot pathway (Klepper, 1983). Water and minerals taken up by the lateral root must pass across this junction in monocot roots which have no secondary growth. Many dicot roots have secondary growth so that smooth elbows of xylem tissue are laid down to carry the xylem sap smoothly from the lateral root up the parent root in a continuous linear series of xylem elements (Klepper, 1983). Presumably, this structure offers less resistance to flow than does the monocot junction.

The materials brought into a plant root system by different lateral roots are mixed, to some extent at least, in the parent root xylem where water and

Table 10-2. Morphological features of primary root systems on seedlings of four species grown for up to 12 d under laboratory conditions. From MacLeod and Thompson (1979).

Species	Broadbean	Pen	Maize	Bean
Distance of primordia from tip, cm†	2.3	1.6	3.2	2.1
Distance of youngest lateral from tip, cm	8.2	5.5	11.4	5.6
Primordia formed,				
no. cm^{-1}	5.4	3.9	8.4	6.8
no. d^{-1}	11.0	7.8	25.2	12.4
Laterals, no. cm^{-1}	6.2	4.7	7.7	6.8
Time from primordia formation to appearance of lateral, d	3.1	3.1	3.2	2.4

† Visible under dissecting microscope.

minerals can be slowly exchanged by movement through tangential vessel walls and by anastomoses which often occur in stems (Esau, 1965). However, in many plants such mixing is minimal so that water taken up by roots on one side of the plant primarily supplies that same side of the canopy (Kolesnikov, 1971). In these plants, certain areas of leaf are "targetted" for delivery of materials from particular roots.

E. Timing of Branching

The timing of branch appearance is controlled primarily by tissue maturation; i.e., root branch initials will not differentiate prior to a set point in tissue maturation. Field and laboratory observations both suggest that there is a window of time during which lateral roots grow out from a particular segment of root. Generally, when root systems are extracted from soil or from potting medium, the proximal lateral roots are progressively older, longer, and more branched than are the distal ones. Rarely do we see a mixture of old, branched laterals and young, fresh laterals. Therefore, there appears to be a limited time window for a root segment to produce lateral roots. For controlled-environment growing conditions, this means that lateral roots will appear at a specific time after the root hairs grow out and that the distance between the root tip and the point of appearance of the newest lateral is generally constant for the conditions selected. For field conditions, a combination of time and temperature must be used to time the period between initial growth of a segment of root and the appearance of lateral roots.

F. Branch Density

The linear frequency or density of lateral roots (number per centimeter) also varies with species and culture conditions. There is an effect of nutrients (Hackett, 1972), with especially marked decreases under severe nutrient deficiencies (Tennant, 1976a). Legumes have about 5 roots per centimeter, maize

8 (MacLeod & Thompson, 1979), radish (*Raphanus sativus* L.) 5 to 6 (Blakely et al. 1982), and barley (*Hordeum vulgare* L.) 1.7 to 3.7 (Hackett & Bartlett, 1971) or 5 to 6 (May et al., 1965). Apparently, there are significant effects of environment, even under optimized conditions in the laboratory, on linear frequency. The number of lateral roots may be correlated with the diameter of the parent axis (Sasaki et al., 1983). Unfortunately, most data are from laboratory experiments concerned only with the earliest root of seedlings under experimental conditions, which are often far removed from those found in the field.

In a recent field experiment (Belford et al., 1986), intact root systems of winter wheat were extracted from the field at different times in the growing season. The individual root axes were identified (Klepper et al., 1984) and examined for degree and linear frequency of branching. Those axes which developed branches after spring fertilizer was applied showed an increase in the linear frequency of branching (Belford et al., 1987).

The spatial arrangement of root branches on the parent root appears to be nonrandom. Lateral roots generally form opposite the xylem poles in dicots and can often be seen to occur in ranks (Charlton, 1983). In a study of five species, Mallory et al. (1970) found that the degree of regularity is inversely related to the number of protoxylem poles. Riopel (1969) found a nonrandom dispersed pattern of lateral roots in six monocot species. As Mallory et al. (1970) pointed out, these patterns are often difficult to discern because not all the lateral primordia develop into lateral roots and the pattern is presumably set by the initiation of primordia.

G. Effect of Environment on Root Anatomy

The anatomy of roots is not immutable. Environmental factors can have marked impacts on tissues. For example, changes in temperature can affect root diameter and length (Abbas Al-Ani & Hay, 1983), the diameter of the stele, and the number of metaxylem vessels in maize (Beauchamp & Lathwell, 1966), and the region of the bean root where differentiated tissues are seen (Brouwer & Hoogland, 1964). Mechanical impedance tends to reduce root length but increase root diameter (Logsdon et al., 1987). This increase in diameter is primarily due to an increase in radial size of cortical cells in soybean (*Glycine max* L.) (Peterson & Barber, 1981) and in barley (Wilson et al., 1977). In dry bean roots, mechanical impedance increases branching and decreases tissue porosity. Lack of O_2 also changes root anatomy by causing an increase in the tissue porosity through the formation of large intercellular spaces in cortical tissues. Unfortunately, most of our information on root anatomy has been based on sections of roots grown in artificial media. Soil environments can have profound influences on the physiologically relevant features of roots.

II. ROOT ELONGATION

A. Physics of the Elongation Zone

The cell enlargement process was modelled by Lockhart (1965) as an increase in volume caused by the influx of water into cells. Because of the constraining bands in root cell walls, most of the change in cell volume goes into increasing cell length. Most modern theories of tissue growth are based on Lockhart's cell growth concepts, with appropriate changes for tissue geometry and complexity (Cosgrove, 1986). Lockhart's concept was that increases in cell volume occur when cell turgor pressure exceeds the plastic yield limit of the wall. This concept can be stated as

$$\frac{1}{v}\frac{dv}{dt} = \phi\,(P - Y)$$ [1]

where

v = cell volume (m^3)
t = time (s)
P = turgor pressure (MPa)
Y = minimum turgor pressure for expansion (MPa)
ϕ = wall extensibility (S^{-1} MPa^{-1})

For roots growing in soil, this equation can be modified to include the constraint of the soil to root penetration and to include the fact that volume change is primarily length change in the root elongation zone.

$$\frac{1}{L}\frac{dL}{dt} = \phi\,(P - Y - M)$$ [2]

where M is the external pressure of the soil as it resists deformation by the root (Greacen & Oh, 1972; Veen, 1982). This formula says that the rate of change of root length (dL/dt) per unit of length (L) depends on two things: (i) the extensibility of the cell walls (Y), which is controlled primarily by endogenous plant biochemical factors (Goss et al., 1987) and (ii) the net pressure operating to expand cell volume. This net pressure is positively related to the turgor pressure (P) inside the root cells and is reduced by the restraints to cell expansion from both the friction within the walls of the expanding tissue (Y) and the friction between the moving root surfaces and the surrounding soil (M).

The exact mechanism by which soil resistance to penetration (M) slows root elongation is not yet known. A number of root properties change in response to the resistance. The first change that occurs in a root that encounters resistance is that forward motion of the tip ceases causing a reaction pressure on the cells proximal to the elongation zone. The root becomes thicker. Roots also become thicker when growth is slowed osmotically (Ciamporova & Luxova, 1976). Also, root hairs may become denser in the root region above the elongation zone (Eavis, 1967). These changes result in better basal anchorage of the root and a broader area for the exertion of pressure. Roots have a mechanical advantage in penetration because of the conical shape of the root cap. Force exerted by the tip (*root cap*) is spread over a small area, whereas the same force is spread over a large area in the basal part of the root. The mechanical advantage is that of a wedge and equals the ratio of the area of the base to area of the tip (Fig. 10–4). Thus, the root response to resistance is such that greater force is exerted by the root anatomical changes brought about by the resistance.

For a typical root in moist, loose, friable soil, the length of the elongation zone (L) is of the order of 5 mm and elongation rates (dL/dt) are about 1 mm h^{-1}. A value for ϕ of 0.24 MPa h^{-1} has been obtained for pea (*Pisum sativum* L.) internodes (Cosgrove, 1985). Using this value for roots would result in a net pressure for elongation of about 0.8 MPa. This net pressure might arise, for example, from a cell osmotic pressure of 1.2 MPa and a water potential of −0.1 MPa, to give a net cellular turgor pressure (P) of 1.1 MPa with a 0.2 MPa yield threshold (Y) and 0.1 MPa root-soil frictional resistance term (M). Unfortunately, these values for properties of the growing root in soil are extremely difficult to measure and they all may vary with changes in soil environment or with endogenous plant factors. For a particular soil type and plant species, an empirical approach (Whisler et al., 1986) can estimate root elongation rates. In this approach, the relative rate of root elongation is calculated from the penetration resistance of the soil which in turn is calculated from empirical relationships between penetration resistance and soil bulk density and water content for the particular soil type.

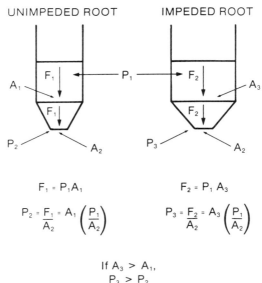

Fig. 10–4. Diagram showing how the mechanical advantage of roots is affected by thin shape.

B. Physiology of the Elongation Zone

To maintain high rates of cell elongation, roots need adequate supplies of energy, which requires O_2 for oxidative phosphorylation. If O_2 is below about 3% concentration in the soil atmosphere or if compaction and/or waterlogging causes the O_2 diffusion rate in soil to fall below a critical threshold, root elongation will slow and eventually cease. The presence of a water film around the root itself is very restrictive to O_2 supply rate since the rate of O_2 diffusion through water is 10^4 slower than through air. Some O_2 can be supplied from the shoot in those roots that develop aerenchyma. Production of these air-filled spaces is increased in many species in response to waterlogging.

Low O_2 supplies can have other effects as well. For example, Eavis (1972b) found that reductions in O_2 supply caused roots to be thicker. He found that, from 21 to 8% O_2 concentrations, the rate of cell division remained the same; decreases in elongation resulted from impacts on cell enlargement. However, cell division dropped off at 3% O_2. Asady et al. (1985) found that accumulations of toxic anaerobic metabolites in bean xylem increased as soil O_2 diffusion rates decreased in a soil compaction experiment. Considerable work has also been done on the $CH_2 = CH_2$ metabolism of anaerobic roots (Bradford & Yang, 1981). For example, tomato plant roots increase their export of ethylene precursors to shoots so that shoots manufacture the ethylene which brings about the well-known epinasty characteristic of flooded plants (Bradford & Yang, 1980).

Biosynthesis of new cell materials such as membranes and wall materials is one of the growth processes that requires large amounts of O_2. Large amounts of carbohydrates are also required as a base material for these syntheses. This photosynthate is delivered to the growing root tip via the phloem, which is mature relatively close to the root apex. Carbohydrates released from phloem must move across the tissues of the root, presumably by diffusion processes. Little is known about partitioning processes which allocate materials to the different growth points of the root system; nor do we have good measurements of the storage of carbohydrates within growing root systems. Nevertheless, it is logical to assume that root elongation rate can be limited by the rate of supply of photosynthate.

Another energy-requiring process is the active and selective accumulation of ions from the surrounding soil solution. Under laboratory conditions at least, there is a decrease in root surface pH which coincides with the elongation zone, probably associated with proton efflux from the growing tissues (Pilet et al., 1983). Ions taken up into the root cells serve, along with sucrose and other sugars, to maintain the osmotic potential in the cells of the elongating zone. This osmotic potential is necessary to maintain the P term in Eq. [2]. Osmotic values of elongation zones generally range from about 0.8 to 1.2 MPa (Taylor & Ratliff, 1969a).

One ion crucial to root growth is Ca, which is required in the elongation process (Burström, 1983). With insufficient Ca, roots are thickened, distorted, and have poor development of laterals (Simpson et al., 1977).

To sustain rapid root elongation in soils, there must also be a rapid radial influx of water from the surrounding soil solution. This water movement takes place down a water potential gradient. The control of this process is still being elucidated, but the wall-yielding process is thought to be of central concern in controlling the rate of elongation (Cosgrove 1987; Goss et al., 1987).

Root temperature has an important effect on root growth. Most species show a temperature optimum which is often lower than the optimum temperature for shoots. Temperature affects rates of metabolic processes, the rate of diffusion of materials within the growing tip, rates of ion uptake from surrounding media, rates of cell division, and rates of cell enlargement. Hydraulic conductivity of plant membranes decreases with decreasing temperature, making it more and more difficult to move water into elongating cells as root temperature declines.

C. Soil-Root Relations in Elongation

Root growth rate is influenced greatly by soil factors including soil structure, strength (compressibility), water content, temperature, porosity, gaseous diffusivity, pH, toxicity, and fertility. In reference to Eq. [2], the main requirements that must be met for elongation are (i) an adequate delivery rate of carbohydrates to the root tip via the phloem, both for osmoticum and for metabolic use (impacts primarily P and ϕ), (ii) soil space to move into (impacts M), (iii) adequate supplies of inorganic ions for root uptake both to maintain electrical balance and to provide osmoticum (impacts P, possibly ϕ), (iv) water uptake to maintain turgor as growth proceeds, and (v) a sufficient supply of chemical energy, presumably from ATP, to carry out metabolic processes (impacts P, Y). Table 10–3 shows which of the requirements are influenced by the soil properties listed in the table. Most of the soil properties influence more than one requirement. Some of these soil properties change rapidly and others change slowly. For example, soil structure, pH, and porosity may not change much during a crop growth season, but water content and temperature change on a weekly or hourly basis.

The requirement for space to permit addition of new root material is often not met in agricultural soils. Traffic pans, plow pans, and other adverse subsoil conditions limit pore space (high bulk density) and cause excessive soil strength (low compressibility) from compaction of subsoils. Both compaction and cementation limit root extension by increasing the M term in Eq. [2].

We tend to think of soils as being homogeneous in any particular horizon. If we were to imagine ourselves shrunk to the size of a plant root-tip, we would have an entirely different view of soil. It would appear full of cracks, worm holes, old root channels, and voids between soil particles or between soil aggregates. Some of these voids would be larger than the root tip and some would be smaller. The large pores provide channels of easy root penetration of soil and may have profound effects upon the shape

Table 10-3. Primary influences which indicifual soil properties have on root elongation requirements.

Soil property	Primary effects			
	Ions	Space	Water	Chemical energy
Structure	--	X	X	--
Porosity	--	X	X	--
Water content	--	X	X	--
Strength	--	X	--	--
Gaseous diffusivity	X	--	X	X
Temperature	X	--	X	X
pH	X	--	--	--
Fertility	X	--	--	--

of a root system because of the ease with which roots can penetrate (Wang et al., 1986). Roots tend to grow along ped surfaces in soils, even when the ped has a relatively low resistance to penetration (Whiteley & Dexter, 1983). Roots growing into a large void have been observed in time-lapse cinematography to circumnutate until the root reencounters soil material whereupon the root proceeds to grow touching the newly found soil surface; in other words, roots grow along the surfaces of cracks, not in cracks (M.G. Huck, 1988, personal communication). These easy routes cause the roots to grow around rather than through soils that have cracks and large voids (Russell, 1977). Vertically oriented cracks, which correspond to the geotropic orientation of the root, give the most rapid penetration. The proportion of roots that enter a vertical crack decreases with decreasing width of the crack and with approach to perpendicular of the root path to the direction of the crack (Dexter, 1986a, b).

When roots penetrate soil materials, they compress the cylinder of soil immediately around themselves (Dexter, 1987). Thus, the elastic modulus or compressibility of the soil is an important factor in root growth. This factor is usually measured as soil resistance or mechanical impedance with a penetrometer which monitors the force required to press a root-shaped probe through the soil. Penetrometers require more force than comparably shaped roots. The mechanical advantage provided by the cone-shaped root cap, the sloughing of root cap cells, the presence of lubricating mucilages at the tips of roots, the presence of localized weak spots in soil, and the circumnutation and flexibility of growing tips all allow roots to penetrate soils with penetrometer resistances greater than the osmotic potentials of their elongating cells. For example, Whiteley et al. (1981) found that the penetrometer had to exert a pressure 5.1 times greater than a root tip to penetrate a fine sandy loam soil.

Two factors that strongly influence soil resistance to penetration in agricultural soils are soil compaction (increase in bulk density) and soil water content (Taylor & Ratliff, 1969b; Blanchar et al., 1978). Both of these can be further confounded by concomitant effects on O_2 diffusion rate (Eavis, 1972a).

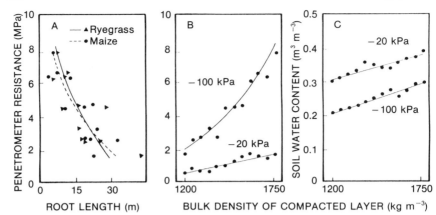

Fig. 10–5. Relation of penetrometer resistance (*A*) and volumetric water content (*B*) to the bulk density of the soil at water potentials of −20 and −100 kPa and relationship of root length for maize and ryegrass at −100 kPa soil water potential as related to penetrometer resistance. Redrawn from Shierlaw and Alston (1984).

Increases in bulk density decrease the soil pore space and thus its compressibility (Cornish et al., 1984b). Figure 10–5 shows the relationship between mechanical resistance and bulk density for a sandy loam soil. Soil texture influences the sensitivity of roots to changes in bulk density. For example, Jones (1983) found that at near-optimum soil water contents, the percentage silt + clay or the percentage clay both were good predictors of the bulk density at which rooting was restricted to 20% of maximum. Viewed from a statistical perspective, the critical bulk density for a taprooted species which has only one vertical axis is considerably less than it is for a cereal which has several seminal axes and thus several opportunities to penetrate a volume of soil (Dexter, 1986b).

As soil water content increases, soil strength declines when bulk density is held constant (Taylor & Ratliff, 1969b). This change occurs because the water in soil acts as a lubricant to the friction between soil particles as they are moved against one another.

Poor aeration is often a confounding factor when compaction decreases available pore space and when soil water content increases to the point where pore space is filled with water (Eavis & Payne, 1969; Warnaars & Eavis, 1972). Since the rate of diffusion of O_2 through air is 10 000 times as fast as it is through water (Lemon, 1962), the O_2 diffusion rate decreases as soil water content increases. Aeration problems are generally more common in soils with a high fraction of clays.

III. SOIL MANAGEMENT INFLUENCES
ON ROOT DISTRIBUTION

The phenotypic expression of the root system of a particular cultivar depends to a great extent upon the soil conditions left by antecedent history of a field. This section will discuss the rooting patterns of crops where soil physical and chemical properties brought about primarily by cultural practices limit the exploration of soil volume by crop root systems.

A. Tillage Against Mechanical Impedance

One of the most common field soil problems encountered in agricultural fields is the presence of high-strength layers. These layers can be brought about by the compaction of soil from the traffic of heavy vehicles over the field, by the smearing of wet soil by a plow share or other implement, or by other processes such as slaking or puddling which may reorient and consolidate soil particles and increase their strength. High-strength layers may also be a naturally occurring part of the profile as, for example, in caliche layers, fragipans, or other cemented soil zones. Models have been developed of the operation of traffic to compact soil. Changes in bulk density can be predicted using fundamental soil physical parameters (Gupta & Larson, 1982; Smith, 1987; Gupta & Allmaras, 1987). These bulk density changes usually have significant effects on soil strength.

Generally, when soil strength exceeds 2.0 MPa (Taylor et al., 1966), root growth is severely restricted. However, presence of worm holes and other biopores can permit root elongation through high-strength layers. For example, Ehlers et al. (1983) showed that the limiting soil strength for root growth was 3.6 MPa in the tiller layer of a loess soil but increased to 4.6 to 5.1 MPa both in an untilled surface soil and in the subsoil of both the tilled and untilled treatments. They found upon excavation that roots were preferentially following old biopores which provided localized low-strength channels in an otherwise impenetrable soil. Absence of tillage promotes the development of high populations of soil invertebrates, including earthworms (Ellis et al., 1977; Edwards & Lofty, 1978).

If left untreated, these pans can severely retard root exploration of deep soil. Roots that are unable to penetrate the pan will grow along it or will be slowed in their growth. Lateral branching may be increased when the axis is stopped. The net result is a shallow, less effective root system (Finney & Knight, 1973).

High-strength soil can be disrupted by deep tillage (e.g., subsoiling, ripping) to provide a more favorable matrix for root growth in the part of the profile where tillage can have an influence. Generally these tillages increase crop yields, especially in drier regions, but there are many soils where benefits are minimal (Unger, 1979). Where tillage is beneficial, it acts by increasing the volume of soil that can be accessed by the crop root system (Vepraskas & Miner, 1986). For example, a silt loam soil in Belgium with a plow pan which provided a penetrometer resistance of more than 3.0 MPa was subsoiled at 60 cm prior to planting winter barley (*Hordeum vulgare* L.). At flowering, total root weight at 50 to 75 cm was 325% of the control, the concentration of mineral nutrients in the barley shoots was significantly increased, and grain yields were greater (Ide et al., 1984).

B. Soil Water Status

As soil dries, its strength increases due to the closer appression of soil particles as films of water are reduced. Root growth rate slows well before the −1.5 MPa soil water potential (permanent wilting point). For example, Newman (1966) found a marked reduction in flax (*Linum usitatissimum* L.) root growth at −0.7 MPa soil water potential in a potting soil (fertilized sand-loam-peat mixture). Cotton (*Gossypium hirsutum* L.) root length densities decreased (decay rate exceeded elongation rate) in a fine sandy loam soil at about −0.1 MPa soil water potential in a rhizotron experiment (Taylor & Klepper, 1974) (Fig. 10–6).

The general sequence of events in the root system development of a crop involves the proliferation of roots in moist parts of the soil profile, the gradual use of water from moist areas to dry the soil down to a point where it no longer supports root elongation (about −0.7 MPa), and finally a drying of the same soil down to the permanent wilting point (about −1.5 MPa). Some growth of roots can take place in soil at or near the permanent wilting point if other parts of the root system are in moist soil (Portas & Taylor, 1976). In a loess soil, Ehlers et al. (1980) found that oat (*Avena sativa* L.) roots ceased elongation at a soil water potential of −1.9 MPa but approached zero at −0.8 MPa.

Generally, root length densities under well-watered conditions exponentially decline with depth but when water is withheld and the upper part of

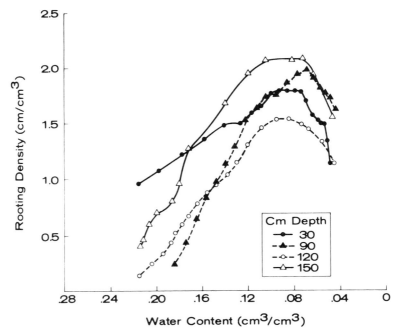

Fig. 10–6. Root length density of cotton in different horizontal layers of soil in a rhizotron as related to soil water content in the layer. Redrawn from Taylor and Klepper (1974).

the profile becomes dry, there is a proliferation of roots in the deeper, moister part of the profile. For example, Fig. 10–7 shows cotton root length densities in a well-watered profile compared to a profile that was allowed to dry (Klepper et al., 1973). Note that the total amount of root material did not change greatly. The loss of roots from decay in the upper part of the profile which was allowed to dry was about the same as the gain of roots deep in the profile. Similar results were shown for sorghum (*Sorghum bicolor* L.) by Blum and Arkin (1984) who found that irrigation promoted shallow root growth and a highly skewed distribution of roots with high concentrations in shallow soil. Hoogenboom et al. (1987) found that soybean produced most of its roots in the top 0.6 m of the profile while roots of plants not irrigated were deeper. In this species, total root length was increased slightly by lack of irrigation (Huck et al., 1986). Similar increases have been observed in other crops (Jodari-Karimi et al., 1983).

Surface soil moisture is especially important for monocots which initiate a series of axes which must penetrate surface soils to reach deeper soil. Dry surface soils can completely stop these axes from growth (Blum & Ritchie, 1984).

C. Soil Texture and Aggregation

Generally roots proliferate best in medium to coarse-textured soils. In Australia, Tennant (1976b) found that wheat roots penetrated deeper in a deep sand soil or a sandy loam soil than in clay soils. For a silt loam soil, Donald et al. (1987) found that root systems were smaller on coarser

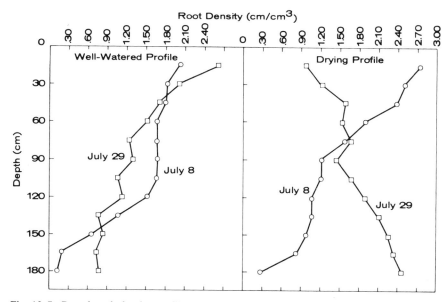

Fig. 10–7. Root length density profiles on two dates of well-watered cotton plants compared to plants in unirrigated soil. Redrawn from Klepper et al. (1973).

aggregates than on finer ones. The difference was primarily due to a decrease in the length of secondary lateral roots in the pots containing large (6.4–12.8 mm) aggregates. Effects of soil texture on root growth patterns may result from factors directly associated with texture or they may relate to other differences such as water relationships arising from texture.

D. Fertility

Impacts of increased soil fertility on root growth directly are difficult to separate from the indirect impacts caused by the fact that shoots are more vigorous with better fertility and provide better nutrition to roots. Roots generally respond to increased fertility by increasing the root length, usually through longer and/or more numerous branches. For example, Maizlish et al. (1980) found that with increased N, corn root axis numbers increased and first order laterals increased in both number (slightly) and length (greatly). In grasses, effects of differences in nutritional levels are most obvious and readily measured with respect to higher order laterals and the nodal system which generally reflects plant response better than the seminal system (Tennant, 1976a). For example, Christie (1975) found that P deficiency decreased the elongation rate of laterals on the seminal root system and decreased both the length and the number of nodal roots in two grasses native to semiarid Queensland. Belford et al. (1987) found that increasing N fertilizer amounts increased branching of nodal roots when applied to winter wheat during the tillering phase.

E. Residue Distribution and Mulching

Surface mulches can increase rooting densities in surface soil (Barber, 1971) probably because of the effects of the mulch on conserving moisture, and in some cases at least, because of cooler soil temperatures under mulch (Chaudhary & Prihar, 1974). Barber (1971) showed that removal of residues depressed root growth in the top 10 cm of the profile. Depression in root growth associated with surface residues in the plant row have been reported (Allmaras et al., 1973; Rickman & Klepper, 1980). These reductions are thought to be associated with toxic leachates or increased microbial activity associated with residues (Elliott et al., 1987).

F. Soil Aeration

The required rate of O_2 delivery to the root surface may not be met in soils that are waterlogged, in high density soils, and/or in soils with high levels of microbiological activity. Aeration problems can be especially acute in clay soils. Air-filled porosity is affected by both soil bulk density and water content. In addition, high bulk density layers can restrict the infiltration of water and create perched water or temporary waterlogging. Decreases in aeration decrease the rate of root elongation (Ellis & Barnes, 1980) and can cause significant yield reductions (Hardjoamidjojo et al., 1982; Meyer et al.,

1985). Damage to roots and to yield depends on the timing of the flooding, the temperature during flooding, and the frequency and duration of flooding (Belford et al., 1985). Cannell et al. (1985) showed that surface flooding restricted root growth in the soils above 50 cm but root elongation continued in deeper horizons. When reduced aeration and mechanical impedance are both imposed, root growth can be more severely restricted than by either factor separately (Gill & Miller, 1956).

G. Planting Geometry

Increases in planting density cause root systems of crop plants to compete sooner for water and minerals. Optimum row spacing and within-row plant distances are specific to site, season, and crop (Klepper et al., 1983). In one experiment on loess soil in western Iowa, Böhm (1977) found that soybean plants on 25-cm rows had about three times the length of root present compared to fields planted on 100-cm rows, but root distribution in the profile was similar for the two row spacings with about three-fourths of the roots in the top 10 cm of the profile and the other one-fourth distributed down to 50 cm at 44 d after planting (Taylor & Klepper, 1978).

Increased plant populations and narrow rows did not influence corn root distributions, but cotton plants in a low-strength soil tended to show higher root length densities at lower profile depths when they were planted densely (Grimes et al., 1975).

H. Planting Date

The main impact of planting date is on the soil temperature regime that the root system encounters, especially early in the growing season. Use of thermal time instead of calendar time is often helpful in comparing material from different planting dates. For example, Barraclough and Leigh (1984) found that winter wheat root growth between December and June was reasonably linear when plotted against accumulated thermal time, which was calculated as the sum of daily average air temperatures from sowing with a 0 °C base temperature.

I. Agricultural Chemicals

Some soil-applied herbicides and fungicides affect root growth either by persisting in the soil from application to previous crops or by direct injury from current application. The injury may be indirect as when application of chlorsulfuron {2-chloro-N-[(4-methoxy-6-methyl-1,3,5-triazin-2-yl) aminocarbonyl] benzenesulfoamide} causes an increase in *Rhizoctonia* damage to cereals (Rovira & McDonald, 1986) or it may result from the direct effects of the chemical on roots. Currey and Teem (1977) discuss root injury for 10 classes of herbicides. Symptoms include such factors as reductions in root elongation rate, clubbing or callusing of root tips, necrosis, suppression of root branching, and disruption of mitosis.

Excessive fertilizer and salinity in soils can also impact root growth. If concentrated fertilizers are banded too close to the seed, the high local osmotic potential of the soil solution may damage seedling roots with resultant impacts on emergence and stand establishment (Klepper et al., 1983). Saline soils are often caused by frequent shallow irrigation of slowly permeable soils in arid environments with the resulting accumulation of salt in the root zone from rapid evaporation of water from the soil surface. The principal effect of salinity is to decrease root elongation but there is an effect on cell division if the salinity gets high enough. For example, at -1.2 MPa osmotic potential, onion (*Allium cepa* L.) roots grow at a rate of 37% of control roots at 0 MPa osmotic potential; new cell production is reduced to 77% of the control and cell size is reduced to 45% of the control (Gonzalez-Bernaldez et al., 1968). Roots are capable of osmotic adjustment, especially under field conditions (Turner et al., 1987). In a growth chamber experiment, Oosterhuis and Wullschleger (1987) found that cotton responded to water stress by increasing the osmotic value in leaves 0.41 MPa and in roots by 0.19 MPa. The osmotic adjustment in roots was a larger percentage (46%) of the control osmotic value than for leaves (22%) because of the lower osmotic value in root tissues.

J. Acid Soils

Where high precipitation has leached soils extensively, acidity is often a problem. Although liming can correct this problem in surface soils, roots still encounter acidity in subsoils. The primary difficulty brought about by acid soil is the solubilization of Al which is toxic to roots. Calcium deficiency can also occur (Adams & Moore, 1983). Further problems can arise from lack of P and from release of heavy metals other than Al (Wong & Bradshaw, 1982). The limiting pH below which most plants begin to show adverse effects is about 5.0, but varies with the crop.

IV. BIOLOGICAL FACTORS IN ROOT SYSTEM DEVELOPMENT

Unless adverse soil factors intervene, the pattern of root system development depends primarily on the plant species. Each crop has a characteristic sequence of below-ground phenological events analogous to the familiar sequence of events in canopy phenology.

Root system types are of three basic patterns (Fig. 10–8): diffuse or fibrous, characteristic of most monocotyledonous crops, and either taprooted or with a modified taproot, characteristic of most dicotyledonous crops.

A. Geotropic Effects

Root systems grow downwards but not all individual roots do this. Primarily the taproots of dicots and the main seminal axes of cereals show *positive orthogeotropism* (grow vertically downward) while lateral roots

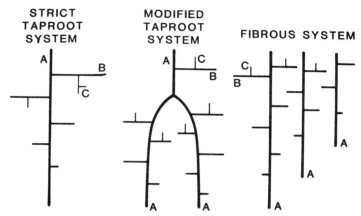

Fig. 10–8. Diagram showing the three basic root system types (left to right): (1) many axes with branches characteristic of cereals, (2) tap-rooted and (3) with a modified taproot where several strong laterals penetrate vertically into the soil.

generally grow perpendicular to the parent root and may, in fact, even grow upwards (Jackson and Barlow, 1981). The crown root axes of cereals generally grow for a short distance at an oblique angle (are *plagiogeotropic*) before becoming orthogeotropic deeper in the profile. Temperature can affect the geotropic response. For example, Mosher and Miller (1972) showed that the direction of growth of corn radicles could be changed by changes in soil temperature. Also the angle of inclination of soybean lateral roots is controlled by soil temperature (Kaspar et al., 1981).

B. Monocotyledons

These plants generally produce roots from stem nodes. In the grass family, these nodal roots consist of seminal roots, which arise from primordia present in the embryo, and crown roots, which arise from the lower vegetative nodes (Klepper et al., 1984). Generally, for the grass family, crown roots begin development by the three-leaf stage (Newman & Moser, 1988) and all support prior to this stage must come from the seminal system.

For wheat, the seminal root system consists of the primary root and up to five roots associated with the two nodes in the seed; it originates at the depth of planting. Nodal roots from the wheat coleoptilar node may also originate from the planting depth (Peterson et al., 1982). The foliar nodes in the crown give rise to a crown root system which arises from successively higher nodes with the passage of time (Klepper et al., 1984; Gregory, 1983). Generally there are 2 to 4 axes per node (Klepper, 1987).

Figure 10–9 illustrates for cereals a naming system which identifies nodal axes of monocots by referencing each axis to its node of origin and by indicating a direction of growth with respect to the other plant organs associated with the node. In wheat, each of the lower nodes produces a pair of roots (for the *n*th node, *n*A and *n*B) on either side of the stem at right angles to the bud associated with that node. Later in development, the plant can produce another pair of roots (nX and nY) at right angles to the first pair (Fig. 10–9).

In monocots, the unique relationship between stem nodes and crown root axes permits correlations between shoot and root development to be derived. For example, Fig. 10–10 shows the degree of branching found on individually identified root axes and the number of leaves on the main stem as a function of phyllochrons after emergence for wheat plants (Klepper et al., 1984). A *phyllochron* is the unit of time between equivalent growth stages of successive leaves, e.g., between a visible leaf number of n and $n + 1$. This diagram shows that each identified axis produces branches about 2.5 phyllochrons after it first appears. Since each phyllochron requires approximately 100 growing degree days (Rickman et al., 1985), then the appearance and branching of wheat root axes can also be predicted from cumulative growing degree days. Cumulative growing degree days is calculated by summing the average air temperatures in degrees C for each day with all negative means taking a value of zero. A plant sample 600 growing degree days after emergence would be expected to have six main stem leaves, well-branched seminal root axes, and a crown root system with branching only on the axes associated with the first foliar node (Fig. 10–10). Similar information has been developed for rice plants (Yamazaki & Harada, 1982) and for corn (Picard et al., 1985) and the techniques would presumably apply for all cereals.

The development of a monocot root system, then, can be described in terms of the production of a sequence of root axes and their subsequent branching. The axes from higher nodes tend to grow out to the side before going downward so that they explore ever-widening rings or cylinders of soil. Monocots do not undergo secondary growth. The early seminal system is

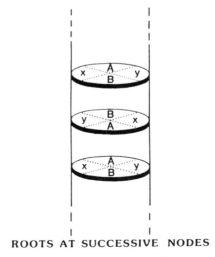

ROOTS AT SUCCESSIVE NODES

Fig. 10-9. Cereal naming system showing association of wheat axes with stem nodes. After Klepper et al. (1984).

not persistent if it is subjected to pathogen attack or if it is torn by frost-heaving. It commonly atrophies in corn (Maizlish et al., 1980). It can, however, persist until grain-filling in wheat (Rickman et al., 1985). The roots from the crown are usually important in extraction of water and minerals from soil during the late vegetative and the reproductive periods.

Early in cereal root system development, the growth rates of the main axes are slowed as branch roots are initiated and begin to grow (May et al., 1967; Brouwer, 1981). Generally the axes extend in depth at the rate of about 1 cm d^{-1}. Rooting depths over 1 m are the rule, not the exception, for cereals. For example, Weaver et al. (1924) reported winter wheat root depths of 5 to 7 ft (1.5–2.1 m) in NE. The length of root material produced by a cereal crop is also high. Values over 15 km m^{-2} of land surface for corn (Mengel & Barber, 1974) and over 10 km m^{-2} for wheat (Böhm, 1978) have been reported.

When the number of axes is restricted in cereals, compensatory growth occurs in other axes so that more branches are produced and they tend to grow longer (Jordan et al., 1979). Also, when growth is suppressed in one part of the root system, there is compensatory growth in the rest of the system.

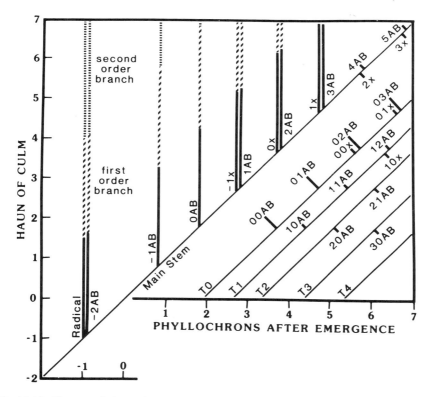

Fig. 10–10. Diagram relating mainstem cereal production of leaves, tillers, and root axes showing the branching times of named axes relative to shoot development. From Klepper et al. (1984).

C. Dicotyledons

These plants are generally taprooted so that the entire root system develops from the radicle and its branches. Adventitious roots produced from stems can provide part of the root system; this is especially common in tomato. Some species such as cotton have a strong woody taproot which grows vertically to considerable depth in the profile. Others have a shallow taproot with several strong lateral roots which take over the downward penetration of the profile. For example, Mitchell and Russell (1971) found that soybean root systems had strong lateral roots which grew out from the upper part of the taproot, grew horizontally for 35 to 40 cm and then turned downward to penetrate to considerable depth in the profile.

Dicots may be deep rooted. Kaspar et al. (1978) found that all seven soybean cultivars studied in a rhizotron experiment reached the compartment bottom at 2.2 m. As a general rule, dicots produce less root length in a growing season than cereals do. For example, Barber (1978) found that soybean root density was about 20% of the root density of corn grown in the same field. He found the maximum soybean root length to be 2.4 km m^{-2} of land surface on a Chalmers silty clay loam soil (fine-silty, mixed, mesic Typic Haplaquolls). This value was nearly the same as the 2.5 km m^{-2} found by Sivakumar et al. (1977) on an Ida silt loam [fine-silty, mixed (calcareous), mesic Typic Udorthents].

Most dicot roots are capable of secondary growth and develop complex vascular relationships among parent roots and their branches. The secondary xylem formed on the upper side of branch roots is generally laid down as a smooth elbow of continuous xylem tissue whereas the xylem on the lower side appears not to be continuous (Klepper, 1987).

D. Genetic Effects

Roots differ between different plant cultivars but have received much less attention than shoots. A recent review (O'Toole & Bland, 1987) lists numerous morphological features of roots studied for genotypic variation, including dry weight, diameter, length, branching, pulling resistance, elongation rate, orientation, horizontal and vertical spread, xylem vessel numbers and radii, and root/shoot ratios. Most of these parameters have been shown to differ among genotypes and some of them have been shown to be heritable. For example, Richards and Passioura (1981) showed that root vessel diameter in wheat has a narrow-sense heritability of 52%.

For most traits, especially under field conditions, it is difficult to separate shoot effects on root development from genetic effects because of the correlation between shoot vigor and root system size and vigor. Another problem is the overriding influence of soil structure and physical properties on rooting pattern.

E. Root-Shoot Relationship

Roots and shoots are interdependent. Limitations for one impose limitations on the other. Heavy grazing, for example, can decrease the root mass and length of grass swards of Caucasian bluestem (*Bothriochloa caucasica*) (Svejcar & Christiansen, 1987). Likewise, severe restriction of the root system depresses shoot growth (Peterson et al., 1984). The time of production of individual root axes from the crown is dependent on shoot development in cereals (Porter et al., 1986).

Thus there are feedback mechanisms for maintaining shoots and roots in relation to one another. This interrelationship can influence crop rooting patterns in field soil. For example, water stress generally reduces shoot expansion at a milder level of stress than it influences photosynthesis. Under mild water stress, roots still receive photosynthate from shoots and elongate new root material wherever soil conditions are favorable. Similar rules apply to nutrient deficiencies in general: they affect shoots more than and sooner than roots. Eventually, however, when stresses become severe, root growth can be impacted by effects of shoot stresses on carbohydrate supply. Such factors as shading of plants by weeds can decrease carbohydrates available for root proliferation. Also, in some species there is a decrease in root growth rate when growth of reproductive structures causes a diversion of assimilate away from the below-ground plant. For example, total root length of corn increased until silking (80 d), remained constant for 14 d and then declined (Mengel & Barber, 1974). Similar observations have been made on other species where measurements have shown either a cessation or a slowing of root growth rate after pollination.

F. Pathogens

Roots can be directly attacked by plant pathogens or they can be indirectly affected by the influence of shoot pathogens on shoot physiology. For example, powdery mildew (*Erysiphe graminis* f. sp. hordei Marchal) causes reductions in photosynthetic leaf area directly, but indirectly affects root dry weight and length, primarily through effects on laterals (Walters & Ayres, 1981).

The direct effects of soil-borne pathogens on roots are primarily from suppression of root elongation rate (e.g., inhibitory *Pseudomonads* as described by Frederickson and Elliott, 1985), from actual consumption of root material (e.g., by cutworms or other larvae), distortion of root tissues to make them less effective (e.g., by rootknotting types of nematodes), and, the most usual case, the rotting of roots as is seen in *Pythium* or *Rhizoctonia*.

G. Root Senescence

Although many of the roots of crop plants persist throughout the season, some branch roots senesce and decay during the season. This root "turnover" is a normal part of root system development. We do not know what the physiological and anatomical distinctions are between long-lived and ephemeral branches. In dicots, some branches undergo secondary growth and become a part of the framework of the root system; others do not undergo secondary growth and senesce. In cereal axes, the cortex is normally short lived (Deacon & Mitchell, 1985), but the stele and endodermis persist. Fusseder (1987) found that corn laterals with dead tips could remain functional and that the cortex of the main seminal root was still alive at the late grain-filling stage.

V. WATER MOVEMENT IN THE SOIL-PLANT CONTINUUM

Extraction of water from soil by plant roots takes place as part of a larger picture. Figure 10–11 shows the general relationship of transpiration and water absorption (Kramer, 1937). Transpired water is lost from leaves and must be resupplied by water from the root system. Figure 10–11 shows that there is a time lag in the response of root absorption to water loss from leaves. This lag means that there is a decrease in plant water content during the morning hours and a rehydration of plant tissues in the afternoon and evening. This loss of plant water content provides the plant water potential decreases which move water into the plant root system from the surrounding soil.

Water in the root system is hydraulically connected through the low-resistance xylem pathway, where water is held under tension at least during periods of rapid transpiration. In dicots which have taprooted or modified

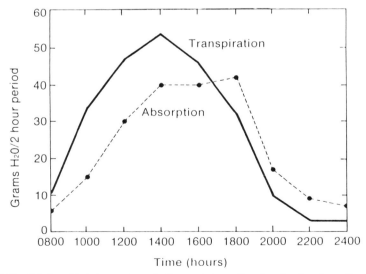

Fig. 10–11. Relationship between transpiration and absorption showing the absorption lag that causes plant water potential to decrease and creates the potential gradient for water entry into roots.

taprooted systems, water from all of the roots is channelled to the central taproot eventually. This central structure permits ready lateral exchange of water within the xylem of the taproot. In cereals that produce many axes, water is channelled to the crown where anastamoses permit some exchange (Klepper, 1983). Figure 10–12 shows this difference in taprooted and fibrous-rooted systems. The taprooted system is wired, using an electrical analogue, in series down the central taproot; the cereal system has many axes wired in parallel. The cereal system can be considered as a group of taproots (axes) all originating from the same plant crown at different times in the crop growing season. The primary difference in the two systems is the restriction to transfer between different parts of the root system at depth for the cereal. This same restriction applies to the modified taproot system common in some dicots.

Roots proliferate most readily in wet, but well-aerated, soil. They then absorb water from the newly explored soil volume because the water poten-

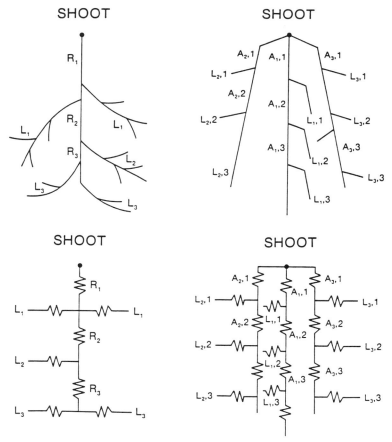

Fig. 10–12. Diagramatic representation of the "plumbing" differences in diffuse- and tap-rooted systems.

tials in these wet parts of the profile are higher than the water potentials of the root, at least during the daily transpiring period. While the newest root material takes up water rapidly from wet soil, older roots, which are usually in soil dried down by root extraction of moisture, take up water more slowly both because they are in drier soil and because they are older and have fewer newly expanded permeable root lengths available for uptake. This section concerns water movement into and through root systems from field soils.

A. Ohm's Law Analogues

Water movement in the soil-root system occurs almost exclusively as liquid flow, which is generally analyzed using an analogy to Ohm's law, where flow is directly proportional to driving potential and inversely proportional to resistance. This analogy can be used to describe a steady-state flow situation where the flux along a continuum is divided into a series of fluxes which remain constant even as the resistance of the material to flow changes over the continuum. These fluxes are then set equal to the water potential difference between set points in the system. For a plant in uniform soil, this equation can be stated as

$$\text{Water flux} = \frac{\psi r - \psi s}{Rrs} = \frac{\psi p - \psi r}{Rrp} \qquad [3]$$

where ψr is water potential (MPa) at the root surface, ψs is water potential of soil, ψp is shoot water potential at the soil line, Rrs is resistance of the soil cylinder surrounding the root and of the root-soil contact area to water transfer from bulk soil to the outermost cell layer of the root system, and Rrp is the resistance of the root and vascular system to water transfer. Care must be taken when equating fluxes across plant parts to ensure that all of the fluxes and potentials are considered for the entire plant (Richter, 1973).

The Ohm's law analogy suffers from some short comings when used to analyze water movement under field conditions. For example, the resistances are not constant; they change with flow rate. The capacitance of the compartments is neglected in calculating fluxes. Also, the most convenient points for referencing the fluxes (i.e., bulk soil, soil-root interface, root xylem) are not necessarily points that provide easy measurement of either water potentials or fluxes for validation. This discussion will follow work of Taylor and Klepper (1978) to illustrate how Ohm's law had been used to analyze water movement in the soil-root system.

Water taken up by root systems comes from soil at different depths with different water contents and antecedent root growth histories. Horizontal layers are usually assumed for partitioning water uptake over the profile. Then each separate layer can be addressed using an Ohm's law approach. Water flow into the root system of each layer is directly proportional to the length of root present in the layer and the difference in water potential between bulk soil in the layer and the shoot, and inversely proportional to the resistance to water flow between bulk soil and the shoot. Stated in equation form, the water uptake (Ui) from a uniformly rooted layer of soil of volume (Vi) is

$$Ui = (Vi)(Lvi)(qi)(\psi si - \psi ri) \qquad [4]$$

where

Vi = the volume of soil in the layer
Lvi = the root length density
qi = the average root water uptake per unit length of root or unit of driving potential
ψsi = soil water potential, and
ψri = root xylem water potential at midpoint of layer.

Horizontal layers are interconnected by the vertically growing roots (taproots and axes, generally) as shown in Fig. 10-13. Taylor and Klepper (1978)

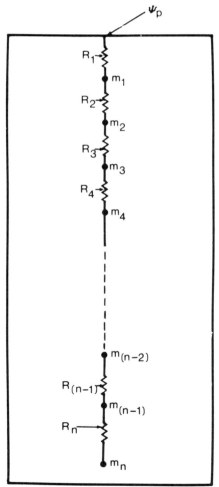

Fig. 10-13. Diagramatic representation of the way in which resistances and potential drops can be visualized in a root system.

estimated ψri from ψp (the xylem water potential at the soil line), the loss in potential due to elevation (ψzi), and the loss in potential due to friction between moving water and xylem vessel walls ($\Sigma \psi fi$). Thus, Eq. [4] can be rewritten as

$$Ui = (Vi)(Lvi)(qi)(\psi si - \psi p + \psi zi + \Sigma \psi fi).$$

In this equation, $Vi,\ Lvi,\ qi,\ \psi si,\ \psi p,$ and ψzi can be measured. The value of $\Sigma \psi fi$ can be estimated for a given plant by using an iterative technique (Klepper et al., 1983). In this approach, both axial and radial root resistances have been included: the axial in the ψfi term and the radial in the qi term.

B. Axial Resistances

Axial resistances to water flow occur within root vascular tissues as water moves from deep in the soil profile to shoots. Under most agronomic situations, axial resistances are rather small, but where the number or diameter of the xylem vessels delivering water to the soil surface is reduced, or where anatomical constrictions occur, (Luxova, 1986), axial resistances can become significant. An excellent example occurs in cereals where crown root axes are prevented from growing as, for example, when surface soils are dry (Passioura, 1972). Since cereal roots do not undergo secondary growth, seminal root xylem vessels must then suffice to supply shoots with water and can develop significant axial resistances (Meyer et al., 1978; Cornish et al., 1984a). Since most dicot roots are capable of secondary growth, axial resistances are often not significant (Rowse & Goodman, 1981). Occasionally, restrictive soil layers can impose excessive vascular constriction on dicots and cause axial resistances to be high (Taylor et al., 1964). Also, if cavitation occurs in root xylem (Byrne et al., 1977), it could cause an increase in axial resistance.

Axial resistances calculated from theory using the Poiseuille equation are generally less than the resistances measured in plant xylem (Tyree & Zimmerman, 1971; Petty, 1978). The comparatively low measured conductivities could be accounted for by the resistance to flow of scalariform endwalls. In addition, presence of boundary conditions different from those specified for the Poiseuille equation (Giordano et al., 1978), especially lateral exchange of water or solutes through porous vessel walls, may contribute to such differences (Klepper, 1983). Allowance must also be made for changes in vessel diameters caused by pressure differences across xylem walls (Greacen et al., 1976) and from point to point in the system. For example, metaxylem vessel radii increase with depth in seminal roots of wheat (Meyer & Alston, 1978). Finally, presence of tyloses that block vessels may preclude their participation in conduction.

In summary, then, conductivity of xylem tissue depends on the number, radii, and structures of conducting cells comprising the tissue, and flow depends on the hydraulic potential differences that drive flow. Anatomical obstruction is sufficient in xylem elements and vessels are sufficiently narrow to allow development of significant axial resistance, especially if some restriction is placed on the number of conducting vessels.

C. Radial Resistance

Resistances to water flow in root systems may account for the majority of the resistance to water flow through the plant (Black, 1979; Boyer, 1971). These resistances are not necessarily a constant, as implied in the Ohm's law treatment, but instead may decrease with increasing flow rates (Meyer & Ritchie, 1980). This behavior may arise in part because of the coupling of solute and water flux, especially at low flow rates (Fiscus, 1975) or because of the increase in permeability of the root system at higher flow rates (Aston & Lawlor, 1979). This increase in permeability could arise from an increase in the length of root used in absorption, from an increase in the possible pathways for water entry across a given piece of root or from a change in the permeability of a critical membrane (Powell, 1978). The shoot environment can also influence how leaf water potential changes as transpiration increases because of the effects that shoot environments have on root elongation rate and, therefore, on the presence of highly permeable new lengths of root (Bunce, 1978). It is also possible that the response of a system to a change in flow rate will depend on the presence or absence of dry surface soil layers (Landsberg & Fowkes, 1978; Seaton & Landsberg, 1978).

The radial resistance of root systems to water flow arises because water that crosses root tissues from epidermis to xylem may traverse either cell wall material (apoplast) or cytoplasmic material (symplast) up to the endodermis, but is forced to pass through cell membranes at least at the endodermal layer (Kramer, 1969). Consequently this layer of root tissues has received considerable attention with respect to its water transport properties (Clarkson & Sanderson, 1974). Figure 10-14 shows the rate of water uptake by intact root segments along a barley seminal axis at two humidities. The more mature tissues transmit considerably less water than does the region of the root 4 to 8 cm behind the root tip. The influence of lowered humidity (increased transpiration rate) is to increase water flow through this less permeable part of the root. The uptake rate appears to be controlled by the incidence of state III (with both suberin lamellae and tertiary wall formation internal to these lamellae) endodermis cells (Sanderson, 1983).

Conductance values for bean, soybean, and maize (Steudel et al., 1987) under laboratory conditions and for cotton in rhizotron experiments (Taylor & Klepper, 1975) are of the order of 2×10^{-7} ms^{-1} MPa^{-1}. These values correspond to hydraulic conductivities of the order of 10^{-6} cm d^{-1} (Taylor & Klepper, 1975). In older suberized pine roots, hydraulic conductance was about 40% of the conductance in young pine roots (Sands et al., 1982).

The involvement of membranes in the pathway for water movement through roots implies that there will be strong temperature effects (Dalton & Gardner, 1978). Such effects are found. For example, Markhart et al. (1979) showed much stronger effects of temperatures in soybean and broccoli root permeability than could be accounted for by the effects of temperature on water viscosity. These strong temperature effects result from impacts of temperature on radial, rather than axial, resistance (Fiscus, 1983).

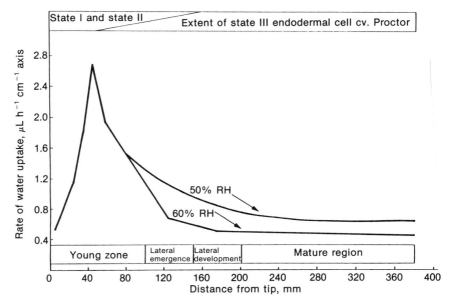

Fig. 10–14. The rate of water uptake by intact segments along the length of the seminal axes of barley, at two humidity regimes, in relation to the development of the endodermis. From Sanderson (1983).

Anaerobiosis can also influence root conductance of water. For example, Holder and Brown (1980) found that water uptake of bean roots decreased as soil O_2 concentrations were lowered. The mechanism for this effect is not clear.

Tissue dehydration can impact radial resistances to water flow. For example, Nobel and Sanderson (1984) found a 10-fold decrease in radial water flow in roots of desert succulents subjected to dehydration at 50% relative humidity.

D. Perirhizal Resistance

In addition to the resistances to water flow of radial and longitudinal movement through root systems, there are resistances associated with the root-soil interface and with the rhizosphere soil. The resistances, called *perirhizal* (around the root), are difficult to measure and have generally been inferred from circumstantial evidence. Two general types are considered: (i) a draw-down resistance, diurnally imposed by the rapid loss of water from the soil immediately adjacent to the root and (ii), a contact resistance which increases as the root surface has less contact with the liquid water in the soil.

The draw down resistance results from the fact that the hydraulic conductivity of soil decreases as it dries. If roots extract water rapidly from the cylinder of soil immediately adjacent to them, the replacement of that water from "bulk" soil at a distance from the root depends on the conductivity of the rhizosphere soil for water. There would be a diurnal pattern where the soil near the root would be dried out during the day and would recover at night. This phenomenon was analyzed by Gardner (1960), who predicted that substantial perirhizal resistance could occur when flow rates per unit length of root were high in soils that were close to the permanent wilting point. Cowan (1965) showed how the occurrence of significant soil resistance depends on the root length density. When root length density is low, flow rates per unit length of root are high and significant resistances occur at relatively high soil water contents. With dense root populations, the average flow rate is small and diurnal draw-down resistances are minimized. Soil texture is an important factor in the probability of having substantial perirhizal resistance (Sykes & Loomis, 1967).

This perirhizal resistance has also been described as a contact resistance (Herkelrath et al., 1977). Contact resistance increases as water retreats from large pores into smaller and smaller capillary areas in the soil and decreases the fraction of the root length actually wetted. The contact resistance would cause root water uptake rates to respond differently to a reduction in soil water potential arising from matric potential than to a reduction arising from reductions in soil water potential due to osmotic factors (Schleiff, 1986).

A contact factor could also arise when roots shrink diurnally (Huck et al., 1970), as a result of long-term decreases in root water potential (Cole & Alston, 1974), or from asymmetrical arrangement of roots in pores (Tinker, 1976). Shrinking of the root creates a gap between the root and the soil where water cannot move in liquid form. Vapor movement is much slower and a localized high resistance to water transfer would occur. Faiz (1973) demonstrated that these perirhizal gaps could apparently be temporarily eliminated in an experiment where he caused a transient increase in water uptake rate by squeezing a plastic bag containing the root system and soil, and presumbaly closed the air gaps without affecting cortical pressure potentials sufficiently to induce flow (Weatherley, 1976).

Root hairs might help in preventing air gaps since they are able to grow into very small pores and can effectively "glue" themselves to soil particles with exuded muscilages. The hairs can thus maintain contact with soil water better than the rest of the epidermal surface. The amount of water taken up by the root hairs is not clear, although it has been shown that they can absorb water (Rosene, 1943). Newman (1976) argued that they are less effective conduits of soil water than the soil itself.

E. Crop Water Uptake Patterns

Seasonal crop water uptake patterns depend on the precipitation and infiltration history of the field, long-term redistribution of water in the soil profile, and the crop root growth and water extraction history. All of these factors are influenced by the structure, texture, and management history of the soil (Hamblin, 1986).

The general water extraction pattern shows greater uptake from surface layers early in the season with progressive deepening of the uptake bulge as the season progresses. This pattern is especially evident in plots that receive

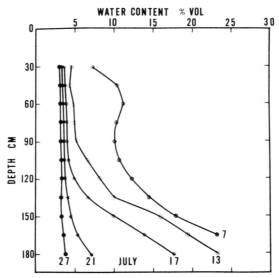

Fig. 10-15. Volumetric soil water content with depth for several dates during a drying cycle for cotton growing in a loamy fine sand (Klepper et al., 1973).

no precipitation or irrigation during the season. For example, Fig. 10-15 shows the water content remaining in the soil profile on several dates in July for nonirrigated soil containing roots of mature cotton plants at the Auburn rhizotron (Klepper et al., 1973). As upper soil layers were depleted, water extraction was shifted to deeper and deeper layers. This pattern is accentuated by the fact that the surface layers are generally permeated with roots earlier in the season than are the deeper layers.

Under continually well-watered conditions, the majority of the water is extracted from the upper part of the profile because the plant maintains high populations of roots in the upper profile and because these surface soils are wet. Because of the catenary arrangement of soil layers (Fig. 10-16), the upper layers supply the water first and low xylem water potentials are not transmitted to the xylem of deeper roots when demand is already satisfied in the surface layers.

The length of root present in a layer, the water content and potential of the layer, and the diurnal water potential levels in the plant xylem determine the amount of water extracted by roots in any given soil layer in a general sense (Taylor & Klepper, 1975). However, there are numerous reports showing that a few roots in moist soil can provide amounts of water disproportionate to their number. For example, Gregory et al. (1978) showed that when only 3% of the root dry weight was below 1 m, 19% of the water was taken up from below 1 m. Similar results have been shown for maize (Sharp & Davies, 1985) and soybean (Willatt & Taylor, 1978).

The actual maximum depth of rooting and the resultant depth of water extraction depend on crop species. For example, Hamblin and Tennant (1987) found that pea roots penetrated less deep than cereal roots and that water

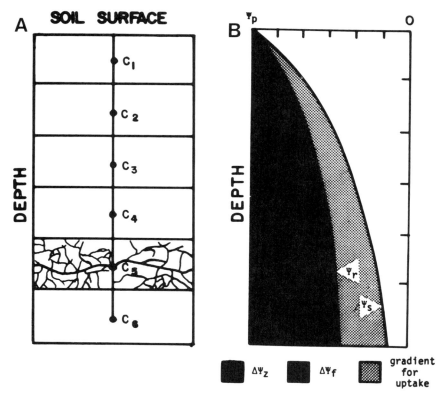

Fig. 10–16. Diagrams of depth relationships in model. (*A*) Relationships between absorbing and transporting roots. Water is extracted and transmitted to the mid-plane of a layer by absorbing roots; it is moved to the surface by transporting roots. (*B*) Potential relationships with flow-dependent axial resistance in transport roots.

uptake rates were also shallower for peas. Cultivars within a species can differ in rooting depth and in rates of exploration of deep soil as well (O'Toole & Bland, 1987).

Models describing crop water uptake rate as a function of depth and of time in the crop season have been written (Hillel et al., 1976; Taylor & Klepper, 1978; Ritchie, 1985; Hoogenboom & Huck, 1986; Whisler et al., 1986). Few of these models have validated dynamic root growth and decay algorithms which permit the root system to change over time and allow root growth and water uptake to interact dynamically. Effects of root age are rarely incorporated in an explicit way and only occasionally do authors address the problems associated with application of the model to range of environments and soil types. Few experimental data are available for assessing axial resistances and for validating models that include axial resistances explicitly. All of these areas need more research to improve our ability to predict the root growth and water uptake patterns of crops over a whole growing season.

REFERENCES

Abbas Al-Ani, M.K., and R.K.M. Hay. 1983. The influence of growing temperature on the growth and morphology of cereal seedling root systems. J. Exp. Bot. 34(149):1720–1730.

Adams, F., and B.L. Moore. 1983. Chemical factors affecting root growth in subsoil horizons of coastal plain soils. Soil Sci. Soc. Am. J. 47:99–102.

Allmaras, R.R., A.L. Black, and R.W. Rickman. 1973. Tillage, soil environment, and root growth. p. 62–86. In Proc. Natl. Conservation Tillage Conference, Des Moines, IA. 28–30 March. Soil Conserv. Soc. Am., Ankeny, IA.

Asady, G.H., A.J.M. Smucker, and M.W. Adams. 1985. Seedling test for the quantitative measurement of root tolerances to compacted soil. Crop Sci. 25:802–804.

Aston, M.J., and D.W. Lawlor. 1979. The relationship between transpiration, root water uptake, and leaf water potential. J. Exp. Bot. 30(114):169–181.

Barber, S.A. 1971. Effect of tillage practice on corn (Zea mays L.) root distribution and morphology. Agron. J. 63:724–728.

Barber, S.A. 1978. Growth and nutrient uptake of soybean roots under field conditions. Agron. J. 70:457–461.

Barraclough, P.B., and R.A. Leigh. 1984. The growth and activity of winter wheat roots in the field: The effect of sowing date and soil type on root growth of high-yielding crops. J. Agric. Sci. (Camb.)103:59–74.

Beauchamp, E., and D.J. Lathwell. 1966. Root-zone temperature effects on the vascular development of adventitious roots in Zea mays. Bot. Gaz. 127(2–3):153–158.

Belford, R.K., R.Q. Cannell, and R.J. Thompson. 1985. Effects of single and multiple water loggings on the growth and yield of winter wheat on clay soil. J. Sci. Food Agric. 36:142–156.

Belford, R.K., B. Klepper, and R.W. Rickman. 1987. Studies of intact shoot-root systems of field-grown winter wheat. II. Root and shoot development patterns as related to nitrogen fertilizer. Agron. J. 79:310–319.

Belford, R.K., R.W. Rickman, B. Klepper, and R.R. Allmaras. 1986. Studies of intact shoot-root systems of field-grown winter wheat. I. Sampling techniques. Agron. J. 78:757–760.

Black, C.R. 1979. A quantitative study of the resistances to transpirational water movement in sunflower (Helianthus annuus L.). J. Exp. Bot. 30(118):947–953.

Blakely, L.M., M. Durham, T.A. Evans, and R.M. Blakely. 1982. Experimental studies on lateral root formation in radish seedling roots. I. General methods, developmental stages, and spontaneous formation of laterals. Bot. Gaz. 143(3):341–352.

Blanchar, R.W., C.R. Edmonds, and J.M. Bradford. 1978. Root growth in cores formed from fragipan and B2 horizons of Hobson soil. Soil Sci. Soc. Am. J. 42:437–440.

Blum, A., and G.F. Arkin. 1984. Sorghum root growth and water use as affected by water supply and growth duration. Field Crops Res. 9:131–142.

Blum, A., and J.T. Ritchie. 1984. Effect of soil surface water content on sorghum root distribution in soil. Field Crops Res. 8:169–176.

Böhm, W. 1977. Development of soybean root systems as affected by plant spacing. Z. Acker Planzenbau 144:103–112.

Böhm, W. 1978. Untersuchengen zur Wurzelentwicklung bei Winterweizen. Z. Acker-und Pflanzenbau 147:264–269.

Boyer, J.S. 1971. Resistances to water transport in soybean, bean, and sunflower. Crop Sci. 11:403–407.

Bradford, K.J., and S.F. Yang. 1980. Xylem transport of 1- aminocyclopropane-1-carboxylic acid, an ethylene precursor, in water logged tomato plants. Plant Physiol. 65:322–326.

Bradford, K.J., and S.F. Yang. 1981. Physiological responses of plants to waterlogging. HortScience 16(1):25–30.

Brouwer, R. 1962. Influence of temperature of the root medium on the growth of seedlings of various crop plants. JAARB. IBS. 1962:11–18.

Brouwer, R. 1981. Coordination of growth phenomena within a root system of intact maize plants. Plant Soil 63:65–72.

Brouwer, R., and A. Hoogland. 1964. Responses of bean plants to root temperatures. II. Anatomical aspects. JAARB ISB 23–36.

Bunce, J.A. 1978. Effects of shoot environment on apparent root resistance to water flow in whole soybean and cotton plants. J. Exp. Bot. 29(110)595–601.

Burström, H.G. 1983. Calcium, strontium, and root growth. Z. Pflanzenphysiol. 109:91–93.

Byrne, G.F., J.E. Begg, and G.K. Hansen. 1977. Cavitation and resistance to water flow in plant roots. Agric. Meteor. 18:21–25.

Cannell, R.Q., R.K. Belford, P.S. Blackwell, G. Govi, and R.J. Thompson. 1985. Effects of waterlogging on soil aeration and on root and shoot growth and yield of winter oats (*Avena sativa* L.). Plant Soil 85:361-373.

Caradus, J.R. 1981. Effect of root hair length on white clover growth over a range of soil phosphorus levels. N.Z. J. Agric. Res. 24:353-358.

Charlton, W.A. 1983. Patterns of distribution of lateral root primordia. Ann. Bot. 51:417-427.

Chaudhary, M.R., and S.S. Prihar. 1974. Root development and growth response of corn following mulching, cultivation, or interrow compaction. Agron. J. 66:350-351.

Christie, E.K. 1975. Physiological responses of semiarid grasses. II. The pattern of root growth in relation to external phosphorus concentration. Aust. J. Agric. Res. 26:437-446.

Ciamporva, M., and M. Luxova. 1976. The effect of polyethylene glycol-induced water stress on the maize root apex. Biol. Plant. (Praha) 18(3):173-178.

Clarkson, D.T., and J. Sanderson. 1974. The endodermis and its development in barley roots as related to radial migration of ions and water. p. 87-100. *In* J. Kolek (ed.) Structure and function of primary root tissue, Veda, Publ. House Slovak Acad. Sci., Bratislavia.

Cole, P.J., and A.M. Alston. 1974. Effect of transient dehydration on absorption of chloride by wheat roots. Plant Soil 40:243-247.

Cornish, P.S., J.R. McWilliam, and H.B. So. 1984a. Root morphology, water uptake, growth and survival of seedlings of ryegrass and phalaris. Aust. J. Agric. Res. 35:479-492.

Cornish, P.S., H.B. So., and J.R. McWilliam. 1984b. Effects of soil bulk density and water regimen on root growth and uptake of phosphorus by ryegrass. Aust. J. Agric. Res. 35:631-644.

Cosgrove, D.J. 1985. Cell wall yield properties of growing tissue. Evaluation by in vitro stress relaxation. Plant Physiol. 78:347-356.

Cosgrove, D.J. 1986. Biophysical control of plant cell growth. Annu. Rev. Plant Physiol. 37:377-405.

Cosgrove, D.J. 1987. Wall relaxation and the driving forces for cell expansive growth. Plant Physiol. 84:561-564.

Cowan, I.R. 1965. Transport of water in the soil-plant-atmosphere system. J. Appl. Ecol. 2:221-239.

Currey, W.L., and D.H. Teem. 1977. Herbicides and root development. Soil Crop Sci. Soc. Fla. Proc. 36:23-28.

Dalton, F.N., and W.R. Gardner. 1978. Temperature dependence of water uptake by plant roots. Agron. J. 70:404.

Deacon, J.W., and R.T. Mitchell. 1985. Comparison of rates of natural senescence of the root cortex of wheat (with and without mildew infection), barley, oats, and rye. Plant Soil 84:129-131.

Dexter, A.R. 1986a. Model experiments on the behavior of roots at the interface between a tilled seed-bed and a compacted subsoil. I. Effects of seed-bed aggregate size and sub-soil strength on wheat roots. Plant Soil 95:123-133.

Dexter, A.R. 1986b. Model experiments on the behavior of roots at the interface between a tilled seed-bed and a compacted subsoil. II. Entry of pea and wheat roots into sub-soil cracks. Plant Soil 95:135-147.

Dexter, A.R. 1987. Compression of soil around roots. Plant Soil 97:401-406.

Donald, R.G., B.D. Kay, and M.H. Miller. 1987. The effect of soil aggregate size on early shoot and root growth of maize (*Zea mays* L.). Plant Soil 103:251-259.

Eavis, B.W. 1967. Mechanical impedance and root growth. Pap. 4 (F/39). Agricultural engineering symp., Inst. Agric. Eng., 14 Sept. 1967, Silsoe, England.

Eavis, B.W. 1972a. Soil physical conditions affecting seedling root growth. I. Mechanical impedance, aeration and moisture availability as influenced by bulk density and moisture levels in a sandy loam soil. Plant Soil 36:613-622.

Eavis, B.W. 1972b. Soil physical conditions affecting seedling root growth. III. Comparisons between root growth in poorly aerated soil and at different oxygen partial pressures. Plant Soil 37:151-158.

Eavis, B.W., and D. Payne. 1969. Soil physical conditions and root growth. p. 315-338. *In* W.J. Whittington (ed.) Root growth. Butterworths, London.

Edwards, C.A., and J.R. Lofty. 1978. The influence of arthropods and earthworms upon root growth of direct drilled cereals. J. Appl. Ecol. 15:789-795.

Ehlers, W., B.K. Khosla, U. Kopke, R. Stulpnagel, W. Bohm, and K. Baeumer. 1980. Tillage effects on root development, water uptake and growth of oats. Soil Till. Res. 1:19-34.

Ehlers, W., U. Kopke, F. Hesse, and W. Bohm. 1983. Penetration resistance and root growth of oats in tilled and untilled loess soil. Soil Till. Res. 3:261-275.

Elliott, L.F., H. Bolton Jr., H.F. Stroo, and R.I. Papendick. 1987. Some effects of residue management on rhizosphere biology. p. 67–80. *In* L.F. Elliott (ed.) STEEP-conservation concepts and accomplishments. Washington State Univ. Press, Pullman.

Ellis, F.B., and B.T. Barnes. 1980. Growth and development of root systems of winter cereals grown after different tillage methods including direct drilling. Plant Soil 55:283–295.

Ellis, F.B., J.G. Elliott, B.T. Barnes, and K.R. Howse. 1977. Comparison of direct drilling, reduced cultivation and ploughing on the growth of cereals. 2. Spring barley on a sandy loam soil: soil physical conditions and root growth. J. Agric. Sci. 89:631–642.

Esau, K. 1965. Plant anatomy. 2nd ed. John Wiley and Sons, New York.

Ewens, M., and R.A. Leigh. 1985. The effect of nutrient solution composition on the length of root hairs of wheat (*Triticum aestivum* L.). J. Exp. Bot. 36(166):713–724.

Faiz, S.M.A. 1973. Soil-root-water relations. Ph.D. thesis. Univ. Aberdeen, Scotland, U.K.

Feldman. L.J. 1977. The generation and elaboration of primary vascular tissue patterns in roots of Zea. Bot. Gaz. 138(4):393–401.

Finney, J.R., and B.A.G. Knight. 1973. The effect of soil physical conditions produced by various cultivation systems on the root development of winter wheat. J. Agric. Sci. 80:435–442.

Fiscus, E.L. 1975. The interaction between osmotic- and pressure-induced water flow in plants roots. Plant Physiol. 55:917–922.

Fiscus, E.L. 1983. Water transport and balance within the plant: Resistance to water flow in roots. p. 183–194. *In* H.M. Taylor et al. (ed.) Limitations to efficient water use in crop production. ASA, CSSA, and SSSA, Madison, WI.

Foehse, D., and A. Jungk. 1983. Influence of phosphate and nitrate supply on root hair formation of rape, spinach and tomato plants. Plant Soil 74:359–368.

Frederickson, J.K., and L.F. Elliott. 1985. Colonization of winter wheat roots by inhibitory rhizobacteria. Soil Sci. Soc. Am. J. 49:1172–1177.

Fusseder, A., 1987. The longevity and activity of the primary root of maize. Plant Soil 101:257–265.

Gardner, W.R. 1960. Dynamic aspects of water availability to plants. Soil Sci. 89:63–73.

Gill, W.R., and R.D. Miller. 1956. A method for study of the influence of mechanical impedance and aeration on the growth of seedling roots. Soil Sci. Soc. Am. Proc. 20:154–157.

Giordano, R., A. Salleo, S. Salleo, and F. Wanderlingh. 1978. Flow in xylem vessels and Poiseuille's law. Can. J. Bot. 56:333–338.

Gonzalez-Bernaldez, F., J.F. Lopéz-Sáez,, and G. Garcia-Ferrero. 1968. Effect of osmotic pressure on root growth, cell cycle and cell elongation. Protoplasma 65:255–262.

Goss, M.J., A.R. Dexter, and M. Evans. 1987. Mechanics of root elongation and the effects of 3,5-diiodo-4-hydroxybenzoic acid (DIHB). Plant Soil 99:211–218.

Gracean, E.L., and J.S. Oh. 1972. Physics of root growth. Nature (London) New Biol. 235(53):24–25.

Greacen, E.L., P. Ponsana, and K.P. Barley. 1976. Resistance to water flow in the roots of cereals. p. 86–100. *In* O.L. Lange (ed.) Water and plant life. Ecological studies. Analysis and Synthesis. Vol. 19. Springer-Verlag, Berlin.

Gregory, P.J. 1983. Response to temperature in a stand of pearl millet (*Pennisetum typhoides* S. & H.) 3. Root development. J. Exp. Bot. 34:744–756.

Gregory, P.J., M. McGowan, and P.V. Biscoe. 1978. Water relations of winter wheat. 2. Soil water relations. J. Agric. Sci. 91:103–116.

Grimes, D.W., R.J. Miller, and P.L. Wiley. 1975. Cotton and corn root development in two field soils of different strength characteristics. Agron. J. 67:519–523.

Gupta, S., and R.R. Allmaras. 1987. Models to assess the susceptibility of soils to excessive compaction. Adv. Soil Sci. 6:65–100.

Gupta, S.C., and W.E. Larson. 1982. Modeling soil mechanical behavior during tillage. p. 151–178. *In* Predicting tillage effects on soil physical properties and processes. ASA and SSSA, Madison, WI.

Hackett, C. 1972. A method of applying nutrients locally to roots under controlled conditions, and some morphological effects of locally applied nitrate on the branching of wheat roots. Aust. J. Biol. Sci. 25:1169–1180.

Hackett, C., and B.O. Bartlett. 1971. A study of the root system of barley. III. Branching pattern. New Phytol. 70:409–413.

Hamblin, A.P. 1986. The influence of soil structure on water movement, crop root growth, and water uptake. Adv. Agron. 38:95–157.

Hamblin, A., and D. Tennant. 1987. Root length density and water uptake in cereals and grain legumes: How well are they correlated? Aust. J. Agric. Res. 38:513–527.

Hardjoamidjojo, S., R.W. Skaggs, and G.O. Schwab. 1982. Corn yield response to excessive soil water conditions. Trans. ASAE 25:922–927.

Herkelrath, W.N., E.E. Miller, and W.R. Gardner. 1977. Water uptake by plants: II. The roots contact model. Soil Sci. Soc. Am. J. 41:1039–1043.

Hillel, D., H. Talpaz, and H. van Keulen. 1976. A macroscopic-scale model of water uptake by a non-uniform root system and of water and salt movement in the soil profile. Soil Sci. 121:242–255.

Holder, C.B., and K.W. Brown. 1980. The relationship between oxygen and water uptake by roots of intact bean plants (*Phaseolus vulgaris* L.). Soil Sci. Soc. Am. J. 44:21–25.

Hoogenboom, G., and M.G. Huck. 1986. ROOTSIMU V4.0 A dynamic simulation of root growth, water uptake, and biomass partitioning in a soil-plant-atmosphere continuum: Update and documentation. Alabama Agric. Exp. Stn. Agron. Soil Dep. Ser. No. 109.

Hoogenboom, Gerrit, M.G. Huck, and C.M. Peterson. 1987. Root growth rate of soybean as affected by drought stress. Agron. J. 79:607–614.

Huck, M.G., B. Klepper, and H.M. Taylor. 1970. Diurnal variations in root diameter. Plant Physiol. 45:529–532.

Huck, M.G., C.M. Peterson, G.H. Hoogenboom, and C.D. Busch. 1986. Distribution of dry matter between shoots and roots of irrigated and nonirrigated determinate soybeans. Agron. J. 78:807–813.

Ide, G.,G. Hofman, C. Ossemerct, and M. Van Ruymbeke. 1984. Root-growth response of winter barley to subsoiling. Soil Till. Res. 4:419–431.

Jackson, M.B., and P.W. Barlow. 1981. Root geotropism and the role of growth regulators from the cap: A re-examination. Plant Cell Environ. 4:107–123.

Jaunin, F., and R.M. Hofer. 1986. Root hair formation and elongation of primary maize roots. Physiol. Plantarum 68:653–656.

Jodari-Karimi, F., V. Watson, H. Hodges, and F. Whisler. 1983. Root distribution and water use efficiency of alfalfa as influenced by depth of irrigation. Agron. J. 75:1207–1211.

Jones, C.A. 1983. Effect of soil texture on critical bulk densities for root growth. Soil Sci. Soc. Am. J. 47:1208–1211.

Jordan, W.R., M. McCrary, and F.R. Miller. 1979. Compensatory growth in the crown root system of sorghum. Agron. J. 71:803–806.

Kaspar, T.C., C.D. Stanley, and H.M. Taylor. 1978. Soybean root growth during the reproductive stages of development. Agron. J. 70:1105–1107.

Kaspar, T.C., D.G. Woolley, and H.M. Taylor. 1981. Temperature effect on the inclination of lateral roots of soybean. Agron. J. 73:383–386.

Klepper, B. 1983. Managing root systems for efficient water use: Axial resistances to flow in root systems—anatomical considerations. p. 115–125. *In* Limitations to efficient water use in crop production. ASA, CSSA, and SSSA, Madison, WI.

Klepper, B. 1987. Origin, branching and distribution of root systems. p. 103–124. *In* J.V. Lake et al. (ed.) Root development and function—Effects of the physical environment. Cambridge Univ. Press, New York.

Klepper, B., R.K. Belford, and R.W. Rickman. 1984. Root and shoot development in winter wheat. Agron. J. 76:117–122.

Klepper, B., R.W. Rickman, and H.M. Taylor. 1983. Farm management and the function of field crop root systems. Agric. Water Manage. 7:115–141.

Klepper, B., H.M. Taylor, M.G. Huck, and E.L. Fiscus. 1973. Water relations and growth of cotton in drying soil. Agron. J. 65:307–310.

Kolesnikov, V. 1971. L. Aksenova (Trans.) The root system of fruit plants. Mir Publ., Moscow.

Kramer, P.J. 1937. The relation between rate of transpiration and rate of absorption of water in plants. Am. J. Bot. 24:10–15.

Kramer, P.J. 1969. Plant and soil water relationships: A modern synthesis. McGraw-Hill, New York.

Landsberg, J.J., and N.D. Fowkes. 1978. Water movement through plant roots. Ann. Bot. 42:493–508.

Lemon, E.R. 1962. Soil aeration and plant root relations. I. Theory. Agron. J. 54:167–170.

Lockhart, J.A. 1965. An analysis of irreversible plant cell elongation. J. Theor. Biol. 8:264–275.

Logsdon, S.D., R.B. Reneau, Jr., and J.C. Parker. 1987. Corn seedling root growth as influenced by soil physical properties. Agron. J. 79:221–224.

Luxova, M. 1986. The hydraulic safety zone at the base of barley roots. Planta 169:465–470.

Mackay, A.D., and S.A. Barber. 1985. Effect of soil moisture and phosphate level on root hair growth of corn roots. Plant Soil 86:321–331.

MacLeod, R.D., and A. Thompson. 1979. Development of lateral root primordia in *Vicia faba, Pisum staivum, Zea mays* and *Phaseolus vulgaris:* Rates of primordium formation and cell doubling times. Ann. Bot. 44:435–449.

Maizlish, N.A., D.D. Fritton, and W.A. Kendall. 1980. Root morphology and early development of maize at varying levels of nitrogen. Agron. J. 72:25–31.

Mallory, T.E., Su-Hwa Chiang, E.G. Cutter, and E.M. Gifford, Jr. 1970. Sequence and pattern of lateral root formation in five selected species. Am. J. Bot. 57(7):800–809.

Markhart, A.H., E.L. Fiscus, A.W. Naylor, and P.J. Kramer. 1979. Effect of temperature on water and ion transport in soybean and broccoli systems. Plant Physiol. 64:83–87.

May, L.H., F.H. Chapman, and D. Aspinall. 1965. Quantitative studies of root development. I. The influence of nutrient concentration. Aust. J. Biol. Sci. 18:25–35.

May, L.H., F.H. Randles, D. Aspinall, and L.G. Paleg. 1967. Quantitative studies of root development. II. Growth in the early stages of development. Aust. J. Biol. Sci. 20:273–283.

Mengel, D.B., and S.A. Barber. 1974. Development and distribution of the corn root system under field conditions. Agron. J. 66:341–344.

Meyer, W.S., and A.M. Alston. 1978. Wheat responses to seminal root geometry and subsoil water. Agron. J. 70:981–986.

Meyer, W.S., H.D. Barrs, R.C.G. Smith, N.S. White, A.D. Heritage, and D.L. Short. 1985. Effect of irrigation on soil oxygen status and root and shoot growth of wheat in a clay soil. Aust. J. Agric. Res. 36:171–185.

Meyer, W.S., E.L. Greacen, and A.M. Alston. 1978. Resistance to water flow in the seminal roots of wheat. J. Exp. Bot. 29(113):1451–1461.

Meyer, W.S., and J.T. Ritchie. 1980. Resistance to water flow in the sorghum plant. Plant Physiol. 65:33–39.

Mitchell, R.L., and W.J. Russell. 1971. Root development and rooting patterns of soybean [*Glycine max* (L.) Merrill] evaluated under field conditions. Agron. J. 63:313–319.

Mosher, P.N., and M.H. Miller. 1972. Influence of soil temperature on the geotropic response of corn roots (*Zea mays* L.). Agron. J. 64:459–462.

Newman, E.I. 1966. Relationship between root growth of flax (*Linum usitatissimum*) and soil water potential. New Phytol. 65:273–283.

Newman, E.I. 1976. Water movement through root systems. Phil. Trans. R. Soc. London. B. 273:463–478.

Newman, P.R., and L.E. Moser. 1988. Seedling root development and morphology of cool-season and warm-season forage grasses. Crop Sci. 28:148–151.

Nobel, P.S., and J. Sanderson. 1984. Rectifier-like activities of roots of two desert succulents. J. Exp. Bot. 35(154):727–737.

Oades, J.M. 1978. Mucilages at the root surface. J. Soil Sci. 29:1–16.

Oosterhuis, D.M., and S.D. Wullschleger. 1987. Osmotic adjustment in cotton (*Gossypium hirsutum* L.) leaves and roots in response to water stress. Plant Physiol. 84:1154–1157.

O'Toole, J.C., and W.L. Bland. 1987. Genotypic variation in crop plant root systems. p. 91–145. *In* N.C. Brady (ed.) Advances in agronomy, Vol. 41. Academic Press, New York.

Passioura, J.B. 1972. The effect of root geometry on the yield of wheat growing on stored water. Aust. J. Agric. Res. 23:745–752.

Peterson, C.M., B. Klepper, F.V. Pumphrey, and R.W. Rickman. 1984. Restricted rooting decreases tillering and growth of winter wheat. Agron. J. 76:861–863.

Peterson, C.M., B. Klepper, and R.W. Rickman. 1982. Tiller development at the coleoptilar node in winter wheat. Agron. J. 74:781–784.

Peterson, W.R., and S.A. Barber. 1981. Soybean root morphology and K uptake. Agron. J. 73:316–319.

Petty, J.A. 1978. Fluid flow through the vessels of birch wood. J. Exp. Bot. 29:1463–1469.

Picard, D., M.O. Jordan, and R. Trendel. 1985. Rythme d'apparition des racines primaires du mais (*Zea mays* L.) I. Etude detaillee pour une variete en un lieu donne. Agron. 5(8):667–676.

Pilet, P.E., and A. Senn. 1980. Root growth gradients: A critical analysis. Z. Pflanzenphysiol. 99:121–130.

Pilet, P.E., J.M. Versel, and G. Mayor. 1983. Growth distribution and surface pH patterns along maize roots. Planta 158:398–402.

Portas, C.A.M., and H.M. Taylor. 1976. Growth and survival of young plant roots in dry soil. Soil Sci. 121:170–175.

Porter, J.R., B. Klepper, and R.K. Belford. 1986. A model (WHITROOT) which synchronizes root growth and development with shoot development for winter wheat. Plant Soil 92:133–145.

Powell, D.B.B. 1978. Regulation of plant water potential by membranes of the endodermis in young roots. Plant Cell Environ. 1:69–76.

Richards, R.A., and J.B. Passioura. 1981. Seminal root morphology and water use of wheat. II. Genetic variation. Crop Sci. 21:253–256.

Richter, H. 1973. Frictional potential losses and total water potential in plants: A re-evaluation. J. Exp. Bot. 24(83):983–994.

Rickman, R.W., and B.L. Klepper. 1980. Wet season aeration problems beneath surface mulches in dryland winter wheat production. Agron. J. 72:733–736.

Rickman, R.W., B. Klepper, and R.K. Belford. 1985. Developmental relationships among roots, leaves, and tillers in winter wheat. p. 83–98. In W. Day and R.K. Adkins (ed.) Wheat growth modelling. Plenum, New York.

Riopel, J.L. 1969. Regulations of lateral root positions. Bot. Gaz. 130(2):80–83.

Ritchie, J.T. 1985. A user-oriented model of the soil water balance in wheat. p. 293–305. In W. Day and R.K. Atkin (ed.) Wheat growth and modelling. Plenum, New York.

Rosene, H.F. 1943. Quantitative measurement of the velocity of water absorption in individual root hairs by a microtechnique. Plant Physiol. 18:588–607.

Rougier, M., and A. Chaboud. 1985. Mucilages secreted by roots and their biological function. Is. J. Bot. 34:129–146.

Rowse, H.R., and D. Goodman. 1981. Axial resistance to water movement in broad bean (Vicia faba) roots. J. Exp. Bot. 32(18):591–598.

Rovira, A.D., and H.J. McDonald. 1986. Effects of the herbicide chlorsulfuron on Rhizoctonia bare patch and take-all of barley and wheat. Phytopathology 76:876–882.

Russell, R. Scott. 1977. Plant root systems: Their function and interaction with the soil. McGraw-Hill, London.

Salim, K.M., and H. Oryem-Origa. 1981. The relationship between cell size and cell number and the distance from the root tip to the region of phloem initiation in groundnuts (Arachis hypogea L. var. Bukene Red). J. Exp. Bot. 32(129):813–820.

Sanderson, J. 1983. Water uptake by different regions of the barley roots. Pathways of radial flow in relation to development of the endodermis. J. Exp. Bot. 34(140):240–253.

Sands, R., E.L. Fiscus, and C.P.P. Reid. 1982. Hydraulic properties of pine and bean roots with varying degrees of suberization, vascular differentiation and mycorrhizal infection. Aust. J. Plant Physiol. 9:559–569.

Sasaki, O., K. Yamazaki, and S. Kawata. 1983. The relationship between the growth of crown roots and their branching habit in rice plants. Jpn. J. Crop Sci. 52(1):1–6.

Schleiff, U. 1986. Water uptake by barley roots as affected by the osmotic and matric potential in the rhizosphere. Plant Soil 94:143–146.

Seaton, K.A., and J.J. Landsberg. 1978. Resistance to water movement through wheat root systems. Aust. J. Agric. Res. 29:913–924.

Sharp, R.E., and W.J. Davies. 1985. Root growth and water uptake by maize plants in drying soil. J. Exp. Bot. 36(170):1441–1456.

Shierlaw, J., and A.M. Alston. 1984. Effect of soil compaction on root growth and uptake of phosphorus. Plant Soil 77:15–28.

Simpson, J.R., A. Pinkerton, and J. Lazdovskis. 1977. Effects of subsoil calcium on the root growth of some lucerne genotypes (Medicago sativa L.) in acidic soil profiles. Aust. J. Agric. Res. 28:629–638.

Sivakumar, M.V.K., H.M. Taylor, and R.H. Shaw. 1977. Top and root relations of field-grown soybeans. Agron. J. 69:470–473.

Smith, D.L.O. 1987. Measurement, interpretation and modelling of soil compaction. Soil Use Manage. 3(3):87–93.

Steudle, E., R. Oren, and E.D. Schulze. 1987. Water transport of maize roots. Measurement of hydraulic conductivity, solute permeability, and of reflection coefficients of excised roots using the root pressure probe. Plant Physiol. 84:1220–1232.

Svejcar, T., and S. Christiansen. 1987. The influence of grazing pressure on rooting dynamics of Caucasian bluestem. J. Range Manage. 40(3):224–227.

Sykes, D.J., and W.E. Loomis. 1967. Plant and soil factors in permanent wilting percentages and field capacity storage. Soil Sci. 104:163–173.

Tanaka, N. 1979. Studies on the growth of root hairs in rice plants and their microfibrous network structure with a scanning electron microscope. Agric. Bull. Saga Univ. No. 46:1–8.

Taylor, H.M., and B. Klepper. 1974. Water relations of cotton. I. Root growth and water use as related to top growth and soil water content. Agron. J. 66:584–588.

Taylor, H.M., and B. Klepper. 1975. Water uptake by cotton root systems: An examination of assumptions in the single root model. Soil Sci. 120(1):57–67.

Taylor, H.M., and B. Klepper. 1978. The role of rooting characteristics in the supply of water to plants. Adv. Agron. 30:99–128.

Taylor, H.M., L.F. Locke, and J.E. Box. 1964. Pans in southern Great Plains soils. III. Their effects on yield of cotton and grain sorghum. Agron. J. 56:542–545.

Taylor, H.M., and L.F. Ratliff. 1969a. Root growth pressures of cotton, peas, and peanut. Agron. J. 61:398–402.

Taylor, H.M., and L.F. Ratliff. 1969b. Root elongation rate of cotton and peanuts as a function of soil strength and soil water content. Soil Sci. 108:113–119.

Taylor, H.M., G.M. Robertson, and J.J. Parker, Jr. 1966. Soil strength—root penetration relations for medium- to coarse-textured soil materials. Soil Sci. 102:18–22.

Tennant, D. 1976a. Root growth of wheat. I. Early patterns of multiplication and extension of wheat roots including effects of levels of nitrogen, phosphorus and potassium. Aust. J. Agric. Res. 27:183–196.

Tennant, D. 1976b. Wheat root penetration and total available water on a range of soil types. Aust. J. Exp. Agric. Anim. Husb. 16:570–577.

Tinker, P.B. 1976. Transport of water to plant roots in soil. Phil Trans. R. Soc. London B. 273:445–461.

Turner, N.C., W.R. Stern, and P. Evans. 1987. Water relations and osmotic adjustments of leaves and roots of lupins in response to water deficits. Crop Sci. 27:977–983.

Tyree, M.T., and M.H. Zimmerman. 1971. Theory and practice of measuring transport coefficients and sap flow in the xylem of red maple stems (Acer rubrum). J. Exp. Bot. 22:1–18.

Unger, P.W. 1979. Effects of deep tillage and profile modification on soil properties, root growth, and crop yields in the United States and Canada. Geoderma 22:275–295.

Veen, B.W. 1982. The influence of mechanical impedance on the growth of maize roots. Plant Soil 66:101–109.

Vepraskas, M.J., and G.S. Miner. 1986. Effects of subsoiling and mechanical impedance on tobacco root growth. Soil Sci. Soc. Am. J. 50:423–427.

Versel, J.M., and G. Mayor. 1985. Gradients in maize roots: Local elongation and pH. Planta 164:96–100.

Walters, D.R., and P.G. Ayres. 1981. Growth and branching pattern of roots of barley infected with powdery mildew. Ann. Bot. 47:159–162.

Wang, J., J.D. Hesketh, and J.T. Woolley. 1986. Preexisting channels and soybean rooting patterns. Soil Sci. 141(6):432–437.

Warmbrodt, R.D. 1985. Studies on the root of Hordeum vulgare L.—ultrastructure of the seminal root with special reference to the phloem. Am. J. Bot. 72(3):414–432.

Warnaars, B.C., and B.W. Eavis. 1972. Soil physical conditions affecting seedling root growth. II. Mechanical impedance, aeration and moisture availability as influenced by grain-size distribution and moisture content in silica sands. Plant Soil 36:623–634.

Weatherley, P.E. 1976. Introduction: Water movement through plants. Phil. Trans. R. Soc. London B. 273:435–444.

Weaver, J.E., J. Kramer, and M. Reed. 1924. Development of root and shoot of winter wheat under field environment. Ecology 5:26–50.

Whisler, F.D., B. Acock, D.N. Baker, R.E. Rye, H.F. Hodges, J.R. Lambert, H.E. Lemmon, J.M. McKinion, and V.R. Reddy. 1986. Crop simulation models in agronomic systems. Adv. Agron. 40:141–208.

Whiteley, G.M., and A.R. Dexter. 1983. Behavior of roots in cracks between soil peds. Plant Soil 74:153–162.

Whiteley, G.M., W.H. Utomo, and A.R. Dexter. 1981. A comparison of penetrometer pressures and the pressures exerted by roots. Plant Soil 61:351–364.

Willatt, S.T., and H.M. Taylor. 1978. Water uptake by soya-bean roots as affected by their depth and by soil water content. J. Agric. Sci. 90:205–213.

Wilson, A.J., A.W. Robards, and M.J. Goss. 1977. Effects of mechanical impedance on root growth in barley, Hordeum vulgare L. II. Effects on cell development in seminal roots. J. Exp. Bot. 28(106):1216–1227.

Wong, M.H., and A.D. Bradshaw. 1982. A comparison of the toxicity of heavy metals, using root elongation of rye grass, Lolium perenne. New Phytol. 91:255–261.

Yamazaki, K., and J. Harada. 1982. The root system formation and its possible bearings on grain yield in rice plants. Jpn. Agric. Res. Q. 15:153–160.

11 Plant Responses to Water Deficits[1]

M. B. KIRKHAM

Kansas State University
Manhattan, KS

Plants must regulate their water status to survive drought. Under equilibrium conditions, the state of water at a particular point in a plant may be written in terms of the various components of the potential energy (Kirkham et al., 1969) as follows:

$$\Psi = \Psi_s + \Psi_p + \Psi_m + \Psi_g \qquad [1]$$

where Ψ is the water potential, Ψ_s is the osmotic (solute)-potential component, Ψ_p is the pressure (turgor)-potential component, Ψ_m is the matric component due to capillary or adsorption forces such as those in the cell wall, and Ψ_g is the component due to gravity. The matric potential and the gravitational potential are usually neglected [however, see Kirkham (1983b) for cases when gravity is important], and Eq. [1] reduces to the classical equation of plant physiology, as follows:

$$DPD = OP - TP \qquad [2]$$

where the diffusion pressure deficit (DPD) is a measure of the water potential, OP is the osmotic pressure, and TP is the turgor pressure. The signs in Eq. [1] of the water potential, Ψ, and the osmotic potential, Ψ_s, are negative, while in Eq. [2] the signs of the DPD and the OP are positive. Turgor potential, Ψ_p in Eq. [1], and TP in Eq. [2], are positive or zero. Equation [2] is seldom employed in research today. Only Eq. [1] is used now.

In this chapter, we consider the effects of water deficits on water potential, osmotic potential, and turgor potential. In addition, because stomata are the main pathways through which water is lost from a plant, we discuss the effects of water deficits on stomata. In the first edition of this monograph, Vaadia and Waisel (1967) cited the need for more research on water deficits and plant hormones. Much work is being done in this area, especially with abscisic acid (ABA) and ethylene, which we review. Finally, we look at the effects of drought on growth, because the definition of *drought* ("a shortage of precipitation sufficient to adversely affect crop production"

---

[1] Contribution No. 88-521-B from the Evapotranspiration Laboratory, Dep. of Agronomy, Kansas State Univ., Manhattan, KS 66506.

(Rosenberg, 1980, p. 2) is based on growth. The chapter focuses on recent literature (i.e., mainly 1987), as previous reviews have documented earlier literature (e.g., Hale & Orcutt, 1987; International Rice Research Institute, 1982; Jarvis & Mansfield, 1981; Kozlowski, 1983; Paleg & Aspinall, 1981; Parsons, 1982; Schulze et al., 1987; Simpson, 1981; Teare & Peet, 1983; Turner & Kramer, 1980). We do not consider roots, because they are dealt with in the preceding chapter. Also, we do not consider climatology or methods to measure plant-water deficits. We do, however, examine *drought-resistant* plants. They are defined as plants with "relatively less growth inhibition under stress," compared to other plants (Blum & Sullivan, 1986).

I. POTENTIALS

A. Water Potential

Water stress decreases leaf water potential. The amount of decrease depends upon the drought resistance of the plant. Under drought, drought-resistant plants maintain higher water potentials than do drought-sensitive plants. For example, Goyal (1987) found that, under water stress, a drought-resistant cultivar of rice (*Oryza sativa* L. 'N-22') had a higher leaf water potential than a drought-sensitive cultivar ('Jaya'). Kirkham (1988, 1989) got similar results for sorghum [*Sorghum bicolor* (L.) Moench 'KS-9' and 'IA-25'] and winter wheat (*Triticum aestivum* L. 'KanKing' and 'Ponca'). Bannister (1986) measured water potential of many species during a drought in New Zealand. Some species avoided drought and maintained a high water potential, even though adjacent species were wilted. The most sensitive species showed wilting at the highest water potentials (about -1.0 MPa) (1 MPa = 10 bars = 9.87 atm), and the most resistant at the lowest (about -4.5 MPa). In Florida, Lorens et al. (1987a) found that the primary difference between two hybrids of corn (*Zea mays* L.), one drought sensitive ('Pioneer 3192') and one drought resistant ('Pioneer 3165'), was the ability of Hybrid 3165 to avoid desiccation by maintaining higher potentials than Hybrid 3192. Hybrid 3165 also yielded more grain under drought than Hybrid 3192 (Lorens et al., 1987b). They concluded that the greater growth rate, biomass, and grain yield of Hybrid 3165 during severe water stress, compared to Hybrid 3192, probably resulted from its higher leaf water potential and turgor potential.

B. Osmotic Potential and Turgor Potential

Plants can maintain their turgor potential by *osmotic adjustment* (lowering of osmotic potential due to solute accumulation). However, species vary in amount of osmotic adjustment. For example, McCree and Richardson (1987) compared cowpea [*Vigna unguiculata* (L.) Walp 'California Blackeye No. 5'] and sugarbeet (*Beta vulgaris* L. 'Mono-Hy-D2'). During drought, sugarbeet adjusted osmotically by 0.8 MPa and reached a daytime minimum leaf water potential of -2.6 MPa. Cowpea adjusted by half as much, but maintained a higher water potential (-1.2 MPa) by stomatal closure. Water loss rates were similar in the two species. The results suggested that osmotic adjustment was as effective as stomatal closure for survival under drought. In addition to osmotic adjustment and stomatal closure, changes in the elasticity of the cell wall may be important in maintaining turgor, too. Joly and Zaerr (1987) saw that drought induced changes in the bulk modulus of elasticity in seedlings of Douglas fir [*Pseudotsuga menziesii* (Mirb.) Franco]. They said that regulation of turgor in Douglas fir was mediated by structural factors associated with the cell walls, rather than by osmotic adjustment.

Not only do different species vary in osmotic adjustment, but also different cultivars of the same species vary. Gupta and Berkowitz (1987) subjected two cultivars of spring wheat, Condor and Cappelle Desprez, to a 9-d stress. Under water stress, leaf osmotic potential declined to a greater degree in Condor plants, but photosynthesis was 106.5% higher in Condor than in Cappelle Desprez. They hypothesized that there was an association between protoplast- (and presumably chloroplast-) volume reduction at low water potential and low-water-potential inhibition of photosynthesis. In Condor, osmotic adjustment allowed for maintenance of relatively greater volume at low water potential, thus reducing low-water-potential inhibition of photosynthesis.

Organs of the same plant differ in the amount of osmotic adjustment. Oosterhuis and Wullschleger (1987) studied the osmotic adjustment of leaves and roots of water-stressed cotton (*Gossypium hirsutum* L. 'Deltapine 41'). The osmotic adjustment of roots was 0.19 MPa, compared to 0.41 MPa in leaves. Turner et al. (1986) also observed similar values (0.5 to 0.6 MPa) for leaves of 'Deltapine 16' cotton. Turner et al. (1987) measured osmotic adjustment of leaves and roots of lupines in response to water deficits under both greenhouse and field conditions. In the greenhouse, leaves of *Lupinus pilosus* Murr. (common name not given) and *L. atlanticus* Gladst. (a naturalized, bitter line from western Australia) had the greatest degree of osmotic adjustment (0.4 to 0.5 MPa), and those of *L. augustifolius* L. (a narrowleaved lupine) had the least (0.1 MPa). No osmotic adjustment was observed in the roots in the greenhouse. However, in the field the leaves of *L. augustifolius* adjusted osmotically by 0.2 to 0.3 MPa and roots by about 0.4 MPa. The osmotic adjustment in both leaves and roots was accompanied by a decrease in the turgid weight/dry weight ratio. The authors suggested that changes in this ratio may provide a simple screening method for osmotic adjustment in lupines.

Osmotic adjustment may be important for short-term survival under drought, but not for long-term survival. Toft et al. (1987) studied *Eustachys paspaloides* (Vahl) Lanza & Mattei, a tropical, perennial grass on the Serengeti Plain of east Africa, under conditions of water stress. Osmotic adjustment (0.2 MPa) occurred in tillers of water-stressed plants. The delay in permanent wilting afforded by lowered osmotic potential was about 1 d. This indicated that osmotic adjustment would not reduce the effect of drought on growth for a long time. It may, however, be of value during short droughts and limit dehydration of growing points. Initiation of tillers appears to be less sensitive to stress than does subsequent vegetative growth (Davidson & Chevalier, 1987).

C. Osmolytes

Water-stressed plants must lower solute potential without damaging metabolic functions (Hanson et al., 1986). This may be achieved by accumulating benign osmolytes, such as betaine (glycinebetaine) in the cytoplasm. High concentrations of betaine have been found not to interfere with metabolic activities and may be under genetic control. Grumet and Hanson (1986) altered genetically the betaine levels in spring barley (*Hordeum vulgare* L.) by creating two F_4 populations with different betaine levels. Although selected only for differing betaine levels, the parents and populations differed also in solute potential. The high-betaine genotypes maintained a solute potential 1.0 MPa lower than the low-betaine genotypes. Furthermore, in both populations and parents, betaine level was linearly related to solute potential. These observations suggested that betaine level and osmotic potential were under genetic control and that betaine accumulation is a component of osmotic adjustment in barley.

Proline is thought to be another osmolyte that enables water-stressed plants to lower their osmotic potential. Stewart and Voetberg (1987) compared proline accumulation in wilted leaves from the wilty tomato (*Lycopersicon esculentum* Mill.) mutant, *flacca*, to that in the wild type, the cultivar Rheinlands Ruhm. The rate of proline accumulation was faster in *flacca* than in the wild type, because the wilty mutant wilted faster. They showed that proline accumulation in tomato leaves was induced by wilting. Sheriff et al. (1986) examined *Macroptilium atropurpureum* 'Siratro' (authority not given) and *Galactia striata* (cultivar and authority not given), two twining pasture legumes. Siratro is drought sensitive, and *G. striata* is drought resistant. Leaves of *G. striata* exhibited osmotic adjustment of up to 2.3 MPa and accumulated proline, but in leaves of Siratro, osmotic adjustment was < 0.26 MPa and no proline accumulated.

Even though betaine and proline are thought to be important osmolytes that accumulate under drought, other substances, such as sucrose (Drossopoulos et al., 1987), K (Pier & Berkowitz, 1987; Tanguilig et al., 1987), and pinitol (Sheriff et al., 1986), have been observed to increase, too. Uptake may be dependent on pH (Rygol et al., 1987).

II. STOMATA

The intercellular (internal) CO_2 concentration is defined as follows (Farquhar et al., 1978):

$$c_i = c - 1.6 \, (A/g) \tag{3}$$

where c_i is the CO_2 concentration inside the leaf, c is the ambient CO_2 concentration, A is the net rate of assimilation (photosynthesis), g is the stomatal conductance (the reciprocal of g is the stomatal resistance), and the factor 1.6 is the ratio of the diffusivities of water vapor and CO_2 in air. The equation is based on Ohm's law. The value of c_i is calculated and not measured directly.

If c_i is reduced under water deficit as a result of stomatal closure, the condition is called *stomatal inhibition of photosynthesis* (Johnson et al., 1987). If, however, c_i remains relatively stable, reduced stomatal conductance is not considered to be the major cause of reduced photosynthesis, and the condition is called *nonstomatal inhibition of photosynthesis*. The commercially available infrared CO_2 gas analyzers (e.g., those made by The Analytical Development Company, Ltd., Hoddesdon, Hertfordshire, England, and by Li-Cor, Inc., Lincoln, NE, USA) automatically calculate c_i. Therefore, c_i is now being reported in many experiments, along with discussions of stomatal and nonstomatal inhibition of photosynthesis.

As for osmotic adjustment, species vary when one considers stomatal and nonstomatal inhibition of photosynthesis. For example, Johnson et al. (1987) compared photosynthesis of a desert annual [*Triticum kotschyi* (Boiss.) Bowden] and winter wheat (*T. aestivum* L. 'TAM W-101') over a range of leaf-water potentials from -0.50 to -2.9 MPa. *Triticum kotschyi* had a higher stomatal conductance and a higher photosynthetic rate at a given water potential than did *T. aestivum*. The higher stomatal conductance of *T. kotschyi* under well-watered conditions was associated with a higher internal CO_2 concentration (c_i). Under water deficits, however, c_i did not differ significantly between the two species. *Triticum kotschyi* had a higher photosynthetic rate under water deficits than did *T. aestivum*. The results showed that stomatal conductance played an important role in the high photosynthetic rate of *T. kotschyi* under well-watered conditions; but, under water deficits, nonstomatal factors were involved in maintaining its higher photosynthetic rate.

Cox and Jolliff (1987) compared sunflower (*Helianthus annuus* L. '894') and soybean [*Glycine max* (L.) Merr. 'S09-90'] under dryland conditions in Oregon. Although increased stomatal resistance of dryland sunflower did not occur until 50 d into a soil-drying cycle, a 15% reduction in the photosynthetic rate was observed 22 d into the cycle. The data suggested that nonstomatal effects were most responsible for the reduced photosynthetic rate in sunflower. Stomatal closure, which occurred early in the soil-drying cycle of dryland soybean, was accompanied by a 50% reduction in photosynthetic rate. A significant correlation ($r = -0.80$) was observed between the photosynthetic rate and stomatal resistance of soybean, which suggested that stomatal closure was responsible for the reduced photosynthetic rate in dryland soybean. Thus, the results indicated that reduced stomatal conductance was the major cause of reduced photosynthesis in soybean, but not sunflower, which showed nonstomatal inhibition of photosynthesis.

The severity of the stress also appears to be important in determining whether or not nonstomatal factors affect photosynthesis. During early stages of stress, stomatal closure is often the cause of reduced photosynthesis. But during prolonged stress, nonstomatal factors also inhibit photosynthesis. For example, Cornic et al. (1987) studied photosynthesis of leaves of kidney bean (*Phaseolus vulgaris* L. 'Rangally') submitted to either a rapid (4-d) or a slow (8-d) drought cycle. The maximum daily net photosynthesis decline during the first 3 d of the rapid drought cycle was due to stomatal closure, but on the last day of the rapid drought cycle and during the slow drought cycle, it was due both to stomatal closure and inhibition of mesophyll activity (nonstomatal inhibition of photosynthesis).

If plants are not stressed long enough, nonstomatal inhibition of photosynthesis may not be apparent. Lopez et al. (1988) subjected pigeonpea [*Cajanus cajan* (L.) Huth. 'US-10'] to an 8-d drought, followed by an 18-d recovery after rewatering. Differences in stomatal conductance and photosynthetic rate between unwatered and control plants were first detected at 5 and 8 d after withholding water, respectively. As a result of these changes in stomatal conductance and photosynthetic rate, the CO_2 in the intercellular air spaces of the leaves decreased 5 to 8 d after withholding water, and, therefore, during this period they observed stomatal inhibition of photosynthesis. They were unable to determine if nonstomatal inhibition of photosynthesis occurred, which they did observe in an earlier experiment (Lopez et al., 1987). Lopez et al. (1988) did note that pigeonpea recovered slowly from drought (i.e., it took 18 d). Pigeonpea is an unusual grain legume, compared to others [cowpea, black gram (*Vigna mungo*), soybean], because it has a low epidermal conductance (Sinclair & Ludlow, 1986). This conductance determines water loss from leaves when stomata are closed. The low epidermal conductance of pigeonpea may be an adaptation that allows it to survive droughts longer than do other grain legumes (Sinclair & Ludlow, 1986).

These and other experiments [e.g., Hutmacher & Krieg (1983) for cotton (*Gossypium hirsutum* L.); Krieg & Hutmacher (1986) for sorghum; Tsunodo & Fukoshima (1986) for rice; Kirschbaum (1987) for *Eucalyptus*] indicate that, in many cases, nonstomatal factors determine photosynthetic rate under water deficit, particularly if stress is extended. Abscisic acid, a hormone that increases when plants are water stressed (see section III in this chapter), may be one such factor that affects photosynthesis. Bunce (1987) measured photosynthesis and stomatal conductance in velvet-leaf (*Abutilon theophrasti* Medik.), princes-feather (*Amaranthus hypochondriacus* L.), and soybean after injecting ABA into petioles of attached leaves. Response to ABA was compared with response to water stress. Abscisic acid reduced photosynthetic rate and stomatal conductance in proportion to the amount applied. In no case was the relative decrease in substomatal CO_2 concentration as large as the relative decrease in photosynthesis. This indicated that nonstomatal inhibition of photosynthesis occurred. The data suggested that the reductions of both photosynthesis and stomatal conductance by water stress were mediated by ABA.

III. ABSCISIC ACID

Abscisic acid rises rapidly after water stress is imposed. It closes stomata. The stomatal response to ABA constitutes one of the most sensitive hormonal reactions known in plants (Raghavendra & Reddy, 1987). For example, a 10^{-19} mol/m^3 solution induces closure in "wilty" mutants of plants. They are unable to remain turgid in normal growing conditions and have been found to be incapable of closing stomata because of a deficiency of ABA (Mansfield, 1983). The mutants include the "droopy" mutant of potato (*Solanum tuberosum* L.) (Quarrie, 1982) and the *flacca* mutant of tomato (Tal et al., 1979). A daily application of ABA restores them to normal physiological conditions. Sensitive techniques, such as those used in immunoassays, are needed to detect the tiny quantities that cause the stomatal aperture to narrow (Weiler, 1984).

Zhang et al. (1987) suggested that ABA can move from the roots to the epidermis of leaves and restrict stomatal aperture, even when leaf water potentials and turgor potentials remain constant. They split roots of dayflower (*Commelina communis* L.) between drying soil and well-watered soil. Increased ABA content of the epidermis of the leaves and stomatal closure coincided with increased ABA content of the roots in the drying soil. Roots of the same plant in moist soil did not show increases in ABA content. Jones et al. (1987) found that ABA at 25 mmol/m^3 increased turgor potential of roots as well as the osmolality of the cell sap of the roots. However, root growth was inhibited at ABA concentrations that caused a turgor-potential increase. Their results showed that, although ABA can affect root-cell-turgor potential, it does not result in increased root growth.

Abscisic acid affects the exchange of H ions and K ions across cell membranes, which is thought to regulate guard-cell turgor. Dehydration causes an increase in pH which then enhances the release of ABA from mesophyll cells (Hartung et al., 1988). Abscisic acid inhibits K uptake into guard cells which is necessary for stomatal opening. Temperature also appears to affect ABA accumulation (Eamus, 1987; Radin & Hendrix, 1986). Heat stress causes ABA to accumulate. Tropical species wilt when the plants are chilled, because the stomata remain open, perhaps due to a lack of ABA. Addition of ABA ameliorates the injury from wilting.

Drought-resistant and drought-sensitive plants respond differently to ABA. Kirkham (1983a) found that, under ample water supply, ABA decreased water and osmotic potentials of a drought-sensitive cultivar of winter wheat (Ponca), but had no effect on these potentials in a drought-resistant cultivar (KanKing). Under water-deprived conditions, ABA increased water and osmotic potentials of Ponca, but did not change these potentials in KanKing. The overall effect of ABA was to decrease the differences in the water and osmotic potentials between the two cultivars.

Studies have been done to relate ABA levels in plants to drought resistance, but results are mixed. Rose (1985) found that ABA levels in a drought-resistant cultivar of winter wheat (KanKing) did not differ from those

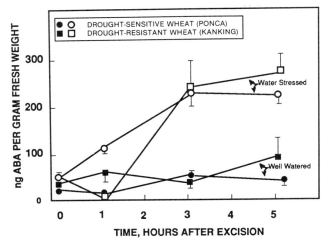

Fig. 11-1. Abscisic acid content of excised leaves of a drought-resistant and a drought-sensitive cultivar of winter wheat under well-watered and water-stressed conditions. Vertical bars = ±SD; half of the bars drawn for clarity. Adapted from Rose (1986).

of a drought-sensitive cultivar of winter wheat (Ponca), both under well-watered and water-stressed conditions (Fig. 11-1). Terry et al. (1988) saw that, under dry conditions, a drought-sensitive cultivar of coleus (*Coleus blumei* Benth. 'Buckley Supreme') accumulated more ABA than did a drought-resistant cultivar of coleus ('Marty'). However, in maize, Pekić and Quarrie (1987) found the opposite. Lines of maize that were most drought resistant accumulated the most ABA. Lines differed in ABA levels, despite similar water potentials. Chumakovski (1986) also found that an increase in ABA content in five cultivars of spring wheat during water stress was directly correlated with the drought-resistance of the cultivars. Since plants cope with drought in various ways, it is unlikely that a particular ABA response (e.g., ABA accumulation) will be the same for all drought-resistant plants (Walton, 1980).

IV. ETHYLENE

Ethylene ($CH_2 = CH_2$) is a gaseous hormone. It has been known for a long time that a gas has injurious effects on plants. In 1879, Fahnestock listed the damage that plants suffered when a gas leaked into some Philadelphia greenhouses. Abscission was the most common response, and it was severe. The gas was identified later (1920s) to be ethylene (Thimann, 1972). Besides causing abscission, ethylene has been shown to decrease plant height, increase stem thickness, decrease leaf length, increase leaf width, and increase the angle between the leaf and stem (Kirkham, 1985).

Ethylene production in plants often increases following damage, wounding, or stress from a variety of sources (Abeles & Abeles, 1972). This increase

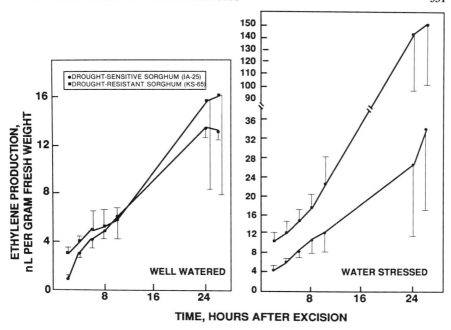

Fig. 11-2. Ethylene production by excised leaves of a drought-resistant and a drought-sensitive cultivar of sorghum under well-watered and water-stressed conditions. For vertical bars, see legend of Fig. 11-1. From Zhang and Kirkham (unpublished).

has been called *wound ethylene* or *stress ethylene*. Many papers have been published indicating that the causes of stress ethylene are varied and include chemicals; insects; temperature extremes; irradiation; disease; and mechanical effects such as incision, pressure, and abrasion (Abeles & Abeles, 1972).

Water deficit also causes an increase in ethylene production. This was noted by Jordan et al. (1972) and McMichael et al. (1972). The latter authors found enhancement of ethylene production by cotton petioles as leaf water potentials decreased to -2.0 to -2.5 MPa. Jordan et al. (1972) saw that abscission of cotyledonary leaves from cotton seedlings occurred following relief from water stress. The amount of abscission was related to the magnitude of the plant water deficit. Leaf abscission was promoted by exogenous ethylene and was enhanced in seedlings subjected to water stress. Treatment with ethylene raised the threshold plant water potential required to induced abscission from -1.7 to -0.7 MPa.

Ethylene does not have the dramatic effect on stomata that ABA does (i.e., closure). Ethylene has been shown to decrease, increase, or have no effect on stomatal resistance (Kirkham, 1983c). However, ethylene does appear to have a consistent effect on osmotic potential. Application of ethylene always lowers osmotic potential (Eisinger et al., 1983; Kirkham, 1983c, 1985; Miyamoto & Kamisaka, 1987). As stated before, plants can maintain their turgor by osmotic adjustment.

Recent results suggest that drought resistance and ethylene production may be related. Zhang and Kirkham (1988) found that, under drought, sorghum ('Dahuangqiao' and 'Dasheyan') produced more ethylene than did maize ('72:HLR 43-1' and 'Mexicana'). Sorghum is considered to be more drought resistant than maize. In addition, cultivars of the same species vary in ethylene production. Zhang and Kirkham (unpublished) found that a drought-resistant cultivar of sorghum (KS-65) produced more ethylene under water stress than did a drought-sensitive cultivar of sorghum (IA-25) (Fig. 11-2). However, Stumpff and Johnson (1987) observed opposite results with loblolly pine (*Pinus taeda* L.). Ethylene production in needles from a mesic area (Virginia coastal plain source; female parent clone R-523) and from a dry area (east Texas drought-hardy source; female parent clone CR1-8) both increased with initial stress (down to about −1.2 MPa in the needles). But only in the Virginia source did ethylene production increase with increasing stress (needle water potential of −3.0 MPa). That is, the mesic source produced more ethylene under severe drought than did the xeric source. More research needs to be done to see if drought-resistant plants produce more ethylene under stress than do drought-sensitive ones.

V. GROWTH

A. Leaves

Upon the development of water stress, restriction of leaf expansion growth is one of the first symptoms of stress. Expansive growth is the process most sensitive to water stress (Hsiao & Jing, 1987; Hsiao et al., 1985). The relative rate of volume (*V*) enlargement of the cell may be expressed as follows (Hsiao et al., 1985):

$$\frac{1}{V}\frac{dV}{dt} = \phi\,(P - Y) \qquad [4]$$

where ϕ is the volumetric wall extensibility, P is the turgor pressure in the cell, and Y is the yield threshold pressure, below which the force on the wall is insufficient to cause wall relaxation. The left-hand side of Eq. [4] can be equated to the growth rate in three dimensions (equal to the relative rate of volume increase) (Hsiao et al., 1985). Thus, in the nondifferential form, we get (Van Volkenburgh & Cleland, 1986)

$$GR = \phi\,(P - Y) \qquad [5]$$

where GR is the rate of cell enlargement.

Even though the mathematics of growth with no water stress is being studied (e.g., see Bertaud & Gandar, 1986), little work is being done concerning the mathematics of growth under water deficit. This is an area that needs more research. What is known is reviewed by Hsiao and Jing (1987) and Hsiao et al. (1985). Nevertheless, many recent studies do document that drought stress reduces leaf growth (Guralnick & Ting, 1987; McIntyre, 1987; Sammis et al., 1986; Sobrado & Turner, 1986). Decreasing leaf area during dry weather is a way that plants can reduce water loss by transpiration (El-Sharkawy & Cock, 1987). In fact, artificial reduction of leaf area has been shown to be beneficial under drought. Shanahan and Nielsen (1987) applied growth retardants (antigibberellins) to maize ('Pioneer 3732' and 'Dekalb DK-524'). The compounds reduced vegetative growth, which was advantageous for production under water stress, but was counterproductive under well-watered conditions.

At the beginning of a dry period, plants diminish leaf area by producing fewer and smaller leaves and by shedding older leaves. For example, Huda et al. (1987) in India observed that two sorghum cultivars ('CSH 8' and 'M 35-1') each produced one leaf less under nonirrigated conditions compared to irrigated conditions. In Colombia, El-Sharkawy and Cock (1987) found that varieties of cassava (*Manihot esculenta* Crantz), which yielded well under both well-watered and water-stressed conditions, had a higher than optimal leaf-area index. Their leaf-area index was about 5.0. Optimal levels were 2.5 to 3.5. They suggested that a variety with more than optimal leaf-area index could reduce leaf area under stress with a minimal effect on yield. In Alabama, Hoogenboom et al. (1987) saw that, during periods of moisture stress, leaves of soybean ('Braxton') formed on nonirrigated plants were smaller than leaves at comparable nodes on irrigated plants (0.0075 vs. 0.010 m^2). After rain, shoots of nonirrigated plants grew more rapidly than those of irrigated plants (0.025 vs. 0.020 m/d), and most of the growth reduction during the previous stressed period was regained.

The reduction in leaf growth due to water stress varies among species. Tanguilig et al. (1987) in the Philippines compared the response of rice ('IR36'), maize ('DMR-2'), and soybean ('Clark 63') to soil water stress. Leaf elongation, leaf water potential, and transpiration rate declined earlier in rice than in maize or soybean. They concluded that the ability of maize and soybean to grow better than rice under water stress was due to their ability to maintain turgor by osmotic adjustment.

Growth appears to be reduced more by water deficit than photosynthesis. In Arizona, Allen et al. (1987) found that photosynthesis of guayule (*Parthenium argentatum* A. Gray 'N565 II'), a drought-resistant, rubber-producing shrub native to the desert Southwest of the USA, was not affected by a 70-d drought, even though height and width of the plants were reduced by the stress. Similarly, in Australia, Turner et al. (1986) found that leaf expansion of cotton ('Deltapine 16') was more sensitive to soil and leaf water deficits than was photosynthesis. Photosynthetic rate and stomatal conductance decreased progressively with leaf water potential between -1.9 and -3.5 MPA. Leaf size decreased when the predawn leaf water potential decreased below -0.8 MPa. Water deficits also reduced the number of floral buds, bolls retained, and final yield.

B. Reproductive Structures

Reproductive stages of growth are often the most sensitive to drought, and, if a water deficit occurs then, yield is depressed more than if it occurs at other periods (Chang, 1968, p. 215). For example, Singh et al. (1987) in India found that cessation of irrigation at flowering of chickpea (*Cicer arientinum* L. 'H-355') induced a rapid decrease in photosynthesis and reduced grain yield by 33% due to a decrease in the number of pods set. Grain yield was linearly related to pod number, leaf water potential, and photosynthetic rate. In Texas, Eck et al. (1987) saw that soybean ('Douglas') was vulnerable to drought during seed development. Soybean stressed between early flowering and the beginning of pod development, and then rewatered, had a yield that was 9 to 13% below that of the well-watered control. However, stress imposed at the beginning of bean development, and then relieved, reduced yields from 15 to 46%.

Reproductive stages are not always the sensitive stages. For example, Sionit et al. (1987) stressed the agricultural weed, goosegrass [*Eleusine indica* (L.) Gaertn.], at the vegetative stage, reproductive stage, and during both stages. Plants stressed during flowering had a smaller decline in biomass during stress, and a higher rate of growth after rewatering, compared to plants stressed during vegetative growth and flowering or stressed only during vegetative growth. The lack of sensitivity of goosegrass to water stress at flowering may be an adaptation that allows it to out-compete crop plants.

Even different stages of grain filling are differentially sensitive to drought. Ouattar et al. (1987a) found that kernel growth of maize ('DRA 400') was more sensitive to water deficits during endosperm cell division than during the period of rapid starch deposition. However, they also observed (Ouattar et al., 1987b) that kernel growth was relatively unaffected by a water deficit that completely inhibited photosynthesis. Kernel growth was maintained by remobilization of assimilates from nongrain parts, especially the stem. High stem moisture content, despite severe leaf dehydration, may have facilitated translocation from the stem to the grain. Grain-water status (water potential, osmotic potential, turgor potential) was independent of leaf-water status.

C. Leaf Angle and Leaf Color

Not only does leaf area decrease with drought, but leaf angle can change, too. In South Africa, Oosterhuis (1986) observed that, with increasing water stress, leaflets of soybean ('Bragg') inverted (angle greater than 90° to the horizontal), so that the more reflective abaxial leaflet surface was exposed. Well-watered soybean leaflets were horizontal. Inversion of the soybean leaflets was related to leaf-water potential, stomatal resistance, and soil-water content. All of these measurements have been used as indicators of crop-water stress. Determination of leaflet-angle appears to be a promising and easy method for scheduling soybean irrigations (Wright & Berliner, 1986). Oosterhuis et al. (1987) also found a relation between leaf-water potential and crop color. A decrease in leaf-water potential of pea (*Pisum sativum* L. 'Puget') was associated with the green-to-blue color change that develops when pea is water stressed. The authors said that crop color could be an early, identifiable indicator of the onset of water stress.

D. Morphological Changes

Morphological changes occur, along with growth reduction, during water stress. Čiamporová (1987) reported that endoplasmic reticulum and mitochondria of root cells of maize ['A15 (VIR)'] were the cell components most sensitive to water stress. Their structure changed after 5 min of water stress, induced by polyethylene glycol 4000 with an osmotic potential of -1.56 MPa. After 8 h of stress, the condensation of nuclear chromatin was apparent. The rapid changes in the structure of the endoplasmic reticulum and mito-

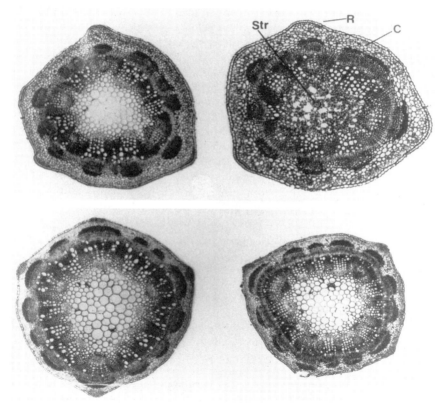

Fig. 11-3. Transverse sections of stems of a xeric alfalfa, *Medicago prostrata* (*top*), and of a mesic alfalfa, *M. sativa* (*bottom*). Ridges, xylem, and starch granules from irrigated (*left*) and nonirrigated (*right*) plants are shown. *Medicago prostrata* × 50 (*left*) and × 75 (*right*). *Medicago sativa* × 40 (*left* and *right*). R = ridge; C = cortex; Str = starch granule. Adapted from Suksayretrup (1986).

chondria coincided with the rapid decrease of water potential of the roots and the immediate cessation of growth.

Hampton et al. (1987) observed that, within the cotton germplasm, there were morphological features associated with drought resistance. These features included thick cuticles, high trichome densities, thick leaves, and thick epidermal cell walls. Significant differences were also seen between stomatal densities on both leaf surfaces (adaxial and abaxial), but the differences did not correlate with differences in stomatal resistance or transpiration. Thick leaves may be an adaptive advantage under drought because they could hold more water. Búrquez (1987) found a strong correlation between leaf thickness and relative water content in rape (*Brassica napus* L. 'Maris Haplona'), green bean (*Phaseolus vulgaris* L. 'Canadian Wonder'), snapweed (*Impatiens parviflora* DC), and four-o'clock (*Mirabilis jalapa* L.).

Suksayretrup (1986) studied the morphology, under well-watered conditions in a greenhouse, of *Medicago prostrata* Jacq., an alfalfa that grows in dry regions, and of *M. sativa* L., the normal hay-type of alfalfa that grows in mesic regions. Many xeromorphic characters were identified in *M. prostrata*. Its leaves had thick cuticles and cell walls, especially on the abaxial surface. The adaxial leaf surface had heavy wax plates, and the abaxial leaf surface had amorphous wax. Glandular hairs were distributed over the plant and were particularly dense on the abaxial side of the leaves. Its leaves were small and thick with compact veins and isolateral mesophyll. *Medicago prostrata* had a thicker cortex than did *M. sativa*. A thick cortex in the stem of xerophytic plants like *M. prostrata* may lessen loss of water from the vascular tissues and thus prevent damage to the shoot. The plants also were grown under irrigated and nonirrigated conditions in the field. As a result of the wider cortex, *M. prostrata* had a narrower pith than did *M. sativa* under both well-watered and water-stressed conditions (Fig. 11–3). Under stress, *M. prostrata* had wider and better developed xylem vessels than did *M. sativa*. Also, under stress, starch granules developed in the pith of *M. prostrata*, but not in the pith of *M. sativa* (Fig. 11–3).

E. Cell Walls

Cell walls change when plants are water stressed. Bozarth et al. (1987) investigated proteins in cell walls of soybean ('Williams') seedlings subjected to low water potentials by withholding water from the vermiculite in which they grew. The vermiculite of the control plants and water-stressed plants had water potentials of −0.1 and −3.0 MPa, respectively. A 28-kDa protein increased in the cell walls as plant-water potential decreased. [A dalton is a unit of mass equal to one-twelfth the mass of an atom of C_{12}. One Da equals $N^{-1}g = 1.663 \times 10^{-24}g$, where N is Avogadro's number (CBE Style Manual Committee, 1978, p. 182).] In contrast, a 70-kDa protein in the cell walls appeared to decrease at low water potentials. The results suggested that water stress and protein changes in the cell walls are related. Singh et al. (1987) found that a 26-kDa protein (osmotin) accumulated in tobacco (*Nicotiana tabacum* L. 'Wisconsin 38') cells adapted to grow under osmotic stress. However, the osmotin was concentrated in the vacuole, not the cell walls.

During water stress, deformation of the cell wall also takes place. Pearce and Beckett (1987) studied cells in leaves of well-watered and drought-stressed barley ('Mazurka') seedlings by scanning electron microscopy. Dehydration reduced cell volume. Folds occurred in the walls of the cells that were just wilting. In severely stressed and damaged plants, a range of cell shapes and deformations were evident.

Water potential, stomata, hormones, and growth of drought-resistant and drought-sensitive cultivars under dry conditions have been compared. However, there appear to be no studies of the cell walls of drought-resistant and drought-sensitive cultivars under drought. This might be an important area for investigation, to help determine why drought-resistant plants survive better under drought.

VI. SUMMARY

Water deficits decrease the water potential of plants. Under dry conditions, drought-resistant plants often have higher water potentials than do drought-sensitive plants. Osmotic adjustment enables plants to maintain turgor during drought. Osmotic adjustment of leaves can be either more or less than that of roots. Species and cultivars vary in osmotic adjustment. A plant that can keep a higher turgor potential by osmotic adjustment is more likely to survive drought than a plant that has little osmotic adjustment. Betaine and proline are two osmolytes that accumulate under drought and permit a plant to lower its osmotic potential without damaging metabolic functions. Accumulation of osmolytes differs among species and cultivars and may be under genetic control.

Both stomatal and nonstomatal inhibition of photosynthesis occurs under drought. The relative amount of the two types of inhibition is affected by the severity of the stress. Nonstomatal factors often become more important as stress is prolonged.

Concentration of the plant hormone, abscisic acid, rises rapidly after water stress is imposed. It closes stomata. Several studies indicate that drought-resistant cultivars accumulate more ABA than do drought-sensitive cultivars, but this is not a consistent finding. Water deficit causes an increase in production of ethylene, a gaseous plant hormone. Ethylene does not have the dramatic effect on stomata (i.e., closure) that ABA has, but ethylene does appear always to lower osmotic potential. Recent work suggests that drought-resistant plants produce more ethylene than do drought-sensitive plants, but more research is needed to confirm this observation.

Drought reduces growth. Growth, such as leaf elongation, decreases at a higher water potential than that necessary to decrease photosynthesis. If drought occurs during reproductive stages of growth, yield is often more reduced than if drought occurs at other stages of growth. Different stages of grain filling are differentially sensitive to drought. For example, endosperm cell division is more sensitive to water deficit than starch deposition. Kernel growth is decreased at a lower water potential than photosynthesis. Morphological features associated with drought resistance include the following: (i) erect leaf angle; (ii) wax accumulation; (iii) high trichome density; (iv) wider and better developed xylem vessels; (v) thick cuticle; (vi) thick leaf; (vii) thick cortex; and (viii) thick cell wall. Changes in cell-wall proteins and cell-wall shapes also occur under water deficit. Future studies might focus on cell walls of drought-resistant and drought-sensitive cultivars, since characteristics of cell walls appear to be important for survival under drought.

REFERENCES

Abeles, A.L., and F.B. Abeles. 1972. Biochemical pathway of stress-induced ethylene. Plant Physiol. 50:496–498.

Allen, S.G., F.S. Nakayama, D.A. Dierig, and B.A. Rasnick. 1987. Plant water relations, photosynthesis, and rubber content of young guayule plants during water stress. Agron. J. 79:1030–1035.

Bannister, P. 1986. Observations on water potential and drought resistance of trees and shrubs after a period of summer drought around Dunedin, New Zealand. N.Z. J. Bot. 24:387–392.

Bertaud, D.S., and P.W. Gandar. 1986. A simulation model for cell growth and proliferation in root apices. I. Structure of model and comparisons with observed data. Ann. Bot. 58:285–301.

Blum, A., and C.Y. Sullivan. 1986. The comparative drought resistance of landraces of sorghum and millet from dry and humid regions. Ann. Bot. 57:835–846.

Bozarth, C.S., J.E. Mullet, and J.S. Boyer. 1987. Cell wall proteins at low water potentials. Plant Physiol. 85:261–267.

Bunce, J.A. 1987. Species-specific responses to water stress of gas exchange parameters mimicked by applied abscisic acid. Can. J. Bot. 65:103–106.

Búrquez, A. 1987. Leaf thickness and water deficit in plants: A tool for field studies. J. Exp. Bot. 38:109–114.

CBE Style Manual Committee. 1978. Council for Biology Editors style manual: Guide for authors, editors, and publishers in the biological sciences. Council of Biology Editors. Am. Inst. Biol. Sci., Arlington, VA.

Chang, J.-H. 1968. Climate and agriculture. An ecological survey. Aldine Publ. Co., Chicago.

Chumakovskii, N.N. 1986. Stem height and drought resistance in relation to leaf content of abscisic acid and ethylene in wheat plants. Sov. Plant Physiol. 33 (no. 3, part 2):401–407.

Čiamporová, M. 1987. The development of structural changes in epidermal cells of maize roots during water stress. Biol. Plant. 29:290–294.

Cornic, G., I. Papgeorgiou, and C. Louason. 1987. Effect of a rapid and a slow drought cycle followed by rehydration on stomatal and non-stomatal components of leaf photosynthesis in Phaseolus vulgaris L. J. Plant Physiol. 126:309–318.

Cox, W.J., and G.D. Jolliff. 1987. Crop-water relations of sunflower and soybean under irrigated and dryland conditions. Crop Sci. 27:553–557.

Davidson, D.J., and P.M. Chevalier. 1987. Influence of polyethylene glycol-induced water deficits on tiller production in spring wheat. Crop Sci. 27:1185–1187.

Drossopoulos, J.B., A.J. Karamanos, and C.A. Niavis. 1987. Changes in ethanol soluble carbohydrates during the development of two wheat cultivars subjected to different degrees of water stress. Ann. Bot. 59:173–180.

Eamus, D. 1987. Influence of preconditioning upon the changes in leaf conductance and leaf water potential of soybean, induced by chilling, water stress and abscisic acid. Aust. J. Plant Physiol. 14:331–339.

Eck, H.V., A.C. Mathers, and J.T. Musick. 1987. Plant water stress at various growth stages and growth and yield of soybeans. Field Crop Res. 17:1–16.

Eisinger, W., L.J. Croner, and L. Taiz. 1983. Ethylene-induced lateral expansion in etiolated pea stems. Kinetics, cell wall synthesis, and osmotic potential. Plant Physiol. 73:407–412.

El-Sharkawy, M.A., and J.H. Cock. 1987. Response of cassava to water stress. Plant Soil 100:345–360.

Farquhar, G.D., D.R. Dubbe, and K. Raschke. 1978. Gain of the feedback loop involving carbon dioxide and stomata. Theory and measurement. Plant Physiol. 62:406–412.

Goyal, A. 1987. Effects of water stress on glycolate metabolism in the leaves of rice seedlings (*Oryza sativa*). Physiol. Plant. 69:289–294.

Grumet, R., and A.D. Hanson. 1986. Genetic evidence for an osmoregulatory function of glycine-betaine accumulation in barley. Aust. J. Plant Physiol. 13:353–364.

Gupta, A.S., and G.A. Berkowitz. 1987. Osmotic adjustment, symplast volume, and non-stomatally mediated water stress inhibition of photosynthesis in wheat. Plant Physiol. 85:1040–1047.

Guralnick, L.J., and I.P. Ting. 1987. Physiological changes in *Portulacaria afra* (L.) Jacq. during a summer drought and rewatering. Plant Physiol. 85:481–486.

Hale, M.G., and D.M. Orcutt. 1987. The physiology of plants under stress. John Wiley and Sons, New York.

Hampton, R.E., D.M. Oosterhuis, J.M. Stewart, and K.S. Kim. 1987. Anatomical differences in cotton related to drought tolerance. Ark. Farm Res. 36(6):4.

Hanson, A.D., N.E. Hoffman, and C. Samper. 1986. Identifying and manipulating metabolic stress-resistance traits. HortScience 21:1313–1317.

Hartung, W., J.W. Radin, and D.L. Hendrix. 1988. Abscisic acid movement into the apoplastic solution of water-stressed cotton leaves. Role of apoplastic pH. Plant Physiol. 86:909–913.

Hoogenboom, G., C.M. Peterson, and M.G. Huck. 1987. Shoot growth rate of soybean as affected by drought stress. Agron. J. 79:598–607.

Hsiao, T.C., and J. Jing. 1987. Leaf and root expansive growth in response to water deficits. p. 180–192. *In* D.J. Cosgrove and D.P. Knievel (ed.) Physiology of cell expansion during plant growth. Am. Soc. Plant Physiol., Rockville, MD.

Hsiao, T.C., W.K. Silk, and J. Jing. 1985. Leaf growth and water deficits: Biophysical effects. p. 239–266. *In* N.R. Baker et al. (ed.) Control of leaf growth. Cambridge Univ. Press, Cambridge.

Huda, A.K.S., M.V.K. Sivakumar, Y.V. Sri Rama, J.G. Sekaran, and S.M. Virmani. 1987. Observed and simulated responses of two sorghum cultivars to different water regimes. Field Crop Res. 16:323–335.

Hutmacher, R.B., and D.R. Krieg. 1983. Photosynthetic rate control in cotton. Stomatal and nonstomatal factors. Plant Physiol. 73:658–661.

International Rice Research Institute. 1982. Drought resistance in crops with emphasis on rice. Int. Rice Res. Inst., Los Baños, Laguna, Philippines.

Jarvis, P.G., and T.A. Mansfield (ed.). 1981. Stomatal physiology. Cambridge Univ. Press, Cambridge.

Johnson, R.C., D.W. Mornhinweg, D.M. Ferris, and J.J. Heitholt. 1987. Leaf photosynthesis and conductance of selected *Triticum* species at different water potentials. Plant Physiol. 83:1014–1017.

Joly, R.J., and J.B. Zaerr. 1987. Alteration of cell-wall water content and elasticity in Douglas-fir during periods of water deficit. Plant Physiol. 83:418–422.

Jones, H., R.A. Leigh, A.D. Tomos, and R.G. Wyn Jones. 1987. The effect of abscisic acid on cell turgor pressures, solute content and growth of wheat roots. Planta 170:257–262.

Jordan, W.R., P.W. Morgan, and T.L. Davenport. 1972. Water stress enhances ethylene-mediated leaf abscission in cotton. Plant Physiol. 22:185–196.

Kirkham, M.B. 1983a. Effect of ABA on the water relations of winter-wheat cultivars varying in drought resistance. Physiol. Plant. 59:153–157.

Krikham, M.B. 1983b. Physical model of water in a split-root system. Plant Soil 75:153–168.

Kirkham, M.B. 1983c. Effect of ethephon on the water status of a drought-resistant and a drought-sensitive cultivar of winter wheat. Z. Pflanzenphysiol. 112:103–112.

Kirkham, M.B. 1985. Effect of ethephon on growth and water status of *Striga*-susceptible genotypes of pearl millet. Field Crop Res. 11:219–231.

Kirkham, M.B. 1988. Hydraulic resistance of two sorghums varying in drought resistance. Plant Soil 105:19–24.

Kirkham, M.B. 1989. Growth and water relations of two wheat cultivars grown separately and together. Biol. Agric. Hortic. 6:35–46.

Kirkham, M.B., W.R. Gardner, and G.C. Gerloff. 1969. Leaf water potential of differentially salinized plants. Plant Physiol. 44:1378–1382.

Kirschbaum, M.U.F. 1987. Water stress in *Eucalyptus pauciflora*: Comparison of effects on stomatal conductance with effects on the mesophyll capacity for photosynthesis, and investigation of a possible involvement of photoinhibition. Planta 171:466–473.

Kozlowski, T.T. (ed.). 1983. Water deficits and plant growth. Vol. 7. Additional woody crop plants. Academic Press, New York.

Krieg, D.R., and R.B. Hutmacher. 1986. Photosynthetic rate control in sorghum: Stomatal and nonstomatal factors. Crop Sci. 26:112–117.

Lopez, F.B., T.L. Setter and C.R. McDavid. 1987. Carbon dioxide and light responses of photosynthesis in cowpea and pigeonpea during water deficit and recovery. Plant Physiol. 85:990–995.

Lopez, F.B., T.L. Setter, and C.R. McDavid. 1988. Photosynthesis and water vapor exchange of pigeonpea leaves in response to water deficit and recovery. Crop Sci. 28:141–145.

Lorens, G.F., J.M. Bennett, and L.B. Loggale. 1987a. Differences in drought resistance between two corn hybrids. I. Water relations and root length density. Agron. J. 79:802–807.

Lorens, G.F., J.M. Bennett, and L.B. Loggale. 1987b. Differences in drought resistance between two corn hybrids. II. Component analysis and growth rates. Agron. J. 79:808–813.

Mansfield, T.A. 1983. Movements of stomata. Sci. Prog. 68:519–542.

McCree, K.J., and S.G. Richardson. 1987. Stomatal closure vs. osmotic adjustment: A comparison of stress responses. Crop Sci. 27:539–543.

McIntyre, G.I. 1987. The role of water in the regulation of plant development. Can. J. Bot. 65:1287–1298.

McMichael, B.L., W.R. Jordan, and R.D. Powell. 1972. An effect of water stress on ethylene production by intact cotton petioles. Plant Physiol. 49:658–660.

Miyamoto, K., and S. Kamisaka. 1987. Effect of water stress and ethylene on osmoregulation in the subhook region of pea epicotyls. Biochem. Physiol. Pflanzen 182:41–48.

Oosterhuis, D.M. 1986. Soybean leaf movements as an indicator of drought stress. Oilseeds News (December) 1986:19–20.

Oosterhuis, D.M., F. Le Maire, and C. Le Maire. 1987. Leaf water potential and crop color change in water-stressed peas. HortScience 22:429–431.

Oosterhuis, D.M., and S.D. Wullschleger. 1987. Osmotic adjustment in cotton (*Gossypium hirsutum* L.) leaves and roots in response to water stress. Plant Physiol. 84:1154–1157.

Ouattar, S., R.J. Jones, and R.K. Crookston. 1987a. Effect of water deficit during grainfilling on the pattern of maize kernel growth and development. Crop Sci. 27:726–730.

Ouattar, S., R.J. Jones, R.K. Crookston, and M. Kajeiou. 1987b. Effect of drought on water relations of developing maize kernels. Crop Sci. 27:730–735.

Paleg, L.G., and D. Aspinall (ed.). 1981. The physiology and biochemistry of drought resistance in plants. Academic Press, Sydney.

Parsons, L.R. 1982. Plant responses to water stress. p. 175–192. *In* M.N. Christiansen and C.F. Lewis (ed.) Breeding plants for less favorable environments. John Wiley and Sons, New York.

Pearce, R.S., and A. Beckett. 1987. Cell shape in leaves of drought-stressed barley examined by low temperature scanning electron microscopy. Ann. Bot. 59:191–195.

Pekić, S., and S.A. Quarrie. 1987. Abscisic acid accumulation in lines of maize differing in drought resistance: A comparison of intact and detached leaves. J. Plant Physiol. 127:203–217.

Pier, P.A., and G.A. Berkowitz. 1987. Modulation of water stress effects on photosynthesis by altered leaf K^{+1}. Plant Physiol. 85:655–661.

Quarrie, S.A. 1982. Droopy: A wilty mutant of potato deficient in abscisic acid. Plant Cell Environ. 5:23–26.

Radin, J.W., and D.L. Hendrix. 1986. Accumulation and turnover of abscisic acid in osmotically stressed cotton leaf tissue in relation to temperature. Plant Sci. 45:37–42.

Raghavendra, A.S., and K.B. Reddy. 1987. Action of proline on stomata differs from that of abscisic acid, G-substances, or methyl jasmonate. Plant Physiol. 83:732–734.

Rose, E. 1985. Water relations of winter wheat. I. Genotypic differences in ethylene and abscisic acid production during drought stress. II. Effect of a dwarfing gene on root growth, shoot growth, and water uptake in the field. Ph.D. diss. Kansas State Univ., Manhattan (Diss. Abstr. 85-10242).

Rosenberg, N.J. (ed.). 1980. Drought in the Great Plains: Research on impacts and strategies. Water Resourc. Publ., Littleton, CO.

Rygol, J., K. Winter, and U. Zimmermann. 1987. The relationship between turgor pressure and titratable acidity in mesophyll cells of intact leaves of a Crassulacean-acid-metabolism plant, *Kalanchoe daigremontiana* Hamet et Perr. Planta 172:287–493.

Sammis, T.W., S. Williams, D. Smeal, and C.E. Kallsen. 1986. Effect of soil moisture stress on leaf area index, evapotranspiration and modeled soil evaporation and transpiration. Trans. ASAE 29:956–961.

Schulze, E.-D., N.C. Turner, T. Gollan, and K.A. Shackel. 1987. Stomatal responses to air humidity and to soil drought. p. 311–321. *In* E. Zeiger et al. (ed.) Stomatal function. Stanford Univ. Press, Stanford, CA.

Shanahan, J.F. and D.C. Nielsen. 1987. Influence of growth retardants (anti-gibberellins) on corn vegetative growth, water use, and grain yield under different levels of water stress. Agron. J. 79:103–109.

Sheriff, D.W., M.J. Fisher, G. Rusitzka, and C.W. Ford. 1986. Physiological reactions to an imposed drought by two twining pasture legumes: *Macroptilium atropurpureum* (desiccation sensitive) and *Galactia striata* (desiccation insensitive). Aust. J. Plant Physiol. 13:431–445.

Simpson, G.M. 1981. Water stress on plants. Praeger Publ., New York.

Sinclair, T.R., and M.M. Ludlow. 1986. Influence of soil water supply on the plant water balance of four tropical grain legumes. Aust. J. Plant Physiol. 13:329–341.

Singh, D.P., P. Singh, H.C. Sharma, and N.C. Turner. 1987. Influence of water deficits on the water relations, canopy gas exchange, and yield of chickpea (*Cicer arietinum*). Field Crop Res. 16:231–241.

Singh, N.K., C.A. Bracker, P.M. Hasegawa, A.K. Handa, S. Buckel, M.A. Hermodson, E. Pfankoch, F.E. Regnier, and R.A. Bressan. 1987. Characterization of osmotin. A thaumatin-like protein associated with osmotic adaptation in plant cells. Plant Physiol. 85:529–536.

Sionit, N., D.T. Patterson, R.D. Coffin, and D.A. Mortenson. 1987. Water relations and growth of the weed, goosegrass (*Eleusine indica*), under drought stress. Field Crops Res. 17:163–173.

Sobrado, M.A., and N.C. Turner. 1986. Photosynthesis, dry matter accumulation and distribution in the wild sunflower *Helianthus petiolaris* and the cultivated sunflower *Helianthus annuus* as influenced by water deficits. Oecologia 69:181–187.

Stewart, C.R., and G. Voetberg. 1987. Abscisic acid accumulation is not required for proline accumulation in wilted leaves. Plant Physiol. 83:747–749.

Stumpff, N.J., and J.D. Johnson. 1987. Ethylene production by loblolly pine seedlings associated with water stress. Physiol. Plant. 69:167–172.

Suksayretrup, K. 1986. Xeroptism in *Medicago*. Ph.D. diss. Kansas State Univ., Manhattan Diss. Abstr. 86-17133.

Tal, M., D. Imber, A. Erez, and E. Epstein. 1979. Abnormal stomatal behavior and hormonal imbalance in *flacca*, a wilty mutant of tomato. Plant Physiol. 63:1044–1048.

Tanguilig, V.C., E.B. Yambao, J.C. O'Toole, and S.K. De Datta. 1987. Water stress effects on leaf elongation, leaf water potential, transpiration, and nutrient uptake or rice, maize, and soybean. Plant Soil 103:155–168.

Teare, I.D., and M.M. Peet (ed.). 1983. Crop-water relations. John Wiley and Sons, New York.

Terry, P.H., D.T. Krizek, and R.M. Mirecki. 1988. Genotypic variation in coleus in the ability to accumulate abscisic acid in response to water deficit. Physiol. Plant. 72:441–449.

Thimann, K.V. 1972. Ethylene. p. 213–221. *In* F.C. Stewart (ed.) Plant physiology. A treatise. Vol. VIB. Physiology of development. The hormones. Academic Press, New York.

Toft, N.L., S.J. McNaughton, and N.J. Georgiadis. 1987. Effects of water stress and simulated grazing on leaf elongation and water relations of an east African grass, *Eustachys paspaloides*. Aust. J. Plant Physiol. 14:211–226.

Tsunoda, S., and M.T. Fukoshima. 1986. Leaf properties related to the photosynthetic response to drought in upland and lowland rice varieties. Ann. Bot. 58:531–539.

Turner, N.C., A.B. Hearn, J.E. Begg, and G.A. Constable. 1986. Cotton (*Gossypium hirsutum* L.): Physiological and morphological responses to water deficits and their relationship to yield. Field Crop Res. 14:153–170.

Turner, N.C., and P.J. Kramer (ed.). 1980. Adaptation of plants to water and high temperature stress. John Wiley and Sons, New York.

Turner, N.C., W.R. Stern, and P. Evans. 1987. Water relations and osmotic adjustment of leaves and roots of lupins in response to water deficits. Crop Sci. 27:977–983.

Vaadia, Y., and Y. Waisel. 1967. Physiological processes as affected by water balance. p. 354–372. *In* R.M. Hagan et al. (ed.) Irrigation of agricultural lands. ASA, Madison, WI.

Van Volkenburgh, E., and R.E. Cleland. 1986. Wall yield threshold and effective turgor in growing bean leaves. Planta 167:37–43.

Walton, D.C. 1980. Biochemistry and physiology of abscisic acid. Annu. Rev. Plant Physiol. 31:453–489.

Weiler, E.W. 1984. Immunoassay of plant growth regulators. Annu. Rev. Plant Physiol. 35:85–95.

Wright, A.D., and P.R. Berliner. 1986. The use of soybean leaflet angle data for irrigation scheduling. Irrig. Sci. 7:245–248.

Zhang, J., and M.B. Krikham. 1988. Ethylene production by maize and sorghum. Plant Physiol. 86(suppl.):156.

Zhang, J., U. Schurr, and W.J. Davies. 1987. Control of stomatal behaviour by abscisic acid which apparently originates in the roots. J. Exp. Bot. 38:1174–1181.

12

Theoretical Considerations in the Description of Evaporation and Transpiration

THOMAS R. SINCLAIR

USDA-ARS
Gainesville, Florida

Equations defining the processes of evaporation and transpiration are readily available in many text books. Interestingly, there are several approaches used to define each of these processes, but many times a full consideration of the theoretical basis on which they are derived does not accompany these presentations. In particular, discussions concerning the simplifying assumptions and empiricisms required to derive commonly used equations from fundamental theoretical relationships are often omitted. The intent of this chapter is to examine the origins of some of the common relationships describing evaporation and transpiration so that the relative merits of the various equations can be assessed.

This chapter is divided into two major sections. The first section reviews the relationships used to define evaporation from an open surface of pure, liquid water. Two processes are examined: the liquid-vapor equilibrium at the liquid surface, and the vapor flux from near the liquid surface. The second major section examines the more complicated situations of transpiration when water is vaporized inside a leaf, and consequently, the path of vapor flux is much more complex. Consideration is given to the additional restrictions on vapor diffusion resulting from individual leaf morphology and from the structure of the whole leaf canopy.

I. EVAPORATION

A. Vapor Flux Density

A key process in evaporation is the movement of water vapor from adjacent to a liquid surface to the bulk atmosphere. However, the appropriate formulation of this process is still not altogether clear. Since temperature gradients almost always exist in natural evaporating systems, it seems an analysis based on the thermodynamics of irreversible processes is appropriate. The basis for expressing all the thermodynamic flows in a thermodynamically irreversible system is a set of phenomenological equations generally referred to as the Onsager equations (Katchalsky & Curran, 1967). The appropriate Onsager expression for vapor flow in a coupled system with a water vapor and temperature gradient is

$$J = L_v \, \Delta\mu_v + L_{vT} \, \Delta T \qquad [1]$$

where

J = vapor flux density (mol cm^{-2} s^{-1}),
L_v = Onsager coefficient for vapor,
L_{vT} = Onsager vapor-temperature coupling coefficient,
$\Delta\mu_v$ = vapor chemical potential difference across the diffusion path (J mol^{-1}),
ΔT = temperature difference across the diffusion path.

A premise of the Onsager equation is that there is reciprocity among the coupling coefficients. In Eq. [1], the coupling coefficient (L_{vT}) defining the effect of the temperature gradient on water vapor flow would be equivalent to the coefficient defining the effect of the water vapor gradient on heat flow, L_{vT}. Cary (1963) experimentally tested Onsager's equation for coupled thermal and water vapor flux and confirmed equivalency in the coupling coefficients. However, the value of the coupling coefficient is dependent on both vapor pressure and temperature, making its numerical evaluation difficult.

Since under many natural evaporative conditions both the temperature gradients and the coupling coefficients are thought to be relatively small, the effect of the coupled thermal flux is invariably ignored. Consequently, Eq. [1] is reduced to the more familiar expression of flux density of an individual molecular species under reversible, steady-state conditions (Nobel, 1983).

$$J = m \, \rho_v \left(-\frac{\partial\mu}{\partial x} \right) \qquad [2]$$

where

m = mobility coefficient (cm^2 mol J^{-1} s^{-1}),
ρ_v = local vapor concentration or density (mol cm^{-3}),
μ = chemical potential of vapor (J mol^{-1}),
x = distance (cm).

Equation [2] defines the evaporation rate as dependent on the vapor chemical potential gradient.

The use of Eq. [2] to solve for vapor flux density looks to be quite formidable due not only to its dependence on the chemical potential gradient, but also due to its dependence on the local vapor concentration, ρ_v. However, derivations from Eq. [2] lead to more tractable relationships. The vapor potential gradient term is eliminated by defining the components of the vapor chemical potential (Nobel, 1983)

$$\mu = \mu^* + RT \ln \frac{P_v}{P_v^*} + \epsilon gZ \qquad [3]$$

where

μ^* = reference vapor potential,
R = ideal gas constant (8.314 J mol^{-1} K^{-1}),
T = temperature (K),
P_v = vapor partial pressure (J cm^{-3}),
P_v^* = saturation vapor pressure,
ϵ = molecular weight of water (18 g mol^{-1}),
g = gravitational acceleration (980 cm s^{-1}),
Z = height above reference (cm).

Assuming a constant temperature and a negligible height difference along the vapor gradient, and differentiating with respect to x, Eq. [3] becomes

$$-\frac{\partial \mu}{\partial x} = -RT \frac{\partial \left(\ln \frac{P_v}{P_v^*} \right)}{\partial x} \qquad [4]$$

Since isothermal conditions were assumed, P_v^* is constant and Eq. [4] gives the vapor potential gradient

$$-\frac{\partial \mu}{\partial x} = -\frac{RT}{P_v} \frac{\partial P_v}{\partial x} . \qquad [5]$$

To eliminate the local vapor concentration from Eq. [2], ρ_v can be defined by the ideal gas law

$$\rho_v = \frac{n}{V} = \frac{P_v}{RT} \qquad [6]$$

where

n = moles,
V = volume (cm^3).

Substituting Eq. [5] and [6] into Eq. [2], the derived expression for evaporation rate is

$$J = -m \frac{P_v}{RT} \left(\frac{RT}{P_v} \frac{\partial P_v}{\partial x} \right) \qquad [7a]$$

or

$$J = -m \frac{\partial P_v}{\partial x} \qquad [7b]$$

Equation [7] can be further simplified into a finite difference equation when the atmosphere adjacent to the liquid surface is saturated with vapor (P_v^*).

$$J = \frac{m}{\Delta x} (P_v^* - P_v).$$ [8]

The next step in deriving a more recognizable evaporation equation is to evaluate the mobility coefficient. Be again assuming an ideal gas, the mobility coefficient is defined as

$$m = \frac{D}{RT} = \frac{D\rho_a}{P_a}$$ [9]

where
  D = molecular diffusion coefficient of water vapor (0.24 cm^2 s^{-1} at 20 °C),
  ρ_a = density of air (mol cm^{-3}),
  P_a = pressure of air (J cm^{-3}).

Consequently, on a molar basis the evaporation rate can be derived by substituting Eq. [9] into Eq. [8] and obtaining

$$J = \frac{D}{\Delta x RT} (P_v^* - P_v)$$ [10a]

or

$$J = \frac{D\rho_a}{\Delta x P_a} (P_v^* - P_v).$$ [10b]

A useful variant of Eq. [10] is to express the vapor pressure as a fraction of the total pressure. That is,

$$J = \frac{D\rho_a}{\Delta x} \frac{(P_v^* - P_v)}{P_a} = \frac{D\rho_a}{\Delta x} \Delta N_v$$ [10c]

where N_v is the mole fraction of water vapor. Since the values of D and ρ_a change in an inverse manner with respect to each other in response to changes in either temperature or pressure, a conductance coefficient derived from D and ρ_a in Eq. [10c] would be less environmentally sensitive (Nobel, 1983).

Evaporation rate on a mass basis, E (g cm^{-2} s^{-1}), can be derived simply by inserting into Eq. [10] the molecular weight of water (ϵ = 18 g mol^{-1}).

$$E = \frac{D\epsilon}{\Delta x RT} (P_v^* - P_v)$$ [11a]

or

$$E = \frac{D}{\Delta x} \frac{\epsilon \rho_a}{P_a} (P_v^* - P_v).$$ [11b]

The term $D/\Delta x$ has units of cm s^{-1} and can be viewed as the conductance of vapor during evaporation.

Sometimes the evaporation equation is written as a function of water vapor density (ρ_v). This derivation can be obtained from Eq. [10] and [11] by again assuming the ideal gas law and isothermal conditions. Since this latter assumption is rarely satisfied in the natural environment, the use of the derived expressions should be treated with caution. In any case, with these assumptions the vapor pressure can be defined in terms of a density (Eq. [6]), and Eq. [10a] and [11a] become, respectively,

$$J = \frac{D}{\Delta x} (\rho_v^* - \rho_v)$$ [12]

and

$$E = \frac{D}{\Delta x} \epsilon(\rho_v^* - \rho_v).$$ [13]

The above derived expressions (Eq. [10] and [11]) defining evaporation contain three variables: (i) saturated vapor pressure, (ii) atmospheric vapor pressure, and (iii) a conductance coefficient. In the section I.B. of this chapter, the basis for determining saturated vapor pressure is examined and in section I.C. the evaluation of the conductance coefficient for conditions other than still air is considered.

B. Liquid-Vapor Equilibrium

In the previous section it was assumed that the vapor immediately adjacent to the pure, water surface was in equilibrium with the liquid. Consequently, immediately adjacent to the liquid a saturated vapor pressure existed and the value of the saturated vapor pressure was dependent on the temperature. The derivation of an expression defining saturated vapor pressure as a function of temperature originates from classical, equilibrium thermodynamic arguments.

It must be assumed that the liquid-vapor system is subjected to isothermal and isobaric conditions, and the two phases are in equilibrium. Consequently, from the Gibbs equation it can be shown that

$$\frac{dP_v^*}{dT} = \frac{S_v - S_l}{V_v - V_l}$$ [14]

where
  S_v = entropy of the vapor ($J \ mol^{-1} \ K^{-1}$),
  S_l = entropy of the liquid,
  V_v = molar volume of vapor ($cm^3 \ mol^{-1}$),
  V_l = molar volume of liquid.

The numerator on the right-hand side of Eq. [14] is the entropy change for the two phases and can be expressed as a function of the heat of vaporization, H_v ($J \ mol^{-1}$).

$$S_v - S_l = \frac{H_v}{T} \qquad [15]$$

Since any change in the molar volume of liquid would be small relative to the molar volume of vapor, substitution of Eq. [15] into Eq. [14] yields the Clausius-Clapeyron equation

$$\frac{dP_v^*}{dT} = \frac{H_v}{V_v T} . \qquad [16]$$

Solving for the vapor molar volume from the ideal gas law and substituting into Eq. [16]

$$\frac{dP_v^*}{dT} = \frac{H_v P_v}{RT^2} \qquad [17]$$

Integration of Eq. [17] yields

$$\ln (P_v^*) = -\frac{H_v}{RT} + c_1 \qquad [18a]$$

or

$$P_v^* = \exp \left(-\frac{H_v}{RT} + c_1 \right) \qquad [18b]$$

where c_1 is a constant of integration.

Since Eq. [18] was derived for constant conditions, it really represents a state equation rather than a function equation defining saturated vapor pressure over a broad range of temperatures. Water vapor in air is not an ideal gas and both H_v and c_1 are temperature dependent. No expression as simple as Eq. [18] accurately describes saturated vapor pressure over a broad range of temperatures.

Considerable experimental effort has been devoted to developing an empirical expression for saturated vapor pressure. It appears the most universally accepted standard for vapor pressure in the range of -105 to $100\,°C$ was presented by Goff and Gratch (1946). Using a combination of theoretical arguments and empirical evidence, they presented the following equation (converted to be dimensionally consistent with the units used in this chapter, J cm^{-3}):

$$P_v^* = 7.95357 \times 10^6 \exp \left\{ -18.19728 \left(\frac{T_s}{T} \right) \right.$$

$$+ 5.02808 \ln \left(\frac{T_s}{T} \right) - 70242.1852 \exp \left[\frac{26.12052}{T_s/T} \right]$$

$$\left. + 58.06919 \exp \left[-8.03945 \left(\frac{T_s}{T} \right) \right] \right\} \qquad [19]$$

where $T_s = 373.16$ K.

Obviously Eq. [19] is complex, and before the era of computers it was very difficult to use. Subsequently, there were a number of efforts to develop simpler, alternative equations which accurately reproduced Eq. [19], especially in the temperature range of 0 to 50 °C. One of the more accurate of these simplified expressions was presented by Murray (1967) and was shown to result in no more than a 0.06% deviation from the Goff and Gratch equation in the 0 to 50 °C range (Hull, 1974).

$$P_v^* = 6.1078 \times 10^{-4} \exp \left(\frac{17.2694 \, T - 4717.31}{T - 35.86} \right) . \qquad [20]$$

Even though the form of Eq. [20] parallels the result of the derivation from the Clausius-Clapeyron equation, it is also apparent that considerable empiricism and approximation were used to produce Eq. [20]. While the use of Eq. [20] seems justified, especially in naturally evaporating environments, it should be remembered that clear constraints on the temperature range and accuracy of Eq. [20] do exist.

C. Conductance

Equations [10] and [11] defining the evaporation rate had the coefficient $D/\Delta x$ with units cm s^{-1}. That is, evaporation as defined in these equations is dependent on a term called the *vapor conductance in still air*, which in turn is defined by the diffusion coefficient of the vapor and the distance of diffusion. However, evaporation in the natural environment rarely occurs in still air. Rather, air movement over a surface results in a boundary layer of momentum, heat, and/or mass exchange that directly influences the conductance. Expressions of conductance under a variety of conditions have been developed that are analogous to the conductance in the original evaporation equations (Brutsaert, 1982).

The simplest equations are derived from the boundary layer and its impact on conductance for flat plates of finite length. Gebhart (1961) presented the solution for the conductance to mass transfer above a single, flat surface of uniform temperature subjected to laminar flow as

$$h = 0.664 \frac{K}{L} Re^{1/2} Sc^{1/3} \qquad [21]$$

where

h = vapor boundary layer conductance (cm s^{-1}),
K = fluid thermal diffusivity (0.215 cm^2 s^{-1} at 20 °C),
L = length of flat surface (cm),
Re = Reynolds number (uL/K_v, where u = wind speed, cm s^{-1}, and K_v = kinematic viscosity, 0.15 cm^2 s^{-1} at 20 °C),
Sc = Schmidt number (0.63 for water vapor).

Combining the constants when the temperature is 20 °C, Eq. [21] predicts

$$h = 0.316 \sqrt{\frac{u}{L}} . \qquad [22]$$

Unfortunately, laminar flow is not common in natural environments so the straightforward analysis that led to Eq. [22] is generally not applicable. The derivation of conductance in the presence of turbulent flow is complicated, involving a number of assumptions (Brutsaert, 1967). Gebhart (1961) presented the empirical equation for conductance of turbulent boundary layers as

$$h = 0.032 \, u^{0.8} \, L^{-0.2} . \qquad [23]$$

Consequently, the conductance in turbulent flow is more sensitive to windspeed and less sensitive to the dimensions of the surface than in laminar flow. Brutsaert and Yu (1968) found for square water surfaces of area 0.09 to 5.9 m^2 subjected to natural winds, that the numerical terms in Eq. [23] needed to be adjusted somewhat for the best empirical fit.

In many evaporative situations the conductance for a surface of large extent is desired. In this case, it is assumed there is a uniform boundary layer over the entire surface and momentum transfer theory can be used to derive a conductance. It is assumed that the momentum conductance in neutral conditions (vertical temperature gradient is equal to the dry adiabatic lapse rate) is directly proportional to the shearing stress and inversely proportional to the wind speed (Monteith, 1973).

$$h = \frac{\tau}{\rho_a u} = \frac{u k^2}{[\ln (Z/Z_o)]^2} \cdot \quad [24]$$

where

τ = shearing stress,
u = wind speed at reference height (Z),
k = von Karman's constant (0.41),
Z_o = roughness length of the surface (cm).

Under natural evaporative situations, neutral conditions are not common so adjustments for atmospheric buoyancy are required (Tanner, 1968). While approaches for accounting for nonneutral conditions have been presented (e.g., Mahrt & Ek, 1984), fairly elaborate experimental observations are required to incorporate the buoyancy terms. To complicate matters further, it is likely under many conditions that the momentum and mass transfer boundary layers are different so the momentum conductance must be converted to a conductance for vapor transfer.

Due to these uncertainties in theoretically calculating the conductance for a large surface, a number of empirical formulas have been developed. Penman (1948) offered an empirical relationship making conductance linearly dependent on the wind speed measured at 2 m above the surface. Sweers (1976) reviewed many of the conductance formulas and similarly concluded that conductance could be expressed as a linear function of windspeed measured at some reference height. The conductance above large water surfaces in the units used in this chapter (cm s^{-1}) was predicted best by

$$h = 5.7 + 0.04 \, u_3 \quad [25]$$

where u_3 = windspeed at 3 m (cm s^{-1}), and above land surfaces was

$$h = 7.0 + 0.03 \, u_{10} \quad [26]$$

where u_{10} = windspeed at 10 m (cm s^{-1}).

Obviously, equations for evaporation in natural environments rely heavily on empirical formula to define the conductance. The theoretical analyses show that there are many factors and assumptions which can make the extrapolation of these formula to new situations unreliable. Accurate estimations of the conductance term are a major difficulty in describing the evaporation process.

II. TRANSPIRATION

Transpiration is a specialized situation for evaporation in that the vaporization of water occurs inside a leaf. Therefore, the leaf environment generally makes it even more unlikely that the assumptions and simplifications of the previous discussions are satisfied. As an example, isothermal conditions rarely exist in the leaf environment and three-dimensional temperature variations within leaves are quite likely. Obviously, considerable caution needs to be taken as the equations defining transpiration are developed.

The consideration of transpiration is also more complicated because the diffusion pathway is more complex. Vapor must diffuse from the site of vaporization inside the leaf, through the leaf stomata, through the leaf boundary layer, and finally through the canopy boundary layer. Transpiration will be discussed first at the leaf level and then at the canopy level.

A. Leaf Transpiration

Since water in the leaves contains solutes, the description of the chemical potential (Eq. [3]) must be reconsidered in view of the existence of an osmotic term and a hydrostatic pressure term. The presence of solutes in the liquid lowers the chemical potential of the water, while the hydrostatic pressure may be positive and raise it. The combined effect of osmotic and hydrostatic pressure can be expressed as a water potential (Ψ), which is related to the vapor pressure over the liquid (P_v)

$$\Psi = \frac{RT}{V_w'} \ln \frac{P_v}{P_v^*} \qquad [27]$$

where V_w' is the partial molal volume of vapor ($cm^3 \ mol^{-1}$). Equation [27], in fact, provides the basis by which thermocouple psychrometry is used to measure water potentials.

Rearrangement of Eq. [27] allows the effect of leaf water potential on the vapor pressure to be assessed

$$\frac{P_v}{P_v^*} = \left(\exp \frac{\Psi V_w'}{RT} \right) . \qquad [28]$$

In a severely stressed leaf where Ψ might be $-3 \ J \ cm^{-3}$, it can be calculated from Eq. [28] that P_v/P_v^* equals 0.978. Consequently, in the usual physiological range of water potentials there is relatively little error introduced by assuming saturated vapor pressure for pure water adjacent to the sites of vaporization.

While the saturated vapor pressure can then be calculated directly from leaf temperature (by Eq. [20], for example), it is no small task to determine leaf temperature. The thermal mass of most leaves is usually quite small so that alterations in the energy fluxes to and from the leaf can result in rapid changes in leaf temperature. Changes in the radiant energy, sensible heat flux, and/or transpiration itself can result directly in changes in leaf temperature so that the saturated vapor pressure is altered. The coupling of the saturated vapor pressure to the energy balance of the leaves makes transpiration responsive to a number of variables in the physical environment. Consequently, all the approximations and empiricisms required to calculate leaf energy balances are inherent in attempts to calculate transpiration rates based on leaf temperatures.

A further complication in describing leaf transpiration is to identify the actual sites within leaves where vaporization occurs. Since the vapor exits the leaf through stomata by means of a vapor pressure gradient, those surfaces nearest the stomata pore would seemingly have the greatest vaporization rate. In fact, several experimental studies have shown that the inner walls of the guard cells and subsidiary cells are the major sites of evaporation in leaves (Meidner, 1975; Byott & Sheriff, 1976; Aston & Jones, 1976). In a theoretical analysis of vapor diffusion inside leaves, Rand (1977) showed that 90% of the vaporization occurs within 8 μm of the stomatal pore. Tyree and Yianoulis (1980) came to a similar conclusion in a more recent analysis.

However, temperature gradients within a leaf might alter the sites of evaporation. For example, in an experiment by Sheriff (1977) where a leaf was subjected to high illuminance from one side, there was vapor distillation from the warm mesophyll cells to the inner epidermal walls of the nonilluminated side. Consequently, temperature gradients across a leaf could influence the distribution of the evaporation sites.

In any case, under many conditions it is expected that vapor conductance inside leaves is quite large ($D/\Delta x = 0.24/(8 \times 10^{-4}) = 300$ cm s^{-1}). Therefore, the main restriction to vapor loss from leaves is through the stomatal pore. Studies on the effects of the stomatal pore on vapor diffusion have been undertaken for nearly 100 yr. Parlange and Waggoner (1970) reviewed and analyzed the physical restrictions to gas diffusion caused by the stomatal pore. They confirmed that the conductance through an individual pore is closely related to the cross-sectional area of the pore aperature and the pore depth, such that

$$k_p \approx \frac{\pi \, abD}{d} \qquad [29]$$

where

k_p = individual pore conductance (cm^3 s^{-1}),
a,b = semilength of major and minor axis of pore aperature, respectively (cm),
d = pore depth (cm).

Parlange and Waggoner also considered the end effects of gas diffusion external to the pore and concluded this conductance to be

$$k_e = \frac{2\pi aD}{\ln\left(4\,\dfrac{a}{b}\right)} . \qquad [30]$$

For cases where the interstomatal distance is three times greater than the length of the major axis, there was negligible interaction of diffusion among stomata. Therefore, the conductance per unit leaf area attributable to stomata for a single leaf surface (h_s) can be obtained by combining Eq. [29] and [30]) and multiplying by the stomatal density (n, number cm^{-2}).

$$h_s = \frac{\pi\, ab\, Dn}{d + \dfrac{b}{2} \ln\left(\dfrac{4a}{b}\right)} \tag{31}$$

From Eq. [31] it can be calculated in the case where $a = 8 \times 10^{-4}$ cm, $b = 3 \times 10^{-4}$ cm, $d = 10 \times 10^{-4}$ cm, and $n = 10 \times 10^3$ stomata cm^{-2} that the stomatal conductance is equal to 1.3 cm s^{-1}. If these variables were equivalent for both sides of a leaf, then the stomatal conductance for the whole leaf is calculated as 2.5 cm s^{-1}.

Seemingly, Eq. [31] provides an extremely useful relationship for explicitly defining stomata conductance for transpiration. However, it turns out that the semilength of the stomatal pore width (b) is a very dynamic variable with b having the potential to change rapidly between values from zero to approximately 4×10^{-4} cm. The change in the value of b is under physiological control and obviously provides a key mechanism for the regulation of transpiratory water loss. The value of b is related to the turgor pressures in both the stomata guard cells and their neighboring cells (Cooke et al., 1976). As of yet, no fundamental relationships have been discovered relating the turgor pressures of these cells to environmental and physiological input conditions (Cowan, 1977). Many empirical equations have been produced in an attempt to predict stomatal conductance using such factors as bulk leaf water potential, light, temperature, and humidity. But all such relationships remain empirical and allow little opportunity for extrapolation beyond the localized experimental conditions. Most recently, bulk leaf turgor has been used as an independent variable predicting stomatal conductance. While this variable would seemingly be of use, it turns out to be neither independent of transpiration rate nor a predictor of the turgor in the two cell-types regulating stomatal conductance (Shackel & Brinckmann, 1985; Sinclair & Ludlow, 1985). Bennett et al. (1987) found none of the variables defining bulk leaf water relations accurately predicted the stomatal conductance of field-grown soybean (*Glycine max* L.) and maize (*Zea mays* L.) leaves.

Leaf transpiration is also influenced by the vapor conductance through the boundary layer around leaves. The analysis of the conductance through the boundary layer over a flat plate (Eq. [22]) offers a theoretical background for estimating the boundary layer conductance, h_{bl}. The combined h_{bl} for both sides of a leaf in a laminar flow is predicted from Eq. [22] as

$$h_{bl} = 0.63 \sqrt{\frac{u}{L}} . \tag{32}$$

However, experimental evidence suggests the above equation underestimates the natural boundary layer conductance. The turbulent nature of the boundary layer around many leaves probably accounts for much of the increased conductance. Monteith (1964) proposed that the leaf boundary layer conductance for both sides of the leaf is approximately expressed as

$$h_{bl} = 0.77 \sqrt{\frac{u}{L}} .$$ [33]

Consequently, for a leaf of 8-cm width subjected to a windspeed of 100 cm s^{-1}, the h_{bl} is calculated as 2.7 cm s^{-1}, which is nearly equivalent to the stomatal conductance estimated from Eq. [31].

Since both stomatal and boundary layer conductances influence leaf transpiration rates, the effects of these two conductances are combined. Leaf transpiration rate (J cm^{-2} s^{-1}) is thus defined as

$$J H_v = \frac{h_s h_{bl}}{h_s + h_{bl}} (P_L^* - P_v)$$ [34]

where P_L^* is the saturated vapor pressure at leaf temperature.

B. Canopy Transpiration

In most natural systems the transpiration rate of the entire leaf canopy is desired rather than that of individual leaves. Consideration of the whole canopy greatly complicates matters because individual leaves within the canopy vary in their stomatal conductance, boundary layer conductance, temperature, and distance from the reference humidity. To determine accurately the canopy transpiration rate, all these factors influencing leaf energy budgets and physiology need to be known and the individual leaf transpiration rates summed.

Sinclair et al. (1976) did a computer study to seek simplified methods for estimating canopy transpiration. Their reference predictions were made by dividing the canopy into 15 horizontal layers and splitting each layer into 20 leaf elements having differing temperatures and intercepted solar radiation. Stomatal conductance was calculated as an empirical function of intercepted solar radiation. They found results comparable to the 300 leaf-element model were obtained by using only a two leaf-element model. Discrepancies were less than 10% at moderate and low transpiration rates but larger differences occurred at high transpiration rates.

The two leaf elements used by Sinclair et al. (1976) were defined as being those leaves exposed either to direct solar radiation or to shaded conditions. The amount of leaf area in each element was calculated from an exponential radiation interception model. The stomatal conductance and leaf boundary conductance were calculated separately for each element. It was assumed the leaf elements transpired in parallel at a mean source height in the canopy. Consequently, additional conductances were included to account for vapor transfer within the canopy from the mean source height to the zero-plane displacement height, and then through the canopy boundary layer to the reference height. The zero plane displacement height (d_o) is the height where the extrapolated logarithmic windspeed profile above the canopy goes to zero. The value of d_o must be empirically derived but it is commonly found to be 0.6 to 0.8 of the canopy height (Monteith, 1973). Equation [24] was used to calculate canopy boundary layer conductance by including the zero-plane displacement height

$$h = \frac{uk^2}{\left(\ln \dfrac{Z - d_o}{Z_o} \right)^2} \qquad [35]$$

It should be recognized that Eq. [35] is for a rather idealized canopy boundary layer. Such factors as atmospheric buoyancy and the effect of windspeed on d_o and Z_o complicate the evaluation of Eq. [35]. Bluff-body effects due to the canopy may also affect the conductance to vapor transfer (Stewart & Thom, 1973). However, Thom and Oliver (1977) proposed for regional evaluations of water loss by natural surfaces that the boundary layer conductance was well approximated by the following modifications of Eq. [35]

$$h = \frac{126 \, (1 + 0.0054 \, u) \, k^2}{\left(\ln \dfrac{Z - d_o}{Z_o} \right)^2} \qquad [36]$$

A further simplification of the calculation of canopy transpiration was presented by Tanner and Sinclair (1983). They used the two-leaf element approach assuming a canopy with a leaf area index > 3 so that nearly all the solar radiation was intercepted by the leaf canopy. Tanner and Sinclair derived a formulation whereby transpiration rate was expressed as a function of atmospheric vapor pressure deficit rather than leaf vapor pressure deficit.

$$T = \frac{\epsilon \rho_a}{P_a} \frac{(P_v^* - P_v)_a}{(1/h_s + 1/h_{bl})_d} L_T \qquad [37]$$

where

T = transpiration rate (g cm^{-2} s^{-1}),
a = subscript denoting vapor pressures are for the atmosphere,
d = subscript denoting conductance for leaves intercepting direct solar radiation,
L_T = "effective" transpiring leaf area index (approximately 2.2).

The size of the deviation of leaf vapor pressure deficit from atmospheric vapor pressure deficit was incorporated into the variable L_T. The value of L_T was explicitly defined in the derivation of Tanner and Sinclair (1983) and is dependent on the ratio of conductances between leaves in direct radiation and those in the shade, and on the leaf temperatures. They found L_T was generally about 2.2 but could deviate substantially under arid or humid conditions.

The obvious difficulty in calculating transpiration with the above approaches is that they necessitate assumptions about leaf temperature and an evaluation of the stomatal conductance and leaf boundary layer conductance. Such evaluations, especially for stomatal conductance, rely on empirical relationships whose extrapolative capabilities are questionable. An alternative is to determine evapotranspiration rates based on equations derived from an energy budget.

Penman (1948) derived a commonly used evaporation equation from the energy balance for a water surface. It was assumed that the radiation budget at the water surface was balanced by the energy flux density resulting from evaporation plus sensible heat flux density.

$$R_n = J H_v + S \qquad [38]$$

where

R_n = net radiation (J cm^{-2} s^{-1}),
S = sensible heat flux density (J cm^{-2} s^{-1}).

By assuming the conductances in the atmosphere for heat and water vapor are equivalent, the following ratio is obtained

$$\frac{S}{J H_v} = \gamma \frac{(T_w - T_a)}{(P_w^* - P_v)}, \qquad [39]$$

where

γ = psychrometric constant (0.66 $\times$ 10^{-4} J cm^{-3} C^{-1}),
T_w = water temperature (C),
T_a = air temperature (C),
P_w^* = saturated vapor pressure at T_w (J cm^{-3}),
P_v = vapor pressure of air (J cm^{-3}).

To simplify the derivation, the change in vapor pressure with respect to a change in temperature is calculated from the air temperature (see Eq. [17]). That is,

$$\Delta = \frac{P_w^* - P_v^*}{T_w - T_a} \qquad [40]$$

Substitutions of Eq. [39] and [40] into Eq. [38] result in

$$\Delta \cdot R_n = J H_V \left(\Delta + \gamma - \gamma \frac{(P_v^* - P_v)}{(P_w^* - P_v)} \right), \qquad [41a]$$

or

$$J H_V (\Delta + \gamma) = \Delta \cdot R_n + J H_V \gamma \frac{(P_v^* - P_v)}{(P_w^* - P_v)} \qquad [41b]$$

Substituting into Eq. [41] the definition of $J\,H_V$ derived from Eq. [10] results in the Penman equation,

$$J\,H_V = \frac{\Delta \cdot R_n + \gamma h\,(P_v^* - P_v)}{\Delta + \gamma}.$$ [42]

As discussed previously, the atmospheric conductance (h) used by Penman was an empirical function of windspeed.

Monteith (1965) sought to extend Penman's energy budget approach describing evaporation from a water surface to the case of transpiration rate for a leaf canopy. Implicit in Monteith's derivation was that the bulk of incident radiation is intercepted by the leaf canopy so that the partitioning of energy at the soil surface is relatively insignificant. The initial assumption in the Monteith derivation is identical to Penman's in that the conductances in the atmosphere for heat and water vapor are equal. However, in the case of a leaf canopy it is not readily apparent that such an assumption is valid. The inference of such an assumption is that the sinks and/or sources for water vapor and heat are distributed identically throughout the canopy. While it is clear such an equality in distribution does not exist, Monteith argued the overall effect of the canopy was to produce mean source heights for water vapor and heat that are approximately equal. Consequently, the ratio given in Eq. [39] was assumed.

The next crucial difference between the derivation of Monteith and Penman was in the evaluation of Eq. [41b]. An expression directly defining $J\,H_V$ on the right-hand side of Eq. [41b] is required. The value $J\,H_V$ is defined by using an expression similar to that defining transpiration rate for a single leaf (Eq. [34]). It is assumed that boundary layer and stomatal conductances can be replaced by analogous conductances for the whole canopy. The atmospheric conductance (h) used in the Penman equation replaces h_{bl} used in the individual leaf equation. The leaf stomatal conductance is replaced by a conductance indicative of vapor transfer through stomata for the entire canopy (h_c). With these substitutions into Eq. [34] and subsequently solving Eq. [41], the Monteith formulation is derived.

$$J\,H_v = \frac{\Delta\,R_n + \gamma h\,(P_v^* - P_v)}{\Delta + \gamma \left(\dfrac{h + h_c}{h_c}\right)}.$$ [43]

The major problem in the use of Eq. [43] remains the difficulty in evaluating h_c. Techniques for integrating individual leaf stomatal conductances into a single representation value of h_c have not been proven. Furthermore, the prediction of individual leaf stomatal conductance is empirical, as discussed previously, so that the basic input information to calculate h_c remains ambiguous. In practice, Eq. [43] is generally rearranged to solve for h_c so that some insights about this variable might be obtained. The physical and physiological data requirements are seemingly too great for the Monteith equation to be used as a practical predictor of canopy transpiration rates.

In attempts to minimize the amount of information required to estimate evaporation from canopies, formulas that are purely empirical have been developed. Several formulas have been developed predicting transpiration based on temperature (Thornthwaite, 1948; Blaney & Criddle, 1950). Jensen and Haise (1963) developed an empirical relationship whereby evaporation was dependent on both temperature and radiation. Priestly and Taylor (1972) used a form of the Penman equation to predict evaporation based on net radiation and temperature data. All these formulas suffer, of course, from questions about their use under conditions differing from those existing when and where the formulas were developed. Drought conditions, for example, where stomatal conductance limits transpiration would surely cause the empirical formulations to be in error. The practical problems of actually using these empirical formulas, as well as that of the energy balance approach, (Penman-Monteith) are discussed more fully in subsequent chapters in this section.

III. CONCLUSION

There is substantial theoretical background for deriving from first principles the equations defining evaporation and transpiration. Fundamental thermodynamic relationships and mechanics can be used to formulate these equations. Unfortunately, it was also found that all these derivations are dependent on a number of simplifying assumptions that are rarely, if ever, met under the natural conditions of interest. Equilibrium and isothermal conditions are but two assumptions that are required by these theoretical derivations.

Nevertheless, the theoretical derivations offer substantial guidance in formulating the empirical relationships required to define evaporation and transpiration. The descriptions of saturated vapor pressure as a function of temperature and of boundary layers conductance as a function of windspeed have benefited from detailed theoretical studies of these processes. It is clear from the derivations that the variables ultimately controlling evaporation and transpiration rates are the: (i) saturated vapor pressure at the temperature of the vaporizing surface, (ii) ambient vapor pressure, and (iii) vapor conductance.

Vapor conductance remains the largest source of empiricism in defining evaporation and transpiration. Vapor transfer in the boundary layers over vaporizing surface is still not fully understood for natural environments. These surfaces are aerodynamically ill defined, and the natural boundary layers are highly variable including both free convection and unique turbulences. While there has been a reasonably widespread confirmation of the expressions for the boundary conductances around individual leaves and over surfaces of large extent, these equations remain somewhat empirical and are subject to revision as the understanding of boundary layer mechanics improves.

Transpiration remains especially difficult to define because of its dependence on stomatal conductance. Considerable progress has been made in recent years in understanding the mechanics of stomatal regulation of vapor transfer. Similarly, knowledge of guard cell physiology has been greatly expanded. Yet a quantitative relationship between stomatal physiology and vapor conductance remains at the level of regression analysis. No fundamental descriptors exist for predicting the effect of environmental changes on stomatal conductance. As seen by the equations presented earlier, transpiration is difficult to predict accurately without a full understanding of stomatal behavior. Transpiration equations remain at an empirical level until stomatal conductance can be accurately predicted.

ACKNOWLEDGMENT

The author gratefully acknowledges the helpful discussions and suggestions of Dr. L.H. Allen, Jr., USDA-ARS, Gainesville, FL.

REFERENCES

Aston, M.J., and M.M. Jones. 1976. A study of the transpiration surfaces of *Avena sterilis* L. var. Algerian leaves using monosilic acid as a tracer for water movement. Planta 130:121–129.

Bennett, J.M., T.R. Sinclair, R.C. Muchow, and S.R. Costello. 1987. Dependence of stomatal conductance on leaf water potential, turgor potential, and relative water content in field-grown soybean and maize. Crop Sci. 27:984–990.

Blaney, H.F., and W.D. Criddle. 1950. Determining water requirements in irrigated areas from climatological and irrigation data. USDA Soil Conserv. Serv. Tech. Rep. N. 96.

Brutsaert, W. 1967. Evaporation from a very small water surface at ground level: Three-dimensional turbulent diffusion without convection. J. Geophys. Res. 72:5631–5639.

Brutsaert, W. 1982. Evaporation into the atmosphere. P. Reidel Publ. Co., Dordrecht, Netherlands.

Brutsaert, W., and S.L. Yu. 1968. Mass transfer aspects of pan evaporation. J. Appl. Meteorol. 7:563–566.

Byott, G.S., and D.W. Sheriff. 1976. Water movement into and through *Tradescantia virginiana* (L.) leaves. II. Liquid flow pathways and evaporative sites. J. Exp. Bot. 27:634–639.

Cary, J.W. 1963. Onsager's relation and the non-isothermal diffusion of water vapor. J. Phys. Chem. 67:126–129.

Cooke, J.R., J.G. DeBaerdemaeker, R.H. Rand, and H.A. Mang. 1976. A finite element shell analysis of guard cell deformations. Trans. ASAE 19:1107–1121.

Cowan, I.R. 1977. Stomatal behavior and environment. Adv. Bot. Res. 4:117–228.

Gebhart, B. 1961. Heat transfer. McGraw-Hill Book Co., New York.

Goff, J.A., and S. Gratch. 1946. Low-pressure properties of water from − 160 to 212 F. Trans. Am. Soc. Heat. Ventilat. Eng. 52:95–122.

Hull, A.N. 1974. Comments on "A simple but accurate formula for the saturation vapor pressure over liquid water." J. Appl. Meteorol. 13:607.

Jensen, M.E., and H.R. Haise. 1963. Estimating evapotranspiration from solar radiation. Irrig. Drain. Div. Am. Soc. Civil Eng. 89:15–41.

Katchalsky, A., and P.F. Curran. 1967. Nonequilibrium thermodynamics in biophysics. Harvard Univ. Press, Cambridge, MA.

Mahrt, L., and M. Ek. 1984. The influence of atmospheric stability on potential evaporation. J. Clim. Appl. Meteorol. 23:222–234.

Meidner, H. 1975. Water supply, evaporation, and vapour diffusion in leaves. J. Exp. Bot. 26:666–673.

Monteith, J.L. 1965. Evaporation and environment. p. 305–234. *In* The state and movement of water in living organisms. XIX Symp. Soc. Exp. Biol. Academic Press, New York.

Monteith, J.L. 1973. Principles of environmental physics. American Elsevier Publ. Co., New York.

Murray, F.W. 1967. On the computation of saturation vapor pressure. J. Appl. Meteorol. 6:203–204.

Nobel, P.S. 1983. Biophysical plant physiology and ecology. W.H. Freeman, San Francisco.

Parlange, J., and P.E. Waggoner. 1970. Stomatal dimensions and resistance to diffusion. Plant Physiol. 46:337–342.

Penman, H.L. 1948. Natural evaporation from open water, bare soil and grass. R. Soc. London Proc. Ser. A. 193:120–145.

Priestley, C.H.B., and R.J. Taylor. 1972. On the assessment of surface heat flux and evaporation using large scale parameters. Mon. Weather Rev. 100:81–92.

Rand, R.H. 1977. Gaseous diffusion in the leaf interior. Trans. ASAE 20:701–704.

Shackel, K.A., and E. Brinckmann. 1985. *In situ* measurement of epidermal cell turgor, leaf water potential, and gas exchange in *Tradescantia virginiana* L. Plant Physiol. 78:66–70.

Sheriff, D.W. 1977. Evaporation sites and distillation in leaves. Ann. Bot. 41:1081–1082.

Sinclair, T.R., and M.M. Ludlow. 1985. Who taught plants thermodynamics? The unfulfilled potential of plant water potential. Aust. J. Plant Physiol. 12:213–217.

Sinclair, T.R., C.E. Murphy, and K.R. Knoerr. 1976. Development and evaluation of simplified models for simulating vegetative photosynthesis and transpiration. J. Appl. Ecol. 13:813–829.

Stewart, J.B., and A.S. Thom. 1973. Energy budgets in pine forest. Q. J. R. Meteorol. Soc. 99:154–170.

Sweers, H.E. 1976. A nomogram to estimate the heat-exchange coefficient at the air-water interface as a function of wind speed and temperature; A critical survey of some literature. J. Hydrol. 30:375–401.

Tanner, C.B. 1968. Evaporation of water from plants and soil. p. 73–106. *In* T.T. Kozlowski (ed.) Water deficits and plant growth. Vol. I. Academic Press, New York.

Tanner, C.B., and T.R. Sinclair. 1983. Efficient water use in crop production: Research or re-search? p. 1–27. *In* H.M. Taylor et al. (ed.) Limitations to efficient water use in crop production. ASA, CSSA, and SSSA, Madison, WI.

Thom, A.S., and H.R. Oliver. 1977. On Penman's equation for estimating regional evaporation. Q. J. R. Meteorol. Soc. 103:345–357.

Thornthwaite, C.W. 1948. An approach toward a rational classification of climate. Geogr. Rev. 38:55–94.

Tyree, M.T., and P. Yianoulis. 1980. The site of water evaporation from sub-stomatal cavities, liquid path resistances and hydroactive stomatal closure. Ann. Bot. 46:175–193.

13

Soil and Plant Factors Affecting Evaporation

J. T. RITCHIE AND B. S. JOHNSON

Michigan State University
East Lansing, Michigan

Evaporation of water from a crop is composed of evaporation from the soil surface and evaporation from the plant tissue. The proportion of these two components can vary greatly in a few days time. When the crop leaf area index is low, soil evaporation can be a large part of the total evaporation, especially if the soil surface is wetted frequently. Because most crops have incomplete cover throughout a significant part of a growing season, the estimation of the water loss from soil and plants is vital for evaluating irrigation water demands. Accurate evaluation of total evaporation for a growing crop with less than full cover requires that evaporation from the soil and vegetation be considered separately.

In this chapter we will define the factors needed to estimate evaporation from plant and soil surfaces by providing models of these processes and describing information necessary to implement the models. We will also provide a basis for a reasonable approximation for leaf area index, information needed in practically all models that calculate soil and plant evaporation separately, and show how these approximations can be used to calculate a crop coefficient during partial cover for use in traditional methods of scheduling irrigation.

Before proceeding, we wish to clarify a departure from use of the term *evapotranspiration*, partly because we will discuss evaporation from soil and plant surfaces separately and because the term is unnecessary. We subscribe to Monteith's (1985) statement that evapotranspiration is "inappropriate because its components are not strictly congruous; the word *transpiration* implies a flux of vapor whereas the primary meaning of evaporation is a change of phase from liquid to gas." When it is necessary to describe the loss of water by evaporation from the soil (E_s) and plants (E_p) as they occur together, the term *total evaporation* (E_T) will be used and can be interpreted the same as the more traditional evapotranspiration.

I. MEASUREMENT OF SOIL AND PLANT EVAPORATION

There has been an increased interest in direct, separate measurement of E_s and E_p in recent years. Methods have principally focused around the use of micro-lysimeters for measuring E_s in bare soil (Boast & Robertson, 1982) or for soil beneath canopies (Walker, 1983; Shawcroft & Gardner, 1983; Lascano et al., 1987; Klocke et al., 1986; Martin et al., 1985). These *micro-lysimeters* consist of undisturbed cores of soil 10 to 30 cm deep and 10 to 30 cm diam. taken from below the crop canopy at several times during drying cycles. A core is removed from the field, the bottom is sealed, and the core is then replaced in its original position. It is removed from that position occasionally and weighed on a balance to determine the rate of water loss. When the no-flow bottom boundary condition is thought to cause evaporation to deviate from reality, the cores are discarded and new ones taken. Results from cores have been shown to agree well with calculations made with validated models (Lascano & van Bavel, 1986; Shawcroft & Gardner, 1983) when precautions are observed to keep the micro-lysimeter soil water and temperature conditions reasonably similar to adjacent field soil.

Shawcroft and Gardner (1983) regressed soil water content in the surface 5 cm with micro-lysimeter-measured E_s and found a fair correlation for dry soil water content when E_s was low. Although they realized the relationship was unique to their soils, they suggested that the E_s could be reasonably inferred from such an experimentally established relationship. Ritchie and Burnett (1971) correlated lysimeter-measured E_s values with water contents in the surface 3 cm and used the relationship to infer E_s beneath sorghum (*Sorghum bicolor* (L.) Moench] and cotton (*Gossypium hirsutum* L.) canopies on the same soil type to estimate E_p from lysimeter-measured E_T values.

A technique designed to measure E_s below canopies during the energy-limited stage when the soil surface is wet was described by Arkin et al. (1974). Measured water flow through a thin evaporation plate with a known surface area was used as a direct indicator of evaporation rates. Flow rates could be recorded manually or automatically. Field tests under bare soil conditions indicated the resolution of the system was equal to that of a precision-weighing lysimeter and that the system gave reasonable accuracy on an hourly basis. As with the coring technique, the water content conditions of the plate must be near that of the surrounding soil for the results to be accurate.

Measurement of E_p has been done principally through use of lysimeters in which E_s was prevented or held to a minimum (Harrold et al., 1959; Briggs & Shantz, 1913). Unfortunately, the technique provides an unnatural condition because E_s is prevented at the surface of the lysimeter but not in the surrounding area. Thus, use of such information would be questionable for prediction purposes because of the unnatural microclimate near the soil surface unless the entire surface has near-zero E_s. When lysimeter values are available for E_T, and E_s measurements are made with evaporation plates or micro-lysimeters, it is possible to make reasonable approximations for E_p = E_T − E_s (Ritchie, 1985, Table 1).

Denmead (1973) used an aerodynamic approach to estimate E_s and E_p by measuring vapor pressure and temperature profiles within and over a wheat canopy. The measurements indicated that E_s and E_p were essentially independent except where E_p is limited by soil water supply.

Denmead (1984) reviewed plant physiological methods including chamber systems and tracer techniques that are available to measure E_T or the E_p component of E_T. In the case of chamber systems, the objective is to enclose a volume of air surrounding transpiring plants so that changes in ambient humidity could be monitored. Although temperature and humidity are difficult to control in the chambers, Reicosky and Peters (1977) circumvented this problem by implementing a tractor-mounted chamber that was placed over the crop for only 1 min. Fans induced rapid circulation within the chamber and a fast response psychrometer was used to detect the rise in humidity. Chamber systems usually provide measurements of E_T except where they are operated at an excess pressure or when the soil surface is covered for the measurement of E_p through suppression of E_s (Davis & Ludlow, 1981). As with covered soil in lysimeters, the result may not be representative of field conditions.

The E_p of individual plants can be estimated through the use of tracer techniques which facilitate the measurement of flow rates in water-conducting tissue of the stem. The heat pulse method and heat balance method are examples of this approach (Baker & van Bavel, 1987). The *heat pulse method* consists of determining the time required for a discrete heat input to travel the length of a stem segment bounded by an inserted heater and sensor. The *heat balance method* is a modification of this approach as stem flow measurement is based on the balance of heat into and out of a stem segment, rather than heat pulse velocity. These methods have been employed predominantly in large plants (i.e., trees). However, the heat balance technique as proposed by Sakuratani (1981, 1984) for application to herbaceous plants has been extended to the measurement of E_p in soybean grown in pots (Sakuratani, 1987) and to cotton and sunflower (*Helianthus annuus* L.) grown in pots (Baker & van Bavel, 1987). Use of this technique on monocotyledon plants with leaf sheaths surrounding the stem may present a greater challenge. Furthermore, the heat pulse and heat balance methods have yet to be demonstrated as a useful tool for E_p measurement under field conditions where the temperature variation is great. There is also the biometric difficulty of extrapolating from one, or a few plants to the crop as a whole.

II. PLANT AND SOIL EVAPORATION INTERACTION

For many years, researchers have recognized that E_p can be influenced by local sources of sensible heat from nonevaporating surfaces such as dry soil surrounding transpiring plants. Marlatt (1961) found that with screen-level meteorological conditions being similar, E_T from orchard grass (*Dactylis glomerata* L.) with an adequate root-zone water supply was as great with 50% cover as with full cover. In the study, the soil between rows was dry at the surface, thus contributing little to the E_T. As a consequence, E_p per unit covered area was about double that of a full-cover crop. This phenomenon, called the *clothesline effect* (Tanner, 1957), occurs when hot, dry air from between rows passes through plants in the rows, increasing E_p. The vapor pressure deficit of the air must be considerably greater near the surface when the soil is dry in a partial crop canopy than is measured with standard weather instruments some distance from the point where it is being influenced by the microclimate. Hanks et al. (1971) found that about 64% of the sensible heat flux from a dry soil surface below a sorghum crop with a leaf area index (LAI) of 1.2 was used for E_p.

Ritchie and Burnett (1971) quantified the influence of partial cover on E_T and found that LAI of sorghum and cotton were more generally related to E_p as fractions of the potential evaporation rate (E_o) than ground cover or plant dry weight. The empiricisms used to estimate E_p from crops with an adequate supply of water in the root zone usually make E_p a function of E_o, LAI or plant cover. Generally, E_p equals E_o, when LAI approaches about 2.5 and when the soil surface is dry (Ritchie, 1983). Any increase in LAI beyond about 2.5 causes little increase in E_T for most row crops with rows spaced < 1 m apart.

The relationship between E_p/E_o reported in the model of Ritchie (1972) was obtained under conditions when the soil surface between rows was dry most of the time. Thus, the clothesline effect should have been significant, at least for low LAI. A relatively large difference between E_p when the soil surface was dry as compared with E_p when it was wet became apparent when separately calculating E_s and E_p for a partial-cover condition when using the fitted equation for E_p in the Ritchie model. When the soil surface was wet, the sum of E_s and E_p always exceeded calculated E_o. The E_p value

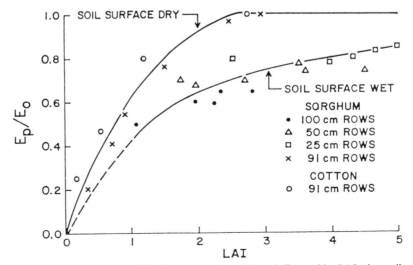

Fig. 13-1. Plant evaporation (E_p) relative to estimated E_o as influenced by LAI when soil water in the root zone is nonlimiting for dry and wet soil surfaces. Data are from Ritchie (1971), Ritchie and Burnett (1971), and Adams et al. (1976).

had to be reduced in such cases for the model to agree with lysimeter-measured E_T.

Later interest in this problem led to independent measurements of E_s from wet soil by using special evaporation plates designed to measure E_s under a canopy (Arkin et al., 1974) as described earlier in this chapter. Using the plates, Adams et al. (1976) found that for LAI values as high as 5, 15% of the E_o was E_s. The lower curve in Fig. 13-1 is a plot of E_p/E_o calculated from the Adams et al. data (their Table 2 and Fig. 5), assuming E_p is the difference between E_o and measured values for E_s. Values of E_o were taken with evaporation plates placed in the center of a large, bare area that was wet at the surface.

The upper curve of Fig. 13-1 shows E_p/E_o values obtained from lysimeter E_T data of Ritchie (1971) and LAI data of Ritchie and Burnett (1971). Measured E_p/E_o values were selected for the upper curve using only days when the soil surface was dry following a relatively long period without rainfall. Values of E_o for these measurements were estimated from 24-h net radiation, which usually produced approximate values of E for Temple, TX, at that time of the year. Although E_o for the wet and dry soil surface comparison were measured using different estimates and in different years, both E_o estimates were close to 6 mm d^{-1} during July and August when measurements were made. Cotton showed evidence of being influenced by the clothesline effect somewhat more than sorghum at a given LAI when the soil surface was dry. In the wet surface tests with sorghum, there was a demonstrated difference in E_p for a given LAI as influenced by row spacing because of different amounts of radiation interception per unit leaf area for the different canopy configurations (Adams et al., 1976).

In another field experiment using the E_s measurement techniques of Arkin et al. (1974) and a weighing lysimeter, the E_s and E_p of a maize (*Zea mays* L.) crop were measured during a 4-d period from 3 to 6 June soon after tasseling when the LAI was about 2.7 (Ritchie, 1985). There had been 26-d without rain at Temple, TX, and the surface was quite dry. The E_T was almost equal to calculated E_o, however, because of the large amount of extractable water held in the clay soil. On 3 June the soil surface was dry, but on 4 June, a plastic cover was placed over the soil between plants to ensure soil evaporation was negligible. The plastic was covered with a thin layer of air-dry soil. Results for the 2 d were almost identical (Table 13-1). On the third day, about 3000 m^2 of land was sprinkler-irrigated in the lysimeter area of the field with 28 mm of water. On the fourth day, the evaporation

Table 13-1. Soil evaporation (E_s) and plant evaporation (E_p) from a weighing lysimeter for maize grown at Temple, TX, for wet and dry soil surface conditions. (LAI was 2.7, and the potential evaporation rate (E_o) is the Priestly-Taylor estimate).

Soil surface condition	Date	E_o	E_T	E_s	E_p
			mm d^{-1}		
Dry	3 June	6.5	6.1	0	6.1
Covered and dry	4 June	7.2	6.4	0	6.4
Wet	6 June	6.9	7.3	2.5	4.9

plates were used to measure E_s below the canopy in the lysimeter, making possible the reasonable separation of E_s and E_p on that day. Although the crop had an LAI of 2.7, daily E_s amounted to 2.5 mm and E_p was 4.9 mm (Table 13–1). These directly measured results agree well with the generalized relationships shown in Fig. 13–1 developed for grain sorghum and cotton. Similar results for wet soil conditions were reported by Walker (1984) for maize canopies with LAI values of 3 to 4. He found that E_s on some days exceeded the radiant energy supply reaching the soil surface, indicating a downward flux of sensible heat from the canopy. The plant-environmental simulation model CUPID reported by Norman and Campbell (1983) was shown to produce a greatly reduced E_p on a day with high E_s when compared with a similar day with low E_s (their Fig. 7 and 8). The models of Shuttleworth and Wallace (1985) and Choudry and Monteith (1988) also treat this problem conceptually.

III. APPROACHES TO MODELING SOIL AND PLANT EVAPORATION

Several types of mathematical models are available for calculating E_s and E_p. Because models of these processes and others of interest in crop production have different purposes and are approached from different levels of organization, it is possible to categorize them for discussion. They can be described as deterministic or stochastic, mechanistic or functional, and rate or capacity types (Addiscott & Wagenet, 1985). *Deterministic models* produce a unique outcome for a given set of events. However, due to the spatial variability of the mediating processes there will be a certain degree of uncertainty associated with the results. *Stochastic models* have been developed to accommodate this spatial variability and to quantify the degree of uncertainty. Stochastic models produce an uncertain outcome because they include one or more parameters that are random variables with an associated probability distribution. Rice and Jackson (1985) used remote sensing techniques to show that E_s can vary substantially in space, especially during the transition from the energy-limiting phase to the soil-limiting phase. While stochastic models are gaining in popularity, they have been applied little in modeling E_T.

Most models used for estimating E_s and E_p are deterministic and can be further categorized as mechanistic or functional. *Mechanistic models* for E_T are usually based on dynamic rate concepts. They incorporate basic mechanisms of processes such as Darcy's law or Fourier's law and the appropriate continuity equations for water and heat flux, respectively. *Functional models* are usually based on capacity factors and treat processes in a more simplified manner, reducing the amount of input required.

Perhaps the most important difference between mechanistic and functional models is their usefulness as either research or management tools. Mechanistic models are useful primarily as research tools for better understanding of an integrated system, and are usually not used by nonauthors due to their complexity. The functional models have modest input requirements making them useful for management purposes. Because of their simplicity, functional models are more widely used and independently validated than their mechanistic counterparts.

Every model of the plant-soil-atmosphere system, whether mechanistic or functional, uses some level of empiricism to reduce the need for input information. Thus it may be somewhat difficult to distinguish between mechanistic and functional models. In the following section, examples of mechanistic and functional models used to calculate E_s and E_p are compared to provide information on how various soil and plant factors affect E_T.

A. Soil Evaporation Models

An abundance of mechanistic models have been reported in which the general equation of water flow is used as a basis to calculate E_s. Some of these models predict evaporative losses of water from a bare soil (Rose, 1968; Gardner & Gardner, 1969; van Bavel & Hillel, 1976; Hillel, 1977; Hillel & Talpaz, 1977; Lascano & van Bavel, 1983, 1986) while others facilitate the separate calculation of E_s and E_p in the presence of a crop (Fedes et al., 1978; Norman & Campbell, 1983; Huck & Hillel, 1983; Lascano et al., 1987). Functional models for calculation of E_s are less evident in the literature. A model developed by Ritchie (1972) calculates E_s from a bare, partial, or full cover crop surface. Other functional models with similar levels of detail have been used successfully to calculate E_s and E_p separately (Hanks, 1974; Kanemasu et al., 1976; Tanner & Jury, 1976).

Two papers by Lascano and van Bavel (1983, 1986) are used here to illustrate the salient features of mechanistic models and to provide a comparison with the functional model of Ritchie (1972). These papers were chosen because they contain excellent validation data and sufficient detail to facilitate comparison of their approach with a functional model. Lascano and van Bavel (1983) described a mechanistic model, CONSERVB, which was used to simulate profiles of water and temperature for a bare soil. Evaporation rates estimated by CONSERVB were verified in a subsequent paper by comparing them with values measured using micro-lysimeters (Lascano & van Bavel, 1986). The model explicitly accounts for long-wave radiation from the soil surface; the stability of the atmosphere in the boundary layer; and the effect of drying on the albedo, vapor pressure, and the thermal and hydraulic properties of the soil surface. Numerical integration techniques are used in the simultaneous solution of water flow and heat flow.

A functional model developed by Ritchie (1972) calculates E_s from a bare soil on a daily basis. When a crop is present, E_s and E_p are calculated separately but only the soil evaporation component of the model will be described in this section and used in the comparison that follows. Concepts for this model were originally developed by Black et al. (1969). Potential soil evaporation (E_{os}) is first calculated from an estimate of E_o. Although the following relationships did not appear explicitly in Ritchie (1972), they are used in the crop model, CERES-Maize (Jones et al., 1986):

$$E_{os} = E_o [1.0 - 0.43(LAI); \qquad LAI < 1 \qquad\qquad [1]$$

$$E_{os} = (E_o/1.1)e^{-0.4(LAI)}; \qquad LAI \geq 1. \qquad\qquad [2]$$

In this case, a modified Priestly-Taylor equation is used to calculate daily values of E_o although other equations could be used. The potential rates for E_o and E_{os} are equivalent in this example when applied to bare soil conditions (i.e., LAI = 0).

The remaining calculations of Ritchie (1972) are based on the premise that soil evaporation takes place in two stages: (i) the constant rate stage and (ii) the falling rate stage (Philip, 1957). During the constant rate stage (stage 1), evaporation occurs at the potential rate until the upper limit of stage 1 evaporation (U) is reached. Thus, U is reached more rapidly under conditions of high E_o than under low E_o. When measured in the field, the upper limit of stage 1 drying varies from about 5 mm in sands and heavy shrinking clays to about 14 mm in clay loams. These limits apply only to field soils where a shallow water table is not present, where drainage is not greatly impeded, and where evaporation and redistribution commence immediately following wetting. Where drainage is slow or when laboratory cores are used to determine U, values can be much higher. Sadeghi et al. (1984) reported U values in excess of 35 mm for a soil with a lower horizon fragipan which restricted water flow. Data of Bond and Willis (1970) and Jaafar et al. (1978) demonstrate that values for U obtained from experiments involving laboratory columns are considerably larger than those found in normal, deep soil. Jalota and Prihar (1986) demonstrated that when evaporation commenced after 2 d of redistribution time, the generality that cumulative E_s increases with increasing E_o may not hold true.

During stage 2 drying, E_s from below a crop canopy is assumed to be identical to evaporation from a bare soil because E_s in this stage is more dependent on the hydraulic properties of the soil and less dependent on the available energy. Cumulative evaporation during stage 2 drying of an initially wet, deep soil can be expressed by the following equation (Black et al., 1969):

$$E_s = \alpha t^{1/2} \qquad\qquad [3]$$

where α can be determined experimentally from cumulative evaporation data for a single drying cycle on a given soil. Several direct measurements of the coefficient α in a diversity of soils have consistently resulted in values of about 3.5 mm d$^{-1/2}$.

Inputs for the mechanistic and functional model are compared in Table 13–2. This shows that inputs differ in amount and complexity. Hourly weather data were used in all simulation runs for CONSERVB (Lascano & van Bavel, 1983, 1986). The simulation procedure (Runge-Kutta, fourth order) involves numerical integration with variable time steps. This is common and necessary where numerical techniques are used to solve partial differential equations. The time steps must be sufficiently small to minimize error but large enough to maximize the efficiency of the computations. Zaradny (1978) solved

Table 13-2. Comparison of inputs and initial values required for a mechanistic and functional model used to simulate soil evaporation.

Type of input	Mechanistic[†]	Functional[‡]
	hourly	daily
Weather		
Time dependent:	Global radiation	Global radiation
	Air temperature	Maximum, minimum air
	Rainfall/irrigation	temperature
	Dewpoint temperature	Rainfall/irrigation
	Wind speed	
Constants:	Height of measurement	None
	Roughness parameter	
Soil		
Functions:	Water retention curve	None
	K vs. θ	
	Soil albedo vs. θ	
Constants:	Ksat	Soil albedo
	Porosity	Upper limit of stage 1
		drying (U)
Number of soil layers[§]:	13	Not critical
Depth of first layer:	5 mm	Not critical
Initial values[¶]	θ vs. depth	SUMES1, SUMES2
	Temeprature vs. depth	

† From Lascano and van Bavel (1983, 1986).
‡ From Jones et al. (1986), based on Ritchie (1972).
§ The number of soil layers (13) and depth of the first layer (5 mm) were the values used by the authors but are not necessarily the values required for use of the mechanistic model.
¶ SUMES1 and SUMES2 refer to cumulative stages 1 and 2 drying in millimeters, respectively. These variables were initialized at zero in this example.

the Richard's equation numerically and found that the desired time step is directly related to the nodal spacing and is inversely proportional to flux.

Soil inputs for the mechanistic approach require the rate parameter hydraulic conductivity, K (Table 13-2). Unique relationships between K and water content (θ) and between matric potential (ψ_m) and θ are assumed, although K and ψ_m depend not only on θ but also on the past wetting history of the soil (Kirkham & Powers, 1972). Inputs for such models are often estimated rather than measured because their measurement is so time consuming. Furthermore, measured values for "rate" parameters (e.g., K) may not adequately describe the system as they are more spatially variable than "capacity" factors such as U (Addiscott & Wegenet, 1985). The $K(\theta)$ function used for CONSERVB was estimated from measured values of saturated hydraulic conductivity by using the calculation procedure of Jackson (1972). Results produced by mechanistic models, or any model that uses estimated inputs, can be no better than the assumptions used and the accuracy of these inputs.

Table 13-2 also shows that numerous soil layers are needed to desribe the profile for the mechanistic model. This multiplies the quantity of soil input data required because the "constant" and "functional" soil inputs must

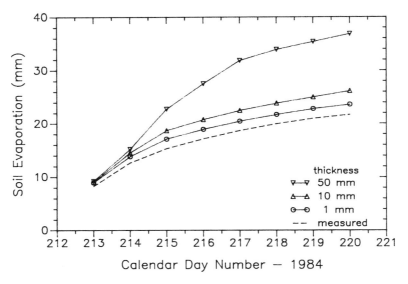

Fig. 13-2. Measured and simulated cumulative soil evaporation for various surface layer thicknesses during one drying cycle in 1984 at Lubbock, TX. From Lascano and van Bavel (1986).

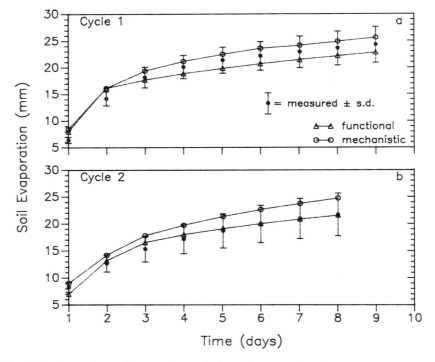

Fig. 13-3. Comparison of measured cumulative soil evaporation with simulated amounts produced by two evaporation models. Measured data are from Lascano and van Bavel (1986).

be specified for each depth. In addition, the depth of the first soil layer must be small to minimize discretization errors. Discretization errors result from the numerical solution of flow equations at a set of discrete nodal points rather than through a continuum. Figure 13–2 shows the accuracy of estimation for CONSERVB is sensitive to the assigned thickness of the surface layer. A surface layer thickness of 5 mm or less was necessary to produce simulated results that agreed with measurements and to minimize the computer time needed to run the simulations. Reynolds and Walker (1984) used discretization analysis to evaluate the sensitivity of a numerical evaporation model similar to CONSERVB. In their case, a nodal spacing of 2 mm was required in the first 10 mm of soil to obtain accurate estimates of evaporative flux and the near-surface water content profile. The functional evaporation model of Ritchie (1972) is not subject to discretization error. In fact, the number and depth of soil layers has no direct effect on the evaporation estimates. They become important only in the more comprehensive water balance models that include plant evaporation and root water uptake at several depths.

Results of each model are compared with measured E_s in Fig. 13–3. Measured evaporation amounts and simulated values for the mechanistic model were taken from Table 3 in Lascano and van Bavel (1986). Daily global radiation and maximum and minimum air temperatures, reported in the same study, were used as inputs to produce the curves for the functional model. The coefficient α in Eq. [3] was assumed to be 3.5 mm d$^{-1/2}$ and a value of 13 mm was used for the upper limit of stage 1 drying. This value for U was obtained from the first drying cycle curve of measured cumulative E_s by determining where there was a distinct reduction in the rate of cumulative evaporation after surface wetting. Figure 13–3 shows that both modeling approaches estimated E_s reasonably well. At the end of each drying cycle the standard deviation for measured cumulative evaporation was 3.4 and 4 mm, respectively. Measured E_s and values simulated by CONSERVB differed by about 1 mm after the first drying cycle and about 3 mm after the second. For this study, the functional model produced results with approximately the same accuracy.

B. Plant Evaporation Models

As with modeling E_s, there are also mechanistic and functional models for E_p. Two mechanistic-type models for E_p to be discussed include the Penman-Monteith model (Monteith, 1973) and the ENWATBAL model (Lascano et al., 1987). They are somewhat contrasting type models in terms of detail and input but represent models with reasonable theoretical bases. Both models usually operate on a 1-h time step. The Penman-Monteith equation is considerably simpler than ENWATBAL, but the former is only intended to calculate E_p or E_T. The Penman-Monteith formula has been used extensively to estimate E_T in recent years. This equation was derived from the well-known Penman equation by adding a term for canopy resistance to water and vapor flow to account for stomatal control of E_p.

The Penman-Monteith equation will not be specifically presented here since it appears in many references (e.g., McNaughton & Jarvis, 1984). To properly calculate E_p, hourly input or estimates from empirical functions are needed for humidity, temperature, net radiation, soil heat flux, heat storage rate in the canopy air column, transport resistance from the leaf surface to instrument height (r_a), and the resistance of the crop (r_c). Assumptions involved in the derivation and application of the Penman-Monteith formula are discussed by Monteith (1981). A critical simplifying assumption is that the canopy is treated as if it were one "big leaf." McNaughton and Jarvis (1984) quote several references that indicate this simplification is reasonable. They also indicate that the equation has been commonly used for the analysis of experimental results to determine E_p if r_c is known and to determine r_c if E_p is known. When needed, values of r_c must come from an empirical model or suitably aggregated porometry data to calculate E_p.

Denmead (1984) suggested that the Penman-Monteith formula should be as useful for prediction as it is for measurement because it correctly describes the dependence of the evaporation rate for an individual leaf on the prevailing atmospheric conditions. However, use of the Penman-Monteith formula to predict E_T in applications such as irrigation scheduling is a problem because the required meteorological inputs are difficult to obtain. Meteorological data from above the crop at some reference height are needed but the data are usually only available at a weather station some distance away. In addition, the value of the surface resistance is a complex function of many climatological and biological factors (Monteith, 1985). All the meteorological variables that must be known are dependent to some extent on the properties of the vegetation. Thus they must be estimated on the basis of previous experience or calculated from models of the exchange processes (McNaughton & Jarvis, 1984).

Concepts similar to the Penman-Monteith formula have been incorporated in several mechanistic models that calculate E_T and the soil water balance. One such model, ENWATBAL, described by Lascano et al. (1987) calculates the water and energy balance for both the soil surface and crop canopy. Theories regarding E_s, E_p, infiltration, and root water uptake are combined in ENWATBAL to calculate the water balance of a soil profile that is divided into horizontal layers. The model requires input information on LAI, and root depth and density throughout the season as plant inputs to the model. The values of LAI are used in empirical relationships to calculate soil surface resistance, shortwave absorbance of the crop and soil, and the view factor from the soil to the sky. Although soil and weather inputs are similar to those mentioned earlier for the E_s model CONSERVB, daily weather data were disaggregated into the hourly values needed to drive the ENWATBAL model. Empirical relationships were used for the disaggregation to produce 24 estimated values from one or two measured values.

The model ENWATBAL was used to simulate E_s and E_p for cotton grown at Lubbock, TX. Profile soil water content was measured periodically during the growing season using the neutron probe method. This information was used to infer cumulative E_T between measurements.

Micro-lysimeters were used to measure daily E_s under the canopy during two drying cycles.

The functional model of Ritchie (1972), as used in CERES-Maize, was used to calculate E_s and E_p for comparison with the results from ENWATBAL. The functional model calculates E_s as described in the previous section. Although the CERES-Maize model calculates E_p for conditions where soil water limits E_p, the equations given below for E_p are those that were used for the comparison because the crop had been irrigated at the beginning of the test period. Where soil water is nonlimiting, the functional model calculates E_p using the relationships:

$$E_p = E_o (1.0 - e^{-\text{LAI}}), \quad \text{LAI} \leq 3.0 \qquad [4]$$

$$E_p = E_o, \quad \text{LAI} > 3.0. \qquad [5]$$

If $E_s + E_p > E_o$,

$$E_p = E_o - E_s. \qquad [6]$$

The conditional Eq. [6] is necessary because values for E_s and E_p are calculated independently and their sum can exceed the potential rate on a given day because Eq. [4] and [5] are for estimating E_p when the soil surface is dry. When E_s is relatively large, E_p is smaller because of the more humid microclimate surrounding the plants. This condition was discussed earlier in the section describing E_s and E_p interactions. Here daily values for LAI provide a direct variable to estimate E_T.

Figure 13–4 shows measured daily E_s for one drying cycle, and values for E_s and E_T simulated by the ENWATBAL model. The LAI increased from about 1.0 to 1.2 during that part of the season. Because E_T was not measured on a daily basis during the time period shown, only the error associated with cumulative E_T measurement is given. Also shown in Fig. 13–4 are simulated values for E_s and E_T produced using the functional approach described in Eq. [1] to [6]. Measured values for E_s under a partial cover of cotton compared favorably with simulated values, as was the case in Fig. 13–3 where E_s for a bare soil was simulated. However, cumulative E_T predicted by the two models differed substantially as E_T calculated by the functional model exceeded E_T for the mechanistic model by about 20 mm for the 10-d period.

The lack of accurate daily measurements of E_T during the drying cycle precludes evaluation of which of the two estimates is most nearly correct. The simulated E_T values produced by the ENWATBAL model were consistently lower than the predicted E_T values for the functional model during the period from calendar day 190 to 220. The functional model accounts for the clothesline effect on E_p when the soil surface is dry, as it was during most days in this time, and as a result, may account for the higher E_T. However, the mechanistic model should also account for the drier soil surface because it uses predicted surface temperature to calculate vapor pressure gradients at canopy level.

The time required for computer simulation is not evident in the comparison given in Fig. 13–4 but is nevertheless an important consideration where use of a model for management decision making is desired. The mechanistic model for E_T calculation, ENWATBAL (Lascano et al., 1987), uses rectangular integration, a time step of 10 s, and a soil depth increment at the surface of 5 mm. Simulations required 3.5 min of CPU time per simulation day on a mini-computer (Micro Vax II) or approximately 6 h to simulate E_s and E_p for a 100-d season. This contrasts to the functional evaporation model of Ritchie (1972) as implemented in the daily incrementing crop growth model CERES-Maize (Jones et al., 1986) which calculates a single season simulation run in less than 5 s of CPU time on a mini-computer (HP-9000-550). The latter computation time also includes functional models of crop development, growth, yield, and N balance for the season. Using modern micro-computers (IBM PC-AT) with a math coprocessor, the CERES-Maize model takes about 20 s to run one season of water balance, N balance, and crop growth. This favorably lends the model to multi-season simulations for use in prediction and control decision support.

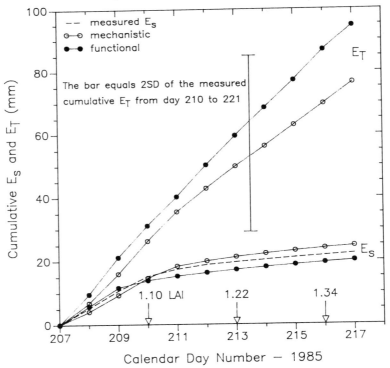

Fig. 13–4. Cumulative E_s and E_T curves produced by a mechanistic model (Lascano et al., 1987) and a functional model (Ritchie, 1972; Jones et al., 1986). Measured values for E_s and the error associated with measured values for E_T are from Lubbock, TX, during 1985 (Lascano et al., 1987).

We have identified several problems that may preclude the use of a mechanistic model for prediction purposes and for supporting farm decisions such as irrigation scheduling. On the other hand, functional models are quite widely used in some aspects of farm decision making because they require few inputs, most of which are readily obtainable. As we will point out in section IV, functional models of E_T have indeed improved in accuracy over the years as experience in using them has been gained. We feel certain that further improvements toward developing more rational, generalizable functional models will be made and they will become increasingly useful for prediction and control. Mechanistic models may provide the information needed to derive some of the empiricisms upon which functional models are based.

IV. APPLICATION OF SOIL AND PLANT EVAPORATION MODELS TO IRRIGATION SCHEDULING

Irrigation scheduling involves preventing the soil water deficit from falling below some threshold level for a particular crop and soil situation. This may involve estimating the earliest date to permit efficient irrigation or the latest date to avoid the detrimental effects of water stress on the crop. Irrigation scheduling also involves estimating the amount of water to be applied, especially where the intent is to eliminate the soil water deficit with each application of water. The irrigation requirement at any time during the growing season is roughly equal to the soil water depletion since the previous application of water. This quantity can be determined by periodic water content measurements, a method which is laborious and may be subject to error due to spatially varying soil conditions. Methods of determining irrigation requirements using estimated E_T have become increasingly popular in recent years. These methods permit us to make reasonable estimates of daily crop-water use with available information on the climate, crop, and soil conditions.

The functional model that calculates E_s and E_p, discussed in earlier sections of this chapter, has been shown to provide reasonably accurate estimates of E_T. However, this or other models that separate E_s and E_p have not been widely used for E_T prediction in irrigation scheduling, possibly because the required inputs of LAI are usually not known or are considered too impractical to measure. As a consequence, Kanemasu et al. (1983) suggested that the lack of technique for estimating or predicting leaf area has been a major problem associated with water balance, and other "hydrological" methods of irrigation scheduling. Irrigation scheduling procedures have used a simpler, but less generalizable functional approach to predict E_T. This approach involves determination of a time-varying set of crop coefficients to estimate the irrigation requirements for a given crop from a meteorologically based E_o estimate. The development and nature of crop coefficients used to estimate crop E_T have been adequately discussed by Wright (1981, 1982, 1985). The general E_T crop coefficient is defined by

$$K_c = E_T/E_o \qquad [7]$$

and is used in the following form:

$$E_T = (K_c)(E_o) \qquad [8]$$

where K_c is the dimensionless crop coefficient for a particular crop at a given growth stage. In the western USA, many irrigation scientists have used an alfalfa (*Medicago sativa* L.) reference E_T (E_{tr}) or a grass reference E_T (E_{to}) in place of E_o (Burman et al., 1983). The Penman combination approach has been recommended for estimation of E_{tr} or E_{to} where sufficient data are available (Wright, 1985). As indicated by Eq. [7], K_c can be determined by simultaneously measuring the daily crop E_T and the climatic parameters needed to calculate E_o. A crop coefficient curve represents the seasonal variation of the empirically derived K_c values. Once this curve has been developed for a given location, daily crop E_T can be estimated using Eq. [8] with daily climatic data, as long as the same procedure for calculating E_o is used as was used to estimate K_c.

The variation in time of season of the crop coefficient in Eq. [8] has not proven to be generalizable because it is often management, site, and weather specific. Values of K_c may be management specific as a result of differences in planting date, plant population, and row spacing. The values may be site specific because of large-scale soil spatial variability and may not be reproducible from one year to the next for a given location because weather sequences are usually not reproducible from year to year. Crop coefficients are dependent on weather because air temperature, radiation, and frequency of rainfall affect E_s and E_p directly and temperature influences the rate of crop development. Hanks (1985) compared crop coefficients measured at Logan, UT, and Davis, CA, for the same crop. The crop coefficient curves differed markedly for the two locations, especially early in the season. He attributed this site specificity to the dependence of E_s on the rainfall frequency and amount or irrigation regime when plant cover is low and suggested that crop coefficients vary from year to year for the same reason.

Wright and Jensen (1978) recognized this limitation of the crop coefficient procedure and developed a crop curve that was based only on conditions where the soil surface is dry. Their model accounts for increased evaporation when the soil surface is wet and effectively reduced the estimated E_T of snap bean (*Phaseolus vulgaris* L.) during leaf area development. Wright (1981) defined these modified crop coefficients, designed to represent conditions where the soil surface is dry but water is readily available, as "basal E_T crop coefficients" (K_{cb}). Estimates of an adjusted crop coefficient in terms of K_{cb} were accomplished through use of the following equation:

$$K_a = K_{cb} + (1 - K_{cb})[1 - (t/td)^{1/2}](f_w) \qquad [9]$$

where K_a is the adjusted crop coefficient, t is the number of days after major rain or irrigation, t_d is the usual number of days for the soil surface to dry, and f_w is the relative proportion of the soil surface originally wetted. Most of the adjustment takes place in the first few days after wetting the soil. The usefulness of this adjustment procedure may be limited to arid regions where evaporative conditions are relatively uniform during the season and t_d may indeed be constant for a given soil. In humid regions, it may be necessary to accommodate the unpredictable temperature and radiation conditions by considering the constant rate stage of soil evaporation and its upper limit U. Clearly, however, Eq. [9] helps to diminish the year-to-year variation of the crop coefficients caused by varying frequency of rainfall and irrigation.

Burman et al. (1983) identified several factors that affect K_c, including soil surface wetness, and suggested that life cycle of the crop (i.e., growth stage) is one of the most important. Because of the influence of growth stage on E_T, especially with respect to the development of plant cover, the horizontal axis or time scale of the crop coefficient curve becomes an important consideration. A normalized time base consisting of percentage time from planting to full cover and elapsed days after full cover has been commonly used with estimates of K_c (or K_a) because large differences in planting dates often have no effect on the date that full cover is reached within a region (Wright & Jensen, 1978). As pointed out by Neale and Bausch (1985), this normalized time base does not always allow matching the crop coefficient in the irrigation scheduling model with actual crop growth. This is especially true when abnormal climatic conditions occur as in a cold, wet spring when E_T could be overestimated.

Rather than percentage time between planting and full cover or elapsed days after full cover as the normalized time base, Wright (1985) proposed that it would be better to have a means of relating crop coefficients more directly to crop development. The desired refinement may be achieved through models that relate crop development directly to climatic and growing conditions (e.g., plant density). Hanks (1985) offered an alternate approach by suggesting that some of the difficulties associated with crop coefficients could be dealt with more easily if crop coefficients for E_p were used instead of crop coefficients for E_T. He stated that these coefficients should also be more transferable from one location to another.

V. A NEW GENERALIZED CROP COEFFICIENT CONCEPT FOR USE DURING PARTIAL CANOPY COVER

We believe it is possible to provide an accurate calculation of K_c that is crop specific but not site, year, or management specific. The proposed approach provides a reasonable estimate of LAI and uses the functional model described in this chapter to calculate E_s and E_p. This procedure incorporates the effects of temperature on leaf appearance and expansion growth and accounts for the influence of varying plant populations on E_T. The new coefficient proposed here is given by

$$K_{s+p} = (E_s + E_p)/E_o \qquad [10]$$

where K_{s+p} is a daily crop coefficient based on separate calculation of E_s and E_p, and E_o is the potential rate of evaporation from a crop surface. The calculations of E_s, E_p, and E_o were described earlier in this chapter. The value of K_{s+p} is always ≤ 1 because the sum of E_s and E_p, by definition, never exceeds E_o.

Adoption of Eq. [10] for irrigation scheduling requires prediction of daily values for LAI in terms of readily obtainable inputs. When the water supply to roots is nonlimiting, the development of leaf area on an individual plant is influenced primarily by temperature. The area increase of individual leaves depends on temperature but it is the appearance of successive leaves or nodes on a main stem that has a more direct influence on the increase in plant leaf area. Hesketh et al. (1973) found a linear relationship between the rate of appearance of emerged soybean leaves on the main stem and temperature. The relationship obtained was expressed by

$$d\text{PI}/dt = 0.018\ (T - T_b) \qquad [11]$$

where the plastochron index (PI) is the integer count of the number of emerged leaves plus a decimal fraction representing the progress of the emerging leaf toward full emergence, T is the daily mean temperature, and T_b is the base temperature below which leaf development essentially ceases. This relationship has since been shown to hold under several tested circumstances. Sinclair (1984) found good agreement between observed leaf numbers for several soybean cultivars and those simulated using Eq. [11] by setting T_b to a value ranging from 9 to 11 °C, depending on the cultivar.

The interval between appearance of successive leaves when expressed in thermal time (degree-days) has been called the *phyllochron* (P) (Klepper et al., 1982; Bauer et al., 1984). *Daily thermal time* (DTT) is expressed by the normal degree-day calculation:

$$\text{DTT} = [(T_{max} + T_{min})/2.0 - T_b] \qquad [12]$$

where DTT are degree-days, T_{max} and T_{min} are maximum and minimum daily temperatures, and T_b is the base temperature as in Eq. [11]. Because leaf appearance rate does not increase indefinitely with increasing daily mean temperature, the calculation of DTT could be improved by accounting for the critical temperature (T_c) above which the generation of leaf area ceases. For maize, a temperature of 44 °C has been used as the T_c value for prediction of LAI (Coelho & Dale, 1980) and for calculation of phenological changes (Jones et al., 1986). The P for a crop species is usually fairly constant when the same value of T_b is used. Baker et al. (1986) reported a P value of 106 degree-days per leaf for winter wheat (*Triticum aestivum* L.) in Kansas after double ridge formation. The reciprocal of the coefficient in Eq. [11] gives a P value of about 56 degree-days per leaf for soybean. A value for P of about 40 degree-days per leaf applies to several maize cultivars grown in a diversity of climates (Ritchie, unpublished data).

The constancy of the relationship between leaf appearance rate and thermal time and the dependence of plant leaf area on the number of emerged leaves for a given species provides a fundamentally sound basis for estimating the daily LAI for partial-cover conditions until plant competition begins to dominate the rate of new leaf growth. Sinclair (1984) developed a three-parameter allometric model that related plant leaf area to leaf number for several soybean cultivars with leaf area < 1000 cm^2 plant^{-1}. We found that a flexible growth function described by Richards (1959, 1969) fits the Sinclair (1984) soybean leaf area well in addition to leaf area expansion for other crops tested. This function, called the *Gompertz function*, is the only one of five growth functions described by Richards (1969) that allows us to limit predicted plant leaf area to small values for the appropriate length of time. The Gompertz function is expressed by

$$A = A_o e^{-be^{-kt}} \qquad [13]$$

where A represents the leaf area (cm^2 plant^{-1}) at time t, the final size A_o (cm^2 plant^{-1}) is approached asymptotically for large values of t, and b and k are constants that determine the position and the spread of the curve along the time axis, respectively. The Gompertz function produces an asymmetric sigmoid curve that has been used in animal and population studies but has not been applied extensively to the growth of higher plants (Richards, 1969). Amer and Williams (1957) used this function to study the growth in the area of single *Pelargonium* leaves. More recently, Baker et al. (1975) applied Eq. [13] to maize leaf area development and evaluated the influence of environmental conditions on the parameters of the Gompertz function, especially the coefficient k.

Because environmental factors such as temperature influence the rate of leaf appearance and expansion growth, Eq. [13] can be made as general as possible for all crops by replacing time with the variable *TI*, where *TI* is cumulative thermal time after emergence divided by the phyllochron (ΣDTT$/P$). We fit the Gompertz function to leaf area expansion curves for several crops during most of the vegetative growth period except where competition for light becomes important at LAI values near 3. The first step in the fitting procedure involves linearizing Eq. [13].

$$\log_e \log_e (A/A_o) = \log_e b - k(TI) \qquad [14]$$

The left side of Eq. [14], when plotted vs. "plant development time" *TI*, has a slope k and an intercept $\log_e b$ which gives the constant b when exponentiated.

The maximum or final leaf area A_o of an individual plant is difficult to predict in advance as a generality because the value is quite dependent on plant population, species, genotype and other factors that affect plant size, such as nutrient availability. We have observed that during early plant development the genotype and population have relatively small influence on individual plant leaf area within a species when plant-to-plant competition for light is small. We also found that when using actual A_0 values for fitting the coefficients in Eq. [13], there is a consistent bias of the leaf area that falls below a linearized fit of the $\log_e \log_e (A_0/A)$ vs. *TI* for *TI* values between 1 and 6 for maize. By choosing values for A_0 that are considerably larger than the measured values, the leaf area curve corresponding to low leaf number is better. A value for A_0 of about five times the average plant maximum leaf area is usually suitable to use for estimating LAI for partial cover conditions. Of course, the equation with the biased A_0 will not produce accurate values of leaf area for large values of *TI*. Our concern, however, is only to provide reasonable approximations of leaf area until a LAI of 2.5 to 3 is reached because E_s and E_p calculations are critically influenced by LAI in this range.

Figure 13-5 illustrates an example where Eq. [14] was used to determine the parameters for the Gompertz function. The leaf area data are from Dale et al. (1980) where LAI of maize hybrid PAG 2X-29 was measured for early and late plantings of two populations. Figure 13-5 was fit only from the early planting, at a population of 62 000 plants ha^{-1}. No attempt was made to fit Eq. [14] to all of the data in Fig. 13-5 as we were most interested in using the Gompertz function to simulate leaf area development from emergence to a time when LAI becomes approximately 3.

Equation [13] and the parameters given in Fig. 13-5 were used to simulate the development of leaf area for maize at a different plant population, also from the data of Dale et al. (1980). Because the units of Eq. [13] are cm^2 plant^{-1}, LAI at each point in time was calculated by converting the units of Eq. [13] to m^2 plant^{-1} and multiplying by the appropriate plant density in plants m^{-2}. The results are given in Fig. 13-6 where simulated

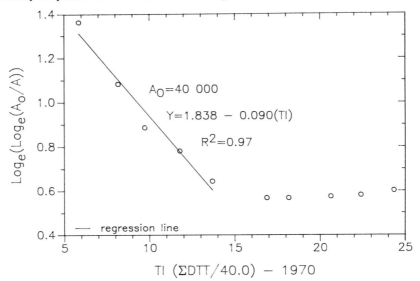

Fig. 13-5. Determination of the Gompertz function parameters for maize. Data are from Dale et al. (1980) for the early plantings of hybrid PAG 2X-29 during 1970 at a population of 62 000 plants ha^{-1}.

Table 13–3. Base temperature (T_b), phyllochron (P), and the limiting leaf area (A_o) used in the linear regression of $\log_e\log_e (A_o/A)$ on leaf number $(\Sigma DTT/P)$.

Crop	T_b	P	A_o	b†	k	R^2	Reference
	C	C d	cm^2				
Maize	8.0	40.0	40 000	6.28	0.090	0.97	Dale et al., 1980‡
Soybean	9–11	56.0	3 500	5.43	0.152	0.95	Sinclair, 1984
Sorghum	8.0	49.0	5 000	11.95	0.18	NA	Ritchie, unpublished data
Sweet sorghum	8.0	49.0	10 000	7.21	0.14	0.99	Shih et al., 1981
Millet	8.0	43.0	2 100	9.68	0.13	NA	Ritchie, unpublished data

† Parameter b is the exponent of the intercept obtained from the regression of $\log_e\log_e (A_o/A)$ on leaf number.
‡ "Early planting" of cultivar PAG SX-29 at a population of 62 000 plants ha^{-1} in 1970.

LAI is compared with measured values for maize at a population of 43 000 plants ha^{-1} (4.3 plants m^{-1}). The measured values for LAI used to fit Eq. [14], and those reported in Fig. 13–6, represent independent measurements of LAI for maize at different populations. Simulated LAI compared favorably with the observed data until LAI reached a value of 3. This suggests that leaf area development per plant was relatively independent of plant densities up to a LAI of about 4 and that the parameters obtained for Eq. [13] should be applicable to a range of plant populations.

Table 13–3 gives the parameters for Eq. [13] for several crops, as well as the source of the corresponding leaf area data. The experiments cited in Table 13–3 share the following features: (i) soil water deficits did not limit

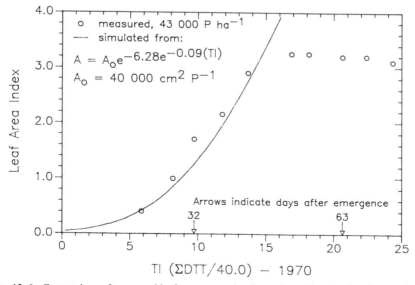

Fig. 13–6. Comparison of measured leaf area expansion for maize with a simulated expansion curve produced by the Gompertz function. Data are from Dale et al. (1980) for the early planting of hybrid PAG 2X-29 during 1970 at a population of 43 000 plants ha^{-1}.

the rate of leaf area development; (ii) leaf area per plant was given, or LAI and plant populations were reported; (iii) daily maximum and minimum temperatures were availale for the site; and (iv) the date of emergence was known. The parameters reported in Table 13–3 differ markedly between crop species. Values of T_b should be constant for a given crop species and preliminary observations indicate that values of P are usually constant for a given crop. Winter cereals are an exception as P has been shown to vary considerably with sowing dates and latitude of the sites (Baker & Gallagher, 1983; Baker et al., 1986). Our experience in using the Gompertz function has revealed that the parameters will vary somewhat between cultivars of a particular crop and for circumstances where the temperature is exceptionally high or low with respect to the usual temperature range for plant development. However, it should not be necessary to fit Eq. [13] to each cultivar of a crop. Errors produced by Eq. [13] are always relatively small at low LAI where the actual E_T is most dependent on wetness of the soil surface (Fig. 13–1).

As an illustration of the utility of the crop coefficient procedure described in this section, crop curves for maize were calculated for 2 yr that differed in the amount and frequency of early season rainfall at Flint, MI. An emergence date of 20 May was used in each simulation. Determination of daily crop coefficients (K_{s+p}) involved calculation of daily values for LAI using Eq. [13] and the parameters reported in Table 13–3 for maize. The model of Ritchie (1972), as used in the CERES-Maize crop simulation model, was used to calculate E_o, E_s, and E_p. Soil inputs used to calculate E_s (Table 13–2) were 0.13 for the soil albedo and 9 mm for the upper limit of stage 1 evaporation U. Daily E_p is a function of E_o in this model except when soil water deficits reduce E_p. These conditions are usually satisfied in the case of ir-

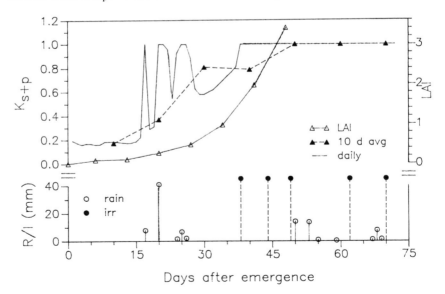

Fig. 13–7. Daily rainfall and irrigation, simulated LAI, daily and 10-d average simulated crop coefficients for maize at Flint, MI, during 1966.

rigated crops. They were provided for in these simulations by arbitrarily defining a profile with 150 mm of total plant-extractable soil water (PESW), by calculating the water balance on a daily basis, and irrigating 45 mm at 50% soil water depletion (PESW = 75 mm) to increase PESW to 120 mm.

Simulated crop coefficients (K_{s+p}) for maize at Flint, MI, are given in Fig. 13-7 and 13-8 for 1966 and 1973, respectively, in addition to simulated LAI. The curves constructed using the daily estimates of K_{s+p} were directly proportional to LAI during the early part of the season but only between periods of rainfall or irrigation. These crop coefficients approached 1 on only three occasions prior to the establishment of an LAI of 3 in 1966. This year was characterized by infrequent rainfall during May and June. The daily values for K_{s+p} were less consistent during the wet year of 1973. They reached 1 more frequently and remained at 1 for several days at a time during particularly wet periods. The crop coefficients resemble those reported by Wright (1981, 1985) where Eq. [9] was used to estimate K_a in terms of the empirically derived values for K_{cb}. However, an extensive amount of research using weighing lysimeters was required to determine K_{cb} and the result would still be somewhat site specific because of the possibility of seasonal temperature differences, and management specific because of the possibility of different plant populations or planting dates. The methodology used to produce the simulated crop curves for Flint, MI, required only daily weather data (Table 13-2), reasonable estimates for the parameters in Eq. [8], and measured or estimated values for soil albedo and the upper limit of stage 1 soil evaporation.

The dependence of the crop coefficients on the prevailing climate is even more evident when curves representing 10-d averages for 1966 and 1973 are

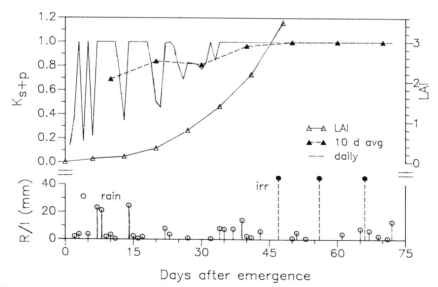

Fig. 13-8. Daily rainfall and irrigation, simulated LAI, daily and 10-d average simulated crop coefficients for maize at Flint, MI, during 1973.

compared (Figs. 13–7 and 13–8). The mean curves differed by a factor of 4.0 for the first 10 d after emergence, averaging 0.17 in 1966 and 0.69 in 1973. The means for the second 10-d period after emergence differed by a factor of 2.3 as the mean K_{s+p} in 1973 again exceeded the 1966 mean. Figures 13–7 and 13–8 demonstrate that no single crop curve can be used to accurately determine irrigation requirements for a crop because crop curves are not reproducible from one year to the next. However, the model used here accounts for the influence of temperature on the rate of leaf area development and the combined effects of climate and LAI on E_s and E_p. Furthermore, it is not management specific as varying plant population is easily incorporated into the calculations of K_{s+p} through its influence on LAI. To illustrate the effects of plant population on crop coefficients produced by this model, values for K_{s+p} were calculated for maize at populations ranging from 32 000 to 62 000 plants ha^{-1} (3.2–6.2 plants m^{-2}). Weather data used for these calculations were from Flint, MI, in 1973. All other inputs were the same as those used to produce the crop coefficients in Fig. 13–7 and 13–8. The results are in Fig. 13–9 where simulated crop curves consisting of 10-d average K_{s+p} are shown along with arrows indicating when simulated LAI reached 3 for each plant density. Decreasing the plant density reduced the rate of leaf area development and diminished the 10-d average K_{s+p} during a relatively dry period in 1973. This example suggests that the K_{s+p} should be a more transferrable technology than the crop coefficients currently used. They also facilitate more accurate calculation of irrigation requirements, especially in humid climates where early season rainfall frequency is highly variable.

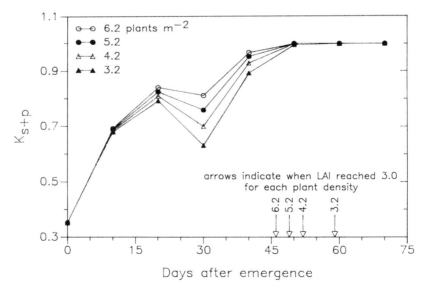

Fig. 13–9. Influence of plant density on 10-d average simulated crop coefficients for maize at Flint, MI, during 1973.

VI. SUMMARY

As we look toward refinement of irrigation technology to improve water conservation, farm profitability, and production stability, the accurate prediction of water loss by evaporation is imperative. Many crops of agronomic importance provide only partial cover for the soil surface during a part of the growing season. Thus the partitioning of the evaporative water loss between soil and plants can vary considerably depending on the LAI and the soil surface wetness. Accurate estimation of E_T during times when LAI is low requires the separate analysis of E_s and E_p because E_T under such circumstances can be considerably lower than E_o.

We have discussed methods that have been used to measure E_s and E_p separately and then presented examples of both mechanistic and functional models that can be used to reasonably calculate the two components separately. The models need, as a minimum, temperature, solar radiation, and rainfall input from weather stations; soil parameters related to upward water flow to the soil surface; and vegetation characteristics as expressed by LAI.

Since plant leaf area is a dynamic quantity that changes throughout the season with leaf growth and death, evaporation models needing LAI as an input have not been widely used for irrigation scheduling. We have presented a model in the form of a modified Gompertz equation that can be used along with plant population to reasonably predict LAI over the early part of the growing season before the LAI reaches a value of 2 to 3. The predicted LAI provides the necessary information to calculate a crop coefficient, K_{s+p}, which becomes the multiplier for E_o to determine E_T for use in functional irrigation scheduling models.

In this chapter, we have not discussed the influence of soil water deficits on E_p, because it is discussed in other chapters of this book.

REFERENCES

Adams, J.E., G.F. Arkin, and J.T. Ritchie. 1976. Influence of row spacing and straw mulch on first stage drying. Soil Sci. Soc. Am. Proc. 40:436–442.

Addiscott, T.M., and R.J. Wagenet. 1985. Concepts of solute leaching in soils: A review of modelling approaches. J. Soil Sci. 36:411–424.

Amer, F.A., and W.T. Williams. 1957. Leaf-area growth in *Pelargonium zonale*. Ann. Bot. 21:339–342.

Arkin, G.F., J.T. Ritchie, and J.E. Adams. 1974. A method for measuring first-stage soil water evaporation in the field. Soil Sci. Soc. Am. Proc. 38:951–954.

Baker, C.K., and J.N. Gallagher. 1983. The development of winter wheat in the field. 2. The control of primordium initiation rate by temperature and photoperiod. J. Agric. Sci. 101:337–344.

Baker, C.H., R.D. Horrocks, and C.E. Goering. 1975. Use of the Gompertz function for predicting corn leaf area. Trans. ASAE 18:323–326, 330.

Baker, J.M., and C.H.M. van Bavel. 1987. Measurement of mass flow of water in the stems of herbaceous plants. Plant Cell Environ. 10:777–782.

Baker, J.T., P.J. Pinter, Jr., R.J. Reginato, and E.T. Kanemasu. 1986. Effects of temperature on leaf appearance in spring and winter wheat cultivars. Agron. J. 78:605–613.

Bauer, A., A.B. Frank, and A.L. Black. 1984. Estimation of spring wheat leaf growth rates and anthesis from air temperature. Agron. J. 76:829–835.

Black, T.A., C.B. Tanner, and W.R. Gardner. 1969. Evapotranspiration from a snap bean crop. Agron. J. 62:66–69.

Boast, C.W., and T.M. Robertson. 1982. A "micro-lysimeter" methods for determining evaporation from bare soil: Description and laboratory evaluation. Soil Sci. Soc. Am. J. 46:689–696.

Bond, J.J., and W.O. Willis. 1970. Soil evaporation: First stage drying as influenced by surface residue and evaporation potential. Soil Sci. Soc. Am. Proc. 33:445–448.

Briggs, L.J., and H.L. Shantz. 1913. The water requirement of plants: I. Investigations in the Great Plains in 1910 and 1911. USDA Bur. Plant Indus. Bull. 284.

Burman, R.D., R.H. Cuenca, and A. Weiss. 1983. Techniques for estimating irrigation water requirements. p. 335–394. In D. Hillel (ed.) Advances in irrigation. Vol. 2. Academic Press, New York.

Choudhury, B.J., and J.L. Monteith. 1988. A four-layer model for the heat budget of homogeneous land surfaces. Q. J. R. Meteorol. Soc. 114:373–398.

Coelho, D.T., and R.F. Dale. 1980. An energy-crop growth variable and temperature factor for predicting corn growth and development: Planting to silking. Agron. J. 72:503–510.

Dale, R.F., D.T. Coelho, and K.P. Gallo. 1980. Prediction of daily green leaf area index for corn. Agron. J. 72:999–1005.

Davis, R., and M.M. Ludlow. 1981. A field gas exchange measuring system for pasture canopies. CSIRO Sust. Div. Trop. Crops Past. Trop. Agron. Tech. Mem. 28.

Denmead, O.T. 1973. Relative significance of soil and plant evaporation in estimating evapotranspiration. p. 505–511. In R.O. Slatyer (ed.) Plant response to climatic factors. UNESCO, Paris.

Denmead, O.T. 1984. Plant physiological methods for studying evapotranspiration: Problems of telling the forest from the trees. Agric. Water Manage. 8:167–189.

Feddes, R.A., P.J. Kowalik, and H. Zaradny. 1978. Simulation of field water use and crop yield. PUDOC, Wageningen, Netherlands.

Gardner, H.R., and W.R. Gardner. 1969. Relation of water application to evaporation and storage of soil water. Soil Sci. Soc. Am. Proc. 33:192–196.

Hanks, R.J. 1974. Model for predicting plant yield as influenced by water use. Agron. J. 66:660–665.

Hanks, R.J. 1985. Crop coefficients for transpiration. p. 431–438. In Advances in evapotranspiration. Proc. Natl. Conf. on Advances in Evapotranspiration, Chicago. 16–17 Dec. ASAE, St. Joseph, MI.

Hanks, R.J., L.H. Allen, and H.R. Gardner. 1971. Advection and evapotranspiration of wide-row sorghum in the central Great Plains. Agron. J. 52:520–527.

Harrold, L.L., D.B. Peters, F.R. Dreibelbis, and J.L. McGuinness. 1959. Transpiration evaluation of corn grown on a plastic-covered lysimeter. Soil Sci. Soc. Am. Proc. 23(2):174–178.

Hesketh, J.D., D.L. Myhre, and C.R. Willey. 1973. Temperature control of time intervals between vegetative and reproductive events in soybeans. Crop Sci. 13:250–254.

Hillel, D. 1977. Computer-simulation of soil water dynamics: A compendium of recent work. Int. Dev. Res. Ctr., Ottawa, Canada.

Hillel, D., and H. Talpaz. 1977. Simulation of soil water dynamics in layered soils. Soil Sci. 123:54–62.

Huck, M.G., and D. Hillel. 1983. A model of root growth and water uptake accounting for photosynthesis, respiration, transpiration, and soil hydraulics. p. 273–333. In D. Hillel (ed.) Advances in irrigation. Vol. 2. Academic Press, New York.

Jaafar, M.N., E.T. Kanemasu, and W.L. Powers. 1978. Estimating soil factors for nine Kansas soils used in an evapotranspiration model. Trans. Kans. Acad. Sci. 81:57–63.

Jackson, R.D. 1972. On the calculation of hydraulic conductivity. Soil Sci. Soc. Am. Proc. 36:380–382.

Jalota, S.K., and S.S. Prihar. 1986. Effects of atmospheric evaporativity, soil type and redistribution time on evaporation from bare soil. Aust. J. Soil Res. 24:357–366.

Jones, C.A., J.T. Ritchie, J.R. Kiniry, and D.C. Godwin. 1986. Subroutine structure. p. 49–112. In C.A. Jones and J.R. Kiniry (ed.) CERES-Maize: A simulation model of maize growth and development. Texas A&M Univ. Press, College Station.

Kanemasu, E.T., J.L. Steiner, A.W. Biere, F.D. Worman, and J.F. Stone. 1983. Irrigation in the Great Plains. Agric. Water Manage. 7:157–178.

Kanemasu, E.T., L.R. Stone, and W.L. Powers. 1976. Evapotranspiration model tested for soybean and sorghum. Agron. J. 68:569–572.

Kirkham, D., and W.L. Powers. 1972. Advanced soil physics. John Wiley and Sons, New York.

Klepper, B., R.W. Rickman, and C.M. Peterson. 1982. Quantitative characterization of vegetative development in small grain cereals. Agron. J. 74:789–792.

Klocke, N.L., G.W. Hergert, and R. Todd. 1986. Soil evaporation and evapotranspiration from fully, limited and non-irrigated corn. Paper 86-2524. Chicago. 16–19 Dec. ASAE, St. Joseph, MI.

Lascano, R.J., and C.H.M. van Bavel. 1983. Experimental verification of a model to predict soil moisture and temperature profiles. Soil Sci. Soc. Am. J. 47:441–448.

Lascano, R.J., and C.H.M. van Bavel. 1986. Simulation and measurement of evaporation from a bare soil. Soil Sci. Soc. Am. J. 50:1127–1132.

Lascano, R.J., C.H.M. van Bavel, J.L. Hatfield, and D.R. Upchurch. 1987. Energy and water balance of a sparse crop: Simulated and measured soil and crop evaporation. Soil Sci. Soc. Am. J. 51:1113–1121.

Marlatt, W.E. 1961. The interaction of microclimate, plant cover, and soil moisture content affecting evapotranspiration rates. Colorado State Univ. Dep. Atmos. Sci. Tech. Pap. 23. Fort Collins.

Martin, D.L., N.L. Klocke, and D.L. DeHann. 1985. Measuring evaporation using mini-lysimeters. p. 231–240. In Advances in evapotranspiration. Proc. Natl. Conf. on Advances in Evapotranspiration, Chicago. 16–17 Dec. ASAE, St. Joseph, MI.

McNaughton, K.G., and P.G. Jarvis. 1984. Using the Penman-Monteith equation predictively. Agric. Water Manage. 8:263–278.

Monteith, J.L. 1973. Principles of environmental physics. Edward Arnold, London.

Monteith, J.L. 1981. Evaporation and surface temperature. Q. J. R. Meteorol. Soc. 107:1–27.

Monteith, J.L. 1985. Evaporation from land surfaces: Progress in analysis and prediction since 1948. p. 4–12. In Advances in evapotranspiration. Proc. Natl. Conf. on Advances in Evapotranspiration, Chicago. 16–17 Dec. ASAE, St. Joseph, MI.

Neale, C.M.U., and W.C. Bausch. 1985. Reflectance-based crop coefficients for use in irrigation scheduling models. p. 250–258. In Advances in evapotranspiration. Proc. Natl. Conf. on Advances in Evapotranspiration, Chicago. 16–17 Dec. ASAE, St. Joseph, MI.

Norman, J.M., and G.C. Campbell. 1983. Application of a plant-environment model to problems in irrigation. p. 155–188. In D. Hillel (ed.) Advances in irrigation. Vol. 2. Academic Press, New York.

Philip, J.R. 1957. Evaporation and moisture and heat fields in the soil. J. Meteorol. 14:354–366.

Reicocsky, D.C., and D.B. Peters. 1977. A portable chamber for rapid evapotranspiration measurements on field plots. Agron. J. 69:729–732.

Reynolds, W.D., and G.K. Walker. 1984. Development and validation of a numerical model simulating evaporation from short cores. Soil Sci. Soc. Am. J. 48:960–969.

Rice, R.C., and R.D. Jackson. 1985. Spatial distribution of evaporation from bare soil. p. 447–453. In Advances in evapotranspiration. Proc. Natl. Conf. on Advances in Evapotranspiration, Chicago. 16–17 Dec. ASAE, St. Joseph, MI.

Richards, F.J. 1959. A flexible growth function for empirical use. J. Exp. Bot. 10:290–300.

Richards, F.J. 1969. The quantitative analysis of growth. p. 3–76. In F.C. Stewart (ed.) Plant physiology. Vol. VA. Analysis of growth: Behavior of plants and their organs. Academic Press, New York.

Ritchie, J.T. 1971. Dryland evaporative flux in a subhumid climate: I. Micrometeorological influences. Agron. J. 63:51–55.

Ritchie, J.T. 1972. Model for predicting evaporation from a row crop with incomplete cover. Water Resour. Res. 8:1204–1213.

Ritchie, J.T. 1983. Efficient water use in crop production: Discussion on the generality of relations between biomass production and evapotranspiration. p. 29–44. In H.M. Taylor et al. (ed.) Limitations to efficient water use in crop production. ASA, CSSA, and SSSA, Madison, WI.

Ritchie, J.T. 1985. Evapotranspiration empiricisms for minimizing risk in rainfed agriculture. p. 139–150. In Advances in evapotranspiration. Proc. Natl. Conf. on Advances in Evapotranspiration, Chicago. 16–17 Dec. ASAE, St. Joseph, MI.

Ritchie, J.T., and E. Burnett. 1971. Dryland evaporative flux in a subhumid climate: II. Plant influences. Agron. J. 63:56–62.

Rose, C.W. 1968. Evaporation from bare soil under high radiation conditions. Trans. Int. Congr. Soil Sci. 9th, Adelaide. I:57–66.

Sadeghi, A.M., H.D. Scott, and J.H. Ferguson. 1984. Estimating evaporation: A comparison between Penman, Idso-Jackson and zero-flux methods. Agric. For. Meteorol. 33:225–234.

Sakuratani, T. 1981. A heat balance for measuring water flux in the stem of intact plants. J. Agric. Meteorol. 37:9–17.

Sakuratani, T. 1984. Improvement of the probe for measuring water flow rate in intact plants with the stem heat balance method. J. Agric. Meteorol. 40:273–277.

Sakuratani, T. 1987. Studies on evapotranspiration from crops. (2) Separate estimation of transpiration and evapotranspiration from a soybean field without water shortage. J. Agric. Meteorol. Soc. 111:839–855.

Shawcroft, R.W., and H.R. Gardner. 1983. Direct evaporation from soil under a row crop canopy. Agric. Meteorol. 28:229–238.

Shih, S.F., G.J. Gascho, and G.S. Rahi. 1981. Modeling biomass production of sweet sorghum. Agron. J. 73:1027–1032.

Shuttleworth, W.J., and J.S. Wallace. 1985. Evaporation from sparse crops—an energy combination theory. Q. J. R. Meteorol. Soc. 111:839–855.

Sinclair, T.R. 1984. Leaf area development in field-grown soybeans. Agron. J. 76:141–146.

Tanner, C.B. 1957. Factors affecting evaporation from plants and soils. J. Soil Water Conserv. 12:221–227.

Tanner, C.B., and W.A. Jury. 1976. Estimating evaporation and transpiration from a row crop during incomplete cover. Agron. J. 68:239–243.

van Bavel, C.H.M., and D. Hillel. 1976. Calculating potential and actual evaporation from a bare soil surface by simulation of concurrent flow of water and heat. Agric. Meteorol. 17:453–476.

Walker, G.K. 1983. Measurement of evaporation from soil beneath crop canopies. Can. J. Soil Sci. 63:137–141.

Walker, G.K. 1984. Evaporation from wet soil surfaces beneath plant canopies. Agric. For. Meteorol. 33:259–264.

Wright, J.L. 1981. Crop coefficients for estimates of daily crop evapotranspiration. p. 18–26. *In* Irrigation scheduling for water and energy conservation in the 80's. Proc. Irrigation Scheduling Conf., Chicago. 14–15 Dec. ASAE, St. Joseph, MI.

Wright, J.L. 1982. New evapotranspiration crop coefficients. Am. Soc. Civil Eng. J. Irrig. Drain. Div. 108(IA1):57–74.

Wright, J.L. 1985. Evapotranspiration and irrigation water requirements. p. 105–113. *In* Advances in evapotranspiration. Proc. Natl. Conf. on Advances in Evapotranspiration, Chicago. 16–17 Dec. ASAE, St. Joseph, MI.

Wright, J.L., and M.E. Jensen. 1978. Development and evaluation of evapotranspiration models for irrigation scheduling. Trans. ASAE 21:88–96.

Zaradny, H. 1978. Boundary conditions in modeling water flow in unsaturated soils. Soil Sci. 125:75–82.

14 Relationships Between Crop Production and Transpiration, Evapotranspiration, and Irrigation

TERRY A. HOWELL

USDA-ARS
Bushland, Texas

Crops consume water in the process of transpiration, and water evaporates from the soil. These processes are defined collectively as *evapotranspiration* (Thornthwaite, 1948). Only the transpiration portion of evapotranspiration directly influences crop production (de Wit, 1958). Although soil evaporation can be reduced, it is practically impossible to totally eliminate soil water evaporation, even with expensive plastic or artificial mulches (Klocke et al., 1985). There has been a long history dating back to the late 17th century (Woodward, 1699) of efforts to determine water use by crops and vegetation. The necessary amount of irrigation water for crop production has been of interest to investigators at least since the 19th century in the USA (Mead, 1887).

The purpose of this chapter is to summarize the voluminous information on the relationship between crop production and evapotranspiration. Soil water evaporation, deep percolation, runoff, and soil water recharge can result from irrigation but may not directly increase crop production. Previous chapters have discussed the physical and biological limitations to the evapotranspiration process, and this chapter will focus on the crop production associated with transpiration and evapotranspiration with an extension to the relationship between crop production and applied irrigation water. Past reviews of this subject are Doorenbos and Kassam (1979), Hanks and Rasmussen (1982), Taylor et al. (1983), Stanhill, 1986), and van Keulen and Wolf (1986).

The quantity of irrigation water necessary for crop production has been historically important, particularly in the arid western USA. The water right granted to an irrigator as a result of prior appropriation or adjudication was called the *duty of water* (Powers, 1922). The term duty of water was widely used throughout the late 19th century and is still in use (Lety & Vaux, 1984). The duty of water was the amount of water required to be diverted to irrigate a crop area sufficiently to produce an economic yield. The term *consumptive use* (ASCE, 1930) was used beginning in the early 20th century and was defined as the evapotranspiration of the crop and has largely replaced the term duty of water in legal institutions. Early research defined the terms *water requirement* and *transpiration ratio* (Briggs & Shantz, 1913a, 1913b) to mean the ratio of the amount of transpiration (usually expressed in units of mass) to the production of crop dry matter (usually expressed in units of mass and excluding the root mass). Transpiration ratio was also known as the transpiration coefficient (Maximov, 1929). Viets (1962) defined *water use efficiency* as the ratio of the crop production to evapotranspiration. Water use efficiency has become a widely used agronomic term implying the yield (photosynthesis, biological, or economic) per unit of water (transpiration, evapotranspiration, or applied water). Sinclair et al. (1984) classified the water use efficiency terminology for several production measurements (photosynthesis to yield) for different time scales.

The agronomic or physiological characterization of water use efficiency is defined differently than the engineering definition in which water use efficiency means the ratio between the amount of water stored in the crop root zone to the amount delivered for irrigation (Bos & Nugteren, 1978). The engineering characterization of water use efficiency is normally expressed as a volume percentage. This chapter is associated with the agronomic view of water use efficiency as contrasted with the engineering view. When the agronomic value of water use efficiency is increased, the engineering value of water use efficiency will likely be improved, although maybe not in a direct proportion.

Clearly, the relationship between crop production and the amount of irrigation water applied to the crop is important to agronomists, engineers, economists, and water resource planners. This importance is currently accentuated due to competition among users, declining groundwater reserves, various legal institutions, and degradations in water quality. The relationship between crop production and irrigation applications is not unique and is often not clearly defined.

Crop production models with resource and management inputs (as input-output models) have been widely used, particularly by agricultural economists, and called *production functions* (Hexem & Heady, 1978; Vaux & Pruitt, 1983). These production functions have permitted analyses of resource problems, usually in terms of one or two inputs. Agricultural production depends on many resource or managerial inputs, in addition to irrigation or rainfall, that may not be properly characterized in such one- or two-dimensional systems. The relationship of crop production to irrigation also depends on the salinity of the soil and irrigation water, the uniformity of the irrigation applications, the spatial variability of the soil physical properties, specific crop variety characteristics, and crop cultural practices (e.g., weed and pest control, fertility, plant population, row spacing, and planting date).

I. ANALYTICAL CONCEPTS

The analytical framework for describing the effects of irrigation on crop yield is complex. This chapter will discuss the effects of many parameters on crop yield through their effects on several processes. These processes (mainly assimilation, transpiration, evapotranspiration, etc.) are discussed in other chapters in this monograph. The framework for this chapter is based on understanding the dynamic nature of the following relationships:

$$ET = f (Q, \theta\ C, W, M), \tag{1}$$

$$T = f (ET, C, W, M), \tag{2}$$

$$A = f (T, C, W, M) \tag{3}$$

$$P = f (A, C, M) \tag{4}$$

$$\text{and } Y = f (P, C, M) \tag{5}$$

where f represents a functional relationship between many specific production vectors, ET is evapotranspiration, Q is irrigation, θ is various soil vectors (water content, nutrient content, salinity, etc.), C is various crop vectors (species, diffusion resistances, CO_2 compensation point, and partitioning), W is various weather vectors (solar radiation, air temperature, vapor pressure deficit, rainfall, etc.), M is various miscellaneous vectors (diseases, critical water deficit periods, insects, agronomic culture, etc.), T is transpiration, A is assimilation, P is dry matter production, and Y is economic yield. (These and other symbols used in this chapter are listed and explained in the Appendix.) The development of the complete functional relationships in this framework would be extremely difficult, if not impossible. Even when simple relationships between the various vectors have been developed, the integration of all the factors related to crop yield remains complex.

If the costs of production are neglected for the moment, the goal of most agronomic systems is to produce the most yield subject to the physical and chemical limitations expressed by the above equations. Improving engineering water use efficiency relates to maximizing ET from Q. Improving agronomic water use efficiency ($Y\ ET^{-1}$) relates to maximizing the yield partitioning (harvest index, yield structural components, etc.), minimizing the transpiration ratio ($T\ A^{-1}$), and maximizing the partitioning of transpiration from evapotranspiration (Viets, 1962; Tanner & Sinclair, 1983; Cooper et al., 1987). This section will discuss this analytical framework within the current level of understanding for the interactions of the specific production vectors.

A. Assimilation-Transpiration Relationships

The mean rate of leaf transpiration is given as

$$T = (W_s - W_a) (\Sigma r)^{-1} \tag{6}$$

where T is the transpiration rate in kg (H_2O) m^{-2} s^{-1}, W_a is the atmospheric water vapor concentration in kg (H_2O) m^{-3}, W_s is the substomatal water vapor concentration inside the leaf in kg (H_2O) m^{-3}, and Σr is the sum of all the water vapor diffusion resistances in s m^{-1} from the atmosphere to the substomatal cavity (normally, the resistance terms considered are r_a, atmospheric boundary layer diffusion resistance; and r_s, leaf stomatal diffusion resistance). The mean rate of leaf assimilation (net or apparent photosynthesis) is given as

$$A = (C_a - C_s) \, (\Sigma r')^{-1} \tag{7}$$

where A is the CO_2 assimilation rate in kg (CO_2) m^{-2} s^{-1}, C_a is the atmospheric CO_2 concentration in kg m^{-3}, C_s is the substomatal CO_2 concentration (compensation point) in kg m^{-3}, and $\Sigma r'$ is the sum of all the diffusion resistances in s m^{-1} for CO_2 from the atmosphere to the substomatal cavity inside the leaf (normally, the resistance terms considered are r'_a, the atmospheric diffusion resistance; r'_s, leaf diffusion resistance; and r'_m, cell diffusion resistance).

Penman and Schofield (1951) were the first to analytically examine these relationships based mainly on diffusion approaches and using resistances based on "unstressed" or "potential" crop conditions. They realized similarities in the equations (Eq. [6] and [7]) and analyzed the ratio of the two processes—transpiration ratio in terms of carbohydrate, CH_2O, and water—with several assumptions for the resistance terms and proposed the following equation:

$$T \, A^{-1} = 1.18 \, 10^6 \, (e_s^* - e) \, P_b^{-1} \, (\rho - \rho_s)^{-1} \tag{8}$$

where e_s^* is the saturated vapor pressure at the leaf temperature in kPa, e is the atmospheric vapor pressure in kPa, P_b is the atmospheric pressure in kPa, ρ is the atmospheric CO_2 concentration in mg (CO_2) kg^{-1}, and ρ_s is the substomatal CO_2 concentration in mg (CO_2) kg^{-1}. They estimated the transpiration ratio for sugarbeets (*Beta vulgaris* L.) to be 25 when ρ_s was assumed to be 0 and using the mean atmospheric vapor pressure deficit ($e^* - e$, where e^* is the saturated vapor pressure at air temperature in kPa) as 0.667 kPa for summertime conditions in Great Britain. Since their estimate of the transpiration ratio was about seven times too small as compared to measured values, they concluded that the internal CO_2 concentration must be larger than 0. The transpiration ratio equation proposed by Penman and Schofield (1951), in which the transpiration ratio is directly proportional to the vapor pressure deficit, agreed with the experimental studies conducted by Kisselbach (1916) using container studies in greenhouses maintained at several humidities.

De Wit (1958), using similar arguments, postulated that the transpiration ratio should be linearly related to free water evaporation (potential evaporation) at high solar irradiance levels, almost constant at intermediate solar irradiance levels, and increased as solar irradiance declines to lower levels. Figure 14-1 illustrates his conceptual relationships between the transpiration ratio ($T \, A^{-1}$), transpiration (T), and assimilation (A) as influenced by the solar irradiance (R) or potential evaporation (E_o) (Penman, 1948). De Wit proposed that the transpiration ratio would not be greatly affected by temperature, water deficits, or mutual shading.

Following the logic of Penman and Schofield (1951), Bierhuizen and Slatyer (1965) used improved concepts of plant resistances to H_2O vapor flux and net CO_2 flux to estimate the transpiration ratio as

$$T A^{-1} = 6.0 \ 10^5 \ (e_s^* - e) \ (\Sigma r') \ P_b^{-1} \ (\rho - \rho_s)^{-1} \ (\Sigma r)^{-1}. \qquad [9]$$

If the ratio of diffusion resistances for CO_2 to H_2O from the atmosphere to the inside of the leaf is about 2, then Eq. [9] is equivalent to Eq. [8] (Penman and Schofield's). Bierhuizen and Slatyer found that the ratio of these resistances for well-watered cotton (*Gossypium hirsutum* L.) leaves varied from 2 to 8 and depended on both ventilation and irradiance. They reported that Eq. [9] represented a wide range of conditions for different irradiances, ventilation, atmospheric air temperature and humidities, and atmospheric CO_2 concentrations.

These relationships for transpiration ratio indicate that the three main factors are: (i) the vapor pressure gradient from the leaf to the air, (ii) the CO_2 gradient from the atmosphere to the leaf, and (iii) the diffusion resistances for both CO_2 and water. The first factor is mainly an atmospherically controlled variable, although the surface temperature of the leaf will actively respond to atmospheric parameters (e.g., mainly radiation and vapor pressure deficit). The last two factors are clearly related to plant-controlled parameters. These parameters are both genetically determined and environmentally responsive.

Gifford (1974) described the main photosynthetic differences between C_3 and C_4 plants as: (i) in the C_4 photosynthetic pathway, the primary carboxylating enzyme has about twice the affinity for CO_2 as in the C_3 photosynthetic pathway; (ii) C_3 plants have photorespiration (respiration which

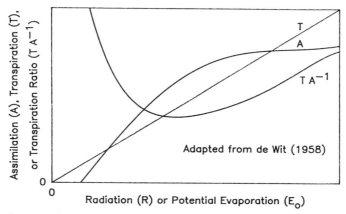

Fig. 14–1. Conceptual relationships between net assimilation (A), transpiration (T), and transpiration ratio (T A^{-1}) for leaves or plants and radiation (R) or potential evaporation (E_o) (de Wit, 1958).

Table 14-1. Transpiration ratio [T A $^{-1}$, kg (H_2O) per kg (CH_2O)] for C_3 and C_4 species plants with regulating and nonregulating stomata at three levels of atmospheric humidity as computed by van Keulen and van Laar (1986) using the model of de Wit (1978) and the ratio of CH_2O to CO_2 as 0.68.

	C_3 species			C_4 species		
	Atmospheric humidity, %					
Stomatal regulation	25	50	75	25	50	75
Regulating	123	103	82	84	66	51
Nonregulating	221	171	132	176	140	106

occurs simultaneously with photosynthesis in the light) which requires O_2, while this process does not occur in C_4 plants; and (iii) leaves of C_4 type plants maintain about one-half the intercellular CO_2 level compared to C_3 plants. Downton (1975) and Raghavendra and Das (1978) provided lists of C_4 photosynthetic pathway species. Since the assimilation rate will generally be larger in C_4 plants due to the higher affinity for CO_2 of the carboxylating enzyme [this is mainly true in higher light intensities (Goudriaan & van Laar, 1978)], the transpiration ratio of C_3 plants will be greater than the transpiration ratio of C_4 plants.

Raschke (1975) proposed that the stomatal control by plants when water was not limiting could be characterized as: (i) regulating when the internal CO_2 concentration is kept within narrow limits and (ii) nonregulating when the internal CO_2 concentration is not controlled by the plant. Goudriaan and van Laar (1978) demonstrated examples of both situations. Van Keulen and van Laar (1986), using the model described in de Wit (1978), computed values of the transpiration ratio for C_3 and C_4 crops of both stomatal regulation types (R—regulating and NR—nonregulating) for three atmospheric humidity levels (Table 14-1). These estimates demonstrate the range of transpiration ratio values that might be found and the complex nature of the relationship between assimilation and transpiration.

The close coupling between photosynthesis and transpiration is obvious since CO_2 and H_2O simultaneously move through the stomata. The diffusive conductance of the stomata opening imposes a major control on the rates of both processes, although the internal CO_2 concentration and the external H_2O vapor concentration determine the magnitude of the respective gradients. However, changes in stomatal resistance may not necessarily affect transpiration and assimilation similarly (Cowan & Troughton, 1971).

Cowan (1977) and Cowan and Farquhar (1977) proposed that plants dynamically adjust their stomatal resistance to maintain an optimum balance between assimilation and transpiration. For a given daily transpiration rate, the resistance adjusts to provide the maximum daily assimilation, and for a given daily assimilation rate, the daily transpiration rate is minimized. This concept was clarified in subsequent work (Cowan, 1982; Schulze & Hall, 1982) and defined such that

$$\partial T \, (\partial A)^{-1} = \lambda \qquad [10]$$

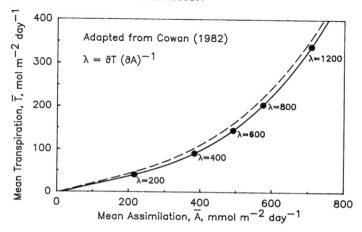

Fig. 14–2. Hypothetical combinations of the average rate of assimilation $(\overline{A})$ and transpiration $(\overline{T})$ per unit leaf area per unit time of a plant during a day. Each point on the broken curve corresponds to a particular constant stomatal resistance. Each point on the full curve corresponds to a particular variation of stomatal resistance that is optimal, in the sense that no other variation could lead to a smaller $\overline{T}$ at the same time $\overline{A}$, or a larger $\overline{A}$ at the same $\overline{T}$ (Cowan, 1982).

where λ is a constant Lagrange multiplier. Figure 14–2 illustrates this concept for daily assimilation $(\overline{A})$ and daily transpiration $(\overline{T})$ from Cowan (1982). This figure shows the daily temporally and spatially averaged (spatial averaging is over the foliage surface) values of CO_2 assimilation $(\overline{A})$ and transpiration $(\overline{T})$ where the particular day represents a single point on the curve. Cowan (1982) stated that "each point on the curve (Fig. 14–2) that bounds this region represents a unique variation of stomatal aperture that could not have been bettered—in the sense that no other variation could have led to a smaller $\overline{T}$ at the same $\overline{A}$, or a larger $\overline{A}$ at the same $\overline{T}$." This optimal stomatal control theory has been verified from experimental results by Farquhar et al. (1980) over a range in ambient temperatures and humidities, by Hall and Schulze (1980) in the laboratory, and by Field et al. (1982) with field data. However, Cowan (1986) stated that "the paradigm of optimality can be no more than an approximation to the truth." Farquhar and Sharkey (1982) discussed that this theory requires both feedback control for the internal CO_2 concentration and feedforward control for humidity and radiation influences that require a close correspondence between assimilation and conductance (Wong et al., 1979).

The relationship between assimilation and transpiration is well founded because of the coupling between CO_2 influx and water efflux from the leaf. Stanhill (1986) reviewed many methods to decouple transpiration and photosynthesis in order to decrease the transpiration ratio. He discussed several means to reduce transpiration while maintaining photosynthesis, e.g., increased cuticular and boundary layer resistances, chemical antitranspirants, plant breeding (Crassulacean acid metabolism [CAM] pathways), selective spectral modification of radiation, as well as means to increase dry matter production while not increasing transpiration, e.g., plant breeding (C_4 and CAM metabolism pathways compared to C_3 pathways). Clearly, the transpiration ratio is associated with the plant species, the plant and atmospheric diffusion resistances for both CO_2 and water, and various environmental parameters (most notable are solar radiation, air temperature, and vapor pressure deficit).

B. Dry Matter-Transpiration Relationships

The interpretation of the relation between dry matter production and assimilation is difficult since aboveground dry matter is usually measured and not total dry matter. This inconsistency can lead to some incorrect conclusions and requires close attention to how various components of water use efficiency are expressed (Sinclair et al., 1984). Fischer and Turner (1978) summarized many reports of the similarities between the transpiration ratio within C metabolism pathways. They reported transpiration ratios of 667, 303, and 50 kg of water per kilogram of dry matter for C_3, C_4, and CAM plant species, respectively. Stanhill (1986) grouped the transpiration ratios from the container studies from Akron, CO (Shantz & Piemeisel, 1927), by CO_2 metabolism groups. He reported that the 51 C_3 species had transpiration ratios of 640 $\pm$ 165 kg of water per kilogram dry matter, and the 14 C_4 species had transpiration ratios of 320 $\pm$ 43 kg of water per kilogram dry matter.

Briggs and Shantz (1913a) investigated the effect of the environmental conditions on durum wheat (*Triticum turgidum* L. var. durum) and sorghum [*Sorghum bicolor* (L.) Moench] dry matter production and transpiration in Akron and Dalhart, TX. They reported that transpiration ratio for wheat was approximately proportional to pan evaporation for the growing seasons for Akron and Dalhart but that the transpiration ratio for sorghum was nearly constant (1–5% increase at Dalhart compared to Akron for 2 yr) at both locations in spite of a 10 to 14% higher pan evaporation at Dalhart compared to Akron for the 2 yr (1910 and 1911).

De Wit (1958) reanalyzed the early container experiments and found that crop dry matter production was linearly related to the ratio of transpiration to pan evaporation (a sunken pan 1.83 m diam., 0.61 m deep, with the water level maintained at soil level, later called a "BPI [Bureau of Plant Industry] sunken pan") for climates with bright growing season sunshine. De Wit expressed the relationship as

$$P = m_e \, T \, E_e^{-1} \qquad [11]$$

where P is the crop dry matter production in kg container^{-1}, m_e is a crop specific proportionality coefficient in mm d^{-1} (the subscript e on m and E refer to the data recorded within screened plot enclosures), T is the transpiration in kg container^{-1}, and E_e is the pan evaporation in mm d^{-1} averaged over the growing period. De Wit computed values of m for sorghum, wheat (*Triticum durum* Desf.), and alfalfa (*Medicago sativa* L.) from data of Briggs and Shantz (1913a, 1914), Shantz and Piemiesel (1927), and Dillman (1931) which were 0.0252, 0.0139, and 0.00662 mm d^{-1}, respectively. Figure 14-3 illustrates the relationships for these three crops between the container production and the ratio of the container transpiration to BPI pan evaporation as reported by de Wit (1958). De Wit reported that the standard errors of the lines through the origin were 0.025, 0.015, and 0.020 kg for the sorghum, wheat, and alfalfa, respectively. He concluded that the relationship (Eq. [11]) was accurate except when the production was small. For conditions outside of the screened plots, de Wit used a relation between transpiration inside and outside the screened plots and a relation between BPI pan evaporation and E_o to adjust the m values to 0.0207, 0.0115, and 0.0055 mm d^{-1} for the sorghum, wheat, and alfalfa, respectively. He found that similar cultivars within several species from experiments and environments had similar values of m; and he concluded that for a first approximation, one could assume that m was a constant, depending only on the crop species. However, de Wit found that in several experiments in the Netherlands (he noted that the precision of these experiments for this purpose was not as good as those conducted earlier in the USA), production was more accurately estimated by the equation

$$P = n T \qquad [12]$$

where n is a crop specific coefficient of proportionality in units of kg kg^{-1}. He determined that the value of n from the experiments in the more temperate climates was indeed different from $m E_o^{-1}$. De Wit analyzed several fertility experiments conducted in containers and determined that m and n were reduced when production was seriously limited by nutrient availability. He proposed that m and n should be independent of nutrient status if the production was mainly limited by other factors. De Wit found that the value of n was consistent in many experiments in which the crops were allowed to

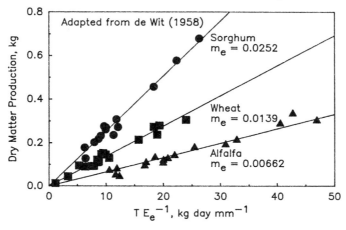

Fig. 14-3. Relationships between production and the ratio of transpiration to pan evaporation for sorghum, wheat, and alfalfa (de Wit, 1958).

deplete various amounts of soil water so long as the containers were not overly irrigated (aeration problems) or greatly underirrigated (greatly stressed). He also extrapolated his methods to analyze the relationship between production and irrigation water application, albeit de Wit acknowledged the many pitfalls of this step.

Arkley (1963) postulated that crop growth and transpiration were related but that advection would distort the relationship. He reanalyzed the data summarized largely by de Wit (1958) and determined that crop production, when fertility was constant or adequate as estimated by Eq. [11] and [12], could be unified with the equation

$$P = k_a \, T \, (100 - H)^{-1} \qquad [13]$$

where k_a is a crop specific coefficient in units of percentage, and H is the mean daily relative atmospheric humidity in percentage during the growing season. Equation [13], thus, effectively provided a means to use the relationships presented by de Wit (1958) in various climatic conditions (mainly advection differences). Arkley estimated that daytime atmospheric relative humidity values should be more meaningful in Eq. [13] than daily averaged atmospheric relative humidity values. He also investigated the relationship between crop dry matter production and the ratio of transpiration to vapor pressure deficit arranged in the following form:

$$P = k_a' \, T \, (e^* - e)^{-1} \qquad [14]$$

where k_a' is the crop specific coefficient in units of kPa. Since $H = 100 \, e \, (e^*)^{-1}$, the crop coefficients are related by the following:

$$k_a' = 0.01 \, e^* \, k_a. \qquad [15]$$

Since e^* is temperature dependent, the crop specific coefficient, k_a', would be temperature dependent also. Arkley recommended that Eq. [13] contained the necessary temperature dependency within the H term. This contradicted the theory proposed by Penman and Schofield (1951) that potential assimilation was inversely proportional to atmospheric vapor pressure deficit ($e^* - e$).

Hanks et al. (1969) reported that dry matter was linearly related to evapotranspiration for wheat (*T. aestivum* L.), millet (*Panicum miliaceum* L.), oat (*Avena sativa* L.), and grain sorghum at Akron, CO, in both lysimeters and field plots. They estimated soil evaporation by several techniques and subtracted it from total evapotranspiration to obtain an estimate of transpiration. They reported m values (Eq. [11]) of 125 kg ha^{-1} d^{-1}) for winter wheat [note that these units are derived for P in units of kg ha^{-1}, T in units of mm, and E_o in units of mm d^{-1}], 94 to 223 kg ha^{-1} d^{-1} for oat, 132 to 167 kg ha^{-1} d^{-1} for millet, and 141 kg ha^{-1} d^{-1} for grain sorghum. Hanks (1974) concluded that for studying only the effects of limited water on crop production, the de Wit (1958) or Arkley (1963) relationships

(Eq. [11] and [13]) could be simplified since m and E_o would be constant for a given crop in a given year to the following:

$$P P_m^{-1} = T T_m^{-1} \qquad [16]$$

where P_m is maximum or potential dry matter production in kg ha^{-1}, T is transpiration for the growing season in mm, and T_m is the maximum or potential transpiration when soil water does not limit transpiration or yield. Hanks (1974) demonstrated validation of Eq. [16] for yield prediction with model estimates of transpiration for corn (*Zea mays* L.) and grain sorghum.

Tanner and Sinclair (1983) researched the relationship developed by Bierhuizen and Slatyer (1965) (Eq. [9]) in order to determine if current simplified relationships for transpiration and assimilation would lead to an expression for the transpiration ratio of crops that would be consistent with observed differences in the transpiration ratio among species. They developed an equation for transpiration ratio from a crop which is

$$T P^{-1} = 1.5 \ 10^4 \ \rho_a \ \epsilon \ L_T \ B' \ (e^* - e) \ (a \ b \ c \ P_b \ C_a \ \text{LAI}_D)^{-1} \qquad [17]$$

where T is in mm d^{-1}, P is in kg ha^{-1} d^{-1}, a is the molecular weight of hexose to carbon dioxide ($CH_2O \ CO_2^{-1}$, 0.68), b is a factor for the conversion of CH_2O to biomass which ranges from 0.33 to 0.83 (Penning de Vries, 1975) c is the CO_2 factor $[(\rho - \rho_s) \ \rho^{-1}$, where ρ_s is the intercellular CO_2 concentration in the leaf in mg kg^{-1} and ρ is the atmospheric CO_2 concentration in mg kg$^{-1}]$ which is approximately constant for a crop with values of 0.3 for C_3 crops and 0.7 for C_4 crops (Wong et al., 1979), C_a is the atmospheric CO_2 density in kg m^{-3}, LAI_D is the leaf area index of leaves directly exposed to incident radiation, ρ_a is the air density in kg m^{-3}, ϵ is the molecular weight of water vapor to air (0.622), L_T is the effective transpiration leaf area index, and B' is a correction term for the shaded and nonshaded leaf area which is approximately 1 ± 0.2 when LAI > 3. They proposed that dry matter production could be estimated by

$$P = 1.0 \ 10^4 \int [k_d \ T \ (e^* - e)^{-1}] \ dt \qquad [18]$$

where dt is days and k_d is a crop-specific coefficient determined as

$$k_d = (0.667 \ a \ b \ c \ P_b \ C_a \ \text{LAI}_D) \ (\rho_a \ \epsilon \ L_T \ B')^{-1} \qquad [19]$$

where k_d is in kPa. They reported consistent agreements between the predicted value of k_d and experimentally derived values (Table 14–2) from several experiments where the necessary data were measured, except for potato (*Solanum tuberosum* L.). Tanner (1981) discussed this difference in detail, but did not find the source for the difference. Tanner and Sinclair (1983) emphasized the following points in their review of the use of Eq. [18]:

1. Since many of the factors in Eq. [18] and [19] are correlated, great care is necessary in applying the equation to experimental data. In particu-

Table 14–2. Comparison of computed estimates for k_d by Eq. [19] and experimental measurements (from Tanner & Sinclair, 1983).

Crop	Location	Source	Experimental k_d	Computed k_d
			———— kPa ————	
Corn	Davis, GA	Stewart et al., 1977	0.0100	0.0118
Sorghum	Manhattan, KS	Teare et al., 1973	0.0138	0.0118
Potato	Madison, WI	Tanner, 1981	0.0065	0.0055
Potato	Netherlands	Rijtema & Endrodi, 1970	0.0015	0.0055
Alfalfa	Madison, WI	Tanner & Sinclair, 1983	0.0043	0.0050
Soybean	Manhattan, KS	Teare et al., 1973	0.0040	0.0041

lar, they emphasized that $\int T[B'\,(e^* - e)^{-1}]$ dt would not equal T_T $[\overline{B'}$ $(e^* - e)^{-1}]$ where T_T is the season total transpiration in mm and equal to $\int T$ dt.

2. The partitioning of sunlit and shaded leaves is necessary to estimate the effects of the environment on leaf temperature to improve the equation. Leaf temperature in low air temperature environments might be warmer than air, while sunlit leaf temperatures in warmer temperature environments might be cooler than the surrounding air.

3. Since the daytime values of transpiration and assimilation should be the most important, the daytime vapor pressure deficit would be more sensitive than the daily mean vapor pressure deficit.

4. Improvements in the transpiration ratio through breeding must result in modifications to the c factor in Eq. [17] where $c = [1 - (\rho_s\,\rho^{-1})]$. They cited as support for their position the similarity of k_d values for corn from 1912 to 1975, although they recognized that partitioning of dry matter into grain has improved through breeding.

Although Tanner and Sinclair (1983) indicated that the potential to increase k_d through breeding was limited, Farquhar and Richards (1984), demonstrated a screening technique that showed differences in ρ_s among wheat genotypes. Richards (1987) reviewed the potential to use the differences in the transpiration ratio (or k_d) in breeding programs.

The crop dry matter production relationship to transpiration is quantitatively similar to the relationships between assimilation and transpiration. The main factors affecting the relationship are the CO_2 metabolism pathway and environmental factors (e.g., vapor pressure deficit, potential evaporation, and air and leaf temperature). The crop and atmospheric diffusion resistances that affect both photosynthesis and transpiration are also important but are more difficult to quantify at the crop level.

C. Economic Yield-Evapotranspiration Relationship

The previous discussions have considered only assimilation and dry matter production. Economic production is normally only a portion of the total dry matter production of a crop. In many cases, the quality of this portion of the crop production significantly affects its economic value. Transpiration, likewise, is a portion of the total water supply provided to produce a specific crop. Additionally, transpiration is practically impossible to measure on a field level, and even evapotranspiration is sometimes difficult to measure. Therefore, many times economic evaluation of irrigation systems or irrigation management is made on the basis of economic yield and applied water. The ratio of crop yield (economic yield) to applied irrigation water has often been termed water use efficiency, but this term is confusingly applied and does not always correctly express how applied irrigation water impacts crop productivity.

1. Economic Yield-Dry Matter Production Relationships

The economic product of a crop can be the dry matter as in forages; but, more likely, it is either the grain, fiber, seed, fruit, root, tuber, or some other plant component. Since the economic product is included in the total plant dry matter, it is logical to quantify the partitioning of the economic production in terms of the total crop or plant dry matter. Many experiments are not concerned with the dry matter production, only the economic production; therefore, the partitioning between economic yield and total dry matter yield is not often reported. Even when dry matter production is determined, the root component of the dry matter yield is rarely measured. The ratio of economic yield to aboveground dry matter yield is termed the *harvest index* (Donald & Hamblin, 1976) and is useful in characterizing a wide range of agronomic experiments. The harvest index is defined as

$$H_i = Y (P_s + Y)^{-1} \qquad [20]$$

where H_i is the harvest index (dimensionless), Y is the economic yield (dry basis) in kg ha^{-1}, and P_s is the stover yield in kg ha^{-1}. Defining aboveground dry matter (P_a) as the sum of P_s and Y and the total crop dry matter production (P_t) as sum of P_a and the root dry matter production (P_r), relationships between economic yield and dry matter yield can be expressed as follows:

$$Y = H_i P_a \qquad [21a]$$

and

$$Y = H_i (P_t - P_r) \qquad [21b]$$

where P_a is aboveground dry matter in kg ha^{-1}, P_t is total dry matter production in kg ha^{-1} ($P_t = P_s + Y + P_r$), and P_r is root dry matter production in kg ha^{-1}. Since P_a includes Y, it is evident that Eq. [21a] should be indicative of the high degree of self correlation that must exist between Y and P_a.

De Wit (1958) indicated that the slope of the linear relationship between grain yield and aboveground dry matter yield of Kubanka wheat (*T. turgidum* L. var. *durum*) from several container studies in the USA was 0.36 when the relationship was forced through the origin. De Wit (1958) also reported that the relationship between grain yield of Kubanka wheat and aboveground dry matter yield from field studies of the early USDA dryland research sites in the Great Plains of the USA (Cole & Mathews, 1923) was linear with a slope of 0.42 and passed through the origin, assuming that 15% of the dry matter was left in the field as stubble. The linear regression between the reported grain yield (assuming 0.773 kg L^{-1} for wheat) and total reported dry matter yield (total of grain and straw yields) from Cole and Mathews' data, as shown in Fig. 14-4, was

$$Y = 0.404 \ (P_s + Y) - 0.178 \ (r^2 = 0.876, \ N = 83) \qquad [22]$$

where Y and P_s are expressed in units of Mg ha^{-1}. The intercept (-0.178 Mg ha^{-1}) was significantly different from zero ($P < 0.05$) and the slope was 0.382 when the relationship was forced through the origin. This relationship is remarkably similar to one determined by Aase and Siddoway (1981) for spring wheat where the intercept was -0.298 and the slope was 0.478 ($r^2 = 0.946$) based on data from 20 different experiments. The following statement by Cole and Mathews (1923) seems appropriate to describe this relationship:

> . . . when affected by drought the wheat crop seems to spend its last energy in producing grain, and that if there is any chance at all, it will produce some yield of grain. This study indicates that a high yield of straw means a high yield of grain. There have been a few cases when exceptionally favorable weather enables wheat to fill so well that the yield of grain was out of proportion to the yield of straw. These years are very infrequent, and as a whole the yield of grain and straw are nearly proportional.

Speath et al. (1984) reported that the harvest index of soybean [*Glycine max* L. (Merr.)] was a conservative characteristic within specific cultivars. Snyder and Carlson (1984) reviewed the harvest index in relation to improved

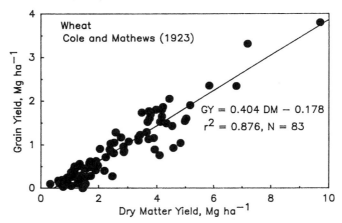

Fig. 14-4. Relationship between grain yield and aboveground dry matter yield of Kubanka wheat from dryland field studies in the Great Plains of the USA with data from Cole and Mathews (1923).

economic yields of crops through plant breeding and also discussed both environmental and biological factors that might affect the harvest index.

Slabbers et al. (1979) investigated the relationship between grain yield (economic yield) and dry matter yields for grain sorghum and corn. They reported the following linear regression equations:

$$\text{Sorghum:} \quad Y = 0.58 \, (P_s + Y) - 1.26 \quad (r^2 = 0.941) \qquad [23]$$

$$\text{Corn:} \quad Y = 0.49 \, (P_s + Y) - 1.21 \quad (r_2 = 0.865) \qquad [24]$$

where Y and P_s are in units of Mg ha^{-1} (note that it was not explicitly stated whether the grain was dry or at standard water content). These equations accounted for 94 and 86% of the variation in yield of sorghum and corn, respectively, when tested against independent data. Figure 14–5 illustrates the relationship described by Eq. [23] for sorghum and the harvest index [$Y (P_s + Y)^{-1}$]. Figure 14–5 demonstrates the importance of the intercept in the relationship between economic yield and aboveground dry matter yield of crops. The economic yield would be better estimated by an equation that accounts for the dry matter yield threshold. At the higher levels of dry matter yields, the harvest index become conservative, as shown in Fig. 14–5.

The relationship between economic yield and dry matter yield has been widely used in many procedures to estimate the effects of crop water use on crop economic yield (e.g., Slabbers et al., 1979; Doorenbos & Kassam, 1979; van Keulen & Wolf, 1986). Generally, the relationship between economic yield and dry matter yield is based on a concept utilizing the harvest index as a constant (Eq. [21a]). The adjusted harvest index (H_{ia}) defined as

$$H_{ia} = Y (P_s + Y - P_r - P_o)^{-1} \qquad [25]$$

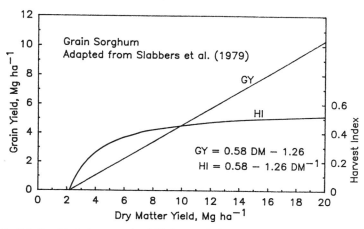

Fig. 14–5. Relationship between grain yield, harvest index, and aboveground dry matter yield of grain sorghum based on the regression equation presented by Slabbers et al. (1979).

where P_o is the *dry matter yield threshold* (amount of total stover production necessary to produce the first increment of economic yield) could improve the accuracy of the relationship between economic yield and dry matter yield. If P_r is neglected and the above examples are used, P_o for wheat, sorghum, and corn could be estimated as 442, 2178, and 2465 kg ha^{-1}, respectively, with the resulting values of 0.40, 0.58, and 0.49 for the adjusted harvest indices, respectively. The relationship between economic yield and dry matter yield might be more accurately determined if the partitioning components of economic yield, such as seed number, seed mass, etc., could be estimated.

2. Transpiration-Evapotranspiration Relationships

The measurement of transpiration in the field is complex (Klocke et al., 1985) and subject to many errors. Even the field measurement of evapotranspiration can be rather complex and difficult in many situations where the drainage from the root zone, water uptake from saturated zones, and runon and runoff from the area are difficult to measure, both temporally and spatially. Generally, soil water balance techniques are used to measure seasonal evapotranspiration from crops. Precise field soil water balance measurements are usually possible only when using lysimeters to precisely define the water movement across the lower soil boundary. Several micrometeorological techniques can be used to measure the energy balance and evapotranspiration from crops as described in other chapters in this Monograph.

Most often, the transpiration is estimated from evapotranspiration measurements using (i) subtraction of an estimate of E_s (usually, E_s is assumed to be the intercept of the P-ET linear regression), which is most often taken to be a seasonal constant from the measured seasonal ET (Hanks et al., 1969); (ii) daily water balance simulation using empirical functions to separately calculate T from daily calculations (or measurements) of ET using measured plant parameters such as leaf area index or ground cover (Ritchie, 1972; Tanner, 1981; Howell et al., 1984; Hanks, 1985); or (iii) measuring E_s and subtracting it from measurements of ET (Lascano et al., 1987). All of these measurement techniques yield indirect estimates of transpiration. Direct plant measurements of water movement rates have been made using the heat-pulse velocity technique (Bloodworth et al., 1955), but the estimation of transpiration flux remains difficult because of volume calibration difficulties as well as sampling limitations. However, newer techniques and improvements in heat-pulse instrumentation appear to greatly solve calibration problems (Sakuratani, 1984) or even eliminate them (Baker & van Bavel, 1987).

The relationship first proposed by Ritchie and Burnett (1971) or variations (Tanner & Jury, 1976; Kanemasu et al., 1976; Al-Khafaf et al., 1978) have been widely used to estimate field transpiration of "unstressed" crops. The Ritchie and Burnett relationship developed for cotton and grain sorghum is

$$T = E_o \left[-0.21 + 0.70 \, (LAI)^{1/2} \right] \qquad [26]$$

where T is transpiration in mm d^{-1}, LAI is the leaf area index, and E_o is "potential" evapotranspiration in mm d^{-1}. It is interesting to note that the data from Ritchie and Burnett (1971) illustrate that the ratio of transpiration to "potential" evapotranspiration ($T E_o^{-1}$) was more closely related to aboveground dry matter than to either leaf area index or ground cover for cotton and grain sorghum and two row spacings of grain sorghum. The data for $T E_o^{-1}$ and LAI for Ritchie and Burnett (1971) can be closely approximated by the simpler equation

$$T = E_o \,[1 - \text{EXP}(-0.8 \text{ LAI})] \qquad [27]$$

where EXP represents the exponential function $[\text{EXP}(x) = e^x]$ with little loss in accuracy. Ritchie (1983) discussed the bias in Eq. [26] and [27] due to the advective influences that enhance transpiration with dry soil surface conditions. He presented a curve for wet soil surface conditions that would suggest that the exponential coefficient in Eq. [27] (-0.8) would be reduced to about -0.38 to -0.40 when the soil surface was wet.

The effects of reduced soil water contents (or, in effect, soil water potential) on transpiration are more difficult to precisely estimate. Transpiration under soil water deficits is strongly influenced by the crop rooting depths, rooting densities, soil hydraulic properties, and the evaporative demand. Campbell and Campbell (1982) illustrated the influences of rooting density, soil water potential, and evaporative demand on crop water uptake from the soil by using the Ohms-law electrical analogy to simulate water flow from the soil through the plant to the atmosphere.

The relationship between transpiration and evapotranspiration is not clearly defined in most cases. Various model forms have been used to estimate transpiration from "potential" or "maximum" evapotranspiration estimates. These relationships are often site as well as crop specific.

3. Economic Yield-Evapotranspiration Relationships

All of the previous discussion illustrate the estimation of dry matter by using transpiration as the independent variable. The estimation of transpiration from the total evapotranspiration is difficult (Hanks & Rasmussen, 1982). Since total evapotranspiration (ET) is the process most closely related to transpiration that can be measured in the field, many approaches have been based on the economic yield relationship to ET.

Cole and Mathews (1923) and Mathews and Brown (1938) investigated grain yield for winter wheat and sorghum across the southern Great Plains in the USA in relation to precipitation, the practice of fallowing, and effects of growing conditions (soils and locations). They used linear regression techniques to evaluate the function

$$Y = b \text{ ET} + a \qquad [28]$$

where Y is grain yield in kg ha^{-1}, ET is the estimated growing season evapotranspiration in mm, a and b are regression coefficients in units of kg ha^{-1} and kg ha^{-1} mm^{-1}, respectively. They estimated ET as growing season precipitation plus soil water depletion from seeding until harvest. They found a to be negative as a result of soil water evaporation (Hanks, 1974) (note that the soil evaporation can be approximated by the ratio $a\,b^{-1}$). The equation determined by Mathews and Brown (1938) for winter wheat at Garden City and Colby, KS, was

$$Y = 5.19 \text{ ET} - 972 \qquad (r^2 = 0.561, N = 81) \tag{29}$$

where Y is in kg ha^{-1} and ET is in mm. The average error in wheat yield estimation over 20 yr at Colby, KS, was 98 kg ha^{-1}. They tested their model with data from three other USDA dryland stations in Texas and Oklahoma. The results indicated that the model explained slightly $>50\%$ of the variance in the yield data from the additional three sites. In addition, the model was not biased in that the intercept (-117 kg ha^{-1}) was not different from zero ($P < 0.05$) and the slope was not different from 1.0 ($P < 0.05$). However, the standard error of the model was 430 kg ha^{-1} when tested against estimated yields at these three sites. By current standards, this model would seem to be quite applicable for the intended purpose of estimating dryland winter wheat production in the southern Great Plains, although the model is rather site specific. The slope from the Mathews and Brown (1938) regression equation (5.19 kg ha^{-1} mm^{-1}) compares well to 6.38 kg ha^{-1} mm^{-1} from a later dryland wheat study at Bushland, TX (Johnson & Davis, 1980). These empirical models of crop production-evapotranspiration are widely used for many agronomic, engineering, and/or economic purposes but are widely criticized for site specificity, effects of specific periods of water stress effects, the lack of climatic influences, and empiricisms that do not increase the understanding of the fundamental relationships between production and water use. Many debates have occurred regarding whether the empirical relationship between economic yield of a crop and crop ET was linear, quadratic, or some other function (Barrett & Skogerboe, 1980).

Many crop production-evapotranspiration models have evolved to predict economic crop yield. Various techniques were used to address the crop yield response in relation to ET and to ET deficits in specific crop growth stages. Jensen (1968) proposed two models of crop yield in relation to ET during specific crop growth stages: (i) for determinate crops and (ii) for indeterminate crops. His model for determinate crops is

$$Y\,Y_{\text{m}}^{-1} = \prod_{i=1}^{n} [\text{ET ET}_{\text{m}}^{-1}]_i^{\lambda_i} \tag{30}$$

where Y_{m} is maximum or potential grain yield in kg ha^{-1} with water not limiting production, ET_{m} is the crop water use in mm with water not limiting production, λ_i is the relative sensitivity factor (dimensionless) of the crop to water deficits in growth stage i, and n is the number of growth stages. The right side of Eq. [30] is a product. Jensen gave λ values of 0.5, 1.5, and

0.5 for three periods of grain sorghum as emergence to boot, boot to milk stage, and milk to harvest, respectively. His indeterminate crop yield model was of the form

$$Y \, Y_m^{-1} = [\sum_{i=1}^{n} \lambda_i \, (ET)_i] \, [\sum_{i=1}^{n} \lambda_i \, (ET_m)_i]^{-1} \qquad [31]$$

Jensen stated that the primary difference between Eq. [30] and [31] was that for indeterminate crops, the effects of water stress on yield during specific growth stages are independent of other growth stages. Stewart et al. (1977) and Doorenbos and Kassam (1979) proposed to estimate crop yield in relation to ET as

$$Y \, Y_m^{-1} = \{ 1 - B[1 - (ET)(ET_m)^{-1}] \} \qquad [32]$$

where B is the yield response factor (dimensionless). Hanks and Rasmussen (1982) determined that soil evaporation and maximum transpiration could be estimated from Eq. [32] as follows:

$$E = ET_m \, (1 - B^{-1}) \qquad [33]$$

$$T_m = ET_m \, B^{-1} \qquad [34]$$

where E is soil water evaporation in mm, and T_m is maximum transpiration in mm. The value of B should be >1 if the intercept of the yield and ET line (or curve) is negative. Hanks (1983) discussed that the B values <1 reported by Doorenbos and Kassam (1979) and Stewart et al. (1977), probably resulted from use of limited data and incorrect estimates of ET_m and Y_m. Hanks also proposed that the values of B and m (de Wit, 1958) were related as

$$m = Y_m \, E_o \, B \, ET_m^{-1} \qquad [35]$$

where E_o is the mean daily growing season potential ET in mm d^{-1}. Hanks and Rasmussen (1982) and Hanks (1983) reviewed additional models that have been used to relate crop yields to water use which, in general, are basically some variation or combination of the above models. Although the discussion is limited in the literature, the yield response of the crop is considered to be constant at ET $\geq$ ET$_m$ (meaning that soil evaporation is larger than the minimum E necessary to produce Y_m).

   To summarize this section, an example analysis might be enlightening and useful in illustrating these concepts. A winter wheat study (Jensen & Sletten, 1965) that has been used in the literature for this purpose (Ritchie, 1983) was chosen. The study consisted of six water treatments and six fertility treatments over 3 yr. Dry matter yields were not reported, but straw-to-grain ratios were reported for selected treatments in 1955–56. Stover yield for the 1955–56 data was computed as the product of the straw-to-grain ratio times the reported grain yields. Total aboveground dry matter production for the 1955–56 data was computed as the sum of the stover and grain yields. The source of the data did not report the water contents of the grain or the straw, but the combined water contents are likely <8% (wet basis) and no correction was applied for water in the estimated dry matter or grain. The ET was measured by soil moisture sampling, and the plots were level borders so no runoff occurred. The reported data from the M_1 through M_6 water treatments and the F_2, F_4, and F_5 fertility treatments (34 kg ha^{-1} P_2O_5 each with 0 kg ha^{-1}, 90 kg ha^{-1}, and 135 kg ha^{-1} of N, respectively) were utilized in this analysis since these were the only treatments with published data for the straw-to-grain ratio. The grain yield was highly correlated to the estimated total aboveground dry matter as illustrated by the regression equation

$$Y = 0.388 \, (P_s + Y) - 0.05 \qquad (r^2 = 0.976, \, N = 18) \qquad [36]$$

where Y and P_s are in Mg ha^{-1}. Both the estimated aboveground dry matter yield $(P_s + Y)$ and grain yield (Y) were highly correlated to the measured ET as expressed by the regression equation

$$(P_s + Y) = 0.0169 \, \text{ET} - 5.00 \qquad (r^2 = 0.870, \, N = 18) \qquad [37]$$

$$Y = 0.00648 \, \text{ET} - 1.96 \qquad (r^2 = 0.846, \, N = 18) \qquad [38]$$

where Y and P_s are in Mg ha^{-1}, and ET is in mm. The relative reduction in both estimated aboveground dry matter yield and grain yield was also highly correlated to the relative reduction in ET as expressed by Eq. [32] with coefficients of determination of 0.878 and 0.711, respectively. Figure 14-6 shows the relationship between the relative decrease in total dry matter yield in relation to the relative ET decrease for the 1955–56 season for the six water and three fertility treatments. Maximum dry matter was 9.10 Mg ha^{-1}, and Y_m was 3.37 Mg ha^{-1}, while ET_m was taken as 864 mm for the 1955–56 sea-

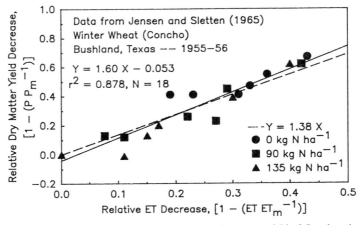

Fig. 14-6. Relationship between the relative decrease in dry matter yield of Concho winter wheat at Bushland, TX, in 1956 (Jensen & Sletten, 1965) and the relative decrease in ET.

son. The slopes of these relationships when forced through the origin to determine B values for Eq. [32] were 1.38 and 1.32 for dry matter and grain, respectively. These B values, along with E_o determined from pan evaporation data reported by Jensen and Sletten (1965) and the values of Y_m and ET_m, were used to compute m values (de Wit, 1958) which were 111 kg ha^{-1} d^{-1} and 39 kg ha^{-1} d^{-1} for dry matter and grain, respectively. The estimated dry matter m value is close to that value reported by de Wit (1958) for wheat of 115, the range of values for wheat of 110 to 140 reported by Fischer and Turner (1978) (note that their m's include root biomass), and 125 for wheat reported by Hanks et al. (1969). The proportionality factor, B, was similar to the seasonal value for wheat of 1.0 to 1.15 reported by Doorenbos and Kassam (1979), although my value is larger. The soil evaporation component of ET_m was approximately 25% based on the value of B, and the transpiration component of ET_m was about 75%. Although these data contain some definite trends that illustrate fertility interactions with irrigation (Ritchie, 1983), in general, the data can be adequately represented by functions similar to Eq. [11] and [32] over a relatively wide range in both fertilizer applications and irrigation water management.

The information regarding economic yield of crops and evapotranspiration can be summarized as illustrated in Fig. 14–7. It should be kept in mind that these functional relationships are only applicable to small plots where (i) the soil is relatively uniform, (ii) all water applications (both rainfall and irrigation) are applied uniformly, (iii) severe water deficits during critical crop growth periods are avoided, (iv) salinity (of either the soil or irrigation water) does not limit production, and (v) fertility and cultural management techniques do not limit production. Figure 14–7 is illustrative of the functions in the above discussion for wheat grown in a variety of locations. Aboveground dry matter yields are linearly related to ET from a point which is about 20 to 25% of ET_m up to the point P_m, ET_m. It would be conceivable that ET could exceed ET_m if the soil surface was kept wet

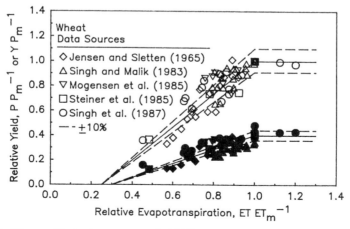

Fig. 14–7. Diagram illustrating concepts of yield-ET relationships for wheat. The open symbols represent dry matter production and the filled symbols represent grain production.

from frequent rains or irrigations that increased soil evaporation. Economic yield (in this example, wheat grain yield) would also increase linearly from a point at approximately 25 to 30% of ET_m up to the point $Y_m P_m^{-1}$, ET_m. Although Fig. 14–7 is considered a gross simplification, it provides a discussion framework and is, in fact, realistic, as demonstrated by the data illustrated in the figure for wheat from Singh and Malik (1983), Steiner et al. (1985), and Mogensen et al. (1985). Similar diagrams could be developed with the minimum inputs of P_m, ET_m, B (or m or k_d), and H_i. Procedures to develop this information are readily found in Doorenbos and Kassam (1979) and van Keulen and Wolf (1986). Such diagrams can provide production envelops that can represent the upper limits to expected production. Specific yields might not equal the estimated production limits because of various limitations such as disease, pests, fertility, critical period water stresses, salinity, nonuniform irrigation applications, etc.

II. EFFECTS OF OTHER ENVIRONMENTAL AND MANAGEMENT FACTORS

The previous section has summarized concepts used to describe the relationship between crop assimilation-transpiration, crop dry matter production-transpiration, crop economic production-dry matter production, crop economic production-evapotranspiration. This section will discuss the effects of other factors, i.e., evaporative demand, fertility, salinity, critical periods of water deficits, soil variability, and irrigation application uniformity.

A. Evaporative Demand Effects

Evaporative demand clearly affects the relationships between assimilation-transpiration, crop dry matter-transpiration, and yield-evapotranspiration. The evaporative demand influences are quantified through the vapor pressure deficit in several of the equations presented in the previous section or the environmental factors that influence ET_m. Basically, the evaporative demand affects the partitioning between soil evaporation and transpiration which depends on the surface soil wetness and the amount of crop development (Ritchie, 1983) and affects the transpiration ratio directly. Clearly, the addition of extra advective energy to drive the evaporative process will not result in increased crop productivity. Crop dry matter production depends on the amount of photosynthetically active radiation (PAR) absorbed by the crop (Monteith, 1977). Canopy light interception depends on many factors, with the canopy architecture and leaf area index being the important crop factors and the distribution of PAR between direct beam and diffuse components being important radiation parameters. As discussed by Ritchie (1983), both absorbed PAR and ET are driven largely by radiation, but ET is also increased by advective influences (vapor pressure deficit and wind) and nearly maximized at leaf area indices approaching 3, while absorbed PAR continues to increase with leaf area indices exceeding 4 to 5.

B. Fertility Effects

Crop productivity is strongly influenced by nutrition and water availability. Viets (1962) investigated these interactions in terms of water use efficiency for crops with unlimited water supplies. When the water supply to a crop is fixed, any management factor that increases production, such as fertilizers, weed control, disease control, planting density, planting geometry, will increase the water use efficiency. The crop production-evapotranspiration relationships presented in the previous section assume that crop nutrition is adequate and nonlimiting to production.

Crop nutrition through fertilizer applications does not greatly affect crop water use unless significant effects on leaf area development are present. Even in the cases where fertilizers increase leaf area index, generally, the time distribution of the effects results in minor changes in actual crop water use unless the nutrient deficiency is very extreme. Nutrient uptake is largely determined by the nutrient demand to meet sink requirements in the crop materials but can be limited by nutrient status in the soil or by water-limited conditions in soil layers where crop nutrients are available and crop rooting is sufficient for nutrient uptake but where root water uptake is limited by the soil water potential. Crop fertility management can have both positive and negative effects on crop productivity when the water supply is fixed and/or limited (Black, 1966; Viets, 1966). Fertilizer applications to a crop with limited available water could result in early depletion of the limited soil water and the development of severe water deficits during later critical crop development stages, possibly reducing yield and water use efficiency. With sufficient available soil water and a nutrient deficient soil, nutrient additions should increase dry matter and economic yield, thereby increasing the water use efficiency. Jones et al. (1986) reported that neither water nor N stress affected the value of k_d for corn. Viets (1972) concluded that nutrient and water uptake were largely independent processes in crop roots and that plants do not need a constant supply of nutrients.

Rhoads (1984) has summarized the literature dealing with water and N responses of crops which indicates that when N was limiting yield, the water use efficiency was improved sometimes as much as 41% when higher rates of N were applied. Figure 14-8 illustrates the interactions of N and water applications on aboveground dry matter and grain yield of corn (Stapleton et al., 1983). Hanks et al. (1983) stated that production surfaces similar to that illustrated in Fig. 14-8 are site specific. The response surfaces shown in Fig. 14-8 illustrate the points: (i) the yield response to irrigation applications will increase with increasing N applications until N no longer limits yield for the amount of irrigation applied; (ii) there is a broad range of N fertilizer applications that result in approximately similar yields (the data in Fig. 14-8 show small dry matter or grain yield increases, regardless of the irrigation level, as N application is increased from 150 to 300 kg ha^{-1}); and (iii) N and irrigation applications affect the relationship between economic yield and dry matter yield similarly (no differential effect on harvest index). Thus, these empirical production surfaces are of little general usefulness because of the site-specific nature of the soil fertilizer interactions, local rain-fall patterns, etc.; but they are useful examples of the interaction of irriga-tion and nutrients.

The relationship between crop yield and nutrient requirement is fully explored by van Keulen (1986a) based on graphical procedures suggested by de Wit (1953). Figure 14–9 illustrates this graphical analysis procedure using an irrigated corn fertilizer uptake study (Stapleton et al., 1983). Both dry

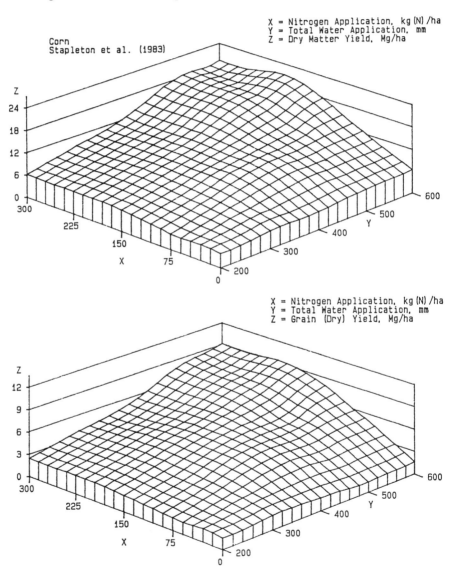

Fig. 14–8. Response surfaces of corn dry matter and grain yield (dry) to irrigation applications and rainfall and to N applications. Data from Stapleton et al. (1983).

matter yield and grain were affected by the applied water and N fertilizer (Fig. 14-8). Part A of Fig. 14-9 shows the corn grain yield as affected by the fertilizer and four irrigation levels, W1 through W4. Part D of Fig. 14-9 shows the efficiency of the N recovery by the corn crop as affected by the different water levels. The high N recovery, even exceeding 100% by W4, may be due to N in the irrigation water that was not measured or otherwise to errors in determining the yields or N concentrations. Nevertheless, increased irrigation, and presumed increased ET, greatly increased the N use by the crop and resulted in improved growth and yields. Interestingly, the water and fertility levels did not affect the relationship between grain yield and aboveground dry matter which was highly linear with a coefficient of determination of 0.982. Although this study did not report the crop water use, it demonstrates the following concepts discussed by van Keulen (1986a): (i) grain yield is approximately proportional to N uptake at lower levels of N uptake, which leads to consistent minimum N concentrations in the plant material; (ii) at higher N uptake levels the yield response is nonlinear, reflecting increasing N concentrations in the economic yield products, resulting in lower N use efficiency but probably greater protein content in the case of grains; (iii) at some point on the N application curve, yield response plateaus, indicating the limitation of some other parameter (water, light, temperature, salinity, other nutrients, etc.); and (iv) the yield-N uptake curve will extend to the point where the plant has reached the point of maximum N concentration in its tissues throughout its life cycle. The intercept of the N uptake-

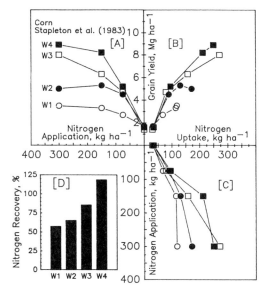

Fig. 14-9. The relation between (A) corn grain yield and water and N applications; (B) grain yield relation to nitrogen uptake; (C) N uptake and N application; and (D) the efficiency of nitrogen recovery as affected by the irrigation applications (W1 dry to W4 wet) for sprinkler-irrigated corn at Kaysville, UT (Stapleton et al., 1983). The graphs were developed using the procedure outlined by van Keulen (1986a).

Table 14-3. Minimum concentrations (g kg^{-1}) for the major plant nutrients in the economic and the crop residue portions of crop yield for several types of crops (van Keulen, 1986b).

Crop type	Economic yield component			Residue yield component		
	N	P	K	N	P	K
				g kg^{-1}			
Grains	10	1.1	3	4	0.5	8
Oil seeds	15.5	4.5	5.5	3.4	0.7	8
Root crops	8	1.3	12	12	1.1	3.3
Tuber crops	4.5	0.5	5	15	1.9	5

N application curve (Part C of Fig. 14-9) indicates the mineralized N available from the soil which is not greatly affected by the different irrigation treatments but would vary from year to year depending on water and temperature levels, crop rotations, and other management factors. The slope of N uptake-N application represents the fertilizer recovery efficiency which depends on the type of fertilizer material, application methods, application timing, environmental factors, etc., and would normally be < 80% (van Keulen, 1986a). These procedures can be extended to other nutrients as well as nutrient interactions. Van Keulen (1986b) summarized the minimum concentrations of major plant nutrients for several crops, as given in Table 14-3. The data from Table 14-3—when combined with estimates of the soil available nutrients contributing to crop uptake, and fertilizer recovery efficiency and potential dry matter yield (or estimated economic yield along with an estimated value for the harvest index)—permit the minimum fertilizer applications to be estimated (Stanford & Legg, 1984) that are necessary to produce the estimated yield level. Much information, including the previous cropping history and organic amendments, is required to precisely estimate the fertilizer requirements of a crop.

C. Salinity Effects

Salinity (soluble salts or specific ions) present in the soil or in the irrigation water solution can significantly affect crop yield as well as the relationship between crop yield and evapotranspiration. Basically, the plant transpires pure water and only pure water evaporates from the soil, leaving the soluble salts within the soil solution. These processes change both the osmotic and matric potentials within the soil profile. Childs and Hanks (1975) demonstrated that these two components of total soil water potential are additive in terms of their effects on crop transpiration and should be additive in terms of their effect on crop production. However, as discussed by Bresler and Hoffman (1986), the interpretation or prediction of the interaction of salinity and irrigation water quantity on crop production is complex, depending on the transient nature of the soil salinity patterns.

Maas and Hoffman (1977) summarized much of the existing literature on salt tolerance of crops in terms of yield response to salinity as shown in

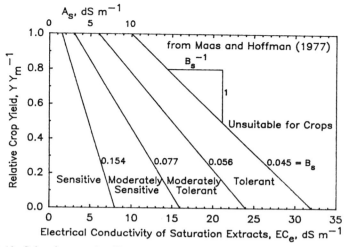

Fig. 14-10. Salt tolerance classification of crops (Maas & Hoffman, 1977).

Fig. 14-10. Recently, Maas (1986) updated the previous summary of data on crop salt tolerance. The salt tolerance of crops is defined as

$$Y\,Y_m^{-1} = 1 - B_s\,(EC_e - A_s) \qquad \text{[for } EC_e \geq A_s] \qquad \text{[39a]}$$

$$Y\,Y_m^{-1} = 1 \qquad\qquad\qquad \text{[for } EC_e \leq A_s] \qquad \text{[39b]}$$

where A_s is the salinity threshold in dS m^{-1}, B_s is the sensitivity of the crop to salinity above the threshold level in m dS^{-1}, and EC_e is the electrical conductivity of saturation extract in dS m^{-1}. The saturated extract electrical conductivity of soils is generally considered to be about one-half of the electrical conductivity of the actual soil solution for mineral soils. These relationships are applicable where Cl is the main ion affecting yield. The main difficulty in using these descriptions of salt tolerance of crops is that the electrical conductivity of the soil is dynamic depending on the salinity of the applied irrigation water and/or the received precipitation (Meiri, 1984), crop water extraction profiles (Raats, 1974; Hoffman & van Genuchten, 1983), the initial salinity profiles, and the soil chemical reactions (particularly for irrigation waters high in sulfates and carbonates) (Rhoades & Merrill, 1976).

Feinerman et al. (1984) developed procedures to compute corn yield in relation to applied water, salinity, and application uniformity using the salt tolerance concepts of Maas and Hoffman (1977) and using both steady-state soil salinity and transient cases based on the model of Bresler (1967). Lety et al. (1985) included effects of plant adjustments to the root zone salinity such that even with limited irrigation applications using saline water, leaching (drainage from the root zone) could occur if plant water use was reduced due to the salinity of the soil water solution. Lety and Dinar (1986) presented yield relationships for several crops based on their previously developed procedures. Solomon (1985) developed procedures similar to Lety et al. (1985) to predict crop production from water and salinity relationships and included two functions for water: (i) for the increasing production side of the curve and (ii) for describing the aeration effects (and/or nutrient leaching) on crop production due to excessive irrigation.

Bresler and Hoffman (1986) demonstrated for a variety of crops that crop yield (both dry matter and economic yield) was related to applied irrigation water and the water salinity by using steady-state and transient models. The major effect of soil salinity was to reduce plant water uptake as determined by the root zone water potential (total of matric and osmotic potentials). They illustrated the difficulties of predicting the dynamic characteristics of leaching using the steady-state model. Bresler (1987) postulated that the transient model could explain both crop yield response to irrigation quantity and water quality as well as specific crop salt tolerances based on the limiting (or lowest possible) value for total plant root potential.

The effects of irrigation water salinity on crop yields are illustrated in Fig. 14-11 based on the concepts developed by Bresler (1987), Bresler and Hoffman (1986), and Letey et al. (1985). With nonsaline irrigation water, the crop yield response is similar to that previously described; as the irrigation water salinity increases to moderate levels, the yield declines almost in proportion to the salinity level (Fig. 14-11). But as the irrigation water salinity continues to increase, the generally linear lines become pronounced curves. Bresler (1987) emphasized that this relationship would not be applicable to all conditions due to aeration or leaching of plant nutrients as irrigation applications became excessive to overcome the irrigation water salinity.

Bresler and Hoffman (1986) analyzed both dry matter production (above ground) and economic yield components from a variety of experiments dealing with salinity. They reported that economic yield was highly correlated to dry matter yield ($r^2 = 0.99$), and thus, the relationship between these yield components was not differentially affected by salinity. Hanks et al. (1978) reported that both grain yield and dry matter yield of corn were linearly

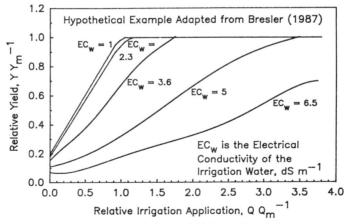

Fig. 14-11. Example estimates of relative crop yield in relation to relative water application and irrigation water salinity. Adapted from Bresler (1987).

related to evapotranspiration over a wide range of water and salinity treatments, implying that grain yield was also linearly related to dry matter yields. Figure 14–12 illustrates this relationship for corn using the data from four locations for 2 yr at each location, with four to six different salinity and irrigation levels applied at each location as well as several corn varieties and several irrigation deficit period schedules (Stewart et al., 1977). The relationship between grain yield (dry) and aboveground dry matter is similar to that proposed by Slabbers et al. (1979) and accounts for more than 84% of the variance in the grain yields (note when the 1975 data shown in Fig. 14–12 for Fort Collins and Yuma were deleted due to an early freeze and extreme heat stress at the respective locations in that year, the coefficient of determination increased to 0.90). Similar results were obtained when the relationships between grain sorghum yield and dry matter yield and seed cotton yield and dry matter were examined using data from Maas et al. (1986) and Russo and Bakker (1987), respectively, from a variety of water and salinity treatments. The cotton yield (seed cotton or lint) in relationship to dry matter yield is slightly nonlinear, however, as verified with data from Davis (1983).

D. Effects of Water Deficits at Critical Crop Growth Periods

Water deficits at critical crop development stages have been reported to adversely affect crop yields (Hagan et al., 1959). The effects of water deficits and/or irrigation additions at specific crop growth stages were summarized by Salter and Goode (1967) for many types of crops (Table 14–4). In general, crop water deficits during floral initiation or anthesis have been reported to have the greatest effects on crop economic or grain yields through reductions in seed or grain numbers, while water deficits after anthesis through grain filling generally reduce seed or grain mass. Doorenbos and Kassam (1979) provided summary information regarding effects of critical periods of water deficits on crop production. One of the major problems

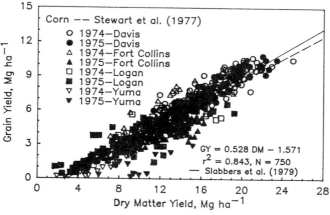

Fig. 14–12. Relationship between corn grain yield (dry) and aboveground dry matter. Data from Stewart et al. (1977).

Table 14-4. Summary of the most critical growth periods for water deficits of selected crops on their production (Salter & Goode, 1967).

Crop	Most critical periods for water deficits
Wheat	Shooting (elongation of internodes)
	Earing (emergence of the ear from the boot)
Corn	Flowering
	Early grain formation
Barley	Shooting
	Earing
Oat	Heading
	Flowering
Rye	Flowering
	Early grain formation
Sorghum	Booting (end of shooting stage just prior to the emergence of the head)
	Heading
Rice	Heading
	Flowering
Cereal summary	Main effects of water deficits at critical periods appear to mainly affect the number of grains with some effects on other yield components like tiller number and grain mass.
Peas	Flowering
	Pod filling
Soybean and other beans	Flowering
	Pod set
Peanuts	Flowering
	Seed development
Annual legume summary	Differing results were reported but generally indicated that flowering and pod development were the most critical periods.
Tomato, pepper, and cucumber	Start of fruit set onwards
Annual fruit crop summary	Basically, all annual fruit crops are sensitive to water deficits at the time that the first fruits start to develop
Cotton	Flowering
	Boll development
Flax	Vegetative growth (fiber production)
	Flowering (seed production)
Safflower	Rosetting
	Flowering
	Seed filling
Sunflower	Heading
	Grain filling
Fiber and seed crop summary	Generally, like the cereals, these crops show critical periods of development near to or at flowering.
Sugar beet	No critical stages (root production)
	Flowering (seed production)
Carrot	No critical stages
Turnip	Seedling (leaf and root production)
	Prior to harvest (root production)
Cabbage	Head formation
Cauliflower and broccoli	All stages (curd production)

(continued on next page)

Table 14-4. Continued.

Crop	Most critical periods for water deficits
Biennial crop summary	Basically, the biennial crops have critical periods near to the time that the storage organ begins to develop.
Potato	Tuber initiation through maturity
Onions	Flowering (seed production)
Tuber and bulb crop summary	Generally, water deficits before tuber initiation reduce the number of tubers, and water deficits during tuber formation reduce tuber size.

involved in identifying critical periods for crops in relation to water deficits is quantifying the degree of imposed crop water deficit (or crop water stress). Many indices of crop water deficit have been used to quantify the deficit (Hiler & Clark, 1971). Several example indices include the following:

$$(ET \ ET_m^{-1})_i \qquad\qquad \text{(Jensen, 1968)} \qquad\qquad [40]$$

$$[1 - (ET \ ET_m^{-1})]_i \qquad\qquad \text{(Hiler \& Clark, 1971)} \qquad\qquad [41]$$

$$\{1 - [1 - (ET \ ET_m^{-1})]^2\}_i \qquad \text{(Minhas et al., 1974)} \qquad\qquad [42]$$

where $(ET)_i$ is crop water use during specific crop growth period i, and $(ET_m)_i$ is the crop water use during period i without any imposed crop water deficits. These indices are used in various forms of production functions (see Eq. [30], [31], [32a], etc.) in which either additive or multiplicative functions are developed (note that generally these functions are applied to grain or economic yields and not to dry matter yields). Singh et al. (1987) reported that multiple-period models did not predict wheat yields any more consistently or accurately than simpler yield-ET models. Wenda and Hanks (1981) reported similar results for corn.

The ET deficit $[(ET_m - ET) \ ET_m^{-1}]$ experienced in a specific crop growth stage will seldom exceed 0.5 unless the soil water deficit is large when the growth stage is initiated. Also, the field measurement of the ET deficit is subject to large potential errors in many cases since it may be small and of short duration. Little experimental evidence has been reported that illustrates differential effects of water deficits at specific growth stages on the transpiration ratio as normalized by E_o or $(e^* - e)$. Asrar et al. (1984) did report that ET P^{-1} for wheat declined following anthesis not in proportion to decreases in $(e^* - e)$, but ET P^{-1} was not affected by planting density for two genotypes. If crop water deficits at specific growth stages differentially affect economic crop yields, the water deficits would have to also affect the relationship between economic yield and dry matter yield (harvest index). This topic has not been widely studied. Figure 14-12 illustrates a summary of several corn irrigation experiments where ET deficits were intentionally created in specific crop growth stages, yet no major effects of the water deficits on the partitioning of grain from the aboveground dry matter production were evident (See also Hanks et al., 1978.), although considerable variation is present in the data illustrated. However, it is apparent that certain environmental parameters can produce significantly different results, like the 1975 data from Fort Collins and Yuma as reportd by Stewart et al. (1977), and that certain cultivars may be different (DeLoughery & Crookston, 1979). Consistent relationships between economic yield and dry matter yield have been determined based on data reported for grain sorghum using specific periods of osmotic stress by Maas et al. (1986) ($r^2 = 0.914$ for two varieties of grain sorghum and three growth stages), for navy beans (*Phaseolus vulgaris* L.) by Gunton and Evenson (1980) ($r^2 = 0.977$ for two growth stages and their combination), for wheat by Singh et al. (1987) ($r^2 = 0.950$ for two locations, Germany and India, with two different varieties and seven stress periods), for dwarf wheat in India by Singh and Malik (1983) ($r^2 = 0.848$ for three different stress periods), for spring wheat in Denmark by Mogensen et al. (1983) ($r^2 = 0.937$ with seven different stress periods), for cowpea [*Vigna unguiculata* (L.) Walp] by Ziska and Hall (1983) ($r^2 = 0.904$ for six irrigation levels with two fertility levels), and lima bean (*Phaseolus lunatus* L.) by Ziska et al. (1985) ($r^2 = 0.766$ for three levels of soil water depletion and two different stress periods).

E. Effects of Soil Spatial Variability and Irrigation Uniformity

With the current level of soil and irrigation science, the effects of soil spatial properties and irrigation application variability can not be easily evaluated independently; however, their individual and/or combined effects can be estimated for specific situations. As the foregoing discussions have indicated, crop production is greatly influenced by soil water availability, with irrigation having the major effect on soil water levels. Soil properties greatly affect the processes of infiltration, root development, plant water extraction, chemical reactions, redistribution of profile soil water, and water holding capacity. Each of these processes, in turn, can affect crop production through the availability of soil water to the crop. Warrick and Nielsen (1980), Russo and Bresler (1981), and Trickler (1981) have reported that the distribution of water infiltration rates and/or hydraulic conductivity is highly skewed and likely log-normally distributed. The distribution of water held within the soil profile will likely be less skewed and more normally distributed (Cassel & Bauer, 1975; Russo & Bresler, 1981). Peck (1983) discussed soil spatial variability and its effects on water and solute transport within fields. For irrigation systems that depend on the soil for distribution (surface methods, furrow, border, flood, etc.), the distribution of soil hydraulic properties will directly affect the distribution of infiltrated water. However, the soil will redistribute the infiltrated water such that the resulting soil water distribution may be more uniform than the infiltration distribution for nonuniform irrigation applications (Hart, 1972). If the application distribution is perfectly uniform, however, then the resulting soil water storage could be less uniform due to soil variability. For pressurized irrigation distribution systems (e.g., sprinkler, drip, trickle), the distribution of the soil hydraulic properties will not greatly affect the infiltration distribution, which will depend mainly on the water distribution from the irrigation system. Both irrigation application uniformity and soil water storage uniformity will affect the mean production from a field. The previous sections have assumed that all applied water is evenly distributed and that the water in the soil is uniformly available to the crop.

Solomon (1984) reviewed the irrigation uniformity parameters that have been used by engineers to describe the application distribution from irrigation systems such as the following:

$$UC = 1 - (\delta \, \overline{Q}^{-1}) \qquad \text{(Christiansen, 1942)} \qquad [43]$$

$$DU = \overline{Q}_{lq} \, \overline{Q}^{-1} \qquad \text{(Criddle et al., 1956)} \qquad [44]$$

where UC is called a uniformity coefficient, DU was originally called a pattern efficiency and later called the distribution uniformity, δ is the mean deviation of the application ($\overline{|Q_i - \overline{Q}|}$, where Q_i is an individual observation in mm), $\overline{Q}$ is the mean application in mm, and $\overline{Q}_{lq}$ is mean of the lowest one-quarter of the applications in mm. Warrick (1983) examined these two functions and reported their characterizations in terms of the coefficient of variation (CV, $\sigma \, \overline{Q}^{-1}$, where σ is the standard deviation of the applications in mm) and several types of application population distributions (normal, log normal, specialized power, beta, and gamma). For most of the distributions (particularly for CV ≤ 0.5), Warrick (1983) proposed that the uniformity coefficients could be estimated as follows:

$$UC = 1 - 0.8 \, CV \qquad [45]$$

$$DU = 1 - 1.3 \, CV. \qquad [46]$$

Since the spatial variability of resulting irrigation intake and the resulting spatial variability of crop water use is complex, most early research on the effects of irrigation uniformity simply considered yield effects related to irrigation uniformity. Zaslavsky and Buras (1967) used a Taylor series expansion of the yield-water relationship to determine the yield as:

$$\overline{Y} = \overline{Y(Q)} + 0.5 \, [(\partial^2 Y) \, (\partial Q^2)^{-1} \, (\overline{Q})] \, \sigma_q^2 \qquad [47]$$

where $\overline{Y}$ is the mean yield in kg ha^{-1}, $Y(\overline{Q})$ is the yield in kg ha^{-1} from the yield-water function at the mean water application, $[(\partial^2 Y) \, (\partial Q^2)^{-1} \, (\overline{Q})]$ is the second derivative in kg ha^{-1} mm^{-2} of the yield-water function evaluated at the mean water application, and σ_q^2 is the variance in mm^2 of the irrigation application. Equation [47] assumes that the higher order terms of the Taylor series are negligible, and it requires a twice differentiable yield-applied water function. Varlev (1976) investigated the interactions of irrigation uniformity and irrigation quantity using similar concepts and focused on "infiltrated water" contrasted to "applied water" (this implicitly brings into the function the soil spatial variability), and discussed the trade-offs between irrigation uniformity and water applications to achieve optimum yield levels. Seginer (1978) estimated that the marginal economic value of an increment of irrigation uniformity (UC) would be approximately one-half of the maximum income per unit land area (Y_m times the commodity price). Seginer (1983) extended his previous analysis to include the economic evaluation of irrigation uniformity in terms of land and water constraints. Warrick and Gardner (1983) examined the problem of soil spatial and irrigation application variability on yield and irrigation water use efficiency (yield per unit applied water). Unlike Varlev (1976) and Seginer (1978), they attempted to combine the irrigation application variability and soil spatial variability distributions using the joint probability distributions which can be determined by convolution or Monte Carlo methods. Figure 14–13 shows an example from Warrick and Gardner (1983) using a log-normal irrigation distribution and a uniform distribution of available soil water for CVs of 0, 0.5, 1, and 1.5. This example indicates that any variation in irrigation application (and/or equivalent soil water infiltration) will tend to skew the yield-water function. The effect of soil infiltration and irrigation application variability is similar to the effects of increasing irrigation water salinity on crop yields (Fig. 14–11). This skewing of the linear crop yield-ET lines obtained from small plots (where properties are uniform) into curves for fields (where natural variation might be large) may be one factor to help explain the curved nature of other yield-ET and yield-applied water functions widely found in the literature. Slight irrigation or soil infiltration variability can produce nonlinear yield-ET functions even though the basic yield function is exactly linear. Additional discussions of the effects of irrigation or soil infiltration variability on crop yield are found in Stern and Bresler (1983), Solomon (1984), Feinerman et al. (1984), Lety et al. (1984), and Lety (1985).

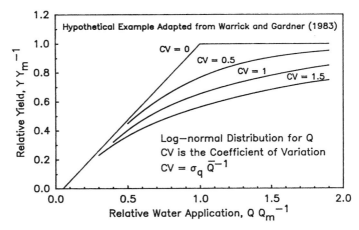

Fig. 14–13. Hypothetical relationship between crop yield and applied irrigation water for several appliation variabilities (expressed in terms of the CV, coefficient of variation) for log-normally distributed irrigations and uniform soil water variability (Warrick & Gardner, 1983).

III. CROP PRODUCTION-IRRIGATION RELATIONSHIPS

The relationships between net assimilation and transpiration, dry matter production and transpiration, and economic crop production and crop water use have been discussed along with descriptions of the effects of evaporative demand, fertility, salinity, and soil and irrigation variability. The basic relationship between net assimilation and transpiration depends on the photosynthetic pathway, plant diffusion resistances, and specific environmental factors. Dry matter production is controlled by the same factors that affect net assimilation, although the interpretation of field data is difficult because of the inability to measure field transpiration as well as the usual omission of the measurement of root dry matter. Economic production can be estimated from the dry matter production in most instances. The relationships between economic yield and evapotranspiration are basically empirical and depend largely on the above described factors as well as the evaporative demand. Severe water deficits in certain critical crop development periods can interfere with the crop development, in particular reproductive processes, and reduce economic crop yield, but the effects may be difficult to precisely define due to the interactions of water deficits between several crop growth stages. Nutrition, if inadequate, can limit production as well as reduce the efficiency of crop water use. Salinity, soil spatial variability, and irrigation application variability all act similarly to skew the relationship between crop production and water application and will generally reduce the mean crop production on a field basis.

The interpretation or the prediction of the effects of irrigation on crop production is complex. Obviously, the effects of irrigation on crop production must be accurately predicted to permit economic analyses of irrigation systems, irrigation management, and water resource allocation decisions. The relation of irrigation to crop production is essentially site specific. Yaron and Bresler (1983) and Vaux and Pruitt (1983) provide excellent interpretations and reviews along with discussions of the limitations of productions functions for economic evaluation of irrigation water applications.

The graphical presentation of the relationship between crop production and the field water supply presented by Stewart and Hagan (1973) illustrates the concepts discussed in this chapter (Fig. 14-14). They defined the field water supply (FWS) to be the sum of the soil water in the profile at planting that will become available to the crop during the season (ASW), the gross seasonal irrigations (Q) (also would include preplant irrigations if not included in ASW), and the rainfall (R) received during the season. This example shows a case where the sum of ASW and R is 250 mm and where 1150 mm of irrigation water is needed to be applied (with the implied application efficiency, application uniformity, inherent soil variability, and irrigation water salinity) to obtain maximum crop production, P_m and Y_m. The example P_m and Y_m are 24 and 11 Mg ha^{-1} for aboveground dry matter and grain yield (dry basis), respectively. Note that the points (P, Q_o) and (Y, Q_o) represent the yields without irrigation (dryland). The slope of the dry matter production line (S_p) is determined by the species of the crop (basically, the k_d value) and

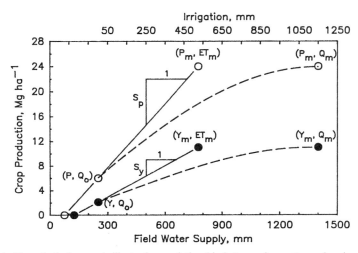

Fig. 14–14. Hypothetical example illustrating a relationship between dry matter and grain production and field water supply.

the environment (evaporative demand as characterized by either potential ET or vapor pressure deficit). The slope of the grain yield line (S_y) is determined by the partitioning between dry matter and economic yield and S_p. The deviation of the dashed curves from the lines represents the combined effects of the irrigation hydrology (runoff, deep percolation, soil water recharge, spray evaporation, drift, etc.) with the effects of the irrigation water salinity, irrigation application uniformity, and the spatial variability of the soil physical parameters. Vaux and Pruitt (1983) discussed these concepts in detail and reported that relationships like those illustrated in Fig. 14–14 closely resembled the results reported by a number of investigators. Martin et al. (1984) developed procedures to estimate the contribution of irrigation to evapotranspiration, which was then related with production functions to grain yield. The relationship between grain yield and field water supply shown for the specific example in Fig. 14–14 (which is site, crop, and irrigation specific) demonstrates the following conditions: (i) maximum water use efficiency ($Y\,ET^{-1}$) occurs at the point (Y_m, ET_m); (ii) maximum irrigation water use efficiency ($Y\,Q^{-1}$) occurs at a value of Q of about 600 mm for this example, which is considerably less than the 1150 mm necessary to produce maximum grain yield (this value can be graphically determined by the tangent on the curve to a line constructed through the origin); and (iii) assuming a constant water cost, the maximum net profit will normally occur at a value of FWS exceeding ET_m but $< Q_m$ (unless water is free) and will decrease as the water price increases for fixed land but increase with higher fixed production costs (Yaron & Bresler, 1983). Generally, the net profit will be rather insensitive to a relatively broad range in applied water (likely to be ± 25–50 mm in this example) and, therefore, the grower would likely choose the higher irrigation applications (if sufficient water is available) within this range to avoid risk associated with critical water deficits as well as the other intangible factors.

Future research into crop production as affected by irrigation will not benefit from developing empirical production functions, except where the basic knowledge of the irrigation uniformity or soil variability or crop-soil rooting interactions is deficient. Procedures are available to estimate P_m and ET_m as well as the relationships between P and T and Y and P (e.g., Doorenbos & Kassam, 1979; Doorenbos & Pruitt, 1977; Tanner & Sinclair, 1983; Feddes, 1986; van Kuelen & Wolf, 1986). Additional improvements in the quantification of the effects of water deficits at critical crop growth stages are needed. As the development of comprehensive crop growth models increases, the irrigation economic analyses and real-time irrigation decisions (Swaney et al., 1983) can be accomplished with expert systems (Lemon, 1986) that rely on crop simulation. Few current crop simulation models contain the sophistication to deal with all of the simultaneous problems related to water, salinity, fertility, insects, diseases, soil chemical and physical limitations, and irrigation dynamics—as well as the environmental variability—but the future for their application to irrigation management problems appears promising.

APPENDIX

Symbols

A = Assimilation flux density, kg (CH_2O) m^{-2} d^{-1}, mmol (C_2O) m^{-2} d^{-1}, or kg (CO_2) m^{-2} s^{-1}

A_s = Salinity-yield threshold, dS m^{-1}

B = Crop-specific coefficient

B_s = Sensitivity factor for salinity, m dS^{-1}

B' = Correction factor for leaf shading

C = Crop vector

C_a = Atmospheric CO_2 concentration, kg m^{-3}

C_s = CO_2 concentration inside substomatal cavity, kg m^{-3}

CV = Coefficient of variation

DU = Distribution uniformity

E = Soil water evaporation, mm or mm d^{-1}

E_e = Seasonal mean pan evaporation within screened enclosure, mm d^{-1}

E_o = Potential ET, mm d^{-1}

EC_e = Electrical conductivity of saturated extract, dS m^{-1}

ET = Evapotranspiration, mm or mm d^{-1}

ET_m = Maximum ET without water deficits, mm

H = Seasonal mean daily relative humidity, %

H_i = Harvest index

H_{ia} = Adjusted harvest index

LAI_D = Sunlit leaf area index

L_T = Effective transpiration leaf area index

M = Miscellaneous vector
P = Crop dry matter production, kg per container, kg ha^{-1}, or Mg ha^{-1}
P_a = Aboveground dry matter yield, Mg ha^{-1}
P_b = Barometric pressure, kPa
P_m = Maximum dry matter production without water deficits, kg ha^{-1} or Mg ha^{-1}
P_o = Dry matter production required to initiate economic production, Mg ha^{-1}
P_r = Root dry matter yield, Mg ha^{-1}
P_s = Stover dry matter yield, Mg ha^{-1}
P_t = Total dry matter yield, Mg ha^{-1}
Q = Irrigation application, mm
$\overline{Q}$ = Mean irrigation application, mm
Q_m = Irrigation application required to produce the maximum yield, mm
R = Incident solar radiation, W m^{-2}
S_p = Dry matter yield-evapotranspiration slope, Mg ha^{-1} mm^{-1}
S_y = Economic yield-evapotranspiration slope, Mg ha^{-1} mm^{-1}
T = Transpiration, kg m^{-2} d^{-1}, mmol m^{-2} s^{-1}, kg per container, or mm d^{-1}
T_m = Seasonal transpiration without water deficits, mm
T_T = Seasonal transpiration, mm
UC = Christiansen's uniformity coefficient
W = Weather vector
W_a = Atmospheric water vapor concentrations, kg m^{-3}
W_s = Atmospheric water vapor concentrations inside substomatal cavity, kg m^{-3}
Y = Economic yield, kg ha^{-1} or Mg ha^{-1}
Y_m = Maximum economic yield without water deficits, kg ha^{-1}
a = Molecular mass ratio of CH_2O to CO_2, 0.68
b = Conversion factor for CH_2O to biomass, 0.33 to 0.83
c = CO_2 gradient factor $[(\rho - \rho_s)\,\rho^{-1}]$
e = Vapor pressure, kPa
e_s^* = Saturated vapor pressure at the crop temperature, kPa
e^* = Saturated vapor pressure at the air temperature, kPa
k, k_a', k_d = Crop-specific coefficients, kPa
k_a = Crop-specific coefficient, % (RH)
m = Crop-specific coefficient, mm d^{-1} or kg ha^{-1} d^{-1}
m_e = Crop-specific coefficient within screened enclosure, mm d^{-1}
n = Crop-specific coefficient, kg kg^{-1}
r = Leaf diffusion resistance to H_2O, s m^{-1}
r' = Leaf diffusion resistance to CO_2, s m^{-1}
δ = Mean irrigation application deviation, mm
ρ = Ambient CO_2 concentration, mg kg^{-1}
σ = Standard deviation of irrigation application, mm
ρ_s = CO_2 concentration inside substomatal cavity, mg kg^{-1}
ρ_a = Density of air, kg m^{-3}
ϵ = Molecular mass ratio of water vapor to air, 0.622
λ = Constant (Lagrange multiplier)
l_i = Crop-specific coefficient for growth stage i
θ = Soil vector

REFERENCES

Aase, K., and F.H. Siddoway. 1981. Spring wheat yield estimates from spectral reflectance measurements. Inst. Electr. Electron. Eng. Trans. Geosci. Remote Sens. GE-19:78–84.

Al-Khafaf, S., P.J. Wierenga, and B.C. Williams. 1978. Evaporative flux from irrigated cotton as related to leaf area index, soil water, and evaporative demand. Agron. J. 70:912–917.

American Society of Civil Engineers. 1930. Consumptive use of water in irrigation. Progress report of the Duty of Water Committee of the Irrigation Division. Trans. Am. Soc. Civ. Eng. 94:1349–1399.

Arkley, R.J. 1963. Relationship between plant growth and transpiration. Hilgardia 34:559–584.

Asrar, G., L.E. Hipps, and E.T. Kanemasu. 1984. Assessing solar energy and water use efficiencies in winter wheat: A case study. Agric. For. Meteorol. 31:47–58.

Baker, J.M., and C.H.M. van Bavel. 1987. Measurement of mass flow of water in stems of herbaceous plants. Plant Cell Environ. 10:779–782.

Barrett, J.W.H., and G.Y. Skogerboe. 1980. Crop production functions and the allocation and use of irrigation water. Agric. Water Manage. 3:53–64.

Bierhuizen, J.F., and R.O. Slatyer. 1965. Effect of atmospheric concentration of water vapour and CO_2 in determining transpiration-photosynthesis relationships of cotton leaves. Agric. Meteorol. 2:259–270.

Black, C.A. 1966. Crop yields in relation to water supply and soil fertility. p. 177–206. In W.H. Pierre et al. (ed.) Plant environment and efficient water use. ASA and SSSA, Madison, WI.

Bloodworth, M.E., J.B. Page, and W.R. Cowley. 1955. A thermoelectric method for determining the rate of water movement in plants. Soil Sci. Soc. Am. Proc. 19:411–414.

Bos, M.G., and J. Nugteren. 1978. On irrigation efficiencies. Int. Inst. Land Reclam. Improve. Publ. 19 (2nd ed.). Wageningen, Netherlands.

Bresler, E. 1967. A model for tracing salt distribution in the soil profile and estimating the efficient combination of water quality under varying field conditions. Soil Sci. 104:227–233.

Bresler, E. 1987. Application of a conceptual model to irrigation water requirement and salt tolerance of crops. Soil Sci. Soc. Am. J. 51:788–793.

Bresler, E., and G.J. Hoffman. 1986. Irrigation management for soil salinity control: Theories and tests. Soil Sci. Soc. Am. J. 50:1552–1560.

Briggs, L.J., and H.L. Shantz. 1913a. The water requirement of plants: I. Investigations in the Great Plains in 1910 and 1911. USDA Bur. Plant Ind. Bull. 284.

Briggs, L.J., and H.L. Shantz. 1913b. The water requirement of plants: II. A review of the literature. USDA Bur. Plant Ind. Bull. 285.

Briggs, L.J., and H.L. Shantz. 1914. Relative water requirements of plants. J. Agric. Res. 3:1–65.

Campbell, G.S., and M.D. Campbell. 1982. Irrigation scheduling using soil moisture measurements: Theory and practice. p. 25–42. In D. Hillel (ed.) Advances in irrigation. Vol. 1. Academic Press, New York.

Cassel, D.K., and A. Bauer. 1975. Spatial variability in soils below depth of tillage. Soil Sci. Soc. Am. J. 39:247–250.

Childs, S.W., and R.J. Hanks. 1975. Model of soil salinity effects on crop growth. SSSA Proc. 39:617–622.

Christiansen, J.E. 1942. Irrigation by sprinkling. Agric. Exp. Stn. Bull. 670. Univ. Calif., Berkeley.

Cole, J.S., and O.R. Mathews. 1923. Use of water by spring wheat on the Great Plains. USDA Bur. Plant Ind. Bull. 1004.

Cooper, P.J.M., P.J. Gregory, D. Tully, and H.C. Harris. 1987. Improving water use efficiency of annual crops in rainfed farming systems of west Asia and North Africa. Exp. Agric. 23:113–158.

Cowan, I.R. 1977. Stomatal behavior and environment. Adv. Bot. Res. 4:1176–1227.

Cowan, I.R. 1982. Regulation of water use in relation to carbon gain in higher plants. p. 589–613. In O.L. Lange et al. (ed.) Physiological plant ecology, I. Vol. 12A. Encyclopedia of plant physiology. Springer-Verlag, Berlin.

Cowan, I.R. 1986. Economics of carbon fixation in higher plants. p. 133–170. In T.J. Givnish (ed.) On the economy of plant form and function. Cambridge Univ. Press, Cambridge, Great Britain.

Cowan, I.R., and G.D. Farquhar. 1977. Stomatal function in relation to leaf metabolism and environment. Symp. Soc. Exp. Biol. 31:471–505.

Cowan, I.R., and J.H. Troughton. 1971. The relative role of stomata in transpiration and assimilation. Planta 97:325–336.

Criddle, W.D., S. Davis, C.H. Pair, and D.G. Shockley. 1956. Methods for evaluating irrigation systems. U.S. Dep. Agric. Agric. Handb. 82. Washington, DC.

Davis, K.R. 1983. Trickle irrigation of cotton in California. p. 34–38. *In* Proc. Western Cotton Prod. Conf., Las Cruces, NM. 9–11 Aug.

DeLoughery, R.L., and R.K. Crookston. 1979. Harvest index of corn affected by population density, maturity rating, and environment. Agron. J. 72:999–1005.

de Wit, C.T. 1953. A physical theory on placement of fertilizers. Versl. Landbouskd. Onderz. 59.4, Staatsdrukkerij, 's-Gravenhage, Netherlands.

de Wit, C.T. 1958. Transpiration and crop yields. Instituut Voor Biologisch en Scheikundig Onderzoek van Landbouwgewassen, Versl. Landbouskd. Onderz. 64.6, Wageningen, Netherlands.

de Wit, C.T. 1978. Simulation of assimilation, respiration and transpiration of crops. Simulation Monogr. PUDOC, Wageningen, Netherlands.

Dillman, A.C. 1931. Water requirements of certain crop plants and weeds in the northern Great Plains. J. Agric. Res. 42:187–238.

Donald, C.M., and J. Hamblin. 1976. The biological yield and harvest index of cereals as agronomic and plant breeding criteria. Adv. Agron. 28:361–405.

Doorenbos, J., and A.H. Kassam. 1979. Yield response to water. FAO Irrig. Drain. Pap. 33. Rome.

Doorenbos, J., and W.O. Pruitt. 1977. Crop water requirements. FAO Irrig. Drain. Pap. 24. Rome.

Downton, W.J.S. 1975. The occurrence of C_4 photosynthesis among plants. Photosynthetica 9:96–105.

Farquhar, G.D., and R.A. Richards. 1984. Isotopic composition of plant carbon correlates with water-use efficiency of wheat genotypes. Austr. J. Plant Physiol. 11:539–552.

Farquhar, G.D., E.-D. Schulze, and M. Kuppers. 1980. Responses to humidity by stomata of *Nicotiana glauca* L. and *Corylus avellana* L. are consistent with the optimization of carbon dioxide uptake with respect to water loss. Aust. J. Plant Physiol. 7:315–327.

Farquhar, G.D., and T.D. Sharkey. 1982. Stomatal conductance and photosynthesis. Annu. Rev. Plant Physiol. 33:317–345.

Feddes, R.A. 1986. Modelling and simulation in hydrologic systems related to agricultural development: State of the art. p. 1–16. *In* Proc. Water Manage. Agric. Dev., Greece.

Feinerman, E., K.C. Knapp, and J. Lety. 1984. Salinity and uniformity of water infiltration as factors in yield and economically optimal water application. Soil Sci. Soc. Am. J. 48:477–481.

Field, C., J.A. Berry, and H.A. Mooney. 1982. A portable system for measuring carbon-dioxide and water vapor exchange of leaves. Plant Cell Environ. 5:179–186.

Fischer, R.A., and N.C. Turner. 1978. Plant productivity in the arid and semiarid zones. Annu. Rev. Plant Physiol. 29:277–317.

Gifford, R.M. 1974. A comparison of potential photosynthesis, productivity, and yield of plant species with differing photosynthetic metabolism. Aust. J. Plant Physiol. 1:107–117.

Goudriaan, J., and H.H. van Laar. 1978. Relations between leaf resistance, CO_2-concentration, and CO_2-assimilation in maize, beans, lalang grass, and sunflower. Photosynthetica 12:241–249.

Gunton, J.L., and J.P. Evenson. 1980. Moisture stress in navy beans. I. Effect of withholding irrigation at different phenological stages on growth and yield. Irrig. Sci. 2:49–58.

Hagan, R.M., Y. Vaadia, and M.B. Russel. 1959. Interpretation of plant responses to soil moisture regimes. Adv. Agron. 11:97–113.

Hall, A.E., and E.-D. Schulze. 1980. Stomatal responses to environment and a possible interrelationship between stomatal effects on transpiration and CO_2 assimilation. Plant Cell Environ. 3:467–474.

Hanks, R.J. 1974. Model for predicting plant yield as influenced by water use. Agron. J. 66:660–665.

Hanks, R.J. 1983. Yield and water-use relationships: An overview. p. 393–411. *In* H.M. Taylor et al. (ed.) limitations to efficient water use in crop production, ASA, CSSA, and SSSA, Madison, WI.

Hanks, R.J. 1985. Crop coefficients for transpiration. p. 431–438. *In* Advances in evapotranspiration. ASAE, St. Joseph, MI.

Hanks, R.J., G.L. Ashcroft, V.P. Rasmussen, and G.D. Wilson. 1978. Corn production as influenced by irrigation and salinity—Utah studies. Irrig. Sci. 1:47–59.

Hanks, R.J., H.R. Gardner, and R.L. Florian. 1969. Plant growth-evapotranspiration relations for several crops in the central Great Plains. Agron. J. 61:30–34.

Hanks, R.J., D.W. James, and D.W. Watts. 1983. Irrigation management and crop production as related to nitrate mobility. p. 141–151. *In* Chemical mobility and reactivity in soil systems. ASA and SSSA, Madison, WI.

Hanks, R.J., and V.P. Rasmussen. 1982. Predicting crop production as related to plant water stress. Adv. Agron. 35:193–215.

Hart, W.E. 1972. Subsurface distribution of nonuniformly applied surface waters. Trans. ASAE 15:656–661, 666.

Hexem, R.W., and E.O. Heady. 1978. Water production functions and irrigated agriculture. Iowa State Univ. Press, Ames.

Hiler, E.A., and R.N. Clark. 1971. Stress day index to characterize effects of water stress on crop yields. Trans. ASAE 14:757–761.

Hoffman, G.J., and M.Th. van Genuchten. 1983. Soil properties and efficient water use: Water management for salinity control. p. 73–85. *In* H.M. Taylor et al. (ed.) Limitations to efficient water use in crop production. ASA, CSSA, and SSSA, Madison, WI.

Howell, T.A., K.R. Davis, R.L. McCormick, H. Yamada, V.T. Walhood, and D.W. Meek. 1984. Water use efficiency of narrow row cotton. Irrig. Sci. 5:195–214.

Jensen, M.E. 1968. Water consumption by agricultural plants. p. 1–22. *In* T.T. Kozlowski (ed.) Water deficits and plant growth, Vol. II. Academic Press, New York.

Jensen, M.E., and W.H. Sletten. 1965. Evapotranspiration and soil moisture-fertilizer interrelations with irrigated winter wheat in the southern High Plains. USDA-ARS, Conserv. Res. Rep. U.S. Gov. Printing Office, Washington, DC.

Johnson, W.C., and R.G. Davis. 1980. Yield-water relationships of summer-fallowed winter wheat. Agric. Res. Results, South. Ser. (USDA) 5, ARR-S-5.

Jones, J.W., B. Zur, and J.M. Bennett. 1986. Interactive effects of water and nitrogen stresses on carbon and water vapor exchange of corn canopies. Agric. For. Meteorol. 38:113–126.

Kanemasu, E.T., L.R. Stone, and W.L. Powers. 1976. Evapotranspiration model tested for soybean and sorghum. Agron. J. 68:569–572.

Kisselbach, T.A. 1916. Transpiration as a factor in crop production. Nebr. Agric. Exp. Stn. Res. Bull. 6.

Klocke, N.L., D.F. Heermann, and H.R. Duke. 1985. Measurement of evaporation and transpiration with lysimeters. Trans. ASAE 28:183–189, 192.

Lascano, R.J., C.H.M. van Bavel, J.L. Hatfield, and D.R. Upchurch. 1987. Energy and water balance of a sparse crop: Simulated and measured soil and crop evaporation. Soil Sci. Soc. Am. J. 51:1113–1121.

Lemon, H. 1986. Comax: An expert system for cotton crop management. Science 233:19–33.

Lety, J. 1985. Irrigation uniformity as related to optimum crop production—Additional research is needed. Irrig. Sci. 6:253–263.

Lety, J., A. Dinar, and K.C. Knapp. 1985. Crop-water production function model for saline irrigation waters. Soil Sci. Soc. Am. J. 49:1005–1009.

Lety, J., and A. Dinar. 1986. Simulated crop-water functions for several crops when irrigated with saline waters. Hilgardia 54:1–32.

Lety, J., and H.J. Vaux, Jr. 1984. Water duties for California agriculture. Rep. Calif. State Water Resour. Control Board, Sacramento.

Lety, J., H.J. Vaux, Jr., and E. Feinerman. 1984. Optimum crop water application as affected by uniformity of water infiltration. Agron. J. 76:435–441.

Maas, E.V. 1986. Salt tolerance of plants. Appl. Agric. Res. 1:12–26.

Maas, E.V., and G.J. Hoffman. 1977. Crop salt tolerance—Current assessment. J. Irrig. Drain. Div. Am. Soc. Civ. Eng. 103(IR2):115–134.

Maas, E.V., J.A. Poss, and G.J. Hoffman. 1986. Salinity sensitivity of sorghum at three growth stages. Irrig. Sci. 7:1–11.

Martin, D.L., D.G. Watts, and J.R. Gilley. 1984. Model and production function for irrigation management. J. Irrig. Drain. Eng. 110:149–164.

Mathews, O.R., and L.A. Brown. 1938. Winter wheat and sorghum production in the southern Great Plains under limited rainfall. USDA Circ. 477. U.S. Gov. Print. Office, Washington, DC.

Maximov, N.A. 1929. The plant in relation to water. George Allen and Unwin, London.

Mead, E. 1887. Report of experiments in irrigation and meteorology. Colorado Agric. Exp. Stn. Bull. 1.

Meiri, A. 1984. Plant response to salinity: Experimental methodology and application to the field. p. 284–297. In I. Shainberg and J. Shalhavet (ed.) Soil salinity and irrigation, Ecol. Studies 51. Springer-Verlag, Heidelberg, West Germany.

Minhas, B.S., K.S. Parikh, and T.N. Srinivasan. 1974. Toward the structure of a production function for wheat yields with dated inputs of irrigation water. Water Resour. Res. 10:383–393.

Mogensen, V.O., H.E. Jensen, and M.A. Rab. 1985. Grain yield, yield components, drought sensitivity, and water use efficiency of spring wheat subjected to water stress at various growth stages. Irrig. Sci. 6:131–140.

Monteith, J.L. 1977. Climate and efficiency of crop production in Britain. Phil. Trans. R. Soc. London Ser. B 281:277–294.

Peck, A.J. 1983. Field variability of soil physical properties. p. 189–221. In D. Hillel (ed.) Advances in irrigation. Vol. 2. Academic Press, New York.

Penman, H.L. 1948. Natural evaporation from open water, bare soil, and grass. Proc. R. Soc. London A193:120–145.

Penman, H.L., and R.K. Schofield. 1951. Some physical aspects of assimilation and transpiration. Symp. Soc. Exp. Biol. 5:115–129.

Penning de Vries, F.W.T. 1975. Use of assimilates in higher plants. p. 459–480. In J.P. Cooper (ed.) Photosynthesis and productivity in different environments. Cambridge Univ. Press, Cambridge, Great Britain.

Powers, W.L. 1922. Some ways of increasing the duty of irrigation water. Soil Sci. 14:377–382.

Raats, P.A.C. 1974. Steady flows of water and salt in uniform soil profiles with plant roots. Soil Sci. Soc. Am. Proc. 38:717–722.

Raghavendra, A.S., and V.S.R. Das. 1978. The occurrence of C4 photosynthesis: A supplementary list of C_4 plants reported during late 1974–mid-1977. Photosynthetica 12:200–208.

Raschke, K. 1975. Stomatal action. Annu. Rev. Plant Physiol. 26:309–340.

Rhoads, F.M. 1984. Nitrogen or water stress: Their interrelationships. p. 207–317. In R.D. Hauck (ed.-in-chief) Nitrogen in crop production. ASA, CSSA, and SSSA, Madison, WI.

Rhoades, J.D., and S.D. Merrill. 1976. Assessing the suitability of water for irrigation: Theoretical and empirical approaches. FAO Soils Bull. 31:69–109.

Richards, R.A. 1987. Physiology and the breeding of winter-grown cereals for dry areas. p. 133–140. In J.P. Srivastava et al. (ed.) Drought tolerance in winter cereals. John Wiley, Chichester.

Rijtema, P.E., and G. Endrodi. 1970. Calculation of production of potatoes. Neth. J. Agric. Sci. 18:26–36.

Ritchie, J.T. 1972. Model for predicting evaporation from a row crop with incomplete cover. Water Resour. Res. 8:1204–1212.

Ritchie, J.T. 1983. Efficient water use in crop production: Discussion on the generality of relations between biomass production and evapotranspiration. p. 29–44. In H.M. Taylor et al. (ed.) Limitations to efficient water use in crop production. ASA, CSSA, and SSSA, Madison, WI.

Ritchie, J.T., and E. Burnett. 1971. Dryland evaporative flux in a subhumid climate. II. Plant influences. Agron. J. 63:56–62.

Russo, D., and D. Bakker. 1987. Crop-water production functions for sweet corn and cotton irrigated with saline waters. Soil Sci. Soc. Am. J. 51:1554–1562.

Russo, D., and E. Bresler. 1981. Soil hydraulic properties as stochastic processes: I. An analysis of field spatial variability. Soil Sci. Soc. Am. J. 45:682–687.

Sakuratani, T. 1984. Improvement of the probe for measuring water flow rate in intact plants with the stem heat balance method. J. Agric. Meteorol. 40:273–277.

Salter, P.J., and J.E. Goode. 1967. Crop responses to water at different stages of growth. Commonw. Bur. Hortic. Plant. Crops (G.B.) Res. Rev. 2. Farnham Royal, G.B.

Schulze, E.-D., and A.E. Hall. 1982. Stomatal responses, water loss and CO_2 assimilation rates of plants in contrasting environments. p. 181–230. In O.L. Lange et al. (ed.) Physiological plant ecology, II. Vol. 12B. Encyclopedia of Plant Physiology. Springer-Verlag, Berlin.

Seginer, I. 1978. A note on the economic significance of uniform water application. Irrig. Sci. 1:19–25.

Seginer, I. 1983. Irrigation uniformity effect on land and water allocation. Trans. ASAE 26:116–122.

Shantz, H.L., and L.N. Piemeisel. 1927. The water requirements of plants at Akron, CO. J. Agric. Res. 34:1093–1190.

Sinclair, T.R., C.B. Tanner, and J.M. Bennett. 1984. Water-use efficiency in crop production. BioScience 34:36–40.

Singh, T., and D.S. Malik. 1983. Effect of water stress at three growth stages on the yield and water-use efficiency of dwarf wheat. Irrig. Sci. 4:239–245.

Singh, P., H. Wolkewitz, and R. Kumar. 1987. Comparative performance of different crop production functions for wheat (*Triticum aestivum* L.). Irrig. Sci. 8:273–290.

Slabbers, P.J., V.S. Herrendorf, and M. Stapper. 1979. Evaluation of simplified water-crop yield models. Agric. Water Manage. 2:95–129.

Snyder, F.W., and G.E. Carlson. 1984. Selecting for partitioning of photosynthetic products in crops. Adv. Agron. 37:47–72.

Solomon, K.H. 1984. Yield related interpretations of irrigation uniformity and efficiency measures. Irrig. Sci. 5:161–172.

Solomon, K.H. 1985. Water-salinity-production functions. Trans. ASAE 28:1975–1980.

Spaeth, S.C., H.C. Randall, T.R. Sinclair, and J.S. Vendeland. 1984. Stability of soybean harvest index. Agron. J. 76:482–486.

Stanford, G., and J.O. Legg. 1984. Nitrogen and yield potential. p. 263–272. *In* R.D. Hauck (ed.-in-chief) Nitrogen in crop production. ASA, CSSA, and SSSA, Madison, WI.

Stanhill, G. 1986. Water use efficiency. Adv. Agron. 39:53–85.

Stapleton, A.R.A., R.J. Wagenet, and D.L. Turner. 1983. Corn growth and nitrogen uptake under irrigated, fertilized conditions. Irrig. Sci. 4:1–15.

Steiner, J.L., R.C.G. Smith, W.S. Meyer, and J.A. Adeney. 1985. Water use, foliage temperature and yield of irrigated wheat in south-eastern Australia. Aust. J. Agric. Res. 36:1–11.

Stern, J., and E. Bresler. 1983. Nonuniform sprinkler irrigation and crop yield. Irrig. Sci. 4:17–29.

Stewart, J.I., and R.M. Hagan. 1973. Functions to predict effects of crop water deficits. J. Irrig. Drain. Div. Am. Soc.Civ. Eng. 99:421–439.

Stewart, J.I., R.M. Hagan, W.O. Pruitt, R.J. Hanks, J.P. Riley, R.E. Danielson, W.T. Franklin, and E.B. Jackson. 1977. Optimizing crop production through control of water and salinity levels in the soil. Utah Water Res. Lab. Publ. PRWG151-1.

Swaney, D.P., J.W. Mishoe, J.W. Jones, and W.G. Bogges. 1983. Using crop models for management: Impact of weather characteristics on irrigation decisions in soybeans. Trans. ASAE 26:1809–1814.

Tanner, C.B. 1981. Transpiration efficiency of potato. Agron. J. 73:59–64.

Tanner, C.B., and W.A. Jury. 1976. Estimating evaporation and transpiration from a row crop during incomplete cover. Agron. J. 68:239–243.

Tanner, C.B., and T.R. Sinclair. 1983. Efficient water use in crop production: Research or re-search? p. 1–27. *In* H.M. Taylor et al. (ed.) Limitations to efficient water use in crop production. ASA, CSSA, and SSSA, Madison, WI.

Taylor, H.M., W.R. Jordan, and T.R. Sinclair (ed.). 1983. Limitations to efficient wateruse in crop production. ASA, CSSA, and SSSA, Madison, WI.

Teare, I.D., E.T. Kanemasu, W.L. Powers, and H.S. Jacobs. 1973. Water-use efficiency and its relation to crop canopy area, stomatal regulation, and root distribution. Agron. J. 65:207–211.

Thornthwaite, C.W. 1948. An approach towards a rational classification of climate. Geog. Rev. 38:55–94.

Trickler, A.S. 1981. Spatial and temporal patterns of infiltration. J. Hydrol. 49:261–277.

van Keulen, H. 1986a. Crop yield and nutrient requirements. p. 155–181. *In* H. van Keulen and J. Wolf (ed.) Modelling of agricultural production: Weather, soils, and crops. Ctr. Agric. Publ. Doc., Wageningen, Netherlands.

van Keulen, H. 1986b. Plant data. p. 235–247. *In* H. van Keulen and J. Wolf (ed.) Modelling of agricultural production: Weather, soils, and crops. Ctr. Agric. Publ. Doc., Wageningen, Netherlands.

van Keulen, H., and H.H. van Laar. 1986. The relation between water use and crop production. p. 117–129. *In* H. van Keulen and J. Wolf (ed.) Modelling of agricultural production: Weather, soils, and crops. Ctr. Agric. Publ. Doc., Wageningen, Netherlands.

van Keulen, H., and J. Wolf (ed.). 1986. Modelling of agricultural production: Weather, soils and crops. Ctr. Agric. Publ. Doc., Wageningen, Netherlands.

Varlev, I. 1976. Evaluation of nonuniformity in irrigation and yield. J. Irrig. Drain. Div. Am. Soc. Civ. Eng. 102:149–164.

Vaux, H.J., Jr., and W.O. Pruitt. 1983. Crop-water production functions. p. 61–97. *In* D. Hillel (ed.) Advances in irrigation. Vol. 2. Academic Press, New York.

Viets, F.G., Jr. 1962. Fertilizers and the efficient use of water. Adv. Agron. 14:223–264.

Viets, F.G. 1966. Increasing water use efficiency by soil management. p. 259–274. *In* W.H. Pierre et al. (ed.) Plant environment and efficient water use. ASA and SSSA, Madison, WI.

Viets, F.G. 1972. Water deficits and nutrient availability. p. 217–239. *In* T.T. Kozlowski (ed.) Water deficits and plant growth. Vol. III. Academic Press, New York.

Warrick, A.W. 1983. Interrelationships of irrigation uniformity terms. J. Irrig. Drain. Eng. 109:317–332.

Warrick, A.W., and W.R. Gardner. 1983. Crop yield as affected by spatial variations of soil and irrigation. Water Resour. Res. 19:181–186.

Warrick, A.W., and D.R. Nielsen. 1980. Spatial variability of soil physical properties in the field. p. 319–344. *In* D. Hillel (ed.) Applications of soil physics. Academic Press, New York.

Wenda, W.I., and R.J. Hanks. 1981. Corn yield and evapotranspiration under simulated drought. Irrig. Sci. 2:193–204.

Wong, S.C., I.R. Cowan, and G.D.Farquhar. 1979. Stomatal conductance correlates with photosynthetic capacity. Nature (London) 282:424–426.

Woodward, J. 1699. Some thoughts and experiments concerning vegetation. Phil. Trans. R. Soc. London 21:193–227.

Yaron, D., and E. Bresler. 1983. Economic analysis of on-farm irrigation using response functions of crops. p. 223–255. *In* D. Hillel (ed.) Advances in irrigation, Vol. 2. Academic Press, New York.

Zaslavsky, D., and N. Buras. 1967. Crop yield response to nonuniform application of irrigation water. Trans. ASAE 10:196–198, 200.

Ziska, L.H., and A.E. Hall. 1983. Seed yields and water use of cowpeas [*Vigna unguiculata* (L.) Walp.] subjected to planned-water deficit irrigation. Irrig. Sci. 3:237–245.

Ziska, L.H., A.E. Hall, and R.M. Hoover. 1985. Irrigation management methods for reducing water use of cowpea [*Vigna unguiculata* (L.) Walp.] and lima bean (*Phaseolus lunatus* L.) while maintaining seed yield at maximum levels. Irrig. Sci. 6:223–239.

15   Methods of Estimating Evapotranspiration

J. L. HATFIELD

USDA-ARS
Ames, Iowa

Evapotranspiration (ET) research has progressed greatly in the 20 yr since Tanner (1967) prepared a similar chapter for an earlier monograph. Evaporation from natural surfaces continues to occupy a great deal of research effort in order to characterize the rate of water loss from the soil and plants. Evapotranspiration is a key component in the growth of plants and the impacts of water deficit can be clearly seen on yield and biomass reductions. Monteith (1985) stated that the progress made to our understanding of ET in the past 30 yr has been large and that information is being practically applied today. The references on evaporation research are voluminous and to include them all would be well beyond the scope of this chapter. Rosenberg (1968) and Jensen (1973) provided comprehensive reviews of the current state of ET research at that time. This chapter will build upon their reviews in an attempt to present the current theories and techniques.

The process of evaporation (E) is both physical and dynamic. The physical process has been described in chapter 14 of this book and is driven by meteorological and soil water conditions. We must fully understand all of these factors if we are to utilize the information. In the utilization of data on E there is no consensus about the required accuracy on E but are generally assumed to be to the same value as the best lysimeter: 0.1 mm d^{-1}. The accuracy requirements needed for irrigation management or hydrology have not been made and would assist in the evaluation of ET models.

Monteith (1985) suggested that the term ET should be replaced by the more appropriate term, E, and then denoted by plant or soil, E_p or E_s, respectively. He defined TE as the total evaporation. As an advance toward science it would be desirable to foster such a change; however, the current literature has adopted the term ET and a change may be perceived as a typographical error rather than a scientific advance. Therefore, ET is used in this chapter.

Methods of estimating ET fall within three general categories, direct and indirect methods, and simulation models of the soil water balance. This chapter will overview the current state of knowledge in each of these areas and the limitations in each method. The symbols for the terms are defined in the Appendix.

I. DIRECT MEASUREMENT METHODS

Evapotranspiration is directly inferred from the residual of the soil water balance after all other terms have been measured and is given as

$$ET = SW + P + I - D - R. \quad [1]$$

This method has been successfully used in a number of hydrologic scale studies where entire drainage basins have been studied. In small field studies this method has been used to determine the soil water use by crops for periods of 7 to 10 d. Soil water measurement methods are described in detail in chapter 6 on Measurement of Soil Water Content and Potential. The hydrologic balance method is described briefly in order to provide a comparison with other techniques.

A. Hydrologic Balance

The soil water balance, as described in Eq. [1], requires accurate measurement of all terms and, if the estimate is to be expanded to an entire field, then a measure of the spatial variability of the measurements is necessary. Rouse and Wilson (1972) performed a detailed analysis of the errors involved in the water balance approach. They concluded that this method is acceptable at intervals of 4 d if the actual ET is high, there is no precipitation, and a number of sites are averaged to give a mean. In practice, the current limits of accuracy of $P, I, D,$ and R from Eq. [1] are considered to be 0.2 mm d^{-1}. The procedures for calculating the spatial relationships between samples have been described by Vieira et al. (1983). These procedures aid in defining the number of samples needed to be obtained from a field and the distribution of sampling sites. An example of this approach was given by Russo (1986a, b). He analyzed the effects of differential irrigation amounts and salinity on the crop yields in a heterogenous field. This approach provides a framework for examining ET over large production fields and is an example of the recent advances in combining the soil properties with a soil water balance to evaluate a management practice.

In Eq. [1] it is necessary to determine the change in soil water content over the effective rooting depth or another predefined depth with sufficient accuracy to allow a reasonable estimate of ET. The method of choice for taking these measurements depends upon the depth of soil to be measured: Near the soil surface, and deeper in the soil profile.

1. Surface Soil Water Content

The traditional method of obtaining soil water contents in the upper soil surface, 0 to 200 mm, has been to use gravimetric samples. This procedure is labor intensive and subject to large sample variability. Neutron probes have been used for measuring soil water content in the upper 0 to 100 mm, but the measurements have exhibited large variability. However, neutron probes can be calibrated for this region (Haverkamp et al., 1984). Even with proper calibration and use there are large errors in the data obtained with the neutron probe near the surface.

One new method for obtaining soil evaporation has been the use of the "micro-lysimeter" as first proposed by Boast and Robertson (1982) and described in detail by Boast (1986). Several researchers have used microlysimeters to obtain estimates of soil evaporation (Shawcroft & Gardner, 1983; Walker, 1983). Microlysimeter equipment consists of a thin-walled cylinder pushed into the soil. The cylinder is then removed, the bottom sealed with a rubber stopper, weighed to determine the total mass, and reinstalled in the field with the top flush with the surrounding soil. Typically, the cylinder is again removed and reweighed with the difference in mass being the evaporation loss from the soil surface. Boast and Robertson (1982) and Boast (1986) found the error for a cylinder 70 mm in length to be 0.5 mm for a 1- to 2-d period. Later, Reynolds and Walker (1984) showed through a careful theoretical analysis that the microlysimeter could provide accurate data if detail was provided for the upper 2 mm. Boast and Robertson (1982) stated that although the microlysimeter technique is still labor intensive, it requires little equipment. They proposed that this method would offer an acceptable technique where the spatial resolution of larger, traditional lysimeters is impractical (e.g., evaporation as a function of distance between crop rows, partial cover) where the cost of large lysimeters is prohibitive or where lysimeters are not available. Microlysimeters may provide an acceptable method for evaluating models such as those proposed by Ritchie (1972).

2. Soil Profile

Neutron probes or neutron scattering devices have become accepted techniques for the measurement of soil water content throughout the soil profile. As a technique for obtaining estimates of ET, this procedure remains widely used, particularly when consumptive water use for intervals > 7 d are required. The neutron probe still remains a labor-intensive procedure and Sinclair and Williams (1979) discussed in detail the errors associated with using a neutron probe. Haverkamp et al. (1984), after a detailed analysis, found that the major component of the total variance was the calibration. They showed that profiles of water contents were best determined through the use of Simpsons rule for integration compared to the classical trapezoidal method. The error in estimating soil water content compared to gravimetric samples was < 6.4 mm or 1.6% when Simpsons rule for integration was used.

Lascano et al. (1986) proposed a method for field calibration of the neutron probe using a two-probe, gamma-density gauge as a reference. This procedure, compared to gravimetric samples, was accurate within 1% of the total lower soil profile (0.30–1.50 m) water content. Nakayama and Reginato (1982) had earlier proposed a technique of using a master probe to calibrate other neutron probes for field studies. The master probe would be calibrated against gravimetric samples. All of these procedures would appear to provide suitable measures of soil water content for use in Eq. [1] for routine studies.

Rouse and Wilson (1972) compared the water budget approach with an energy budget calculation of ET for a 25-d period during July and concluded that the accuracy was dependent upon the time span. They stated that, on a daily basis, the soil water budget cannot estimate ET as accurately as an energy budget calculated from measurements of net radiation, Bowen ratio, and soil heat flux. In their study, good agreement was found when there was no precipitation and a 4-d period was totaled. They also cautioned that multiple soil water sites would be necessary to obtain this agreement, even at 4 d. If a single site was used and precipitation occurred, a longer time period for accurate soil water measurement would be necessary.

3. Limitations

Direct measures of soil water content in the hydrologic balance are limited by the following factors:

1. Labor-intensity.
2. Spatial dependence.
3. Temporal resolution.
4. Separate techniques for upper soil profile 0–0.30 m and the lower soil profile 0.30–2.00 m.
5. Definition of the rooting zone of crops.

Each of these factors place constraints on the use of direct measurements, however, these are not insurmountable and the direct measures will continue to be widely used for evapotranspiration studies.

4. Advantages

These are the advantages of using direct measure of the soil water content:

1. Relatively low cost.
2. Low training and operating costs.
3. Multiple site capability.
4. Data easily processed.
5. Accuracy acceptable for many applications.

Direct measures of the soil water content can be easily used and the advantages have outweighed the limitations.

B. Lysimeters

In his review, Tanner (1967) stated that "lysimetry is the only hydrological method in which the experimenter has complete knowledge of all of the terms in the hydrological balance (Eq. [1])." Lysimeters and the technology associated with them have developed rapidly in the past 20 yr;

however, the basic considerations proposed by Pelton (1961) and van Bavel (1961) are still followed today. The information collected from lysimeters has proven valuable in advancing the knowledge base for calibration and derivation of ET equations.

Lysimeters are not without problems. Pruitt and Lourence (1985) stated that "the availability of a highly-sensitive lysimeter can lead to over-confidence. The most vital requirement, however, related to how well the loss of water from the enclosed system represents field losses." The considerations proposed by Pelton (1961), van Bavel (1961), and Pruitt and Lourence (1985) include the following:

1. Large area/rim ratio.
2. Uniformity of soil mass to surroundings, e.g., soil water profile, temperature, and bulk density.
3. Forced drainage or deep construction.
4. Uniformity of crop stand compared to the surrounding field.
5. Differential growth because of different soil conditions, e.g., fertility or compaction.
6. Differences in maturity due to differential soil water use.
7. Destruction of the narrow row crop, e.g., small grains, when servicing the lysimeter.
8. Adequate fetch.

These differences can create errors in obtaining precise values of ET even under conditions assumed to be at or near potential water use rates.

1. Lysimeter Installations

a. Weighing Lysimeters. Lysimeters range in exposed area as well as depth depending on the installation. The Coshocton, OH, and the Davis, CA, lysimeters represented rapid advances in lysimeter technology when they were installed. The Coshocton lysimeter was constructed for a monolithic volume of soil 2 m deep encased by concrete walls and weighed mechanically (Harrold & Dreibelbis, 1953). These lysimeters were constructed with sufficient depth to avoid the problem of an unnatural soil water profile. Pruitt and Angus (1960) avoided the drainage problem by forcing drainage through suction devices installed in the bottom of the lysimeter. The change in weight was measured with a truck scale.

The depth of a lysimeter is dependent upon the intended purpose of the study. The Coshocton lysimeters were constructed to be deep enought to allow for normal root development under rainfed conditions. The Davis lysimeters were intended for irrigated conditions that would maintain a soil water content in the 0.9-m profile near maximum available soil water. Mottram and deJager (1973) developed a load cell-based, weighing lysimeter with a depth of only 0.4 m because their application included perennial ryegrass (*Lolium perenne* L.) in which almost all of the roots were found in the upper 0.4 m of soil in a natural profile. Ritchie and Burnett (1968) proposed that load cells would provide several advantages over mechanical systems through increased accuracy and sensitivity, continuous recording capability, and relatively low cost. Van Bavel and Myers (1962) had earlier developed a 3-Mg lysimeter with dimensions of 1 × 1 × 1.6 m weighed with a strain gauge load cell to allow for precise hourly observations.

Green et al. (1974) and Green and Bruwer (1979) adapted a load cell lysimeter for citrus trees. Gifford et al. (1982) described six lysimeters each containing an individual Monterey pine (*Pinus radiata* D. Don) tree in a 0.5-ha area of trees. Their design placed the beam and counter balance mechanism above ground rather than buried below the lysimeter, however they did not report comparison data between lysimeters. Fritschen et al. (1973) placed a 28-m Douglas fir [*Pseudotsuga menziesii* (Mirbel) Franco] tree in a weighing lysimeter. All of these lysimeters are relatively shallow, with a depth not exceeding 1.8 m. Green and Bruwer (1979) located three 3.7 × 3.7 × 1.8 m 60-Mg weighing lysimeters adjacent to each other in a large orchard of Valencia orange (*Citrus sinensis* L.) trees. When they compared among lysimeters similarly treated there was <4% difference over an entire year. These data suggest that it is possible to achieve good agreement between lysimeters when all are located within the same environment. Edwards (1986) developed four lysimeters for different trees (*Pinus radiata, Populus fleuo, Salix matsudana,* and *Eucalyptus fastigatu*) to examine the water use by the different species. His design used an above-ground cable system attached to a counterbalanced arm with a load cell. Even though the weighing mechanism was above ground, the effect of wind on this system was minimal. He gave the sensitivity in grams per tree since isolated trees are difficult to express in terms of millimeters (Edwards, 1986).

Technology has allowed the development and installation of larger lysimeter systems (Dugas et al. 1985; Howell et al. 1985b; Marek et al., 1987). The lysimeter designed by Howell et al. (1985b) and located at Fresno, CA, has a surface area of 4 m^2 with a soil profile depth of 2 m. It contains four neutron access tubs, 15 soil water electrical conductivity sensors, 15 soil water matric potential sensors, and 3 soil heat flux plates along with soil temperature sensors. They installed one lysimeter under a grass sod and another in a cultivated field. Each lysimeter contains its own data acquisition system for sampling all sensors and communicating via telephone lines with a central computer located 80 km away. Both a free water and vacuum drainage system were included. The weighing mechanism used a flexure balance with a load cell and requires no counterbalance mechanism. The lysimeter is designed in five components (i) inner tank, (ii) outer tank bottom, (iii) outer tank top, (iv) access tunnel, and (v) scale. The inner tank was filled with disturbed soil packed to the bulk density of the surrounding field. Good agreement was found between the field and the lysimeter after several wetting cycles for the soil water content. After irrigations and settling times the lysimeter-measured soil water profiles were in good agreement with the lysimeter field profiles.

To avoid disturbed soil in lysimeters, soil monoliths taken in situ have been proposed by Brown et al. (1974). This procedure was used by Dugas et al. (1985) in installing their lysimeter at Temple, TX. They pushed a 1.5 × 2.0 × 2.68-m lysimeter into the soil and inverted the box for installation of the drainage system. This weighing mechanism consists of a scale and load cell with a counter balance system. Their lysimeter is unique in that after the monolith was installed they added soil water psychrometers to monitor matric potential with depth and 36 horizontal, clear acrylic tubes for observing root growth within the soil profile. An interesting feature of this system is that the lysimeter was installed in a field whose soil has a high shrink-swell characteristic and is different from the lysimeter soil.

Currently, four large lysimeters of the type at Fresno are being installed at Bushland, TX. These lysimeters measure 3 × 3 × 2.3 m and are monolithic profiles (Marek et al., 1987). The monoliths are being taken from the exact location of the lysimeter installation and have potential accuracy of better than 0.1 mm d^{-1}. The procedure for pulling down the lysimeter to extract the large monolith is described by Schneider et al. (1987). In this procedure the piers used to pull down the inner box form the foundation for the scale mechanism.

b. Hydraulic and Other Lysimeters. Hydraulic pillow or floating lysimeters such as those described by Hanks and Shawcroft (1965) have continued to be used, but with the increased sensitivity and reduced cost of load cells, little new construction has been done. Kruse and Neale (1985) developed a hydraulic lysimeter with a manometer. They tested it under laboratory conditions and concluded that temperature variations were the largest source of error in the system. They found that the errors would not be cumulative in the field and for periods > 10 d the error would be <4% of the mean.

Other lysimeters often consist of drums made of various materials, e.g., concrete or steel, which are weighed with a derrick system. These have limited utility in obtaining data at less than a daily resolution and are subject to many errors. This type of installation, is used when several treatments are imposed on adjacent lysimeters, does not allow comparison to surrounding fields.

2. Accuracy

Given that lysimeters are a combination mechanical and electrical system, there are limitations to the accuracy in which water loss from the soil volume can be measured. Harrold and Dreibelbis (1953) reported a precision of 0.25 mm while Pruitt and Angus (1960) reported an accuracy of 0.03 mm on their large weighing lysimeter. Van Bavel and Myers (1962) developed their lysimeter systems to have an accuracy of 0.025 mm. These accuracy limits have not been exceeded by the newer facilities reported by Dugas et al. (1985) and Howell et al. (1985b). Dugas et al. (1985) reported a sensitivity and resolution of 0.02 mm while Howell et al. (1985b) found a resolution between 0.02 and 0.05 mm. Both groups stated this degree of resolution would permit accurate measurement of hourly ET fluxes.

Factors that affect the accuracy of the lysimeter include the wind and the surface area to depth ratio of the lysimeter. Wind effects are minimized by decreasing the wall gap to lysimeter perimeter ratio. Dugas et al. (1985) used a flexible gasket to seal between the inner and outer tank; however, they still found wind interference with windspeeds above 5 m s^{-1}. They were able to minimize the wind effect by rapid sampling (5-s interval) and computing a 10-min. average of the readings. Howell et al. (1985b) found no change in sensitivity with windspeeds up to 6 m s^{-1}. Problems with wind are further complicated with tall crops, e.g., trees; Gifford et al. (1982) reported little noise and good sensitivity with windspeed up to 5 m s^{-1}. However, higher windspeeds caused the signal to be too noisy for interpretation even when integrated over hourly intervals.

Surface area-to-depth ratio affects the accuracy of lysimeter readings. Most lysimeters have optimized this parameter for the intended application. The overall accuracy of most lysimeters is considered excellent with values of 0.1 mm attainable on a daily basis. These values are obtainable with careful management and operation.

3. Size and Depth Considerations

Van Bavel (1961) defined the criteria followed today for lysimeter depth and drainage consideration if soil water extraction is to represent field situations. The depth, however, depends upon the application. If the application is to evaluate ET fluxes under conditions not limited by water, then lysimeter depths of 1 m can be effectively used. However, if the application involves a more realistic simulation of field water-extraction patterns, then a depth of 2 m with a drainage system should be used.

Pruitt and Lourence (1985) showed the problem with shallow lysimeters under less than optimum water conditions. They found hastened rate of maturity in wheat (*Triticum aestivum* L.) and increased plant stress in the lysimeter compared to the surrounding field. This difference can also occur throughout the season if the soil water extraction profiles are different between lysimeter and field, particularly for studies designed to follow crop response to soil water depletion.

4. Surrounding Area and Field Size

There are no set criteria for the size of field in which a lysimeter should be placed and most installations are located within areas with fetch of 100 m or more in the direction of prevailing wind. To be correct in the evaluation of ET formulations, the downwind distance should be sufficient to allow for a constant flux layer with the fetch: (z-d) ratio of 100:1. The surrounding area should be treated as similarly to the lysimeter as possible in order to place the crop within a microclimate indicative of its soil water status, e.g., a drying lysimeter surrounded by an irrigated crop would be placed in a situation of reduced atmospheric demand because of the evaporation from the surrounding crop (Tanner, 1967). Conversely, a well-watered lysimeter surrounded by a dry field would be subject to an oasis effect due to advection. These conditions can be easily corrected with proper lysimeter and field management.

Any ET values reported from lysimeters should include descriptions of the management practices and field surroundings. There have not been any studies on the effect of the size of surrounding fields on the ET fluxes from lysimeters. However, studies of this type may become necessary as lysimeters are used less to determine maximum rate of water loss and more to determine how crops use water during soil water depletion cycles.

5. Calibration

Although calibration procedures for lysimeters are relatively simple, they are critical for the proper operation of lysimeter installations. Calibration procedures involve the incremental addition and removal of known mass to the lysimeter while the soil surface is covered to prevent E. These tests define the accuracy and sensitivity of the scale to changes in mass. Howell et al. (1985b) described in detail their procedure for calibration. Lysimeters should be calibrated at least yearly to ensure proper operation.

6. Costs

Dugas et al. (1985) and Howell et al. (1985b) reported lysimeter construction costs. The hardware cost for each lysimeter at Fresno exceeded $16 000 (Howell et al., 1985b) while the Temple lysimeter was approximately $14 000 (Dugas et al., 1985). Howell et al. (1985b) stated that this cost is comparable to that reported by Armijo et al. (1972) for a large lysimeter installed in a grassland site. It is safe to assume that the technology reported in the most recent installations will continue to be used over the next several years, so the costs reported will be appropriate.

However, hardware cost do not include the cost of any sensors or labor for fabrication, installation, or maintenance. Above the initial labor cost, which may exceed the hardware cost by twice, it should be realized that additional labor costs will be incurred because the reliability of the system is directly proportional to its maintenance.

7. Limitations

Lysimeters provide extremely accurate and sensitive measures of ET if they are properly maintained and operated. The primary limitation to lysimeters is their cost, which limits their wide use and multiple installation. Similar installations at different sites are not available at the present time; however, this has not been a hindrance to evaluating different models as shown by Hatfield et al. (1984). Lysimeter technology will continue to be refined, but accuracies and sensitivities improved beyond what are available today may not be possible or necessary.

C. Portable Chambers

Comparison of ET among various treatments is not possible with lysimeters because few sites have multiple lysimeters. Reicosky and Peters (1977) proposed and developed a portable chamber technique as an alternative to measuring ET with lysimeters. Their chamber, $1.83 \times 2.03 \times 1.37$

m (L × W × H) covered with 0.127-mm Mylar film, was mounted on a hydraulic system of a tractor for placement over field plots. In operation the chamber is lowered over the plot for a short interval and the change in water vapor concentration is measured within the chamber.

Reicosky and Peters (1977) used wet-bulb and dry-bulb measurements of temperature to measure the change in water vapor density. The equation describing ET for this chamber is

$$\text{ET} = \frac{(\rho_{vf} - \rho_{vi})}{(t_f - t_i)} \frac{V}{A_s}$$
[2]

Reicosky et al. (1983) compared alfalfa (*Medicago sativa* L.) ET from the portable chamber and from a lysimeter. The lysimeter provided continuous ET measurements while the chamber provided a discontinuous measurement. When the chamber data were extrapolated to hourly values, the agreement throughout a day was within 1% of the lysimeter values. The largest differences occurred between 1100 and 1200 when the lysimeter exhibited erratic behavior. Integrated daily total of ET from the lysimeter was 8.0 mm while that of the chamber was 7.7 mm, which is exceptionally good agreement.

The portable chamber provides a method of measuring ET on different treatments or at different locations within a field for use in simulation model validation. Reicosky (1985) described the incorporation of CO_2 measurements on his tractor-mounted system. Baker and Musgrave (1964) used a portable field chamber to measure photosynthesis in corn (*Zea mays* L.). Zur and Jones (1984) described the results of incorporating CO_2 and water vapor measurements in a single chamber to evaluate water use efficiency. In a later study Jones et al. (1986) used a similar chamber to evaluate the instantaneous changes in water use efficiency of irrigation and fertilizer treatments. The ET data collected from the chambers provide only an instantaneous value, however, are useful for comparing different genetic strains or crop management practices. This chamber can be easily moved by two individuals and can be taken to any location in a field without destroying the crop with the tractor mounted system. Each of these systems, hand-carried or tractor-mounted, provide another measure of actual ET over crop surfaces.

The use of chambers over crops creates an artificial environment. Even though the solar radiation is only slightly modified, the windspeed is substantially reduced. The use of thin mylar (0.1 mm) has good thermal infrared transmissivity and reduced the temperature increase. These chambers are over a plot for <2 min.; however, as shown by Reicosky et al. (1983), the agreement to lysimeter data is extremely good. The primary limitation is the lack of continuity which would be serious under partially cloudy conditions.

D. Heat Flow Measurement

Velocity of water flow in the xylem of plants was first proposed by Bloodworth et al. (1956) as a method of obtaining transpiration. Since the introduction of that technique, there have been continued refinements and adaptation to many species. The primary application of the heat flow technique has been on trees where other methods are not feasible. Swenson (1972) found a correlation of 0.98 between calculated transpiration with the heat pulse and measured transpiration with a lysimeter. These studies have been improved by Cohen et al. (1981) and Lassoie et al. (1977) for use on trees.

The technique is being applied to cotton (*Gossypium hirsutum* L.) and peanuts (*Arachis hypogaea* L.) in order to measure transpiration independently. Sakuratani (1981, 1984) described a heat balance technique that is fairly simple to use and requires only a calibration of the thermal conductivity of the stem. This technique appears to have promise as a useful procedure in a number of crop species but has not been fully tested. Cohen et al. (1987) have improved the unit for cotton which allows for improved resolution in the heat flow in the small stems. These units can remain attached for long periods of time. Continued research will improve these procedures.

II. INDIRECT MEASUREMENT METHODS

Discussion of the indirect methods of measuring ET immediately brings to mind the theoretical and empirical methods of estimating ET. Jensen (1973), in compiling methods of estimating consumptive use of water by crops, provided a complete list of the available equations used at that time. Brutsaert (1982) provided a detailed explanation of the history of E research and various meteorological approaches; refer to this treatise for a complete survey of the limitations of different approaches.

A. Actual vs. Potential Evapotranspiration

Potential E (PE) or potential ET (PET) was a concept first proposed by Thornthwaite (1948) to classify climates. The present day definition has been taken as that proposed by Penman (1956) where PET is "the amount of water transpired in a unit time by a short, green crop, completely shading the ground, of uniform height, and never short of water." It is implicit in this definition that the area is sufficiently large to avoid advective effects. Rijtema (1965) provided a comprehensive analysis of actual ET for various surfaces. Brutsaert (1982) argued that PE is an ambiguous term and is best referred to as potential evaporation since it refers to a large uniform surface, which is sufficiently wet, causing the air in contact with the surface to be fully saturated. Brutsaert also cautioned that PE is often calculated from meteorological data obtained from nonpotential conditions. This note of caution should be borne in mind when using the PE concept. Morton (1983) developed a complementary relationship between a real ET and PET, which he suggests will overcome the concerns raised by Brutsaert. Morton has shown this concept to be realistic for large watersheds. Further research will define its limitations.

The problems with PE have caused the development of the concept of reference crop evapotranspiration. *Reference crop evapotranspiration,* ET_r, is taken to mean ET from a crop grown in an environment to which the ET equations have been calibrated to represent water loss under nonlimiting soil water conditions. One example of this approach is the estimation of ET, by the Blaney-Criddle method as proposed by Doorenbos and Pruitt (1977). Calculation of reference crop evapotranspiration is a procedure commonly used in irrigation management programs and is related to empirical methods or evaporation pans. *Actual evapotranspiration,* ET_a, is the amount of water lost from the plant and soil to the atmosphere given the current meteorological conditions, soil water availability, and plant growth stage. The literature is crowded with papers relating ET_p to ET_a for different crops and soils, and space limitations prohibit a complete list of the available references. Doorenbos and Pruitt (1977) and Jensen (1973) discussed the variation among species and its effect on irrigation management and assessment of crop water requirements. Baier (1969) summarized five types of relationships between available soil water and the ratio of ET_a/PE. Since the relationships are a function of the soil water content, it may be more realistic to develop procedures which estimate ET_a directly rather than from a series of empirical relationships. Sadler and Camp (1986) compiled the available data on crop water use for the southeastern USA which list the sources of data for the major crops in this region.

B. Energy Balance

Micrometeorological or empirical methods involve partitioning of energy at the earth's surface which can be described in a simplistic form as

$$Rn - G = H + LE \tag{3}$$

or expanded to separate the net radiation terms as

$$(1 - a)S_t + Ld = H + LE + G + \epsilon\sigma T_s. \tag{4}$$

These equations, as written and used, are subject to a number of constraints: steady-state conditions, a homogeneous surface, and one-dimensional vertical transport with no horizontal gradients. This form of the equation neglects the photosynthesis, heat storage within the canopy, and advective terms which are important but not easily measured. The energy exchanges for a surface are best considered three-dimensionally as suggested by Thom (1975). In this manner, the storage of energy within the vegetation layer and the effects of advection can be treated and one can visualize the process more completely. Thom (1975) described this approach in detail and with a complete mathematical treatment. In the following discussion, the various energy balance approaches are reported and the accompanying results to date will be described. In development and evaluation, these various approaches have been compared against lysimeters as the standard.

Evapotranspiration estimated from all micrometeorological methods (e.g., profile, Bowen ratio, aerodynamic, or combination) depends on explicit assumptions about vertical fluxes and gradients. Since these assumptions are common to several methods, they are discussed first.

1. Surface Aerodynamic Properties and Stability

Wind flow over a surface causes momentum exchange and can be described by a logarithmic wind profile. Energy exchanges occur within the turbulent boundary layer above the crop. Within this region and under conditions of novertical temperature gradients, there is a simple logarithmic profile of windspeed with height which can be stated as

$$u(z) = \frac{u^*}{k} \ln \left(\frac{z - d}{z_o} \right).$$ [5]

The implicit assumption in Eq. [5] is that $z > h$. Values of z_o and d are dependent on crop height, crop density, and windspeed.

Estimation of z_o and d from the logarithmic wind profile requires an accurate measure of the windspeed with height. Computer algorithms can be developed to calculate z_o and d; however, the value of z_o is dependent upon d. Thus, an iterative procedure is necessary to arrive at these parameters. The requirement of neutral conditions is an absolute necessity for this type of analysis. Stanhill (1969) proposed that

$$z_o = 0.13 \ h$$ [6]

and

$$d = 0.65 \ h$$ [7]

would be applicable for a wide range of conditions and these parameters are routinely used for a number of crops in ET models.

Hatfield (1988) and Hatfield et al. (1985) showed that under neutral conditions for cotton with less than full ground cover, the values of z_o and d were much different than those available from the simple model of Eq. [6] and [7]. The ratio of z_o/h was 0.60 when the crop was small (ground cover < 15%) and decreased as the crop developed. In the irrigated plots the ratio of z_o/h approached 0.15 toward the end of the season. In the dryland canopies, the ratio did not decrease below 0.25 (Hatfield, 1988). Shaw and Pereira (1982), after an analysis of roughness lengths, displacement heights, and canopy foliage distributions, concluded that z_o and d were functions of vegetation density. Their simulations showed that z_o initially increased with vegetation density and then decreased while d increased monotonically with increasing vegetation density. Raupach and Legg (1984) had earlier shown through wind tunnel studies, that as the surface becomes more rough, the flux-gradient relationships become less reliable. These data suggest that limitations of the logarithmic profile occur under conditions of rough surfaces, incomplete cover, and, particularly, when the atmosphere is unstable. Continued research will provide new insights into these complexities.

Corrections for the temperature gradients within the atmosphere above the crop are necessary because of the bouyancy effect on the rate of momentum exchange. The effects of a non-neutral temperature condition on the momentum transfer has been investigated by a number of researchers (Businger et al., 1971; Dyer & Hicks, 1970; Pruitt et al., 1973). To fully understand the stability corrections it is first necessary to show the flux gradients. For momentum, heat, and water vapor these are

$$\tau = -\rho k_m \frac{\delta u}{\delta z} = -\rho k_m \frac{u^*}{kz} \phi_m , \qquad [8]$$

$$H = -\rho \text{Cp} \, k_h \frac{\delta T}{\delta z} = -\rho \text{Cp} \, k_h \frac{T^*}{kz} \phi h, \qquad [9]$$

$$E = -\frac{\rho e}{P} k_v \frac{\delta e}{\delta z} = -\frac{\rho e}{P} k_v \frac{e^*}{kz} \phi_v. \qquad [10]$$

An assumption often used is that $k_m = k_h = k_v$ over a wide range of conditions. Motha et al. (1979) found that under conditions of strong sensible heat advection that $k_h > k_v$. These differences between k_h and k_v occurred during the afternoon hours over a well-watered alfalfa canopy. Warhaft (1976) had earlier suggested that k_h would be larger than k_v when the gradients of temperature and water vapor were of opposite direction. Verma et al. (1978) found, under advective conditions, that $k_h > k_v$ because the sensible heat transfer was toward to surface. The changes in k_m relative to k_h and k_v would cause errors, particularly in situations of strong temperature gradients or within canopies (Raupach & Legg, 1984). Over nonhomogenous surfaces there are separate sources and sinks of heat, momentum, and water vapor negating the assumption of homogeneous surface and placing a constraint on the limits of Eq. [8] to [10].

The transport coefficients must be modified to account for stability. Gradients of windspeed, temperature, and water vapor occur with height above a natural surface. The temperature gradient can either enhance or detract from the rate of mixing because of only momentum transfer. Mahrt and Ek (1984) showed that atmosphere stability was a primary factor in the rate of potential evaporation and had to be considered for realistic values to be obtained. Two approaches have been used: The Monin-Obukhov length, L, and the Richardson number Ri. The Monin-Obukhov length is expressed as

$$L = -\frac{u^{*3} \, T\rho \text{Cp}}{kgH} \qquad [11]$$

and the Richardson number given as

$$Ri = \frac{g/T \ (\delta T/\delta z)}{(\delta u/\delta z)^2} \qquad [12]$$

or

$$Ri = \frac{g}{T} \frac{(T_z - T_i)(Z_2 - Z_1)}{(u_2 - u_1)^2} \qquad [13]$$

when two levels of temperature and windspeed are known. If surface temperature is measured, then Ri can be calculated from

$$Ri = (g/T) \ (z - d - z_0) \ [T(z) - T_s]/u(z)^2 \qquad [13a]$$

which will become important as procedures that use T_s are introduced.

Under stable conditions when $\phi m = \phi h = \phi v$ then

$$\phi_m = \phi_h = \phi_v = 1. + 5 \ (z - d)/L \qquad [14]$$

appears to be the best description (Dyer, 1974), while under unstable conditions

$$\phi_m = [1 - 16(z - d)/L]^{-1/4} \qquad [15]$$

and

$$\phi_h = \phi_v = [1 - 16(z - d)/L]^{-1/2}. \qquad [16]$$

If we assume that these equations hold, then

$$Ri = \frac{(z - d)}{L} \frac{\phi_h}{\phi_m{}^2} \qquad [17]$$

and we can freely substitute Ri directly into Eq. [15] to [16]. Thom (1975) agreed that Eq. [17] would be applicable under stable conditions and

$$Ri = \frac{(z - d)}{L} \phi_m \qquad [18]$$

would apply under stable conditions.

Equations [8] to [10], when integrated for windspeed, can be expanded as

$$u = \frac{u^*}{k} \ln \frac{(z - d)}{z_0} - \psi_m \qquad [19]$$

in which ψ_m are also familiar functions related to Eq. [14] to [16]. Dyer and Hicks (1970) gave empirical relationships for ψ_m, ψ_h, and ψ_v for unstable conditions which are useful in this treatise because many individuals only work with the forms of Eq. [19].

These relationships for ψ_m for unstable conditions are

$$\psi_m = 0.032 + 0.448 \ln [-(z - d)/L] - 0.132 \ln [-(z - d)/L]^2 \qquad [20]$$

and

$$\psi_h = \psi_v = 0.598 + 0.390 \ln [-(z - d)/L]$$
$$- 0.090 \ln [-(z - d)/L]^2. \qquad [21]$$

In all of these situations we can rearrange equations to calculate $u*$ as

$$u* = \frac{k[u(z) - 5(z - d - z_0)] g/T[(T_2 - T_1)/(u_2 - u_1)]}{\ln [(z - d)/z_0]} \qquad [22]$$

as given by Thom (1975). This is important for individuals who wish to use this approach in calculating the stability corrections.

The equations for ϕ_m or ϕ_h are used to calculate the rate of exchange of momentum, heat, or water vapor above the crop surface. In many approaches these are expressed as an aerodynamic resistance rather than a transfer coefficient. The aerodynamic resistance to momentum can be expressed as

$$r_{am} = u(z)/u*^2. \qquad [23]$$

According to the argument of equivalence of k_h, k_m, and k_v, r_{ah} and r_{av} can be calculated as

$$r_{ah} = r_{av} = [\ln(z - d)/z_h - \psi_h] [\ln(z - d)/z_a - \psi_m]/k^2u \qquad [24]$$

as shown by Kanemasu et al. (1979). Then z_h and z_0 are related for most agricultural situations as

$$\ln (z_0/z_h) = 2. \qquad [25]$$

The utility of this approach has not been fully exploited or evaluated in ET equations. Thom (1975) suggested that the aerodynamic resistance to momentum will always be less than the resistance to water vapor or sensible heat. Stewart and Thom (1973) referred to this as the excess resistance and formulated it for water vapor as

$$r_{bv} = r_{av} - r_{am} \qquad [26]$$

and for sensible heat as

$$r_{bh} = r_{ah} - r_{am}.$$ [27]

From Eq. [25] we can rearrange to obtain $r_{av, h}$ as

$$r_{av, h} = \frac{1}{ku^*} \ln \frac{(z - d)}{z_{v, h}}$$ [28]

with $z_{v, h}$ the roughness lengths for water vapor and sensible heat, respectively. Thom (1975) formulated $r_{bv, h}$ as

$$r_{bv, h} = \frac{1}{ku^*} \frac{\ln z_o}{z_{v, h}}$$ [29]

which leads to the conclusion that $z_{v, h}$ is about 0.2 of z_o. The treatment of the excess resistance provides a clearer understanding of the coupling between an evaporating surface and the atmosphere.

2. Profile Techniques

To utilize the profile techniques, measurements of windspeed, temperature, and water vapor are needed at a minimum of two heights. In the finite difference form with stability corrections this is given by

$$E/\lambda = \frac{(\rho\epsilon/p)k^2(\phi v/\phi_m) (e_2 - e_1) (u_2 - u_1)}{[\ln(z_2 - d)/(z_1 - d)]^2}.$$ [30]

Inspection of Eq. [30] reveals that in order to obtain good estimates of E, accurate estimates of windspeed and water vapor content are required. The instrumentation requirements and calibration techniques are discussed in detail by Fritschen and Gay (1979). The mass transfer approach has given way to the more complete approach of energy exchanges. Using an approach-based on aerodynamic theory, Penman and Long (1976) found that the results were less than desirable, partially due to their study location in the British Isles. During long periods of calm or near-clam winds, their approach did not work at all. Verma and Rosenberg (1975), after a comparison of lysimetric, energy balance, and aerodynamic methods, concluded that aerodynamic methods (Eq. [30]) can be effectively used provided that the proper stability corrections are used. The stability corrections available have prompted the use of the more complete forms of ET equations, e.g., combination approaches.

3. Turbulence Methods/Eddy Correlation

The term eddy correlation immediately brings to mind complicated, expensive instrumentation which is used in only the most sophisticated micrometeorological experiments. The theory of turbulence exchanges above a surface is relatively simple. Estimation of ET above a surface is directly calculated from

$$ET = L \overline{w' q'}. \qquad [31]$$

Kanemasu et al. (1979) summarized several problems with the practical application of Eq. [31], which include expensive, fragile instruments, and accuracy requirements for measurement of humidity fluctuations. Recent technological advances have improved the reliability of eddy correlation systems and reduced the instrumentation cost. Campbell and Unsworth (1979) and Shuttleworth et al. (1982) each described the development of a one-dimensional sonic anemometer. A fast response hygrometer has been developed by Campbell and Tanner (1985) which may provide a sensor capable of measuring humidity with same accuracy and sample rate as vertical windspeed.

Tanner et al. (1985) compared results from five one-dimensional eddy correlation systems to lysimeter and energy balance ET values at Temple, TX, and found that: (i) turbulence measurements tend to underestimate the fluxes, (ii) systematic errors occur within systems, (iii) closure between individual systems was within 25 to 30%, and (iv) the lysimeter consistently exceeded energy balance closure. Energy balance closure was defined as

$$closure = R_n - G - H - LE = 0. \qquad [32]$$

They concluded that a lysimeter would not be a necessary prerequisite for evaluating the eddy correlation method since the energy balance closure was adequate for most ET applications. Baldocchi et al. (1985) in an experiment involving eddy correlation over an oak-hickory forest, found that the lack of closure (Eq. [32]) may be due to the imprecision in estimating G and topographic effects on R_n in the upwind direction of fetch. A comprehensive analysis of the reasons for the failure to close the energy balance were not all evident but further research should improve our understanding and use of this technology.

The one-dimensional system may not be adequate and three-dimensional systems may be required. A simple system described by Coppin and Taylor (1983) used a pathlength of 0.1 m. This shorter pathlength allows for placement of the systems closer to the crop surface, thus reducing the fetch requirement which, in turn, may improve the utility of the technique. However, this has yet to be evaluated. Campbell and Unsworth (1979) also used this pathlength in their design. Wind tunnel studies by Coppin and Taylor (1983) found an accuracy to within 10% of measured flux values.

Biltott and Gaynor (1986), in comparing a three-dimensional sonic with a 0.25-m pathlength to a one-dimensional sonic with a 0.1-m pathlength, found good agreement in the vertical fluxes over a wide range of conditions. There was, however, large disagreement between instruments when large thermal changes occurred. These results suggest that a one-dimensional system may not be adequate under certain atmospheric conditions.

Eddy correlation systems may become practical tools for ET research. They may provide the instrumentation capable of calibrating other ET approaches under conditions where lysimeters do not exist. The technological problems will continue to limit their widespread application.

4. Combination Methods

Combination methods by definition imply the use of more than one method to arrive at ET estimates. These approaches involve a solution of the energy balance given as

$$R_n = H + LE + G, \qquad [33]$$

where R_n and G are measured directly and $H + LE$ are determined from various methods. Three widely used techniques to determine H and LE are the Bowen ratio, Penman, and surface energy balance. Bowen (1926) proposed a ratio of sensible to latent heat exchanges as an indicator of the rate of energy exchange at the surface. Evapotranspiration can be easily estimated from the following equation:

$$\lambda ET = \frac{R_n - G}{[(1 + \rho CP)/L_e] \, (k_h/k_v) \, (\Delta T/\Delta e)} . \qquad [34]$$

Fritschen (1965) determined the accuracy requirements for the sensors to measure temperature and humidity differences, and these guidelines are generally still followed.

Fritschen (1965) found that in comparison to lysimeters, the errors with the Bowen ratio were <5% for short-period calculations. To date, there have been several improvements in the overall system, including equalizing sensor bias automatically; however, the overall accuracy requirements have not changed. The attractive, simplifying assumption is that $k_h = k_v$, which may not always apply over crops that are not uniform in cover, e.g., partial canopies and under advection.

b. Penman and Other Aerodynamic Methods. The simplicity of the Bowen ratio approach lies with the avoidance of the measurement of the aerodynamic properties. There have been many forms of combination equations proposed in the past 40 yr, the first of which was Penman's (1948). Van Bavel (1966) showed the utility of an alternative approach. Monteith (1981) reviewed the changes in the combination approach to accommodate aerodynamic and canopy resistance. It is this latter form which has been used as the beginning for this discussion

$$ET = \frac{\Delta(R_n - G) + \rho Cp[e^*(z) - e(z)]/r_{ah}}{\Delta \times \gamma (r_{av} + r_s)/r_{ah}} \qquad [35]$$

Thom (1975) discussed this approach in detail and examined the problems associated with Eq. [35]. Van Bavel's (1966) derivation of a combination equation for ET_p and the comparison against lysimeter data for alfalfa suggest an alternative to Eq. [35]. He stated that the output of this model was not hampered by considering daily rather than hourly meteorological data. His treatment provides a good analysis of the combination method and its limitations. Buchan (1984), in an analysis of the errors associated with the bulk aerodynamic terms, concluded that the intra-diurnal fluctuation in windspeed caused errors in the estimation of E and surface temperature. He suggested that these diurnal variations need to be corrected for in a combination model when applied to a bare soil. This form of Eq. [35] proposed by Penman allows for the introduction of a surface resistance, r_s. If r_s is not included, free water E is assumed and an approximation of PE is obtained. Monteith (1965) introduced this concept into the treatment of ET from canopies. The values for r_{av} and r_{ah} are calculated via the aerodynamic approach given previously in the discussion on aerodynamic properties (Eq. [28]).

Black et al. (1970) used Eq. [35] to estimate ET for a snap bean (*Phaseolus vulgaris* L.) crop and found agreement to within 4% of a lysimeter. They also stated that better estimates of water vapor flux for the canopy are needed plus improved estimates of soil water E. The results of Black et al. (1970) are interesting because of the good model performance under conditions of less than full cover with a large amount of exposed soil. These results would suggest that the criteria for horizontal homogeneity may not be as critical as often assumed. Their model did, however, include a soil E term and a modification of Eq. [35] to the following:

$$LE = \frac{\Delta(R_n - G) + \rho Cp[e^*(z) - e(z)]/\lambda r_{av} - \Delta + \gamma \lambda E_s}{\Delta + \lambda (r_s + r_{av})r_{av}} + \lambda E_s. \qquad [36]$$

Brun et al. (1972) and Szeicz et al. (1973) found good agreement between the model- and lysimeter-obtained ET for a range of canopies, e.g., snap bean, soybean (*Glycine max* L.), and grain sorghum [*Sorghum bicolor* (L.) Moench]. These models provide a good estimation of the ET rate without the complexities of multilayer models.

In applying these models to crop canopies, the parameterization of canopy resistance has been the most difficult. Singh and Szeicz (1980) described a forest canopy resistance to water vapor transport as

$$r_c = r_s/LAI \qquad [37]$$

following the approach by Monteith (1973). Cowen (1968) and Thom (1975) both discussed the theoretical aspects in considering r_c as a bulk stomatal resistance. This analysis was based on the differences in aerodynamic resistance between momentum and sensible heat (Eq. [29]). Shuttleworth (1976) also analyzed the importance of r_c and he showed that r_c is a physiological parameter. Finnigan (1985) after evaluating Eq. [37], concluded that Eq. [37] is not entirely correct. In Eq. [37] it is assumed that the sum of the resistances, r_s, is the total resistance, r_c. However, these individual resistances exhibit a discrepancy which may be due to light response of individual stomata. Szeicz and Long (1969) showed that canopy resistance changes with soil water content, and van Bavel (1967) found that alfalfa canopy resistance became measurable as the ET rate declined below the ETp rate. Hatfield (1985) found that the canopy resistance of well-watered wheat was approximately 20 s m^{-1} and increased linearly with decreasing soil water availability. Szeicz and Long (1969) found a value of 25 s m^{-1} for alfalfa under well-watered conditions. In a controlled environment study on St. Augustine grass [*Stenotaphrum secundatum* (Walt.) Ktze.], Johns et al. (1983) found that the internal resistances were less than the external (aerodynamic) resistances. Their methodology provides a technique applicable to the computation of the resistances that could be applied to various field crops.

Chen (1985) expanded the technique of Monteith (1981) to obtain canopy resistance from profiles of temperature and humidity above the canopy. In this analysis, as in Eq. [37], canopy resistance is assumed to be approximately the bulk stomatal resistance. Several assumptions, as shown by Thom (1975), have to be considered and the introduction of an excess resistance is necessary. This complication has been avoided by developing a technique with a direct measure of surface temperature rather than extrapolating to an uncertain value of surface temperature. In a theoretical analysis of the coupling between the vegetation and the atmosphere, Shuttleworth (1976) showed that the canopy resistance had a physiological meaning but complications arose when the leaves of the canopy were wet with either dew or rain.

Webb (1984) suggested a combination approach that combines the Bowen ratio with the Penman approach. He found that the Bowen ratio-bulk aerodynamic method was more sensitive to surface characteristics, e.g., the aerodynamic resistance and canopy resistance, than the Penman equation alone. This procedure allows data to be collected in short time intervals (20 min.) which may be useful under conditions of drying soil, wet foliage, and partly cloudy skies.

Methods that have been applied directly to potential ET estimation have involved a combination of theoretical and empirical approximations. These are of the form

$$LE = \frac{(\Delta/\gamma)\,(R_n - G) + \rho Cp\,P/\epsilon_v(e_s - e_a)\,u[\ln(z - d/z_o)]^{-2}}{1 + \Delta/\gamma} \qquad [38]$$

with humidity and wind data collected at only one height. Saxton (1975) performed a sensitivity analysis on Eq. [38] and concluded that Δ, R_n, $(e_s - e_a)$, u, and z_o are all interrelated and that ET is sensitive to changes in any one of these parameters.

In most applications the u and $\ln(z - d)/z_o$ terms are replaced by a "wind function" term. Penman (1963) proposed that the wind function would be approximated by 0.26 (1 + $u/160$) with wind passage at 2 m. Doorenbos and Pruitt (1977) modified the wind function to be 0.27 (1 + $u/100$). Batchelor (1984) found there was a 23% difference between these two empirical approaches for rice (*Oryza sativa* L.) with no problem with the basic physics of the combination equation. He suggested that measurements of the parameters included in the wind function specific to a given set of conditions be used and not generalized values.

c. Surface Temperature. Equations that have combined a direct measure of the surface temperature have appeal because the extrapolation of surface temperature from the air temperature profile is no longer necessary. The energy balance partitioning would still place it within the context of a combination method; however, ET is solved as a residual from the other components. This approach is expressed as

$$LE = Rn - G - \rho Cp \, (T_s - T_a)/r_{ah}. \tag{39}$$

Technological advances in the measurement of surface temperature have allowed for this approach to become more viable and have been the focus of many studies involving remote sensing techniques. Initially, the T_s term was measured via thermometers attached to the leaf and recently surface temperature has been measured with infrared thermometry. Jackson (1982) has provided an in–depth review of the changes in technology.

Earlier, Brown (1974) and Stone and Horton (1974) evaluated this approach and began to introduce the concept of using remotely obtained surface temperature. Brown (1974) proposed that the inclusion of surface temperature eliminated the need to know canopy resistance in an actual ET model.

Heilman and Kanemasu (1976) and Verma et al. (1976) used thermocouples to obtain canopy temperatures. Heilman and Kanemasu (1976) found differences between Eq. [39] and measured ET from lysimeters of 3 and 7% for soybean and sorghum canopies, respectively. Under warm, windy conditions the agreement decreased to 15% for the sorghum and 4% for the soybean. Verma et al. (1976) also using thermocouples, found the errors between lysimeter and Eq. [39] were within 10% over a series of days. They found that errors in measuring T_s were more critical than errors in r_{ah}, especially under nonadvective conditions. Hatfield et al. (1984) evaluated Eq. [39] over a range of locations and crops and found the agreement to within 10% of the lysimeter ET over these sites. Later, Hatfield and Wanjura (1985) used this approach to estimate ET throughout a growing season for nonirrigated cotton and found the cumulative ET for a 100-d period to be within 5% of the cumulative ET estimated from a soil-water balance model. The agreement of these results are encouraging because of the length of season covered and the wide range of environmental conditions.

The problem in applying Eq. [39] has been the estimation of r_{ah} without detailed measurements of the wind profile to estimate thermal stability function. Stewart and Thom (1973), Hatfield et al. (1983), and Verma et al. (1976) showed the need for stability correction in Eq. [39] and the application by Hatfield and Wanjura (1985) used the Richardson number with surface temperature. Choudhury et al. (1986) analyzed the aerodynamic resistance for conditions ranging from stable to unstable and showed a procedure for calculating r_{ah}. In their study, the errors averaged 47 Jm^{-2} s^{-1} and the slope of the fit between measured and calculated ET values was 0.975.

The utility of remotely sensed data has been the extension to a regional scale (Schieldge, 1978). Soer (1980) proposed that remotely sensed crop temperatures could be used to estimate ET over nonuniform areas. He used a simulation model to estimate daily ET totals from the near-instantaneous measures of surface temperature. Price (1982) extended the surface temperature to estimate latent heat and sensible heat at scales of 1 km. Hatfield (1983) reviewed the various approaches of remote sensing applicable to ET. Gurney and Camillo (1984) developed a generalized model for combining remotely sensed data into ET models. Nieuwenhaus et al. (1985) applied the surface temperature approach with a simulation model and proposed that this procedure would allow for hydrologic mapping of areas in excess of 100 km^2. Equation [39] requires a combination of remotely sensed and ground-based meteorological data to estimate ET over large areas. Reginato et al. (1985) addressed this problem with surface temperature data obtained from aircraft. Their results are encouraging in terms of the possibilities of this technique. Jackson (1985) reviewed the past research in evaluating ET at local and regional scales using remotely sensed data. The remote sensing approach will continue to develop as the technology of obtaining and processing remotely sensed data improves.

5. Limitations

The limitations within the energy balance approaches are mostly associated with instrumentation and the collection of data. Accurate measurements are needed of several parameters, which has prompted the development for empirical approximations rather than the more generalized theoretical approach. However, comparisons against lysimeters have shown the energy balance approaches to be both accurate and reliable.

Current research involving simulation models and the use of high-order closure models such as those proposed by Paw et al. (1985) may offer insight into the evaporation process. As this procedure is developed, it is possible that new avenues or parameterizations of ET will result.

One limitation to the ET process and the reliability of the energy balance approach occurs under the condition of advection. Hanks et al. (1971) found errors of over 45% when warm, dry air moved over an irrigated plot. The volumetric energy balance allows for the inclusion of terms quantifying the horizontal gradients of temperature and water vapor. Lang (1973) developed a procedure using a two-dimensional energy balance where the effect of advection could be corrected. The first approximation was to use a single height measurement of horizontal temperature and humidity gradients. Lang et al. (1974) later reported that measurements made at the center of a uniform flooded rice field would be an adequate approximation for the entire field without any correction for advection. Rosenberg (1969) concluded from studies on well-watered alfalfa in Nebraska that advection of sensible heat would contribute to energy used in ET. With the current availability of computers and simulation programs, the problem with advection may be overcome with the approach proposed by McNaughton (1976a, b). This approach has been expanded and evaluated by Lang et al. (1983) to include a pressure term but they concluded that any more improvement will require better measures of the pressure gradients. Although advection has been considered important, there is little information available from large-scale studies. One such study was that conducted by Davenport and Hudson (1967a, b). They found that advection was a large component of ET in desert environments where cropped fields were interspersed among fallow fields. As the wind moved over the irrigated fields, the mean temperature decreased and humidity increased, which reduced the Penman estimates.

Use of energy balance equations requires that instrumentation be placed within a fully adjusted boundary layer over the crop. The single layer equations have explicit assumptions of no horizontal gradients and of a uniform surface. Both of these conditions are invalidated during investigations of row crops; however, the agreement between the ET models and lysimeters by Black et al. (1970) would tend to support the single layer model. Placement of instrumentation within fields with a fetch-to-measurement height ratio of 100:1 would be considered adequate and is generally followed in experimental studies.

C. Statistical Methods

Statistical or empirical methods have been derived to develop relationships that can be used with routine meteorological data. These equations predict either PE or ET_r because few of them consider any restriction to the E process. Jensen (1973) lists several empirical approaches in his treatise on consumptive use of water. Doorenbos and Pruitt (1977) listed 31 different methods to estimate crop water requirements. In this paper it is not the intent to repeat what is well described by these reports, but rather to discuss some of the limitations of these methods. Jensen (1973) provided a complete list of the coefficients derived for the various empirical methods. The empirical methods can be classified into radiation, temperature, and combination methods.

1. Radiation Methods

Jensen (1966) found that empirical methods involving either net or solar radiation take on the following general forms:

$$ET_p = C\left(1 + \frac{G + H}{R_n}\right)R_n \qquad [40]$$

or

$$ET_p = C\left(1 - \frac{L}{S_t} + \frac{G + H}{S_t}\right)S_t. \qquad [41]$$

Direct regression was used between the measured parameters and ET_p. The most-often used equation is that of Jensen and Haise (1963) with the SI unit form

$$ET_p = 0.025 \, [T - (-3)]S_t. \qquad [42]$$

They obtained these coefficients from regression of temperature, radiation, and ET_p over full-cover crops in the western USA. This type of approach has given way to the more complete energy balance approach of Priestley-Taylor or modified Penman.

2. Temperature Methods

The most-referred-to temperature methods are those proposed by Thornthwaite (1948) and Blaney and Criddle (1950). The original intent was to provide monthly estimates of ET_p rather than daily estimates, for which the methods are often used today. As shown by Tanner (1967), the generalized form of these relationships are

$$ET_p = C_1 \, d_1 \, T(C_2 - C_3 \, H). \qquad [43]$$

The Blaney-Criddle method proposed to use monthly mean temperatures calculated from daily maximum and minimum temperatures and daylength available from tables such as those given later by Doorenbos and Pruitt (1977). Normal relative humidity is obtained from climatic summaries for the location. The form of the Blaney-Criddle method is

$$ET_p = Cd \, T \qquad [44]$$

and although values for constants are available, the equation is best calibrated locally, with either a short, well-watered grass or alfalfa.

3. Modified Combination Methods

The original Penman equation was modified to replace the aerodynamic resistance term with a wind function. The exact value for the wind function is dependent upon location and Doorenbos and Pruitt (1977) showed that the vapor pressure gradient term is also locally dependent when daily averages are used. They presented the computation procedure in detail when using the modified Penman equation to estimate ET_p. Messem (1975) proposed a rapid method for the calculation of the Penman model with data obtained at weekly intervals. The agreement between the weekly and daily values for the March to October period for 7 yr was better than 95%. However, with the development of rapid computer algorithms, this type of nomogram approach is no longer necessary. Merva and Fernandez (1985) simplified the Penman model to estimate ET_p in humid regions. They replaced the saturated vapor pressure term $[e^*(z)]$ with a vapor pressure calculated from the minimum temperature. The windspeed term was found to have little influence when vapor pressures were large and they generalized this term as $(1 + 0.0062\ u)$ with windspeed in km d^{-1}. Solar radiation could be estimated from mean values for the location. Net radiation was given as

$$S_t(1 - \alpha)(0.18 + 0.55S) - \sigma T_a^4(0.56 - 0.291\ e_d^{1/2})(0.1 + 0.9S) \qquad [45]$$

which is empirically fit to their data in comparison to pan evaporation rates.

Other terms in the Penman equation have been empirically fit to field data. Grant (1975) derived empirical fits between r_a, crop height, and windspeed that are specific to the barley (*Hordeum vulgare* L.) crop of his experiment and of the form:

$$r_a = 1/u\ (0.003 + 2.4 \times 10^{-4}h + 1.9 \times 10^{-6}\ h^2). \qquad [46]$$

The empirical equation for r_s was found for a range of soil water conditions to be

$$\frac{1}{r_s} = \frac{I}{1.0\ [1 + 0.5\ (D' - 1)]} + \frac{1 - 0.69^{LAI}}{0.2\ (1.0 + 0.06\ SW)} . \qquad [47]$$

Again, this equation could be specific to his experiment but the form may provide a useful parameterization for other researchers. This approach relates actual to potential ET and has also been used by Shepherd (1973) for potato (*Solanum tuberosum* L.) in which the r_s term was modified by the leaf water content of the upper leaves.

The Priestley-Taylor model, as developed by Priestley and Taylor (1972), assumes that the vapor pressure deficits at the surface and in the air are equal, which may occur under large areas with nonadvective conditions and a wet surface. Their model is expressed in the form

$$\lambda E = \frac{\alpha'\Delta}{\Delta + \gamma}\ (R_n - G). \qquad [48]$$

The constant α' is referred to as a Priestly-Taylor constant and varies from 1.08 to 1.34. Jury and Tanner (1975) reported values of 1.28 for daily ET of irrigated potato while Kanemasu et al. (1976) found α' values of 1.28 and 1.45 for sorghum and soybean, respectively. Bailey and Davies (1981), in evaluating α' for a growing season of soybean, concluded that a value of 1.26 would apply to a freely evaporating surface with a LAI > 1. They suggested a linear decrease in α' with decreasing soil water availability. Davies and Allen (1973) proposed a modification for soil water availability of the form:

$$\alpha'' = \alpha'[1 - \exp(-10.563 \frac{Sw}{Sw_f})]. \qquad [49]$$

They suggested that surface soil water content could be used to modify α' for less than potential conditions.

Clothier et al. (1982) found the Priestley-Taylor method to be similar to the Penman method, with errors to 15 to 20% for daily ET values. They found a value of 1.21 for α' and stated that the Priestley-Taylor was preferable because of the simplicity of inputs. Gunston and Batchelor (1983) agreed that the Priestley-Taylor method is more desirable operationally compared to the Penman model because of a reduced number of inputs. They examined 30 tropical stations for wet and dry months. Wet months were defined as mean monthly rainfall > monthly mean Penman ET while dry months had monthly mean rainfall < Penman ET/2. They concluded that both models estimated reference crop ET when the meteorological data came from a grassed meteorological station in humid area. The models estimated ET_p or ET_r reliably when the monthly rainfall exceeded ET; however, the relationship became less reliable under conditions of low humidities and high wind speeds, for which the α' value of 1.26 was too small.

In describing empirical models, the implicit assumption is that LE is a function of R_n and highly coupled. McNaughton and Jarvis (1983) proposed a coupling coefficient as

$$\Omega = [1 + \lambda/(\lambda + s)] (r_c/r_{av})^{-1}. \qquad [50]$$

The coefficient has values between 0 and 1, with low values related to canopy resistance and vapor pressure deficit and not related to net radiation. High values of Ω show a strong coupling between LE and R_n. They stated that for grass surfaces values would be near 0.8. This type of analysis may provide a more thorough understanding of canopy response to the soil water supply.

Samani and Pessarakli (1986) found that the Hargreaves and Samani (1982) equation expressed as

$$ET_r = 0.0162 - S_t (T + 17.8) \qquad [51]$$

best estimated reference ET for alfalfa in arid areas compared to six other empirical methods. Hargreaves (1974) equation was originally developed as a function of air temperature. In Eq. [50], McNaughton and Jarvis estimated solar radiation from the following relationship:

$$S_t = k_T \times S \times (T_{max} - T_{min})^{1/2} \qquad [52]$$

of which k_T would be a locally derived coefficient. This equation has not been evaluated extensively over a range of environmental conditions.

4. Limitations

The approaches described as empirical or statistical methods estimate PET or ET_r. The primary limitation of any of these approaches is related to the statistical equations that were fitted to the available data. Another limitation is the availability of data with which to drive the models. Saeed (1986) gave the coefficients derived for several equations against pan evaporation and ET for a well-watered alfalfa crop. He found that different models performed better at various times of the year, indicating that the empirical coefficients are not only site specific but environmentally specific. All empirical approaches require large sets of data to obtain reasonable empirical constants applicable over a number of years. Persaud and Chang (1985) proposed using a time series analysis of solar radiation and temperature to provide input for the Jensen-Haise model. They stated that this method may provide a method of simulating data for locations with minimal or no climatic records. Their estimates of ET_p agreed closely to the daily pan evaporation data in the San Joaquin Valley of California.

Any empirical approach is constrained by the representativeness of the original data. These methods may work well for selected sites but should be used with caution beyond the original limits. The primary limitation of the empirical method is that the equations only estimate PET or ET_r. To arrive at ET_a for a crop requires an adjustment of the form

$$ET = k_c \times (ET_p \text{ or } ET_r). \qquad [53]$$

Doorenbos and Pruitt (1977) listed these for most agricultural crops that were developed under conditions of nonlimiting soil water. The crop coefficient, k_c, is crop specific and varies with stage of crop development, amount of ground cover, and soil water availability. Sammis et al. (1985) proposed that k_c values could be developed from growing-degree-days for several crops: alfalfa, cotton, corn, and sorghum. These results still vary considerably, with coefficients of determinations of 0.70; however, they stated that this procedure was better than estimating k_c from day of year only.

The k_c can vary and can include modifications such as those suggested by Grant (1975) or used by Slabbers (1980) and Wallace et al. (1981). Slabbers (1980) used an empirical fit between a critical leaf water potential, and ET_p and the fraction of soil water available to modify the ET_p rate. He showed good agreement between this model and ET_a obtained from lysimeters for a number of crops. He stated that this approach still needs verification but is simpler than obtaining canopy resistance required for the Penman model. Wallace et al. (1981), however, used a soil water balance method and the Penman model to calculate the change in surface resistance throughout a season for five crops. In turn, these relationships could be used in subsequent years to estimate ET_a.

Heilman et al. (1982) proposed that k_c could be estimated using remote sensing techniques, based on the premise that k_c was a linear function of ground cover. Ground cover could be estimated from a remotely sensed vegetative index such as the perpendicular vegetative index. Neale and Bausch (1985) compared the reflectance-derived crop coefficient with computed k_c values for corn from well-watered alfalfa and found that reflectance could provide a useful tool. They suggested that the reflectance data could provide an input to an irrigation scheduling model at each update. The use of reflectance data has not been tested in an irrigation management program.

Crop coefficients must be used with caution, since k_c should be a function of the leaf area, not of the date; most forms use the latter because it is more easily obtained after planting. Doorenbos and Pruitt (1977) gave specifics on the use of crop coefficient.

Eagleman (1971) adjusted ET_p with soil water availability to estimate ET_a over large areas. The coefficients were derived from a number of data sets from the USA and agreement with an independent soil water balance was within 5%. However, this method requires a measure of the available soil water.

D. Evaporation Pans

4. Uses

The ease of use, simplicity of data, and low cost have prompted the wide adaptation of the evaporation pan. The literature abounds in references to the use of evaporation pans and the development of crop coefficients for the estimate of ET_p from pan data. Pruitt (1966) and Jensen (1973) provided summaries of the use of evaporation pans. Pruitt (1966) described in detail the adjustments that need to be made in pan evaporation data to account for the surrounding area, windspeed, and humidity of the air. Doorenbos and Pruitt (1977) used this approach in developing their guidelines for determining crop water use. Pan evaporation data is so commonly used today that it is one of the more routine methods. Thom et al. (1981) made a comprehensive analysis between the Penman model and pan evaporation and concluded that pan evaporation was adequately described by a combination equation with an adjustment in the psychrometric constant and wind function for the pan. Their adjustment provides a meteorological quantification of the adjustments given in tables by Doorenbos and Pruitt (1977).

Pan evaporation data has been used to develop curves for potential consumptive use of water by crops. Relationships among meteorological parameters and pan evaporation have been developed for locations around the world. One example of this type of approach is that developed by Iruthayaraj and Morachan (1978). For the Monsoon and summer season in India they found different relationships between pan evaporation data and meteorological data, i.e., windspeed at 1.2 m, relative humidity, hours of sunshine per day, air temperature, and solar radiation. Their empirical approach is an extension to that proposed by Doorenbos and Pruitt (1977) and would only apply to that location and for a U.S. Weather Bureau Class A pan.

Technology has also been introduced into the evaporation pan with several techniques proposed to automate the readings in order to make it compatible with data acquisition systems. Phene and Campbell (1975), Asrar et al. (1982), and McKinion and Trent (1985) all described techniques for automating a Class A evaporation pan. These methods employ either strain gages, pressure transducers, or linear variable distance transducers to measure the level of water in the evaporation pan. The automation of pans has been proposed as input for irrigation control (Phene & Campbell, 1975).

2. Limitations

Limitations to the use of evaporation pans are related to the environment in which they are located. In dry and wet environments, the pan serves as a source of available water for wildlife, and screens are necessary to protect the pan. Howell et al. (1985a), in comparing the evaporation from screened and uncovered evaporation pans, found that the screen reduced the rate by 10%. They concluded this effect was due to reduced radiation rather than aerodynamic effects. In wet, humid environments rainfall may not be accurately measured. Thom et al. (1981) reported a 8.5% loss of rainfall by splash out of the pan. Adjustments must be made to account for the surroundings. Pruitt (1966), after detailed experiments on evaporation pans, showed how the effect of surroundings, windspeed, and humidity influenced the pan evaporation rate. The coefficients that he developed can be applied to evaporation pans to adjust for environmental changes. These limitations can be overcome but need to be considered in collecting and using the data.

III. SOIL-PLANT-ATMOSPHERE MODELS

The evolution of computer languages and computer science has prompted the increased development of computer simulation models of ET. These models are rapidly being developed and evaluated for a variety of crops and conditions. Limited space in this chapter prevents listing all of the available computer models, which would provide a survey of the current state of knowledge. The current models provide a mechanism of estimating ET_a and also the seapration of soil water E and plant transpiration. These physically based models have a minimum of empiricisms.

Ritchie (1972) first proposed that soil water E could be separated from total ET. This model is currently used as a subroutine in many more complex models. The structure of this model includes the physics of the soil water E as a function of surface soil water content and the amount of crop cover. Tanner and Jury (1976) refined this procedure to accommodate changes in α' of the Priestley-Taylor equation. Van Bavel and Hillel (1976) combined heat and water flow in a theoretical analysis to estimate E from a bare soil. Their numerical simulation procedure provides a useful tool for understanding the process of soil E. Boesten and Stroosnijder (1986) developed a single parametric model for soil water E which requires knowledge of only the soil hydraulic properties. The model does not relate cumulative soil water E to time but rather to cumulative PET.

De Jong (1981) reviewed the available soil water models to provide a comparison of the theory encompassed in each model. An example of the type of advanced soil water model currently available is that described by Hoogenboom and Huck (1986) which includes root growth, water uptake, and biomass partitioning.

More complete models, which include soil and plant influences, generally follow the electrical analog of resistances. Rose et al. (1976) referred to this type as a lumped parameter model which includes empirical relationships between the resistances to water flow and leaf water potential. Shuttleworth and Wallace (1985) used a similar approach to estimate E from sparse crops. Their approach involves a combination of energy partitioning and resistance to water flow by considering crop interception of radiation. Lascano et al. (1987), followed the concept of van Bavel et al. (1984), combined with the approach of Shuttleworth and Wallace (1985), to simulate the water balance for a dryland crop. They found that for a 74-d period the model was within 1 standard deviation of the measured ET obtained from a soil water balance.

Saxton et al. (1974) found that a soil-plant-atmosphere model based on the combination method of estimating ET could reliably estimate soil water balance for a watershed. Rosenthal et al. (1977) used the Priestley-Taylor model to estimate ET for corn and found that the model needed daily inputs of leaf area, solar radiation, precipitation, and maximum and minimum air temperatures. The model estimates were within 6% of the soil water balance.

Computer simulation models currently in use require some local calibration to account for the crop or soil characteristics. Belmans et al. (1983) and de Jong and Zentner (1985) presented analyses of models where calibrations are minimal. This trend will continue as the need for more general responses increases.

At the present time, a more thorough examination is needed of the input parameters and how they may be estimated. Camillo and Gurney (1984) provided a sensitivity analysis of an ET model much like that of Saxton (1975), but used Monte Carlo methods to simulate the daily meteorological data. Another approach is that proposed by Cordova and Bras (1981) where probability is used to describe the inputs. This may be particularly useful in areas where limited irrigation is practiced.

Computer simulation models of actual ET will provide a more complete description of the interactions between the soil-plant-atmosphere components. They will provide a clearer understanding of the physics of energy exchanges.

IV. CONCLUSION

Research on ET methods has progressed rapidly in the past 20 yr. We expect that the same amount or more understanding will be gained during the next 10 yr. The information on ET is being used each day in application ranging from irrigation scheduling to watershed hydrology. This is not to suggest, however, that we completely understand the ET process as we try to utilize the various techniques to provide daily or hourly values. There is no agreement as to which method provides the best ET estimate in comparison to the lysimeters. As we attempt to utilize ET information in management decisions, we will gain a fuller understanding of the limitations.

Research needs to continue on improving the ET methods, however, what we have available today should be fully used and evaluated. These data derived from well-conceived and evaluated formuli will be invaluable in the irrigation management of crops.

APPENDIX

Symbols

a = Albedo
α = Priestley-Taylor coefficient
A_s = Area of soil (m^2)
C = Constants in a formula
Cp = Heat Capacity of air (J kg^{-1}°C^{-1})
d = Displacement height (m)
Δ = Slope of saturation water vapor pressure curve (kPa°C^{-1})
D = Drainage (mm)
e = Ratio of molecular weight water to air (0.622)
$e(z)$ = Water vapor pressure at height z (kPa)
$e*(z)$ = Saturated water vapor pressure at height z (kPa)
ϵ = Emissivity
E = Evaporation (J m^{-2} s^{-1})
E_p = Evaporation from plant canopy (J m^{-2} s^{-1})
E_s = Evaporation from soil surface (J m^{-2} s^{-1})
ET = Evapotranspiration (J m^{-2} s^{-1}) or (mm d^{-1})
ET_a = Actual evapotranspiration (J m^{-2} s^{-1})
ET_p or PET = Potential evapotranspiration (J m^{-2} s^{-1})
ET_r = Reference crop evapotranspiration (J m^{-2} s^{-1})
γ = Psycometric constant
G = Soil heat flux (J m^{-2} s^{-1})
h = Crop height (m)
H = Sensible heat flux (J m^{-2} s^{-1})
I = Irrigation (mm)
k = von Karmans constant (0.4)
k_c = Crop coefficient (dimensionless)
L = Monin-Obukhov length
Ld = Longwave radiation(J m^{-2} s^{-1})
LAI = Leaf area index (m^2/m^2)
λ = Latent heat of vaporization
P = Atmospheric pressure (kPa)
P = Precipitation (mm)

PE = Potential evaporation (J m^{-2} s^{-1})

ϕ_m, ϕ_h, ϕ_v = Stability functions for momentum, heat, and water vapor, respectively

q' = Deviation of absolute humidity from mean

r_{bh} = Excess resistance for sensible heat (s m^{-1})

r_{bv} = Excess resistance for water vapor (s m^{-1})

r_c = Canopy resistance (s m^{-1})

r_s = Stomatal resistance (s m^{-1})

R = Runoff (mm)

Ri = Richardson number

R_n = Net radiation (J m^{-2} s^{-1})

ρ = Density of air (kg m^{-3}

ρv = Water vapor density

S_t = Solar radiation (J m^{-2} s^{-1})

SW = Soil water content (mm or m^3/m^3)

ϕ = Stefan-Boltzman constant

t_i = Time

T_a = Air temperature (°C)

T_c = Canopy temperature (°C)

T_s = Surface temperature (°C)

$T(z)$ = Air temperature at height z (°C)

ψ_m, ψ_h, ψ_v = Diabatic corrections for momentum, heat, and water vapor, respectively

u = Windspeed (m s^{-1})

$u(z)$ = Windspeed at a specific height (m s^{-1})

$u*$ = Friction velocity

V = Volume (m^3)

w' = Deviation of vertical windspeed from mean

z = Height above surface (m)

z_o = Roughness length (m)

REFERENCES

Armijo, J.D., G.A. Twitchell, R.D. Burman, and J.R. Munn. 1972. A large, undisturbed, weighing lysimeter for grassland studies. Trans. ASAE 15:827-830.

Asrar, G., R.J. Kunze, and D.E. Linuitt. 1982. Automating a class A evaporation pan for semi-continuous recording. Agric. Meteorol. 25:275-281.

Baier, W. 1969. Concepts of soil moisture availability and their effect on soil moisture estimates from a meteorological budget. Agric. Meteorol. 6:165-177.

Bailey, W.G., and J.A. Davies. 1981. Evaporation from soybean. Boundary-Layer Meteorol. 20:417-428.

Baker, D.N., and R.B. Musgrave. 1964. Photosynthesis under field conditions. V. Further plant chamber studies of the effect of light on corn (*Zea mays* L.). Crop Sci. 4:127-131.

Baldocchi, D.D., D.R. Matt, R.T. McMillen, and B.A. Hutchinson. 1985. Evapotranspiration from an oak-hickory forest. p. 414-422. *In* Advances in evapotranspiration. ASAE, St. Joseph, MI.

Batchelor, C.H. 1984. The accuracy of evapotranspiration estimated with the FAO modified Penman equation. Irrig. Sci. 5:223-233.

Belmans, C., J.G. Wesseling, and R.A. Feddes. 1983. Simulation model of the water balance of a cropped soil: SWATRE. J. Hydrol. 63:271-286.

Biltott, C.A., and J.E. Gaynor. 1986. Comparison of two types of sonic anemometers and fast response thermometers. p. 173-176. *In* Proc. 66th AMS Annu. Meet. Miami, FL. 13-17 Jan. Am. Meteorol. Soc. Boston, MA.

Black, T.A., C.B. Tanner, and W.R. Gardner. 1970. Evapotranspiration from a snap bean crop. Agron. J. 63:66–69.

Blaney, H.F., and W.D. Criddle. 1950. Determining water requirements in irrigated areas from climatological and irrigation data. SCS-TP 96. USDA-SCS, U.S. Gov. Print. Office, Washington, DC.

Bloodworth, M.E., J.B. Page, and W.R. Cowley. 1956. Some applications of the thermoelectric method for measuring water flow rates in plants. Agron. J. 48:222–228.

Boast, C.W. 1986. Evaporation from bare soil measured with high spatial resolution. *In* A. Klute (ed.) Methods of soil analysis. Part I. 2nd ed. Agronomy 9:889–900.

Boast, C.W., and T.M. Robertson. 1982. A "micro-lysimeter" method for determining evaporation from bare soil: Description and laboratory evaluation. Soil Sci. Soc. Am. J. 46:689–696.

Boesten, J.J., T.I., and L. Stroosmijder. 1986. Simple model for daily evaporation from fallow tilled soil under spring conditions in a temperate climate. Neth. J. Agric. Sci. 34:75–90.

Bowen, I.S. 1926. The ratio of heat losses by conduction and by evaporation from any water surface. Phys. Rev. 27:779–787.

Brown, K.W. 1974. Calculations of evapotranspiration from crop surface temperature. Agric. Meteorol. 14:199–209.

Brown, K.W., C.J. Gerard, B.W. Hipp, and J.T. Ritchie. 1974. A procedure for placing large undisturbed monoliths in lysimeters. Soil Sci. Soc. Am. J. 38:981–983.

Brun, L.J., E.T. Kanemasu, and W.L. Powers. 1972. Evapotranspiration from soybean and sorghum fields. Agron. J. 64:145–148.

Brutsaert, W.H. 1982. Evaporation into the atmosphere. Theory, history, and applications. D. Reidel Publ. Co., Boston.

Buchan, G.D. 1984. Correlation errors due to time-averaging of aerodynamic terms in combination estimates of evaporation and surface temperature. Agric. Meteorol. 32:23–30.

Businger, J.A., J.C. Wyngaard, Y. Izumi, and E.F. Bradley. 1971. Flux-profile relationships in the atmospheric surface layer. J. Atmos. Sci. 28:181–189.

Camillo, P.J., and R.J. Gurney. 1984. A sensitivity analysis of a numerical model for estimating evapotranspiration. Water Resourc. Res. 20:105–112.

Campbell, G.S., and B.D. Tanner. 1985. A krypton hygrometer for measurement of atmospheric water vapor concentration. p. 609–612. *In* Moisture and humidity, 1985. Instr. Soc. Am., Research Triangle Park, NC.

Campbell, G.S., and M.H. Unsworth. 1979. An inexpensive sonic anemometer for eddy correlation. J. Appl. Meteorol. 18:1072–1077.

Chen, J. 1985. A graphical extrapolation method to determine canopy resistance from measured temperature and humidity profiles above a crop canopy. Agric. For. Meteorol. 33:291–297.

Choudhury, B., R.J. Reginato, and S.B. Idso. 1986. An analysis of infrared temperature observations over wheat and calculation of latent heat flux. Agric. For. Meteorol. 37:75–88.

Clothier, B.E., J.P. Kerr, and J.S. Talbut, and D.R. Scotter. 1982. Measured and estimated evapotranspiration from well-watered crops. N.Z. J. Agric. Res. 25:301–307.

Cohen, Y., M. Fuchs, V. Falkenburg, and S. Moreshet. 1987. Calibrated heat pulse method for determining water uptake in cotton. Agron. J. 80:398–402.

Cohen, Y., M. Fuchs, and G.C. Green. 1981. Improving the heat pulse method for determination sap flow in trees. Plant Cell Environ. 4:391–397.

Coppin, P.A., and K.J. Taylor. 1983. A three-component sonic anemometer/thermometer system for general micrometerological research. Boundary-Layer Meteorol. 27:27–42.

Cordova, J.R., and R.L. Bras. 1981. Physically based probabilistic models of infiltration, soil moisture, and actual evapotranspiration. Water Resourc. Res. 17:93–106.

Cowen, I.R. 1968. Mass, heat and momentum exchange between stands of plants and their environment. Q. J. R. Meteorol. Soc. 94:523–544.

Davenport, D.C., and J.P. Hudson. 1967a. Local advection over crops and fallow. I. Changes in evaporation rates along a 17-Km transact in the Sudan Gezira. Agric. Meteorol. 4:339–352.

Davenport, D.C., and J.P. Hudson. 1967b. Local advection over crops and fallow. II. Meteorological observations and Penman estimates along a 17-km transect in the Sudan Gezira. Agric. Meteorol. 4:405–414.

Davies, J.A., and C.D. Allen. 1973. Equilibrium, potential and actual evaporation from cropped surfaces in southern Ontario. J. Appl. Meteorol. 12:649–657.

deJong, R. 1981. Soil water models. A review. Agric. Canada. Publ. LRRI 123. Ministry of Agric., Ontario.

deJong, R., and R.P. Zentner. 1985. Assessment of the SPAW model for semi-arid growing conditions with minimal local calibration. Agric. Water Manage. 10:31–46.

Doorenbos, J., and W.O. Pruitt. 1977. Guidelines for predicting crop water requirements. FAO Pap. 24. (Revised) Rome.

Dugas, W.A., D.R. Upchurch, and J.T. Ritchie. 1985. A weighing lysimeter for evapotranspiration and root measurements. Agron. J. 77:821–825.

Dyer, A.J. 1974. A review of flux-profile relationship. Boundary-Layer Meteorol. 7:363–372.

Dyer, A.J., and B.B. Hicks. 1970. Flux-gradient relationships in constant flux layer. Q. J. R. Meteorol. Soc. 96:715–721.

Eagleman, J.R. 1971. An experimentally derived model for actual evapotranspiration. Agric. Meteorol. 8:385–394.

Edwards, W.R.N. 1986. Precision weighing lysimetry for trees. Tree Physiol. 1:127–144.

Finnigan, J.J. 1985. Turbulent transport in flexible plant canopies. p. 443–480. In B.A. Hutchinson and B.B. Hicks (ed.) The forest-atmosphere interaction. D. Reidel Publ. Co., New York.

Fritschen, L.J. 1965. Accuracy of evapotranspiration determinations by the Bowen ratio method. I.A.S.H. Bull. 10:38–48.

Fritschen, L.J., L. Cox., and R. Kinerson. 1973. A 28-meter Douglas-fir tree in a weighing lysimeter. For. Sci. 19:256–261.

Fritschen, L.J., and L. Gay. 1979. Environmental Instrumentation. Springer-Verlag, New York, New York.

Gifford, H.H., D. Whitehead, R.S. Thomas, and D.S. Jackson. 1982. Design of a new weighing lysimeter for measuring water use by individual trees. N.Z. J. For. Sci. 12:448–456.

Grant, D.R. 1975. Comparison of evaporation from barley with Penman estimates. Agric. Meteorol. 15:49–60.

Green, G.C., W.P. Burger, and J. Conradie. 1974. Lysimetric determination of citrus tree evapotranspiration. Agrochemophysica 6:35–42.

Green, G.C., and W. Bruwer. 1979. An improved weighing lysimeter facility for citrus evapotranspiration studies. Water SA 5:189–195.

Gunston, H., and C.H. Batchelor. 1983. A comparison of the Priestley-Taylor and Penman methods for estimating reference crop evapotranspiration in tropical countries. Agric. Water Manage. 6:65–77.

Gurney, R.J., and P.J. Camillo. 1984. Modeling daily evapotranspiration using remote sensed data. J. Hydrol. 69:305–324.

Hanks, R.J., L.H. Allen, and H.R. Gardner. 1971. Advection and transpiration of wide-row sorghum in the central Great Plains. Agron. J. 63:520–527.

Hanks, R.J., and R.W. Shawcroft. 1965. An economical lysimeter for evapotranspiration studies. Agron. J. 57:634–636.

Hargreaves, G.H. 1974. Estimation of potential and crop evapotranspiration. Trans. ASAE 17:701–704.

Hargreaves, G.H., and Z.A. Samani. 1982. Estimating potential evapotranspiration. J. Irrig. Drain. Div. Am. Soc. Civ. Eng. 108:223–230.

Harrold, L.L., and F.R. Dreibelbis. 1953. Water use by crops as determined by weighing monolith lysimeters. Soil Sci. Soc. Am. Proc. 17:70–74.

Hatfield, J.L. 1983. Evapotranspiration obtained from remote sensing methods. p. 395–416. In D.E. Hillel (ed.) Advances in irrigation. Academic Press, New York.

Hatfield, J.L. 1985. Wheat canopy resistance determined by energy balance techniques. Agron. J. 77:279–283.

Hatfield, J.L. 1988. Priorities in ET Research: Evolving methods. Trans. ASAE 31:490–495.

Hatfield, J.L., A. Perrier, and R.D. Jackson. 1983. Estimation of evapotranspiration at one-time-of-day using remotely sensed surface temperature. Agric. Water Manage. 7:341–350.

Hatfield, J.L., R.J. Reginato, and S.B. Idso. 1984. Evaluation of canopy temperature-evapotranspiration over various crops. Agric. Meteorol. 32:41–53.

Hatfield, J.L., and D.F. Wanjura. 1985. Actual evapotranspiration from dryland agriculture. p.151–158. In Advances in evapotranspiration. ASAE, St. Joseph, MI.

Hatfield, J.L., D.F. Wanjura, and G.L. Barker. 1985. Canopy temperature response to water stress under partial canopy. Trans. ASAE 28:1607–1611.

Haverkamp, R., M. Vauclin, and G. Vachaud. 1984. Error analysis in estimating soil water content from neutron probe measurements. I. Local standpoint. Soil Sci. 137:78–90.

Heilman, J.L., W.G. Heilman, and D.G. Moore. 1982. Evaluating the crop coefficient using spectral reflectances. Agron. J. 74:967–971.

Heilman, J.L., and E.T. Kanemasu. 1976. An evaluation of a resistance form of the energy balance to estimate evapotranspiration. Agron. J. 68:607–611.

Hoogenboom, G., and M.G. Huck. 1986. Rootsim V4.0. A dynamic simulation of root growth, water uptake, and biomass partitioning in a soil-plant-atmosphere continuum: Update and documentation. Auburn Univ. Res. Bull. 109.

Howell, T.A., C.J. Fine, D.W. Meek, and R.J. Miller. 1985a. Evaporation from screened Class A pans in a semi-arid climate. Agric. Meteorol. 29:111–124.

Howell, T.A., R.L. McCormick, and C.J. Phene. 1985b. Design and installation of large weighing lysimeters. Trans. ASAE 28:106–112, 117.

Iruthayaraj, M.R., and Y.B. Morachan. 1978. Relationship between evaporation from different evaporimeters and meteorological parameters. Agric. Meteorol. 19:93–100.

Jackson, R.D. 1982. Canopy temperature and crop water stress. p. 43–85. In D.E. Hillel (ed.) Advances in irrigation. Academic Press, New York.

Jackson, R.D. 1985. Evaluating evapotranspiration of local and regional scales. Proc. Inst. Electrical Electronics Eng. 73:1086–1096.

Jensen, M.E. 1966. Empirical methods of estimating or predicting evapotranspiration using radiation. p. 57–61, 64. In Proc. Conf. on evapotranspiration. ASAE, St. Joseph, MI.

Jensen, M.E. (ed.). 1973. Consumptive use of water and irrigation water requirements. Am. Soc. Civ. Eng., New York.

Jensen, M.E., and H.R. Haise. 1963. Estimating evapotranspiration from solar radiation. J. Irrig. Drain. Div. Am. Soc. Civ. Eng. 89:15–41.

Johns, D., J.B. Beard, and C.H.M. van Bavel. 1983. Resistances to evapotranspiration from a St. Augustine grass turf canopy. Agron. J. 75:419–422.

Jones, J.W., B. Zur, and J.M. Bennett. 1986. Interactive effects of water and nitrogen stresses on carbon and water vapor exchange of corn canopies. Agric. For. Meteorol. 38:113–126.

Jury, W.A., and C.B. Tanner. 1975. Advection modification of the Priestley and Taylor evapotranspiration formula. Agron. J. 67:840–842.

Kanemasu, E.T., L.R. Stone, and W.L. Powers. 1976. Evapotranspiration model tested for soybean and sorghum. Agron. J. 68:569–572.

Kanemasu, E.T., M.L. Wesley, B.B. Hicks, and J.L. Heilman. 1979. Techniques for calculating energy and mass fluxes. p. 156–182. In B.J. Barfield and J.F. Gerber. (ed.) Modification of the aerial environment of crops. ASAE, St. Joseph, MI.

Kruse, E.G., and C.M.U. Neale. 1985. Sources of error in hydraulic weighing lysimeter data. p. 70–78. In Advances in evapotranspiration. ASAE, St. Joseph, MI.

Lang, A.R.G. 1973. Measurement of evapotranspiration in the presence of advection by means of a modified energy balance procedure. Agric. Meteorol. 12:75–81.

Lang, A.R.G., G.N. Evans, and P.Y. Ho. 1974. The influence of local advection on evapotranspiration from irrigated rice in a semi-arid region. Agric. Meteorol. 13:5–13.

Lang, A.R.G., K.G. McNaughton, C. Fozu, E.F. Bradley, and E. Ohtaki. 1983. An experimental appraisal of the terms in the heat and moisture flux equations for local advection. Boundary-Layer Meteorol. 25:89–102.

Lascano, R.J., J.L. Hatfield, and C.H.M. van Bavel. 1986. Field calibration of neutron meters using a two-probe, gamma density gauge. Soil Sci. 141:442–477.

Lascano, R.J., C.H.M. van Bavel, J.L. Hatfield, and D.R. Upchurch. 1987. Simulation and measurement of water use by cotton in a semiarid climate. Soil Sci. Soc. Am. J. 51:1113–1121.

Lassoie, J.P., D.R. Scott, and R.M. Fritschen. 1977. Transpiration studies in Douglas-fir using the heat pulse technique. Forest Sci. 23:377–390.

Marek, T.H., A.D. Schneider, T.A. Howell, and L.L. Ebeling. 1987. Design and construction of large weighing monolith lysimeters. Trans. ASAE 31:477–489.

Mahrt, L., and M. Ek. 1984. The influence of atmospheric stability on potential evaporation. J. Clim. Appl. Meteorol. 23:222–234.

McKinion, J.M., and A. Trent. 1985. Automation of a Class A evaporation pan. Trans. ASAE 28:169–171.

McNaughton, K.G. 1976a. Evaporation and advection. I. Evaporation from extensive homogeneous surfaces. Q. J. R. Meteorol. Soc. 102:181–191.

McNaughton, K.G. 1976b. Evaporation and advection. II. Evaporation downward of a boundary separating regions having different surface resistances and available energies. Q. J. R. Meteorol. Soc. 102:193–202.

McNaughton, K.G., and P.G. Jarvis. 1983. Predicting effects of vegetation changes on transpiration and evaporation. p. 1–47. In T. Kozlowski (ed.) Water deficits and plant growth. Academic Press, New York.

Merva, G., and A. Fernandez. 1985. Simplified application of Penman's equation for humid regions. Trans. ASAE 28:819–825.

Messem, A.B., 1975. A rapid method for the determination of potential transpiration derived by the Penman combination model. Agric. Meteorol. 14:369–384.

Monteith, J.L. 1965. Evaporation and environment. p. 205–234. In G.E. Fogg (ed.) The state and movement of water in living organisms. Proc. Symp. Soc. Exp. Biol. Vol. 19. Academic Press, New York.

Monteith, J.L. 1973. Principles of environmental physics. Arnold Press, London.

Monteith, J.L. 1981. Evaporation and surface temperature. Q. J. R. Meteorol. Soc. 107:1–27.

Monteith, J.L. 1985. Evaporation from land surfaces: Progress in analysis and prediction since 1948. p. 1–12. In Advances in evapotranspiration. ASAE, St. Joseph, MI.

Morton, F.I. 1983. Operational estimates of a real evapotranspiration and their significance to the science and practice of hydrology. J. Hydrol. 66:1–76.

Motha, R.P., S.B. Verma, and N.J. Rosenberg. 1979. Exchange coefficients under sensible heat advection determined by eddy correlation. Agric. Meteorol. 20:273–280.

Mottram, R., and J.M. de Jager. 1973. A sensitive recording lysimeter. Agrochemophysica 5:9–14.

Nakayama, F.S., and R.J. Reginato. 1982. Simplifying neutron moisture meter calibration. Soil Sci. 133:48–52.

Neale, C.M.U., and W.C. Bausch. 1985. Reflectance-based crop coefficients for use in irrigation scheduling models. p. 250–258. In Advances in evapotranspiration. ASAE, St. Joseph, MI.

Nieuwenhaus, G.J.A., E.H. Smidt, and H.A.M. Thunnissen. 1985. Estimation of regional evapotranspiration of arable crops from thermal infrared images. Int. J. Remote Sens. 6:1319–1334.

Paw, U., K.T., R.H. Shaw, and J.P. Meyers. 1985. Evapotranspiration as modeled by a higher order closure scheme. p. 43–50 In Advances in evapotranspiration. ASAE, St. Joseph, MI.

Pelton, W.L. 1961. The use of lysimetric methods to measure evapotranspiration. p. 106–134. In Proc. 2nd Hydrol. Symp. Queens Printer, Ottawa.

Penman, H.L. 1948. Natural evaporation from open water, bare soil, and grass. Proc. R. Soc. London 193:120–146.

Penman, H.L. 1956. Evaporation: An introductory survey. Neth. J. Agric. Sci. 4:9–29.

Penman, H.L. 1963. Vegetation and hydrology. Tech. Commun. Common. Bur. Soils 53. London, England.

Penman, H.L., and I.F. Long. 1976. Profiles and evaporation. Q. J. R. Meteorol. Soc. 102:841–855.

Persaud, N., and A.C. Chang. 1985. Time series analysis of daily solar radiation and air temperature measurements for use in computing potential evapotranspiration. Trans. ASAE 28:462–470.

Phene, C.J., and R.B. Campbell. 1975. Automating pan evaporation measurements for irrigation control. Agric. Meteorol. 15:181–191.

Price, J.L. 1982. On the use of satellite data to infer surface fluxes at meteorological scales. J. Appl. Meteorol. 21:1111–1122.

Priestley, C.H.B., and R.J. Taylor. 1972. On the assessment of surface heat flux and evaporation using large scale parameters. Mon. Weather Rev. 100:81–92.

Pruitt, W.O. 1966. Empirical method of estimating evapotranspiration using primary evaporation pans. p. 57–61. In Proc. Conf. on Evapotranspiration. ASAE, St. Joseph, MI.

Pruitt, W.O., and D.E. Angus. 1960. Large weighing lysimeter for measuring evapotranspiration. Trans. ASAE 3:13–18.

Pruitt, W.O., and E.T. Lourence. 1985. Experiences in lysimetry for ET and surface drag measurements. p. 51–69. In Advances in evapotranspiration. ASAE, St. Joseph, MI.

Pruitt, W.O., D.L. Morgan, and F.J. Lourence. 1973. Momentum and mass transfer in the surface boundary layer. Q. J. R. Meteorol. Soc. 99:370–386.

Raupach, M.R., and B.J. Legg. 1984. The uses and limitations of flux-gradient relationships in micrometeorology. Agric. Water Manage. 8:119–131.

Reginato, R.J., R.D. Jackson, and P.J. Pinter, Jr. 1985. Evapotranspiration calculated from remote multispectral and ground station meteorological data. Remote Sens. Environ. 18:75–89.

Reicosky, D.C. 1985. Advances in evapotranspiration measured using portable field chambers. p. 79–86. *In* Advances in evapotranspiration. ASAE, St. Joseph, MI.

Reicosky, D.C., and D.B. Peters. 1977. A portable chamber for rapid evapotranspiration measurements on field plots. Agron. J. 69:729–732.

Reicosky, D.C., B.S. Sharratt, J.E. Ljungkull, and D.G. Baker. 1983. Comparison of alfalfa evapotranspiration measured by a weighing lysimeter and a portable chamber. Agric. Meteorol. 28:205–211.

Reynolds, W.D., and G.K. Walker. 1984. Development and validation of a numerical model simulating evaporation from short cores. Soil Sci. Soc. Am. J. 48:960–969.

Rijtema, P.E. 1965. An analysis of actual evapotranspiration. Versl. Landbouwk. Onderz. 659:1–107.

Ritchie, J.T. 1972. Model for predicting evaporation from a row crop with incomplete cover. Water Resourc. Res. 8:1204–1213.

Ritchie, J.T., and E. Burnett. 1968. A precision weighing lysimeter for row crop water use studies. Agron. J. 60:545–549.

Rose, C.W., G.F. Byrne, and G.K. Hansen. 1976. Water transport from soil through plant to atmosphere: A lumped parameter model. Agric. Meteorol. 16:171–184.

Rosenberg, N.J. 1968. Evapotranspiration: Review of Research. Nebr. Agric. Exp. Stn. Bull. MP20.

Rosenberg, N.J. 1969. Advection contribution of energy utilized in evapotranspiration in the east central Great Plains (USA). Agric. Meteorol. 6:179–184.

Rosenthal, W.D., E.T. Kanemasu, and R.J. Raney, and L.R. Stone. 1977. Evaluation of an evapotranspiration model for corn. Agron. J. 69:461–464.

Rouse, W.R., and R.G. Wilson. 1972. A test of the potential accuracy of the water-budget approach to estimating evapotranspiration. Agric. Meteorol. 9:421–446.

Russo, D. 1986a. A stochastic approach to the crop yield-irrigation relationships in heterogeneous soils. I. Analysis of the field spatial variability. Soil Sci. Soc. Am. J. 50:736–745.

Russo, D. 1986b. A stochastic approach to the crop yield-irrigation relationships in heterogeneous soils. II. Application to irrigation management. Soil Sci. Soc. Am. J. 50:745–751.

Sadler, E.T., and C.R. Camp. 1986. Crop water use data available from the southeastern USA. Trans. ASAE 29:1070–1079.

Saeed, M. 1986. The estimation of evapotranspiration by some equations under hot and arid conditions. Trans. ASAE 29:434–438.

Sakuratani, T. 1981. A heat balance method for measuring water flux in the stem of intact plants. J. Agric. Meteorol. 37:9–17.

Sakuratani, T. 1984. Improvements of the probe for measuring water flow rate in intact plants with a stem heat balance method. J. Agric. Meteorol. 40:273–277.

Samani, Z.A., and M. Pessarakli. 1986. Estimating potential crop evapotranspiration with minimum data in Arizona. Trans. ASAE 29:522–524.

Sammis, T.W., C.L. Mapel, and D.G. Lugg, R.N. Lansford, and J.T. McGuckin. 1985. Evapotranspiration crop coefficients predicted using growing-degree-days. Trans. ASAE 28:773–780.

Saxton, K.E. 1975. Sensitivity analysis of the combination evapotranspiration equation. Agric. Meteorol. 15:343–353.

Saxton, K.E., H.P. Johnson, and R.H. Shaw. 1974. Watershed evapotranspiration estimated by the combination method. Trans. ASAE 17:668–672.

Schieldge, J.P. 1978. On estimating the sensible heat flux over land. Agric. Meteorol. 19:315–328.

Schneider, A.D., T.H. Marck, L.L. Ebeling, T.A. Howell, and J.L. Steiner. 1987. Hydraulic pulldown procedure for collecting large soil monoliths. Trans. ASAE 31:1092–1097.

Shaw, R.H., and A.R. Pereira. 1982. Aerodynamic roughness of a plant canopy: A numerical experiment. Agric. Meteorol. 26:51–65.

Shawcroft, R.W., and H.R. Gardner. 1983. Direct evaporation from soil under a row crop canopy. Agric. Meteorol. 28:29–238.

Shepherd, W. 1973. Plant moisture status effects in estimation of evaporation by a combination methods. Agric. Meteorol. 11:213–222.

Shuttleworth, W.J. 1976. A one-dimensional theoretical description of the vegetation-atmosphere interaction. Boundary-Layer Meteorol. 10:273–302.

Shuttleworth, W.J., D.D. McNeil, and C.J. Moore. 1982. A switched continuous-wave sonic anemometer for measuring surface heat fluxes. Boundary-Layer Meteorol. 23:425–448.

Shuttleworth, W.J., and J.S. Wallace. 1985. Evaporation from sparse crops—an energy combination theory. Q. J. R. Meteorol. Soc. 11:839–855.

Sinclair, D.F., and J. Williams. 1979. Components of variance involved in estimating soil water content and water content changes using a neutron moisture meter. Aust. J. Soil Res. 17:237–247.

Singh, B., and G. Szeicz. 1980. Predicting the canopy resistance of a mixed hardwood forest. Agric. Meteorol. 21:49–58.

Slabbers, P.J. 1980. Practical prediction of actual evapotranspiration. Irrig. Sci. 1:185–196.

Soer, G.J.R. 1980. Estimation of regional evapotranspiration and soil moisture conditions using remotely sensed crop surface temperatures. Remote Sens. Environ. 9:27–45.

Stanhill, G. 1969. A simple instrument for the field measurement of turbulent diffusion flux. J. Appl. Meteorol. 8:508–512.

Stewart, R.B., and A.S. Thom. 1973. Energy budgets in pine forest. Q. J. R. Meteorol. Soc. 99:154–170.

Stone, L.R., and M.L. Horton. 1974. Estimating evapotranspiration using canopy temperature: Field evaluation. Agron. J. 66:450–454.

Swenson, R.H. 1972. Water transpired by trees is indicated by heat pulse velocity. Agric. Meteorol. 10:277–281.

Szeicz, G., and I.F. Long. 1969. Surface resistance of crop canopies. Water Resour. Res. 5:622–633.

Szeicz, G., C.H.M. van Bavel, and S. Takami. 1973. Stomatal factor in the water use and dry matter production by sorghum. Agric. Meteorol. 12:361–389.

Tanner, B.D., M.J. Tanner, W.A. Dugas, E.C. Campbell, and B.L. Bland. 1985. Evaluation of an operational eddy correlation system for evapotranspiration measurements. p. 87–99. *In* Advances in evapotranspiration. ASAE, St. Joseph, MI.

Tanner, C.B. 1967. Measurement of evapotranspiration. *In* Irrigation of agricultural lands. R.M. Hagen et al. (ed.) Agronomy 11:534–574.

Tanner, C.B., and W.A. Jury. 1976. Estimating evaporation and transpiration from a row crop with incomplete cover. Agron. J. 68:239–243.

Thom, A.S. 1975. Momentum, mass, and heat exchange. p. 57–109. *In* J.L. Monteith (ed.) Vegetation and the atmosphere. Academic Press, New York.

Thom, A.S., J.L. Thony, and M. Vauclin. 1981. On the proper employment of evaporation pans and atmometers in estimating potential transpiration. Q. J. R. Meteorol. Soc. 107:711–736.

Thornthwaite, C.W. 1948. An approach toward the rationale classification of climate. Geogr. Rev. 38:55–94.

van Bavel, C.H.M. 1961. Lysimeter measurements of evaportranspiration rates in the eastern United States. Soil Sci. Soc. Am. Proc. 25:138–141.

van Bavel, C.H.M. 1966. Potential evaporation: The combination concept and its experimental verification. Water Resour. Res. 2:455–467.

van Bavel, C.H.M. 1967. Changes in canopy resistance to water loss from alfalfa induced by soil water depletion. Agric. Meteorol. 4:165–176.

van Bavel, C.H.M., and D.E. Hillel. 1976. Calculating potential and actual evaporation from a bare soil surface by simulation of concurrent flow of water and heat. Agric. Meteorol. 17:453–476.

van Bavel, C.H.M., R.J. Lascano, and L. Stroosnijder. 1984. Test and analysis of a model of water use by sorghum. Soil Sci. 137:443–456.

van Bavel, C.H.M., and L.E. Myers. 1962. An automatic weighing lysimeter. Agric. Eng. 43:580–583, 587–588.

Verma, S.B., and N.J. Rosenberg. 1975. Accuracy of lysimetric, energy balance, and stability-corrected aerodynamic methods of estimating above-canopy flux of CO_2. Agron. J. 67:699–704.

Verma, S.B., N.J. Rosenberg, and B.L. Blad. 1978. Turbulent exchange coefficients for sensible heat and water vapor under advective conditions. J. Appl. Meteorol. 17:330–338.

Verma, S.B., N.J. Rosenberg, B.L. Blad, and M.U. Baradas. 1976. Resistance energy balance method for predicting evapotranspiration: Determination of boundary layer resistance and evaluation of error effects. Agron. J. 68:776–782.

Vieira, S.R., J.L. Hatfield, D.R. Nielsen, and J.W. Biggar. 1983. Geostatistical theory and application to variability of some agronomic properties. Hilgardia 51:1–75.

Walker, G.K. 1983. Measurement of evaporation from soil beneath crop canopies. Can. J. Plant Sci. 63:137–141.

Wallace, J.S., C.H. Batchelor, and M.G. Hodnett. 1981. Crop evaporation and surface conductance calculated using soil moisture data from Central India. Agric. Meteorol. 25:83–96.

Warhaft, Z. 1976. Heat and moisture flux in the stratified boundary layer. Q. J. R. Meteorol. Soc. 102:703–704.

Webb, E.K. 1984. Evaluation of evapotranspiration and canopy resistance: An alternative combination method. Agric. Water Manage. 8:151–166.

Zur, B., and J.W. Jones. 1984. Diurnal changes in the instantaneous water use efficiency of a soybean crop. Agric. For. Meteorol. 33:41–51.

16

Comparison of Irrigation Systems

E. GORDON KRUSE

USDA-ARS
Fort Collins, Colorado

DALE A. BUCKS

USDA-ARS
Beltsville, Maryland

R. D. VON BERNUTH

University of Tennessee
Knoxville, Tennessee

Modern irrigation methods can be divided into five categories: surface, sprinkler, micro (drip/trickle), subirrigation, and hybrid systems. In *surface irrigation*, the irrigation water supply is introduced at one edge of a field and flows across the soil surface by gravity, infiltrating into the soil while the stream advances across or is ponded on the field. *Sprinkler irrigation* systems are those where water is supplied in a pressurized network and emitted from sprinkler heads mounted on either fixed or moving supports. *Micro irrigation* includes the methods commonly known as drip or trickle irrigation, and other low pressure systems. Water is often distributed in plastic conduits and emitted through drippers, tricklers, bubblers, small misters, foggers, or sprayers. On well-designed sprinkler and micro systems, the water infiltrates the soil at or very near the point where it is applied. There is minimal redistribution of water by flow on the soil surface. *Subirrigation* systems provide water to crops by controlling the water-table level so that crop roots can reach the capillary fringe above the water table and extract all their water needs from it. Hybrid irrigation methods are those systems that don't neatly fall within the categories of the other four methods.

In this chapter, the on-farm irrigation methods will be described and compared in more detail. Systems commonly in use in the 1980s will be treated in greatest detail, with minimal attention to those older systems that do not require a great deal of care in design and that frequently perform at low levels of efficiency.

The authors wish to emphasize that surface, sprinkler, micro, and other irrigation methods all have their place in present-day agriculture. Each can be efficient in use of water, in labor requirements, and in financial feasibility if properly matched to the site to be irrigated and managed. Similarly, even the best designed, inherently efficient system will perform poorly if management of the system is inadequate.

Several reference works will be useful to readers interested in the design, evaluation, or management of any type of on-farm irrigation system. Some are listed here to avoid the need to mention them repeatedly in the following sections of this chapter. Design and efficiency concepts and useful design charts and tables are available in the Irrigation section of the *Soil Conservation Service National Engineering Handbook* (USDA, 1967, 1974, 1983a, b, 1984b). Merriam and Keller (1978) gave detailed procedures for evaluating the efficiencies of common irrigation systems. An overview of all aspects of on-farm irrigation is contained in Jensen (1980). Proceedings of *Specialty Conferences of the Irrigation and Drainage Division of the American Society of Civil Engineers* and Symposia of the *Soil and Water Division of the American Society of Agricultural Engineers* also contain much useful information that is not widely available.

I. DESCRIPTION OF ON-FARM IRRIGATION METHODS

A. Surface Irrigation

Surface irrigation systems range from wild flooding, where little if any land preparation precedes irrigation, to basin systems where the soil surface is precisely leveled. Graded surface irrigation systems are those where the field surface has a slope. The energy gradient causing advance of water over the field is a combination of the soil surface slope and the water surface gradient of the advancing stream. In the various types of level basin systems, the field surface has no slope and stream advance is a result of the water surface gradient only.

The uniformity of water infiltration on a surface-irrigated field is dependent upon the variations in the infiltration characteristics of the soil and on variations in the *intake opportunity time*—that time that a unit area of soil surface is inundated. A rigorous analysis of surface irrigation systems (e.g., mathematical simulation of irrigation stream movement) often considers the first phase of the irrigation as the time during which the stream advances from the field turnout to the downstream end of the field where runoff potentially begins (Bassett et al., 1980). Following advance, there is a water storage or ponding phase while the irrigation stream is still being introduced at the upstream end of the field. During ponding, part of the irrigation stream is running off the lower end of graded fields but depth of water stored on the soil surface is increasing. In the third or depletion phase, inflow is stopped and water depth at the upslope of the field begins to decrease. During depletion the surface water may be redistributing itself at other points in the field. Finally, in the recession phase free water disappears from the upstream end of the field and the recession front travels across the field until the surface is no longer inundated. For fields where there is no slope, the time of recession is zero in theory, and in practice may be small.

To obtain uniform intake opportunity time across a field, the advance and recession times should be equalized. In other words, plotted curves of advance and recession distance with time should be parallel so that all sections of the fields are inundated for nearly equal times (Fig. 16–1a). If this cannot be accomplished, as on level fields, the advance time should be as short as possible so that it does not represent a significantly large portion of the total intake opportunity time.

With a more-or-less constant, shallow depth of water on the soil surface, such as occurs in surface irrigation, the variation of infiltration rate with time can be described by the infiltration rate curve in Fig. 16–1b. The second curve shows the accumulated depth of infiltration with time, the integral of the area under the infiltration rate curve. Since infiltration rate decreases with intake opportunity time, it is apparent that variations in infiltrated depth are not as great as variations in infiltration opportunity time. Thus, if the soil characteristics are uniform across the field, the final depth of infiltrated water may be more uniform than the intake opportunity time variation would suggest.

1. Surface Irrigation Methods

a. Flooding and Contour Ditches

Some of the oldest irrigation systems in the USA can be described as wild flooding. The soil surface topography is not appreciably modified for flood irrigation. Irrigators, based on their experience, bring water to the high points of fields in ditches, then release it to cover the field in a somewhat haphazard manner. If the irrigation stream has to run any great distance it will tend to flow to the lower spots, causing overirrigation, and leave the ridges or high areas too dry. In some cases, irrigation runs are made shorter by introducing water in a series of more or less parallel ditches, breaking larger fields into smaller ones, and allowing more uniform water distribution. Uniformity of water coverage is sometimes improved by small, hand-shoveled dikes and ditches through the field.

b. Border Dike

In the *border dike* method, earthen dikes are constructed parallel to the direction of maximum land slope. Border strips between dikes are essentially level in a direction perpendicular to the dikes. Water introduced at the upstream end of such a strip tends to spread across the full width as it advances down the field. Hay or small grains are sometimes planted perpendicular to the slope so the closely spaced plants in each drill row will encourage water to distribute itself across the width of the strip. As in all graded surface irrigation systems, water that does not infiltrate will flow from the field as surface runoff.

Irrigation stream sizes need to be relatively large in order that the advance phase will be short enough to permit uniform infiltration. Border dikes must be high enough to contain this stream. Construction should allow for

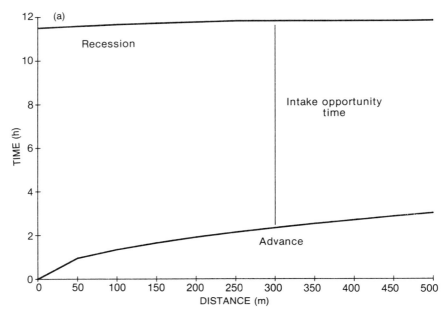

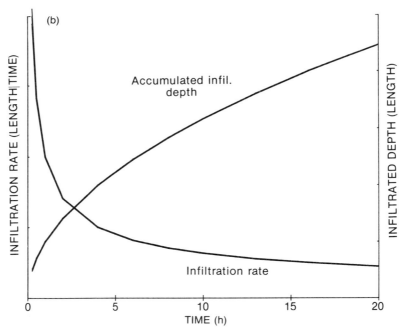

Fig. 16-1. (a) Typical advance and recession curves for graded furrow irrigation and (b) trends of infiltration rates and accumulated depths, with time for typical irrigated soils.

initial soil settlement and also for the leveling action of livestock or farm machinery travel over the dikes. When border dike irrigation is used for perennial crops such as alfalfa (*Medicago sativa* L.), many irrigators periodically reshape the dikes with border disks or other implements. A plane soil surface prior to the time of dike construction allows the highest efficiency and the best water infiltration uniformity with these systems.

Border dikes are sometimes closely spaced to minimize the amount of earth moving to prepare the soil surface. The small spacings can cause difficulty in the operation of farm equipment between the borders. Border dike designers should take into consideration the widths of hay swathers, combines, or other harvesting equipment that will be used for all crops anticipated.

c. Graded Furrow

Graded furrow irrigation is commonly used for annual row crops as well as some tree and vine crops. Prior to the first irrigation, furrows to convey the water across the field are formed between the crop rows. Water is introduced to each furrow at the upper end of the field by siphon tubes or cutouts in a ditch, or from gated pipe outlets.

Furrow irrigated crops are often planted parallel to the maximum field slope. However, if irrigation streams are erosive at this slope, the rows can be planted at an angle to the slope. Furrow slope is then decreased, but there will be cross slope perpendicular to the furrows. The irrigator must take care that water does not break between furrows, causing poor water distribution and accelerated soil erosion. This is an especially severe problem on soils that crack extensively between irrigations.

With graded furrows, a finite intake opportunity time at the lower end of the field is necessary to refill the root zone with water. Surface runoff occurs during this time and may amount to as much as half the total water applied. Furrow irrigation is often regarded as inefficient because of this runoff. However, in most situations this water is not lost but is available for reuse either on the same field or farm, by a neighboring farm, or by downstream water diverters from the river system to which this flow returns.

d. Corrugation

A system similar to graded furrow irrigation, called *corrugation irrigation*, is sometimes used for close-growing crops where the soil surface was not carefully smoothed to prevent slope variations before the crop was established. Small, shallow furrows, or corrugates, help in distributing the irrigation water uniformly over the soil surface. The small flow rate in the corrugates can help to minimize erosion on erosive soils. Special implements are available to form the corrugates. For some soils and crops, corrugates can be formed or reshaped after the crop is established.

e. Level Basin

Level basin is an irrigation method where the soil surface is leveled—a leveled basin has no slope in any direction. Close-growing crops such as small grains, hay, or grasses are planted on the basin surface. Water is introduced from one or more turnouts along one edge of the basin. Even though there is no soil surface slope, the water surface gradient causes the water to advance across the basin. When advance is complete, water begins to pond because the basin is surrounded on all sides by dikes. No runoff can occur. With proper management, water is introduced into the basin only until the volume necessary to apply a predetermined irrigation depth has been furnished. During the ponding and depletion phases, water distributes itself over the basin surface. Water depth is then quite uniform until infiltration is complete. Peri et al. (1979) gave detailed procedures for evaluating and improving basin irrigation systems. With level basin irrigation, small variations in elevation, on the order of 1 or 2 cm, will affect the uniformity of water infiltration by a similar amount. Large application depths will reduce this variability. Precise leveling is important (Erie & Dedrick, 1979; Dedrick et al., 1982; Ross & Swanson, 1957).

A variation of level basin irrigation, often called *basin-furrow irrigation*, can be used for row crops. Planting and furrowing are similar to graded furrow systems. Row alignment is generally at right angles to the edge of the basin where water is supplied. Prior to the first irrigation, a shallow, level distribution channel is constructed to intersect the upstream end of all furrows. Water from the turnouts then flows into this header channel and supplies all furrows more or less simultaneously. Since the rate of advance down all furrows will never be identical, many irrigators construct a similar distribution channel at the downstream end of the furrows. The first water to reach the downstream channel is directed to those furrows where the advance was slower and begins to flow "upstream," equalizing the intake opportunity time for all portions of the basin.

2. Applicability of Surface Irrigation Methods

All crops can be irrigated by some surface irrigation method. Surface irrigation is best suited to fine to medium textured soils. Coarse-textured or fine-textured soils that crack may have high intake rates, causing excessive deep percolation at the upstream end of a field before sufficient water has advanced to or infiltrated at the far end of the field. Nonuniform infiltration can be corrected to some extent by shortening the lengths of run, but such a solution requires a greater investment in water distribution channels and frequently hampers cultural operations. On soils of low intake, excessive runoff may occur when graded surface irrigation systems are used. With level basin or basin-furrow irrigation methods, there will be no runoff. However, water may stand on the soil surface for long periods before infiltration is complete. Some crops, such as beans (*Phaseolus vulgaris* L.), are sensitive to such standing water, perhaps because it inhibits root aeration. Stands of alfalfa can also be killed if water stands for prolonged periods in hot weather.

A major initial expense for surface irrigation systems is preparation of field topography so that irrigation can proceed efficiently with uniform water distribution. Lands that can be prepared for surface irrigation at low cost, therefore, are those lands where the original slope approximates the final design slope.

The depth of cut required to obtain the desired land surface and the depth of soil profile development on the field also need to be considered. Excessive cuts may remove so much topsoil that several years will be required before productivity of the cut areas can be restored. This condition can be avoided, again at extra cost, by storing the topsoil, moving subsoil to obtain the desired topography, and then replacing the topsoil uniformly on the field surface.

The surface irrigation designer must estimate the necessary seasonal water supply based on the evapotranspiration of the crop to be grown plus any leaching requirement necessary to maintain salt balance in the soil root zone. This volume then needs to be adjusted upward, based on the field water application efficiency. Since runoff from graded surface irrigation systems often decreases water application efficiencies into the range of 0.40 to 0.70, the seasonal water requirement may be large compared to other irrigation methods. If a tailwater recycling system is available, the runoff can be reapplied to the field, reducing the necessary primary supply. For level basin and basin-furrow systems, the field water application efficiency value can be 0.85 or greater and the necessary water supply may not be much larger than the estimated crop need.

The required rate of flow can be low for furrow applications if only a few furrows are supplied at any one time. The rate of flow to each furrow must be large enough so that advance is completed in a small percentage (generally 25% or less) of the intake opportunity time to assure uniform infiltration. Bishop (1962) presented relationships between soil characteristics, length of run, and advance time that are necessary for acceptably uniform furrow irrigation. For the border dike method, at least enough water must be available to supply one border strip efficiently. The largest water flow rate will be required for the level basins and basin-furrow methods, especially where basins are large and intake rates are medium to high. In such cases a large stream size is necessary to get quick coverage of the basin surface. Basins should be designed to operate efficiently with the irrigation supply flow rate available. Since the application efficiency will be high, however, the high rate of inflow needs to be supplied for only the short time necessary to supply enough water to refill the soil root zone.

In the western USA surface irrigation systems are often supplied by gravity flow through canals from streams or reservoirs. In such cases no external energy is required for the irrigation system. Runoff from such systems may return to the supplying stream and be available for diversion by downstream users, again with no external energy requirement. As irrigated areas have expanded since the mid-20th century, more ground-water supplies have been developed. For groundwater, the pumping energy needed is only that to lift water to the high point of the irrigated field and distribution is again by gravity flow. The energy required for pumping is increased if field water application efficiency is low.

Water quality is a lesser consideration for surface systems than for sprinkler or micro systems. However, several quality factors are important. The greatest cause of poor quality water in the USA is sediment. Proper elevation and location of on-farm turnouts relative to distribution canals can reduce sediment concentrations in the farm supply.

If sediment remains in suspension until applied to the field, many irrigators, especially those with high intake soils, find it beneficial. As water infiltrates, sediment particles tend to seal the larger soil pores, reducing the infiltration capacity and allowing faster advance of the irrigation streams over the field surface.

Trash, principally plant remnants, is a problem for many irrigation systems. Trash can plug turnouts, measuring devices, siphon tubes, or the gates in gated pipe. As more and more surface irrigation systems are automated, trash problems become more serious because the systems are unattended for prolonged periods of time. Mechanical equipment to remove trash is available for on-farm ditches (Kemper et al., 1986). Good weed control, especially on ditch banks, will minimize trash problems to a significant extent. Weed control will also reduce the number of weed seeds carried onto irrigated fields in the irrigation water.

Dissolved solids also need to be considered in determining the suitability of water for surface irrigation. All irrigators should be alert to increases in root zone salinity caused by saline water application. See chapter 35 (Tanji and Hanson) for a discussion of salinity. Van Schilfgaarde et al. (1974) provided a thorough discussion of saline irrigation water management. Maas and Hoffman (1977) presented salt tolerance values for many irrigated crops. Irrigators should also ascertain that Na, as measured by the sodium absorption ratio (SAR), is not so high as to reduce the permeability of their soils. Guidance on allowable SAR's and salinity is available in *USDA Handbook 60* (U.S. Salinity Lab. Staff, 1954) and *ASCE Manual* (Tanji, 1990).

Significant amounts of plant nutrients, especially nitrates, may be present in the irrigation supply. In some irrigated areas the NO_3 content of groundwater used for irrigation is adequate to supply an appreciable portion of the crops' seasonal requirements. Fertilizer application rates should be adjusted accordingly.

3. Automation

Development of equipment and techniques to automatically apply water to surface irrigation fields began in the 1960s. Many types of automation are now available.

The principal justification for automating surface irrigation systems is to save irrigation labor. There may be other benefits, however. In order for an automated system to operate for prolonged lengths of time without the operator's attention, the land forming, diking, furrowing, runoff channels, etc., must be well enough designed and constructed that no attention is necessary during an irrigation. Thus, a field to be automated will frequently have to be rehabilitated first.

The simplest types of semiautomatic system may consist of check gates controlled by spring-wound or electronic clocks in the farm ditch. Gates can be made from sheet metal, plywood, or fabrics in wood, metal, or concrete structures. Other useful structures for automating surface irrigation, including float gates, center-of-pressure gates, pneumatic valves, cylinder-operated gates, and cablegation, were described by Haise et al. (1980), Haise and Fischbach (1970), Humphreys (1967, 1969, 1971, 1975), Haise and Kruse (1966, 1969), Humphreys et al. (1970), and Kemper et al. (1981, 1985).

While automated systems for surface irrigation have been available for some 20 yr, malfunctions can and do occur. The owner of an automated system should not assume that the system can be set up and forgotten and all will work properly. Rather, a regular schedule of maintenance is needed. Also, automated systems should be designed to "fail safe." Thus, if any component of the system should not function as intended, there should be a back-up mechanism to keep excessive pressures from pipelines or to keep open ditches from overtopping and spilling water where damage could be done.

4. Efficiencies

Many types of efficiency have been defined for surface irrigation and other irrigation systems. Different writers have given different definitions for the same efficiency term. Therefore, anyone considering the efficiencies of irrigation systems needs to define terms carefully and be sure that the use of all terms is clearly understood. The effectiveness of an irrigation system cannot be described with any single efficiency term. For example, an effective irrigation system will store most of the applied water in the soil root zone where it is available to the crop (high water application efficiency); each irrigation will replace nearly all the soil water deficit in the soil root zone (high storage efficiency) and water will be applied uniformly to all parts of the field being irrigated (high uniformity coefficient).

Field water application efficiencies, defined as the average depth of water stored in the root zone divided by the average depth of water applied to a field, for surface irrigation often appear to be low compared to sprinkler or micro irrigation systems. One reason for this is that a significant fraction of the applied water runs off of most graded surface-irrigated fields. Tailwater recovery systems can salvage this runoff, causing an effective water application efficiency much higher than the calculated value. If an irrigator does not salvage his tailwater, it frequently returns to water courses where it is reused for agricultural, municipal, or industrial purposes. In such cases the water is not lost and the effectiveness of the system is better than the water application efficiency value might indicate. It follows that the effectiveness of surface irrigation can compare well with that of sprinkler or micro irrigation systems if calculated on a regional basis, or even for an entire river system. Runoff from surface irrigation may, of course, not always be recoverable. The quality of runoff will frequently be poorer than the initial supply, with increased sediment, nutrient, and/or pesticide content. Some return flows may not immediately enter streams where they can be rediverted, but may flow to swamps, to saline water bodies, or areas where the water is consumed by phreatophytes rather than by economically beneficial vegetation.

Another reason for relatively low application efficiencies of surface irrigation systems is soil variability and resulting nonuniform infiltration capacities. On fields with such conditions, infiltration depth will be highly variable even when the irrigation system is designed and managed to provide uniform intake opportunity times. If enough water is applied to satisfy soil water deficits in areas of low infiltration, deep percolation will occur from the high intake areas.

5. Management Skills and Technologies

The most modern, best-designed surface irrigation system, or any irrigation method, can apply water efficiently only if the system is properly managed. Accurate water measurement devices are an essential part of surface irrigation systems in order for proper management to be possible. Good management involves the application of the proper amounts of water at the proper times. Chapter 17 of this book describes irrigation scheduling techniques.

However, management of surface irrigation involves more than applying the proper amount of water at the proper time. Water should be applied uniformly and can vary from one irrigation to the next because of changes in the soil intake characteristics. Thus, an irrigator may need to adjust flow rates or application times to adjust for soil or crop changes.

The physical design of an irrigation system can ease management decisions. For instance, a system that collects and reuses tailwater may allow the irrigator to apply larger than optimal furrow streams, allowing rapid advance of water and thus uniform intake opportunity time along the full length of the furrow. Systems such as level basin irrigation, where rate of stream advance is not critical as long as it is not excessively slow, may also simplify management decisions.

Surge irrigation, a process patented by Utah State University Foundation, has become popular recently to facilitate rapid advance of water across the field. Surging involves the intermittent release of water into an irrigation furrow (Stringham & Keller, 1979; Bishop et al., 1981; Coolidge et al., 1982). On some soils a given volume of applied water will advance to the end of a furrow more quickly under surge than with constant inflow, allowing more uniform intake opportunity times along the furrow.

B. Sprinkler Irrigation

Sprinkler irrigation is the application of water to the soil using a device that directs a stream of water through the air onto the soil. Water is delivered to the sprinkler device through a pressurized pipeline. Sprinkler systems are human's attempt to duplicate natural rainfall; water sprayed from a pressurized pipeline into the air breaks into drops which fall to the earth like rain. The size of the drops, the uniformity with which they fall, and the rate at which they fall are all affected by the design of the system. Consequently, the design of a sprinkler system is important to its overall success.

1. Types of Systems

Sprinkler systems are generally categorized by how they are used in the field. Usually a sprinkler system is designed so that only a portion of the field is being irrigated at one time, and if the individual sprinklers are stationary for the duration of the time that they are in use, the system is called a *set system*. The design of a set system is fundamentally different from that of a system in which the sprinklers are moved constantly. A set system should be designed so that the application rate from the system does not exceed the infiltration capacity of the soil; otherwise runoff can occur. When continuously moving sprinklers are used, the application rate can exceed the infiltration capacity without excess runoff due to the short duration of application.

The reason that sprinkler systems are designed to irrigate only a portion of the field at a time is due to individual sprinkler performance characteristics. In order to obtain reasonable uniformity of application, most sprinkler systems must apply water at a rate much greater than the crop use rate. Furthermore, it is much more economical to design systems that do not apply water over the entire field simultaneously, and continuous wetting is probably not desirable. This leads to cycling or moving the individual sprinklers.

a. Set Systems

Set systems are those in which the sprinklers are placed on a fixed grid or spacing. There may be enough sprinklers to cover all the irrigated area, in which case the sprinklers are not moved, or there may be only enough sprinklers to cover a part of the area, and the sprinklers must be portable. A portable system is usually made of lines of sprinklers, and an entire line is moved from one set to the next. The simplest of the portable systems are moved by hand. Typical pipe lengths are 6 m (20 ft) to 12 m (40 ft) and typical diameters are from 25 mm (1 in.) to 127 mm (5 in.).

Set systems are often categorized by the materials used in the pipe-lines. Aluminum is the most commonly used material for pipelines that are moved. If the lines are not moved during the irrigation season the system is called a *solid set system*. Hand-moved set systems and solid set systems can be removed from the field to facilitate tillage and harvesting. Solid set systems made of polyvinyl chloride (PVC) pipe are usually buried and therefore not portable. Care must be taken during field operations in order to avoid damage to the underground pipeline and the individual sprinkler risers. A significant advantage of solid set systems is that they can be used to modify the crop environment; they can be used to cool the crop during hot periods or to prevent damage due to subfreezing conditions.

b. Mobile Systems

Mobile sprinkler systems make up a second major category. The *center pivot* is the most widely used of the mobile systems. A center pivot consists of a pipeline mounted on a series of wheeled towers. The entire pipeline rotates about a fixed end through which the water is fed. A similar system which moves in a straight line and is fed water from a ditch, series of hydrants, or a flexible hose is known as a *lateral- or linear-move system*. Sprinkler heads or sprayers apply water from the moving pipe. Depth of water applied is usually quite uniform, but application rates under continuously moving systems are usually much higher than with set-type systems.

Manual assistance is required in order to complete the irrigation cycle with some types of mobile systems. The traveling gun must be manually moved at the end of each run and the boom sprinkler must be moved from one land to another.

There are other systems that are movable, often requiring labor for movement but that are actually set systems. The wheel line (side roll) system and the tow line system are both mobile and set systems but should be designed as set systems. Sketches of several system types are shown in Fig. 16–2.

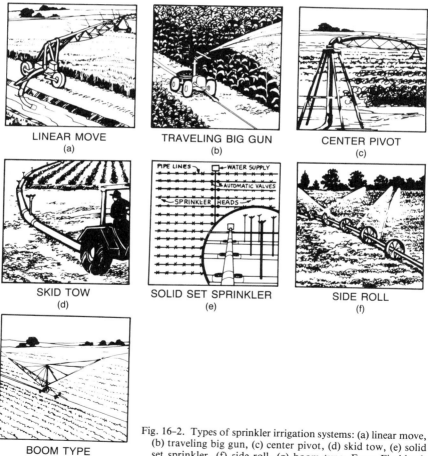

LINEAR MOVE
(a)

TRAVELING BIG GUN
(b)

CENTER PIVOT
(c)

SKID TOW
(d)

SOLID SET SPRINKLER
(e)

SIDE ROLL
(f)

BOOM TYPE
(g)

Fig. 16–2. Types of sprinkler irrigation systems: (a) linear move, (b) traveling big gun, (c) center pivot, (d) skid tow, (e) solid set sprinkler, (f) side roll, (g) boom type. From Fischbach (1988).

2. Selection of Sprinkler System Type

Harrison (1982) and USDA (1984a) listed factors involved in selection of the proper type of sprinkler system. It is most important to choose a system that will apply water at a rate and in quantities that are compatible with the infiltration capacity and water-holding capacity of the soil. However, it is also important to consider the crop to be grown, the climate, the topography, the system performance (uniformity and efficiency), and the initial and operating costs of the system.

A set-type system can usually be designed so that the application rate of the system matches the infiltration capacity of the soil, by balancing the individual sprinkler flow rate and the spacing of the sprinklers. The minimum spacing of the sprinklers is limited by the cost of the system and the maximum spacing is limited by the distribution uniformity of the system. Sprinkler spacing guidelines can be found in Christiansen (1942), Strong (1972), and Jensen (1980). Designs are compromises of cost vs. uniformity.

In the design of the system, consideration is given to soil conditions—both in determining the application rate and in the duration of irrigation. The *irrigation cycle time* (time between irrigations) is determined by the total system capacity, the total soil water holding capacity, and the water use rate of the crop. The rate at which the crop will use water depends upon the crop and the climate, as discussed in chapter 13 (Ritchie and Johnson) of this book.

The evenness in the depth of irrigation water application is known as *irrigation uniformity*. It is almost impossible to apply exactly the same depth of water to all parts of the field. The continuously moving types of systems generally apply water more uniformly than the set-type systems, but moving systems apply the required depth so quickly that application rates during portions of the application time can exceed the infiltration capacity of the soil. Water then flows over the soil surface and the uniformity of infiltrated depths is less than the uniformity of application to the soil surface.

Pressure differences may occur among sprinklers due to elevation or hydraulic differences. Pressure regulators can be used to reduce pressure to a more uniform level. It is also possible to operate sprinklers at reduced pressures using droplet-size control nozzles, but there is an accompanying wetted diameter reduction with most of these nozzles which may reduce application uniformity. Operating sprinklers at excessively low pressure produces larger droplets that in turn impart more energy to the soil and can lead to infiltration reduction.

Irrigation efficiency terms can be used to describe how much of the water applied is made available for plant use. If water is applied unevenly or in excess, or if runoff occurs, the efficiency of the system will be degraded.

A key factor in the profitability of irrigation is the cost of installing and operating an irrigation system. While the initial investment in the system is important, operating costs cannot be forgotten. Operating costs include the cost of pumping water (which will be higher for higher pressure systems), the cost of pumping water that is used inefficiently, and the cost of nutrients lost with deep-percolating water or runoff.

3. Special Uses for Sprinklers

a. Crop Cooling

Sprinklers may be useful for purposes other than for meeting crop water requirements. Two of the most important uses are for crop cooling and frost protection. Crop cooling is accomplished when water is sprayed on the crop during periods of high ambient temperatures. While some cooling results from the sensible heat transfer to the water (the water is cooler than the crop), most of the cooling comes from evaporation of water. Because it takes about 2450 J/g to convert water from liquid to vapor, that energy must be removed from the surrounding environment. The result is a lowering of ambient temperatures. The cooling effect of water caused by sensible heat exchange is only about 4.2 J/g per °C. In areas where the relative humidity is not high, the effect of evaporative cooling can be significant. Sprinklers have been successfully used in lowering the ambient temperatures on crops such as onion (*Allium cepa* L.) and apple (*Malus sylvestris* Mill.). Wright et al. (1981) used sprinklers to reduce the temperature of onion florets as much as 15 °C below ambient. Unrath (1972a, b) used sprinklers to reduce the temperatures on red delicious apple and significantly increased fruit quality.

b. Frost Protection

Sprinklers have been widely used to prevent damage to sensitive parts of plants by low ambient temperatures. Such frost protection is most common in areas where damage occurs from radiative frost. Radiative frosts may occur on clear, calm nights, usually following the passage of a cold front within the previous 1 or 2 d. Radiative frosts are differentiated from windborne frosts which occur when a cold air mass is blown through the area. In order for frost protection by sprinkling to be effective, the air mass must be relatively still. The sensitive parts of the crop are protected by the heat released when water freezes into ice. The *latent heat of fusion* (heat released when water freezes) of water is 335 J/g. Although water used in sprinkling must be above the temperature at which damage will occur to a plant, only 4.2 J/g of heat is released for each degree of water is cooled. Most of the benefit from sprinkling for frost protection occurs as a result of the sprinkled water freezing. However, because most plants do not suffer damage until the temperature falls below −2 °C (28 F) or so, the latent heat of fusion released actually serves to protect the sensitive parts of the plant. Sprinkling for frost protection under windy conditions is usually ineffective. General guidelines, including sprinkler cycling times, were given by Fry and Gray (1971) and Fry (1977). Hansen et al. (1979) listed more detailed guidelines that allow consideration for wind velocity, and Perry et al. (1982) listed guidelines considering relative humidity. Hamer (1986) developed an extensive model of frost protection and concluded that under reasonable conditions, a rate of 2.5 mm/h (0.1 in./h) would protect most crops.

4. Water Quality Considerations

Irrigation management is complicated when the water supply is high in dissolved salts or suspended solids. The tolerance of the crop to salts in the water supply must be considered and is an integral part of the system design. The most effective method to prevent harmful accumulations of salt is leaching. Leaching is discussed in chapter 35 of this book, by Hansen et al. (1979), and Tanji (1990).

Sprinkling poor-quality water on the foliage or fruit of plants may cause undesirable effects. Some fruits, e.g., apple and cherry (*Prunus* sp.), can be damaged by deposits on the fruits. Deposits on the foliage of some plants may have adverse effects on the life of the plant (Maas, 1985).

5. Disease as Influenced by Sprinkling

Irrigation can create microenvironments favorable to pathogens when conditions were previously unfavorable (Fry, 1982). Rotem and Palti (1969) presented an extensive review of the relationship of irrigation and plant diseases. They cited numerous conditions where the plant environment had been altered by irrigation, giving rise to diseases that would not otherwise have been factors. They also pointed out situations where the presence of irrigation reduced plant susceptibility to disease. Two specific examples of plant diseases are noteworthy. Blad et al. (1978) studied the effects of irrigation on the incidence of white mold (*Sclerotinia sclerotiorum*) in dry edible bean (*Phaseolus vulgaris* L.) on the Great Plains. They concluded that more frequent irrigation led to a more dense plant canopy and a much higher incidence of white mold. Avoiding excess irrigation was suggested. Sugar and Lombard (1981) studied the incidence of pear scab caused by *Venturia pirina* in pear (*Pyrus communis* L.) trees irrigated overtree and undertree. Those irrigated overtree had a much higher incidence of scab than the others, even after normal preventative chemical application had taken place.

6. Advantages and Disadvantages of Sprinklers

The water control achieved with sprinkler irrigation is a major advantage. Well-designed sprinkler systems can deliver water efficiently and uniformly over a wide range of conditions. They can be adapted to virtually any soil so as to eliminate runoff and minimize erosion, and are highly adaptable to terrain with so much slope or soils so shallow that leveling is infeasible. Application rates can be designed for low-permeability soils.

The application of chemicals through sprinkler systems is a common practice. The systems are especially well adapted to that use because they can apply small depths of water uniformly. Consideration must be given to the effect of the chemical upon the foliage of the crop because much of the water sprinkled will be intercepted by the plant, a desirable condition for the application of many chemicals. Sprinklers can also be used to modify the environment of the crop. Where a late spring frost can cause bud damage, the use of the sprinkler system to protect against such damage is a major consideration. Sprinklers can also be used to reduce potentially damaging high temperatures or to sustain a low temperature and delay the onset of budding.

There are characteristics of sprinkler systems that can be disadvantageous. Water must be delivered to the sprinklers under pressure, and unless sufficient pressure is developed by gravity, the cost of energy to pressurize the system must be considered. Wind can degrade the distribution uniformity of a sprinkler system. Some crops are susceptible to disease when the foliage is wetted by a sprinkler, and that can be a significant disadvantage.

Sprinkler systems can allow the designer broad flexibility in design of application rate, but the use of mobile systems restricts that flexibility. Because the application rate under a mobile system is governed only by the system flow rate and the sprinkler pattern, application rates under some mobile systems (e.g., center pivots) are too high for consideration on low infiltration-capacity soils.

7. Efficiencies and Uniformities

The efficiency and uniformity of a sprinkler irrigation system are inseparable. Well-managed sprinkler systems generally have high water application efficiencies, but the efficiency with which that water is made available to the plant depends upon the uniformity of its application. The uniformity of a sprinkler system is determined primarily by the designer, and is fundamentally a function of the sprinklers selected and the spacing of those sprinklers. The designer must consider the climate, the soil, and the terrain when choosing the sprinkler and the spacing. A well-designed sprinkler system can have water application efficiencies of 0.90, but that is rarely attained. Efficiencies above 0.75 can be attained, especially with mobile systems. Uniformity Coefficients (Christiansen, 1942) of 80 are generally targeted by the designer, and a good mobile system should have a Uniformity Coefficient of about 90.

8. Automation

The adaptability of a sprinkler system to automation depends on the type of system. Some mechanically moved and stationary systems can be totally automated. These systems do not require labor for the movement of the system, and hence an irrigation cycle can be automatically initiated and terminated, with the automatic injection of chemicals during the cycle as necessary. As a practical matter, few systems are operated fully automatically.

Portable systems can only be partially automated. The injection of chemicals and termination of the cycle can be automated, but the cycle must then be interrupted until the laborer moves the system. Automatic termination of the cycle is common in many types of portable systems.

9. Management Skills and Technologies

Successful operation of any irrigation system, especially complex systems, requires management skill and the use of appropriate technology. Well-designed irrigation systems, inherently capable of high efficiencies will, if poorly managed, rapidly become inefficient and unprofitable.

Beyond the fundamentals of maintaining an irrigation system in good working order, a good manager must make water available to the crop so that water stress does not restrict growth. The manager must be cognizant of the crop's daily water use, rain received, and the amount and timing of the application of irrigation water (see chapter 17 by Heermann et al. in this book; Fischbach, 1980; Jensen, 1980). In order to facilitate irrigation management, the manager must make use of a variety of technologies. The basic technology is that of irrigation scheduling, which is based on a current accounting of the water available in the soil. There are a variety of methods and sensors available for monitoring or estimating soil water, ranging from simple to sophisticated (see chapter 6 by Campbell & Mulla in this book), and there are numerous techniques for estimating crop water use. Many of the methods used involve the use of computers, and there are a few fully automated irrigation systems in use today where all of the inputs are processed, controlled, and monitored by a computer. The appropriate technology varies with personnel and location.

C. Micro Irrigation

The term micro irrigation encompasses several methods or concepts, chief of which are drip/trickle, subsurface, bubbler, and spray irrigation (ASAE, 1988). In microirrigation systems, water is delivered through a network of plastic lateral lines that are fitted with emitters that dissipate the pressure through narrow nozzles or long flow paths and discharge water at only a few liters per hour to each unit of field area. The area that can be watered from each emission point is, therefore, limited by the water's horizontal flow. The limitation on the wetted soil volume can be overcome by choosing an application rate and volume of application that will meet both the evapotranspiration demand of the crop and the infiltration and water holding characteristic of the soil.

Emitter types are usually characterized by their mode of operation, such as laminar, turbulent, orifice, or compensating emitter. Laminar, turbulent, and orifice emitters can effectively be used on level or uniformly sloping terrain, whereas compensating emitters are best adapted to undulating fields. Operating pressure at the emission devices along the lateral lines is typically low, ranging from 35 to 140 kPa (5–20 psi). However, additional water pressure of about 245 to 310 kPa (35–45 psi) is usually required at the control station to overcome losses through filtration and other regulating equipment (von Bernuth & Solomon, 1986).

1. Types of Systems

a. Drip/Trickle

Drip/trickle irrigation is the slow application of water through small emitter openings at or near the soil surface. Rates of discharge for wide-spaced individual applicators are generally < 12 L/h (3 gal/h). For close-spaced outlets along a lateral tube or for porous tubing, these rates are usually < 12

L/h per m of tubing length (1 gal/h per ft). The primary reasons for using surface drip/trickle over other micro systems are the ease of system installation, of field inspection, and of changing and cleaning emitters, plus the possibility of checking soil surface wetting patterns and monitoring individual emitter discharge rates. Drip/trickle systems have been installed on row, vine, and orchard crops. For row crops, drip/trickle systems are capable of maintaining high soil-water contents in the crop root zone and moving salts away from the seed row for plant germination and establishment. On vine crops, drip/trickle systems can be placed above the soil surface on the first trellis wire to expedite cultural operations. On orchard crops, the number and spacing of emitters can be altered as the trees mature.

b. Subsurface

Subsurface irrigation is the application of water below the soil surface through emitters that have rates of discharge generally in the same range as those for drip/trickle irrigation. Subsurface irrigation is not to be confused with *subirrigation*, a method of irrigating the root zone through water table control. Lately, subsurface systems have gained wider acceptance on small fruit, row, and vegetable crops. A subsurface system, in comparison with surface drip/trickle systems, eliminates the need to anchor the lateral lines at the beginning or to remove them at the end of the growing season, reduces interference with cultivation or other cultural practices, and possibly results in a longer operational life. In addition, subsurface irrigation is becoming more recognized as an efficient method for applying fertilizer, fungicides, insecticides, and other chemicals precisely within the crop root zone. Plugging of subsurface emitters, the greatest single disadvantage of such systems, can be difficult to detect.

c. Bubbler

Bubbler irrigation is the application of a small stream or fountain of water to a pond on the soil surface from an emitter having a rate of discharge greater than that for drip or subsurface irrigation, but usually less than 225 L/h (60 gal/h). The applicator discharge rate exceeds the soil's infiltration capacity, and a small basin is required to control the distribution of water. Bubbler systems are being used primarily on tree crops planted on flat or moderate slopes with minimal soil erosion problems. Benefits of bubbler over other micro systems include decreased filtration, maintenance or repair, and slightly lower energy requirements. However, larger lateral lines are usually required with bubbler systems to reduce the pressure loss associated with the higher discharge rates.

d. Spray

Spray irrigation is the application of water to the soil surface by a small spray or mist. In this method water is distributed through the air, whereas drip, bubbler, and subsurface irrigation use the soil as the primary means of distribution. Emitter discharge rates are usually lower than 100 L/h (25 gal/h). Spray systems are used to irrigate orchard or other widely spaced crops. Spray systems can be vulnerable to high wind and evaporation losses, particularly when tree crops are young with a limited canopy. However, spray and bubbler systems normally have minimal filtration and other maintenance problems, provided the emitters are properly designed and manufactured.

2. Advantages and Disadvantages

A recent survey conducted by the International Commission on Irrigation and Drainage (Abbott, 1984) indicated that the main reasons for choosing micro irrigation were as follows: labor and water were expensive, the water supply was limited, the water was saline (although periodic leaching water was still required), use of other methods was difficult (e.g., hillside orchards), or landscape or greenhouse irrigation was required. A recent list of advantages with micro irrigation includes increased beneficial use of available water; enhanced plant growth, quality, and yield; reduced salinity hazards to plants; improved fertilizer and other chemical application; limited weed growth; decreased energy requirements; and improved cultural practices. Potential disadvantages compared to other methods include persistent maintenance requirements, salt accumulation near plants, restricted plant root development, and high initial costs (Bresler, 1977; Bucks et al., 1982).

Most crops are adaptable to micro irrigation. However, since initial costs of these systems are high, crops and land of the highest value are best suited to micro irrigation. The present list of major crops irrigated by such systems include avocado (*Persea americana* Mill.), grape (*Vitis vinifera* L.), citrus, cotton (*Gossypium hirsutum* L.), stone fruit (*Prunus* sp.), strawberry (*Fragaria* sp.), sugarcane (*Saccharum officinarum* L.), and vegetables. Micro irrigation systems are found on all soil types, but sandy soils predominate. Micro systems can be used on steep slopes where cultivation is the limiting factor (e.g., avocado in California). Possible reasons for increased plant growth and yield under some field conditions are related to improved water distribution along the row, to reduced effects of variabilities in texture and water-holding capacity of heterogeneous soils, and to increased water uptake efficiencies from the more concentrated root systems close to the emitters.

Micro irrigation is not generally affected by weather since the water is applied directly to the soil. Wind typically has little effect on water distribution or losses, although extreme hot, dry winds can increase surface evaporation. Nearly all the rainfall that does not run off can be considered effectively available for crop production with skillful irrigation scheduling. In the past, micro systems have had the unfounded reputation of substantially reducing evaporation losses. Recent research studies have shown that unless micro systems are applying water below the soil surface without moistening the surface, little saving of water can be expected on row crops where the wetted surface is exposed to direct sunlight (Bucks & Davis, 1986).

Since micro irrigation encourages a small application rate of water over a longer period of time in comparison to other irrigation methods, low to medium discharging groundwater wells are usable. If well capacity is not adequate to meet peak crop evapotranspiration rates, the irrigator will have to depend on timely rainfall or run the risk of periodic crop water stress. For a surface water supply, the delivery of water to the farm will need to be on a nearly continuous basis. A water storage facility on the farm may be necessary. Some evidence suggests that waters of higher salinity can be used with microirrigation than can be used with other irrigation methods without reducing yields. Minimal salinity hazard to plants by micro irrigation can be attributed to keeping salts in the soil water more diluted by high-frequency water applications, eliminating leaf damage caused by foliar salt absorption, and moving salts beyond the active root zone.

Micro irrigation offers maximum flexibility in chemigation. Frequent or nearly continuous application of plant nutrients, insecticides, fungicides, or other chemical amendments along with the irrigation water is feasible and in most cases beneficial for crop production (Rolston et al., 1986). Various explanations have been given for increased efficiency of chemigation with micro irrigation, such as decreased amounts of chemical application because the chemical is applied only to the root zone, improved timing of nutrients because the more frequent applications can match plant requirements at various stages, and improved distribution of chemical amendments with minimal losses below the root zone or on the soil surface.

The lower water discharge rates and higher operating pressures (to maintain water distribution uniformity) dictate an emitter design with small opening, which can create a major clogging problem with micro irrigation systems. Solutions to clogging have included development of emitters that require less maintenance and filtration, or treatment of water before it reaches the emitters. An irrigation water classification has been proposed by Bucks et al. (1979) to predict potential clogging and to indicate the extent of water filtration, chemical water treatment, field inspection, and pipeline flushing required to minimize clogging problems.

3. Efficiencies

The designer of a micro irrigation system must consider the emitter type, emitter uniformity, hydraulics, topography, water distribution uniformity, crop salt tolerance, water quality, soil salinity, fertilizer requirements, cultural practices, crop evapotranspiration rate, and other site-specific variables. When all conditions are optimum, field water application efficiencies for micro irrigation systems of 0.90 to 0.95 are attainable but seldom maintainable. In addition to high application efficiencies, nearly all the infiltrated rainfall can be effectively used with the light, frequent irrigation schedule of a micro system. On the other hand, a large rainfall soon after a surface or sprinkler irrigation will probably result in water drainage below the root zone.

Distribution uniformity of a micro irrigation system is primarily a function of the design, equipment selection, installation, adjustment, and maintenance. Bralts (1986) developed a simplified procedure for determining the statistical distribution uniformity based on the highest one-sixth and lowest one-sixth of times needed for emitters to fill a container (such times are directly related to the variation in emitter discharge rates). By using a statistical treatment, factors such as emitter manufacturing variation, lateral line friction, elevational differences, and emitter plugging are considered in the final uniformity estimate.

In a survey performed between 1981 and 1986 on 60 micro irrigation systems in California (Aslan et al., 1986), the average water distribution uniformity was 73% with individual systems ranging from a low of 24% to a high of 96%. The problems were divided as follows: 42% system operation, 33% environmental factors, and 25% design and installation. The most common design and installation problems were traced to improper operating pressures, mixing of emitter types or discharge rates, placement of emitters, and emitter clogging.

4. Automation

With frequent irrigations, the control of the soil-water-root environment is critically dependent upon the irrigation manager, either human or computer. Any disruption or disturbance to the irrigation schedule can quickly create a detrimental water or oxygen stress on the crop. A full-automated micro irrigation system can be extremely flexible and capable of responding to small and rapid changes in soil water content, plant water stress, or evapotranspiration (Phene, 1986). An automated micro irrigation system can use feedback sensors to monitor on a real-time basis important functions such as: flow rate, water pressure, wind speed, air temperature, pan evaporation, solar radiation, and rainfall. Various types of soil water, plant water stress, or crop canopy temperature sensors can be used to initiate or override a system controller. A controller issues commands to operate water valves, booster pumps, and chemical injectors; to clean filters; etc., according to the modified or corrected irrigation cycle. Either simplified or sophisticated automation equipment can be easily adapted to a micro system. Also, the frequent irrigations mean that the automation will be used on a continuous basis, such that a significant portion of management labor can be saved.

5. Management Skills and Technologies

Management of a micro irrigation system is unique in many respects. Irrigation by small, frequent amounts is quite different from sprinkler and surface irrigation methods where large, infrequent applications are normally applied. With micro irrigation, precise information on crop evapotranspiration rates on a daily basis is required to determine adequately the irrigation amount (Howell et al., 1986). Also, more attention must be placed on system maintenance (James & Shannon, 1986). On the other hand, many micro irrigation users indicate that labor for system operation is reduced because the micro system can be fully automated and that water control has been improved such that crop production risks are also reduced. As technologies advance, micro irrigation offers opportunities for applying irrigation water efficiently where other methods can have difficulties. As a management tool, a micro irrigation system can improve the level of water management or conservation. However, most micro systems are not well adapted to the poor or uneducated user.

D. Subirrigation

Subirrigation provides water to crops by controlling the water table level. Crop roots can then reach the capillary fringe above the water table and extract their water needs from it. Subirrigation should not be confused with subsurface irrigation, a method of micro irrigation where water is supplied by conduits beneath the soil surface (see section I. C in this chapter). Subirrigation has been in use for many years in the USA. In some situations, the high water table has been maintained by natural conditions and farmers have taken advantage of it, needing to do little or nothing to manipulate the system to their advantage. In other cases, irrigators have constructed systems of open ditches to affect the level of the adjacent water table.

Subirrigation systems have been used largely for perennial crops. In areas of low rainfall, evaporation of water from the capillary fringe often has left salt concentrations near the soil surface; in many such systems, only highly salt-tolerant plants of low economic value are currently produced. Fox et al. (1956), in discussing design of subirrigation systems, pointed out that certain natural conditions must exist—either an impermeable layer or a permanent water table at a shallow depth so that seepage losses will not be excessive. In addition, the surface topography must be nearly level so that depth to the water table is reasonably uniform over the entire field.

Since the early 1970s, subirrigation systems based on more rational design methods have been used in the southeastern USA. Typically, these areas have relatively high rainfall, on the order of 1300 mm/yr (50 in./yr), but droughts occasionally occur or rainfall is scarce during critical plant growth stages. Because of the rainfall, tile lines or open ditches are necessary and often already in place for draining excess water. The subirrigation systems make use of these to raise the groundwater level during water stress periods. Skaggs et al. (1972) quoted the advantages of such a combined subirrigation-drainage system as having low labor and maintenance requirements, no delays in cultural practices because of irrigation, and little or no nutrient leaching from the root zone. They recognized the potential hazard of salt buildup, but concluded that periodic high rainfall conditions will provide the necessary leaching. An additional potential problem is deterioration of soil structure when the water table is maintained close to the soil surface for long intervals. During drought periods, a supplemental water source is necessary to supply the drainage facilities that effect the water table control.

E. Other Irrigation Systems

Several modern irrigation methods do not fit the descriptions of surface, sprinkler, micro or subsurface irrigation, but may combine characteristics of two or more such systems. These methods have been developed since 1975 in response to special requirements for conserving water, conserving energy, or protecting water quality.

1. Types of Systems

a. Low Energy Precision Applicator

The *low energy precision applicator* (LEPA) system was developed by Lyle and Bordovsky (1979, 1981). Early systems were modifications of center pivot or linear move sprinkler systems. The sprinkler heads on such systems were replaced by closely spaced drop pipes that release water at pressures of 7 to 35 kPa (1–5 psi) just a few centimeters above the soil surface. Water can be applied at a high enough rate to meet the crop requirements because dikes are constructed at 1 to 2 m (3–6 ft) intervals in the furrows between crop rows and all water applied accumulates in small ponds behind the dikes until it infiltrates. The dikes also serve to trap all but the largest, most intense rainfalls on the irrigated fields. Irrigation application efficiency is high because surface runoff is prevented, and the irrigation water requirement can be reduced because of the improved use of rainfall on such systems.

b. Closed Conduit Gravity System

A closed conduit gravity system (Rawlins, 1977) was developed specifically for irrigation of orchards. A buried corrugated plastic pipe distributes water between each two rows of trees. From this pipe a much smaller (< 10 mm inside diam.) tube brings water to each individual tree. The end of this tube is elevated several centimeters above the ground surface, but is left open with no emitter or other fitting. To obtain a uniform discharge rate at each tree, Rawlins has specified a procedure for adjusting the relative elevation of the open end of each tube. Since this system generally applies water at rates greater than the soil intake capacity, small dikes similar to those used for bubbler systems are formed around each tree to hold applied water until it infiltrates.

II. COMPARISONS OF DIFFERENT IRRIGATION SYSTEMS

A number of research studies have been reported that compare one or more aspects of different irrigation methods. Results of these studies can be quite instructive to a person selecting a system. However, the specific conditions of the study, the site, soils, topography, climate, etc., have to be examined closely to assure that the study results will be applicable at locations other than where the study was conducted. Results of several such studies are cited in the following sections.

Field water application efficiency is defined as the average depth of water stored in the root zone divided by the average depth of water applied to a field. Water application efficiency is reduced when portions of the applied water run off of the field, percolate below the effective root zone of the crop, evaporate during the application, or are carried by wind beyond field boundaries.

Many different definitions of irrigation efficiencies have been published. Anyone using these terms needs to be certain that they are clearly defined and understood. The definitions used in this chapter correspond to those published by the ASCE Committee on On-Farm Irrigation (Kruse, 1978). Other treatments of efficiencies are given by Hansen et al. (1979, chapter 8), Bos (1985), Hall (1960), Hansen (1960), Hart et al. (1979), and Jensen (1967).

If the depth of irrigation applied is much less than the soil water depletion of the root zone, the water application efficiency can be near 1.0. This does not mean that the irrigation is satisfactory, however, because the plants may soon suffer damaging water stress. The uniformity of the water distribution, and the percentage of the root zone water-holding capacity that is refilled are also important in evaluating the effectiveness of an irrigation. In some special cases it may be important to minimize surface runoff, for instance where excess sediment might be lost from erosive soils or where discharges of low-quality tail-water are restricted by law. In other cases it may be more important to limit deep percolation as in the upper Colorado River basin where deep percolating irrigation waters carry dissolved salts from shallow, geologic formations to the Colorado River.

Low values of water application efficiency are often cited by those seeking to imply that water is used wastefully in irrigated agriculture. However, low water application efficiency should only be of concern if the water is available at high cost, especially where it has to be pumped; when the quality of return flows from irrigated fields is significantly degraded; or when the return flows are not available for reuse at points downstream in the drainage basin. In many areas of the western USA the irrigation supply is by surface diversion from streams and the return flows run to the same stream system with little deterioration in quality. In such cases, low water application efficiencies may be of little concern.

Till and Bos (1985) discussed the factors that contribute to nonuniformity in sprinkler, micro, and surface irrigation, and how these factors affect the application efficiency. The authors recommend that uniformity of water application be taken into account when evaluating the field application efficiency. Uniformity considerations include the percent of field area that needs a completely adequate irrigation and the negative benefits caused by underapplication in other areas of the field.

Ranges of possible water application efficiencies for several surface, sprinkler, and micro irrigation methods are given in the *USDA-SCS National Engineering Handbook* (USDA, 1974, 1983a, b, 1984b). The range in values depends on soil type, the depth of irrigation applied, and to a lesser extent, conditions of climate and topography. With poor management any of the systems could operate at lower efficiency than those cited.

A. Uniformity of Application

In general, water distribution in pipelines gives the irrigator greater control over the water supply than the use of open ditches; thus, sprinkler and drip systems often afford better water control than surface irrigation systems. However, many surface irrigation systems now distribute water to different points on the farm with low pressure pipelines, giving good control up to the point of release. With surface irrigation systems, the condition of the soil surface and crop controls uniformity to a much greater extent than the hydraulic characteristics of the irrigation equipment. Variations in the intake opportunity time at various points on the field surface, in the infiltration capacity of the soil, and in flow resistance provided by soil roughness and vegetation cause nonuniformities in infiltrated depth of water. The irrigator can overcome some of these by providing a proper stream size and application time.

The uniformity of water distribution under micro irrigation systems is generally not affected by the soil condition. Variations in uniformity are caused principally by various degrees of clogging of emitter openings as well as by the variation in these orifices that resulted from the manufacturing process. Since clogging is typically a function of time in use, uniformity of micro irrigation can vary widely with the age of a system. Variations in elevation of the field surface may also have effects on the pressure differential and the flow from different emitters.

With sprinkler irrigation, the system design becomes important because proper attention to the water distribution pattern from the individual sprinkler heads and the necessary overlap of those patterns is important. The design must also consider variations in field topography as it affects sprinkler pressures and discharge. Finally, wind distortion of sprinkler patterns contributes significantly to the final uniformity of water application. Vories and von Bernuth (1986) presented a detailed analysis of sprinkler performance in wind.

B. Runoff

Runoff of water from irrigated fields is an important consideration, both as it affects the water application efficiency and because of the possible poor quality of this water as it enters other water bodies for reuse. Water application rates under micro irrigation are so low that runoff seldom, if ever, occurs.

When small sprinklers or sprayers are used on some low intake soils, water may redistribute on the surface, but will seldom accumulate in volumes great enough to leave the field. Sprinkler systems used on sloping lands or on row crops where a significant portion of the soil surface has no vegetative cover, may cause runoff in significant amounts. Recent efforts to save energy by reducing sprinkler operating pressure, or by replacing larger sprinkler heads with spray heads that wet a smaller diameter, tend to increase application rates and runoff. Von Bernuth and Gilley (1985) presented an analysis of the effect of reduced operating pressures and the subsequent increase in application rates under center pivot irrigation.

Except for level basin or basin-furrow systems, runoff will occur with surface irrigation. Management techniques that can minimize runoff volume include terminating water inflow at the proper time on irrigated borders, cutting back the irrigation stream size, or practicing surge irrigation on graded furrows.

C. Energy Requirement

A complete analysis of the energy requirement for an irrigation system will require, first of all, examining the energy required to manufacture the different components of the system and the energy required for any trenching, ditch construction, or earth moving in system preparation or installation. These energies, then, need to be amortized over the anticipated useful system life. In addition, there will be an annual energy requirement for most systems, generally for pumping water from wells, streams, or reservoirs to the field being irrigated and for pressurizing the water if the system requires it.

Manufacturing energies are typically high for micro or sprinkler irrigation systems where the water is conveyed in pressurized conduits to a point close to its application to the soil surface. Site preparation energy will tend to be highest for level basin systems if considerable earth moving is required to form the level field surface and prepare the necessary dikes and berms.

The pumping energy to lift water to the field surface would be nearly the same for all irrigation methods except for differences in water application efficiency (Batty et al., 1975). Systems with a low water application efficiency will require the pumping of more water to achieve a given seasonal net application depth than those having a higher efficiency. In a small but significant number of situations, water can be transmitted by pipeline from a stream, canal, or reservoir at a sufficient elevation above the irrigated field so that pressure is adequate for operation of sprinklers or micro irrigation and no pumping is required.

Hagan and Roberts (1981) compared several types of surface, sprinkler, and micro irrigation installations for a hypothetical cotton field in central California. No single system provided lowest energy requirements, lowest water requirements and lowest total costs for the assumed conditions.

Massey et al. (1983) simulated water and energy requirements for sprinkler and subirrigation in North Carolina over a 27-yr period. Where water supply was pumped from wells, subirrigation sometimes used more energy than sprinklers and sometimes less, depending on site conditions. For a surface water supply, subirrigation required only 6 to 9% of the energy of sprinklers.

Allen and Brockway (1984) reviewed energy efficient modifications of new or existing surface and sprinkler systems and concluded that energy savings were possible for both systems.

D. Labor Requirement

Thorfinnson et al. (1955) compared labor and other costs for gated pipe, siphon tube, and hand-move sprinkler systems. They found six-fold variations in equipment costs per unit area and 50% variations in labor requirements. With all costs considered, including labor, surface irrigation with siphon tubes was least costly and sprinkler irrigation about six times as expensive.

Irrigation systems now in use have large variations in the labor requirements. Many types of systems are highly automated so that little, if any, labor is required for operation. For these systems a significant amount of labor may be required for the initial system installation and this should be considered as one of the factors in system selection. Also, it is important that the irrigator devote the necessary amount of time to monitoring automated systems for malfunctions and for maintaining all components of the system in proper order for reliable irrigation. Irrigation specialists in state extension service offices may have more up-to-date information on systems in current use.

E. Water Use Efficiency

Water use efficiency (WUE) (Haise et al., 1960) is a term that relates crop production to water used. Thus, if one irrigation method produces more desirable soil water and micro climate conditions for growth of a given crop than another method, while using the same amount of water, the first will have the higher water use efficiency. In this chapter, WUE will be considered as the usable crop yield divided by the crop evapotranspiration, which may be supplied by either irrigation or precipitation.

Bernstein and Francois (1973), Bielorai (1982), Sepaskhah et al. (1976), Ravelo et al. (1977), and Freeman et al. (1976) measured water use efficiencies for selected crops and irrigation methods. A common conclusion is that methods that can reduce water applications by improving application efficiencies, or increase yields by decreasing crop water stress, such as micro systems, tend to raise values of WUE.

F. Salinity Management

Salinity management can be an important factor in the choice of an irrigation system. When the water supply is saline, for instance, and crops are salt sensitive, micro systems can maintain a higher soil water regime, and therefore, a more dilute salt concentration near the emerging seedling for improved stand establishment. Also, the salts tend to be moved to the outer limits of the soil volume wetted by each emitter. After a season of water application by such a method, however, future crops may need to be carefully located so that seeds are sown in the nonsaline locations near the soil surface, not in those areas where the large salt accumulation has occurred. Special management may be necessary where crops are grown under furrow irrigation with saline soils or a saline water supply. In this case, salts tend to accumulate in the center of the furrow ridge where evaporation occurs after each irrigation, but where there is little if any leaching. Many irrigators in the southwestern USA plant their crops on the sides of the ridges near the water line so that the young plants can be established in soils where salt concentrations are lower.

Saline water supplies may also limit the use of sprinkers if the saline water on crop leaves would be harmful. Bernstein and Francois (1973) noticed poor production of peppers (*Capsicum annuum* L.) sprinkled with saline water. They attributed the problem less to foliar damage from the sprinkled water than to flushing of accumulated salts from the soil surface to the crop root zone with each irrigation.

G. Cropping Limitations

In general, any of the major irrigation methods can be adapted for use on any irrigated crop. However, the topography, soil type, and crop itself can sometimes affect the selection of an irrigation system. Surface irrigation systems are seldom adaptable to slopes in excess of 3% or sandy soils unless lengths of run are extremely short. Although both surface and sprinkler irrigation can apply water to annual and perennial crops, a micro irrigation system is needed to apply water under a plastic mulch to specialized vegetable or small fruit cropping systems [e.g., tomato (*Lycopersicon esculentum* Mill.) or strawberry]. Similarly, a furrow or micro irrigation system can be more suited to applying water on tall crops such as sugarcane where wind drift can be a major problem. Usually, more conventional crops, such as corn (*Zea mays* L.), wheat (*Triticum aestivum* L.), alfalfa, or grass are better irrigated by surface or sprinkler than by micro irrigation. Steiner et al. (1983) examined the microclimate under frequent center pivot-irrigated and less frequent surface-irrigated corn. In a hot, dry year, mean daily leaf temperatures were reduced by 1 °C by the sprinkler method and soil temperatures were also lower under sprinkler irrigation.

H. Economics

An economic analysis of any irrigation system should consider the whole, rather than the individual parts of the system. Land and water cost can be the two most important factors in any economic comparison of different irrigation technologies. Other factors affecting costs (labor, tillage, weed control, fertilization, and harvesting) and profits (higher yield, earlier ripening, price, and product quality) need to be considered in a complete economic analysis for the selection of the best irrigation method for each location.

Initial costs for surface irrigation systems vary widely. Wild flooding systems may require no land surface preparation. The system is constructed by making a few unlined ditches and providing simple wooden or canvas water control structures. On graded furrow, border dike, and the level basin

Table 16-1. Key features and cost estimates for several irrigation systems.

Category	Relative required water	Required labor	Initial cost	Pumping cost	Soil adapta-bility†	Terrain adapta-bility‡	Special features§	Field adapta-bility¶	Chemigation applica-ble#
		h/ha	$/ha	$/(ha yr)		%			
				Surface					
Wild flooding	1.4	30	700	35	L,C	<10	--	NL	N
Border dike	1.2	12	1150††	25	L,C	<2	SFC	NL	N
Graded furrow	1.2	35	1000	25	L,C	<3	SFC	NL	N
Corrugation	1.2	3	900	25	L,C	<3	SFC*	NL	Y
Level basin	1.0	3	1400	15	L,C	<1	SFC*	NL	Y
				Sprinkler					
Movable set									
Hand lines	1.0	30	1000	50	All	15	--	R	N
Wheel lines	1.0	15	1200	50	All	10	--	R	N
Tow lines	1.0	15	1200	50	All	10	--	R	N
Stationary set									
PVC solid set	1.0	5	2650	50	All	NL	FC,C	NL	Y
Aluminum solid set	1.0	7	2550	50	All	NL	FC,C	NL	Y
Mobile									
Center pivot	1.0	3	1050	45	S,L	15	--	C‡‡	Y
Lateral	1.0	3	1150	45	S,L	15	--	R	Y
Labor assisted									
Wheel lines	1.0	15	1200	50	All	10	--	R	N
Tow lines	1.0	15	1200	50	All	10	--	R	N
Traveler	1.1	20	1000	60	S,L	10	--	N	N
				Micro					
Drip/trickle	0.9	10	1850	35	All	NL	--	NL	Y
Subsurface	0.8	10	1950	35	All	NL	--	NL	Y
Bubbler/spray	1.0	7	2300	40	All	30	SFC	NL	Y

† S: Sand, L: Loam, C: Clay.
‡ Maximum % slope (NL: No limit).
§ FC: Frost control, C: Cooling, SFC: Some frost protection is possible.
¶ C: Limited to circular shapes, R: Limited to rectangular shapes, NL: No limit.
# Y: Yes, N: No or limited adaptability to chemigation.
†† Includes $700/ha for moving 1500 m³/ha soil.
‡‡ Some center pivots are available with adaptations to accommodate noncircular field shapes.

type systems, earth moving is usually necessary to create a field surface with appropriate topography for uniform water distribution. Cost of the field preparation will increase with the volume of soil to be moved. Earth moving costs are especially great for level basin systems, unless the field has low slope initially.

Initial costs of sprinkler systems also vary considerably. The hand-moved, portable set system is the least expensive, and the stationary set system most expensive, with automatic and labor-assisted mechanical move systems typically somewhere in between. Spacing and selection of sprinklers will affect costs, and designs often must consider trade-offs of initial costs vs. water application uniformity.

Micro irrigation systems are usually costly and can require different management skills and farming practices than other irrigation methods. Initial costs of micro systems can range from \$1500/ha to \$3000/ha. In a study of drip/trickle, subsurface, and furrow irrigation for cotton in the southwestern USA (Wilson et al., 1984), operation costs were within 5%, but annual fixed costs per unit of land area were 2.3 times higher for micro than surface irrigation. Implications were that microirrigation of cotton could become profitable only where yields were increased by at least 1.2 bales/ha.

Key factors that affect costs of all irrigation systems are as follows: water requirement, labor requirement, initial costs, pumping costs, adaptability to soil types, adaptability to terrain, special features (such as frost control or evaporative cooling), adaptability to field shapes, and utility for applying chemicals. Based on some of the estimates of Turner and Andersen (1980) and Fereres et al. (1978), some of the key features and costs that one could expect for an average or typical irrigation system in terms of 1989 U.S. dollars are shown in Table 16-1.

III. SUMMARY

In this chapter the authors have described the major types of irrigation methods in use in the USA during the late 20th century. The principal features of each system have been described along with their applications, the factors affecting performance, and the types of management that are required for effective use. This chapter has dealt mainly with the physical features of different irrigation methods. It should be stressed again, however, that to perform effectively, even a well-designed system that conforms properly to the crop, climate and site conditions is not enough. The system must also be well managed.

All major types of irrigation systems—surface, sprinkler, and micro—have a place in today's agriculture. While potential for high efficiency may be greater with some systems than with others, when the economics, including labor, capital costs, and the cost of the water supply are considered, no one method will consistently be most practical.

REFERENCES

Abbott, J.S. (chair). 1984. Micro irrigation—world wide usage. Working group on Micro Irrigation, Int. Comm. Irrig. Drain. Bull. 33(1):4-9.

Allen, R.G., and C.E. Brockway. 1984. Concepts for energy-efficient irrigation system design. J. Irrig. Drain. Div. Am. Soc. Civ. Eng. 110(2):99-106.

American Society of Agricultural Engineers. 1988. Design and installation of microirrigation systems. p. 536-539. *In* Standards 1988. EP405. ASAE, St. Joseph, MI.

Aslan, S., D. Ackley, and P. Willey. 1986. A summary of the drip irrigation problems of the Coachella Valley. USDA Soil Conserv. Serv., Coachella Valley Resour. Conserv. Distr., Escondido, CA.

Bassett, D.L., D.D. Fangmeier, and T. Strelkoff. 1980. Hydraulics of surface irrigation. p. 447-498. *In* M. Jensen (ed.) Design and operation of farm irrigation systems. ASAE, St. Joseph, MI.

Batty, J.C., S.H. Hamad, and J. Keller. 1975. Energy inputs to irrigation. J. Irrig. Drain. Div. Am. Soc. Civ. Eng. 101(IR4):293-307.

Bernstein, L., and L.E. Francois. 1973. Comparisons of drip, furrow, and sprinkler irrigation. Soil Sci. 115(1):733-786.

Bielorai, H. 1982. The effect of partial wetting of the root zone on yield and water use efficiency in a drip- and sprinkler-irrigated mature grapefruit grove. Irrig. Sci. 3:89-100.

Bishop, A.A. 1962. Relation of intake rate to length of run in surface irrigation. p. 282-288. Trans. Am. Soc. Civ. Eng. Part III. Vol. 127.

Bishop, A.A., W.R. Walker, N.L. Allen, and G.J. Poole. 1981. Furrow advance rates under surge flow systems. J. Irrig. Drain. Div. Am. Soc. Civ. Eng. 107(IR3):257-264.

Blad, B.L., J.R. Steadman, and A. Weiss. 1978. Canopy structure and irrigation influence white mold disease and microclimate of dry edible beans. Phytopathology 68:1431-1437.

Bos, M.G. 1985. Summary of ICID definitions on irrigation efficiency. Int. Comm. Irrig. Drain. Bull. 34(1):28-31.

Bralts, V.F. 1986. Field performance and evaluation. p. 216-240. *In* F.S. Nakayama and D.A. Bucks (ed.) Trickle irrigation for crop production. Elsevier Sci. Publ., Netherlands.

Bresler, E. 1977. Trickle-drip irrigation: principles and applications to soil-water management. Adv. Agron. 29:343-393.

Bucks, D.A., and S. Davis. 1986. Historical development. p. 1-52. *In* F.S. Nakayama and D.A. Bucks (ed.) Trickle irrigation for crop production. Elsevier Sci. Publ., Netherlands.

Bucks, D.A., F.S. Nakayama, and R.G. Gilbert. 1979. Trickle irrigation water quality and preventive maintenance. Agric. Water Manage. 2:149-162.

Bucks, D.A., F.S. Nakayama, and A.W. Warrick. 1982. Principles, practices, and potentialities of trickle (drip) irrigation. p. 219-298. *In* D. Hillel (ed.) Advances in irrigation. Vol. 1. Academic Press, New York.

Christiansen, J.E. 1942. Irrigation by sprinkling. Univ. Calif. Bull. 670.

Coolidge, P.S., W.R. Walker, and A.A. Bishop. 1982. Advance and runoff-surge flow furrow irrigation. J. Irrig. Drain. Div. Am. Soc. Civ. Eng. 108(IR1):35-42.

Dedrick, A.R., L.J. Erie, and A.J. Clemmens. 1982. Level-basin irrigation. p. 105-145. *In* D. Hillel (ed.) Advances in irrigation. Vol. 1. Academic Press, New York.

Erie, L.J., and A.R. Dedrick. 1979. Level-basin irrigation: Method for conserving water and labor. USDA Farm Bull. 2261.

Fereres, E., J.L. Meyer, F.K. Aljibury, H. Shulbach, A.W. Marsh, and A.D. Reed. 1978. Irrigation costs. Univ. Calif. Ext. Leafl. 2875.

Fischbach, P.E. (ed.). 1980. Irrigation scheduling handbook. Nebr. Coop. Ext. Serv.

Fischbach, P.E. (ed.). 1988. Irrigation scheduling-management handbook. 5th ed. Agric. Eng. Dep., Inst. Agric. Nat. Resour., Univ. Nebraska, Lincoln.

Fox, R.L., J.T. Phelan, and W.D. Criddle. 1956. Design of subirrigation systems. Agric. Eng. 37(2):103-108.

Freeman, B.M., J. Blackwell, and K.V. Garzoli. 1976. Irrigation frequency and total water application with trickle and furrow systems. Agric. Water Manage. 1(1):21-31.

Fry, A.W. 1977. Frost protection by sprinkling. P/N D31441. Rain Bird Sprinkler Mfg. Corp., Glendora, CA.

Fry, A.W., and A.S. Gray. 1971. Sprinkler irrigation handbook. 10th ed. Rain Bird Sprinkler Mfg. Corp., Glendora, CA.

Fry, W.E. 1982. Principles of plant disease management. Academic Press, New York.

Hagan, R.M., and E.B. Roberts. 1981. Energy, water and cost trade-offs in irrigation system selection and management. Trans. ASAE 24:1539–1545.

Haise, H.R., and P.E. Fischbach. 1970. Auto-mechanization of pipe distribution systems. p. M1–M15. *In* Proc. Nat. Irrig. Symp., Lincoln, NE. 10–13 Nov.

Haise, H.R., and E.G. Kruse. 1966. Pneumatic valves for automation of irrigation systems. p. S1–S8. *In* Proc. 6th Congr. Int. Comm. Irrig. Drain. New Delhi, Jan. Spec. Sess. Rep. 1.

Haise, H.R., and E.G. Kruse. 1969. Automation of surface irrigation systems. J. Irrig. Drain. Div. Am. Soc. Civ. Eng. 95(IR4):503–516.

Haise, H.R., E.G. Kruse, M.L. Payne, and H.R. Duke. 1980. Automation of surface irrigation. 15 years of USDA research and development at Fort Collins, Colorado. USDA Prod. Res. Rep. 179.

Haise, H.R., F.G. Viets, Jr., and J.S. Robins. 1960. Efficiency of water use related to nutrient supply. Int. Congr. Soil Sci., Trans. 7th (Madison, WI) I:663–671.

Hall, W.A. 1960. Performance parameters of irrigation systems. Trans. ASAE 13(1):75, 76, 81.

Hamer, P.J.C. 1986. The heat balance of apple buds and blossoms. Part II. The water requirements for frost protection by overhead sprinkler irrigation. Agric. Forest Meteorol. 37:159–174.

Hansen, V.E. 1960. New concepts in irrigation efficiency. Trans. ASAE 3(1):55–57.

Hansen, V.E., O.W. Israelsen, and G.E. Stringham. 1979. Irrigation principles and practices. 4th ed. John Wiley and Sons, New York.

Harrison, K.A. 1982. Factors to consider in selecting an irrigation system. Georgia Coop. Ext. Serv. Bull. 882.

Hart, W.E., G. Peri, and G.V. Skogerboe. 1979. Irrigation performance: An evaluation. J. Irrig. Drain. Div. Am. Soc. Civ. Eng. 105(IR3):275–288.

Howell, T.A., D.A. Bucks, D.A. Goldhamer, and J.M. Lima. 1986. Irrigation scheduling. p. 241–279. *In* F.S. Nakayama and D.A. Bucks (ed.) Trickle irrigation for crop production. Elsevier Sci. Publ., Netherlands.

Humphreys, A.S. 1967. Control structures for automatic surface irrigation systems. Trans. ASAE 10(1):21–23, 27.

Humphreys, A.S. 1969. Mechanical structures for farm irrigation. J. Irrig. Drain. Div. Am. Soc. Civ. Eng. 95(IR4):463–479.

Humphreys, A.S. 1971. Automatic furrow irrigation systems. Trans. ASAE 14(3):466–470.

Humphreys, A.S. 1975. Automated valves for surface irrigation pipelines. J. Irrig. Drain. Div. Am. Soc. Civ. Eng. 101(IR2):95–109.

Humphreys, A.S., J.E. Garton, and E.G. Kruse. 1970. Auto mechanization of open channel distribution systems. p. L1–L20. *In* Proc. Nat. Irrig. Symp., Lincoln, NE. 10–13 Nov.

James, L.G., and W.M. Shannon. 1986. Flow measurement and system maintenance. p. 280–316. *In* F.S. Nakayama and D.A. Bucks (ed.) Trickle irrigation for crop production. Elsevier Sci. Publ., Netherlands.

Jensen, M.E. 1967. Evaluating irrigation efficiency. J. Irrig. Drain. Div. Am. Soc. Civ. Eng. 93(IR1):83–98.

Jensen, M.E. (ed.). 1980. Design and operation of farm irrigation systems. ASAE, St. Joseph, MI.

Kemper, W.D., J.A. Bondurant, and T.J. Trout. 1986. Irrigation trash screens pay. J. Soil Water Conserv. 41:17–20.

Kemper, W.D., W.H. Heinemann, D.C. Kincaid, and R.V. Worstell. 1981. Cablegation: I. Cable controlled plugs in perforated supply pipes for automatic furrow irrigation. Trans. ASAE 24(6):1526–1532.

Kemper, W.D., D.C. Kincaid, R.V. Worstell, W.H. Heinemann, T.J. Trout, and J.E. Chapman. 1985. Cablegation systems for irrigation: Description, design, installation, and performance. USDA-ARS-21.

Kruse, E.G. 1978. Describing irrigation efficiency and uniformity. J. Irrig. Drain. Div. Am. Soc. Civ. Eng. 104(IR1):35–41.

Lyle, W.M., and J.P. Bordovsky. 1979. Traveling low energy precision irrigator. p. 121–131. *In* Irrigation and drainage in the nineteen-eighties. Proc. Specialty Conf. Irrig. Drain. Div., Albuquerque, NM. 17–20 July. Am. Soc. Civ. Eng., New York.

Lyle, W.M., and J.P. Bordovsky. 1981. Low energy precision application (LEPA) irrigation system. Trans. ASAE 24(5):1241–1245.

Maas, E.V. 1985. Crop tolerance to saline sprinkling water. Plant Soil 89:273–284.

Maas, E.V., and G.J. Hoffmann. 1977. Crop salt tolerance-current assessment. J. Irrig. Drain. Div. Am. Soc. Civ. Eng. 103(IR2):115–134.

Massey, F.C., R.W. Skaggs, and R.E. Sneed. 1983. Energy and water requirements for subirrigation vs. sprinkler irrigation. Trans. ASAE 26:126–133.

Merriam, J.L., and J. Keller. 1978. Farm irrigation system evaluation: A guide for management. Utah State Univ., Logan.

Peri, G., G.V. Skogerboe, and D.I. Horum. 1979. Evaluation and improvement of basin irrigation. Water Manage. Tech. Rep. 49B. Water Manage. Res. Proj., Colorado State Univ., Fort Collins.

Perry, K.B., C.T. Morrow, A.R. Jarrett, and J.D. Martsoff. 1982. Evaluation of sprinkler application rate models used in frost protection. HortScience 17(6):884–885.

Phene, C.J. 1986. Automation. p. 188–215. In F.S. Nakayama and D.A. Bucks (ed.) Trickle irrigation for crop production. Elsevier Sci. Publ., Netherlands.

Ravelo, C.J., E.A. Hiler, and T.A. Howell. 1977. Trickle and sprinkler irrigation of grain sorghum. Trans. ASAE 20:96–99.

Rawlins, S.L. 1977. Uniform irrigation with a low-head bubbler system. Agric. Water Manage. 1:167–178.

Rolston, D.E., R.J. Miller, and H. Schulbach. 1986. Fertilization. p. 317–344. In F.S. Nakayama and D.A. Bucks (ed.) Trickle irrigation for crop production. Elsevier Sci. Publ., Netherlands.

Ross, P.E., and N.P. Swanson. 1957. Level basin irrigation. J. Soil Water Conserv. 12(5):209–214.

Rotem, J., and J. Palti. 1969. Irrigation and plant diseases. Annu. Rev. Phytopathol. 7:267–288.

Sepaskhah, A.R., S.A. Sichani, and B. Bahrani. 1976. Subsurface and furrow irrigation evaluation for bean production. Trans. ASAE 19:1089–1092.

Skaggs, R.W., G.J. Kriz, and R. Bernal. 1972. Irrigation through subsurface drains. J. Irrig. Drain. Div. Am. Soc. Civ. Eng. 98(IR3):363–373.

Steiner, J.L., E.T. Kanemasu, and D. Hasza. 1983. Micro climatic and crop responses to center pivot sprinkler and to surface irrigation. Irrig. Sci. 4:201–214.

Stringham, G.E., and J. Keller. 1979. Surge flow for automatic irrigation. p. 132–142. In Proc. 1979. Specialty Conf., Irrig. Drain. Div., Albuquerque, NM. 17–20 July. Am. Soc. Civ. Eng., New York.

Strong, W.C. 1972. Sprinkler irrigation manual. General Sprinkler Corp., Fresno, CA.

Sugar, D., and P.B. Lombard. 1981. Pear scab influenced by sprinkler irrigation above the tree or at ground level. Plant Dis. 65(12):980.

Tanji, K.K. (ed.). 1990. Agricultural salinity assessment and management. Am. Soc.Civ. Eng., New York.

Thorfinnson, T.S., M. Hund, and A.W. Epp. 1955. Cost of distribution of irrigation water by different methods. Univ. of Nebraska Agric. Exp. Stn. Bull. 432.

Till, M.R., and M.G. Bos. 1985. The influence of uniformity and leaching on the field application efficiency. Int. Comm. Irrig. Drain. Bull. 34(1):28–35, 60.

Turner, J.H., and C.L. Anderson (ed.). 1980. Planning for an irrigation system. 2nd ed. Am. Assoc. for Voc. Instructional Materials, Athens, GA.

U.S. Salinity Laboratory Staff, L.A. Richards (ed.). 1954. Diagnosis and improvement of saline and alkali soils. USDA Handb. 60. Washington, DC.

U.S. Department of Agriculture. 1967. Planning farm irrigation systems. p. 3-1 to 3-92, section 15 (Irrigation). In SCS national engineering handbook. U.S. Gov. Print. Office, Washington, DC.

U.S. Department of Agriculture. 1974. Border irrigation. p. 4-1 to 4-50, section 15 (Irrigation). In SCS national engineering handbook. U.S. Gov. Print. Office, Washington, DC.

U.S. Department of Agriculture. 1983a. Furrow irrigation. p. 5-4 to 5-73, section 15 (Irrigation). In SCS national engineering handbook. 2nd ed. U.S. Gov. Print. Office, Washington, DC.

U.S. Department of Agriculture. 1983b. Sprinkle irrigation. p. 11-1 to 11-121, section 15 (Irrigation). In SCS national engineering handbook. 2nd ed. U.S. Gov. Print. Office, Washington, DC.

U.S. Department of Agriculture. 1984a. Georgia irrigation guide. Soil Conservation Service.

U.S. Department of Agriculture. 1984b. Trickle irrigation. p. 7-1 to 7-129, section 15 (Irrigation). In SCS national engineering handbook. U.S. Gov. Print. Office, Washington, DC.

Unrath, C.R. 1972a. The evaporative cooling effects of overtree sprinkler irrigation on "red delicious" apples. J. Soc. Hortic. Sci. 97(1):55–58.

Unrath, C.R. 1972b. The quality of "red delicious" apples as affected by overtree sprinkler irrigation. J. Soc. Hortic. Sci. 97(1):58–59.

van Schilfgaarde, J., L. Bernstein, J.D. Rhoades, and S.L. Rawlins. 1974. Irrigation management for salt control. J. Irrig. Drain. Div. Am. Soc. Civ. Eng. 100(IR3):321–338.

von Bernuth, R.D., and J.R. Gilley. 1985. Evaluation of center pivot application packages considering droplet induced infiltration reduction. Trans. ASAE 28:1940–1946.

von Bernuth, R.D., and K.H. Solomon. 1986. Emitter construction. p. 27–52. In F.S. Nakayama and D.A. Bucks (ed.) Trickle irrigation for crop production. Elsevier Sci. Publ., Netherlands.

Vories, E.D., and R.D. von Bernuth. 1986. Single nozzle sprinkler performance in wind. Trans. ASAE 29:1325–1330.

Wilson, P., H. Ayer, and C. Snider. 1984. Drip irrigation for cotton: Implications for farm profits. Nat. Resour. Econ. Div. Econ. Res. Serv. USDA Rep. No. 517. Washington, DC.

Wright, J.L., J.L. Stevens, and M.J. Brown. 1981. Controlled cooling of onion umbels by periodic sprinkling. Agron. J. 73(3):481–490.

17 Irrigation Scheduling Controls and Techniques

D. F. HEERMANN

USDA-ARS
Fort Collins, Colorado

D. L. MARTIN

University of Nebraska
Lincoln, Nebraska

R. D. JACKSON

USDA-ARS
Temple, Arizona

E. C. STEGMAN

North Dakota State University
Fargo, North Dakota

Irrigation systems for agricultural crops should be properly designed, installed, and managed to achieve high efficiencies. *Irrigation scheduling*, a key element of proper management, is the accurate forecasting of water application (in timing and amount) for optimal crop production. Intelligent scheduling requires knowledge of soil water initially available to the plant and expected change in levels for individual fields over the succeeding 5 to 10 d. This knowledge enables estimating the earliest date at which the next irrigation should be applied for efficient irrigation with the particular system. Simultaneously, an estimate of the probable latest date for the next irrigation, before water stress affects crop production, could also be made. The irrigator then has the latitude to schedule an irrigation between these dates and accommodate cultural and harvesting requirements for the particular crop. This range of dates also allows the irrigator to develop a strategy for using the rainfall expected to occur before the next irrigation. An irrigation system can only be efficient when it is both scheduled properly and operated to apply the desired amount of water efficiently. A breakdown in either of these can result in poor management.

Operation of an irrigated farm involves planning and decision making activities both before and during the growing season. For example, irrigation scheduling may be integrated with electrical load management to minimize electrical supply costs. Other considerations include salinity control, retaining adequate soil moisture storage capacity for expected rainfall, providing adequate soil moisture for planting, optimizing soil moisture for the harvesting of root crops, minimizing leaching of crop nutrients, and maximizing profit when water or energy supplies are limited. In addition, producers may have varying management objectives depending on physical and financial resources.

I. MANAGEMENT OBJECTIVES

A farmer makes many annual and long-term decisions in producing agricultural crops. The type of crops, cultural practices, fertility levels, and an appropriate irrigation system must all be selected. Each decision impacts production costs and potential for realizing profit. Many decisions and plans are made annually (purchasing of seeds, fertilizers, and supplies), while other decisions are made less frequently (such as purchase of capital equipment and installation of irrigation systems). However, the irrigator is faced with making more frequent decisions regarding the time and amount of water to apply. Maximizing net returns is highly dependent on minimizing production costs for the selected crops. Many irrigators recognize the value of installing irrigation systems that require minimum energy. Many irrigators, however, fail to recognize the value of improved technical abilities either through self education or by contracting with consultants to improve their agronomic and irrigation scheduling methods. Improved scheduling can either

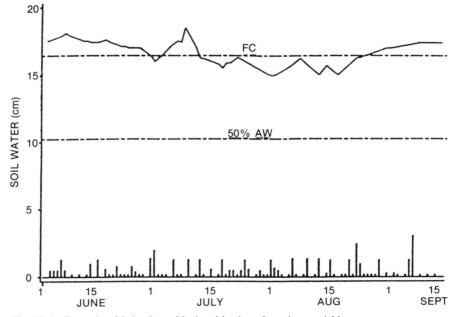

Fig. 17–1. Example of irrigation with the objective of maximum yields.

reduce irrigation costs and/or increase crop quantity or quality. The general goal of most producers is to maximize yield (Fig. 17–1). For example, local contests usually recognize managers who attain the highest yield. Each manager, however, must set his own goals and specific managment objectives for an entire agricultural production system. These goals influence the type of management and scheduling strategies that an irrigator might implement, and may not imply maximum absolute yield, but rather maximum economic yield or economic return (Fig. 17–2).

The maximum yield objective usually requires that irrigations be scheduled to maintain a nonstressing soil water status for much of the growing season. The schedules must also prevent overirrigation which can leach fertilizers and reduce the root zone aeration below optimal levels. This yield objective is often best implemented with irrigation systems that can irrigate frequently and apply water efficiently.

Another management strategy is to assure adequate water supplies at critical growth stages. This management strategy may not produce maximum yields but often contributes to maximum water use efficiency. For example, the time of flowering and grain fill are critical stages for grain crops. Forage crops are generally not irrigated on the basis of critical growth stages.

Irrigation schedules for surface irrigation systems are often designed to minimize labor costs. This is particularly true with systems where there is significant labor associated with changing water from one set to another. The general scheduling objective is to deplete the soil water to mild stress levels, but avoid yield reduction. The strategy minimizes the number of irri-

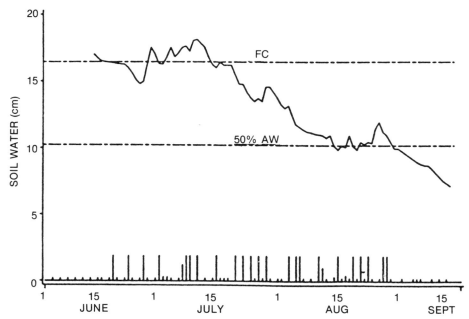

Fig. 17–2. Illustrated the saving of water in the later part of the season where maximum yield is not the management objective.

gations per season and results in larger irrigation amounts per irrigation which allows longer set times that may better match available labor schedules. The frequency of irrigation may vary from 1- to 2-wk intervals.

Irrigation management may include the objective of decreasing the energy costs for pumping irrigation water. Many irrigators are converting their sprinkler systems from high pressure to low pressure. Pressure requirements can also be reduced by selecting surface or trickle irrigation systems. A fact often overlooked in estimating energy costs is that irrigation system efficiency with reduced pressures affects total water pumped. Even though hourly operational costs decrease with reduced pressure, the total water volume pumped may increase because of low efficiency. Thus, the total cost is a function of the pressure and the total volume required for crop production and may increase with lower pressure systems.

An irrigation manager's major objective is to maximize net returns. Often the available capital limits the degree of irrigation system modernization that may decrease operating costs to more economical levels. Furthermore, as competition for the available water resource with urban and industrial users increases or the supply is limited, irrigations may need to be scheduled to maximize net return based on available water supplies. Maximizing net return is both appropriate under limited water conditions and is desirable as the level of management increases. Application of the latest technology to control irrigation applications, to sense crop water stress, and to better estimate crop performance may help in maximizing returns.

II. IRRIGATION SCHEDULING

Many techniques and technologies can forecast the date and amount of irrigation water to apply. The appropriate technique or technology is a function of the irrigation water supply, technical abilities of the irrigator, irrigation system, crop value, crop response to irrigation, cost of implementing technology, and personal preference. Historically, many irrigation schedules were based on the available water supply and the irrigated area. Where the water supply is a direct diversion from an adjacent stream and the distribution system is jointly shared by a number of irrigators, water is often supplied on a rotational basis for fixed periods. Irrigators generally apply all available water during the delivery time to the entire irrigable area. The irrigator selects crops based on production experience and known water requirements. Usually irrigations are scheduled to prevent losses of crop production due to water stress.

Significant irrigation development has recently occurred in areas where water is pumped from ground water reservoirs or rivers (i.e., along the Columbia and Snake Rivers). Many of these areas are irrigated with center pivot irrigation systems. The management of these systems differs significantly from that in older irrigated areas where water is supplied by direct gravity diversion from streams and reservoirs. These farmers, many who had no previous irrigation experience, have developed their own management styles. Some irrigators run the center pivot systems continuously throughout the growing season. This continuous operation frequently results in significant overirrigation, particularly early and late in the irrigation season. The false assumption has been made that excess water acts like an insurance policy. However, irrigators' attitudes are changing as educational programs and the popular press encourage improving management for resource conservation and economic return. The following sections illustrate tools and techniques available for improving irrigation scheduling.

A. Water Budgets

An irrigation scheduling program using meteorological data to calculate water use and maintain a water budget was developed by Jensen (1969), Jensen and Heermann (1970), and Jensen et al. (1970, 1971) for the USDA. Many variations of the water balance program have been developed and are in use. In general, the water budget approach allows an irrigator to maintain a current balance of plant-available water in the soil profile. The water balance equation applied to each field for each day is

$$D_{p_i} = D_{p_{i-1}} + K_{c_i} \times E_{tp_i} + E_{tr_i} - (R_i - RO_i) + W_{d_i} \qquad [1]$$

where
D_{p_i} = the depletion on day i,
K_{c_i} = crop coefficient (a function of crop development),
E_{tp_i} = the reference evapotranspiration,
E_{tr_i} = additional soil evaporation following an irrigation or rain,
R_i = the sum of effective rainfall and net irrigation on day i,
RO_i = surface runoff,
W_{d_i} = the drainage below root zone or upward flow $(-)$ from the groundwater.

The surface runoff is often not input separately but considered in determining the effective rainfall. With well-drained soil and deep water tables, the upward flow from the groundwater is assumed to be zero. If the calculated depletion becomes negative due to large rainfalls or irrigation, the depletion is set to zero and the excess water is assumed to drain below the root zone or run off. Lundstrom et al. (1981) developed a checkbook method which provides inexperienced irrigators with a simple technique for estimating crop water use and determining when to irrigate. The checkbook system is based on the Jensen et al. (1971) scheduling model. However, rather than estimate reference evapotranspiration (ET) from daily mean air temperature and daily solar radiation, the checkbook system utilizes daily maximum air temperatures and long-term monthly averages of solar radiation. This simplification allows development of tables of daily water use as a function of maximum daily temperature and stage of crop growth. The water balance (checkbook sheet) is maintained by the irrigator to avoid either excessive soil water depletions or overirrigation. The checkbook scheduling method is a system for individual use on the farm for improved water management. It is easy to understand and use, while still maintaining an acceptable level of accuracy. Techniques have been developed which do not require real-time climatic data for the estimation of ET. They use either a constant ET for the season (Woodruff et al., 1972) or a historical climatic average for a given location (Fereres et al., 1980). Depending upon the accuracy required and climatic variation in the area, these techniques can provide reasonable guidelines for managing irrigation systems.

Computer-based scheduling models are more sophisticated but have not been as readily accepted as the checkbook system. The checkbook method is an excellent way to introduce irrigators to the concept of water budgeting. The same concepts can be expanded and used with increased accuracy and ease through the use of modern computer technology. Computerized scheduling is also more useful when fields are irrigated frequently or where there are a large number of fields to schedule.

B. Computers

Jensen and associates (Jensen, 1969; Jensen & Heermann, 1970; Jensen et al., 1971) developed irrigation scheduling models for operation on centralized main-frame computers and accessed through terminals by researchers, consulting services, and other government agencies. This operating procedure results in a built-in delay from program execution to user access.

Kincaid and Heermann (1974) developed an irrigation scheduling program for a desktop programmable calculator. The model demonstrated the advantage of having local control and up-to-date information for scheduling individual fields to improve water management. The first adaptation was rather crude compared to the present scheduling models used on modern personal computers. Many consulting firms and researchers have written individualized irigation scheduling programs for personal computers. The interest and acceptance level of irrigation scheduling technology by irrigators is increasing significantly. Practically all of the computer irrigation scheduling programs maintain a water budget based on meteorological data, provide forecasts of water requirements, and estimate the appropriate time and amount of future irrigations.

C. Field-data Requirements

The successful use of a soil-water budget for irrigation scheduling requires accurate and timely data from the individual fields. These include the amount of rainfall and irrigation applied, meteorological data to estimate ET, soil water status, and crop condition.

The amount and frequency of meteorological data input depends on the accuracy required. In most real-time irrigation scheduling applications meteorological data are collected with automatic weather stations for a general region or area. However, some approaches have used the long-term average climatic data for scheduling irrigations.

More detailed methods for estimating ET are discussed in Section V of this book. Method selection by the irrigator often depends on the available data for use in calculating ET. The Penman (1948, 1963) equation is a combination equation that has generally been accepted as a scientifically sound formulation for estimating reference ET. This equation is expressed as a combined function of radiation, maximum and minimum temperature, vapor pressure, and wind run.

The requirement for vapor pressure and wind-run data has discouraged many from using the Penman combination equation. Instead, equations that are a function of solar radiation and temperature have been used. The Priestly-Taylor (1972) and Jensen-Haise (1963) methods estimate potential ET as a function of solar radiation and temperature. These two formulations have been most successful in humid areas where the advective energy is either relatively small or nearly constant. However, the combination equation provides better accuracy in estimating potential ET in arid climates.

Evaporation pans also provide estimates of potential ET, and several investigators (Doorenbos & Pruitt, 1975; Jensen et al., 1961) have reported satisfactory results. However, siting of evaporation pans is critical. The relatively large volume of Class A pans acts as a heat sink which causes a lag in response to daily fluctuation of potential ET. For longer periods (5–10 d), pan evaporation estimates can be quite acceptable.

The location of weather stations to measure the necessary meteorological data is important for accurate estimates of ET. Stations must be located in an environment that is representative of the irrigated area being scheduled. Solar radiation measurements require avoidance of sensor shading or sunlight reflection from adjacent structures. Temperature, vapor pressure, and wind should be sensed above an irrigated crop and measured where there is sufficient upwind fetch. Most recording weather stations collect daily values of maximum and minimum temperature, average vapor pressure, total wind run, and solar radiation. Automatic weather stations provide hourly or more frequent capability to monitor wind, vapor pressure, temperature, and radiation data. Additional research is needed to develop accurate equations which use frequently sampled meteorological data if it is desired to manage irrigation systems with frequent application (i.e., drip irrigation).

Equipment is available to automatically monitor weather stations and store data directly into a personal computer. This allows automatic processing and/or calculations of reference ET or actual crop ET values for a number of individual fields. The same monitoring equipment can automatically measure rainfall and monitor irrigation systems for determining the total water applied. Irrigation scheduling technology is more readily accepted and implemented by irrigators with the use of automatic monitoring and prediction capabilities.

1. Rainfall and Irrigation

The measurement of rainfall and irrigation amounts is essential to maintain water budgets for irrigation scheduling. Rainfall should be measured for each scheduled field because amounts may be highly variable, even within 1 km. Without these individual measurements, the accuracy of a water budget is significantly decreased. Irrigation amounts must also be measured. An irrigation system evaluation should also include the uniformity and water application efficiencies (water available for ET per water delivered in the field). Incorrect timing or application depths of irrigations can reduce the irrigation efficiency. It is the water application available for ET for the entire field or management unit that must be input to the water budget equation.

Only effective precipitation is input to the water budget as rainfall. The amount of precipitation that is effective is dependent on the particular water budget technique. The total rainfall less runoff is entered into the water budget equation. If the water budget equation includes a term for estimating the soil water evaporation, the addition of small rainfall amounts into the water budget will increase the soil water evaporation terms. Rainfall < 5 mm should not be entered if the water budget does not account for soil evaporation.

2. Soil Water Status

Periodic observations of the soil water status can be used to adjust the calculated soil water depletion. Accumulated errors in estimating ET or measuring the water added by irrigation and rainfall can cause significant differences between the calculated and actual soil water status. Irrigation system uniformity and water application efficiency affect the application depth and cause variability in actual soil water status in the field. A ± 5% error in measuring irrigation volumes is typical. This, coupled with uncertain water application efficiency, can easily lead to ± 10% error in determining the effective irrigation amount. Weekly to monthly sampling intervals are usually sufficient to update water budgets.

Accurate ET estimates are also important in maintaining water budgets. When ET errors are suspected, it is important to check the accuracy of the instruments measuring the temperature, vapor pressure, wind run, and radiation. Vapor pressure is probably the most difficult parameter to measure on a continuous basis with currently available sensors. Vapor pressure sensors are particularly sensitive to age and dirt. The ET equations and input data will introduce some random error and in certain cases the error may be systematic. Systematic errors are common if conditions for the climatic data are different from the irrigated area.

Soil water measurement is an excellent technique for evaluating the adequacy of an irrigation system. Measurements at the head and lower ends of surface-irrigated fields can determine under and over irrigated areas. Measurements before and after irrigations can determine the range of soil water variations. If the soil profile is not being refilled, this should be considered when scheduling successive irrigations. Soil sampling by itself does not provide a forecast of the time and amount of the next irrigation. Sampling can be used in conjunction with forecasts of water use, or frequent sampling results can be projected to estimate future depletion levels.

A soil probe is a simple and cost effective tool commonly used for measuring soil water status. Use of the probe requires experience to accurately determine the available water of extracted soil samples. Klocke and Fischbach (1984) developed a guide to instruct the user to estimate the soil water by appearance and feel. Many consulting services providing on-farm irrigation scheduling recommendations use this approach to determine the current soil water status. It is simple and requires no sensor or tube access sites in the field.

Tensiometers and gypsum blocks have long been available for monitoring soil water status. These sensors require labor, both for reading and sensor maintenance. Sensor, meter, or gauge calibrations are often required for accurate soil water measurements. These sensors are used as a tool for irrigation scheduling. Irrigator acceptance has not been widespread. The acceptance of these sensors has largely been limited to where high cash value crops are grown and/or where water is expensive, where rainfall is a significant scheduling factor, or high water tables exist.

The neutron probe is another tool for measuring soil water status. This instrument is expensive but provides the opportunity of repeated soil water measurements at the same location within the field. The neutron probe must be calibrated to give the total volume of water per unit depth in the soil profile. Consulting services, government agencies, and irrigators have used the neutron probe as a direct tool for irrigation scheduling.

The typical procedure for scheduling with soil water measurements in lieu of a water budget is to make periodic (3–7 d) measurements and to extrapolate the measured soil water depletion to project the time of the next irrigation. Measurements of total available water can also be used as inputs to water budget programs to periodically correct the estimates of soil water depletion. The frequency of soil water measurement can be decreased when used in conjunction with a water balance model.

3. Plant Water Status

Irrigation scheduling methods generally concentrate on the water status of the root zone profile. Direct measurements of plant water status can also be used to schedule irrigations. Pressure chambers have been used, particularly by researchers, for determining the current water status (usually leaf water potentials) in a plant. The time of day for these measurements becomes particularly important, with the lowest water potentials occurring during the peak of the day when the potential ET rate is highest. This technique of measuring plant water status, however, is best suited for researchers and has seen less acceptance by irrigation managers, or even consulting services. The remote-sensing of canopy temperature, which will be discussed in later sections can be made quite simply with commercially available instruments. These data can be used as feedback into an irrigation scheduling water budget program.

III. FORECASTING SCHEDULES

Water budget methods using real-time climatic data can provide current estimates of the soil water status. An irrigation can be initiated when the current estimated depletion reaches a management allowable depletion. Management of an irrigation system with this technique requires that water be available upon demand and that the system is large enough to instantly irrigate the entire area. It is generally more desirable to forecast a schedule of future irrigation dates and amounts for the fields that are being managed by an irrigator. The water budget equation can be extended into the future by forecasting ET rates and rainfall amounts. A new schedule may be made every day, after each irrigation, or even at periods spanning several irrigations.

A. Forecasting Evapotranspiration

One forecast option is to assume that the future daily ET will be equal to the average ET for the previous 3 to 7 d. The actual ET rate for the period before the next irrigation may or may not be equal to that just preceding the current date. In general, the crop coefficient increases during the crop development stage, then gradually decreases as the crop matures. For many crops, the actual ET follows the same trend. A better ET estimator is the climatic average reference ET and the projected crop coefficients based on expected crop development.

Heermann and Jensen (1970) demonstrated that forecasting the next irrigation varies when using the running 3-d average ET. They advised an update with real-time climatic data every 3 to 5 d for the forecast of the next irrigation date to converge to the required date.

Climatic average ET data provide a more stable estimator of future ET rates than short-term averages. Where climatic data are available, Jensen and Heermann (1970) proposed a simple exponential equation fitted to historical seasonal average data. Heermann et al. (1984), recognizing the 3 to 5 d persistence in weather, developed a technique to adjust the forecasted long-term average ET rates with the current observed reference ET. The procedure assumes the difference between the observed reference ET and the reference ET based upon the long-term average decays exponentially. This technique assumes the ET will persist for a short time and approach the long-term average within 5 d.

Real-time weather forecasts may be used for forecasting future ET. However, forecasts are commonly given only for temperature and precipitation probability. Limited forecasts are made for solar radiation and wind run. The forecasts for wind are generally in a broad range of expected values rather than the expected average amount which is required to calculate reference ET.

B. Forecasting Rain

Local forecasts for rain are generally made as a probability of occurrence—not amount—in a general area. In many irrigated areas, thundershowers are highly variable from location to location. This makes it extremely difficult for the irrigator to use rainfall forecasts in irrigation scheduling. The most typical way of including rainfall forecasts in irrigation scheduling programs is to forecast the expected date of the next irrigation with different rainfall amounts. The irrigator can then evaluate the effect of various rainfall amounts on the actual operation of the irrigation system.

Heermann et al. (1976) developed irrigation scheduling concepts and procedures for center pivot irigation systems. They forecasted an earliest and latest date for the next irrigation. With this range of dates for the next irrigation, the manager can adjust the actual schedule toward the later date if rainfall is forecast with a high probability.

The availability of historical normal precipitation amounts also can be of value in irrigation management. If the probability of rainfall is high, it may be desirable to manage the irrigation system so as not to refill the soil water profile but maintain a reservoir in the soil profile for expected rain. These management decisions are highly dependent on the probability of rainfall, the soil water holding capacity, and ability of the irrigation system to apply water. For example, when irrigating sandy soils in Colorado, it is generally not recommended to maintain a reservoir for rain because of the low probabilities of rainfall and the limited number of days with which the soil water storage can satisfy the expected ET demand.

IV. REMOTE SENSING

Techniques based on remotely sensed data have been developed that are useful for irrigation scheduling. By integrating measurements over relatively large areas, remote sensing techniques reduce the problem of variability associated with point sampling. Furthermore, entire fields can be sampled in a short time. Both reflected and emitted radiation provide information about plant canopies that is useful in irrigation scheduling. Reflected radiation can be used to infer crop coefficients. Canopy temperature can be estimated from measurements of emitted radiation and used in conjunction with air temperature and vapor pressure data to provide useful information concerning plant water stress.

A. Crop Coefficient Estimation

Bausch and Neale (1987) developed an algorithm that uses reflected radiation in a vegetation index to directly estimate the crop coefficient. This technique provides a real-time estimate of the crop coefficient for use in the water budget program. The conventional crop coefficient is assumed to be a unique function between planting to effective cover, and effective cover to harvest. This is generally acceptable for average climatic conditions. However, when the climate is either cooler or hotter than normal, the actual crop coefficient will deviate from the conventional value. The estimation of the crop coefficient via remote sensing allows adjustments to be made periodically which can account for differences in climatic conditions as well as affects that disease and fertility may have on crop development. Preliminary research results indicate that irrigation scheduling may be improved with the use of a remotely sensed crop coefficient. Experiments are currently underway to test the use of such a real-time crop coefficient for corn (*Zea mays* L.) in a scheduling program.

B. Canopy Temperature

The measurement of canopy temperatures offers a means of using the plant itself to indicate when irrigations are needed. This approach is based on the assumption that, as water becomes limiting, transpiration is reduced causing less evaporative cooling of leaves and increases plant temperatures in relation to air temperature. Although leaf temperatures have been measured for over 100 yr, only during the last 10 yr has the measurement of canopy temperatures become routine. This is a result of the development of small, portable infrared thermometers that measure radiation emitted from all parts of the canopy within the field of view of the instrument. Hand-held infrared thermometers have become ubiquitous tools that allow rapid, quantitative, field measurements of plant water stress. Numerous reports have appeared on this subject, many of which have been reviewed by Jackson (1982) and more recently by Idso et al. (1986).

An energy balance at a surface can be expressed as

$$R_n = G + H + \lambda E, \tag{2}$$

where

R_n = the net radiation (W m^{-2}),
G = the heat flux into the soil (W m^{-2}),
H = the sensible heat flux (W m^{-2}) into the air above the surface, and
λE = the latent heat flux (W m^{-2}).

The difference between the canopy and the air temperatures can be related to the vapor pressure deficit of the air ($e_a^* - e_a$), the net radiation, and the aerodynamic and crop resistances (Jackson, 1982) as

$$T_c - T_a = \frac{r_a R_n}{\rho C_p} \cdot \frac{\gamma(1 + r_c/r_a)}{\Delta + \gamma(1 + r_c/r_a)} - \frac{e_a^* - e_a}{\Delta + \gamma(1 + r_c/r_a)} \tag{3}$$

where

ρ = the density of air (kg m^{-3}),
C_p = the heat capacity of air (J kg^{-1} °C^{-1}),
T_c = the canopy temperature (°C),
T_a = the air temperature (°C),
e_a^* = the saturated vapor pressure of the air (Pa) at T_c,
e_a = the vapor pressure of the air (Pa),
γ = the psychometric constant (Pa °C^{-1}),
r_a = the aerodynamic resistance (s m^{-1}), and
r_c = the canopy resistance to water loss (s m^{-1}).

Under constant environmental conditions, the temperature difference ($T_c - T_a$) will vary with the canopy resistance r_c, and thus r_c determines the upper and lower limits of the temperature difference. The upper limit of $T_c - T_a$ can be found by allowing the canopy resistance r_c to increase without bound. As $r_c \to \infty$, Eq. [3] reduces to

$$(T_c - T_a)_{ul} = r_a R_n / \rho C_p,$$ [4]

the case of a nontranspiring crop.

The lower bound, found by setting $r_c = 0$ in Eq. [3], is

$$(T_c - T_a)_{\ell\ell} = \frac{r_a R_n}{\rho C_p} \cdot \frac{\gamma}{\Delta + \gamma} - \frac{e_a^* - e_a}{\Delta + \gamma},$$ [5]

the case for a wet canopy acting as a free water surface. It is assumed that the aerodynamic resistances in Eq. [4] and [5] are identical, although this assumption is not strictly valid (Choudhury et al., 1986). Theoretically, Eq. [4] and [5] form the bounds for all canopy-air temperature differences. However, the temperature difference for most well-watered crops will be greater than the lower limit because most crops exhibit some resistance to water flow, even when water is not limiting. For these crops, the lower limit should be modified by replacing γ in Eq. [5] with $\gamma^* = \gamma(1 + r_{cp}/r_a)$, where r_{cp} is the canopy resistance at potential transpiration.

Equation [5] shows that the lower limit is linearly related to the vapor pressure deficit. This linearity was determined experimentally by Ehrler (1973), and later in much greater detail by Idso et al. (1981) and Idso (1982). Idso (1982) proposed that a unique lower limit of ΔT (or non-water stressed baseline), exists for each plant species. Numerous articles have supported this view. In recent years, however, it has been demonstrated that the baseline is not entirely independent of net radiation, but may vary with season (Bucks et al., 1985). This result is not surprising in the light of Eq. [3] to [5]. However, on a daily basis, variation in the radiation term is minimized because temperature data are usually taken within a few hours of solar noon and usually on sunny days. Therefore the empirical non-water-stressed baselines have enjoyed some success.

The upper limit of ΔT, however, is difficult to measure experimentally. Some results have been reported by O'Toole and Hatfield (1983), and by Jackson (1982). A frequently used upper limit is 5 °C as calculated using Eq. [4] with $r_a = 10$ s m^{-1} and $R_n = 600$ W m^{-2}.

If Eq. [3] and [5] are used, a crop water stress index can be defined as

$$\text{CWSI} = \frac{(T_c - T_a) - (T_c - T_a)_{\ell\ell}}{(T_c - T_a)_{ul} - (T_c - T_a)_{\ell\ell}}$$ [6]

where $(T_c - T_a)$ is the measured temperature difference. The CWSI was shown to be equal to one minus the ratio of actual ET to potential ET by Jackson (1982).

The quantification of water stress using the CWSI provides a way to schedule irrigations based on plant measurements. Although much work needs to be done, available results indicate that crops having a CWSI value >0.25 may have incurred yield-reducing stress. Irrigating at about that value should minimize stress and maximize yields. A number of research projects have as their objective the determination of the optimum value of the CWSI for maximum yields with efficient use of water.

C. Future Applications

Successful research on estimating the crop coefficient by use of spectral vegetation indices should lead to the use of satellite data to evaluate crop coefficients over large agricultural areas. Irrigation districts could obtain satellite imagery, calculate the crop coefficients for each field, and provide the information to the farm manager. Crop coefficients updated frequently during the growing season should improve irrigation efficiency.

Although hand-held infrared thermometers can provide canopy temperature inputs to the CWSI, their use in a large operational sense is limited. To cover numerous fields would take time and labor that may not be available. The logical extension of this technique is to use satellites to obtain canopy temperatures for numerous fields over a large area, and calculate a crop water index for each field. However, a number of problems will delay this development. Some of the infrared radiation emitted from the surface of the earth is absorbed by water vapor in the atmosphere before it reaches the satellite. Therefore, accurate surface temperatures can only be obtained if readings are corrected for the atmospheric absorption. Furthermore, only one satellite is currently in orbit that can provide surface temperatures at a scale sufficient for agricultural needs. This satellite obtains imagery for a particular scene once every 16 d, weather permitting. Cloud cover frequently prevents data acquisition, causing the time between data collection to be too infrequent for operational use. Nevertheless, future satellites, additional research and improved logistics may provide ways to surmount these problems, and large area crop water stress assessments may be possible.

V. INTEGRATION OF IRRIGATION SCHEDULING WITH ENERGY MANAGEMENT

Irrigation scheduling can assist in energy management where significant energy is used to pump irrigation water. Pumping increases the operating cost and provides economic incentives to limit the amount of water pumped. It is often desirable to select the irrigation system that has the highest application efficiency to minimize the amount of water pumped. Another option is to reduce the pressure requirement for the selected irrigation system. Often a tradeoff exists between the most energy efficient system and the system with the highest water application efficiency. The two constraints may not exist in the same system. The advantage of implementing the best possible irrigation scheduling program is to limit the water application to just meet the crop demands.

The control of peak electrical demands is an important consideration where electrical energy is used to pump the water. Stetson et al. (1975) reported that many power suppliers offer rates for interruptible power which can significantly decrease the cost of energy for irrigation. Irrigation scheduling methods can serve an important role in electrical load control programs to limit crop losses due to water stress. Most scheduling/load control programs successfully decrease peak energy demands significantly by shifting demands to nonpeak periods, preventing interruption without inducing significant crop stress.

Heermann et al. (1984) developed an integrated system for irrigation scheduling and power control during peak electrical use periods. The integrated system provided four functions: (i) monitoring the system operation, (ii) control of irrigation systems, (iii) irrigation scheduling, and (iv) electrical load control. The monitoring and control of individual irrigation systems provided the irrigator assurance that systems were operating when energy was available. It minimized labor and travel time in monitoring and turning systems on and off. The power suppliers monitored the electric demand and sent radio signals to a computer-based controller at the farm headquarters when it was necessary to interrupt load. The computer-based system then stopped irrigation of individual units based on a priority for all irrigation systems under control. An algorithm for prioritizing the individual fields was added to the irrigation scheduling program. Individual priorities were based on the current soil water status, the irrigation system capacity, the water holding capacity of the soil, and the value of the crop. The irrigator was given the option to exercise personal preferences in the prioritization. This preference could include such things as the reliability of the irrigation systems on various fields or making allowances for cultural operations, fertilizer, or chemical applications. The irrigator could change the priority for the entire season or for a given day.

The integrated system was interfaced to an automatic weather station by a radio telemetry system. In this way, the irrigation schedule for each of the fields was updated daily, providing current recommended schedules for the next irrigation. This system also monitored the amount of irrigation water applied and the rainfall on each field.

VI. OPTIMIZING TECHNIQUES

Irrigation schedules based strictly on a water budget and forecasted ET often recommend the same irrigation date for several fields. When crops on different fields begin growing at the same time or when rainfall refills the soil profile, the crop water use may be nearly identical. In such cases several irrigated fields may require irrigation at the same time, which is not physically possible with many systems. Trava et al. (1977) and Pleban et al. (1983)

developed optimization programs that minimize the cost of labor within the constraints of the irrigation system and available water. The optimization programs were constrained so that water never limited crop production. An irrigation scheduling program was used to predict the depth of water required to refill the profiles for each day within a forecast period. An irrigation system has a minimum application depth that can be efficiently applied and was a constraint on the earliest irrigation date to prevent unintentional leaching. The maximum allowed depletion is the most water a crop can extract from the soil before stress occurs and usually determines the latest date for irrigation. The optimization programs by Trava et al. (1977) and Pleban et al. (1983) develop a schedule to irrigate all fields within the timing constraints and to minimize the labor costs for implementing the schedule.

A. Simulation for Scheduling Strategies

A water budget simulation approach provides techniques for evaluating various management-allowed depletions, frequencies of irrigation, and system capacities. Several individuals have used simulation programs to develop irrigation schedules and management strategies. Simulation models can be integrated with crop production functions to determine the expected returns from various irrigation schedules. For example, Ahmad (1987) evaluated the value of having water on demand in adequate amounts as compared to the losses experienced from limited water. The results provided the economic value for developing additional water supplies and improving irrigation systems. Simulation models offer a means to provide decision-making information that goes well beyond traditional

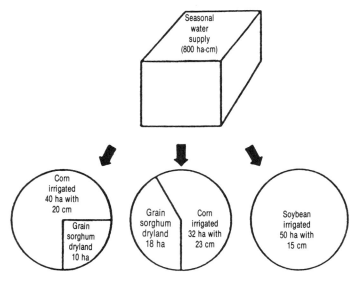

Fig. 17-3. Schematic of a hypothetical situation where an irrigator must decide how to allocate the available water supply (800 ha-cm) over the irrigable area (50 ha).

scheduling. However, the parameters for these models must be readily available and well defined. The simulation model must also be calibrated and verified.

The supply of irrigation water can be limited in several ways. The seasonal volume of water available to irrigate a tract of land may be less than required for maximum yield. With surface irrigation projects, the reservoir storage and inflows may be less than required to fully irrigate the contracted area. In several areas of the USA, groundwater supplies are declining because of extensive development of irrigation. In some of these areas, allocation systems have been established to control the rate of groundwater decline.

Stewart and Hagan (1973) termed the case where seasonal water availability limits production as the *water-limiting condition*. In this case, the irrigator is faced with decisions that don't occur when water is readily available. Irrigation management with seasonal limitations involves three general decisions: (i) how much area to irrigate, (ii) which crops to plant, and (iii) how to distribute the available supply over the irrigable area during the season. The schematic shown in Fig. 17-3 illustrates the type of decision that an irrigator must make before the season begins. Traditional scheduling methods can be useful in answering some of these questions by considering various water supply and ET patterns during the season, but the crop response to irrigation needs to be known. Characteristics of the irrigation system must also be considered to develop realistic schedules. Research has been conducted for various water supply constraints such as by Hall and Butcher (1968) and by Dudley et al. (1971a, b). Recently, simulation models have been combined with optimization programs to provide schedules that include the physical constraints of the irrigation system and crop (Martin, 1984). Thus, analyses beyond conventional scheduling are required to develop management techniques for water-limiting conditions.

Traditionally, land is the limiting resource and the economic objective is to irrigate until the marginal net return from applying a unit of irrigation equals the marginal cost of applying a unit of water. The marginal net return and the result from such an analysis can be expressed as

$$\partial N_r / \partial I = (V_c \, \partial Y / \partial I - C_w) \, 10 \qquad [7]$$

where

N_r = net return ($ ha^{-1})
I = irrigation depth (mm), and
$\partial N_r / \partial I$ = marginal net return ($/m^3).
V_c = adjusted value of the crop ($ kg^{-1}),
Y = crop yield (kg ha^{-1})
C_w = marginal cost of irrigation water ($/m^3).

The crop value can be adjusted to include yield-related production costs. The irrigation water cost can also be increased to include variable costs related to operation of an irrigation system such as labor, repairs, and maintenance. The optimal irrigation depth for land-limiting irrigation is often near the maximum yield depth. Therefore traditional scheduling procedures have been very useful when water is generally available.

The economic criteria for optimizing the use of a limited water supply is different than for scheduling an unlimited source. When water is limiting, the economic criteria is to maximize the average net return per unit of water used. The *average net return* is defined as the net return from applying an amount of irrigation, minus the net return from dryland, all divided by the depth of water applied. Mathematically the average net return is given by

$$(N_I - N_{rd})/I \hspace{3cm} [8]$$

where N_{rd} is the dryland net return. It can be shown that the optimum is attained where the marginal net return equals the average net return.

The optimal cropping pattern and the associated irrigation depth for water-limiting conditions depends upon the profitability of the dryland crop. Hypothetical examples are shown in Fig. 17–4 for a water supply of 1200

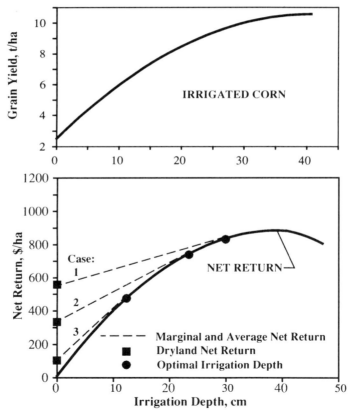

Fig. 17–4. Illustration of differences that occur in optimal strategy for land limiting and two water limiting cases. Analysis is for the water limiting conditions shown in Fig. 17–3 where 80 000 m³ water is available and the maximum irrigated area is 50 ha.

ha-cm and a field with a maximum irrigated area of 50 ha. The yield of corn as a function of the depth of irrigation is shown in the upper graph. In the lower graph the net return from irrigated corn as a function of the irrigation depth is shown as a solid curve. The net return of three dryland crops is shown by the squares along the ordinate. The marginal net return is the slope of the net return curve at the specified irrigation depth. The irrigation depths where the marginal and average net return are equal are shown in Fig. 17–4 for each condition. When the dryland net return is $550 ha^{-1} the optimal irrigation depth is approximately 30 cm, which means that 40 ha of corn could be irrigated. If the dryland net return was $325 ha^{-1} the optimal irrigation depth would be 24 cm, which would allow the entire 50-ha field to be irrigated. If the dryland net return was $100 ha^{-1} the optimal land-limiting depth would be 12 cm, which corresponds to 100 ha. However, the maximum land area that can be irrigated is 50 ha. In the third condition, the optimal irrigated area would be the maximum field area (50 ha) with a depth of 24 cm. These examples show that when the crop is profitable compared to a crop grown in soil irrigated with small depths, the irrigated area should be reduced from the maximum possible. If the dryland crop is not profitable, the limited water supply should be spread uniformly over the entire irrigable area.

Planning for the optimal irrigated area also depends upon the factors considered in the net return calculation. Stewart and Hagan (1973) and Barrett and Skogerbee (1980) evaluated several specific conditions to determine the optimal irrigated area and depth for water-limiting conditions. Recently, Martin et al. (1989) developed a general method to predict the optimal irrigated area and depth of irrigation. They showed that several parameters are involved in the decision, including the efficiency of the irrigation system, the cost of preparing land for irrigation, the cost of water, and the yield response and price expected for the irrigated and dryland crops. Their results indicate that the optimal policy varies from irrigating for maximum yield on a small area, to spreading the available water over the entire irrigable area. One cannot make a general statement about the appropriate management strategy for deficit irrigation; it will be unique to each situation. It is also essential to determine the breakeven condition of irrigation compared to dryland conditions when developing management plans for water-limiting conditions.

Martin and van Brocklin (1985) developed methods to determine the area to irrigate and type of crops to plant for a multiseasonal water allocation system. They showed that the management objective of the producer drastically affected the optimal irrigated area and timing of water use. Gollehon (1987) developed a long-term planning model to allocate limited water over a 20-yr planning horizon, including the ability to borrow and carry over water to alternate years and the ability to abandon irrigation and liquidate irrigation systems during the planning horizon. Once perfected, management tools such as these can be used to help producers plan irrigation and other farming operations when the water supply limits irrigation.

Once the decision has been made to irrigate a specific area of a field and to plant the remaining area to a dryland crop, the optimal time during the season to irrigate needs to be determined. Howell and Hiler (1975) and Martin (1984) used combined simulation and optimization models to determine the proper timing of irrigations. Martin (1984) included the characteristics of the irrigation model to describe system constraints and showed that the form of the yield model had a significant effect upon the timing of irrigation. This indicates that improvements in crop simulation models may be required to develop deficit irrigation schedules. With existing knowledge, some results can be generalized regarding the timing of deficit irrigation. If a critical period exists for a crop where large yield reductions occur for mild stress, then that period may be the best time to irrigate if the rainfall pattern and soil water storage are such that stress is likely to occur. However, withholding too much water until late in the season can reduce effective seasonal water use if ample rainfall is received late in the season.

Generally, the vegetative period of grain crops is the least sensitive stage to stress. New root growth provides soil water if the root zone was not initially depleted. Water stress during the reproductive period is usually the most damaging and is the best time to assure water is available from rain or irrigation. Early season irrigation promotes vigorous plant growth which allows the crop to use the maximum amount of soil water during the season. Late season irrigations may not be used completely by the time the crop matures. However, enough water must be available late in the season to produce the marketable product. Early season irrigation also results in more evaporation that is not productive to yield formation. Irrigation scheduling techniques that consider the comparative worth of irrigating now to later, or not at all, are being developed. These techniques involve combined optimization and simulation models that describe the performance of the irrigation system and response of the crop to water stress. Future methods need to include uncertainties introduced because of the inability to predict weather.

Water can also be limited by the supply rate, such as would occur with low-capacity wells or underdesigned irrigation systems. If irrigation is withheld too long with these systems, stress cannot be relieved. If stress occurs during a critical growth stage, it may limit the effectiveness of subsequent applications. However, storing too much soil water early in the season leaves the irrigator vulnerable to wasted water if rain occurs. Scheduling when the supply rate of the irrigation system is limited has received little attention. Such scheduling is important, especially when groundwater levels decline. In this case, irrigation well yields would also decrease. Some scheduling models consider capacity directly, but to the authors' knowledge, these models have not been applied to water-limiting conditions.

With deficit irrigation, the risk of lost profit increases greatly as the water supply diminishes. Therefore, choosing the wrong management strategy can result in substantial losses. For water-limiting conditions, Martin et al. (1989) showed that a net return plateau existed where profit would not change substantially when the area irrigated varied over a substantial range. However, if the producer decided to water for maximum yield on a reduced area, or to spread the water supply over the entire area, the net return decreased substantially. Irrigators must be prepared for risk and should evaluate the magnitude of risk that results when specific management strategies are used for their particular irrigation and financial conditions. This analysis requires a substantial increase of technology above that normally employed by most irrigators, but offers a dramatic economic benefit. Future methods must consider these factors directly in scheduling.

VII. SCHEDULING ADAPTATIONS

The type of irrigation system and local climatic conditions can also influence scheduling. The water-budgeting technique is suitable for use with all types of irrigation systems and environments but the management decisions may differ either for the timing or depth of application.

A. Scheduling of Humid Area Irrigations

Annual irrigation amounts are generally less in the humid areas of the USA than in the arid areas. However, even though there may be fewer decisions in terms of when to irrigate, the actual decisions become much more complex. Higher probabilities of rainfall must be considered in the decision-making process. Increased rainfall can lead to significant deep percolation losses and fertilizer leaching. Irrigation systems should usually be operated to partially refill the profile, with each irrigation retaining a reservoir for rain. The biggest problem comes when irrigating soils with low water-holding capacities which do not provide much margin for error in either applying irrigation water or managing the system to maximize the effectiveness of rainfall.

Many humid areas have sufficient annual precipitation for crop production but do not have the proper seasonal distribution. Irrigation is necessary to supplement the rainfall during short periods of drought. Irrigation scheduling may be simplified to account for the rainfall and irrigate when water is deficient. In simplified accounting, daily ET is often assumed constant throughout the irrigation season, which generally begins when the crop is near effective cover. Thus, large variations are not expected. Formulas for estimating ET may be simplified to functions of temperature, or at most functions of temperature and radiation. Because rainfall frequently refills the root zone soil profile, fewer long periods of accurate ET estimates are needed because accumulated errors are reset to zero after heavy rains.

Often, humid region irrigation systems are designed with a smaller capacity than required for extended drought periods for the area irrigated, or systems are moved from one field to another. These factors complicate the scheduling process. Irrigations, for example, must begin earlier than would otherwise be necessary for the first field to allow the complete coverage of the total area served by a single irrigation system. Delays in starting the irrigation system can result in crop stress on the latter areas to be irrigated if drought occurs. The entire area or many fields to be irrigated may require water at the same time following a rainfall that refilled the profile. It is more difficult to establish a staggered schedule than is often the case in the arid west where rainfall rarely confounds the schedule. If rainfall occurs during an irrigation, many producers will discontinue the irrigation, which complicates accounting in soil water balances by increasing the variability of stored soil water in the field.

B. Surface Irrigation

The scheduling of most surface irrigation systems requires the forecasting of the time when the management-allowed depletion is reached. The selected depletion is usually the maximum amount that will not significantly decrease crop production. For many crops it is assumed that 50% of the available water capacity in the root zone depth can be depleted before each irrigation. Figure 17–5 is an example recommendation for a surface irrigation system. Most surface systems are then operated to completely refill the soil profile. It is important to measure the water applied to assure that the profile is refilled. Refilling the profile also minimizes the seasonal labor input to most surface irrigation systems. Additional water may be applied to satisfy the leaching requirements if salinity is a problem.

When surface irrigation systems are automated, smaller management-allowed depletions are frequently used. However, because the infiltration rate controls the application depth, it is difficult to apply <25 mm with most surface irrigation systems and still attain a reasonable water application ef-

Fig. 17–5. Example of computer output sent to farmer with recommended irrigation dates and amounts for a surface irrigation system (updated 11 May).

Field Number	Depletion mm	Optimum Depletion mm	Day to Irrigate	Apply mm
Corn				
1	5.1	22	June 2	56
2	5.1	12	May 21	25
3	9.9	18	May 27	48
4	0.0	18	May 29	51
Beets				
1	5.1	11	May 22	25
2	5.1	7	May 14	10
3	5.1	10	May 19	31
4	9.9	13	May 25	56
Beans				
1	5.1	19	May 27	38
2	5.1	11	May 17	18
3	9.9	16	May 23	36
4	0.0	16	May 24	36
Alfalfa				
20	5.6	130	June 12	173
2	72.6	130	May 31	173
3	37.3	76	May 27	112
4	5.6	76	June 2	109

ficiency. Dead level basins and surge irrigation systems are often used to minimize the application depth. These systems are usually operated to apply a given depth with varied frequency based on the actual ET requirements. However, this cannot always be rigorously followed when the water supply is delivered on a rotation or preset schedule. With these constraints, the actual depth of application is also variable.

It should be emphasized that the water delivered to the field should be measured. Unknown amounts applied can result in either under- or over-irrigation. Excessive irrigation amounts not only lead to increased deep percolation (and fertilizer/pesticide leaching) or surface runoff, but may increase the time between irrigations if water supplies limit the timing of the next irrigation. This delay in successive irrigations could result in crop stress if water supply is limited.

C. Sprinkler Irrigation

The scheduling of sprinkler irrigation systems depends on the type of system. Hand-moved or tow line systems are frequently operated to apply larger amounts, usually > 100 mm. Sprinkler systems provide easily controlled applications and deliver more uniform applications since they are less dependent on variable infiltration rates. Where sprinkler irrigation systems are used for crop germination, application amounts are typically small.

Solid set, continuous move laterals, or center pivot sprinkler systems are often used to apply small irrigation amounts. The application rates are quite high at the outer end of center pivot systems. Center pivots are usually run at one speed to apply a constant depth of water in each revolution. The frequency of operation is scheduled to satisfy the crop ET requirements. Irrigation scheduling for center pivot, linear move, and solid set systems typically provides a window of time for the next water application. This window is bracketed by a "no earlier than" date and a "no later than" date. The "no earlier than" date is the date at which the soil water profile would be depleted in an amount equal to the desired irrigation depth. The "no later than" date would be the date to start the irrigation system so as to irrigate all portions of the field before the soil water is depleted below some management-determined level. For example, this depletion level could be 50 or 60% of the total available water holding capacity. Figure 17-6 is an example of an irrigation schedule for a center pivot system. The recommended application with the center pivot system is typically the maximum amount of water that can be applied without surface runoff. Maximizing each application increases the interval between irrigations and reduces the amount of soil water evaporation that occurs each time the soil is wetted. The system can also be operated to partially refill the soil water profile and maintain a reservoir to effectively use rainfall that may come immediately following an irrigation. The size of the reservoir is a direct function of the available soil water-holding capacity. On very sandy soils in arid areas with low water-holding capacities, it is generally not advisable to maintain a reservoir for

Fig. 17-6. Example of computer output submitted to farmer indicating recommended weekly
irrigation schedule for corn.

REGION CROOK

FARM-CONDON   NUMBER 2   CORN   DATE: Aug. 10

| | | | | Depletion Where System | |
Day	Water Used	Irrigation and Rains	Irrigation Dates	Starts	Stops
Aug. 4	3.8	0.00		3.8	3.8
Aug. 5	5.6	0.00		9.1	9.1
Aug. 6	6.6	20.3	Started	0.0	16.0
Aug. 7	6.8	0.00		6.9	2.3
Aug. 8	4.8	0.00		11.4	7.1
Aug. 9	6.4	20.3	Started	0.0	13.5
Aug. 10	7.1	0.00		7.1	6.3

Maximum useful Rain and Irrigation Amounts

Larger amounts will be lost

Date	Amount
Aug. 11	14.7
Aug. 12	22.4
Aug. 13	29.7
Aug. 14	36.8
Aug. 15	42.7
Aug. 16	48.3
Aug. 17	53.6
Aug. 18	58.9
Aug. 19	64.3

If the system applies 20 mm and makes a revolution in 51 hours, the recommended
starting times are—

Amount of Rain	Start	No Later Than
No Rain	Aug. 13	Aug. 19
6	Aug. 14	Aug. 20
12	Aug. 15	Aug. 22
25	Aug. 17	Aug. 24

Assume the system was started Aug. 13, the next starting times are—

Amount of Rain	Start	No Later Than
No Rain	Aug. 16	Aug. 22
6	Aug. 17	Aug. 23
12	Aug. 18	Aug. 24
25	Aug. 20	Aug. 27

Probable rainfall next week is 7.1 mm

Note: All water amounts are expressed in mm in the above tables.

rain. In these cases, it is more important to maintain a full soil profile and
minimize the depletion so as to provide some safety should an irrigation system malfunction. Properly designed center pivot systems, which minimize
surface runoff, generally do not have sufficient capacity to catch up during
the peak use period. Linear move and solid set systems have nearly constant
application rates throughout the entire system and can be designed with higher
average rates. It is important to maintain the systems in excellent operating
condition and not allow the soil water depletion to increase to levels that
could decrease crop production.

D. Trickle Irrigation

Trickle systems are typically scheduled at high frequencies with small application depths. When the systems are installed below the soil surface, the soil evaporation is minimized. This decreases the water requirements somewhat early in the season as the crops are developing. Once the crop has reached effective cover, research has demonstrated that there is little difference in crop water use as a function of the irrigation system. Most trickle irrigation systems are installed where the water supply is sufficient and irrigations can be scheduled on demand. The typical irrigation schedule is to replace ET as it is used.

VIII. SUMMARY

Irrigation scheduling is a key element to proper management of irrigation systems. The goal is to apply the correct amount of water at the right time to meet management objectives. The objective may be to maximize water efficiency, crop production, or economic return or to minimize irrigation costs. Irrigation schedules can be as simple as a rotation based on water availability or as complex as scheduling trickle irrigation systems to deliver water on demand. Field observation of soil water status, plant water status, water budgets, and combinations of these can be used to schedule irrigations. The water budget can be based on real time meteorological data or average climatic data. Modern computer technology and instrumentation is gaining more wide-spread adoption by current irrigators. The computer technology can be used for real time scheduling, optimization, and seasonal planning of irrigation strategies. Limited water availability is demanding more sophisticated technology and remote sensing is being developed to provide instant assessment of crop conditions for establishing irrigation schedules. Irrigation scheduling must be integrated with the total management, including energy, labor and cultural practices.

REFERENCES

Ahmad, S. 1987. Management strategies for scheduling irrigations in Pakistan. Unpubl. Ph.D. diss. Colorado State Univ., Fort Collins.

Barrett, J.W., and G.V. Skogerboe. 1980. Crop production functions and the allocation and use of irrigation water. Agric. Water Manage. 3:53–64.

Bausch, W.C., and C.M.U. Neale. 1987. Crop coefficients derived from reflected canopy radiation: A concept. Trans. ASAE 30(3):703–709.

Bucks, D.A., F.S. Nakayama, O.F. French, W.W. Legard, and W.L. Alexander. 1985. Irrigated guayule—Evapotranspiration and plant water stress. Agric. Water Manage. 10:61–79.

Choudhury, B.J., R.J. Reginato, and S.B. Idso. 1986. An analysis of infrared temperature observations over wheat and calculation of latent heat flux. Agric. For. Meteorol. 37:75–88.

Doorenbos, J., and W.O. Pruitt. 1975. Crop water requirements. FAO Irrig. Drain. Pap. 24. FAO, Rome.

Dudley, N.J., D.T. Howell, and W.F. Musgrave. 1971a. Optimal intraseasonal irrigation water allocation. Water Resour. Res. 7(4):770–778.

Dudley, N.J., D.T. Howell, and W.F. Musgrave. 1971b. Irrigation planning 2: Choosing optimal acreages within a season. Water Resour. Res. 7:(5)1051–1063.

Ehrler, WL. 1973. Cotton leaf temperatures as related to soil water depletion and meteorological factors. Agron. J. 65:404–409.

Fereres, E., R.E. Goldfein, W.O. Pruitt, D.W. Henderson, and R.M. Hagen. 1981. The irrigation management program: A new approach to computer assisted irrigation scheduling. p. 202–207. In Proc. ASAE Irrigation Scheduling Conference, Chicago. 14–15 Dec. Publ. 23-81. ASAE, St. Joseph, MI.

Gollehon, N.R. 1987. Methodology and strategies for multi-season farm-level irrigation decisions under limited water conditions. Ph.D. diss. Univ. of Nebraska, Lincoln (Diss. Abstr. 88-03751).

Hall, W.A., and W.S. Butcher. 1968. Optimal timing of irrigation. J. Irrig. Drain. Div. Am. Soc. Civ. Eng. 94(IR2):267–275.

Heermann, D.F., G.W. Buchleiter, and H.R. Duke. 1984. An integrated water-energy management system—Implementation. Trans. ASAE 27(5):1424–1429.

Heermann, D.F., H.R. Haise, and R.H. Mickelson. 1976. Scheduling center pivot sprinkler irrigation systems for corn production in eastern Colorado. Trans. ASAE 19(2):284–287.

Heermann, D.F., and M.E. Jensen. 1970. Adapting meteorological approaches in irrigation scheduling to high rainfall areas. p. 1–10. In Natl. Irrig. Symp. Papers. ASAE, St. Joseph, MI.

Howell, T.A., and E.A. Hiler. 1975. Optimization of water use efficiency under high frequency irrigation—I. Evapotransporation and yield relationship. Trans. ASAE 18(5):873–878.

Idso, S.B. 1982. Non-water-stressed baselines: A key to measuring and interpreting plant water stress. Agric. Meteorol. 27:59–70.

Idso, S.B., K.L. Clawson, and M.G. Anderson. 1986. Foliage temperature: Effects on environmental factors with implications for plant water stress assessment and the CO_2/climate connection. Water Resour. 22:1702–1716.

Idso, S.B., R.D. Jackson, P.J. Pinter, Jr., R.J. Reginato, and J.L. Hatfield. 1981. Normalizing the stress degree day for environmental variability. Agric. Meteorol. 24:45–55.

Jackson, R.D. 1982. Canopy temperature and crop water stress. p. 43–85. In D. Hillel (ed.) Advances in irrigation. Vol. 1. Academic Press, New York.

Jensen, M.E. 1969. Scheduling irrigations with computers. J. Soil Water Conserv. 24(8):193–195.

Jensen, M.E., and H.R. Haise. 1963. Estimating evapotranspiration from solar radiation. Am. Soc. Civ. Eng. Proc. 89(IR4):15–41.

Jensen, M.E., and D.F. Heermann. 1970. Meteorological approaches to irrigation scheduling. p. NN1–NN10. In Natl. Irrig. Symp. Papers. ASAE, St. Joseph, MI.

Jensen, M.C., J.E. Middleton, and W.O. Pruitt. 1961. Scheduling irrigation from pan evaporation. Wash. Agric. Exp. Stn. Circ. 386.

Jensen, M.E., D.C.N. Robb, and C.E. Franzoy. 1970. Scheduling irrigations using climate-crop-soil data. J. Irrig. Drain. Div. Am. Soc. Civ. Eng. 96(IR1):25–38.

Jensen, M.E., J.L. Wright, and B.J. Pratt. 1971. Estimating soil moisture depletion from climate, crop and soil data. Trans. ASAE 14(5):954–959.

Kincaid, D.C., and D.F. Heermann. 1974. Scheduling irrigations using a programmable calculator. USDA Bull. ARS-NC-12.

Klocke, N.L., and P.E. Fischbach. 1984. Estimating soil moisture by appearance and feel. Coop. Ext. Serv., Instit. of Agric. and National Resour., Univ. of Nebraska, Lincoln, Nebraska Guide B-12.

Lundstrom, D.R., E.C. Stegman, and H.L. Werner. 1981. Irrigation scheduling by the checkbook method. p. 187–193. In Proc. ASAE Irrigation Scheduling Conf., Chicago. 14–15 Dec. ASAE, St. Joseph, MI.

Martin, D.L. 1984. Using crop yield models in optimal irrigation scheduling. Ph.D. diss. Colorado State Univ., Fort Collins (Diss. Abstr. 84-27861).

Martin, D.L., and J. van Brocklin. 1985. The risk and return with deficit irrigation. Am. Soc.Civ. Eng. Tech. Pap. No. 85-2594. Presented at ASAE winter meeting, Chicago. 17–20 Dec.

Martin, D.L., J.R. Gilley, and R.J. Supalla. 1989. Evaluation of irrigation planning decisions. J. Irrig. Drain. Div. Am. Soc. Civ. Eng. 115(IR1):58–77.

O'Toole, J.C., and J.L. Hatfield. 1983. Effect of wind on the crop water stress index derived by infared thermometry. Agron. J. 75:811–817.

Penman, H.L. 1948. Natural evaporation from open water, bare soil and grass. Proc. R. Soc. A. No. 1032 193:120–145.

Penman, H.L. 1963. Vegetation and hydrology. Tech. Commun. Commonw. Bur. Soils No. 53. Harpenden, England.

Pleban, S., J.W. Labadie, and D.F. Heermann. 1983. Optimal short term irrigation schedules. Trans. ASAE 26(1):141–147.

Priestley, C.H.B., and R.J. Taylor. 1972. On the assessment of surface heat flux and evaporation using large-scale parameters. Mon. Weather Rev. 100(2):81–92.

Stetson, L.E., D.G. Watts, F.C. Corey, and I.D. Nelson. 1975. Irrigation system management for reducing peak electrical demands. Trans. ASAE 18(2):303–306, 311.

Stewart, J.I., and R.M. Hagan. 1973. Functions to predict effects of crop water deficits. J. Irrig. Drain. Div. Am. Soc. Civ. Eng. 99(IR4):421–439.

Trava, J., D.F. Heermann, and J.W. Labadie. 1977. Optimal on-farm allocation of irrigation water. Trans. ASAE 20(1):85–88, 95.

Woodruff, C.M., M.R. Petersen, D.H. Schnarre, and C.F. Cromwell. 1972. Irrigation scheduling with planned soil moisture depletion. Pap. 72-772 presented at ASAE winter meeting, Chicago. 12–15 Dec. ASAE, St. Joseph, MI.

18   Alfalfa

J. C. GUITJENS

University of Nevada-Reno
Reno, Nevada

The U.S. Census for Agriculture shows that 26% (2.2×10^4 km^2) of the land area in alfalfa (*Medicago sativa* L.) was irrigated. Irrigated alfalfa yielded an average of 0.9 kg m^{-2} as compared to 0.6 kg m^{-2} for nonirrigation (USDC, 1984). Some areas of California and Arizona had annual field yields of 1.45 kg m^{-2} from eight harvests (Heichel, 1983). At higher elevations (1200 m) is northwest Nevada, a field yield of 0.9 kg m^{-2} from four harvests was considered reasonable (Nevada Agricultural Statistics, 1984). Other areas produced only one growth cycle per season. Yields twice as large as reported field yields have been obtained on experimental plots (Guitjens, 1982; Sammis et al., 1985).

Understanding the plant-soil-water system creates an opportunity for management of water application in response to soil moisture depletion from evapotranspiration by alfalfa plants. Evapotranspiration models can provide reliable estimates. Amounts of readily available soil water in reserve and depleted can be obtained by measuring soil moisture. With irrigation scheduling, timing and amount of water application can be controlled to achieve yield goals while practicing water conservation by preventing unnecessary drainage losses.

Since water demands and competition for shares of limited supplies continue to grow, especially in arid and semiarid areas, water conservation has received considerable attention and new advances have been made. This review relied heavily on research published in the past 10 to 15 yr. Other reviews concerning alfalfa also remain important references (e.g., Hanson, 1972; Heichel, 1983). The same is true for subjects in irrigation (e.g., Hagan et al., 1967; Jensen, 1980), evapotranspiration modeling (e.g., Jensen, 1973; Doorenbos & Pruitt, 1977), and soil-water-plant relations (e.g., Kozlowski, 1968; Doorenbos & Kassam, 1979; Taylor et al., 1983; Teare & Peet, 1983).

I. CROP DEVELOPMENT

A. Vegetative Development

Prior to seeding perennial alfalfa, a seed bed is prepared. Fertilization to prevent a deficiency of essential elements usually occurs at this time. The first irrigation provides soil moisture for seed germination and seedling de-

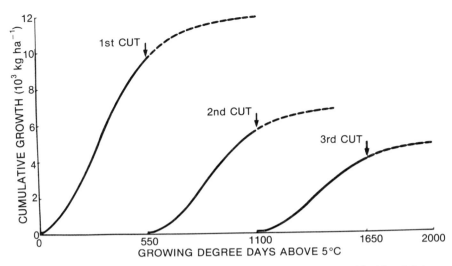

Fig. 18-1. Cumulative growth as a function of growing-degree-days above 5 °C. After Selirio and Brown (1979).

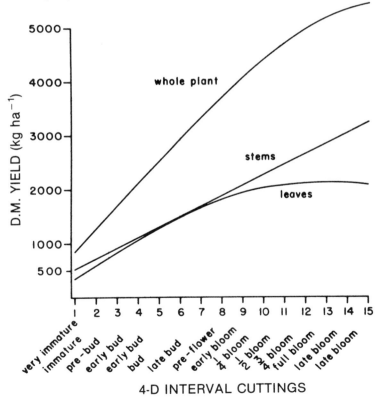

Fig. 18-2. Dry matter yield of alfalfa through an 8-wk growing period (Kilcher & Heinrichs, 1974).

velopment (Bula & Massengale, 1972). Subsequent irrigation is necessary to support initial root and plant growth. Light irrigations are desirable to reduce deep drainage losses.

A new late-summer seeding should occur early enough to allow for adequate root and shoot development or the new plants won't survive frosts during winter dormancy. Bula and Massengale (1972) observed that germination will increase with increases in temperature, and that daily mean air and soil temperatures of 25 °C are optimal. Since evapotranspiration continues during dormancy (Goodrich, 1986), the roots should be well enough developed to give them a soil volume in which an adequate supply of soil moisture is stored.

The number of growth cycles during a growing season varies with geographic location. Metochis (1980) reported nine harvests on the island of Cyprus in the Mediterranean Sea and Guitjens (1982) four harvests in northwest Nevada. Yields vary among growth cycles. Optimal yields in northwest Nevada could be 9000, 4500, 4500, and 2500 kg ha^{-1} for four successive growth cycles, respectively. This covers a growing season from mid-March to October.

B. Scheduling harvest

Forage alfalfa is always in a vegetative state. Only seed alfalfa is allowed to set seed and dry up before harvesting the seed. Growth starts from the crown and continues until the plant is ready to be harvested at about 10% bloom. At this point the growth cycle is stopped by cutting off the plants near the ground. Regrowth from the crown initiates the next growth cycle. Indicators such as new regrowth at the crown or growing-degree-day accumulation to harvest (Doorenbos & Kassam, 1979; Fick, 1984; Selirio & Brown, 1979) offered alternatives to the conventional approach of 10% bloom. Growing-degree-day accumulation (Growing-degree-days above 5 °C) has been used as an index for growth from start of regrowth to maturity. Selirio and Brown (1979) used 550 degree-days as a harvest index as shown by three idealized growth functions in Fig. 18–1. Figure 18–2 demonstrates that harvesting prior to early bloom favors a higher proportion of leaves.

Production functions, yield = f(evapotranspiration), may be used to set yield goals on the basis of available or anticipated water supplies for irrigation (Guitjens, 1982; Guitjens et al., 1984). Since a deficit in evapotranspiration will create a proportional yield deficit (Doorenbos & Kassam, 1979; Guitjens, 1982; Guitjens et al., 1984; Hanks, 1974; Sammis, 1981), growers can to some extent control yield through water management. Less is known about management practices required to reach potential productivity, i.e., the highest yield that may be obtained (Fick, 1984; Selirio & Brown, 1979).

II. EVAPOTRANSPIRATION

A. Geographical and Environmental Effects

Evapotranspiration requires an energy input which is provided by solar radiation. Radiation received at the earth's surface depends on latitude and is greatly influenced by cloud cover. Advective inputs from a sea, ocean, or continental land mass and elevation above sea level will influence total energy available for evapotranspiration. Minimum temperature for growth will delineate the effective growing season.

1. Daily Evapotranspiration

Evapotranspiration by an alfalfa crop should not be moderated by an insufficient supply of soil moisture unless water supplies are insufficient. Before soil moisture availability limits evapotranspiration, irrigation should replenish depleted soil water. A fully developed stand during the peak of the growing season may consume water at around 10 mm d^{-1}. Short-term evapotranspiration predictions are necessary if irrigation scheduling is used for setting the next irrigation date and the amount of water to be applied.

2. Seasonal Evapotranspiration

Seasonal evapotranspiration is used in planning for water supplies. If the area of irrigated alfalfa within a farm unit or an irrigation district is multiplied by the seasonal water consumption, the volume of water required can be determined after allowing for certain inefficiencies in water conveyance and water distribution across a field.

$$\text{Water diverted into a field} = \frac{\text{evapotranspiration}}{\text{irrigation efficiency}}. \quad [1]$$

$$\text{Water delivered to a farm head gate} = \frac{\text{diverted water}}{\text{conveyance efficiency}}. \quad [2]$$

3. Annual Evapotranspiration

Annual evapotranspiration covers the seasonal amount during the period of active plant growth and also recognizes that evapotranspiration occurs during winter dormancy (Goodrich, 1986). An annual water balance should be considered because input by precipitation and output by evapotranspiration during dormancy influence soil moisture conditions at the start of a growing season.

b. Crop Growth Stage vs. Evapotranspiration

Evapotranspiration rate appears to correlate with rate of CO_2 exchange through the stomates and, hence, to actual and potential yield. Research has generally indicated that any deficit in evapotranspiration will create a yield reduction (Doorenbos & Kassam, 1979; Guitjens, 1982; Guitjens et al., 1984;

Hanks, 1974; Sammis, 1981). Figure 18-3 illustrates the functional relationship between yield deficit and evapotranspiration deficit.

Many evapotranspiration models are available that require meteorological information (Doorenbos & Pruitt, 1977; Jensen, 1973; Blad, 1983). These models usually estimate potential or reference evapotranspiration independent of any crop or referenced to a specific crop, respectively. The product of one of these evapotranspiration estimates and a suitable crop coefficient gives an estimate of actual evapotranspiration.

Alfalfa evapotranspiration varies with the stage of plant development and the regrowth time since the previous harvest. After harvest and prior to the next irrigation, evapotranspiration from a stubble field is at a minimum since both transpiration and evaporation are greatly reduced. As plants grow, evapotranspiration increases until it reaches a maximum at full plant development. Full evapotranspiration is usually associated with a leaf area index of about 3 (Stegman et al., 1980). Several crop coefficients are needed to link potential and reference evapotranspiration to estimated evapotranspiration. Table 18-1 gives an example of estimating reference evapotranspiration (ET_o) from mean relative humidity, wind speed, and pan evaporation. Crop coefficient (K_{crop}) links ET_o to estimated evapotranspiration (ET_{crop}).

Other methods for estimating either evapotranspiration or soil moisture depletion involve measurement of soil moisture content (Gear et al., 1977), leaf water potential (Heichel, 1983), or canopy temperature (Doorenbos & Kassam, 1979; Idso et al., 1981a; Idso et al., 1981b; Jackson et al., 1981).

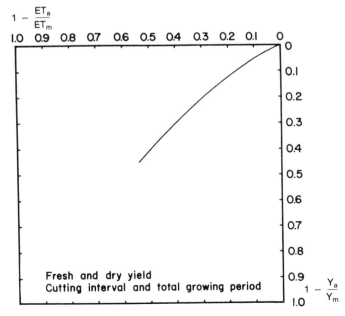

Fig. 18-3. Relative yield decrease (Y_a/Y_m) as a function of relative evapotranspiration deficit (ET_a/ET_m) (a = actual, m = maximum). After Doorenbos and Kassam (1979).

Table 18-1. FAO Class A pan evapotranspiration model. After Guitjens et al. (1984).

Week ending	RH†	Wind	K_{pan}‡	E_{pan}	ET_o	K_{crop}§	ET_{crop}
1981	%	km d^{-1}		———mm———			mm
25 Mar.	60	146	0.85	26.4	22.5	0.50	11.3
1 Apr.	60	251	0.80	31.2	25.0	0.50	12.5
8 Apr.	56	190	0.80	42.7	34.1	0.50	17.1
15 Apr.	59	142	0.85	47.2	40.2	0.60	24.1
22 Apr.	62	119	0.85	33.0	28.1	0.70	19.6
29 Apr.	55	147	0.85	45.7	38.9	0.80	31.1
1 May	59	69	0.85	14.0	11.9	0.80	9.5
8 May	59	162	0.85	59.7	50.7	0.80	40.6
15 May	60	137	0.85	53.3	45.3	1.15	52.1
22 May	62	146	0.85	33.0	28.1	1.15	32.3
29 May	60	86	0.85	42.4	36.1	1.15	41.5
1 June				First harvest			
5 June	57	113	0.85	57.2	48.6	0.52	25.3
12 June	59	263	0.80	87.6	70.1	0.95	66.6
19 June	58	166	0.85	68.6	58.3	0.70	40.8
26 June	60	102	0.85	68.3	58.1	0.95	55.2
3 July	64	98	0.85	66.8	56.8	1.15	65.3
10 July	56	116	0.85	70.1	59.6	0.40	23.8
10 July				Second harvest			
17 July	53	120	0.85	80.0	68.0	0.40	27.2
24 July	58	95	0.95	75.7	64.3	0.40	25.7
31 July	61	103	0.85	70.4	59.8	0.70	41.9
7 Aug.	61	70	0.85	60.2	51.2	1.15	58.8
14 Aug.	58	75	0.85	56.4	47.9	1.05	50.3
21 Aug.	63	76	0.85	49.5	42.1	0.50	21.1
23 Aug.				Third harvest			
28 Aug.	57	74	0.85	64.3	54.6	0.50	27.3
4 Sept.	59	103	0.85	64.8	55.1	0.50	27.5
11 Sept.	63	79	0.85	50.8	43.2	0.70	30.2
18 Sept.	64	81	0.85	52.8	44.9	1.00	44.9
25 Sept.	63	92	0.85	41.4	35.2	1.15	40.5
2 Oct.	64	103	0.85	39.6	33.7	1.15	38.7
9 Oct.	59	99	0.85	31.5	26.8	1.15	30.8
16 Oct.	69	95	0.85	13.7	11.7	1.15	13.4
22 Oct.	60	38	0.85	20.3	17.3	0.50	8.6
22 Oct.				Fourth harvest			
29 Oct.	58	122	0.85	25.1	21.4	0.50	10.7
5 Nov.	61	60	0.85	19.1	16.2	0.50	8.1

† Mean relative humidity.
‡ Doorenbos and Pruitt (1977).
§ Except for the values 0.4 and 1.15 which were taken from Doorenbos and Pruitt (1977), values were judged from interpolation.

III. STRESS RESPONSE

A. Effects on Crop Development

Various features of the alfalfa plant are shown in Fig. 18-4. Alfalfa stems show less elongation and leaves remain smaller under water stress (Bauder et al., 1978; Brown & Tanner, 1981; Brown & Tanner, 1983; Sharrat et al.,

1983; Vough & Marten, 1971). Water stress results in reduced evapotranspiration. Since the rate of CO_2 exchange through the stomates depends on stomatal conductance, an evapotranspiration deficit usually results in reduced dry matter production. Figure 18–5 illustrates how irrigation and stress affect leaf area per stem, stem length, relative growth rate ratios of stem and leaf, leaf area, and internode length. Figure 18–6 shows corresponding soil moisture pressures, leaf water pressures, and cumulative evapotranspiration. A drop in soil moisture pressure to -35 kPa at a depth of 0.30 m caused the leaf water pressure to decrease to -810 kPa in the afternoon and growth

Fig. 18–4. The alfalfa plant. (A) An extensive root system and several young shoots growing from the crown; (B) flowering branch from subsequent growth; (C) three views of an enlarged flower; (D) seed pods, four seeds, and enlargement of single seed. (Drawings by Regina O. Hughes, Scientific Illustrator, USDA-ARS) (Hanson, 1972).

was significantly reduced. A soil moisture pressure > -10kPa at 0.9 and
1.50 m did not provide enough water to the plant to compensate for the lack
of water at 0.30 m. Little growth occurred with soil water pressure below
-1013 kPa.

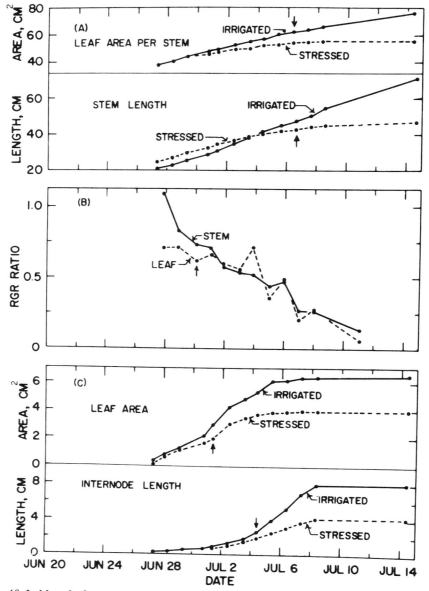

Fig. 18-5. Mean leaf area per stem and stem length (A), stem and leaf relative growth rate
ratios (B), and area and length of leaves and internodes which initiated 27 June 1980 (C)
for alfalfa during a stress period. Arrows on curves indicate when stressed alfalfa parameters
became smaller ($P < 0.02$). After Brown and Tanner (1983).

Whitfield et al. (1986) studied alfalfa growth and dry matter production on heavy soils irrigated weekly and biweekly. Weekly irrigation produced significantly more dry matter than the biweekly treatment. With a 25% lower leaf area index under biweekly irrigation, the smaller leaf area may have intercepted less light. Weekly irrigation gave a significantly greater leaf weight but a significantly smaller leaf vs. stem ratio because of an even greater stem dry matter. Nitrogen concentration was initially greater in the leaves than in the stems but it decreased as plant development progressed.

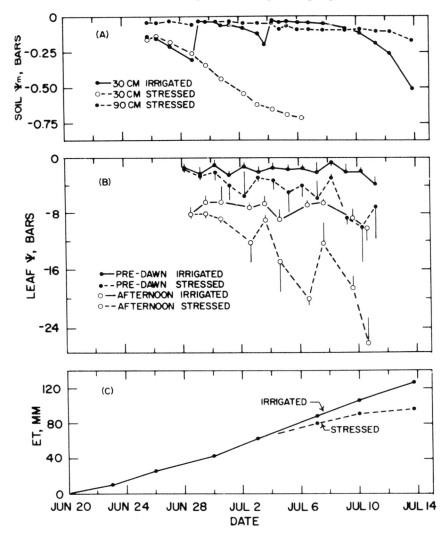

Fig. 18–6. Soil moisture pressure 30 cm in irrigated and stressed plots and 90 cm in stressed plots (A), predawn and afternoon leaf water potentials with $S_{\bar{x}}$ as error bars (B), and cumulative evapotranspiration of irrigated and stressed alfalfa. Data are from the initial 1980 regrowth period (Brown & Tanner, 1983). Note: 1 bar = 101.3 kPa.

If soil conditions and water table location permit, alfalfa plants will develop a deep root system. The opportunity for water uptake from considerable depth gives alfalfa the reputation of being drought resistant. A water table at the bottom of the root zone may also supply water to the lower roots. Plant survival and, perhaps, some growth in the absence of irrigation or rainfall may be due to the deeper water supplies. As a lack of available soil moisture persists and evapotranspiration remains in a deficit condition, plants will go dormant and top growth will be minimal.

B. Stress Duration and Intensity

Brown and Tanner (1983) found that stem length and leaf area developed in a linear manner with time when alfalfa is well watered. Under stressed conditions growth of stems and leaves decreased and eventually stopped. Leaf water pressures of -213 and -850 kPa at predawn and in the afternoon, respectively, resulted in significantly less growth than under leaf water pressures of -101.3 and -648 kPa. With leaf water pressures below -1013 kPa in the afternoon, there was little stem and leaf growth; remaining photosynthetic products went to the crown and roots. Heichel (1983) concluded that the observed stomatal closure at -1013 kPa of leaf water pressure and -405 kPa of soil moisture pressure does not resolve the lack of understanding how the alfalfa plant exercises control over its internal water pressures. Yields have returned to normal levels when stressed plants were returned to nonstressed conditions (Perry & Larson, 1974).

C. Environmental Stress

Evenson (1979) observed that top and root growth respond in opposite directions to crown temperature. Top growth increases whereas root growth decreases as crown temperature rises. Maximum total plant weight occurred at 36 °C but maximum top growth required a crown temperature of 40 °C. When the evapotranspiration demand exceeds the plant's ability to transport water through its system, stress occurs. This phenomenon, called *summer slump* (Metochis & Orphanos, 1981; Stegman et al., 1980), requires high summer temperatures. Although water stress reduces yield (Donovan & Meek, 1983), and severe, continued stress results in a total loss of yield, plants have the ability to recover when the stress condition is over.

D. Yield Loss

Brown and Tanner (1983) reported that water stress in the last half of a growth cycle did not affect stem density and leaf percentage as a portion of the total dry weight. However, stem density decreased by a significant 23% when the stress occurred during the first 14 d of regrowth. Once stem number has been reduced initially, full irrigation later in the same growth cycle may not produce as much growth as with a normal number of stems. This gives the first 14 d after cutting a greater plant sensitivity to water supply.

Growth in response to water stress during specific stages of plant development could provide important information needed in planning for optimal yield in a situation with limited water supply. Hiler and Howell (1983) and Jones (1983) addressed this issue. They discussed a sequence of crop deficits during the growing season in response to a sequence of evapotranspiration deficits with the objective to find an optimal sequence which will minimize yield loss. Less optimal sequences would have excessive evapotranspiration realtive to yield response.

IV. GROWTH AND SOIL WATER

A. Rate of Growth

Forage alfalfa remains in a vegetative state because it is usually harvested at early bloom or before. Between early regrowth and midbloom, total dry matter increases almost linearly with time (Fig. 18–2). At the beginning of plant growth and when regrowth starts after cutting, evapotranspiration is at a minimum. As growth progresses, evapotranspiration increases until the maximum rate has been reached. Wright (1987) reported a 75% drop in crop evapotranspiration during the first week after cutting at harvest time; it may take 21 d after the first and second harvests in Idaho before maximum evapotranspiration has been restored by a full canopy.

1. Soil Physical Properties

Heavy soils with slow infiltration properties are prone to water logging (Whitfield et al., 1986). A low hydraulic conductivity often means poor aeration (Peterson, 1972). Phytophthora root rot occurs under poor drainage. Deep and well-drained soils that are slightly alkaline are best suited for growing alfalfa, while acid conditions are unfavorable (Lowe et al., 1972). Shallow soils provide less volume for root development and plants will experience stress sooner. Drought resistance depends on deep root development (Bula & Massengale, 1972; June & Larson, 1972; Metochis, 1980). Compaction may limit growth because entry of root hairs is impeded and aeration may be insufficient (Christian, 1977). Meek et al., (1980) found losses in alfalfa stands from 15 to 55% when O_2 levels in a silty clay loam soil dropped to 6% or less for more than 24 h. Bornstein et al. (1984) showed a linear increase in O_2 diffusion rate when soil water content decreases in fine-textured soil. This relationship was repeated every time a soil drained after wetting. They found that with drainage to water table depths of 0.15, 0.45, and 0.75 m yield increased linearly with drainage to the deeper water table.

2. Soil Water Depletion

Although soil moisture is considered to be available in the range between field capacity and permanent wilting point, it is not "equally" available (Selirio & Brown, 1979). The pressure range between field capacity and permanent wilting point lies between -0.03 and -1.5 MPa. Water ceases to be readily available below a soil moisture pressure of about -0.2 MPa (van Bavel, 1967). Undiminished vegetative growth requires that water is readily available.

A water extraction model estimates the relative uptake of soil water with soil depth (e.g., the 40-30-20-10% extraction pattern for four equal increments of the total root depth). Readily available water represents about half of the available water. For the 40-30-20-10% model, this represents a 70% extraction from the upper half of the root zone. Extraction patterns are related to root density. Bula and Massengale (1972) gave a 46-26-18-10% extraction pattern over four equal depth increments of 0.6 m. Other patterns may be quite different (Hansen et al., 1980).

A logarithmic decrease in root density with increasing depth means that the root mass is concentrated in the top layer of the soil (Heichel, 1983). Abdul-Jabbar et al. (1982) investigated root pattern relative the moisture level by varying water application from 280 to 1530 mm with a line-source sprinkler. Excluding the tap root, root mass correlated linearly with yield and curvilinearly with evapotranspiration. Nearly two thirds of the root mass was concentrated in the top 450 mm of soil. The maximum root mass correlated with an average soil depth of 1.6 m. Moisture levels exceeding 1000 mm produced a significantly smaller root mass.

V. PLANT NUTRITION

Application of the alfalfa production function, yield = f(evapotranspiration), implies that water consumption is an effective limiting factor. As regrowth from the crown proceeds and leaf area expands, evapotranspiration increases gradually until the maximum rate has been reached. Beyond this point a further increase in leaf area index will not affect evapotranspiration. However, at a given evapotranspiration rate, yield can differ because of soil fertility (Ritchie, 1983). When inadequate fertility delays plant growth, leaf area expansion takes place at a slower rate (Bourget & Carson, 1962). Consequently, the time when maximum evapotranspiration starts occurs later than under full fertility. The response to N fertilization of grain crops and grasses is a good example, but symbiotic bacteria of inoculated alfalfa will fix the required N from the atmosphere.

Phosphorous applications generally give the greatest yield response under most conditions. Its natural supply depends on the soil. One thousand kg of alfalfa hay removes about 2.5 kg of P. The availability of P is pH dependent. At lower pH levels, P is less available. Potassium is required in larger amounts than P. Potassium may have a greater influence on high yield levels of high quality. The monograph *Alfalfa Science and Technology* (Hanson, 1972) provides more details about fertility. Fertilizer recommendations are usually made on the basis of information obtained from analyzing soil samples.

Kilcher and Heinrichs (1974) found that N content of leaves was greater than that of stems, that both decreased as plants developed, but that the decrease proceeded more rapidly in stems. The same was true for P but stems showed an even greater rate of decrease. Whereas digestibility of leaves decreased at a constant rate, that of stems dropped rapidly. Snaydon (1972) reported a 35% increase in P concentration when irrigation was increased from 10% of pan evaporation to 90% of pan evaporation; N content of the combined plant was not effected. Earlier, Bourget and Carson (1962) found an increase in N content at lower water levels and when no P and K fertilizers were applied. They found that P and K applications increased yield and water-use efficiency at four different levels of available water. The crop's mineral contents of P and K increased under fertilization, Mg and Ca contents decreased, but N showed little change. Water availability had little influence on P and K contents but the effect on N, Ca, and Mg was variable.

Alfalfa grown under moisture stress will yield less, but it usually has a higher leaf percentage giving it a better quality. Harvesting at an earlier growth stage will also improve hay quality but dry matter yield will be less. Meyer and Jones (1962) found that hay harvested in early bud stage had superior feed value over alfalfa cut in flowing stage. Lignin content had increased with plant maturity and protein content had decreased. Supplemental irrigation and harvesting at bud or early bloom increased protein content from 17.5 to 20.5% on the sandy soils of Wisconsin (Groskopp et al., 1963). Higher day and night temperatures, combined with moisture stress, reduced both yield and quality (Vough & Marten, 1971).

Plant survival may require periodic cuttings at a stage between early to late bloom to stimulate an increase in root mass. Root mass and storage of carbohydrate decline with early cutting and influence regrowth afterwards (Heichel, 1983).

Alfalfa is moderately tolerant to salinity (Hoffman et al., 1983; Maas & Hoffman, 1977). It will not suffer any yield reduction at an electriacl conductivity of soil water extract (EC_e) of 2.0 ds m^{-1} and if the electrical conductivity of the irrigation water (EC_i) is not more than 1.3 ds m^{-1}. EC_e and EC_i values of 3.4 and 2.2 ds m^{-1}, respectively, will reduce yield by 10% and values of 8.8 and 5.9 ds m^{-1} by 50%. Under drip and high frequency irrigation (nearly daily) EC_e may be as high as 16 ds m^{-1} before a yield reduction occurs (Ayers & Wescot, 1976).

VI. CULTURAL PRACTICES

A comprehensive review of cultural practices can be found in the monograph *Alfalfa Science and Technology* (Hanson, 1972). Plowing and disking are standard cultivation practices. Calcium may be incorporated to raise the pH above 6.8. Phosphorus is also applied before cultivation to allow for a mixing with the soil. If K is required it may be applied at the same time. Seeds should be planted, usually by drilling, in a firm seed bed at a depth of 5 to 10 mm in fine-texture soils and at a depth of 10 to 40 mm in course soils of more arid climates. Preferred seeding time varies geographically from early spring to late summer. Air and soil temperatures near 25 °C are optimal, and readily available soil moisture favors growth. Seeding rate is about 11 kg ha^{-1}. During the first 10 d after emergence, the seedling depends on stored reserves in the seed for survival. The number of harvests depends on the length of the growing season and weather conditions as demonstrated by the concept degree-days above 5 °C before harvest. In addition to cultural practices, weed control with herbicides can be effective on mature alfalfa stands. There are many known diseases that may infect alfalfa but not all varieties are equally susceptible. Insect control may also be required. Local observations and recommendations are most effective.

VII. WATER MANAGEMENT

Water is essential for hydration of plant tissues. The plant provides a pathway for water moving from the soil into the atmosphere and dry matter production of alfalfa is a simple function of transpiration (Turner & Burch, 1983). Evapotranspiration occurs in response to an atmospheric demand for water (Blad, 1983) but the actual rate can be modified by inadequate soil moisture. The energy gradient involved in water flow extends from the soil matrix to the stomates of the plant. A decreasing soil moisture pressure will eventually affect water transport through the plant and, therefore, plant growth.

Soil pores can retain water and they provide a reservoir for water storage (Stegman et al., 1980; Thien, 1983). The soil pores need periodic replenishment, preferably before excessive depletion adversely affects plant growth. Irrigation refills empty soil pores. Ideally, soil moisture pressure should not

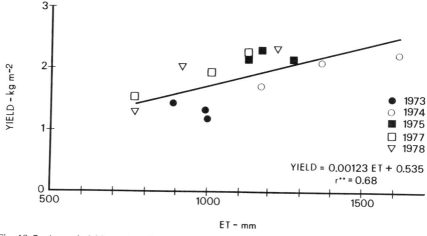

Fig. 18-7. Annual yield as a function of annual evapotranspiration; 1973 to 1978 lysimeter data. After Guitjens et al. (1984).

drop much below −200 kPa. Higher moisture levels in the soil also reduce adverse chemical effects such as specific ion toxicity. Excessively wet soils may have poor aeration.

Production functions for annual yield and the yield of each growth cycle correlate yield to evapotranspiration. Figure 18–7 shows this relationship for annual yield from three lysimeters at a single location in northwest Nevada; the biomass yield is expressed as having 12% moisture content. Figure 18–8 gives results from New Mexico for experiments with lysimeters and line-source sprinklers. The line-source water application decreases with lateral distance from the sprinkler line and yield follows a similar pattern. Production functions of four growth cycles in northwest Nevada are shown in Fig. 18–9. The figure includes also the production function obtained by combining data of four growth cycles. Successive harvests in northwest Nevada (Fig. 18–9) decreased in yield and the observed yield ranges were variable.

Irrigation scheduling is a procedure that considers when to irrigate relative to the reserve of readily available soil moisture and how much water to apply to restore depleted soil moisture. Usually, an evapotranspiration model is involved to estimate water transport into the atmosphere. If effective winter precipitation has not been sufficient to recharge empty soil moisture storage in the root zone, an alfalfa stand should be irrigated as soon as possible around the time the plants are breaking dormancy. Once recharged, irrigation scheduling can maintain the soil moisture content at required levels. Depending on expected water supplies for evapotranspira-

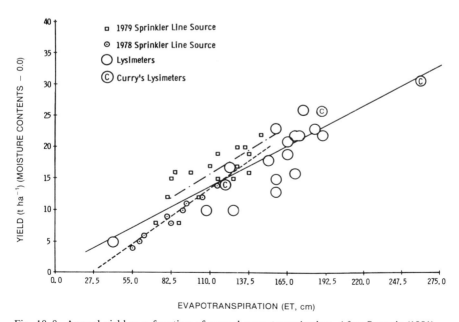

Fig. 18–8. Annual yield as a function of annual evapotranspiration. After Sammis (1981).

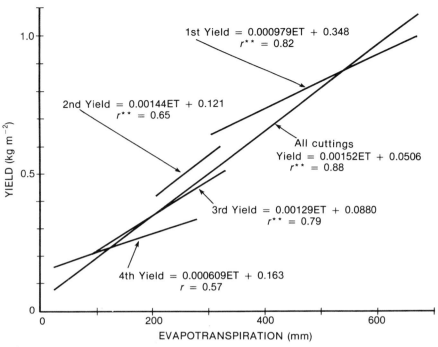

Fig. 18-9. Yield as a function of evapotranspiration by growth cycle; 1973 to 1978 lysimeter data. After Guitjens et al. (1984).

tion, yield goals can be estimated prior to the irrigation season and prior to each growth cycle; for example, planning irrigation to maintain maximum soil moisture levels which eliminate the chance of any evapotranspiration deficit could be a choice when water supply is unlimited. Harvests may create some unique challenges. Irrigation is discontinued some time before cutting. Drying of alfalfa in the field requires several days, and irrigation is postponed until the hay bales have been removed. The time interval between the last irrigation before cutting and the first irrigation of the new growth cycle may be long enough to create a soil moisture deficiency.

Whether or not to irrigate after the last harvest in the fall depends on the chance of getting winter precipitation. Evidence exists to indicate that winter evapotranspiration contributes to dry matter production harvested with the first cutting of the new growing season (Goodrich, 1986). A fall irrigation would reduce the chance of having a soil moisture deficit during dormancy.

A. Scheduling

Irrigation management uses scheduling to decide when to irrigate and how much water to apply. Since reduced evapotranspiration begins when the supply of soil moisture becomes deficient, water balance information is the basis for irrigation scheduling.

$$\mathrm{ET} = \mathrm{IRRIG.} + \mathrm{PRECIP.} - \mathrm{DRAIN.} - (-\Delta\mathrm{SM}) \qquad [3]$$

where ET = evapotranspiration, IRRIG. = irrigation, PRECIP. = precipitation, DRAIN. = drainage, and $-\Delta\mathrm{SM}$ = decrease in soil moisture. Recording precipitation and metering irrigation should be routine procedures. Drainage includes surface runoff and deep percolation below the root zone. Quantification of the latter is difficult in the field but in the confines of a lysimeter it involves metering or volumetric measurement. To obtain changes in soil moisture, periodic measurement of soil water content is recommended. The neutron meter can provide such information.

Since an inadequate supply of readily available soil moisture reduces evapotranspiration and dry matter production, irrigation should take place before this condition develops. Amount of application depends on depleted soil water in the root zone and on the irrigation efficiency. A decision to irrigate earlier than strictly needed means a reduction in irrigation requirement because of a smaller volume of empty pores. Additional considerations may also influence scheduling decisions: system capacity, irrigation duration, required advanced notice to the irrigation district, and other management choices affecting the crop (Reginato & Howe, 1985). The plant's need for water will dominate here.

Scheduling decisions rely on modeling of evapotranspiration. Meteorological parameters provide adequate information for making dependable estimates of atmospheric demand for water. Jensen (1973), Doorenbos and Pruitt (1977), Burman et al. (1980) and Blad (1983) presented comprehensive treatments. Evapotranspiration of a reference crop or potential evapotranspiration is calculated first. Estimation of actual evapotranspiration comes next; it is usually the product of crop coefficient and reference evapotranspiration. Crop coefficients for alfalfa vary with stage of crop development; at first regrowth, the value is at a minimum and it increases until a full stand has developed (Table 18–1).

The approach of estimating evapotranspiration directly from the atmospheric demand for water circumvents the necessity of repeatedly quantifying all components of Eq. [3]. Soil water holding capacity has to be documented at least once. Periodic neutron meter measurements can give moisture contents between dry and wet conditions, and provide information for determining optimal and deficient soil water ranges. Characterization of the soil as a water reservoir (Thien, 1983) is important but also subject to considerable spatial variation. Local soil surveys may also provide the desired information.

Carter and Conway (1984) developed a scheduling procedure for users of the lower Colorado region on the basis of soil moisture measured with the neutron meter. It helped to reduce irrigation diversions and drainage return flows to the Colorado River. Calibration and soil variation may contribute to errors in neutron meter measurements. Haverkamp et al. (1984) and Vauclin et al. (1984) gave an error analysis in terms of water content and number of observations required. Calibration appears to be the major source of error in local measurements and spatial variability of the soil in a heterogeneous field.

Table 18–2. Pan coefficient (K_p) for Class A pan for different groundcover and levels of mean relative humidity and 24-h wind (after Doorenbos and Pruitt, 1977).

Wind km d^{-1}	Windward side distance of green crop m	Case A: Pan placed in short green cropped area			Case B:† Pan placed in dry fallow area		
		RH mean, %			RH mean, %		
		Low <40	Medium 40–70	High >70	Low <40	Medium 40–70	High >70
Light <175	1	0.55	0.65	0.75	0.7	0.8	0.85
	10	0.65	0.75	0.85	0.6	0.7	0.8
	100	0.7	0.8	0.85	0.55	0.65	0.75
	1000	0.75	0.85	0.85	0.5	0.6	0.7
Moderate 175–425	1	0.5	0.6	0.65	0.65	0.75	0.8
	10	0.6	0.7	0.75	0.55	0.65	0.7
	100	0.65	0.75	0.8	0.5	0.6	0.65
	1000	0.7	0.8	0.8	0.45	0.55	0.6
Strong 425–700	1	0.45	0.5	0.6	0.6	0.65	0.7
	10	0.55	0.6	0.65	0.5	0.55	0.65
	100	0.6	0.65	0.7	0.45	0.5	0.6
	1000	0.65	0.7	0.75	0.4	0.45	0.55
Very strong >700	1	0.4	0.45	0.5	0.5	0.6	0.65
	10	0.45	0.55	0.6	0.45	0.5	0.55
	100	0.5	0.6	0.65	0.4	0.45	0.5
	1000	0.55	0.6	0.65	0.35	0.4	0.45

† For extensive areas of bare fallow soils and no agricultural development, reduce K_{pan} by 20% under hot, windy conditions; by 5 to 10% for moderate wind, temperature, and humidity conditions.

B. Evapotranspiration Modeling

Evapotranspiration modeling has taken several approaches. Energy balance gives the net amount of energy available for water evaporation. Mass transfer is the basis for an aerodynamic approach; a wind speed function and a vapor pressure gradient between the crop canopy and the overlying air influence vapor transport. The combination equation includes both energy balance and mass transfer, giving it the most complete physical description of evapotranspiration. Other more empirical methods consider temperature, relative humidity, or evaporation from a Class A pan. Every method requires calibration against actual evapotranspiration.

1. Models

a. Class A pan. Doorenbos and Pruitt (1977) reported values of pan coefficient (K_p) on the basis of mean relative humidity, wind speed, and windward side-distance of a green crop (Table 18–2). They considered two pan locations, one in a short, green crop and the other in a dry, fallow area. Reference evapotranspiration (ET_o) was defined as

$$ET_o = K_p \times E_p \qquad [4]$$

where pan evaporation has been measured for a desired time increment. The reference crop is a short, green grass. With alfalfa as the reference crop $K_{alfalfa} = 1.0$ at full crop development. This compares with an equivalent $K_{alfalfa} = 1.15$ for short, green grass reference. (Wright, 1982)

Alfalfa evapotranspiration ($ET_{alfalfa}$) is estimated from the product

$$ET_{alfalfa} = K_{alfalfa} \times ET_o \qquad [5]$$

where $K_{alfalfa}$ = crop coefficient for alfalfa. Alfalfa coefficients should be determined in plots in which plants are not stressed. Otherwise, the coefficients will be too small and lead to underestimation of evapotranspiration.

Rashedi (1983) and Rashedi and Guitjens (1984) presented alfalfa coefficients that apply to weekly increments of four growth cycles of a growing season in northwest Nevada. The best-fit function of the 12-wk growth cycle prior to the first harvest is:

$$K_{crop} = 0.31 \text{ (week No.)}^{0.55} \qquad [6]$$

and of three subsequent 5-wk growth cycles:

$$K_{crop} = 0.12 \text{ (week No.)}^{0.66}. \qquad [7]$$

Since values of $K_{alfalfa}$ are equal to the ratio of measured $ET_{alfalfa}$ and calculated ET_o, lysimeter research has played an important role in providing dependable values of $ET_{alfalfa}$. Relative humidity, wind speed, and evaporation from a Class A pan are uncomplicated measurements. Simplicity makes the Class A method attractive.

b. Penman method. The Penman equation incorporates an energy balance and a description of aerodynamic vapor transport

$$ET_r = \frac{\Delta}{\Delta + \gamma} (R_n + G) + \left(\frac{\gamma}{\Delta + \gamma}\right) [15.36 \ W_f \ (e_a - e_d)] \qquad [8]$$

where

ET_r = reference crop evapotranspiration, W m^{-2},
Δ = slope of vapor pressure − temperature curve, Pa C^{-1},
γ = psychometric constant, Pa C^{-1},
R_n = net radiation, W m^{-2},
G = soil heat flux to the surface, W m^{-2},
W_f = dimensionless wind function,
$e_a - e_d$ = average daily vapor pressure deficit, Pa, and
15.36 = constant, W m^{-2} per Pa.

Burman et al. (1980) gave a complete explanation and derivations of all auxilliary equations. They gave also crop coefficients for alfalfa to be used with ET_r after converting watts per square meter to millimeters. The Penman method has been routinely used in several states as the basis for estimating evapotranspiration.

The user of a specific model must select the appropriate set of alfalfa coefficients to convert reference evapotranspiration to estimated evapotranspiration at a certain stage of plant growth in a given growth cycle. The reason is the differences in estimation of reference evapotranspiration among models. Furthermore, changes in model reference evapotranspiration with geographic location may not be strictly a response to the new values of a model's variables. A variation in unexplained influences, lumped together in the alfalfa coefficients, makes the use of a locally developed alfalfa coefficient a recommended practice.

Sammis et al. (1985) developed a statistically significant correlation between crop coefficient and growing-degree-days (G), the summation of average daily temperature minus 5 °C

$$K_{crop} = 0.405 + 1.11 \times 10^{-3} G - 4.25 \times 10^{-7} G^2 + 3.56 \times 10^{-1} (G^3) \qquad [9]$$

based on combined data from four New Mexico locations that had seasonal yields ranging from 1.47 to 2.34 kg m^{-2}. The crop coefficients are for the modified Penman equation referenced to alfalfa ($K_{crop} = ET_{measured}/ET_{Penman}$).

c. Other models. Jensen (1973), Doorenbos and Pruitt (1977), Burman et al. (1980), and Blad (1983) gave comprehensive reviews of other models. Those based on solar energy and aerodynamic principles account for the most complete physical description of evapotranspiration but data requirements are more complex. All models give the equivalent of some kind of reference evapotranspiration but the term *potential evapotranspiration* is also widely accepted. Each model requires a unique set of alfalfa coefficients for estimating actual evapotranspiration.

To make different models estimate the same value for reference evapotranspiration, Doorenbos and Pruitt (1977) made several adjustments to four models by adding the influence of additional variables: mean relative humidity, sunshine hours, and daytime wind speed to the Blaney-Criddle model (based on temperature and daytime hours); mean relative humidity and daytime wind speed to a radiation model (based on solar radiation and a weighing factor for temperature and altitude); maximum relative humidity, daytime wind speed, and relative daytime vs. nighttime wind to the Penman combination model (based on temperature, net radiation, wind speed, and vapor pressure difference); and mean relative humidity, wind speed, and upwind side-distance of cropped area to pan evaporation measurements. Pruitt et al. (1986) reported 11 zones of similar evaporative demand in California and they prepared maps of daily reference evapotranspiration during each month of the year. This approach simplifies the need for model-specific alfalfa coefficients but it may not resolve the need for local coefficients.

Since escape of water vapor through stomates and entry of CO_2 are related, there is an interest in quantifying transpiration separately from evapotranspiration. Presumably, if a relatively larger portion of total evapotranspiration is used in transpiration, water use will be more efficient in terms of crop dry matter production. However, in practice, no attempt has been made to apply this distinction. Furthermore, the evaporation component depends on the degree of wetness at the soil surface when plants are not fully shading the ground. A wet soil surface not completely covered by the crop canopy will evaporate more water than a dry soil surface. Following rainfall or irrigation, evaporation will reach a maximum and decrease thereafter. Wright (1982) reported basal or minimum crop coefficients which apply to conditions that exist when evaporation is at a minimum but available soil moisture for transpiration remains adequate. Figure 18–10 shows the

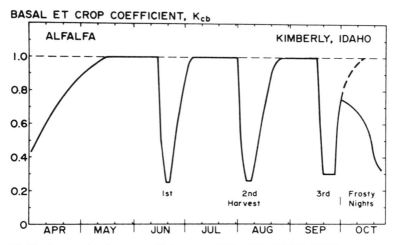

Fig. 18–10. Average basal crop coefficient for alfalfa. After Wright (1982).

basal crop coefficients; alfalfa is the reference crop and Eq. [8] the evapotranspiration model.

Measurement of canopy temperature with an infrared thermometer has also been used to evaluate stress conditions (Idso et al., 1977; 1981a, b; Jackson et al., 1981; Jackson 1982; Reginato, 1983). This approach involves a crop water stress index (CWSI):

$$CWSI = 1 - \frac{ET}{PET}. \qquad [10]$$

Abdul-Jabbar et al. (1985) showed a linear decline in alfalfa yield with a change in CWSI from zero to one (Fig. 18–11). They recommended further research to determine the critical value before which irrigation should occur.

2. Uptake of Shallow Groundwater

In the water balance equation, drainage below the root zone appears usually as a negative quantity (Eq. [3]). However, areas affected by a high water table may offer an opportunity to apply water management techniques that will induce plant uptake of water from the shallow groundwater supply. Besides reducing irrigation demand, drainage needs can be reduced.

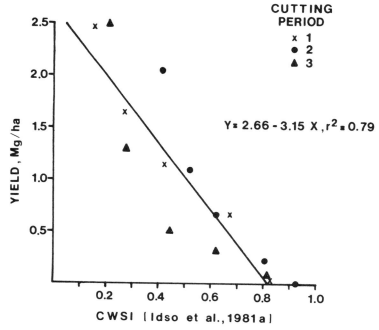

Fig. 18–11. Yield as a function of mean Crop Water Stress Index. After Abdul-Jabbar et al. (1985).

Hanks et al. (1977) noted that a water table at 2.3-m depth apparently supplied an adequate amount of water to the lower root zone of alfalfa plants to meet the demand of evapotranspiration. Earlier, Tovey (1969) found little difference in seasonal consumptive use when irrigated alfalfa grown in lysimeters had static water tables of 0.6, 1.2, and 2.4 m, respectively, but for similar nonirrigated treatments the seasonal consumptive use decreased linearly with water table depth. In Tovey's work, a 0.6-m water table depth shows little difference in yield between the irrigated (16 000 kg ha^{-1}) and nonirrigated (16 400 kg ha^{-1}) treatments. Guitjens (1984) reported significant yield declines under nonirrigation and an unstable water table. Chiodi (1986) questioned whether water use from the shallow groundwater may have occurred during deficit irrigation. Where irrigation supplements rainfall during the growing season Benz et al. (1983, 1984) showed that increased irrigation depressed the shallow groundwater's contribution to evapotranspiration. Increased supplemental irrigation decreased the contribution to as low as zero, whereas less supplemental irrigation raised the contribution to as much as 50% without experiencing a significant reduction in yield. An optimum water table depth of 1.5 m required supplemental irrigation to produce a yield increase, whereas a constant water table at a depth of 0.46 m showed excessive evapotranspiration and lower yield.

Favorable aspects of successful plant water uptake from in-situ groundwater include a stable water table, shallow water table depth, periodic rainfall, and acceptable groundwater quality. Exact management procedures are unclear. More information is also needed that compares results relative to potential yields.

3. Limited Water Environments

Income optimization with a limited water supply may require under-irrigation. The irrigation program could first aim at producing maximum yield, and irrigation could be stopped once the available supply has been used. For example, the growth cycles that produce the first two yields in Fig. 18–9 could be irrigated for yield near the upper limit of the production functions whereas the last two growth cycles receive no irrigation water. Nonirrigated growth cycles would not be harvested. Alternatively, all growth cycles could be underirrigated, thereby reducing the yield of each harvest. The result would be similar to the yield from the line-source sprinkled field in Fig. 18–8; the yield drops off with increasing distance from the line source. Both cases may give the same annual yield but labor and operation costs may differ and affect net income.

Going a step further, adequate irrigation may be replaced with deficit irrigation during less sensitive stages of plant growth after stems and leaves have developed. Apparently, initial regrowth after cutting occurs in a sensitive period that would benefit from a full water supply for evapotranspiration (Fig. 18–5). More information is needed about plant sensitivity before it can be considered in routine water management decisions.

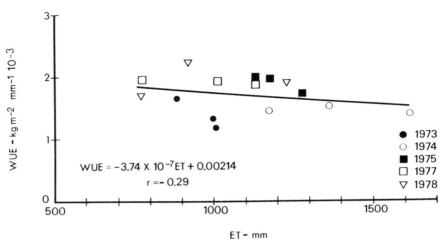

Fig. 18-12. Annual water use efficiency as a function of annual evapotranspiration; 1973 to 1978 lysimeter data. After Guitjens et al. (1984).

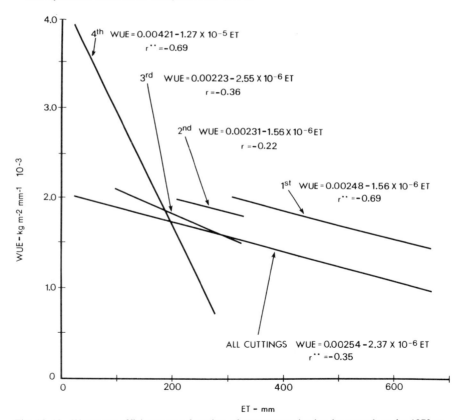

Fig. 18-13. Water use efficiency as a function of evapotranspiration by growth cycle; 1973 to 1978 lysimeter data. After Guitjens et al. (1984).

C. Water Use Efficiency

Taylor and Ashcroft (1972) defined *water use efficiency* as the ratio kilograms of dry matter and cubic meters of evapotranspiration. Guitjens (1982) and Guitjens et al. (1984) showed a nonsignificant, linear trend between declining annual water use efficiency and increasing annual water consumption (Fig. 18–12); the average annual water use efficiency was 1.73×10^{-3} kg m^{-2} per mm, with dry weight rated at 12% moisture content for simulating field weight. The decline differs among growth cycles of the irrigation season; Fig. 18–13 shows significant correlations for first and fourth cuttings. Fourth-cutting results are dramatically different but the last growth cycle contributes only about 10% of the annual yield. For all cuttings combined, only 12% of the data explains the correlation; the average water use efficiency was 1.88×10^{-3} kg m^{-2} per mm at a dry weight moisture content of 12%. Earlier Peterson (1972) found a water use efficiency of 1.47 $\times$ 10^{-3} kg m^{-2} per mm (presumably at 0% moisture content).

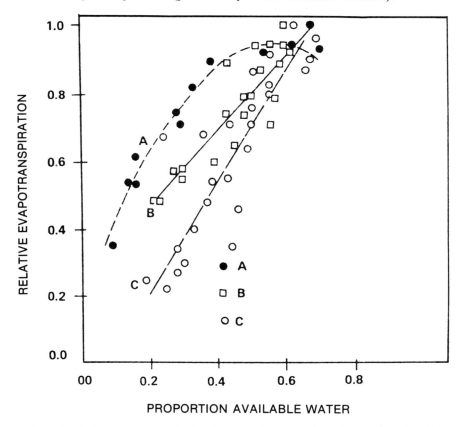

Fig. 18–14. Relative evapotranspiration (ET/ET$_{max}$) as a function of proportion of available soil water. A = sandy loam (ET$_{Penman}$ = 6.3 mm d^{-1}), B = clay loam (ET$_{Penman}$ = 6.1 mm d^{-1}), and C = clay loam (ET$_{Penman}$ = 7.6 mm d^{-1}). After Abdul-Jabbar et al. (1983).

Applied water use efficiency is the ratio kilograms of dry matter and cubic meters of applied water; applied water replaces evapotranspiration in water use efficiency. The trend between water use efficiency and evapotranspiration will be the same as in Fig. 18-12 and 18-13 as long as reduced evapotranspiration limits dry matter production because of underirrigation (Stewart et al., 1983) and additional increments of applied water correlate with further increases in evapotranspiration. Thereafter, additional water applications will produce more surface runoff or deep seepage (Jensen et al., 1980; Jensen, 1986). As drainage increases, yield as a function of applied water will become curvilinear. Combinations of soils, field layouts, and irrigation systems will create different irrigation efficiencies (Sammis, 1981), making applied water use efficiencies inconsistent among locations.

This point is also supported by studies comparing relative evapotranspiration (ET/ET_{max}) and proportion of available soil water (Abdul-Jabbar et al., 1983). On two out of three soils, relative evapotranspiration increased linearly with increases in proportion of available soil water. On a clay loam this trend continued until 68% of available water was reached but for sandy loam the percentage was smaller (Fig. 18-14). In addition to soil differences, evaporative demand also had an influence on the rate of increase in relative evapotranspiration. Variable water applications with the line-source sprinkler created the differences in the proportion of available soil water.

Hanks (1983) reviewed early research in water use efficiency with alfalfa grown in containers; it showed a good, linear relationship between yield in grams and transpiration in kilograms (a slope of 1.17 g kg^{-1}) and even a

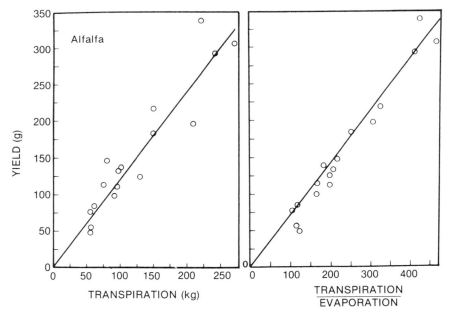

Fig. 18-15. Yield as a function of transpiration and the ratio of transpiration and evaporation. After Hanks (1983).

better linear relationship between yield in grams and the ratio of transpiration and free water evaporation. The inclusion of evaporation removes the influence of differences among years (Fig. 18-15).

Literature values for water use efficiency vary considerably. Metochis and Orphanos (1981) in Cypress gave values ranging from 2.25 to 2.85 kg m^{-3} when annual evapotranspiration is considered. Their values were much lower during the "summer slump" in July and August with values of 0.50 to 0.65 kg m^{-3}. Donovan and Meek (1983) in the Imperial Valley of southern California gave values ranging from 0.98 to 1.09 kg m^{-3} without significant differences between dry and wet water treatments. Also using annual evapotranspiration, Guitjens (1982) reported values of 0.95 to 1.57 kg m^{-3} for northwest Nevada; the values were not significantly different and covered a range of annual evapotranspirations. Values based on seasonal evapotranspiration in North Dakota ranged from 1.21 to 2.31 kg m^{-3}, the latter under the highest recorded summer precipitation (Bauder et al., 1978). For specific growth cycles in Minnesota, Carter and Sheaffer (1983) gave values of 1.31 to 3.01 kg m^{-3} for a third growth cycle and values of 0.97 to 3.01 kg m^{-3} for a fourth growth cycle; their values come from different water treatments but without a trend for dry or wet conditions.

Considering the slope of the linear production function yield = f (evapotranspiration), data reported by Sammis (1981) show values of 1.2 kg m^{-3} for Las Cruces, NM, 1.4 kg m^{-3} for Reno, NV, 1.6 kg m^{-3} for North Dakota, 1.3 kg m^{-3} for Nebraska, and a composite value for all locations of 1.2 kg m^{-3}. Donovan and Meek (1983) gave a similar value of 1.2 kg m^{-3} for Imperial Valley, CA and Guitjens et al. (1984) gave a value of 1.1 kg m^{-3} for Fallon, NV. Las Cruces, Imperial Valley, and Fallon data involve annual evapotranspiration, whereas Reno, North Dakota, and Nebraska omit evapotranspiration during dormancy.

VIII. OTHER ASPECTS

A. Economic Decisions

1. Economic Models

Although water management can effectively conserve water by exercising control over water losses, economic considerations may further influence irrigation decisions. Growers who produce other crops in addition to alfalfa may choose to change water allocations among crops when water supplies are limited; changing crop prices may make certain crops more attractive. Underirrigation will reduce alfalfa yield, but when water applications return to normal at a later time, production appears to return to former levels. Growers have also the choice to limit irrigation to more productive growth cycles or to irrigate only a portion of the area in alfalfa. The losses in yield may outweigh the economic value of the water saved.

Setting production goals prior to irrigation and practicing water conservation through irrigation scheduling are realistic possibilities. Economic optimization recognizes differences in yield response to water deficit of incremental growth stages and requires more advanced management skills and the use of growth models. Yaron and Bresler (1983) and Raju et al. (1983) presented the fundamentals of saving water by changing irrigation dates with the objective of minimizing yield decreases. The illustrations do not involve alfalfa.

B. Irrigation Systems

The most common methods for irrigating alfalfa are flood and sprinkler irrigation. To allow for a uniform water depth on the surface of a field, the land surface has to be shaped for flood irrigation. With sprinklers, the irrigation system distributes the water evenly even when the land surface is uneven. Most flooding systems give little control over depth of application. Consequently, the irrigation interval should be decreased when evapotranspiration increases. Sprinklers give more flexibility. They can be operated at variable or fixed intervals. A fixed interval but a variable sprinkling duration can apply the same amount of water as a variable interval but a fixed duration.

Flood irrigation systems can be of several types. With the border method water spreads out between two levees by gravity flow. Prior to seeding, the cultivated borders should be laser-leveled to create a uniform slope for water flow down the field and zero side-slope between the levees. When excessive sideslope makes leveling of the ground surface impractical, a corrugated land surface can prevent side flow while the water flows down the field through the corrugates. Basin irrigation requires a horizontal land surface. Other flooding systems are less common for alfalfa irrigation. Hart et al. (1980) gave detailed information for the design and operation of flood systems.

Sprinkler irrigation systems can be divided into stationary and moving systems. A stationary system can be permanent or movable, and sprinkling duration controls the depth of application. Moving systems apply water as they travel across the field, either in a circle or along a straight path, and their speed of travel controls the depth of application. Addink et al. (1980) discussed the design and operation.

C. Yield Goals

Annual production functions, the annual yield's dependence on evapotranspiration, are useful for setting seasonal yield goals. Production functions of individual growth cycles serve the same purpose for each harvest. With abundant water supplies careful irrigation scheduling gives growers an opportunity to apply water for the upper yield level. When water supplies are limited, some periods of plant stress are inevitable. A choice must be made between allowing water stress before irrigation and irrigation for full production followed by no irrigation. Perhaps a grower decides not to irrigate his last one or two growth cycles but concentrates on full growth of earlier growth cycles. Such a choice could reduce harvest costs. The author has unpublished data to show that three consecutive years without irrigation during the last one or two growth cycles did not reduce yield in the fourth year when full irrigation was restored.

D. Production Efficiency

Water use efficiency provides a measure of dry matter production per unit volume of evapotranspiration. As discussed, values from the literature are inconsistent and the dependence of water use efficiency on evapotranspiration has not been clearly demonstrated. There is also a difference between dry matter produced and dry matter harvested because of harvesting losses. In hay production they may consist of 20 to 25% of the dry matter produced. Alfalfa dried in the field leaves visible signs of depressed regrowth where the windrow covered plants. Such interference may affect uniformity in regrowth and evapotranspiration. Consequently, it may also affect the degree of water conservation provided by irrigation scheduling.

IX. SUMMARY

The linear dependence of alfalfa yield on evapotranspiration illustrates the importance of meeting evapotranspiration demand through timely irrigation. If a lack of irrigation reduced evapotranspiration, yield decreased. Since evapotranspiration rate varies from first regrowth to a fully developed stand, variable crop coefficients should account for the change in water consumption with stage of crop development during each growth cycle. A combination of a suitable model for estimating reference evapotranspiration and appropriate crop coefficients can provide reliable evapotranspiration estimates. However, it may be more difficult to select crop coefficients that are suitable for a given geographic location than it would be to adopt a model for calculating reference evapotranspiration.

Relative uptake of water by alfalfa decreases with soil depth because root distribution dictates patterns of water extraction in the soil. If two-thirds of the root mass is concentrated in the top 450 mm of soil, irrigation should supply adequate amounts of water to the soil depth with the greatest concentration of roots. The water balance can be used to determine soil moisture depletion and the required amount of water application at time of irrigation.

REFERENCES

Abdul-Jabbar, A.S., D.G. Lugg, T.W. Sammis, and L.W. Gay. 1985. Relationship between crop water stress index and alfalfa yield and evapotranspiration. Trans. ASAE 28:454-461.

Abdul-Jabbar, A.S., T.W. Sammis, and D.G. Lugg. 1982. Effect of moisture level on the root pattern of alfalfa. Irrig. Sci. 3:197-207.

Abdul-Jabbar, A.S., T.W. Sammis, D.G. Lugg, C.E. Kallsen, and D. Smeal. 1983. Water use by alfalfa, maize, and barley as influenced by available soil water. Agric. Water Manage. 6:351-363.

Addink, J.W., J. Keller, C.H. Pair, R.E. Sneed, and J.W. Wolfe. 1980. Design and operation of sprinkler systems. p. 621-660. In M.E. Jensen (ed.) Design and operation of farm irrigation systems. ASAE Monogr. 3. ASAE, St. Joseph, MI.

Ayers, R.S., and D.W. Wescot. 1976. Water quality for agriculture. Irrig. Drain. Pap. 29. FAO, Rome.

Bauder, J.W., A. Bauer, J.M. Raminez, and D.K. Cassel. 1978. Alfalfa water use and production on dryland and irrigated sandy loam. Agron. J. 70:95–99.

Benz, L.C., E.J. Doering, and G.A. Reichmann. 1984. Water-table contribution to alfalfa evapotranspiration and yields in sandy soils. Trans. ASAE 27:1307–1312.

Benz, L.C., G.A. Reichmann, and E.J. Doering. 1983. Drainage requirements for alfalfa grown on sandy soils. Trans. ASAE 26(1):161–164, 166.

Blad, B.L. 1983. Atmospheric demands for water. p. 1–44. In I.D. Teare et al. (ed.) Crop-water relations. John Wiley and Sons, New York.

Bornstein, J., G.R. Benoit, F.R. Scott, P.R. Hepler, and W.E. Hedstrom. 1984. Alfalfa growth and soil oxygen diffusion as influenced by depth to water table. Soil Sci. Soc. Am. J. 48:1165–1169.

Bourget, S.J., and R.B. Carson. 1962. Effects of soil moisture stress on yield, water-use efficiency and mineral composition of oats and alfalfa grown at two fertility levels. Can. J. Soil Sci. 42:7–12.

Brown, P.W., and C.B. Tanner. 1981. Alfalfa water potential measurement: A comparison of the pressure chamber and leaf dew-point hygrometers. Crop Sci. 21:240–244.

Brown, P.W., and C.B. Tanner. 1983. Alfalfa stem and leaf growth during water stress. Agron. 75:799–805.

Bula, R.J., and M.A. Massengale. 1972. Environmental physiology. In C.H. Hanson (ed.) Alfalfa science and technology. Agronomy 15:167–184.

Burman, R.D., P.R. Nixon, J.L. Wright, and W.O. Pruitt. 1980. Water requirements. p. 189–232. In M.E. Jensen (ed.) Design and operation of farm irrigation systems. ASAE Monogr. 3. ASAE, St. Joseph, MI.

Carter, P.R., and C.C. Sheaffer. 1983. Alfalfa response to soil water deficits. I. Growth, forage quality, yield, water use and water-use efficiency. Crop Sci. 23:669–675.

Carter, V.H., and S. Conway. 1984. Predicting irrigation dates and amounts with a neutron moisture gage. Trans. Twelfth Int. Congr. Irrig. Drain. I(A):1007–1020.

Chiodi, M.T. 1986. Shallow groundwater reuse options in alfalfa irrigation. Unpubl. M.S. thesis. Univ. of Nevada, Reno.

Christian, K.R. 1977. Effects of environment on the growth of alfalfa. Adv. Agron. 29:183–227.

Donovan, T.J., and B.D. Meek. 1983. Alfalfa responses to irrigation treatment and environment. Agron. J. 75:461–464.

Doorenbos, J., and A.H. Kassam. 1979. Yield response to water. Irrig. Drain. Pap. 33. FAO, Rome.

Doorenbos, J., and W.O. Pruitt. 1977. Crop water requirements. Irrig. Drain. Pap. 24. FAO, Rome.

Evenson, P.D. 1979. Optimum crown temperature for maximum alfalfa growth. Agron. J. 71:798–800.

Fick, G.W. 1984. Simple simulation models for yield prediction applied to alfalfa in the northeast. Agron. J. 76:235–239.

Gear, R.D., A.S. Dransfield, and M.D. Campbell. 1977. Irrigation scheduling with neutron probe. J. Irrig. Drain. Div. Am. Soc. Civ. Eng. 103(IR3):291–298.

Goodrich, M.T. 1986. Dormant season evapotranspiration in alfalfa. Unpubl. M.S. thesis, Univ. of Nevada, Reno.

Groskopp, M.O., J.M. Sund, and J.T. Murdock. 1963. Irrigated alfalfa in central Wisconsin. Wisc. Agric. Exp. Stn. Bull. 558.

Guitjens, J.C. 1982. Models of alfalfa yield and evapotranspiration. J. Irrig. Drain. Div: Am. Soc.Civ. Eng. 108(IR3):212–222.

Guitjens, J.C. 1984. Alfalfa yield variance under scheduled irrigation. p. 323–329. In Proc. Specialty Conf. Irrig. Drain. Div. 24–26 July. Am. Soc. Civ. Eng. Flagstaff, AZ.

Guitjens, J.C., P-S. Tsui, J.M. Connor, and D.F. Thran. 1984. Toward total water management. Twelfth Int. Congr. Irrig. Drain. Vol. 1(A), Q38:169–184.

Hagan, R.M., H.R. Haise, and T.W. Edminster. (ed.). 1967. Irrigation of agricultural lands. Agronomy 11.

Hanson, C.H. (ed.). 1972. Alfalfa science and technology. Agronomy 15.

Hanks, R.J. 1974. Model for predicting plant yield as influenced by water use. Agron. J. 66:660–665.

Hanks, R.J. 1983. Yield and water-use relationships: An overview. In H.M. Taylor et al. (ed.) Limitations to efficient water use in crop production. ASA, CSSA, and SSSA, Madison, WI.

Hanks, R.J., T.E. Sullivan, and V.E. Hunsaker. 1977. Corn and alfalfa production as influenced by irrigation and salinity. Soil Sci. Soc. Am. J. 41:606–610.

Hansen, V.E., O.W. Israelsen, and G.E. Stringham. 1980. Irrigation principles and practices. John Wiley and Sons, New York.

Hart, W.E., H.G. Collins, G. Woodward, and A.S. Humpherys. 1980. Design and operation of gravity or surface systems. p. 501–580. *In* M.E. Jensen (ed.) Design and operation of farm irrigation system. ASAE Monogr. 3. ASAE, St. Joseph, MI.

Haverkamp, R., M. Vauclin, and G. Vachaud. 1984. Error analysis in estimating soil water content from neutron probe measurements: 1. Local standpoint. Soil Sci. 137:78–90.

Heichel, G.H. 1983. Alfalfa. p. 127–155. *In* I.D. Tear et al. (ed.) Crop-water relations. John Wiley and Sons, New York.

Hiler, E.A., and T.A. Howell. 1983. Irrigation options to avoid critical stress: An overview. p. 479–497. *In* H.M. Taylor et al. (ed.) Limitations to efficient water use in crop production. ASA, CSSA, and SSSA, Madison, WI.

Hoffman, G.J., R.J. Ayers, E.J. Doering, and B.L. McNeal. 1983. Salinity in irrigated agriculture. p. 135–185. *In* M.E. Jensen. (ed.) Design and operation of farm irrigation systems. ASAE Monogr. 3. ASAE, St. Joseph, MI.

Idso, S.B., R.D. Jackson, P.J. Pinter, R.J. Reginato, and J.L. Hatfield. 1981a. Normalizing the stress-degree-day parameter for environmental variability. Agric. Meteorol. 24:45–55.

Idso, S.B., R.J. Jackson, and R.J. Reginato. 1977. Remote sensing of crop yields. Science 196:19–25.

Idso, S.b., R.J. Jackson, D.C. Reicosky, and J.L. Hatfield. 1981b. Determining soil-induced plant water potential depression in alfalfa by means of infrared thermometry. Agron. J. 73:826–830.

Jackson, R.D. 1982. Canopy temperature and crop water stress. p. 43–85. *In* D. Hillel (ed.) Advances in irrigation. Vol. 1. Academic Press, New York.

Jackson, R.D., S.B. Idso, R.J. Reginato, and P.J. Pinter. 1981. Canopy temperature as a crop water stress indicator. Water Resour. Res. 17:1133–1138.

Jensen, M.E. 1973. Consumptive use of water and irrigation water requirements. Am. Soc. Civ. Eng., New York.

Jensen, M.E. (ed.). 1980. Design and operation of farm irrigation systems. ASAE Monogr. 3. ASAE, St. Joseph, MI.

Jensen, M.E. 1986. Design and performance of irrigation and drainage systems. Int. Comm. Irrig. Drain. Bull. 34(2)1–10.

Jensen, M.E., D.S. Harrison, H.C. Korven, and F.E. Robinson. 1980. The role of irrigation in food and fiber production. p. 15–41. *In* M.E. Jensen (ed.) Design and operation of farm irrigation systems. ASAE Monogr. 3. ASAE, St. Joseph, MI.

Jones, J.W. 1983. Irrigation options to avoid critical stress: Optimization of on-farm water allocation to crops. p. 507–516. *In* H.M. Taylor et al. (ed.) Limitations to efficient water use in crop production. ASA, CSSA, and SSSA, Madison, WI.

Jung, G.A., and K.L. Larson. 1972. Cold, drought, and heat tolerance. *In* C.H. Hanson (ed.) Alfalfa science and technology. Agronomy 15:185–209.

Kilcher, M.R., and D.H. Heinrichs. 1974. Contribution of stems and leaves to the yield and nutrient level of irrigated alfalfa of different stages of development. Can. J. Plant Sci. 54:739–742.

Kozlowski, T.T. (ed.). 1968. Water deficit and plant growth. Vol. II. Academic Press, New York.

Lowe, C.C., V.L. Marble, and M.D. Rumbaugh. 1972. Adaptation, varieties, and usage. *In* C.H. Hanson (ed) Alfalfa science and technology. Agronomy 15:391–413.

Maas, E.V., and G.J. Hoffman. 1977. Crop tolerance—current assessment. J. Irrig. Drain. Div. Am. Soc. Civ. Eng. 103(IR2):115–134.

Meek, B.D., T.J. Donovan, and L.E. Graham. 1980. Summertime flooding effects on alfalfa mortality, soil oxygen concentration, and metric potential in a silty clay loam soil. Soil Sci. Soc. Am. J. 44:433–435.

Metochis, C. 1980. Irrigation of lucerne under semi-arid conditions. Irrig. Sci. 1:247–252.

Metochis, C., and P.I. Orphanos. 1981. Alfalfa yield and water use when forced into dormancy by withholding water during the summer. Agron. J. 73:1048–1050.

Meyer, J.H., and L.G. Jones. 1962. Controlling alfalfa quality. Univ. California Agric. Exp. Stn. Bull. 784.

Nevada Agricultural Statistics. 1984. Nevada Crop and Livestock Reporting Service, Reno.

Perry, L.J., and K.L. Larson. 1974. Influence of drought on tillering and internode number and length in alfalfa. Crop Sci. 14:693–696.

Peterson, H.B. 1972. Water relationships and irrigation. *In* C.H. Hanson (ed.) Alfalfa science and technology. Agronomy 15:469–480.

Pruitt, W.O., E. Fereres, K. Kaita, and R.L. Snyder. 1986. Reference evapotranspiration (ET_o) for California. Univ. Calif. Agric. Exp. Stn. Bull. 1922.

Raju, K.S., E.S. Lee, A.W. Biere, and E.T. Kanemasu. 1983. Irrigation scheduling based on a dynamic crop response model. p. 257–271. *In* D. Hillel (ed.) Advances in irrigation. Vol. 2. Academic Press, New York.

Rashedi, N. 1983. Evapotranspiration crop coefficients for alfalfa at Fallon, Nevada. Unpubl. M.S. thesis. Univ. of Nevada, Reno.

Rashedi, N., and J.C. Guitjens. 1984. Alfalfa crop coefficients for NW Nevada. p. 227–283. *In* Proc. Specialty Conf. Irrig. Drain Div. 24–26 July. Am. Soc. Civ. Eng. Flagstaff, AZ.

Reginato, R.J. 1983. Field quantification of crop water stress. Trans. ASAE 26(3)772–775, 781.

Reginato, R.J., and J. Howe. 1985. Irrigation scheduling using crop indicators. J. Irrig. Drain. Div. Am. Soc. Civ. Eng. 111(IR2):125–133.

Ritchie, J.T. 1983. Efficient water use in crop production: Discussion on the generality of relations between biomass production and evapotranspiration. p. 29–44. *In* H.M. Taylor et al. Limitations to efficient water use in crop production. ASA, CSSA, and SSSA, Madison, WI.

Sammis, T.W. 1981. Yield of alfalfa and cotton as influenced by irrigation. Agron. J. 73:323–329.

Sammis. T.W., C.L. Mapel, D.G. Lugg, R.R. Lansford, and J.T. McGuckin. 1985. Evapotranspiration crop coefficients predicted using growing-degree-days. Trans. ASAE 28(3):773–780.

Selirio, I.S., and D.M. Brown. 1979. Soil moisture-based simulation of forage yield. Agric. Meteorol. 20:99–114.

Sharratt, B.S., D.C. Reicosky, S.B. Idso, and D.G. Baker. 1983. Relationships between leaf water potential, canopy temperature, and evapotranspiration in irrigated and nonirrigated alfalfa. Agron. J. 75:891–894.

Snaydon, R.W. 1972. The effect of total water supply, and of frequency of application, upon lucerne. Aust. J. Agric. Res. 23:253–256.

Stegman, E.C., J.C. Musick, and J.I. Stewart. 1980. Irrigation management. p. 763–816. *In* M.E. Jensen. Design and operation of farm irrigation system. ASAE Monogr. 3. ASAE, St. Joseph, MI.

Stewart, B.A., J.J. Musick, and D.A. Dusek. 1983. Yield and water use efficiency of grain sorghum in a limited irrigation-dryland farming system. Agron. J. 75:629–634.

Taylor, H.M., W.R. Jordan, and T.R. Sinclair (ed.) 1983. Limitations to efficient water use in crop production. ASA, CSSA, and SSSA, Madison, WI.

Taylor, S.A., and G.L. Ashcroft. 1972. Physical edaphology. Freeman and Co., San Francisco.

Teare, E.D., and M.M. Peet. 1983. Crop-water relations. John Wiley and Sons, New York.

Thien, S.J. 1983. The soil as a water reservoir. p. 45–72. *In* I.D. Teare et al. (ed.) Crop-water relations. John Wiley and Sons, New York.

Tovey, R. 1969. Alfalfa water table investigations. J. Irrig. Drain. Div. Am. Soc. Civ. Eng. 95(IR4):525–535.

Turner, N.C., and G.T. Burch. 1983. The role of water in plants. p. 73–126. *In* T.D. Teare et al. (ed.) Crop-water relations. John Wiley and Sons, New York.

U.S. Department of Commerce. 1984. Census of agriculture. Vol. 1, Part 51. U.S. Gov. Print. Office, Washington, DC.

van Bavel, C.H.M. 1967. Changes in canopy resistance to water loss from alfalfa induced by soil water depletion. Agric. Meteorol. 4:165–176.

Vauclin, M., R. Haverkamp, and G. Vachaud. 1984. Error analysis in estimating soil water content from neutron probe measurements: 2. spatial standpoint. Soil Sci. 137:141–148.

Vough, L.R., and G.C. Marten. 1971. Influence of soil moisture and ambient temperature on yield and quality of alfalfa forage. Agron. J. 63:40–42.

Whitfield, D.M., G.C. Wright, O.A. Gyles, and A.J. Taylor. 1986. Growth of lucerne (*Medicago sativa* L.) in response to frequency of irrigation and gypsum application on a heavy clay soil. Irrig. Sci. 7:37–52.

Wright, J.L. 1982. New evapotranspiration crop coefficients. J. Irrig. Drain Div. A. Soc. Civ. Eng. 108(IR1):57–74.

Wright, J.L. 1988. Daily and seasonal evapotranspiration and yield of irrigated alfalfa in southern Idaho. Agron. J. 80:662–669.

Yaron, D., and E. Bresler. 1983. Economic analysis of on-farm irrigation using response functions of crops. p. 223–255. *In* D. Hillel (ed.) Advances in irrigation. Vol. 2. Academic Press, New York.

19   Corn[1]

F. M. RHOADS

University of Florida
Quincy, Florida

J. M. BENNETT

University of Florida
Gainesville, Florida

Efficient water management for corn (*Zea mays* L.) production is dependent on an understanding of the growth and development of the crop. Considerable research has focused on the response of corn growth and productivity as a function of the growth phases during which water stress was imposed. Although several authors have described the growth and development of corn in detail (Leng, 1951; Duncan, 1975; Ritchie & Hanway, 1982), a brief summary of the developmental phases of corn is necessary before discussing the responses of the crop to water stress.

I. CROP DEVELOPMENT

A. Vegetative Growth

Following seed germination and seedling emergence, corn typically initiates and expands 20 to 21 leaves during a period which may range from 60 to 65 d (Ritchie & Hanway, 1982). Hybrids grown in Indiana have been shown to average 69 d from planting to silking and, although the time to maturity increased from near 100 to 110 d in the 1950s to 120 to 135 d in the 1970s, little change in the time required for completion of vegetative growth was apparent (McGarrahan & Dale, 1984). Vegetative development is dependent on many crop, soil, and environmental factors including variety, temperature, soil water, and radiation. Generally, a shorter vegetative growth period is associated with higher temperatures (Watts, 1974; Dale et al., 1980), while water deficits lengthen the vegetative growth period.

---

[1] Contribution of the Dep. of Soil Sci. and Agron., Inst. of Food and Agric. Sci., Univ. of Florida. Florida Agric. Exp. Stn. Journal Series No. 8160.

Development of adequate leaf area necessary for interception and utilization of incident radiation is important and has been shown to be closely related to final grain yield (Williams et al., 1965; Eik & Hanway, 1966; Williams et al., 1968; Scarsbrook & Doss, 1973; Prior & Russell, 1976). Even mild water deficits during vegetative growth reduce leaf area expansion (Acevedo et al., 1971; Van Volkenburgh & Boyer, 1984), while more severe stresses are generally required before the number of leaves produced is affected. Although the initiation of vegetative structures is complete relatively early during crop growth, leaf appearance and expansion continue until near tassel emergence. Appearance of new leaves may be as rapid as one every 3 d (Ritchie & Hanway, 1982). Maximum leaf area index (LAI) typically ranges between 3 and 5 for crops grown under optimal conditions (Dale et al., 1980; Bennett & Hammond, 1983). Although optimal leaf area indices have been difficult to accurately determine (Prior & Russell, 1976), light interception generally increases up to LAI of about 3.5 (Eik & Hanway, 1966). Once maximum leaf area has been achieved, stresses that occur during reproductive growth may cause loss of green leaf area through senescence of leaves. Maintenance of leaf area for maximum photosynthetic capacity during grain filling is essential for maximum yields (Alessi & Power, 1975).

B. Root Growth

Extraction of water and nutrients from the soil profile is dependent on the development of an adequate root system. Although some research has been conducted to allow a general characterization of the depth and development of the root system of corn, additional research is needed to more fully describe the development with time and in response to various soil physical factors and environmental growing conditions. A clear picture of root development through the growing season is currently difficult to define based on current literature.

At physiological maturity, Mayaki et al. (1976) observed that approximately 64 and 92% of the root dry matter was confined to the upper 0.30 and 0.90 m of soil, respectively, for irrigated corn growing on a Muir silt loam soil (fine-silty, mixed, mesic Cumulic Haplustolls). Other studies have shown that root dry matter in the upper 0.30 m of sandy soil profiles with varying depths to the water table ranged between 69 and 97% of the total amount of roots observed (Follett et al., 1974). Such observations suggest that the root system of corn is highly dependent on the depth of the soil profile and the fluctuation of the depth of a water table. Although numerous studies show that most of the roots of corn are confined to the upper soil layers, Allmaras et al. (1975a, b) demonstrated that the maximum rooting depth (1.45 m) for corn was slightly deeper than that observed for soybean (*Glycine max* L.) when rooting of the two crops were compared on a Nicollet sandy loam soil (fine-loamy, mixed, mesic Aquic Hapludolls) in Minnesota. However, Mayaki et al. (1976) observed slightly less root dry matter and depth for corn when compared to both soybean and grain sorghum (*Sorghum bicolor* L. Moench).

Maximum rooting depths of well-watered corn are commonly between 1.2 and 1.5 m (Allmaras et al., 1975b; Mayaki et al., 1976; Robertson et al., 1980). Most of the water requirements of corn have been shown to be supplied by root uptake in the upper 1.0 m of sandy soils (Follet et al., 1974). Both water and nutrient uptake patterns are related to the extent and distribution of the root system (Mackay & Barber, 1985; Sharp & Davies, 1985).

Although water deficits clearly affect both above- and below-ground plant growth, the relative effect on the root and shoot seems to be related to the severity, and likely to the timing, of the stress. With mild water deficits, an increase in the absolute weight of roots produced has been observed (Sharp & Davies, 1979; Dwyer & Stewart, 1985), suggesting preferential allocation of photosynthate to the roots at expense of the shoots. However, with more severe stresses, the total amount of roots is often decreased with conflicting reports as to the effect on the depth of the root system. For example, in controlled environment studies, Sharp and Davies (1985) reported that withholding water reduced root mass in the upper soil profile while stimulating deeper rooting and soil water extraction at a greater depth. However, in two field studies (Mayaki et al., 1976; Robertson et al., 1980), total root mass was decreased by water stress, while rooting depth was increased in one study and decreased in the other, presumably as a result of differences in soil factors or the timing and intensity of the stress periods. Although little information is available concerning production of new roots and senescence of older roots during stress periods, some enhanced root senescence undoubtedly occurs with severe stress (Dwyer & Stewart, 1985). Generally, root-to-shoot ratios increase with stress (Barlow et al., 1976; Sharp & Davies, 1979, 1985; Dwyer & Stewart, 1985), although occasional reports of an opposite response are found (Saini & Chow, 1982).

C. Reproductive Development

The corn plant changes from vegetative to reproductive development when the apical meristem within the stem begins to elongate (Duncan, 1975). Before the tassel is differentiated, early initials of ear shoots appear as buds at the axils of the lower leaves. By the time the plant has six collared leaves, the growing point is near the soil surface and some ear shoots are visible if the plant is dissected (Ritchie & Hanway, 1982). Although a number of ear shoots have been initiated by the time the plant has nine collared leaves, generally only one or two ear shoots will ultimately develop into an ear on commercial corn hybrids. By the 12th collared leaf stage, the size of the ear is being established. At about 1 wk before silking, the potential number of kernels per row has been determined. Rapid development of the ear shoot begins near the time the tassel emerges (Duncan, 1975). When the plant has produced about 17 collared leaves, upper ear shoots begin to emerge and the tip of the tassel becomes visible (Ritchie & Hanway, 1982). Stevens et al. (1986) observed the reproductive development of B73 × MO17 corn with respect to emerged leaf stages and found tassel initiation occurred as the fourth leaf emerged, ear initiation coincided with emergence of the sixth leaf, and pistillate and silk formation occurred as the 11th leaf emerged. Full plant height is attained near silking which occurs 2 to 3 d after tasseling. Stresses imposed near the silking period have been shown to have dramatic effects on final grain yields. For example, Prine (1971) demonstrated that the period of time during and shortly after silking was sensitive to reduced light intensity. Numerous other studies, which will be discussed in a later section, have demonstrated that the pollination and early seed establishment periods are quite sensitive to water deficits.

After pollination, seeds develop through blister, milk, dough, and dent stages before reaching physiological maturity (Ritchie & Hanway, 1982). Although development through all phenological stages of growth is highly dependent on temperature and other environmental factors, blister stage generally occurs 10 to 14 d after silking (DAS) followed by dough (24–28 DAS), dent (35–42 DAS), and physiological maturity (55–65 DAS) for a typical corn hybrid.

Seed dry weight increases and seed moisture percentage decreases as the grain filling process progresses. Filling of grain is dependent on the current supply of photosynthate as well as remobilization of previously stored assimilate from other plant parts such as the leaves and stem. Grain yield is dependent on both the rate and duration of the seed filling period. The average duration from silking to physiological maturity for corn hybrids grown in Indiana increased from near 50 d in the 1950s to near 60 d in 1980 (McGarrahan & Dale, 1984), demonstrating that corn breeders have selected for slightly longer periods of grain filling. Both the rate and duration of grain filling are affected by genetic as well as environmental factors. Johnson and Tanner (1972) demonstrated that at least 90% of the increase in grain dry weight was linearly accumulated beginning about 2.5 wk after silking and continuing until 90% of the final grain weight was attained. Remobilization of assimilates from the stalk to the developing grain is an important process which tends to maintain a rather linear increase in seed dry weight despite varying environmental conditions which result in differing levels of daily photosynthate (Duncan et al., 1965; Hume & Campbell, 1972).

By physiological maturity, adapted corn hybrids may have *harvest indices* (ratio of seed weight to total biomass) near 0.50. DeLoughery and Crookston (1979) observed an average harvest index of 0.32 over a number of environmental conditions, but found maximum harvest indices of 0.50 at each location. Harvest indices of 0.45 to 0.48 have also been observed in north Florida (Lorens et al., 1987).

II. EVAPOTRANSPIRATION

A. Geographical and Environmental Effects

The literature abounds with data of daily and seasonal water use of corn grown in various geographic regions and environmental conditions. Although numerous factors influence the rate at which water is lost from a corn canopy, evapotranspiration (ET) is most strongly dependent on environmental factors such as water vapor pressure and radiation, as well as crop factors which include the crop water status and the crop ground cover which is a function of leaf area. Seasonal water use is also closely related to the duration of leaf area. Soil physical properties and soil water contents directly affect evaporation from the soil and indirectly regulate crop transpiration through their influence on crop water status.

1. Daily Evapotranspiration

Soil evaporation is the major component of total ET during the early stages of crop growth. However, after leaf area begins to increase, crop transpiration gradually becomes the dominant component of ET. Daily water use rates of corn increase in parallel with increases in leaf area and light interception, and generally peak near complete closure of the crop canopy. Reddy et al. (1982) observed that maximum rates of daily ET increased through vegetative growth and peaked between 7 and 8 mm d^{-1} after full canopy closure before declining later towards maturity as leaves began to senesce. Similar results have been observed in north Florida. Higher plant populations result in more rapid increases in leaf area during early vegetative growth and thus cause higher rates of water loss. Under semi-arid dryland cropping environments, higher plant populations may result in increased water extraction during vegetative growth, leaving less water available for extraction during the more critical reproductive phases (Alessi & Power, 1976). Research in the early 1960s at Thorsby, AL, demonstrated that water loss from a corn canopy increased with crop development to a maximum of 7.6 mm d^{-1} at the dough stage before declining until maturity (Doss et al., 1962). Later studies at the same location revealed maximum daily water use rates averageing 6 mm d^{-1} during July and early August (Doss et al., 1970). Similar maximum daily ET rates of 5 to 7 mm d^{-1} were observed in a cooler environment in southern Alberta, Canada (Krogman et al., 1980). However, in a more semiarid environment at Clovis, NM, potential daily water use rates ranging from 7.5 to 9.8 mm d^{-1} were computed for corn (Abdul-Jabbar et al., 1983).

2. Seasonal Evapotranspiration

Not unlike daily ET rates discussed in the above section, seasonal corn ET has been extensively studied and is clearly a function of the length of the growing season and the environment in which the crop is grown. For example, simulated seasonal ET for corn grown in Minnesota has been computed to be as low as 375 mm (Morey et al., 1980), while seasonal ET at Bushland, TX, has been measured as high as 964 mm (Eck, 1986). Two studies in Kansas have shown that measured seasonal ET of corn ranges from 600 to 650 mm (Mayaki et al., 1976; Rosenthal et al., 1977), while data from northern Utah reveal seasonal ET for corn ranging from 500 to 600 mm (Retta & Hanks, 1980). Stewart et al. (1975) measured seasonal ET of near 650 mm at Davis, CA.

In the cooler environment of southern Alberta, Canada, Krogman et al. (1980) reported a seasonal ET of 436 mm for an early corn hybrid. Most of the research to examine water use rates of corn have historically been conducted in the more semiarid, western regions of the USA. However, increases in acreage of irrigated corn in the southeastern USA over the past 20 yr, coupled with the increasing knowledge of how to manage the sandy soils for maximum yields, have led to renewed interest in efficiently managing water for corn production. Although fewer studies of corn water use have been conducted in the more humid regions of the southeastern USA (Sadler & Camp, 1986), recent evidence has shown water use of well-watered corn to be similar at several southeastern locations. Average computed seasonal ET for corn grown for 5 yr in Georgia was 430 mm (Hook, 1985) while ET of corn grown in north Florida ranged from 430 to 440 mm (Hammond, 1981). Similarly, Doty (1980) computed seasonal corn ET to average 450 mm over a 3-yr period at Florence, SC. The lower seasonal ET compared to the western regions of the USA reflects the lower evaporative demands in the more humid environments. It has been clearly demonstrated that high corn yields can be produced with less water required for ET in the more humid environments, resulting in increased water use efficiency (WUE). Irrigation management, however, becomes a much more complex factor in the southeastern USA because of the erratic and sometimes intense rainfalls and the lower water-holding capacities of the sandy-textured soils.

B. Yield-Evapotranspiration Relationships

Using both yield and ET data obtained from well-watered, maximum-yield corn crops, Musick and Dusek (1980) compared seasonal *evapotranspiration water use efficiencies* (ET-WUE, defined as total grain yield divided by total seasonal ET) obtained in their studies to data from various other geographic locations. Seasonal ET-WUE for well-watered corn ranged from 0.012 Mg ha^{-1} per mm at Bushland, TX (Musick & Dusek, 1980), to as high as 0.022 Mg ha^{-1} per mm for corn grown in the southern Negev region of Israel (Hillel & Guron, 1973). Stewart et al. (1975) observed maximum ET-WUE of 0.017 at Davis, CA. Although maximum corn yields differ only slightly from semiarid to more humid environments, seasonal ET rates are considerably different. As a result of the lower seasonal ET in the humid regions, ET-WUE's as high as 0.030 Mg ha^{-1} per mm have been reported for corn grown in the southeastern USA (Hook, 1985). Clearly, ET-WUE is a function of the environmental conditions under which the experiment was conducted.

Methods of improving either irrigation or evapotranspiration WUE have been considered for numerous years. Obviously, water deficits that result in stomatal closure reduce plant transpiration and seasonal rates of water loss. Although predictions of ET rates by use of empirical equations have been widely used, similar predictions of ET under conditions of limiting soil moisture are more difficult (Ritchie, 1973). As a result, direct measurements of ET and corn dry matter and grain yields in field environments where various irrigation treatments have been imposed and subsequent analysis of the resultant ET-WUE relationships have been the focus of numerous studies during the 1970s and the early 1980s. Neither time nor space will allow thorough discussion of all such studies, but an attempt to summarize the important conclusions is included in the following paragraphs.

When water deficits are imposed somewhat uniformly throughout the growing season and result in lower seasonal ET, it is clear from the literature that corn yield is linearly related to the amount of ET (e.g., Hillel & Guron, 1973; Alessi & Power, 1976; Stewart et al., 1977; Morey et al., 1980; Musick & Dusek, 1980; Wenda & Hanks, 1981; Stegman, 1982; Abdul-Jabbar et al., 1983). For example, Wenda and Hanks (1981) observed a linear relationship between both corn dry matter and grain yield and ET. Linear regressions performed on ET-yield data resulted in r^2 values of 0.95 and 0.87 for corn dry matter and grain yields vs. ET, respectively. A linear relationship between grain yield and corn water use with an r^2 value of 0.91 was reported over ET's ranging from 400 to near 1000 mm (Abdul-Jabbar et al., 1983). However, the slope of the ET-yield relationship is dependent on many factors including the environmental conditions and the crop growth phase during which the stress is imposed. It should be emphasized that when water stress is imposed, ET-WUE generally declines when compared to that of the well-watered crop. Although many attempts have been made to improve ET-WUE by imposing water deficits uniformly through crop development or only at less sensitive crop growth stages, results overwhelmingly suggest that water deficits reduce ET-WUE. Only a few reports of increased ET-WUE with stress are evident in the literature. Reductions in single leaf WUE efficiencies for C_4 crop plants have been predicted from a theoretical analysis (Sinclair et al., 1975) and such conclusions have been consistently substantiated with field data from numerous experiments. Musick and Dusek (1980) observed 67 and 94% reductions in ET-WUE when water deficits were imposed and recommended that limited irrigation of corn was not a feasible practice on the Texas High Plains because of the sensitivity of corn to water stress. A similar conclusion was reached by Eck (1986).

The magnitude of the yield reductions associated with a given reduction in ET is dependent on the environment and the crop growth stages when stress is imposed. Sensitivity of corn to stress at various growth stages will be discussed in a later section.

III. WATER STRESS RESPONSE

A. Effects on Crop Development

Water stress imposed during vegetative development reduces expansive growth of stems and leaves and results in reduced plant height and lower LAI (Bennett & Hammond, 1983). Leaf expansion of corn is extremely sensitive to water deficits, with reductions in leaf expansion occurring well before leaf photosynthesis is reduced (Boyer, 1970; Acevedo et al., 1971).

Reduced internode lengths and plant heights are obvious stress symptoms associated with even mild water deficits which occur during vegetative development. Despite the sensitivity of leaf area expansion to water stress, the final number of leaves produced is generally unaffected by mild water deficits even though leaf appearance may be somewhat delayed (Bennett & Hammond, 1983). Except following severe water deficits, leaf production and expansion resume quickly after relief of stress. Musick and Dusek (1980) concluded that, if necessary, planned water deficits should occur only during early vegetative growth. Such timing would allow leaf production and expansion to occur after relief of the stress. Water deficits imposed during vegetative growth have their primary effect on reducing the size of the assimilatory structure.

Water deficits during silking, tasseling, and pollination are especially detrimental to yield and may result in the delay of silking (Barnes & Woolley, 1969; Hall et al., 1980), reduced silk elongation (Herrero & Johnson, 1981), and inhibition of pollination. Stresses imposed shortly before or after silking considerably reduce seed numbers (Musick & Dusek, 1980). Stress imposed later during the grain-filling period may cause increased leaf senescence, a shorter duration of the seed-filling period, increased lodging, and lower individual seed weights. The primary effect of water stress during the grain-filling period is a reduction in current photosynthate supply which is critical for optimum seed filling.

B. Stress Duration, Intensity, and Timing

The duration and intensity of stress is dependent on the environmental conditions, water-holding capacity of the soil, and crop growth stage during which water deficits occur. Several reports indicate that application of multiple stress cycles on corn do not result in additive yield reductions, suggesting that the crop may "condition" during stress events (Stewart et al., 1975; Harder et al., 1982). However, little information is available to suggest a mechanism for the "conditioning" response. Similarly, little research has been done to compare crop response to rapid imposition of stress which occurs on sandy soils with more gradual development of water stress which is associated with clayier soil types.

Research to determine crop growth stages of corn that are most sensitive to water deficits extends back a number of years. Short periods of wilting (1–2 d) during tasseling and pollination have been shown to result in large grain-yield reductions (Robins & Domingo, 1953). The extreme sensitivity of corn to water deficits during the pollination and seed development period has been substantiated by numerous other experiments (Claasen and Shaw, 1970; Stewart et al., 1975; Harder et al., 1982; Stegman, 1982). For example, Musick and Dusek (1980) computed the slope of ET-yield relationships frcm data presented by Robins and Domingo (1953) and concluded that the stresses they imposed during tasseling reduced grain yields by 0.049 Mg ha^{-1} per mm ET reduction, while water stress imposed during grain filling reduced yields by only 0.011 Mg ha^{-1} per mm ET. Generally, the sensitivity of crop growth stages to water deficits has been shown to decline in the following order: flowering and pollination > grain filling > vegetative stress (Musick & Dusek, 1980).

More recent work has focused on factors responsible for the extreme sensitivity of corn grain yield to water deficits imposed during flowering, pollination, and early seed set. Moss and Downey (1971) observed numerous abnormal embryo sacs in ears from corn plants that had been stressed during embryo sac formation. Even though heat stress was shown to affect both ear receptivity and pollen viability, water stress alone did not reduce pollen viability (Schoper et al., 1986) or in vitro pollen tube germination (Herrero & Johnson, 1981). Severe stress during the pollination period may reduce elongation and emergence of silks, even though pollen shedding continues normally. Herrero and Johnson (1981) concluded that water stress imposed at anthesis had a larger effect on the female compared to the male flower. Plant water deficits have also been shown to affect the water potential of silks more than that of the pollen (Westgate & Boyer, 1986a). Using reciprocal crosses with pollen and/or silks from water stressed plants, Westgate and Boyer (1986b) observed that the water potential of the silk was a determining factor regulating seed establishment. Somewhat surprisingly, they reported that the pollen tube grew within water-stressed silks, the egg sac was fertilized, but the embryo, endosperm, and seed coat did not develop beyond 2 to 3 d after fertilization. As a result, Westgate and Boyer (1986b) concluded that the failure of grain production as a result of water stress during flowering may be attributed to factors that arrest embryo development after pollination has occurred. Additional research will hopefully reveal more about the causal factors involved in the sensitivity of seed set to stress.

Even though stress during reproductive growth has been shown to reduce yields more than similar stresses imposed earlier during vegetative growth, it should be emphasized that relatively mild water deficits during vegetative growth have been shown to reduce ET-WUE (Eck, 1986) as well as significantly reduce grain yields (Bennett & Hammond, 1983; Eck, 1984). While the period surrounding silking and tasseling is commonly identified as the period most susceptible to water stress, it should be recognized that research suggests that grain yields are often reduced with crop water deficits that occur at any growth stage. Although much data exist in the literature describing the effects of water deficits on corn productivity, most studies have focused on relatively severe stress periods which cause large reductions in yield. With emphasis on high-input, high-yield production in the USA, it is suggested that additional information describing the effects of mild water deficits on corn growth, development, and yield should receive a high priority. Although it is clear that severe stresses reduce yields, improved precision of irrigation management practices requires additional information which could result from the imposition of mild water deficits in future studies.

C. Yield Losses

Several studies showing a linear relation between corn yield and seasonal ET are cited in section II of this chapter. Considering this commonly observed linear response, a general relation between yield loss and plant water stress can be derived from the slope of yield vs. ET and the reduction in ET as a result of water stress. Part of the apparent increased sensitivity of yield often associated with water stress during pollination and seed development may be associated with the higher rate of water use during these growth stages. A lower water demand during vegetative growth requires a longer water stress period as compared with reproductive growth to reduce yield by the same amount. For example, stress during vegetative growth with a daily ET of 4 mm d^{-1} would have to be applied twice as long to achieve the same yield loss as stress during reproductive growth with a daily ET of 8 mm d^{-1}. For illustration, if zero ET during stress is assumed, a 10-d stress period during reproductive growth would reduce ET by 80 mm and ET-WUE by 0.004 Mg ha^{-1} per mm, while a 10-d stress during vegetative growth would reduce ET by only 40 mm and ET-WUE by only 0.002 Mg ha^{-1} per mm. These calculations are based on the following assumptions: maximum seasonal ET = 480 mm and yield = $-10.5 + 0.045$ ET (Hook, 1985). These results illustrate the difficulty in imposing similar ET reductions during different growth stages to adequately classify growth stage sensitivities to water deficits.

Excessive water can reduce yield as well as water deficits. Irrigation and drainage both were shown to increase yields during a 5-yr study conducted in southern Illinois on clay pan soils (Walker et al., 1982). Irrigation and drainage alone increased corn yields by 0.8 and 2.4 Mg ha^{-1}, respectively, while irrigation and drainage in combination increased yields by 4.8 Mg ha^{-1}.

After analysis of data reported in the literature, Musick and Dusek (1980) concluded that "yield reductions associated with reduced water application are less severe (i) on deep soils having high water storage capacities where root extension into moist subsoil limits the severity of stress, (ii) where the severity of the stress at any one time is limited by distributing moderate stress periods throughout the season . . ., and (iii) where lower evaporative demand climate limits the severity of water stress."

IV. GROWTH AND SOIL WATER

A. Rate of Growth

1. Crop Growth Rates

After developing through a lag phase of growth before a complete leaf canopy is produced, corn achieves a near linear crop growth rate until approaching maturity, provided the environment is relatively stable and adequate nutrients and water are available. Soil water deficits may significantly reduce crop growth rates of corn as illustrated by data from Lorens et al. (1987) who measured average crop growth rates of 32.5 g m^{-2} per d for an irrigated treatment and 22.9 g m^{-2} per d for a rain-fed treatment in Florida during the linear phases of growth. A similar crop growth rate of 34.0

g m^{-2} per d was observed for irrigated corn grown in north Florida (Valle Melendez, 1981). Obviously, the reduction in growth rate will be dependent on the severity and duration of the water stress period.

2. Kernel and Seed Growth Rates

Similar to total crop growth rates, the seed accumulates dry matter linearly through much of the grain-filling period. Despite variations in environmental conditions, the rate of seed growth is remarkably linear unless severe crop stress is imposed during the period of grain filling. The rather constant seed growth rates are maintained by remobilization of photosynthate from other plant parts to the seed during periods when the current supply of photosynthate is insufficient to meet the demand of the growing seeds (Duncan et al., 1965). Water deficits have been shown to decrease the rate of the linear increase in seed weight and may significantly reduce the duration of the seed-filling period. For example, seed growth rates averaged over two corn hybrids were increased from 21.4 to 30.5 g m^{-2} per d with irrigation in a study conducted in north Florida (Lorens et al., 1987).

Individual kernel growth rates ranging from 6.2 to 9.7 mg kernel^{-1} per d have been reported for 20 inbred corn lines (Carter & Poneleit, 1973), while Jones and Simmons (1983) observed kernel growth rates (averaged over a 3-yr study) of 6.5 mg kernel^{-1} per d. Rates as high as 10 mg kernel^{-1} per d have been reported (Poneleit & Egli, 1979). Irrigation increased kernel growth rates from 7.9 to 9.9 mg kernel^{-1} per d (Lorens et al., 1987).

B. Growth Stage Responses

1. Vegetative

Vegetative growth provides the leaf area and root system necessary for maintaining the crop throughout the remainder of the growing season. Expansion of the root system through the soil profile continually provides new sources of nutrients and water for the developing crop. Photosynthate stored during vegetative growth has been shown to provide nearly 10% of the dry matter that eventually accumulates in the grain (Simmons & Jones, 1985), and under stressful environments this contribution undoubtedly increases. Presilking assimilates may be quite important for optimum grain development.

2. Reproductive

By silking, the root system has completed most of its development and the crop extracts water and nutrients from a fixed soil volume throughout the grain-filling period. Seed yield is dependent on pollination, seed set, and dry matter accumulation in the seeds. McPherson and Boyer (1977) imposed water deficits that reduced photosynthesis to near zero during the entire period of grain filling. Grain yields were reduced by only 47 to 69% of the well-watered control as a result of remobilization of assimilates accumulated before initiation of the stress period. They concluded that grain yield was dependent on total dry matter accumulation during the entire season rather than that accumulated only during grain filling. Similar results were reported by Jurgens et al. (1978). With water stress imposed at silk emergence, early grain fill, and mid-grain fill, yield losses resulted from a cessation of silk and ear development, decreased seed size and number, and decreased seed size, respectively (Westgate & Boyer, 1985). When stress was applied during silk emergence, there was little remobilization from other plant parts and carbohydrates remained at low levels until maturity, suggesting that carbohydrate supply at the time of silk emergence may be critical to yield.

V. PLANT NUTRITION

A. Seasonal Uptake

Nutrient uptake must be adequate for full expression of plant response to irrigation. Nitrogen is the nutrient that most often limits corn yield. Uptake of N determined from plant samples taken shortly after silking was significantly ($P < 0.01$) correlated with grain yield of irrigated corn during a 2-yr test in which rainfall varied widely between years in Florida (Rhoads & Stanley, 1984). Nitrogen uptake accounted for 98% of the variation in yield from yield vs. uptake data in the linear (0–3 kg of N per 1000 plants) portion of the response function. Fertilizer N in excess of 3 kg per 1000 plants (220 kg N ha^{-1} for 7.4 plants m^{-2}) resulted in no yield increase. Both yield and N uptake were dependent on amount of fertilizer N applied. Eck (1984) reported that the adequate level of fertilizer N for nonstressed irrigated corn was 210 kg ha^{-1} on a Pullman clay loam (fine, mixed, thermic Torrertic Paleustolls) in Texas. At lower N rates, N deficiency limited yields more than water stress. With adequate N, water was the main yield-limiting factor. In Iowa, N content of grain increased while yield was reduced by 30% by water stress in corn during July and August (Harder et al., 1982). Another Texas study (Onken et al., 1985) considered the importance of residual NO_3^- in the soil prior to fertilization. They used a power function to predict the N requirements of irrigated corn for various levels of yield and residual NO_3^-. The required fertilizer N predicted for 12.0 Mg of grain ha^{-1} was 220 kg ha^{-1} with a soil NO_3^--N level of 20 kg ha^{-1} and 170 kg ha^{-1} with 70 kg NO_3^--N ha^{-1}. The soil had a clay loam texture and NO_3^- was measured in the 0- to 0.15-m depth. Fertilizer N requirements for irrigated corn on sandy soils in Florida were estimated to be 180 kg ha^{-1} for a 12.0 Mg ha^{-1} grain yield and 270 kg ha^{-1} for a 14.4 Mg ha^{-1} yield (Rhoads & Stanley, 1984).

Phosphorus uptake during the growing season is highly dependent on nutrient balance since a deficiency of other elements can limit total uptake even though P content may be high in plants with reduced dry matter yield. Seasonal P uptake for maximum yield of irrigated corn in Florida averaged 52 kg ha^{-1} during five growing seasons, with a range of 36 to 76 kg ha^{-1} (F.M. Rhoads, 1983, unpublished data). Seasonal P uptake in mature shoots of irrigated corn in South Carolina was 51 kg ha^{-1} for 13.0 Mg ha^{-1} grain yield (Karlen et al., 1985). Application of 22 kg P ha^{-1} produced 95% of maximum irrigated corn yield in Nebraska with soil test P (Bray & Kurtz) before fertilizer application of 5.4 mg kg^{-1} (Rehm et al., 1983). Ear leaf P was 2.56 g kg^{-1} for 95% of maximum grain production and 2.20 for 95% of maximum silage production. Irrigated corn in Florida contained 45 kg P ha^{-1} in the shoots shortly after silking for a grain yield of 14.5 Mg ha^{-1} while yield was no higher for P uptake of 56 kg ha^{-1} (Rhoads & Stanley, 1984). Reduced P uptake at low temperature is primarily because of reduced root growth rate (Mackay & Barber, 1984). Parameters describing the movement of P to the root surface and P uptake at the root surface appear to play only minor roles in causing reduced P uptake by corn at low temperatures.

Seasonal K uptake in mature shoots of irrigated corn in South Carolina was 257 kg ha^{-1} for yields of 13.0 to 14.0 Mg ha^{-1} (Karlen et al., 1985). There was no response to K fertilizer applied to corn in Nebraska on an irrigated sandy soil where soil test K was 88 mg kg^{-1} (approximately 175 kg ha^{-1}) (Rehm & Sorensen, 1985). Potassium contained in shoots of irrigated corn in Florida was 238 kg ha^{-1} for a grain yield of 9.7 Mg ha^{-1}, and 250 and 290 kg ha^{-1} for grain yields of 12.7 and 14.5 Mg ha^{-1}, respectively (Rhoads & Stanley, 1981, 1984).

Calcium contained in shoots of high-yielding irrigated corn was 33 kg ha^{-1} in South Carolina (Karlen et al., 1985) and 27 kg ha^{-1} in Florida (Rhoads, 1981a). Magnesium content at these locations was 39 and 52 kg ha^{-1}, respectively. No response to Mg fertilization was observed for irrigated corn in Nebraska with exchangeable soil Mg of 54 mg kg^{-1} (Rehm & Sorensen, 1985). Sulfur applied to 30-d-old corn seedlings appearing S deficient increased S in plant tissue at 55 d but had no effect on yield (Mitchell & Gallaher, 1980). However, Mg fertilization increased grain yield of short season corn by 23% but had no effect on full-season corn. Seasonal uptake of Zn by irrigated corn in South Carolina ranged from 0.4 to 0.5 kg ha^{-1} (Karlen et al., 1985). Maximum corn herbage yield occurred at a Zn application rate of 4.5 mg kg^{-1} and Mehlich I extractable Zn of 6.0 mg kg^{-1} (Blue et al., 1982).

These data suggest that nutrient requirements of irrigated corn are not extremely high in comparison to nonirrigated corn because it appears that grain yields > 12.0 Mg ha^{-1} can be produced with available N–P–K levels in the soil of 250, 50, and 275 kg ha^{-1}, respectively. Efficient irrigation management maximizes nutrient utilization efficiency and increases yield potential of corn. However, overirrigation can result in leaching of nutrients and contribute to groundwater contamination in some regions.

B. Rate of Uptake

A knowledge of nutrient uptake rates is helpful when planning nutrient application schedules for irrigated corn. Generally, nutrient uptake rates are proportional to plant growth rate. In most cases, highest uptake rates occur

Table 19-1. Nutrient content of irrigated corn (grain yield of 14.6 Mg ha^{-1}) at each of four growth stages (Rhoads, 1981a).

Weeks after emergence	Growth stage	Nutrient content					
		N	P	K	Ca	Mg	Zn
		kg ha^{-1}					
4	18-leaf	31	3	25	1	3	0.06
7	12-leaf	110	16	159	7	18	0.26
9	Silk	199	32	312	17	30	0.26
14	Mature	348	60	357	27	52	0.31

between the eight-leaf and silking stages of growth. However, for some nutrients the uptake rate continues to be quite high during grain fill while others are taken up at a small fraction of the presilking rate.

Nutrient uptake rates can be calculated from the data in Table 19-1 for several nutrients. For example, N and P were accumulated during the grain-filling period at 67 and 70% of the presilking rate while K uptake during grain filling was only 12% of the presilking rate. Potassium uptake by corn on a Grossarenic Paleudult in Florida ceased after silking, although three applications of K fertilizer were made during the filling period (Rhoads & Stanley, 1981).

For satisfactory plant development the soil must be able to supply at least 0.9 g N plant^{-1} per wk and 0.1 g P plant^{-1} per wk (Bar-Yosef & Kafkafi, 1972). This is equivalent to 54 kg N ha^{-1} per wk and 6 kg P ha^{-1} per wk for 60 000 plants ha^{-1}. Maximum N uptake rates for irrigated corn in Florida were 38.5 kg ha^{-1} per wk for 60 000 plants ha^{-1} (F.M. Rhoads, 1978, unpublished data) and 58.1 kg ha^{-1} per wk for 146 000 plants ha^{-1} (Rhoads, 1981a). Nitrogen uptake appears to be more limiting in Florida than P uptake because the ratio of P to N was 0.18 while only 0.11 in Israel. As more data become available, similar calculations can be made for other nutrients. Nutrient supply to irrigated corn depends upon nutrient availability and root plus shoot growth. Therefore, soil test data must be interpreted in terms of uptake rate as well as total amount available.

C. Scheduling Nutrient Application

Nutrients that are relatively immobile in the soil can be applied most economically prior to planting. The application rates of these nutrients should be based on soil test results and grower experience. Phosphorus, Ca, Mg, Zn, Fe, and Mn are not easily leached from most soils. Potassium does not leach readily in fine-textured soils but significant losses of K can occur in sandy soils. Generally, Ca and Mg should be applied as liming material between growing seasons.

Fertilizer application through irrigation systems has become popular in many regions of the USA. Information is available on injection formulas and equipment needed to inject fertilizer materials into irrigation systems (Harrison & Rhoads, 1978). The use of tractor-mounted fertilizer applicators is eliminated and use of the irrigation facility is increased while irrigating. Since nutrient application through irrigation systems is essentially a broadcast procedure, sidebanding of mobile nutrients in humid regions may be best accomplished while plants are small. Nutrients most successfully applied through irrigation systems are N, S, B, and K on sandy soils. It is more efficient to apply K preplant on fine-textured soils.

All N should be applied preplant on fine-textured soils, since no yield advantage was observed in Kansas for split applications or in season application over preplant application (Anderson et al., 1982). This conclusion is also supported by research conducted by Russelle et al. (1981) in Nebraska. However, on coarse-textured sandy soils, results are quite different. Corn responded to method of N application in Nebraska on a soil with a loamy fine sand surface and a fine sand subsoil (Rehm & Wiese, 1975). However, there was no response on a soil with a loamy fine sand surface, a silt loam layer at 0.5 to 0.7 m depth and fine sand below the silt loam layer. Highest yield was produced with 182 kg N ha^{-1} sidedress and 70 kg N ha^{-1} applied through the irrigation system in three equal applications. In Florida, biweekly applications of fertilizer increased grain yield of irrigated corn on a sand by 39% above that obtained with a preplant and two sidedress applications (Rhoads et al., 1978).

The importance of adequate nutrients during the seedling to eight-leaf stage was shown by delaying fertilizer application until 6 wk after planting in Florida (Stanley & Rhoads, 1977). Yield was reduced by 20% as a result of the delay, although more than 300 kg N and more than 200 kg K ha^{-1} were applied along with about 100 kg P ha^{-1} in three equal applications. Nitrogen applied at 200 kg ha^{-1} produced 98% of maximum yield obtained with 336 kg N ha^{-1}.

Research results and experience suggest that mobile nutrients should be scheduled as follows for worst case leaching conditions: 17% of total nutrients at each of six applications, starting when plants emerge and at 20, 30, 40, 50, and 60 d after emergence. Results of research in Florida suggest that 3 kg N 1000 plants^{-1} is adequate for nonprolific cultivars on coarse-textured soils (Rhoads & Stanley, 1984).

D. Nutrient Content Vs. Grain Yield

The ratio of grain to nutrient content of mature corn shoots gives some idea of nutrient requirements for a specific yield goal. It is also an index of nutrient utilization efficiency (NUE). When the ratio is high, all other growth factors are near optimum relative to the nutrient in question. Low values indicate that some other factor is limiting yield. The ratio of grain to P uptake of irrigated corn ranged from 254:1 to 329:1 in Florida (Rhoads & Stanley, 1981) while the ratio of grain to K uptake ranged from 43:1 to 51:1. Grain yield divided by N uptake of corn ranged in value from 41:1 to 85:1 in a 2-yr study in Florida (Rhoads & Stanley, 1984). There was a tendency for values to be constant between 1 and 3 kg of fertilizer N 1000 plants^{-1} and to decrease at N levels above that required for maximum yield. Grain to N content ratios ranged from 48:1 to 78:1 in a Texas study (Eck, 1984). There was a tendency for values to be constant after maximum yield was attained in unstressed plants. However, N content did not increase at higher fertilizer N levels after yield reached the maximum. Plants continued to increase in N content with increasing fertilizer N after maximum yield was attained in the Florida study.

Regressing yield on nutrient application rates only may fail to identify nutrient availability as a limiting factor, whereas a regression of yield on uptake would properly identify the problem. If maximum yield occurs before maximum nutrient uptake, then availability of added nutrients is not a yield-limiting factor. Some other factor such as genetic potential or lack of another nutrient has become limiting. This kind of data should be helpful in identifying unknown yield-limiting factors as well as verifying that certain factors are not limiting. Water stress does not limit nutrient availability with proper irrigation practices.

Grain to mature shoot nutrient ratios of 60:1 for N, 250:1 for P, and 50:1 for K would require uptake levels of 200, 48, and 240 kg ha^{-1}, respectively, for N, P, and K to produce 12.0 Mg ha^{-1} of grain.

VI. CULTURAL PRACTICES

A. Effects of Tillage on Nutrient and Water Availability

Variation in tillage depth and intensity can increase or decrease rooting depth and rooting volume. Nutrient and water availability are directly related to the extent of the plant root system with or without irrigation. In arid regions, increasing rooting depth in soils with high water-holding capacity can significantly increase yield in the absence of irrigation and make scheduling more flexible where irrigation is practiced. Tillage influences water intake of the soil with surface irrigation. Increased rooting depth can enhance recovery of mobile nutrients in humid regions where leaching is likely to occur.

1. Subsoiling

Subsoiling without irrigation on a soil with a tillage pan produced a mean corn yield of about 7.6 Mg ha^{-1} during a 3-yr test in North Carolina (Cassel & Edwards, 1985). Irrigation alone produced a mean yield of about 8.1 Mg ha^{-1} while subsoiling plus irrigation produced the highest 3-yr mean yield (9.8 Mg ha^{-1}). The 3-yr mean yield without subsoiling or irrigation was 2.6 Mg ha^{-1}.

Yield increases from irrigation ranged from 0.5 to 1.5 Mg ha^{-1} in a year with 568 mm of rain during the growing season in Florida (Rhoads & Mansell, 1986). The average response to subsoiling was 1.4 Mg ha^{-1} with and without irrigation. Irrigated plots produced about 10 times as much grain as nonirrigated plots when rainfall was 262 mm but there was no response to subsoiling. Yield of nonirrigated plots for the dry year was 1.2 Mg ha^{-1} and 10.2 Mg ha^{-1} for the irrigated plots. Therefore, subsoiling in row increased yield only when rainfall was sufficient to cause downward movement of nutrients. Subsoiling on soils with low water-holding capacity appears to be more effective when stress periods are short and frequent than when stress is severe over long periods (F.M. Rhoads, 1984, unpublished data).

2. Moldboard Plowing

Ohio researchers concluded that corn yields were remarkably insensitive to tillage (Van Doren & Triplett, 1976). On a Mollic Ochraqualf soil, a no-tillage treatment averaged 0.98 Mg ha^{-1} or 13% lower yield than the average of plowed treatments for 10 yr of continuous corn. Yields were the same for all tillage treatments in other rotations. No-tillage averaged 0.75 Mg ha^{-1} or 10% greater yield than the average of plowed treatments on a Typic Fragiudalf soil for 11 yr of continuous corn or corn-soybean rotation. In an earlier test, mulch cover produced three times as great a yield effect as any single tillage variable (Van Doren & Triplett, 1973).

Corn yield in Florida was higher on a Typic Paleudult for a moldboard plowing treatment (12.9 Mg ha^{-1}) than for conventional tillage (9.9 Mg ha^{-1}) or minimum tillage plus subsoiling in row (10.9 Mg ha^{-1}) with no irrigation during a year (1985) with favorable rainfall distribution (F.M. Rhoads, 1985, unpublished data). However, in a dry year (1986) the yield from a moldboard plowing treatment was no better than from in-row subsoiling. There was no yield response to tillage under irrigation in either 1985 or 1986.

3. Chiseling

Chisel plowing is often used in the Midwest to incorporate fertilizer and crop residue deeper in the soil profile. This is one of the key factors to which the record corn yield of 23.2 Mg ha^{-1} (370 bu acre^{-1}) in Illinois was attributed (Potash and Phosphate Institute, 1985–1986). There was no difference in yield of corn between subsoiling and chisel plowing in South Carolina during 3 yr with no irrigation (Camp et al., 1984). However, in-row subsoiling was better than chiseling two out of 3 yr with irrigation on Arenic and Typic Paleudults.

Researchers in Kentucky reported that higher grain yields were associated with a greater depth to fragipan on a Zanesville soil (fine-silty, mixed, mesic Typic Fragiudalfs) (Frye et al., 1983). However, on a deep well-drained Maury silt loam (fine, mixed, mesic Typic Paleudults), grain yield and P, Ca, and Mg uptake were unaffected by tillage (Blevins et al., 1986). Potassium uptake was strongly related to the surface stratification of K in the no-till soil environment.

B. Planting Rate, Population

Planting rate is generally 10 to 15% greater than the desired population to minimize stand loss due to reduced germination or weather conditions.

Researchers in South Carolina reported near-maximum yields with a corn population of 7.1 plants m^{-2} (Karlen & Camp, 1985). They also reported that a twin row configuration (inter-row spacings of 0.30 and 0.96 m) increased grain yield an average of 0.64 Mg ha^{-1} (10 bu acre^{-1}). Maximum yields in Florida were obtained with 7.2 plants m^{-2} in 0.45-m rows (Stanley & Rhoads, 1975). Available equipment cannot harvest corn in rows 0.45 m apart; however, twin rows spaced 0.30 m apart on 0.90 m centers can be harvested with available equipment. Twin rows consistently produced higher yields than single rows spaced 0.90 m apart but not as high as 0.45-m row spacing. The average yield increase for twin rows over single rows was 1.5 Mg ha^{-1}. Suggested practices for producers are a population range of 7 to 8 plants m^{-2} and a row spacing of 0.76 m or twin rows with 0.30-m spacing on 0.90-m centers. Lower plant populations may improve yield in environments with limited water, and higher plant populations may be desirable in northern-most regions of the Corn Belt.

C. Hybrids and Varieties

Yields in the State of Florida field corn variety tests ranged from 8.8 Mg ha^{-1} (140 bu acre^{-1}) to 15.6 Mg ha^{-1} (248 bu acre^{-1}); however, 45 to 63 varieties produced more than 12.5 Mg ha^{-1} (Horner et al., 1985). Average yields over 3 yr ranged from 9.3 to 12.8 Mg ha^{-1} for 24 hybrids in irrigated tests. Nonirrigated yields ranged from 2.6 to 5.6 Mg ha^{-1} (42–89 bu acre^{-1}). High-yielding hybrids are available but care must be exercised in choosing hybrids because some do not have the yield potential to be profitable for irrigated corn production.

Corn hybrids are generally classified into three maturity groups, i.e., early season, mid-season, and full season. The geographic range of adaptability must also be considered. Early season hybrids mature before mid- or late-season hybrids as implied by the term "early." There are also prolific and nonprolific hybrids. Prolific varieties produce more than one ear per plant.

Under irrigation, early hybrids responded more to spacing variables and produced higher yields than full-season hybrids (Stanley & Rhoads, 1975). Full-season hybrids reached a yield peak at lower populations than early maturing hybrids. When population remained constant, both types produced highest yields at a spacing more nearly equidistant.

The number of ears per plant decreases as population increases for prolific hybrids, and ear size decreases with population increase for nonprolific hybrids (Rhoads & Stanley, 1979). Fortunately, ear size does not change at the same rate as population. Yield potential for nonprolific hybrids is about the same as for prolific hybrids when population is adjusted to give the same ear number ha^{-1}.

VII. IRRIGATION MANAGEMENT

A. Scheduling

The main objective of irrigation scheduling is to manage irrigations for greatest effectiveness. Water must be applied frequently enough to avoid plant water stress and in amounts adequate to recharge the soil to the depth required by local conditions. Overirrigation during the growing season may eliminate crop water stress but will also lessen irrigation efficiency and yield response to other management practices such as fertilization, planting date, population, and weed control. Inadequate irrigation, on the other hand, results in crop water stress and less yield response to other management factors, even though irrigation efficiency remains high because runoff and deep percolation are reduced.

1. Plow Layer Management

One approach to irrigation management is to recharge only the plow layer with each irrigation (Rhoads, 1981b). This method is especially suited to humid regions because it leaves part of the root zone unrecharged to reduce percolation loss when rain occurs soon after irrigation.

Many soils have extreme differences in water-release characteristics between the plow layer and the subsoil. In a soil with a loamy fine sand plow layer and a sandy clay loam subsoil, about 20 mm of water can be released from the plow layer in the soil-water pressure range of -5 to -20 kPa while little is released from the subsoil. The more favorable water condition in the plow layer promotes more rapid root growth in that horizon. Corn plants have limited root systems in the early stages of growth and are completely contained within the plow layer. The use of tensiometers to monitor soil water extraction by corn roots in a soil with a tillage pan showed that plants were 50 d old before water was extracted below the 0.3-m depth.

Table 19-2. Estimated water use of corn during various stages of growth and irrigation frequency for a loamy fine sand plow layer scheduled for irrigation at a soil-water pressure of -20 kPa.†

Growth stage	Water use, mm d^{-1}	Days after planting	Days between irrigations
Seedling	1.5	0–20	20
0.12–0.25 m	2.3	20–30	13
0.25–0.50 m	3.8	30–40	8
0.50–1.25 m	5.1	40–50	6
1.25–2.00 m	5.3	50–60	6
200 m–silking	6.4	60–70	5
Silking–grainfill	8.4	70–100	4
Grainfill	6.4	100–110	5
Maturity	5.8	110–120	5

† Recharge of 30.5 mm calculated from release characteristic and 0.3-m depth.

The plow layer is easily defined (by depth of plowing) and is uniform due to the mixing action of tillage. Research in Alabama indicated that 72 to 88% of water used by corn came from the soil layer above 0.5 m (Weatherly & Dane, 1979). The amount of water to apply per irrigation can be determined from the water-release characteristic of the plow layer and the allowable depletion level of available water (see Section III of this monograph for more details). Irrigation frequency is easily determined by dividing the amount of water per application by the daily water use rate of the crop or by monitoring soil water content of the plow layer. Water use rate increases as crop growth continues (Table 19-2). Therefore, irrigation frequency increases as the crop grows toward maturity.

2. Pan Evaporation, Water Budget, and Tensiometers

Evaporation pans are used to estimate evapotranspiration or total water use by the crop. Irrigation can be scheduled from crop water use data and amount of water required to recharge the root zone at a selected depletion level. Pan evaporation data can also be used in water budget scheduling techniques. Use of evaporation pans is discussed in chapter 15 of this monograph. Water budget irrigation scheduling is covered in more detail in chapter 17. Tensiometers are discussed in more detail in chapter 6. Yield response of corn to various irrigation scheduling techniques and comparison between scheduling techniques will be discussed in the following paragraphs.

Two irrigation scheduling techniques were compared in a 3-yr study in North Carolina (Cassel et al., 1985). A computer-based water balance model was compared with scheduling by tensiometers. Tensiometers were placed 0.23 m from crop rows and sensor cups were at a depth of 0.25 m. Irrigation was applied when the mean soil water pressure (SWP) of the tensiometers from four plots reached −60 kPa in 1979 and −40 kPa in 1980 and 1981. The average amount of water applied per year was 219 mm for the computer-based water balance model and 179 mm for tensiometer scheduling. The amount of water applied per irrigation was usually 20 to 30 mm. Average yields were 2.58 Mg ha^{-1} for the rain-fed treatment, 7.73 Mg ha^{-1} for the water balance model, and 8.14 Mg ha^{-1} for the tensiometer irrigation schedule. Yield difference between the two irrigation scheduling techniques was not significant ($P < 0.05$). Research in South Carolina also revealed no difference in yield between a computer-based water balance model and tensiometers for scheduling irrigation of corn (Karlen & Camp, 1985; Camp et al., 1985).

Yield response to irrigation was different between soils for the deep placement of control tensiometers in a 5-yr study on corn in Georgia (Hook, 1985). Both deep (0.45–0.60 m) and shallow (0.15–0.30 m) placement of tensiometers to trigger irrigation was effective on the deep sand soil; however, when rainfall during the vegetative growth period was limited, the deep placement resulted in periods of plant stress due to dry topsoil and lower crop yields. Deep placement of control tensiometers resulted in fewer irrigations in years with more rainfall during the vegetative period. Available water-holding capacity was greater for the soil with a sandy clay loam subsoil. Timeliness was less critical on this soil and deep placement of tensiometers did not result in fewer irrigations. A linear response for yield vs. SWP (-20 to -60 kPa) was observed in Florida where irrigation was scheduled with tensiometers in the row with sensor cups at 0.15-m depth (Rhoads & Stanley, 1973).

3. Limited Water Environment

As previously discussed in this chapter a strong linear correlation between ET and corn dry matter and grain yield has been demonstrated in many studies. Yield vs. ET equations have a negative value for the Y intercept indicating that a threshold ET must be reached before any grain can be produced. The threshold ET range was 200 to 340 mm in Utah studies (Wenda & Hanks, 1981) and 230 to 240 mm in Georgia studies (Hook, 1985). Such data are useful for irrigation planning in limited water environments. For example, if sufficient water is stored in the soil profile from rainfall to meet or exceed the threshold ET for grain production, irrigation water use efficiency (IWUE) will be maximized. Whereas, if total ET is supplied by irrigation, IWUE is minimized.

The effect of withholding irrigation on grain yield of corn in limited water environments to conserve water is dependent on atmospheric demand. Post-silking water stress reduced yield as much as 33% in rain shelter plots at Ames, Iowa (Harder et al., 1982). Variation in plant water stress between years was attributed to differences in mean daily temperatures in Texas (Eck, 1984). Yields were reduced an average of 1.2% of nonstressed yield per day of stress imposed during grain filling. Models that predict ET should be effective in predicting yield response to irrigation management where other growth factors (fertilizer and weed control) are not limiting.

B. Irrigation Water Use Efficiencies

The yield increase per unit of irrigation water applied is useful information for irrigation planners. Apparent irrigation water use efficiency (AIWUE) can be calculated by

$$\text{AIWUE} = (Y_{irr} - Y_{dry})/W_{irr} \qquad [1]$$

where Y_{irr} and Y_{dry} are the grain yields of irrigated and dryland treatments, respectively, and W_{irr} is amount (mm) of irrigation water applied (Cassel & Edwards, 1985). These researchers reported AIWUE values with sprinkler irrigation in North Carolina ranging from 0.003 to 0.036 Mg ha^{-1} per mm. Values varied with years and tillage methods. Values of AIWUE calculated from furrow irrigation data of Sorensen et al. (1980) ranged from 0.014 to 0.020 Mg ha^{-1} per mm. The maximum AIWUE from the Utah data (Sorensen et al., 1980) was about 0.55 of the one found in North Carolina. However, there are many possible sources of variation including method of irrigation plus soil and climatic differences. Values of AIWUE have been found to vary with soil types, years, and irrigation treatments in Florida (Rhoads & Stanley, 1975; F.M. Rhoads, 1986, unpublished data).

Seasonal irrigation water use efficiencies are commonly less than seasonal ET water use efficiencies. The average slope of yield vs. adjusted ET (AET) in Georgia was 0.045 Mg ha^{-1} per mm, while the maximum slope was 0.085 Mg ha^{-1} per mm (Hook, 1985). A maximum yield of 21 Mg ha^{-1} (this yield has been achieved both in research plots and on farm tests) was predicted from the equation

$$Y = -20.5 + 0.085 \text{ AET} \qquad [2]$$

where Y = corn grain yield in Mg ha^{-1} and AET = adjusted evapotranspiration of 488 mm. The maximum AIWUE calculated from the Georgia data was 0.036 Mg ha^{-1} per mm. The highest reported AIWUE (0.046 Mg ha^{-1} per mm) found in the literature is from Florida (Robertson et al., 1981). Their highest yield was 7.5 Mg ha^{-1}, while much higher yields have been reported with AIWUE values in the 0.020 to 0.030 Mg ha^{-1} per mm range.

VIII. OTHER ASPECTS

A. Economic Decisions

The first decision required of irrigation planners is whether or not to install a crop-watering system. Increased yield in response to irrigation must have a higher value than the cost of irrigation. Yield increase required to pay irrigation costs will vary with installation costs, energy costs, and corn prices. A study in Illinois indicated that a yield response to irrigation of 4.5 to 5.6 Mg ha^{-1} (72–90 bu acre^{-1}) was required to be financially feasible (Schoney & Massie, 1981). Soil type and rainfall distribution influence corn yield response to irrigation. A North Dakota economic analysis indicated that supplemental irrigation on soils with water-holding capacities > 250 mm is not profitable (Wilson & Eidman, 1983). Rainfall probability is the deciding factor for installing irrigation systems on fine-textured soils. However, dry periods of 2 wk or more can cause significant yield losses on sandy soils in humid regions. Extended droughts can cause corn production on sandy soils to be near zero. For example, in a 1986 water management experiment in Florida, corn yield on a sand without irrigation was 0.6 Mg ha^{-1} (9 bu acre^{-1}). However, with irrigation the yield was 13.9 Mg ha^{-1} (221 bu acre^{-1}) (F.M. Rhoads, 1986, unpublished data).

A production budget similar to one published by Eason and Rhoads (1982) should be prepared before the production period begins in order to estimate costs, yields, and returns. Costs must be known before the producer can effectively calculate an asking price based on the desired profit margin above the break-even price.

There are several types of irrigation systems available including center pivot, lateral move, cable tow, and furrow distribution. The choice of systems will depend on field size, shape, and topography. An economic analysis of irrigation data from Florida showed that net returns per hectare were greater with a cable tow system when using a 5-d irrigation cycle rather than a 3-d cycle (Eason & Rhoads, 1983). However, increasing the length of the irrigation cycle to 7 d reduced net returns per hectare by a factor of 0.10. More hectares were irrigated by cable tow with longer irrigation cycles. A 3-d cycle was more profitable with a center pivot because of the fixed land area irrigated, regardless of length of irrigation cycle.

B. Yield Goals

Record yields of above 19.9 Mg ha^{-1} (300 bu acre^{-1}) have been reported from both research plots and farm trials. However, the majority of irrigators should expect grain yields of irrigated corn to be in the range of 11.0 to 15.0 Mg ha^{-1} (175–240 bu acre^{-1}) based on a survey of the literature. Response to irrigation will vary from year to year but in years with favorable rainfall (unirrigated yields of 9.4–12.5 Mg ha^{-1}), yield increases in response to irrigation in the range of 2.5 to 3.8 Mg ha^{-1} (40–60 bu acre^{-1}) have been observed in Florida (F.M. Rhoads, 1985, unpublished data).

C. Production Efficiency

There are other inputs besides water that increase production costs of irrigated corn. Most prominent are fertilizer and seed. Higher levels of N–P–K and plant population will be needed to produce the larger yield expected with irrigation. These additional costs cannot be ignored when evaluating production efficiency associated with irrigation management.

IX. SUMMARY

Maintaining efficient management of water resources while maximizing corn productivity is best accomplished through a better understanding of the many interactions among the phases of crop development, soil physical properties, and water and nutrient characteristics—as well as the environmental factors that affect crop water use and yield formation. Both seasonal and daily ET are dependent on environmental factors and on crop characteristics, including the crop water status and light interception. Daily rates of ET increase in parallel with increases in leaf area and light interception and generally peak near complete closure of the crop canopy. Seasonal ET's of corn have ranged from 375 mm in Minnesota to as high as 964 mm in Texas, with intermediate values of 420 to 440 mm in the more humid southeastern USA.

Expansive growth is quite sensitive to water deficits imposed during vegetative growth, and reductions in development of the photosynthetic factory as well as the root system may result. Water deficits imposed during silking and tasseling are generally assumed to be most detrimental to corn yields, although stresses at any growth stage may significantly reduce grain yields. Stresses during reproductive growth often result in increased leaf senescence and reductions in both the rate and duration of seed filling.

Numerous studies have demonstrated a near-linear response of yield to seasonal ET, especially when stresses are imposed somewhat uniformly throughout the growing season. Other studies have shown that the yield reduction per unit reduction in ET is greater when stress is imposed near flowering compared to stresses imposed during vegetative stress. The magnitude of the yield reductions associated with a given reduction in ET is dependent on the environment and the crop growth stage when stress is imposed. Although many irrigation strategies have been imposed in attempts to improve the WUE of corn, results generally suggest that stress imposed at any growth stage reduces ET-WUE.

Efficient irrigation management must be coupled with adequate nutrient availability in the soil. Rates of uptake of major nutrients in response to the developmental stages of corn are well documented and can be helpful for efficient scheduling of nutrient applications. Grain yields of >12.0 Mg ha^{-1} can be produced with available N–P–K levels in the soil of 250, 50, and 275 kg ha^{-1}, respectively, and plant populations of near seven plants m^{-2}. Irrigation maximizes nutrient utilization efficiency and increases the yield potential. Locally adapted, high-yielding hybrids should be selected to ensure that genetic potential does not limit yield when corn is grown with sufficient water and nutrients.

Irrigation scheduling to maintain ample water in the plow layer, and utilization of tensiometers and computer-based crop models have been shown to be useful irrigation-scheduling techniques in humid areas. The response of corn grain yield to supplemental irrigation depends on rainfall amount and seasonal distribution. The decision to install a crop-watering system for corn should be based on careful consideration of projected costs and returns.

Corn irrigation has been the subject of many studies which have led to improved management strategies, although the complexity of the numerous factors involved suggests that additional research will be required for continued progress. Although research has clearly demonstrated that severe water stresses reduce corn yields, future improvements in irrigation management practices may likely result from information gained through studies that focus on the response of the crop to only minimal stress levels.

REFERENCES

Abdul-Jabbar, A.S., J.W. Sammis, D.G. Lugg, C.E. Kallsen, and D. Smeal. 1983. Water use by alfalfa, maize, and barley as influenced by available soil water. Agric. Water Manage. 6:351–363.

Acevedo, E., T.C. Hsiao, and D.W. Henderson. 1971. Immediate and subsequent growth responses of maize leaves to changes in water status. Plant Physiol. 48:631–636.

Alessi, J., and J.F. Power. 1975. Effect of plant spacing on phenological development of early and midseason corn hybrids in a semiarid region. Crop Sci. 15:179–182.

Alessi, R.J., and J.F. Power. 1976. Water use by dryland corn as affected by maturity class and plant spacing. Agron. J. 68:547–550.

Allmaras, R.R., W.W. Nelson, and W.B. Voorhees. 1975a. Soybean and corn rooting in southwestern Minnesota. I. Water uptake sink. Soil Sci. Soc. Am. Proc. 39:764–771.

Allmaras, R.R., W.W. Nelson, and W.B. Voorhees. 1975b. Soybean and corn rooting in southwestern Minnesota. II. Root distributions and related water flow. Soil Sci. Soc. Am. Proc. 39:771–777.

Anderson, C.K., L.R. Stone, and L.S. Murphy. 1982. Corn yield as influenced by in-season application of nitrogen with limited irrigation. Agron. J. 74:396–401.

Barlow, E.W.R., L. Boersma, and J.L. Young. 1976. Root temperature and soil water potential effects on growth and soluble carbohydrate concentration of corn seedlings. Crop Sci. 16:59–62.

Barnes, D.L., and D.G. Woolley. 1969. Effect of moisture stress at different stages of growth. I. Comparison of single-eared and a two-eared corn hybrid. Agron. J. 61:788–790.

Bar-Yosef, B., and U. Kafkafi. 1972. Rates of growth and nutrient uptake of irrigated corn as affected by N and P fertilization. Soil Sci. Soc. Am. Proc. 36:931–935.

Bennett, J.M., and L.C. Hammond. 1983. Grain yields of several corn hybrids in response to water stresses imposed during vegetative growth stages. Soil Crop Sci. Soc. Fla. Proc. 42:107–111.

Blevins, R.L., J.H. Grove, and B.K. Kitur. 1986. Nutrient uptake of corn grown using moldboard plow or no-tillage soil management. Commun. Soil Sci. Plant Anal. 17:401–417.

Blue, W.G., E.O. Jacome, and J.L. Afre. 1982. Corn response to zinc and magnesium in a Florida entisol. Soil Crop Sci. Soc. Fla. Proc. 41:209–213.

Boyer, J.S. 1970. Leaf enlargement and metabolic rates in corn, soybean, and sunflower at various leaf water potentials. Plant Physiol. 46:233–235.

Camp, C.R., G.D. Christenbury, and C.W. Doty. 1984. Tillage effects on crop yield in coastal plain soils. Trans. ASAE 27:1729–1733.

Camp, C.R., D.L. Karlen, and J.R. Lambert. 1985. Irrigation scheduling and row configuration for corn in the southeastern coastal plain. Trans. ASAE 28:1159–1165.

Carter, M.W., and C.G. Poneleit. 1973. Black layer maturity and filling period duration among inbred lines of corn (Zea mays L.). Crop Sci. 13:436–439.

Cassel, D.K., and E.C. Edwards. 1985. Effects of subsoiling and irrigation on corn production. Soil Sci. Soc. Am. J. 49:996–1001.

Cassel, D.K., C.K. Martin, and J.R. Lambert. 1985. Corn irrigation scheduling in humid regions on sandy soils with tillage pans. Agron. J. 77:851–855.

Claasen, M.M., and R.H. Shaw. 1970. Water deficit effects on corn. I. Vegetative components. Agron. J. 62:649–652.

Dale, R.F., D.T. Coelho, and K.P. Gallo. 1980. Prediction of daily green leaf area index for corn. Agron. J. 72:999–1005.

DeLoughery, R.L., and R.K. Crookston. 1979. Harvest index of corn affected by population density, maturity rating, and environment. Agron. J. 72:577–580.

Doss, B.D., O.L. Bennett, and D.A. Ashley. 1962. Evapotranspiration by irrigated corn. Agron. J. 54:497–498.

Doss, B.D., C.C. King, and R.M. Patterson. 1970. Yield components and water use by silage corn with irrigation, plastic mulch, nitrogen fertilizer, and plant spacing. Agron. J. 62:541–543.

Doty, C.W. 1980. Crop water supplied by controlled and reversible drainage. Trans. ASAE 23:1122–1126.

Duncan, W.G. 1975. Maize. p. 23–50. In L.T. Evans (ed.) Crop physiology. Cambridge Univ. Press, England.

Duncan, W.G., A.L. Hatfield, and J.L. Ragland. 1965. The growth and yield of corn. II. Daily growth of corn kernels. Agron. J. 57:221–223.

Dwyer, L.M., and D.W. Stewart. 1985. Water extraction patterns and development of plant water deficits in corn. Can. J. Plant Sci. 65:921–933.

Eason, M.A., and F.M. Rhoads. 1982. Production and marketing of irrigated corn. Univ. Florida Agric. Res. Educ. Cent. Res. Rep. NF-82-2.

Eason, M.A., and F.M. Rhoads. 1983. Economics of irrigation scheduling for field corn in north Florida. Univ. Florida Agric. Res. Educ. Cent. Res. Rep. NF-83-1.

Eck, H.V. 1984. Irrigated corn yield response to nitrogen and water. Agron. J. 76:421–428.

Eck, H.V. 1986. Effects of water deficits on yield, yield components and water use efficiency of irrigated corn. Agron. J. 78:1035–1040.

Eik, K., and J.J. Hanway. 1966. Leaf area in relation to yield of corn grain. Agron. J. 58:16–18.

Follett, R.F., R.R. Allmaras, and G.A. Reichman. 1974. Distribution of corn roots in a sandy soil with a declining water table. Agron. J. 66:288–292.

Frye, W.W., L.W. Murdock, and R.L. Blevins. 1983. Corn yield-fragipan depth relations on a Zanesville soil. Soil Sci. Soc. Am. J. 47:1043–1045.

Hall, A.J., H.D. Ginzo, J.H. Lerncoff, and A. Soriano. 1980. Influence of drought during pollen-shedding on flowering, growth, and yield of maize. J. Agron. Crop Sci. 149:287–298.

Hammond, L.C. 1981. Irrigation efficiency and controlled root-zone wetting in deep sands. Univ. Florida Water Resour. Res. Publ. 52.

Harder, H.J., R.E. Carlson, and R.H. Shaw. 1982. Yield, yield components, and nutrient content of corn grain as influenced by post-silking moisture stress. Agron. J. 74:275–278.

Harrison, D.S., and F.M. Rhoads. 1978. Application of fertilizer through center pivot and self-propelled gun sprinkler irrigation systems. Univ. Florida. Agric. Eng., Ext. Rep. 78-1.

Herrero, M.P., and R.R. Johnson. 1981. Drought stress and its effects on maize reproductive systems. Crop Sci. 21:105–110.

Hillel, D., and Y. Guron. 1973. Relation between evapotranspiration rate and maize yield. Water Resour. Res. 9:743–748.

Hook, J.E. 1985. Irrigated corn management for the coastal plain: Irrigation scheduling and response to soil water and evaporative demand. Univ. Georgia Agric. Exp. Stn. Res. Bull. 335.

Horner, E.S., F.G. Martin, H.A. Peacock, J.R. Rich, and R.L. Stanley, Jr. 1985. Hybrid field corn variety tests in central, north, and west Florida. Univ. Florida. Inst. Food Agric. Sci. Agron. Res. Rep. AY 86-1.

Hume, D.J., and D.K. Campbell. 1972. Accumulation and translocation of soluble solids in corn stalks. Can. J. Plant Sci. 52:363–368.

Johnson, D.R., and J.W. Tanner. 1972. Calculation of the rate and duration of grain filling in corn (Zea mays L.). Crop Sci. 12:485–486.

Jones, R.J., and S.R. Simmons. 1983. Effect of an altered source sink ratio on growth of maize kernels. Crop Sci. 23:129–134.

Jurgens, S.K., R.R. Johnson, and J.S. Boyer. 1978. Dry matter production and translocation in maize subjected to drought during grain fill. Agron. J. 70:678–682.

Karlen, D.L., and C.R. Camp. 1985. Row spacing, plant population, and water management effects on corn in the Atlantic coastal plain. Agron. J. 77:393–398.

Karlen, D.L., C.R. Camp, and J.P. Zublena. 1985. Plant density, distribution, and fertilizer effects on yield and quality of irrigated corn silage. Commun. Soil Sci. Plant Anal. 16:55–70.

Krogman, K.K., M.D. MacDonald, and E.H. Hobbs. 1980. Response of silage and grain corn to irrigation and N fertilizer. Can. J. Plant Sci. 60:445–451.

Leng, E.R. 1951. Time relationships in tassel development of inbred and hybrid corn. Agron. J. 43:445–449.

Lorens, G.F., J.M. Bennett, and L.B. Loggale. 1987. Differences in drought resistance between two corn hybrids. II. Component analysis and growth rates. Agron. J. 79:808–813.

Mackay, A.D., and S.A. Barber. 1984. Soil temperature effects on root growth and phosphorus uptake by corn. Soil Sci. Soc. Am. J. 48:818–823.

Mackay, A.D., and S.A. Barber. 1985. Soil moisture effects on root growth and phosphorus uptake by corn. Agron. J. 77:519–523.

Mayaki, W.C., L.R. Stone, and I.D. Teare. 1976. Irrigated and nonirrigated soybean, corn, and grain sorghum root systems. Agron. J. 68:532–538.

McGarrahan, J.P., and R.F. Dale. 1984. A trend toward a longer grain-filling period for corn: A case study in Indiana. Agron. J. 76:518–522.

McPherson, H.G., and J.S. Boyer. 1977. Regulation of grain yield by photosynthesis in maize subjected to a water deficiency. Agron. J. 69:714–718.

Mitchell, C.C., Jr., and R.N. Gallaher. 1980. Sulfur fertilization of corn seedlings. Soil Crop Sci. Soc. Fla. Proc. 39:40–44.

Morey, R.V., J.R. Gilley, F.G. Bergsrud, and L.R. Dirkzwager. 1980. Yield response of corn related to soil moisture. Trans. ASAE 23:1165–1170.

Moss, G.I., and L.A. Downey. 1971. Influence of drought stress on female gametophyte development in corn (Zea mays L.) and subsequent grain yield. Crop Sci. 11:368–372.

Musick, J.T., and D.A. Dusek. 1980. Irrigated corn yield response to water. Trans. ASAE 23:92–98.

Onken, A.B., R.L. Matheson, and D.M. Nesmith. 1985. Fertilizer nitrogen and residual nitrate-nitrogen effects on irrigated corn yield. Soil Sci. Soc. Am. J. 49:134–139.

Poneleit, C.G., and D.B. Egli. 1979. Kernel growth rate and duration in maize as affected by plant density and genotype. Crop Sci. 19:385–388.

Potash and Phosphate Institute. 1985–1986. Here's how Herman Warsaw produced 370 bu/A corn yield. p. 12–13. *In* Better Crops (winter).

Prine, G.M. 1971. A critical period for ear development in maize. Crop Sci. 11:782–786.

Prior, C.L., and W.A. Russell. 1976. Leaf area index and grain yield for nonprolific single crosses of maize. Crop Sci. 16:304–305.

Reddy, M.D., I. Krishnamurthy, K.A. Reddy, and A. Venkatachari. 1982. Consumptive use and daily evapotranspiration of corn under different levels of nitrogen and moisture regimes. Plant Soil 56:143–147.

Rehm, G.W., and R.C. Sorensen. 1985. Effects of potassium and magnesium applied for corn grown on an irrigated sandy soil. Soil Sci. Soc. Am. J. 49:1446–1450.

Rehm, G.W., R.C. Sorensen, and R.A. Wiese. 1983. Application of phosphorus, potassium, and zinc to corn grown for grain or silage: Nutrient concentration and uptake. Soil Sci. Soc. Am. J. 47:697–700.

Rehm, G.W., and R.A. Wiese. 1975. Effect of method of nitrogen application on corn (*Zea mays* L.) grown on irrigated sandy soils. Soil Sci. Soc. Am. Proc. 39:1217–1220.

Retta, A., and R.J. Hanks. 1980. Corn and alfalfa production as influenced by limited irrigation. Irrig. Sci. 1:135–147.

Rhoads, F.M. 1981a. Nutrient management for irrigated agronomic crops. p. 13–19. *In* Proc. Natl. Symp. Chemigation. 20–21 Aug. 1981. Rural Development Center, Tifton, GA.

Rhoads, F.M. 1981b. Plow layer soil water management and program fertilization on Florida Ultisols. Soil Crop Sci. Soc. Fla. Proc. 40:12–16.

Rhoads, F.M., and R.S. Mansell. 1986. Effect of tillage and water management on soil-water depletion and grain yield of corn on a sandy soil. Univ. Fla. Inst. Food Agric. Sci. Res. Educ. Cent. Res. Rep. 86-1.

Rhoads, F.M., R.S. Mansell, and L.C. Hammond. 1978. Inflence of water and fertilizer management on yield and water-input efficiency of corn. Agron. J. 70:305–308.

Rhoads, F.M., and R.L. Stanley, Jr. 1973. Response of three corn hybrids to low levels of soil moisture tension in the plow layer. Agron. J. 65:315–318.

Rhoads, F.M., and R.L. Stanley, Jr. 1975. Response of corn (*Zea mays* L.) grown on soils of three textural classes to plow layer water management. Soil Crop Sci. Soc. Fla. Proc. 34:1–3.

Rhoads, F.M., and R.L. Stanley, Jr. 1979. Effect of population and fertility on nutrient uptake and yield components of irrigated corn. Soil Crop Sci. Soc. Fla. Proc. 38:78–81.

Rhoads, F.M., and R.L. Stanley, Jr. 1981. Fertilizer scheduling, yield, and nutrient uptake of irrigated corn. Agron. J. 73:971–974.

Rhoads, F.M., and R.L. Stanley, Jr. 1984. Yield and nutrient utilization efficiency of irrigated corn. Agron. J. 76:219–223.

Ritchie, J.T. 1973. Influence of soil water status and meteorological conditions on evaporation from a corn canopy. Agron. J. 65:893–897.

Ritchie, S.W., and J.J. Hanway. 1982. How a corn plant develops. Iowa State Univ. Coop. Ext. Serv. Spec. Rep. 48.

Robertson, W.K., L.C. Hammond, J.T. Johnson, and K.J. Boote. 1980. Effects of plant-water stress on root distribution of corn, soybeans, and peanuts in sandy soil. Agron. J. 72:548–550.

Robertson, W.K., L.C. Hammond, G.M. Prine, and F.G. Martin. 1981. Response of corn cultivars on sandy soil to irrigation, row-spacing, plant population, and nitrogen. Soil Crop Sci. Soc. Fla. Proc. 40:101–105.

Robins, J.S., and C.E. Domingo. 1953. Some effect of severe soil moisture deficits at specific growth stages in corn. Agron. J. 45:618–621.

Rosenthal, W.D., E.T. Kanemasu, R.J. Raney, and L.R. Stone. 1977. Evaluation of an evapotranspiration model for corn. Agron. J. 69:461–464.

Russelle, M.P., E.J. Deibert, R.D. Hauck, M. Stevanovic, and R.A. Olson. 1981. Effects of water and nitrogen management on yield and ^{15}N-depleted fertilizer use efficiency of irrigated corn. Soil Sci. Soc. Am. J. 45:553–558.

Sadler, E.J., and C.R. Camp. 1986. Crop water use data available from the southeastern USA. Trans. ASAE 29:1070–1079.

Saini, G.R., and T.L. Chow. 1982. Effect of compact subsoil and water stress on root and shoot activity of corn (*Zea mays* L.) and alfalfa (*Medicago sativa* L.) in a growth chamber. Plant Soil 66:291–298.

Scarsbrook, C.E., and B.D. Doss. 1973. Leaf area index and radiation as related to corn yield. Agron. J. 65:459–461.

Schoney, R.A., and L.R. Massie. 1981. An investment feasibility analysis of a traveling gun irrigation system in the humid Midwest. North Centr. J. Agric. Econ. 3:53–61.

Schoper, J.B., R.J. Lambert, and B.L. Vasilas. 1986. Maize pollen viability and ear receptivity under water and high temperature stress. Crop Sci. 26:1029–1033.

Sharp, R.E., and W.J. Davies. 1979. Solute regulation and growth by roots and shoots of water-stressed maize plants. Planta 147:43–49.

Sharp, R.E., and W.J. Davies. 1985. Root growth and water uptake by maize plants in drying soils. J. Exp.Bot. 36:1441–1456.

Simmons, S.R., and R.J. Jones. 1985. Contributions of pre-silking assimilate to grain yield in maize. Crop Sci. 25:1004–1006.

Sinclair, T.R., G.E. Bingham, E.R. Lemon, and L.H. Allen, Jr. 1975. Water use efficiency of field-grown maize during moisture stress. Plant Physiol. 56:245–249.

Sorensen, V.M., R.J. Hanks, and R.L. Cartee. 1980. Cultivation during early season and irrigation influences on corn production. Agron. J. 72:266–270.

Stanley, R.L., Jr., and F.M. Rhoads. 1975. Response of corn (Zea mays L.) to population and spacing with plow-layer soil water management. Soil Crop Sci. Soc. Fla. Proc. 34:127–130.

Stanley, R.L., Jr., and F.M. Rhoads. 1977. Effect of time, rate, and increment of applied fertilizer on nutrient uptake and yield of corn (Zea mays L.). Soil Crop Sci. Soc. Fla. Proc. 36:181–184.

Stegman, E.C. 1982. Corn grain yield as influenced by timing of evapotranspiration deficits. Irrig. Sci. 3:75–87.

Stevens, S.J., E.J. Stevens, K.W. Lee, A.D. Flowerday, and C.O. Gardner. 1986. Organogenesis of the staminate and pistillate inflorescences of pop and dent corns: Relationships to leaf stages. Crop Sci. 26:712–718.

Stewart, J.I., R.J. Hanks, R.E. Danielson, E.B. Jackson, W.O. Pruitt, W.T. Franklin, J.P. Riley, and R.M. Hagan. 1977. Optimizing crop production through control of water and salinity levels in the soil. Utah Water Resour. Lab. Rep. PRWG151-1.

Stewart, J.I., R.D. Misra, W.O. Pruitt, and R.M. Hagan. 1975. Irrigating corn and grain sorghum with a deficient water supply. Trans. ASAE 18:270–280.

Valle Melendez, R.R. 1981. Physiological aspects of maize (Zea mays L.) yield. Ph.D. diss., Univ. of Florida, Gainesville. (Diss. Abstr. 81-24461).

Van Doren, D.M., Jr., and G.B. Triplett, Jr. 1973. Mulch and tillage relationships in corn culture. Soil Sci. Soc. Am. Proc. 37:766–769.

Van Doren, D.M., Jr., and G.B. Triplett, Jr. 1976. Influence of long-term tillage, crop rotation, and soil type combinations on corn yield. Soil Sci. Soc. Am. J. 40:100–105.

Van Volkenburgh, E., and J.S. Boyer. 1984. Inhibitory effects of water deficit on maize leaf elongation. Plant Physiol. 77:190–194.

Walker, P.N., M.D. Thorne, E.C. Benham, and S.K. Sipp. 1982. Yield response of corn and soybeans to irrigation and drainage on a claypan soil. Trans. ASAE 25:1617–1621.

Watts, W.R. 1974. Leaf extension in Zea mays. III. Field measurements of leaf extension in response to temperature and leaf water potential. J. Exp. Bot. 25:1085–1096.

Weatherly, A.B., and J.H. Dane. 1979. Effect of tillage on soil-water movement during corn growth. Soil Sci. Soc. Am. J. 43:1222–1225.

Wenda, W.I., and R.J. Hanks. 1981. Corn yield and evapotranspiration under simulated drought conditions. Irrig. Sci. 2:193–204.

Westgate, M.E., and J.S. Boyer. 1985. Carbohydrate reserves and reproductive development at low leaf water potentials in maize. Crop Sci. 25:762–769.

Westgate, M.E., and J.S. Boyer. 1986a. Silk and pollen water potentials in maize. Crop Sci. 26:947–951.

Westgate, M.E., and J.S. Boyer. 1986b. Reproduction at low silk and pollen water potentials in maize. Crop Sci. 26:951–956.

Williams, W.A., R.S. Loomis, W.G. Duncan, A. Dovrat, and F. Nunez A. 1968. Canopy architecture at various population densities and the growth and grain yield of corn. Crop Sci. 8:303–308.

Williams, W.A., R.S. Loomis, and C.R. Lepley. 1965. Vegetative growth of corn as affected by population density. I. Productivity in relation to interception of solar radiation. Crop Sci. 5:211–215.

Wilson, P.N., and V.R. Eidman. 1983. The financial profitability of irrigating fine textured soils in the upper Corn Belt. North Centr. J. Agric. Econ. 5:103–110.

20   Wheat[1]

J. T. MUSICK

USDA-ARS
Bushland, Texas

K. B. PORTER

TAES-TAMU
Amarillo, Texas

Wheat (*Triticum aestivum* L. and *T. turgidum* L. var. *durum*) is the world's major crop source of calories and protein (Thorne, 1977). It is grown over a wide range of precipitation and temperature conditions, mostly in the 25 to 50° lat range. Wheat production has expanded into the lower latitudes to less than 15° as a cool-season crop (Khalifa et al., 1977) and to about 60° lat in the Northern Hemisphere as a spring-planted, warm-season crop grown at the lower elevations.

Irrigation is widely practiced in some of the major production regions in the Northern Hemisphere and to only a limited extent in the Southern Hemisphere, primarily southeast Australia (Smith et al., 1983). Most of the wheat in the major production regions of India, Pakistan, and the People's Republic of China (PRC) is grown under irrigation. Irrigation has played a major part in the PRC becoming the world's number one wheat-producing country. In the USSR, the world's second largest producer, wheat is mostly grown under rain-fed conditions. The USA is the third largest producing country, and wheat is the third largest irrigated crop after corn (*Zea mays* L.) and alfalfa (*Medicago sativa* L.). Irrigated yields in 1984 averaged 4.6 Mg/ha (USDC, 1984). Wheat is principally a cool-season, temperate-zone crop. However, it is grown during the cool season at higher elevations in some tropical zone countries. The only major crop production regions where it is not grown are in the hot, low-lying tropics.

In India and Pakistan, wheat is usually grown during the dry season and irrigation is the major water source. It is also grown as an irrigated crop in the arid regions of Arizona, California, Washington, and Idaho and in the irrigated valleys of northwest Mexico. In semiarid regions such as the Pacific Northwest and Great Plains states, where wheat is widely grown as a dryland crop, it is also grown under limited irrigation in conjunction with seasonal rainfall. Supplemental or limited irrigation of wheat is also practiced in some semiarid regions of India and the Middle East.

---

[1] Contribution from the USDA-ARS Conservation and Production Research Laboratory, P.O. Drawer 10, Bushland, TX 79012; and Texas Agric. Exp. Stn., Bushland, TX 79012.

In 1979, irrigated wheat in India occupied 14.7 million ha, about two-thirds of the cultivated area (Sinha et al., 1985). In the USA, almost all of the irrigated wheat is grown in 17 western states, where, in 1984, it totaled 1.82 million of the 23.9 million ha (USDC, 1984). Wheat irrigation was widely dispersed in the Central and Southern Plains states and in the northwest and southwest states. Except for 67 000 ha in Montana, irrigation was relatively insignificant in the northern Plains states. The largest production area was in Texas (363 000 ha), followed by California and Kansas (290 000 ha each), Idaho (266 000 ha), and Washington (145 000 ha).

A U.S. Wheat Studies Delegation that visited the PRC in May 1976 reported that 80 to 85% of the wheat was irrigated (Johnson & Beemer, 1977). Wheat was being grown on 20% of the arable land, with production in all provinces, ranging from 18 to 50°N lat. Wheat in the PRC has been stable in area since 1978 at 44 to 45 million ha (PRC, 1986). In 1978, national yields averaged 1.84 Mg/ha, and soil fertility was a major limitation to irrigated yields. From 1978 to 1984, fertilizer consumption for all crops increased from 9 to 18 Tg (Brown, 1986); and by 1984, national wheat yields averaged 2.97 Mg/ha (PRC, 1986). The increased yields were accomplished without further expansion in irrigated area.

I. GROWTH AND DEVELOPMENT

A. Vegetative-Reproductive

Extensive literature exists on the growth and development of the wheat plant. The Feekes scale of growth stages as reported by Large (1954) has been widely used in the USA, while in Europe, it has been largely replaced by the Zadoks decimal code (Zadoks et al., 1974; Tottman, 1987). Further developments in classification of growth and development were described by Haun (1973) and Klepper et al. (1982, 1984). A modern standard reference on development of wheat and barley is Kirby and Appleyard (1984). A simple but physiologically sound three-stage growth system, used in temperature studies of wheat by Warrington et al. (1977) and discussed by Eastin et al. (1983) and Evans and Wardlaw (1976), will be used in this review. Wheat growth and development physiology related to the three phases of vegetative, reproductive, and grain growth is described as follows:

GS1—Emergence to floral initiation (FI). Leaves and roots grow until the apex changes from initiating leaves to initiating the inflorescence as described in photographic sequence by Bonnet (1966) and George (1982). This period can be long for early planted winter wheat cultivars that require vernalization for FI or short for late fall- or spring-planted spring wheat cultivars that do not have a vernalization requirement.

GS2—Floral initiation to postanthesis beginning of grain growth. This is an exponential reproductive growth period in which green leaf

Area index (LAI) increases and peaks 2 to 3 wk before anthesis, root growth increases on a weight basis until anthesis, and the inflorescence grows and develops the potential grain numbers. Pollination occurs at anthesis, and grain growth begins. Culm growth continues only a few days after anthesis. Environmental stresses may severely limit potential grain numbers and yield potential.

GS3—Grain growth to physiological maturity. Grain growth begins and proceeds, following a few days lag after anthesis, as a linear process with time except for a few days at the beginning and end of the filling process. During grain growth, green LAI progressively declines, root growth continues at a reduced rate where conditions are favorable, and some previously stored assimilates are relocated to the grain. Environmental stresses primarily reduce yields by reducing the duration of grain filling and grain weight. Physiological maturity, in general, correlates with loss of head greenness and can be more specifically estimated by the water content of the grain as indicated by the absence of a wet endosperm surface when a kernel is pulled apart at the crease.

Some wheat genotypes require long photoperiods for floral development to anthesis, while others are photoperiod-neutral. Winter types require a chilling period (vernalization) for floral initiation, while spring types have little or no vernalization response. However, substantial variation occurs among both winter and spring types in vernalization requirement and photoperiod response. Spring types are late fall planted in the lower latitudes and spring planted in the higher latitudes where they cannot survive the cold temperatures.

The effects of environment on growing season duration of spring wheats are illustrated in Fig. 20–1 for locations in India with latitudes ranging from 11 to 32 °N (Sinha et al., 1985). The growing season duration to maturity of 100 to 150 d was associated with a range of irrigated grain yields from 1.6 to 5.0 Mg/ha, and yields and days to maturity were highly correlated. In these tests, relatively high irrigated yield levels of 4 to 5 Mg/ha from research plots were obtained only at locations where the latitudes exceeded 25 °, corresponding to about the southern boundary of the major production region in India. Increasing altitude reduces temperatures and increases length of growing season, thus increasing yields. In Zimbabwe, Cackett and Wall (1971) reported irrigated yields of 6.5 to 8.5 Mg/ha in the highveld and 4.0 to 5.5 Mg/ha in the lowveld, at elevations differing about 1100 m.

Wiegand and Cuellar (1981) determined from field plot studies at 26 °N lat that for each °C increase in mean air temperature during grain filling, duration of grain filling was shortened by 3.1 d and final kernel weight was reduced by 2.8 mg. The data by Wiegand and Cuellar (1981) compare with 2.8 d and 1.5 mg per kernel per °C from eight previous studies reported in the literature. In the low-latitude environments (below about 35 °N), spring wheat is late fall planted and matures in mid to late spring. The crop is grown

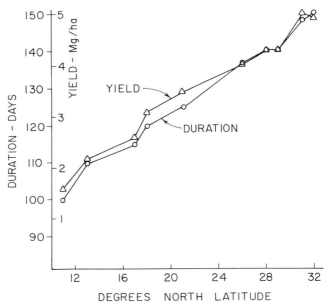

Fig. 20-1. Growing season duration of spring wheat and grain yield of irrigated wheat in India for locations having latitude ranging from 11 to 32 °N (adapted from Sinha et al., 1985).

during the cool season and escapes some of the adverse effects of increasing late spring temperatures.

At the higher latitudes, where spring wheat has insufficient cold tolerance for winter survival, spring planting is practiced and harvest occurs in late summer. In these climatic regions of long days and moderate air temperatures, solar radiation has a dominant climatic influence on yields. Under the long-day environment, improved cultivars produce high yields given adequate water, fertility, and good management. Fall-planted irrigated winter wheat is grown mostly in the central and western USA from about 32 ° to 44 °N lat, with production extending to 46 to 48 °N lat at the relatively low to moderate elevations of the Columbia Basin, the Palouse region of Washington, and north-central Montana.

Winter wheat cultivars planted in the USA in September or October mature from late May in central Texas to August in Washington. In the mid-latitude region of continental climate with relatively hot summers, winter wheat is better adapted than spring wheat and normally produces higher yields. Low temperature episodes can cause freeze damage when the wheat is not covered with snow. As winter wheat resumes growth in late winter and early spring, it loses cold conditioning, and low temperatures can cause death of green leaf area and damage to the elongating culm and developing apex (George, 1982). Later frost events can result in head sterility. Heavily fertilized irrigated wheat, with its lusher growth and higher N content, is more likely to be damaged by frost (Fischer, 1981a).

B. Roots

The root system of wheat was carefully observed and recorded by Weaver (1926), and root growth was received by MacKey (1973). Spring wheat cultivars commonly develop roots to about 1-m depth, while winter wheats, with their longer vegetative growth period, develop roots to about 1.5 m. In favorable rooting environments, winter wheat roots can develop to depths in excess of 3 m (Kmoch et al., 1957; Cholick et al., 1977). Studies involving semidwarf cultivars have shown no relationship between plant height and rooting depths or soil water depletion (Cholick et al., 1977, Holbrook & Welsh, 1980).

The root system terminology of seminal and nodal is used in this discussion. *Seminal roots*, three to seven in number, develop from primordia in the seed. Due to their early start, they dominate rooting during the seedling stage and continue a decreasing dominance until about mid-season, when nodal roots (crown or adventitious) become increasingly important (MacKey, 1973). *Nodal roots* develop from nodes in the crown in sequence as the main culm and tillers develop, with the earlier developed roots being longer and more branched than later ones. The *main culm roots* develop first and tend to dominate the nodal root system. Later-developing *tiller nodal roots* are short in length and senesce when tillers senesce.

Seminal roots are deeper, finer, have a higher order of branching than nodal roots, and tend to have a near-linear distribution with depth. They tend to have a direct downward orientation (Belford et al., 1987), remain active throughout the season, and provide the greater fraction of the roots in the lower profile; thus, their importance increases under conditions of major soil water deficits.

Nodal roots are exponentially distributed with depth; after about mid-season, they comprise the major weight of the root system in the upper profile. Klepper et al. (1984) reported that root length densities of 6 cm/cm^3 were common in surface soils, while densities of 1 cm/cm^3 were found at the 0.5- to 1.0-m depth. This distribution change with depth is associated mostly with changes in nodal root densities.

Nodal roots have an oblique outward development pattern that reorients in the downward direction, which increases the extent of lateral rooting. MacKey (1973) indicated that early developed nodal roots increased effectiveness in water and nutrient absorption compared with the later developing ones which are thicker, shorter, and less branched. He indicated that early development of nodal roots appears to have a restrictive effect on the growth of seminal roots, and the nodal root system tends to become dominant in rooting activity after about mid-season.

Nodal root growth normally begins about 25 to 30 d after plant emergence unless growth is prevented by cool temperatures. Dry surface soil at the crown can prevent the initial extension of nodal roots, and major surface soil drying during tillering can slow or prevent new nodal root growth from tiller nodes. For fall-planted wheat that emerges and grows during a dry season, irrigation may be essential for providing surface soil water to permit normal extension of nodal roots. In the southern High Plains, dry winters at Bushland, TX, have resulted in persistence of dry surface soil and delay in nodal root extension from November until the jointing irrigation in April. In contrast, the senior author has observed nodal roots in soil core samples to depths of 1.5 m by jointing in April. On this soil (Richfield clay loam; fine, montmorillonitic mesic family of Aridic Argiustolls) with a deep silt subsoil favorable for root development (1.25 Mg/m^3 bulk density), irrigated wheat depleted soil water to <15% available before yield reductions occurred (Musick et al., 1963). Early deep development of the nodal root system is believed to have contributed to the ability of the plants to deplete soil water from the lower profile.

Irrigation research in India has shown that crown (nodal) root initiation stage is the most critical for scheduling irrigation of late fall-planted spring wheat grown on sandy soils in environments where plants have a relatively short duration of vegetative growth. The crown root stage is critical in environments where significant water deficits can develop due to limited rooting and the water deficits are expressed through reduced vegetative growth (tillers, leaf area, and dry matter accumulation). Also, delayed irrigation in this environment fails to compensate for the lost growth in the relatively short growing season (Fischer, 1981b). Misra and Choudhary (1985) found a highly significant ($R^2 = 0.92$) curvilinear relationship between root density by weight and grain yield. Under dryland conditions, Black (1970b) found an association between nodal roots per plant and grain yields.

Gregory et al. (1978) have discussed the partitioning of plant assimilates between roots and tops under water deficit conditions. They indicated that during fall growth, the plant allocates about one-fourth of its dry weight accumulation to roots; the allocation to roots increases to about one-third during winters as cool temperatures limit vegetative growth; it rapidly declines during spring to about 0.09 at anthesis as developing leaves, culms, and the growing head become increasingly stronger sinks for assimilate. The root system reaches its maximum weight at anthesis. During grain filling, as older roots die and new root activity slows in the lower profile, partition of assimilates to roots is about 0.07. At harvest, the root system constitutes about 10% of the plant dry matter.

If severe late-season stress develops, the plant's investment of assimilates in developing a large early season root system may be detrimental to late-season soil water reserves and grain yield (Passioura, 1977). However, on irrigated land, the soil profile is normally fully wet to the potential rooting depth; a large and, thus, deeper root system may be effective in exploring additional water resources for plant use. This extensive profile wetting to lower depths may not be the case under dryland farming conditions with limited rainfall.

Death of tillers results in death of tiller nodal roots (Gregory et al., 1978). However, failure of tiller nodal roots to develop may not be the cause for failure of tiller development or senescence of developed tillers. In the absence of nodal roots providing nutrients and water uptake to tillers, death of late-developed tillers is more likely to be related to reduced assimilate from shading, water stress, or low competitive partitioning from the main plant during rapid growth.

C. Yield and Yield Components

1. Yield

Wheat yields in the USA increased by 32% during a 20-yr period through 1980 (Schmidt, 1984). The genetic yield potential of experimental lines compared with long-term check cultivars increased from 25 to 46% during the 20-yr period. About half of the increase in wheat yields was attributed to breeding for improved cultivars. In addition to improved cultivars, increased yields can be attributed to improved irrigation, fertilizer use, and cultural management. In the semiarid climate of the southern High Plains, the annual rate of irrigated wheat yield increase from 1968 to 1986 was double that of dryland wheat (58 vs. 26 kg/ha per yr, J.T. Musick, 1987, unpublished data). The importance of irrigation in increasing wheat yields is further emphasized by the contribution to the "green revolution" of expanded irrigation of wheat in India (4.8 million ha in 1966 compared with 14.7 million ha in 1979) (Sinha et al., 1985).

Irrigated wheat yields in the irrigated valleys of northwest Mexico have increased dramatically from 1.0- to 1.3-Mg/ha range prior to 1950 to almost 5 Mg/ha in 1975 (Waddington et al., 1986). The yield improvement associated with release of semidwarf cultivars was discussed by Dalrymple (1986).

Grain yields of winter wheat in the Netherlands have increased from 3.8 Mg/ha in 1950 to 6.7 Mg/ha in 1982 (Spiertz & Vos, 1985). The yield improvements were attributed to modern cultivars that have greater resistance to lodging and greater efficiency in using larger amounts of fertilizer N, to higher harvest index, and to improved management. After about 1970, use of systemic fungicides, insecticides, and split applications of N further increased yields. Similar increased yields have occurred in other northwestern European countries (Vlassak & Verstraeten, 1985). Yields in excess of 10 Mg/ha have occurred in some production environments, but these high yields have not been repeatable in different climatic seasons. A record yield of 14.0 Mg/ha was reported in the state of Washington (Stanford & Legg, 1984).

The release of semidwarf cultivars with increased harvest index was primarily a one-time progress in incremental yield improvement, while release of newer cultivars has contributed to increased yields without further reduction in plant height (Dalrymple, 1986). Further releases of high-yielding cultivars of bread (Waddington et al., 1986) and durum (Waddington et al., 1987) wheats in northwest Mexico have shown genetic yield improvement associated with increased aboveground biomass, increased seed numbers per unit area (primarily by increasing seed per head), and increased total grain-filling rate. Yields of durum cultivars have been improved in association with shortening the time to anthesis and lengthening the grain-filling period. From this perspective, genetic yield improvement is being expressed through growth and development processes other than height reduction, and irrigation will continue to play an important role in realizing the higher genetic yield potentials in semiarid and arid environments.

In semiarid regions where seasonal precipitation is a significant contribution to water requirements, wheat is widely grown under limited irrigation with attainment of moderate yield levels. Under such conditions, yields obtained by farmers are normally much lower than those obtained from research station tests that are irrigated and managed for high yields. In India, state-irrigated wheat yields as a percentage of research station yields averaged 58% in Punjab and 30% in Madhya Pradesh and Utter Pradesh (Sinha et al., 1985). In the Murrumbridgee Irrigation Area (MIA) of southeastern New South Wales, Australia, irrigated wheat yields obtained by farmers averaged 2.6 Mg/ha, while yields on research plots managed for high yields averaged 6 to 8 Mg/ha (Smith et al., 1983; Steiner et al., 1985). Smith et al. (1983) compared irrigated wheat yields obtained in the MIA with those obtained in Yolo County, California, where irrigated wheat yields averaged 6.4 Mg/ha. Low irrigated yields in the MIA were attributed to high clay soils having low permeability, transient water logging following flood irrigation, and limited rooting depth (Smith et al., 1983).

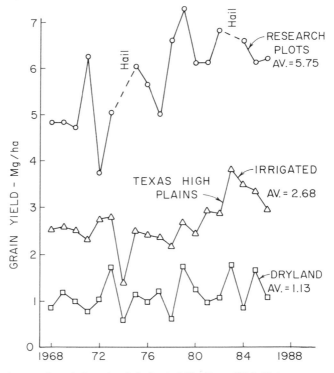

Fig. 20–2. Average farm irrigated and dryland yields, Texas High Plains, compared with the average yields of the five high-yielding varieties each year in irrigated wheat improvement research tests at Bushland, TX. Farm yields are from annual reports, Texas Agricultural Statistics Service.

Irrigated yields in the Texas High Plains are compared with dryland yields in Fig. 20-2. These irrigated yields in semiarid regions of limited ground-water supplies and significant seasonal rainfall average 47% of the average research plot yields by the five highest-yielding cultivars grown each year.

2. Yield Components

Yield component analysis involves viewing yield as the product of plants per unit area, heads per plant, spikelets per head, grain per spikelet, and weight per grain. The number of grain per head can be influenced by water deficits until a few days after anthesis, when the grain numbers are firmly set. Because of the plastic nature of the plant and associated negative correlations among the yield components, a simplified approach is combining the four grain number components into number per unit area. Thus, a yield component analysis can logically involve only two components: grain number per unit area and grain weight. The grain numbers component establishes the yield potential that is realized if conditions after anthesis are favorable for grain filling.

Grain-filling termination is influenced by availability of assimilates and/or nutrients (Gallagher & Biscoe, 1978). Water deficits after anthesis largely limit yields by reducing the photosynthetic rate and duration. As photosynthesis continues to decline during the latter phase of grain filling, relocation of preanthesis dry matter helps to sustain the grain-filling process. Grain weight often exceeds postanthesis dry matter increase, sometimes by as much as 50% under water deficits (Gallagher & Biscoe, 1978). The photosynthesis-assimilate relocation process influencing grain filling is less sensitive to water deficits than growth (leaf area and dry matter accumulation) and development processes that influence seed numbers per m^2.

The historic increases in wheat yields have involved improved cultivars that have increased spikelets per head, grain per spikelet, and higher grain weights and larger leaves (Evans & Dunstone, 1970). In an evolutionary perspective, higher grain weights have resulted from both faster and longer grain-filling periods (Sofield et al., 1977). While unit area photosynthesis rates of leaves have not increased in higher-yielding cultivars (Evans, 1981), modern production practices have resulted in high photosynthetic surface area (LAI of 4-8) that provides assimilate capacity to support both high grain numbers and high grain-filling rates associated with high yields.

3. Harvest Index

The above-ground components of the wheat plant were determined by weight as 40% grain, 33% culm, 10% chaff, 9% leaves, and 8% vegetative tillers (Singh & Stoskopf, 1971). Gifford and Evans (1981) indicated the best wheats now have *harvest indices* (HI, the grain fraction of aboveground dry matter by weight) around 0.50. A review of the literature indicates that the HI of modern high-yielding cultivars is mostly in the range of 0.38 to 0.50. The highest HI found in the literature was 0.60 for early-planted spring wheat cultivar 'Twin' in Utah (Hanks & Sorensen, 1984), while the lowest values of slightly < 0.2 were found for winter wheat at Manhattan, KS (Asrar & Kanemasu, 1985), for wheat grown under major water deficits in southern Iran (Poostchi et al., 1972), and for wheat grown under severe stress after anthesis in Australia (Passioura, 1977).

Dry matter accumulation during head development proceding anthesis influences grain numbers (Fischer, 1985a); the more preanthesis dry matter produced, the more grain there is to be filled and the more critical the postanthesis water supply and dry matter accumulation becomes in preventing grain shrivelling and a low HI. Passioura (1977) found a linear relationship between HI and the percent of the seasonal water use after anthesis which ranged from HI of 0.2 when water use after anthesis was less than 5% of the season total to about 0.5 when water use after anthesis was 25 to 30% of the total. The results demonstrate grain filling as a critical stage for water deficit effects on HI.

Line source sprinkler irrigation studies involving water deficits normally are managed to distribute the deficits and, thus, distribute dry matter production throughout the season. Under this method of deficit management, Hanks and Sorensen (1984) found that HI was not affected until seasonal ET was reduced by over one-half. When ET was reduced to 33% (from 534–176 mm), HI for spring wheat varieties grown in Utah were reduced from 0.51 to 0.44.

Hanks and Sorensen (1984) determined that a range of five planting densities for spring wheat did not affect HI. However, high planting densities in winter wheat increases early season biomass and reduces HI. Darwinkel (1978) found that increasing planting density from 100 to 800 plants/m^2 decreased HI from 0.44 to 0.38. As higher planting densities increased heads per m^2 from 500 to 900, HI decreased from 0.47 to 0.43. When heads increased beyond 800/m^2, the decline in grain yields was associated with the lower HI.

A relatively wide range in planting dates for spring wheat in Utah, early April to mid-May, did not affect HI. However, a further delay in planting to late May reduced HI from 0.49 to 0.44 because of reduced grain yield (Hanks & Sorensen, 1984). Very early planting of winter wheat reduces HI by increasing early season biomass production.

Donald and Hamblin (1976), in a review of biological yield and HI, concluded that the application of N fertilizer to cereals commonly gives an increase in biological yield that is proportionately greater than the increase in grain yields and the HI declines. Data from 12 yr of tests at the Rothamsted Experiment Station indicated that increasing N application from 0 to 200 kg/ha decreased HI from 0.36 to 0.30. Data quoted from McNeal et al. (1971) indicated that increasing N application from 0 to 90 kg/ha for spring wheat in Montana decreased HI from 0.43 to 0.39.

Evans (1980) stated that the yield increase of wheat through 1976 came from the HI, which rose to about 0.50. Gifford et al. (1984) presented data for eight major winter wheat cultivars released in England over a period of 70 yr. Grain yield increased from 5 to 7 Mg/ha, while HI increased from 0.35 to 0.50, with the increased yields being associated with the increased HI.

Syme (1970) tested nine cultivars in which the HI ranged from 0.24 to 0.39 and found that increasing HI accounted for 92% of the yield variance. He indicated that the close correlation ($r = 0.96$) of HI with yield suggests improvement in this ratio may be associated with most of the varietal increase in wheat yield in Australia. A significant contribution to increased HI has been the development of semidwarf and dwarf cultivars. Singh and Stoskopf (1971) found that dwarf compared with tall cultivars increased HI from 0.38 to 0.42.

Austin et al. (1980) concluded that the newer, high-yielding cultivars were shorter and had lower culm weights per m^2 than older cultivars but similar maximum leaf areas and leaf weights per m^2. Total dry matter production was similar, and the increase in grain yield due to cultivar improvement was associated mainly with greater HI. They estimated the potential through breeding to increase HI from present values of about 0.50 to 0.60 and predicted further genetic gain in yield will depend on exploiting genetic variation in biomass production.

Other authors indicate that increased biomass, in addition to increased HI, has contributed to increased yields. MacKey (1973) presented wheat productivity data for 50 yr of winter wheat breeding in Sweden through 1955, in which grain yields were increased from 3.9 to 5.5 Mg/ha, and for 60 yr of spring wheat breeding (through 1965), in which grain yields were increased from 3.3 to 4.3 Mg/ha. The HI increased from 0.34 to 0.41 for both winter and spring types and was associated with about one-half the yield increase in winter wheat and about two-thirds of the yield increase in spring wheats.

More recent tests by Waddington et al. (1986, 1987) indicated that the increase in bread wheat yields for cultivars released since 1950 (5.5–7.8 Mg/h) and in durum wheat cultivars released since 1960 (3.7–8.4 Mg/ha) were associated with both increased HI and aboveground biomass. Harvest indices of bread wheats increased and peaked in 1968 with the release of 'Sonalika 68' (HI = 0.47), while HI of durum wheat cultivars peaked in 1975 with the release of 'Mexicali 75' (HI = 0.46). Continued increased yields from more recent releases have been entirely associated with increased biomass and seed numbers per head. The dramatic increase in irrigated durum wheat yields from 3.7 to 8.4 Mg/ha was associated with an increase in HI from 0.24 to 0.42 and an increase in aboveground biomass from 13.7 to 17.7 Mg/ha.

Dry matter (DM) at anthesis generates a sink that represents a given yield potential, as expressed through seed numbers per m^2. In modern short wheats, anthesis DM is approximately 1.5 times grain yield, while with tall cultivars, it is 2.0 or more. The favorable ratio by the short wheats may be further improved by nontillering uniculm types that offer prospects of increased yields. Atsmon and Jacobs (1977) determined HI of 0.41 for a tillered cultivar compared with 0.54 for a uniculm line that had 13% higher grain yield per plant. Darwinkel (1978) found that late-formed shoots at low plant densities reduced HI from 0.45 to 0.36.

In summary, the wheat plant conservatively adjusts grain numbers to preanthesis water deficit-induced reductions in dry matter to the extent that

HI is affected little, while HI is sensitive to postanthesis water deficits. Irrigation has played an important role in the understanding of HI relationships of wheat in some climatic environments by eliminating postanthesis water deficits.

4. Quality

Common wheats are classified for commercial purposes as being hard or soft, either red or white, and spring or winter. Durum wheat is considered separately in some areas but not in others (Shellenberger, 1978). Zeleny (1978) discussed the botanical, physical, and chemical criteria of wheat quality; more recently, Finney et al. (1987) discussed milling properties and bread-making properties of hard (bread) wheat, soft wheat, and milling and pasta-processing properties of durum wheats. The quality of wheat is usually determined by its suitability for a particular end use (Zeleny, 1978). Finney et al. (1987) described actual quality of a wheat as the summation effect of soil, climate, and seed stock on the wheat plant and kernel components.

The major environmental factors influencing quality (water, soil fertility, and temperature during grain filling) were discussed by Evans and Wardlaw (1976). Solar radiation has little effect on protein, since grain protein and starch accumulation are affected similarly by reduced radiation. Although soil water and fertility may influence many quality criteria, their effect on the quantity and quality of protein in wheat grain may be of the greatest significance.

Grain filling of wheats proceeds at a linear rate with time from about 15 to 85% of final grain weight (Gallagher & Biscoe, 1978). Under conditions of relatively high N where uptake continues through grain filling, both protein and starch content of the grain increase linearly until near physiological maturity. When soil nitrates are severely depleted by the time of anthesis and little uptake occurs during grain filling, most of the N accumulation in the grain is relocated from leaves and culms. Where low N is combined with adequate irrigation, leaf senescence is delayed, relocation of previously assimilated N to grain filling is slowed, and protein content of grain is reduced. Low protein contents, indicating N deficiency, are mostly in the range of 8.5 to 10.5%. Water deficits during grain filling accelerate leaf senescence, and relocation of N contributes to increasing protein at a time when wheat starch accumulation has slowed. Thus, low yields from water deficits during grain filling are generally accompanied by high grain protein contents. High protein contents are mostly in the range of 12 to 16%.

Finney et al. (1963) found that wheat samples harvested in 1962 from irrigated regional performance trials at a number of locations in the southern Great Plains had shorter mix times and lower corrected loaf volumes for protein content than samples from comparable dryland trials. Although these differences, in some cases, could be accounted for by differences in protein content of the samples, a generally longer fruiting period of irrigated wheat was believed to have contributed to the shorter mix times. Certain irrigated samples, particularly in 1972, appeared to have undergone protein degradation, as evidenced by mixing times and, in certain instances, by loaf volumes that were materially less than those of comparable dryland samples. Other irrigated wheats in 1958 and 1959 had fruiting periods that were not much longer than those for dryland samples and had mixing times that were equal to or longer than those of dryland wheats.

Loaf volume of bread wheats is linearly related to protein content of the flour. Where loaf volume was not limited by the effects of high field environmental temperatures, protein content accounted for 95% of the variability in volume (Finney & Fryer, 1958). Although high wheat protein content (up to 17–18%) normally contributes to increased gluten strength, Tipples et al. (1977) found that in several instances, irrigated hard red spring wheat with protein content of over 17% showed a marked weakening of physical dough characteristics and a deterioration of baking quality.

High temperatures (above 32 °C) during the latter part of the grain-filling period reduce protein content (Smika & Greb, 1973) and loaf volume (Finney & Fryer, 1958). Finney and Fryer (1958) found that varieties with long dough mixing times were more tolerant to detrimental effects of high temperatures on loaf volume. When considering temperature effects in subhumid eastern Kansas compared with semiarid regions of the Great Plains, Finney and Fryer (1958) indicated that the continuation of water use through grain filling reduces the detrimental effects of high temperatures on loaf volume. Musick et al. (1963) found that adequate irrigation increased loaf volume per unit of grain protein when compared with treatments that experienced water deficits. Canopy temperatures are normally cooler than air temperatures (Howell et al., 1986; Steiner et al., 1985) and improved protein quality may be associated with transpirational cooling. Evans et al. (1975) indicated that increasing temperature during grain filling has a similar effect to water stress on increasing grain protein. However, Smika and Greb (1973) found that the late grain-filling effect was curvilinear, with protein increasing as temperature increased from 22 to about 30 °C and then decreasing in the 32 to 37 °C range.

Irrigation, by reducing water deficits and increasing yields, strongly interacts with N fertility in influencing protein content and quality (Fernandez-G. & Laird, 1959; Musick et al., 1963; Jensen & Sletten, 1965; Bole & Dubetz, 1986; Eck, 1988). In studies in the southern High Plains, Jensen and Sletten (1965) found that lower rates of applied N increased yields with little effect on protein, while higher N rates had a reduced effect on yields per unit of N and an increased effect on protein content. Low N rates that increase relative yield in the range of 0.2 to 0.5 have been shown to reduce grain protein (Fernandez-G. & Laird, 1959; Strong, 1981).

The interrelationships of yield and protein content of wheat grown under adequate irrigation and involving water deficits during GS2 and GS3, along with N rates ranging from deficient to excessive for maximum yields, were reported by Musick et al. (1963) and Eck (1988). Where yields were reduced to about 0.5 by severe stress in the study by Eck (1988), applied N did not affect yields but markedly increased protein content within a range of 10.5 to 16%. As water deficits were reduced, applied N increased yields with only a modest increase in protein content until maximum yields were attained for the water level involved. After attainment of maximum yields, the increase in protein averaged 1% per 42 kg/ha of applied N to a maximum protein content of 16%.

The results by Eck (1988) were obtained on a slowly permeable clay loam that normally experiences low profile drainage and N leaching from irrigation. Ritter and Manger (1985) stated that NO_3 leaching, which can be substantial on some soils where all the N is applied before planting, is directly related to profile drainage volume and is associated with low irrigation application efficiencies. They recommended applying only enough N fertilizer to meet crop requirements for a realistic yield goal. However, some additional N application may be desirable to increase the protein content and improve baking quality of the irrigated hard red wheats. Spratt and Gasser (1970) found that N applied to spring wheat at boot stage increased dry matter much less than application at planting, but caused a greater increase in the concentration of N in the grain.

For quality of pastry flour, the soft white wheats should contain less than 10.5% protein (Bole & Dubetz, 1986). Adequate irrigation is required through grain filling to prevent excessive protein content. They indicated that excessive protein in soft wheats can be prevented by limiting fertilizer N plus soil NO_3 N to 30 kg/ha for each Mg/ha increase in target yield.

Finney et al. (1987) indicated that sufficient protein is necessary in durum wheat; generally, semolina protein of 11 to 12% (14% moisture basis) is sufficient for gluten strength. They also indicated that as with other classes of wheat, the protein content of durum is controlled by fertilizer, environment, and heredity and that the protein content of durum generally ranges over 13% in the whole grain.

Inherent differences in protein content occur among cultivars of all classes of wheat, but differences in protein of wheat of a given class are closely associated with cultural practices and environmental factors directly or as they influence yield. Johnson (1978) stated that protein levels of hard red winter wheat in the Great Plains began a steady downhill trend in 1964 as farmers became actively concerned with high-yielding wheat cultivars. In many cases, inherently high-yielding cultivars may sacrifice protein content for yield. However, this trend need not continue, for there are cultivars like the hard red winter wheat cultivar Lancota, numerous experimental lines, and germplasm releases that have the potential for elevated protein, even at high yield levels (Johnson, 1978). In addition, Stein et al.(1987) indicated that the use of the rapid near-infrared screening procedure for protein will permit simultaneous selection for both grain protein and yield.

Good quality, high-yielding irrigated wheat of all classes, as related to intended end use, is possible if proper attention is given to irrigation practices, soil fertility levels, other cultural practices, and cultivar selection.

II. EVAPOTRANSPIRATION

Evapotranspiration (ET) is a complex energy-driven process, but in simple terms, it is primarily influenced by the evaporative demand of the climate, the nature and extent of vegetative ground cover, the soil and plant water status, and the length of the growing season. Evapotranspiration and methods of measurements are discussed by Ritchie and Johnson (chapter 13 in this book) and by Hatfield (chapter 15 in this book). Evapotranspiration is widely used in determining water requirements, scheduling irrigations, and assessing crop growth and yield response to water and water deficits. In this section, we discuss a review of ET data by adequately irrigated wheat obtained from publications of field and lysimeter studies.

Papers from 26 studies containing seasonal ET data from sites in the USA, Britain, Zambia, Israel, India, and Australia were reviewed. The lowest seasonal ET values of 300 to 360 mm were measured from both short growing seasons in India (Prashar & Singh, 1963; Rao & Bhardwaj, 1981; Reedy & Bhardwaj, 1982) and a cool growing season at Copenhagen, Denmark (Mogensen et al., 1985). The highest seasonal ET of 818 mm was reported from Australia (Cooper, 1980).

Data from the different climatic environments indicated, in general, that for fall-planted spring wheat, the longer the growing season, the higher the seasonal ET. For the relatively short growing seasons, ET of 300 to 350 mm were measured in the India studies and in Zambia (Bunyolo et al., 1985). In the southwestern USA, where growing seasons are longer and development extends into periods of high temperatures, seasonal ET was measured in the 650 to 700 mm range in Arizona (Erie et al., 1973, 1982) and in the Imperial Valley, California (Ehlig & LeMert, 1976). Maximum daily ET rates are strongly influenced by climatic environment and average 2.5 to 3 mm/d in Britain (Monteith & Scott, 1982); 3 to 5 mm/d in the shorter seasons and, thus, in cooler environments of India and Israel where maturity occurred before high temperatures developed (Prashar & Singh, 1963; Shimshi et al., 1973); and 8 to 9 mm/d in the warmer environments during grain filling in the southwestern USA (Erie et al., 1982; Ehlig & LeMert, 1976). The maximum daily values for fall-planted spring wheat grown in the southwestern USA were similar to values for winter wheat grown in the central and southern High Plains (Jensen & Musick, 1960; Jensen & Sletten, 1965; Musick et al., 1963; Shawcroft & Croissant, 1986).

In the High Plains, seasonal ET has been measured in irrigation studies ranging in N latitude from 35 (Bushland, TX) to 38° (Garden City, KS) and 40° (Akron, CO). Seasonal ET at the three locations averaged 710 mm at Bushland (Jensen & Sletten, 1965; Schneider et al., 1969; Musick et al., 1984; Eck, 1988), 610 mm at Garden City (Musick et al., 1963), and 503 mm at Akron (Shawcroft, 1983). As latitude increased by 5° from Bushland to Akron, planting dates were advanced by about 3 wk, while anthesis dates were delayed by 3 wk. Thus, for winter wheats, lengthening the growing season by increasing latitude reduces seasonal ET in association with increased length of winter dormancy and lower temperatures during GS1 and GS2. Temperatures during GS3 in the irrigated production regions of the western High Plains are not appreciably different unless associated with differences in elevation; peak ET rates, as shown by studies at Bushland, Garden City, and Akron were similar (Jensen & Musick, 1960; Shawcroft & Croissant, 1986).

Daily ET rates for late fall-planted winter wheat were measured in lysimeter studies at Kimberly, ID, and compared with alfalfa as a reference crop (Wright, 1982). The ratio of wheat to alfalfa ET (basal crop coefficient) for a late fall planting increased to 1.0 after about 4 wk of spring growth. Normal planted winter wheat in Idaho that tillers extensively in the fall develops effective ground cover for a basal crop coefficient of 1.0 after about 2 wk of spring growth (Wright, 1982).

A ratio of daily measured lysimeter ET to Class A pan evaporation at Brawley, CA, indicated that fall-planted spring wheat reached the potential rate about 60 d after irrigation for emergence (Elhig & LeMert, 1976). Evapotranspiration continued at the potential rate of 0.8 of Class A pan evaporation for about 80 d before declining rapidly with loss of green vegetation.

Green LAI peaks before anthesis and gradually declines during grain filling, while ET increases due to increasing climatic evaporative demand, until it peaks at about milk stage. Seasonal ET curves suggest that the emergence of wheat heads and their prominence in interception of incoming solar radiation apparently does not result in reduced ET rates below potential. The ET decline below potential becomes pronounced during late grain filling as the loss of greenness accelerates approaching senescence. Wright (1982) indicated senesced wheat at maturity reaches a basal coefficient of 0.15 (a similar value to other senesced crops). Ehlig and LeMert (1976) indicated a low ET ratio to pan evaporation of 0.2 after surface drying following irrigation for emergence and a ratio of 0.15 after complete senescence. The lower value after maturity suggests the senesced canopy reduces soil evaporation.

III. WATER DEFICITS

A. Plant Water Deficits

Water relations of wheat were reviewed by Jones (1977) and Kirkham and Kanemasu (1982). According to Turner (1986), plants can be classified into thre categories for drought resistance. Wheat is classified in the category of "drought tolerance with low plant water potential." This drought resistance category has the following adaptation mechanisms: maintenance of turgor, osmotic adjustment, increase in cell elasticity, decrease in cell size, desiccation tolerance, and cell protoplasmic tolerance. Wheat is not sensitive to stomatal closure (Morgan, 1977), and Meyer and Green (1980) indicated that ET reduction under severe stress may result from increased soil-root resistance rather than stomatal control of water loss.

The drought tolerance of wheat reduces critical stage sensitivty for yield compared with many other crops and permits irrigation management involving a wide range of allowable water deficits. Wheat genotypes can range widely in osmotic adjustment (Morgan, 1977). However, in the southern High Plains, six commonly grown cultivars maintained leaf turgor pressures above 0.6 MPa under a wide range of water deficits, indicating marked osmotic adjustment (A.C. Mathers, 1985, personal communication). Minimum turgor pressures under severe stress were more than double those obtained from tests with corn, sorghum [*Sorghum bicolor* (L.) Moench], soybean [*Glycine max* (L.) Merr], and sunflower (*Helianthus annuus* L.). Osmotic adjustment may tend to be a characteristic of wheat varieties grown in semiarid environments since the varieties have been selected for performance under both irrigated and dryland conditions.

Winter wheat is grown in cooler environments than summer crops, and the normally slow development of water deficits in a field environment may contribute to the marked osmotic adjustment found in wheat compared with summer crops that experience more rapid stress development. When leaf water potentials of nonstressed wheat were in the range of -1.6 to -2.0 MPa, severely stressed wheat had potentials of -3.0 MPa by heading and to -4.0 MPa or lower during grain filling before leaf death occurred (Ehrler et al., 1978; Fischer & Sanchez, 1979; Sojka et al., 1981). Osmotic adjustment is influenced by solute accumulation in cells, and osmotic potentials vary in diurnal cycles similar to leaf water potentials (Morgan, 1984). When deficits are terminated by irrigation, cell metabolic processes may use the solutes and cause a loss of osmotic adjustment within a few days. Osmotic adjustment during stress contributes to maintaining turgor in developing heads prior to anthesis and in the developing root systems (Turner, 1986).

Papendick et al. (1971) found that winter wheat has the ability to deplete soil water on Ritzville loam (coarse-silty, mixed, mesic Calciorthidic Haploxerolls) to -4.0 MPa or lower and that the lower limit of soil water potential cannot be precisely defined. This remarkable ability of wheat to thoroughly deplete the profile of available water is likely associated with osmotic adjustment of roots, and Fischer (1980) suggested the maximum extent of root zone soil drying may be more dependent on plant water potential gradients than on rooting density.

Since plant water deficit values are sensitive to evaporative demand, their use in irrigation scheduling may require greater knowledge and sophistication than the use of soil water deficits. With experience, plant water deficit effects can be assessed visually in a qualitative way. Visual appearance of stress effects on reduced growth are extensively used in conjunction with development stages and the calendar for irrigation scheduling in the Great Plains.

B. Soil Water Deficits

The use of soil water deficits has a long history in irrigation scheduling and is receiving renewed emphasis in soil-water-plant relations studies. In reassessing plant adaptation to water deficits, Turner (1986) concluded that a reduction in leaf area appears to be largely affected by soil water status and root hydration and is a response rather than an adaptation to water deficits. Meyer and Green (1980), using lysimeters in South Africa, found that expansive growth declined when soil water in a 1.0-m profile was depleted below 33% available. To prevent the reduction in expansive growth, which was also influenced by temperature and evaporative demand, they recommended irrigation scheduling based on 50% available soil water (ASW) depletion. The decline in ET did not occur until depletion was below 20% ASW.

Musick et al. (1976) found that in the southern High Plains, soil water depletion to the −1.5-MPa potential, 1.2-m profile depth, resulted in a 20% yield reduction in relatively dry seasons and no reduction in a season when profile drying occurred in conjunction with periodic rewetting of the surface soil by rainfall. These studies, conducted on Pullman clay loam (fine, mixed, thermic family of Torrertic Paleustolls), indicated the ability of wheat under prolonged stress to deplete profile water to about 50 mm or 35% below the −0.03- to −1.5-MPa field capacity to wilting point water content range. When Jensen and Sletten (1965) weighted the water potentials of the root zone by 4, 3, 2, and 1 to correspond to an approximate 40, 30, 20, and 10% depletion pattern by rooting depth increments for irrigated crops, ASW depletion to −0.4 MPa did not affect yields, but depletion to −0.9-MPa potential reduced yields by 17%.

Plant water deficits can be taken as predawn water potentials that approach root zone soil water potentials. Sojka et al. (1981) found that a decline in predawn potential to −0.7 MPa on a clay soil in northwest Mexico resulted in 40% yield reduction. When severe deficits were terminated by irrigation, wheat rapidly recovered to normal water potentials of nonstressed wheat.

Simulation modeling can be useful in assessing optimal irrigation policy. An optimized simulation study by Yaron et al. (1973) indicated depletion of ASW to 13% for a loess (silt) profile to 1.5-m depth. The results were similar to results obtained from field tests on a deep loess profile by Musick et al. (1963) in southwest Kansas. Depletion of ASW (1.8-m depth) to 10% during milk stage reduced yields by 6% and increased water-use efficiency (WUE).

The recommendation by Meyer and Green (1980) that irrigation be scheduled by 50% ASW depletion to prevent reduction in expansive growth may not apply to water-yield optimization of winter wheat when some reduction in expansive growth is allowable. Before the introduction of the stiff straw semidwarf cultivars in the Great Plains, reduction in expansive growth during early GS2 by delaying irrigation was commonly practiced to limit height and lodging. Spring wheat is more susceptible than winter wheat to early GS2 soil water deficits because of a less developed root system (Gales, 1983). Growth of the root system increases the allowable soil water depletion before deficits become limiting to growth.

IV. IRRIGATION MANAGEMENT

A. Irrigation Water Requirements

Irrigation water requirements consist of initial application for crop establishment and recharging the soil profile, and seasonal irrigations to meet ET demands. Seasonal irrigation requirements can be predicted based on water budget procedures using available soil water storage, allowable depletion level, ET prediction, and irrigation application efficiency (see chapter 17 in this book). Irrigation water requirements include losses associated with application efficiencies. In the Texas High Plains, furrow application efficiencies were estimated from irrigation inventories as mostly in the range of 50 to 80% without considering tailwater reuse and were higher on the slowly permeable clays. Sprinkler application efficiencies were measured from 223 field evaluations of center pivot systems as mostly in the 80 to 90% range and were higher under lower windspeeds (Musick et al., 1988).

Several studies have related ET and irrigation requirements for wheat to the National Weather Service Class A pan evaporation (E_{pan}) (Miller & Hang, 1982; Agarwal & Yadav, 1978; Bunyolo et al., 1985; Choudhary & Kumar, 1980; Jalota et al., 1980; Prihar et al., 1976; Singh, 1978). Evapotranspiration averaged approximately 0.8 of E_{pan} after full ground cover (Ehlig & LeMert, 1976; Shimshi et al., 1981).

Wheat normally develops full ground cover before seasonal irrigations are needed. Irrigation studies in India indicate that application based on 0.75 to 0.8 of E_{pan} provided adequate water, while Miller and Hang (1982) found that daily sprinkler application based on 0.95 E_{pan} replacement on a relatively high water-storage soil was excessive and caused yield reduction. A daily deficit application of about 0.4 E_{pan} resulted in maximum yield of winter wheat in a moderate evaporative demand climate at Prosser, WA, when the irrigation season was started with a wet soil profile. The study by Miller and Hang (1982) demonstrates the success of allowing depletion of profile soil water while avoiding critical water deficits. Critical deficits were minimized by the daily deficit irrigations that allowed plant conditioning to slowly developing water deficits.

In India, where the common practice is to apply five seasonal irrigations based on stage of development, a scheduling procedure based on water application as 0.75 E_{pan} permitted deleting early season irrigations and reducing irrigation water requirements (Prihar et al., 1976). This study demonstrated the value of an irrigation scheduling method that adjusts water application to climatic evaporative demand and allowable soil water depletion, particularly for wheat grown on high water-storage soils.

1. Preseason Irrigation

Irrigation management has emphasized beginning the growing season with a wet soil profile. In regions where rainfall is inadequate for soil profile wetting, emphasis has been given to an irrigation before planting or for crop emergence when planted into dry soil. In areas with dry seasons, an irrigation may be applied before planting to facilitate tillage and seedbed preparation and to control crop volunteer plants and weeds (Echert et al., 1978). In the study by Echert et al. (1978), seeding into dry soil and irrigation for emergence successfully replaced the preplant and the first postplant irrigation and the reduced application greatly reduced drainage losses to groundwater. In areas of dry summers and late fall and winter rainfall, seeding into dry soil and irrigating for emergence results in planting on an optimum date for growth and yield and is an efficient water management practice.

In the Great Plains, irrigated wheat is mostly planted as continuous cropping, after summer fallow, or, to a lesser extent, after harvest of a summer crop such as corn. The priority for water supplies from wells is usually given to summer row crops, and preplant irrigation is not commonly practiced because of conflicts in water demands. Germination results from seeding into moist soil, from rainfall after planting, or from an emergence irrigation. A common practice is to apply the initial irrigation sometime after emergence to rewet the soil profile. When soil water conditions are adequate for a period of growth after emergence, the initial irrigation may be delayed or deleted. The use of summer fallow for soil profile water storage at planting reduces the need for an initial irrigation before a period of substanital water use by the crop.

Because of primary tillage effects on increasing water intake, application amount by surface methods during an initial preplant or emergence irrigation frequently is about double that normally applied in subsequent seasonal irrigations (Jensen & Sletten, 1965). Due to surface sealing and runoff problems, large sprinkler applications are not made on bare soil, and preplant or emergence irrigation to recharge the soil profile is not normally practiced.

2. Seasonal Irrigation

Irrigation of wheat is widely practiced in regions where seasonal irrigation by surface methods ranges from only one application to as many as six or seven, depending on seasonal ET requirements, initial soil water storage, seasonal rainfall, application amounts, and yield goals. Where irrigation is practiced as a supplement to rainfall, such as in Israel, the common practice is to initially irrigate after seeding into dry soil for timely stand establishment and to apply a second irrigation for grain filling, depending on April rainfall (Shimshi et al., 1973). When wheat is grown as a dry season crop, an adequate irrigation level that does not limit ET and yields can be four to five applications in India (Lal, 1985); three to four applications in Arizona (Erie et al., 1973); five to seven applications in the Imperial Valley, California (Ehlig & LeMert, 1976); six to seven applications in northwest Mexico (Fischer et al., 1977); and seven applications in southeast Australia (Cooper, 1980). In many of these studies where frequent irrigations were applied, high yields attained were in the 7 to 8 Mg/ha range. On slowly permeable clays, the practice of frequent irrigation may not appreciably increase irrigation water requirements because low permeability limits water intake and application amounts during irrigation.

In the semiarid Great Plains, precipitation frequencies decline in the fall as the crop is established, increase with spring growth, and peak during grain filling. Irrigated wheat production is mostly concentrated on the fine-textured soils that are moderately deep and relatively high in water storage. Tests at Garden City, KS (Musick et al., 1963; Hooker et al., 1983), and at Akron, CO (Shawcroft & Croissant, 1986), indicate that when an initial irrigation is used to wet the profile, seasonal irrigation is not needed before boot stage, and irrigation that rewets the profile at this stage provides the irrigation water requirements for high yields. Precipitation during grain filling reduces and frequently eliminates the need for a second spring irrigation. In the central Great Plains, where sprinkler irrigation is practiced on relatively high water-storage soils, restricting water application to the boot to heading period reduces the risk of lodging (R.W. Shawcroft, personal communication).

In the Texas High Plains, wheat is mostly grown on slowly permeable clay loams that are surface irrigated. Three to four seasonal irrigations are required to meet ET demand and produce high yields (Jensen & Sletten, 1965; Schneider et al., 1969; Musick et al., 1984). When fine-textured soils such as Pullman clay loam are fully wet in the fall, a common irrigation schedule is to apply the first irrigation by jointing stage, the second during boot stage, and the third during grain filling. In a 5-yr test, Musick et al. (1984) found that during 2 yr when spring rainfall was above normal, no yield response occurred from an irrigation during grain filling. When wheat was irrigated in the fall on Sherm clay loam (fine, mixed, mesic family of Torrertic Paleustolls) at Etter in the Texas High Plains, yield response occurred from irrigation during jointing in April but not from irrigation during tillering in March (Shipley & Regier, 1972b). In south-central Washington on Ritzville fine sandy loam (coarse-silty, mixed, mesic family of Calciorthidic Haploxeralfs), Robins and Domingo (1962) found that when wheat was initially irrigated to begin the season with a wet profile, further irrigation was not needed until boot stage. Delaying irrigation past early vegetative growth reduced height and lodging with little or no reduction in grain yield. In the climatic environment of dry summers in the Pacific Northwest, irrigation of spring wheat during grain filling was necessary for high yields (Robins & Domingo, 1962).

B. Critical Stages

With wheat, a crop that has excellent drought resistance, development stages are not as critical as for other crops that are more sensitive to critical-stage water deficits. However, the critical stage terminology is extensively used in the literature for irrigation of wheat to indicate that some stages are more critical than others. Sensitivity to water deficits most frequently relates to physiological and morphological responses that reduce seed numbers per m^2. However, studies have reported critical stage effects that range from crown root initiation (tillering) to grain filling.

As a cool-season crop, wheat is mostly grown in relatively low to moderate evaporative demand climates. Under these conditions, stress normally develops slowly in a field environment, and adaptive response can limit yield reductions. Since critical stage effects are normally more pronounced through their effect on the grain yield component of grain number per m^2, they are more likely to occur during preanthesis. However, both the evaporative demand and the rapidity of water deficits increase as the season progresses; some studies have reported the critical stage effect as continuing well into grain filling (Misra et al., 1969; Schneider et al., 1969; Shipley & Regier, 1972b).

In controlled environment studies, Fischer (1970) reported the most critical stage as 10 d before anthesis. In later field studies of wheat grown in dry seasons in northwest Mexico, Fischer et al. (1977) reported the most sensitive stage as 25 d before to 20 d after head emergence (to about one-half grain filling). Milk stage, which occurs at about one-half grain filling, is the most sensitive stage for hot, dry wind effects of shrivelled grain and is prob-

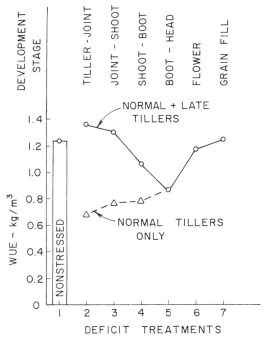

Fig. 20-3. Development stage effects of water deficit periods from tillering to grain filling of spring wheat on water-use efficiency (WUE), including stress-recovery compensation by late-developing tillers through boot stage (Copenhagen, Denmark, 55°N) (adapted from Mogensen et al., 1985).

ably a combined effect of high temperature and lowered plant water potential (Fischer, 1980). Sensitivity to hot, dry winds on grain shrivelling is reduced by high soluble sugars (5–7%) in plant culms at anthesis. Late grain filling is less sensitive to water deficits, primarily because of relocation of preanthesis assimilates to grain filling. Major depletion of profile soil water storage after milk stage is commonly practiced and contributes to efficient water use. Also, in the Great Plains, increasing spring rainfall often limits the extent of water deficits during grain filling.

Studies that report crown root initiation (CRI) as the most critical stage involve fall seeding of spring wheat on sandy soils in the Indian subcontinent (Agarwal & Yadav, 1978; Choudhary & Kumar, 1980; Gajri & Prihar, 1983; Lal, 1985; Singh et al., 1979a). In studies on a clay soil, Fischer et al. (1977) found that crown roots become established from irrigation at planting and that delay in irrigation past CRI from 27 to 50 d did not affect yield, while Srivastava and Bansal (1975) found that a 2-wk delay on mixed black soils did not affect yield. Although the CRI irrigation enhanced rooting in sandy soils (Singh et al., 1979a), the yield response is more likely associated with increased tillering (Gajri & Prihar, 1983) and head numbers per m^2 (Agarwal & Yadav, 1978). In this short growing season environment, the adjustment in increased seed per head failed to compensate for the water stress effects on reduced tillering. In studies at Copenhagen, Denmark (55 °N lat), where spring wheat is grown as a summer crop, the tillering to jointing stage was determined by Mogensen et al. (1985) as the most critical if late developing stress-recovery tillers were excluded. Development stage effects of water deficit periods from tillering through grain filling on WUE are illustrated in Fig. 20–3. The compensation effect of late stress-recovery tillers declined to zero by boot stage, and boot to heading was determined as the most critical stage on yields and WUE.

In the southwestern USA, an initial irrigation (preplant or emergence) provides soil water storage that limits the development of critical water deficits during tillering; the jointing stage becomes critical (Day & Intalap, 1970; Ehlig & LeMert, 1976). In the study by Day and Intalap (1970), a stress period during jointing reduced seed numbers per m^2 by 45%. Stress during flowering and grain filling were only modestly less critical than during jointing, indicating only moderate differences in development stages to stress sensitivity.

Winter wheat grown in the Great Plains has a relatively long period for tillers to develop prior to FI, and tillers and potential head numbers are not usually critical yield components. The critical stage is reduced grain numbers per m^2, both as a combined influence of fewer heads (tiller senescence) and fewer seeds per head. This stage has been indicated as critical for irrigation by Musick et al. (1963), Shipley and Regier (1972b), Robins and Domingo (1962), and Schneider et al. (1969). In areas having a Mediterranean climate, irrigation may be needed only for timely establishment of the crop near the end of the dry season and to insure grain filling after the rains stop (Shimshi & Kafkafi, 1987); thus, grain filling becomes a critical stage for irrigation.

C. Water-Yield Relationships

Water-yield relationships have been determined between ET and yield and between irrigation water and yields. Relationships between ET and yield have been determined as both linear (Hunsaker & Bucks, 1987; Steiner et al., 1985) and curvilinear (Ehlig & LeMert, 1976; Musick et al., 1963; regression analyses of data by Sharratt et al., 1980). In addition, linear relationships have been determined from ET deficits during selected growth stages by Schneider et al. (1969) and during postanthesis grain filling by Aggarwal et al. (1986). In the high seasonal ET range, a decline in the yield response can be influenced by ET data calculated from soil water depletion that included some profile drainage (Sharratt et al., 1980).

In the high yield range, dry matter may be more responsive to increased ET than grain yield. Shipley and Regier (1972b) found that irrigation during early GS2 increased straw yields by 24% but had no effect on grain yields, since grain numbers per m^2 were not increased. When water deficits were limited to late GS2 and GS3 at Akron, CO, the ET-yield relationship was linear. However, irrigation applied during early GS2 increased ET but reduced grain yield, causing the ET-yield relationship to become curvilinear (Shawcroft, 1983). In the studies where the ET-grain yield relationship is curvilinear, the response decreases with increasing ET (Ehlig & LeMert, 1976; Musick et al., 1963).

Yield relationships to applied irrigation that exclude surface runoff are curvilinear diminishing return functions when the range in water applied is due to differences in scheduling (Echert et al., 1978; Hunsaker & Bucks, 1987;

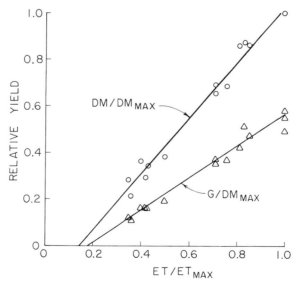

Fig. 20–4. Relative yield relationships of spring wheat dry matter (DM) to nonstressed DM_{max} and grain yield to DM_{max} over a range of water deficits (relative ET/ET_{max} of 0.35–1.0) from a sprinkler line gradient test that varied application depths only (adapted from Sharratt et al., 1980).

Shimshi, 1979). The exception may be water applied by the sprinkler line gradient method, which can be managed to vary only the amount applied without varying the timing (Sharratt et al., 1980). When profile drainage is avoided and profile storage is mostly depleted by maturity, water applied is utilized as ET and the water applied-yield relationships can become linear. An example of linear relationships of dry matter and grain yield, with ET expressed on a relative basis of ET_{max} and dry matter and grain yield expressed on a relative basis of DM_{max}, is illustrated in Fig. 20-4. The relationships in Fig. 20-4, calculated from sprinkler line gradient data by Sharratt et al. (1980), illustrate the grain yield threshold (about 0.2 ET_{max}) before the first yield increment and the stable HI when the grain yield is visually shown as a fraction of the dry matter.

D. Water-Use Efficiency

Water-use efficiency is defined as grain yield per unit of ET and is expressed as kg/m^3. Water-use efficiency of fall-planted irrigated spring wheat frequently is in the range of 1.0 to 1.2 kg/m^3 (Aggarwal et al., 1986; Bunyolo et al., 1985; Cooper, 1980; Lal, 1985; Shimshi & Kafkafi, 1978; Singh & Dastane, 1971). Some irrigation studies have resulted in a higher range of WUE, such as 1.5 to 1.9 kg/m^3 by Rao and Bhardwaj (1981), 1.4 to 1.5 kg/m^3 by Ehlig and LeMert (1976), and 1.2 to 1.6 kg/m^3 by Fischer (1970). Several studies have indicated that WUE values are higher under deficit conditions, especially when irrigation is applied in relation to critical stages (Ehlig & LeMert, 1976; Fischer, 1970; Lal, 1985; Miller, 1977; Rao & Bhardwaj, 1981; Schneider et al., 1969; Singh et al., 1979b). Effects of development stage deficits on WUE that indicate a critical stage effect during boot to heading are illustrated in Fig. 20-3.

Although irrigated wheat can be grown over a relatively wide range of water deficits and yields, the use of irrigation to manage water deficits normally prevents the reduction of WUE. However, in dryland cropping where water deficits cannot be controlled by irrigation, the severe deficits that occur in the southern High Plains can considerably lower WUE. In 25 yr of dryland wheat after summer fallow at Bushland, TX, WUE averaged 0.35 kg/m^3, for average yields of 1.1 Mg/ha (O.R. Jones, 1987, personal communication). Water-use efficiency of dryland wheat averaged about one-half the WUE of irrigated wheat grown over a wide range of water deficits with yields of 3 to 6 Mg/ha (Musick et al., 1984).

Water-use efficiency values are sensitive to yield levels and have substantially improved with the release of higher-yielding cultivars. In the southern High Plains, WUE increased from 0.44 kg/m^3 with 'Concho' grown in the late 1950s (Jensen & Sletten, 1965) to 0.54 kg/m^3 with 'Tascosa' grown in the late 1960s (Schneider et al., 1969) to 0.94 kg/m^3 with semidwarf cultivars grown during 1979 to 1982 (Musick et al., 1984). Irrigation tests in the Great Plains, Washington, and Utah have indicated WUE values to be mostly in the 0.8 to 1.0 kg/m^3 range (Miller, 1977; Musick et al., 1984; Sharratt et al., 1980; Shawcroft, 1983). Where the ET-grain yield relationship is linear, the response slope reflects a higher WUE value than for the maximum ET-yield value. For example, Steiner et al. (1985) determined by linear regression a response slope value of 1.6 kg/m^3 compared with WUE of 1.3 kg/m^3 for maximum yield.

The WUE of seasonal irrigations (IWUE) applied to level plots without surface runoff is usually similar to or only slightly lower than seasonal WUE values (Musick et al., 1984). However, IWUE values can be quite low when irrigations are applied that result in low yield response, and low IWUE values are more likely to occur in the higher water application and yield range.

V. PLANT NUTRITION

A. Fertilizer Requirements and Nutrient Uptake

Fertilizer requirements and nutrient uptake are the supply and demand sides of an equation that is influenced by many other factors. For N, the supply side includes soil-available N at planting and a somewhat uncertain quantity of N mineralized during the growing season. Also, some irrigation waters provide a significant source of N. Factors such as soil water, climatic conditions, cultural practices, and plant growth and development affect the uptake of N; the total plant need for N is somewhat difficult to accurately predict.

Irrigated wheat is mostly grown in semiarid to arid regions where soils are not deficient in K; thus, studies have emphasized N and P deficiencies and fertilizer requirements. Nitrogen is the plant nutrient that is normally most limiting and has been given the most emphasis in research.

Uptake efficiencies of applied N can be influenced by many variables and range from about 40 to 80%. When the N supply is variable over a wide range from severely depleted to luxurious, plant N concentration relative to dry weight is also variable (Viets, 1965). In a water-fertilizer study involving applied N rates of 0, 70, 140, and 210 kg/ha, Eck (1988) found near maximum grain yields of 6 Mg/ha with 140 kg/ha of applied N, which resulted in above-ground N uptake of 150 kg/ha. The 210 kg/ha rate failed to further increase grain yield but increased N uptake to 174 kg/ha.

Estimates of nutrient uptake for wheat yields of 6.7 Mg/ha by the Soil Improvement Committee, Fertilizer Association (1975) were 196, 39, and 130 kg/ha for N, P, and K, respectively. Several publications reporting results of irrigation-fertility studies indicate N fertilizer requirements for maximum yields with adequate irrigation to be in the range of 80 to 140 kg/ha, while maximum applied rates were as high as 220 kg/ha.

Phosphorus fertilizer requirements for maximum yields were mostly in the 20 to 40 kg/ha of applied P. Phosphorus fertilizer uptake is most efficient when band-applied near the seed at planting. A large broadcast application can be adequate for multiple crops, but total uptake efficiencies may be reduced. Land leveling is common in surface irrigation. Since P levels are frequently higher in the surface soil, removal of surface soil can result in P deficiency.

The ability of soils to supply N requirements of irrigated wheat varies widely, ranging from no more than 5% in some desert soils that are low in organic carbon to some productive soils high in organic carbon that can supply adequate N. In the water-fertility study by Eck (1988) on Pullman clay loam (1.5% organic carbon in 0.3-m depth), growing season mineralization of N supplied 60 kg/ha for irrigated wheat.

Wheat requires approximately 30 kg/ha of available soil N per Mg of grain produced (Tucker & Murdock, 1984). In the southern Great Plains, where irrigated wheat is grazed during the fall and winter and then managed for grain production, additional applied N of about 20 kg/ha is needed per additional Mg/ha of forage production available for grazing (Tucker & Murdock, 1984).

Under irrigation, applied N rates for the southern Plains and intermountain states range as high as 180 kg/ha (Tucker & Murdock, 1984; Westfall, 1984). On the slowly permeable clays with low leaching losses, N is mostly applied before planting as anhydrous NH_3. Split applications are more common on the moderately permeable soils that experience deep percolation of water and leaching of N. On these soils, a common practice is to apply one-half before planting and the remaining one-half during early spring tillering before rapid growth begins. Wheel track damage from ground-driven application equipment is less during tillering than later during culm elongation (jointing). When tractor passes after jointing are required for multiple applications of N or other chemicals, unplanted rows as traffic zones can minimize damage to the growing crop.

Under sprinkler irrigation, N solutions can be injected into irrigation water and applied in single or multiple applications. Injection of N into furrow irrigation water is not commonly practiced because of nonuniform application and losses in tailwater runoff. Soil tests are recommended for determining residual nitrates before planting and for estimating fertilizer requirements. Plant analysis can be useful in assessing N deficiency. Both total plant N uptake (aboveground) and NO_3 concentration in cell sap may be used. However, reliable, rapid methods of measuring tissue N status are lacking. Nitrogen may be applied late in the season to increase grain protein content and to correct late-season N deficiency associated with excessive irrigation and N leaching.

High levels of crop residues at planting in conservation tillage systems increase the need for applied N. Saffigna et al. (1982) found that 5 kg/ha of applied N was immobilized per t of incorporated residues. Frequent irrigation that maintains moist soil may increase N uptake compared with less frequent irrigation that allows the surface soil layer to remain dry for extended periods of time (Strong, 1981).

B. Growth and Yield Responses

Plants growing on soils that are deficient in N respond to fall N application by increasing seminal rooting (Richman et al., 1985). Since the seminal root system develops early, fall-applied N increased rooting to a much greater extent than spring-applied N. Applied N increases tiller development and associated nodal root development (Black, 1970a; Woodruff, 1980) and reduces tiller abortion during jointing (Power & Allesi, 1978; Rickman et al., 1985).

Nutrient deficiencies cause interruption of the tillering process by preventing elongation of auxilliary buds and by reducing the growth rate of the youngest tillers (Malse, 1985). The slowing of growth spreads progressively to older tillers, then to the main culm. For younger tillers, reduction in growth rate can be rapid and lead to senescence.

Black (1970a) found that early formation of nodal roots was necessary for tillers to develop and produce heads and that the increased heads per m^2 accounted for 97% of the yield response to applied P. Woodruff (1980) found that banded P increased nodal roots and tillers and P uptake to a greater extent than broadcast P. Nodal root development is necessary for adequate N and P uptake by tillers. Deficiencies of N and P can cause cessation of growth (Malse, 1985). Dryness of the surface soil reduces nodal root elongation and N uptake by tillers, thus increasing tiller senescence. Tiller senescence can be identified by yellowing of the terminal leaf.

Except in extreme conditions, nutrient deficiencies do not affect morphological development. Grain yields are reduced through reductions in growth rate and dry matter accumulation and are expressed through the reduction in grain numbers per unit area (both reduced head numbers and grain per head).

Nitrogen deficiencies cause protein deficiencies in leaves, which reduce photosynthesis, leaf area expansion, and dry matter accumulation and ac-

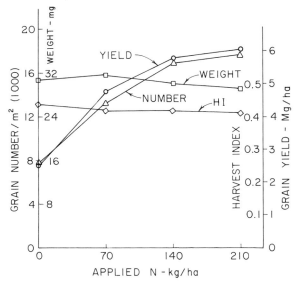

Fig. 20-5. Effect of N application on grain yield and harvest index (HI) and on the yield components of grain numbers/m^2 and weight for adequately irrigated winter wheat, Bushland, TX (adapted from Eck, 1988).

celerate senescence. It has been shown that N application increases the N content of leaves and delays senescence, which, in the absence of water deficits and high temperatures, extends the grain-filling period (Spiertz & Ellen, 1978). After anthesis, N compounds are relocated from vegetative parts and roots to filling grain (Gregory et al., 1981). Under adequate irrigation and fertility, N available for relocation can be estimated from the difference between the N content at anthesis (about 150 kg/ha) and the N residue in straw and chaff (about 60 kg/ha).

Late-season lodging is a problem associated with excess N and adequate irrigation. Excess N increases luxuriant growth and susceptibility to lodging, while lodging reduces photosynthesis and grain-filling rates and cause shrivelling of grain. The irrigated semidwarf cultivars with stiff straw have greatly reduced lodging problems associated with excess N. This has permitted the use of higher water and fertility levels for increased yields.

Increased yields from applied N are entirely associated with the grain yield component of grain numbers per unit area. In some studies, a fertilizer response that increased grain numbers per m^2 was associated with a slight to modest decrease in grain weight (Shimshi & Kafkafi, 1978; Eck, 1988). Also, the partitioning of dry matter into grain is affected little by N fertilization.

The yield components' responses to applied N are illustrated from data by Eck (1988) (Fig. 20–5). Yield response to applied N under adequate irrigation was about equally accounted for by increased head density and increased grain numbers per head. However, the yield component response to applied N may be cultivar-specific. The high early-season uptake of N and the sensitivity of tiller and floral development to N deficiency suggests

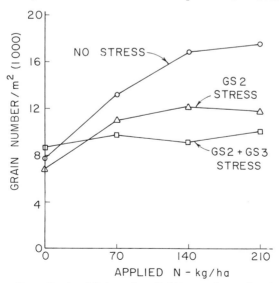

Fig. 20–6. Interactions of water deficits and applied N on grain numbers/m^2 for winter wheat at Bushland, TX (adapted from Eck, 1988). Yields were highly correlated with grain numbers/m^2.

that N application should not be delayed past early GS2. If deficiencies develop during rapid growth, further application may be needed and, under sprinkler irrigation, can be readily applied with the water. Late season-applied N can be effective in increasing grain protein content.

The interaction effects of water and N deficits are primarily expressed through the yield component of grain numbers per m^2. Water and applied N treatments from the study by Eck (1988) illustrate the interaction (Fig. 20–6). A reduction in yields by about one-half due to severe water deficits (during GS2 + GS3) or by a severe N deficiency where none was applied, both had a similar effect in reducing grain numbers per m^2. As the deficits in both water and applied N were conjunctively reduced, grain numbers per m^2 and yields were proportionately increased. Water deficit effects on growth and development normally occur during mid to late season, while major N deficiencies normally develop early and persist. Since water deficits are likely to increase in severity late in the season, they affect both grain numbers and grain weight, while early developing N deficiencies appear to only affect the grain numbers per m^2.

Economic models that use production functions and marginal analysis (additional yield from one additional input unit) are useful in optimization of water and fertilizer inputs. Applied water and N response are normally expressed as curvilinear diminishing return functions. Kloster and Whittlesey (1971) concluded that since the marginal yield response to water and N declined with increasing input, management for maximum yields is inconsistent with profit maximization and economic efficiency in allocating resources. Resource optimization and maximizing net returns are discussed in chapter 17 of this book.

VI. CULTURAL PRACTICES

A. Planting Dates

In most climatic regions in the Northern Hemisphere, winter wheat is planted from about mid September to mid- or late October. In the central and southern Great Plains, where irrigated winter wheat is planted for both grazing and grain production, planting dates are advanced to late August. Very early planting without grazing can cause excessive vegetative growth and increases seasonal water use and the risks of abortive loss of the advanced culms. Loss of culms can result from over-winter lodging of lush growth. In addition, early-planted winter wheat is more likely to be affected by diseases such as leaf rust, viruses, and root and crown rots.

Early planting advances spring development, which increases the vulnerability to freeze damage with culm elongation and elevation of the head above the soil surface. George (1982) indicates −5 °C as a critical temperature for 50% culm sterility when developing heads were 50 to 100 mm above the soil surface.

Late-planted winter wheat can encounter cold temperatures that limit tillers, roots, and leaf area growth which can be associated with reduced yields (Thill et al., 1978). The delay in tillering until temperatures warm in late winter and early spring limits the extent of tillering prior to cessation with FI (Porter, 1985), and delayed planting may necessitate the use of higher seeding rates for adequate head density and yield potential. At higher elevations in the lower latitudes, spring types should be planted late enough in the fall to avoid freeze damage prior to and during the sensitive flowering period.

At lower elevations in the lower latitudes, early planting is desired for development of GS2 and GS3 during the cooler months (Midmore et al., 1984). In these environments, growth continues during the winter and later planting can shift GS2 and GS3 to periods of increasing temperatures that reduce grain numbers per m^2 and grain-filling duration. In higher latitude environments, spring wheat is normally planted as soon after winter as conditions permit (early April through early to mid-May). Delayed planting past mid-May and associated delayed maturity can increase the risk of adverse weather conditions during harvest. In the southern part of the Great Plains winter wheat region, where fall-planted spring wheat may not survive the winter, spring wheat can be planted in February or March, but yields are usually lower than for fall-planted winter wheat.

Fischer (1981b) presented data from studies of fall-planted wheat in five regions that indicated yield reductions per week of delayed planting past the optimum as latitude at location sites declined from 40 °N lat to 18 °S lat. For two locations in India, the yield reductions were 10% per wk at 18 °N lat and 7% at 28 °N lat. For two locations in Australia, the reductions were 6% per wk at 31 °S lat and 4% at 35 °S lat; and for one location in Turkey, 5% at 40 °N lat. About a 4-wk delay in planting at Pullman, WA (about 47 °N lat), from an early September date and at Bushland, TX (35 °N lat), from a mid-October date reduced grain yields by 6% per wk (Thill et al., 1978; Musick & Dusek, 1980). For spring-wheat grown at Prosser, WA, a delay in planting from early March to mid-April reduced yields by 4% per wk (Nelson & Roberts, 1961). Delayed planting of spring wheat reduced yields by 7% per wk at Sidney, MT, and by 7.7% per wk at Logan, UT, when irrigation was adequate for high yields (Sharratt et al., 1980). These results indicate a 4 to 7% per wk yield reduction reported by Fischer (1981b) for delayed planting past the optimum date can be applied across a wide range in latitudes.

A line source sprinkler irrigation system was used in Utah to establish an ET range from about one-third of ET_{max} to ET_{max} for high yields. When seasonal ET was reduced to less than one-half of ET_{max}, planting date did not affect yields. The interaction of planting date response to water deficits indicates that the higher the yield level, the greater the yield loss from delayed planting. A 5-wk delay in planting (16 Apr.–21 May) reduced WUE from 1.14 to 0.61 kg/m^3 and IWUE from 0.94 to 0.60 kg/m^3. These results by Sharratt et al. (1980) indicate that delayed planting that reduces yields also reduces WUE.

Fall planting of spring wheat has expanded into the lower latitudes, and optimum planting dates for cultivars are important for optimizing anthesis dates to benefit from increasing solar radiation (day length) during grain filling while limiting the adverse effect of increasing temperatures (Fischer, 1985b). Climatic environmental optimization of planting dates can be addressed by a computer simulation model, as discussed by Fischer (1985b), where climatic data for a representative range of seasons can be used to estimate yield probabilities for a sequence of planting dates.

B. Row Spacing and Planting Rates

Irrigated wheat is planted in rows spaced 0.15 to 0.25 m, while dryland wheat in semiarid environments is planted in rows spaced 0.25 to 0.35 m. Wheat has relatively short leaves and lateral root extension, which necessitates the use of narrow rows for efficient absorption of solar radiaiton and soil water resources. When grown under high yield potentials, narrow row spacing increases yields (Joseph et al., 1985).

Under sprinkler irrigation, wheat is planted into a relatively smooth surface soil (without bed-furrows). Under surface irrigation with grades less than about 1%, the common irrigation practice in the Great Plains is to use beds and furrows, with furrows having relatively large flow capacities to irrigate field lengths of 400 to 800 m. A common bed-furrow spacing is 1 m (spacings of 0.75 and 1.5 m are less common) with planting in the direction of the bed-furrows. With 1-m bed-furrow spacing, four 0.25- or five 0.2-m-spaced rows are planted on each bed-furrow with single or double disk opener drills. With the single disk drills, the disks are set facing the bed to maintain existing bed-furrows during the planting operation.

On steeper sloping soils where small flow rates are applied to small furrows (rills or corrugations) to minimize erosion by irrigation water, furrows are spaced closer, mostly 0.5 to 0.75 m. In the USA, large planting units are used on smooth surface soils and small irrigation furrows are installed after planting. Furrow placement after planting reduces plant density (Nelson & Roberts, 1961) and delays canopy closure over the small furrows. This practice is more common in the northwestern states where surface-irrigated soils have greater slopes and more erosive textures. In the southwestern states, a common practice is to surface irrigate leveled blocks of about 5 to 20 ha that are either planted flat or on relatively large bed-furrows. Having bed-furrows can improve the uniformity distribution of irrigation water. However, planting of spring wheat in furrows can slow plant development and delay maturity and harvest. In this region of dry summers, delayed harvest is not a weather-related problem.

Planting rates for irrigated winter wheat are mostly in the range of 60 to 100 kg/ha, plant densities are mostly 120 to 200 per m^2, and head densities are mostly 500 to 800 per m^2. With reduced tillering associated with genotype or environment, planting rates are substantially increased and densities can exceed 300 per m^2. Spring wheats produce proportionately less grain on tiller heads than winter wheat, and seeding rates are normally about 50% higher (Reitz, 1976), while head densities tend to be 20 to 30% lower.

The compensation of tillering can result in similar grain yields over a relatively wide range in planting rates (Nelson & Roberts, 1963; Hooker et al., 1983; Shipley & Regier, 1972b; Joseph et al., 1985); however, higher planting rates normally increase vegetative growth and are more important for grazing (Shipley & Regier, 1972a). Higher plant densities also increase lodging potential (Nelson & Roberts, 1963). However, high plant densities may be needed for yield potential of late-planted winter wheat that fails to tiller in the fall and develops limited tillering during renewed growth in late winter before new tiller cessation after FI. Low planting rates combined with limited fall tillering can reduce yields. Asrar and Kanemasu (1985) found that reducing planting rate from 67 to 24 kg/ha for two wheat cultivars planted on a normal date for Manhattan, KS (3 October), reduced average LAI after jointing by 20%, grain yields by 27%, seasonal ET by 17%, and WUE by 9%.

Tiller senescence is accelerated by preanthesis soil-water deficits, and the sensitivity of planting rates to water deficits is relatively low. Because water deficits cause adjustments in tillers that produce heads and also in grain numbers per head, planting rates may not need adjusting for a range in irrigation water levels. Uniculm cultivars require higher planting rates and plant densities for adequate head densities since heads per plant may not appreciably exceed 1.0.

C. Grazing

On irrigated land, animal grazing of winter wheat forage is normally a planned management strategy that involves early planting and increased application of water and fertilizer. In the central and southern High Plains (Kansas, Oklahoma, northwest Texas, and small areas in eastern Colorado and New Mexico), winter wheat is managed for both grazing by stocker cattle and for grain production, with the grazing period beginning in November and terminating in February or March. Wheat that is planted on the optimum dates for grain production produces limited vegetative growth during GS1 and can support only moderate grazing.

Various aspects of wheat pasture grazing were reviewed in the *Proceedings of the National Wheat Pasture Symposium* (Horn, 1984), and wheat forage and grazing have been discussed in earlier state experiment station bulletins (Anderson, 1956; Holt et al., 1969; Malm et al., 1973). Because of winter dormancy and the limited spring growing period before the need for cattle removal for grain production, the major vegetative growth period that produces forage for grazing occurs in the fall. When winter wheat is grown for both forage and grain production, it is planted about 3 to 6 wk earlier than the optimum time for grain production. This earlier planting allows increased forage production but increases the need for fall irrigation. One additional surface irrigation is usually required for sustaining the longer period of fall growth.

During the longer warm fall growing period, winter wheat plants grow vegetatively (without stems) during GS1 to the 0.2- to 0.3-m height, main culm and tiller density increases to 2000 to 3000/m^2, and 2 to 4 Mg/ha of dry forage is produced. Growth mostly stops when mean air temperatures drop below 4 °C, normally in early December (Winter & Thompson, 1987).

In the central and southern Great Plains, irrigated wheat planted for grazing and grain production usually begins seasonal growth in early September with a wet soil. The first seasonal irrigation is normally applied by late October prior to the beginning of the grazing period in early November. Wheat is mostly grown on the fine-textured soils having relatively high water-storage capacity. When wheat is planted near the optimum date for grain production only, the fall irrigation is normally not applied. Thus, wheat managed for grazing plus grain production requires about one additional surface irrigation of about 100 mm, and the longer growing season increases seasonal ET by a proportional amount. October is the transition rainfall period to the dry winter months; in some years, the fall irrigation can be important for nodal root growth and anchoring the plant prior to beginning the grazing period.

Winter and Thompson (1987) found the need to remove cattle by the beginning of spring growth (before FI) to avoid yield reductions in high-yielding cultivars (4–6 Mg/ha). This work has emphasized the importance of limiting the duration of forage removal to GS1. As wheat develops during GS2, the increases in leaf area and biomass are important for grain yield potential; grazing harvest is probably a direct tradeoff with loss of grain yield potential. In earlier work with older cultivars, when irrigated wheat was managed for 3- to 4-Mg/ha grain yields, cattle grazing was allowed to continue for about 3 to 4 wk of renewed vegetative growth and removed prior to the extension of the developing head above the soil surface when primary culms were subject to removal by cattle (Reitz, 1976). The dry matter produced during 3 to 4 wk of slow, early spring growth following winter dormancy constitutes only about 10% of the grazing dry matter normally produced during the 8 to 10 wk of more rapid fall growth.

Where management and environmental conditions combine to produce excessive LAI and biomass during fall growth, fall grazing may be desirable to prevent yield reductions from overwinter lodging of lush growth and abortive loss of lodged culms. Shipley and Regier (1972a) found that early planted nongrazed wheat yielded 28% less than wheat planted on the same date and grazed from 12 November to 20 March. Severe grazing that results in almost complete removal of green leaf area of primary culms can result in overwinter abortive loss of primary culms and reduced grain yields produced on secondary culms. Moderate grazing of early planted wheat in the absence of disease problems with grazing termination near the end of GS1 results in normal or only slightly reduced grain yields compared with planting on an optimum date for grain production only.

VII. SUMMARY

Wheat ranks third in irrigated crop area in the USA (after corn and alfalfa), with average yields in 1984 of 4.6 Mg/ha. Adequate irrigation for high yields is mostly practiced in arid regions, while limited irrigation is widely practiced in semiarid climates where the crop is also grown without irrigation.

Wheat has excellent drought resistance and falls in the crop category of "drought tolerant with low water potential" (Turner, 1986). In field environments where stress develops slowly, cultivars have shown marked ability for osmotic adjustment to water deficits and rapid recovery following irrigation. Wheat has an extensive root system that permits rooting depth to exceed 1.0 m for spring wheats and 1.5 m for winter wheats in favorable rooting environments. It has a wide range of allowable water deficits for scheduling irrigations for efficient water use.

Irrigation tests, conducted in a wide range of climatic environments, indicated a yield range under adequate irrigation of mostly 4 to 8 Mg/ha, seasonal ET of about 350 to 700 mm, peak daily ET during grain filling of 3 to 9 mm/d, and seasonal WUE of 0.8 to 1.6 kg/m^3. Water-use efficiency of applied irrigation is largely similar to seasonal WUE except when reduced by application losses and nondepleted profile storage.

Wheat is only moderately sensitive to critical-stage plant water deficits compared with many other field crops. Deficits are normally the most sensitive during a period of about 10 to 20 d preceding and continuing through anthesis that reduces grain numbers per m^2. Development during grain filling is normally less sensitive, perhaps associated with plant ability to relocate preanthesis assimilate to filling grain as late-season water stress reduces photosynthesis. The preanthesis stress sensitivity is related to deficit effects on reduced biomass accumulation (head weight in particular) that reduce grain numbers per m^2. Crown root initiation and tillering, as well as jointing, have been reported as most critical for spring wheat in dry environments of sandy soil and appreciable evaporative demand and are probably associated with limited early-stage rooting for water extraction. Grain filling has been reported as critical in climates that experience periods of hot, dry winds—particularly during milk stage—that cause grain shrivelling.

Harvest index is mostly in the range 0.38 to 0.50 for high-yielding irrigated cultivars. During preanthesis, potential grain numbers are adjusted to biomass accumulation, and HI is affected little by water deficits. Once grain numbers become fixed, HI becomes sensitive to water deficits during grain filling.

Evapotranspiration-yield relationships indicate a threshold of about 0.2 ET_{max} for the first grain increment. Evapotranspiration-yield relationships are mostly linear over a wide range of water deficits, while WUE relationships are curvilinear. Yield relationships to applied irrigation are mostly curvilinear, diminishing-return functions. Similar response functions for applied N permit optimization analysis of marginal return inputs of water and N.

Although ET-yield relationships are mostly linear, they can become curvilinear when irrigation timing results in one or more applications that fail to increase yields and when profile drainage may be included in ET calculations at the high application levels. In the winter wheat region of the Great Plains, irrigation during early vegetative growth may increase straw more than grain yields. Also, increasing frequency of spring rainfall frequently limits yield response to irrigation during grain filling, thus reducing IWUE. Wheat can be planted over a wide range of dates. However, a delay past the optimum period reduced irrigated yields by 4 to 7% per wk over a wide range of latitudes. Early planting is important for producing early-season forage for grazing by cattle in the central and southern Great Plains. For irrigated wheat managed for both grazing and grain production, early termination of grazing (before GS2) is more critical when managed for high yields. Nitrogen fertility is increasingly important for irrigated wheat managed for high yields and acceptable grain quality.

REFERENCES

Agarwal, S.K., and S.K. Yadav. 1978. Effect of nitrogen and irrigation levels on the growth and yield of wheat. Indian J. Agron. 23:137–143.

Aggarwal, P.K., A.K. Singh, G.S. Chaturvedi, and S.K. Sinha. 1986. Performance of wheat and triticale cultivars in a variable soil-water environment. Field Crops Res. 13:301–315.

Anderson, K.L. 1956. Winter wheat pasture in Kansas. Kans. Agric. Exp. Stn. Bull. 345.

Asrar, G., and E.T. Kanemasu. 1985. Seasonal distribution of water use and photosynthetic efficiencies in winter wheat. Proc. Int. Conf. Crop Water Requirements, Paris. 11–14 Sept.

Atsmon, D., and E. Jacobs. 1977. A newly bred 'Gigas' form of bread wheat (*Triticum aestivum* L.): Morphological features and thermo-photoperiodic responses. Crop Sci. 17:31–35.

Austin, R.B., J. Bingham, R.D. Blackwell, L.T. Evans, M.A. Ford, C.L. Morgan, and M. Taylor. 1980. Genetic improvement in wheat yields since 1900 and associated physiological changes. J. Agric. Sci. Camb. 94:675–689.

Belford, R.K., B. Klepper, and R.W. Rickman. 1987. Studies of intact shoot-root systems of field-grown winter wheat. II. Root and shoot development patterns as related to nitrogen fertilizer. Agron. J. 79:310–319.

Black, A.L. 1970a. Adventitious roots, tillers, and grain yields of spring wheat as influenced by N–P fertilization. Agron. J. 62:32–36.

Black, A.L. 1970b. Soil water and soil temperature influences on dryland winter wheat. Agron. J. 62:797–801.

Bole, J.B., and S. Dubetz. 1986. Effect of irrigation and nitrogen fertilizer on the yield and protein content of soft white spring wheat. Can. J. Plant Sci. 66:281–289.

Bonnet, O.T. 1966. Inflorescences of maize, wheat, rye, barley, and oats: Their initiation and development. Illinois Agric. Exp. Stn. Bull. 721.

Brown, L.R. 1986. A generation of deficits. p. 1–21. *In* State of the world. A World Watch Institute report on progress toward a sustainable society. W.W. Norton & Co., New York.

Bunyolo, A., K. Munyinda, and R.E. Karamanos. 1985. The effect of water and nitrogen on wheat yield on a Zambian soil. II. Evaluation of irrigation schedules. Commun. Soil Sci. Plant Anal. 16:43–53.

Cackett, K.E., and P.C. Wall. 1971. The effect of altitude and season length on the growth and yield of wheat (*Triticum aestivum* L.) in Rhodesia. Rhod. J. Agric. Res. 9:107–120.

Cholick, F.A., J.R. Welsh, and C.V. Cole. 1977. Rooting patterns of semidwarf and tall winter wheat cultivars under dryland field conditions. Crop Sci. 17:637–639.

Choudhary, P.N., and V. Kumar. 1980. The sensitivity of growth and yield of dwarf wheat to water stress at three growth stages. Irrig. Sci. 1:223–231.

Cooper, J.L. 1980. The effect of nitrogen fertilizer and irrigation frequency on a semidwarf wheat in southeast Australia. I. Growth and yield. 2. Water use. Aust. J. Exp. Agric. Anim. Husb. 20:359–369.

Dalrymple, D.C. 1986. Development and spread of high-yielding wheat varieties in developing countries. Bur. Sci. Tech., Agency Int. Dev., Washington, DC.

Darwinkel, A. 1978. Patterns of tillering and grain production of winter wheat at a wide range of plant densities. Neth. J. Agric. Sci. 26:383–398.

Day, A.D., and S. Intalap. 1970. Some effects of soil moisture stress on the growth of wheat (*Triticum aestivum* L. em Thell.). Agron. J. 62:27–29.

Donald, C.M., and J. Hamblin. 1976. The biological yield and harvest index of cereals as agronomic and plant breeding criteria. Adv. Agron. 28:361-405.

Eastin, J.D., T.E. Dickson, D.R. Kreig, and A.B. Maunder. 1983. Crop physiology in dryland agriculture. *In* H.E. Dregne and W.O. Willis (ed.) Dryland agriculture. Agronomy 23:333-364.

Echert, J.B., N.M. Chaudhry, and S.A. Qureshi. 1978. Water and nutrient response of semidwarf wheat under improved management in Pakistan: Agronomic and economic implications. Agron. J. 70:77-80.

Eck, H.V. 1988. Winter wheat yield response to nitrogen and irrigation. Agron. J. 80:902-908.

Ehlig, C.F., and R.D. LeMert. 1976. Water use and productivity of wheat under five irrigation treatments. Soil Sci. Soc. Am. J. 40:750-755.

Ehrler, W.L., S.B. Idso, R.D. Jackson, and R.J. Reginato. 1978. Diurnal changes in plant water potential and canopy temperature of wheat as affected by drought. Agron. J. 70:999-1004.

Erie, L.J., D.A. Bucks, and O.F. French. 1973. Consumptive use and irrigation management for high-yielding wheats in central Arizona. Prog. Agric. Ariz. XXV(2):14-15.

Erie, L.J., O.F. French, D.A. Bucks, and K. Harris. 1982. Consumptive use of water by major corps in the southwestern United States. U.S., Agric. Res. Serv. Conserv. Res. Rep. 29.

Evans, L.T. 1980. The natural history of crop yields. Am. Sci. 68:388-397.

Evans, L.T. 1981. Yield improvement in wheat: Empirical or analytical? p. 203-222. *In* L.T. Evans and W.J. Peacock (eds.) Wheat science—Today and tomorrow. Cambridge Univ. Press, New York.

Evans, L.T., and R.L. Dunstone. 1970. Some physiological aspects of evolution in wheat. Aust. J. Biol. Sci. 23:725-741.

Evans, L.T., and I.F. Wardlaw. 1976. Aspects of comparative physiology of grain yields in cereals. Adv. Agron. 28:301-359.

Evans, L.T., I.F. Wardlaw, and R.A. Fischer. 1975. Wheat. p. 101-149. *In* L.T. Evans (ed.) Crop physiology, some case histories. Cambridge Univ. Press, London.

Fernandez-G., R., and R.J. Laird. 1959. Yield and protein content of wheat in central Mexico as affected by available soil moisture and nitrogen fertilization. Agron. J. 51:33-36.

Finney, K.F., and H.C. Fryer. 1958. Effect on loaf volume of high temperatures during the fruiting period of wheat. Agron. J. 50:28-34.

Finney, K.F., G.L. Rubenthaler, L.C. Bolte, R.C. Hoseney, M.D. Shogren, C.M. Barmhart, and J.A. Shellenberger. 1963. Quality characteristics of hard winter wheat varieties grown in the southern, central, and northern Great Plains of the United States, 1962 crop. U.S. Dep. Agric. Hard Winter Wheat Qual. Lab. Rep. CR-10-64.

Finney, K.F., W.T. Yamazaki, V.L. Youngs, and G.L. Rubenthaler. 1987. Quality of hard, soft, and durum wheats. *In* G.E. Heyne (ed.) Wheat and wheat improvement. Agronomy 13:677-748.

Fischer, R.A. 1970. The effects of water stress at various stages of development on yield processes in wheat. Proc. symp. plant responses to climatic factors, Uppsala, Sweden. 15-20 Sept.

Fischer, R.A. 1980. Influences of water stress on crop yield in semiarid regions. p. 323-339. *In* N.C. Turner and P.J. Kramer (ed.) Adaptation of plants to water and high temperature stress. John Wiley and Sons, New York.

Fischer, R.A. 1981a. Optimizing the use of water and nitrogen through breeding of crops. Plant Soil 58:249-278.

Fischer, R.A. 1981b. Developments in wheat agronomy. p. 249-269. *In* L.T. Evans and W.J. Peacock (ed.) Wheat science—Today and tomorrow. Cambridge Univ. Press, New York.

Fischer, R.A. 1985a. Number of kernels in wheat crops and the influence of solar radiation and temperature. J. Agric. Sci. Camb. 105:447-461.

Fischer, R.A. 1985b. The role of crop simulation in wheat agronomy. p. 237-255. *In* W. Day and R.K. Atkin (ed.) NATO ASI Series A: Life Sciences Vol. 86. Plenum Press, New York.

Fischer, R.A., J.H. Lindt, and A. Glave. 1977. Irrigation of dwarf wheats in the Yaqui Valley of Mexico. Expl. Agric. 13:353-367.

Fischer, R.A., and W. Sanchez. 1979. Drought resistance in spring wheat cultivars. II. Effect on plant water relations. Aust. J. Agric. Res. 30:801-814.

Gajri, P.R., and S.S. Prihar. 1983. Effect of small irrigation amounts on the yield of wheat. Agric. Water Manage. 6:31-41.

Gales, K. 1983. Yield variation of wheat and barley in Britain in relation to crop growth and soil conditions—A review. J. Sci. Food Agric. 34:1085-1104.

Gallagher, J.N., and P.V. Biscoe. 1978. A physiological analysis of cereal yield. II. Partitioning of dry matter. Agric. Prog. 53:51-70.

George, D.W. 1982. The growing point of fall-sown wheat: A useful measure of physiological development. Crop Sci. 22:235–239.

Gifford, R.M., and L.T. Evans. 1981. Photosynthesis, carbon partitioning, and yield. Annu. Rev. Plant Physiol. 32:485–509.

Gifford, R.M., J.H. Thorne, W.D. Hitz, and R.T. Giaquinta. 1984. Crop productivity and photosynthate partitioning. Science 225:801–808.

Gregory, P.J., B. Marshall, and P.V. Biscoe. 1981. Nutrient relations of winter wheat. 3. Nitrogen uptake photosynthesis of flag leaves, and translocation of nitrogen to grain. J. Agric. Sci. Camb. 96:539–547.

Gregory, P.J., M. McGowan, P.V. Biscoe, and B. Hunter. 1978. Water relations of winter wheat. I. Growth of the root system. J. Agric. Sci. Camb. 91:91–102.

Hanks, R.J., and R.B. Sorensen. 1984. Harvest index as influenced in spring wheat by water stress. p. 205–209. In W. Day and R.K. Atkins (ed.) Wheat growth and modeling. NATO ASI Series A: Life Sciences Vol. 86. Plenum Press, New York.

Haun, J.R. 1973. Visual quantification of wheat development. Agron. J. 65:116–119.

Holbrook, F.S., and J.R. Welsh. 1980. Soil water use by semidwarf and tall winter wheat cultivars under dryland field conditions. Crop Sci. 20:244–246.

Holt, E.C., M.J. Norris, and J.A. Lancaster. 1969. Production and management of small grains for forage. Texas Agric. Exp. Stn. B-1082.

Hooker, M.L., S.H. Mohiuddin, and E.T. Kanemasu. 1983. The effect of irrigation timing on yield and yield components of winter wheat. Can. J. Plant Sci. 63:815–823.

Horn, G.W. 1984. Proc. national wheat pasture symp. Oklahoma Agric. Exp. Stn. Misc. Publ. 115.

Howell, T.A., J.T. Musick, and J.A. Tolk. 1986. Canopy temperature of irrigated winter wheat. Trans. Am. Soc. Agric. Eng. 29:1692–1698.

Hunsaker, D.J., and D.A. Bucks. 1987. Wheat yield variability in level basins. Trans. Am. Soc. Agric. Eng. 30:1099–1104.

Jalota, S.K., S.S. Prihar, B.A. Sandhu, and K.L. Khera. 1980. Yield, water use, and root distribution of wheat as affected by presowing and postsowing irrigation. Agric. Water Manage. 2:289–297.

Jensen, M.E., and J.T. Musick. 1960. The effects of irrigation treatments on evapotranspiration and production of sorghum and wheat in the southern Great Plains. Int. Congr. Soil Sci., Trans. 7th (Madison, WI) Vol. 1:386–393.

Jensen, M.E., and W.H. Sletten. 1965. Evapotranspiration and soil moisture-fertilizer interrelations with irrigated winter wheat in the southern High Plains. U.S. Dept. Agric. Conserv. Res. Rep. 4.

Johnson, V.A. 1978. Protein in hard red winter wheat. Baker's Digest 52(April):22–28, 67.

Johnson, V.A., and H.L. Beemer, Jr. 1977. Wheat in the People's Republic of China. Committee on scholarly communications with the People's Republic of China Rep. Natl. Acad. Sci., Washington, DC.

Jones, H.G. 1977. Aspects of the water relations of spring wheat (*Triticum aestivum* L.) in response to induced drought. J. Agric. Sci. Camb. 88:267–282.

Joseph, K.D.S.M., M.M. Alley, D.E. Brann, and W.D. Gravelle. 1985. Row spacing and seeding rate effects of yield and yield components of soft red winter wheat. Agron. J. 77:211–214.

Khalifa, M.A., M.A. Akasha, and M.B. Said. 1977. Growth and N uptake as affected by sowing date and nitrogen in irrigated semiarid conditions. J. Agric. Sci. Camb. 89:35–42.

Kirby, E.J.M., and M. Appleyard. 1984. Cereal development guide. 2nd ed. Nat. Agric. Cent. Arable Unit, Stoneleigh, England.

Kirkham, M.B., and E.T. Kanemasu. 1982. Wheat. p. 481–520. In I.D. Teare and M.M. Peet (ed.) Crop water relations. John Wiley and Sons, New York.

Klepper, B., R.K. Belford, and R.W. Rickman. 1984. Root and shoot development in winter wheat. Agron. J. 76:117–122.

Klepper, B., R.W. Rickman, and C.M. Peterson. 1982. Quantitative characterization of vegetative development in small cereal grains. Agron. J. 74:789–792.

Kloster, L.D., and K. Whittlesey. 1971. Production function analysis of irrigation water and nitrogen fertilizer in wheat production. Washington Agric. Exp. Stn. Bull. 746.

Kmoch, H.C., R.E. Raming, R.L. Fox, and F.E. Koehler. 1957. Root development of winter wheat as influenced by soil moisture and nitrogen fertilization. Agron. J. 49:20–25.

Lal, R.B. 1985. Irrigation requirement of dwarf durum and aestivum wheat varieties. Indian J. Agron. 30:207–213.

Large, E.C. 1954. Growth stages in cereals—Illustrations of the Feekes scale. Plant Pathol. 3:128-129.

MacKey, J. 1973. The wheat root. p. 827-849. *In* Proc. 45th Int. Wheat Genetics Symp., Columbia, MO.

Malm, N.R., J.S. Arlege, and C.E. Barnes. 1973. Forage production from winter small grains in southwestern New Mexico. New Mexico Agric. Exp. Stn. Bull. 602.

Malse, J. 1985. Competition among tillers in winter wheat: Consequences for growth and development of the crop. p. 55-58. *In* W. Day and R.K. Atkin (ed.) Wheat growth and modeling. NATO ASI Series A: Life Sciences Vol. 86. Plenum Press, New York.

McNeal, F.M., M.A. Berg, P.L. Brown, and C.F. McGuire. 1971. Productivity and quality response of five spring wheat genotypes, *Triticum aestivum* L., to nitrogen fertilizer. Agron. J. 63:908-910.

Meyer, W.S., and G.C. Green. 1980. Water use by wheat and plant indicators of available soil water. Agron. J. 72:253-256.

Midmore, D.J., P.M. Cartwright, and R.A. Fischer. 1984. Wheat in tropical environments. II. Growth and grain yield. Field Crops Res. 8:207-227.

Miller, D.E. 1977. Deficit high-frequency irrigation of sugarbeets, wheat, and beans. p. 269-282. *In* Proc. Conference Water Management for irrigation and drainage. Am. Soc. Civ. Eng., Reno, NV. 20-22 July 1977.

Miller, D.E., and A.N. Hang. 1982. Deficit, high-frequency sprinkler irrigation of wheat. Soil Sci. Soc. Am. J. 46:386-389.

Misra, R.D., E.C. Sharma, B.C. Wright, and V.P. Singh. 1969. Critical stages in irrigation and irrigation requirement of wheat variety of 'Larma Jojo'. Indian J. Agric. Sci. 39:898-906.

Misra, R.K., and T.N. Chaudhary. 1985. Effect of a limited water input on root growth, water use, and grain yield of wheat. Field Crops Res. 10:125-134.

Mogensen, V.O., H.E. Jensen, and M.A. Rab. 1985. Grain yield, yield components, drought sensitivty, and water use efficiencies of spring wheat subjected to water stress at various growth stages. Irrig. Sci. 6:131-140.

Monteith, J.L., and R.K. Scott. 1982. Weather and yield variation of crops. p. 127-153. *In* K. Blaxter and L. Fowden (ed.) Food, nutrition, and climate. Applied Sci. Publ., London.

Morgan, J.M. 1977. Changes in diffusive conductance and water potential of wheat plants before and after anthesis. Aust. J. Plant Physiol. 4:75-86.

Morgan, J.M. 1984. Osmoregulation and water stress in higher plants. Annu. Rev. Plant Physiol. 35:299-319.

Musick, J.T., and D.A. Dusek. 1980. Planting date and water deficit effects on development and yield of irrigated winter wheat. Agron. J. 72:45-52.

Musick, J.T., D.A. Dusek, and A.C. Mathers. 1984. Irrigation water management of wheat. ASAE Paper 84-2094. Am. Soc. Agric. Eng., St. Joseph, MI.

Musick, J.T., D.W. Grimes, and G.M. Herron. 1963. Water management, consumptive use, and nitrogen fertilization of irrigated winter wheat in western Kansas. U.S. Dep. Agric. Prod. Res. Rep. 75.

Musick, J.T., L.L. New, and D.A. Dusek. 1976. Soil water depletion-yield relationships of irrigated sorghum, wheat, and soybeans. Trans. ASAE 19:489-493.

Musick, J.T., F.B. Pringle, and J.D. Walker. 1988. Sprinkler and furrow irrigation trends—Texas High Plains. Appl. Eng. Agric. 4:46-52.

Nelson, C.E., and S. Roberts. 1961. Cultural practices affecting stand and their relationships with tillering, lodging, and yield of rill-irrigated wheat. Washington Agric. Exp. Stn. Bull. 629.

Nelson, C.E., and S. Roberts. 1963. Spring wheat variety, population, and planting date experiment. Washington Agric. Exp. Stn. Circ. 419.

Papendick, R.I., V.L. Cochran, and W.M. Woody. 1971. Soil water potential and water content profiles with wheat under low spring and summer rainfall. Agron. J. 63:731-734.

Passioura, J.B. 1977. Grain yield, harvest index, and water use of wheat. J. Aust. Inst. Agric. Sci. 43:117-120.

People's Republic of China. 1986. The rural, economic, and social statistics of China. Minist. Agric. Anim. Husb. Fish. Inf. Dep. Bull.

Poostchi, I., I. Revohani, and K. Razmi. 1972. Influence of levels of spring irrigation and fertility on yield of winter wheat (*Triticum aestivum* L.) under semiarid conditions. Agron. J. 64:438-440.

Porter, J.R. 1985. Approaches to modeling canopy development in wheat. p. 69–81. *In* W. Day and R.K. Atkin (ed.) Wheat growth and modeling. NATO ASI Series A: Life Sciences Vol. 86. Plenum Press, New York.

Power, J.F., and J. Alessi. 1978. Tiller development and yield of standard and semidwarf spring wheat varieties as affected by nitrogen fertilizer. J. Agric. Sci. Camb. 90:97–108.

Prashar, C.R.K., and M. Singh. 1963. Soil-moisture studies and the effects of varying levels of irrigation and fertilizers on wheat under intensive system of cropping. Indian J. Agric. Sci. 33:75–93.

Prihar, S.S., K.L. Khera, E.S. Sandhu, and B.S. Sandhu. 1976. Comparison of irrigation schedules based on pan evaporation and growth stages in winter wheat. Agron. J. 68:650–653.

Rao, Y.G., and R.B.L. Bhardwaj. 1981. Consumptive use of water, growth and yield of *aestivum* and durum wheat varieties at varying levels of nitrogen under limited and adequate irrigation situations. Indian J. Agron. 26:243–250.

Reedy, A.S., and R.B.L. Bhardwaj. 1982. Water use studies in wheat as influenced by levels of nitrogen and phosphorus under limited and adequate irrigation. Indian J. Agron. 27:22–27.

Reitz, L.P. 1976. Wheat in the United States. U.S. Dep. Agric. Inf. Bull. 386. Washington, DC.

Rickman, R.W., B. Klepper, and R.K. Belford. 1985. Developmental relationships among roots, leaves, and tillers in wheat. p. 83–98. *In* W. Day and R.K. Atkin (ed.) Wheat growth and modeling. NATO ASI Series A: Life Sciences Vol. 86. Plenum Press, New York.

Ritter, W., and E.A. Manger. 1985. Effect of irrigation efficiencies on nitrate leaching losses. J. Irrig. Drain. Div. Am. Soc. Civ. Eng. 111(3):230–239.

Robins, J.S., and C.E. Domingo. 1962. Moisture and nitrogen effects on irrigated spring wheat. Agron. J. 54:135–138.

Saffigna, P.G., A.L. Cogle, W.M. Strong, and S.A. Waring. 1982. The effect of carbonaceous residues on ^{15}N fertilizer nitrogen transformations in the field. p. 83–87. *In* I.E. Gabally and J.R. Freney (ed.) The cycling of carbon, nitrogen, sulfur, and phosphorus in terrestrial and aquatic ecosystems. Aust. Acad. Sci., Canberra.

Schmidt, J.W. 1984. Genetic contributions to yield gains in wheat. p. 89–101. *In* W.R. Fehr (ed.) Genetic contribution to yield gains of five major crop plants. CSSA Spec. Publ. 7. ASA and CSSA, Madison, WI.

Schneider, A.D., J.T. Musick, and D.A. Dusek. 1969. Efficient wheat irrigation with limited water. Trans. ASAE 12:23–26.

Sharratt, B.S., R.J. Hanks, and J.K. Aase. 1980. Environmental factors associated with yield differences between seeding dates of spring wheat. Utah Agric. Exp. Stn. Res. Rep. 92.

Shawcroft, R.W. 1983. Limited irrigation may drop yield, up profit. Colo. Rancher Farmer 37(4):35–38.

Shawcroft, R.W., and R. Croissant. 1986. Irrigation of winter wheat in Colorado. Colorado State Univ. Coop. Ext. Serv. Action 556.

Shellenberger, J.A. 1978. Production and utilization of wheat. p. 1–18. *In* Y. Pomeranz (ed.) Wheat chemistry and technology. Am. Assoc. Cereal Chem., St. Paul.

Shimshi, D. 1979. Leaf permeability as an index of water relations, CO_2 uptake, and yield of irrigated wheat. Irrig. Sci. 1:107–111.

Shimshi, D., H. Bielorai, and A. Mantell. 1973. Irrigation of field crops. p. 369–372. *In* B. Yaron et al. (ed.) Arid zone irrigation. Ecol. Stud. 5. Springer-Verlag New York, New York.

Shimshi, D., S. Gairon, J. Rubin, M. Khilfa, and Y. Khilmi. 1981. Field crops: Wheat. p. 7–15. *In* J. Shalhevet et al. (ed.) Irrigation of field and orchard crops under semiarid conditions. Int. Irrig. Inf. Cent. (Israel) Publ. 1.

Shimshi, D., and U. Kafkafi. 1978. The effect of supplemental irrigation and nitrogen fertilization on wheat (*Triticum aestivum* L.). Irrig. Sci. 1:27–38.

Shipley, J., and C. Regier. 1972a. Optimum forage production and the economic alternatives associated with grazing irrigated wheat, Texas High Plains. Texas Agric. Exp. Stn. MP-1068.

Shipley, J., and C. Regier. 1972b. Irrigated wheat yields with limited irrigation and three seeding rates, northern High Plains of Texas. Texas Agric. Exp. Stn. PR-3031.

Singh, A. 1978. Response of late sown wheat to N fertilization and irrigation management in western Rajasthan. Indian J. Agron. 23:44–48.

Singh, K.D., and N.C. Stoskopf. 1971. Harvest index in cereals. Agron. J. 634:224–226.

Singh, N.P., and N.G. Dastane. 1971. Effects of moisture regimes and nitrogen levels on growth and yield of dwarf wheat varieties. Indian J. Agric. Sci. 41:952–958.

Singh, N.T., R. Singh, P.S. Mahajan, and A.C. Vig. 1979a. Influence of supplemental irrigation and presowing soil water storage on wheat. Agron. J. 71:483–486.

Singh, N.T., A.C. Vig, R. Singh, and M.R. Choudhary. 1979b. Influence of different levels of irrigation and nitrogen on yield and nutrient uptake of wheat. Agron. J. 71:401–404.

Sinha, S.K., P.K. Agarwal, and R.E. Chopra. 1985. Irrigation in India: A physiological and phenological approach to water management in grain crops. Adv. Irrig. 3:129–212. Academic Press, New York.

Smika, D.E., and B.W. Greb. 1973. Protein content of winter wheat grain as related to soil and climatic factors in the semiarid central Great Plains. Agron. J. 65:433–436.

Smith, R.C.G., W.K. Mason, W.S. Meyer, and H.D. Barrs. 1983. Irrigation in Australia: Development and prospects. Adv. Irrig. 2:99–153. Academic Press, New York.

Sofield, K., L.T. Evans, M.G. Cook, and I.F. Wardlaw. 1977. Factors influencing the rate and duration of grain filling in wheat. Aust. J. Plant Physiol. 3:785–797.

Soil Improvement Committee, Fertilizer Association. 1975. Essential plant nutrients. p. 54–75. *In* Western fertilizer handbook. 5th ed. Interstate Printing and Publishers, Danville, IL.

Sojka, R.E., L.H. Stolzy, and R.A. Fischer. 1981. Seasonal drought response of selected wheat cultivars. Agron. J. 73:838–845.

Spiertz, J.H.J., and J. Ellen. 1978. Effects of nitrogen on crop development and grain growth of winter wheat in relation to assimilation and utilization of assimilates and nutrients. Neth. J. Agric. Sci. 26:210–231.

Spiertz, J.H.J., and J. Vos. 1985. Grain growth of wheat and its limitation by carbohydrate and nitrogen supply. p. 129–141. *In* W.Day and R.K. Atkin (ed.) Wheat growth and modeling. NATO ASI Series A: Life Sciences Vol. 86. Plenum Press, New York.

Spratt, E.D., and J.K.R. Gasser. 1970. Effects of fertilizer-nitrogen and water supply on distribution of dry matter and nitrogen between different parts of wheat. Can. J. Plant Sci. 50:613–625.

Srivastava, S.P., and O.N. Bansal. 1975. Effect of delay in first irrigation, methods of N application, and cycocel in wheat on mixed black soils. Indian J. Agron. 20:381–382.

Stanford, G., and J.O. Legg. 1984. Nitrogen and yield potential. p. 263–272. *In* R.D. Hauck (ed.) Nitrogen in crop production. ASA, CSSA, SSSA, Madison, WI.

Stein, I.S., D.L. Wetzel, and R.G. Sears. 1987. NIR protein screening in KSU wheat breeding program. p. 81. *In* Agronomy Abstracts, ASA, Madison, WI.

Steiner, J.L., R.C.G. Smith, W.S. Meyer, and J.A. Adeney. 1985. Water use, foliage temperature, and yield of irrigated wheat in southeastern Australia. Aust. J. Agric. Res. 36:1–11.

Strong, W.M. 1981. Nitrogen requirements of irrigated wheat on the Darling Downs. Aust. J. Exp. Agric. Anim. Husb. 21:424–431.

Syme, J.R.1970. A high-yielding Mexican semidwarf wheat and the relationship of yield to harvest index and other varietal characteristics. Aust. J. Exp. Agric. Anim. Husb. 10:350–353.

Thill, D.C., R.E. Witters, and R.I. Papendick. 1978. Interactions of early- and late-planting winter wheat and their environment. Agron. J. 70:1041–1047.

Thorne, M.D. 1977. Agronomists and food: Contributions and challenges. ASA Spec. Publ. 30. ASA, Madison, WI.

Tipples, K.H., S. Dubetz, and G.N. Irvine. 1977. Neepawa wheat grown under irrigation. II. Milling and baking quality. Can. J. Plant Sci. 57:337–350.

Tottman, D.R. 1987. The decimal code for the growth stages of cereals, with illustrations. Ann. Appl. Biol. 110:441–454.

Tucker, B.B., and L.W. Murdock. 1984. Nitrogen use in the south-central states. p. 734–749. *In* R.D. Hauck (ed.) Nitrogen in crop production. ASA, CSSA, and SSSA, Madison, WI.

Turner, N.C. 1986. Adaptation to water deficits: A changing perspective. Aust. J. Plant Physiol. 13:175–190.

U.S. Department of Commerce, Bureau of Census. 1984. Farm and ranch irrigation survey. Spec. Rep. Ser. AG-84-SR-1. U.S. Gov. Print. Office, Washington, DC.

Viets, F.G., Jr. 1965. The plant's need for and use of nitrogen. *In* W.V. Bartholomew and F.E. Clark (ed.) Soil nitrogen. Agronomy 10:503–549.

Vlassak, K., and L.M.J. Verstraeten. 1985. Nitrogen nutrition of winter wheat. p. 217–236. *In* W. Day and R.K. Atkin (ed.) Growth and modeling. NATO ASI Series A: Life Sciences Vol. 86. Plenum Press, New York.

Waddington, S.R., M. Osmanzai, M. Yoshida, and J.K. Ransom. 1987. The yield potential of durum wheat released in Mexico between 1960 and 1984. J. Agric. Sci. Camb. 108:469–477.

Waddington, S.R., J.K. Ransom, M. Osmanzai, and D.A. Saunders. 1986. Improvement in the yield potential of bread wheat adapted to northwest Mexico. Crop Sci. 26:698–703.

Warrington, I.J., R.L. Dunstone, and L.M. Green. 1977. Temperature effects at three development stages on the yield of the wheat ear. Aust. J. Agric. Res. 28:11–27.

Weaver, J.E. 1926. Root development of field crops. McGraw-Hill Book Co., New York.

Westfall, D.G. 1984. Management of nitrogen in the mountain states. p. 750–763. In R.D. Hauck (ed.) Nitrogen in crop production. ASA, CSSA, and SSSA, Madison, WI.

Wiegand, C.L., and J.A. Cuellar. 1981. Duration of grain filling and kernel weight of wheat as affected by temperature. Crop Sci. 21:95–101.

Winter, S.J., and E.K. Thompson. 1987. Grazing duration effects on wheat growth and grain yield. Agron. J. 79:110–114.

Woodruff, D.R. 1980. Influence of nodal root production on wheat yields under conditions of limited soil water and phosphorus supply. Queensl. J. Agric. Anim. Sci. 37:53–62.

Wright, J.L. 1982. New evapotranspiration crop coefficients. Proc. Am. Soc. Civ. Eng. 108(IR2):57–71.

Yaron, D., G. Strateener, D. Shimshi, and M. Weisbrod. 1973. Wheat response to soil moisture and the optimal irrigation policy under conditions of unstable rainfall. Water Resour. Res. 9:1145–1154.

Zeleny, L. 1978. Criteria of wheat quality. p. 19–49. In Y. Pomeranz (ed.) Wheat chemistry and technology. 2nd ed. Am. Assoc. Cereal Chem., St. Paul.

Zadoks, J.C., T.T. Chang, and C.F. Konzak. 1974. A decimal code for the growth stages of cereals. Weed Res. 14:415–421.

21   Soybean[1]

D. C. REICOSKY

USDA-ARS
Morris, Minnesota

L. G. HEATHERLY

USDA-ARS
Stoneville, Mississippi

Soybean [*Glycine max* (L.) Merr.] is a dominant world crop for vegetable oil and protein for animal and human consumption. Interest in soybean production has increased because soybean has a fairly wide range of adaption involving a wide array of climatic, soil, and growth conditions. Most soybean crops in the USA are produced under rainfed conditions; however, a significant portion is now taking place under irrigation to stabilize production from year to year. The principal area of soybean production is the North Central region (Fig. 21–1) with two other significant production areas in the Mississippi Delta and the mid-Atlantic Coastal Plains.

Availability of adequate rainfall imposes definite limitations on areas suitable for soybean production throughout the USA. As natural rainfall and soil properties interact to make less stored water available to plants, the importance of irrigation to maintain stable production of soybean increases. Figure 21–1 shows an abrupt drop in the soybean hectarage west of a line running through eastern Texas, through the Dakotas, primarily because of inadequate rainfall and irrigation resources. It is in this area that the relative amount of irrigated soybean production increases. Throughout the southeastern Coastal Plains, there is a relatively large percentage of irrigated soybean as a result of soils with low water-holding capacities or shallow rooting depths.

I. CROP DEVELOPMENT

A. Crop Growth and Canopy Development

1. Classification, Development, and General Water Use

Soybean water use is not uniform throughout the growing season and is strongly dependent on evaporative demand and crop growth and canopy development. A typical seasonal water use pattern that consists of both soil

---

[1] Contribution from the North Central Soil Conservation Research Laboratory and Soybean Production Research Unit, USDA-ARS, Morris, MN, and Stoneville, MS.

evaporation and plant transpiration for rainfed soybean grown in the midwestern USA is shown schematically in Fig. 21–2. The soybean classification schemes of Fehr et al. (1971) and Fehr and Caviness (1977) show the approximate time of vegetative and reproductive stages of growth. The schematic only provides a qualitative picture of soybean water use. Specific quantitative examples of water use may be more informative but can be cultivar, year, and site dependent to make generalization difficult. For example, Clawson et al. (1986) showed a 31-mm difference in cumulative ET and 1-mm/d difference in peak daily ET because of pubescence on the 'Harosoy' isoline. They reported a daily peak ET of 5.8 mm/d using a water balance method and Kanemasu et al. (1976) reported 8.8 mm/d for 'Calland' using a lysimeter. Generally, soybean water use is low during the germination and seedling stages, with a large portion of water loss due to soil evaporation. However, as the plant develops from V3 to V6 (Fig. 21–2), there is a rapid increase in water use. Maximum water use occurs during R1 to R6 reproductive stages when full canopy and rooting volume have been attained. Once plants have started to mature and pods are filled, there is a rapid decrease in water use associated with leaf and root senescence at the end of the growing season and a concurrent reduction in evaporative demand. The length of season represented in Fig. 21–2 is approximate and can range from 120 to 160 d after planting, depending on geographic location and genotype.

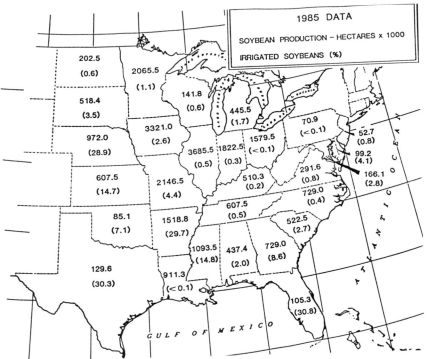

Fig. 21–1. Soybean production area and percentage of soybeans irrigated in the USA (Source: 1986 Irrig. J. 36:21–29 and Crop Reporting, SRS, USDA).

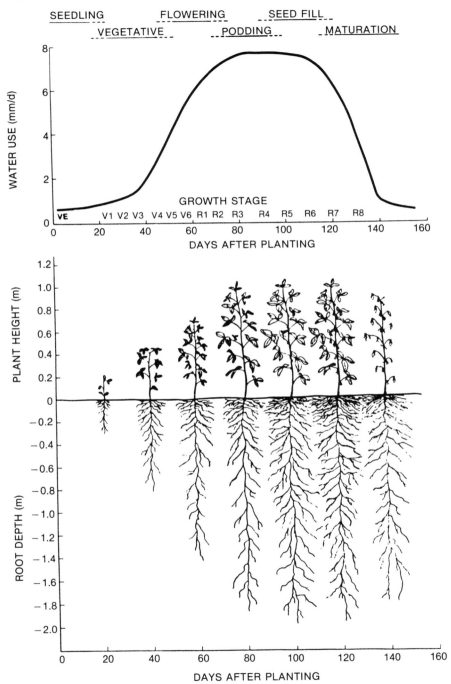

Fig. 21-2. Schematic representation of soybean water use, growth stages, plant height, and root depth during the growing season.

2. Plant Height

Plant height of soybean is a product of the number of nodes and the internode length. The number of nodes is determined by 35 d after planting (Johnson et al., 1960). Drought stress will rarely have occurred by this time, and irrigation will generally have little or no effect on the number of nodes (Huck et al., 1986; Kadhem et al., 1985a). Therefore, internode length, especially of later-formed nodes, will be the primary determinant of plant height, and environmental stresses such as drought that occur during internode elongation can affect height.

Determinate cultivars reach maximum height at about R3, but most of this has been attained by R1. Even though flowering generally indicates cessation of vegetative growth, a cluster of four to six unexpanded leaves and unelongated internodes present at the stem apex complete canopy height development at about R3. Indeterminate cultivars can greatly increase in height after R1, but this stage will occur earlier (sooner after emergence) in adapted indeterminate cultivars than in adapted determinate cultivars (Kadhem et al., 1985a).

Because determinate cultivars produce little increase in height after R1, irrigation has little or no effect on plant height if drought stress begins at R1 or later (Doss & Thurlow, 1974; Hunt et al., 1981; Kadhem et al., 1985a; Specht et al., 1986). With indeterminate cultivars, because R1 occurs earlier and vegetative growth continues (Kadhem et al., 1985a), irrigation initiated at R1 would not have the same effect. Specht et al. (1986) measured an average linear increase in height of 8.5 and 6.8 cm, respectively, per 10 cm of applied water in 1983 and 1984 ($r^2 = 0.99$). Main stem elongation increased until about R4, and was thus stimulated by irrigation. Lodging increased linearly by an average of 0.67 and 0.51 units (1–5 scale) in 1983 and 1984, respectively, per 10 cm of applied water.

The amount of available water in most soils planted to soybean in the USA will support adequate height development up to R1 in determinate and later in indeterminate cultivars (Fig. 21–2). If irrigation is started during the vegetative phase, height still may not be increased if water is not limiting growth during this time (Huck et al., 1986). Irrigation that increases plant height can indirectly lead to yield reduction if the increased height contributes to lodging, especially in indeterminate cultivars (Kadhem et al., 1985a).

There is little or no correlation between height of soybean and seed yield. Yields of 3500 kg ha^{-1} have been reported where plants were as short as 60 cm (Kadhem et al., 1985a; Heatherly, 1983). Irrigation should be managed to ensure that plant height is sufficient for harvest, but not great enough to contribute to lodging.

3. Leaf Area Index

Leaf area development parallels height development in soybean. Water is the primary component necessary for cell expansion, and both leaf expansion and internode elongation result from cell expansion. Therefore, conditions that relate to plant height increases, as described above, also relate to leaf area increases.

Expansion of leaves is sensitive to shortages of available water (Boyer, 1970a; Bunce, 1978; Pandey et al., 1984). Extreme drought stress also results in shorter leaf area duration (Pandey et al., 1984). Therefore, leaf area index (LAI) is susceptible to varying soil water supply and drought stress since it results from both increase in leaf area and maintenance of accumulated leaf area (Constable & Hearn, 1978; Scott & Batchelor, 1979).

Irrigated determinate soybean will typically produce leaf area similar to that of nonirrigated soybean (Huck et al., 1986; Scott & Batchelor, 1979) if moisture is not limiting prior to reproductive development. However, maintenance of leaf area accumulated before irrigation was started can result in a higher LAI during reproductive development when drought stress usually occurs (Constable & Hearn, 1978; Scott & Batchelor, 1979).

Increased leaf area resulting from irrigation during vegetative growth can be detrimental if increased transpirational demand exceeds the capability of soil to supply adequate water later in the growing season. Therefore, a key to irrigation management for soybean is to avoid irrigation during the vegetative phase that could promote leaf area increases beyond levels needed for optimum photosynthesis and canopy closure. Leaf area indexes greater than the optimum can lower irrigation efficiency by promoting excessive transpiration and soil water depletion, especially since maximum leaf area will be present during months when rainfall is usually lowest and atmospheric demand is highest.

4. Dry Matter Accumulation

Dry matter accumulation in soybean is a result of leaf, root, and stem growth during the vegetative phase, and a combination of pod and seed growth concurrent with shifts in leaf, stem, and root mass during the reproductive phase. Irrigation has the most profound effect on dry matter accumulation during the reproductive phase (Hunt et al., 1981; Scott & Batchelor, 1979) because this is the period when soil water supply and rainfall are usually lowest and atmospheric demand is highest. Irrigation will also moderate the process of leaf senescence with increasing maturity (Pandey et al., 1984); i.e., adequate water can ensure that the maturing process will proceed with maximum translocation of photosynthate from leaf to seed (Huck et al., 1986). Water deficits during any portion of the reproductive phase can reduce photosynthesis (Boyer, 1970a, b) and the efficiency of translocation and result in lower seed dry matter accumulation, even though leaf and stem dry matter accumulation may have been at acceptable levels earlier in the season (Scott & Batchelor, 1979).

Drought stress has significantly reduced dry matter accumulation of all aboveground tissue measured at harvest (Ashley & Ethridge, 1978; Huck et al., 1986; Pandey et al., 1984; Scott & Batchelor, 1979; Mayaki et al., 1976). Harvest index, however, was consistently higher in an irrigated than in a nonirrigated treatment. Drought stress increased total root length, but root weight was significantly greater when soybean was irrigated (Huck et al., 1986). Pandey et al. (1984) found highly significant relationships between seed yield and leaf area duration, LAI, and shoot dry weight ($r^2 = 0.95$, 0.92, and 0.99, respectively) of soybean when grown under increasing drought.

5. Growth after Flowering

The majority of dry matter accumulation between R1 and R4 occurs in stems (Scott & Batchelor, 1979; Scott et al., 1983) of determinate cultivars. After R4, dry matter increases are derived from rapid accumulation in pods and seeds. Leaf mass or weight is also being reduced by translocation of photosynthate to the seed at about R4. The accelerated demise of vegetative tissue caused by drought stress will result in irreversible loss of yield potential if water is not added before this stage.

Increase in total accumulated dry matter generally occurs through R7. Leaves, petioles, and stems continue to increase dry matter accumulation through R5, and then start to decrease (Scott & Batchelor, 1979). Part of this decrease is attributable to dry matter transfer to the filling seed (Constable & Hearn, 1978).

Meckel et al. (1984) found that individual seed growth rate was not affected by drought stress during the reproductive phase in any year of a 3-yr study. The duration of seed fill was more sensitive to drought stress, but response was variable and nonsignificant in two of the three yr. Ramseur et al. (1984) found that seed growth rate was increased and effective filling period was decreased by irrigation, with no difference in single seed weight between irrigated and nonirrigated treatments. These reports support the conclusion that one effect of severe drought stress on seed yield is reduced number of seed (Boerma & Ashley, 1982; Constable & Hearn, 1980).

6. Leaf Pubescence and Reflectivity

One aspect of canopy development of recent interest has been the effect of leaf pubescence on water use by soybean. Theoretically, increased reflectance should reduce evapotranspiration (ET) and affect management decisions for irrigation scheduling and the adaption of soybean to more arid climates. Ghorashy et al. (1971a, b) found that soybean leaf pubescence increases reflectivity slightly in the visible band, but more significantly in the near infrared band. They also found that pubescence decreases transpiration because of decreased radiation absorption and because of an increased boundary layer resistance. Ghorashy et al. (1971b) found no effect of pubescence on photosynthesis in soybean. Agronomic trials of Hartung et al. (1980) and microclimate-physiology studies of Baldocchi et al. (1983) support the idea that pubescent soybean can be introduced into arid regions and can affect total water use (Clawson et al., 1986). Evapotranspiration was reduced overall by about 7% on a densely pubescent isoline and water use efficiency was greater in the pubescent isoline than in the nonpubescent isoline. The major effect was through altering the net radiation distribution, with deeper penetration into the canopy of the pubescent isoline (Nielsen et al., 1984). While some beneficial effects of pubescence on transpiration have been shown under field conditions with high radiant energy, effects on yield have been inconclusive. Hartung et al. (1980) showed that pubescent isolines tended to outyield normal isolines. However, caution is needed in interpreting these results because Garay and Wilhelm (1983) found differences in the root systems for two isolines differing in pubescence. The 'Harosoy' dense pubescent isoline had a greater root density, penetrated deeper into the soil and extracted more water during drought than did the normal pubescent isoline. However, the rate of water extraction on a per unit length basis was greater for the normal pubescent isoline. Under midsummer drought, the root distribution was characterized by low root density in the dry surface layer and maximum root proliferation in the wetter, deeper soil layers.

B. Root Growth and Water Absorption

1. Root Development and Depth

Soybean develops an extensive root system that is weakly taprooted with secondary and tertiary roots and adventitious roots originating from the hypocotyl. Lateral roots emerge shortly after germination, grow and branch profusely in upper soil layers until the plants are approximately 50-d old, and then start a downward growth pattern. Soil physical conditions strongly influence growth and development of root systems, but plant genetic factors and shoot growth and development also control root extension. Under soil conditions that are conducive to deep root penetration, about five or six of the primary laterals extend downward and eventually reach about the same depth as that of the taproot, which can be as much as 1.3 to 1.8 m below the soil surface (Allmaras et al., 1975b; Jung & Taylor, 1984; Mason et al., 1982; Mayaki et al., 1976). Most research has indicated that soybean needs a deep vigorous root system during drought stress to avoid affects on final yield. Increased rooting depth should increase the total quantity of water available for extraction during the growing season if all other conditions are equal throughout the season (Jones & Zur, 1984).

Research characterizing vertical distribution of soybean root dry weight and root length has shown that most roots are located in the upper 0.3 m of soil (Allmaras et al., 1975b; Mitchell & Russell, 1971; Raper & Barber, 1970). Mitchell and Russell (1971) showed that 90% or more of the roots were located in the upper 0.15 m of a Clarion loam-Nicollet clay loam soil (fine-loamy, mixed, mesic Typic Hapludoll and fine-loamy, mixed, mesic Aquic Hapludoll, respectively). Raper and Barber (1970), using widely spaced plants, analyzed root samples from cylindrical soil cores radiating from the plant center and showed that more than 83% of the root weight was contained in the upper 0.15 m and more than 90% was in the upper 0.3 m of a Chalmers silt loam (fine-silty, mixed, mesic Typic Haplaquoll) profile. Their data also showed that more than 70% of the root weight was located in the central 0.15-m cylinder of soil around the plant. The data of Mayaki et al.

(1976) on a Muir silt loam (fine-silty, mixed, mesic Cumulic Haplustoll) showed a greater downward distribution of roots with only about 70% of the root dry weight in the upper 0.3 m of the soil. Apparent differences in the soil profile could result in this difference in root distribution. Extension of a root system can be limited by soil physical characteristics, but soybean roots have the potential to extend 0.5 m or more horizontally and over 2 m vertically.

Irrigation can impact partitioning of dry matter between shoot and roots. Huck et al. (1986) determined the relative effect of water stress upon partitioning of dry matter between roots and shoots of soybean during both vegetative and reproductive growth periods in a rhizotron. Water stress significantly reduced total shoot and seed weight, but increased total root length. Irrigated plants had fewer roots that were generally shallower and more evenly distributed than nonirrigated plants. Although total weight of roots recovered was greater in irrigated treatments, weight of roots recovered from below 0.2 m was greater for the nonirrigated treatment, which agreed with the root length. Statistical significance of the difference between root growth in irrigated and nonirrigated treatments increased at successively deeper layers in the profile. The effect of water deficits on root-shoot relationhsip was evident, but the effect of water deficits on phenological development of the shoot was minimal. Similar increases in root growth of nonirrigated soybean was observed by Brown et al. (1985) during a dry year.

2. Root Extension

Growth and development of soybean root systems are strongly influenced by genotype. Taylor et al. (1978) and Kaspar et al. (1978) showed that taproot elongation rates of cultivars varied by as much as 13 mm/d. After 27 d, roots of the cultivar with the slowest rate had averaged 835 mm long, whereas length of the fastest growing cultivar averaged 1173 mm, a difference of 12.5 mm/d. Only small fractions of the increase in root elongation rate were associated with individual seed weight at planting and top weight at harvest time. Kaspar et al. (1978) determined root extension rates for seven soybean cultivars during reproductive development. Roots extended deeper at average rates of 11 to 18 mm/d during vegetative development, 33 to 81 mm/d as plants progressed from R1 to R3, and 36 mm/d from R4 to R5. These cultivars were indeterminate in growth habit and allowed root growth during reproductive periods. Results of Kaspar et al. (1978) suggest substantial downward root extension during pod filling, especially during dry years. Selection of the proper cultivar for increased rooting depth for maximum response to irrigation should consider both rate and duration of taproot extension.

There are limited data available on the effect of irrigation on root distribution of soybeans. Mayaki et al. (1976) noted substantial differences in plant height of irrigated and nonirrigated soybean as they matured, but essentially no difference in depth of rooting as a function of time. The actual rooting front progressed to a depth of about 1.6 m at an approximate rate of 20 mm/d. Both irrigated and nonirrigated soybean showed essentially the same rate of penetration. The depth of the root penetration measured under field conditions is similar to that measured in special plant containers by Taylor et al. (1978) with uniform soil conditions. Based on model predictions of Jones and Zur (1984), the rate of root extension is most critical in enabling the plant to cope with stress. Simulations indicate that root density and root permeability have only a minor effect on water extraction by the plant.

3. Root Water Extraction

Soil water extraction patterns have been used by some investigators to indicate depth and activity of soybean roots. A few published data sets indicate a poor correlation between root distribution and water uptake. Reicosky et al. (1972) showed that approximately 90% of the water extracted by a soybean root system was accomplished by about 15% of the root length in the presence of a water table. Similar behavior in profile soil water depletion was observed in the field by Stone et al. (1976) for both irrigated and nonirrigated soybean. Depletion of water by roots reached a depth of 1.5 m by R1 (flowering) with similar extraction patterns on both the irrigated and nonirrigated plots. However, differences in depletion patterns on the irrigated and nonirrigated plots were observed at early pod development. Roots of irrigated plants were more effective in depleting the water from the upper soil layers per unit of root dry weight than were those of nonirrigated plants. Ability to extract water on a unit dry weight basis increased with increase in soil depth in both irrigated and nonirrigated plots. A similar trend observed by Arya et al. (1975) showed the depletion rate from shallower depths of nonirrigated plots was less than on irrigated plots, primarily due to water availability and unsaturated conductivity. Most of the water extracted from nonirrigated profiles as a drying cycle advanced came from the lower portion of the root zone. This is in general agreement with results of Allmaras et al. (1975a) and Jung and Taylor (1984), who found that maximum soybean rooting depth coincided with maximum depth of water extraction.

A limited number of studies on the effect of irrigation on total root dry matter (Mayaki et al., 1976) and on root length (Robertson et al., 1980) showed little difference between irrigated and nonirrigated soybean. Roots of nonirrigated soybean profilerated more laterally and in the lower portion of the soil profile than did those of irrigated plants, but the differences were relatively small in relation to total root mass. Effect of intermittent drought on top-root ratio reported by Sivakumar et al. (1977) showed that leaf area continued to increase as root length decreased. Decrease in root length density was because of root death at all levels in the profile where the soil water potential Ψ_s was < -200 kPa. Although total quantity of roots appears not to be drastically affected by irrigation in many instances, this does not mean that the rate of water removal is not affected by soil water availability.

II. EVAPOTRANSPIRATION

A. Water Requirement and Consumptive Use

1. Evaporative Demand and Crop Water Use

Water use rates throughout the growing season are important to determine periods of maximum water use; however, seasonal water use can be more important in determining in which areas soybean is likely to need irrigation. Carter and Hartwig (1963) indicated seasonal water use for soybean can range from between 508 to 762 mm. In Missouri, Whitt and van Bavel (1955) found that seasonal water requirements ranged from 330 to 584 mm, with a maximum water use rate of 7.6 mm/d in July and August. In Kansas, Herpich (1963) found that total water requirements ranged between 508 and 610 mm with peak rates comparable to those found in Missouri. One year's data obtained by Kanemasu et al. (1976) showed that lysimetrically determined ET was 651 mm for a full season soybean crop and that this was approximately 13% more than a sorghum [*Sorghum bicolor* (L.) Moench] crop grown under similar conditions. In Nebraska, Somerhalder and Schleusener (1960) reported 457 to 635 mm of water are required for optimum growth

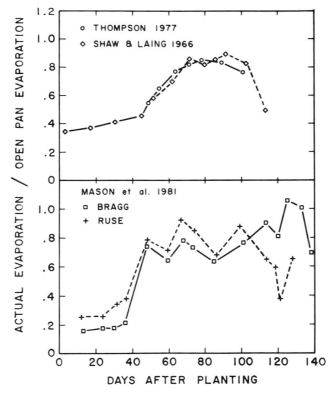

Fig. 21-3. Ratio of soybean evaporation to open pan evaporation during the growing season.

and yield. Musick et al. (1976) showed that soybean grown in Texas is very sensitive to soil water deficit and requires as much as 660 mm of water to complete the growing season. The largest reported value for soybean seasonal water use was 1575 mm for subirrigated soybean with the water table controlled at 0.6 m (Doss & Pearson, 1972). Doorenbos and Pruitt (1977) indicated that seasonal ET for soybean can range from 450 to 825 mm of water where the growing season ranges from 100 d in warm low altitude areas up to 190 d in higher altitudes. Even in northern climates, Hobbs and Muendel (1983) and Brun et al. (1985) showed a water use peak late in July and early August with a maximum of 7.0 and 7.2 mm/d, respectively. These water requirements for soybean indicate a fairly broad range that is highly dependent on rainfall distribution and total evaporative demand.

The seasonal pattern of crop water use relates to probability of developing drought stress. Water use has been expressed as a ratio of ET to open pan evaporation (EP) which represents an estimate of the atmospheric demand for water. As a soybean canopy develops, the ratio between ET and EP increases as shown in Fig. 21–3 (Shaw & Laing, 1966). The ratio is relatively low up to approximately 40 d after planting, and then shows a sharp increase with a maximum ratio between 60 and 100 d after planting. Similar data sets for two locations in Australia are shown for comparison (Mason et al., 1981; Thompson, 1977). In all three cases, results were similar with the ratio of ET to EP increasing 40 d after planting. A fairly high ratio is maintained up to about 120 d after planting, which parallels the general crop growth shown in Fig. 21–2.

Reicosky et al. (1982) investigated the diurnal relationship between ET and the leaf water potential (Ψ_l) for field grown soybean as affected by irrigation on an Ida silt loam (fine-silty, mixed [calcareous], mesic Typic Udorthents) in Iowa. The Ψ_l-ET relationship for nonirrigated soybean exhibited hysteresis while irrigated plants did not. The simplified Ohms Law did not adequately describe the relationship of plant water flux to resistance and potential gradients under field conditions because of the dynamic nature of field environment, which results in nonsteady state flow in the plant.

B. Irrigation Efficiency

For maximum economic benefits to accrue from irrigation, seed yield must be maximized with minimal amount of added water. This is not always an attainable objective because of weather and cost of irrigation. An economical source of water may be available, but stresses other than that caused by drought may be limiting plant performance. Conversely, an excellent yield potential with irrigation may be attainable, but a cheap, accessible source of water may not be available.

The ultimate goal of irrigation is to utilize added water efficiently and on soybean hectarage that can give the greatest seed yield increase from added water. *Irrigation efficiency*, defined as the increase in seed yield per unit of irrigation water applied, can be calculated if the amount of added water plus actual yield increase from irrigation are known. From data in Table 21–1,

Table 21-1. Effect of irrigation at different growth stages on seed yield and irrigation efficiency of soybean cultivars.

Cultivar or maturity group[†]	Planting date	Irrigation treatment[‡]	Irrigation water added	Irrigated yield	Yield increase above nonirrigated	Irrigation efficiency[§]
			cm	———kg/ha———		kg (ha cm)
GA (USA)—Ashley and Ethridge (1978)						
Hampton 266A	May 1972	VE	23.4	2130	1740	74
(VIII)		FL	18.3	2265	1875	102
		SD	12.5	1940	1550	124
Ransom (VIII)	May 1973	VE	19.6	3650	1605	82
		FL	13.2	3575	1530	116
		SD	9.9	3615	1570	159
Coker 102 (VIII)	May 1973	VE	19.6	2670	1465	75
		FL	13.2	2520	1315	100
		SD	9.9	2150	945	95
Hampton 266A	May 1974	VE	16.5	2010	80	5
		FL	9.9	1945	15	2
Ransom	May 1974	VE	16.5	2925	555	34
		FL	9.9	2985	615	62
KS (USA)—Brady et al. (1974b)						
Calland (III)	May 1972	VE	49.0	3550	470	10
		FL	44.4	3315	235	5
		PS	31.2	3820	740	24
Calland	May 1973	VE	40.6	3295	565	14
		FL	33.8	3335	605	18
		PS	19.6	3280	550	28
ND (USA)—Cassel et al. (1978)						
SRF100 (I)	June 1972	FL	38.1	1875	810	19
I	May 1973	FL	32.4	3105	2895	90
I	May 1974	FL	31.8	2305	1955	61
NSW (AUST)—Constable and Hearn (1978)						
Ruse (V)	Dec. 1975	VE	NA[¶]	2735	1030	--
		FL	NA	2530	825	--
Bragg (VII)	Dec. 1975	VE	NA	2945	1035	--
		FL	NA	3095	1185	--
NSW (AUST)—Constable and Hearn (1980)						
Ruse	Dec. 1976	VE	NA	2610	890	--
		FL	NA	2600	880	--
Bragg	Dec. 1976	VE	NA	3040	1025	--
		FL	NA	3130	1110	--
Ruse	Dec. 1977	VE	NA	3385	1970	--
		FL	NA	3175	1760	--
Bragg	Dec. 1977	VE	NA	3595	2000	--
		FL	NA	3425	1830	--
AL (USA)—Doss and Thurlow (1974)						
Bragg	June 1968	VE	30.2	2570	800	26
Bragg	June 1969	VE	24.8	3240	1120	45
Bragg	June 1970	VE	18.4	2520	160	9
Hampton 266	June 1968	VE	30.2	2070	300	10
Hampton 266	June 1970	VE	18.4	2160	200	11

(continued on next page)

Table 21-1. Continued.

Cultivar or maturity group†	Planting date	Irrigation treatment‡	Irrigation water added	Irrigated yield	Yield increase above nonirrigated	Irrigation efficiency§
			cm	⸺kg/ha⸺		kg/(ha cm)
AL (USA)—Doss et al. (1974)						
Bragg	May–June 1970	VE	37.7	2790	180	5
		FL	28.7	2670	60	2
Bragg	May–June 1971	VE	37.7	3250	1190	31
		FL	28.7	3120	1060	37
		SD	10.8	2170	110	10
Bragg	May–June 1972	VE	37.7	2940	1060	28
		FL	28.7	2770	890	31
		SD	10.8	1420	−460	0
MS (USA)—Heatherly (1983)						
Bedford	June 1979	PS	15.2	2670	−80	0
		SD	8.6	2580	−170	0
Tracy (VI)	June 1979	PS	17.4	3380	10	1
		SD	14.7	3300	−70	0
Bragg	June 1979	PS	9.0	3210	40	4
		SD	9.0	3630	460	51
Bedford (V)	May 1980	FL	48.0	2000	1270	26
		PS	48.9	1750	1020	21
		SD	35.0	1110	380	11
Tracy	May 1980	VE	60.0	2890	1740	29
		FL	63.8	2810	1660	26
		SD	29.1	1470	320	11
Bragg	May 1980	VE	70.0	3560	2240	32
		FL	58.0	3220	1900	33
		SD	29.4	1790	470	16
MS (USA)—Heatherly (1984)						
Forrest (V)	June 1979	SD	7.7	3625	115	15
	May 1980	FL	30.9	2980	1855	60
		PS	23.0	2480	1355	59
		SD	17.2	2450	1325	77
MS (USA)—Heatherly and Elmore (1986)						
Bedford	May 1980	FL	52.9	2730	1740	33
	June 1980	FL	38.8	3150	1995	51
Bragg	May 1980	FL	69.1	3525	2195	32
	June 1980	FL	46.9	2980	1460	31
Bedford	May 1981	FL	31.8	2780	1800	56
	June 1981	FL	19.6	2375	1325	68
Braxton (VII)	May 1981	FL	41.9	3275	2245	53
	June 1981	FL	29.4	2940	1245	42
Bedford	May 1982	FL	29.4	2245	1270	44
	June 1982	FL	29.4	1670	790	27
Braxton	May 1982	FL	39.2	2720	1710	44
	June 1982	FL	29.4	2345	1145	39
SC (USA)—Hunt et al. (1981)						
Bragg	June 1978	FL	20.5	3090	1815	88

(continued on next page)

Table 21-1. Continued.

Cultivar or maturity group†	Planting date	Irri- gation treat- ment‡	Irrigation water added	Irrigated yield	Yield increase above nonirrigated	Irrigation efficiency§
			cm	kg/ha		kg/(ha cm)
AR (USA)—Jung and Scott (1980)						
Forrest	May 1978	FL	27.0	3455	2550	94
SC (USA)—Karlen et al. (1982)						
Coker 338 (VIII)	June 1978	FL	20.2	3070	1480	73
Bragg	June 1978	FL	20.2	2890	1340	66
Ransom	June 1978	FL	20.2	3510	1490	74
Coker 338	May 1979	FL	24.5	1730	290	11
Bragg	May 1979	FL	24.5	1960	290	12
Ransom	May 1979	FL	24.5	2320	580	24
NE (USA)—Korte et al. (1983a)						
II, III, IV	May 1977	F	NA	3000	0	--
		P	NA	3250	250	--
		S	NA	3030	30	--
II, III, IV	May 1978	F	NA	2650	−80	--
		P	NA	3150	580	--
		S	NA	3540	970	--
II, III, IV	May 1979	F	NA	2700	90	--
		P	NA	3020	410	--
		S	NA	2760	150	--
NC (USA)—Mahler and Wollum (1981)						
Ransom	May 1979	FL	28.3	2070	775	27
NC (USA)—Martin et al. (1979)						
Ransom	May 1977	VE	28.0	2065	335	12
		FL	15.2	2125	395	26
		PS	3.0	1685	−45	0
SC (USA)—Ramseur et al. (1984)						
Braxton	May 1980	VE	38.8	2740	1395	36
		FL	23.8	2645	1300	55
Braxton	May 1981	VE	39.9	2845	1415	35
		FL	25.4	2655	1225	48
SC (USA)—Reicosky and Deaton (1979)						
Davis (VI)	May 1975	VE	NA	2340	350	--
McNair 800 (VIII)	May 1975	VE	NA	2405	365	--
NE (USA)—Specht et al. (1986)						
Corsoy 79	May 1983	FL	36.7	4050	2450	67
	May 1984	FL	31.1	3250	1590	51
Century	May 1983	FL	36.7	4020	2890	79
	May 1984	FL	31.1	3450	1780	57
Fremont	May 1983	FL	36.7	3870	2760	75
	May 1984	FL	31.1	3450	1940	62
Mead	May 1983	FL	36.7	3770	2640	72

(continued on next page)

Table 21-1. Continued.

Cultivar or maturity group†	Planting date	Irrigation treatment‡	Irrigation water added	Irrigated yield	Yield increase above nonirrigated	Irrigation efficiency§
			cm	kg/ha	kg/ha	kg/(ha cm)
	May 1984	FL	31.1	3450	1590	51
Pella	May 1983	FL	36.7	3790	2910	79
	May 1984	FL	31.1	3360	1390	45
Will	May 1983	FL	36.7	3690	2290	62
	May 1984	FL	31.1	3380	1990	64
A3127	May 1983	FL	36.7	4130	3160	86
	May 1984	FL	31.1	3520	1710	55
Hobbit	May 1983	FL	36.7	3450	2350	64
	May 1984	FL	31.1	3630	2010	65

† Where only maturity group or groups are given, tabular values are averages of more than one cultivar in a group or groups.
‡ VE, FL, PS, SD = irrigation started before flowering, at beginning of flowering or R1, at beginning of podset or R3, or at beginning of podfill or R5, respectively, and continued as needed until near maturity. F, P, S = irrigation at R1 only, at R3 only, or at R5 only.
§ Values are kg/ha yield increase from irrigation divided by total amount of irrigation water in cm added to that treatment.
¶ NA means that the amount of irrigation water was not given or could not be accurately calculated.

some general conclusions can be made: (i) efficiency of irrigation was usually less when irrigation was started before R1 because no significant yield increase was realized from additional irrigation water; (ii) in years where soybean had not been irreparably damaged by drought stress prior to the R3 to R5 period, efficiency of irrigation initiated at R3 was as great as or greater than that initiated at R1 because relatively less water was required per unit increase in seed yield when watering was delayed; and (iii) effects of year, cultivar, location, and their interactions are evident because of weather-related stresses that irrigation alone could not overcome, and also because of the varying effect that different ET requirements had on a specific cultivar among different years.

Specht et al. (1986) suggested that cultivars that exhibit a relatively small change in seed yield per unit of applied water could be described as being less sensitive to drought, whereas those cultivars that exhibit a larger response in seed yield per unit of applied water could be described as being more responsive to irrigation. Cultivars that are less susceptible to drought-induced reductions in seed yield might be adapted to rain-fed culture where periods of drought stress are common, whereas cultivars that are more responsive to increasing levels of irrigation might be more specifically adapted to irrigation culture. Regardless of varying irrigation efficiencies among cultivars, selection of cultivar and hectarage for irrigation application should also be based on net economic return and not just response to applied water. It is this economic efficiency that will dictate optimum use of a limited supply of irrigation water and the method of application (Crabtree et al., 1985).

III. WATER STRESS (ADAPTION AND RECOVERY)

A. Plant Water Stress—Plant Water Relations

1. Water Relations and Growth

The primary purpose of irrigation is to maintain an optimum plant water status throughout the entire growing season because soil water deficit has a direct effect on plant tissue growth and seed yield. Huck et al. (1983) and Silvius et al. (1977) demonstrated that stomatal closure and reduced photosynthetic activity occurred whenever transpiration from soybean leaves exceeded the absorptive capacity of the root system. Boyer (1970a) reported that leaf enlargement rate of soybean was dramatically reduced as Ψ_l decreased from -0.2 to -0.4 MPa. Enlargement rate progressively decreased with a decrease in Ψ_l and nearly stopped at -1.2 MPa. Other investigations have shown that stem elongation and leaf enlargement are the first processes reduced by plant water deficit. Prolonged periods of stress apparently result in a reduced rate of cell division in meristematic tissue (Meyer & Boyer, 1972).

Based on growth chamber studies, Boyer (1970a) found dry weight accumulation rates are much less sensitive to plant water deficit than those indicated for leaf expansion. However, during the summer growth period, the midday Ψ_l in the field is often less than that required to reduce leaf expansion rates even when there is abundant soil water available. Brady et al. (1974a), Reicosky and Deaton (1979), and Jones et al. (1983) showed that the diurnal patterns of Ψ_l in irrigated and nonirrigated soybean are related to the evaporative demand. The amplitude of diurnal variation can be as much as 1.3 to 1.4 MPa, with differences in minimum Ψ_l of irrigated and nonirrigated plots being relatively small. Reicosky and Deaton (1979) noted larger differences in ET between irrigated and nonirrigated plots that suggested substantial stress for nonirrigated soybean. They found small differences in midday Ψ_l for two soybean cultivars, whereas ET from a nonirrigated treatment was about two-thirds the irrigated treatment 25 and 32 d into the drought. Nonirrigated plants exhibited severe wilt symptoms during drought and soil water potential (Ψ_s) data indicated significant water extraction at 1.53 m on nonirrigated treatments. These results and those of Mayaki et al. (1976), Huck et al. (1986), Zur and Jones (1984), and Zur et al. (1983) point out the importance of root distribution and subsoil water in partially meeting the evaporative demand during drought.

The importance of root systems in determining plant-water status was demonstrated by Sullivan and Brun (1975) who studied genetic differences in root stocks and their effect on midday water deficit. Using a grafting technique, they established the effect of genetic differences of root stocks on plant water status under midday water deficit. Boyer et al. (1980) showed cultivars with low yields generally experience larger water deficits than those with high yields. Largest water deficits were found in oldest cultivars and were present in all the fully illuminated leaves. Rooting densities of two cultivars showing extremes of midday water deficits indicated that 'Wayne' (with the least water deficit) had the greater root density and 'Richland' (with the largest deficit) had the smallest root density. Old cultivars experienced more water stress that resulted in limited yields, while the newer cultivars had higher yields, apparently from improved root systems.

2. Physiological Activity

Plant response to soil water deficits is closely related to Ψ_l and evaporative demand. Both Ψ_l measured with pressure chambers and stomatal resistance (r_l) measured with porometers have been used to characterize plant water status under field conditions. Jung and Scott (1980) found Ψ_l of nonirrigated soybean decreased earlier and increased later in the afternoon than that in irrigated soybean. The stomates of nonirrigated soybean were partially closed during the daylight hours and closed earlier in the afternoon than those of the irrigated plants. Brady et al. (1975) showed a definite relationship for two consecutive growing seasons between Ψ_s and r_l. Adaxial r_l increased rapidly when Ψ_s decreased to near -0.4 MPa. They noted a distinct difference in r_l between irrigated and nonirrigated soybean.

Sojka and Parsons (1983) studied the effect of canopy microclimate on soybean water status. They observed no row spacing effects on Ψ_l for any of the varieties, but did note that midday Ψ_l for 'Coker 338' was lower than that for 'Davis.' Leaf water potential was highly correlated with leaf vapor pressure deficit. However, they did notice the parallel leaf diffusive resistance was not highly correlated with any of the canopy microclimate or water stress parameters. Carlson et al. (1979) showed that Ψ_l and leaf conductance are related but other factors affect this relationship. Two cultivars acted differently under the same environmental conditions, and as a result, they could not find a unique linear relationship between Ψ_l and stomatal conductance. Atmospheric demand, time of day, and some type of drought stress hardening contributed to the variation. Part of the variation found in field measurements (Sojka & Karlen, 1984) could also be related to the lack of standard technique for covering the leaves while measuring Ψ_l (Meyer & Reicosky, 1985) and may have been due to errors in measurements of leaf temperature using various stomatal porometers (Meyer et al., 1985).

3. Photosynthesis

One of the key physiological processes affected by water deficits is photosynthesis. Boyer (1970b) reported that a decline in apparent photosynthesis of young soybean leaves occurred below -1.1 MPa Ψ_l. The apparent photosynthetic rate was half maximum rate when Ψ_l was -1.7 MPa. Boyer attributed reduction in apparent photosynthesis to an increase in r_l and a decrease in diffusion of CO_2 to the fixation site. Ghorashy et al. (1971a) found that apparent photosynthetic rates decreased linearly as Ψ_l decreased. This decrease was less during the flowering stage than during pod filling stage for three isolines differing in pubescence density. They found no significant difference between regression coefficients of the three isolines when stress was applied at flowering, suggesting no difference between dense pubescent and normal pubescent leaves. In all isolines, apparent photosynthesis was more sensitive to drought stress during pod filling than during flowering.

Ghorashay et al. (1971a) found that soybean leaves remained fully turgid to about -1.1 MPa and at this Ψ_1, photosynthesis was not seriously affected. However, at -2.0 MPa leaves were wilted and relative apparent photosynthesis decreased to below 10% of well-watered plants. They concluded that an inherent limitation of soybean roots to supply water rapidly enough to meet transpiration demand resulted in substantial water stress.

It is commonly accepted that chamber grown plants differ from field grown plants in their response to drought stress. Turner et al. (1978) showed field-grown soybean plants do not show the same decrease in photosynthesis as those grown in chambers until Ψ_1 reached -1.5 to -1.7 MPa. The field environment conditioned the plants to withstand greater drought stress than plants grown under controlled environmental conditions. They proposed restricted rooting volume and lower evaporative demand under growth-chamber conditions could account for differences in response to field-grown plants. Christy and Porter (1982) concluded from field data that soybean yield is the result of a continuing readjustment of sink load, seed number, and seed size to concurrent photosynthesis throughout reproductive stages.

IV. GROWTH AND SOIL WATER

A. Soil Water Effects on Crop Growth

1. Soil Water Deficit and Growth

One of the most sensitive indicators of soil water deficit is leaf growth (Boyer, 1970a; Heatherly et al., 1977; Sivakumar & Shaw, 1978). This sensitivity is so great that most leaf growth occurs at night when Ψ_1 is high and evaporative demand low (Boyer, 1968). Even with well-watered plants, turgor during midday is insufficient to sustain significant leaf growth (Boyer et al., 1980; Jung & Scott, 1980; Sojka et al., 1977). Stem growth follows the same pattern. Pod growth occurs later in the growing season when soil moisture supply and rainfall are less, and thus is dependent on supplemental water, although its sensitivity to soil moisture deficit is no greater than that of leaf growth.

Irrigation during any phase of soybean growth and development will alter growth potential if drought stress is significant (Ashley & Ethridge, 1978; Ramseur et al., 1985; Scott & Batchelor, 1979). If the plant is vegetaive, then leaf and stem growth rate can be increased (Pandey et al., 1984; Scott & Batchelor, 1979). If the plant is reproductive (or in the case of indeterminates, both vegetative and reproductive), then number of pods and seeds can be increased if irrigation occurs during early reproductive development (Ashley & Ethridge, 1978; Egli et al., 1983; Kadhem et al., 1985b; Ramseur et al., 1984). The majority of yield increase from irrigation of soybean where drought stress is severe but alleviated during the reproductive period comes from this increase in number of seed rather than from increase in weight of individual seeds that may occur if irrigation is applied only during late reproductive development (Ashley & Ethridge, 1978; Egli et al., 1983; Heatherly, 1983; Huck et al., 1986; Ramseur et al., 1984).

Mayaki et al. (1976) evaluated seasonal dry weight accumulation of both roots and aboveground portions of plants under irrigated and nonirrigated conditions. The total amount of roots was not affected appreciably by irrigation even though greater lateral and deeper root development was observed in the nonirrigated plots. Aboveground components were adversely affected by soil water deficit with LAI of nonirrigated plants about 62% of irrigated plants. Fruit weight of nonirrigated plants was about 74% of irrigated plants near physiological maturity. Scott and Batchelor (1979) found similar differences in plant growth due to soil water regime during two dry seasons. Differences in weight were initially found during late vegetative and early bloom stages for both leaves and stems. Generally, irrigated soybean accumulated greater amounts of dry matter and leaf area and had higher seed yields than did nonirrigated soybean. However, they did note the lack of response to irrigation during a wetter than normal growing season.

2. Soil Type and Growth

Soil properties (texture, depth, structure, and hydraulic conductivity) control availability of water to plants. Sandy soils contain limited water, but almost all of it is readily available for plant use. Clay soils, on the other hand, contain a large amount of water, but a portion of it is unavailable for plant uptake, and that water available is only slowly available. Soils that have a large proportion of silt have the greatest amount of available water, and it is those soils that will support greatest growth and biomass accumulation. Those soils will also give least response to irrigation given periodic rainfall input.

Variable water retention properties can cause growth and development of soybean to be drastically different among soil types (Heatherly & Russell, 1979; Scott & Batchelor, 1979). However, this growth difference is not necessarily manifested in seed yield difference (Heatherly & Russell, 1979). Canopy size developed on a soil that supports rapid and large growth can be detrimental when late-season rainfall is limiting because transpirational surface of a larger canopy may exceed the soil water availability. This can result in severe drought stress that requires frequent and large irrigations if yield potential is to be maintained.

Largest yield increases from adequate irrigation have been achieved on soils that have lower available water-holding capacities (Boerma & Ashley, 1982; Cassel et al., 1978; Heatherly, 1983; Hunt et al., 1981; Karlen et al., 1982a; Mahler & Wollum, 1981; Ramseur et al., 1984). This is because of relatively low nonirrigated yields from these soils rather than abnormally high yields with irrigation. Therefore, it appears that irrigation water should be allocated to soybean grown on soil types that have a low yield in a nonirrigated environment. This assumes, of course, that other limiting factors besides water can be economically alleviated.

3. Critical Demand Periods

Increasing need for water as soybean development progresses is shown in Fig. 21-2. Most soil types, assuming periodic rainfall, can supply the water necessary to meet ET demands and support adequate growth during the vegetative phase. Exceptions to this are those that have a shallow rooting depth (Griffin et al., 1985) or low water-holding capacity.

Numerous studies (Ashley & Ethridge, 1978; Brady et al., 1974b; Constable & Hearn, 1978; Doss et al., 1974; Heatherly, 1983, 1984; Martin et al., 1979; Ramseur et al., 1984; Sojka et al., 1977) have investigated soybean yield response to both full-season irrigation (water applied as needed during both the vegetative and reproductive phases of development, designated VE) vs. irrigation during reproductive development only (water applied as needed from R1 to near maturity, designated FL). In all cited cases (Table 21-1), VE irrigation produced no appreciable yield advantage above that realized from FL irrigation. Also, irrigation efficiency, defined here as increase in seed yield per hectare due to irrigation [kg/(ha cm)], was usually higher for FL treatment than for VE treatments. Thus, irrigation of monoculture soybean prior to R1 appears to be of little benefit, even though demand for water is increasing through R1.

Several research efforts have sought to establish a critical period for irrigation of soybean. Sionit and Kramer (1977) found that stress applied during either pod formation or podfill resulted in greater yield reductions than stress applied during flower induction or flowering. This finding is supported by Doss et al. (1974), who obtained lowest seed yield when they withheld irrigation during podfill. In both cases, adequate water had been supplied to plants up to the induced stress. Stress applied during podfill produced the smallest seed, but did not reduce the total number of seed below the level produced by plants that were irrigated during all stages of reproductive development (Sionit & Kramer, 1977). Obviously, number of seeds that are set is maintained during drought stress that occurs after seed formation; however, seed size cannot be maintained where number of seed is at a level equaling that in nonstressed soybean. Korte et al. (1983a) found that one irrigation at pod elongation (R3–R4) or seed enlargement (R5–R6) significantly enhanced 3-yr average yields above both the nonirrigated check and the treatment which received a single irrigation at flowering (R1–R2) (Table 21-1). Most of the 3-yr effect, however, was the result of large differences in 1978 compared to those in both 1977 and 1979 which resulted in a significant year by irrigation interaction. Smallest seed in this study were obtained from the treatment irrigated only at flowering (Korte et al., 1983b), but seed per plant were similar for all irrigation treatments. Again, number of seed established as a result of adequate water during flowering was not affected by stress during later stages, but seed size was reduced. Brown et al. (1985) found the effect of drought stress at R2–R3 vs. R4–R5 was variable, with R4 stress being more critical in a drier year. Both weight and number of seed were less with R5 stress. Stress at R2 reduced number of seed, but not seed weight.

Doss et al. (1974) concluded that the podfill stage of soybean is the most critical period for irrigation to be used to obtain maximum seed yield. This would appear to be supported by studies of Sionit and Kramer (1977) and Korte et al. (1983a). However, in the former two cases, all of the results are from situations where water was available from either rainfall or irrigation during all other phases of growth leading up to the designated stress stage. This would allow plants to set maximum fruit load allowed by available moisture. If water was suddenly cut off (as was the case), then plants would be unable to maintain the yield potential established earlier, and the result would be smaller seed, as found by Sionit and Kramer (1977) and Korte et al. (1983b). Conversely, plants stressed only during flowering produced significantly fewer seed than those stressed during podfill, but the seed were equal in size to those of well-watered control plants (Sionit & Kramer, 1977). Yield from this treatment (stressed during flowering only) was between yield of the other two. Mahler and Wollum (1981) found that nonirrigated soybean actually had larger seed, and they attributed this to the fewer seeds in the nonirrigated treatment benefitting from late-season rainfall during podfill.

Delaying initiation of irrigation until beginning of podset (PS) or beginning of podfill or seed development (SD) in years when rainfall was limited during early reproductive stages resulted in seed yields that were lower than those realized from either VE or FL irrigation (Table 21-1), but equal to yield from a nonirrigated treatment in most cases (Ashley & Ethridge, 1978; Griffin et al., 1985; Heatherly, 1983, 1984; Martin et al., 1979). In a year where rainfall was adequate during all reproductive phases except podfill (Heatherly, 1983) (Table 21-1), 'Bragg' produced significantly more yield from one irrigation at this stage than was produced by either the PS or nonirrigated treatments. Earlier-maturing cultivars that were essentially through filling pods before this late-season stress occurred did not show the same response. Obviously, stage of development of a particular cultivar in relation to the occurrence of a period of limited rainfall will have a dominant effect on the cultivar's response to irrigation.

In some cases, irrigation of soybean at any stage of development did not significantly increase or only slightly increased yields above the level of a nonirrigated treatment (Table 21-1). This resulted from rainfall or soil moisture supply being sufficient to produce yields of 2500 kg/ha or greater (Brady et al., 1974a; Doss et al., 1974; Heatherly, 1983, 1984; Huck et al., 1986; Scott & Batchelor, 1979), lodging or disease precluded a response (Ashley & Ethridge, 1978; Doss & Thurlow, 1974), not enough irrigation water was applied (Martin et al., 1979), nonirrigated plants extracted enough subsoil water to maintain a similar yield level (Reicosky & Deaton, 1979), irrigation was terminated before pods were filled (Doss & Thurlow, 1974), or pod abortion caused by inadequate late-season photosynthate supply occurred (Taylor et al., 1982). Excessive soil water caused by unexpected rainfall following irrigation that results in low O content, reduced N fixation and stomatal closure can affect soybean response to irrigation (Bennett & Albrecht, 1984).

Based on the preponderance of data (Table 21-1), irrigation during the entire reproductive phase in the abscence of rainfall is the most desirable. This is because: (i) development occurs during the portion of the growing season when rainfall amount is usually below the ET requirement and soil water supply has been significantly diminished because of plant extraction up to this time; (ii) leaf area is at or near maximum levels; and (iii) components of seed yield are being determined. Drought stress during any portion of this phase can limit yield by limiting the contribution of one or more yield components. For maximum yield in dry years, water must be available for irrigation to be started at beginning of flowering, and in the absence of rainfall, continued until seed are fully developed (Thompson, 1977; Heatherly, 1983; Ashley & Ethridge, 1978). Less irrigation than this will give smaller though relatively efficient yield increases (Table 21-1). If only a limited amount of water is available (not enough for full reproductive phase irrigation), reported data support the practice of using this water for irrigation during podfill. However, this latter practice will not produce the maximum yield that may be required to maximize net return (unless adequate rainfall has been received prior to this time).

4. Flooding and Soil Aeration Effects

Soybean roots require adequate O for growth and maintenance respiration. Stanley et al. (1980) investigated the effects of temporary water tables imposed for 7 d at three stages of growth on soybean root and shoot response. Water tables were raised to within 0.45 and 0.90 m of the soil surface for 7 d in a rhizotron at three stages: (i) during vegetative growth prior to flowering, (ii) immediatey following flowering, and (iii) after the pods had set. When water tables were imposed before flowering, roots below the water table level stopped growing downward but remained alive and resumed downward growth after the water table was lowered. When the water tables were imposed either during the flowering or at the pod filling stage, however, roots below the water table became discolored and tended to decompose after the water level was lowered. No new roots grew into soil volumes that had been previously below the water table even though new root growth occurred above this depth.

Response of soybean top growth to flooding can be different from that of the root system. Bennett and Albrecht (1984), using a sandy soil in a greenhouse study, imposed both flooding and drought stress treatments for 14 d. Flooding had little effect on Ψ_1, nodule water potential, and r_1. Nitrogenase activities of nodules after removal from the flooded environment were similar to the controls at the end of the experiment. However, a slight leaf yellowing and altered root system were noted. Effects of the drought stress treatments were more dramatic with a decrease in N_2 fixation observed slightly before Ψ_1 decreased and r_1 increased.

Researchers in Queensland, Australia, observed that continuous water on the soil surface increased shoot dry matter and seed yield when compared with other irrigated controls. This recent work has shown soybean can produce higher yields under "wet soil culture" (Hunter et al., 1980; Troedson et al., 1985; Nathanson et al., 1984) when compared to conventional methods. Briefly, the seed is conventionally planted in raised beds and furrow irrigated by continuous low-volume water flow when the first trifoliate appears. The water level was maintained near the surface of the soil bed throughout the season. After an initial acclimation period when the soybean exhibited chlorotic symptoms characteristic of N stress, the plants recovered and yielded an average of 22% more for 14 varieties than conventionally grown soybean that yielded 3444 kg/ha in a field study (Trodeson et al., 1985). The yield increase for one variety was as large as 68% and is presently attributed to complete absence of soil moisture stress (no stomatal closure) and root proliferation and nodule dry matter accumulation in the saturated zone just above the water table. While this method requires specific soil conditions, the soybean response is intriguing. The practical consequence of this work awaits further research where water use is quantified and O_2 status is characterized.

V. PLANT NUTRITION

A. Fertility Irrigation Relations

Soybean, like other crops, grows and produces well only when it has an adequate supply of water and essential mineral nutrients. Fertilization and irrigation should work in concert for maximum crop production. One would expect considerable research on the interaction between irrigation and fertility levels, however, DeMooy and Sutherland (1981) indicated the opposite is true and that the recent work combining foliar nutrient applications and irrigation still leaves much to be done. Mineral nutrition of soybean has recently been reviewed by Barber (1985), Borkert and Barber (1985a, b), de Mooy et al. (1973), Kamprath (1974), Kurtz (1976), and Peaslee et al. (1985).

Little information is available on effects of irrigation on the placement of nutrients on soybean yield. Lutz and Jones (1975) evaluated effects of irrigation on placement of lime, P, and K and micronutrients on yield and composition of soybean on a Davidson clay loam (clayey, kaolinitic, thermic Rhodic Paleudults [oxidic]) in Virginia. Soybean yields were increased each year with irrigation, with an average annual increase for 3 yr of 514 kg/ha or 22%. Yields were unaffected by P and K treatments during the first 2 yr, but in the third year yields were lower where P and K had not been applied. Fertility treatments did not affect seed oil or protein content. Irrigation had little influence on oil and protein content of soybean seed.

As part of a larger study to determine why narrow-row soybean often outyields wide-row soybean, Bennie et al. (1982) evaluated the concentration, accumulation, and translocation of 12 elements in soybean tissue as affected by irrigation. They found no significant effect of row spacing or irrigation level on the concentration of the 12 elements in the mature seeds. Row spacing had no significant effect (at the 5% level) on leaf concentration of several elements immediately before the start of pod set. Irrigation significantly increased (at the 5% level) leaf concentrations of N, P, K, B, and Cu, but significantly decreased leaf concentrations of Ca, Mg, and Fe. Some of these effects were due to increased biomass caused by irrigation, but some effects were caused by a higher uptake rate associated with irrigation. These small differences due to irrigation occurred despite the fact that total biomass production during the season was increased by irrigation but seed yield was not (Taylor et al., 1982).

Nitrogen provides a major management challenge on moderately coarse-textured soils with low inherent fertility because of the mobility of NO_3, but is usually not much of a concern for soybean. Cassel et al. (1978) reported results of an irrigation N study on a Maddock sandy loam (sandy, mixed Udorthentic Haploboroll) soil in southeastern North Dakota. They established four water levels for irrigation and four subplots for application of N fertilizer. Because of low rainfall, irrigation level had a significant effect on seed grain yield in each of the 3 yr. Soybean yields were increased by the application of N only in the first year of this study when seed was not inoculated. Analysis of the NO_3-N in the soil profile at different dates during the 3-yr period showed that under heavy fertilization, NO_3-N moved below the crop rooting zone when excess water was applied.

Soil and plant water status primarily influences nutrient absorption and accumulation because of its effect on plant growth. De Mooy et al. (1973) stated that nutrient accumulation by soybean was affected in proportion to the length of effective drought during the growing season and the portion of the root system located in water-depleted soil. Cation absorption is especially dependent on irrigation to maintain high soil water status. Oliver and Barber (1966) showed K transport is dependent on diffusion, and Ca and Mg are dependent on mass flow. The uptake patterns for these cations showed different responses to drought stress, presumably because the transport mechanisms in the soil differed (Karlen et al., 1982a, b). Uptake of K was more responsive to soil water status than Ca or Mg and has also been shown to be influenced by soybean root morphology (Raper & Barber, 1970). Irrigation did not significantly change the cation concentration within the soybean plant and differences in cation accumulation over the full season were primarily caused by increased plant growth. Potassium uptake under nonirrigated conditions fluctuated with the soil water availability, whereas K uptake with irrigation was relatively constant until physiological maturity. Both Ca and Mg uptake were relatively constant regardless of whether or not irrigation was applied until physiological maturity.

Karlen et al. (1982a) showed that irrigation appeared to enhance P diffusion to plant roots and increased P concentration at some sampling times, but rarely influenced the Fe, Mn, or Zn concentration or accumulation. Seed yields were not limited by P or micronutrient accumulation. Phosphorus concentrations within irrigated plants were generally greater than in nonirrigated plants. Irrigation increased P accumulation by increasing plant growth and perhaps by enhancing P diffusion to the plant roots.

B. Soil Water Availability and Dinitrogen Fixation

Only in recent years has detailed information on how soil water availability affects soybean N_2 fixation become available. Sprent (1971) concluded that both N_2 fixation and respiration of detached soybean nodules nearly ceased when nodule fresh weight dropped below 80% of the fully turgid weight and that N_2 fixation was possibly related to nodule water content between 80 and 100%. In other experiments, N_2 fixation by nodules attached to soybean roots was also reduced in water-stressed plants (Sprent, 1972b) or by osmotically induced root zone drought stress (Sprent, 1972a). Apparently, the outer cortex of the nodule shows structural changes in cytoplasmic constituents under moderate stress, with more severe stress needed to affect nodule activity.

Pankhurst and Sprent (1975) associated the reduction in nodule activity under drought stress with changes in the O_2 diffusion and respiration rates. Leaf water potential was found to be closely associated with rate of N_2 fixation (Huang et al., 1975a, b; Patterson et al., 1979). Huang et al. (1975a, b) noted a close association between rates of photosynthesis and transpiration and N_2 fixation with decreasing Ψ_l, but respiration remained fairly constant at all water-stress levels. These observations led to conclusions that photosynthate translocation or some direct effect on the nodule other than respiration was most likely the cause of decreased N_2 fixation as the Ψ_s decreased.

Sinclair et al. (1985) and Denison et al. (1983a, b) confirmed the O_2 response of drought-stressed soybean under field conditions. They used continuous flow root chambers, where partial pressure of O_2 was increased in the short term response and acetylene reduction rate was measured. A close correlation between nodule conductance for O_2 and acetylene-reduction rate at all stages of the drying cycle demonstrated the existence of an O_2 diffusion barrier that limits the N_2 fixation rate. Nodule permeability decreases in response to the dehydrating conditions associated with limited soil water. The drought study showed the flux density of acetylene was linearly related to nodule permeability regardless of the severity of stress or magnitude of the nodule permeability. These results confirm the importance of optimum water management and proper irrigation scheduling for maximum N_2 fixation.

Matheny and Hunt (1983) studied the impact of irrigation on accumulation of both soil and symbiotically fixed N in nodulating and nonnodulating determinate soybean on a Norfolk loamy sand (fine-loamy, siliceous, thermic Typic Paleudults). In a year with above-normal rainfall, irrigation did not significantly affect the seed yield in the nonnodulating variety, but irrigation resulted in a two-fold increase in the following year with a major drought. They concluded irrigation and rainfall patterns greatly affect the N accumulation and that the percentage of the N supplied from fixation may be as high as 90%. The difference in the yield between the nodulating and nonnodulating soybean was nearly 1000 kg/ha greater under irrigated than nonirrigated conditions.

Nitrogen redistribution in soybean as affected by water deficit was reported by Egli et al. (1983). Their data showed the contribution of redistributed N to seed N, and maturity was more closely related to total amount of N available rather than on the ability of plants to fix N or take up N from the soil. Plant weight at R5 and yield were affected by drought stress treatments, which had only a limited effect on N concentration in the plant tissue. The portion of seed N that came from redistribution varied among treatments and there was no consistent relationship between level or timing of drought stress and contribution from redistributed N to seed N. Similar results were reported by Hunt et al. (1983).

C. Salinity

Salinity is normally not a problem in the production of soybean in the Midwest, but can be a problem associated with irrigation in other parts of the world (van Schilfgaarde, 1979). Soybean growth decreased linearly with increasing salinity in a greenhouse study (Bernstein & Ogata, 1966). However, under field conditions, soybean was more tolerant of the same salinity levels, apparently due to a N interaction. They suggest available soil N in the field may decrease plant dependence on N_2 fixation and probably accounts for the capability of field crops to yield 70% of the maximum yield at a salinity level that completely inhibited nodulation in the greenhouse. Nodule number declined appreciably only at the highest salinity level, dropping to 46% of the nonsaline control. Only the dry matter percent of nodules tended to parallel inhibitory effects of salinity on growth of inoculated soybean.

Soybean has been found to be extremely sensitive to salinity at low inorganic P concentrations in solution (Grattan & Maas, 1984, 1985). The interactive effects of salinity and inorganic P on soybean growth and foliar injury were evaluated using several different cultivars selected according to their potential to exclude or accumulate leaf P and Cl from saline solutions. Increased inorganic P had no effect on growth under nonsaline conditions, but significantly reduced the shoot and root growth of 'Clark' and 'Kanrich' in the presence of salinity. In contrast they observed no foliar injury in 'Lee.' The striking interaction between salinity and P requires an awareness of critical factors in solution culture studies. Grattan and Maas (1985) identified the root control mechanism in soybean response to salinity stress using reciprocally grafted soybean to evaluate the role of scion and root stock on leaf P and Cl accumulation and foliar injury. Foliar injury was controlled predominately by genotype of the root stock and was correlated with leaf P and Cl concentration. Scions on Lee root stocks remained healthy during the entire experiment, and leaf P and Cl were maintained below critical levels. Although the P and Cl accumulations were controlled predominantly by genotype of root stock, they identified another possible mechanism for controlling P and Cl concentrations.

VI. CULTURAL PRACTICES AND IRRIGATION

A. Plant Density (or Plant Spacing)

Theoretically, soybean plant arrangement should be designed to give a complete canopy as quickly as possible and to give each individual plant equal space for optimum light interception, growth (height and leaf area), and development. Soybean is unique in that it exhibits both determinant and indeterminant growth characters that may respond differently to row spacing and population and in establishing early canopy cover. Early canopy closure will supposedly reduce soil evaporation significantly, and the optimum space per plant will allow maximum genotypic potential to be expressed for each individual plant. Early canopy closure will result in some erosion and weed control that may not translate directly to yield. With irrigation, this maximum potential is thought to increase even more.

To reach this optimum plant arrangement and early canopy closure, both inter- and intrarow spacing can be manipulated. Interrow spacing of soybean has been evaluated for years; however, much of this effort has dealt with dryland environments. Taylor (1980), using indeterminant Wayne soybean in Iowa with no irrigation, found that 25-cm rows outyielded 100-cm rows (3480–2970 kg/ha) only in a season with the greatest seasonal water supply. During a season with an intermediate water supply (avg. yield = 2480 kg/ha) and one classified as a drought year (avg. yield = 785 kg/ha), no difference in yield was measured among row spacings of 25, 50, 75, and 100 cm. Results from this study also showed that early canopy closure produced greater early season water loss from the soil profile through transpiration than did later canopy closure with prolonged soil exposure.

Research on the effect of row spacing using indeterminant cultivars in conjunction with irrigation is minimal. Hartung et al. (1980) measured a nonsignificant 110 kg/ha increase from 18-cm rows vs. 81-cm rows when indeterminate cultivars were irrigated in Nebraska. Taylor et al. (1982) measured a 300 to 400 kg/ha increase in yield of Wayne soybean when grown in 25 vs. 100-cm rows and either irrigated or not irrigated. Irrigation had no effect on seed yield in the Iowa study. On a sandy soil, Reicosky et al. (1985) showed a much larger response to irrigation, but a similar response to row spacing as Taylor et al. (1982). The narrow row spacing had higher ET earlier in the season, depleting the stored water available on the nonirrigated treatment during pod fill. Basnet et al. (1974) found that irrigated soybean seed yield increased as within row spacing increased from 3.8 to 6.4 cm in 46-cm rows and when within-row spacing decreased from 6.4 to 3.8 cm in 92-cm rows. The effect of row width alone was nonsignificant in 1969 and significant in 1970, with wide rows having a 300-kg/ha higher yield. Yields of the five indeterminate cultivars irrigated in this study all exceeded 3000 kg/ha.

The limited research on row spacing and irrigation of determinant cultivars has yielded mixed results. Doss and Thurlow (1974), using determinant cultivars Bragg and Hampton 266 in Alabama, reported a slight yield advantage for a 60-cm vs. a 90-cm row spacing when irrigated. A nonirrigated treatment had no yield difference between the two-row spacings. Heatherly (1984), using 'Bedford,' 'Tracy,' and Bragg in Mississippi, measured slight and sometimes significant yield increases from both irrigated and nonirrigated 50-cm vs. 100-cm row spacings. Boerma and Ashley (1982) found that irrigation of soybean planted in June or early July in 51-cm and 91-cm rows provided an 11% yield advantage for narrow rows, but nonirrigated narrow rows showed only a 3% yield advantage. The average increase from irrigation was 87%. When both irrigation and row spacing are considered, proper irrigation is vastly more important; i.e., much greater yield responses can be achieved with irrigation of any row spacing than can be achieved by changing row spacing in the absence of adequate water.

Intrarow spacing appears to be even less critical than row width when irrigation is used. Ramseur et al. (1984) reported similar yields for determinant 'Braxton' when intrarow spacings ranged from 43 to 305 mm in 0.91-m rows both with and without irrigation. Average irrigated yields were above 2600 kg/ha for 2 yr. Doss and Thurlow (1974) measured no significant effect on yield when plant densities of 11, 22, and 33 plants/m were used and either irrigated or not irrigated.

The subtle difference in response of determinant and indeterminant soybean to irrigation-row spacing interactions requires caution in making blanket recommendations. Soybean appears flexible in its response to row spacing and irrigation until lodging and subsequent yield losses become a factor. While the relative yield increases for determinant cultivars have not been as large as for indeterminant cultivars, available evidence supports using narrow rows and intermediate populations for maximum yield potential with irrigated soybean.

B. Double Cropping

Double cropping of soybean is generally restricted to wheat (*Triticum aestivum* L.) and soybean grown in rotation in a 1-yr period. It is a practice that has widespread acceptance and usage in the southeastern USA, and between 25 and 40% of the soybean hectarage is double cropped in this region (Boerma and Ashley, 1982).

Crabtree and Rupp (1980) found that yields of nonirrigated double cropped soybean and wheat were about 400 and 300 kg/ha below the yields of nonirrigated monocrop soybean and wheat, respectively. Both monocrop and doublecrop soybean were planted in late June or early July. This same trend also occurs when monocrop soybean is planted at normal time vs. later planting date for doublecropped soybean and when soybean in both plantings was irrigated (Wesley et al., 1988). Data of Boerma and Ashley (1982) and Heatherly and Elmore (1986) indicate that this yield reduction for the doublecropped soybean results from a later planting date and not from wheat causing soybean yield reduction. This effect may not always be evident when dryland soybean is grown monocrop vs. doublecrop following small grain; i.e., a dryland doublecrop soybean planting can yield just as well as a dryland monocrop planting if rainfall is deficient and there is no irrigation to ensure that monocrop planting will reach maximum yield potential based on an earlier planting date. If rainfall patterns and stored soil water are ideal, then monocrop planting should outyield double crop planting based solely on earlier planting date.

C. Planting Date

Planting date as a factor in soybean production has been studied for many years. It is generally acknowledged that delayed planting beyond an accepted optimum period can result in reduced potential seed yield of nonirrigated soybean, but it has been difficult to determine a predictable reduction caused by a specific delay. Delays in planting until late May or early June have not generally resulted in yield reductions in the southern USA (Pendelton & Hartwig, 1973). Weather and rainfall patterns following planting on any date can have a profound positive or negative effect on the performance of soybean.

Effect of irrigation on soybean planted on varying dates is easier to assess because the capability of supplemental watering allows maximum potential expression of seed yield in the absence of drought stress. The majority of data shows that soybean planted at the accepted optimum time (Heatherly & Elmore, 1986) or at the earliest possible time after this (Boerma & Ashley, 1982) for a particular area will give optimum yields. Irrigation using the best available guidelines will produce a greater yield and a greater yield increase than when planted at some later time and irrigated similarly. This has been shown to be true for a difference in planting date of as little as 16 d (12 May vs. 28 May) (Heatherly & Elmore, 1986), and for planting dates as late as late June, July, and August (Boerma & Ashley, 1982). Thus, management of irrigated soybean hectarage with regard to planting date is obvious: plant at the earliest acceptable time for an individual cultivar and location to provide opportunity for the maximum potential seed yield. When both irrigated and nonirrigated hectarage is involved, irrigated hectarage should be planted first to ensure its seeding at or near the optimum period.

D. Tillage Methods

Tillage should be used for (i) killing weed vegetation that will compete for resources needed by a developing soybean crop; (ii) incorporating chemicals, weeds, and crop residues that may interfere with the crop; (iii) altering soil properties that may restrict optimum root development and proliferation; and (iv) preparing an adequate seedbed. If none of the above needs exist, then tillage is wasted input.

Preplant tillage that delays planting of soybean to be irrigated can actually result in less than maximum potential yield because of delayed planting (Heatherly & Elmore, 1983, 1986). An array of herbicides that can eliminate weed vegetation at planting is available. Increased costs that may be incurred by replacing tillage with judicious use of herbicides can be more than offset by potentially higher yields attained from earlier-planted crop (Heatherly et al., 1986). This concept is especially useful where soybean can be adequately planted in an undisturbed seedbed. Thus, elimination of preplant tillage can indirectly lead to higher yield potential with irrigation.

Subsoiling or deep tillage of various sorts is often used to enhance soybean root proliferation, especially in the southeastern USA. It can increase the amount of stored water available for plant growth (Martin et al., 1979; Reicosky & Deaton, 1979) as well as increase infiltration capability of compacted soils. Campbell et al. (1984) found that subsoiling doublecropped "full-season" soybean was beneficial in a dry season when suppressed early biomass production conserved stored soil water for the reproductive phase. With a cover crop using the stored soil water early, "late-season" soybean had lower yield with this form of conservation tillage. Martin et al. (1979) found that an irrigated, nonsubsoiled treatment in North Carolina produced a yield (2100 kg/ha) below a nonirrigated, subsoiled treatment (2500 kg/ha). Irrigation initiated before flowering (V8) in subsoiled treatment resulted in a yield (2200 kg/ha) significantly below that attained with both nonirrigated, subsoiled treatment and a subsoiled treatment with irrigation initiated at flowering (2600 kg/ha).

Use of deep tillage will not guarantee alleviation of drought stress even though rooting depth may be enhanced. In situations where irrigation is available, deep tillage may be eliminated in favor of the less risky supplemental watering. Of course, the economic feasibility of replacing tillage with irrigation must be considered, but the lower risk factor involved with irrigation vs. deep tillage in areas where summer rainfall is deficient cannot be overlooked.

E.   Cultivar Selection

The majority of the research conducted with irrigation of soybean has incorporated the use of more than one cultivar. These cultivars have often been from more than one maturity group; therefore, performance of cultivars can often be more accurately defined as performance of individual maturity groups as represented by selected cultivars from those groups.

Whenever cultivars from the same or different maturity groups have been evaluated with and without irrigation, differences generally occur. Lodging is sometimes a factor (Kadhem et al., 1985a, b); i.e., a cultivar that lodges significantly when irrigated will usually yield less, even though its potential yield may have been similar to a cultivar that did not lodge. In reality, then, a difference in yield response to irrigation did not occur, but rather a difference in lodging that resulted from irrigation.

Response to irrigation of cultivars that are of different maturity is often a confounded effect. Since such cultivars are at different reproductive stages on any given date, environmental stresses such as extreme temperature that irrigation alone could not alleviate would have a different effect on the cultivars. Also, earlier-maturing cultivars may be too far advanced to be affected by alleviation of environmental stresses that can occur later in a growing season (Griffin et al., 1985; Heatherly, 1983; Snyder et al., 1982;

Specht et al., 1986). Thus, differential responses of these different cultivars to irrigation may actually be related to the differing stage of development and not to the difference in cultivar per se.

Adapted cultivars from the same maturity group may perform similarly when irrigated (Brown et al., 1985). Similarity of stature and rate or time of reproductive development in relation to drought stress is probably more important than genotype per se. Mederski and Jeffers (1973) measured large and significant differences in both yield and change in yield among cultivars within maturity groups when cultivars were subjected to low and high soil moisture stress. The difference in change in yield seemed to be correlated with yield produced at low stress level; i.e., yield difference among cultivars was relatively greater in low stress environment. However, less adapted cultivars used in this study gave the smallest increase in yield with low soil moisture stress; therefore, the confounding of adaptation may have been present.

REFERENCES

Allmaras, R.R., W.W. Nelson, and W.B. Voorhees. 1975a. Soybean and corn rooting in southwestern Minnesota. I. Water uptake sink. Soil Sci. Soc. Am. Proc. 39:764–771.

Allmaras, R.R., W.W. Nelson, and W.B. Voorhees. 1975b. Soybean and corn rooting in southwestern Minnesota. II. Root distribution and related water flow. Soil Sci. Soc. Am. Proc. 39:771–777.

Arya, L.M., G.R. Blake, and D.A. Farrell. 1975. A field study of soil water depletion patterns in presence of growing soybean roots: Rooting characteristics and root extraction of soil water. Soil Sci. Soc. Am. Proc. 39:437–444.

Ashley, D.A., and W.J. Ethridge. 1978. Irrigation effects on vegetative and reproductive development of three soybean cultivars. Agron. J. 70:467–471.

Baldocchi, D.D., S.B. Verma, N.J. Rosenberg, B.L. Blad, A. Garay, and J.E. Specht. 1983. Influence of leaf pubescence on the mass and energy exchange between soybean canopies and the atmosphere. Agron. J. 75:543–548.

Barber, S.A. 1985. Fertilizer rates and placement effects on nutrient uptake by soybeans. p. 1007–1015. In R.M. Shibles (ed.) World soybean research conference III. Westview Press, Boulder, CO.

Basnet, B., E.L. Mader, and C.D. Nickell. 1974. Influence of between and within-row spacing on agronomic characteristics of irrigated soybeans. Agron. J. 66:657–659.

Bennett, J.M., and S.L. Albrecht. 1984. Drought and flooding effects on N_2 fixation, water relations and diffusive resistance of soybeans. Agron. J. 76:735–740.

Bennie, A.T.P., W.K. Mason, and H.M. Taylor. 1982. Responses of soybeans to two-row spacings and two soil water levels. III. Concentration, accumulation and translocation of 12 elements. Field Crops Res. 5:31–43.

Bernstein, L., and G. Ogata. 1966. Effects of salinity on nodulation, nitrogen fixation and growth of soybeans and alfalfa. Agron. J. 58:201–203.

Boerma, H.R., and D.A. Ashley. 1982. Irrigation, row spacing, and genotype effects on late and ultra-late planted soybeans. Agron. J. 74:995–999.

Borkert, C.M., and S.A. Barber. 1985a. Soybean shoot and root growth and phosphorus concentration as affected by phosphorus placement. Soil Sci. Soc. Am. J. 49:152–155.

Borkert, C.M., and S.A. Barber. 1985b. Predicting the most efficient phosphorus placement for soybeans. Soil Sci. Soc. Am. J. 49:901–904.

Boyer, J.S. 1968. Relationship of water potential to growth of leaves. Plant Physiol. 43:1056–1062.

Boyer, J.S. 1970a. Leaf enlargement and metabolic rates in corn, soybeans, and sunflowers at various water potentials. Plant Physiol. 46:233–235.

Boyer, J.S. 1970b. Differing sensitivity of photosynthesis to low water potentials in corn and soybean leaves. Plant Physiol. 46:236–239.

Boyer, J.S., R.R. Johnson, and S.G. Saupe. 1980. Afternoon water deficits and grain yields of old and new soybean cultivars. Agron. J. 72:981–986.

Brady, R.A., S.M. Goltz, W.L. Powers, and E.T. Kanemasu. 1975. Relation of soil water potential to stomatal resistance of soybean. Agron. J. 67:97–99.

Brady, R.A., W.L. Powers, L.R. Stone, and S.M. Goltz. 1974a. Relation of soybean leaf water potential to soil water potential. Agron. J. 66:795–798.

Brady, R.A., L.R. Stone, C.D. Nickell, and W.L. Powers. 1974b. Water conservation through proper timing of soybean irrigation. J. Soil Water Conserv. 29:266–268.

Brown, E.A., C.E. Caviness, and D.A. Brown. 1985. Response of selected soybean cultivars to soil moisture deficit. Agron. J. 77:274–278.

Brun, L.J., L. Prunty, J.K. Larsen, and J.W. Enz. 1985. Evapotranspiration and soil water relationships for spring wheat and soybean. Soil Sci. 139:547–552.

Bunce, J.A. 1978. Effects of water stress on leaf expansion, net photosynthesis, and vegetative growth of soybeans and cotton. Can. J. Bot. 56:1492–1498.

Campbell, R.B., R.E. Sojka, and D.L. Karlen. 1984. Conservation tillage for soybean in the U.S. southeastern Coastal Plain. Soil Tillage Res. 4:531–541.

Carlson, R.E., N.N. Momen, O. Arjmand, and R.H. Shaw. 1979. Leaf conductance and leaf water potential relationships for two soybean cultivars grown under controlled irrigation. Agron. J. 71:321–325.

Carter, J.L., and E.E. Hartwig. 1963. The management of soybeans. p. 161–226. In A.J. Norman (ed.) The soybean. Academic Press, New York.

Cassel, D.K., A. Bauer, and D.A. Whited. 1978. Management of irrigated soybeans on a moderately coarse textured soil in the upper Midwest. Agron. J. 70:100–104.

Christy, A.L., and C.A. Porter. 1982. Canopy photosynthesis and yield in soybean. p. 499–511. In Govindjee (ed.) Photosynthesis: Development, carbon metabolism, and plant productivity. Vol. II. Academic Press, New York.

Clawson, K.L., J.E. Specht, B.L. Blad, and A.F. Garay. 1986. Water use efficiency of soybean pubescence density of isolines—a calculation procedure for estimating daily values. Agron. J. 78:483–487.

Constable, G.A., and A.B. Hearn. 1978. Agronomic and physiological responses of soybean and sorghum crops to water deficits. I. Growth and development and yield. Aust. J. Plant Physiol. 5:159–167.

Constable, G.A., and R.B. Hearn. 1980. Irrigation for crops in a sub-humid environment. I. The effect of irrigation of growth and yield of soybeans. Irrig. Sci. 2:1–12.

Crabtree, R.J., and R.N. Rupp. 1980. Double and monocropped wheat and soybeans under different tillage and row spacings. Agron. J. 72:445–448.

Crabtree, R.J., A.A. Yassin, I. Kargougou, and R.W. McNew. 1985. Effects of alternate furrow irrigation: Water conservation on the yields of two soybean cultivars. Agric. Water Manage. 10:253–264.

de Mooy, C.J., J. Pesek, and E. Spaldon. 1973. Mineral nutrition. In B.E. Caldwell (ed.) Soybeans—Improvement, production, and uses. Agronomy 16:267–352.

de Mooy, C.J., and P.L. Sutherland. 1981. Soil fertility requirements of soybeans with reference to irrigation. p. 25–37. In W.H. Judy and J.A. Jackobs (ed.) Proceedings of the irrigated soybean production in arid and semi-arid regions. Univ. of Illinois at Urbana-Champaign.

Denison, R.F., T.R. Sinclair, R.W. Zobel, M.M. Johnson, and G.M. Drake. 1983a. A nondestructive field assay for soybean nitrogen fixation by acetylene reduction. Plant Soil 70:173–182.

Denison, R.F., P.R. Weisz, and T.R. Sinclair. 1983b. Analysis of acetylene reduction rates of soybean nodules at low acetylene concentrations. Plant Soil 73:648–651.

Doorenbos, J., and W.O. Pruit. 1977. Guidelines for predicting crop water requirements. FAO Irrig. Drain. Pap. No. 24. FAO, Rome.

Doss, B.D., and R.W. Pearson. 1972. Response of soybeans to subirrigation. Soil Sci. 114:264–267.

Doss, B.D., R.W. Pearson, and H.T. Rogers. 1974. Effect of soil-water stress at various growth stages on soybean yield. Agron. J. 66:297–299.

Doss, B.D., and D.L. Thurlow. 1974. Irrigation, row width, and plant population in relation to growth characteristics of two soybean varieties. Agron. J. 66:620–623.

Egli, D.B., L. Meckel, R.E. Phillips, D. Radcliffe, and J.E. Leggett. 1983. Moisture stress and nitrogen redistribution in soybean. Agron. J. 75:1027–1031.

Fehr, W.R., C.E. Caviness, D.T. Burmood, and J.S. Pennington. 1971. Stage of development description for soybeans, Glycine max (L.). Crop Sci. 11:929–931.

Fehr, W.R., and C.E. Caviness. 1977. Stages of soybean development. Iowa Agric. Exp. Stn. Spec. Rep. 80.

Garay, A.F., and W.W. Wilhelm. 1983. Root system characteristics of two soybean isolines undergoing water stress conditions. Agron. J. 75:973–977.

Ghorashy, S.R., J.W. Pendleton, R.L. Bernard, and M.E. Bauer. 1971a. Effect of leaf pubescence on transpiration, photosynthetic rate, and seed yield of three isogenetic lines of soybeans. Crop Sci. 11:426–427.

Ghorashy, S.R., J.W. Pendleton, D.B. Peters, J.S. Boyer, and J.E. Beuerlein. 1971b. Internal water stress and apparent photosynthesis with soybeans differing in pubescence. Agron. J. 63:674–676.

Grattan, S.R., and E.V. Maas. 1984. Interactive effects of salinity and substrait phosphate on soybean. Agron. J. 76:668–676.

Grattan, S.R., and E.V. Maas. 1985. Root control of leaf phosphorus and chlorine accumulation in soybeans under salinity stress. Agron. J. 77:890–895.

Griffin, J.L., R.W. Taylor, R.J. Habetz, and R.F. Regan. 1985. Response of solid-seeded soybeans to flood irrigation. I. Application timing. Agron. J. 77:551–554.

Hartung, R.C., J.E. Specht, and J.H. Williams. 1980. Agronomic performance of selected soybean morphological variance in irrigated culture with two-row spacings. Crop Sci. 20:604–609.

Heatherly, L.G. 1983. Response of soybean cultivars to irrigation of a clay soil. Agron. J. 75:859–864.

Heatherly, L.G. 1984. Soybean response to irrigation of Mississippi River Delta soils. USDA-ARS ARS-18. U.S. Gov. Print. Office, Washington, DC.

Heatherly, L.G., and C.D. Elmore. 1983. Response of soybeans (Glycine max) to planting in untilled, weedy seedbed on clay soil. Weed Sci. 31:93–99.

Heatherly, L.G., and C.D. Elmore. 1986. Irrigation and planting date effects on soybeans grown on clay soil. Agron. J. 78:576–580.

Heatherly, L.G., J.A. Musick, and J.G. Hamill. 1986. Economic analysis of stale seedbed concept of soybean production on clay soil. Mississippi Agric. For. Exp. Stn. Bull. 944.

Heatherly, L.G., and W.J. Russell. 1979. Vegetative development of soybeans grown of different soil types. Field Crops Res. 2:135–143.

Heatherly, L.G., W.J. Russell, and T.M. Hinckley. 1977. Water relations and growth of soybeans in drying soil. Crop Sci. 17:381–386.

Herpich, R.L. 1963. Irrigating soybeans. Kansas State Univ. Agric. Ext. Land Reclamation 12.

Hobbs, E.H., and H.H. Muendel. 1983. Water requirements of irrigated soybeans in southern Alberta. Can. J. Plant Sci. 63:855–860.

Huang, C.Y., J.S. Boyer, and L.N. Vanderhoef. 1975a. Acetylene reduction (nitrogen fixation) and metabolic activities of soybean having various leaf and nodule water potentials. Plant Physiol. 56:222–227.

Huang, C.Y., J.S. Boyer, and L.N. Vanderhoef. 1975b. Limitation of acetylene reduction (nitrogen fixation) by photosynthesis in soybean having low water potentials. Plant Physiol. 56:228–232.

Huck, M.G., K. Ishihara, C.M. Peterson, and T. Ushijima. 1983. Soybean adaption to water stress at selected stages of growth. Plant Physiol. 73:422–427.

Huck, M.G., C.M. Peterson, G. Hoogenboom, and C.D. Bush. 1986. Distribution of dry matter between shoots and roots of irrigated and nonirrigated determinant soybeans. Agron. J. 78:807–813.

Hunt, P.G., T.A. Matheny, A.G. Wollum II, D.C. Reicosky, R.E. Sojka, and R.B. Campbell. 1983. Effect of irrigation and Rhizobium japonicum strain 110 upon yield and nitrogen accumulation and distribution of determinate soybeans. Commun. Soil Sci. Plant Anal. 14:223–238.

Hunt, P.G., A.G. Wollum II, and T.A. Matheny. 1981. Effects of soil water on Rhizobium japonicum infection, nitrogen accumulation, and yield in Bragg soybeans. Agron. J. 73:501–505.

Hunter, M.N., P.L.M. de Jabrun, and D.E. Blyth. 1980. Response of nine soybean lines to soil moisture conditions close to saturation. Aust. J. Exp. Agric. Anim. Husb. 20:339–345.

Johnson, H.W., H.A. Borthwick, and R.C. Leffel. 1960. Effects of photoperiod and time of planting on rates of development of the soybean in various states of the life cycle. Bot. Gaz. 122:77–95.

Jones, J.W., B. Zur, and K.J. Boote. 1983. Field evaluation of a water relations model for soybean. II. Diurnal fluctuations. Agron. J. 75:281–286.

Jones, J.W., and B. Zur. 1984. Simulation of possible adaptive mechanisms in crops subjected to water stress. Irrig. Sci. 5:251–264.

Jung, P.K., and H.D. Scott. 1980. Leaf water potential, stomatal resistance and temperature relations in field grown soybeans. Agron. J. 72:986–990.

Jung, Y.S., and H.M. Taylor. 1984. Differences in water uptake rates of soybean roots associated with time and depth. Soil Sci. 137:341–350.

Kadhem, F.A., J.E. Specht, and J.H. Williams. 1985a. Soybean irrigation serially timed during stages R1 to R6. I. Agronomic responses. Agron. J. 77:291–298.

Kadhem, F.A., J.E. Specht, and J.H. Williams. 1985b. Soybean irrigation serially timed during stages R1 to R6. II. Yield component responses. Agron. J. 77:299–304.

Kamprath, E.J. 1974. Nutrition in relation to soybean fertilization. p. 28–32. In Soybean production—Marketing and use. TVA, Muscle Shoals, AL.

Kanemasu, E.T., L.R. Stone, and W.L. Powers. 1976. Evapotranspiration model tested for soybean and sorghum. Agron. J. 68:569–572.

Karlen, D.L., P.G. Hunt, and T.A. Matheny. 1982a. Accumulation and distribution of phosphorus, iron, manganese and zinc by selected determinant soybean cultivars grown with and without irrigation. Agron. J. 74:297–303.

Karlen, D.L., P.G. Hunt, and T.A. Matheny. 1982b. Accumulation and distribution of potassium, calcium and magnesium by selected determinant soybean cultivars grown with and without irrigation. Agron. J. 74:347–354.

Kaspar, T.C., C.D. Stanley, and H.M. Taylor. 1978. Soybean root growth during reproductive stages of development. Agron. J. 70:1105–1107.

Korte, L.L., J.E. Specht, J.H. Williams, and R.C. Sorensen. 1983b. Irrigation of soybean genotypes during reproductive ontogeny. II. Yield component responses. Crop Sci. 23:528–533.

Korte, L.L., J.H. Williams, J.E. Specht, and R.C. Sorenson. 1983a. Irrigation of soybean genotypes during reproductive ontogeny. I. Agronomic responses. Crop Sci. 23:521–527.

Kurtz, L.T. 1976. Fertilizer needs of the soybean. p. 85–100. In L.D. Hill (ed.) World soybean research. Proc. World Soybean Res. Conf. Interstate Printers and Publ., Danville, IL.

Lutz, Jr., J.A., and G.D. Jones. 1975. Effect of irrigation, lime and fertility treatments on the yield and chemical composition of soybeans. Agron. J. 67:523–526.

Mahler, R.L., and A.G. Wollum, II. 1981. The influence of irrigation and Rhizobium japonicum strains on yields of soybeans grown in a Lakeland sand. Agron. J. 73:647–651.

Martin, C.K., D.K. Cassel, and E.J. Kamprath. 1979. Irrigation and tillage effects on soybean yield in a Coastal Plain soil. Agron. J. 71:592–594.

Mason, W.K., G.A. Constable, and R.C.G. Smith. 1981. Irrigation for crops in a subhumid environment. II. The water requirement of soybeans. Irrig. Sci. 2:13–22.

Mason, W.K., H.R. Rowse, A.T.P. Bennie, T.C. Kaspar, and H.M. Taylor. 1982. Responses of soybeans to two-row spacings, and two soil water levels. II. Water use, root growth and plant water status. Field Crops Res. 5:15–29.

Matheny, T.A., and P.G. Hunt. 1983. Effects of irrigation on accumulation of soil and symbiotically fixed nitrogen by soybean grown on a Norfolk loamy sand. Agron. J. 75:719–722.

Mayaki, W.C., I.D. Teare, and L.R. Stone. 1976. Top and root growth of irrigated and nonirrigated soybeans. Crop Sci. 16:92–94.

Meckel, L., D.B. Egli, R.E. Phillips, D. Radcliffe, and J.E. Leggett. 1984. Effect of moisture stress on seed growth in soybeans. Agron. J. 76:647–650.

Mederski, H.J., and D.L. Jeffers. 1973. Yield response of soybean varieties grown at two soil moisture stress levels. Agron. J. 65:410–412.

Meyer, R.F., and J.S. Boyer. 1972. Sensitivity of cell division and cell elongation to low water potentials in soybean hypocotyls. Planta 108:77–87.

Meyer, W.S., and D.C. Reicosky. 1985. Enclosing leaves for water potential measurement and its effect on interpreting soil-induced water stress. Agric. Forest Meteorol. 35:187–192.

Meyer, W.S., D.C. Reicosky, and N.L. Schaefer. 1985. Errors in field measurement of leaf diffusive conductance associated with leaf temperature. Agric. Forest Meteorol. 35:55–64.

Mitchell, R.L., and W.J. Russell. 1971. Root development and rooting patterns of soybean evaluated under field conditions. Agron. J. 63:313–316.

Musick, J.T., L.L. New, and D.A. Dusek. 1976. Soil water depletion-yield relationships of irrigated sorghum, wheat and soybeans. Trans. ASAE 19:489–493.

Nathanson, K., R.J. Lawn, P.L.M. de Jabrun, and D.E. Byth. 1984. Growth, nodulation and nitrogen accumulation by soybean in saturated soil culture. Field Crops Res. 8:73–92.

Nielsen, D.C., B.L. Blad, S.B. Verma, N.J. Rosenberg, and J.E. Specht. 1984. Influence of soybean pubescent type on radiation balance. Agron. J. 76:924–929.

Oliver, S., and S.A. Barber. 1966. An evaluation of the mechanisms governing the supply of calcium, magnesium, potassium and sodium in soybean roots (*Glycine max* L.). Soil Sci. Soc. Am. Proc. 30:80–86.

Pandey, R.K., W.A.T. Herrera, A.N. Villegas, and J.W. Pendleton. 1984. Drought response of grain legumes under irrigation gradient: III. Plant growth. Agron. J. 76:557–560.

Pankhurst, C.E., and J.I. Sprent. 1975. Effects of water stress on the respiratory and nitrogen fixing activity of soybean root nodules. J. Exp. Bot. 26:287–304.

Patterson, R.P., C.D. Raper, and H.D. Gross. 1979. Growth and specific nodule activity of soybean during application and recovery of a leaf moisture stress. Plant Physiol. 64:551–556.

Peaslee, D.E., B.F. Hicks, Jr., and D.B. Egli. 1985. Soil test levels of potassium yields and seed size in soybean cultivars. Commun. Soil Sci. Plant Anal. 16:899–907.

Pendleton, J.W., and E.E. Hartwig. 1973. Management. *In* B.E. Caldwell (ed.) Soybeans: Improvement, production, and uses. Agronomy 16:211–237.

Ramseur, E.L., V.L. Quisenberry, S.U. Wallace, and J.H. Palmer. 1984. Yield and yield components of 'Braxton' soybeans as influenced by irrigation and intrarow spacing. Agron. J. 76:442–446.

Ramseur, E.L., S.U. Wallace, and V.L. Quisenberry. 1985. Growth of 'Braxton' soybeans as influenced by irrigation and intrarow spacing. Agron. J. 77:163–168.

Raper, C.D., Jr., and S.A. Barber. 1970. Rooting systems of soybeans. II. Physiological effectiveness as nutrient absorption surfaces. Agron. J. 62:585–588.

Reicosky, D.C., and D.E. Deaton. 1979. Soybean water extraction, leaf water potential and evapotranspiration during drought. Agron. J. 71:45–50.

Reicosky, D.C., T.C. Kaspar, and H.M. Taylor. 1982. Diurnal relationships between evapotranspiration and leaf water potential of field grown soybeans. Agron. J. 74:667–673.

Reicosky, D.C., R.J. Millington, A. Klute, and D.B. Peters. 1972. Patterns of water uptake and root distribution of soybeans (*Glycine max*) in the presence of a water table. Agron. J. 64:292–297.

Reicosky, D.C., D.D. Warnes, and S.D. Evans. 1985. Soybean evapotranspiration, leaf water potential and foliage temperature as affected row spacing and irrigation. Field Crops Res. 10:37–48.

Robertson, W.K., L.C. Hammond, J.T. Johnson, and K.J. Boote. 1980. Effects of plant-water stress on root distribution of corn, soybeans and peanuts in sandy soil. Agron. J. 72:548–550.

Scott, H.D., and J.T. Batchelor. 1979. Dry weight and leaf area production rates of irrigated determinant soybeans. Agron. J. 71:776–782.

Scott, H.D., R.E. Sojka, D.L. Karlen, F.B. Arnold, V.L. Quisenberry, and C.W. Doty. 1983. Bragg soybeans grown on a southern Coastal Plains soil. I. Dry matter distribution, nodal growth analysis, and sample variability. J. Plant Nutr. 6(2):133–162.

Shaw, R.H., and D.R. Laing. 1966. Moisture stress and plant response. p. 73–94. *In* W.H. Pierre et al. (ed.) Plant environment and efficient water use. ASA and SSSA, Madison, WI.

Silvius, J.E., R.R. Johnson, and D.B. Peters. 1977. Effects of water stress on carbon assimilation and distribution in soybean plants at different stages of development. Crop Sci. 17:713–716.

Sinclair, T.R., P.R. Weisz, and R.F. Denison. 1985. Oxygen limitation to nitrogen fixation in soybean nodules. p. 797–806. *In* R. Shibles (ed.) Proceedings of the world soybean research conference III. Westview Press, Boulder, CO.

Sionit, N., and P.J. Kramer. 1977. Effect of water stress during different stages of growth of soybeans. Agron. J. 69:274–278.

Sivakumar, M.V.K., and R.H. Shaw. 1978. Relative evaluation of water stress indicators of soybeans. Agron. J. 70:619–623.

Sivakumar, M.V.K., H.M. Taylor, and R.H. Shaw. 1977. Top and root relations of field grown soybeans. Agron. J. 69:470–473.

Snyder, R.L., R.E. Carlson, and R.H. Shaw. 1982. Yield of the indeterminant soybeans in response to multiple periods of soil water stress during reproduction. Agron. J. 74:855–859.

Sojka, R.E., and D.L. Karlen. 1984. Measurement variability in soybean water status and soil nutrient extraction in a row spacing study in the U.S. southeastern Coastal Plains. Commun. Soil Sci. Plant. Anal. 15:1111–1154.

Sojka, R.E., and J.E. Parsons. 1983. Soybean water status and canopy microclimate relationships at four row spacings. Agron. J. 75:961–968.

Sojka, R.E., H.D. Scott, J.A. Ferguson, and E.M. Rutledge. 1977. Relation of plant water status to soybean growth. Soil Sci. 123:182–187.

Somerhalder, B.P., and P.E. Schleusener. 1960. Irrigation can increase soybean production. Nebraska Agric. Exp. Stn. Q. 7(1):16–17.

Specht, J.E., J.H. Williams, and C.J. Weidenbenner. 1986. Differential responses of soybean genotypes subjected to a seasonal soil water gradient. Crop Sci. 26:922–934.

Sprent, J.I. 1971. The effect of water stress on nitrogen-fixing root nodules. I. Effect of the physiology of detached soybean nodules. New Phytol. 70:9–17.

Sprent, J.I. 1972a. Effects of water stress on nitrogen-fixing root nodules. II. Effects on the fine structure of detached soybean nodules. New Phytol. 71:443–450.

Sprent, J.I. 1972b. The effects of water stress on nitrogen-fixing root nodules. IV. Effects on whole plants of *Vicia faba* and *Glycine max*. New Phytol. 71:603–611.

Stanley, C.D., T.C. Kaspar, and H.M. Taylor. 1980. Soybean response to temporary water tables. Agron. J. 72:343–346.

Stone, L.R., I.D. Teare, C.D. Nickell, and W.C. Mayaki. 1976. Soybean root development and soil water depletion. Agron. J. 68:677–680.

Sullivan, T.P., and W.A. Brun. 1975. Effect of root genotype on shoot water relations in the soybeans. Crop Sci. 15:319–322.

Taylor, H.M. 1980. Soybean growth and yield as affected by row spacing and by seasonal water supply. Agron. J. 72:543–547.

Taylor, H.M., E. Burnett, and G.D. Booth. 1978. Taproot elongation rates of soybeans. Z. Acker. Pflanzenbau. 146:33–39.

Taylor, H.M., W.K. Mason, A.T.P. Bennie, and H.R. Rowse. 1982. Responses of soybeans to two row spacings and two soil water levels. I. An analysis of biomass accumulation, canopy development, solar radiation interception and components of seed yield. Field Crops Res. 5:1–14.

Thompson, J.A. 1977. Effect of irrigation termination on yield of soybeans in southern New South Wales. Aust. J. Exp. Agric. Anim. Husb. 17:156–160.

Troedson, R.J., R.J. Lawn, D.E. Byth, and G.L. Wilson. 1985. Saturated soil culture—An innovative water management option for soybeans in the tropics and subtropic. p. 171–180. *In* S. Shanmugasundaram and E.W. Sulzberger (ed.) Soybean in tropical and subtropical cropping systems. Asian Vegetable Res. and Develop. Ctr., Shanhua, Taiwan.

Turner, N.C., J.E. Begg, H.M. Rawson, S.D. English, and A.B. Hearn. 1978. Agronomic and Physiological responses of soybean and sorghum crops to water deficits. III. Components of leaf water potential, leaf conductance, $^{14}CO_2$ photosynthesis and adaption to water deficits. Aust. J. Plant Physiol. 6:169–177.

van Schilfgaarde, J. 1979. Salinity management for soybean production. p. 105–110. *In* W.H. Judy and J.A. Jackobs (ed.) Proceedings of the irrigated soybean production and arid and semi-arid regions. Univ. Illinois Urbana-Champaign.

Wesley, R.A., L.G. Heatherly, H.C. Pringle III, and G.R. Tupper. 1988. Seedbed tillage and irrigation effects on yield of mono- and doublecrop soybean and wheat on a silt loam. Agron. J. 80:139–143.

Whitt, D.M., and C.H.M. van Bavel. 1955. Irrigation of tobacco, peanuts and soybeans. p. 376–381. *In* Water. USDA Yearbook. U.S. Gov. Print. Off., Washington, DC.

Zur, B., and J.W. Jones. 1984. Diurnal changes in the instantaneous water use efficiency of a soybean crop. Agric. For. Meteorol. 33:41–51.

Zur, B., J.W. Jones, and K.J. Boote. 1983. Field evaluation of a water relations model for soybean. I. Validity of some basic assumptions. Agron. J. 75:272–280.

22 Peanut

K. J. BOOTE

University of Florida
Gainesville, Florida

D. L. KETRING

USDA-ARS
Stillwater, Oklahoma

Irrigation of peanut (*Arachis hypogaea* L.) has become a popular production practice in the USA to ensure the yield potential of the crop. A 1978 survey in Georgia indicates that about 45% of the allotted peanut acreage was irrigated, with new installations increasing steadily (Henning et al., 1979). The increase in irrigation in the humid southeastern USA has occurred because of the high value of the crop and the desire to stabilize and ensure production against rainfall-deficit periods even though total seasonal rainfall may be adequate. Irrigation is equally important in the dryland areas of the southwestern USA, where optimization of water resources is even more critical because of declining water tables and increasing costs of pumping. Irrigation and peanut response to drought are also of interest worldwide, since more than half of the world's peanut acreage is grown in the semiarid tropics and is frequently subjected to low rainfall and water deficits. India, for example, is the largest producer of peanuts, having nearly one-third of the world's total acreage of peanuts, mostly grown in water-deficit situations. African countries, such as Sudan, Senegal, and Nigeria each have acreage as great as the USA, but realize much lower yields because of drought limitations.

In this chapter we will discuss peanut growth and development, factors affecting evapotranspiration (ET), yield response to ET and irrigation, water relations during drought, growth and soil water deficit, genotypic differences in drought tolerance, and approaches for scheduling irrigation.

I. PEANUT GROWTH AND DEVELOPMENT

A. Crop Growth Stages and Life Cycle

Knowledge of the crop growth stages of peanut should aid scheduling irrigation and interpreting impacts of drought periods, especially relative to formation and maturation of the underground fruits.

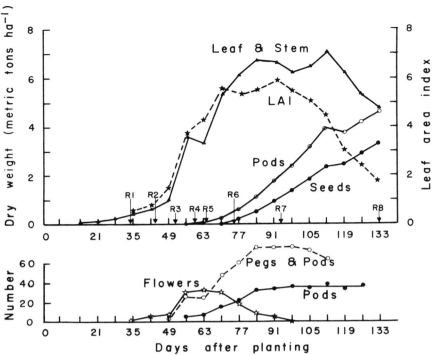

Fig. 22-1. Dry matter distribution and reproductive development of Florunner peanut relative to reproductive growth stages. Adapted from growth analysis of McGraw (1977) and growth stages of Boote (1982). Reproductive stages are R1 (beginning bloom), R2 (beginning peg), R3 (beginning pod), R4 (first full-sized pod), R5 (beginning seed), R6 (full seed), R7 (beginning maturity), and R8 (harvest maturity). Reprinted from Boote et al. (1982).

Vegetative and reproductive growth stages for peanut have been published by Boote (1982). The vegetative stages are based on the number of fully developed leaves expressed on the main axis of the plant. Reproductive stages depend on visually observable events in the ontogeny of reproductive structures. Because of peanut's indeterminate growth, vegetative and reproductive stages occur concurrently and should be observed simultaneously. The phenological events are affected by temperature, soil water content, and genotype. Crop growth stages of a 'Florunner' peanut crop are shown in Fig. 22-1 to illustrate their relationship to leaf area development, dry matter accumulation, pod addition, and seed growth of the crop. The shorter season spanish and valencia market types differ mainly in having a shorter interval between the reproductive stages and having a shorter period of dry matter accumulation and pod dry matter accumulation. Wynne and Coffelt (1982) provide a description of growth habit (erect vs. spreading), branching type (alternate or sequential), maturity, and traditional market types (virginia, spanish, and valencia) of peanut.

B. Germination and Seedling Growth

When peanut seeds are planted in warm, moist soil at about 5-cm depth, seedling emergence occurs in 5 to 7 d. Over the next 7 d, leaf, stem, and root growth continue substantially at the expense of seed reserves of lipid and protein.

The seedling growth process is sensitive to soil water status, particularly if the seed is not planted in moist soil or is planted too shallow. Peanut seeds require a seed moisture level >35% for germination to proceed (Mixon, 1971). Producers without irrigation plant only after sufficient rainfall is received on previously prepared fields, whereas those with irrigation may plant after an irrigation event. Considering the high value of the commodity in the USA, many producers have invested in irrigation systems to ensure planting on optimum dates, germination and emergence, and herbicide activation.

C. Root Growth Pattern

Root growth is the dominant activity during early germination and seedling development. The radicle emerges after 1 or 2 d of germination. Within 5 to 6 d, the taproot may grow 10 to 16 cm deep and develop a number of lateral roots (Yarbrough, 1949). Also by 5 to 6 d, the hypocotyl elongates, the cotyledons reach the soil surface, and the first few leaves begin unfolding. Once the seedling has achieved this stage, the plant can survive rather severe water deficits, although shoot growth may be severely restricted.

The time course of root growth can be described in terms of increases in root dry matter accumulation, rooting depth, or root length density. Results from potted plant studies conducted outdoors by Maliro (1987) showed that dry matter allocation to roots for Florunner peanut is dominant early in the life cycle, but becomes less rapid as the crop increases in age. Root mass sampled by excavation in a field study showed that roots in the top 15 cm of soil comprised 37% of the crop dry weight at 21 d after planting, but only 1.5% at harvest (McCloud, 1974). Although the latter study ignored roots below 15 cm, studies conducted by Lenka and Misra (1973) demonstrated that the majority (77%) of the root weight is in the top 15 cm of soil, partly because of the large taproot.

Total rooting depth and the pattern of root-length distribution are important to nutrient and water uptake. Seasonal progression of total root length increase was followed by Gregory and Reddy (1982) for cv. Robut 33-1 grown on a sandy clay loam topsoil overlying a clay subsoil (medium deep alfisol). Total root length followed a sigmoid growth curve over time and peaked at 25 cm root length per cm^2 land surface area by 68 d after planting for this short-season cultivar. In general, time series studies of root dry matter (McCloud, 1974) or total root length density (Gregory & Reddy, 1982) show that root growth rate of peanut slows considerably and nearly ceases as the crop progresses into the phase of rapid seed fill.

D. Leaf and Stem Growth

Growth of leaves and stems begins slowly while roots are being established. During early vegetative development, the partitioning of dry matter to leaves and roots is relatively high, and that to stems is low. Computed partitioning of assimilate among leaves and stems shows a gradual shift from leaf to stem as the plant grows from emergence to mid-pod fill (Boote, 1981, unpublished data). Dry matter accumulation in leaves and stems continues until about 12 wk in Florunner, but may cease sooner in early cultivars. In some highly indeterminate cultivars, leaf and stem growth may continue to maturity, even at the expense of reduced pod growth (Duncan et al., 1978; Pixley, 1985).

E. Seasonal Pattern of Leaf Area Index

Early leaf growth is slow, but eventually leaf area index (LAI) begins to accumulate more rapidly resulting in the familiar sigmoid-shaped curve for seasonal leaf area development (Fig. 22-1). The LAI may be between 1 and 2 as late as 6 wk after planting, depending on planting date, temperature, and planting density. During early growth the crop is quite drought tolerant, because the LAI and soil water extraction are low and the crop responds to stress by slowing leaf area expansion. Maximum LAI for Florunner peanut can reach values of 6 to 7 by 10 to 14 wk. Thereafter, the decline in the LAI is primarily a function of disease control (Pixley, 1985), because peanut is an indeterminate, nonsenescing crop that retains most of its LAI until maturity. With good disease control, LAI may only decline 1 to 2 LAI units.

F. Pattern of Dry Matter Accumulation

Dry matter accumulation is initially slow during crop establishment, but increases as LAI and canopy cover begin to intercept more of the available solar radiation (McGraw, 1977). Beginning about 7 wk after planting, dry matter accumulation becomes rapid and begins its linear phase that can last well into the middle of seed fill. Measured crop growth rate of peanut under optimum conditions is about 20 g m^{-2} per d (Ketring et al., 1982a; Duncan et al., 1978; Maliro, 1987). Crop growth rate is about 40% greater for Florunner peanut than for 'Bragg' soybean [*Glycine max* (L.) Merr.] when grown side by side in the same irrigated experiments (Duncan et al., 1978; Maliro, 1987). Its higher crop growth rate is one of the reasons for the high productivity of peanut. Peanut maintains its LAI and dry matter accumulation better throughout pod fill and has a longer pod-fill period than soybean.

G. Flowering and Podset

Flowering of peanut begins about 25 to 35 d after planting, depending on the lateness of the cultivar and the thermal environment under which the plants are grown (Fortanier, 1957; Leong & Ong, 1983; Ong, 1986). Peanut has a self-fertilized flower that withers soon after fertilization. After 5 to 7 d, the base of the fertilized ovary (Gynophore) begins to elongate geotropically toward the soil (R2, pegging stage). After growing into the soil 4 to 5 cm, the ovary tip begins to swell (R3 stage) and turns horizontally away from the base of the plant. The time from R1 to R3 may require 15 to 20 d (Boote, 1982), after which the pod begins to expand rapidly until it reaches dimensions characteristic of the cultivar. The R4 stage (first full pod stage) is defined as the time when 50% of the plants have achieved one fully expanded pod.

The flowering period of peanut generally extends from about 5 to 13 wk after planting (Duncan et al., 1978; McGraw, 1977). The pegging period continues from 7 to 13 wk after planting, although there is a 1- to 2-wk lag from peg penetration into the soil until rapid pod expansion is visible (Fig. 22-1). For a crop under optimum stable environment, the rate of addition of gynophores and rate of appearance of full-sized pods occur at a nearly constant rate (constant number per day) (Duncan et al., 1978; Boote, 1981, unpublished data; Maliro, 1987).

H. Seed Growth and Maturation

The seed growth phase begins after the R5 growth stage, when seed cotyledon growth is visible in at least one pod on 50% of the plants. Because of the slow rate of pod addition, many more pods must be added and seeds initiated in them to achieve a full pod load. At this point the linear seed growth phase begins. In the example in Fig. 22-1, linear seed dry weight accumulation commenced about 12 wk after planting. Thereafter, seed growth occurred at a constant rate until shortly before maturity. Ketring et al. (1982a) indicated that pod (plus seed) growth rate in the linear phase averaged 8.3 $\pm$ 2.1 g m^{-2} per d over 24 experiments they examined. Pod growth rates can differ among cultivars (Duncan et al., 1978; McGraw, 1979; Pixley, 1985), and are affected by the temperature of the fruiting zone (Dreyer et al., 1981). Pod growth rate frequently appears to slow down during late pod fill.

I. Yield, Quality, and Maturation

Predicting pod maturation and the optimum harvesting date is difficult for peanut because of its slow rate of pod addition and because its vegetative canopy remains green and active if protected from diseases. A number of methods for estimating pod maturation have been reviewed by Sanders et al. (1982). These include the *shell-out* and the *hull-scrape* methods. For the shell-out method, the pods are opened and the percentage of pods with tan to brown coloration inside the hull and pink to dark pink seed coats are determined. Unless diseases or pod losses mandate earlier harvest, harvest should be scheduled when 60 to 80% of the pods (except soft watery immature pods) demonstrate this condition, and when the shelling percentage (seed weight $\times$ 100%/pod wt.) reaches values characteristic of the cultivar. The hull-scrape method, now called the *pod maturity profile* (Williams & Drexler, 1981), is based on color of the pod mesocarp observable by scraping off the outside of the shell (the exocarp). Physiologically, the coloration of the pod mesocarp is caused by the same factors that cause the inside of the hull to develop coloration or cause the testae to develop coloration. The advantage of the pod maturity profile is that one does not need to open the pods, and that various classes of pods from immature to mature are separated on a color-coded counting board. By using the pod maturity profile, producers can project ahead to the optimum harvest date. The method has been adopted in many areas in the southern USA.

II. EVAPOTRANSPIRATION

A. Environmental, Soil, Plant, and Cultural Effects on Evapotranspiration

1. Daily Evapotranspiration Pattern Throughout the Growing Season

The seasonal trend in daily water use and ET of peanut has been measured under controlled lysimetric conditions by Kassam et al. (1975), Stansell et al. (1976), Vivekanandan and Gunasena (1976), and Dancette and Forest (1986). In these studies, rainfall was excluded and water applications for peanut growth were carefully controlled to prevent or allow measurement of percolation. Daily water use (ET) was calculated from soil water profiles and irrigation water applied.

Seasonal ET patterns for peanut cultivars, Florigiant, Florunner, and Tifspan were determined by Stansell et al. (1976). Daily water use was low during early vegetative growth, but increased as the canopy developed and closed (Fig. 22–2). Maximum daily water use rates were about 0.5 to 0.6 cm d^{-1} as reported by Kassam et al. (1975), Stansell et al. (1976), Vivekanan-

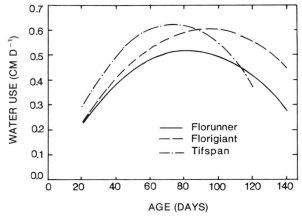

Fig. 22–2. Water use by three peanut cultivars as a function of age when grown under optimum soil water supply at Tifton, GA. Reprinted from Stansell et al. (1976).

dan and Gunasena (1976), and Dancette and Forest (1986). Under high evaporative demand in Israel, daily water use of Improved Virginia Bunch averaged 0.69 cm d^{-1} during 53 to 83 d after planting (Mantell & Goldin, 1964). In Stansell's study, the times to maximum daily ET were 70, 80, and 95 d after planting for Tifspan, Florunner, and Florigiant cultivars, respectively. These delays to peak ET are consistent with cultivar differences in maturity and time to maximum LAI. The period of maximum ET for spanish peanut was reported at 67 to 77 d after planting by Kassam et al. (1975), and at 55 to 80 d by Vivekanandan and Gunasena (1976). Goldberg et al. (1967) observed peak ET to potential ET ratios (ET/ET$_o$) for Improved Virginia Bunch peanut at 70 to 110 d after planting. Ishag et al. (1985) reported peak ET rates of 0.7 to 0.8 cm d^{-1} at 75 to 85 d after planting for the Ashford semispreading virginia-type cultivar in Sudan. In Oklahoma, Harp et al. (1986) and Grosz (1986) reported midseason ET rates generally approaching 0.8 cm d^{-1}, although rates up to 1.0 cm d^{-1} were observed for short 3- to 4-d periods when evaporative demand was high.

2. Plant Effects on Evapotranspiration

The seasonal pattern in ET of peanut is substantially related to the pattern of canopy development and establishment of LAI. In absence of any crop ground cover, average evaporation from the soil surface has been reported to be 30 to 40% of pan evaporation (Goldberg et al., 1967; Kassam et al., 1975; Ishag et al., 1985; Dancette & Forest, 1986). Actual soil evaporation is dependent on soil type and the fraction of time the surface is recently wetted by rain or irrigation. The ratio ET/ET$_0$ was reported to increase nearly linearly from 0.3 to 1.0 as the percentage ground cover increased from 0 to 100% and as the LAI increased during peanut growth (Goldberg et al., 1967; Kassam et al., 1975). Kassam et al. (1975) reported that peak ET occurred shortly before peak LAI was achieved. After full foliage development and ground cover, daily ET for all reported peanut crops gradually declined from the maximum until the plants reached maturity. This decline may be due in part to plant senescence (both loss in LAI and leaf conductance) and to seasonal decrease in evaporative demand. Harp et al. (1986) estimated the crop coefficient (K_c) for peanut in Oklahoma from measured ET of irrigated peanut divided by measured ET of a full canopy reference crop (alfalfa [*Medicago sativa* L.]). Their K_c curve approximated the previously described LAI and ET curves, peaking at slightly above 1.0 at 70 to 80 d and declining slowly for the remainder of the season. Figure 22–3 shows similar results for the ratio (ET/ET$_0$) estimated for a peanut crop grown in a rainy season in Nigeria (Yayock & Owonubi, 1986). Dancette and Forest (1986), in Senegal, reported K_c for short-season cultivars peaked slightly above 1.0 between 50 to 70 d after planting. The 90-d cultivar had the lowest K_c (<0.90) and peaked earlier (50 d), compared to 105- or 120-d cultivars which had higher and later peaks in their K_c.

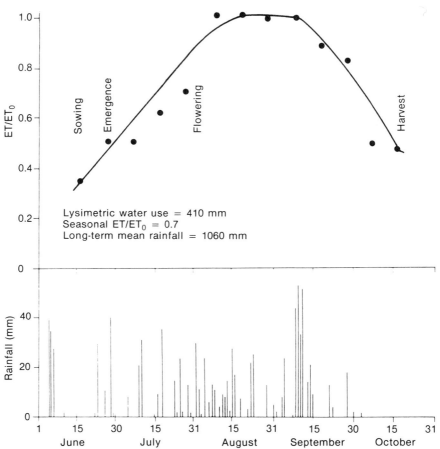

Fig. 22–3. Ratio of evapotranspiration to potential evapotranspiration (ET/ET$_o$) for a rainfed peanut crop at Samaru, Nigeria, 1973. Adapted from Yayock and Owonubi (1986).

3. Effects of Soil Water Deficit on Evapotranspiration

Soil water deficit may reduce both evaporation and transpiration. Wormer and Ochs (1959) found that the transpiration rate of peanut remained nearly constant as soil water was reduced until two-thirds of the available soil water had been depleted. Vivekanandan and Gunasena (1976) reported that maximum daily ET was 0.61, 0.48, and 0.38 cm d^{-1} for a crop grown in wet (-33 kPa), intermediate (-55 kPa), and dry (-73 kPa) soil water potential treatments, respectively. Mantell and Goldin (1964) observed that seasonal consumptive water use was higher for plots receiving the most frequent irrigation, although errors from deep percolation cannot be excluded in their data. Daily ET was reported as 0.56, 0.65, 0.61, 0.72, and 0.75 cm d^{-1} for peanut irrigated to field capacity at 40-, 30-, 21-, 14-, and 10-d intervals, respectively.

Effects of soil water deficit on ET can be mediated by stomatal closure and restricted LAI growth; nevertheless soil water supply to maintain evaporation from the soil may also affect daily ET without reducing daily transpiration. There is evidence that the ratio of ET/ET_0 during the season can be increased by more frequent irrigations (Mantell & Goldin, 1964; Goldberg et al., 1967). Increased soil water supply increased the ET/ET_0 ratio, but these increases were not always associated with increased yield. Apparently, some of the additional water may be evaporated from the soil or transpired without increasing the rate of dry matter production by the crop. Ritchie (1974) and Ritchie et al. (1976) described a method for predicting ET of vegetated areas, which considers evaporation from the soil separately from transpiration. The method uses a two-stage evaporation from the soil (energy-limiting phase, followed by a soil water supply-limiting phase). During the energy-limiting phase, the energy available for evaporation is a function of crop LAI.

4. Climatic Factors Influencing Evapotranspiration

The diurnal pattern of transpiration and evapotranspiration for a closed canopy of nonstressed peanuts is a normal-shaped curve influenced by diurnal evaporative demand and radiant energy load (Slayter, 1955; Stone et al., 1976; McCauley et al., 1978). If soil water supply becomes deficient, the daily transpiration curve is reduced and becomes progressively skewed to a peak earlier in the morning (Slayter, 1955).

Solar radiation is the primary energy source which warms the air, soil, and crop, thus providing heat for vaporization. Air temperature and relative humidity of the air are also important. Latent heat loss is dependent on the vapor pressure gradient from the leaf to the air, which is controlled by leaf temperature, air temperature, and air relative humidity. Sensible heat loss from leaves is driven by the temperature gradient from the leaf to the air. If leaf temperature is greater than air temperature, sensible heat is transferred to the air, cooling the leaf. As shown by data of Stone et al. (1976), peanut may frequently extract heat energy from the air (where air temperature is warmer than leaf temperature) and use that energy for evapotranspiration.

Vapor pressure deficit (VPD) has a major effect on the rate of latent heat loss from foliage. Simmonds and Ong (1987) showed that transpiration rate per unit fractional light interception increased linearly as mean daytime VPD was increased in the range of zero to 1.7 kPa. Any restriction due to reduced stomatal conductance in drier air is apparently offset by the steeper gradient in vapor pressure from the leaf to the air.

Wind and air turbulence in and above the crop canopy is the primary transport mechanism for vapor removal. So long as water vapor is continuously removed, evaporation can continue at the energy-limited rate. If vapor is continuously removed, ET may continue for several hours after sunset with energy coming from sensible heat stored in the air, crop, and soil. Conversely, under calm conditions, evapotranspiration ceases after sunset as net radiation becomes negative. The role of wind in supplying advected sensible heat energy for evaporation is important in arid regions and in the southwestern U.S. peanut region. Erickson et al. (1986), in Oklahoma, reported that effects on stomatal closure occurred when daily advective energy (a function of VPD and wind velocity) exceeded 8.5 MJ m^{-2}. In arid regions, the method for computing ET should include effects of advective energy via the average daily VPD and daily wind run. The modified Penman equation (Burman et al., 1980) is one such approach that considers the average daily VPD and a wind function. Harp et al. (1986), using measured daily ET of peanut over two seasons, calibrated the wind function as a linear regression vs. daily wind run. During high advective energy days, maximum measured ET in Oklahoma was up to 1.0 cm d^{-1} (Grosz, 1986; Harp et al., 1986).

5. Influence of Cultural Practices on Evapotranspiration

Dense planting of peanuts (22.2 vs. 5.6 plants m^{-2}) resulted in more rapid LAI establishment, higher daily ET, and 16% higher yield (Ishag et al., 1985). Because of the higher daily ET, the more densely planted crop required eight irrigations instead of seven. Bhan and Misra (1970) observed slightly higher ET and less percolation losses for peanuts in 30- or 45-cm rows as compared to 60-cm rows. Applications of N and P tended to increase ET and water use efficiency (WUE). Increased ET would be expected if greater LAI were established. Studies in the southwestern U.S. peanut area by Stone et al. (1976), Chin Choy et al. (1977), and McCauley et al. (1978) indicated interactions of row spacing (30- vs. 90-cm rows) and row direction on daily and diurnal patterns of ET on different evaporative demand days. On high evaporative demand days, stomatal resistance of narrow-row plants increased earlier in the day and was higher for a given water potential than for wide-row plants (Abdul-Jabbar, 1978; Stone et al., 1985). Erickson et al. (1986) concluded that the days when differential stomatal action occurred had one thing in common: cumulative daily advective energy (a function of VPD and wind velocity) exceeded 8.5 MJ m^{-2}. Unfortunately this series of experiments was not done at equal planting density per unit land area; hence narrow-row peanut produced two- to threefold higher LAI because the plant density per m^2 was threefold greater. Our interpretation of their data is that stomata closed sooner for narrow-row plants on the high advective demand days because more total LAI was available to extract advected energy for the air. In either case, narrow-row peanuts produced greater pod yields (Chin Choy et al. 1977, 1982) and resulted in higher water use efficiency.

B. Yield Response to Evapotranspiration, Water Use, Irrigation, and Rainfall

About 60 cm of water is required for satisfactory peanut yields if the rainfall or irrigation is well distributed (Mantell & Goldin, 1964; Gorbet & Rhoads, 1975; Stansell et al., 1976; Pallas et al., 1977; Hammond et al., 1978; Nageswara Rao et al., 1985; Stansell & Pallas, 1985). Timing and amounts of rainfall or irrigation are obviously important. For a crop grown on sandy soils, as in the coastal plains of the USA, irrigation may be necessary at approximately 7-d intervals to prevent water deficit during a prolonged rain-free period. For a soil with greater water-holding capacity, rainfall or irrigation interval can be longer. In Israel, on fine-textured sandy clay or clay loam soils, Mantell and Goldin (1964) obtained high yields (5800 kg ha^{-1}) on plots irrigated at 30-d intervals. The seasonal water requirement is also dependent on evaporative demand. In the semiarid Southwest USA, total seasonal water required (rain plus irrigation) to produce a peanut crop ranges from 51 to 71 cm (20–28 in.) depending on potential ET (Oklahoma and Texas Coop. Agric.Ext. Serv., 1988, 1985). In irrigation studies in Caddo County, Oklahoma, fields with the highest and lowest yields received 90 and 41 cm of applied seasonal water (rain plus irrigation) (Grosz, 1986). The soils were a sandy loam and loamy fine sand, respectively. The high-yield field was maintained above 50% available water by sprinkler irrigation for most of the growing season.

Yield response to water use (ET) has been measured under carefully controlled lysimetric conditions. Figure 22–4 shows results from experiments by Stansell et al. (1976), Pallas et al. (1977), and Hammond et al. (1978)—all conducted in field lysimeters during the normal mid-summer growing season in the southeastern USA. Figure 22–5 shows yield response to water applied for short-season cultivars grown under variable field conditions, some of which may have suffered periods of excess rainfall and apparent percolation. The highest yield of short-season or full-season peanuts occurred when

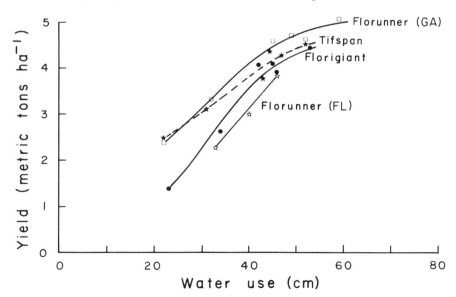

Fig. 22–4. Peanut yield response to water use in controlled experiments. Data for Florunner (squares), Tifspan (solid stars), and Florigiant (solid dots) at Tifton, GA, from Stansell et al. (1976) and Pallas et al. (1977). Data for Florunner (open stars) at Gainesville, FL, from Hammond et al. (1978). Reprinted from Boote et al. (1982).

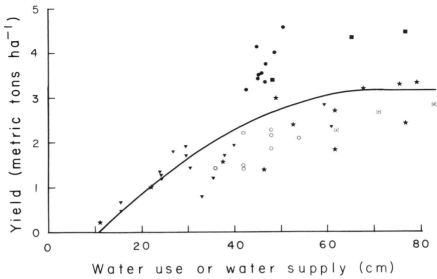

Fig. 22–5. Yield response of short-season peanuts to water use or water supply (rainfall plus irrigation) in a variety of field experiments. Data from Matlock et al., 1961 (closed star); Subramanian et al., 1975 (open circle); Narasimham et al., 1977 (closed square); Reddi and Reddy, 1977 (closed circle); Lenka and Misra, 1973 (open square); and Bockelee-Morvan et al., 1974 (closed triangle). Reprinted from Boote et al. (1982).

seasonal water use was 60 cm or more. The yield response was curvilinear between 40 and 60 cm, but yield rapidly dropped to zero as water use declined from 40 to 10 cm. In the linear range of yield response to water use (up to 50 cm), yield increased 93 kg ha^{-1} for each additional cm of water evapotranspired (Fig. 22–4). Under field conditions where < 50 cm of rainfall plus irrigation was received (Fig. 22–5), yield response was 76 kg ha^{-1} for each additional centimeter of rainfall or irrigation received.

Thus, 60 cm of water appears to be the minimum water supply for optimum yield if seasonal distribution is good. If distribution is poor or potential ET is high, even more water must be supplied. Those may be two of the reasons that the line on Fig. 22–5 accounts for only 50% of the yield variation in response to water supply. Other possible reasons for yield variability include diseases, insects, nematodes, rooting depth, soil water-holding capacity, evaporative demand, and growth stage at which water deficits occurred. Nageswara Rao et al. (1985) reported that yield response to seasonal ET is highly dependent on the stage of growth at which the water deficits occurred. In a narrow range of nearly the same seasonal ET (50–60 cm), there were several different lines describing yield response to ET, each for a different stage of growth when water deficit occurred. Data of Stansell and Pallas (1985) also indicated different yields obtained at the same seasonal water use, also caused by timing of water deficit. Yield was reduced most for a given seasonal water use, when deficit occurred at 70 to 105 d. We suggest that a modeling approach (given later) may predict yield response to uneven rainfall and irrigation applications as well as integrate the effects of water deficits during different crop growth stages, effects of evaporative demand, soil water-holding capacity, and rooting depth.

III. WATER RELATIONS OF PEANUT DURING DROUGHT AND RECOVERY

A. Leaf Water, Osmotic, and Turgor Potential

Leaf water potential of frequently irrigated peanuts is generally above -1.2 to -1.3 MPa for crops grown in the humid southeastern USA (Allen et al., 1976; Bennett et al., 1984; Pallas et al. 1977, 1979; Rodrigues, 1984) although lower values have been reported in semiarid regions (Erickson & Ketring, 1985). The water potential for water-stressed crops has been reported to be as low as -3.0 to -4.5 MPa (Bhagsari et al., 1976; Erickson & Ketring, 1985; Pallas et al. 1977, 1979).

Figure 22–6 illustrates how leaf water potential, osmotic potential, and canopy temperature change throughout a sunny day for an irrigated and rainfed peanut crop in the semiarid southwestern USA (Erickson & Ketring, 1985). Minimum values of water potential are reached between 1300 and 1400 h. The low water potentials for the rainfed crop indicate considerable water stress, as does the large diurnal swing in osmotic potential.

Figure 22–7 illustrates leaf water potential changes throughout the day for treatments exposed to different VPD and growing on stored water (Ong et al., 1985). Lower leaf water potentials were attained as higher VPDs were imposed. The depression of water potential was not as great as observed by Erickson and Ketring, presumably because solar irradiance and advective energy load was less in the greenhouse environment.

One way to evaluate the onset of water stress during a drying cycle is to compare water relations parameters meaured only at mid-day during successive days until stress is relieved by water application. Bennett et al. (1984) measured leaf water status of Florunner peanut growing during a 28-d drying cycle in the humid southeastern USA. Mid-day leaf water potential of the irrigated crop remained between -0.6 and -1.1 MPa, and turgor potential fluctuated between 0.3 and 0.6 MPa (Fig. 22–8). Water and turgor potentials of the nonirrigated crop became distinctly lower by 11 d after withholding water and fell as low as -2.0 and 0 MPa, respectively. We suggest that the more arid environment in the southwestern USA is one of the reasons that mid-day water potentials reported by Erickson and Ketring (1985) were lower than those reported by Bennett et al. (1984).

B. Stomatal Resistance and Plant Water Status During Drought

As the water and turgor potentials continue to decline during a drought cycle, stomatal closure is induced. As shown by Bennett et al. (1984), R_s of peanut leaves began to increase as water and turgor potentials declined below -1.4 and 0.1 MPa, respectively (Fig. 22–9). In the data of Rodrigues (1984), the point of the initial increase in R_s of peanut occurred at water and turgor potentials of about -1.2 and 0.3 MPa, respectively. Bennett et al. (1984) observed a curvilinear relationship of leaf turgor potential to water

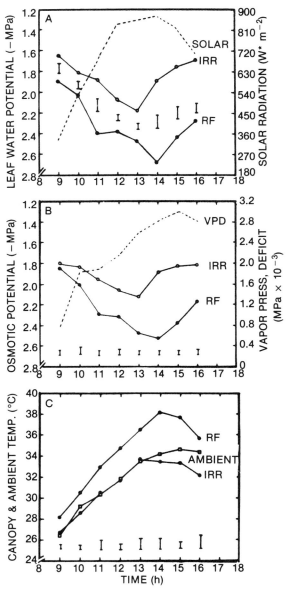

Fig. 22–6. Diurnal patterns of (A) leaf water potential, (B) leaf osmotic potential, and (C) canopy temperature of irrigated (IRR) and rainfed (RF) peanut plants, and solar radiation, vapor pressure deficit (VPD), and ambient temperature on 9 Aug. 1983. Reprinted from Erickson and Ketring (1985).

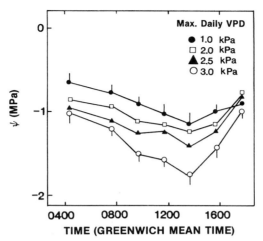

Fig. 22–7. Diurnal patterns in water potential of expanding leaflets for Robut 33-1 plants grown in a greenhouse under different vapor pressure deficit (VPD) treatments. Symbols indicate approximate maximum midday VPD: 1.0, 2.0, 2.5, and 3.0 kPa, respectively. Adapted from Ong et al. (1985).

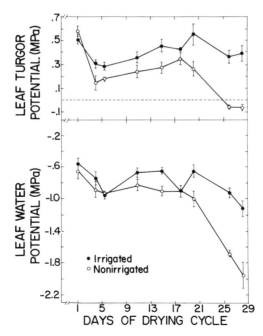

Fig. 22–8. Leaf turgor potential and leaf water potential of irrigated and nonirrigated Florunner plants sampled at midday during a 28-d drying cycle in the field. Reprinted from Bennett et al. (1984).

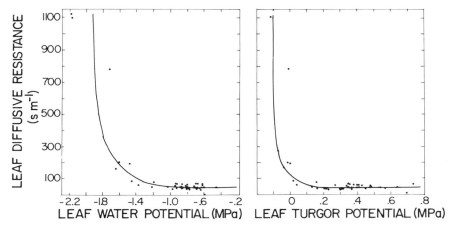

Fig. 22–9. Leaf stomatal diffusive resistance in relationship to leaf water potential or leaf turgor potential, measured on Florunner peanut leaves at midday during successive days during a field drying cycle. Reprinted from Bennett et al. (1984).

potential as leaf water potential declined (Fig. 22–10). Zero turgor was achieved when leaf water potential had fallen to −1.6 MPa at which point leaf relative water content (RWC) had fallen to 87%. These values for zero turgor point vs. water potential and RWC are close to those reported earlier by Bennett et al. (1981).

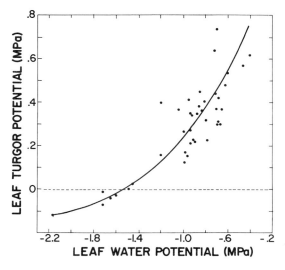

Fig. 22–10. The relationship of leaf turgor potential to leaf water potential for Florunner peanut plants sampled at midday during a field drying cycle. Reprinted from Bennett et al. (1984).

IV. GROWTH AND SOIL WATER DEFICIT

A. Root Growth Patterns and Water Extraction During Drought

Peanut rooting depth over time can be estimated based on the presumption that soil water extraction indicates root presence. Figure 22–11 shows results of studies conducted on Florunner peanut at Gainesville, FL, on a Lake fine sand soil (hyperthermic, coated Typic Quartzipsamments) (Boote et al., 1982). After a 30-d lag phase of crop establishment, the apparent root depth progression (based on front of water extraction) progressed steadily at 2.2 to 2.8 cm d^{-1} until 130 d after planting. These observations were confirmed by periodic root core observations. On a somewhat heavier soil [Teller sandy loam (fine-loamy, mixed, thermic Udic Argiustolls)] in a cooler early season climate (Oklahoma), the early lag phase was longer and the rate of depth increase was slower, 1.6 to 2.1 cm d^{-1} (Huang, 1985). Williams et al. (1986) reported that the water extraction front progressed downward at 1.0 to 1.2 cm d^{-1} for 120 d for four cultivars grown on an alfisol, sandy clay loam topsoil overlying a clay subsoil, during the somewhat cool post-rainy season in India. Simmonds and Ong (1987) reported that the rate of increase in water extraction depth was greater (2.0 vs. 1.5 cm d^{-1}) for plants exposed to a greater vapor pressure deficit (2.5 kPa) compared to a smaller

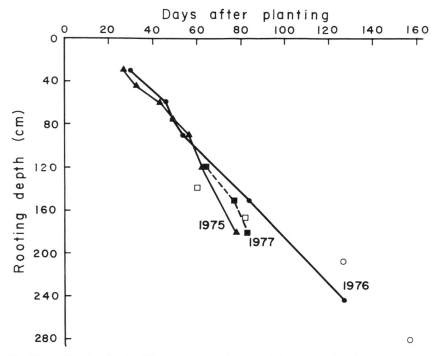

Fig. 22–11. Rooting depth of Florunner peanut in a deep Lake fine sand during several seasons at Gainesville, FL. Rooting observations in soil cores (open symbols). Reprinted from Boote et al. (1982).

VPD (1.0 kPa). Thus, the rate of root penetration into the soil appears to depend on the soil type, temperature conditions, and the aerial environment (VPD).

Information on peanut root length distribution vs. depth can be valuable to indicate the potential soil layers from which the crop can extract soil water. Root length density of peanut in a fine sandy soil approaches 1 to 2 cm cm^{-3} of soil volume in the topsoil and is typically 20 to 40% of that density in the deeper layers (Robertson et al., 1980; Gregory & Reddy, 1982; Rodrigues, 1984; DeVries, 1986). The depth and distribution of root length is dependent on soil type. In a fine sand soil, Robertson reported only 55% of the roots in the top 30 cm of soil, whereas on a clay loam soil, Gregory and Reddy (1982) reported that 80% of the total root length was in the top 30 cm. In sandy soils, root depths to 200 cm were observed by Hammond et al. (1978) and Robertson et al. (1980). On a sandy to clay loam soil, Stern (1968) reported root depths to 130 cm, and Gregory and Reddy (1982) reported depths to only 70 cm on a sandy clay loam topsoil overlying a clay subsoil (medium deep alfisol).

Differential root growth under drought stress may be an important adaptive mechanism in peanut. Greater rooting depth for water-stressed peanut than irrigated peanut has been reported (Lin et al., 1963; Lenka & Misra, 1973; Narasimham et al., 1977). Allen et al. (1976) observed continued rapid downward movement of the water extraction front for drought-stressed Florunner, even though top growth appeared to stop. Simmonds and Ong (1987) found that Robut 33-1 grown in drier air, had a more rapid extraction of water from depth and a more rapid rate of increase in the depth of the water extraction front than plants grown in more humid air. Root length profiles reported by Rodrigues (1984) and DeVries (1986) illustrate the ability of Florunner peanut to increase root length density considerably in deeper layers (60–150 cm) of a deep fine sand soil during drought periods. Florunner exhibited greater capacity for deep rooting, particularly with the onset of drought stress, when compared to a drought-susceptible peanut, 'Pearl Early Runner' (Rodrigues, 1984), or to soybean or pigeonpea [*Cajanus cajan* L. (Mills)] (DeVries, 1986). Pandey et al. (1984b), in a comparison of four grain legumes under line source irrigation, reported that peanut cv. Kidang had greater root length density at 40 to 80 cm depth at 55 d than did soybean or mungbean (*Vigna radiata* L.), especially when grown under drought stress. In their opinion, this characteristic appeared to be a major adaptive mechanism for peanut's drought tolerance.

The fraction of water extracted from the various soil layers depends on root length density in the respective zones and the pattern of water application to the soil. With frequent rains or irrigations, more of the water will be extracted from the upper soil layers. During peak rooting and water use periods in rain-free conditions in Israel and with a 10-d irrigation frequency, peanut extracted 48% of its water from the upper 30 cm, 23% from 30 to 60 cm, 15% from 60 to 90 cm, 9% from 90 to 120 cm, and 5% from 120–150 cm (Mantell & Goldin, 1964). In humid rainfed regions, the fraction from upper layers could be even higher. When Mantell and Goldin lengthened the irrigation interval (more water applied per irrigation), a greater fraction of water was extracted from lower layers.

B. Shoot Growth Processes and Soil Water Deficit

1. Crop Development

Crop development rate can be delayed by soil water deficit, but effects on expansive growth, dry matter accumulation, and pod formation are more important. Vegetative growth stage progression (number of nodes produced on the main axis) is delayed by soil water deficit (Boote & Hammond, 1981). Ong et al. (1985) found that rate of leaf production on the main stem was reduced from 0.3 to 0.23 leaves d^{-1} as soil water deficit increased.

There have been few studies of soil water deficit on the timing of reproductive development in peanut. Il'ina (1958) reported that floral initiation was delayed 7 d and flowering was inhibited where the soil was maintained at or dryer than 35% of field capacity. Under most field situations, the start of flowering is not delayed much by drought stress. Nevertheless, the subsequent events in reproductive growth—peg elongation (R2), beginning of pod expansion (R3), achievement of first full pod expansion (R4), and beginning of seed growth (R5)—are all turgor-dependent processes that are progressively inhibited by insufficient plant turgor and lack of assimilates caused by soil water deficit. These stages can also be delayed by lack of soil water in the fruiting zone and insufficient Ca uptake. Limited data (Boote & Hammond, 1981; Williams et al., 1986) indicated delays in these stages under prolonged soil water deficit. If the drought delays early reproductive growth, maturity will be delayed proportionately.

2. Photosynthesis, Stomatal Conductance, and Leaf Display

Relatively few studies have been conducted on the effects of soil water deficit on photosynthesis and stomatal conductance of peanut leaves, although dry matter accumulation is clearly reduced by prolonged water deficit (Stansell et al., 1976; Vivekanandan & Gunasena, 1976; Pallas et al., 1979; Nageswara Rao et al. 1985; Sivakumar & Sarma, 1986). As drought progresses and plant water status is reduced, stomatal closure is induced. The main effect of soil water deficit on leaf carbon exchange rate (CER) is exerted through stomatal closure as documented by simultaneous measurements of leaf CER, transpiration, and stomatal diffusive resistance (R_s) (Bhagsari et al., 1976). Nevertheless, the long-term effect of soil water deficit on canopy assimilation must consider the additional effects of drought on leaf area expansion. The latter is more sensitive to soil water deficit than is stomatal closure (Black et al., 1985).

With potted plants, Bhagsari et al. (1976) reported decreased leaf CER, decreased transpiration and increased R_s within 3 d of withholding water. In field environments with deeply rooted peanut, drought stress develops more slowly, causing a more gradual increase in R_s (Allen et al., 1976; Pallas et al., 1977; Rodrigues, 1984; Sivakumar & Sarma, 1986). Allen et al. (1976)

measured R_s of peanut leaves under increasing water stress over a 21-d period in the field. By the seventh day of the drying cycle, there was a slight increase in R_s, indicating that stomatal closure was just beginning. Ten days into the drying cycle, R_s was significantly higher in the stressed plants. Only after most of the available soil water was depleted in this sandy soil (after 17–18 d of withholding water), did R_s increase dramatically and plant water and turgor potential begin to drop even more rapidly (Allen et al., 1976; Bennett et al., 1984).

Chen and Chang (1974) found no difference in CER at water regimes of 40 to 80% field capacity, except for a slight reduction in CER at 40% field capacity during later reproductive growth stages. Iyama and Murata (1961) suggested that peanut maintained higher leaf water content during soil water deficits than did soybean or small grains. They reported that CER of peanut first declined when soil water was depleted to 32% of field capacity compared to CER declines for soybean and small grains at 41–45% of field capacity. Wormer and Ochs (1959) reported that stomatal closure did not begin until two-thirds of the available soil water was depleted, i.e., at 33% of field capacity.

Atmospheric vapor pressure deficit has been shown to influence R_s and photosynthesis of peanut (Black & Squire, 1979; Erickson et al., 1986; Stone et al., 1985; Tsuno, 1975) and rate of dry matter accumulation (Ong et al., 1987). Ong et al. (1987) observed that the rate of dry matter accumulation of peanut was reduced as VPD increased. Dry matter production per unit of intercepted solar radiation was also decreased at high VPDs or if average daytime leaf water potential (from 1000–1600 h) fell lower than -1.0 MPa. Dry matter produced per unit of water transpired fell from 5.2 to 1.5 g kg^{-1} as mean daily VPD increased from 1.0 to 2.0 kPa. In semiarid regions where VPD is frequently 2.0 kPa or greater, field-grown peanut produces 1.7 to 1.9 g of shoot dry matter per kg of water transpired (Kassam et al., 1975; Nageswara Rao et al., 1985). Tsuno (1975) reported that CER of relatively young peanut plants was insensitive to short-term changes in VPD; however, as the plant became older, CER was decreased in response to increasing the VPD (lower relative humidity). Tsuno (1975) concluded that CER was limited in older peanut plants because of the reduced ability of the root system to supply water to the leaves. The ratio of root weight to leaf area was less in the older plants.

Stomatal resistance of irrigated peanuts tends to be higher at equivalent water potential values on high evaporative demand days than on either low or moderate evaporative demand days (Stone et al., 1985; Erickson et al., 1986). The R_s in the presence of low VPD was routinely as low as 0.33 to 0.5 s cm^{-1} (Black & Squire, 1979; Black et al., 1985). However, under the same irradiance (600 W m^{-2}), R_s increased to 1 s cm^{-1} as VPD increased from 1 to 3 kPa. Stomatal sensitivity to changes in VPD was greatly reduced or absent in nonirrigated plants in which R_s was already high. Covering some leaves of nonirrigated plants decreased R_s of remaining leaves and partially restored stomatal sensitivity to VPD.

During periods of severe water deficit and high radiant energy, peanut leaflets have been observed to fold from their more horizontal (and partly sun-tracking positons) to an orientation nearly parallel to the incident radiation (Il'ina, 1958; Allen et al., 1976; Bhagsari et al., 1976). The vertical folding together of the compound leaflets minimizes leaflet interception of direct-beam radiation. This minimizes leaf heat load and temperature at a time when stomata are nearly closed and transpirational cooling is limited. Canopy temperature may rise as much as 8 to 9 °C if stomata close and transpirational cooling is reduced 90% or more (Sivakumar & Sarma, 1986). Leaflet folding could be considered a heat stress avoidance strategy, but does not seem to occur until stomata are nearly closed (Allen et al., 1976).

3. Vegetative and Expansive Growth

Water deficit reduces dry matter production of vegetative components (Fourrier & Prevot, 1958; Ochs & Wormer, 1959; Su et al., 1964; Lenka & Misra, 1973; Stansell et al., 1976; Vivekanandan & Gunasena, 1976; Pallas et al., 1979; Ong, 1984; Sivakumar & Sarma, 1986; Ong et al., 1987) as well as crop growth rate (Slatyer, 1955; Pandey et al., 1984b). Soil water deficit inhibits leaf expansion and stem elongation through its reduction of relative turgidity (Slatyer, 1955; Allen et al., 1976; Vivekanandan & Gunasena, 1976) or leaf turgor potential (Rodrigues, 1984; Ong et al., 1985). Slayter (1955) reported that the rate of dry matter accumulation by peanut was first reduced when relative turgidity dropped below 90%. The point of reduced growth coincided with decreased crop water use.

Water deficits have been shown to reduce the rate of daily leaf production (Ochs & Wormer, 1959; Billaz & Ochs, 1961; Vivekanandan & Gunasena, 1976; Boote & Hammond, 1981; Ong et al., 1985). Rate of leaf appearance on the main stem is reduced progressively as soil water deficit increases (Boote & Hammond, 1981; Ong et al., 1985), although the total number of leaves is generally reduced more than number of leaves on the main axis, thus indicating reduced branching (Ong et al., 1985). Based on reductions in LAI, leaf size is reduced even more by soil water deficit than is number of leaves (Ong et al., 1985). Vivekanandan and Gunasena (1976) grew peanut throughout a life cycle in lysimeters controlled at −33, −55, and −73 kPa of soil water potential. Differences in LAI's ensued and persisted throughout the season, with each treatment achieving maximum LAI of 6.25, 4.75, and 3.81, respectively. Leaf longevity and leaf area duration were reduced by decreasing soil water potential. Pandey et al. (1984b) reported that leaf area expansion rate, leaf area duration, and LAI were progressively reduced as soil water deficit was intensified.

Leaf area expansion is dependent on leaf turgor, temperature, and assimilate supply for growth, all factors which are affected by drought. Ong et al. (1985) observed that rate of leaf expansion during 0900 to 1600 h was linearly related to leaf turgor potential in the range of 1.0 to 0 MPa, or as leaf water potential dropped from −0.5 to −1.5 MPa. During predawn hours (0400–0700 h), the relationship of leaf expansion to turgor potential was not as good for the driest treatments, suggesting other limitations such as as-similate shortages induced by water deficit. Ketring (1984b) found that long durations of moderately high daily temperature (up to 35 °C) reduced leaf area as well as mature seed yield of spanish-type peanut. Increasing the day-time temperature from 30 to 35 °C caused both area per leaf and specific leaf area to be reduced.

Leaf and stem morphology are altered by water stress. Main axis and cotyledonary branches are shorter for water-stressed peanut plants (Lin et al., 1963; Su et al., 1964; Gorbet & Rhoads, 1975; Boote & Hammond, 1981). Soil water deficit reduces internode length more drastically than node number (Ochs & Wormer, 1959), although rate of node development is also reduced (Boote & Hammond, 1981). Continuous soil water deficit causes fewer and smaller leaves which have smaller and more compact cells (Il'ina, 1958; Lin et al., 1963) and greater specific leaf weight (Pandey et al., 1984b). The xero-morphic leaf structure was retained even after adequate water was supplied, although new leaf development would apparently be normal.

Greater partitioning to roots (greater root/shoot ratio, and lower leaf area per unit root weight) occurs in response to greater VPD, especially for plants growing on stored soil water (Ong et al., 1987). After growing 30 d since planting on stored soil water, the root/shoot ratio of Robut 33-1 was 0.10 and 0.18 on plants exposed to maximum midday VPDs of 1.0 and 2.5 kPa, respectively. The corresponding ratios for leaf area to root mass were 1277 and 752 $cm^2 g^{-1}$.

4. Flowering

The effect of soil water deficit on flowering and pod formation depends on the timing and severity of the water deficit. More importantly, reduc-tions in flower numbers per se may not reduce pod yield because the peanut plant produces many more flowers than mature fruits (Smith, 1954; Su et al., 1964). The peanut plant can compensate for reduced flower numbers resulting from early water deficit (Lin et al., 1963; Ono et al., 1974; Boote et al., 1976) by producing a flush of flowers and fruits when the water stress is relieved (Billaz & Ochs, 1961; Pallas et al., 1979; Sivakumar & Sarma, 1986) or by maturing a higher percentage of the fruits (Vivekanandan & Gunasena, 1976).

Moderate soil water deficit has been shown to delay flowering by 1 to 2 d and to reduce the total number of flowers (Lin et al., 1963; Lenka & Misra, 1973). Continuous severe soil water deficit (soil maintained at or dryer than 35% of field capacity) was shown to delay floral initiation by 7 d and caused flowering to be inhibited (Il'ina, 1958). Flowering is most severely affected by water stress at or just before peak flowering (Fourrier & Prevot, 1958; Billaz & Ochs, 1961; Su et al., 1964). Billaz and Ochs (1961) found that water deficit (50–80 d after planting on a short-season peanut) reduced flowering and pegging and produced a greater yield reduction than stress at any other growth period.

What are the mechanisms by which water stress reduces flower numbers? Possible causes include reduced photosynthate supply, reduced turgor, and low relative humidities. Low relative humidity, which often accompanies drought, has been shown to reduce the rate of flowering (Lee et al., 1972; Ong et al., 1987). Bolhuis et al. (1965) reported that low humidities increased the occurrence of flowers with short styles (probably due to reduced turgor). This abnormality lowered the rate of fertilization and varied with cultivar. Not only flowering, but number of pegs (Ong et al., 1987) and the rate of peg elongation (Lee et al., 1972) are reduced by low humidity. This effect could be an important impediment to pod formation in arid environments (with less dew and lower humidity).

5. Pegging and Pod Formation

The effect of soil water deficit during pegging and pod formation is primarily to reduce the number of pods formed while scarcely affecting weight per pod (Matlock et al., 1961; Skelton & Shear, 1971; Underwood et al., 1971; Lenka & Misra, 1973; Ono et al., 1974; Boote et al., 1976; Vivekanandan & Gunasena, 1976; Pallas et al., 1979). High VPDs have also been reported to reduce numbers of flowers and pegs formed (Lee et al., 1972; Ong et al., 1987).

Because of the subterranean fruiting habit of peanut, soil water content has two distinct effects on peg penetration and pod formation. First, soil water availability in the root zone affects plant growth, photosynthesis, and transport of water and assimilates to the developing fruit. Secondly, soil water availability in the fruiting zone (top 4–5 cm of soil) has independent additional effects on peg penetration, pod formation, and pod uptake of Ca and water. These two effects commonly occur together during drought periods, but there are significant periods during drying cycles when the fruiting zone is deficit in soil water, yet the root zone has adequate water for maintaining plant turgor and growth.

Soil water status in the top centimeter or two of soil is critical to peg entrance into the soil. Cox (1962), Underwood et al.(1971), and Boote et al. (1976) observed that pegs frequently failed to penetrate effectively into air-dry soil, thus preventing fruit growth. Boote et al. (1976) reported that within 4 d of water withholding in the field, the soil surface became too dry for peg entrance. Skelton and Shear (1971) reported that adequate root zone moisture will keep pegs alive until pegging zone moisture content is sufficient to allow penetration and initiation of pod development.

The transition from pegs to pods has been identified by Underwood et al. (1971) and Ono et al. (1974) as a critical period during which fruit zone soil water availability limits mature pod numbers. Once pegs are in the soil, adequate moisture and darkness are needed for pod development (Zamski & Ziv, 1976). Underwood et al. (1971) reported that if pegs penetrated to 1 to 1.5 cm depth, some pod development occurred but that development was slower near the surface. By introducing pegs to adequate depth in soil with various water contents, they found that the plant could offset fruit zone soil-water potentials down to -1500 kPa, but not those of air-dry soil. Ono et al. (1974) observed that adequate pegging zone moisture was critical for development of pegs into pods and that adequate soil water in the root zone could not compensate for lack of pegging zone water for the first 30 d of peg development. After 30 d of adequate pegging zone moisture, fruits could continue normal growth in dry soil if roots had adequate moisture. 30-day-old fruits are usually fully expanded, have a rigid shell, and have begun seed growth (Schenk, 1961; Boote, 1976, unpublished data). Stern (1968) proposed that seed growth could continue after full pod expansion with root-supplied water even if surface moisture was inadequate. Bhagsari et al. (1976) reported that leaves and immature fruits on water-stressed plants have similar water potentials. Thus, the underground location confers little protection from drought, except that the xylem connection between plant and fruits would maintain reasonable fruit water status if the roots had access to sufficient soil water. This situation would not, however, allow adequate Ca uptake from soil by developing fruits.

Failure of pegs to penetrate and develop pods in a dry pegging zone may also result from high soil temperature (Ono et al., 1974), greater strength of dry soil (Underwood et al., 1971), lack of turgor if roots are also in dry soil (Allen et al., 1976; Bhagsari et al., 1976; Boote et al., 1976), and lack of Ca uptake by developing fruits (Skelton & Shear, 1971).

A 6 to 9 °C increase in canopy temperature (Sivakumar & Sarma, 1986) and a 3 to 4 °C increase in fruiting zone temperature (Sanders et al., 1985a) during severe soil water deficit can certainly have adverse effects on pegging and pod formation. The optimum temperature for pod growth is about 23 °C (Cox, 1979), which is close to the mean fruiting zone temperature reported during pod fill for an irrigated crop in Florida (Dreyer et al., 1981). Warming the fruiting zone by 3, 7, or 10 °C above the mean 23 °C was sufficient to reduce number of fruits, fruit size, filling period, and pod yield (Dreyer et al., 1981). Cox (1979) also found that pod growth rate, pod size, and filling period were reduced as temperature was increased above 23 °C. Ketring (1984b) reported that increasing the daytime growth temperature from 30 to 35 °C caused a significant reduction in number of subterranean pegs and a significant reduction in mature seed yield and seed size.

6. Seed Growth, Maturation, Germinability, and Quality

Early stages of pod and seed formation appear to be highly sensitive to soil water and Ca availability in the fruiting zone, and deficits of either water or Ca may result in seed abortion (pops and single locules), poorly formed embryos and the "hollow heart" condition. After each individual pod has grown for about 30 d, formed a rigid shell, and initiated seed growth, then seed growth appears to be fairly insensitive to soil water status of the fruiting zone (Ono et al., 1974; Stern, 1968). The specific effect of drought on pod yield will depend, first, on the timing of water deficit relative to achieving a sufficient number of pods of this minimum size and, secondly, on availability of photoassimilate to fill the existing pods.

Water deficits during seed fill have been reported to reduce weight per seed (Gorbet & Rhoads, 1975; Varnell et al., 1976; Pallas et al., 1977, 1979; Pandey et al., 1984a) and weight per pod (Underwood et al., 1971; Lenka & Misra, 1973). The shelling percentage or percent sound mature kernels (SMK) is frequently several units lower for drought-stressed peanuts (Matlock et al., 1961; Su et al., 1964; Gorbet & Rhoads, 1975; Stansell et al., 1976; Varnell et al., 1976; Vivekanandan & Gunasena, 1976; Pallas et al., 1977). There is often a decrease in the percentage extra large kernels (Gorbet & Rhoads, 1975; Stansell et al., 1976) and an increase in the percentage of other kernels (OK) (Matlock et al., 1961; Stansell et al., 1976; Pallas et al., 1979).

In addition to reduced market quality, soil water deficit may reduce seed germination, thereby affecting the quality of seed to be used for next season's planting (Cox et al., 1976; Pallas et al., 1977, 1979). Pallas et al. (1977) observed significantly lower germination of Florigiant, Florunner, and Tifspan seed when the seed-producing plants encountered water stress sufficient to cause wilting without overnight recovery. The problem was greater for the larger-seeded Florigiant. Cox et al. (1976) demonstrated that drought-depressed germination of Florigiant seed was associated with insufficient seed Ca concentration. Germinability was reduced 17% for every 100 mg Ca kg^{-1} below the critical seed level of 420 mg Ca kg^{-1}. Either irrigation or addition of $CaSO_4$ increased seed Ca and improved germination, but $CaSO_4$ application was more effective than irrigation in improving germinability and seed quality.

The specific effect of water deficit on seed growth, size, and maturation depends substantially on the timing of the water deficit. Early season droughts can delay pegging and pod formation, yet pod addition may resume normally when moisture becomes available (Boote et al., 1976; Vivekanandan & Gunasena, 1976; Boote & Hammond, 1981). Billaz (1962) observed that water deficit from planting to 67 d delayed the period of rapid fruit growth about 10 d and decreased final yield. Boote and Hammond (1981) reported that drought between 40 and 80 d after planting delayed the period of rapid fruit addition and delayed pod maturation of Florunner by 11 d. Under this situation and assuming late season moisture is adequate, the crop should be given additional time to mature before harvest; otherwise the weight per pod, seed size, and the shelling percentage will be reduced. Stansell and Pallas (1985) reported that drought from 36 to 105 d, followed by a 30 to 40 d period of adequate irrigation, caused a large number of immature pods at harvest (at 135–148 d) and reduced percent sound mature kernels of Florunner to 33%. Water deficit from 36 to 70 d or from 71 to 105 d also caused an increase in immature pods and reduced shelling percentage to 73.4 and 69.7%, respectively, compared to 76.5% for the irrigated check. Water deficit periods during mid-season can cause difficulties at harvest because early added pods may continue to maturity, yet a late flush of pods may also be set. In such a situation, two or more flushes of pod formation may occur. If the producer delays harvest to wait for the second crop of pods to mature, he or she risks losing the first flush of pods to deterioration of peg attachments and pod abscission. A 35-d water deficit during late pod fill (105–140 d) caused the least yield reduction, and actually increased shelling percentage because the late drought eliminated the addition of young, immature pods (Stansell & Pallas, 1985).

7. Effects of Water Deficit on Harvest Index and Partitioning to Vegetation vs. Fruits

Excessive rainfall or irrigation may promote vegetative growth at the expense of reproductive growth. High soil water potential has been reported to cause greater LAI and excessive vegetative growth, but no increase in pod yield (Vivekanandan & Gunasena, 1976). The ratio of pods to vegetation was lowest for the most frequently watered lysimeter. Gorbet and Rhoads (1975) reported that irrigation (to maintain soil water potential above -40 kPa) increased mainstem length from 34 to 52 cm and cotyledonary branch length from 51 to 70 cm as compared to a nonirrigated control. They also reported yield enhancement from application of Kylar, a vegetative growth retardant (succinic acid 2,2 dimethyl-hydrazide), primarily in wet years or frequently irrigated plots. Desa et al. (1984) differentially applied irrigation water (IW) to peanuts at 0.5, 0.7, 0.9, and 1.1 fraction of cumulative pan evaporation (CPE) during a dry season. The ratio of pods to vegetation was highest for the 0.7 IW/CPE treatment and was reduced by irrigating at lower or higher IW/CPE ratios, even though yield was highest for the 0.9 IW/CPE treatment and vegetative growth was maximum at the 1.1 IW/CPE treatment. An increased ratio of pods to vegetative growth under small periodic water deficits may be a natural and important mechanism of peanut adaptation to droughty conditions, except where pod formation is considerably restricted by the drought or where the drought is of long duration during reproductive growth. Ong (1986) found that the rate of peg production was less sensitive to declining plant water potential than was leaf area expansion.

The particular influence of water deficit on partitioning to vegetative growth or fruits depends on the timing and severity of the water deficit relative to fruit set. Moderate degrees of drought during early vegetative growth reduce vegetative growth, yet allow good pod yield and thus result in higher pod harvest indices (Nageswara Rao et al., 1985). In fact, Nageswara Rao et al. (1985) reported that moderate water deficits from planting to the start of peg initiation (0–51 d) had no effect on total biomass, but increased pod yield by 12 to 19%, primarily via the effect on pod harvest index. When drought was relieved at 51 d, pod addition was more rapid and pod growth entered the linear phase sooner (Sivakumar & Sarma, 1986). By contrast, water deficit during pod formation (50–80 d) was reported by Billaz and Ochs (1961) to reduce flowering, pod formation, and final yield more than water deficits occurring at any other growth stage. This treatment resulted in less partitioning to fruits and more to leaves during the subsequent seed fill period (80–120 d) during which the plants were irrigated. Billaz and Ochs (1961) showed that the degree of partitioning to leaves during the seed growth phase (80–120 d) of spanish peanut was inversely proportional to the number of fruits formed earlier. If pod formation was nearly complete prior to imposing water deficit, then partitioning to fruits was increased by water deficit (re-supplying water only at 100% soil water depletion) (Ochs & Wormer, 1959).

Pod harvest index is progressively reduced by longer droughts and for droughts later in the life cycle of Florunner (Pallas et al., 1979) and Robut 33-1 peanut (Nageswara Rao et al., 1985). In the study of Nageswara Rao et al. (1985), the ratio was 0.50 for fully irrigated peanuts, as high as 0.57 for stress during 0 to 51 d, and as low as 0.24 for prolonged drought during the seed-fill phase. This shift in pod harvest index was caused by the timing of water deficit relative to the crop life cycle. Where soil water deficit was continuous from planting to maturity (under line source irrigation), harvest index was unaffected by moderate water deficit, but was reduced as greater deficits were imposed (Pandey et al., 1984a).

V. INTERACTIVE EFFECTS OF SOIL WATER DEFICIT AND SOIL CHARACTERISTICS ON PLANT NUTRITION

A. Interaction of Soil Water in Fruiting Zone and Ca Uptake

Insufficient soil water in the fruiting zone can depress Ca uptake by de-veloping pods and cause more unfilled pods (pops), fewer double-loculed pods, and lower Ca concentrations in the hull and seed (Skelton & Shear, 1971; Cox et al., 1976; Gillier, 1969). Typical symptoms of Ca deficiency in seeds include "hollow heart" and "concealed damage" (poor embryo or plumule development), which are especially prevalent after droughts during the pod initiation and formation period (Gillier, 1969).

The Ca requirement for developing peanut fruits is related to the sub-terranean nature of the fruits. Calcium is an element that is not significantly transported in the phloem. For crops with aerial fruits that transpire some-what, Ca is passively supplied via the xylem transpiration stream. Peanut fruits, being underground, do not transpire significantly and, therefore, do not receive sufficient Ca from xylem flow into the fruit (Bledsoe et al., 1949; Harris & Bledsoe, 1951; Skelton & Shear, 1971). Under normally moist fruiting zone conditions, xylem water probably flows from fruits to tran-spiring leaves in response to hydrostatic pressure gradient. Wiersum (1951) demonstrated dye uptake by fruits and movement to nearby stem nodes which suggests water uptake by fruits. Thus water uptake by fruits appears to be a major factor in Ca uptake by fruits. Bledsoe et al. (1949) and Slack and Morrill (1972) conclusively demonstrated Ca uptake by developing peanut fruits and even observed some Ca movement from fruits to the foliage.

Calcium uptake by peanut fruits is, therefore, dependent on: (i) adequate Ca in soil solution in the fruiting zone, (ii) adequate soil water in the fruiting zone during the pod expansion period to allow water flow from soil to fruit, and (iii) the absorbing surface area of the peanut fruit relative to its seed requirement for Ca. Application of $CaSO_4$ to the pegging zone is a common cultural practice in peanut-producing regions to ensure adequate Ca, especially if the soil is low in available Ca, irrigation is not available, or the cultivar is a large-podded type. Cox et al. (1976) demonstrated that application of $CaSO_4$ to the pegging zone enhanced Ca uptake by fruits and, especially in dry years, caused a significant increase in yield and grade. In wet years or in irrigated fields, adequate water in the fruiting zone sufficiently increased the availability of soil Ca to the developing fruits such that yield and quality showed little response to $CaSO_4$. Larger-podded Virginia cultivars more frequently demonstrate Ca-related quality problems under drought stress (Stansell et al., 1976; Pallas et al., 1977). They require higher fruiting zone Ca (Slack & Morrill, 1972) and respond more to $CaSO_4$ applications (Walker, 1975). The Virginia-sized fruits require more soil Ca and/or more adequate soil moisture than smaller-fruited peanut for equal Ca uptake and equal fruit quality.

B. Interactions of Water Deficit with Soil Bulk Density and Soil Strength

Increasing the bulk density of a sandy clay loam soil in the range of 1.3 to 1.7 g cm^{-3} was shown to progressively reduce LAI, biomass accumulation, filled pod number, and harvest index (Venkaiah, 1985). The effect of increased bulk density on LAI and biomass accumulation is presumed to act via restricted root penetration and less water extraction for photoassimilate production. By contrast, the effect on harvest index may result from a direct effect of bulk density and soil strength on pegging as suggested by Underwood et al. (1971). LAI, biomass, filled pod number, and harvest index were reduced as the bulk density and resistance for this soil increased above 1.5 g cm^{-3} and about 300 kPa, respectively.

C. Dinitrogen Fixation and Nodulation

There is limited information on the effects of water stress on nodulation and N_2 fixation of peanut. Lenka and Misra (1973) reported fewer nodules per plant (240 vs. 553) for peanuts irrigated at 75% soil water depletion (SWD) compared those irrigated at 25% SWD. By contrast, Shimshi et al. (1967) reported no effects of irrigation frequency (7, 14, or 21 d) on nodule number, nodule weight, or pod yields at the end of the season for peanut grown on a deep loess soil. Peanut nodulation in response to inoculation on Rhizobium-free soil was not inhibited by repeated drying of the top 5 to 10 cm of soil between 2- or 3-wk irrigation events, although more frequent irrigation tended to give somewhat better early nodulation. Irrigation in the moderate climate of Ontario also caused no effect on nodule number or nodule weight per plant, although $N_2(C_2H_2$ reduction) fixation and yield of spanish peanut was increased (Reddy, 1978; Reddy & Tanner, 1980; Reddy et al., 1981). Irrigation increased the percentage of total plant N coming from N_2 fixation, partly because more fertilizer N was leached from the soil. DeVries (1986) observed that N_2 fixation (and plant turgor) of Florunner peanut was sustained much longer during a severe drought period than was N_2 fixation (and plant turgor) of soybean or pigeonpea, when grown on a deep sandy soil in Florida. Deeper rooting and deeper water extraction were factors contributing to Florunner's better tolerance to drought, but nodule activity of peanut also appeared more tolerant to prolonged drying of the topsoil. Total accumulation of N by peanut was less affected by the drought than was total N accumulation by soybean or pigeonpea.

D. Interactions of Water Deficit with Soil Nutrient Status

Calcium deficiency, which is possible in acid soil environments, may make peanut more susceptible to drought stress. Chari et al. (1986) grew peanut plants in nutrient culture at different Ca levels and then subjected the plants to water deficit. For water-stressed plants, as the Ca concentration of tissue increased, the relative water content was enhanced and proline (amino acid) accumulation less. The authors found that Ca-deficient treatments exhibited greater electrolyte leakage from leaf tissues when exposed to water stress. Calcium presence is generally presumed to enhance membrane stability.

Soils in many developing countries are deficient in P. Studies of irrigation management and P fertilization of P-deficient soils generally indicate additive effects of both P and irrigation water; i.e., there is a yield response to irrigation at each P treatment level (Rao & Singh, 1985).

VI. GENOTYPIC VARIATION IN DROUGHT TOLERANCE

Drought-tolerant peanut cultivars are needed in semiarid regions. Researchers in the southwestern USA, West Africa, and India have attempted to screen germplasm for drought tolerance defined in the following broad categories: (i) escape from drought (earliness), (ii) avoid droughts (by deeper rooting to maintain plant water status), or (iii) resist droughts (by withstanding low water potential).

Breeding for earliness as a way to escape drought is an easily understood objective, especially where the goal is to fit a cultivar into a short rainy season in the semiarid tropics (Africa or India). The only compromise is that yield potential in good years is sacrificed.

In regions with long rainy periods but with periodic droughts, screening for deeper rooting has been attempted as a way to maintain plant water status throughout dry periods. Selecting cultivars with more extensive root systems should allow more soil water extraction from greater soil volumes than for plants having limited root systems. Advantages of more extensive rooting and water extraction should include greater drought avoidance and ability to maintain leaf area during periods of drought. Ketring et al. (1982b, 1984a) found genetic diversity for root and shoot growth traits among peanut germplasm and suggested the potential for selecting parental germplasm to develop cultivars for drought-prone environments. Among Virginia-type cultivars, Florunner and a related Univ. of Florida breeding line had the longest taproots, followed closely by Early Runner, Early Bunch, and Flor-igiant. Dixie Runner, the earliest release from the Florida breeding program, had the shortest taproot among the Virginia types. It is of interest that genetic improvement for yield in the successive releases of Dixie Runner, Early Runner, Florunner, and Early Bunch from the Florida breeding program (Duncan et al., 1978) may be related to improved rooting traits as well.Rodrigues (1984), comparing four peanut genotypes, found that wilt susceptibility of the Pearl Early Runner cultivar during a field drought period was associated with less extensive root length density and less water extraction in deeper soil layers (60–150 cm) as compared to Florunner, Early Bunch, and PI 383426. In irrigated tube culture in the greenhouse, Pearl Early Runner had the slowest rate of root depth progression, the lowest ratio of root mass to shoot mass, and the lowest ratio of root length or mass to leaf area, all factors contributory to early wilting. Williams et al. (1986), in a comparison of four cultivars, found that Robut 33-1 had a somewhat more rapid rate of increase in its depth of water extraction than the other cultivars.

In the third category of selecting genotypes that can survive and produce even at low water potential, Gautreau (1978) in West Africa, separated a collection of peanut germplasm into three groups having low, intermediate, and high leaf water potentials which correspond inversely to their yield levels. The low-yielding cultivars had higher water potentials and the higher-yielding, drought-tolerant cultivars had lower, more negative water potentials. Theoretically, a more negative water potential should allow drought-tolerant cultivars to have a steeper water potential gradient to drive root uptake of water. This would be a drought-resisting or drought-withstanding trait. As reviewed by Elston and Bunting (1978), other varieties lacking this trait may also be adapted to dry areas because of low LAI, short season (escape or avoidance), or because of indeterminacy and recovery mechanisms.

Breeding for drought tolerance may conflict with breeding for high yield potential in adequate water environments. Williams et al. (1986) found that sensitivity of a genotype to drought increases with its yield potential. Genotypes of similar maturity had small differences in total water use, up to 30% difference in dry matter produced per unit water transpired, and nearly twofold differences in pod harvest index.

VII. IRRIGATION MANAGEMENT

A. Irrigation Management Relative to Peanut Growth Stages

To effectively irrigate peanut, one must consider the current stage of growth and development of the crop. For example, water extraction depth is influenced by rooting extension, and crop ET is influenced by canopy cover and LAI. Furthermore, pegging and pod formation have additional requirements for adequate moisture in the fruiting zone.

Water deficits that occur only during early vegetative growth cause minor reductions in growth and yield (Il'ina, 1958; Su et al., 1964; Stansell et al.,

1976; Reddi & Reddy, 1977; Pallas et al., 1979; Nageswara Rao et al., 1985). Likewise, less frequent irrigation and irrigation at greater SWD has its least detrimental effects if applied prior to pegging and pod formation (Subramanian et al., 1975; Reddi & Reddy, 1977). As seen in Fig. 22–1, beginning peg (R2 stage) and beginning pod (R3 stage) occur at about 6 and 7 wk after sowing, respectively. Thus water deficit during the first 50 d after sowing should reduce primarily vegetative growth, since few flowers and pegs are present. Since vegetative growth is frequently excessive in wet years (Gorbet & Rhoads, 1975; Vivekanandan & Gunasena, 1976) and pod harvest index may be improved by moderate stresses prior to 50 d (Nageswara Rao et al., 1985), we would expect minor effects on final yield if moderate water deficit is relieved by 50 d after sowing. Stresses during vegetative growth are less damaging because ground cover and LAI are incomplete. Thus, less water is consumed for ET, and irrigation amounts and frequency can be reduced during this phase. Moreover, early vegetative growth may continue by using stored soil water as root extension progresses. Obviously, irrigation should be used to ensure germination and emergence and to relieve extreme stress if irrigation is available.

Water deficits are more detrimental to yield if they occur while pods are being added and formed. In terms of peanut growth stages, this period starts when the peg tips first begin to swell (R3-beginning pod) and ends when a full pod load has been established and vegetative growth begins to slow (Fig. 22–1). For Florunner peanut, this period continues from about 50 to 90 d after sowing. For short-season cultivars, pod formation starts a few days sooner, and ends about 10 d sooner. The greater drought sensitivity of the pod formation and addition phase is partly related to the fact that the crop reaches its peak water use and ET during this time. Additionally, this is the period during which a pod load is set. Soil water deficit can reduce pod addition in one of several ways. First, insufficient soil water in the fruiting zone reduces peg penetration and pod formation (Cox, 1962; Underwood et al., 1971; Ono et al., 1974; Boote et al., 1976). Secondly, water deficit in the entire rooting zone may reduce both plant turgor and photosynthesis. Insufficient turgor reduces peg elongation and pod expansion whereas reduced photosynthesis would limit the number of pods added and thereby also limit the final yield potential. The period 65 to 86 d after sowing were shown by Hang et al. (1984) to be the most sensitive period in terms of yield reduction caused by shade-induced reductions in photosynthetic supply. Drought-induced reduction in photosynthesis would likely have its most critical effect on pod addition during this growth phase.

There is some disagreement in the literature as to whether the pod set and formation period is the phase most sensitive to water deficit, or whether the period of seed growth and filling is most sensitive in terms of yield reduction. Stansell and Pallas (1985) reported that drought from 71 to 105 d was more damaging to seed yield of Florunner than drought at 36 to 70 d or at 106 to 140 d. The answer to the above question is related to the normal life cycle of the cultivar. For the full-season Florunner cultivar, the 71- to 105-d period includes much of the period for adding a full pod load, and only part of the linear seed growth phase. By contrast, 71 to 105 d would be substantially into the seed-fill phase for a short-season cultivar.

After a full pod load is set (by about 90 d for Florunner), water deficit reduces yield initially by causing smaller and younger fruits to terminate growth, and eventually by reducing growth rate of older fruit. Water deficit from 105 to 140 d on Florunner caused less yield reduction than 35-d deficit periods from 36 to 70 d or from 71 to 105 d (Stansell & Pallas, 1985). Late-season drought caused termination of young pods or prevented the addition of new young pods, thus causing a higher shell-out (SMK). A 70-d drought during late season was less detrimental to yield than a 70-d drought at mid-season (36–105 d). Thus irrigation can be managed more conservatively during the last 30 d. Furthermore, excessive and too frequent irrigation or rainfall during the latter part of seed fill (last 30 or so days) can result in yield loss, especially on heavier soils or those with restricted drainage. Decreased yields, probably associated with increased disease, weakening of peg attachments, and loss of detached pods, were related to more frequent irrigations on a clay soil (Mantell & Goldin, 1964) and to frequent irrigations to maintain soil water potential above -40 kPa in loamy sand soils (Gorbet & Rhoads, 1975; Stansell et al., 1976).

B. Methods for Scheduling Irrigation

The goal of irrigation is to supply the crop rooting zone with sufficient water at appropriate times in order to prevent soil water deficit from reducing growth and yield. There are several approaches to accomplish this goal. One is to measure soil water status with appropriate instrumentation and schedule irrigation at a soil water potential above which the crop will not be stressed. This approach requires careful measurements of soil water potential, but is less dependent on information of rainfall, irrigation amounts, soil water-holding capacity, or good estimates of daily crop ET. The second approach is essentially a bookkeeping approach in which crop water use (ET) is estimated from calibrated data or from weather inputs or as a fraction of cumulative pan evaporation, and irrigation is scheduled as needed to resupply water lost by ET. Soil water potential measurements are not required for this approach, but estimates of effective rooting zone, soil water-holding capacity, rainfall, irrigation amounts, and daily crop ET are needed to gauge when the available soil water will deplete enough to cause crop stress and fail to supply evapotranspirational demand. Scheduling irrigation by use of crop simulation models is a special form of the bookkeeping approach because model simulation of crop LAI and root growth allows dynamic simulation of the crop water use coefficient and changing rooting depth and root length proliferation. Dynamic computation of ET_o is done using current weather information and methods such as the modified Penman equation or Priestly-Taylor equilibrium method.

1. Scheduling Irrigation Based on Soil Water Potential

In this approach, irrigation is scheduled when soil water potential reaches a defined value at the 15- to 30-cm soil depth (Matlock et al., 1961; Hiler & Clark, 1971; Gorbet & Rhoads, 1975; Stansell et al., 1976; Hammond et al., 1978). This system theoretically should be an excellent approach because the soil water sensors employed for this technique estimate the soil water potential in the zone of greatest rooting density, and thereby can be used to minimize low soil water potentials in the rooting zone. Secondly, the use of soil water sensors eliminates the need to record rainfall and pan evaporation. One of the problems with irrigation scheduling by soil water potential is that soils have different soil water retention curves. At -30 kPa, most of the available water may be depleted from a sandy soil; yet considerable available water will remain in a clay loam. Thus, irrigation must be scheduled at a water potential determined for the given soil type and crop. Other problems include the accuracy, reliability, and calibration of sensors for measuring soil water potential, as well as spatial variability of soil and time required for producers to observe the sensors.

A number of researchers in the southeastern USA have proposed the use of soil water potential at 15- to 30-cm depth for scheduling irrigation of peanut. Gorbet and Rhoads (1975) observed that yield response to irrigation at -60 kPa was as good as irrigating at -30 or -40 kPa on a sandy loam soil. Yield was only slightly less if irrigation was delayed until plants showed a noticeable wilt. In wet years, yield was reduced by frequent irrigation to maintain soil water potential above -30 kPa. Rochester et al. (1978) in Alabama reported satisfactory yield responses for irrigation initiated when soil water potential at 15 or 30 cm dropped to -20 to -60 kPa. On more sandy soils (loamy sand or a fine sand), Stansell et al. (1976) and Hammond et al. (1978) obtained highest yields when irrigation was scheduled when soil water potential at 15- or 30-cm depth dropped to -20 kPa. Maximum harvestable pod yields in the study of Stansell et al. (1976) were 4460, 5080, and 4540 kg ha^{-1} for Florigiant, Florunner, and Tifspan, respectively, when irrigated to a depth of 60 cm when water potential in the 0- to 30-cm zone of loamy sand reached -20 kPa. Yields of Florunner and Tifspan were lower, but not significantly so if irrigation was scheduled at -60 kPa in the 0- to 30-cm zone. Yield was significantly reduced if irrigation was delayed until water potential reached -1500 kPa or the plants wilted to the stage of no overnight recovery.

Water use efficiency of Florunner peanut was lowest for the highest-yielding treatment and usually highest for the driest treatment (Stansell et al., 1976), although Hammond et al. (1978) reported highest WUE for the most frequent irrigation treatment. These studies and one by Pallas et al. (1979) suggest that aiming only for high WUE is a poor and variable indicator of optimum economic use of water for yield. On sandy loams and loamy sands in the southeastern USA, optimum economic yield and water use appear to result from irrigation scheduling when soil water potential reaches -60 kPa or when peanut plants show temporary, but noticeable wilt at midday. Irrigation when soil water potential reaches -100 or -200 kPa has usually allowed satisfactory growth and yield of spanish-type peanuts grown on finer-textured soils in Oklahoma (Matlock et al., 1961). Excessive or frequent irrigations in a wet year or during the latter half of the season have been reported to reduce harvestable yields, partly because of increased disease development and accelerated pod losses (Gorbet & Rhoads, 1975; Wright et al., 1986).

2. Scheduling Irrigation Based on Soil Water Depletion

This approach is similar to the one above, except that soil water status is defined in terms of volumetric or gravimetric water content. This method allows estimation of soil water content by other methods such as neutron moderation, gamma transmission, or gravimetric techniques. It can also be determined from soil water potential measurements, because a given soil water depletion (SWD) can be defined from a soil water release curve vs. soil water potential for the particular soil. For this method, irrigation is scheduled when the soil water in a given layer (usually 0 to 30 cm) reaches a given percentage of SWD. Lenka and Misra (1973) and Narasimham et al. (1977) irrigated semierect short-season cultivars at 75, 50, or 25% SWD during the dry season in India. Irrigation at 25% SWD gave the highest yield, but lowest WUE. Irrigation at 75% SWD gave high WUE, but significantly lower yield. Scheduling irrigation at 50% SWD was a good general compromise which gave good WUE without losing too much of the yield response possible from irrigation. Stage-specific irrigation treatments were studied by Reddi and Reddy (1977) on cv. AH1192 and TMV2 during a nearly rainless winter season in India. During the period from sowing to pegging, irrigating at 50% SWD was an acceptable strategy which mostly maintained yield potential and WUE; however, irrigation had to be scheduled more frequently, at 25% SWD, during the phases of "pegging to pod formation," and from pod formation to harvest in order to obtain highest yields and WUE. Irrigating at 25% SWD for the entire season gave the highest yield of 4680 kg ha^{-1} and highest WUE of 92 kg ha^{-1} per cm.

3. Scheduling Irrigation Based on Cumulative Pan Evaporation

In this bookkeeping approach, sufficient irrigation water (IW) is applied to replace a given fraction of CPE less rainfall received since the last irrigation. This approach has been used for full-season cultivars on sandy soils under arid conditions in Israel (Goldberg et al., 1967) and on short-season cultivars grown on sandy loam soils under arid conditions in India (Subramanian et al., 1975; Khan & Datta, 1982; Pahalwan & Tripathi, 1984). It probably does not work well in humid rainfall regions because of uncertainty of deep drainage. A number of researchers (Goldberg et al., 1967; Subramanian et al., 1975; Kahn & Datta, 1982; Pahalwan & Tripathi, 1984) have evaluated irrigation scheduling at 0.45, 0.5, 0.6, 0.7, 0.75, 0.9, and 1.05 fraction of CPE between irrigations and at different growth phases. Their consensus is that irrigating at 0.90 CPE is best during full canopy development, although Khan and Datta (1982) found higher yield and WUE if irrigation was scheduled at 0.75 IW/CPE. Since their growing season had considerable rainfall, the conservative irrigation would allow storage in the soil for unexpected following rains and probably made more effective use of those rains. Irrigation at 0.5 or 0.6 CPE was not detrimental during early growth (up to pod formation), and caused an increase in WUE if followed by irrigation at 0.9 CPE during later stages (Subramanian et al., 1975; Pahalwan & Tripathi, 1984). An underlying problem is that the CPE method fails to consider crop aspects, i.e., crop coefficient. The ratio of ET/ET_o is relatively low during early growth because of insufficient ground cover. The more conservative irrigation (0.5 of CPE) up to the pod formation stage resulted in greater contribution of soil profile water to crop water use (Pahalwan & Tripathi, 1984). Irrigating at 0.9 CPE was essential during pegging to pod formation, but irrigating at 0.7 CPE during pod formation to maturity only slightly reduced yields while maintaining high WUE. Irrigating at 0.90 of CPE works well for closed canopies in arid regions because ET/ET_o under such conditions is nearly 0.90, but a different approach may be necessary when plants are small or in humid regions to maximize use of rainfall.

4. Bookkeeping of a Water Account

A bookkeeping method was suggested by Stansell (1975) in which available water-holding capacity of the top 60 cm of soil is considered the total water account. Rainfall and irrigations are considered deposits to the water account and daily ET as withdrawals from the account. Rainfall or irrigations in excess of the capacity of the account are lost to deep drainage. Daily crop water use rates (Fig. 22–2) have been developed from field data and account for plant age, ground cover, and canopy decline at the end of the season (Mantell & Goldin, 1964; Stansell et al., 1976). These rates can be used to estimate daily profile water depletion. After some preset depletion, irrigation is applied to replace water that has been used. Since previous research has shown that water withdrawals of more than 50% of available soil water decreased yield, the peanut crop is usually allowed to withdraw one-half of the total available water before irrigation is scheduled.

5. Modeling the Crop Water Budget and Integrating Crop Response to Evapotranspiration and Weather

Irrigation scheduling with dynamic crop growth simulators may be possible in the future. For example, the PNUTGRO crop growth and yield simulator (Boote et al., 1988) contains the same type of water account described above and considers rainfall and irrigation receipts, deep drainage, and daily ET. The PNUTGRO and similar simulators have several advantages over the simple bookkeeping approach:

1. They dynamically increase rooting depth over time and proliferate root length density in the various soil layers and thus reflect the changes in the size of the water account over time as the crop grows.

2. They dynamically simulate the increase in crop leaf area index over time as well as its senescence decline and thus automatically generate an appropriate crop water use coefficient (K_c).

3. They dynamically simulate ET from daily weather inputs and thus are not dependent on a standard fixed calibration. This is particularly an advantage in situations where early season water deficit causes low LAI, and the model thus simulates a low effective crop water use coefficient. Standard fixed calibrations of K_c or ET over time do not allow such feedback loops. The requirement for daily solar radiation and weather information is a disadvantage, however.

4. The models should be able to address stage specific water requirements because such features are usually implicitly coded into the computer coding for development, growth, pod formation, and yield development of the crop.

5. They predict the growth, yield, and maturity date, thus providing information on yield potential and a projected date to cease irrigations. The PNUTGRO model is described by Boote et al. (1986) and is available for use (Boote et al., 1988).

To illustrate the use of PNUTGRO for evaluating responses to water application, simulations were conducted on the Florunner cultivar using an initially dry soil water profile and increasing the total amount of water received during the season from zero application until the point at which maximum pod yield was obtained. As ET increased, yield increased slowly at first, then linearly, and achieved a maximum yield of 6 Mg ha^{-1} with 55 cm of seasonal ET, at which point water application fully satisfied the climatic potential ET (Fig. 22–12). The simulated seasonal water requirement for maximum Florunner yield is similar to that measured by Stansell et al. (1976).

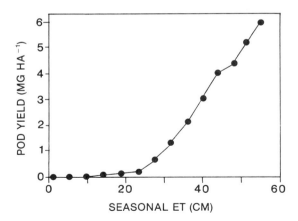

Fig. 22–12. Simulated yield response of Florunner peanut to seasonal evapotranspiration for 1 May planting at Gainesville, FL, using the PNUTGRO crop growth model. Initial soil profile was depleted of water, and water applications increased systematically until full ET demand was met.

C. Irrigation Management Relative to Diseases and Pests

Irrigation applications may have positive or negative effects on disease, insect, and weed control. For example, Backman et al. (1978) showed that irrigation increased leafspot diseases and damage from *Sclerotium rolfsii*. Wright et al. (1986) reported increased leafspot, increased pod rot, and increased damage from Sclerotina blight (caused by *Sclerotina minor*) on irrigated vs. rainfed plots in Virginia. Use of irrigation may, therefore, increase the need for better disease control practices or result in economic losses from disease. Wright et al. (1986) observed large yield response to irrigation in one drought year of a 4-yr study. In other years with better rainfall, irrigation caused yield losses (presumably because of diseases). Thus, irrigation applications may need to be timed more conservatively, unless the drought is prolonged.

In some cases, irrigation can also be useful to improve disease and pest control. For example, irrigation may improve weed control by insuring activation of pre-emergence herbicides. Irrigation or adequate rainfall are commonly believed to reduce the incidence of lesser cornstalk borer [*Elasmopalpus lignosellus* (Zeller)], since this pest is a common dry-weather pest (Smith & Barfield, 1982). Irrigation may be of use to reduce aflatoxin contamination of peanut pods (toxin produced by *Aspergillus favus* Link). Aflatoxin contamination of pods is more prevalent when peanuts are exposed to drought conditions, especially when drought occurs during pod-filling and maturation phases (Pettit et al., 1971; Wilson & Stansell, 1983; Diener et al., 1982). Conditions most favorable for invasion by *A. favus* and alfatoxin contamination include elevation of fruiting zone temperature to 28 to 30°C and soil water deficit during the 50-d period prior to harvest (Sanders et al., 1985b). Irrigation is frequently effective in reducing the amount of aflatoxin, partly because it reduces soil temperature. The critical period for irrigation to prevent aflatoxin is from 50 to 20 d before harvest (Sanders et al., 1985b; Wilson & Stansell, 1983). Water deficits during the last 20 d did not enhance aflatoxin damage.

VIII. A SYSTEMS APPROACH TO IRRIGATION: HARDWARE, ECONOMICS, MODELS, PESTS, AND DISEASE

The decision to irrigate peanut involves more than just alleviating drought stress. The decision should consider water supply, feasible irrigation equipment, labor and energy conservation, capital availability, economic return, water infiltration rates, the potentials of using the same system for other crops in a multiple cropping rotation, and the benefits from using the irrigation system for doing other cultural and pest control practices. For example, what is the economic benefit if irrigation allows the producer to select the optimum planting date for high yield? What is the economic benefit of the use of irrigation to ensure effective application and activation of pre-emergence herbicides for weed control? To control lesser cornstalk borer?

Or to reduce aflatoxin damage to pods? On the other hand, there may be economic losses because irrigation has been shown to increase leafspot diseases and damage from *S. rolfsii* (Backman et al., 1978).

Irrigation hardware, technology, water supply requirements, and system deployment for peanut irrigation are similar to those for other crops, and readers are referred to earlier chapters in this monograph that cover those topics. Most of the peanut irrigation in the USA is done with overhead sprinklers, usually with center pivot units. The water supply is commonly from deep wells although impounded surface water is used in some regions. Aspects of irrigation system design and operation are discussed by Davidson et al. (1973), Keese et al. (1975), and Samples (1981). For additional region-specific assistance or information on irrigation, readers are referred to peanut production guides available in their own states.

What is the future for peanut irrigation in the USA? Because of the high value of the crop in the USA, we anticipate growers will continue to invest in irrigation systems until the majority of growers who have access to water have installed systems. There are exciting potentials for use of irrigation systems to apply nutrients, herbicides, and pesticides for control of soil-borne insects, diseases, and nematodes. Such uses require uniform application capabilities and flexible application rates. Future irrigation scheduling may involve combinations of the following: automated soil water measurements, automated rainfall and weather measurements, microprocessor-controlled irrigation systems, modeling of crop growth and water flow processes to schedule irrigation and to predict response to irrigation, and expert systems approaches.

REFERENCES

Abdul-Jabbar, A.S. 1978. Leaf diffusive resistance of peanuts as influenced by environment and row spacing. M.S. thesis, Oklahoma State Univ., Stillwater.

Allen, L.H., Jr., K.J. Boote, and L.C. Hammond. 1976. Peanut stomatal diffusion resistance affected by soil water and solar radiation. Proc. Soil Crop Sci. Soc. Fla. 35:42–46.

Backman, P.A., E.W. Rochester, and J.M. Hammond. 1978. Effects of irrigation on peanut disease. Proc. Am. Peanut Res. Educ. Assoc. 10:68 (Abstr.).

Bennett, J.M., K.J. Boote, and L.C. Hammond. 1981. Alterations in the components of peanut leaf water potential during desiccation. J. Exp. Bot. 32:1035–1043.

Bennett, J.M., K.J. Boote, and L.C. Hammond. 1984. Relationships among water potential components, relative water content, and stomatal resistance of field-grown peanut leaves. Peanut Sci. 11:31–35.

Bhagsari, A.S., R.H. Brown, and J.S. Schepers. 1976. Effect of moisture stress on photosynthesis and some related physiological characteristics in peanuts. Crops Sci. 16:712–715.

Bhan, S., and D.K. Misra. 1970. Water utilization by groundnut (*Arachis hypogaea* L.) as influenced due to variety, plant population and soil fertility level under arid zone conditions. Indian J. Agron. 15:258–263.

Billaz, R. 1962. Comparison de quatre varietes d'arachide pour leur resistance a la secheresse. Oleagineau 17:35–39.

Billaz, R., and R. Ochs. 1961. Stades de sensibilite de l'arachide a la secheresse. Oleagineux 16:605–611.

Black, C.R., and G.R. Squire. 1979. Effects of atmospheric saturation deficit on the stomatal conductance of pearl millet (*Pennisetum typhoides* S. and H.) groundnut (*Arachis hypogaea* L.). J. Exp. Bot. 30:935–945.

Black, C.R., D.Y. Tang, C.K. Ong, A. Solon, and L.P. Simmonds. 1985. Effects of soil moisture stress on water relations and water use of groundnut stands. New Phytol. 100:313-328.

Bledsoe, R.W., C.L. Comar, and H.C. Harris. 1949. Absorption of radioactive calcium by the peanut fruit. Science 109:329-330.

Bockelee-Morvan, A., J. Gautreau, J.C. Mortreuil, and O. Roussel. 1974. Results obtained with drought-resistant groundnut varieties in West Africa. Oleagineux 29:309-314.

Bolhuis, G.G., H.D. Frinking, J. Leenwaugh, R.G. Rens, and G. Staritsky. 1965. Occurrence of flowers with short style in the groundnut (*Arachis hypogaea*). Oleagineux 20:293-296.

Boote, K.J. 1982. Growth stages of peanut (*Arachis hypogaea* L.). Peanut Sci. 9:35-40.

Boote, K.J., and L.C. Hammond. 1981. Effect of drought on vegetative and reproductive development of peanut. Proc. Am. Peanut Res. Educ. Soc. 13:86 (Abstr.).

Boote, K.J., J.W. Jones, G. Hoogenboom, G.G. Wilkerson, and S.S. Jagtap. 1988. PNUTGRO V1.01, Peanut crop growth simulation model, user's guide. Florida Agric. Exp. Stn. J. 8420.

Boote, K.J., J.W. Jones, J.W. Mishoe, and G.G. Wilkerson. 1986. Modeling growth and yield of groundnut. p. 243-255. *In* Agrometeorology of groundnut. Proc. Int. Symp. ICRISAT Sahelian Ctr., Niamey, Niger. 21-25 Aug. 1985. Int. Crops Res. Inst. for the Semi-Arid Tropics, Patencheru, A.P. India.

Boote, K.J., J.R. Stansell, A.M. Schubert, and J.F. Stone. 1982. Irrigation, water use, and water relations. p. 164-205. *In* H.E. Pattee and C.T. Young (ed.) Peanut science and technology. Am. Peanut Res. and Educ. Soc. Yoakum, TX.

Boote, K.J., R.J. Varnell, and W.G. Duncan. 1976. Relationships of size, osmotic concentration, and sugar concentration of peanut pods to soil water. Proc. Soil Crop Sci. Soc. Fla. 35:47-50.

Burman, R.D., P.R. Nixon, J.L. Wright, and W.O. Pruitt. 1980. Water requirements. p. 189-232. *In* M.E. Jensen (ed.) Design and operation of farm irrigation systems. ASAE, St. Joseph, MI.

Chari, M., K. Gupta, T.G. Prasad, K.S. Krishna Sastry, and M. Udaya Kumar. 1986. Enhancement of water status by calcium pretreatment in groundnut and cowpea plants subjected to moisture stress. Plant Soil 91:109-114.

Chen, C.Y., and H.S. Chang. 1974. A study of the relations of several physiological functions of crops to the leaf water potential, soil moisture stress and growth stage. Corn, kaoliang, soybeans, peanut and sweet potato. J. Agric. Assoc. China 88:1-16.

Chin Choy, E.W., J.F. Stone, and J.E. Garton. 1977. Row spacing and direction effects on water uptake characteristics of peanuts. J. Soil Sci. Soc. Am. 41:428-432.

Chin Choy, E.W., J.F. Stone, R.S. Matlock, and G.N. McCauley. 1982. Plant population and irrigation effects on spanish peanuts (*Arachis hypogaea* L.). Peanut Sci. 9:73-76.

Cox, F.R. 1962. The effect of plant population, various fertilizers, and soil moisture on the grade and yield of peanuts. North Carolina State Univ., Raleigh. Diss. Abstr. Int. 22:3326-B.

Cox, F.R. 1979. Effect of temperature treatment on peanut vegetative and fruit growth. Peanut Sci. 6:14-17.

Cox, F.R., G.A. Sullivan, and C.K. Martin. 1976. Effect of calcium and irrigation treatments on peanut yield, grade and seed quality. Peanut Sci. 3:81-85.

Dancette, C., and F. Forest. 1986. Water requirements of groundnuts in the semi-arid tropics. p. 69-82. *In* Agrometeorology of groundnut. Proc. Int. Symp. ICRISAT Sahelian Ctr., Niamey, Niger. 21-25 Aug. 1981. Int. Crops. Res. Inst. for the Semi-Arid Tropics, Patencheru, A.P. India.

Davidson, J.M., J.E. Garton, R.S. Matlock, D. Schwab, J.F. Stone, and L.D. Tripp. 1973. Irrigation and water use. p. 361-382. *In* Peanuts: Culture and uses. Am. Peanut Res. and Educ. Assoc., Stillwater, OK.

Desa, N.D., R.S. Josh, and K.R. Patel. 1984. Response of summer groundnut to various levels of irrigation on clayey soils. Madras Agric. J. 71:617-620.

DeVries, J.D. 1986. Water relations and nitrogen fixation of three grain legumes during water stress. M.S. thesis. Univ. Florida, Gainesville.

Diener, U.L., R.E. Pettit, and R.J. Cole. 1982. Aflatoxins and other mycotoxins in peanuts. p. 486-519. *In* H.E. Pattee and C.T. Young (ed.) Peanut science and technology. Am. Peanut Res. and Educ. Soc., Yoakum, TX.

Dreyer, J., W.G. Duncan, and D.E. McCloud. 1981. Fruit temperature, growth rates, and yield of peanuts. Crop Sci. 21:686-688.

Duncan, W.G., D.E. McCloud, R.L. McGraw, and K.J. Boote. 1978. Physiological aspects of peanut yield improvement. Crop Sci. 18:1015-1020.

Elston,J., and A.H. Bunting. 1978. Water relations of legume crops. p. 37–42. *In* R.J. Summerfield and A.H. Bunting (ed.) Advances in legume science. Royal Botanic Gardens, Kew, England.

Erickson, P.E., and D.L. Ketring. 1985. Evaluation of peanut genotypes for resistance to water stress in situ. Crop Sci. 25:870–876.

Erickson, P.I., J.F. Stone, and J.E. Garton. 1986. Critical evaporative demands for differential stomatal action in peanut grown in narrow and wide row spacings. Agron. J. 78:254–258.

Fortanier, E.J. 1957. Control of flowering in *Arachis hypogaea* L. Meded. Landbouwhogesch. Wageningen 57:1–116.

Fourrier, P., and P. Prevot. 1958. Influence sur l'arachide de la pluviosite, de la fumure minerale et de trempage des graines. Oleagineux 13:805–809.

Gautreau, J. 1978. Niveaux de potentiels foliaires intervarietaux et adaptation de l'arachide a la secheresse au Senegal. Oleagineux 32:323–332.

Gillier, P. 1969. Effects secondaires de la secheresse sur l'arachide. Oleagineux 24:79–81.

Goldberg, S.D., B. Gornat, and D. Sadan. 1967. Relation between water consumption of peanuts and Class A pan evaporation during the growing season. Soil Sci. 104:289–296.

Gorbet, D.W., and F.M. Rhoads. 1975. Response of two peanut cultivars to irrigation and kylar. Agron. J. 67:373–376.

Gregory, P.J., and M.S. Reddy. 1982. Root growth in an intercrop of pearl millet/groundnut. Field Crops Res. 5:241–252.

Grosz, G.D. 1986. Simulation of peanut growth in Oklahoma. M.S. thesis. Oklahoma State Univ., Stillwater.

Hammond, L.C., K.J. Boote, R.J. Varnell, and W.K. Robertson. 1978. Water use and yield of peanuts on a well-drained sandy soil. Proc. Am. Peanut Res. Educ. Assoc. 10:73 (Abstr.).

Hang, A.N., D.E. McCloud, K.J. Boote, and W.G. Duncan. 1984. Shade effects on growth, partitioning, and yield components of peanuts. Crop Sci. 24:109–115.

Harp, S.L., R.L. Elliott, D.P. Schwab, G.D. Grosz, and M.A. Kizer. 1986. Irrigation scheduling program. Project completion report, Corporation Commission of Oklahoma. Agric. Eng. Dep., Oklahoma State Univ., Stillwater.

Harris, H.C., and R.W. Bledsoe. 1951. Physiology and mineral nutrition. p. 89–121. *In* The peanut: The unpredictable legume. Natl. Fertilizer Assoc., Washington, DC.

Henning, R.J., J.F. McGill, L.E. Samples, C. Swann, S.S. Thompson, and H. Womack. 1979. Growing peanuts in Georgia: A package approach. Univ. Georgia Coop. Ext. Serv. Bull. 640.

Hiler, E.A., and R.N. Clark. 1971. Stress day index to characterize effects of water stress on crop yields. Trans. ASAE. 14:757–761.

Huang, M.T. 1985. Physiological aspects of drought and heat tolerance of peanut (*Arachis hypogaea* L.). Ph.D. diss. Oklahoma State Univ., Stillwater (Diss Abstr. Int. 46:4072-B).

Il'ina, A.I. 1958. Definition of the periods of high sensitivity of peanut plants to soil moisture. Soviet Plant Physiol. 5:253–258. Translated to French in Oleagineux 14:89–92.

Ishag, H.M., Osman A. Fadl, H.S. Adam, and A.K. Osman. 1985. Growth and water relations of groundnuts (*Arachis hypogaea*) in two contrasting years in the irrigated Gezira. Exp. Agric. 21:403–408.

Iyama, J., and Y. Murata. 1961. Studies on the photosynthesis in upland field crops. 2. Relationships between the soil moisture and photosynthesis of some upland crops and rice plant (*In* Japanese, English summary.) Proc. Crop Sci. Soc. Jpn. 29:350–352.

Kassam, A.H., J.M. Kowal, and C. Harkness. 1975. Water use and growth of groundnut at Samaru, Northern Nigeria. Trop. Agric. (Guilford) 52:105–112.

Keese, C.W., J.S. Denton, E.A. Hiler, and J.S. Newman. 1975. Irrigation practices p. 42–47. *In* Peanut production in texas. Texas Agric. Exp. Stn. RM 3.

Ketring, D.L. 1984a. Root diversity among peanut genotypes. Crop Sci. 24:229–232.

Ketring, D.L. 1984b. Temperature effects on vegetative and reproductive development of peanut. Crop Sci. 24:877–882.

Ketring, D.L., R.H. Brown, G.A. Sullivan, and B.B. Johnson. 1982a. Growth physiology. p. 411–457. *In* H.E. Pattee and C.T. Young (ed.) Peanut science and technology. Am. Peanut Res. Educ. Soc., Yoakum, TX.

Ketring, D.L., W.R. Jordan, O.D. Smith, and C.E. Simpson. 1982b. Genetic variability in root and shoot growth characteristics of peanut. Peanut Sci. 9:68–72.

Khan, A.R., and B. Datta. 1982. Scheduling of irrigation for summer peanut. Peanut Sci. 9:10–13.

Lee, T.A., Jr., D.L. Ketring, and R.D. Powell. 1972. Flowering and growth response of peanut plants (*Arachis hypogaea* L. var. Starr) at two levels of relative humidity. Plant Physiol. 49:190–193.

Lenka, D., and P.K. Misra. 1973. Response of groundnut (*Arachis hypogaea* L.) to irrigation. Indian J. Agron. 18:492–497.

Leong, S.K., and C.K. Ong. 1983. The influence of temperature and soil water deficit on the development and morphology of groundnut (*Arachis hypogaea* L.). J. Exp. Bot. 34:1551–1561.

Lin, H., C.C. Chen, and C.Y. Lin. 1963. Study of drought resistance in the virginia and spanish types of peanut. J. Agric. Assoc. China 43:40–51.

Maliro, C.E. 1987. Physiolgical aspects of yield among four legume crops under two water regimes. M.S. thesis. Univ. Florida, Gainesville.

Mantell, A., and E. Goldin. 1964. The influence of irrigation frequency and intensity on the yield and quality of peanuts (*Arachis hypogaea*). Israel J. Agric. Res. 14:203–210.

Matlock, R.S., J.E. Garton, and J.F. Stone. 1961. Peanut irrigation studies in Oklahoma, 1956–1959. Oklahoma Agric. Exp. Stn. Bull. B-580.

McCauley, G.N., J.F. Stone, and E.W. ChinChoy. 1978. Evapotranspiration reduction by field geometry effects in peanuts and grain sorghum. Agric. Meteorol. 19:295–304.

McCloud, D.E. 1974. Growth analysis of high yielding peanuts. Proc. Soil Crop Sci. Soc. Fla. 33:24–26.

McGraw, R.L. 1977. Yield dynamics of Florunner peanuts (*Arachis hypogaea* L.). M.S. thesis. Univ. Florida, Gainesville.

McGraw, R.L. 1979. Yield physiology of peanuts (*Arachis hypogaea* L.). Ph.D. diss. Univ. Florida, Gainesville (Diss. Abstr. 41:0430-B).

Mixon, A.C. 1971. Moisture requirements for seed germination of peanuts. Agron. J. 63:336–338.

Nageswara Rao, R.C.N., S. Singh, M.V.K. Sivakumar, K.L. Srivastava, and J.H. Williams. 1985. Effect of water deficit at different growth phases of peanut. I. Yield responses. Agron. J. 77:782–786.

Narasimham, R.L., I.V. Subba Rao, and M. Singa Rao. 1977. Effect of moisture stress on response of groundnut to phosphate fertilization. Indian J. Agric. Sci. 47:573–576.

Ochs, R., and T.M. Wormer. Influence de l'alimentation en eau sur la croissance de l'arachide. Oleagineaux 14:281–291.

Oklahoma Cooperative Extension Service. 1988. Peanut production guide for Oklahoma. Circ. E-608. Oklahoma State Univ., Stillwater.

Ono, Y., K. Nakayama, and M. Kubota. 1974. Effects of soil temperature and soil moisture in podding zone on pod development of peanut plants. Proc. Crop Sci. Soc. Jpn. 43:247–251.

Ong, C.K. 1984. The influence of temperature and water deficits on the partitioning of dry matter in groundnut (*Arachis hypogaea* L.). J. Exp. Bot. 35:746–755.

Ong, C.K. 1986. Agroclimatological factors affecting phenology of groundnut. p. 115–125. *In* Agrometeorology of groundnut. Proc. Int. Symp. ICRISAT Sahelian Ctr., Niamey, Niger. 21–25 Aug. 1985. Int. Crops Res. Inst. for the Semi-Arid Tropics, Patencheru, A.P. India.

Ong, C.K., C.R. Black, L.P. Simmonds, and R.A. Saffell. 1985. Influence of saturation deficit on leaf production and expansion in stands of groundnut (*Arachis hypogaea* L.) grown without irrigation. Ann. Bot. 56:523–536.

Ong, C.K., L.P. Simmonds, and R.B. Matthews. 1987. Responses to saturation deficit in a stand of groundnut (*Arachis hypogaea* L.). 2. Growth and development. Ann. Bot. 59:121–128.

Pahalwan, D.K., and R.S. Tripathi. 1984. Irrigation scheduling based on evaporation and crop water requirement for summer peanuts. Peanut Sci. 11:4–6.

Pallas, J.E., Jr., J.R. Stansell, and R.R. Bruce. 1977. Peanut seed germination as related to soil water regime during pod development. Agron. J. 69:381–383.

Pallas, J.E., Jr., J.R. Stansell, and T.J. Koske. 1979. Effects of drought on Florunner peanuts. Agron. J. 71:853–858.

Pandey, R.K., W.A.T. Herrera, and J.W. Pendleton. 1984a. Drought response of grain legumes under irrigation gradient: I. Yield and yield components. Agron. J. 76:549–553.

Pandey, R.K., W.A.T. Herrera, A.N. Villegas, and J.W. Pendleton. 1984b. Drought response of grain legumes under irrigation gradient: III. Plant growth. Agron. J. 76:557–560.

Pettit, R.E., R.A. Taber, H.W. Schroeder, and A.L. Harrison. 1971. Influence of fungicides and irrigation practice on aflatoxin in peanuts before digging. Appl. Microbiol. 22:629–634.

Pixley, K.V. 1985. Physiological and epidemiological characteristics of leafspot resistance in four peanut genotypes. M.S. thesis. Univ. Florida, Gainesville.

Rao, K.V., and N.P. Singh. 1985. Influence of irrigation and phosphorus on pod yield and oil yield of groundnut. Indian J. Agron. 30:139–141.

Reddi, G.H.S., and M.N. Reddy. 1977. Efficient use of irrigation water for wheat and groundnut. Mysore J. Agric. Sci. 11:22–27.

Reddy, V.M. 1978. The effects of irrigation, inoculants, and fertilizer nitrogen on nitrogen fixation and yield of peanuts (*Arachis hypogaea* L.). M.S. thesis. Univ. Guelph, Ont. Canada.

Reddy, V.M., and J.W. Tanner. 1980. The effects of irrigation, inoculants and fertilizer nitrogen on peanuts (*Arachis hypogaea* L.). I. Nitrogen fixation. Peanut Sci. 7:114–119.

Reddy, V.M., J.W. Tanner, R.C. Roy, and J.M. Elliot. 1981. The effects of irrigation, inoculants and fertilizer nitrogen on peanuts (*Arachis hypogaea* L.). II. Yield. Peanut Sci. 8:125–128.

Ritchie, J.T. 1974. Evaluating irrigation needs for southeastern USA. p. 262–279. *In* Proceedings contribution of irrigation and drainage to world food supply. ASCE, Biloxi, MS.

Ritchie, J.T., E.D. Rhoades, and C.W. Richardson. 1976. Calculating evapotranspiration from native grassland watersheds. Trans. ASAE 19:1098–1103.

Robertson, W.K., L.C. Hammond, J.T. Johnson, and K.J. Boote. 1980. Effects of plant-water stress on root distribution of corn, soybeans, and peanuts in sandy soil. Agron. J. 72:548–550.

Rochester, E.W., P.A. Backman, S.C. Young, and J.M. Hammond. 1978. Irrigation policies for peanut production. Alabama Agric. Exp. Stn. Circ. 241.

Rodrigues, T.J.D. 1984. Drought resistance mechanisms among peanut genotypes. Ph.D. diss. Univ. Florida, Gainesville (Diss. Abstr. 45:2746-B).

Samples, L.E. 1981. Peanut irrigation in Georgia. Univ. Georgia Circ. 685.

Sanders, T.H., P.D. Blankenship, R.J. Cole, and R.A. Hill. 1985a. Temperature relationships of peanut leaf canopy, stem, and fruit in soil of varying temperature and moisture. Peanut Sci. 12:86–89.

Sanders, T.H., R.J. Cole, P.D. Blankenship, and R.A. Hill. 1985b. Relation of environmental stress duration to *Aspergillus flavus* invasion and aflatoxin production in preharvest peanuts. Peanut Sci. 12:90–93.

Sanders, T.H., A.M. Schubert, and H.E. Pattee. 1982. Maturity methodology and postharvest physiology. p. 624–654. *In* H.E. Pattee and C.T. Young (ed.) Peanut science and technology. Am. Peanut Res. and Educ. Soc., Yoakum, TX.

Schenk, R.U. 1961. Development of the peanut fruit. Georgia Agric. Exp. Stn. Tech. Bull. New Ser. 22.

Shimshi, D., J. Schiffmann, Y. Kost, H. Bielorai, and Y. Alper. 1967. Effect of soil moisture regime on nodulation of inoculated peanuts. Agron. J. 59:397–400.

Simmonds, L.P., and C.K. Ong. 1987. Responses to saturation deficit in a stand of groundnut (*Arachis hypogaea* L.). 1. Water Use. Ann. Bot. 59:113–119.

Sivakumar, M.V.K., and P.S. Sarma. 1986. Studies on water relations of groundnut. p. 83–98. *In* Agrometeorology of groundnut. Proc. Int. Symp. ICRISAT Sahelian Ctr., Niamey, Niger. 21–25 Aug. 1985. Int. Crops Res. Inst. for the Semi-Arid Tropics, Patencheru, A.P., India.

Skelton, B.J., and G.M. Shear. 1971. Calcium translocation in the peanut (*Arachis hypogaea* L.). Agron. J. 63:409–412.

Slack, T.E., and L.G. Morrill. 1972. A comparison of a large-seeded (NC2) and a small-seeded (Starr) peanut (*Arachis hypogaea* L.) cultivar as affected by levels of calcium added to the fruit zone. Proc. Soil Sci. Soc. Am. 36:87–90.

Slatyer, R.O. 1955. Studies on the water relations of crop plants grown under natural rainfall in northern Australia. Aust. J. Agric. Res. 6:365–377.

Smith, B.W. 1954. *Arachis hypogaea*. Reproductive efficiency. Am. J. Bot. 41:607–616.

Smith, J.W., Jr., and C.S. Barfield. 1982. Management of preharvest insects. p. 250–325. *In* H.E. Pattee and C.T. Young (ed.) Peanut science and technology. Am. Peanut Res. and Educ. Soc., Yoakum, TX.

Stansell, J.R. 1975. Peanut irrigation. A science rather than an art. Southeast. Peanut Farmer 13:12–20.

Stansell, J.R., and J.E. Pallas, Jr. 1985. Yield and quality response of Florunner peanut to applied drought at several growth stages. Peanut Sci. 12:64–70.

Stansell, J.R., J.L. Shepherd, J.E. Pallas, Jr., R.R. Bruce, N.A. Minton, D.K. Bell, and L.W. Morgan. 1976. Peanut responses to soil water variables in the Southeast. Peanut Sci. 3:44–48.

Stern, W.R. 1968. The influence of sowing date on the yield of peanut in a short summer rainfall environment. Aust. J. Exp. Agric. Anim. Husb. 8:594–598.

Stone, J.F., P.I. Erickson, and A.S. Abdul-Jabbar. 1985. Stomatal closure behavior induced by row spacing and evaporative demand in irrigated peanuts. Agron. J. 77:197–202.

Stone, J.F., G.N. McCauley, E.W. ChinChoy, and H.E. Reeves. 1976. Evapotranspiration reduction by field geometry effects. Final Tech. Compl. Rep., Oklahoma Water Resourc. Res. Inst., Stillwater.

Su, K.C., T.R. Chen, S.C. Hsu, and M.T. Tseng. 1964. Studies on the processing of water absorption and economized irrigation of peanuts. (In Chinese, English summary.) J. Agric. Assoc. China 45:31–40.

Subramanian, S., S.D. Sundarsingh, K.P. Ramaswamy, S.P. Packiaraj, and K. Rajagopalan. 1975. Effect of moisture stress at different growth stages of groundnut. Madras Agric. J. 62:587–588.

Texas Agricultural Extension Service. 1985. Peanut production guide for Texas. B-1514. Texas A&M Univ., College Station.

Tsuno, T. 1975. The influence of transpiration upon the photosynthesis in several crop plants. Proc. Crop Sci. Soc. Jpn. 44:44–53.

Underwood, C.V., H.M. Taylor, and C.S. Hoveland. 1971. Soil physical factors affecting peanut pod development. Agron. J. 63:953–954.

Varnell, R.J., H. Mwandemere, W.K. Robertson, and K.J. Boote. 1976. Peanut yields affected by soil water, no-till, and gypsum. Proc. Soil Crop Sci. Soc. Fla. 35:56–59.

Venkaiah, K. 1985. Effect of bulk density on growth and yield of groundnut. Indian J. Agron. 30:278–280.

Vivekanandan, A.S., and H.P.M. Gunasena. 1976. Lysimetric studies on the effect of soil moisture tension on the growth and yield of maize (*Zea mays* L.) and groundnut (*Arachis hypogaea* L.). Beitr. Trop. Landwirtsch. Veterinaermed. 14:369–378.

Walker, M.. 1975. Calcium requirements for peanut. Commun. Soil Sci. Plant Anal. 6:299–313.

Wiersum, L.K. 1951. Water transport in the xylem as related to calcium uptake by groundnuts (*Arachis hypogaea* L.). Plant Soil 3:160–169.

Williams, E.J., and J.S. Drexler. 1981. A non-destructive method for determining peanut pod maturity. Peanut Sci. 8:134–141.

Williams, J.H., R.C. Nageswara Rao, R. Matthews, and D. Harris. 1986. Responses of groundnut genotypes to drought. p. 99–111. *In* Agrometeorology of groundnut. Proc. Int. Symp. ICRISAT Sahelian Ctr., Niamey, Niger. 21–25 Aug. 1985. Int. Crops Res. Inst. for the Semi-Arid Tropics, Patancheru, A.P. India.

Wilson, D.M., and J.R. Stansell. 1983. Effect of irrigation regimes on aflatoxin contamination of peanut pods. Peanut Sci. 10:54–56.

Wormer, T.M., and R. Ochs. 1959. Humidite du sol, ouverture des stomates et transpiration du palmier a huile et l'arachide. Oleagineux 14:571–580.

Wright, F.S., D.M. Porter, N.L. Powell, and B.B. Ross. 1986. Irrigation and tillage effects on peanut yield in Virginia. Peanut Sci. 13:89–92.

Wynne, J.C., and T.A. Coffelt. 1982. Genetics of *Arachis hypogaea* L. p. 50–94. *In* H.E. Pattee and C.T. Young (ed.) Peanut science and technology. Am. Peanut Res. and Educ. Soc., Yoakum, TX.

Yarbrough, J.A. 1949. *Arachis hypogaea*. The seedling, its cotyledons, hypocotyl, and roots. Am. J. Bot. 36:758–772.

Yayock, J.Y. and J.J. Owonubi. 1986. Weather-sensitive agricultural operations in groundnut production: The Nigeria situation. p. 213–226. *In* Agrometeorology of groundnut. Proc. Int. Symp. ICRISAT Sahelian Ctr., Niamey, Niger. 21–25 Aug. 1985. Int. Crops Res. Inst. for the Semi-Arid Tropics, Patancheru, A.P., India.

Zamski, E., and M. Ziv. 1976. Pod formation and its geotropic orientation in the peanut, *Arachis hypogaea* L., in relation to light and mechanical stimulus. Ann. Bot. 40:631–636.

23   Sorghum

DANIEL R. KRIEG

Texas Tech University
Lubbock, Texas

ROBERT J. LASCANO

Texas A&M University
Lubbock, Texas

Sorghum (*Sorghum bicolor* L. Moench) originated in, is well adapted to, and is primarily grown in the semiarid regions of the world. It has long been recognized as being drought tolerant compared with other major cereal crops and therefore is well suited to semiarid conditions characterized by: (i) insufficient water supply to meet the evaporative demand, (ii) uneven seasonal distribution of precipitation, and (iii) high year-to-year variation in rainfall and surface water supplies.

The land area devoted to sorghum and the per hectare grain yields of the major sorghum-producing countries of the world are listed in Table 23-1. Yield variation among countries reflects the effects of adoption of new production technology as well as the impact of environmental constraints on yield.

Leng (1982) recently analyzed yield trends of the major cereal crops over a 40-yr period (1945–1985) for various regions of the world. Sorghum grain yields have remained static or slightly declined for Africa, remained static with a tendency to increase for Asia, and increased rapidly between 1950 and 1970 but have remained relatively static since then for Mexico and the USA. A comparison of yield trends for the major cereal crops grown in the USA indicates that only maize (*Zea mays* L.) has exhibited a steady increase in mean grain yield over this same period (Fig. 23-1). Maize yield improvement reflects a combination of genetic gain and utilization of improved management systems. Grain yield of sorghum increased dramatically with the introduction of hybrids in the mid-1950s. Yields more than doubled from 1955 to 1970 but remained essentially static from 1970 to 1985. A large year-to-year variation exists, emphasizing the location of sorghum in marginal production environments. Wheat yields steadily increased over this same 40-yr period but at a relatively slow rate of gain. The static yield of sorghum and wheat can largely be explained by the concentrated production in the Great Plains states where rainfall is usually inadequate to meet the evaporative demand and plant water stress is the major yield-limiting factor for all rainfed

Table 23-1. Land area used and per-hectare yields of the major sorghum-producing areas of the world (FAO, 1981–1983.)

Location	Land area	Grain yield	World area	World production
	1000 ha	kg ha^{-1}	%	
World	47 565	1 433	--	--
India	16 403	706	34.5	17.0
Africa	15 528	673	32.6	15.3
USA	5 108	3 598	10.7	27.0
S. America	3 052	3 349	6.4	15.0
PRC	2 734	3 175	5.8	12.7
Mexico	1 646	3 540	3.5	8.5
Australia	668	1 754	1.4	1.7
USSR	141	915	0.3	0.2
			95.2%	97.4%

crop production. The four Great Plains states of Texas, Oklahoma, Kansas, and Nebraska comprise more than 80% of the total U.S. sorghum production (Table 23–2). However, only 17% of the U.S. sorghum production occurs under irrigation, with almost 90% of the irrigated sorghum area existing in the Great Plains (USDA, 1986).

Grain yields measured on the Texas High Plains, where the majority of the state's sorghum production exists, reveals the significance of supplemental irrigation in this production area (Fig. 23–2). Average irrigated

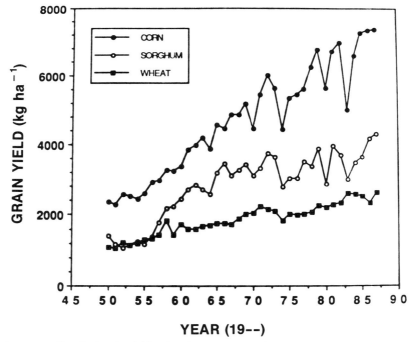

Fig. 23-1. Historical grain yields of major cereal crops in the USA (USDA, 1986).

Table 23-2. Cropland and sorghum distribution in the Great Plains states relative to total U.S. cropland and sorghum production (USDA, 1986).

Location	Crop production area			Sorghum production area		
	Total	Irrigated	Irr./Total	Total	Irrigated	Irr./Total
	——— 10⁶ ha ———		%	——— 10⁶ ha ———		%
USA	173.98	22.27	12.8	5.76	0.99	17.3
Nebraska	9.46	3.64	38.5	0.75	0.09	12.0
Kansas	12.89	1.59	12.3	1.45	0.29	20.0
Oklahoma	4.97	0.40	8.0	0.21	0.04	19.0
Texas	11.80	3.59	30.4	2.19	0.44	20.0

yields are two to three times the mean rainfed (dryland) yields. In very dry years (e.g., 1974, 1980, and 1983) the yield advantage of irrigated sorghum production exceeds the dryland yields by more than fourfold.

The genetic potential for grain yield is at least two and probably closer to five times the average yield realized at the farm level. Small plot grain yields in excess of 20 000 kg ha^{-1} have been reported (Boyer, 1982). Sorghum performance tests conducted annually by the various State Agricultural Experiment Station and Extension Service personnel routinely include hybrids that yield in excess of 10 000 kg ha^{-1} grain under normal management. Grain yield results from the hybrid sorghum perform tests conducted at Lubbock, TX (Pietsch et al., 1981–1985) reveal that the average grain yields from the trials are 40 to 60% greater than the irrigated yields from the county statistics in the area (Table 23–3). Secondly, maximum yields in the trials approach 10 000 kg ha^{-1} in nearly every year, implying that the genetic

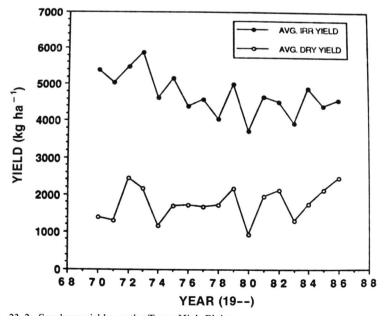

Fig. 23-2. Sorghum yields on the Texas High Plains.

Table 23-3. Yield of irrigated sorghum performance tests on the Texas High Plains (Pietsch et al., 1981-1985).

Year	Number of entries	Grain yield		High plains Average
		Average	Maximum	
			kg ha^{-1}	
1981	61	6 829	8 509	3 951
1982	65	8 378	9 465	3 452
1983	45	6 914	8 370	2 992
1984	36	9 845	10 872	4 072
1985	64	8 164	9 428	3 384

potential for yield is not being challenged in current production systems, but rather inadequate production resources are responsible for the resultant yields. The stability of yield in U.S. sorghum production during the past 15 yr (Fig. 23-1) reflects environmental constraints or management limitations to realization of more of the inherent genetic yield potential.

I. YIELD COMPONENT DEVELOPMENT

To capitalize on more of the genetic yield potential, it is imperative to understand the relative importance of each component to total grain yield. It is then important to understand how that component responds to various environmental factors such as water supply, radiant energy, and thermal energy. The yield of sorghum, similar to other seed-producing plants, is largely determined by the number of seed produced per unit land area with only a minor contribution from the weight or size variation of the individual seed (Table 23-4).

The seed number component is comprised of the number of harvested panicles per unit land area and the number of seed per panicle (Table 23-4) with the seed per panicle component being the more critical of the two. The number of harvested panicles per unit land area is largely a function of seeding rate and seedling survival. Plant population is more critical to yield attainment under irrigated conditions than under dryland or limited water supply conditions, as indicated by the magnitude of the regression coefficient and the partial R^2 (Table 23-4). However, sorghum, like most *Graminae* species, is capable of producing basal tillers under certain conditions which can alter the number of panicles produced relative to the seeding rate. In general, conditions that favor the accumulation of assimilate in excess of the growth demand by the main shoot favor tiller initiation and survival, and include low initial plant density, high radiation levels, and relatively cool air temperatures (Downes, 1968, 1972). If photosynthate production is only adequate to support the growth of the main shoot, tiller production will be arrested and tiller survival is low unless the tillers are essentially self-supporting. Tillering can be beneficial or detrimental depending on the water supply (Blum & Naveh, 1976; Hedge et al., 1976; Heslehurst, 1983; Karachi & Rudich, 1966; Myers & Foale 1981; Steiner, 1986; Stickler & Wearden,

Table 23–4. Multiple regression analyses of yield and yield components of sorghum. Data were collected over a 5-yr period using 16 hybrids grown under irrigated and dryland conditions each year (D.R. Krieg, 1976, unpublished data).

Analyses	Partial R^2	Total R^2
Pooled		
Grain yield (kg ha^{-1}) = -7516 + 3.79 (seed panicle^{-1})	0.59	
+ 0.028 (panicle ha^{-1})	0.28	
+ 130.1 (g 10^3 seed^{-1})	0.07	
		0.94
Irrigated		
Grain yield (kg ha^{-1}) = -9967.6 + 3.92 (seed panicle^{-1})	0.40	
+ 0.030 (panicle ha^{-1})	0.32	
+ 178.73 (g 10^3 seed^{-1})	0.23	
		0.95
Dryland		
Grain yield (kg ha^{-1}) = -4309.4 + 3.12 (seed panicle^{-1})	0.57	
+ 0.015 (panicle ha^{-1})	0.12	
+ 96.79 (g 10^3 seed^{-1})	0.12	
		0.81

1965). Under irrigated conditions, tillering can compensate for low primary plant density, however, the yield from tiller panicles is usually less than from primary panicles. Under rainfed conditions, restricted basal tillering and control of resultant leaf area is especially critical. Water use rates by the crop must be controlled to extend the stored soil water supply through the critical growth stages and to minimize plant water stress. Maximum allowable water use rates will depend upon the soil water storage capacity, the evaporative demand, and the probability of rainfall during critical developmental periods. Axillary tillering can be beneficial under conditions where the primary panicles have been severely damaged by early season stress and rains in mid- to late-season have relieved the stress. Yields will be low and maturity highly erratic but at least some grain will be produced from axillary tillers.

Within a given plant population, grain yields are primarily determined by the number of seed per panicle (Table 23–4) which is largely determined by the conditions that exist during the middle one-third of the life cycle occurring from panicle initiation until flowering (Fig. 23–3). The panicle differentiation period (commonly referred to as growth stage 2, GS-2) begins with the transition of the apical meristem from a vegetative to a reproductive structure and terminates with flowering. Various components of the total panicle differentiation and development process exhibit differing sensitivities to environmental stress conditions, resulting in variations in seed number per panicle. (Eastin, 1981; Krieg, 1983; Jordan & Sullivan, 1982). The floret differentiation process which occurs 15 to 22 d after panicle initiation appears to be the single most sensitive process to water or heat stress. Eastin et al. (1975) reported yield reductions of 25 to 36% from heat stress in sorghum held 5 °C above "near optimum" air temperature at night during GS-1 and GS-2. Ogunlela (1979) found that by raising the night temperature

DEVELOPMENTAL SEQUENCE OF SORGHUM

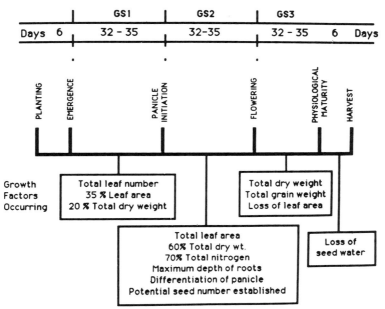

Fig. 23-3. Sorghum developmental sequence.

just 5 °C above ambient, a 28% reduction in seed number and a 30% reduction in yield occurred with the most sensitive period being 2 to 3 wk after panicle initiation. Reduction in assimilate supply due to shading, independent of water or heat stress, resulted in similar reductions in seed number when the stress was imposed during the floret differentiation period (Bennett, 1975). Blasted or sterile florets are the visible symptoms of stress during this period of development (Rosenow et al., 1983). Stress during the panicles' exsertion period (boot stage) can result in reduced seed numbers at the base of the panicle because of incomplete exsertion from the flag leaf sheath, and thus, lack of opportunity for pollination (Lewis et al., 1974). Plant density and genetic variation are other factors affecting the degree of panicle exsertion due to water stress. High plant populations are more likely to exhibit poor exsertion characteristics within a given hybrid. Management systems must be designed to ensure that the critical panicle development period is allowed to occur with the least likelihood for environmental stress, if maximum yield is to be attained. (Hiler & Howell, 1983).

Seed size (g 1000 kernels^{-1}) varies among genotypes, ranging from <20 g 1000 kernels^{-1} to over 30 g 1000 kernels^{-1} and contributes 10 to 30% to total grain yield variation (Table 23-4). Within a genotype, the variation in seed weight is relatively small (<20%) because of environmental conditions prevalent during the grain-filling period. Seed size normally in-

Table 23-5. Relative contribution of preanthesis assimilate to grain weight of senescent and nonsenescent hybrids in response to water supply (Harden, 1985).

Genotype	Dry matter		Nitrogen	
	Irr.	Dry	Irr.	Dry
		%		
Senescent	9.9	45.0	48.5	75.0
Nonsenescent	2.7	17.7	20.8	35.0

creases when seed numbers have been reduced by stress prior to flowering, however, the increased size rarely offsets the reduced seed numbers, resulting in yield losses. Significant reductions in seed size occur only when severe conditions such as an early freeze, insect infestation, or disease infections such as charcoal rot or *Fusarium* spp. exist prior to physiological maturity of the grain.

Grain mass, which represents approximately 40 to 45% of the total phytomass at physiological maturity, is largely derived from photosynthate produced during GS-3, with <20% of the total grain mass derived from assimilate stored in the plant prior to anthesis under well-watered conditions (Fischer & Wilson, 1971; Harden, 1985). Water stress during GS-3 results in a greater contribution of preanthesis assimilate to final grain mass with differences in the magnitude of the contribution among hybrids due to genetically controlled leaf senescence characteristics (Table 23–5). A larger percentage of the grain N is derived from preanthesis assimilate than is grain dry matter (Table 23–5). Grain mass accumulation uses a declining leaf area and a deteriorating photosynthetic system per unit leaf area. The rate of decline in leaf area and photosynthetic rate is subject to genetic variation (Phipps, 1987; Kidambi, 1987) and is exaggerated by environmental stresses, especially water stress.

It is apparent that grain yield per unit land area is largely dependent upon the number of seed produced. Plant population is largely influenced by seeding rate with modification because of degree of tillering. The seed per panicle component then becomes the major affector of grain yield. Seed per panicle is largely influenced by environmental conditions prevalent during the period from panicle initiation to flowering with differing degrees of sensitivity of various developmental processes during this entire time span. The reader is referred to excellent descriptions of sorghum growth and development provided by Vanderlip and Reeves (1972), Eastin (1981), and Wilson and Eastin (1982) and to the relationships between growth and development and plant water relations (Krieg, 1983). A thorough understanding of the yield components responsible for yield variation, and the developmental period in the life of the plant when each component is being established is essential to develop management systems to use existing environmental resources with the greatest efficiency and to minimize the opportunity for significant plant stress.

II. GRAIN YIELD—EVAPOTRANSPIRATION RELATIONSHIPS

The relationship between grain yield or total phytomass and seasonal evapotranspiration (ET) is essentially linear for sorghum, as has been shown for other crops (Garrity et al., 1982b; Hanks, 1983; Inuyama et al., 1976; Kanemasu, 1983; Stewart et al., 1975; Stone et al., 1978). Stewart (1985) summarized yield responses of sorghum to seasonal ET for the southern Great Plains (Bushland, TX) over a period of 26 yr and developed the relationship shown in Fig. 23–4. A minimum of about 150 mm ET was required to produce any grain, with yield increasing at a rate of 15.5 kg ha^{-1} mm^{-1} above the minimum requirement.

A major concern of sorghum research during the past 10 to 15 yr has been the question "Can the slope of the yield-ET relationship be altered by water management and/or genetics?" The most comprehensive report addressing this question is that of Garrity et al. (1982b) in which they evaluated yield-ET relationships of three sorghum hybrids as a function of water supply during each of the three major growth stages. A modified line-source sprinkler irrigation system was used such that the water supply during each growth stage could be varied from 0 to 100% ET replacement. Their GGG treatment involved gradient irrigation from 0 to 100% ET replacement across all three growth stages. The GGI and GII treatments represented a gradient during GS-1 and either a gradient (GGI) or full irrigation (GII) during GS-2, with complete replacement of ET during GS-3. During the course of the 2-yr experiment, they observed a curvilinear, rather than a linear, response between grain yield and ET replacement, with maximum grain yield being at-

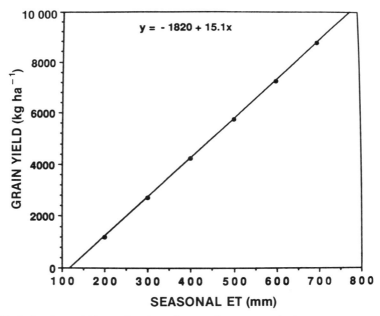

Fig. 23–4. Sorghum yield as a function of seasonal evapotranspiration.

Table 23-6. Grain yield as a function of total water supply and growth stage sensitivity (adapted from Garrity et al., 1982a, b).

Water supply	Irrigation treatment			
	GGG	GGI	GII	IIG
mm		kg ha^{-1}		
400	5 438†	5 068	4 613	4 830
500	7 085	7 451	8 073	7 016
600	8 732	9 835	11 237	9 236

† Mean of three hybrids.

tained with 80% ET replacement rather than 100%. The yield reduction between 80 and 100% ET replacement was due to fewer seed per panicle. Lack of adequate aeration was ruled out as a causative factor due to the sandy texture of the soil. Similarly, leaching of the nutrient supply in the 100% ET was ruled out because the 80% ET replacement reflected less water available than was being replaced, implying drainage was not occurring. No explanation for the curvilinear response was provided—just emphasizing the fact that water supply is but one factor required in the total production picture. As the water supply limitation is removed some other environmental component becomes the limiting factor to continued productivity. When the linear portion of grain yield-ET relationship was used, a mean slope of 16.5 kg ha^{-1} mm^{-1} was obtained, which compares favorably with that reported by Stewart (1985) using long-term data. However, the slope differed among the three hybrids ranging from 11.8 to 19.2 kg ha^{-1} mm^{-1} when the water supply gradient was imposed over the entire life cycle (GGG) and differed drastically due to growth stage-water supply interactions. The largest slopes were associated with ample water supplies during GS-2 and GS-3 as compared to those obtained from a gradient in water supply across the entire life cycle (GGG).

Mean yield data for the three hybrids as a function of the water supply-growth stage relationship are depicted in Table 23-6. A minimum water supply of 150 to 200 mm was required to produce a harvestable yield, similar to the results of Stewart's (1985) analyses. When the total water supply was approximately 400 mm the GGG and GGI treatments were superior for the three hybrids. These results reflect the importance of a continuous water supply even though not totally adequate to replace ET. The crop suffers short-term stress periods but does have a chance to recover. The GGI treatment reflects the need for water during grain filling to minimize lodging. When the water supply exceeded 400 mm, the GII and GGI treatments resulted in the greatest grain yield per unit additional water supply. The grain return was 16.0 kg ha^{-1} mm^{-1} for the GGG vs. 24.0 kg ha^{-1} mm^{-1} for the GGI and GII treatments. The GII treatment had the greatest rate of increase when the water supply exceeded 500 mm with a production rate of 32.0 kg ha^{-1} mm^{-1}. The yield response to water during GS-2 and GS-3 emphasize the critical nature of these periods in the life of the plant in terms of seed production.

When the total water supply is not adequate to meet the evaporative demand, management decisions of when to apply the water make large differences in the yield response. It is apparent that GS-2 is extremely sensitive to environmental stress, resulting in large reductions in the seed number component of yield. Managing the water supply to minimize the opportunity for plant water stress during this critical growth stage is essential for maximum water use efficiency. Water supply during this critical growth stage can be ensured by choice of planting date coupled with rainfall probabilities, or irrigation scheduling dependent upon soil water supply-evaporative demand relationships.

Significant genetic variation also existed among the three hybrids in grain yield per unit water supply (Garrity et al., 1982a). For instance, at the crossover point of 450 mm total water, the early maturity class hybrid, NB505, produced about 5000 kg ha^{-1}, whereas the medium-maturity hybrids RS626 and NC + 55X produced grain yields in excess of 6500 kg ha^{-1} with RS626 slightly exceeding NC + 55X. The fact that significant genetic variation existed in the grain yield-ET relationships points out the need to select the best-adapted hybrids (maturity class) and seeding rate to match the available water supply to maximize both yield and water use efficiency.

The results of this type of grain yield response to water supply and timing of irrigation are extremely important in water management for maximum yield within the limits of the available water supply. Although the yield-ET relationships can change due to atmospheric demand differences, the response to water supply at various growth stages will be consistent. Depending upon the total water resource available, various strategies need to be implemented to maximize yield per unit available water as evidenced in the different responses of the timing of stress occurrence. It is obvious that ample water supplies during GS-2 are the most critical.

III. WATER USE—EVAPOTRANSPIRATION RATES

The preceding chapters of this treatise define the interaction of environmental factors determining evaporative demand, plant factors interacting with the environment to determine transpiration rates, and soil factors that determine total water supply available to the crop. Daily water use rates of a sorghum crop are dependent upon leaf area development and distribution up to complete canopy closure which occurs between leaf area indices (LAI's) of 2 to 3 depending on distance between rows (Ritchie, 1972). Prior to complete canopy closure, a significant proportion of the total water loss can be from soil evaporation depending upon the degree of surface wetness. Leaf area development is relatively slow during GS-1 with only 25 to 30% of maximum leaf area present at the end of GS-1 (Fig. 23-3). However, leaf area development is extremely rapid during GS-2, reaching a maximum at the time of flowering. Maximum leaf area is maintained for 10 to 14 d into grain filling and then leaf senscence begins reducing existing LAI. The rate and extent of senescence is dependent upon both genotype (senescent vs. nonsenescent types) and degree of water stress.

The daily rate of water use, therefore, is dependent upon evaporative demand, LAI, soil evaporation, and soil water supply (Chaudri & Kanemasu, 1985; Kanemasu et al., 1976; Steiner, 1986) and is a dynamic function well defined in several chapters of this book.

Our purpose in this section is to demonstrate the use of a water balance model hereafter referred to as ENWATBAL to quantitatively describe ET of a field-grown sorghum crop subject to varying soil water supplies. van Bavel and Ahmed (1976) initially proposed the model and it was subsequently tested by van Bavel et al. (1984) for a sorghum crop in a subhumid environment. It was demonstrated that calculated ET over a 50-d period was not significantly different from the measured values and soil water content profiles could be accurately calculated. Lascano et al. (1987) subsequently modified the model to include soil heat flux and tested the model over a 74-d period in the life of a cotton (*Gossypium hirsutum* L.) crop at Lubbock, TX (a semiarid environment). The predicted values for ET, soil evaporation, and soil profiles of temperature and water content were not significantly different from measured values. The distinguishing characteristics of this model are that the energy and water balances of the entire soil profile are calculated separately from the crop canopy, making it possible to find the two components of ET. (Lascano & van Bavel, 1986).

Inputs to the ENWATBAL model include soil and crop variables and daily weather data. Soil inputs are the soil-water retention curve, the unsaturated hydraulic conductivity, and the number and thickness of the soil layers. Crop inputs are the relation between leaf conductance and leaf water potential, root distribution as a function of time and depth, and LAI as a function of time. Weather inputs are daily solar radiation, daily maximum and minimum air and dewpoint temperatures, daily windspeed, and the quantity of rain or irrigation. In addition, initial values of the water and temperature profile must be known at the start of the simulation.

To illustrate the use of the ENWATBAL model, the components of the water balance equation were simulated for a sorghum crop for the climatic conditions at Brownfield, TX, which is typical of the semiarid southern Great Plains. The simulated period was between 3 July 1985 (calendar day 184) and 5 Sept. 1985 (calendar day 248). The simulated period encompassed all of GS-2 and essentially all of GS-3. Input and initial conditions for the simulation were taken from unpublished data of (D.R. Krieg) generated during the 1985 growing season at this location.

The soil at the site is classified in the Amarillo series (fine-loamy, mixed, thermic Aridic Paleustalfs). Sorghum [cv (ATx 623) × (RTx 430)] was planted in 0.67-m rows with a population density of 280,000 plants ha^{-1} at emergence. The hydraulic conductivity was calculated using an estimated value of 2×10^{-6} m s^{-1} for the saturated hydraulic conductivity (Jackson, 1972). The weather variables used as model inputs typically were 25 MJ m^{-2} for the total daily solar radiation, 35 °C for the maximum daily air temperature, 19 °C for the daily minimum air temperature, 17 °C for the daily maximum dewpoint temperature, 7 °C for the daily minimum dewpoint temperature, and 3 ms^{-1} for the daily average windspeed. A total of 53 mm of rain fell over the simulation period.

The daily values of LAI used as input were estimated from weekly measurements of leaf area development made over the course of the growing season.

The simulated daily values of crop transpiration based upon measured leaf area index and rain are given in Fig. 23–5a and 23–5b for the irrigated

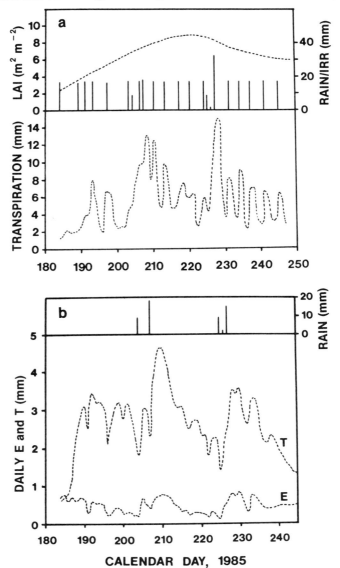

Fig. 23–5. Daily transpiration rate of irrigated (a) and dryland (b) sorghum over a 70-d period beginning 30 d after planting.

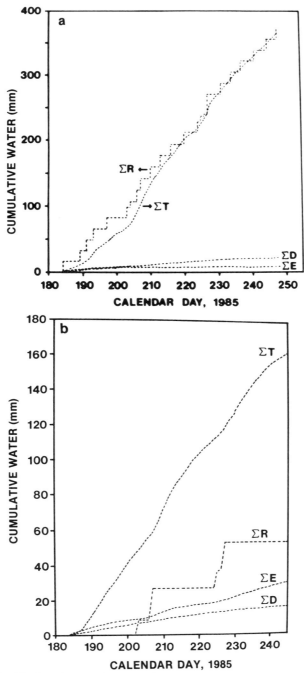

Fig. 23-6. Partitioning of seasonal water use into each component of supply and demand for an irrigated (a) and dryland (b) sorghum crop over a 70-d period beginning 30 d after planting (R = total water supply, T = transpiration, E = evaporation, D = deep drainage).

and dryland treatments, respectively. The daily pattern of water use illustrates the dynamic nature of water use by a sorghum crop under varying levels of soil water content. Under well-irrigated conditions, the calculated daily rate of transpiration ranged from a low of 1.3 mm d^{-1} to a high of 15.0 mm d^{-1} following 60 mm of water added to the profile by a combination of rainfall and irrigation. The daily rate of transpiration for the dryland crop was considerably less. Following a rain of 19.0 mm (i.e., calendar day 207), the daily value of crop transpiration increased from 2.5 to 4.5 mm d^{-1} and then gradually decreased to 1.5 mm d^{-1} until another rainfall event occurred. Under dryland conditions, throughout most of the growing season the crop extracted water from deeper and wetter layers of the soil profile, and was dependent upon the growing root system to mine this water supply.

The different components of the water balance for this sorghum crop for the simulated period are given in Fig. 23–6a and 23–6b for the irrigated and dryland treatments, respectively. For the irrigated treatment, the total amount of crop transpiration was 365 mm from days 184 to 246, soil evaporation 8 mm, drainage below the root zone 23 mm, and rainfall and irrigation 373 mm. This gives a water balance of −23 mm, equal to the sum of the inputs (373 mm) to the sum of the outputs (396 mm). The negative sign indicates that the losses were greater than the gains. In this case the loss of water by ET, mainly due to transpiration, was equal to the amount of water added by irrigation and rainfall. The net loss of 23 mm was by drainage below the root zone in the soil profile. The simulated value of ET was within 5% of the average measured value (355 mm), estimated from the profiles of volumetric water content at the beginning and end of the simulated period.

For the dryland treatment, the total amount of crop transpiration for the simulation period was 160 mm, soil evaporation 30 mm, drainage out of the root zone 16 mm, and rainfall 53 mm. This gives a water balance of −153 mm, equal to the sum of the inputs (53 mm) to the sum of the outputs (206 mm). Again the negative sign indicates that the losses were greater than the gains. The simulated value of ET was within the variance of the average measured value, estimated from the profiles of volumetric water content at the beginning and end of the simulated period.

In summary, it is shown that the components of a water balance equation for a sorghum crop can be calculated using a dynamic simulation model that describes the processes of crop transpiration, soil evaporation, and the hydrological balance of the root zone. The rate of water use by the crop is controlled by a set of interacting factors that cannot be described by a simple formula. Rather it is necessary to maintain and continuously update a calculated water balance that accounts for the daily pattern of weather events, the growth of the crop, and the hydraulic properties of the root zone. The model used employs a mechanistic method that does not require empiricisms derived from locally measured data and is generally applicable.

IV. WATER MANAGEMENT FOR MAXIMUM YIELD
AND WATER USE EFFICIENCY

Sorghum is usually grown under conditions of less than optimum water supply, and crop water stress represents the major yield-limiting factor. Even when irrigation is available, the volume applied to supplement rainfall is usually less than that required for maximum ET and grain yield. Irrigation frequency and replacement volume are dependent upon the soil water volume that can be extracted before plant water stress causes growth rate reductions. The extractable soil volume is a function of soil texture, rooting depth and density, and atmospheric demand. The rooting depth and root length density change with plant growth. In general, approximately 50% of the volumetric water between field capacity and permanent wilting point (-0.03 to -1.5 MPa) can be extracted from the root zone before measurable growth reduction is observed in sorghum (Sweeten & Jordan, 1987). High-frequency, low-volume irrigation usually results in increased loss of water by evaporation both during application and while the soil surface is wet (Howell & Hiler, 1975), especially prior to canopy closure. Greater application efficiency is achieved by applying larger volumes and adjusting the frequency to allow depletion of the non-growth limiting water supply before replacement (Musick & Dusek, 1971). This volume is unique to each soil type, growth stage, and evaporative demand and must be defined accordingly.

It is also quite apparent that stress during certain critical growth stages reduces yield to a greater extent than at other growth stages and managing the water supply becomes paramount in increasing the efficient use of the limited water resource. Numerous options including maturity class of the hybrid, planting date, plant density considerations, and nutrient supply are available to increase the opportunity to maximize grain yield, and thus water use efficiency, within the limits of the water supply.

Two major developmental processes must be considered in choosing a cultivar to match the water supply of the production system. The maturity range (early, medium, or late) determines the duration of the growth period, which coupled with the daily ET, determine the total water requirement for the crop. The duration of the growth period is a function of the thermal environment with each growth stage requiring a certain accumulation of heat units for completion (Arkin et al., 1976). Planting early in the growing season requires more days to complete each growth stage than planting later when the daily heat unit accumulation is higher. For instance, planting in early May at Lubbock, TX, requires 70 to 75 d to reach flowering for a medium maturity hybrid (Table 23–7), whereas planting in early June requires only 60 to 65 d. More importantly, the early June-planted sorghum produces more seed per panicle and less leaf area per plant than the early May-planted sorghum. The combination of less time and less leaf area with greater seed number results in higher grain yields and increased water use efficiency for the June-planted sorghum compared with the early May-planted sorghum.

Table 23–7. Effect of planting date on development and yield of sorghum hybrids on the Texas High Plains.†

Planting date	Grain yield		Seed number		LAI		DTF	
	Irr.	Dry	Irr.	Dry	Irr.	Dry	Irr.	Dry
	— kg ha^{-1} —		Seed panicle^{-1}					
Early May	4820	2100	1100	600	4.4	2.9	72	71
Mid-May	5500	1850	1250	640	3.9	2.6	66	65
Early June	6700	2600	1570	750	3.6	2.4	62	62
Mid-June	5500	2600	1450	700	3.2	1.8	62	62

† Mean of 4 hybrids ranging from early (RS610) to medium late [(ATx 378) × (RTx 7000)] over 2 yr. Plant population was relatively constant at ~25 plants m^{-2}. LAI = Leaf Area Index, DTF = Days to Flower.

In addition to these positive attributes of an early June planting, grain filling occurs when the temperature begins to decline in late August and September. The diminishing heat load reduces the daily maintenance respiration costs and therefore more of the current photosynthate is available for net growth and yield. Delaying planting until mid- to late-June results in grain filling occurring under less than adequate air temperatures (below 10 °C) in the fall and seed weights are reduced because of inadequate daily photosynthate. Therefore choice of a maturity class and planting date to maximize productive potential within the limits of the available water supply can result in increased yield and water use efficiency.

The second major developmental process that must be considered is tillering. Under limited water supply conditions, the rate of water usage must be regulated to maintain the available water supply in accord with the evaporative demand such that the opportunity for significant plant water stress is minimized. Control of the plant population (Bond et al. 1964; Fisher & Wilson 1975; Myers & Foale, 1981; Steiner 1986) and therefore, the existing LAI is essential to match the rate of water usage with the available water supply (stored soil water plus precipitation or irrigation). Tillering is not desirable under these conditions. The degree of tillering is genetically controlled but is also subject to environmental control. Planting when the mean daily air temperature exceeds 20 °C (end of May at Lubbock) reduces the extent of tillering and results in single-culm plants. Under optimum water supply conditions, yields increase with population up to 300 000 plants ha^{-1} and grain yields in excess of 12 000 kg ha^{-1} are possible (Fischer & Wilson, 1975; Hedge et al., 1976; Heslehurst, 1983; Heinrich et al., 1983; Muchow et al., 1982). Under limited water supply, a curvilinear response exists between grain yield and plant population with optimum plant populations dependent upon the total water supply and the timing of rainfall or irrigation events during the growing season (Bond et al., 1964; Plaut et al., 1969).

V. WATER-NUTRIENT INTERACTIONS

Although water supply usually represents that single-most important limitation to grain yield of sorghum, water alone does not control the yield response. An adequate nutrient supply must be available to allow optimum growth rates per unit water used. Nitrogen and P supplies normally represent potential yield limitations in semiarid sorghum production areas when water supply can be managed through irrigation (Eck & Musick, 1979). Based upon water use efficiency results for grain yield and total biomass, one can easily calculate the N and P requirements to maximize productivity per unit available water. Each 1000 kg of grain contains 15 kg of N and 2.5 kg of P (Roy & Wright, 1974). With a harvest index of 0.45, total biomass production would be 2200 kg per 1000 kg of grain. Using a water use efficiency estimate common to semiarid regions of 15.0 kg ha^{-1} mm^{-1} ET$_a$ above a threshold requirement of 150 to 200 mm, the N requirement would be 0.4 to 0.5 kg ha^{-1} mm^{-1} for optimum productivity. The P requirement would be 0.06 kg ha^{-1} mm^{-1}. The total N and P requirements should be provided prior to flowering. It is critical that not only adequate water but also nutrient supplies be available during GS-2 to maximize seed number.

VI. MANAGEMENT OF LIMITED IRRIGATION WATER SUPPPLY

Stewart et al. (1983) developed and tested a limited irrigation-dryland (LID) farming system for the conjunctive use of rainfall and a limited supply of irrigation water such as exists throughout most of the Great Plains sorghum production area. The system uses furrow irrigation for water application. The upper one-half of the field is treated as a fully irrigated section (Fig. 23–7). The next one-fourth is treated as a "tailwater runoff" section under limited irrigation. The lower one-fourth of the field is managed as a complete "rainfed" section. The total water supply to this section is derived from rainfall plus runoff from the wetter sections of the field. Plant populations are adjusted down the row to match the expected water supply within each section. Stewart (1983) even developed and described an automated seeding

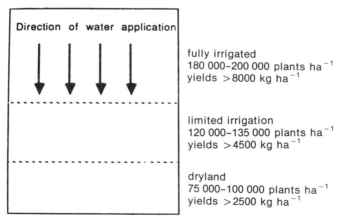

Fig. 23–7. Schematic of limited irrigation-dryland system for sorghum production. From Stewart et al. (1983).

rate selector to modify plant population during the planting process. The system was tested for 3 yr at Amarillo, TX, and successfully increased the utilization efficiency of the irrigation water. Grain yield was increased by 15.4 kg ha^{-1} mm^{-1} of added irrigation water as compared to 9.2 kg ha^{-1} mm^{-1} in the conventional irrigation system. The unique feature of the LID is that it allows an adjustment, during the crop growing season, of the amount of land that can be irrigated with a limited supply of water. During above-average rainfall years, more land can be watered than during below-average years. This water management system should work well for those production areas where furrow irrigation is practiced and the land area exceeds the water supply. The system minimizes the waste of irrigation water due to runoff and deep percolation and more effectively uses the rainfall when it occurs.

VII. USE OF CROP MODELS

The success (profitability) of a crop production system is dependent upon selecting the proper management strategies to maximize the growth rate and productivity within the existing climatic conditions. Developing management strategies using biological-physical based crop simulation models offers the opportunity for making decisions based upon quantitative information rather than trial and error or "educated" guess (Arkin & Dugas 1984; Hargreaves, 1984). A sorghum crop model has existed for about 10 yr and has been refined and revised to be more widely adapted (Arkin et al., 1976; Huda et al., 1984). Use of the model is limited only by the availability of weather data and the appropriate soil physical data specific for each site at which the model is to be used. With further testing and validation, the crop growth model should evolve into a valuable management tool for effective, efficient decision making on a day-to-day basis.

VIII. SUMMARY

Sorghum is noted for its ability to produce grain when subject to water and temperature stresses as compared with other cereals. However, sorghum does have certain developmental periods which are more sensitive to environmental stress than other periods, resulting in significant yield reductions and reduced water use efficiency. Management strategies can be developed to use the existing water resource more efficiently in production of sorghum. These options include choice of hybrid (maturity group), planting period, seeding rate, irrigation schedule, and fertility management. Use of crop simulation models in conjunction with the prevailing weather conditions and soil characteristics can be effectively used to increase production through proper management decisions and applications. Sorghum has the inherent ability to use water efficiently if managed properly.

REFERENCES

Arkin, G.F., and W.A. Dugas, Jr. 1984. Evaluating sorghum production strategies using a crop model. p. 289-295. *In* S.M. Virmani (ed.) Agrometeorology of sorghum and millet in the semi-arid tropics. Proc. Int. Symp. 15-20 Nov. 1982. ICRISAT, Patancheru, A.P. India.

Arkin, G.F., R.L. Vanderlip, and J.T. Ritchie. 1976. A dynamic grain sorghum growth model. Trans. ASAE 19:622-630.

Bennett, J. 1975. Light and water stress effects on sorghum. M.S. thesis. Texas Tech Univ., Lubbock.

Blum, A., and M. Naveh. 1976. Improved water-use efficiency in dryland grain sorghum by promoted plant competition. Agron. J. 68:125-133.

Bond, J.J., T.J. Army, and O.R. Lehman. 1964. Row spacing, plant population and moisture supply as factors in dryland grain sorghum production. Agron. J. 56:3-6.

Boyer, J.S. 1982. Plant productivity and environment. Science 218:443-448.

Chaudri, U.N., and E.T. Kanemasu. 1985. Growth and water use of sorghum [*Sorghum bicolor* (L.) Moench] and pearl millet [*Pennisetum americanum* (L.) Leeke]. Field Crops Res. 10:113-124.

Downes, R.W. 1968. Effect of temperature on tillering of grain sorghum seedlings. Aust. J. Agric. Res. 19:59-64.

Downes, R.W. 1972. Effect of temperature on the phenology and grain yield of *Sorghum bicolor*. Aust. J. Agric. Res. 23:585-594.

Eastin, J.D. 1981. Sorghum development and yield. p. 126-147. *In* S. Yoshida (ed.) Proceedings of the symposium on the potential productivity of field crops under different environments. IRRI, Manila, Philippines.

Eastin, J.D., I. Brooking, and A.O. Taylor. 1975. Temperature influence on sorghum development and yield components. p. 326-334. *In* Physiology of yield and management of sorghum in relation to genetic improvement. Annu. Rep. 8. Univ. Nebraska, Lincoln.

Eck, H.V., and J.T. Musick. 1979. Plant water stress effects on irrigated grain sorghum. II. Effects on nutrients in plant tissues. Crop Sci. 19:592-598.

Fischer, K.S., and G.L. Wilson. 1971. Studies of grain production in *Sorghum vulgare*. I. The contribution of preflowering photosynthesis to grain yield. Aust. J. Agric. Res. 22:33-37.

Fischer, K.S., and G.L. Wilson. 1975. Studies of grain production in *Sorghum bicolor* (L.) Moench. Effect of planting density on growth and yield. Aust. J. Agric. Res. 26:31-41.

Food and Agricultural Organization. 1981. Yearbook of food and agricultural statistics. FAO, Rome.

Garrity, D.P., D.D. Watts, C.Y. Sullivan, and J.R. Gilley. 1982a. Moisture deficits and grain sorghum performance: Effect of genotype and limited irrigation strategy. Agron. J. 74:808-814.

Garrity, D.P., D.D. Watts, C.Y. Sullivan, and J.R. Gilley. 1982b. Moisture deficits and grain sorghum performance: Evapotranspiration-yield relationships. Agron. J. 74:815-820.

Hanks, R.J. 1983. Yield and water use relationships: An overview. p. 393-411. *In* H.M. Taylor et al. (ed.) Limitations to efficient water use in crop production. ASA, CSSA, and SSSA, Madison, WI.

Harden, M.L. 1985. Carbon and nitrogen assimilation during development of sorghum as related to genotype and water stress. Ph.D. diss. Texas Tech Univ., Lubbock (Diss. Abstr. 85-17799).

Hargreaves, G.H. 1984. Developing practical agroclimate models for sorghum and millet. p. 183-188. *In* S.M. Virmani (ed.) Agrometeorology of sorghum and millet in the semi-arid tropics. Proc. Int. Symp. 15-20 Nov. 1980, 1982. ICRISAT, Patancheru, A.P. India.

Hedge, B.R., D.J. Major, D.B. Wilson, and K.K. Krogman. 1976. Effects of row spacing and population density on grain sorghum production in southern Alberta. Can. J. Plant Sci. 56:31-37.

Heinrich, G.M., C.A. Francis, and J.D. Eastin. 1983. Stability of grain sorghum yield components across diverse environments. Crop Sci. 23:209-212.

Heslehurst, M.R. 1983. Effects of population density and planting pattern on yield responses of grain sorghum. Field Crops Res. 7:213-222.

Hiler, E.A., and T.A. Howell. 1983. Irrigation options to avoid critical stress: An overview. p. 479-497. *In* H.M. Taylor et al. (ed.) Limitations to efficient water use in crop production. ASA, CSSA, and SSSA, Madison, WI.

Howell, T.A., and E.A. Hiler. 1975. Optimization of water use efficiency and high frequency irrigation. I. Evapotranspiration and yield relationships. Trans. ASAE 18:873-878.

Huda, A.K.S., M.V.K. Sivakuman, S.M. Virmani, N. Seetharama, S. Singh, and J.G. Sekason. 1984. Modeling the effect of environmental factors on sorghum growth and development. p. 277–287. *In* S.M. Vermani (ed.) Proc. Int. Symp. Agrometeorology of Sorghum and Millet in Semi-arid Tropics. 15–20 Nov. 1982. ICRISAT, Patancheru, A.P. India.

Inuyama, S., J.T. Musick, and D.A. Dusek. 1976. Effect of plant water deficits at various growth stages on growth, grain yield, and leaf water potential of irrigated grain sorghum. Proc. Crop Sci. Soc. Jpn. 45:298–307.

Jackson, R.D. 1972. On the calculation of hydraulic conductivity. Soil Sci. Soc. Am. Proc. 36:360–382.

Jordan, W.R., and C.Y. Sullivan. 1982. Reaction and resistance of grain sorghum to heat and drought. p. 131–142. *In* J.O. Mertin (ed.) Sorghum in the eighties. Proc. Int. Symp. Sorghum. 2–7 Nov. 1981. ICRISAT, Patancheru, A.P. India.

Kanemasu, E.T. 1983. Yield and water use relationships: Some problems of relating grain yield to transpiration. p. 413–417. *In* H.M. Taylor et al. (ed.) Limitations to efficient water use in crop production. ASA, CSSA, and SSSA, Madison, WI.

Kanemasu, E.T., L.R. Stone, and W.L. Powers. 1976. Evapotranspiration model tested for soybeans and sorghum. Agron. J. 68:569–572.

Karachi, Z., and Y. Rudich. 1966. Effects of row width and seedling spacing on yield and its components in grain sorghum grown under dryland conditions. Agron. J. 58:602–604.

Kidambi, S.R. 1987. Genetic control of gas exchange processes affecting water use efficiency of grain sorghum. Ph.D. diss. Texas Tech Univ., Lubbock (Diss. Abstr. 87-17340).

Krieg, D.R. 1983. Sorghum. p. 352–388. *In* I.D. Teare, and M.M. Peet. (ed.) Crop water relations. Wiley Interscience, New York.

Lascano, R.J., and C.H.M. van Bavel. 1986. Simulation and measurements of evaporation from a bare soil. Soil Sci. Soc. Am. J. 50:1127–1133.

Lascano, R.J., C.H.M. van Bavel, J.L. Hatfield, and D.R. Upchurch. 1987. Energy and water balance of a dryland cotton crop: Simulated and measured soil and crop evaporation. Soil Sci. Soc. Am. Proc. 57:1115–1121.

Leng, E.R. 1982. Status of sorghums compared to other cereals. p. 25–32. *In* J.V. Mertin (ed.) Sorghum in the eighties. Proc. Int. Symp. Sorghum. 2–7 Nov. 1981. ICRISAT, Patancheru, A.P. India.

Lewis, R.B., E.A. Hiler, and W.R. Jordan. 1974. Susceptibility of grain sorghum to water deficit at three growth stages. Agron. J. 66:589–591.

Muchow, R.C., D.B. Coates, G.L. Wilson, and M.A. Foale. 1982. Growth and productivity of irrigated *Sorghum bicolor* (L. Moench) in northern Australia. I. Plant density and arrangement on light interception and distribution and grain yield in the hybrid Texas 610 RS in low and medium latitudes. Aust. J. Agric. Res. 33:773–784.

Musick, J.T., and D.A. Dusek. 1971. Grain sorghum response to number, timing and size of irrigations in the southern High Plains. Trans. ASAE 14:401–410.

Myers, R.J.K., and M.A. Foale. 1981. Row spacing and population density in grain sorghum, a simple analysis. Field Crops Res. 4:147–154.

Ogunlela, V.B. 1979. Physiological and agronomic response of a grain sorghum [*Sorghum bicolor* (L.) Moench] hybrid to elevated night temperatures. Ph.D. thesis. Univ. of Nebraska, Lincoln (Diss. Abstr. 80-10871).

Phipps, M.R. 1987. Genetic control of development traits related to productivity of grain sorghum under various soil water supplies. M.S. thesis. Texas Tech Univ., Lubbock.

Pietsch, D.R., F.R. Miller, L. Reyes, R.A. Creelman, D.T. Rosenow, G.C. Peterson, and D.J. Undersander. 1979, 1980, 1981, 1982, 1983, 1984, and 1985. Grain sorghum performance tests in Texas. Texas Agric. Exp. Stn., College Station.

Plaut, Z., A. Blum, and I. Arnon. 1969. Effect of soil moisture regime and row spacing on grain sorghum production. Agron. J. 61:344–347.

Ritchie, J.T. 1972. Model for predicting evaporation from a row crop with incomplete cover. Water Res. Trans. 8:1204–1213.

Rosenow, D.T., J.E. Quisenberry, C.W. Wendt, and L.E. Clark. 1983. Drought tolerant sorghum and cotton germplasm. p. 207–222. *In* J.F. Stone and W.O. Willis (ed.) Plant production and management under drought conditions. Elsevier Science, Netherlands.

Roy, R.N., and B.C. Wright. 1974. Sorghum growth and nutrient uptake in relation to soil fertility. II. N, P, and K uptake pattern by various plant parts. Agron. J. 66:5–9.

Steiner, J.L. 1986. Dryland grain sorghum water use, light interception, and growth responses to planting geometry. Agron. J. 78:720–726.

Stewart, B.A. 1983. An automated seeding rate selector. Agron. J. 75:152–153.

Stewart, B.A. 1985. Limited irrigation-dryland farming system. p. 112–118. *In* Proceedings: Workshop on management of vertisols for improved agricultural production. 18–22 Feb. ICRISAT, Patancheru, A.P. India.

Stewart, B.A., J.T. Musick, and D.A. Dusek. 1983. Yield and water-use efficiency of grain sorghum in a limited irrigation-dryland farming system. Agron. J. 75:629–634.

Stewart, J.I., R.D. Misra, W.D. Pruitt, and R.M. Hagan. 1975. Irrigating corn and grain sorghum with a deficient water supply. Trans. ASAE 18:270–279.

Stickler, F.C., and S. Wearden. 1965. Yield and yield components of grain sorghum as affected by row width and stand density. Agron. J. 57:564–567.

Stone, L.R., R.E. Groin, Jr., and M.A. Dillon. 1978. Corn and grain sorghum yield response to limited irrigation. J. Soil Water Conserv. 33:235–238.

Sweeten, J.M., and W.R. Jordan. 1987. Irrigation water management for the Texas High Plains: A research summary. Texas Water Resour. Inst. Tech Rep. 139. Texas A&M Univ., College Station.

U.S. Department of Agriculture. 1986. Agricultural Statistics. U.S. Gov. Print. Office, Washington, DC.

van Bavel, C.H.M., and J. Ahmed. 1976. Dynamic simulation of water depletion in the root zone. Ecol. Model. 23:189–212.

van Bavel, C.H.M., R.J. Lascano, and L. Stroosnijder. 1984. Test and analysis of a model of water use by sorghum. Soil Sci. 137:443–456.

Vanderlip, R.L., and H.E. Reeves. 1972. Growth stages of sorghum [*Sorghum bicolor* (L.) Moench]. Agron. J. 64:13–16.

Wilson, G.L., and J.D. Eastin. 1982. The plant and its environment. p. 101–110. *In* J.O. Mertin (ed.) Sorghum in the eighties. Proc. Int. Symp. Sorghum. 2–7 Nov. 1981. ICRISAT, Patancheru, A.P. India.

24 Cotton

D. W. GRIMES

University of California
Davis, California

K. M. EL-ZIK

Texas A&M University
College Station, Texas

The majority of the world's cotton (*Gossypium hirsutum* L.) crop is produced with at least part of its water supply provided by irrigation; indeed the agricultural stability of arid and semiarid regions depends on a viable water supply. Even in rain-fed regions of the U.S. Cotton Belt, periods of drought subject the crop to varying water deficits and alter production trends. Cotton grown in California and Arizona requires water delivered by irrigation systems during as much as 85% of the growing season. Water transport to fields and storage in the crop-root zone to meet plant needs requires the best possible technology to achieve a desired resource use efficiency and economic return.

Successful water management requires observation of the water status of soils and plants and an understanding of the often complex, fundamental, and interacting responses of cotton to water, climate, and other production inputs. The purpose of this review is to provide the reader with information that reflects the current state of understanding of cotton plant-water relations and irrigation management.

I. GROWTH AND DEVELOPMENT OF THE COTTON PLANT

The cotton plant is more complex structurally than any other major field crop. Cotton, with its indeterminate growth habit, produces vegetative and reproductive growth simultaneously over a relatively long period (130–180 d). Even with such complexity, growth and development of the cotton plant follow an orderly and predictable sequence where growing conditions are favorable.

A. Seed Germination and Plant Establishment

The seed contains an embryo with two well-developed *cotyledons* (seed leaves) that are protected by a seed coat. Other parts of the embryo are the *epicotyl* (shoot), *hypocotyl,* and *radicle* (root). Imbibition of water for germination is through the chalazal cap. Very early germination activity is supported by soluble carbohydrate reserves. The primary energy and food source during germination and early seedling development is from the embryo's stored lipids and proteins. Seed germination and seedling emergence are favored by high seed quality, a well prepared seed bed with adequate moisture, optimum seed depth that depends on soil type and moisture, high soil O, and soil temperature above 17.8 °C. With favorable conditions, seedlings emerge 5 to 15 d after planting. The root develops more rapidly during germination because the radicle is more massive in the seed than the epicotyl. Before true leaf expansion, the primary root is penetrating deeply into the soil and branch roots are formed.

B. Vegetative Growth

The cotton plant is constantly producing new, specialized cells to form the organs that carry out the functions of growth and reproduction throughout the growing season. The plant established its basic framework—the root, vegetative, and fruiting branches, and partial leaf canopy in 40 to 55 d. A fully developed cotton plant has a prominent, erect mainstem consisting of a series of nodes and internodes. The number and length of internodes, which determine plant height, are influenced by cultivar, environment, and cultural factors such as soil type, moisture availability, and nutrients. The cotyledons are located at the lowest node on opposite sides of the stem. As the seedling grows, the internode above the cotyledons extends and a new node is formed from which the first true leaf unfolds. This process continues at successively higher nodes at 2.5- to 4-d intervals until a fruit load develops. A single leaf forms at each node in a spiral arrangement at progressively higher nodes.

Location of the first fruiting (sympodial) branch is influenced by plant population, temperature, stress, and excessive moisture, coupled with ample soil N, early in the plant growth period. The first fruiting branch usually develops on the fifth to ninth node (El-Zik & Frisbie, 1985). Short-season cultivars may set the first fruiting branch at node four to five, rainbelt cultivars at node six to eight, Acala and other long-season cultivars may not set the first reproductive branch until the ninth node. The higher the first fruiting branch, the longer it will take for the plant to complete fruiting and mature its bolls. If the terminal bud of the main stem is damaged by insects, hail, or other factors, one or more of the branches near the terminal or axillary buds at the higher nodes will assume apical dominance.

C. Root System

The cotton plant has a primary taproot with many laterals. Root length density shows a typical sigmoidal curve, with exponential growth in the early part of the season when the root system is unaffected by soil conditions such as compaction and hardpans. With little or no growth restrictions, the taproot grows rapidly downward, 2.5 cm/d for several weeks after planting (Bassett et al., 1970b). Branching of the taproot begins about the time the seedling straightens up and the cotyledons begin to unfold. Soil type, moisture, aeration, and impeding zones determine how deep taproots penetrate. Although a few will grow as deep as 240 cm, normally about one-half of the total root length is confined to the top 60 cm of soil (Bassett et al., 1970b; Taylor & Klepper, 1978). The lateral roots may extend outward from the taproot to a horizontal length of 2 m (Taylor & Klepper, 1978). The total root length continues to increase as the plant develops until the onset of flowering, about 8 to 10 wk after planting. Toward the end of the season, root length beings to decline as plant height stays the same and older roots die.

Root death during the growing season often involves a major portion of the root biomass. The causes of root death are varied and may involve both biotic and abiotic factors. Soil moisture is a major factor affecting root growth rates and distribution (Grimes & El-Zik, 1982); other variables are soil temperature, soil compaction, lack of O in water-logged soils, soil pH, nutrient supply and availability, and boll load.

D. Fruit Formation

Cotton with its indeterminant growth habit, produces squares (flower buds), flowers (blooms), and bolls over 75 to 105 d. The initiation of squares and continuation of flowering is a function of vegetative growth which produces additional fruiting branches and of formation of additional fruiting sites (nodes) on existing branches. One of the main aspects for profit-

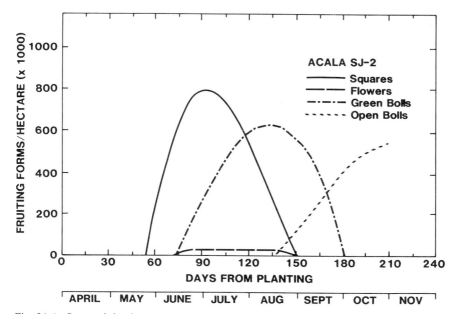

Fig. 24-1. Seasonal development of fruiting forms of Acala SJ-2 cotton in the San Joaquin Valley, California (El-Zik, 1985).

Table 24-1. Phenology† of the cotton plant. Source: El-Zik and Frisbie (1985).

Stage of growth	Area (USA)	Range	Average
		—— no. of days ——	
Planting to emergence		5–20	10
Emergence to square	Southeast‡ (SE)	27–38	32
	High Plains	33–38	35
	West	40–60	50
Square to first bloom		20–27	23
First bloom to peak bloom		26–45	34
Bloom to open boll			
Early and	SE and High Plains	45–55	50
mid-season bloom	West	45–65	58
Late-season bloom	SE and High Plains	55–70	60
	West	65–85	70
Growing season	High Plains	120–150	140
	Southeast	130–170	155
	West	180–210	195

† Based on normal growth rate and no severe adversities, such as insects, diseases, moisture and heat stress, or other environmental setbacks.
‡ Includes Mississippi River Delta and mid-South.

able cotton production is to maintain the balance between vegetative growth and fruit production that is affected by weather conditions, nutrition, and moisture supply. All processes leading to square, flower, boll development, and maturation are temperature dependent.

Seasonal development of fruiting forms of Acala SJ-2 cotton is shown in Fig. 24-1. Fruit formation begins with the appearance of the first square; under normal conditions the first square can be expected between 5 and 8 wk after planting, depending on the cotton growing region and temperature. A decline in squaring or delayed squaring may be due to physiological factors resulting from environmental effects, poor growing conditions, and/or pests. Shedding of some squares, flowers, and small bolls is a natural process, but it is increased by adverse factors such as too much or too little soil moisture, inadequate number of fertilized ovules, insufficient nutrient supply, excessive heat or cold, extended periods of cloudy weather, and damage from insects and diseases (El-Zik, 1980; Guinn, 1982). Fruit abscission also results when the demand for photosynthates by the various plant parts exceeds the supply. This is common when several bolls per plant reach maximum size; natural abscission increases in late season. Guinn (1982) reviewed the causes of square and boll shedding in cotton.

The first flower usually appears 60 to 80 d from planting, or 20 to 27 d (23 d average) after square initiation (Table 24-1). Flower development from squares follows a definite pattern (Dennis & Briggs, 1969; El-Zik, 1980; Mauney, 1983, 1986; Tharp, 1965). In many growing regions of the Northern Hemisphere the effective flowering period occurs from late June or early July to mid-August, and about 60% of the flowers are produced within 110 d of planting.

E. Boll Growth and Maturation

The period of boll growth and maturation begins when the plants are 65 to 95 d old and continues until the last set boll opens at about 140 to 190 d, depending on the growing region. Following anthesis and fertilization of the ovules, the young boll grows rapidly, reaching full size in about 24 d. An additional 24 to 40 d are required for complete maturation of the boll. Seeds attain their full size about 3 wk after fertilization and mature shortly before the boll opens. Boll development is controlled by genetic-environment interactions. The distribution of available carbohydrate to fruiting forms is controlled by the relative strength of the various competing sinks.

Conversion of flowers to bolls that will be retained by the plant is more effective in the early than in the late part of the season. The plant normally sets 80% of its bolls in the first 3 to 6 wk of flowering, depending on the growing region (El-Zik et al., 1980; El-Zik & Frisbie, 1985). Early season flowers require about 45 to 55 d to mature and open, while flowers developing in August and September require between 60 and 80 d. These stages of development vary between cotton-growing regions as shown in Table 24–1. Crop development rates are influenced by cotton cultivar, environment, stresses, insects, and diseases.

Late in the fruiting period, the cotton plant starts to "cut-out"; no new nodes, fruiting branches, or squares are formed. Moisture stress and extreme high or low temperatures may cause premature cut-out, reducing yield and fiber quality.

Each cotton fiber is an extension of an epidermal cell of the seed coat. Fibers attain their full length usually within the first 18 d after flowering and fertilization of the ovule. Thickening of fiber walls starts about 4 d before elongation ceases. Each day, successive layers of cellulose are deposited on the inner surface of the fiber in a spiral fashion. The amount and pattern of cellulose deposition determines fiber quality characteristics such as strength, fineness, and maturity. The genetic makeup of a cultivar largely controls fiber quality traits; however, these are also influenced by environment, management, and pests. Unfavorable growing conditions, including moisture stress, reduce fiber quality.

II. PLANT NUTRITION

Soils used for cotton production vary in native fertility, and the use of fertilizers has played a major role in increasing cotton productivity. Nitrogen, P, and K are required in large quantities and are limiting in many soils or, in some cases, are not available for plant growth even if the element is present in the soil. Micronutrients have critical functions in metabolism. Nutritional stresses affect vegetative and reproductive growth and ultimately yield, and fiber and seed quality. The supply of nutrients functions in the partitioning of vegetative and fruiting growth. Cotton, as with other crops, has the capability to adjust morphogenesis and fruit set to establish a balance between demand for and supplies of water, nutrients, and photosynthate in a way that will insure the development of viable seed for survival.

A. General Nutrient Requirements

Berger (1969) reported that cotton yielding 560 kg/ha lint removes approximately 40 kg/ha of N, 3.5 kg/ha of P, 7.1 kg/ha of K, 4.2 kg/ha of Mg, and 2.9 kg/ha of Ca. Bassett et al. (1970a) found that N contained in-plant material producing 100 units of lint ranged from 9.0 to 13.1 units over 2-yr while Christidis and Harrison (1955) reported a value of 18.7. Similar values (Bassett et al., 1970a) for P, K, Ca, and Mg were, respectively, 1.3 to 1.7, 7.6 to 11.4, 6.2 to 13.1, and 2.4 to 7.2. Required nutrient amounts must be soil supplied or added as supplements.

Since cotton produces vegetative and reproductive growth simultaneously over a relatively long time, its nutritional needs are perhaps more complex than that of any other major field crop. A continuous supply of nutrients is required to sustain morphogenesis. The rate of both nutrient intake and dry matter production increases progressively during the seedling, vegetative, and fruiting periods and peaks near the end of the bloom period (Bassett et al., 1970a). Berger (1969), Hinkle and Brown (1968), Jones and Bardsley (1968), Kamprath and Welch (1968), and Tucker and Tucker (1968) reviewed and discussed the nutritional requirements of the cotton plant, role of nutrients, deficiency symptoms, and methods and timing of fertilization.

B. Specific Nutrient Effects

The ratio of the dry weight of bolls/dry weight of stems and leaves is used to describe fruiting efficiency (fruiting index); the effects of several nutrients on fruiting efficiency were reviewed by Eaton (1955) and Joham (1979). Joham (1986) divided the essential elements for cotton into two broad groups based on fruiting index. The first group consists of N, S, Mo, and Mn; a deficiency of these elements has little or no effect on fruiting index or relative fruitfulness. The second group consists of nutrient elements that seem to have a more direct role in flowering and fruiting, and a deficiency of one or more of these elements will cause a decrease in the fruiting index and relative fruitfulness. Elements which fall within the second group include P, K, Ca, Mg, B, and possibly Zn (Eaton, 1932, 1944; Ergle & Eaton, 1957; Helmy et al., 1960; Joham, 1955; Sorour, 1963). Joham and Rowe (1975) found that Zn had a marked influence on the partitioning of vegetative and fruiting growth that was influenced by temperature.

Calcium, K, and Mg reportedly function in the control of carbohydrate translocation. With deficiencies of each of these elements, carbohydrate movement from the leaves is restricted both in rate and distance moved. Phosphorous and Zn have not been shown to control carbohydrate translocation in cotton, yet, they exert profound effects on fruiting (Joham, 1986). The effects of P are similar to those of light and temperature in promoting fruiting and earliness, and these effects may be mediated through the well-known association of P in energy reactions.

Supplemental N is required in most cotton production systems. Though N nutrition does not alter morphogenetic patterns, it does alter processes which depend upon partitioning of assimilates, i.e., growth rates of leaves and stems, boll shedding, seed and lint weight, root/shoot ratio, earliness, and stomatal response to water stress. The three basic physiological responses of cotton plants to low N fertility are decreased photosynthetic rate and leaf expansion resulting from changes in hydraulic conductivity, and increased stomatal sensitivity to water stress (Radin & Parker, 1979a, b; Mauney et al., 1980).

The cotton seed is the primary site of N, P, Ca, and Mg compartmentation within the developing boll. The primary sink for boll K is the carple wall (bur), and the fiber is the second greatest sink. As with K, Mg appears to be associated with both fiber elongation and secondary wall formation. Leffler (1986) reviewed and discussed mineral compartmentation within the cotton boll, and the nutritional changes that occur during boll development and maturation.

C. Salt Tolerance

Saline soils and irrigation water are common in the semiarid and arid southwestern and western cotton growing regions. Chloride, sulfate and bicarbonate salts of Na, Ca, and Mg contribute in varying degrees to soil and water salinity.

Cotton is classified as a salt-tolerant crop. A threshold soil salinity level at which initial yield decline has been observed is a soil saturation extract conductivity of 7.7 dS m^{-1} (EC$_e$) with 50% reduction in yield observed at an EC$_e$ of 17.0 dS m^{-1} (Ayers, 1977).

Hoffman et al. (1971) reported that transpiration, and thus water requirement, of cotton decreased with increasing concentrations of specific salts in the soil. Boyer (1965) observed a 25% reduction in both photosynthesis and respiration in cotton grown at a NaCl concentration corresponding to an osmotic potential of -0.85 MPa. Increased salinity reduced both photosynthesis and transpiration in a study by Hoffman and Phene (1971) that resulted in lower water use efficiency.

Genotypic differences in salt tolerance and differences at various growth stages were reported among Upland cotton types (Hayward & Wadleigh, 1949; Lauchli et al., 1981). Hayward and Wadleigh (1949) observed that the long staple Egyptian types were somewhat more salt tolerant than the Upland types.

III. EVAPOTRANSPIRATION

A. Total Season Evapotranspiration for Contrasting Regions

Total season cotton evapotranspiration (ET) depends on cultural practices and climatic parameters that include temperature, solar radiation,

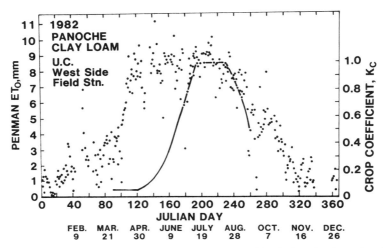

Fig. 24-2. Daily ET_0 at the University of California West Side Field Station in 1982 and a typical K_c curve for cotton in the San Joaquin Valley (Grimes, 1986).

humidity, wind, and length of the growing season. It is not surprising, then, that total season ET exhibits a considerable range in contrasting regions; Doorenbos and Pruitt (1977) reported an approximate range of 55 to 95 cm for nonlimiting soil water conditions for cotton-growing regions of the world. This is also the approximate range in total season ET observed across the Cotton Belt of the USA. Doss et al. (1964) observed a total season ET of approximately 60 cm for cotton in Alabama. At the opposite end of the ET range, total growing season ET was near 105 cm at Mesa and Tempe, AZ (Erie et al., 1982). Total season ET in the lower, desert region of California averages about 95 cm (Peterson et al., 1962) while that of the San Joaquin Valley is about 72 cm (Grimes & El-Zik, 1982). Maximum cotton production was observed with 85 cm ET from a lysimeter in Las Cruces, NM by Sammis (1981).

B. Actual Crop Evapotranspiration—Seasonal Trends

Actual cotton ET (ET_{crop}) at a given time during the growing season is dependent on the effects of climate on reference crop ET (ET_0) and the cotton crop coefficient (K_c); this can be expressed as

$$ET_{crop} = K_c \times ET_o. \qquad \text{Eq. [1]}$$

K_c at a given time is influenced primarily by degree of ground cover since cotton leaf morphology stays constant. Evaporation from the soil surface (E_{soil}) is a part of ET_{crop}, therefore, the frequency of rain or irrigation will also influence K_c in early season before development of full ground cover. Early season K_c values will increase sharply following a rain or irrigation; usually the values are smoothed to reflect average conditions for a region.

Table 24–2. Comparative ET_{crop} for cotton grown in the San Joaquin Valley, California, and at Mesa and Tempe, AZ.

California[†]				Arizona[‡]			
Date	Rate	Date	Rate	Month	Bimonthly rate	Month	Bimonthly rate
	cm/d		cm/d		cm/d		cm/d
25 Apr.	0.03	14 July	0.81	Apr.	0.03	Aug.	0.93
5 May	0.03	18 July–8 Aug.-peak bloom			0.05		0.92
25 May	0.08	13 Aug.	0.81	May	0.11	12 Sept.-last blooms	
2 June–1st square		23 Aug.	0.66		0.21	that usually mature	
4 June	0.19	2 Sept.	0.50	11 June–1st bloom		Sept.	0.74
14 June	0.34	12. Sept.	0.37	June	0.33		0.56
23 June–1st bloom		22 Sept.	0.23		0.56	Oct.	0.37
24 June	0.48	2 Oct.	0.15	18 July–1st bloom peak			0.20
4 July	0.64	12 Oct.	0.09	July	0.76	Nov.	0.10
		Season total = 71.9 cm			0.96	Season total = 104.6 cm	

† Grimes and El-Zik (1982).   ‡ Erie et al. (1982).

The relationship between reference crop ET_0 by a modified Penman equation (Doorenbos & Pruitt, 1977) and K_c for cotton observed in the San Joaquin Valley is ilustrated in Fig. 24–2. Maximum K_c values for the San Joaquin Valley were observed to be slightly lower than those suggested by Doorenbos and Pruitt (1977).

Table 24–2 gives daily rates of ET_{crop} for intermediate (San Joaquin Valley) and severe (Arizona) climatic evaporative demand conditions through a typical growing season in the west and southwest sections of the U.S. Cotton Belt. With a small leaf surface and relatively dry surface soil, ET_{crop} is much below the climatic potential ET_0 in early season. A period of rapid canopy development is observed in June that is accompanied by marked increases in daily ET_{crop}. By mid-July a complete ground cover allows cotton daily ET_{crop} (0.81 and 0.96 cm for San Joaquin Valley and Arizona, respectively) to closely reflect the energy available for evaporating water. Late season ET_{crop} rates are lower, reflecting plant aging and less atmospheric evaporative demand. In practice, late-season water use may be reduced even further by allowing soil water deficits to increase to hasten maturity and provide improved conditions for defoliation. Midseason maximum water use rates occur during peak bloom when a severe water deficit is most harmful to yield (Grimes et al., 1970).

IV. WATER STRESS AND ADAPTATION

A. Osmotic Adjustment

Cotton undergoes an adaptive mechanism, in response to periods of water stress, that provides positive turgor pressure (ψ_p) for some degree of continued growth even at relatively low values of leaf water potential (ψ_l) (Cutler & Rains, 1977; Cutler et al., 1977; Oliveira, 1982). Possible adaptive mechanisms include solute accumulation or osmoregulation, small cell size (more cell walls per unit volume), and greater cell-wall elasticity. Cutler et al. (1977) concluded that such turgor maintenance is due in part to the accumulation of sugars and malate and partly to high cell-wall elasticity. Oliviera (1982), on the other hand, concluded that solute accumulation was the primary adjustment mechanism based on intensive analyses of leaf-moisture-release curves. Figure 24–3 illustrates the various components for osmoregulation during a soil drying cycle (Morgan, 1984).

As much as 1 MPa adjustment of ψ_s (osmotic potential) for whole leaves is commonly reported (Brown et al., 1976). A diurnal fluctuation in ψ_s occurs in cotton leaves with the lowest value lagging about 4 h after minimum ψ_l in the work of Ackerson et al. (1977b) and Cutler et al. (1977). However, a close diurnal parallel between ψ_l and ψ_s was found by Karami et al. (1980).

Cotton genotypes differing in leaf morphology have shown contrasting abilities to osmoregulate. Under severe soil water stress Karami et al. (1980) found that a superokra leaf genotype consistently had the lowest ψ_s; this resulted in 0.2 to 0.3 MPa higher ψ_p throughout the day than a normal leaf genotype.

B. Stomatal Response

The osmotic adjustment of cotton leaves resulting from periods of water deficit conditions imparts a differential sensitivity to stomata for conditioned plants compared with nonstressed plants. Thomas et al. (1976) found stomatal closure initiated at −1.8 MPa for well-watered field grown plants, but plants

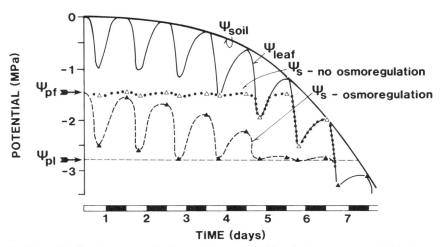

Fig. 24–3. Idealized response of leaf osmotic potential (ψ_s) to leaf water potential (ψ_l) during periods of declining soil water potential (ψ_{soil}) and evaporative demand; full (broken line) and zero (dotted line) turgor maintenance situations are depicted. The ψ_{pf} is the osmotic potential at full turgor and ψ_{pl} is the limit of solute accumulation. Reproduced, with permission, from the Annual Review of Plant Physiology, Vol. 35 © 1984 by Annual Reviews, Inc.

grown in the same environment and subjected to periods of water stress did not show appreciable stomatal resistance until ψ_1 dropped to about -2.8 MPa. Similar observations are reported by a number of investigators.

Stomatal resistances of the top side of cotton leaves are greater than that for the bottom epidermis, due in part to higher stomatal density for the lower leaf surface (Jordan et al., 1975; McMichael & Hesketh, 1982). An additional factor, however, is found in the greater sensitivity of upper surface stomata to lowering ψ_1's and their failure to respond to water stress conditioning. Brown et al. (1976) found the ψ_s of lower surface guard cells to be 0.7 MPa lower than that of upper surface guard cells. Also, a pronounced difference in stomatal sensitivity between young and old leaves exists. Jordan et al. (1975) observed stomatal closure first on the oldest leaves of the cotton plant that progressed to the youngest as soil moisture stress increased; this pattern was independent of radiation effects during the stress period.

A low N status has been reported (Radin et al., 1985) to eliminate the diurnal oscillations of ψ_s in cotton. Plants having adequate N in early season were observed to lose the osmoregulation ability as petiole NO_3-N concentrations declined to low levels later in the season. With low N, stomatal closure was observed at higher ψ_1's. These researchers suggested that this physiological response might be used to increase water use efficiency where a low N regime is used in cases of severely limiting water supplies.

C. Photosynthesis and Transpiration

With adequate soil moisture an increase in evaporative demand is accompanied by a parallel increase in transpirational flux; some lowering of ψ_1 will also occur to increase the soil to leaf water potential gradient (van den Honert, 1948; Ackerson & Krieg, 1977). As a soil dries, increased flow resistances result in additional lowering of the ψ_1 until a level is reached that

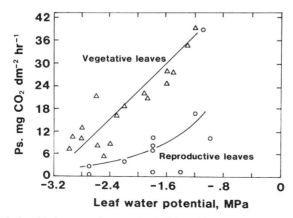

Fig. 24–4. Relationship between photosynthesis (P_s) and leaf water potential for vegetative and reproductive cotton leaves (redrawn from Ackerson et al., 1977b).

triggers some degree of stomatal closure. From the previous discussion it is clear that a critical cotton ψ_l for stomatal response to water stress is dynamic, being different for each leaf position in the canopy, upper and lower leaf surfaces, and contrasting water stress history.

With increased water stress, cotton net photosynthesis generally declines (Fig. 24-4) from a maximum rate of about 40 mg CO_2 dm^{-2} h^{-1} (Ackerson et al., 1977b; McMichael & Hesketh, 1982). For nonosmotically adjusted plants the reduction with declining ψ_l is accompanied by concomitant reduction in transpiration and is under stomatal control as exemplified by the results of Bielorai and Hopmans (1975). With osmotically adjusted plants, field grown and subjected to water stress, photosynthesis still declines linearly with reduced ψ_l, but diffusive resistance may remain relatively low over the range of declining ψ_l (Ackerson et al., 1977a). This type of response has led several researchers (Ackerson et al., 1977a; Hutmacher & Krieg, 1983; Karami et al., 1980) to conclude that photosynthesis is being controlled by non-stomatal factors.

Hutmacher and Krieg (1983) found evidence that nonstomatal photosynthetic rate control varied considerably among contrasting genetic lines. Based on this finding they suggested that genetic selection, to achieve higher photosynthesis per unit leaf conductance, is a viable possibility and should be explored. Karami et al. (1980) found that a superokra genotype was able to maintain a higher ψ_p, because of higher ψ_l and lower ψ_s, that was related to higher photosynthetic rates. These findings indeed suggest that improvements in drought tolerance or water use efficiency should be explored genetically in addition to developing optimum water management strategies.

V. GROWTH AND WATER STATUS IN THE SOIL-PLANT-AIR-CONTINUUM

A. Expansive Growth

Cell growth is the most sensitive process affected by water stress (Hsiao et al., 1976). The ψ_p is essential for irreversible cell expansion and growth and a close correlation between the two has been shown. Although the relationship between ψ_p and ψ_l is nonlinear for cotton (Gardner & Ehlig, 1965), expansive growth was linearly related to ψ_l when relatively long-term measurements (hours or days) of growth were made and are functionally dependent on ψ_l measured at predawn or midday (Cutler & Rains, 1977; Grimes & Yamada, 1982). As pointed out by Cutler and Rains (1977), measurement of both growth and ψ_l in many cases represents integrated values; nevertheless, such measurements are useful for understanding plant-water relations under field conditions and can be used in management.

The diurnal nature of leaf expansion rate is illustrated in Fig. 24-5. (Cutler et al., 1977). Maximum extension rate is observed during the light to dark transition period associated with stomatal closure. For low water stress conditions, extension rate is relatively constant for the dark period into

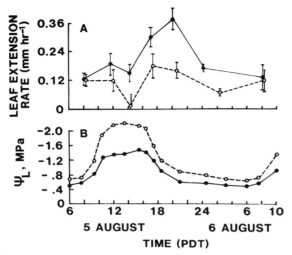

Fig. 24-5. Diurnal variations in elongation rate of expanding leaf blades (A) and ψ_l (B) of cotton with low (●) and high (○) levels of water deficit (Cutler et al., 1977).

light through solar noon. However, under water stress conditions a substantial decline in extension rate corresponds to achievement of minimum ψ_l for the diurnal period. Main stem elongation (Fig. 24-6) is also closely associated with ψ_l and is quite sensitive to water deficit conditions (Grimes & Yamada, 1982). Maximum main stem elongation is achieved at the highest midday measurements of ψ_l and declines linearly with declining ψ_l. Essentially zero extension is observed near -2.5 MPa ψ_l. Cutler and Rains (1977) found leaf growth to be affected by water stress more than main stem growth; however, Fig. 24-5 and 24-6 indicate a response near the same magnitude. Figure 24-7 illustrates the influence of severe water deficits in a semiarid environment

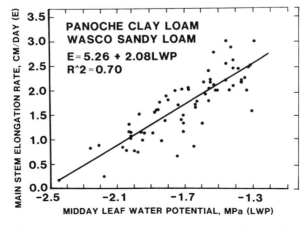

Fig. 24-6. Relation of cotton main stem elongation rate to midday ψ_l (Grimes & Yamada, 1982).

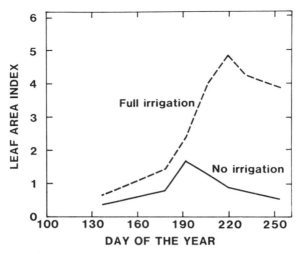

Fig. 24-7. Contrasting leaf area index values of cotton in a semiarid region that is irrigated to fully meet evaporative demand and cotton grown on approximately 33 cm of water stored in the potential root zone at planting. From Grimes (1981, unpublished data).

on canopy development where no effective rainfall was recorded during the growing season, but planting was done with a soil profile near field capacity.

B. Boll and Fiber Growth

Boll and fiber growths are generally observed to be much less sensitive to water stress than is expansive vegetative growth. Radulovich (1984) measured boll diameter and volume for field-grown plants as a function of midday ψ_l to find that boll size was not materially reduced by water stress until ψ_l dropped substantially below -2.2 MPa. Grimes and Yamada (1982) used the techniques of Gipson and Ray (1969) to measure the response of fiber elongation and weight increase to water stress. For field-grown plants neither fiber elongation nor weight increase were affected until midday ψ_l declined to approximately -2.7 MPa, at which time a marked reduction for both parameters was observed. From these responses one would conclude that the boll is a relatively strong photosynthate sink; indeed this has been demonstrated directly by Krieg and Sung (1979). Dry matter accumulation in reproductive plant parts will continue for a period after expansive vegetative growth has stopped.

C. Abscission

Abscission of leaves and fruiting forms in cotton is a complex phenomenon and will receive only limited discussion. Generally it is believed that water deficits alter the normal hormone balance of the abscission zone (Jordan, 1979). Ethylene production is enhanced by water stress and this compound functions as a potent abscission-promoting hormone (Guinn, 1979) by slowing transport, increasing destruction of auxin, and stimulating synthesis of pectinase and cellulase in the abscission zone. Jordan (1979) suggested that abscission, once induced, requires active cell growth near the separation layer; he further suggested that this can explain observed abscission pattern differences between field-grown plants and those grown in small containers. Container-grown plants require stress alleviation by watering before abscission goes to completion whereas field grown plants do not. Jordan (1979) stated that ψ_p required for cellular growth is normally reestablished at night in field-grown plants but not in small container plants because of the rapid water loss in the small soil volume.

Cotton normally produces more fruiting forms than can be carried through to maturity even in the absence of external stresses; therefore, some shedding of fruiting forms, as mentioned earlier, is considered normal. Nevertheless, excessive water stress is associated with changing patterns of reproductive development. Contrasting stages of fruiting forms differ in sensitivity to water stress that will trigger abscission; McMichael (1979) and McMichael and Guinn (1980) summarized the current state of understanding of this phenomenon. As fruiting forms develop from the "pin head" stage to anthesis, a time period of about 21 to 22 d, their sensitivity to water stress decreases (Guinn & Mauney as cited by McMichael, 1979). This was shown indirectly by Grimes et al. (1970) who observed a pronounced depression in flowering rate 3 wk after alleviating a severe water stress interval. Young bolls, until a size of about 2 cm diam. is achieved, are sensitive to abscission due to water stress.

Quantitative data relating ψ_l to the abscission processes is limited; nevertheless, some information is available. McMichael et al. (1973) observed

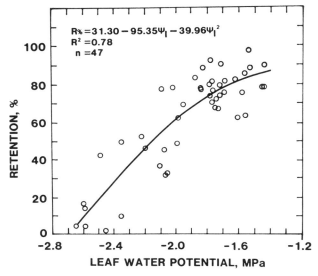

Fig. 24-8. Relationship between ψ_l and retention of young bolls early in the season (redrawn from Guinn & Mauney, 1984b).

a marked increase in <2-cm diam. boll abscission as predawn ψ_l declined below -1.1 MPa in small container-grown plants; leaf abscission increased as ψ_l declined below -0.7 MPa. Guinn and Mauney (1984b) observed a pronounced decline in boll retention as midday ψ_l declined below -1.9 MPa; this is illustrated in Fig. 24–8. Grimes and Yamada (1982) associated high cotton productivity to irrigation scheduling that did not allow midday ψ_l to decline below -1.9 MPa.

VI. IRRIGATION MANAGEMENT

Optimum cotton irrigation scheduling entails an application of fundamental plant-water relations, discussed in earlier sections of this chapter and in other chapters, and the growth rate-photosynthate partitioning patterns of the plant. Vegetative growth is linearly related to water supply over a wide input range (Fig. 24–9); however, harvest index is generally higher at low water supply amounts, reflecting a greater proportion of total biomass in reproductive growth as water stress severity is increased. These characteristics tend to impart a curvilinearity to yield-water supply functions for many circumstances (Grimes et al., 1969a; Grimes & Yamada, 1982; Howell et al., 1984a). The relationship between vegetative and reproductive growth can be manipulated to an advantage in integrated pest management systems (Leigh et al., 1974); this aspect is discussed in more detail in a later section of this chapter.

A. Soil- and Plant-Based Water Status Measurements for Irrigation Scheduling

The criteria used for scheduling irrigations can vary from one situation to another. While scheduling by the calendar and past experience can be done with reasonable success if climatic conditions vary little from one year to

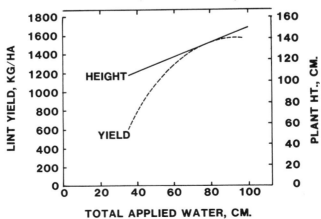

Fig. 24–9. Relation of total applied water, soil-stored at planting plus seasonal irrigation water, to vegetative and reproductive growth characteristics (redrawn from Grimes et al., 1969).

the next, observations on soil water and plant water status will accommodate greater variability in evaporative demand, economic conditions, and water availability.

Soil-based observations of water status are successful for irrigation scheduling, providing additional information is known or can be reasonably estimated. Since the crop is responding to total soil water potential, a measurement reflecting this would translate most directly as a scheduling parameter. Usually, however, a measurement of water content is made that is best used with site-specific information on soil water release characteristics and soil salinity. Also, information is needed on root development; soil compaction or other barriers may slow root extension or prevent root growth into some zones entirely (Grimes et al., 1975). At the other extreme, a shallow or perched water table may contribute substantial amounts of water to the ET demand of a growing crop (Grimes et al., 1984). These conditions can render soil-based measurements of water content difficult to interpret; however, a knowledge of soil water volume is essential to adding the right amount of water to replace that used by crops and to meet any leaching requirement.

Plant-based water status measurements have received much attention in recent years and have the advantage of reflecting the plant-integrated soil and atmospheric environments. Bordovsky et al. (1974) observed improved water use efficiency for cotton when plant-based irrigation scheduling parameters were used instead of soil ψ (tensiometer readings of -0.07 MPa at a 30-cm depth). At present, plant-based measurements are usually in the form of pressure chamber measurements of xylem pressure potentials (Scholander et al., 1965) or infrared thermometer-measured canopy temperature comparisons with ambient air that are "normalized" for climatic conditions (Idso et al., 1981). A crop water stress index (CWSI) is calculated and related to plant responses. Xylem pressure potentials closely estimate ψ_l if properly done and have been related to growth processes, discussed earlier, and seedcotton production (Grimes & Yamada, 1982; Guinn & Mauney, 1984b). Howell et al. (1984b; 1984c), among others, have related cotton productivity to infrared thermometry CWSI values. Plant-based observations of water status are sensitive to both matric and solute components of total soil water potential and to climatic evaporative demand.

Crop simulation models are generating considerable interest in a crop management context as well as being used as agronomic research tools (Whisler et al., 1986). Models vary greatly in their degree of complexity and range from essentially empirical systems to those that are highly mechanistic as increasing causality of processes is incorporated. Usually, models that are more mechanistic will have a broader applications base, but they also are more complex and require more time, computer capacity, and greater detail for input parameters. A dynamic cotton simulation model might predict, for example, the changing number of bolls throughout the season and changing soil water content at definable depth intervals. With continued development and validation, models should experience increased usage as management tools for crop production input optimization that includes irrigation scheduling.

Though the water needs of growing plants are continuous throughout the season, it is convenient to consider irrigation scheduling in distinct phases, namely, preplant or preirrigation, first seasonal irrigation, midseason irrigation(s), and last irrigation (Grimes & El-Zik, 1982; Longenecker & Erie, 1968). This procedure will be followed for this discussion.

B. Preplant Irrigations

Rainfall is seldom adequate in irrigated regions of the Cotton Belt to thoroughly wet the soil profile through the potential rooting depth following soil water depletion of a previous crop. A common and desirable practice is to irrigate before planting to wet through the expected depth of rooting. Davis et al. (1980) found it necessary to apply some preplant irrigation water (19 cm) to achieve maximum cotton yields even when a comparable quantity of postemergence trickle irrigation was applied. Preplant irrigations provide adequate water for germination and early growth; also, if excess salts are present, the timing is optimum for meeting any required leaching.

In some regions early season water is provided by "irrigating up," that is, seeding into dry soil followed by irrigation. Preplant irrigation is generally preferred because soil temperature may be lowered considerably in the irrigation process that slows germination and seedling emergence and creates a favorable soil environment for seedling pathogens. Furthermore, it may be difficult to rewet the whole profile without creating aeration problems by irrigating for emergence. Some soils may crust severely in this process, thereby making emergence and stand establishment difficult.

C. Scheduling Irrigation

1. First Irrigation

When preplant irrigation is done and there is little or no rainfall in the early growing season, the timing of the first irrigation is an important management consideration (Harris & Hawkins, 1942; Levin & Shmueli, 1964) that depends primarily on soil water retention properties and the prevailing climate. Early season growth of cotton affects the plant's response through the entire season; however, considerable disagreement exists in the literature on what constitutes a desirable early season water management scheme. Longenecker and Erie (1968) reviewed much of the work conducted prior to the mid-1960s. Their review placed emphasis on the high positive correlation observed between plant size at first flower and final yield and concluded that minimal stress should be imposed before flowering. They concluded that "all early season management practices in all areas should be designed to encourage vigorous and early plant growth until after first flowers appear, regardless of soil type, available water, or first flowering date." Eaton (1931) and Dunnam et al. (1943) found that removing fruiting forms early in the season stimulated plant growth and caused the plants to develop more fruiting positions. In the long growing season of Arizona, increased early vegetative growth gave higher lint yields, whereas seedcotton production in Mississippi was reduced in proportion to the time that defruiting continued. Good early growth is needed for heavy fruiting, but practices that delay maturity require a longer growing season to enhance seedcotton production. Recent investigations (Grimes et al., 1978; Guinn & Mauney, 1984a; Guinn et al., 1981) show that irrigations that are too early not only lower final yield, but result in reduced efficiency for increasingly scarce water and energy resources.

Guinn et al. (1981) conducted experiments on Avondale clay loam (fine-loamy, mixed [calcareous], hyperthermic Typic Torrifluvents) in Arizona and identified mid-June as an optimum first irrigation date. Both vegetative growth and number of fruiting positions were reduced by a May to June irrigation delay, but both the number of good bolls and boll weight increased to give higher seedcotton production. These workers attributed the detrimental effects of early irrigation to square loss caused by higher Lygus bug (*Lygus hesperus* Knight) populations. Leigh et al. (1974) found higher Lygus bug populations with more vigorously growing plants. A mid-June optimum irrigation date was also observed for a clay loam soil in the San Joaquin Valley, California (Grimes et al., 1978), but the detriment of too early an irrigation was attributed to reduced soil and canopy temperature at a time during the season when temperature is already growth limiting. A midday ψ_l of -1.6 MPa just before irrigation was observed by these workers for the optimum first irrigation. From the previous discussion of this chapter it is recognized that -1.6 MPa ψ_l at midday will slow growth, but will not cause fruiting form abscission. This level of ψ_l is greater than that tolerated at later irrigations, indicating that a less severe stress should be imposed for the first irrigation. While ψ_l is greater for the first than at later irrigations, it should be remembered that it is sufficient to slow growth; also, since evaporative demand is low early in the year, stress levels may be held for longer times than at later dates when evaporative demand is higher.

Because of the root development characteristics of cotton in early season, soil water depletion before the first postplant irrigation is nearly all from the surface 60 cm of soil (Grimes et al., 1978; Levin & Shmueli, 1964). Although site specific, models such as that developed by Grimes et al. (1978) are useful for planning purposes. These workers developed the empirical relation

$$RY = -32.3 + 57.43D^{1/2} - 5.659D - 9.957W + 1.449D^{1/2}W \qquad \text{Eq. [2]}$$

where RY is relative yield, D is days after 30 April, and W is centimeters of water held by the surface 61 cm of soil at planting time at soil ψ's > -1.5 MPa. For production regions having relatively uniform climatic conditions in early season, such a relationship can be used fairly successfully; adjustments for fluctuating evaporative demand can be made by observing soil or

Table 24–3. Optimum dates for the first postplant irrigation derived from Eq. [2] for the San Joaquin Valley, California. From Grimes et al. (1978).

Plant-available water in top 61 cm of soil	D	Calendar date
cm	days after 30 Apr.	
5	33	2 June
7	36	5 June
9	39	8 June
11	42	11 June
13	45	14 June

plant water status or both. Table 24–3 shows the use of this model for average climatic conditions and a normal planting date in the San Joaquin Valley.

2. Midseason Irrigations

Timing of irrigations in midseason is important. Too little water causes water deficits associated with fruiting form abscission and lower yields, while too frequent irrigations may cause excessive vegetative growth not associated with higher yields. Production is optimum with a water status that allows the plant to develop many fruiting positions while avoiding excessive vegetative development. While root systems are usually expanded to maximum development by this time, the potential for ET is high due to climatic conditions and fully developed canopies (see Fig. 24–2).

Critical periods to avoid severe water stress have been identified and the sensitivity of fruiting forms at various stages of development to water stress-induced abscission was discussed in an earlier section. Grimes et al. (1970) showed severe stress during peak flowering to be most detrimental, with the loss of both squares and young bolls contributing to a severe yield loss. Severe stress early in the flowering period was less detrimental since no prevailing bolls were present to abscise. Stress late in the flowering period was also intermediate in yield loss even though both squares and bolls were lost; squares retained in very late season will not have time to mature and contribute to yield.

Guinn and Mauney (1984b) correlated boll retention with active boll load and ψ_1 during the first and second weeks following an irrigation. Abscission rates were low with a small boll load when water stress was absent shortly after irrigation, but rates increased with an increasing boll load. As water stress developed, abscission was related more closely to a declining ψ_1. Similar effects of boll load and water stress on young boll abscission were observed earlier by Stockton et al. (1961).

Longenecker and Erie (1968) cited work that shows an allowable soil moisture depletion of about 65% of the available water in the top 91 cm of soil with optimum productivity maintained for furrow or sprinkler water delivery systems. Grimes and El-Zik (1982) reported that as much as 75% of the available soil water in the root zone of sandy loam soils can be depleted, once the root system is fully developed, without yield loss. On the other hand, only about 60 to 65% of the available water of clay loam soils should be depleted because clay soils hold more water at greater tensions than sandy soils. Neither soil water volume measurements or tensiometric measurements of the matric potential component of soil water take into account the osmotic potential from dissolved solids. This can be appreciable in many areas of the west and southwest regions of the USA and other production regions of the world. The water balance approach to irrigation scheduling can be expanded to include climatic observations for calculating ET_o (see Eq. [1]). This can reduce the required frequency for soil water measurements to occasional periodic checks to verify irrigation amounts, provided that appropriate values of ET_o and K_c are used.

For midseason irrigation scheduling, critical values have been determined for both ψ_l and infrared thermometry-determined CWSI. For midday measured ψ_l, the findings of Guinn and Mauney (1984b) and Grimes and Yamada (1982) are that yield loss can be expected when ψ_l's lower than -1.9 MPa are permitted before an irrigation event. Where water is supplied in frequent small additions, as with drip/trickle systems, production is highest with middayψ_l's maintained at approximately -1.6 to -1.7 MPa (Meron et al., 1984). Howell et al. (1984b) show CWSI to be sensitive to crop osmotic stress as well as to soil water matric potential. These workers, along with Reginato (1983), recommended that CWSI values of about 0.2 to 0.3 should signal the need for an irrigation. A CWSI of 0 represents no stress and 1 represents complete stress. This critical level correlated well with that recommended for ψ_l (Howell et al., 1984c).

Visual plant symptoms have historically been used to schedule cotton irrigations, but are subject to the individual interpretation and skill of the observer. Frequently used visual symptoms that are associated with water stress are darker green leaves, lack of new growth, and transient afternoon wilting (Longenecker & Erie, 1968). A much more quantitative assessment is possible with the techniques and equipment that are now available at an acceptable cost.

A new technology that is emerging for irrigation scheduling involves the use of expert systems or computer systems that perform at the level of human experts. The expert system, COMAX, is reported to be the first linkage with a cotton simulation model, GOSSYM, with the objective of optimizing cotton production by determining irrigation schedules, N requirement and crop maturity date (Lemmon, 1986). Other systems for cotton that are currently under development and testing in the USA include COTFLEX in Texas and CALEX in California (Goddell & Wilson, 1986).

3. Final or Cutoff Irrigation

In late season, greater water stress can be tolerated by the crop because productivity at this development stage is not dependent on retention of fruiting forms that are sensitive to water stress-induced abscission. Boll maturation is less affected by water stress (Grimes & Yamada, 1982), but sufficient water should be available to complete development of bolls that normally have time to mature. Prolonged irrigation delays mature boll opening (Kittock et al., 1983), favors a higher incidence of boll rots, and can make defoliation difficult (Walhood & Yamada, 1972) by increasing late growth or regrowth.

When conventional furrow or sprinkler irrigations replace evapotranspired water in the root zone, a final desired irrigation date within a climatic region can be related to water retention properties of soils (Grimes & Dickens, 1974). In the event irrigations have not replaced all evapotranspired water, or high frequency and low volume irrigation systems are used, scheduling of the last irrigation can be related to actual measurements of plant-available water in the effective rooting soil depth. For San Joaquin Valley, California climatic conditions, this relationship is illustrated by Fig. 24–10. The procedure is similar to that used to estimate when to first irrigate, but differs in that the entire rooting profile provides the base for estimates. Soils able to retain a large amount of plant-available water can achieve near-maximum productivity with a final irrigation near the first of August. Sandy soils of lower water-retaining capabilities will need a last irrigation as late as early September to maintain optimum productivity. Climatic regions differing from that illustrated can use the technique for estimating a desired final irrigation, but it must be emphasized that individual climatic, site-specific relations must be developed. For example, in Arizona and the lower desert region of California, climatic conditions may allow maturity of a late season "top crop" (Longenecker & Erie, 1968) that requires extension of the irrigation season to meet crop water demands. However, even this potential may require modifications, for in these regions in recent years final irrigations are being made early to minimize overwintering populations of pink bollworm (*Pectinophora gossypiella*) and tobacco budworm (*Heliothis virescens*).

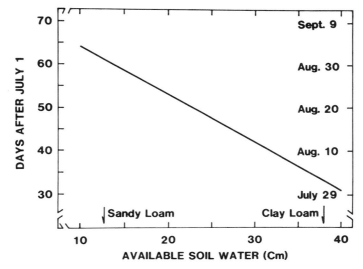

Fig. 24–10. Optimum scheduling of last irrigations in the San Joaquin Valley, California (Grimes & Dickens, 1974).

D. Irrigation in the Rainbelt

Waddle (1984) refers to the cotton-growing region east of 100 °W long. in the USA as chiefly rain-grown cotton. All areas of this region have a summer rain potential and, as pointed out by Waddle (1984), this potential separates irrigation philosophy between west and east. Although rainfall averages about 127 cm annually across the eastern Cotton Belt, periods of water deficit are common in June, July, and August (Thomas, 1987). Thomas (1987) reported that there were 162 000 ha of cotton supplementally irrigated in the Southeast as of 1985. Fundamental plant water relations, of course, are unaffected by region and those concepts discussed earlier in this chapter apply equally well across the Cotton Belt.

Preplant irrigation, generally practiced in the West, is rarely done or needed with rain-grown cotton. However, Bruce and Shipp (1962) and Bruce and Romkens (1965) conducted studies in Mississippi, utilizing plot shelters that closed and opened automatically as rainfall began and ended, that demonstrated the essential nature of an adequate soil water supply during the period from first bloom to 4 to 6 wk past first bloom.

Use of a reliable and quantitative indicator of water status is essential to best use a supplemental irrigation water supply in the Southeast. Waddle (1984) called attention to the fact that reddening of the upper stem areas and darkening of the green in uppermost leaves, once used by growers to indicate water stress, is not as viable with modern cultivars that have heritable reddish tendencies. As in regions totally dependent on irrigation for viable cotton production levels, irrigation scheduling in the Rainbelt depends on the amount of available soil water and the water requirement of the cotton plant at specific growth stages. A soil moisture depletion of 50% in the top 60 cm of the soil profile has shown the highest yield response in the Southeast (Thomas, 1987). Optimum scheduling would be expected from some form of water balance calculations that are coupled with information on soil water status and water retention characteristics, and measurements of plant water status.

With adequate supplemental irrigation, a common grower concern is when to stop irrigating. Waddle (1984) suggested that the last irrigation should not be applied later than 1 to 2 wk before the last effective flower on soils that have a water storage of 20 cm in the root zone. As water storage declines below 20 cm, the last irrigation approaches the last effective flowering date.

E. Water Stress Effects on Fiber Quality

Many fiber and seed quality characteristics are determined primarily by the genetic makeup of the variety. However, unfavorable growing conditions, including plant water stress, may modify the genetic potential of certain fiber characteristics. Bolls that mature late in the season when temperatures are lower require a longer period for fiber growth and development and usually produce lint of lower quality that is also reflective of water stress required to optimize defoliation and harvesting conditions. Fiber elongation and strength are influenced somewhat by water status, but fiber length and micronaire are most consistently affected (Grimes et al., 1969b; El-Zik et al., 1978). Micronaire is generally reduced by water management that delays maturity.

Fiber length is reduced by severe water stress. Grimes and Yamada (1982) showed that fiber elongation and weight increases proceeded relatively unchecked until midday ψ_l's reach -2.5 to 2.8 MPa where both are reduced sharply. Water management systems that restrict yield levels to 50 to 70% of the maximum possible will impose sufficient water stress to significantly shorten fiber.

F. Narrow-row Cotton

Cotton has traditionally been grown in rows that are 97 to 102 cm wide with cultural and harvest equipment designed for this planting configuration. Research conducted during the last 20 yr shows that rows spaced closer than 1 m wide have an increased production potential of as much as 10 to 30% (El-Zik et al., 1982; George et al., 1978). Enhanced production has been most notable for field sites that have some factor limiting productivity that is not uniformly attributable to the entire production region. Frequently cited advantages for narrow-row plantings include increased earliness potential and associated reduced production inputs of chemicals, water, and fertilizers (Howell et al., 1987). The earliness characteristic is frequently used in improved management schemes to control pink bollworm and secondary pests. A primary disadvantage of the narrow-row planting configurations is related to harvesting technology though considerable advances are being made.

Fundamental plant-water relations are unaffected by the narrow-row culture of cotton. However, some aspects of water management are influenced. If narrow-row systems use the full growing season and are not excessively water stressed, little or no reduced water or nutritional resource inputs can be achieved compared with conventional wide-row plantings (Grimes & Dickens, 1977; Howell et al., 1984a, 1987). However, because of the earliness potential of narrow-row plantings, water may be withdrawn earlier in late season to enhance early crop termination or maturity. Such early termination is associated with reduced production potential (Grimes & Dickens, 1977), but yield may still equal that of a full season, wide-row system.

Because the narrow-row system develops leaf surface more rapidly in early season than conventional wide rows, ET is greater in early season (Grimes & Dickens, 1977). Optimum scheduling of irrigations for narrow-row plantings must take into account this earlier water requirement. For example, a soil in the San Joaquin Valley that gives an optimum response to a first season irrigation in mid-June with wide rows should be irrigated at least 10 d earlier with narrow-row plantings.

VII. WATER-YIELD PRODUCTION FUNCTIONS

The production function provides a useful means of analyzing water-productivity relations if it is based on data that utilize proper irrigation scheduling to give the least possible yield reduction from a defined water deficit. A substantial number of water response functions have been developed in western states and several are presented by Hexem and Heady (1978).

As pointed out by Vaux and Pruitt (1983), a variety of independent variables are used to indicate water input; three of greatest importance are ET, applied water, and soil moisture. Evapotranspiration has the greatest rigor and potential for transferability. However, applied water is the controlled variable and, in an economic sense, represents the cost consideration. Soil water status provides a link between ET and applied water.

A. Evapotranspiration-Lint Function

For many crops and growing conditions the relationship between ET and yield is linear up to ET values that result in maximum productivity. This is especially true for crops where the aboveground biomass represents yield. Sammis (1981) reported a linear relation for cotton although longer-term studies (Grimes et al., 1969a; Grimes & El-Zik, 1982) suggested a slight curvature to the function. Considering the nature of cotton reproductive development and plant-water relations, a curvilinear function appears most appropriate for this crop. Figure 24–11 shows a graphic depiction of the square root function

$$Y_L = -3419 + 919.6ET^{1/2} - 42.71ET \qquad \text{Eq. [3]}$$

when Y_L is kg lint/ha and ET is centimeters of evapotranspired water for the growing season. This function represents a transformation of the relative yield (observed yield/maximum yield) $-$ ET function presented by

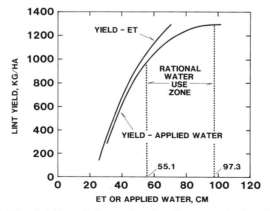

Fig. 24–11. Yield-ET and yield-applied water functions for cotton developed in the San Joaquin Valley, California (redrawn from Grimes & El-Zik, 1982).

Grimes and El-Zik (1982), assuming that maximum yield is 1300 kg lint/ha, a reasonable yield for San Joaquin Valley conditions.

B. Applied Water-Lint Function

As shown by Fig. 24–11, the applied water function progressively departs from the ET function as applied water increases. This results primarily from increased drainage below the root zone and larger amounts of applied water remaining in the soil profile at the end of the growing season at high applied-water input amounts (Grimes et al., 1969a). The yield-applied water function illustrated in Fig. 24–11 is

$$Y_L = -3954 + 1067AW^{1/2} - 54.14AW \qquad \text{Eq. [4]}$$

where Y_L is kg lint/ha and AW is centimeters of applied water. As with the yield-ET function, this is a transformation from a relative yield base assuming a maximum lint yield of 1300 kg/ha.

The limits of a "rational water use zone" are depicted in Fig. 24–11. Applied water to achieve maximum yield (97 cm) is the upper limit while AW required to reach a maximum average product ($AP = Y_L/AW$) is the basis for the lower limit. Heady (1952) presented the theoretical considerations justifying this lower limit and Grimes (1977) discussed the application to a cotton-applied water response. Applied water to maximize profit will always fall within the limits of this AW input zone. For example, if water costs $1.18/cm and cotton lint price is $1.76/kg, Eq. [4] can be used to determine the profit maximizing AW amount for these condition. If traditional production economics are used, AW is used up to an amount that balances the cost of the last centimeter used with the value of the product resulting from this last increment of AW; mathematically, $dY_L/d(AW) = P_w/P_y$ where P_w and P_y are the respective water cost and cotton lint price. For this illustration the profit maximizing AW amount is 94 cm, a value only slightly below the yield-maximizing AW quantity for this relatively high income crop.

C. Transferability

Empirically derived water production functions are usually correct only for the site-specific conditions under which they are developed. However, as pointed out by Vaux and Pruitt (1983), production functions that use relative ET (actual ET/potential ET) and relative yield (actual Y/maximum Y) offer some advantage toward a more generalized function. Sammis (1981) in New Mexico compared his linear production function (Y_L vs. ET) for cotton to a similar function developed in California and concluded that such functions are not transferable. On the other hand, Grimes (1982) used a dimensionless (relative yield vs. relative ET) function developed in California to compare with cotton functions presented by Ayer and Hoyt (1981). The California function was converted first to yield and ET values appropriate for Arizona. Then, by using an observed relation between ET and AW, the function was transformed to a yield vs. AW function for comparison with similar functions developed in Arizona. With this procedure the California and Arizona functions showed reasonably good agreement. This result and that of Misra (1973) and Stewart and Hagan (1973) lend support to the concept of achieving a fair amount of transferability among geographic regions of contrasting soil and climatic conditions.

VIII. WATER MANAGEMENT TO CONTROL PLANT PATHOGENS AND INSECTS

Population growth of insect herbivores and plant pathogens depends, in numerous interacting ways, on environmental conditions such as temperature, humidity, availability of free water, and wind. Of course, it is the microenvironment adjacent to individual organisms that is most relevant. Variations in soil moisture and crop canopy microclimate play an important role in the development and, conversely, in the control of insects, soilborne and aerial pathogens, and nematodes. Soil moisture may directly affect pathogens, their antagonists, or the length of the susceptible period of the host plant. Grimes (1986) and Grimes and El-Zik (1982) demonstrated that water management was an effective tool for manipulating plant growth and development and to control pests. Regulating the amount and timing of irrigations is a precise and practical method of integrated management, with timing of a water application often more important than the amount applied.

It has long been recognized that freshly irrigated cotton provides a lush and attractive host for lygus bugs, boll weevils, and leaf-feeding insects. Fletcher (1941) found a high correlation between percent moisture in the plant terminals and the number of bollworm eggs and larvae, and he reported that increased bollworm damage was associated with increased rainfall during June and July. The timing of irrigations may be altered so that the lush and attractive state of the plant does not coincide with periods of peak insect activity. This can frequently be done without reducing yield potential. In Texas, Slosser (1980) found that irrigations applied to cotton during peak ovipositional activity by *Heliothis zea* increased eggs. In nonstressed irrigated cotton, larval survival was increased. He concluded that *Heliothis* spp. activity in cotton could be reduced by not irrigating during 7- to 10-d before and 3- to 4-d after peak oviposition. Parencia et al. (1962) observed that *Heliothis* infestations in cotton were confined to irrigated acreages in drought years.

Leigh et al. (1974) reported that early July lygus bug populations in nonstressed irrigated plots in California are double that of stressed plots, reflecting differences in microclimate or in plant attractiveness. Guinn et al. (1981) attributed the detrimental effects of an early first irrigation to square loss caused by higher lygus bug populations than those of later irrigated plots. The last irrigation may be applied early to terminate the crop and escape damage from boll-feeding insects and boll-rotting microorganisms, and to minimize the number of overwintering populations of pink bollworm and tobacco budworm (Grimes & El-Zik, 1982).

Seed rot and damping-off of cotton seedlings caused by *Pythium* spp. and *Rhizoctonia solani* can be reduced by maintaining a dry soil surface early in the season. Scheduling the first post-plant irrigation was observed by Grimes and Huisman (1984) and El-Zik (1985) to be closely tied to the severity of foliar Verticillium wilt symptoms. Defoliation was most severe with an early first irrigation on a Panoche clay loam soil (fine-loamy, mixed [calcareous], thermic Typic Torriorthents) with a high *Verticillium dahliae* inoculum density. As the first irrigation was progressively delayed, defoliation severity was diminished and cotton yields increased.

Any modification in prevailing practices in cotton production may create changes, sometimes subtle, in the plant environment. Modifications in the time and amount of irrigation water application have an impact on pest populations and cotton productivity.

IX. SUMMARY

The cotton plant, though complex, is systematically programmed and, under favorable conditions, its growth and development follow an orderly and predictable pattern. Understanding the pattern of development is necessary for planning and implementing efficient management systems.

Fundamental cotton nutrition, plant-water relations, and reaction to saline conditions and toxic ions provide the basic concepts for developing and implementing specific irrigation practices that are directed to site-specific conditions of soil, climate, and pest and disease complexes of the agroecosystem. Irrigation scheduling is best accomplished by monitoring the water status at one or more points in the soil-plant-air continuum and relating defined stresses to stage of cotton plant development. A final decision on irrigation timing may also be contingent on prevailing insect/pathogen pressures during the season or the need to terminate a crop early to provide a defined host-free period. The water, nutrient supply, and planting pattern and density can be managed to achieve a desired canopy architecture that controls, to some extent, the microclimate in the canopy. The microclimate and overall attractiveness of the plant exert significant influences on a number of pests and pathogens; this points to the opportunities and desirability of managing the total cotton production system, not just a single entity.

Irrigation scheduling and some related cultural practices provide a unique opportunity to control the growth and development of cotton. This is especially true in arid and semiarid zones where little rainfall may occur during the growing season. However, application of plant-water relations concepts may well apply to regions where irrigation may be needed only to supplement normal rainfall during occasional periods of drought stress.

REFERENCES

Ackerson, R.C., and D.R. Krieg. 1977. Stomatal and nonstomatal regulation of water use in cotton, corn, and sorghum. Plant Physiol. 60:850–853.

Ackerson, R.C., D.R. Krieg, C.L. Haring, and N. Chang. 1977a. Effects of plant water status on stomatal activity, photosynthesis, and nitrate reductase activity of field grown cotton. Crop Sci. 17:81–84.

Ackerson, R.C., D.R. Krieg, T.D. Miller, and R.E. Zartman. 1977b. Water relations of field grown cotton and sorghum: Temporal and diurnal changes in leaf water, osmotic, and turgor potentials. Crop Sci. 17:76–80.

Ayer, H.W., and P.G. Hoyt. 1981. Crop-water production functions: Economic implications for Arizona. Arizona Agric. Exp. Stn. Tech. Bull. 242.

Ayers, R.S. 1977. Quality of water for irrigation. J. Irrig. Drain. Div. Am. Soc. Civ. Eng. 103(IR2):135–154.

Bassett, D.M., W.D. Anderson,and C.H.E. Werkhoven. 1970a. Dry matter production and nutrient uptake in irrigated cotton (Gossypium hirsutum). Agron. J. 62:299–303.

Bassett, D.M., J.R. Stockton, and W.L. Dickens. 1970b. Root growth of cotton as measured by P^{32} uptake. Agron. J. 60:200–203.

Berger, J. 1969. The world's major fiber crops, their cultivation and manuring. Centre d'Etude de l'Azote, Zurich.

Bielorai, H., and P.A.M. Hopmans. 1975. Recovery of leaf water potential, transpiration, and photosynthesis of cotton during irrigation cycles. Agron. J. 67:629–632.

Bordovsky, D.G., W.R. Jordan, E.A. Hiler, and T.A. Howell. 1974. Choice of irrigation timing indicator for narrow row cotton. Agron. J. 66:88–91.

Boyer, J.S. 1965. Effects of osmotic water stress on metabolic rates of cotton plants with open stomata. Plant Physiol. 40:229–234.

Brown, K.W., W.L. Jordan, and J.C. Thomas. 1976. Water stress induced alterations of the stomatal response to decreases in leaf water potential. Physiol. Plant. 37:1–5.

Bruce, R.R., and M.J.M. Römkens. 1965. Fruiting and growth characteristics of cotton in relation to soil moisture tension. Agron. J. 57:135–140.

Bruce, R.R., and C.D. Shipp. 1962. Cotton fruiting as affected by soil moisture regimes. Agron. J. 54:15–18.

Christidis, B.G., and G.J. Harrison. 1955. Cotton growing problems. McGraw-Hill Book Co., New York.

Cutler, J.M., and D.W. Rains. 1977. Effects of irrigation history on responses of cotton to subsequent water stress. Crop Sci. 17:329–335.

Cutler, J.M., D.W. Rains, and R.S. Loomis. 1977. Role of changes in solute concentration in maintaining favorable water balance in field-grown cotton. Agron. J. 69:773–779.

Davis, K.R., H.I. Nightingale, and C.J. Phene. 1980. Consumptive water requirement of trickle irrigated cotton: I. Water use and plant response. ASAE Proc. Pap. 80-2080.

Dennis, R.E., and R.E. Briggs. 1969. Growth and development of the cotton plant in Arizona. Univ. Arizona Bull. A-64.

Doorenbos, J., and W.O. Pruitt. 1977. Guidelines for predicting crop water requirements. FAO Irrig. Drain. Pap. 24.

Doss, B.D., D.A. Ashley, and O.L. Bennett. 1964. Effect of moisture regime and stage of plant growth on moisture use by cotton. Soil Sci. 98:156–161.

Dunnam, E.W., J.C. Clark, and S.L. Calhoun. 1943. Effect of the removal of squares on yield of upland cotton. J. Econ. Entomol. 36:896–900.

Eaton, F.M. 1931. Early defoliation as a method of increasing cotton yields and the relation of fruitfulness to fiber and boll characters. J. Agric. Res. 42:447–462.

Eaton, F.M. 1932. Boron requirement of cotton. Soil Sci. 34:301–305.

Eaton, F.M. 1944. Deficiency, toxicity, and accumulation of boron in plants. J. Agric. Res. 69:237–277.

Eaton, F.M. 1955. Physiology of the cotton plant. Annu. Rev. Plant Physiol. 6:299–328.

El-Zik, K.M. 1980. The cotton plant—its growth and development. Proc. West. Cotton Prod. Conf. 1980:18–21.

El-Zik, K.M. 1985. Integrated control of Verticillium wilt of cotton. Plant Dis. 69:1025–1032.

El-Zik, K.M., K. Brittan, C. Brooks, R.G. Curley, A.G. George, R.A. Kepner, T.A. Kerby, O.D. McCutcheon, L.K. Stromberg, R.N. Vargas, D. West, and B. Weir. 1982. Effects of row spacing on cotton yield, quality, and plant characteristics. Univ. California Coop. Ext. Serv. Bull. 1903.

El-Zik, K.M., and R.E. Frisbie. 1985. Integrated crop management systems for pest control and plant protection. p. 21–22. *In* N.B. Mandava (ed.) Handbook of natural pesticides: Methods. Vol. I: Theory, practice, and detection. CRC Press, Boca Raton, FL.

El-Zik, K.M., V.T. Walhood, and H. Yamada. 1978. Effects of irrigation scheduling, amount, and plant population on Acala cotton cultivars (*Gossypium hirsutum* L.). p. 204–205. *In* Proc. Beltwide Cotton Prod. Res. Conf. Natl. Cotton. Counc. of Am., Memphis.

El-Zik, K.M., H. Yamada, and V.T. Walhood. 1980. Effect of management on blooming, boll retention, and productivity of upland cotton, *Gossypium hirsutum* L. p. 69–70 *In* Proc. Beltwide Cotton Prod. Res. Conf. Natl. Cotton Counc. of Am., Memphis.

Ergle, D.R., and R.M. Eaton. 1957. Aspects of phosphorus metabolism in the cotton plant. Plant Physiol. 32:106–113.

Erie, L.J., O.F. French, D.A. Bucks, and K. Harris. 1982. Consumptive use of water by major crops in the southwestern United States. USDA-ARS Conserv. Res. Rep. 29.

Fletcher, R.K. 1941. The relation of moisture content of the cotton plant to oviposition by *Heliothis armigera* (Hbn.) and to survival of young larvae. J. Econ. Entomol. 34:856–858.

Gardner, W.R., and C.F. Ehlig. 1965. Physical aspects of the internal water relations of plant leaves. Plant physiol. 40:705–710.

George, A., O.D. McCutcheon, C.R. Brooks, R.G. Curley, K. El-Zik, and R.E. Johnson. 1978. Summary report of narrow row field trials, 1971–1973. Univ. California Coop. Ext. Serv. Spec. Rep. 3205.

Gipson, J.R., and L.L. Ray. 1969. Fiber elongation rates in five varieties of cotton (*Gossypium hirsutum* L.) as influenced by night temperature. Crop Sci. 9:339–341.

Goodell, P.B., and L.T. Wilson. 1986. Computer models for cotton production. Calif. Cotton Prog. Rep. 1986:5–7.

Grimes, D.W. 1977. Physiological response of cotton to water and its impact on economical production. Proc. West. Cotton Prod. Conf. 1977:22–25.

Grimes, D.W. 1982. Water requirements and use patterns of the cotton plant. Proc. West. Cotton Prod. Conf. 1982:27–30.

Grimes, D.W. 1986. Cultural techniques for management of pests in cotton. p. 365–382. *In* R.E. Frisbie, P.L. Adkisson (ed.) Integrated pest management on major agricultural systems. Texas Agric. Exp. Stn. MP-1616.

Grimes, D.W., and W.L. Dickens. 1974. Dating termination of cotton irrigation from soil water-retention characteristics. Agron. J. 66:403–404.

Grimes, D.W., and W.L. Dickens. 1977. Irrigation water management of cotton for a planting configuration and variety conducive to short-season development. Univ. California Water Sci. Eng. Pap. 7003.

Grimes, D.W., W.L. Dickens, and W.D. Anderson. 1969b. Functions for cotton (*Gossypium hirsutum* L.) production from irrigation and nitrogen fertilization variables: II. Yield components and quality characteristics. Agron. J. 61:773–776.

Grimes, D.W., W.L. Dickens, and H. Yamada. 1978. Early-season water management of cotton. Agron. J. 70:1009–1012.

Grimes, D.W., and K.M. El-Zik. 1982. Water management for cotton. Univ. California Div. Agric. Sci. Bull. 1904.

Grimes, D.W., and O.C. Huisman. 1984. Irrigation scheduling and Verticillium wilt interactions in cotton production. p. 88–92. *In* Calif. Plant Soil Conf. Proc. California Chapt. ASA, Madison, WI.

Grimes, D.W., R.J. Miller, and L. Dickens. 1970. Water stress during flowering of cotton. Calif. Agric. 24(3):4–6.

Grimes, D.W., R.J. Miller, and P.L. Wiley. 1975. Cotton and corn root development in two field soils of different strength characteristics. Agron. J. 67:519–523.

Grimes, D.W., R.L. Sharma, and D.W. Henderson. 1984. Developing the resource potential of a shallow water table. Univ. California Water Resour. Cent. Contrib. 188.

Grimes, D.W., and H. Yamada. 1982. Relation of cotton growth and yield to minimum leaf water potential. Crop Sci. 22:134–139.

Grimes, D.W., H. Yamada, and W.L. Dickens. 1969a. Functions for cotton (*Gossypium hirsutum* L.) production from irrigation and nitrogen fertilization variables: I. Yield and evapotranspiration. Agron. J. 61:769–773.

Guinn, D. 1979. Hormonal relations in flowering, fruiting, and cutout. p. 265–276. *In* Proc. Beltwide Cotton Prod. Res. Conf. Natl. Cotton Counc. of Am., Memphis.

Guinn, G. 1982. Causes of square and boll shedding in cotton. U.S. Dep. Agric. Tech. Bull. 1972.

Guinn, G., and J.R. Mauney. 1984a. Fruiting of cotton. I. Effects of moisture status on flowering. Agron. J. 76:90–94.

Guinn, G., and J.R. Mauney. 1984b. Fruiting of cotton. II. Effects of plant moisture status and active boll load on boll retention. Agron. J. 76:94–98.

Guinn, G., J.R. Mauney, and K.E. Fry. 1981. Irrigation scheduling and plant population effects on growth, bloom rates, boll abscission, and yield of cotton. Agron. J. 73:529–534.

Harris, K., and R.S. Hawkins. 1942. Irrigation requirements of cotton on clay loam soils in the Salt River Valley. Arizona Agric. Exp. Stn. Bull. 181.

Hayward, H.E., and C.H. Wadleigh. 1949. Plant growth on saline and alkali soils. Adv. Agron. 1:1–38.

Heady, E.O. 1952. Economics of agricultural production and resource use. Prentice-Hall, Englewood Cliffs, NJ.

Helmy, H., H.E. Joham, and W.C. Hall. 1960. Magnesium nutrition of American upland and Egyptian cottons. Texas Agric. Exp. Stn. MP-411.

Hexem, R.W., and E.O. Heady. 1978. Water production functions for irrigated agriculture. The Iowa State Univ. Press, Ames.

Hinkle, D.A., and A.J. Brown. 1968. Secondary nutrients and micronutrients. p. 281–320. *In* F.C. Elliott et al. (ed.) Advances in production and utilization of quality cotton: Principles and practices. The Iowa State Univ. Press, Ames.

Hoffman, G.J., and C.J. Phene. 1971. Effects of constant salinity levels on water use efficiency of bean and cotton. Trans. ASAE 14:1103–1106.

Hoffman, G.J., S.L. Rawlings, M.J. Garber, and E.M. Cullen. 1971. Water relations and growth of cotton as influenced by salinity and relative humidity. Agron. J. 63:822–826.

Howell, T.A., K.R. Davis, R.L. McCormick, H. Yamada, V.T. Walhood, and D.W. Meek. 1984a. Water use efficiency of narrow row cotton. Irrig. Sci. 5:195–214.

Howell, T.A., J.L. Hatfield, J.D. Rhoades, and M. Meron. 1984b. Response of cotton water stress indicators to soil salinity. Irrig. Sci. 5:25–36.

Howell, T.A., J.L. Hatfield, H. Yamada, and K.R. Davis. 1984c. Evaluation of cotton canopy temperature to detect crop water stress. Trans. ASAE 27(1):84–88.

Howell, T.A., M. Meron, K.R. Davis, C.J. Phene, and H. Yamada. 1987. Water management of trickle and furrow irrigated narrow row cotton in the San Joaquin Valley. ASAE Appl. Eng. 3:222–227.

Hsiao, T.C., E. Acevedo, E. Fereres, and D.W. Henderson. 1976. Stress metabolism: Water stress, growth, and osmotic adjustments. Phil Trans. R. Soc. (London) 273:479–500.

Hutmacher, R.M., and D.R. Krieg. 1983. Photosynthetic rate control in cotton. Plant Physiol. 73:658–661.

Idso, S.B., R.D. Jackson, P.J. Pinter, R.J. Reginato, and J.L. Hatfield. 1981. Normalizing the stress-degree-day parameter for environmental variability. Agric. Meteorol. 24:45–55.

Joham, H.E. 1955. The calcium and potassium nutrition of cotton as influenced by sodium. Plant Physiol. 30:4–10.

Joham, H.E. 1979. The effect of nutrient elements on fruiting efficiency. p. 306–311. *In* Proc. Beltwide Cotton Prod. Res. Conf. Natl. Cotton Counc. of Am., Memphis.

Joham, H.E. 1986. Effects of nutrient elements on fruiting efficiency. p. 79–90. *In* J.R. Mauney and J.M. Stewart (ed.) Cotton physiology. The Cotton Foundation, Memphis.

Joham, H.E., and V. Roe. 1975. Temperature and zinc interactions on cotton growth. Agron. J. 67:313–317.

Jones, U.S., and C.E. Bardsley. 1968. Phosphorous nutrition. p. 213–254. *In* F.C. Elliott et al. (ed.) Advances in production and utilization of quality cotton: Principles and practices. The Iowa State Univ. Press Ames.

Jordan, W.R. 1979. Influence of edaphic parameters on flowering, fruiting, and cutout A. role of plant water deficit. p. 297–301. *In* Proc. Beltwide Cotton Prod. Res. Conf. Natl. Cotton Counc. of Am., Memphis.

Jordan, W.R., K.W. Brown, and J.C. Thomas. 1975. Leaf age as a determinant in stomatal control of water loss from cotton during water stress. Plant Physiol. 56:595–599.

Kamprath, E.J., and C.E. Welch. 1968. Potassium nutrition. p. 256–280. *In* C. Elliott et al. (ed.) Advances in production and utilization of quality cotton: Principles and practices. The Iowa State Univ. Press, Ames.

Karami, E., D.R. Krieg, and J.E. Quisenberry. 1980. Water relations and carbon-14 assimilation of cotton with different leaf morphology. Crop Sci. 20:421–426.

Kittock, D.L., T.J. Henneberry, L.A. Bariola, B.B. Taylor, and W.C. Hofmann. 1983. Cotton boll period response to water stress and pink bollworm. Agron. J. 75:17–20.

Krieg, D.R., and F.J.M. Sung. 1979. Source-sink relations of cotton as affected by water stress during boll development. p. 302–305. In Proc. Beltwide Cotton Prod. Res. Conf. Natl. Cotton. Counc. of Am., Memphis.

Lauchli, A., L.M. Kent, and J.C. Turner. 1981. Physiological responses of cotton genotypes to salinity. Proc. Beltwide Cotton Prod. Res. Conf. 1981:40. Natl. Cotton Counc. of Am., Memphis.

Leffler, H.R. 1986. Mineral compartmentation within the boll. p. 301–309. In J.R. Mauney and J.M. Stewart (ed.) Cotton physiology. The Cotton Foundation, Memphis.

Leigh, T.F., D.W. Grimes, W.L. Dickens, and C.E. Jackson. 1974. Planting pattern, plant population, irrigation, and insect interactions in cotton. Environ. Entomol. 3:492–496.

Lemmon, H. 1986. Comax: An expert system for cotton management. Science 233:29–33.

Levin, I., and E. Shmueli. 1964. The response of cotton to various irrigation regimes in the Hula Valley. Isr. J. Agric. Res. 14(4):211–225.

Longenecker, D.E., and L.J. Erie. 1968. Irrigation water management. p. 322–345. In F.C. Elliott et al. (ed.) Advances in production and utilization of quality cotton: Principles and practices. The Iowa State Univ. Press, Ames.

Mauney, J.R. 1983. Anatomy and morphology of cultivated cottons. In R.J. Kohel and C.F. Lewis (ed.) Cotton. Agronomy 24:59–80.

Mauney, J.R. 1986. Vegetative growth and development of fruiting sites. p. 11–28. In J.R. Mauney and J. Stewart (ed.) Cotton physiology. The Cotton Foundation, Memphis.

Mauney, J.R., G. Guinn, and K.E. Fry. 1980. Analysis of increases of flowers in moisture stressed cotton. p. 38. In Proc. Beltwide Cotton Prod. Res. Conf. Natl. Cotton Counc. of Am., Memphis.

McMichael, B.L. 1979. The influence of plant water stress on flowering and fruiting of cotton. p. 301–302. In Proc. Beltwide Cotton Prod. Res. Conf. Natl. Cotton Counc. of Am., Memphis.

McMichael, B.L., and G. Guinn. 1980. The effects of moisture deficits on square shedding. p. 38. In Proc. Beltwide Cotton Prod. Res. Conf. Natl. Cotton Counc. of Am., Memphis.

McMichael, B.L., and J.D. Hesketh. 1982. Field investigations of the response of cotton to water deficits. Field Crops Res. 5:319–333.

McMichael, B.L., W.R. Jordan, and R.D. Powell. 1973. Abscission processes in cotton: Induction by plant water deficit. Agron. J. 65:202–204.

Meron, M., C.J. Phene, T.A. Howell, K.R. Davis, and D.W. Grimes. 1984. Scheduling drip irrigated narrow row cotton. Proc. Spec. Conf. Irrig. Drain. Div. Am. Soc. Civ. Eng. 1984:314–322.

Misra, R.D. 1973. Responses of corn to different sequences of water stress as measured by evapotranspiration deficits. Ph.D. diss. Univ. California, Davis (Diss. Abstr. 73:32288).

Morgan, J.M. 1984. Osmoregulation and water stress in higher plants. Annu. Rev. Plant Physiol. 35:299–319.

Oliveira, E.C., Jr. 1982. Growth and adaptation of cotton in the field under water deficit conditions. Ph.D. diss. Univ. California, Davis. (Diss. Abstr. 83-04777).

Parencia, C.R., C.B. Cowan, and J.W. Davis. 1962. Relationship of Lepidoptera light-trap collections to cotton field infestations. J. Econ. Entomol. 55:692–695.

Peterson, G.D., Jr., R.L. Cowan, and P.H. Van Schaik. 1962. Cotton production in the desert valleys of California. California Exp. Stn. Circ. 508.

Radin, J.W., J.R. Mauney, and G. Guinn. 1985. Effects of N fertility on plant water relations and stomatal responses to water stress in irrigated cotton. Crop Sci. 25:110–114.

Radin, J.W., and L.L. Parker. 1979a. Water relations of cotton plants under nitrogen deficiency. I. Dependence upon leaf structure. Plant Physiol. 64:495–498.

Radin, J.W., and L.L. Parker. 1979b. Water relations of cotton plants under nitrogen deficiency. II. Environmental interactions on stomata. Plant Physiol. 64:499–501.

Radulovich, R.A. 1984. Reproductive behavior and water relations of cotton. Ph.D. Diss., Univ. California, Davis (Diss. Abstr. 84-25012).

Reginato, R.J. 1983. Field quantification of crop water stress. Trans. ASAE 26(3):772–775, 781.

Sammis, T.W. 1981. Yield of alfalfa and cotton as influenced by irrigation. Agron. J. 73:323–329.

Scholander, P.F., H.T. Hammel, E.D. Bradstreet, and E.A. Nemmingsen. 1965. Sap pressure in vascular plants. Science 148:339–346.

Slosser, J.E. 1980. Irrigation timing for bollworm management in cotton. J. Econ. Entomol. 78:356–359.

Sorour, F.A. 1963. Certain factors affecting manganese and molybdenum accumulation and distribution in the cotton plant. Ph.D. diss., Texas A&M Univ., College Station (Diss. Abstr. 64-01535).

Stewart, J.I., and R.M. Hagan. 1973. Functions to predict effects of crop water deficits. J. Irrig. Drain. Div. Am. Soc. Civ. Eng. 100:179–199.

Stockton, J.R., L.D. Doneen, and V.T. Walhood. 1961. Boll shedding and growth of the cotton plant in relation to irrigation frequency. Agron. J. 53:272–275.

Taylor, H.M., and B. Klepper. 1978. The role of rooting characteristics in the supply of water to plants. Adv. Agron. 30:99–128.

Tharp, W.H. 1965. The cotton plant—how it grows and why its growth varies. USDA Handb. 178. U.S. Gov. Print. Office, Washington, DC.

Thomas, J.C., K.W. Brown, and W.R. Jordan. 1976. Stomatal response to leaf water potential as affected by preconditioning water stress in the field. Agron. J. 68:706–709.

Thomas, J.G. 1987. Altering the crop: Plant requirements and irrigation practices in the east. p. 9–10 In Proc. Beltwide Cotton Prod. Res. Conf. Natl. Cotton Counc. of Am., Memphis.

Tucker, T.C., and B.B. Tucker. 1968. Nitrogen nutrition. p. 183–211. In F.C. Elliot et al. (ed.) Advances in production and utilization of quality cotton: Principles and practices. The Iowa State Univ. Press, Ames.

van den Honert, T.H. 1948. Water transport in plants as a catenary process. Disc. Faraday Soc. 3:146–153.

Vaux, H.J., Jr., and W.O. Pruitt. 1983. Crop-water production functions. p. 61–97. In D. Hellel (ed.) Advances in irrigation. Vol. 2. Academic Press, San Diego.

Walhood, V.T., and H. Yamada. 1972. Varietal characteristics and irrigation practices as harvest aids in narrow row cotton. p. 43–44. In Proc. Beltwide Cotton Prod. Res. Conf. Natl. Cotton Counc. of Am., Memphis.

Whisler, F.D., B. Acock, D.N. Baker, R.E. Fye, H.F. Hodges, J.R. Lambert, H.E. Lemmon, J.M. McKinion, and V.R. Reddy. 1986. Crop simulation models in agronomic systems. Adv. Agron. 40:141–207.

Waddle, B.A. 1984. Cotton growing practices. In R.J. Kohel, and C.F. Lewis (ed.) Cotton. Agronomy 24:233–263.

25   Sunflower[1]

PAUL W. UNGER

USDA-ARS
Bushland, Texas

Sunflower (*Helianthus annuus* L.) is grown in many parts of the world, with the major production countries or regions being the USSR, Argentina, Eastern Europe, USA, People's Republic of China, European Community, and Spain (NDSU, 1985). In the USA, most production occurs in North Dakota, South Dakota, Minnesota, and Texas (NDSU, 1985). Most sunflower is grown without irrigation, but irrigation is important in areas such as the central and southern Great Plains (USA), Spain, Australia, and India. Much of the literature pertaining to irrigation is from these countries.

Where precipitation and the soil water supply are limited, most crops, including sunflower, respond positively to irrigation with respect to growth and yield. For sunflower, the amount and timing of irrigation are important for efficient use of applied water and for maximizing crop yields.

I. CROP DEVELOPMENT

A. Growth Stages

Several systems of classifying sunflower growth stages have been developed (Marc & Palmer, 1981; Robinson, 1971; Schneiter & Miller, 1981; Siddiqui et al., 1975). For each system, time of planting is known and the major stages (emergence, budding, anthesis, and maturity) are readily identifiable. Although subdivisions of growth stages were identified by Schneiter and Miller (1981) and by Siddiqui et al. (1975) and are useful for some purposes, the above classifications generally are adequate with respect to irrigation of sunflower.

The duration of each growth stage varies with cultivar, year, and planting date. While differences among cultivars grown under identical conditions result from inherent characteristics of the cultivars, the same cultivars may develop at different rates in different years and when planted at different dates. The year and planting date effects are largely the result of prevailing environmental conditions, with soil and/or air temperature being especially

---

[1] Contribution from the USDA-ARS, Conserv. and Production Res. Lab., P.O. Drawer 10, Bushland, TX 79012.

---

critical for achene germination and seedling emergence and early growth (Garside, 1984; Robinson, 1971; Unger, 1980, 1986a, b). For later growth stages, the aerial environment (temperature, solar radiation, and daylength) has a greater influence on the duration of growth periods than does the soil environment (Keefer et al., 1976; Robinson, 1971; Unger, 1986a, b). The effect of irrigation per se on duration of later growth stages undoubtedly would be minimal. Irrigation for emergence could lower the soil temperature, which could slow germination, emergence, and early growth.

As for duration of different growth periods, the total time from planting to maturity varies with cultivars, years, and planting dates. For six cultivars planted on seven dates in Minnesota, the average total growing season ranged from 104 to 114 d in three different years. Averaged over years and dates, the growing season ranged from 106 to 113 d for the cultivars; it ranged from 97 to 124 d because of planting dates averaged over years and cultivars (Robinson, 1971). In a 2-yr study in Texas, sunflower required 167 and 93 d from planting to physiological maturity when planted in late March or early July, respectively (Unger, 1986a). Besides planting date, length of most growth periods was significantly correlated with average soil temperature, minimum air temperature, summation of daylight, daylength, and solar radiation. In that study, a relatively high soil water content was maintained by irrigation; hence, an evaluation of irrigation effects was not possible.

B. Plant and Root Development

Sunflower plants grow rather slowly during the initial 30 d after emergence and attain a height of about 30 cm or less during that period. Plant height is about quadrupled during the next 30 d and reaches a maximum height at about 70 to 80 d after emergence. The greatest height response to irrigation occurred when water was applied at budding, and the maximum number of leaves was attained at 45 to 55 d after planting, but there were no consistent trends because irrigations were applied more than 40 d after planting (Unger, 1983). Water stress reduces the number of leaves per plant when it occurs before 20 d after planting (Marc & Palmer, 1976). Leaf areas were not reported for the above studies, but maximum leaf area indices occurred at about 60 to 85 d after planting for different irrigation treatments in other studies (Andhale & Kalbhor, 1980; Connor et al., 1985; Mason et al., 1983; Somasundaram & Iruthayaraj, 1979). For sunflower irrigated at different growth stages, maximum dry matter accumulation in stems and leaves occurred at 65 d after planting; for heads, it occurred at 75 d after planting. Although the maximums occurred at the same time for all treatments, the maximum accumulations differed because of irrigation treatments.

Deep rooting of the sunflower crop is not important where adequate irrigation water is available to maintain a high soil water content. Where water for irrigation is limited, extraction of stored soil water by deep rooting is an important means by which the crop can produce favorable yields. Although root length density in soil may be less for sunflower than for such crops as corn (*Zea mays* L.) and sorghum [*Sorghum bicolor* (L.) Moench]

(Mason et al., 1983), sunflower has been shown to root deeper than these crops in some soils (for example, Pullman series—fine, mixed, thermic Torrertic Paleustoll). In this soil, corn and sorghum usually extract water only to about a 1.2-m depth, even when adequate water for root growth is present at greater depths (Musick et al., 1976; Unger, 1986c; Unger & Jones, 1981). In contrast, sunflower extracted water to depths of 1.8 m or more in Pullman soil (Jones, 1978, 1984; Jones & Unger, 1977; Unger, 1978). Sunflower also extracted soil water to a lower matric potential than corn on Heimdal loam (coarse-loamy, mixed Udic Haploborall) in North Dakota (Stegman & Lemert, 1980). Degree of extraction, however, may be influenced by the soil under consideration because Mason et al. (1983) reported that sunflower extracted less water than sorghum from Marah clay loam in Australia. Besides deep rooting per se, root depth penetration apparently occurs throughout most of the growing season (Jones, 1978), thus providing some water to plants during late reproductive stages. This undoubtedly contributes to the success of limited irrigation for sunflower on soils having relatively high soil water storage potentials (Connor et al., 1985; Unger, 1978, 1982).

C. Harvestable Dry Matter

Most sunflower is grown for edible oil production. Achenes of presently available hybrids contain about 400 to 500 g kg^{-1} of oil, with oleic and linoleic acids comprising about 850 to 900 g kg^{-1} of the oil. In general, oleic acid concentration of the oil decreases and linoleic acid concentration increases with decreases in prevailing temperatures during achene development. Consequently, oil quality, even for the same cultivar, may vary widely, depending on the climatic conditions under which the sunflower is grown (for example, northern or southern USA). Oil high in oleic acid is preferred for cooking or frying snack foods that require storage before consumption. Oil high in linoleic acid, which is polyunsaturated, is preferred for salad oil and margarine (Robertson et al., 1979). After extracting the oil, the remaining meal contains about 300 g kg^{-1} of protein and is suitable for use in livestock rations (Dinusson, 1977; Richardson et al., 1978). Sunflower also is suitable as livestock feed when harvested as silage (Dinusson, 1977), but such use apparently is limited. Other major uses of sunflower are for confectionary (large achene types) and birdseed (small achene types) purposes. Sunflower stalks remaining after achene harvest have been evaluated for use as fuel (LaRue & Pratt, 1977) and for manufacture as particleboard (Gertjejansen, 1976). While feasible, such use is limited.

Achene yields usually are highest when sunflower is adequately irrigated to avoid plant water stress and have exceeded 4 Mg ha^{-1} (Connor et al., 1985; Decau et al., 1973; Delibaltov & Ivanov, 1973; Dimitrov, 1975; Marty et al., 1972; Mikhov, 1974, 1975). Generally, achene yields are below 4 Mg ha^{-1}, with those in the 2- to 3-Mg ha^{-1} range being much more common. Total dry matter yields also vary greatly, with harvest indices (HI's) generally in the 0.30 to 0.40 range (Connor et al., 1985; Karami, 1977; Thompson & Fenton, 1979; Unger, 1983). Connor et al. (1985) reported HI's around

0.50 for some irrigation treatments; values around 0.20 prevailed for some treatments in the studies by Karami (1977) and Rawson and Turner (1983) and for various cultivars in the study by Murphy (1978).

II. EVAPOTRANSPIRATION

Reported values of consumptive water use of evapotranspiration (ET) by sunflower vary widely, ranging from <200 mm (Bhan & Khan, 1980; Jana et al., 1982; Mungse & Bhapkar, 1983) to more than 900 mm (Aranda, 1979; Muriel et al., 1974; Talha, 1976; Unger, 1986a). The wide range in ET, in part, is attributable to the irrigation levels used, climatic regions involved, and length of growing season. A high level of irrigation or rainfall that provides adequate water for nonstress conditions throughout the growing season results in greater ET than where water stress because of limited irrigation or rainfall limits ET. Likewise, ET is greater in a climatic region where the evaporative demand is high than where it is low. The high ET (>900 mm) occurred in Spain, Egypt, and west Texas, whereas the low ET occurred in India. Length of growing season, as affected by planting date, resulted in ET ranging from 605 to 959 mm in west Texas (Unger, 1986a). All sunflower was fully irrigated, but the early planted (March) sunflower developed slower and, therefore, used much more water than the sunflower planted from late May to July, which developed more rapidly.

Some of the low reported ET values also may be because of incomplete accounting of the water used from the soil because, in some reports, the depth of measurement was <1.0 m or not reported. Measurement to <1.0 m may be adequate in soils having a restricting layer in the profile or where water additions (rainfall or irrigations) do not deeply wet the profile. However, sunflower is capable of deep rooting and may have used more water from the profile than the amount reported.

Because of the wide range of reported values, the total water requirement for sunflower is uncertain. However, ET from about 500 to 700 mm resulted in near top yields in many cases (Browne, 1977; Muriel et al., 1974; Schuppan & Thomas, 1976; Somasundaram & Iruthayaraj, 1981; Talha, 1976; Talha & Osman, 1975; Unger, 1982, 1983, 1986a), generally in the drier climatic regions. Amounts in the 300- to 500-mm range were common for more humid regions (Andhale & Kalbhor, 1978; Bauder & Ennen, 1979; Mehrotra et al., 1977; Rawson & Turner, 1983; Subramanian et al., 1979).

The amount of irrigation water required for sunflower is influenced primarily by the amount and distribution of rainfall and plant-available water storage capacity of the soil. The water use efficiency (WUE) of sunflower is influenced, among other factors, by irrigation level (frequency or amount), irrigation timing (relative to growth stage), and planting date. Reported values ranged from 0.09 to 1.37 kg of achenes m^{-3} of water used. In general, WUE increased as irrigation levels decreased. These results suggest that with a decreased level of irrigation, some of the water was applied at critical, highly responsive growth stages for influencing achene yields. Studies wherein irrigations were applied at different growth stages support this suggestion. As compared to a high level of irrigation that required 889 mm total water, irrigations at budding and at full anthesis (one at each stage) required only 219 mm of water. Yields were reduced from 3.5 to 2.3 Mg ha^{-1}, but WUE increased from 0.4 to 1.1 kg of achenes m^{-3} of total water use (Aranda, 1979). In a study by Jana et al. (1982), highest WUE (1.4 kg of achenes m^{-3} of water) was obtained with two irrigations (once each at anthesis and at achene development stages). Also, Unger (1982) obtained high WUE's by irrigating at anthesis or late anthesis growth stages. For fully irrigated sunflower, early (March) planting prolonged the growing season and resulted in high total water use and low WUE. Later-planted (April–early June) sunflower developed more rapidly, used less water, and had higher WUE. Sunflower planted in mid-June or July again had lower WUE, mainly because of lower achene yields (Unger, 1986a).

Based on a report from Spain (Aranda, 1979), WUE for sunflower achene ranged from about 0.4 to 1.1, those for corn grain ranged from about 1.4 to 2.5, and those for seed cotton (*Gossypium hirsutum* L.) ranged from about 0.8 to 0.9 kg m^{-3} of water used. Treatments for the different crops were not identical but were similar in that the crops were irrigated when water in the root zone had been depleted to various levels (based on the soil water-holding capacity). For corn and sunflower, treatments involving irrigations to match ET or at various growth stages or times during the growing season were included also. Although the treatments were not identical and the results, therefore, are not directly comparable, the data indicate that WUE's for cotton and sunflower are in about the same range (0.8–0.9 for cotton and 0.4–1.1 for sunflower) and that WUE's for corn are about two to three times those for cotton and sunflower (1.4–2.5 for corn). Bauder and Ennen (1979) reported similar WUE's for sunflower and corn grown without irrigation in the northern Great Plains (USA), namely, about 0.5 and 1.4 kg m^{-3} for sunflower and corn, respectively. The WUE's of different crops are not directly comparable but are valuable for comparing crops when such information is combined with market prices and production costs.

III. WATER STRESS, ADAPTATION TO STRESS, AND RECOVERY

Water stress apparently has little or no effect on the rate of early sunflower development, namely, the time to budding and anthesis (Bhan, 1977; Marc & Palmer, 1976), but the time to maturity was 2 and 4 d later when the final irrigation was made 16 to 36 d, respectively, after midanthesis than when it was made at midanthesis (Browne, 1977). Stress imposed earlier (before midanthesis) undoubtedly would result in even earlier maturity because of earlier plant drying.

Although water stress has little or no effect on the rate of sunflower development until after anthesis, water stress at critical stages has a marked influence on sunflower growth, achene yield, and achene quality factors. Stress imposed between 10 and 20 d after sowing resulted in 21 to 22 leaves per plant as compared with 29 leaves for nonstressed plants (Marc & Palmer, 1976). Early stress (before budding) also reduced the leaf area per plant (Muriel & Downes, 1974). In this case, plants responded well to subsequent adequate water conditions, but total leaf area never exceeded about 75% of that of the controls.

Besides affecting leaf numbers and areas, early stress reduced plant height (Karami, 1977; Patel & Singh, 1979; Selvaraj et al., 1977; Unger, 1982, 1983) and head diameter (Bhan & Kahn, 1980; Jana et al., 1982; Karami, 1977; Somasundaram & Iruthayaraj, 1981). Alleviating stress by irrigating at anthesis or late-anthesis stages resulted in shorter plants than when stress was alleviated at budding (Unger, 1982, 1983). Early irrigation (four–six leaf stage) followed by irrigations at flower initiation and achene filling tended to result in the largest head diameters, but the trends were not consistent (Jana et al., 1982).

The fact that irrigation level significantly affects sunflower achene yield is well known. However, in most studies, irrigations were made when the available soil water was depleted to a given level or a given potential or at a level to partially or wholly compensate for ET based on pan potential evaporation. Although full irrigation resulted in highest yields, yields generally were near maximum when irrigations were made to maintain soil water content above about the 0.7 to 0.8 available level, based either on measured soil water extraction or estimated from pan evaporation (Bhan, 1977; Bhan & Khan, 1980; Muriel et al., 1974; Pal, 1981; Patel & Singh, 1983; Selvaraj et al., 1977; Somasundaram & Iruthayaraj, 1981; Talha, 1976). While stress may have occurred at certain growth stages in these studies, it would have been coincidental, and it was not reported.

Irrigations to alleviate water stress at various growth stages have been shown to markedly influence achene yields (Aranda, 1979; Browne, 1977; Jadhav & Jadhav, 1978; Jana et al., 1982; Talha & Osman, 1975; Unger, 1982). While maximum yields were obtained with full irrigation, near-maximum yields generally were obtained when irrigations were made to provide adequate water during anthesis and achene filling periods. However, adequate water for good initial plant growth was important for providing a plant capable of responding to later irrigations. Irrigations before or at floral initiation (budding) generally provided for good plant growth (Jana et al., 1982; Talha & Osman, 1975; Unger, 1982). Water stress during achene ripening reduced yields as compared to full irrigation, but the reduction was much less than when stress occurred during anthesis (Browne, 1977; Talha & Osman, 1975). Closely related to yield trends as affected by irrigation at different growth stages were trends for achene numbers per head, achene weight, achene test weight, and achene oil concentration (Aranda, 1979; Browne, 1977; Jana et al., 1982; Talha & Osman, 1975; Unger, 1982). In addition, values for these factors generally increased with increased levels of soil water maintained throughout the growing season. While oil concentration was affected by irrigation level and timing, composition of the oil (linoleic and oleic acid concentration) was little affected by irrigation (Unger, 1982; Unger et al., 1975).

IV. GROWTH AND SOIL WATER

The effect of soil water levels on optimum sunflower growth was covered in general terms in section III of this chapter. Many studies showed that growth decreased as soil water stress increased. However, for most studies, soil water matric potentials associated with different irrigation levels were not reported; but frequent irrigations to replenish the water used by plants undoubtedly maintained the matric potential at above about -200 kPa. This is supported by data of Talha (1976).

At high matric potentials, plants can readily extract water for optimum growth. In addition, such potentials provide favorable conditions for root growth. Deep root growth is a well-established characteristic of sunflower, and it permits sunflower to better withstand water stress than some other crops (Bauder & Ennen, 1979). Continued rooting with depth throughout most of the growing season (Jones, 1978) is another characteristic that enables sunflower to better withstand water stress than some other crops. In this case, some water may be available to the crop for achene filling, thus providing for reasonable yields, even though overall water use may be relatively low.

V. PLANT NUTRITION

Fertilizer requirements for sunflower rise with increased irrigation because of the higher yields obtained. Consequently, fertilizer recommendations should be based on anticipated yields. For optimum utilization of fertilizer resources, recommendations should be based also on soil analyses. Recommendations based on anticipated yields and soil analyses have been developed (Table 25-1) for North Dakota (Dahnke et al., 1981).

The recommendations given in Table 25-1 generally are higher for N and lower for P and K than results obtained in studies at various locations. Studies by Andhale and Kalbhor (1978), Bhosale et al. (1979), Kamel et al. (1980), Mathers and Stewart (1982), and Pal (1981) indicated maximum achene yield with about 75 to 100 kg of applied N ha^{-1}, and Cheng and Zubriski (1978) reported maximum achene yield (4.2 Mg ha^{-1}) with 112 kg applied N ha^{-1}. In contrast, Bhan (1977) obtained maximum yields with 30 to 60 kg N ha^{-1}, apparently because the study was conducted on a "fertile" soil.

The time of N application influences achene numbers, weight, and oil concentration. Adequate N before floret initiation is important for obtaining large numbers of achenes. Single achene weight responded to the N supply after floret initiation (during floret growth), but mainly before anthesis. High N supplies after anthesis depressed oil concentration on an achene dry-weight basis (Steer & Hocking, 1985).

The recommendations of Dahnke et al. (1981) suggest no response to applied P if extractable P in soil is above 33 kg ha^{-1}. Such conclusion could be reached also from the study of Bhan (1977). However, Somasundaram

Table 25-1. Nitrogen, phosphate, and potash recommendations for sunflower. Adapted from Dahnke et al. (1981).

		Broadcast recommendations							
		Phosphorus soil test levels				Potassium soil test levels			
Yield goal	Soil N plus fertilizer N needed†	0-10	10-21	22-32	Over 33	0-111	112-223	224-335	Over 336
		kg ha^{-1}							
1340	73	20	10	0	0	56	37	0	0
1570	78	20	10	0	0	56	37	0	0
1790	90	20	10	0	0	70	46	0	0
2010	101	20	10	0	0	70	46	0	0
2240	112	24	15	0	0	84	56	0	0
2460	123	24	15	10	0	84	56	0	0
2690	134	24	15	10	0	98	65	28	0
2910	146	24	15	10	0	98	65	28	0
3140	162	29	20	15	0	112	74	37	0
3360	179	29	20	15	0	112	74	37	0
3580	196	29	20	15	0	112	74	37	0

† Subtract amount of NO_3-N in top 0.6 m of soil from these figures to determine the amount of N fertilizer to apply. These figures are for soil samples taken between 1 Sept. and 1 Apr.

and Iruthayaraj (1981) obtained responses to applied P up to 60 kg ha^{-1}, even at a relatively low yield level (about 1.6 Mg ha^{-1}). The reasons for these contradictory results is not clear but could be related to such factors as soil texture, organic matter concentration, and pH.

Based on recommendations of Dahnke et al. (1981), K fertilizer required for sunflower depends on the amount of exchangeable K in soil, with greatest amounts required for soils low in K (Table 25-1). However, at an achene yield of about 2.3 Mg ha^{-1}, K uptake by sunflower was 106 kg ha^{-1}, which was greater than that recommended by Dahnke et al. (1981) for that yield level. The high K uptake in the study by Patel and Singh (1983) suggests that the recommendations of Dahnke et al. (1981) may be low because sunflower plant components would contain about 106 kg of K, based on an achene yield of 2.3 Mg ha^{-1} at a 5.9-g kg^{-1} of K concentration; stem, leaf, and head dry matter of 3.7 Mg ha^{-1} at 23.7-g kg^{-1} of K concentration; and root dry matter of 0.9 Mg ha^{-1} at 5.8-g kg^{-1} of K concentration. This total amount is identical to the uptake reported by Patel and Singh (1983) at the same yield level. On soils known to be low in K, these results suggest that sunflower may respond to K amounts in excess of those recommended by Dahnke et al. (1981) and that sunflower growth and yield should be monitored to determine whether low K fertility may be limiting yields.

When they occur, deficiencies of secondary elements—Ca, Mg, and S—usually occur on a regional basis, while those of micronutrients—Fe, B, Mn, Cu, Zn, Mo, and Cl—may be regional but most often occur in parts of fields (Robinson, 1978). No general recommendations for overcoming micronutrient deficiencies are available, but adding $CaCO_3$ to increase soil pH from 4.1 to 5.5 greatly increased sunflower top and root growth. Toxicity of Al was associated with the low pH (Foy et al., 1974). Boron toxicity also was shown to greatly limit sunflower growth and yield (Pathak et al., 1975). In contrast, sunflower growth and yield were not affected by an Fe deficiency that prevented head development or caused death of grain sorghum plants (Mathers et al., 1980).

The salt tolerance of sunflower is rated as medium, with about 3 to 4 g salt L^{-1} of water causing a 50% reduction in yield (Richards, 1954). As salinity of irrigation water increased, plant height, stem diameter, and head diameter decreased (Farah et al., 1984). Decreases in achene oil concentration and oil yield occurred when salt content of irrigation water exceeded 1 g L^{-1} (Raju & Ranganayakulu, 1978).

VI. CULTURAL PRACTICES

A. Soil Preparation and Cropping Practices

Required cultural practices for successful sunflower production generally are similar to those for other large-seeded field crops such as corn and soybean (*Glycine max* L.). Among other factors, cultural practices should: (i) provide a firm, moist seedbed having a favorable temperature for rapid germination and seedling emergence; (ii) permit planting of crops, cultivation, and placement of herbicides, insecticides, and fertilizers, at the proper depth and without undue interference from surface residues; (iii) prevent competition from weeds at planting, emergence, and early seedling growth stages; and (iv) conserve soil and water resources.

Favorable seedbeds for sunflower can be established by a variety of methods ranging from moldboard plowing to no-tillage. In general, moldboard plowing is most effective for loosening soil, incorporating residue, controlling weeds, and mixing nutrients (especially P) with depth in the tillage layer. Moldboard plowing, however, also requires secondary tillage (disking and harrowing) to firm the seedbed, and the resultant bare soil may be highly susceptible to water and wind erosion. Subsoiling improved yields in the southeast USA (Sojka et al., 1986). The no-tillage system may provide excellent erosion control because of surface residues but may cause planting, fertilizer, and herbicide application and cultivating problems. However, sunflower has been successfully no-till planted into wheat (*Triticum aestivum* L.) (Unger, 1981; Unger et al., 1975) and sorghum (Unger, 1984) residues where fertilizer had been applied to the previous crop. Weed control before planting was with paraquat (1,1'-dimethyl-4,4'-bipyridinium ion)[2], glyphosate [N-(phosphono-methyl) glycine], or 2,4-D [(2,4-dichlorophenoxy) acetic acid] (Unger, 1981, 1984). Growing season weed control was provided by surface-applied granular trifluralin (α,α,α-trifluro-2,6-dinitro-N,N-dipropyl-p-toluidine), metolachor [2-chloro-N-(2-ethyl-6-methylphenol)-N-(2-methoxy-

---

[2]Mention of a trade name or product does not constitute a recommendation or endorsement for use by the USDA, nor does it imply registration under FIFRA as amended.

1-methylethyl) acetamide], or with cultivators (rolling or sweep) and hoeing (Unger, 1981, 1984; Unger et al., 1975).

In no-tillage systems, surface residues result in lower soil temperatures, but these have less effect on sunflower than on corn or soybean because corn and soybean require higher temperatures for germination and seedling growth than sunflower. In addition, surface residues conserve water and cool the soil, thus providing more favorable conditions for growth during dry periods (Robinson, 1978).

Growing sunflower in rotation with other crops has beneficial effects but may also have detrimental effects. Beneficial effects result from improved pest control (weeds, volunteer plants, insects, and diseases) and utilization of soil water and nutrient resources. The benefits, however, depend on which crops are grown in rotation with sunflower.

Use of crop rotations improves weed control because of differences in response to herbicides, tillage methods, and time of planting. Weed control with herbicides may be especially beneficial if different types of herbicides can be used for sunflower and the crops grown in rotation with it. For example, trifluralin can be used to control grassy weeds in sunflower, whereas 2,4-D and atrazine [2-chloro-4-ethylamino-6-(isopropylamino)-1,3,5-triazine] can be used to control broadleaf weeds and volunteer sunflower plants. Atrazine also controls volunteer wheat plants. When using herbicides, care must be exercised to assure that herbicide residues do not adversely affect the next crop. Weed and volunteer plant control is possible with rotations when different tillage methods and planting dates can be used to interrupt the normal life cycle of the weeds. Especially beneficial results occur if sunflower, a summer crop, can be grown in rotation with a winter crop such as wheat.

As for weed control, rotations improve insect and disease control, especially if the insects and disease organisms are host plant specific. Some disease organisms, however, remain viable in soil for 5 yr or more and, thus, could infest the sunflower when repeated in the rotation. Consequently, a practical goal of rotations is to reduce the amount of pathogen carryover, not to eliminate it (Robinson, 1978).

The beneficial effects of sunflower in rotation, with respect to soil water and nutrient utilization, result from the deep-rooting habit of sunflower, as compared to shallow-rooted crops. Because of the deeper rooting, sunflower can extract water or nutrients that have moved to depths beyond the reach of other crops. Without use of a deep-rooted crop, the water and nutrients probably would be lost from the system because of continued downward movement. This may be especially true under conditions where full irrigation is practiced. Extracting water to a greater depth also provides a potential for subsequently greater water storage in soil, thus providing an opportunity for improved water use efficiency.

Although deep utilization of water and nutrients may be beneficial to sunflower, this also may be detrimental to other crops, especially to deep-rooted ones because they, likewise, could use the water and nutrients. Consequently, desirable rotations are those in which deep- and shallow-rooted crops are alternated.

In addition to being rotated with other crops, sunflower can also be used in double cropping systems because of its relatively short growing season, provided adequate water is available. In many regions, irrigation would be required. With irrigation, sunflower was double-cropped after winter wheat harvest in the southern Great Plains. Wheat was harvested on 26 June, and sunflower was planted on 28 June. Achene yields were 1.76 and 1.96 Mg ha^{-1} where clean and no-tillage planting, respectively, were used for sunflower (Unger et al., 1975). The relatively short growing season of sunflower also permits double cropping of wheat after sunflower harvest. Sunflower planted during May was harvested in late August or early September, which permitted winter wheat to be planted on a more timely basis in the southern Great Plains than when wheat was planted after grain sorghum (Unger, 1981). Other possibilities for double cropping exist in the south Atlantic Coastal Plain (R.E. Sojka, 1985, personal communication).

B. Fertilization

Fertilizer requirements of sunflower can be met by broadcasting before planting, row placement at or near the achene at planting, or sidedressing after seedling emergence. Foliar application of some nutrients is possible also. The method used would depend on such factors as equipment available, material to be applied, and fertilizer retention capability of the soil.

Broadcast application of NO_3-N fertilizers is as good as row placement because the NO_3 moves with water to plant roots. For P and K, broadcast applications should be about twice that of row placement on soils having low to medium levels of these elements because these elements are adsorbed on clay particles and organic matter and move little with movement of soil water. The usual practice is either to row place all fertilizer at planting (5 cm below and 5 cm to the side of the achene) or to use a combination of row and broadcast application. A common practice is to apply P and K and part of the N in the row and the remaining N by broadcasting before planting or sidedressing after planting (Robinson, 1978).

Besides direct soil application, fertilizer elements, especially N, can be applied through irrigation water and by foliar application. When applied through irrigation water, more uniform application is achieved with sprinkler than with furrow or flood irrigation.

Although some sunflower yield increases have been obtained with foliar application of the primary nutrients (N, P, and K), such method of application generally is considered less practical than soil application (Robinson, 1978). In contrast, foliar applications corrected deficiencies of some micronutrients of sunflower. Data of Semikhnenko et al. (1973), cited by Robinson (1978), indicated that foliar application of Mn and Mo 10 d after anthesis increased yields by 170 to 420 kg ha^{-1} in some years. Application of B, Cu, and Zn had no effect, but the degree of deficiency was not mentioned.

C. Planting Dates

Because sunflower has a relatively short growing season and favorable conditions for growth generally exceed that required for most cultivars at many locations, a wide range of planting dates can be used for this crop. Where the sunflower growing season is affected by low spring and fall temperatures, early planting generally resulted in better growth, achene yields, and oil concentrations than late plantings (Keefer et al., 1976; Murphy, 1978; Robinson, 1971; Unger, 1980). However, extremely early plantings may result in slow plant development (Unger, 1986a) or even frost damage (Murphy, 1978).

Acceptable planting dates for a given location are influenced largely by latitude and elevation. At Bushland, TX, at 35 °N lat at an elevation of 1180 m, planting dates from 21 to 23 March to 12 to 21 June resulted in nonsignificantly different achene yields, but oil concentrations, except for plantings from 1 to 10 May that resulted in a higher concentration, declined from the first (21–23 March) to the last (28 July–1 August) plantings (Unger, 1980). At Redmond, OR (937-m elevation), highest yields were obtained with 28 April and 12 May planting dates (Murphy, 1978). Generally recommended planting date for the northern USA and Canada (40–50 °N lat) is 1 to 20 May (Robinson, 1978). In tropical regions where year-round growth is possible, Garside (1984) still obtained a yield and oil concentration response to sunflower planting dates (February–August). Yields were acceptable (1.8–2.0 Mg ha^{-1}) for plantings from April to July, but for some cultivars, even more restricted planting dates (e.g., May) resulted in highest yields.

Although a relatively wide range of planting dates results in potentially favorable achene yields of sunflower, yields often are greatly reduced by insect or disease outbreaks, which often can be avoided by planting at certain times. For example, sunflower in the southern Great Plains planted after about 10 June had significantly fewer plants infested with larvae of stem weevil (*Cylindrocopturus adspersus* LeConte) and fewer larvae per plant than earlier-planted sunflower (Rogers & Jones, 1979). Planting date also affected infestation by sunflower moths (*Homoeosoma electellum* Hulst), with degree of infestation in Texas generally being lower for late- than for early planted sunflower (Rogers & Jones, 1979). Opposite results were reported for Georgia, and variable results were reported for some other locations (Schulz, 1978). The degree of infestation apparently is related to how well the moth activity coincides with the sunflower growth stage that is most susceptible to infestation by moths. This stage is from anthesis to early achene development, with the time of its occurrence being highly dependent on planting date.

D. Plant Populations

Plant populations have a major effect on sunflower head diameter and weight per achene, with each decreasing with an increase in populations. However, over a wide range of populations, achene yields generally are little or not affected.

In Australia, Thompson and Fenton (1979) used populations from 25 000 to 140 000 plants ha^{-1} for irrigated sunflower planted in rows 76 cm apart. In two experiments, the wide range in population had little effect on yield because individual achene weight and number of achene per head interacted with plant population to give nearly identical yields over the entire range. In a third experiment, achene yields increased significantly as populations increased from about 40 000 to 90 000 plants ha^{-1}.

For several cultivars and irrigation treatments in Texas, between 40 000 and 90 000 plants ha^{-1} gave similar yields, but yields decreased when populations were below 30 000 and above 100 000 plants ha^{-1} (Unger et al., 1975). Similar results were obtained by Radford (1978). The general recommendation of about 60 000 to 70 000 plants ha^{-1} where high yields are expected (Robinson, 1978), as with irrigation, seems appropriate for oil-type sunflower. For large-achened, nonoil cultivars, Robinson (1978) recommended 35 000 plants ha^{-1}.

E. Pest Control

As for any crop, pest control is important for obtaining favorable yields of a high quality sunflower product. Major pests of sunflower include weeds, insects, diseases, birds, and sometimes deer (family Cervidae) and rodents, including rabbits (*Oryctolagus cuniculus*) and hares (order Lagomorpha).

Sunflower yield losses because of weeds vary with weed species, environment, and time of infestation relative to the crop growth stage and have ranged up to 30 or 35% for several weed species (NDSU, 1985). Weeds can be controlled with cultural practices and with herbicides. Culturally, weeds present before or at planting are controlled by disking, harrowing, and sweep plowing, while those emerging after planting can be controlled by harrowing or cultivating. Effective control of weeds in other crops in the rotation, as well as between crops, is beneficial for sunflower.

A number of herbicides can be used for weed control in sunflower (NDSU, 1985). Included are contact herbicides, which can be applied before planting or before sunflower emergence to control existing weeds; preplant or pre-emergence herbicides that interfere with the germination, emergence, or seedling development of weed species; and postemergence herbicides that can be used to control some established weeds in sunflower. In each case, the herbicides must be used at recommended rates and times to avoid damage to surrounding or subsequent crops. The North Dakota State University (NDSU) (1985) publication contains recommended practices for various herbicides that are suitable for sunflower.

As mentioned in previous sections, sunflower insects can be controlled, to some extent, through use of tillage methods and crop rotations to disrupt the normal life cycle of the insects or by using alternate planting dates to shift the most susceptible growth stage of sunflower away from the most active period of the insects. Where problems still occur, some insects can be controlled by timely (preplant or postemergence) insecticides. As for herbicides, insecticides must be used with caution. The NDSU (1985) publication gives information for controlling some insects.

At least 30 diseases affecting sunflower have been identified, but only a few of them are of economic importance (NDSU, 1985). However, in contrast to weeds and insect pests, little can be done to control diseases once they infect the sunflower plants. Therefore, the most effective control practices are to plant disease resistant or tolerant hybrids and to use crop rotations with a minimum of 4 yr between sunflower crops on the same land (NDSU, 1985).

Birds are a problem in sunflower, mainly as the crop approaches maturity. However, problems in crop establishment also have been noted because of birds. Near harvest, birds feed on achenes, with the upright heads providing a convenient perch. Culturally, bird damage can be minimized by planting sunflower as far as possible from bird roosting/nesting areas, by controlling other plants (weeds) that may attract birds, and by planting the sunflower as near as possible to the same time as other sunflower in the same area. Undoubtedly, this is the best and most cost-effective method. Mechanical control consists of using guns, exploders, and frightening devices, but these usually are not effective. Chemical control consists of applying cracked corn treated with Avitrol (4-aminopyridine-HCL), which is a fright-producing repellent (NDSU, 1985).

Birds, at some times, have been observed to dig up newly planted achenes or to feed on germinating achenes and on emerging or newly emerged seedlings. On a large field basis, no effective control measures have been found, but for small research plots, scattering sunflower achenes or grain seeds at the plot perimeters minimized bird damage within the plots (P. Unger, 1980, 1981; unpublished observations).

Rodent damage usually occurs soon after planting and is most severe near fields or pastures having residues that provide protection for the rodents (especially mice, *Mus musculus*). Damage can be minimized by providing rodenticide-treated grain for the rodents to eat.

Rabbits damage sunflower from seedling emergence until plants are about 25-cm tall, with most damage occurring near the edges of fields. Deer damage to sunflower may occur throughout the growing season, with damage being most severe adjacent to wooded or other areas that provide cover for the deer. Scare tactics may provide some protection against deer.

F. Harvesting

Sunflower is physiologically mature and, hence, ready for harvest when the back of the heads have turned yellow and the bracts are turning brown. However, achene moisture content at this stage is about 40%, which is too high for efficient combine harvesting. Achene dry-down can be hastened by applying a chemical desiccant to the plants after physiological maturity. Sunflower can be combine harvested when achene moisture content is about 25%, but the achenes must be dried to about 9.5% moisture for safe storage. Delaying harvest until achene moisture content is below 25% increases the potential for achene losses because of shattering during harvest and because of birds.

Sunflower is harvested with combines that are used for grain crops, but a special header attachment usually is used for more efficient harvesting. Several types of headers are available, with many having provisions for gathering the head, catching shattering achenes, and cutting heads from stalks but minimizing the amount of stalk entering the combine. The heads are gathered, and achenes are caught by pans extending ahead of the cutter bar. The stalks are deflected so that only the heads remain above the cutter bar. After cutting, the heads are moved into the combine with a small reel. In contrast to settings for grain crops, the combine cylinder speed is set as low as possible, the concave setting is set wide open, and the fan is adjusted to a low speed to minimize achene loss and the amount of trash with the achenes (NDSU, 1985).

VII. IRRIGATION MANAGEMENT

The type of irrigation management used for sunflower will depend, among other factors, on amount of water available for irrigation, desired yield level, and concerns for WUE. Decisions regarding each of these factors will determine whether a full irrigation or a limited irrigation approach is used.

A. Full Irrigation

As used for this report, full irrigation means that sufficient irrigation water will be applied as needed so that the sunflower from planting to physiological maturity will not experience visible water stress (plant wilting). However, even under a full irrigation approach, frequency and amount of irrigation may be different, depending on the technique used to determine when to irrigate and how much water to apply.

Under a full irrigation approach, water is applied whenever water use exceeds a predetermined level. This determination may be based on leaf water potential, soil water potential, soil water content, or evaporation from a free water surface (pan evaporation). To maintain turgid leaves, leaf water potential should remain above about −350 kPa (Robinson, 1978). Although leaf water potential measurements directly indicate when irrigation is needed, it generally is easier and more convenient to use soil water or pan evaporation measurements.

Direct measurements of soil water content can be obtained by the *gravimetric method* (i.e., sampling the soil, weighing while wet, oven drying, weighing after drying, and calculating the water content). This method, however, involves a delay from sampling until results are known; consequently, indirect methods of water content determination such as neutron attenuation, electrical resistance, tensiometer, or even the "feel" method (Robinson, 1978) or estimation from pan evaporation often are more desirable. To achieve maximum yields, provided other factors are not limiting, irrigations generally should be made to maintain the soil water content above the 60 to 70% available level (Pal, 1981; Selvaraj et al., 1977; Somasundaram & Iruthayaraj, 1981) or to replenish the soil water supply when the irrigation water to cumulative pan evaporation ratio (IW/CPE) reaches about 0.75 (Bhan, 1977; Bhan & Khan, 1980; Connor et al., 1985; Patel & Singh, 1979, 1983; Robinson, 1978; Somasundaram & Iruthayaraj, 1981; Subramanian et al., 1979). Irrigating to maintain an adequate supply of water also can be made on a time basis, provided the approximate ET rate, soil water-holding capacity, and precipitation amounts are known (Karami, 1977; Talha, 1976; Unger, 1982).

The three basic methods of applying water are furrow, flood, and sprinkler. The application method used will depend on soil conditions (texture, slope, and land preparation), equipment available, and amount of water available for irrigation. On soils with relatively low infiltration rates (clays and clay loams) and slight slopes (generally < 1.0%), and where adequate water is available, furrow irrigation is common. On similar soils, but with leveled surfaces, flood irrigation is possible. Sprinkler irrigation is desirable on sandy or sloping soils and where the water supply is limited. Under these latter conditions, sprinkler application increases the uniformity of application of the available water.

The amount of water applied at a given irrigation depends on soil water-holding capacity, equipment available, and amount of water available for irrigation. The water-holding capacity is mainly dependent on soil texture, depth, and organic matter concentration. In general, sandy, shallow, low organic-matter concentration soils have low water-holding capacities, while the opposite is true for fine-textured (clay and clay loam), deep, high organic-matter concentration soils. The amount added should replenish the root zone to the desired depth but should avoid losses because of deep percolation. Where water at a given time is limited (e.g., by low well outputs), sprinkler irrigation at frequent intervals may permit more effective water application than furrow or flood irrigation. Sprinkler irrigation at frequent intervals also may be most desirable on low water-holding capacity soils.

B. Limited Water Situations

Technically, any irrigation scheme less than full irrigation is a limited water situation. For this report, however, the limited water situations are planned so as to achieve increased WUE (economic yield per unit of water used) rather than maximum yield.

As discussed in section III of this chapter, near-maximum sunflower achene yields were obtained with less than full irrigation when adequate water was supplied by irrigation during critical periods such as anthesis and achene filling, provided that adequate water was available for good initial plant growth to provide a plant capable of responding to later irrigations. By limiting the amount of water applied without reducing yields by a similar amount,

WUE's are increased. Examples of increased WUE's because of limited irrigations were given in section III of this chapter.

Water application techniques under limited water situations generally are similar to those with full irrigation, except that greater care must be exercised to obtain uniform water application. This is especially true for furrow or flood irrigation systems because the soil is drier and may have cracks, which can result in greater than optimum water infiltration at input sites and less than optimum distribution of water in the field. In such cases, sprinkler irrigation may provide for more uniform water distribution.

The amount of water to apply at a given irrigation will depend, basically, on the same factors as for full irrigation. The goal should be to provide sufficient water for plant use yet avoid losses to percolation or runoff. On a clay loam soil, Unger (1982) obtained equally high (not significantly different) irrigation WUE's by applying a growing season irrigation (80–100 mm), either at early or late anthesis, or by applying two growing season irrigations (80–85 mm each) at budding and late anthesis or at anthesis and late anthesis. Aranda (1979) obtained highest WUE when two irrigations totaling 110 mm were applied at budding and full anthesis. Connor et al. (1985) applied between 25 and 137 mm of water, and Rawson and Turner (1983) applied 80 mm of water for various limited irrigation treatments that resulted in highest WUE's. Sprinkler or trickle systems were used by Stegman and Lemert (1980) to apply about 30 to 40 mm of water per irrigation for treatments designed to achieve water stress at various growth stages. Based on achene yields, sunflower was most sensitive to stress during anthesis.

VIII. SUMMARY

Although most sunflower is grown without irrigation, irrigation is an important management option for this crop in some regions. Sunflower, like other crops, responds to good irrigation practices through increased growth and yield, with achene yields >4.0 Mg ha^{-1} having been obtained. However, the relatively high amounts of water needed to achieve high yields may be used less efficiently than when lesser amounts of water are applied. Because sunflower has some growth stages that are more susceptible to stress than others, providing water at those critical stages (generally from budding or anthesis to achene filling) results in more efficient water use with a relatively small decrease in achene yield. Sunflower yields also are influenced by planting dates, land preparation, and fertilization. However, these generally are less critical than irrigation amount and timing for achieving efficient use of water for sunflower production.

REFERENCES

Andhale, R.K., and P.N. Kalbhor. 1978. Effect of irrigational schedules under varying levels of nitrogen on growth, yield, quality, and water use of sunflower (*Helianthus annuus* L.). J. Maharashtra Agric. Univ. 3:200–203.

Andhale, R.K., and P.N. Kalbhor. 1980. Pattern of dry matter accumulation of sunflower as influenced by irrigational schedules under various levels of nitrogen fertilization. J. Maharashtra Agric. Univ. 5:9–14.

Aranda, J.M. 1979. Water use by crops in South-west Spain. p. 177–183. *In* E. Welte (ed.) Proc. Water and Fertilizer Use for Food Production in Arid and Semiarid Zones. 26 Nov.–1 Dec. Garyounis Univ., Benghazi, Libya.

Bauder, J.W., and M.J. Ennen. 1979. Crop water use—how does sunflower rate? Sunflower 5(1):10–11.

Bhan, S. 1977. Studies on the optimum scheduling of irrigation, row spacing, and fertilizer dose for sunflower in Central Tract of Uttar Pradesh. Indian J. Agron. 22:212–216.

Bhan, S., and S.A. Khan. 1980. Effect of frequency and method of irrigation and application of surface mulch on sunflower. Indian J. Agron. 25:645–650.

Bhosale, R.J., B.P. Patil, and S.S. Wadkar. 1979. Effect of graded levels of nitrogen and phosphorus on the yield of sunflower variety Peredovik under Rabi irrigated conditions in Konkan region. Indian J. Agric. Res. 13:164–166.

Browne, C.L. 1977. Effect of date of final irrigation on yield and yield components of sunflowers in a semi-arid environment. Aust. J. Exp. Agric. Anim. Husb. 17:482–488.

Cheng, S.F., and J.S. Zubriski. 1978. Effects of nitrogen fertilizer on production of irrigated sunflower, plant uptake of nitrogen, and water use. p. 400–409. *In* Proc. 8th Int. Sunflower Conf., Minneapolis. 23–27 July. Int. Sunflower Assoc., Toowoomba, Queensland.

Connor, D.J., T.R. Jones, and J.A. Palta. 1985. Response of sunflower to strategies of irrigation. I. Growth, yield, and the efficiency of water-use. Field Crops Res. 10:15–36.

Dahnke, W.C., J.C. Zubriski, and E.H. Vasey. 1981. Fertilizing sunflowers. North Dakota State Univ. Coop. Ext. Serv. Circ. SF-713.

Decau, J., P. Lencrerot, J.R. Marty, J. Puech, and B. Pujol. 1973. The effect of climatic conditions and irrigation on the oil and protein production of three oilseed crops (rape, sunflower, and soybean) in S.W. France. C. R. Seances Acad. Agric. Fr. 59:1464–1474.

Delibaltov, I., and I.M. Ivanov. 1973. The effects of irrigation and fertilizer application on yield and quality of sunflower seed. Rastenievud. Nauki 10:57–68.

Dimitrov, I. 1975. Effect of irrigation and fertilizers on growth, development, and yield of sunflower. Rastenievud. Nauki 12:85–94.

Dinusson, W.E. 1977. Sunflower by-products as feeds. p. 1–2. *In* Proc. Sunflower Forum. 12–13 Jan. Vol. II. Sunflower Assoc. Am., Bismarck, ND.

Farah, M.A., A.M. Daoud, M.A. Barakat, and I.M. Anter. 1984. The effect of the depth to water table and salinity of irrigation water on growth and evapotranspiration of sunflower. Agric. Res. Rev. (1981) 59:199–209.

Foy, C.D., R.G. Orellana, J.W. Schwartz, and A.L. Fleming. 1974. Responses of sunflower genotypes to aluminum in acid soil and nutrient solution. Agron. J. 66:293–296.

Garside, A.L. 1984. Sowing time effects on the development, yield, and oil characteristics of irrigated sunflower (*Helianthus annuus*) in semi-arid tropical Australia. Aust. J. Exp. Agric. Anim. Husb. 24:110–119.

Gertjejansen, R.O. 1976. An evaluation of sunflower stalks for the manufacture of particleboard. p. 30–33. *In* Proc. Sunflower Forum. 12–13 Jan. Sunflower Assoc. Am., Bismarck, ND.

Jadhav, A.S., and S.B. Jadhav. 1978. Irrigation requirement of sunflower in relation to critical growth stages. J. Maharashtra Agric. Univ. 3:66.

Jana, P.K., B. Misra, and P.K. Kar. 1982. Effect of irrigation at different physiological stages of growth on yield attributes, yield, consumptive use, and water use efficiency of sunflower. Indian Agric. 26:39–42.

Jones, O.R. 1978. Management practices for dryland sunflower in the U.S. southern Great Plains. p. 89–98. *In* Proc. 8th Int. Sunflower Conf., Minneapolis. 23–27 July. Int. Sunflower Assoc., Toowoomba, Queensland.

Jones, O.R. 1984. Yield, water-use efficiency, and oil concentration and quality of dryland sunflower grown in the southern High Plains. Agron. J. 76:229–235.

Jones, O.R., and P.W. Unger. 1977. Soil water effects on sunflowers in the southern High Plains. p. 12–16. *In* Proc. Sunflower Forum. 12–13 Jan. Vol. II. Sunflower Assoc. Am., Bismarck, ND.

Kamel, M.S., R. Shabana, A.K. Ahmed, and S.I. El-Mohandes. 1980. Response of an exotic hybrid and a local sunflower cultivar to N application under irrigation in Egypt. J. Agron. Crop Sci. 149:227–234.

Karami, E. 1977. Effect of irrigation and plant population on yield and yield components of sunflower. Indian J. Agric. Sci. 47:15–17.

Keefer, G.D., J.E. McAllister, E.S. Uridge, and B.W. Simpson. 1976. Time of planting effects on development, yield, and oil quality of irrigated sunflower. Aust. J. Exp. Agric. Anim. Husb. 16:417–422.

LaRue, J., and G. Pratt. 1977. Utilization of sunflower stalks. p. 20–23. *In* Proc. Sunflower Forum. 12–13 Jan. Vol. II. Sunflower Assoc. Am., Bismarck, ND.

Marc, J., and J.H. Palmer. 1976. Relationship between water potential and leaf and inflorescence initiation in *Helianthus annuus*. Physiol. Plant. 36:101–104.

Marc, J., and J.H. Palmer. 1981. Photoperiodic sensitivity of inflorescence initiation and development in sunflower. Field Crops Res. 4:155–164.

Marty, J.R., J. Puech, J. Decau, and C. Maertens. 1972. Effects of irrigation on the yield and quality of sunflower. p. 46–53. *In* Proc. 5th Int. Sunflower Conf., Clermont-Ferrand, France. Int. Sunflower Assoc., Toowoomba, Queensland.

Mason, W.K., W.S. Meyer, R.C.G. Smith, and H.D. Barrs. 1983. Water balance of three irrigated crops on fine-textured soils of the Riverine Plain. Aust. J. Agric. Res. 34:183–191.

Mathers, A.C., and B.A. Stewart. 1982. Sunflower nutrient uptake, growth, and yield as affected by nitrogen or manure, and plant population. Agron. J. 74:911–915.

Mathers, A.C., J.D. Thomas, B.A. Stewart, and J.E. Herring. 1980. Manure and inorganic fertilizer effects on sorghum and sunflower growth on iron-deficient soil. Agron. J. 72:1025–1028.

Mehrotra, O.N., M. Pal, and G.S. Singh. 1977. Determination of consumptive use in oilseed crops (sunflower, rai, and linseed) using screened evaporimeters under irrigated conditions. Indian J. Agric. Res. 11:167–172.

Mikhov, I. 1974. On the irrigation regime for sunflower under the conditions of southeastern Bulgaria. Rastenievud. Nauki 11:99–109.

Mikhov, I. 1975. Economic effectiveness of optimum irrigation regime for sunflower given different fertilizer rates. Khidrotekhnika Melioratsii 20:24–27.

Mungse, H.B., and D.G. Bhapkar. 1983. Influence of antitranspirants on the yield and water use of sunflower under varying irrigation schedules. J. Maharashtra Agric. Univ. 8:179–180.

Muriel, J.L., and R.W. Downes. 1974. Effect of periods of moisture stress during various phases of growth of sunflowers in the greenhouse. p. 127–131. *In* Proc. 6th Int. Sunflower Conf., Bucharest, Romania. July. Int. Sunflower Assoc., Toowoomba, Queensland.

Muriel, J.L., R. Gimenez, and J. Berengena. 1974. Yield of sunflower in field plots in response to various watering regimes and to irrigation during critical phases of growth. p. 577–582. *In* Proc. 6th Int. Sunflower Conf., Bucharest, Romania. July. Int. Sunflower Assoc., Toowoomba, Queensland.

Murphy, W.M. 1978. Effects of planting date on seed, oil, and forage yields of irrigated sunflowers. Agron. J. 70:360–362.

Musick, J.T., L.L. New, and D.A. Dusek. 1976. Soil water depletion—yield relationships of irrigated sorghum, wheat, and soybeans. Trans. ASAE 19:489–493.

North Dakota State University. 1985. Sunflower: Production and pest management. North Dakota State Univ. Ext. Bull. 25 (Revised).

Pal, M. 1981. Effect of soil moisture regimes and fertility levels on yield and quality of sunflower varieties. Indian J. Agric. Res. 15:74–78.

Patel, J.C., and R.M. Singh. 1979. Water use and yield of sunflower (*Helianthus annuus* L.) as influenced by irrigation, mulch, and cycocel application. Madras Agric. J. 66:777–782.

Patel, J.C., and R.M. Singh. 1983. Yield and nutrient uptake of sunflower (*Helianthus annuus* L.) as affected by irrigation, mulch, and cycocel. Indian J. Agron. 28:205–210.

Pathak, A.N., R.K. Singh, and R.S. Singh. 1975. Effect of different concentrations of boron in irrigation water on sunflower. J. Indian Soc. Soil Sci. 23:388–390.

Radford, B.J. 1978. Plant population and row spacing for irrigated and rainfed oilseed sunflowers (*Helianthus annuus*) on the Darling Downs. Aust. J. Exp. Agric. Anim. Husb. 18:135–142.

Raju, K.V., and C. Ranganayakulu. 1978. Effect of saline water irrigation on salt distribution in profile and performance of sunflower crop. Indian J. Agric. Res. 12:183–186.

Rawson, H.M., and N.C. Turner. 1983. Irrigation timing and relationships between leaf area and yield in sunflower. Irrig. Sci. 4:167–175.

Richards, L.A. (ed.). 1954. Diagnosis and improvement of saline and alkali soils. USDA Agric. Handb. 60. U.S. Gov. Print. Office, Washington, DC.

Richardson, C.R., R.C. Albin, R.N. Beville, and R.K. Ratcliff. 1978. Nutritional value of sunflower meal as a protein supplement for growing ruminants. p. 576–592. *In* Proc. 8th Int. Sunflower Conf., Minneapolis. 23-27 July. Int. Sunflower Assoc., Toowoomba, Queensland.

Robertson, J.A., W.H. Morrison, III, and R.L. Wilson. 1979. Effects of planting location and temperature on the oil content and fatty acid composition of sunflower seeds. USDA-SEA-ARS Res. ARR-S-3. U.S. Gov. Print. Off., Washington, DC.

Robinson, R.G. 1971. Sunflower phenology—year, variety, and date of planting effects on day and growing degree-day summations. Crop Sci. 11:635–638.

Robinson, R.G. 1978. Production and culture. *In* J.F. Carter (ed.) Sunflower science and technology. Agronomy 19:89–143.

Rogers, C.E., and O.R. Jones. 1979. Effects of planting date and soil water on infestation of sunflower by larvae of *Cylindrocopturus adspersus*. J. Econ. Entomol. 72:529–531.

Schneiter, A.A., and J.F. Miller. 1981. Description of sunflower growth stages. Crop Sci. 21:901–903.

Schulz, J.T. 1978. Insect pests. *In* J.F. Carter (ed.) Sunflower science and technology. Agronomy 19:169–223.

Schuppan, D., and D. Thomas. 1976. Irrigated summer crops in northern Victoria: Maize, millet, and sunflowers. J. Agric. (Melb.) 74:424–426.

Selvaraj, K.V., R. Kaliappa, A. Damodaran, and P.P. Ramasamy. 1977. The effect of different moisture regimen on the yield and yield components of sunflower (*Helianthus annuus* L.). Oils Oilseeds J. 30(1):7–10.

Semikhnenko, P.G., T.E. Guseva, and A.N. Riger. 1973. Effect of trace elements on yield of sunflower grown in the Krasnodar region. Byull. Nauchno-Tekh. Inf. Maslichn. Kult. 1973(3):25–27.

Siddiqui, M.Q., J.F. Brown, and S.J. Allen. 1975. Growth stages of sunflower and intensity indices for white blister and rust. Plant Dis. Rep. 59:7–11.

Sojka, R.E., W.J. Busscher, F.B. Arnold, and D. Gooden. 1986. Sunflower and subsoiling in the southeast Coastal Plains. p. 253. *In* Agronomy abstract. ASA, Madison, WI.

Somasundaram, S., and M.R. Iruthayaraj. 1979. Effect of irrigation and phosphorus levels on growth analysis parameters in sunflower (*Helianthus annuus* L.). Madras Agric. J. 66:783–788.

Somasundaram, S., and M.R. Iruthayaraj. 1981. Effect of irrigation regimes and phosphorus levels on yield and yield attributes in sunflower (*Helianthus annuus* L.). Indian J. Agric. Res. 15:215–222.

Steer, B.T., and P.J. Hocking. 1985. The optimum timing of nitrogen application to irrigated sunflowers. p. 221–226. *In* Proc. 11th Int. Sunflower Conf., Mar Del Plata, Argentina. Int. Sunflower Assoc., Toowoomba, Queensland.

Stegman, E.C., and G.W. Lemert. 1980. Sunflower yield vs. water deficits in major growth periods. ASAE Pap. 80-2569.

Subramanian, S., S.D. Sundarsingh, and K.P. Ramaswami. 1979. Irrigation and manurial requirements of sunflower. Madras Agric. J. 66:115–117.

Talha, M. 1976. A comparison between two systems of irrigation on sunflower production. Egypt. J. Soil Sci. 16:81–92.

Talha, M., and F. Osman. 1975. Effect of soil water stress on water economy and oil composition in sunflower (*Helianthus annuus* L.). J. Agric. Sci. 84:49–56.

Thompson, J.A., and I.G. Fenton. 1979. Influence of plant population on yield and yield components of irrigated sunflower in southern New South Wales. Aust. J. Exp. Agric. Anim. Husb. 19:570–574.

Unger, P.W. 1978. Effect of irrigation frequency and timing on sunflower growth and yield. p. 117–129. *In* Proceedings 8th Int. Sunflower Conf., Minneapolis. July. Int. Sunflower Assoc., Toowoomba, Queensland.

Unger, P.W. 1980. Planting date effects on growth, yield, and oil of irrigated sunflower. Agron. J. 72:914–916.

Unger, P.W. 1981. Tillage effects on wheat and sunflower grown in rotation. Soil Sci. Soc. Am. J. 45:941–945.

Unger, P.W. 1982. Time and frequency of irrigation effects on sunflower production and water use. Soil Sci. Soc. Am. J. 46:1072–1076.

Unger, P.W. 1983. Irrigation effect on sunflower growth, development, and water use. Field Crops Res. 7:181–194.

Unger, P.W. 1984. Tillage and residue effects on wheat, sorghum, and sunflower gown in rotation. Soil Sci. Soc. Am. J. 48:885–891.

Unger, P.W. 1986a. Growth and development of irrigated sunflower in the Texas High Plains. Agron. J. 78:507–515.

Unger, P.W. 1986b. Sunflower development in the Texas High Plains: Environmental effects. Texas Agric. Exp. Stn. Misc. Publ. MP 1598.

Unger, P.W. 1986c. Wheat residue management effects on soil water storage and corn production. Soil Sci. Soc. Am. J. 50:764–770.

Unger, P.W., and O.R. Jones. 1981. Effect of soil water content and a growing season straw mulch on grain sorghum. Soil Sci. Soc.Am. J. 45:129–134.

Unger, P.W., O.R. Jones, and R.R. Allen. 1975. Sunflower experiments at Bushland on the Texas High Plains. Texas Agric. Exp. Stn. Prog. Rep. PR-3304.

26 Sugarbeet

F. J. HILLS

University of California
Davis, California

S. R. WINTER

Texas A&M University
Amarillo, Texas

D. W. HENDERSON

University of California
Davis, California

The sugarbeet (*Beta vulgaris* L.) belongs to a broad species of the family Chenopodeacae that includes red table beet, swiss chard, mangel, and fodder beet. Intense breeding has created the commercial sugarbeet as a plant well suited to the synthesis and storage of sucrose. With favorable treatment and climate, it germinates rapidly from seed, develops large leaves for the efficient capture of sunlight, rapidly extends fibrous roots into the soil for accumulating water and nutrients, and forms a large storage root for accumulating high concentrations of sucrose. Given optimum conditions, the crop may produce 90 to 100 Mg fresh roots ha^{-1} within 24 wk. Commercial production is usually less. For example, California has a wide range of climate and weather patterns suitable for sugarbeet production. Over 5-yr (1980–1984), for different growing regions, root yield averaged from 49 to 77 Mg ha^{-1} and sucrose contents averaged 14.1 to 16.4% sucrose (Hills et al., 1986).

Sugarbeet is adapted to a wide range of climatic conditions, is noted for its tolerance to soil salinity once a stand has been established, and is able to tolerate moderate soil water stress and still produce a profitable crop. It is grown in the central and western USA. States that usually grow over 40 000 ha yr^{-1} are Minnesota, California, Idaho, North Dakota, and Michigan. In the semiarid areas of the western USA, where little rain falls during the growing season, irrigation is crucial for economic yield and is a major cost of production. In these areas, irrigation has a variety of uses. It is used to provide soil water necessary for preparing the seedbed, germinating seed, alleviating soil crusting, and managing soil salinity, as well as supplying the necessary soil water to meet the evapotranspiration (ET) needs of the crop.

Objectives of this chapter are to illustrate how the crop develops, how it responds to soil water stress, and how it should be managed to provide for efficient use of irrigation water.

I. CROP DEVELOPMENT AND EFFECTS OF WATER STRESS

A. Growth and Development

Research by Ulrich (1952, 1954, 1955, 1956) has provided a foundation for understanding the growth and development of the sugarbeet. More recently, plant growth and development and the effects of environmental factors have been reviewed in detail by Loomis et al. (1971) and Wyse (1979). As the seedling emerges, the taproot grows rapidly and may reach 30 cm or more by the time the first leaf is fully developed (Brown & Dunham, 1986). During the first 30 d, growth is confined primarily to top and fibrous roots, and the rate of increase in leaf area and transpiration are slower than most annual crops (Wright, 1982). After about 30 d, both top and storage root growth proceed rapidly. Tops reach near-maximum fresh weight in 60 to 90 d but storage roots continue rapid growth for 20 to 24 wk. Beyond that time, crown (stem) growth becomes an increasingly larger percentage of the commercial "storage root." As the storage root increases in size there is a constant partitioning of sucrose to root dry matter (Wyse, 1979; Ghariani, 1981) but the sucrose content of the root on a fresh weight basis may remain relatively constant. The irrigated (I) crop of Fig. 26–1 illustrates the growth and development of well-watered, well-fertilized sugarbeet from 20 July to 12 October, 76 to 160 d after planting. Cooler temperatures in the fall, particularly night temperatures, coupled with N deficiency usually result in increased root sucrose concentration. This increase is associated with a slowing of root and top growth, allowing the accumulation of sucrose in storage roots rather than being used in growth processes (Ulrich, 1955; Loomis & Nevins, 1963).

There are genetic differences among cultivars for sucrose storage and a general inverse relation exists between high percent sucrose and root yield. This ranges from the mangel with a low root sucrose concentration and high root fresh weight yield to sugarbeet cultivars selected for high sucrose concentrations but with lower fresh weight root yields. For a given cultivar, the final sucrose content of the root is greatly affected by environmental conditions, particularly those prevailing 4 to 8 wk prior to harvest with cooler night temperatures and N deficiency having the most positive influence (Ulrich, 1955).

There are many conditions that lead to decreased sucrose concentration of the growing sugarbeet root. Some of these are: periods of hot days and warm nights; severe defoliation by hail, diseases, and insects; root diseases; and an increase in the uptake of soil N during periods favorable to plant growth. When the growing sugarbeet undergoes a prolonged period of cold temperatures followed by warmer temperatures and longer day lengths, seed stalk production (bolting) takes place. The onset of bolting is usually accom-

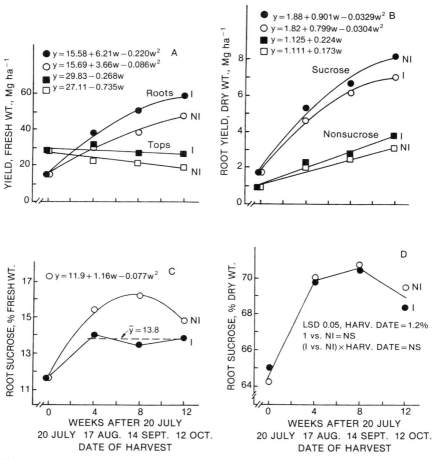

Fig. 26-1. Growth and development of sugarbeet, without and with increasing water stress, on a silt loam soil at Davis, CA, 1982 (Hills, unpublished data). I, irrigated biweekly throughout the season; NI, last irrigated 21 July. Regressions are on weeks (w) following the first harvest on 20 July. For each harvest date, cumulative crop ET was estimated for I from pan E and crop coefficients as 328, 503, 654, and 768 mm, respectively (Pruitt et al., 1987). Crop ET for NI was estimated by neutron probe to 2.74 m soil depth. ET deficit for NI (ETc I − ETc NI) for each harvest date was 0, 91, 180, and 269 mm, respectively.

panied by a reduction in root sucrose concentration. When plants have been stressed for water, a subsequent irrigation or rainfall may reduce root sucrose concentration due to root rehydration and/or increased vegetative growth.

B. Water Stress

1. Effect on Crop Production

In the semiarid areas of the western USA, irrigation is a major cost in crop production. For example, in California's central valleys, about 30% of all preharvest costs are related to irrigation (Hills et al., 1986). It is desirable not to use more water than is necessary to maximize net return. This will also minimize the leaching of NO_3 to groundwater.

Results of an experiment at Davis, CA, on a deep silt loam soil illustrate the response of sugarbeet to continuously increasing soil water stress (Fig. 26-1). In this furrow-irrigated experiment, one set of plots was irrigated every 2 wk to supply 1080 mm of water for the growing season extending from 5 May to 14 October. This amount of water, applied without run-off, was 40% in excess of potential crop ET as estimated from pan evaporation and crop coefficients. Thus, the plants of this treatment should have suffered little or no water stress. Other plots were irrigated for the last time on 21 July, 77 d after planting. These plots received 579 mm water to that date plus 50 mm of rain that fell between 15 and 26 September. Evapotranspiration of the stressed plants was measured by neutron probe to a depth of 2.74 m. Both top and root yield decreased sharply as water stress increased (Fig. 26-1A). Sucrose continued to be partitioned to storage roots, but at a reduced rate, as water stress increased, and root nonsucrose dry matter accumulated at a slower rate in stressed than in nonstressed plants (Fig. 26-1B). There was little effect of water stress on the dry weight concentration of root sucrose (Fig. 26-1D). Roots of stressed and nonstressed plants increased in dry weight sucrose concentration through mid-September, then declined after a late-season rain, probably due to increased NO_3 uptake from the movement of surface accumulations into the root zone by rain. Differences in fresh weight sucrose concentration were largely due to differences in the water content of storage roots (Fig. 26-1C). After water cutoff, as pointed out by Loomis and Haddock (1967), the higher sucrose content of stressed plants is apparently due to a slower accumulation of water relative to sucrose and other root dry matter rather than to any appreciable net loss of water from the root or more rapid accumulation of sucrose.

Jensen and Erie (1971) reviewed earlier work on sugarbeet irrigation and concluded, "it is apparent that sugarbeets can be irrigated at a high soil moisture level with frequent, light irrigations, or they can be allowed to deplete 60 to 70 percent of the available water in the root zone between irrigations if each irrigation refills the amount depleted." The increasing costs of irrigation in the 1970s prompted studies to establish minimum levels of irrigation needed to produce economical yields. These are briefly summarized below.

Miller and Aarstad (1976) in Washington, used daily sprinkler irrigation to apply water based on pan evaporation. From late June to early October, decreasing total water applied to about 68% of crop ET resulted in a 12% loss in root yield that was compensated for by a 2-percentage point increase in sucrose concentration and minimal loss in crop value. Ehlig and LeMert (1979), in California's Imperial Valley, established treatments that provided 23% less to 11% more water than was needed as indicated by a weighing lysimeter located in the experimental area. Total seasonal water use varied from 1195 for the wettest to 900 mm for the driest treatment. Over this range of water use, there was no statistically significant effect on sucrose content or sucrose yield. Root yield, however, was greatest with a seasonal water use of 1036 mm and declined linearly as less water was applied. Miller and Hang (1980), again working with daily deficit irrigation following canopy closure in mid-June to early July, reduced water application rates to 35 to 50% of estimated ET without reducing sugar yields on a loam soil. Declines in root yield were compensated by increases in sucrose concentration. On a sandy soil, however, sugar yields decreased linearly with decreases in applied water. Carter et al. (1980), in Idaho, concluded that very little sucrose yield reduction should occur at a mid-October harvest without irrigation if 200 mm water were available in 1.6 m of soil on 1 August. Winter (1980), in Texas, showed that sugarbeet used available soil water to a depth of 3 m, adapted to an irregular water supply, and still produced reasonable commercial yields. Highest yields, however, used the largest amounts of water with the highest (76.6 Mg ha^{-1} of roots) using the most (1010 mm). Hang and Miller (1986a, b) concluded that with soils initially near the upper limit of available water, daily sprinkler irrigation could be reduced to 40 to 50% of estimated ET for a loam soil but only to 85% of ET in a sandy soil without reducing top, root, or sucrose production. These studies emphasize the ability of sugarbeet to tolerate moderate soil moisture stress without seriously reducing sugar yield.

Other studies, however, have shown that sugarbeet yield is closely related to ET (Ghariani, 1981; Winter, 1988). This is illustrated by Ghariani's work in 1980 at Davis, CA, on a clay loam soil. His experimental treatments were five water levels, applied at weekly intervals throughout the season, to supply approximately 7, 25, 50, 78, and 100% of potential crop ET. Figure 26–2A shows the close relation of sucrose yield to cumulative ET for plots harvested 39 d after the termination of irrigation on 6 September. The data of Fig. 26–2B are for plots harvested only 10 d after the terminal irrigations. Evapotranspiration deficits were determined by subtracting the cumulative ET of each irrigation level from potential ET. Sucrose yield declined 0.12 Mg ha^{-1} per cm of ET deficit. This strongly supports the conclusion of Jensen and Erie (1971) that sugarbeet irrigation should strive to replenish ET, and that the frequency of irrigation and the amount of water applied should be determined by the rapidity and the extent that a given crop uses the available water in its root zone.

2. Allowable Stress at Mid-season

The degree of water stress a beet crop will tolerate without reducing sucrose yield can be estimated from studies similar to the Davis experiment (Fig. 26–1). On 4 August, 2 wk after the terminal irrigation for the plants to be stressed, ET deficit was 59 mm, and about 56% of the available soil water had been removed from the first 0.9 m of soil (Fig. 26–3). At this time, the regressions of Fig. 26–1 indicate a potential crop loss by not irrigating that would exceed the cost of an irrigation. Thus, in this instance, with a crop ET rate averaging 6 mm water d^{-1}, to prevent a loss in sugar yield, stress should not exceed an ET deficit of 60 mm or the removal of more than

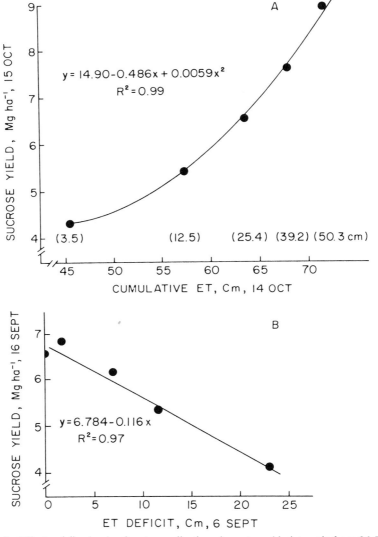

Fig. 26–2. Effects of five levels of water application given at weekly intervals from 26 June to 6 September to sugarbeet on a clay loam soil at Davis, CA in 1980 (Ghariani, 1981): A, total amounts of water applied (parentheses), crop ET, and its relation to sucrose yield 39 d after the final irrigation; B, relation of ET deficit to sucrose yield 10 d following the final irrigation.

about 60% of the available soil water in the first 0.9 m of soil depth. This estimate of allowable stress agrees with the summarization of earlier data by Jensen and Erie (1971). As the growing season progresses, roots will utilize soil water from lower depths, ET declines as temperature declines, and irrigation frequencies can be lengthened or the amounts of water applied reduced without causing water stress that would result in a loss in net crop return.

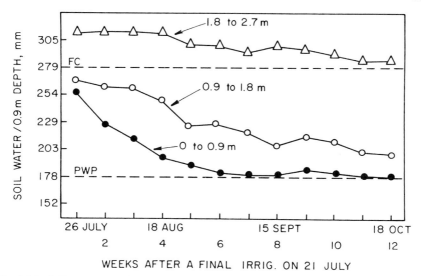

Fig. 26–3. Soil water removal by sugarbeet per 0.9 m of silt loam soil depth for the nonirrigated (NI) treatment of Fig. 26–1. (F.J. Hills, unpublished data, 1982). The FC and PWP are approximate soil water contents at field capacity and permanent wilting percentage.

3. Stress Prior to Harvest

Using soil-stored water to meet ET requirements prior to harvest may save an irrigation, reduce harvest and handling costs by an acceptable reduction in root yield, and thus increase net return. This must be done with caution, however, as prolonging water stress can reduce sugar yield and net monetary return.

In most areas of the irrigated western USA, sugarbeet is purchased on the basis of fresh weight sucrose content and gross monetary return is directly related to gross sugar yield. Limited moisture stress to increase sucrose concentration without reducing or only slightly reducing gross sugar yield can be a profitable practice. The length of the interval between the last irrigation and harvest that can be tolerated will depend on local weather, soil type, soil depth, root distribution, and the extent to which the soil water reservoir is filled at irrigation cutoff. It is not prudent to stop irrigation long enough to allow plants to wilt severely. This is illustrated by the first two trials summarized in Table 26–1. With the cost of a single irrigation about $49 ha^{-1} ($20 acre^{-1}, Hills et al., 1986), the sugar loss at 20 d in the 1963 trial following a terminal irrigation would be profitable in cost saving but stress beyond that would reduce net return. In the 1980 trial, losses in root yield occurred with cutoff periods longer than 4 wk before harvest. Additional irrigations on 4 August and 3 September would have increased net return considerably for the 1 November harvest date. The mineral soils of both these trials were low in organic matter (1–1.5%).

The 1981 trial of Table 26–1 was on an organic soil (about 30% organic matter) of the Sacramento River Delta. This field of sugarbeet had been

Table 26-1. Effect of water stress prior to sugarbeet harvest at three California locations.

| Last irrigation | | Fresh wt. | | Root sucrose | | Sucrose | Observed |
Date	Period to harvest†	Tops	Roots	Conc.	Yield‡	value§	moisture stress
		—Mg ha^{-1}—		%	Mg ha^{-1}	$ ha^{-1}	Visual effect on tops
1963, Kern County (Ferry et al., 1965)							
22 July	10 d	47.3	65.0	13.8	9.04	2451.78	Slight wilt
12 July	20 d	33.4	61.0	14.7	8.91	2416.62	Moderate wilt
2 July	30 d	26.2	58.1	15.0	8.73	2367.71	Severe wilt
Significant contrasts:							
Days linear		**	**	**			
CV, %		23.7	5.1	2.9			Estimated ET deficit, cm
1980, Yolo County (F.J. Hills, 1980, unpublished data)							
3 Oct.	4 wk	33.6	86.3	13.5	11.73	3181.75	0
3 Sept.	8 wk	24.4	81.6	13.8	11.10	3010.49	9.7
4 Aug.	12 wk	25.8	73.3	13.7	10.12	2744.70	35.0
Significant contrasts:							
Weeks linear		NS	**	NS			
CV, %		21.6	6.3	5.3			
1981, Solano County (F.J. Hills, 1981, unpublished data)							
15 Sept.	3 wk	¶	96.6	11.3	10.76	2918.27	3.3
1 Sept.	5 wk	¶	90.1	12.0	11.11	3013.20	4.8
18 Aug.	7 wk	¶	86.6	13.3	11.37	3083.71	11.9
Significant contrasts:							
Weeks linear			*	**			
CV, %			12.5	7.7			

NS, *, ** Not significant and significant at the $P = 0.05$ and 0.01 levels, respectively.
† Respectively, the three experiments were harvested 1 Aug., 1 Nov., and 6 Oct.
‡ Calculated from root yield and percent sucrose predicted from significant regressions. In 1980 the observed mean sugar concentrations were used.
§ Based on a California sugarbeet purchase price of $0.123/lb sucrose.
¶ Data not taken.

sprinkler irrigated regularly at 2-wk intervals through the spring and summer. The experimental treatments were final irrigations 7, 5, and 3 wk prior to a common harvest on 6 October. At harvest, sampling showed the soil to be saturated below 1.22 m and thus the estimated ET deficits given in the table, based on soil sampling to above the saturation level, probably reflect greater stress than actually occurred due to uptake of water just above the water table. Increasing the period from the last irrigation to harvest decreased fresh weight root yield at a rate of 2.6 Mg ha^{-1} per wk but increased fresh weight sucrose concentration 0.5 percentage points per wk, probably in response to increasing N deficiency as soil water decreased. The mean concentration of NO_3-N in dry, recently matured petioles was 4250, 2610, and 1820 mg kg^{-1}, respectively, for 3, 5, and 7 wk from a terminal irrigation to harvest. As observed in other trials on high organic matter soils (Henderson et al., 1968), the drying out of surface soil apparently reduces the mineralization of NO_3, thus reducing its uptake by plants. The changes in N nutrition do not appear to have been prolonged enough to have greatly affected root growth and most of the reduction in root yield was probably due to a reduction of water uptake as a result of increasing stress, as there was little change in dry weight root yield due to the length between the last irrigation and harvest. Sucrose concentration on a dry weight basis, however, increased linearly, 0.93 percentage points wk^{-1} ($r^2 = 0.999$), from 63.1% for 3-wk cutoff to 66.8% for 7 wk. This indicates that the increase in stored sugar was probably due to the onset of a N deficiency that slowed top and, to some extent root growth, resulting in a shift of sucrose from use in growth to storage.

These experiments indicate that while an irrigation may be saved prior to a fall harvest, care should be taken to avoid excessive water stress and that nutritional changes may occur, particularly with soils high in organic matter content, that can influence the profitability of this practice.

4. Effects on Root Quality

Quality usually refers to root fresh weight sucrose content, but often is expressed as purity, which is the percentage of sucrose in juice expressed from roots as a percent of the total soluble solids in the juice. Purity is important to the processor as soluble solids other than sucrose in the expressed sugar juice, particularly soluble N compounds, make it more difficult to recover sucrose in the refining process. Water stress or irrigation amount may affect the sucrose content of sugarbeet roots; however, irrigation effects are frequently difficult to separate from N effects. Sucrose content may be improved, and yield reduced, where excess water leaches N from the soil early in the growing season (Haddock, 1959). Purity is not usually improved by water stress (Bauer et al., 1975; Loomis and Worker, 1963; Reichman et al., 1977) although fresh weight sucrose concentration may increase due to root dehydration (Carter, 1982; Carter et al., 1980; Loomis and Worker, 1963). In Texas, reducing summer irrigation reduced root yield and Na content, but increased K, amino nitrogen content, and sucrose loss to molasses (Winter, 1988). In this case, while reduced irrigation increased sucrose concontration some years, total impurities in the root, and thus, root quality for processing was reduced every year. The apparent increase in sucrose concentration brought on by water stress is not evident if sucrose is expressed on a dry weight basis (Carter, 1982; Fig. 26-1).

II. WATER SCHEDULING

The frequency and amount of irrigation water required by a crop depends on the amount of available water a soil will hold to the effective depth of rooting, the rate of ET, the allowable depletion of available soil water, and the efficiency of irrigation, i.e., the portion of the water applied that is stored in the crop root zone. After planting, about 10 wk of good growing weather are required to develop a root system that can use the available water in the top 0.9 m of soil. In well-drained soils, roots may reach 1.8 m or more but most of the water uptake is from the first 0.9 m. Root penetration may be limited by a high water table, compacted layers, a dry subsoil, or other factors.

It is good practice to have the soil filled to field capacity at the time of planting, or soon after, to provide moist soil for the rapid extension of roots. When sugarbeet is planted during a rapidly warming period it may be necessary to apply three or four light irrigations during the first 10 wk until a deep root system is established. Heavier irrigations are required during midseason when the crop has a complete canopy of leaves and is transpiring at high rates. Each irrigation should be sufficient to replace the water used by the crop as well as take care of losses within the irrigation system. Intervals between irrigations may be as short as 5 d with high rates of ET and soils of low water-holding capacity to as long as 21 d for heavier soils with lower rates of ET. Generally, it is easier to plan farming operations around irrigations scheduled at regular intervals and vary the amount of water applied according to crop demand than to wait until the allowable depletion has been reached which could mean irregular intervals between irrigations. The interval between irrigations can be selected so the allowable depletion would just about be reached during a period of rapid ET. For example, if a soil holds 150 mm available water to a depth of 0.9 m, the allowable depletion is 60%, and peak crop ET is 7 mm d^{-1}, the minimum irrigation interval would be $(150 \times 0.60)/7 = 13$ d. In this case, a 2-wk interval between irrigations may be convenient and should cause little crop loss.

A. Water Balance

Irrigations may be scheduled and amounts of water to apply may be estimated from long-term reference evapotranspiration (ETo) records and crop coefficients (Kc) for specific areas (Pruitt et al., 1987; Snyder et al., 1987a, b). However, current ETo data are increasingly available from local weather stations and, when used with locally developed crop coefficients, give good estimates of crop ET (ETc) during the actual period between irrigations. The estimated crop ET, ETc = ETo × Kc, must be increased sufficiently to take care of losses within the irrigation system. Water losses depend on the efficiency of the irrigation system and usually will mean applying from 20 to 50% more water than indicated by ETc. Table 26–2 gives typical water use by sugarbeet for certain areas of western USA.

It is good practice to check depth of water depletion and penetration by using a soil tube or auger to examine subsurface soil moisture conditions before and after an irrigation. Changes can be made in irrigation procedures to improve efficiency if this appears necessary. Evaluations of depletion and penetration can also be accomplished by installing soil moisture-sensing devices such as tensiometers or gypsum blocks at two or more depths at the beginning, middle, and near the end of the irrigation runs.

Table 26-2. Seasonal water use by sugarbeet in selected semiarid areas.

Location and growing season	Jan.	Feb.	Mar.	Apr.	May	June	July	Aug.	Sept.	Oct.	Nov.	Dec.	Total
							mm						
California (Pruitt et al., 1987)													
Central valley													
1 Apr. to 20 Oct.				24	84	203	243	209	152	72			987
1 Feb. to 15 Sept.		7	20	40	134	226	256	214	72				969
Imperial valley													
1 Oct. to 30 June	82	109	135	214	263	227				54	64	60	1208
Coastal valley													
1 Nov. to 31 Oct.	20	38	74	112	135	142	168	137	107	89	8	13	1043
Idaho (D.C. Larsen, 1988, personnal communication)													
1 Apr. to 31 Oct.				30	56	99	229	213	145	79			851
Texas (Winter, 1988, personal communication)													
15 Mar. to 15 Nov.			38	76	143	213	244	198	147	89	38		1185

B. Other Guides to Irrigation Scheduling

Tensiometers or gypsum blocks placed at 0.5 and 1.0 m of soil depth can be used to assess soil water depletion, schedule irrigations and check on the adequacy of water penetration.

In areas of a field where soil is sandier or water penetration is low, plants tend to wilt sooner than in the rest of the field. Wilting in these areas may be used as indicators of the need for irrigating the entire field. However, certain fungi, nematodes, and root aphids (*Pemphigus populivenae* Fitch) will cause plants to wilt even when the supply of soil water is adequate. Roots should be examined for these agents when patches of wilting occur.

III. INTERACTION OF IRRIGATION WITH OTHER FACTORS

A. Nitrogen Nutrition

Nitrogen nutrition is of primary importance as a determinant of sugar-beet yield and quality. When both N and water are limiting growth from early season on, a greater response to N fertilization will usually occur with irrigation than when soil water is still limiting (Reichman et al., 1986). Loomis and Worker (1963), however, observed similar increases in sucrose concentration and decreases in root yield to late-season soil moisture stress for both high- and low-N plants. With organic soils, soil water contents that restrict plant growth apparently can also severely restrict nitrification and markedly alter the uptake of NO_3-N by plants (Henderson et al., 1968). The foregoing discussion of the third trial of Table 26–1 is another example of this effect. When crops are furrow irrigated and receive little summer rain, NO_3 may accumulate at the surface of planting beds and then be flushed into the root zone by late season rain causing a decrease in root sucrose concentration (Stout, 1964).

B. Stand Establishment

Where rainfall is lacking, no single factor affects seedling emergence as much as does correct, well-timed irrigation. Not only is adequate soil moisture essential for seed germination, but it also helps alleviate soil crusting. Irrigation can also be a factor in managing seed zone soil salinity to which sugarbeet is quite sensitive during germination (Sailsbery et al., 1985).

For furrow irrigation, sugarbeet is usually cultured on raised planting beds. Ideally, fields should be leveled to a grade of 0.05 to 0.15% slope in the direction the irrigation furrows will run. When irrigating, the initial stream of water should reach the end of the row quickly, without overflowing the seed rows at any point. Then the flow should be cut back to maintain water throughout the length of the furrow and minimize runoff until the soil is wetted past the seed line. With warm, dry, and windy weather, evaporation from the soil surface can soon deplete the moisture from around seeds and result in soil crusting. A second irrigation timed to coincide with the beginning of seedling emergence is often needed to replenish soil moisture at the seed level and to soften the soil surface for seedling emergence. A properly operated sprinkler system can assure a high percentage of emergence with reduced water application. Sprinklers can be operated as needed to soften crust, leach salts from around seeds, and maintain optimum soil moisture for germination.

A soil with an initial salinity indicated by a soil-paste conductivity of only 1 or 2 dS m^{-1} can become quite saline (5–10 dS m^{-1}) in the seed zone of single-row beds after one furrow irrigation with good quality water. With such an initial salinity level in single row beds, every other furrow should be irrigated when germinating seed, and beds should be wetted completely across so that salt will be moved by water away from the seed row. When salt in the soil exceeds 4 to 5 dS m^{-1}, sugarbeet can be grown by using specially shaped beds or using sprinkler irrigation during the germination period.

C. Root Disease

Certain of the soil pathogens causing seedling disease known as "damping-off" are favored by high soil moisture. Overirrigation in establishing a stand can increase seedling losses (Leach, 1971). Care should be taken to use no more water than necessary to insure germination and alleviate crusting. Sprinkler irrigation has a distinct advantage over furrow irrigation for reducing the length of time the surface soil remains saturated.

Hot summer temperatures and saturated soil often result in "wet root rot," a disease incited by *Phytophthora* and *Pythium* species. Irrigation practices to avoid prolonged soil saturation are helpful in control of this disease.

IV. CULTURAL PRACTICES IN RELATION TO IRRIGATION SYSTEMS

Cultural practices may vary with different irrigation systems. Furrow irrigation of nearly level soils with long irrigation runs dictates tall beds with wide spacings to reduce lateral water movement. In Texas and California, beds spaced 0.76 m apart, about 13 cm from furrow bottom to top of bed, and one row of beet per bed predominate. Other spacings are 1.0 or 1.5-m beds with two rows of beets on each bed. With sprinkler irrigation, or in some cases with furrow irrigation, narrow rows may be feasible. Some furrow-irrigated areas use a row spacing near 0.5 m or slightly greater with a small water furrow or corrugation.

With well-watered sugarbeet, narrow rows and in-row spacings of 15 to 20 cm increase root yield and sucrose concentration (Dillon & Schmehl, 1971; Hills, 1973). These relationships are not affected by level of irrigation (Winter, 1989). Sugarbeet grown with moderate to severe water stress through the major portion of the growing season still produce more sucrose with narrow rows (0.75 vs. 1.0 m) and in-row spacings of 15 to 20 cm.

Nitrogen may be more difficult to manage with furrow than with sprinkler irrigation. With permeable soil, leaching of NO_3 may be a problem in furrow irrigation, especially on the upper end of the field, if excessive amounts of water are applied. Even with slowly permeable soil, furrow irrigation often results in uneven N distribution associated with uneven water application. Residual NO_3 averaged twice as high on the lower compared to the upper end of 760-m long furrow-irrigated fields in Texas (Winter, 1986). As a result, sucrose content was about 1.5 percentage points higher on the upper end of these fields. In response to this problem, some growers reduce rates of N application to the lower end of long, furrow-irrigated fields.

V. SUMMARY

Sucrose production from sugarbeet depends on maximizing storage root growth over a long growing season. As root growth proceeds, there is a constant partitioning of sucrose to roots, and thus sucrose yield also increases throughout the season. Increases in fresh-weight root sucrose concentration with increasing soil moisture stress are due to slower accumulations of water rather than more rapid accumulation of sucrose. Sugarbeet root and sucrose yields are closely correlated with crop ET. In soils where root development is not restricted, a mid-season depletion of about 60% of the available soil-water to a depth of 0.9 m appears to be allowable before an irrigation to replace the lost water.

Irrigation practices should be directed toward rapid stand establishment, early attainment of a full leaf canopy, and the avoidance of midseason water stress. Depth and frequency of irrigations are dependent on ET, soil texture, and root distribution. Irrigations can be scheduled by the water balance method which depends on access to current or long-time pan-evaporation data, crop coefficients to estimate potential crop ET, and allowable depletion at the site involved.

REFERENCES

Bauer, A., T. Heimbuch, D.K. Cassel, and L. Zimmerman. 1975. Production potential of sugarbeets under irrigation in the West Oakes irrigation district. North Dakota Agric. Exp. Stn. Bull. 498.

Brown, K., and R. Dunham. 1986. The fibrous root system: The forgotten roots of the sugarbeet crop. Br. Sugarbeet Rev. 54(3):22–24.

Carter, J.N. 1982. Effect of nitrogen and irrigation levels, location and year on sucrose concentration of sugarbeets in southern Idaho. J. Am. Soc. Sugar Beet Technol. 21:286–306.

Carter, J.N., M.E. Jensen, and D.J. Traveller. 1980. Effect of mid- to late-season water stress on sugarbeet growth and yield. Agron. J. 72:806–815.

Dillon, M.A., and W.R. Schmehl. 1971. Sugarbeet as influenced by row width, nitrogen fertilization, and planting date. J. Am. Soc. Sugar Beet Technol. 16:585–594.

Ehlig, C.F., and R.D. LeMert. 1979. Water use and yields of sugarbeet over a range from excessive to limited irrigation. Soil Sci. Soc. Am. J. 43:403–407.

Ferry, G.V., F.J. Hills, and R.S. Loomis. 1965. Preharvest water stress for valley sugar beets. Calif. Agric. 19(6):13–14.

Ghariani, S.A. 1981. Impact of variable irrigation water supply on yield-determining parameters and seasonal water-use efficiency of sugarbeets. Ph.D. diss. Univ. of California, Davis.

Haddock, J.L. 1959. Yield, quality and nutrient content of sugar beets as affected by irrigation regime and fertilizers. J. Am. Soc. Sugar Beet Technol. 10:344–355.

Hang, A.N., and D.E. Miller. 1986a. Responses of sugarbeet to deficit, high-frequency sprinkler irrigation. I. Sucrose accumulation, and top and root dry matter production. Agron. J. 78:10–14.

Hang, A.N., and D.E. Miller. 1986b. Responses of sugarbeet to deficit, high-frequency sprinkler irrigation. II. Sugarbeet development and partitioning to root growth. Agron. J. 78:15–18.

Henderson, D.W., F.J. Hills, R.S. Loomis, and E.F. Nourse. 1968. Soil moisture conditions, nutrient uptake and growth of sugarbeets as related to method of irrigation of an organic soil. J. Am. Soc. Sugar Beet Technol. 15:35–48.

Hills, F.J. 1973. Effects of spacing on sugar beets in 30 inch and 14–26 inch rows. J. Am. Soc. Sugar Beet Technol. 17:300–308.

Hills, F.J., S.S. Johnson, and B.A. Goodwin. 1986. The sugarbeet industry in California. Univ. California Agric. Exp. Stn. Bull. 1916.

Jensen, M.E., and L.J. Erie. 1971. Irrigation and water management. p. 189–222. *In* R.T. Johnson et al. (ed.) Advances in sugarbeet production. Iowa State Univ. Press, Ames.

Leach, L.D. 1971. Fungus and bacterial diseases. p. 260–285. *In* R.T. Johnson et al. (ed.) *In* Advances in sugarbeet production. Iowa State Univ. Press, Ames.

Loomis, R.S., and J.L. Haddock. 1967. Sugar, oil, and fiber crops. I. Sugarbeets. *In* R.M. Hagan et al. (ed.) Irrigation of agricultural lands. Agronomy 11:640–648.

Loomis, R.S., and D.J. Nevins. 1963. Interrupted nitrogen nutrition effects on growth, sucrose accumulation and foliar development of the sugarbeet plant. J. Am. Soc. Sugar Beet Technol. 12:309–322.

Loomis, R.S., A. Ulrich, and N. Terry. 1971. Environmental factors. p. 19–48. *In* R.T. Johnson et al. (ed.) Advances in sugarbeet production. Iowa State Univ. Press, Ames.

Loomis, R.S., and G.F. Worker. 1963. Responses of the sugar beet to low soil moisture at two levels of nitrogen nutrition. Agron. J. 55:509–515.

Miller, D.E., and J.S. Aarstad. 1976. Yields and sugar content of sugarbeets as affected by deficit high-frequency irrigation. Agron. J. 68:231–234.

Miller, D.E., and A.N. Hang. 1980. Deficit, high-frequency irrigation of sugarbeets with the line source technique. Soil Sci. Soc. Am. J. 44:1295–1298.

Pruitt, W.O., E. Fereres, K. Kaita, and R.L. Snyder. 1987. Reference evapotranspiration (ETo) for California. Univ. California Agric. Ext. Stn. Bull. 1922.

Reichman, G.A., E.J. Doering, and L.C. Benz. 1986. Water management effects on N-use by corn and sugarbeets. Trans. ASAE 29(1):198–202.

Reichman, G.A., E.J. Doering, L.C. Benz, and R.F. Follett. 1977. Effects of water-table depth and irrigation on sugarbeet yield and quality. J. Am. Soc. Sugar Beet Technol. 19:275–287.

Sailsbery, R.L., F.J. Hills, W.E. Bendixen, R.A. Brendler, and D.W. Henderson. 1985. Sugarbeet crop management series, stand establishment. Univ. California Agric. Exp. Stn. Bull. 1877.

Snyder, R.L., B.J. Lanini, D.A. Shaw, and W.O. Pruitt. 1987a. Using reference evapotranspiration and crop coefficients to estimate crop evapotranspiration for agronomic crops, grasses, and vegetable crops. Univ. California Coop. Ext. Leaf. 21427.

Snyder, R.L., W.O. Pruitt, and D.A. Shaw. 1987b. Determining daily reference evapotranspiration (ETo). Univ. California Coop. Ext. Leaf. 21426.

Stout, M. 1964. Redistribution of nitrate in soils and its effects on sugarbeet nutrition. J. Am. Soc. Sugar Beet Technol. 13:68–80.

Ulrich, A. 1952. The influence of temperature and light factors on the growth and development of sugar beets in controlled climatic environments. Agron. J. 44:66–73.

Ulrich, A. 1954. Growth and development of sugar beet plants at two nitrogen levels in a controlled temperature greenhouse. Proc. Am. Soc. Sugar Beet Technol. 8(2):325–338.

Ulrich, A. 1955. Influence of night temperature and nitrogen nutrition on the growth, sucrose accumulation, and leaf minerals of sugarbeets plants. Plant Physiol. 30:250–257.

Ulrich, A. 1956. The influence of antecedent climates upon the subsequent growth and development of the sugarbeet plant. J. Am. Soc. Sugar Beet Technol. 9:97–109.

Winter, S.R. 1980. Suitability of sugarbeets for limited irrigation in a semi-arid climate. Agron. J. 72:118–123.

Winter, S.R. 1986. Nitrate-nitrogen accumulation in furrow irrigated fields and effects on sugarbeet production. J. Am. Soc. Sugar Beet Technol. 23:174–182.

Winter, S.R. 1988. Influence of seasonal irrigation amount on sugarbeet yield and quality. J. Sugar Beet Res. 25:1–10.

Winter, S.R. 1989. Sugarbeet yield and quality response to irrigation, row width, and stand density. J. Sugar Beet Res. 26:26–33.

Wright, J.L. 1982. New evapotranspiration crop coefficients. Proc. Am. Soc. Civ. Eng. 108:57–74.

Wyse, R. 1979. Parameters controlling sucrose content and yield of sugarbeet roots. J. Am. Soc. Sugarbeet Technol. 20:368–385.

27   Tobacco

MARK J. KING

Virginia Polytechnic Institute and State University
Blackstone, Virginia

The written history of tobacco (*Nicotiana tabacum* L.) began in 1492, when Columbus arrived in the West Indies. Prior to that time it had been used by Indians for many centuries. The colony at Jamestown, VA, was saved from economic collapse by the introduction of tobacco as a cash crop in 1612. Since that time, tobacco has served as both currency and collateral, being used by the Confederacy to guarantee loans from England during the U.S. Civil War. As a cash crop, it remains a leader in the southeastern USA and has been a major export crop for many countries, including the USA.

Botanically, tobacco belongs in the Solanaceae family. There are currently 60 species of *Nicotiana* described (Goodspeed, 1954), but generally, only *N. tabacum* is now used commercially. Tobacco is of tropical origin and is grown from Sweden to Australia (Garner, 1951). It is generally considered to be a mesophytic-type plant (Tso, 1972) containing 85 to 90% water when turgid, and 80 to 85% when wilted (Purseglove, 1968). Tobacco is a determinate plant, which can self-pollinate to produce 2000 to 8000 seeds per capsule (Garner et al., 1936). Purseglove (1968) contains a more extensive review of the botany of tobacco.

There are seven classes of tobacco encompassing 26 closely related grades established by USDA based on location and use (Chaplin et al., 1976). Classes include flue-cured, burley, sun-cured, Oriental, fire-cured, cigar (wrapper, filler, binder), and Maryland. Seedlings are generally produced in small areas called *plant beds,* which are densely seeded and covered with a translucent or transparent cover to maintain high humidity and protect the plants from frost. In some locations, (e.g., Canada) seedlings are produced in greenhouses for field planting. When seedlings are about 15-cm tall, they are transplanted to prepared fields. Seedlings are often planted on continuous soil ridges either mechanically or by hand. A small amount of water is often applied with each plant for establishment of the seedling. The plants generally require 120 frost-free days, depending on type, to mature and ripen in the field.

Tobacco can be grown on many soil types from coarse sands to heavy clays, depending on the type of tobacco (Garner, 1951). Generally, the soils should be well drained; drainage is often enhanced by the use of ridge planting. Soil fertility requirements will vary depending on the tobacco type.

Oriental tobacco is lightly fertilized, with little or no N, while burley tobacco can have as much as 336 kg ha^{-1} of N added to the soil.

Tobacco root systems are mostly fibrous in nature with one or more small taproots. The fibrous portion of the roots (80–90%) will be found in the upper 20 to 35 cm of soil, often within the planting ridge and about 40 cm on either side of the stalk (Hawks & Collins, 1983; Jones et al., 1960). Maximum depth of the roots is usually within 62 cm of the ridge top, though 90-cm depths have been reported in Zimbabwe (H.D. Papenfus, 1981, personal communication). Lack of either sufficient irrigation or normal rainfall will reduce the fibrous root system, but not the taproot development, which penetrates deeper into the soil profile (Papenfus & Quin, 1984).

Upon flowering or when the desired number of leaves are present, the terminal inflorescence is removed (topped) and axillary buds are inhibited from developing either chemically or by physical removal. The plant then accumulates nicotine in the remaining leaves and completes the chemical conversions involved in maturing and ripening. The final product of tobacco production is a leaf having the requisite texture, aroma, tensile strength, and chemical composition desired for that type. Further discussions on desirable leaf quality will follow later in this chapter. For a more complete discussion on tobacco production see Akehurst (1981).

I. IRRIGATION OF TOBACCO

As with most crops, it has long been known that available moisture can affect the final yield and quality of tobacco leaf. One of the first reports was by Darkis et al., (1935), who showed that when tobacco had excess rainfall, it produced a thin, light-colored leaf with a poor balance of chemical constituents. Deficient rainfall produced thick, gummy, dark leaves also having a poor balance of chemical constituents. Weybrew and Woltz (1974) stated that by far the most frequent cause of compositional changes in flue-cured tobacco was due to rainfall. Welch (1964) went so far as to claim 43% of the variation in yield from year to year in Georgia was due to rainfall. Irrigation would seem to be extremely useful to even out the distribution of, or replace, rainfall but in truth it is seldom a complete replacement. This has been widely accepted in tobacco production and was first stated by Garner (1951). There is no question that irrigation can improve yield and quality of a crop in a dry season, but proper scheduling of irrigation is still being debated.

Irrigation of tobacco on a large scale began about 25 to 30 yr ago. Since tobacco is a high-value crop, the initial investment in irrigation equipment may be justified. On a worldwide basis, irrigation application by hand-held sprinkler cans, flooding, overhead sprinklers, or center pivot systems depends primarily on the local cost of labor (Hawks & Collins, 1983; Akehurst, 1981). Most irrigation of U.S. tobacco crops uses overhead sprinklers (Fig. 27–1) or gun sprinklers. In many areas of the world, tobacco is grown in the dry season and irrigation is a necessity. In Australia, South Africa, Pakistan,

Zimbabwe, and Ecuador, for example, tobacco is grown either entirely or partially in the dry season (Akehurst, 1981). Italian burley and tobacco grown in Maryland, though not dry-season crops, may receive more water from irrigation than rainfall (Akehurst, 1981). In Sumatra, the sun-grown cigar wrapper leaves grow larger and thinner if supplemented with irrigation (Akehurst, 1981). In much of the U.S. crop, the need for irrigation depends heavily on rainfall distribution and the tobacco type being grown. At the other extreme is the Oriental and bidi tobacco, which are normally irrigated only once (Akehurst, 1981). In the Xanthi area of Greece, Oriental tobacco produced with a moisture deficit is associated with highly developed aroma (Garner, 1951). Only the large leaf types of Oriental tobacco in eastern Turkey are irrigated once or twice, while the large leaf types produced in the USSR are not irrigated.

In the USA most tobacco is grown under a production control system of allotments usually based on weight or acreage. This system does not always guarantee improved returns from investment in an irrigation system, and the nature of U.S. soils and climate further influences the feasibility of irrigation. The major tobacco-growing region in the USA includes Kentucky, Tennessee, Maryland, Virginia, North Carolina, South Carolina, Georgia, and Florida. Irrigation within each of these states has been studied and provides mixed results based more on the tobacco type and latitude than other factors. In general, the farther north in the flue-cured tobacco growing region of the USA, the less effective irrigation may be in improving yield and quality (Akehurst, 1981). In Florida, substantial yield and grade improvements from

Fig. 27–1. Overhead irrigation of tobacco in Virginia, USA (Courtesy of J.L. Jones, Virginia Polytechnic Institute and State University).

irrigation were noted (Clark & Myers, 1956). In Georgia, substantial yield and quality gains were obtained from irrigation, which was more effective in years when rainfall was inadequate or poorly distributed (Miles, 1957). During 7 out of 28 yr, yield and quality were improved by irrigation in North Carolina (Hawks & Collins, 1983). In Virginia, irrigation can improve yield and quality about 50% of the time. Maryland crops have been shown to benefit from irrigation 9 out of 20 yr (McKee & Street, 1978). Farther inland, work in Tennessee found infrequent and small yield increases and some negative quality effects of irrigation on burley tobacco (Parks et al., 1963; Parks & Safley, 1965). Irrigation can be profitable in some years in Kentucky burley (Welch, 1964) but no yield increases were noted in Ohio burley (Franklin et al., 1964). Farther north in Canada, irrigation of the flue-cured crop is often found to be beneficial to yield, quality and, maturity (Walker & Vickery, 1959), with accelerated maturity being the most important effect.

A. Plant Bed Irrigation

In most tobacco growing regions of the world, seedlings are produced in small, densely seeded areas called *plant beds*. Seed germination is both temperature and moisture dependent, since seeds are not buried but lie in the top 0.5 cm of soil. Beds normally are covered with some translucent material, which serves to maintain high humidity and temperature and to prevent frost injury. In the USA the most commonly used plant bed cover is either plastic, cotton, or synthetic. After the seedlings have germinated and the ambient air temperature has risen to acceptable levels, the covers are removed. When the seedlings are 15 cm in height with several leaves, they are pulled from the bed and transplanted to a prepared field. Work by Bunn and Splinter (1961) found that temperature in the plant bed was more important than soil moisture in determining germination. They also found that seed placed near the soil surface cannot imbibe enough water at -0.152 MPa of soil moisture tension to germinate.

The water requirements of plant beds in the USA are easily satisfied if sufficient cover is provided. A single irrigation before covering the bed may be enough to allow seedling growth until the cover is removed (Akehurst, 1981). In dry areas, the beds must be irrigated more often. In Africa, beds are irrigated two to four times a day until about 1 wk after germination, when only two irrigations per day are used (Akehurst, 1981). At 4 to 5 wk after germination, only once a day irrigation may be enough until just before transplanting. At that time, seedlings that will be planted as a dryland crop will be hardened by withholding irrigation (Akehurst, 1981; Neas, 1955). This common practice in Zimbabwe's early crop allows the seedlings to almost wilt (-0.8 MPa leaf water potential) before rewatering (Papenfus & Quin, 1984). This causes the transpiration to be reduced after transplanting, thus conserving available soil moisture. This practice is used in most dryland crop areas, such as Australia (Hopkinson, 1968), and appears to harden the seedlings for later drought (Papenfus, 1970). Applying the hardening process once results in an increase in soluble carbohydrates, which has been shown to correlate positively with root regeneration (Akehurst, 1981); however, repeated dry/wet periods will cause a decline in carbohydrates.

In more moderate climates, the water needs of a plant bed can usually be met by rainfall (if a water permeable cover is used) or by irrigating with 1.25 of water on a sandy soil when the soil surface appears dry. Irrigation in any situation must be carried out with great care, since the increased humidity of a densely populated plant bed becomes an optimal environment for diseases. Excess seedling irrigation can also inhibit growth (Gatut-Suprijadji, 1984). Irrigation studies in Florida, using buried plastic pipe in plant beds, found that when pipes were deeper than 15 cm, the percent stand established decreased (Rhoads, 1968, 1971) due to insufficient soil moisture. Work in North Carolina found that the optimum depth of the water table beneath a plant bed was 13.5 cm (El-Gheruri, 1980).

When seedlings are transplanted to the prepared field, the soil may be dry with little available moisture near the surface. A common recommendation is to apply a small amount of transplant water with each seedling. This serves two functions: (i) the water creates a hospitable environment in which new roots can develop and (ii) causes the surrounding soil to close around the roots, providing good soil contact. This transplant water is essential in dryland crops (Akehurst, 1981) and is strongly recommended in all tobacco areas of the USA.

The water requirements after transplanting are minimal until the rapid growth phase begins. During the early phase of slow growth in the field, no irrigation is recommended unless the weather is extremely dry. Irrigation of 1.25 cm on a sandy soil, just after transplanting, should maintain the plants until the start of the rapid growth phase.

B. Irrigation During the Rapid Growth Phase

Most research on tobacco irrigation suggests that the best time to irrigate is during the rapid growth phase of the plant. This usually begins at 4 to 6 wk after transplanting and continues until topping of the plant. At the beginning of this phase, the plants are generally at or somewhat less than knee-high. Unless the weather has been very dry up to this point, no irrigation has been needed to establish the crop. Growth during this phase and subsequent leaf quality are addressed in other sections of this chapter.

C. Irrigation During Harvest

The harvest period of tobacco is the time, usually after topping, from the first leaf harvest to the last. This generally follows topping by 30 d, when the whole plant is harvested for curing or up to 6 wk, when only a few leaves are harvested each time (*priming*). The whole plant harvest is generally done for dark-fired, sun-cured, and burley tobacco, while the priming system is used for flue-cured and Oriental tobacco.

When the tobacco is primed, the leaf area for transpiring is constantly being reduced, which reduces the water requirements of the crop. Transpiration of each leaf also decreases with maturity and, subsequently, as the ripening process changes the color from green to yellow in flue-cured tobacco. This is supported by the observation of McCants and Woltz (1967), who noted that topped plants wilted less than untopped plants.

Turner and Incoll (1971) reported that drought delayed ripening when it occurred after topping. The process of ripening requires remobilization of N and decline of chlorophyll content, the latter, which slows during drought stress. After relief of the drought stress, the leaf regreens, which is detrimental to leaf quality (Turner & Incoll, 1971). In contrast, tobacco ripens faster after heavy rain or irrigation (J.L. Jones, 1981, personal communication). Except in extremely hot, dry conditions, enough moisture will already exist in the soil at topping to allow ripening to proceed.

II. IRRIGATION AND CURED LEAF QUALITY

Tobacco leaf quality evaluation varies depending on the type of tobacco and its intended purpose. In simplest terms, leaf quality may be evaluated based on the following characteristics: alkaloid concentration, free sugar concentration, aroma (gum, oil, and resin content), leaf color, elasticity, burn characteristics, and texture. Of these, aroma, elasticity, texture, and color are used in the official grading system for most tobacco types. The other chemical characteristics are accepted on faith at the point of sale and can be adjusted for during manufacturing. Desirable leaf quality depends on the tobacco type. For Oriental tobacco, leaves of strong aroma are desired, while cigar wrapper leaves should be thin and light colored. In flue-cured tobacco, a lemon to orange color with some aroma and good elasticity is important, while burley will have a more brown color. The following discussion describes the overall impact of irrigation on these characteristics across all types of tobacco.

Irrigation during two stages of tobacco growth and development can affect leaf quality; (i) the rapid growth phase and (ii) harvest. The effects are much more pronounced when irrigation occurs during the former stage of growth because irrigation is not strongly recommended during harvest. Water stress during flowering can produce an off-type immature tobacco when cured (McNee et al., 1978). Water stress applied to mature tobacco was shown to have less impact on quality than that imposed on rapidly growing tobacco (Petrie & Arthur, 1943). If no water stress is allowed during the rapid growth phase, the resulting leaf will appear slick and be of general poor quality (Moseley, 1956). Flue-cured tobacco produced under these conditions will usually have neither the desirable color, aroma nor texture (McCants & Woltz, 1967).

Severe water stress or extended dry weather produces leaf that is difficult to cure, brown, stiff with poor chemical balance, and with other poor quality characteristics. Leaves may have rim burn and be small and dark in color, with more aroma and denser leaf structure (Chaplin et al., 1976).

Irrigation generally produces a thinner leaf type of lower density (Akehurst, 1981; Brown et al., 1970). This leaf generally has a higher specific volume, which translates into greater filling capacity (Brown et al., 1970). On the other hand, Walker and Vickery (1959) reported that irrigation had little effect on leaf thickness or filling power. The difference may have been due to how much irrigation was used and at what stage of growth it was applied, which would affect soil N availability. Leaf elasticity and texture are intimately related to leaf density and thickness and will consistently be affected in the same way. Dry weather tobacco has a thicker leaf and lower elasticity, while overwatered tobacco produces a thin leaf of lower elasticity. Texture is improved by some moisture stress, but declines in dry weather tobacco. A compromise in leaf thickness should be sought for good elasticity and texture.

Leaf burn rate is closely related to filling power and can also be affected by available soil moisture. High natural soil-moisture content gave better burning characteristics (van Bavel, 1953), while irrigation gave a greater duration of burn (Street, 1978). On the other hand, Brown et al. (1970) found that irrigation did not affect burn duration, while increasing filling capacity or specific volume. This discrepancy, again, may be due to the degree of moisture deficit allowed in each study.

In general, the concentration of alkaloids, primarily nicotine, in the leaf can be modified by water stress during the rapid growth phase. In flue-cured tobacco, the percent nicotine can vary from 1.88% with abundant water to 2.23% for adequate soil moisture, to 4.14% for insufficient moisture (Hawks & Collins, 1983). In dry weather, nicotine concentration increases under nonirrigated situations (McKee & Street, 1978), probably due to a concentration in lower leaf yields and prolonged synthesis due to delay in exhausting the available soil N (Weybrew & Woltz, 1974). On the other hand, Liu (1978) observed that increasing soil moisture increased total alkaloids and total N in the leaf, which was supported by Chaplin et al. (1976). Irrigation has been shown to reduce the alkaloid concentration of the leaf (Walker & Vickery, 1959; McKee & Street, 1978; Brown et al., 1970; Sficas, 1970). The discrepancy may be becuase of the various levels of available moisture across all of these studies and its impact on N fertility.

The effect of dry weather on sugar concentration in leaves is just the opposite. During dry weather, the sugar concentration declines, as carbohydrate oxidation to acids increases (van Bavel, 1953). Decreasing soil moisture was found to reduce sugar concentration (Liu, 1978). Conversely, if irrigation is used, the sugar concentration rises (Walker & Vickery, 1959; Hawks & Collins, 1983).

The sugar/nicotine ratio of flue-cured tobacco is extremely important for manufacturing. The proper ratio of sugar/nicotine should be in the range of seven to ten. If this ratio is higher or lower, the leaf will be much less desirable and must be blended to compensate for the imbalance. High moisture stress has been shown to affect the relationship of carbohydrates to N compounds (van Bavel, 1953; Weybrew, et al., 1983). Obviously, if dry weather increases nicotine and decreases sugar concentration, the ratio would not be in the desirable range. At the same time, too much rain or irrigation can produce leaves with too little nicotine and too much sugar. The same balance of moisture stress for growth also affects the proper internal chemistry of tobacco.

One study sought to grow tobacco to a prespecified leaf chemistry using both irrigation and soil N applications (Ismail & Long, 1980). They found that although both components were moved in the desired direction, nicotine (which contains N) was more amenable to manipulation than sugar concentration and that the season played a major role, probably making this approach impractical in production. Another study in Virginia, using a line source irrigation system over tobacco fertilized at three rates of N, found no significant interaction of irrigation and applied N on nicotine or sugar levels (M.J. King, 1983, 1984, 1985, unpublished data). This agress well with van Bavel (1953), who found no interaction of fertilizer with water level and Atkinson et al. (1969), who reported no interaction of irrigation with N applications in burley.

Dry weather has been known to increase the aroma or petroleum ether extract of tobacco (van Bavel, 1953; Chaplin et al., 1976). The effects of irrigation on aroma are also consistent over many tobacco types. Aroma is a composite characteristic related to the amount of gum, oil, and resin in the leaf, as well as on the leaf surface. The most aromatic tobacco type, Oriental, is strongly affected by rainfall. In the Turkish (Oriental) tobacco area of Greece, any sporatic rainfall during the growing season was once considered injurious to quality (Garner, 1951). This may have been a result of washing off the gummy deposits on the leaf surface by rain. More recently, Sficas (1970) reported that irrigation in Greece of both the aromatic and neutral types of Oriental tobacco benefited in quality and yield from a moderate supply of water. Again, the balance of available soil moisture appears to be critical for enhancement of yield and quality.

III. GROWTH AND SOIL WATER

Growth of the tobacco plant, in most situations, follows an exponential curve. In one study, only 2.5% of total growth had taken place during the first 21 d post-transplanting (Grizzard et al., 1942). During the next 21 to 28 d the growth rate rises rapidly, and during the last 28 d of a 63-d growth period, 80% of total growth occurs.

It is a well-established concept that soil moisture is intimately associated with plant growth. Leaf area was increased by high soil moisture (Atkinson et al., 1969; Hirata & Sasaki, 1971; Gopalachari et al., 1984; Kume, 1975; Liu, 1977; Matusiewicz & Madziar, 1975; Sasaki & Hirata, 1971). It has also been reported that some degree of soil moisture deficit can increase leaf area (Ligon & Benoit, 1966; Ferguson et al., 1985; Hopkinson, 1968). Farah (1980) found that irrigation had no apparent effect on leaf size. This contradition may be resolved if the relationship of leaf growth to available moisture is considered in terms of the severity of the water stress.

Liu (1978) found that tobacco leaf area development was correlated with leaf water potential ($r = 0.966$). It is accepted that, when leaf water potential begins to decline, the rate of cell expansion being driven by turgor pressure will decline sooner than the rate of cell division. If the moisture deficit is not severe enough to cause a decline in the rate of cell division, but does reduce cell expansion rates, the growth rate after water stress removal may exceed that of well-watered controls (Hopkinson, 1968). Hopkinson (1968) found that water-stressed plants had more leaf area and dry weight, and higher relative growth, relative leaf expansion, and net assimilation rates than controls. Ligon and Benoit (1966) and Liu (1977) found that upon rewatering, the growth rates did not exceed those of controls, except when the controls were approaching maturity. The discrepancy in these studies may be due to the comparison of growth rates of leaves of plants that were at the same chronological age, but due to the water stress, were not at the same developmental stage.

Others have found that in field studies, a period of moisture deficit can increase yields over well-watered plants. Ferguson et al. (1985) reported that withholding water for 28 d post-transplant increased yield by 33%. In many tobacco-growing areas of the world, no irrigation is recommended in the early stages of field growth of both dryland (Papenfus & Baxter, 1969) and humid region tobacco (Jones et al., 1984). Most agree that the increased yield is a direct result of the delay in floral induction, as well as a lengthening of the growth phase of tobacco and production of a more extensive root system.

There are several benefits of irrigating tobacco to accelerate growth. Irrigated tobacco will bloom sooner than nonirrigated or nonwater-stressed tobacco, which shortens the period for disease and insect infestations (Hawks & Collins, 1983). At the same time, this allows the harvesting of the crop earlier to avoid frost injury late in the season, a major concern in Canadian flue-cured production. Also, under irrigation the stalk will be taller (Hawks & Collins, 1983; Hirata & Sasaki, 1971) with leaves spaced farther apart (Hawks & Collins, 1983), which will aid in harvesting. A major drawback to this rapid growth may be the firing of leaf margins (Hirata & Sasaki, 1971) and the production of thin leaf (Tanimoto, 1977).

The balance between effects on cell division and expansion in the short term and the prolongation of leaf initial production and the growth period of the leaf, is critical for producing optimum yield. This alone cannot be the only criteria to determine proper available soil moisture, since the effects on leaf chemistry and quality are equally important in tobacco production. Campbell and Seaborn (1972) found that when tobacco was grown over water tables of varying depth, a 60-cm water table seemed to provide the best balance of soil aeration and water supply. A balanced approach must be taken on irrigation decisions encompassing both the short-term and long-term growth requirements of tobacco.

IV. WATER STRESS RESPONSE

Since tobacco is grown as a dryland crop in some areas and a moisture deficit is desirable in others (where Oriental is grown), the effects of water stress on tobacco have been the subject of much research. McCants and Woltz (1967) suggested that something less than optimal moisture for rapid growth would be better for production of high-quality leaf. Tobacco is considered to be rather drought tolerant and moderate drought will not reduce yields, because smaller leaves have higher leaf density (Garner, 1951).

From a metabolic standpoint, tobacco responds to moisture stress in characteristic ways. During water stress, proline accumulates in the leaves (Boggess et al., 1976). Iwai et al. (1979) suggested that proline accumulation in tobacco was a result of the inhibition of proline oxidation by stress. Boggess and Stewart (1980) found that applications of cycloheximide during water stress did not inhibit proline synthesis, suggesting that proline accumulation does not need protein synthesis. Later, Tanabe et al. (1982) reported that when variegated tobacco leaves were detached and stressed, large amounts of proline accumulated in green but not in white tissues. If exogenous sucrose was supplied to both tissue types, proline accumulation increased, suggesting the need for chloroplasts (Tanabe et al., 1982). This supports the work of Mizuasaki et al. (1964), who noted that proline normally is synthesized from glutamate more rapidly in the light than in the dark. They also reported that proline oxidation did not vary from day to night. Proline was observed to accumulate at different rates depending on leaf position on the stalk. Sano and Kawashima (1982) reported that in detached leaves allowed to wilt for 4 d, proline accumulated more in upper stalk leaves on mature and old plants. Proline was also observed to accumulate more on mature than in young or old leaves. The debate over the utility of proline in a water stress situation continues.

Another metabolic affect of water stress is on NO_3 reductase activity (NRA). In tobacco, NRA is a critical factor in the proper maturing of a leaf. Normally, it increases with available N and only begins to decline as available N is depleted. In flue-cured tobacco this transition from NO_3 reduction to starch accumulation should coincide with topping for proper maturity and ripening of the leaves (Weybrew et al., 1974; Weybrew et al., 1983; Weybrew & Woltz, 1974). Long and Woltz (1977) reported that moisture stress prolonged enzyme activity. Water stress at 26/22 °C (day/night), but not at 22/18 °C inhibited NRA (Crum, 1977). A delay in NRA from water stress would cause NRA to continue beyond topping, thus delaying ripening of the leaves, which besides lowering leaf quality, can present many other problems with disease, pest, and frost injury. The activity of ribulose-1,5-bisphosphate (RUBP)-carboxylase was also reduced during water stress (Long & Woltz, 1977). When water stress was relieved, the activity exceeded that of well-watered controls.

Transpiration in tobacco leaves occurs primarily through stomata on the lower leaf surface. The upper surface transpires at about 10 to 25% of the lower surface. Stomatal density was reported to be higher and transpiration greater in leaves at higher stalk positions (Sasaki & Hirata, 1971). Turner and Incoll (1971) noted that stomatal conductance was small in older leaves even in sunlight. Farquhar et al. (1980) reported evidence that stomata limited transpiration relative to CO_2 assimilation regardless of relative humidity. The regulation of stomata begins at around -0.08 MPa and most stomata are closed at -1.5 MPa (Liu, 1978). Though antitranspirants have been shown to reduce water loss in tobacco by 60% for 2 to 3 d (Kreith et al., 1975), they are not in general use.

Bliss et al. (1957) compared a burley cultivar and a more drought-resistant dark-fired tobacco cultivar. The stomata of both cultivars appeared to respond the same to decreasing turgor potential. The dark cultivar did have higher rates of transpiration than the burley under both field capacity and drought conditions. This may have been a result of the dark cultivar having thicker leaves with more substomatal surface area. Neither relative turgidity nor osmotic potential of the cell sap varied with cultivar. Growth was stopped due to stress in the same manner in both cultivars. The burley removed soil moisture faster, which probably explains the difference in drought resistance. This does not agree with work by Sficas et al. (1961), who stated that leaf area was the most important factor related to drought resistance and that root development was intermediate and absorption capacity least important. Sficas' work was in a growth chamber and the plants were only studied for a relatively short period, while Bliss' study of soil moisture was carried out in the field. Work by Liu (1978) found that soil moisture was more important in the relationship of soil moisture and leaf water potential (0.647) than radiation, temperature, or vapor pressure deficit.

Water relations of tobacco have also been observed to vary with leaf position. Higher leaves had more negative leaf water potential than lower leaves, except at sunrise, when all leaves were about the same (Begg & Turner, 1970; Turner, 1974). This relationship was enhanced as the day progressed from sunrise to 1100 h. In a growth chamber experiment, the ninth leaf from base retained water better than the sixth leaf (Clough & Milthorpe, 1975). Further, the resistance to water flow in the stem was found to increase as the day progressed (Begg & Turner, 1970).

Other work has attempted to screen tobacco cultivars for drought resistance by measuring seed germination in mannitol solutions. Laksminarayana et al., (1979) reported a correlation of $r = 0.7492$ for germination in mannitol and observed field wilting index.

V. DROWNING OF TOBACCO

Tobacco has been described as one of the most susceptible crops to injury from flooding (Kramer & Jackson, 1954). Prevailing cultural practices call for flue-cured tobacco to be grown on well-drained soils and often planted on a ridge to enhance drainage. In areas where irrigation is used to supplement rainfall, an irrigation can be followed by rain, which may cause drowning. This complicates the decision to irrigate when rainfall is imminent.

Tobacco plants are not uniformly susceptible to drowning over the entire growing season. Campbell (1973) noted that plants are more susceptible to drowning at the 12th than at the 17th leaf stage. The difference is that the 12th leaf stage is in early growth phase and the 17th leaf stage is in the late rapid growth phase. Turner and Incoll (1971) observed that plants in the maturing phase were less susceptible to overwatering and that leaf senescence may actually be accelerated by excess water. Tanimoto (1977) reported that high soil moisture in paddy fields caused early senescence of leaves.

Under conditions of excess soil moisture, a tobacco plant will appear to be wilted, as if a soil moisture deficit had developed (Fig. 27-2). The wilting is a direct result of lack of water being transported to the shoot. If tobacco root systems are placed in anaerobic conditions, they will wilt in a fashion similar to field flooding (Williamson & Splinter, 1968; Mizrahi et al., 1972). Harris and van Bavel (1957) found that when CO_2 exceeded O_2 concentration, growth and water absorption were reduced sharply. Willey (1970) observed that near-anaerobic soil conditions reduced water uptake by 50% within 6 h; however, recovery of water uptake was rapid when O_2 was returned to the roots. Root injury did take place within the anaerobic treatments as evidenced by reduced water uptake upon recovery (Willey, 1970). Root injury was noted by Williamson and Splinter (1968) after 24 h of exposure to 1% O_2 and 20% CO_2. Both Willey (1970) and Williamson and Splinter (1968) agreed that the primary cause of root injury was a lack of O_2 and not excess CO_2. Dead root systems were observed to take up 40% as much water as healthy roots on a leaf area basis (Williamson & Splinter,

Fig. 27-2. Drowning of flue-cured tobacco showing characteristic wilting. (Courtesy J.L. Jones, Virginia Polytechnic Institute and State University.)

1968). If plants were flooded for 48 h, yield was reduced by 40% compared to nonflooded tobacco (Campbell, 1973). If plants were flooded for 24 h, the growth was not significantly different from controls; however, the leaf reducing-sugar concentration increased and nicotine decreased (Campbell, 1973).

Another effect of O_2 deficiency was to increase plant and soil ethylene ($CH_2 = CH_2$) content (Hunt et al., 1981). After three d of flooding, the leaf ethylene concentration peaked, then declined to preflood level before soil ethylene appeared. Leaf wilting was associated with appearance of soil ethylene. At wilting, the leaf water potential, stem diameter, and relative water content all decreased. Hunt et al., (1981) suggested that the ethylene from the soil and plant are both involved in reducing root permeability and water uptake.

VI. IRRIGATION SCHEDULING

The decision to irrigate tobacco depends on many factors, not least of which is the appearance of the crop and soil. For dryland tobacco, the decision is not influenced by anticipated rainfall and a simple balance sheet approach will track plant water use and indicate the need for irrigation. In the southeastern USA rainfall, on average, is reasonably well distributed in most growing seasons and irrigation, therefore, is not always economical.

As a general recommendation, Akehurst (1981) suggested irrigation should be based on a visual indication of need, not to exceed 5 cm wk^{-1} between knee-high and flowering. Most agree that only a light irrigation may be needed immediately after transplanting to establish a stand and then water should be withheld for 4 to 6 wk (Chaplin et al., 1976). Mulchi (1985) recommended irrigation should be withheld for 5 to 8 wk after transplanting. During that time, the seedling will make use of available soil water and should not require any further irrigation, if some rain does fall. The plants will establish a good root system and will begin the rapid growth phase at around 5 to 6 wk. At that time, the plants will reach knee-high stage and irrigation may be required to supplement rainfall.

During the rapid growth phase, scheduling of irrigation can be based on several techniques. A low-tech approach, which is generally used in the southeast USA and many foreign countries, requires looking at the plants for signs of wilt before 1100 h or if the soil appears ashy in color (soil color will vary in different soils). If neither condition is met, then irrigation should be delayed another day (Hawks & Collins, 1983). In Virginia, irrigation based on tensiometer readings of 60 to 70 cb (0.06–0.07 MPa) in a sandy loam soil provided too much water and though high yields were attained, the leaf quality declined (M.J. King, 1983, 1984, 1985, unpublished data). Since tensiometers will not accurately read much below this moisture level, their use in irrigation of tobacco is questionable. Another low-tech method would be to keep a water balance record and replace soil water at a rate that would make 2.5 cm water wk^{-1} available to the crop. Walker and Vickery (1959) reported that Thornthwaite evaporation estimates and electrical resistance blocks are suitable for scheduling irrigation in Canada. Other work in India has used a class A pan evaporimeter effectively to increase yield and decrease the number of irrigations on bidi tobacco (Ravindranath & Subrahmanyam, 1978).

After topping, little irrigation is needed for leaf ripening. The plants are not transpiring as heavily, the need for water for cell expansion decreases, leaf cuticle and waxes increase, and at least in flue-cured, the transpiring surface is continually being reduced by individual leaf harvests. A single irrigation at this time may be needed, if there is no precipitation and high temperatures occur. The real danger to the crop at this point is the burning of the leaf margins, which lowers leaf quality and indicates an imbalance of chemical components. The best scheduling technique at this stage would be to observe the plants for signs of severe wilt and irrigate just enough to relieve wilting. Any extensive irrigation may mobilize peripheral soil N allowing it to be taken up by the plant, causing regreening of leaves, and subsequently lowering leaf quality.

VII. EVAPOTRANSPIRATION

Little research has been done on ET in tobacco. It appears that during the first 2 to 5 wk after transplanting, the ET/open pan evaporation (ET/EO) = 0.2 for tobacco in Zimbabwe (Papenfus & Quin, 1984). Jones et al. (1960) suggested that tobacco loses about 0.28 to 0.45 cm water d^{-1}. Over the period from transplanting to final harvest this value changes. Early in the season the demand for moisture is low. During the rapid growth phase, water consumption increases with additional leaves. Orphanos and Metochis (1985) reported that in Cyprus the ET reached its maximum value at 40 to 50 d after transplanting, then declined just before the first harvest, even though soil water was adequate (Fig. 27-3). After topping, the rate of water consumption begins to decline, as the last leaves reach maturity. As the leaves ripen and the color changes from green to yellow, water use declines rapidly based on individual leaf transpiration measurements. The limited amount of published work on the use of ET in tobacco may be partly due to the seasonal variability in the USA.

VIII. TOBACCO NUTRITION AND IRRIGATION

Most tobacco types have fertility requirements specific for their location and desired end product. Breeding programs have developed cultivars that respond to this fertility regime well but would not do well under a vastly different scheme. Generally, N is the major component of any fertility program followed by K and P. Nitrogen is the most influential element in tobacco growth and a primary component of nicotine.

For flue-cured tobacco, the available N in the soil should be depleted at topping of the tobacco. This provides the proper transition from the rapid growth phase to the mature and ripening phase needed for proper leaf quality. Tobacco can be fertilized with NO_3, NH_4 or urea forms of N but the method of choice is usually ammonium nitrate (NH_4NO_3). Nitrate, being mobile in the soil, can be leached by excess water in the well-drained soils in which flue-cured tobacco is usually grown. The interaction of available soil N and precipitation/irrigation is critical. In dryland crops, the amount of water applied can be closely controlled. Under supplemental irrigation and excessive rainfall, however, ripening could begin before maturation has been completed should N be leached below the root zone and the remaining soil N exhausted by the crop. Previous work has shown that N can be leached from the soil by excess irrigation or rainfall during the rapid growth phase (McCants & Woltz, 1967; Terry & McCants, 1979). As insurance against leaching in wet weather, many growers apply extra N at the last cultivation, which is generally before the knee-high stage. If precipitation is low, more irrigation will be required to leach the extra N or leaf quality will be low.

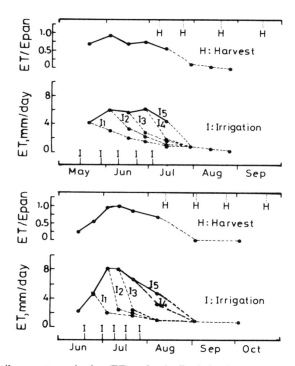

Fig. 27–3. Daily evapotranspiration (ET) under the five irrigation treatments, and ratio of ET to screened pan evaporation (E_{pan}) under the wettest treatment for Exp. 1 (upper graphs) and Exp. 2 (lower graphs). Solid lines indicate ET and ET/E_{pan} under nonlimiting soil moisture conditions. Note that planting in Exp. 1 was on 9 April and in Exp. 2 on 25 May. I_1 to I_5 denote number or irrigation, each of 50 mm in Exp. 1 and 60 mm in Exp. 2. (Orphanos & Metochis, 1985).

In controlled experiments, osmotic stress slightly reduced N absorption (Shoda & Kume, 1978). In an earlier study, N uptake was observed to be higher in stressed plants but protein synthesis was inhibited if stress was continued (Petrie & Arthur, 1943). When stress was relieved, protein synthesis exceeded nonstressed plants, as did growth rate and magnitude. If excess N is available with sufficient moisture, thinner tobacco leaves are produced (McCants & Woltz, 1967).

The effect of interaction of soil N with irrigation is well established (Atkinson et al., 1969; Menser & Street, 1962; Brown et al., 1970). Generally, it was noted that at lower and higher rates of N, irrigation depressed yields but that medium rates of irrigation and N interact to increase yield. Other work has indicated that irrigation (except at excessive rates) does not interact with soil N application rates between 62 and 90 kg ha^{-1} in Virginia on a sandy loam soil (M.J. King, 1983, 1984, 1985 unpublished data).

IX. CHLORIDE, SALINITY, AND NUTRITION

Though many elements are required in tobacco fertility programs, little other work has been done on their interaction with irrigation except for chloride (Cl). Chloride was first established as a micronutrient in tomato (*Lycopersicon esculentum* Mill.) (Broyer et al., 1954), which is in the same botanical family as tobacco (Solanacea). Though the requirement in tobacco is still debated, it may be a moot argument. In most soils in which tobacco is grown, there exists enough Cl or an adequate amount is supplied through standard fertilizers to meet the needs of tobacco. The review by McCants and Woltz (1967) on the role of Cl in tobacco nutrition should be consulted for more detailed information.

Tobacco has been shown to respond to Cl additions. As early as 1929, Moss (1929) found that adding Cl at a rate of 22 to 28 kg ha^{-1} was beneficial to tobacco production. In another study, Peele et al. (1960) applied 0 to 92 kg ha^{-1} of Cl through irrigation water and found no effect on nicotine or sugar concentration, even though leaf Cl concentration ranged from 2 to 3.46%. Elliot (1967) found that under normal Cl availability, the plant will absorb 32.6 kg ha^{-1} of Cl. Hawks and Collins (1983) recommended that no more than 34 kg ha^{-1} of Cl be added to the flue-cured crop from all sources. They calculated that irrigation adds about 0.28 kg ha^{-1} for each mg L^{-1} concentration in the water. Mulchi (1982) found that applications of Cl above 44 kg ha^{-1} produced cured Maryland tobacco with 0.53% Cl, reducing leaf quality and leaf burn.

The available sources of water for irrigating tobacco vary in Cl concentration. In South Carolina, Cl concentration of ponds, wells, and streams tested ranged from 4 to 55 mg L^{-1} and averaged 11 to 12 mg L^{-1} (Peele et al., 1960). In North Carolina, the range may be as high as 2000 mg L^{-1} in the Tidewater region of the state (Hawks & Collins, 1983). In one experiment potted tobacco plants were irrigated with seawater at 10 000 L^{-1} Cl. Growth was depressed, but the effect could be mitigated somewhat by K additions (Kurian et al., 1966). The availability of a source of water high in Cl does not present a problem for tobacco if other steps can be taken, which will be discussed elsewhere in this section.

In general, excess Cl causes the leaves to curl up, be thicker and more brittle, and to appear slick and glabrous with color and aroma alterations (McCants & Woltz, 1967), including a dull or lackluster color and whitish or grayish colored veins (Sabourin, 1967). Wedin and Struckmeyer (1958) found only a slight thickening of the leaves with increasing Cl treatment but others have reported a significant thickening of the leaf (Moss, 1929; McCants & Woltz, 1967). At beneficial rates of Cl application to the soil, the burn characteristics are not affected (Moss, 1929), but at higher rates the duration of burn is significantly reduced (Attoe, 1946; Rhoads, 1972; McCants & Woltz, 1967; Mulchi, 1982). The observed thickening of the leaf may be directly related to the resulting looser arrangement of both palisade and mesophyll cells caused by Cl (McCants & Woltz, 1967). The poor burning properties of the thicker leaves caused by Cl are probably because of their higher hygroscopic or water retention properties (McCants & Woltz, 1967). One recent study found that burn rate increased with irrigation, even though leaf Cl also increased (Orphanos & Metochis, 1985). Their Cl levels in green leaf tissue were all below 2.5% for all irrigation treatments, which would explain the lack of negative effect on leaf burn.

The levels of Cl in leaf tissue acceptable in most tobacco types are generally below 2%. Virginia flue-cured tobacco has been tested at just over 1%, while levels in one South Carolina study ranged from 2 to 3.46% (Peele et al., 1960). In Queensland, Australia, the acceptable limit of Cl is 2% (Akehurst, 1981) and in India a level of 3.06% Cl is unacceptable (Murty, 1964). Irrigation in South Africa can add 75 kg ha^{-1} of Cl, which will raise leaf Cl above 2% (Snyman et al., 1976).

The point at which Cl is detrimental may depend more on other factors than simple leaf concentration. McCants and Woltz (1967) found that the form of soil N played a large role in the relationship of leaf Cl concentration and leaf toxicity symptoms. When nutrient media contained all NO_3 forms of N, a leaf concentration of 8.9% caused no apparent symptoms, but when NH_4 was the primary N source, as little as 0.95% leaf Cl caused toxicity symptons. Though Mulchi (1982) found no interaction of source of N and Cl applied, Fuqua et al. (1976) reported that increasing levels of N decreased the Cl concentration in the leaf. Earlier, Skogley and McCants (1963) found that Cl decreased NO_3 accumulation in the leaf and Fuqua et al. (1974) found that Cl inhibited NO_3 absorption. Overall, there appears to be some competition of Cl and NO_3 and where the irrigation water contains high concentrations of Cl, the NO_3 form of N may mitigate the ill effects of Cl.

Chlorine is not uniformly distributed in the whole plant and tends to concentrate in the lower leaves (McCants & Woltz, 1967). Akehurst (1981) stated that accumulation in the older leaves occurs heavily in the midribs and can concentrate in active growing regions, if Cl is deficient. Elliot (1967) reported that in one study, 42% of the Cl in the whole plant was in the stalk, 33% in the leaves, and 10% in the roots.

The uptake of Cl by Wisconsin cigar tobacco has been well correlated with the Cl content of the top 30 cm of soil, with one-half of Cl absorbed coming from this layer (Myhre et al., 1956). When NaCl was present in the root zone, water uptake was reduced (West and Black, 1978). Dry weather tended to increase Cl uptake (Akehurst, 1981), while plants generally absorbed less Cl during a moist growing season, probably because of the effect of rainfall on Cl distribution in the root zone (McCants & Woltz, 1967).

It has been suggested that Cl uptake could be genetically controlled (Lamprecht & Steenkamp, 1972; Lamprecht & Botha, 1975). As a result, recent work has concentrated on finding salt-tolerant cell cultures that might give rise to salt tolerant tobacco cultivars. Dix and Street (1975) reported finding such a cell line that would grow in the presence of NaCl containing media. Hasegawa et al. (1980) found that the adapted cell line grew almost as well as the unadapted line and Binzel et al. (1985) reported that adaptation led to reduced cell expansion and fresh weight gains but not dry weight gains. Though this work continues, it will probably have a much greater impact on salt tolerance research than on finding a specific new, salt-tolerant tobacco cultivar.

Salination of the root media has some other interesting effects on tobacco. Diffusion resistance was higher in the morning in the presence of NaCl (West & Black, 1978) and transpiration was reduced in plants whose roots had been exposed to a NaCl medium (Mizrahi et al., 1970). Waterlogging at night depressed leaf NaCl levels more than daytime waterlogging by inhibiting NaCl uptake (West and Black, 1978).

Chlorine does seem to increase leaf sugar concentration in some studies but not in others (Woltz et al., 1948). Chlorine can also affect nicotine by decreasing the amount of nitrogenous constituents (Fuqua et al., 1976). This is probably through the effect of Cl on NO_3 uptake discussed elsewhere in this chapter. Other reports have found no effect of Cl on nicotine (McCants & Woltz, 1967; Peele et al., 1960).

On the whole, the problem of Cl toxicity or salination of the soil through irrigation of tobacco appears to be a minor one for most tobacco-growing regions. The soils commonly associated with tobacco production are usually well drained and Cl would not be expected to concentrate over time. There are areas where tobacco is grown that might have problems. Paddy fields with their heavy soils or areas that receive little natural rainfall might be subject to salination, but these represent a small percentage of the total world crop of tobacco.

CONCLUSION

Irrigation of tobacco can be beneficial by increasing yields and leaf quality. However, maintaining the optimum available soil moisture for growth may not provide the best conditions for the development of desirable leaf quality. Tobacco leaf quality is a complex parameter and must be considered equally important with yield in any agronomic decision.

The need for rainfall or irrigation at each phase of tobacco production is generally agreed upon; however, the techniques for scheduling irrigation are still primitive compared to many crops. The restriction lies more with our lack of knowledge on the effects of each irrigation on final leaf quality than on final leaf yield. It appears that further work should pursue this interaction of irrigation, leaf chemistry, and quality.

From this review, it does not appear that Cl toxicity or salination of tobacco soils is a serious problem. The exception would be tobacco grown on heavy soils or areas with little rainfall.

More research is needed to define the minimum irrigation that will produce desired leaf quality and yield and will conserve water resources. Also, more work should be done on both the whole plant and subcellular level on the effects of water stress on tobacco. Only by understanding the biochemistry of quality and growth can we then define the level of irrigation needed for tobacco production.

REFERENCES

Akehurst, B.C. 1981. Tobacco. Longman, New York.

Atkinson, W.O., J.L. Ragland, J.L. Sims, and B.J. Bloomfield. 1969. Nitrogen composition of burley tobacco I. The influence of irrigation on the response of burley tobacco to nitrogen fertilization. Tob. Sci. 13:123–126.

Attoe, O.J. 1946. Leaf-burn of tobacco as influenced by content of potassium, nitrogen, and chlorine. J. Am. Soc. Agron. 38:186–196.

Begg, J.E., and N.C. Turner. 1970. Water potential gradients in field tobacco. Plant Physiol. 46:343–346.

Binzel, M.L., P.M. Hasegawa, A.K. Handa, and R.A. Bressan. 1985. Adaptation of tobacco cells to NaCl. Plant Physiol. 79:118–125.

Bliss, L.C., P.J. Kramer, and F.A. Wolf. 1957. Drought resistance in tobacco. Tob. Sci. 1:120–123.

Boggess, S.F., D. Aspinall, and L.G. Paleg. 1976. Stress metabolism. IX. The significance of end-product inhibition of proline biosynthesis and of compartmentation in relation to stress-induced proline accumulation. Aust. J. Plant Physiol. 3:513–525.

Boggess, S.F., and C.R. Stewart. 1980. The relationship between water stress induced proline accumulation and inhibition of protein synthesis in tobacco leaves. Plant Sci. Lett. 17:245–252.

Brown, G.W., C.G. McKee, and O.E. Street. 1970. Effects of irrigation, nitrogen fertilization, plant population and variety on the physico-chemical properties of Maryland tobacco. Proc. 5th Int. Tob. Sci. Cong., Hamburg, West Germany.

Broyer, T.C., A.B. Carlton, C.M. Johnson, and P.R. Stout. 1954. Chlorine—a micronutrient element for higher plants. Plant Physiol. 29:526–532.

Bunn, J.M., and W.E. Splinter. 1961. The effect of temperature, moisture and light on the germination probability of bright leaf tobacco seed. Tob. Sci. 5:63–66.

Campbell, R.B. 1973. Flue-cured tobacco yield and oxygen content of the soil in lysimeters flooded for various periods. Agron. J. 65:783–786.

Campbell, R.B., and G.T. Seaborn. 1972. Yield of flue-cured tobacco and levels of soil oxygen in lysimeters with different water table depths. Agron. J. 64:730–733.

Chaplin, J.F., A.H. Baumhover, C.E. Bortner, J.M. Carr, T.W. Graham, E.W. Hauser, H.E. Heggestad, J.E. McMurtrey, Jr., J.D. Miles, B.C. Nichols, W.B. Ogden, and H.A. Skoog. 1976. Tobacco production. USDA-ARS Agric. Info. Bull. 245.

Clark, F., and J.M. Myers. 1956. The effects of rates of irrigation, fertilizers and plant spacing on the yield and quality of flue-cured tobacco in Florida. Soil Crop Sci. Soc. Fla. Proc. 16:249–257.

Clough, B.F., and F.L. Milthorpe. 1975. Effects of water deficit on leaf development in tobacco. Aust. J. Plant Physiol. 2:291-300.

Crum, S.D. 1977. Effects of water stress on the interaction of carbohydrate and nitrogen metabolism in burley tobacco. M.S. thesis. North Carolina State Univ., Raleigh.

Darkis, F.R., L.F. Dixon, and P.M. Gross. 1935. Flue-cured tobacco. Factors determining type and seasonal differences. Ind. Eng. Chem. 27:1152-1157.

Dix, P.J., and H.E. Street. 1975. Sodium chloride-resistant cultured cell lines from *Nicotiana sylvestris* and *Capsicum annuum*. Plant Sci. Lett. 5:231-237.

El-Gheruri, S.A. 1980. Sources of plant variation in tobacco seedlings. M.S. thesis. North Carolina State Univ., Raleigh.

Elliot, J.M. 1967. Chemical composition of various parts of flue-cured tobacco plants. Lighter 37:15-18.

Farah, S.M. 1980. Effects of some cultural practices on yield and quality of flue-cured tobacco at Kenana Research Station. J. Agric. Sci. Camb. 95:423-429.

Farquhar, G.D., E.D. Schulze, and M. Kuppers. 1980. Responses to humidity by stomata of *Nicotiana glauca* L. and *Corylus avellana* L. are consistent with the optimization of carbon dioxide uptake with respect to water loss. Aust. J. Plant Physiol. 7:315-327.

Ferguson, K.H., S. Fukai, G.L. Wilson, and M.A. Toleman. 1985. Effects of post-transplant water deficits on leaf development and yield of winter planted tobacco in north Queensland. Aust. J. Agric. Res. 36:51-61.

Franklin, R.E., R.W. Teater, R.B. Curry, and G.O. Schwab. 1964. Nitrogen-irrigation interactions in burley tobacco production. Agron. J. 56:316-362.

Fuqua, B.D., J.E. Leggett, and J.L. Sims. 1974. Accumulation of nitrate and chloride by burley tobacco. Can. J. Plant Sci. 54:167-174.

Fuqua, B.D., J.L. Sims, J.E. Leggett, J.F. Benner, and W.O. Atkinson. 1976. Nitrate and chloride fertilization effects on yield and chemical composition of burley tobacco leaves and smoke. Can. J. Plant Sci. 56:893-899.

Garner, W.W. 1951. The production of tobacco. The Blakiston Co., New York.

Garner, W.W., H.A. Allard, and E.E. Clayton. 1936. Superior germ plasm in tobacco. p. 785-830. *In* USDA Yearbook of agriculture. U.S. Gov. Print. Office, Washington, DC.

Gatut-Suprijadji. 1984. Influence of the quality and the frequency of water sprinkling on the growth of tobacco seedlings. Menara Perkebunan 52:199-203.

Goodspeed, T.H. 1954. The genus *Nicotiana*. Chronica Botanica, Waltham, NA.

Gopalachari, N.C., M.S. Babu, and M.C.M. Reddy. 1984. Effect of drip irrigation on the yield and quality of FCV tobacco. Indian Tob. J. 15:7-12.

Grizzard, A.L., H.R. Davies, and L.R. Kangas. 1942. Time and rate of nutrient absorption by flue-cured tobacco. J. Am. Soc. Agron. 34:327-339.

Harris, D.G., and C.H.M. van Bavel. 1957. Growth, yield and water absorption of tobacco plants as affected by the composition of the root atmosphere. Agron. J. 49:11-14.

Hasegawa, P.M., R.A. Bressan, and A.K. Handa. 1980. Growth characteristics of NaCl-selected and nonselected cells of *Nicotiana tabacum* L. Plant Cell Physiol. 21:1347-1355.

Hawks, S.N., and W.K. Collins. 1983. Principles of flue-cured tobacco production. S.N. Hawks and W.K. Collins, Raleigh, NC.

Hirata, K., and M. Sasaki. 1971. Relationships between soil moisture and water absorption of roots, and leaf water status of tobacco plants. III. Effects of soil moisture changes in the growing stages upon the leaf water status. Utsunomiya Tab. Shikenjo Hokoku 10:27-41.

Hopkinson, J.M. 1968. Effects of early drought and transplanting on the subsequent development of the tobacco plant. Aust. J. Agric. Res. 19:47-57.

Hunt, P.G., R.B. Campbell, R.E. Sojka, and J.E. Parsons. 1981. Flooding-induced soil and plant ethylene accumulation and water status response of field-grown tobacco. Plant Soil 59:427-439.

Ismail, M.N., and R.C. Long. 1980. Growing flue-cured tobacco to prespecified leaf chemistries through cultural manipulations. Tob. Sci. 24:114-118.

Iwai, S., N. Kawashima, and S. Matsuyama. 1979. Effect of water stress on proline catabolism in tobacco leaves. Phytochemistry 18:1155-1157.

Jones, J.L., D.A. Komm, P.J. Semtner, A.J. Lambert, and B.B. Ross. 1984. 1985 Flue-cured tobacco production guide. Virginia Coop. Ext. Ser. Publ. 436-048.

Jones, J.N., G.N. Sparrow, and J.D. Miles. 1960. Principles of tobacco irrigation. ARS-USDA Agric. Info. Bull. 228. U.S. Gov. Print. Office, Washington, DC.

Kramer, P.J., and W.T. Jackson. 1954. Causes of injury to flooded tobacco plants. Plant Physiol. 29:241–245.

Kreith, F., A. Taori, and J.E. Anderson. 1975. Persistance of selected antitranspirants. Water Resour. Res. 11:281–286.

Kume, H. 1975. Influence of soil moisture on the water balances of tobacco plant. II. The influence of soil moisture level on the expansion and water status of tobacco leaves. Seibutsu Kankyo Chosetsu 13:23–27.

Kurian, T., E.R.R. Iyengar, M.R. Narayana, and D.S. Datar. 1966. Effects of sea-water dilutions and its amendments on tobacco. *In* H. Boyko (ed.) Salinity and aridity: New approaches to old problems. W. Junk, The Hague.

Lakshminarayana, R., G.J. Patel, and B.G. Jaisani. 1979. Note on seed germinability in mannitol solution as an index of drought resistance in tobacco. Indian J. Agric. Sci. 49:818–819.

Lamprecht, M.P., and A.H. Botha. 1975. Genetic basis of chloride concentration in flue-cured tobacco. Agroplantae 7:25–30.

Lamprecht, M.P., and C.J. Steenkamp. 1972. Differences in chloride content of flue-cured tobacco cultivars (*Nicotiana tabacum* L.). Agroplantae 4:69–72.

Ligon, J.T., and G.R. Benoit. 1966. Morphological effects of moisture stess on burley tobacco. Agron. J. 58:35–38.

Liu, C.H. 1977. Morphological effects of soil moisture tension on flue-cured tobacco. Taiwan Tob. Wine. Monop. Bur. Tob. Res. Inst. Bull. 7:31–40.

Liu, C.H. 1978. Influence of soil moisture and meteorological factors on the plant water relations and on the yield and quality of flue-cured tobacco. Taiwan Tob. Wine. Monop. Bur. Tob. Res. Inst. Res. Rep. 8:25–37.

Long, R.C., and W.G. Woltz. 1977. Environmental factors affecting the chemical composition of tobacco. Proc. Am. Chem. Soc. Symp. 173:116–163.

Matusiewcz, E., and Z. Madziar. 1975. Influence of soil moisture and pH on the morphological characteristics, transpiration and productivity of the tobacco plant. Coop. Cent. Sci. Res. Relat. Tob. Inf. Bull. (3–4) 6618.

McCants, C.B., and W.G. Woltz. 1967. Growth and mineral nutrition of tobacco. Adv. Agron. 19:211–265.

McKee, C.G., and O.S. Street. 1978. Irrigation studies with Maryland tobacco. Maryland Exp. Stn. Misc. Publ. 931.

McNee, P., L.A. Warrell, and E.W.B. van den Muyzenberg. 1978. Influence of water stress on yield and quality of flue-cured tobacco. Aust. J. Exp. Agric. Anim. Husb. 18:726–731.

Menser, H.A., and O.E. Street. 1962. Effects of air pollution, nitrogen levels, supplemental irrigation, and plant spacing on weather fleck and leaf losses of Maryland tobacco. Tob. Sci. 6:167–171.

Miles, J.D. 1957. The influence of irrigationon flue-cured tobacco in Georgia. Georgia Agric. Exp. Stn. Circ. 8.

Mizrahi, Y., A. Blumenfield, and A.E. Richmond. 1970. Abscisic acid and transpiration in leaves in relation to osmotic root stress. Plant Physiol. 46:169–171.

Mizrahi, Y., A. Blumenfield, and A.E. Richmond. 1972. The role of abscisic acid and salination in the adaptive response of plants to reduced root aeration. Plant Cell Physiol. 13:15–21.

Mizusaki, S., M. Noguchi, and E. Tamaki. 1964. Studies on nitrogen metabolism in tobacco plants. VI. Metabolism of glutamic acid, v-aminobutyric acid, and proline in tobacco leaves. Arch. Biochem. Biophys. 105:599–605.

Moseley,J.M. 1956. Irrigation and bright leaf quality. Irrig. Eng. Maint. (June) 1956:16–17, 48–49.

Moss, E.G. 1929. Nutritional problems of bright tobacco. J. Am. Soc. Agron. 21:137–141.

Mulchi, C.L. 1982. Chloride effects on agronomic, chemical and physical properties of Maryland tobacco. 1. Response to chloride applied to the soil. Tob. Sci. 26:113–116.

Mulchi, C.L. 1985. Environmental factors affecting the growth, chemistry and quality of tobacco. Recent Adv. Tob. Sci. 11:3–46.

Murty, K.S.N. 1964. Influence of irrigation on the chemical composition of flue-cured tobacco. Indian Tob. J. 14:32–36.

Myhre, D.L., O.J. Attoe, and W.B. Ogden. 1956. Chlorine and other constituents in relation to tobacco leaf-burn. Proc. Soil Sci. Soc. Am. 20:547–551.

Neas, I. 1955. Weather factors in relation to tobacco production. Rhod. Tob. 10:10–11.

Orphanos, P.I., and C. Metochis. 1985. Yield and quality of tobacco grown with supplementary irrigation. Agron. J. 77:689–695.

Papenfus, H.D. 1970. The effects of climate and cultural practices on the growth characteristics of flue-cured tobacco. p. 105–116. In Proc. 5th Int. Tob. Sci. Cong., Hamburg, West Germany.

Papenfus, H.D., and D.A. Baxter. 1969. Irrigation of flue-cured tobacco. Tob. Forum Rhod. 33:5–7.

Papenfus, H.D., and F.M. Quin. 1984. Tobacco. In P.R. Goldsworthy and N.M. Fisher (ed.) The physiology of tropical food crops. John Wiley and Sons, New York.

Parks, W.L., B.C. Nichols, R.L. Davis, E.J. Chapman, and J.H. Felts. 1963. Response of burley tobacco to irrigation and nitrogen. Tennessee Agric. Exp. Stn. Bull. 368.

Parks, W.L., and L.M. Safley. 1965. The effect of irrigation and nitrogen upon the yield and quality of dark tobacco. Tennessee Agric. Exp. Stn. Bull. 394.

Peele, T.C., H.J. Webb, and J.F. Bullock. 1960. Chemical composition of irrigation waters in the South Carolina coastal plain and effects of chlorides in irrigation water on the quality of flue-cured tobacco. Agron. J. 52:464–467.

Petrie, A.H.K., and J.I. Arthur. 1943. Physiological ontogeny of the tobacco plant. The effects of varying water supply on the drifts in dry weight and leaf area on various components of the leaves. Aust. J. Exp. Med. Sci. 21:191–200.

Purseglove, J.W. 1968. p. 540–555. Tropical crops. Dicotyledons 2. John Wiley and Sons, New York.

Ravindranath, E., and T.S. Subrahmanyam. 1978. Signaling method of irrigation. Tob. Res. 4:79–80.

Rhoads, F.M. 1968. Subirrigation and fertilization of cigar-wrapper tobacco beds. Tob. Sci. 12:229–231.

Rhoads, F.M. 1971. Effects of pipe depth and fertilizer rates on cigar wrapper tobacco seedling production in a subirrigated plant bed. Proc. Soil Crop Sci. Soc. Fla. 31:155–156.

Rhoads, F.M. 1972. Effect of chloride level in irrigation water on nutrient content, burning quality, yield, leaf quality and tensile strength of cigar wrapper tobacco. Tob. Sci. 16:89–91.

Sabourin, L. 1967. Tobacco quality and combustibility as affected by its chlorine content. 1. The industrial defects of high chlorine tobacco. Coop. Ctr. Sci. Res. Relat. Tob. Info. Bull. 2:14–16.

Sano, M., and N. Kawashima. 1982. Water stress induced proline accumulation at different stalk positions and growth stages of detached tobacco leaves. Agric. Biol. Chem. 46:647–653.

Sasaki, M., and K. Hirata. 1971. Relationships between soil moisture and water absorption of roots, and leaf water status of tobacco plants. II. Effects of soil moisture changes in different layers upon the water absorption of roots and characteristics of above ground parts of tobacco. Utsunomiya Tab. Shikenjo Hokoku 10:15–25.

Sficas, A.G. 1970. Irrigation experiments on Oriental tobacco p. 175 In Proc. 5th Int. Sci. Cong., Hamburg, West Germany.

Sficas, A.G., G.L. Jones, and H. Seltmann. 1961. Factors affecting drought resistance in tobacco. Tob. Sci. 5:39–43.

Shoda, M., and H. Kume. 1978. Influence of water stress control with potassium sulfate and polyethylene glycols on growth and leaf tissue structure of tobacco plants. Utsunomiya Tab. Shikenjo Hokoku 16:87–97.

Skogley, E.O., and C.B. McCants. 1963. Ammonium and chloride influences on growth characteristics of flue-cured tobacco. Proc. Soil Sci. Soc. Am. 27:391–394.

Snyman, H.G., J.G. Coetzee, F.J. Shaw, and H.J. Boshoff. 1976. Influence of fertilizer applied chlorine, chlorine content of irrigation water and leaching of chlorine on the uptake of chlorine by tobacco. Agrochemophysica 8:22–30.

Street, O.E. 1978. History of Maryland tobacco research. Maryland Agric. Exp. Stn. Misc. Publ. 934.

Tanabe, Y., M. Sano, and N. Kawashima. 1982. Changes in free amino acids in white and green tissues of variegated tobacco leaves during water stress. Plant Cell Physiol. 23:1229–1235.

Tanimoto, M. 1977. Studies of the causes of burning up of tobacco leaves in paddy field. V. Effects of soil moisture on growth and senescence of tobacco leaves. Okayama Jpn. Tob. Exp. Stn. Bull. 38:75–82.

Terry, D.L., and C.B. McCants. 1970. Quantitative prediction of leaching in field soils. Soil Sci. Soc. Am. Proc. 34:271–276.

Tso, T.C. 1972. Physiology and biochemistry of tobacco plants. Dowden, Hutchinson and Ross, Stroudsburg, PA.

Turner, N.C. 1974. Stomatal behavior and water status of maize, sorghum, and tobacco under field conditions II. At low soil water potential. Plant Physiol. 53:360–365.

Turner, N.C., and L.D. Incoll. 1971. The vertical distribution of photosynthesis in crops of tobacco and sorghum. J. Appl. Ecol. 8:581–591.

van Bavel, C.H.M. 1953. Chemical composition of tobacco leaves as affected by soil moisture conditions. Agron. J. 45:611–614.

Walker, E.K., and L.S. Vickery. 1959. Some effects of sprinkler irrigation on flue-cured tobacco. Can. J. Plant Sci. 39:164–174.

Wedin, W.F., and B.E. Struckmeyer. 1958. Effects of chloride and sulfate ions on the growth, leaf burn, composition and anatomical structure of tobacco (*Nicotiana tabacum* L.) Plant Physiol. 33:133–139.

Welch, L.F. 1964. Variable rainfall, temperature and yearly tobacco yields. Tob. Sci. 8:17–20.

West, D.W., and J.D.F. Black. 1978. Irrigation timing—its influence on the effects of salinity and waterlogging stresses in tobacco plants. Soil Sci. 125:367–376.

Weybrew, J.A., R.C. Long, C.A. Dunn, and W.G. Woltz. 1974. The biochemical regulation of ripening tobacco leaves. p. 843–847. *In* R.L. Bieleski et al. (ed.) Mechanisms of regulation of plant growth. Bull. 12. Royal Soc. of New Zealand, Wellington.

Weybrew, J.A., W.A. Wan Ismail, and R.C. Long. 1983. The cultural management of flue-cured tobacco quality. Tob. Sci. 27:56–61.

Weybrew, J.A., and W.G. Woltz. 1974. Influence of management and weather.Recent Adv. Tob. Sci. 1:39–49.

Willey, C.R. 1970. Effects of short periods of anaerobic and near-anaerobic conditions on water uptake by tobacco roots. Agron. J. 62:224–229.

Williamson, R.E., and W.E. Splinter. 1968. Effect of gaseous composition of root environment upon root development and growth of *Nicotiana tabacum* L. Agron. J. 60:365–368.

Woltz, W.G., W.A. Reid, and W.E. Colwell. 1948. Sugar and nicotine in cured bright tobacco as related to mineral element composition. Proc. Soil Sci. Soc. Am. 13:385–387.

28 Sugarcane

C. A. JONES

Texas Agricultural Experiment Station,
Formerly USDA-ARS
Temple, Texas

L. T. SANTO

Hawaiian Sugar Planter's Association
Aiea, Hawaii

G. KINGSTON

Bureau of Sugar Experiment Stations
Bundaberg, Queensland, Australia

G. J. GASCHO

University of Georgia
Tifton, Georgia

Sugarcane (*Saccharum* spp. hybrid) is the world's most important sugar crop. It is an erect, robust, tillering, perennial member of the family Poaceae, subfamily Panicoideae, tribe Andropogoneae. It is grown primarily for sugar (sucrose), but molasses, ethyl alcohol, and fiber (bagasse) are important by-products. In 1984 to 1985, world raw sugar production was about 99×10^6 Mg, of which about 63×10^6 Mg were from sugarcane and about 36×10^6 Mg were from sugarbeet (*Beta vulgaris* L.). Average fresh cane and raw sugar yields were 58 and 5.1 Mg ha^{-1}, respectively (USDA, 1985).

Sugarcane is propagated vegetatively by planting stem cuttings (setts) from which axillary buds grow to produce erect primary stalks (main stem). Secondary and tertiary stalks (tillers) are produced at the base of the primary stalk. Sugarcane leaf laminae are 700 to 1200 mm long and up to 100 mm wide. Internodes are up to 250 mm long and 20 to 60 mm diam. About 10 fully expanded leaves are usually present on a stalk, and both lamina and sheath are shed when they senesce. The inflorescence, a panicle (also known as the arrow or tassel), is produced under certain environmental conditions. Genotype, photoperiod, temperature, nutrition, and water stress all affect panicle initiation and growth. Since the stalk ceases to grow and eventually deteriorates after flowering, genotypes that flower readily under field conditions are avoided (Clements, 1980).

When weather conditions permit, sugarcane fields are usually burned to eliminate dead leaves and tillers (trash) and to facilitate harvest. In dry areas of South Africa and in some parts of Australia, cane fields may not be burned, and the heavy mat of trash is left to reduce soil evaporation and to control weeds.

At harvest, the dry weight of a sugarcane tiller usually consists of about 50 to 60% millable cane, 30 to 40% tops (leaves and immature stem), and 10% roots and stubble. Crops grown for more than 1 yr usually have higher percentages of millable cane. Of the millable cane's fresh weight, about 70% is water. Of its dry weight, about 50% is sucrose. Efficient factories can recover about 85% of the sucrose in the cane. Depending on the sucrose content of the cane and its recovery, 8 to 15% of the millable cane harvested is recovered as raw (unrefined) sucrose (Clements, 1980; Jones, 1985).

Commercial sugarcane cultivars are complex hybrids of *S. robustum; S. officinarum,* the "noble" canes; and *S. spontaneum,* a freely tillering wild species used as a source of vigor and disease resistance. The "noble" canes may have been selected from *S. robustum* by Stone Age cultures in New Guinea. They were spread throughout the Pacific and Southeast Asia prior to the arrival of European humans. Natural hybrids of *S. officinarum* and *S. spontaneum* may have resulted in *S. sinense,* the vigorous "thin" canes of northeastern India and southern China. Cane was taken by the Spanish and Portuguese to the New World, to form the basis of sugarcane culture in the 16th century. In the late 18th century, more desirable cultivars of *S. officinarum* were introduced. Modern sugarcane breeding began in Guyana, India, Java, and Barbados at the end of the 19th century due to the discovery that true seeds were viable and the need to obtain increased disease resistance. Initially, only *S. officinarum* was used in breeding programs, but soon interspecific hybrids between *S. officinarum, S. sinense,* and *S. spontaneum* were providing hardier, more disease-resistant cultivars (Jones, 1985).

Sugarcane is adapted to a range of tropical and subtropical climates. It is grown from 37 °N lat in southern Spain to 21 °S lat in the Republic of South Africa. It cannot tolerate freezing temperatures, and growth essentially ceases at mean minimum temperatures below about 12 °C (Ryker & Edgerton, 1931). Maximum photosynthetic rates occur at air temperatures of about 34 °C (Alexander, 1973), and intact plants can survive temperatures in excess of 52 °C (Irvine, 1983). The ideal climate for a 1-yr crop would include at least 16 to 20 wk with mean daytime temperatures of 30 to 35 °C to stimulate growth and 6–8 wk of cooler temperatures prior to harvest to enhance sucrose accumulation (Gascho & Shih, 1982).

Sugarcane is successfully grown under a wide range of temperature, solar radiation, rainfall, and soil conditions. If soil, water, and plant nutrition are adequate, temperature and/or solar radiation can be used to predict cane growth rates (Allen et al., 1978; Das, 1933a, b; Halais, 1935). When water is limiting, rainfall and/or irrigation may be correlated with yields (Jones, 1980; Early, 1974; Thompson, 1976). Sugarcane can be grown on soils as different as Histosols, Andosols, Oxisols, and Vertisols. It will also tolerate soil pH ranging from about 4 to 9, though nutritional problems may occur at the extremes. Though some cultivars tolerate moderate salinity and seasonal flooding, good drainage and salinity management are required for high yields.

Most of the world's sugarcane is harvested once a year, and several ratoon (stubble regrowth) crops usually follow the plant crop. One-year sugarcane is bred and managed to avoid lodging. However, stems of 2-yr crops can reach a length of 10 m or more by growing upward, lodging, then turning upward again. At harvest the 2-yr crop consists of a mat of tangled stems which have lodged several times (Clements, 1980).

I. CROP GROWTH AND DEVELOPMENT

A. Growth Stages

Sugarcane is a perennial crop in which flowering is undesirable, and it often does not occur in commercial fields. Therefore, the growth stages of a commercial sugarcane crop are unrelated to flowering. However, more-or-less distinct periods of vegetative growth can be described. Gascho and Shih (1982) divided sugarcane vegetative growth into four stages: germination and emergence, tillering and canopy establishment, grand growth, and ripening. Duration of the stages is highly dependent on climate. Low temperatures or drought stress can delay germination, emergence, tillering, and canopy development. In subtropical climates, the first two stages may last up to 20 wk. Warm temperatures and plentiful water and nutrients can promote vegetative growth and delay maturation. In addition, where cane is grown for 2 yr the grand growth period can be extended by providing adequate water and nutrients until the crop is 80 to 88 wk old.

B. Root Growth

The early growth of sugarcane roots has been described by Clements (1980), van Dillewijn (1952), and Glover (1967). Under favorable environmental conditions, the axillary buds on setts become active within 3 d of planting, and sett roots begin to grow from the root band at the base of the internode. Glover (1967) reported that sett roots grow at a maximum rate of 24 mm d^{-1} and stop elongating when they are 150 to 250 mm long. They turn dark, decompose rapidly, and disappear within 8 wk after planting. Shoot roots begin to grow from the short basal internodes of the shoot at about the time it emerges from the soil. The first shoot roots are much thicker than the sett roots, their rate of growth is more rapid, they produce few branches, and they penetrate the soil at a steep angle.

Shoot roots produced later are finer and branch more freely than earlier shoot roots. The maximum growth rate of shoot roots is 75 mm d^{-1} for periods of 1 to 2 d or 40 mm d^{-1} when their growth is averaged over 1 wk (Glover, 1967). Wood and Wood (1967) used radioactive P uptake from different depths of a deep sandy soil and concluded that the rooting front reached 0.9 m in 112 d, 1.5 m in 161 d, and 2.1 m in 189 d.

The distribution of roots in the soil is strongly dependent on soil characteristics, cultivars, and soil water content. For example, Paz-Vergara et al. (1980) reported that for 11 furrow-irrigated fields in Peru, the percentage of roots in the 0.30 m horizon is 48 to 68%; from 0.3 to 0.6 m, 16 to 36%; 0.6 to 0.9 m, 3 to 12%; 0.9 o 1.20 m, 4 to 7%; 1.2 to 1.5 m, 1 to 7%; and 1.5 to 1.8 m, 0 to 4%.

Short irrigation intervals which prevent surface soil drying encourage a higher percentage of roots to develop near the soil surface (Baran et al., 1974; Kingston, 1977). However, poor soil aeration restricts root growth. For example, Gosnell (1971) reported that sugarcane roots stop growth 50 to 100 mm above the water table. Though shallow water tables prevent deep rooting, the total weight of roots above the water table may hardly be affected (Juang & Uehara, 1971).

Root systems growing in deep sands tend to be finer, more highly branched, and deeper than those growing in heavy clay soils (Glover, 1968; Lee, 1926c; Thompson, 1976). For example, Glover (1968) found an extensive, fine, well-branched root system to extend more than 140 cm in a sand. In a disturbed clay soil, the thick primary root system was well developed, but secondary branches were poorly developed. In an undistrubed clay soil, even the primary roots were poorly developed below the plow layer.

Genotypic variation in sugarcane root systems is well documented (Dastane, 1957; Evans, 1935, 1936; Lee, 1926a, b; Raheja, 1959). Some cultivars produce roots with a higher degree of branching than other cultivars (Stevenson & McIntosh, 1935), and those producing many tillers normally produce many roots because each new tiller is a source of shoot roots (Stevenson & McIntosh, 1935). Root gravitropism also varies among cultivars, and cultivars with weakly gravitropic (more horizontal) root orientation are more resistant to lodging than those with strongly gravitropic root systems (Mukerji & Alan, 1959; Stevenson & McIntosh, 1935).

Environmental conditions can affect the expression of genotypic differences in root growth. For example, Rostron (1974) reported that the root distributions of two cultivars were similar under good conditions but differed under dry conditions.

C. Shoot Growth

Sugarcane has the capacity to tiller rapidly. Studies in Hawaii (Nickell, 1967), South Africa (Gosnell, 1968), and Australia (Bull & Glaziou, 1975) indicate that stem numbers increase exponentially with time until a maximum of 20 to 30 stalks m^{-2} is reached at 16 to 24 wk. Leaf area index (LAI) increases in a similar exponential manner. When LAI approaches 2.0 to 3.0 at 16 to 24 wk, many younger tillers begin to die, possibly as a result of shading by older tillers, and tiller number normally stabilizes at 10 to 20 stems m^{-2}. A second, smaller flush of new tillers (suckers) may occur during a period of favorable climatic conditions in the second year of a 2-yr crop (Nickell, 1968), but many of these new tillers also die.

Several studies suggest that the maximum LAI for sugarcane is 7 to 8 m^2 leaf m^{-2} (Bull & Tovey, 1974; Irvine & Benda, 1980; Irvine, 1983). This value is only attained in vigorously growing canopies of erect cane. The LAI typically declines as the crop approaches harvest maturity (Bull & Tovey, 1974; Glover, 1972; Gosnell, 1968), especially when N and water stresses or chemical ripeners are used to slow expansion growth and increase sugar storage in the stem.

For crops of 48 wk or shorter duration, leaf area can often be predicted from stem length (Shih & Gascho, 1980b; Irvine, 1983). However, population and row spacing strongly affect tiller numbers and LAI (Gascho & Shih, 1981). For example, full ground cover can be attained by 12 wk at 0.45-m row spacing, but it is delayed until 16 to 20 wk by 1.4- to 1.5-m row spacing (Bull & Glaziou, 1975; Gosnell, 1968). In Louisiana, row spacings of 0.60, 0.91, and 1.81 m produced maximum LAI values of 5.7, 4.8, and 2.8, respectively (Irvine & Benda, 1980).

In much of the world, sugarcane row spacing is about 1.5 m to facilitate mechanization. However, recent work on row spacing in Louisiana (Irving & Benda, 1980) and Florida (Gascho & Shih, 1981; Shih & Gascho, 1980b) suggests narrow row spacing could result in more rapid canopy development, increased light interception, and higher yields in areas with short growing seasons. However, use of narrow row spacing might require modification of harvest mechanization.

D. Dry Matter Accumulation

Maximum dry matter accumulation occurs only with near-optimum temperatues, high solar radiation, complete ground cover, and minimal nutrient and water stresses. Under these conditions, short-term (4–8 wk) production of aboveground biomass can reach 40 to 44 g m^{-2} per d (Irvine, 1983; Shih and Gascho, 1981; Thompson, 1978). Incomplete interception of light during canopy development, low solar radiation, suboptimal temperatures, reduced growth during ripening, and a variety of soil and biotic stresses typically reduce full-season growth rates of commercial fields to 6 to 25 g m^{-2} per d with experimental maxima of 20 to 32 g m^{-2} per d (Irvine, 1983).

E. Ripening

Ripening of sugarcane refers to the gradual increase in percent sucrose on a dry weight basis as harvest approaches. Numerous studies have shown that cool temperatures, high solar radiation, moderate N and/or drought stress, and use of chemical ripeners can stimulate ripening (Clowes & Inman-Bamber, 1980; Lonsdale & Gosnell, 1974; Mason, 1976; Rostron, 1977; Selleck et al., 1974). Therefore, more sucrose is available for translocation to storage tissues in the stem, and the total amount of sucrose in the crop increases. In addition, drought stress and N deficiency reduce the water content of the crop.

Table 28-1. Nutrients present in stalk and leaf of sugarcane in Queensland (Bureau of Sugar Experiment Stations).

Cane yield	Nutrient content									
	N	P	K	Ca	Mg	S	Fe	Mn	Cn	Zn
Mg ha^{-1}					kg ha^{-1}					
89	126	21	203	37	33	29	7	4	0.1	0.4

II. CULTURAL PRACTICES

Cultural practices associated with sugarcane production vary widely as a result of economic, social, climatic, and soil conditions. In areas with high labor costs, essentially all cultural practices are mechanized. In other areas, a great deal of hand labor continues to be used.

Land preparation varies with soil type. Deep plowing, deep ripping, or subsoiling is often used to disrupt compacted layers. Soil may be formed into beds approximately 1.5 m apart to facilitate furrow irrigation, improve surface drainage, or both. However, such beds increase land preparation costs, and, where possible, flat culture is used.

Setts with two or more buds are cut by hand or mechanically, often from special areas maintained to minimize disease infestation. Setts may receive hot water and/or fungicide treatments to further reduce disease problems. They are planted in furrows and are covered with soil. Fertilizer may be applied broadcast, in the furrow, or through the irrigation system. In areas where poor drainage is common, the setts are planted on top of beds. Where furrow irrigation is used, they are often planted in the furrow between beds. Flat culture is practiced whenever possible to reduce land preparation. Ratoon tillage operations are designed to remove postharvest compaction between rows and to control weeds.

Fertilizer requirements of sugarcane are high (Table 28-1), and sugarcane soils vary in natural fertility; therefore, soil testing and/or plant analysis have long been used. Fertilizer products and methods of application vary and are usually determined by economic factors. However, mechanization and fertilizer application via furrow or drip irrigation systems are common. For drip-irrigated sugarcane, frequent (monthly) applications of small amounts of N and K fertilizers in irrigation water are common.

Mechanical and chemical (pre-emergence and postemergence) weed control and both chemical and biological insect control are used. Diseases are controlled through use of resistant germplasm, treatment of stem cuttings with hot water or fungicides, and strict phytosanitary measures.

Cane is usually burned before harvest to reduce the amount of trash and green leaves hauled to the mill. Harvesting may be labor intensive or highly mechanized. Good harvesting systems minimize cane breakage, transport little trash and other extraneous material to the mill, and minimize the time between harvest and cane processing. For many years, manual harvest of unburned cane has been practiced in South Africa. Cane tops and trash are left on the soil surface to conserve moisture and control erosion.

Mechanized harvest of unburned cane is being evaluated in Australia, where it is an integral part of reduced tillage systems designed to reduce costs and conserve soil moisture.

III. EVAPOTRANSPIRATION

A. Potential

The sugar industries in Hawaii, Florida, Australia, South Africa, and Taiwan have historically relied on pan evaporation as an indicator of potential evapotranspiration. However, a few studies (Ekern, 1971; Shih et al., 1977; Thompson & Boyce, 1971) have related evapotranspiration of well-watered sugarcane to theoretical/empirical mathematical estimates of potential evapotranspiration. For example, in South Africa Thompson and Boyce (1971) related potential evapotranspiration measured with weighing lysimeters (E_L, mm d^{-1}) to that estimated according to Penman (1948) (E_P), Penman and Schofield (1951) (E_{PS}), and Monteith (1965) (E_M). In addition, U.S. Weather Bureau Class A pan evaporation (E_{PAN}) was measured. The relationships were developed with 208 daily measurements of E_L during the grand growth period (Thompson & Boyce, 1967).

$$E_L = 0.20 + 1.13\,E_P \qquad\qquad r = 0.93$$

$$E_L = 1.44 + 0.69\,E_{PS} \qquad\qquad r = 0.83$$

$$E_L = 0.33 + 0.90\,E_M \qquad\qquad r = 0.91$$

$$E_L = 0.29 + 0.83\,E_{PAN} \qquad\qquad r = 0.92$$

Thompson and Boyce (1971) concluded that pan evaporation, with its simple instrumentation and calculation requirements, is adequate to estimate potential evapotranspiration of sugarcane for irrigation control.

Several other studies using neutron probes or lysimeters suggest that potential evapotranspiration from a well-developed sugarcane canopy during the grand growth stage is approximately equal to evaporation from a standard U.S. Weather Bureau Class A pan (Ekern, 1971; Fogliata, 1974; Hardy, 1966; Thompson, 1965; Thompson et al., 1963). When calculated on a monthly basis, the ratio of evapotranspiration to pan evaporation usually varies from 0.8 to 1.2, even though ratios as low as 0.63 and as high as 1.59 have been reported (Kingston & Ham, 1975).

Pan size, placement, and maintenance can have dramatic effects on pan evaporation (Ekern, 1971; Shih & Rahi, 1983). For example, Ekern (1971) reported that evaporation from a standard Class A pan was about equal to sugarcane evapotranspiration measured with four weighing lysimeters. However, a pan elevated 1.5 m over sod had 20% more evaporation than the crop, and a pan progressively elevated to match the height of the adjacent cane canopy had 25% more evaporation than the crop. Kingston and

Ham (1975) concluded that there was no consistent advantage in favor of either standard or elevated pans. Either could be used providing the relevant pan calibration factor was applied.

Low ratios of potential evapotranspiration to pan evaporation are often, though not always, found as the crop nears maturity in winter months (Kingston & Ham, 1975; Moberly, 1974; Thompson, 1976) and in ratoon crops (Hardy, 1966; Moberly, 1974; Shih & Gascho, 1980a; Thompson, 1976), presumably due to greater stomatal resistance of slowly growing crops (Thompson, 1986). Lodging can reduce evapotranspiration up to 30% until a uniform canopy is reestablished (Ekern, 1971).

B. Canopy-limited

Transpiration is strongly affected by the amount of solar radiation intercepted by the crop canopy. When the soil surface is dry, actual evapotranspiration is limited by canopy cover. For example, Chang et al. (1965) and Ekern (1971) reported that actual evapotranspiration of adequately watered sugarcane increases from 0.3 to 0.6 of pan evaporation during the first month after planting to 0.8 to 1.0 of pan evaporation at 16 to 20 wk. Kingston (1973) recommends using pan factors of 0.4, 0.6, 0.8, 0.9, and 1.0 for ground cover fractions of 0 to 0.25, 0.25 to 0.50, 0.50 to 0.75, 0.75 to 1.0, and 1.0, respectively.

Doorenbos and Pruitt (1977) provided coefficients to estimate the effects of sugarcane canopy development, ripening, relative humidity, and wind speed on the ratio of potential crop transpiration to potential evapotranspiration.

C. Water-limited

After irrigation there is normally a short period of soil water redistribution when the effects of both gravity and crop water uptake are significant. Within a few days drainage slows and transpiration becomes the dominant process affecting soil water content. If the crop canopy is complete, evapotranspiration proceeds at the potential rate until 60 to 70% of the total plant-extractable water is removed from the soil profile (Koehler et al., 1982; Moberly, 1974). Thereafter, the ratio of evapotranspiration to potential evapotranspiration declines until evapotranspiration ceases in the completely senescent crop.

Soil type strongly affects the amount and distribution of plant-extractable water in the soil profile. For example, Gosnell and Thompson (1965) found that sugarcane extracted water to at least 2.2 m in one of three contrasting soils. However, more water was extracted deep in a sandy soil profile than deep in a sandy loam or a shallow clay loam overlying decomposing shale (Hill, 1966; Thompson et al., 1967). Several other factors, including subsoil Al toxicity, presence of layers with high soil strength, or fragmental or cemented horizons can limit root growth in the subsoil, thereby reducing water available for transpiration, and increasing the susceptibility of the crop to drought stress.

IV. EFFECTS OF WATER STRESS

A. Seedling and Tiller Growth

Sugarcane is most susceptible to drought stress during the first 12 to 16 wk after planting when severe stress can reduce stands and make replanting necessary. During tillering and canopy development young tillers are more susceptible to drought than the main stem or older tillers. Leaves on young tillers begin to roll earlier than those on the primary stem, probably due to the less-developed root systems of the tillers (Clements, 1980).

B. Leaf and Stem Growth

Leaf elongation is sensitive to drought stress, and a reduction in the rate of elongation of the most recently emerged leaf is often used as an indicator of incipient stress. More severe drought stress causes leaves to roll during the day. The rate of lower leaf senescence increases (Inman-Bamber & de Jager, 1986a, b), and severely drought-stressed stems commonly have only four to six green leaves, about half the normal number.

After a tiller begins to produce elongated internodes, the rate of stem elongation and final internode lengths are convenient means of assessing the effects of drought stress. Internodes that elongate during stress are permanently shortened relative to those of well-watered plants, and they serve as a record of the timing, length, and severity of drought stress. Kingston (1977) reported that stem growth began to decrease when the crop had dried the profile to 27 mm below "field capacity." This is consistent with results of Koehler et al. (1982), who showed that for drip-irrigated sugarcane, cane growth rates began to decrease after about 25 mm soil water depletion. Elongation ceased after about 160 mm had been extracted. In contrast, Robinson (1963) found that for furrow-irrigated sugarcane, stem elongation continued at the maximum rate until 65 mm water had been extracted, but it ceased after extraction of only 91 mm. Hill (1966) reported that in a deep sand, stem elongation began to decrease after about 75 mm extraction and ceased after about 96 mm extraction. However, on a clay loam, elongation began to decrease after only 25 to 30 mm extraction and ceased after about 81 mm extraction. Thus, depending on soil characteristics, evaporative demand, and irrigation methods, sugarcane stem elongation begins to decrease after the crop has extracted 25 to 75 mm soil water since the last irrigation. Stem elongation does not cease until leaf senescence is well advanced.

C. Flowering

Flowering reduces cane and sugar production because reproductive stems cease growth, lose sugar content, and eventually die. For most commercial cultivars, panicle initiation occurs within a narrow range of daylength (12 h 30 min–12 h 35 min) (Brett & Harding, 1974; Brett et al., 1975), and it is enhanced by adequate soil water and high temperatures. However, drought stress can inhibit initiation under otherwise inductive environmental conditions. Therefore, prior to periods conducive to floral initiation, controlled drought stress can be used to inhibit initiation until noninductive conditions return (Clements, 1980). Application of ethephon at initiation has dramatically reduced flowering in Sudan (Hardy et al., 1986).

D. Adjustment to Stress

Sugarcane has long been known to adjust to prolonged or repeated drought stress. Ashton (1956) studied the effects of repeated cycles of water stress on photosynthesis of sugarcane grown in containers. He observed that net photosynthesis is reduced less with each successive cycle of stress and that recovery is more rapid in plants that have undergone previous stress. Similar results were recently reported by Inman-Bamber and de Jager (1986b).

Many studies suggest that plants adjust to drought stress by increasing the solute concentration of their cell sap, thereby reducing its osmotic potential. In field-grown sugarcane, reducing sugars accounted for much of the increase in solute concentration (Koehler et al., 1982), while in container-grown plants, sucrose, reducing sugars, and soluble amino acids all increased (P.H. Moore, 1980, personal communication). Koehler et al. (1982) reported a total osmotic adjustment of 0.7 MPa for drought-stressed field-grown sugarcane plants.

Drought stress has long been known to increase nonstructural carbohydrates in sugarcane leaves and stems (Clements & Kubota, 1943). For example, early-morning total sugar concentration of young leaf sheaths falls as low as 5% in rapidly growing, well-watered plants; however, it frequently increases to more than 10% during drought stress. Hartt (1967) concluded that drought stress reduces translocation of photosynthates, but it also causes sucrose content of the stem to increase. This suggests that, as in other C_4 grasses, drought stress increases storage of nonstructural carbohydrates in storage tissues by reducing photosynthate demand more than supply (Jones, 1985).

Drought stress affects several aspects of sugarcane N metabolism. Nitrogen uptake may be reduced both by low NO_3 availability in dry soils and by a decrease in plant demand due to slower growth. A decrease in NO_3 reductase activity is one of the earliest biochemical effects. It is followed by an increase in leaf protein hydrolysis and free amino acid concentrations (Koehler et al., 1982; P.H. Moore & A. Maretzki, 1980, personal communication). Free proline begins to accumulate at approximately the same threshold of leaf water potential as stomatal closure and initial wilting (P.H. Moore & A. Maretzki, 1980, personal communication), and it may increase 16 to 100 fold (Koehler et al., 1982; Rao & Asokan, 1978).

There is a high correlation between leaf sheath water content and leaf N and K concentrations (Clements, 1980; Samuels, 1971; Samuels et al., 1953). All decrease during drought stress, and plant analyses to detect N and K deficiencies are routinely adjusted to take tissue water content into account (Clements, 1980).

V. IRRIGATION MANAGEMENT

Irrigation by flooding, furrow, sprinkler, drip (trickle), and subirrigation by water table adjustment can be used to irrigate sugarcane. Several reviews of sugarcane irrigation practices are available (Finkel, 1983; Gibson, 1974; Gosnell & Pearse, 1971; Humbert, 1968; Leverington & Ridge, 1975; Reynolds & Gibson, 1968; Thompson, 1977).

Thompson and Harding (1986) developed a simulation model to evaluate proposed sugarcane irrigation schemes. Input variables include total available soil water, irrigation efficiency, water application rate, daily rainfall, average monthly pan evaporation, a proposed sequence of planting and harvesting, and monthly estimates of canopy cover. The model simulates the soil water balance and estimates rainfall efficiency, number of days spent irrigating, and the degree of drought stress suffered by the crop.

A. Flood Irrigation

Basin flooding consists of irrigating relatively small plots surrounded by levees. It has been used on impermeable soils in southern Africa and for intercropped rice and sugarcane in Taiwan. In the latter, the cane is often planted on ridges raised 50 to 100 mm above the water level. The cane survives with reduced growth and tillering until the paddies are drained for rice harvesting, after which the sugarcane is irrigated normally (Gosnell & Pearse, 1971).

Humbert (1968) reported that in Los Mochis, Mexico, sugarcane is irrigated in strips 250 m wide and 1000 m long surrounded by levees. The strips are level in the short dimension and have a 2 to 4% grade in the long dimension. Water is released at the upper end of the strip and advances downslope as a sheet. Gosnell and Pearse (1971) described a similar system near Beira, Mozambique, on an alluvial flood plain with low infiltration rate and a slope of 0.1 to 0.4%. The land is divided into bays 50 m wide and 500 to 1000 m long with the slope down the long dimension. Main drains are cut to 1 m depth between the bays and secondary drains cross the bays at 100-m intervals. Cane is grown on ridges 300 to 400 m high to provide internal drainage. When irrigation begins, the drains are closed and water is allowed to spread across and down the bays at a high rate. Within 1 h the water has reached the lower end of the bay. It is allowed to infiltrate for 24 h, after which the drains are opened to allow excess water to leave the field.

B. Furrow Irrigation

Furrow irrigation has been and continues to be used throughout the world for sugarcane irrigation. High labor costs in Hawaii led to a wide variety of furrow irrigation systems to increase the area managed by each irrigator, to improve water distribution over the field, and to recycle irrigation runoff. Furrow systems were developed to irrigate land with up to 45% slopes.

The herringbone system was named for its resemblance to a fish skeleton with a flume as the backbone and furrows spreading at an angle from it. For a typical herringbone layout, a large concrete flume receives water from a supply ditch. The flume has adjustable gates to control water flow into furrows. A typical 70-m furrow has a variable grade, 2% for the first 10 m to reduce overirrigation at the flume, 1.5% for 45 m, and 0% for the last 15 m. The advantages of the herringbone system are a minimum land area in ditches, the ability to conduct water down steep slopes with concrete flumes, and good irrigator performance. The disadvantages are the care needed to obtain good water distribution down the furrow, the need to avoid damage to the permanent concrete flumes at harvest, and the barriers to machinery caused by the flumes (Humbert, 1968; Reynolds & Gibson, 1968).

The level ditch system is used in relatively flat land with ditches on a slope of 0.2 to 0.3% connected by straight furrows between the ditches. Furrow lengths are typically 75 to 100 m. Water flows from one ditch through the furrow to the other ditch. The system has the advantage that no water is wasted at the lower end of the furrow, and distribution along the line is good. However, significant land can be lost to ditches, and erosion may be a problem in sloping fields (Humbert, 1968; Reynolds & Bigson, 1968).

The continuous long line system consists of irrigating continuous furrows up to 300 m long with several lightweight aluminum flumes crossing the furrow at intervals along its length. The flumes run perpendicular to the furrows and rest atop the beds. Water is released into the furrows from the flumes, flows down the furrow, and any water which does not infiltrate simply supplements the water from the next flume down the slope. The system can be used on steep land with long furrows on grade and with flumes running down grade. It produces somewhat better water distribution than the herringbone system on impermeable, poorly aerated soils. In addition, aluminum flumes can be removed to facilitate harvest and land preparation (Humbert, 1968; Reynolds & Gibson, 1968).

Without automation, the herringbone, level ditch, and continuous long line systems allow irrigation of up to 8 ha d^{-1} per irrigator (Norum, 1968). On land slopes greater than 1.5%, the herringbone system provides the best water distribution, but the level ditch system has the lowest yearly cost. At slopes < 1.5%, the level ditch system provides better water distribution at less cost than the herringbone system (Hawaiin Sugar Planters' Association, 1964, unpublished).

In 1965, semiautomatic furrow irrigation methods were introduced in Hawaii to reduce labor costs. Timers were used to open and close flume gates in a predetermined manner, and water sensors were used to close gates when water reached a predetermined position in the furrow (Reynolds & Gibson, 1968). Initial results indicated that a single person could irrigate over 20 ha d^{-1} (Norum, 1968).

C. Sprinkler Irrigation

Sprinkler irrigation is used throughout the world to irrigate sugarcane. Good results have been reported in South Africa (Cleasby, 1959), Mauritius

(Mauritius Sugar Industry Research Institute, 1959), and Hawaii (Humbert, 1968; Norum, 1968; Reynolds & Gibson, 1968) using a variety of permanent, portable, and moving systems. During the 1970s, sprinkler irrigation gradually replaced furrow irrigation in many areas. Its advantages over furrow irrigation include little water loss through percolation, less land required for ditches, reduced land preparation costs due to flat culture, and (sometimes) reduced labor costs (Leverington & Ridge, 1975). Disadvantages include higher energy, capital, and (sometimes) labor costs than furrow systems (Norum, 1968).

In Australia, self-propelled high-volume guns at spacings of about 60 m are the predominant form of sprinkler irrigation on a variety of soil types. Solid-set, low-volume sprinklers have been used to irrigate at 3-d intervals on shallow soils in Swaziland (Durandt & Calder, 1974). Numerous advantages of this system have been reported, though they have not been found on deeper soils elsewhere in southern Africa.

In Hawaii, several types of sprinkler irrigation have been used. Solid-set sprinklers on 18 (crosswind) × 24 m (downwind) spacing with automatic sequencing among laterals have been used. Since the system is fully automatic, it has a low labor requirement and can be operated at night. However, it has a high initial capital cost and is susceptible to damage at harvest. Lateral-ditch roadway systems use trailer-mounted pumps and high-volume guns. They pump directly from irrigation ditches and can easily be towed between sets. Gravity-head systems use a buried mainline downslope from the water source and portable laterals and sprays. They require no pumping. Hose-reel and center-pivot systems have also been used (Norum, 1968; Reynolds & Gibson, 1968).

D. Water Table Management

In many areas sugarcane is grown on nearly level soils in which the water table reaches the root zone during at least part of the year. In Louisiana, sugarcane is grown on low, nearly level land near large bodies of water. High water tables are common, especially during the winter months when sugarcane leaf area and evapotranspiration are low. Response to supplemental irrigation is rare (Carter & Floyd, 1973), probably because the crop can extract much of its water requirement from the water table. A similar situation occurs in western Taiwan, where water tables frequently rise near the soil surface in the rainy season. In the dry season the crop obtains 25 to 50% of its water requirement from a water table at approximately 1.6 m. This dramatically reduces irrigation demands (Yang & Chang, 1976; Chang & Wang, 1983; Hunsigi & Srivastava, 1977).

The most important use of water tables for irrigation occurs on organic soils of the south Florida Everglades. Water from Lake Okeechobee is introduced into a primary canal system. During the dry season, growers pump supplemental water from the primary canal system into a system of farm canals, lateral ditches, and field ditches. Fields are generally 200 × 800 m and are bordered by lateral or field ditches. Lateral movement of water from the field ditches into the field is often facilitated by mole drains 15 cm diam., 70 to 90 cm deep, and 2 to 3 m apart. The organic soils in this region are quite permeable, lateral movement is rapid, and the water table can easily be raised to provide supplemental water during the dry season.

In well-drained sandy soils, artificial barriers have been used to reduce the drainage rate and create a temporary perched water table (Wang et al., 1969). For example, Sumner and Gilfillan (1971) and Erickson et al. (1968) observed dramatically increased nonirrigated sugarcane yields in South Africa when impermeable 2-mm asphalt layers were placed 0.6 m deep in sandy soils. The crop was able to use water perched above this barrier during dry periods; however, the high cost of the layer precluded its commercial use.

If the water table remains near the soil surface, poor aeration may cause setts to decay, reduce ratoon stalk populations (Carter et al., 1985), reduce root growth (Gosnell, 1971; Banath & Monteith, 1966), and cause the production of aerotropic (Srinivasan & Batcha, 1962) or floating (Sartoris & Belcher, 1949) roots. Rudd and Chardron (1977) reported a yield decline of 0.46 Mg cane ha^{-1} for each day the water table rose above 0.5 m.

Louisiana sugarcane farmers plant cane on high (0.30–0.45 m) ridges 1.8 m apart to enhance surface drainage and provide a small volume of aerated soil at almost all times. When winter rainfall is relatively low, these practices provide adequate aeration, but in many years excess rainfall raises water tables and reduces yields (Carter, 1977a; Carter & Floyd, 1975; Carter & Camp, 1983).

A 7-yr lysimeter study on fine-textured Louisiana soils indicated that drainage systems should reduce the water table to 1.2 m within 4 d in order to maximize cane and sugar yields (Carter, 1977b). This maintains high redox potentials in the upper 0.5 m of soil (Carter, 1980), allows good ratoon growth early in the spring, and increases stalk populations. Maintenance of low water tables with drainage systems may also permit more closely spaced rows, more rapid leaf area development, and higher yield potentials than traditional practices (Carter et al., 1985).

Florida sugarcane farmers reverse the flow of water in the system of canals and ditches described previously to remove excess water from their fields and maintain a layer of aerated organic soil. Partial drainage of these soils is necessary for sugarcane production. As a result, the organic matter in these soils has oxidized. Between approximately 1940 and 1980, they have subsided at a rate of 15 to 31 mm yr^{-1} (Shih et al., 1979). The only feasible means of slowing subsidence is to maintain high water tables (Shih & Gascho, 1980a).

Sugarcane cultivars differ in their sensitivity to high water tables (Andries, 1976; Srinivasan & Batcha, 1962). cultivars tolerant to waterlogging respond with production of aerotropic or floating roots (Shah, 1951; Srinivasan & Batcha, 1962). Some Florida cultivars exhibit better growth with water tables at 0.32 m than at 0.61 or 0.84 m; however, the growth of other cultivars is greatly inhibited by the 0.32-m water table (Gascho & Shih, 1979).

E. Drip Irrigation

Hawaii, with its high labor and land costs, has become a leader in drip-irrigated sugarcane. In 1970, the Hawaii sugar industry had 48 000 ha of irrigated sugarcane; 90% was furrow irrigated, and 10% was sprinkler irrigated. The first commercial drip irrigation installations were made in 1971, and by 1973 approximately 1200 ha were drip irrigated (Gibson, 1974). By the end of 1985, there were 46 000 ha of irrigated sugarcane; 80% was drip irrigated, 18% was furrow irrigated, and 2% was sprinkler irrigated. Each furrow or sprinkler irrigator managed an average of 48 ha while each drip irrigator managed an average of 230 ha. With almost no change in irrigated area, the number of irrigation personnel was reduced by half from 1971 to 1985 (Bui & Kinoshita, 1985; Hawaii Sugar Planters' Assoc. 1985, unpublished).

In Hawaii, the advantages of drip irrigation over furrow and sprinkler systems include higher cane and sugar yields, relatively small capital cost, high irrigation efficiency, good irrigator performance, elimination of furrows with their associated costs, continuous supply of irrigation water to the crop, increased fertilizer efficiency, decreased fertilizer application costs, and decreased weed control costs. Drip tube pinching, orifice plugging, and orifice enlargement by ants are the greatest problems encountered; but advances in water filtration, chlorination, tube flushing, tube design, system design, and insecticide use have reduced these problems to acceptable levels.

Drip irrigation can increase yields while reducing irrigation water use. For example, in Taiwan, drip irrigation has increased cane yields by more than 20% while reducing water use by almost 40% (Yang et al., 1978). In Hawaii, for the period 1979 to 1984, drip-irrigated fields produced 22% greater sugar yields ha^{-1} and 26% greater sugar yields $ha^{-1} m^{-1}$ than fields irrigated by furrow or sprinkler. Water distribution uniformity is estimated to be greater than 80% for drip but <35% for furrow (Bui & Kinoshita, 1985; Young, 1985).

Studies of tube placement normally indicate no significant difference between treatments in which a drip tube is placed beside each row and treatments in which a single tube is placed between paired rows (pineapple placement) (Hawaiian Sugar Planters' Association, 1977, unpublished data). With the exception of Hawaii, general adoption of capital-intensive drip irrigation systems has been restricted by low world sugar prices in recent years.

F. Irrigation Frequency

Four methods have been widely used to schedule irrigation: resistance blocks, tensiometers, the water balance, and tissue moisture content. All are based on their ability to predict when incipient drought stress will occur, usually as indicated by a decrease in stem or leaf elongation. Experiments that compare cane and sugar yields from plots irrigated with different frequencies and/or amounts are used to validate these methods. Regardless which method is used to select ideal irrigation schedules, actual irrigation practices are modified in response to availability of water, personnel, and equipment.

In Hawaii, gypsum resistance blocks were first tested as a means of scheduling furrow irrigation in the early 1940s. Extensive experimentation on the design, manufacture, and use of the blocks led to the recommendation that they be installed at two-thirds the depth of maximum root proliferation, which varied among soils. Irrigation was recommended when resistance approached the value at which elongation began to decrease significantly, usually about 5000 ohms or approximately -0.4 MPa at the depth of the block (Humbert, 1968). This is consistent with results from Queensland, where soil water potential was measured with gypsum blocks at 0.23-m depth. Irrigation at -0.4 MPa gave the same yield as irrigation at -0.1 MPa, but only half the water was required (Kingston, 1971).

Tensiometers were tested in Hawaii in the 1930s, and they are still used for irrigation control. They are normally placed 0.45 to 0.60 m deep and are calibrated against cane elongation in a manner similar to that used with gypsum blocks (Humbert, 1968).

Since the 1960s, Class A pan evaporation has been widely used to estimate potential evapotranspiration of sugarcane with a fully developed canopy. Empirical adjustments are used to estimate water use before canopy development is complete, after lodging, and in ratoon crops, all of which use less water than fully developed erect canopies of plant crops (Doorenbos & Pruitt, 1977; Thompson, 1976, 1977). Estimates of water use based on pan evaporation are combined with estimates of water supplied by irrigation and rainfall to predict the soil water balance. Irrigation is scheduled when soil water has decreased enough to cause incipient drought stress (Robinson, 1962; Robinson et al., 1963; Thompson, 1976; Thompson & Collings, 1963). Kingston (1973) compared recommended irrigation dates and crop yields of treatments for which irrigation dates were based on either Class A pan evaporation or gypsum resistance blocks. Over a 4-yr period, irrigation dates based on the two methods were close, and Kingston (1973) concluded that Class A pan evaporation provided the more practical means of irrigation scheduling.

Tissue moisture content has long been used to control irrigation during ripening, when gradual increase in water stress is used to stimulate sugar storage in the stem (Clements, 1980; Humbert, 1968; Thompson, 1977; Yang & Chang, 1978). Clements (1980) described the use to the sheath water content to control ripening. A *ripening log* is used to compare measured and desired sheath water contents during approximately 24 wk prior to harvest of a 2-yr crop. Sheath water content is measured on a periodic basis, and irrigation intervals or amounts are varied to produce a gradual decline of sheath water content, from about 83% at the beginning of ripening to about 73% at harvest. In Hawaii (Clements 1980) and Taiwan (Yang & Chang, 1978) sheath water content has been found to be a good indicator of stem sugar content. Similar methods involving other tissues have been used in Mexico (Humbert, 1968), South Africa (Thompson, 1977), and Zimbabwe (Lonsdale & Gosnell, 1974).

G. Irrigation Amount

Irrigation water is often limited and costly. Therefore, numerous studies have been conducted to determine the optimum amount of irrigation to apply over the course of a crop. Many early studies compared yields of treatments with different irrigation intervals or total numbers of irrigation "rounds." Later, the amount of effective irrigation plus rainfall (water not lost to runoff or deep percolation) was estimated, and its effect on yields was determined.

Early (1974) reviewed the world literature on response of cane and sugar yields to irrigation. He found a wide variety of responses, both linear and nonlinear, between yields and number of irrigations, total irrigation amount, total irrigation plus rainfall, or total effective irrigation plus rainfall. However, the relationships he obtained varied dramatically among years and sites. This is not surprising considering the wide variety of experimental methods, levels of management inputs, and soil conditions represented in the data.

Jones (1980) analyzed several irrigation experiments conducted in Hawaii. After eliminating treatments in which excessive water had reduced yields, presumably due to poor soil aeration and/or leaching of nutrients, he found a linear relationship between *relative water use* (RWU), the ratio of effective water applied to Class A pan evaporation and *relative cane yield* (Y), the ratio of actual yield to maximum yield in the experiment in three experiments.

$$Y = 1.01 \text{ RWU} + 0.03 \quad (R^2 = 0.90, n = 14)$$

Thompson (1976) selected data from studies in South Africa (Boyce, 1969; Thompson & Boyce, 1968, 1971; Thompson & De Robillard, 1968), Australia (Kingston & Ham, 1975), Hawaii (Campbell et al., 1959), and Mauritius (Hardy, 1966) for which evapotranspiration has been estimated with lysimeters or in carefully controlled irrigation rate experiments without excessive water application. For total crop evapotranspiration (ET) ranging from 660 to 3840 mm and cane yields (Y) ranging from 57 to 342 Mg ha^{-1}, he found the following relationship:

$$Y = 0.0969 \text{ ET} - 2.4 \quad (r^2 = 0.90, n = 91)$$

The slope of the relationship is similar to that reported by Jones (1980) for furrow- and sprinkler-irrigated experiments in Hawaii (0.103 Mg cane mm^{-1} effective water).

When irrigation plus rainfall is greater than potential evapotranspiration, anaerobic soil conditions or N losses may reduce crop growth rates. For example, for poorly drained soils in Hawaii, reduced cane yields were observed when effective irrigation plus rainfall were greater than potential evapotranspiration (Jones, 1980; Sylvester, 1984). However, recent work suggests that "excessive" rates of drip irrigation may be beneficial on well-drained soils. For example, in Hawaii, cane yields on an Oxisol were linearly related ($R^2 = 0.98$) to the fraction of pan evaporation (1.7 m above sod) supplied as rainfall plus drip irrigation. Sugar yields were curvilinearly related to the same pan fraction ($R^2 = 0.99$), and there was no indication that yields had reached a maximum at a pan fraction of 1.0 (Santo & Bosshart, 1985a, b; Hawaiian Sugar Planters' Association, 1982, unpublished). In another study of drip irrigation amounts in Hawaii, maximum yields were obtained with a pan fraction of 1.4 on a well-drained Oxisol (Santo & Meinzer, 1986, unpublished data). Batchelor et al. (1985) found that sugarcane roots tend to grow preferentially near drip irrigation tubes, even when rates of irrigation greater than pan evaporation are used. Like Santo and Bosshart (1985a, b), they found that irrigation at rates greater than pan evaporation increased growth of drip-irrigated sugarcane, at least under some conditions. This suggests that in well-drained soils drip irrigation permits adequate aeration of the root zone while providing a limited volume of soil at high water potential.

H. Irrigation Water Quality

Bresler et al. (1982) characterized sugarcane as moderately sensitive to salinity. In general, leaching and addition of Ca are used to maintain good soil structure and low electrical conductivity of the soil water. However, many sugarcane soils in which salinity is a problem are poorly drained. Therefore, adequate drainage must be provided, and irrigation amounts must be adjusted to minimize electrical conductivity in the root zone while maintaining an adequate depth of well-aerated soil.

Most studies suggest that sugarcane yields decline when electrical conductivity of the soil solution exceeds 0.1 to 0.25 S m^{-1}. Serious yield reductions occur at conductivities of 0.4 to 0.8 S m^{-1}, and little cane growth or death occurs above 1.0 S m^{-1} (Bernstein et al., 1966a; Bresler et al., 1982; von der Meden, 1966; Santo, 1981; Valdivia, 1981). In addition, germination and juice quality are reduced by saline conditions (Dev & Bajwa, 1972; Bernstein et al., 1966b; Kingston, 1982; Prothero, 1978). However, several factors, including genotype (Bernstein et al., 1966a, b; Dev & Bajwa, 1972; Syed & El-Swaify, 1972, 1973; Tanimoto, 1969), irrigation frequency (Segovia, 1979), and the presence of a water table or leaching rains (Thomas et al., 1981; Valdivia, 1981) can affect sugarcane response to saline soils and/or saline irrigation waters. For example, Valdivia (1981) reported that sugarcane is much more sensitive to saline soils when the water table is deeper than 2 m than when it occurs between 0.8 and 1.1. m. In Hawaii, drip irrigation promises to improve yields on soils irrigated with saline water. Frequent application of small amounts of irrigation water allow parts of the root zone to be maintained at high water potential while adequate volumes of soil remain well aerated (Segovia, 1979). In some poorly drained soils, drip irrigation with water ranging from 0.1 to 0.8 S m^{-1} required irrigation plus rainfall equivalent to 1.2 to 1.6 of pan evaporation to maintain good cane growth (Santo & Isobe, 1975, unpublished data).

REFERENCES

Alexander, A.G. 1973. Sugarcane physiology. Elsevier, Amsterdam.

Allen, R.J., Jr., G. Kidder, and G.J. Gascho. 1978. Predicting tons of sugarcane per acre using solar radiation, temperature and percent plant cane, 1971 through 1976. Proc. Am. Soc. Sugar Cane Technol. N.S. 7:18–22.

Andreis, H.J. 1976. A water table study on Everglades peat soil. Sugar J. 39:8–12.

Ashton, F.M. 1956. Effects of a series of cycles of alternating low and high soil water contents on the rate of apparent photosynthesis in sugar cane. Plant Physiol. 31:266–274.

Banath, C.L., and N.H. Monteith. 1966. Soil oxygen deficiency and sugar cane root growth. Plant Soil 25:143–149.

Baran, R., D. Bassereau, and N. Gillet. 1974. Measurement of available water and root development on an irrigated sugar cane crop in the Ivory Coast. Proc. Int. Soc. Sugar Cane Technol. 15:726–735.

Batchelor, C.H., G.C. Soopramanien, and J.P. Bell. 1985. Root development of drip irrigated sugarcane. p. 687–695. In Drip/trickle irrigation in action. Vol. II. Proc. Third Int. Drip/Trickle Irrig. Congr. ASAE, St. Joesph, MI.

Bernstein, L., R.A. Clark, L.E. Francios, and M.D. Derderian. 1966a. Salt tolerance of N. Co. varieties of sugar-cane. II. Effects of soil salinity and sprinkling on chemical composition. Agron. J. 58:503–507.

Bernstein, L., L.E. Francois, and R.A. Clark. 1966b. Salt tolerance of N. Co. varieties of sugar cane. I. Sprouting, growth, and yield. Agron. J. 58:489–493.

Boyce, J.P. 1969. First ratoon results of two irrigation experiments at Pongola. Proc. S. Afr. Sugar Technol. Assoc. 43:35–46.

Bressler, E., B.L. McNeal, and D.L. Carter. 1982. Saline and sodic soils. Springer-Verlag, Berlin.

Brett, P.G.C., and R.L. Harding. 1974. Artificial induction of flowering in Natal. Proc. Int. Soc. Sugar Cane Technol. 15:55–66.

Brett, P.G.C., R.L. Harding, and J.G. Parton. 1975. Time and intensity of flowering as influenced by certain temperature and photoperiod treatments. Proc. S. Afr. Sugar Technol. Assoc. 49:202–205.

Bui, W., and C.M. Kinoshita. 1985. Has drip irrigation in Hawaii lived up to its expectations. p. 84–89. In Drip/trickle irrigation in action. Vol. I. Proc. Third Int. Drip/Trickle Irrig. Congr. ASAE, St. Joseph, MI.

Bull, T.A., and K.T. Glaziou. 1975. Sugar cane. p. 51–72. In L.T. Evans (ed.) Crop physiology: Some case histories. Cambridge Univ. Press, London.

Bull, T.A., and D.A. Tovey. 1974. Aspects modelling sugarcane growth by computer simulation. Proc. Int. Soc. Sugar Cane Technol. 15:1021–1032.

Bureau of Sugar Experiment Stations. 1979. Fertilizing sugar cane. Cane Growers' Q. Bull. 42:91–112.

Campbell, R.B., J.-H. Chang, and D.C. Cox. 1959. Evapotranspiration of sugarcane in Hawaii as measured by in-field lysimeters in relation to climate. Proc. Int. Soc. Sugar Cane Technol. 10:637–645.

Carter, C.E. 1977a. Excess water decreases cane and sugar yields. Proc. Am. Soc. Sugar Cane Technol. 6:44–51.

Carter, C.E. 1977b. Drainage parameters for sugarcane in Louisiana. p. 135–138. In Proc. Third Natl. Drain. Symp., Chicago, IL. December, 1976. ASAE, St. Joseph, MI.

Carter, C.E. 1980. Redox potential and sugarcane yield relationship. Trans. ASAE 23:924–927.

Carter, C.E., and C.R. Camp. 1983. Subsurface drainage of an alluvial soil increased sugarcane yields. ASAE 26:426–429.

Carter, C.E., and J.M. Floyd. 1973. Subsurface drainage and irrigation for sugarcane. Trans. ASAE 16:279–281, 284.

Carter, C.E., and J.M. Floyd. 1975. Inhibition of sugarcane yields by high water table during dormant season. Proc. Am. Soc. Sugar Cane Technol. N.S. 4:14–18.

Carter, C.E., J.E. Irvine, V. McDaniel, and J. Dunckelman. 1985. Yield response of sugarcane to stalk density and subsurface drainage treatments. ASAE 28:172–178.

Chang, H.-H., R.B. Campbell, H.W. Brodie, and L.D. Baver. 1965. Evapotranspiration research at the HSPA Experiment Station. Proc. Int. Soc. Sugar Cane Technol. 12:10–24.

Chang, Y.T., and P.L. Wang. 1983. Water consumption and irrigation requirements of sugarcane. p. 15. In Taiwan Sugar Res. Inst. Annu. Rep. 1982–1983. Taiwan Sugar Res. Inst., Tainan, Taiwan.

Cleasby, T.G. 1959. The overhead irrigation of sugarcane in Natal. Proc. Int. Soc. Sugar Cane Technol. 10:621–629.

Clements, H.F. 1980. Sugarcane crop logging and crop control: Principles and practice. Univ. Hawaii Press, Honolulu.

Clements, H.F. and T. Kubota. 1943. The primary index—its meaning and application to crop management with special reference to sugarcane. Hawaiian Plant. Rec. 47:257–297.

Clowes, M. St. J., and N.G. Inman-Bamber, 1980. Effects of moisture regime, amount of nitrogen applied and variety on the ripening response of sugarcane to glyphosates. Proc. S. Afr. Sugar Technol. Assoc. 54:127–133.

Das, U.K. 1933a. Measuring production in terms of temperature. Hawaiian Plant. Rec. 37:32–53.

Das, U.K. 1933b. How to measure effective temperature in terms of day-degrees. Hawaiian Plant. Rec. 37:174–178.

Dastane, N.G. 1957. Evaluation of root efficiency in moisture extraction. J. Aust. Inst. Agric. Sci. 23:223–226.

Dev, G., and M.S. Bajwa. 1972. Studies on salt tolerance of sugarcane. Indian Sugar 22:723–726.

Doorenbos, J., and W.O. Pruitt. 1977. Crop water requirements. FAO Irrig. Drain. Pap. 24 (1977 rev.). FAO, Rome.

Durandt, H.K., and N.S. Calder. 1974. Some observations on short frequent irrigation cycles in Swaziland. Proc. S. Afr. Sugar Technol. Assoc. 48:70–73.

Early, A.C. 1974. The yield response of sugarcane to irrigation in the Philippines. Proc. Int. Soc. Sugar Cane Technol. 15:679–693.

Ekern, P.C. 1971. Use of water by sugarcane in Hawaii measured by hydraulic lysimeters. Proc. Int. Soc. Sugar Cane Technol. 14:805–812.

Erickson, A.E., C.M. Hanson, A.J.M. Smucker, K.Y. Li, L.C. Hsi, T.S. Wang, and R.L. Cook. 1968. Subsurface asphalt barriers for the improvement of sugarcane production and the conservation of water on sand soil. Proc. Int. Soc. Sugar Cane Technol. 13:787–792.

Evans, H. 1935. The root system of the sugar cane. Emp. J. Exp. Agric. 3:351–362.

Evans, H. 1936. The root system of the sugar cane. II. Some typical root systems. Emp. J. Exp. Agric. 4:208–221.

Finkel, H.J. 1983. Irrigation of sugar crops. p. 119–135. In Handbook of irrigation technol. CRC Press, Boca Raton, FL.

Fogliata, F.A. 1974. Sugarcane irrigation in Tucuman. Proc. Int. Soc. Sugar Cane Technol. 15:655–667.

Gascho, G.J., and S.F. Shih. 1979. Varietal response of sugarcane to watertable, I. Lysimeter performance and plant response. Soil Crop Sci. Soc. Flor. Proc. 38:23–27.

Gascho, G.J., and S.F. Shih. 1981. Row spacing effects on biomass and composition of sugarcane in Florida. Proc. Am. Soc. Sugar Cane Technol. N.S. 8:72–76.

Gascho, G.J., and S.F. Shih. 1982. Sugarcane p. 445–479. In I.D. Teare and M.M. Peet (ed.) Crop-water relations. John Wiley and Sons, New York.

Gibson, W. 1974. Hydraulics, mechanics, and economics of subsurface and drip irrigation of Hawaiian sugarcane. Proc. Int. Soc. Sugar Cane Technol. 15:639–648.

Glover, J. 1967. The simultaneous growth of sugarcane roots and tops in relation to soil and climate. Proc. S. Afr. Sugar Technol. Assoc. 41:143–159.

Glover, J. 1968. Further results from the Mount Edgecombe Root Laboratory. Proc. S. Afr. Sugar Technol. Assoc. 42:123–132.

Glover, J. 1972. Practical and theoretical assessments of sugarcane yield potential in Natal. Proc. S. Afr. Sugar Technol. Assoc. 46:138–141.

Gosnell, J.M. 1968. Some effects of increasing age on sugarcane growth. Proc. Int. Soc. Sugar Cane Technol. 13:499–513.

Gosnell, J.M. 1971. Some effects of water-table on the growth of sugarcane. Proc. Int. Soc. Sugar Cane Technol. 14:841–849.

Gosnell, J.M., and T.L. Pearse. 1971. Methods of surface irrigation in sugarcane. Proc. Int. Soc. Sugar Cane Technol. 14:875–885.

Gosnell, J.M. and G.D. Thompson. 1965. Preliminary studies on depth of soil moisture extraction by sugarcane using neutron probe. Proc. S. Afr. Sugar Technol. Assoc. 39:158–165.

Halais, P. 1935. A new index of agricultural climatology. (In French.) Rev. Agric. Maurice 80:44–49.

Hardy, G., H. Dove, and M. Awad. 1986. The use of ethephon for prevention of flowering in sugarcane in Sudan. Proc. Int. Soc. Sugar Cane Technol. 19:305–316.

Hardy, M. 1966. Water consumption of the cane plant. p. 95–101. *In* Annu. Rep. Mauritius Sugar Ind. Res. Inst.

Inman-Bamber, N.G., and J.M. de Jager. 1986a. Effect of water stress on growth, leaf resistance and canopy temperature in field-grown sugarcane. Proc. S. Afr. Sugar Technol. Assoc. 60:156–161.

Inman-Bamber, N.G. and J.M. de Jager. 1986b. The reaction of two varieties of sugarcane to water stress. Field Crops Res. 14:15–28.

Hartt, C.E. 1967. Effect of moisture supply upon translocation and storage of ^{14}C in sugarcane. Plant Physiol. 42:338–346.

Hill, J.N.S. 1966. Availability of soil water to sugarcane in Natal. Proc. S. Afr. Sugar Technol. Assoc. 40:276–282.

Humbert, R.P. 1968. The growing of sugarcane. Elsevier Publ. Co., Amsterdam.

Hunsigi, G., and S.C. Srivastava. 1977. Modulation of ET values of sugarcane because of high water table. Proc. Int. Soc. Sugar Cane Technol. 16:1557–1564.

Irvine, J.E. 1983. Sugarcane. p. 361–381. *In* Biological basis, physical environment, and crop productivity. Int. Rice Res. Inst., Los Baños, Philippines.

Irvine, J.E., and G.T.A. Benda. 1980. Genetic potential and restraints in *Saccharum* as an energy source. p. 1–9. *In* A.G. Alexander (ed.) Alternate uses of sugarcane for development in Puerto Rico. CEER Publ. B-52. Univ. Puerto Rico, San Juan.

Jones, C.A. 1980. A review of evapotranspiration studies in irrigated sugarcane in Hawaii. Hawaiian Plant. Rec. 59:195–214.

Jones, C.A. 1985. C_4 grasses and cereals. Growth, development, and stress response. John Wiley and Sons, New York.

Juang, T.C., and G. Uehara. 1971. Effect of ground-water table and soil compaction on nutrient element uptake and growth of sugarcane. Proc. Int. Soc. Sugar Cane Technol. 14:679–687.

Kingston, G. 1971. An experiment in irrigation scheduling. Further comments. Proc. Queensl. Soc. Sugar Cane Technol. 39:143–151.

Kingston, G. 1973. The potential of 'Class A' pan evaporation data, for scheduling irrigation of sugar cane at Bundaberg. Proc. Queensl. Soc. Sugar Cane Technol. 40:151–157.

Kingston, G. 1977. The influence of accessibility on moisture extraction by sugar cane. Proc. Int. Soc. Sugar Cane Technol. 16:1239–1250.

Kingston, G. 1982. Ash in first expressed cane juice at Rocky Point. I. Factors affecting the inorganic composition of juices. Proc. Aust. Soc. Sugar Cane Technol. 49:11–17.

Kingston, G., and G.J. Ham. 1975. Water requirements and irrigation scheduling of sugar cane in Queensland. Proc. Queensl. Soc. Sugar Cane Technol. 42:57–65.

Koehler, P.H., P.H. Moore, C.A. Jones, A. Dela Cruz, and A. Maretzki. 1982. Response of drip-irrigated sugarcane to drought stress. Agron. J. 74:906–911.

Lee, H.A. 1926a. Progress report on the distribution of cane roots in the soil under plantation conditions. Hawaiian Plant. Rec. 30:511–519.

Lee, H.A. 1926b. A comparison of the root weights and distribution of H109 and D1135 cane varieties. Hawaiian Plant. Rec. 30:520–523.

Lee, H.A. 1926c. The distribution of the roots of sugar cane in the Hawaiian Islands. Plant Physiol. 1:363–378.

Leverington, K.C., and D.R. Ridge. 1975. A review of sugar cane irrigation. Proc. Queensl. Soc. Sugar Cane Technol. 42:37–43.

Lonsdale, J.E., and J.M. Gosnell. 1974. Monitoring maturity of sugarcane during drying-off. Proc. Int. Soc. Sugar Cane Technol. 15:713–725.

Mason, G.F. 1976. Chemical ripening in Trinidad of variety B41227 with glyphosine. p. 130–138. *In* Proc. West Indies Sugar Technol. 094. West Indies Sugar Technol Soc., Edgehill, Barbados.

Mauritius Sugar Industry Research Institute. 1959. p. 1–103. *In* Annu. Mauritius Sugar Ind. Res. Inst., Port Louis, Mauritius.

Moberly, P.K. 1974. The decline in rate of evapotranspiration of fully canopied sugarcane during a winter stress period. Proc. Int. Soc. Sugar Cane Technol. 15:694–700.

Monteith, J.L. 1965. Evaporation and environment. p. 205–234. *In* G.E. Fogg (ed.) The state and movement of water in living organisms. Academic Press, New York.

Mukerji, N., and M. Alan. 1959. Roots as indicator to varietal behavior under different conditions. Indian J. Sugarcane. Res. Dev. 3:131–134.

Nickell, L.G. 1968. Agricultural aspects of transplanting and spacing. p. 147–155. *In* Hawaiian Sugar Technol. Rep. 1967.

Norum, E.M. 1968. The use of high-capacity sprinklers for the irrigation of sugarcane by Kohala Sugar Company. Proc. Int. Soc. Sugar Cane Technol. 13:1443–1463.

Paz-Vergara, J.E., A. Vasquez, W. Iglesias, and J.C. Sevilla. 1980. Root development of the sugarcane cultivars H32-8560 and H57-5174, under normal conditions of cultivation and irrigation in the Chicama Valley. Proc. Int. Soc. Sugar Cane Technol. 17:534–540.

Penman, H.L. 1948. Natural evaporation from open water, bare soil and grass. Proc. R. Soc. (London) Ser. A. 193:120–145.

Penman, H.L., and R.K. Schofield. 1951. Some physical aspects of assimilation and transpiration. Symp. Soc. Exp. Biol. 5:115–129.

Prothero, G. 1978. The effect of saline irrigation water on sugarcane ripening. Proc. Hawaiian Sugar Technol. 37:69–71.

Raheja, P.C. 1959. Performance of sugarcane varieties in relation to ecological habitat. Indian J. Sugarcane Res. Devel. 3:127–130.

Rao, K.C., and S. Asokan. 1978. Studies on free proline association to drought resistance in sugarcane. Sugar J. 40:23–24.

Reynolds, W.N., and W. Gibson. 1968. Sugarcane irrigation in Hawaii. Proc. Int. Soc. Sugar Cane Technol. 13:55–67.

Robinson, F.E. 1962. Using standard Weather Bureau pans for irrigation interval control. Proc. Hawaiian Sugar Technol. 21:105–118.

Robinson, F.E. 1963. Use of neutron meter to establish soil moisture storage and soil moisture withdrawal by sugarcane roots. Proc. Hawaiian Sugar Technol. 22:206–208.

Robinson, F.E., R.E. Campbell, and J. Chang. 1963. Assessing the utility of pan evaporation for controlling irrigation of sugar cane in Hawaii. Agron. J. 55:444–446.

Rostron, H. 1974. Radiant energy interception, root growth, dry matter production and the apparent yield potential of two sugarcane varieties. Proc. Int. Soc. Sugar Cane Technol. 15:1001–1010.

Rostron, H. 1977. A review of chemical ripening of sugarcane with ethrel in southern Africa. Proc. Int. Soc. Sugar Cane Technol. 16:1605–1616.

Rudd, A.V., and C.W.A. Chardon. 1977. The effects of drainage on cane yields as measured by watertable heights in the Macknade mill area. Proc. Queensl. Soc. Sugar Cane Technol. 44:111–117.

Ryker, T.C., and C.W. Edgerton. 1931. Studies on sugar cane roots. Louisiana Agric. Exp. Stn. Bull. 223.

Samuels, G. 1971. Influence of water deficiency and excess on growth and leaf nutrient element content of sugarcane. Proc. Int. Soc. Sugar Cane Technol. 14:653–656.

Samuels, G., B.G. Cap, and I.S. Bangdiwala. 1953. The nitrogen content of sugarcane as influenced by moisture and age. J. Agric. Res. Univ. Puerto Rico 37:1–12.

Santo, L.T. 1981. New method of screening sugarcane cultivars for salt tolerance. p. 65–67. *In* Hawaiian Sugar Technol. Rep. 1980.

Santo, L.T., and R.P. Bosshart. 1985a. Effects of the timing of water deficit on drip-irrigated sugarcane yields. p. A-20. *In* Hawaiian Sugar Technol. Rep. 1984.

Santo, L.T., and R.P. Bosshart. 1985b. Tillage and water effects on yields of drip-irrigated sugarcane on an Oxisol. p. FE-16-18. *In* Hawiian Sugar Technol. Rep. 1984.

Santo, L.T., and M. Isobe. 1975. Drip irrigation with saline water. p. 21. *In* Exp. Stn. Annu. Rep. 1975. Hawaiian Sugar Planters' Assoc.

Sartoris, G.B., and B.A. Belcher. 1949. The effect of flooding on flowering and survival of sugar cane. Sugar 44(1):36–39.

Segovia, R.A.J. 1979. Effects of pH, P, and irrigation frequency on the yields and mineral composition of sugar cane grown under saline conditions. Ph.D. diss. Univ. Hawaii, Honolulu.

Selleck, G.W., K.R. Frost, R.C. Billman, and D.A. Brown. 1974. Sucrose enhancement in field scale sugarcane trials with polaris in Florida, Hawaii, and Louisiana. Proc. Int. Soc. Sugar Cane Technol. 15:938–945.

Shah, R. 1951. Negatively geotropic roots in water-logged canes. Sugar 46(1):39.

Shih, S.F., and G.J. Gascho. 1980a. Water requirement for sugarcane production. Trans. ASAE 23:934–937.

Shih, S.F., and G.J. Gascho. 1980b. Relationships among stalk length, leaf area, and dry biomass of sugarcane. Agron. J. 72:309–313.

Shih, S.F., and G.J. Gascho. 1981. Sugarcane biomass production and nutrient content as related to climate in Florida. Proc. Am. Soc. Sugar Cane Technol. N.S. 8:77–83.

Shih, S.F., J.W. Mishoe, J.W. Jones, and D.L. Myhre. 1979. Subsidence related to land use in Everglades agricultural area. Trans. ASAE 22:560–563, 568.

Shih, S.F., D.L. Myhre, J.W. Mishoe, and G. Kidder. 1977. Water management for sugarcane production in Florida Everglades. Proc. Int. Soc. Sugar Cane Technol. 16:995–1009.

Shih, S.F., and G.S. Rahi. 1983. Pan evaporation as related to sugarcane leaf area index. Proc. Soil Crop Sci. Soc. Fla. 42:80–85.

Srinivasan, K., and M.B.G.R. Batcha. 1962. Performance of clones of *Saccharum* species and allied genera under conditions of waterlogging. Proc. Int. Soc. Sugar Cane Technol. 11:571–577.

Stevenson, G.C., and A.E.S. McIntosh. 1935. Investigations into the root development of the sugar cane in Barbados. (I) Root development in several varieties under one environment. Bull. 5 Br. West Indies Cent. Sugar Cane Breed. Stn., Barbados.

Syed, M.M., and S.A. El-Swaify. 1972. Effect of saline water irrigation on N.Co. 310 and H50-7209 cultivars of sugar-cane. I. Growth parameters. Trop. Agric. (Trinidad) 49:337–346.

Syed, M.M., and S.A. El-Swaify. 1973. Effect of saline water irrigation on N.Co. 310 and H50-7209 cultivars of sugar-cane. II. Chemical composition of plants. Trop. Agric. (Trinidad) 50:45–51.

Sylvester, J.A. 1984. Amounts of water vs. yield on heavy clay soils at Kekaha Sugar Company. p. A32–39. *In* Hawaiian Sugar Technol. Rep. 1985.

Sumner, M.E., and E.C. Gilfillan. 1971. Asphalt barriers to improve productivity of sandy soils—a preliminary assessment. Proc. S. Afr. Sugar Technol. Assoc. 55:165–168.

Tanimoto, T.T. 1969. Differential physiological response of sugar cane varieties to osmotic pressures of saline media. Crop Sci. 9:683–688.

Thomas, J.R., F.G. Salinas, and G.F. Oerther. 1981. Use of saline water for supplemental irrigation of sugarcane. Agron. J. 73:1011–1017.

Thompson, G.D. 1965. The relation of potential evapotranspiration of sugarcane to environmental factors. Proc. Int. Soc. Sugar Cane Technol. 12:3–9.

Thompson, G.D. 1976. Water use by sugarcane. S. Afr. Sugar J. 60:593–600, 627–635.

Thompson, G.D. 1977. Irrigation of sugarcane. S. Afr. Sugar J. 61:126–131, 161–174.

Thompson, G.D. 1978. The production of biomass by sugarcane. Proc. S. Afr. Sugar Technol. Assoc. 52:180–187.

Thompson, G.D. 1986. Agrometeorological and crop measurements in a field of irrigated sugarcane. Mount Edgecombe Res. Rep. 5. S. Afr. Sugar Assoc. Exp. Stn., Mount Edgecombe, South Africa.

Thompson, G.D., and J.P. Boyce. 1967. Daily measurements of potential evapotranspiration from fully canopied sugarcane. Agric. Meteorol. 4:267–279.

Thompson, G.D., and J.P. Boyce. 1968. The plant crop results of two irrigation experiments at Pongola. Proc. S. Afr. Sugar Technol. Assoc. 42:143–157.

Thompson, G.D., and J.P. Boyce. 1971. Comparisons of measured evapotranspiration of sugarcane from large and small lysimeters. Proc. S. Afr. Sugar Technol. Assoc. 45:169–176.

Thompson, G.D., and D.F. Collings. 1963. Supplemental irrigation. S. Afr. Sugar Assoc. Exp. Stn. Bull. 17.

Thompson, G.D., and P.J.M. De Robillard. 1968. Water duty experiments with sugarcane on two soils in Natal. Exp. Agric. 4:295–310.

Thompson, G.D., J.M. Gosnell, and P.J.M. De Robillard. 1967. Responses of sugarcane to supplementary irrigation on two soils in Natal. Exp. Agric. 3:223–238.

Thompson, G.D, and R.L. Harding. 1986. A computerized model for evaluating irrigation schemes for sugarcane. Proc. So. Afr. Sugar. Technol. Assoc. 60:177–182.

Thompson, G.D., C.H.O. Pearson, and T.G. Cleasby. 1963. The estimation of the water requirements of sugarcane in Natal. Proc. S. Afr. Sugar Technol. Assoc. 37:137–141.

U.S. Department of Agriculture. 1985. Agricultural statistics 1985. U.S. Gov. Print. Off., Washington, DC.

Valdivia, V.S. 1981. Advances in Peruvian research on the effects of salinity on sugarcane production. Turrialba 31:237–244.

van Dillewijn, C. 1952. Botany of sugarcane. Chronica Botanica, Waltham, MA.

von der Meden, E.A. 1966. Note on salinity limits for sugarcane in Natal. Proc. S. Afr. Sugar Technol. Assoc. 40:273–275.

Wang, C.C., K.Y. Li, C.C. Yang, and F.W. Ho. 1969. The effect of asphalt barriers on the moisture and nutrients retention in rice and sugarcane fields of deep sands. p. 20–32. *In* Taiwan Sugar Exp. Stn. Annu. Rep. 1968–1969.

Wood, G.H., and R.A. Wood. 1967. The estimation of cane root development and distribution using radiophosphorus. Proc. S. Afr. Sugar Technol. Assoc. 41:160–168.

Yang, S.J., and Y.T. Chang. 1976. Studies on irrigation scheduling by using crop-soil-climatic data. p. 13–14. *In* Taiwan Sugar Res. Inst. Annu. Rep. 1975–1976.

Yang, S.J., and Y.T. Chang. 1978. Forced ripening by properly controlled irrigation. p. 21–22. *In* Taiwan Sugar Res. Inst. Annu. Rep. 1977–1978.

Young, D. 1985. Operation and maintenance of Hawaii systems. p. 90–95. *In* Drip/trickle irrigation in action. Vol. I. Proc. Third Int. Drip/Trickle Irrig. Congr. ASAE, St. Joseph, MI.

29 Potato

JAMES L. WRIGHT

USDA-ARS
Kimberly, Idaho

JEFFREY C. STARK

University of Idaho
Aberdeen, Idaho

Potato (*Solanum tuberosum* L.) is important in the human diet and is by far the most important vegetable crop in terms of quantities produced and consumed worldwide (Li, 1985; Ware & McCollum, 1968). Potato production ranks fourth in the world after rice (*Oryza sativa* L.), wheat (*Triticum aestivum* L.), and maize (*Zea mays* L.). It has fallen in western Europe, but has increased in North America and in developing regions. Over the last 20 yr the rate of increase in potato production has been nearly twice that of total food production (Horton & Sawyer, 1985).

Potato is sensitive to water stress compared to many other common crops (Epstein & Grant, 1973; Singh, 1969). As a result, irrigation has become an essential component of commercial potato production systems in arid and semiarid regions. Supplemental irrigation of potato in humid and subhumid regions is often beneficial (Benoit & Grant, 1980; Bourgoin, 1984) and is becoming increasingly common since even short periods of drought can jeopardize the economic return on a relatively large production investment (Curwen & Massie, 1984; Ritter et al., 1985; Vitosh, 1984). With the advances in irrigation technology during the last 30 yr, there has been a shift in potato production to the arid, temperate regions in North America where the soils and climate produce high-quality tubers. The harvested area of potato in the USA was 526 100 ha in 1984, of which more than 50% was in the irrigated western states (Irrigation Survey, 1986; USDA, 1985). Idaho, Oregon, and Washington accounted for more than 80% of this area and more than 35% of the national area.

There have been considerable research and technological developments in the irrigation of potato since it was included as one of the crops in ASA Monograph 11, *Irrigation of Agricultural Lands* (Robins et al., 1967). We summarize some of this progress and present specific responses, irrigation requirements, and special management practices of potato. The basic principles of irrigation management and crop water requirements are covered in earlier chapters.

I. CROP DEVELOPMENT

The potato is an annual herbaceous dicotyledonous plant with fleshy tubers that arise from underground stems (Smith, 1968). The aboveground stem is erect in early stages of development but later becomes spreading and prostrate or semiprostrate. The tubers have buds or eyes, from which sprouts arise under certain conditions, and are harvested for food and for "seed." The flowers and fruits are not commercially important but are of interest to plant breeders. Although many cultivars are grown, three or four cultivars make up about 70% of the total production in North America, with Russet Burbank the most widely produced cultivar in the irrigated northwest region of the USA (Kleinkopf, 1982).

A. Growth Stages and Tuber Development

The development of the potato can be divided into four major growth stages (Kleinkopf, 1982; Nelson & Hwang, 1975). These growth stages are shown in Fig. 29-1 in relation to typical curves of dry matter partitioning and leaf area index (LAI). Growth stage I includes early vegetative development from planting to tuber initiation, and ranges from 30 to 60 d depending upon varietal differences, cultural practices, and environmental conditions. Stolons begin to develop during this stage but tubers are not present. Stage II, "tuberization," lasts about 10 to 14 d when tubers are formed at the tips of stolons but do not enlarge appreciably. Although most tubers are initiated at this time, additional tubers may form later. The LAI is generally between 1 and 2.

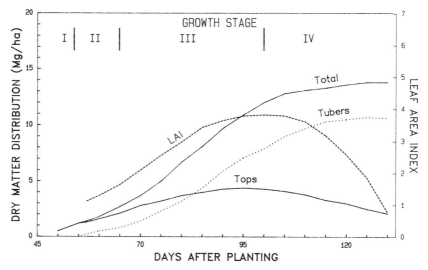

Fig. 29-1. Diagrammatic seasonal progression of dry matter distribution of the tops (stems and leaves), tubers, and total (tops plus tubers) and leaf area index (LAI) of potato, with indicated stages of growth. After Kleinkopf (1982).

Tuber enlargement or "bulking" occurs mostly during growth stage III. Potato tubers are exceptional among plant tissues because of their long period of meristematic development (Plaisted, 1957). Tuber enlargement is usually linear with time for 30 to 60 d under optimal growing conditions. During the latter part of growth stage III, stem and leaf growth continues but growth rates decline. Leaf area index reaches a maximum range of 3.5 to 6.0.

During stage IV, plants begin to senesce and lose leaves. The LAI declines to approximately 1.0, tuber growth rates decrease as the result of reduced leaf area and photosynthetic activity, and the tuber skins mature. Increases in tuber dry weight result primarily from translocation of materials from the tops and roots to the tubers. This maturation period lasts from 10 to 24 d.

The *harvest index* (HI) for potato (the ratio of tuber dry matter to total dry matter, excluding the roots, expressed as a percentage) varies between 60 and 90% depending on growing conditions and the manner of determining it (MacKerron & Heilbroun, 1985). Wolfe et al. (1983) found that by full maturity the HI averaged about 80% for White Rose and Kennebec cultivars over a range of irrigation levels. Similar HI's were indicated in other studies with Norland (Nelson & Hwang, 1975) and Russet Burbank (Kleinkopf et al., 1981).

B. Root Development

Potato has a relatively shallow, fibrous root system with the majority of the roots in the surface 0.3 m (Corey & Blake, 1953; DeRoo & Waggoner, 1961; Lesczynski & Tanner, 1976). Less water is extracted below this depth compared to more deeply rooted crops such as barley (*Hordeum vulgare* L.) and sugarbeets (*Beta vulgaris* L.) (Durrant et al., 1973). The lack of deeper root development has been attributed to the inability of the relatively weak potato root system to penetrate tillage pans or other restrictive layers (Miller & Martin, 1983). Bishop and Grimes (1978) found that soil strengths >1 MPa severely reduced potato root extension as compared to other crops that could extend roots at strengths up to 2 MPa or greater. Rooting depths of 1.2 m or more have been reported for potato under favorable soil conditions (Weaver, 1926; Wolfe et al., 1983). The root system develops rapidly during early growth and achieves maximum development by midseason (Lesczynski & Tanner, 1976). Thereafter, root length, density, and root mass decrease as the plant matures. Fulton (1970) reported that differences in the sensitivity of potato, corn, and tomato (*Lycopersicon esculentum* Mill.) to soil water deficits could not be explained by the distribution of potato roots alone or by maximum rooting depth. He suggested that potato roots have a relatively limited ability to extract soil water.

II. CULTURAL PRACTICES

Potato grows well on a wide variety of soils, but the highest yields of marketable tubers are produced on deep, well-drained, and friable soil (Thornton & Sieczka, 1980). Coarse-textured and loamy soils are preferable but well-drained peat or muck soils are also suitable. Sandy soils will produce high yields if properly fertilized and irrigated.

Seed beds are usually prepared by plowing in the fall or spring. Additional tillage operations are performed with a smoothing implement such as a spring-tooth harrow or a clod buster. On sandy soils, some type of reduced tillage or residue management is becoming common to prevent soil erosion. In many areas, much of the fertilizer is broadcast and incorporated prior to planting. In irrigated areas, particularly those with sandy soils, some of the fertilizer is applied throughout the season through the irrigation system (see section VI of this chapter).

Potato is propagated vegetatively by either whole or cut tubers, called *seed pieces*. Planting date depends on the growing season length and the best marketing time. Preferably, potato should be planted when average soil temperatures at the seed piece depth are 5 to 10 °C (Thornton & Sieczka, 1980).

Seed pieces are usually planted at 15- to 25-cm intervals in rows spaced 76 to 91 cm apart. Planting depth varies from 8 to 14 cm depending on local climatic conditions and soil moisture. The closer intrarow spacings tend to produce small but uniform tubers. Consequently, seed spacing decisions may be based on the intended use of the tubers. Potato crops intended for processing should be grown at relatively wide intrarow spacings to encourage large tubers, while potato crops grown for seed may be closely spaced to produce many small tubers.

After planting, one or more ridging or hilling operations are performed to shape beds, enhance stolon development, and prevent tuber greening or sunburning. These cultivations also control weeds during the first 30 to 40 d after emergence. Thereafter, herbicides and crop shading must be relied upon for weed control. Potato is harvested 12 to 20 wk after planting. Vines are usually mechanically or chemically killed 2 to 3 wk prior to harvest, if not killed by frost, to allow tubers to develop a mature skin. Most potato crops in North America are dug with mechanical harvesters which load the tubers into trucks for transport to storage, packing, or processing facilities.

III. EVAPOTRANSPIRATION AND YIELD

Many studies have considered potato response to irrigation and some data are available on seasonal water use; however, only a few studies report measured evapotranspiration (ET). In some cases, the amount of water applied has been reported (Hang & Miller, 1986; Miller & Martin, 1983; Saffigna et al., 1977a; Sammis, 1980; Shimshi & Susnoschi, 1985), but this may be more or less than actual ET, depending on the irrigation scheduling method and the amount of deep percolation. Daily ET of potato in relation to growth stage and meteorological conditions is needed to develop ET crop coefficients which are used in estimating the soil-water balance.

A. Daily and Seasonal Evapotranspiration

The daily ET of potato varies according to meteorological conditions, surface soil wetness, the stage of growth, and amount of crop cover. Daily ET for a sprinkler-irrigated Russet Burbank potato crop is shown in Fig. 29–2. These measurements were obtained in southern Idaho with a weighing lysimeter located in a 2.5-ha field (Wright, 1982). The soil was a Portneuf silt loam (coarse-silty, mixed, mesic Durixerollic Calciorthid). Plant spacing was 25 cm and row spacing was 91 cm. The field was fertilized the previous fall with 65 kg N ha^{-1}, as NH_4NO_3, and 72 kg P ha^{-1} as triple superphosphate. After planting, liquid N, P, K, and Zn fertilizers were banded beside the seed piece at 145, 39, 104, and 11 kg ha^{-1}, respectively. Potato was planted on day of year (DOY) 116, emerged on DOY 148, and developed a complete canopy by DOY 183. The vines were killed by frost on DOY 263 and harvest was on DOY 278. Soil water potentials were monitored with tensiometers installed at 15, 30, and 45-cm depths. Irrigations were by a solid set sprinkler system and were scheduled to maintain available soil water (ASW) above 65% within the root zone. A total of 16 irrigations were applied during the growing season, plus a preharvest irrigation (see Fig. 29–2). Crop canopy development and yields were on a par with the best commercial potato crops in the region.

Daily ET increased early in the season prior to effective full cover (EFC) when the soil surface was wet following irrigation (see Fig. 29–2, DOY's 145, 156, and 168). After irrigation, daily ET usually returned to the base level in about 5 d as the soil surface dried. When crop cover was sufficient to provide nearly complete shading of the soil, transpiration was the major component of ET. Thus the effects of a wet soil surface were less pronounced

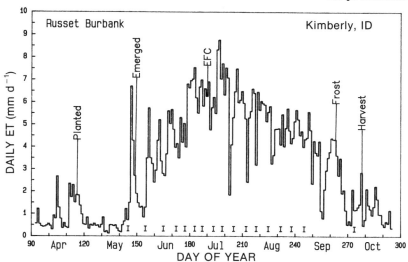

Fig. 29–2. Daily ET of sprinkler-irrigated potato, grown on a silt loam soil, as measured with a weighing lysimeter. Irrigations (I), date of effective full cover (EFC), and other key dates indicated (J.L. Wright, 1972, unpublished data).

than prior to EFC. During June, ET increased as leaf area and transpiration increased and reached near-maximum levels by the end of June just before EFC. The LAI reached 3.5 by this time and peaked at 6.0 in early August. The highest daily ET was 8.5 mm in mid-July when the 7-d mean ET was 7.6 mm d^{-1}. Potato ET gradually decreased during August and early September as evaporative demand decreased and the crop canopy declined, and then rapidly declined to essentially a bare soil evaporation level after vine-killing frosts in mid-September.

Cumulative potato ET calculated from the data of Fig. 29–2 is summarized in Fig. 29–3, along with corresponding summations of Class A pan evaporation (E_{pan}), alfalfa reference ET (E_{tr}), and precipitation. Reference ET was calculated from meteorological data (Wright, 1982), which along with E_{pan} data, were obtained from the U.S. National Weather Service Station about 1 km from the potato lysimeter. Seasonal total potato ET (E_{ta}) for May through September was 604 mm while other corresponding totals in milliliters were: E_{pan}, 1155; E_{tr}, 1066; and precip., 64. Final ratios were $E_{pan}/E_{tr} = 1.08$, $E_{tr}/E_{pan} = 0.92$, $E_{ta}/E_{tr} = 0.57$, and $E_{ta}/E_{pan} = 0.52$. The 64 mm of precipitation was less than normal, but total E_{tr} was near the 7-yr average. Irrigation + precipitation totalled 626 mm, leaving 22 mm for potential drainage.

In comparison with the seasonal potato ET for Kimberly, ID, Tanner (1981) reported that potato ET measured with a lysimeter in the humid Wisconsin area for 92 d, June through August, ranged from 293 to 405 mm during 3 yr of study. This compares with 467 mm in Fig. 29–3 for the same 12 wk. Nkemdirim (1976), using meteorological methods, studied the ET of a potato crop near Calgary, Alberta, Canada, and found midseason daily ET to be about 6 mm. Erie et al. (1965) found that the seasonal water use for

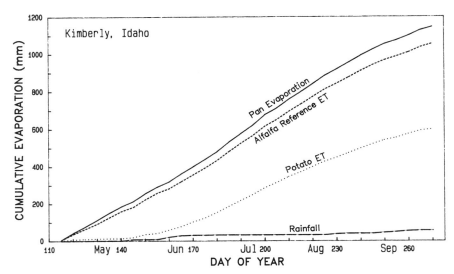

Fig. 29–3. Seasonal progression of cumulative pan evaporation, alfalfa reference ET, potato ET, and rainfall (J.L. Wright, 1972, unpublished data).

potato, from February through June at Mesa, AZ, averaged 617 mm. Data were based on gravimetric soil samples obtained over 5 yr. Reported seasonal water use in irrigated areas of Oregon and Washington ranges from 640 to 700 mm (Hane & Pumphrey, 1984; Hang & Miller, 1986). While potato has generally been considered to require large amounts of water because of the high sensitivity to water stress, this range of seasonal water use is less than for some other field crops such as alfalfa (*Medicago sativa* L.), 600 to 1500 mm; corn, 400 to 750; and sugarbeets, 450 to 850 (Doorenbos & Pruitt, 1977).

B. Evapotranspiration Crop Coefficients

Progress in ET research has permitted development of procedures for estimating the irrigation requirement of a potato crop from information on the meteorological, crop, and soil conditions. One such method uses a reference ET and a crop coefficient approach (Burman et al., 1980; Doorenbos & Pruitt, 1977; Jensen, 1974; Wright, 1981, 1982, 1985; see also chapter 15 in this book). The dimensionless ET crop coefficient, K_c, which can be thought of as relative ET, is given by

$$K_c = E_{tc}/E_{tr} \qquad [1]$$

$$E_{tc} = K_c E_{tr} \qquad [2]$$

where K_c represents conditions for a given growth stage and soil moisture condition, E_{tc} is daily crop ET and E_{tr} is daily reference ET. The distribution of K_c with time throughout the season forms an "ET crop coefficient curve" or "ET crop curve." Crop coefficients are typically derived using Eq. [1] from scientific data while Eq. [2] is used to estimate crop ET for practical purposes when applicable crop coefficients and reference ET data are available.

For estimating daily crop ET, as for irrigation scheduling, the K_c of Eq. [2] can be estimated by:

$$K_c = K_{cb} K_a + K_s \qquad [3]$$

where K_{cb} is a basal crop coefficient, K_a is a dimensionless coefficient dependent on available soil water, and K_s is a coefficient to adjust for increased evaporation from wet surface soil. The basal crop coefficient includes the transpiration component of ET and the small amount of evaporation occurring when the soil surface is visually dry (Wright, 1982). When soil water is not limiting, $K_a = 1$, which is usually the case for irrigated potato; otherwise it is <1. When the soil surface is dry, $K_s = 0$, and when wet, $K_s > 0$, so that $K_c > K_{cb}$ (Jensen et al., 1971; Wright, 1982, 1985).

Several methods are available for estimating E_{tr} depending on data availability and local circumstances (Jensen, 1974; Burman et al., 1980, 1983; Doorenbos & Pruitt, 1977; see also chapter 15 of this book). Alfalfa E_{tr} is often used for arid climates (Jensen et al., 1971; Wright, 1981, 1982; Wright

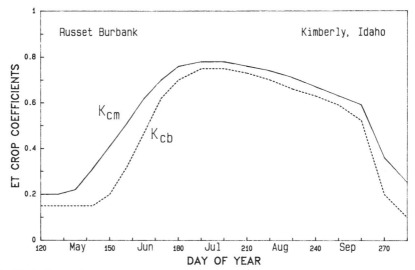

Fig. 29–4. Seasonal ET crop curves of daily basal (K_{cb}) and daily mean (K_{cm}) crop coefficients, derived for use with alfalfa reference ET, for Russet Burbank potato. Adapted from Wright (1981, 1982).

& Jensen, 1972, 1978), while grass reference ET, frequently denoted as E_{to}, and evaporation pan data are also used (Doorenbos & Pruitt, 1977) in the development of crop coefficients.

A mean crop coefficient is sometimes more useful than a basal coefficient to estimate K_c when it is impractical to estimate wet soil effects. The mean crop coefficient, K_{cm}, is > K_{cb} by an amount that depends on the frequency of soil wetting (Wright, 1981). If soil water is not limiting, $K_c = K_{cm}$.

Crop curves of K_{cm} and K_{cb} for sprinkler-irrigated potato are shown in Fig. 29–4. These crop curves were developed from the data of Fig. 29–2 and alfalfa E_{tr} using Eq. [1]. The derived crop coefficients were maximum shortly after the rows closed at EFC. In this case, EFC occurred about 10 July, or 75 d after planting. The curves gradually declined with time after EFC because of plant lodging and natural senescence (Wright, 1982). The crop in the case of Fig. 29–4 remained reasonably healthy, vigorous, and green until vine-killing frosts occurred in mid-September. Time from EFC to harvest was 90 d. The maximum K_{cm} for potato of 0.78 is less than the corresponding K_{cm} of 1.00 for alfalfa, small grains, and sugarbeet, and a K_{cm} of about 0.95 for corn and bean (*Phaseolus vulgaris* L.) (Wright, 1981, 1982). This indicates that with potato the transpiration rates are naturally restricted by the morphological and physiological characteristics of the crop (Enrödi & Rijtema, 1969; Van Loon, 1981). Hane and Pumphrey (1984) found the evaporation pan coefficient for potato to increase from 0.3 at emergence to about 0.8 at maximum leaf area.

For convenience in using the crop coefficients in irrigation scheduling or other procedures, K_{cm} and K_{cb} for potato are listed in Table 29–1, with

Table 29-1. Basal, K_{cb}, and mean, K_{cm}, ET crop coefficients based on alfalfa E_{tr} for irrigated potato grown in an arid temperate region. After Wright (1981, 1982).

Crop coefficient	Percentage time from planting to EFC†										
	0	10	20	30	40	50	60	70	80	90	100
K_{cb}	0.15	0.15	0.15	0.15	0.15	0.20	0.32	0.47	0.62	0.70	0.75
K_{cm}	0.20	0.20	0.20	0.22	0.31	0.41	0.51	0.62	0.70	0.76	0.78
	Elapsed time after EFC, d										
	0	10	20	30	40	50	60	70	80	90	100
K_{cb}	0.75	0.75	0.73	0.70	0.66	0.63	0.59	0.52	0.20	0.10	0.10
K_{cm}	0.78	0.78	0.76	0.74	0.71	0.67	0.63	0.59	0.36	0.25	0.20

† Effective full cover.

time from planting to EFC expressed as a normalized time scale and time after EFC until harvest expressed as elapsed days. Percentage of time during crop development helps adjust for yearly climatic variations because large differences in planting dates usually have little influence on the date of effective full cover for a given area. At Kimberly, ID, K_{cb} for a visibly dry soil surface in April averaged about 0.2, reflecting the combined effects of moisture carry-over from winter precipitation and relatively low evaporative demand. After tillage and with higher evaporative demand in May, K_{cb} decreased to 0.1 or less before crop emergence.

The crop coefficients of Fig. 29-4 and Table 29-1 should be usable in estimating potato ET in areas with a climate similar to that of the area for which they were developed. They should also be usable in other areas with similar climates if verified procedures are used for the reference ET, or if correction factors are used to adjust this reference. The crop curve can be shifted to account for variation in crop development for different locations and years if some crop development characteristics are monitored, such as date of emergence, beginning of rapid growth, and full canopy indicated by the closing of rows. When soil, disease, insect, and other crop and soil management factors cause unusual crop development, crop coefficient curves should be adjusted accordingly. In cases where coarse-textured soils are irrigated frequently, as is possible with center-pivot systems, and the soil surface is moist much of the time, the potato K_{cm} will average about 0.8 throughout the season since the soil evaporation component is high.

C. Evapotranspiration-Yield Relationships

The water use, yield, and *water use efficiency* (WUE), defined here as the ratio of tuber yield to ET, of irrigated potato varies according to climatic and local growing conditions. Results of several studies are summarized in Table 29-2. The yield data are expressed on the basis of total fresh tuber yield and/or No. 1 marketable tuber yield. The ET data for studies 9 and 10 were obtained with weighing lysimeters. All other water use data were calculated from soil water or irrigation data. The data of Table 29-2 show seasonal water use ranges from about 450 mm in central Wisconsin, to 700

Table 29–2. Summary of seasonal water use efficiencies of sprinkler-irrigated potato based on fresh tuber yields and seasonal water use for several locations.

Study	Location	Soil†	Cultivar‡	Water use	Yield		WUE	
				mm	Total	No. 1	Total	No. 1
					— Mg ha⁻¹ —		kg ha⁻¹ per mm	
Hane and Pumphrey, 1984	Hermiston, OR	ls	R.B.	640	--	42	--	66
Hang and Miller, 1986	Patterson, WA	ls	R.B.	700§	70	60	100	86
Hill et al., 1985	Kaysville, UT	l	R.B.	650	51	19	79	29
Hill et al., 1985	Kaysville, UT	l	L.R.	700	51	43	73	62
Hill et al., 1985	Kaysville, UT	l	Ken.	640	57	43	89	68
Hill et al., 1985	Kimberly, ID	sl	R.B.	527	63	46	120	88
Hill et al., 1985	Kimberly, ID	sl	L.R.	532	56	45	106	84
Hill et al., 1985	Kimberly, ID	sl	Ken.	585	68	60	116	103
Hill et al., 1985	Logan, UT	sl	R.B.	548	53	32	97	59
Hill et al., 1985	Logan, UT	sl	L.R.	550	47	41	86	75
Hill et al., 1985	Logan, UT	sl	Ken.	555	56	52	101	94
Sammis, 1980	Las Cruces, NM	cl	Ken.	606	33	--	54	--
Shalhevet et al., 1983	Negev, Israel	T.C.¶	Des.	800	72	62	85	78
Tanner, 1981	Wisconsin	ls	R.B.	450§	50§	--	111	--
J.L. Wright, 1972, unpublished data	Kimberly, ID	sl	R.B.	604	62	50	102	83
Wolfe et al., 1983	Davis, CA	l	Ken.	610	--	36	--	59
Wolfe et al., 1983	Davis, CA	l	W.R.	630	--	34	--	54

† ls = loamy sand; l = loam; sl = sandy loam; cl = clay loam.
‡ R.B. = Russet Burbank; L.R. = Lemhi Russet; Ken. = Kennebec; Des. = Desiree; W.R. = White Rose.
§ Estimated, see text.
¶ Typic Camborthid.

mm in the arid U.S. Pacific Northwest and to 800 mm in Israel. The results show a greater variation in yield than in water use. Yield and water use data for the listed locations indicate that the WUEs for conditions favoring maximum yields range from about 50 to 100 kg ha^{-1} per mm. This gives a water requirement ranging from about 0.50 to 1.0 m^3 kg^{-1} of tuber dry matter. Variations in WUE among locations and years occur for other crops as well (Hanks, 1983). Other results obtained in the 10 studies of Table 29–2 follow.

Potato grown on sandy soil requires irrigation at rates approaching estimated ET if reductions in yield and quality are to be avoided (Hang & Miller, 1986). Hane and Pumphrey (1984) (Study No. 1) found that the yield (Y) of No. 1 and 2 tubers, as related to water use (W), was represented by the production function:

$$Y = 0.142\ W - 28.7 \tag{4}$$

where Y is in megagrams per hectare and W is in millimeters, applicable over the range of 300 to 650 mm. In their study, water application in excess of 650 mm depressed yields of No. 1 tubers according to:

$$Y = 77.0 - 0.015\ W. \tag{5}$$

The 700-mm ET value for Study No. 2 was based on the reported ET of 550 mm for June through August and an estimated additional 150 mm ET that would have occurred before and after the reported period. Miller and Martin (1987a) studied the yield of Russet Burbank, Nooksack, and Lemhi Russet in response to irrigation treatment for 2 yr at the same site as the study of Hang and Miller and obtained mean tuber yields of 57.9, 40.5, and 49.3 Mg ha^{-1}, respectively, for the three cultivars.

Yields listed for the studies of Hill et al. (1985)—Studies No. 3, 4, and 5—were selected for the highest yielding plots obtained with the line source sprinkler scheme. Evapotranspiration was calculated by water balance using neutron meter data for the entire season from time of planting at the end of April, until vine kill, at the end of September. The ET data of Sammis (1980), Study No. 6, were obtained with nonweighing lysimeters that were monitored with a neutron probe. A Kennebec potato crop was planted in February and harvested in July. Surface, trickle, and subsurface irrigated treatments were included in addition to the sprinkler plots. The surface-irrigated plots had the poorest yields. The subsurface irrigated plots yielded 39.5 Mg ha^{-1} total tubers, and water use was assumed to be the same as for the sprinkler treatments.

The study of Shalhevet et al. (1983), Study No. 7, was conducted in the hot climate of the Negev region of Israel. Irrigations applied with a line source sprinkler were based on Class A pan evaporation. Potato was planted in mid-February and harvested in early July. Maximum yields, similar to those reported in Table 29–2 for sprinkler trials, were obtained with drip irrigation treatments. The linear production functions relating total yield (Y), in megagrams per hectare, to seasonal water application depth (W), in millimeters, for sprinkler irrigation were:

$$Y = 0.119 \ W - 23.5, \hspace{3cm} [6]$$

and for drip irrigation:

$$Y = 0.114 \ W - 12.8. \hspace{3cm} [7]$$

The single production function for both irrigation methods describing the yield of marketable tubers was: $Y = 0.128 \ W - 40.0$, applicable for the range of W equal to 0.3 to 1.0 times E_{pan}.

In Study No. 8, Tanner (1981) reported a 3-yr average ET of 357 mm, for June through August, measured with a weighing lysimeter. The total ET of 450 mm shown in Table 29–2 was based on this 12-wk ET plus estimated additional ET before and after the reported period. Tanner actually reported average yields for the 3-yr study on the basis of total tuber dry matter, 10.8 Mg ha^{-1}. A value of 21% solids (Harris, 1978; Kleinkopf et al., 1987) was assumed to estimate the listed tuber fresh yield of 50 Mg ha^{-1} to provide a comparison with the other listed yields. Based on the 12-wk ET of 357 mm, the WUE would be 30.3 kg tuber dry matter ha^{-1} per mm or 140 kg fresh tubers ha^{-1} per mm. Tanner postulated that, considering vapor pressure deficits, WUE should be higher in humid than in arid regions, even though yields may be higher because of increased irradiance and a longer growing season in arid than in humid regions. In comparison with Tanner's results, the lysimetrically measured ET for 12 wk of Study No. 9 at Kimberly, ID, was 485 mm, giving a total tuber dry matter (assuming 21% tuber solids) WUE of 27.4 kg ha^{-1} per mm. The 12-wk results of Hang and Miller (1986) similarly gave a dry matter WUE of 27.3 kg ha^{-1} per mm. Thus, Tanner's WUE was about 10% greater than those for arid regions; however, he did not report quality data so marketable yields are not compared.

The Kimberly, ID, yield data of Study No. 9 were obtained in the same study as the data of Fig. 29–2 and 29–3. The seasonal total ET of this study probably represented a longer time period than many of the other studies which would tend to lower the calculated WUE.

Wolfe et al. (1983), Study No. 10, also used the line source sprinkler to study the growth and yield response of potato to various applied water levels. Potato was planted in late March and harvested in late July. Water use was determined from neutron meter data. They found that yield increased with increasing ET until the point that there was a slight yield depression in the wettest treatment, possibly resulting from poor soil aeration and N deficiency near the end of the season. The final fresh weight yield of marketable White Rose tubers ranged from 4.7 to 33.9 Mg ha^{-1} and from 14.3 to 35.8 Mg ha^{-1} for Kennebec, over the range of dry to well-watered treatments. White Rose had more defects in tuber shape and lower marketable yields under dry soil conditions than Kennebec. Corresponding total seasonal ET ranged from about 325 to 620 mm.

The tuber yields listed in Table 29–2 generally represent maximum yields obtained in the listed studies. Thus, the calculated WUEs would be near optimum. In comparison, Kleinkopf et al. (1981) reported average tuber yields

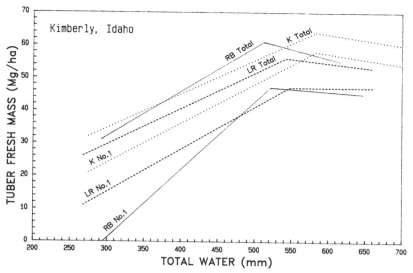

Fig. 29-5. Relation of total and U.S. No. 1 tuber fresh yield to total water use (ET plus excess irrigation) for Russet Burbank (RB), Kennebec (K), and Lemhi Russet (LR) potato. Adapted from Hill et al. (1985).

of 43.1 Mg ha^{-1} obtained in fertility trials in southern Idaho with Russet Burbank. Cho and Iritani (1983) reported commercial yields of 100 Mg ha^{-1} in the Columbia Basin of Washington obtained from fields with 48-ha irrigation circles.

An example of the relationship of tuber yield to total water use (Study No. 4) is presented in Fig. 29-5 for three potato cultivars irrigated with a line source sprinkler system (Hill et al., 1985). Total water refers to ET plus excess irrigation applied after the point of maximum yield. Data are presented for both total and No. 1 yields. The slopes for Kennebec and Lemhi Russet were similar, while Russet Burbank showed the greatest response to irrigation level. Yield was depressed in each case when total water exceeded maximum ET, except for No. 1 Lemhi Russet. Similar depressions were also found in Studies No. 1, 2, 3, 5, 7, and 10 of Table 29-2.

D. Effects of Irrigation Method on Water Use

The effect of irrigation system type on ET is primarily related to the duration of surface soil wetness, especially prior to maximum shading of the soil by the crop. Some of the apparent differences in water requirements between irrigation methods are not due as much to the differences in actual ET but more to the differences in irrigation uniformity, application efficiency, and the drainage of applied water from the root zone. As discussed relative to the ET crop coefficient in section III.B of this chapter, once a complete canopy develops, ET is mainly transpiration which is relatively unaffected by irrigation method. Tanner (1981) found estimated transpiration of sprinkler-irrigated potato, for June through August, to be about 75% of ET in Wisconsin.

Frequent irrigations are required to grow potato on soils with low available water-holding capacities. With sprinkler systems, the surface is wet much of the time and direct evaporation from the soil is higher than with less frequent irrigations. This decreases the evaporative demand on the young plants, and thus no doubt lowers transpiration below what it would be with a dry surface soil, but the net effect is higher ET. With daily light sprinkler irrigations, the K_{cm} of Fig. 29–4 could be about 0.8 from 10 d prior to crop emergence until 10 d prior to vine kill. If so, total water (Fig. 29–5) would increase from about 600 to 750 mm, or about a 25% increase over the less frequent schedule used for the data of Fig. 29–2.

While drip irrigation would seem to provide for decreased soil evaporation, Shalhevet et al. (1983) found no difference in water use between irrigating with a drip system and with sprinklers. Sammis (1980) reported higher WUEs for subsurface irrigation than for sprinkler, trickle, or furrow irrigation. However, the advantage may have resulted from a closer matching of irrigation demand obtained with the subsurface system than with the other systems.

IV. WATER STRESS RESPONSES

Potato response to plant water stress is more complex than for most field crops. While it was previously thought that only modest vegetative and tuber yield depressions resulted from water stress (Corey & Myers, 1955; Robins & Domingo, 1956), potato is now generally considered to be quite sensitive to drought (Harris, 1978; Van Loon, 1981). Many studies have shown that for some drought-sensitive cultivars, even short periods of water stress can cause significant reductions in tuber yield and quality (de Lis et al., 1964; Miller & Martin, 1987a; Nichols & Ruf, 1967; Ruf, 1964; Singh, 1969). The effects of inadequate irrigation are more pronounced on a sandy than a loamy soil (Miller & Martin, 1983). The limited drought tolerance of potato may be due to the effects of a relatively shallow, inefficient root system and the tendency for the stomata to close and expansive growth to decrease in response to mild water deficits (Harris, 1978; Kleinkopf, 1982; Van Loon, 1981).

A. Physiological Effects of Water Stress

Potato stomata begin to close when leaf water potential (Ψ_ℓ) drops to approximately -0.8 to -1.0 MPa (Ackerson et al., 1977a; Campbell et al., 1976; Shimshi et al., 1983). Comparable "threshold" water potentials for field crops such as cotton (*Gossypium hirsutum* L.), wheat, and soybean [*Glycine max* (L.) Merr.] are lower (Carlson et al., 1979; Frank et al., 1973; Thomas et al., 1976). Under optimal soil moisture conditions, potato Ψ_ℓ

usually does not decrease appreciably below -1.0 MPa and stomatal conductance is related primarily to irradiance (Dwelle et al., 1983; Stark, 1987). However, when mild soil water deficits develop, potato stomatal conductance decreases to the extent necessary to prevent Ψ_ℓ from falling below critical levels. At Ψ_ℓ values below -1.0 MPa, conductance decreases markedly in response to falling leaf and turgor water potentials (Ackerson et al., 1977a). Epstein and Grant (1973) reported that when droughted potato plants were irrigated, stomata of some cultivars did not fully reopen until approximately 48 h later. Similarly, Gandar and Tanner (1976) and Ackerson et al. (1977a) observed a relatively slow recovery rate for potato Ψ_ℓ following stress, in contrast with the rapid recovery observed in more drought-tolerant species such as cotton and sorghum (Ackerson et al., 1977b). Preliminary evidence indicates that some cultivars may have superior drought resistance and yield performance in response to differences in stomatal resistance, leaf water retention, epicuticular wax levels, desiccation tolerance, and root growth (Coleman, 1986).

Drought-induced reductions in potato stomatal conductance and Ψ_ℓ are frequently accompanied by reductions in CO_2 assimilation (Ackerson et al., 1977a; Moorby et al., 1975; Munns & Pearson, 1974; Shimshi et al., 1983). Moorby et al. (1975) attributed such reductions in potato photosynthesis primarily to increases in intracellular resistance to CO_2 transport. They did not observe any effect of water stress on the activity of photosynthetic carboxylating enzymes. In contrast, Ackerson et al. (1977a) found that the activities of ribulose-1,5-bisphosphate (RUBP)-carboxylase and phosphoenolpyruvate (PEP)-carboxylase decreased as Ψ_ℓ decreased in response to developing water stress.

Water stress also reduces photoassimilate transport from leaves to tubers (Moorby et al., 1975; Munns & Pearson, 1974). Low leaf water potentials were accompanied by a decrease in assimilate translocation, proportional to the net photosynthesis decline. The results of these studies suggest that there is no direct effect of water stress on the transport mechanism or assimilate distribution throughout the various plant parts. Therefore, tuber growth reductions during drought can be attributed primarily to reductions in photosynthesis.

One of the first physiological processes affected by water stress is the expansive growth of leaves, stems, and roots (Boyer, 1970; Hsiao, 1973). Gandar and Tanner (1976) reported that under greenhouse conditions, potato leaf elongation began to decrease when Ψ_ℓ decreased to approximately -0.3 MPa and stopped completely at -0.5 MPa. Tuber expansion ceased at tuber water potentials of about -0.5 MPa and Ψ_ℓ of -0.7 MPa. By comparison, measured Ψ_ℓ values for irrigated, field-grown potato at midday are usually between -0.7 and -1.1 MPa (Shimshi et al., 1983; Wolfe et al., 1983). Although relationships between water potential and expansive growth likely differ between field and greenhouse conditions, the results of Gandar and Tanner do illustrate the relatively high sensitivity of potato expansive growth to small changes in plant water status.

Kleinkopf (1982) suggested that the tuber may act as a water storage reservoir for the potato plant during periods of drought. As water deficits develop, declining Ψ_ℓ would cause water to move from the tubers to the vines and leaves, thereby reducing or stopping tuber growth. Upon rewatering, tuber growth would resume but the disruption in the normal expansion pattern may result in tuber malformations and secondary growth.

B. Stress Effects on Tuber Growth

Water stress during tuber initiation and early development (growth stage II) generally has the greatest effect on tuber quality, especially with Russet Burbank (de Lis et al., 1964; Miller & Martin, 1987a; Robins & Domingo, 1956). Water deficits interrupt normal patterns of tuber enlargement, increasing the incidence of secondary growth, pointed ends, growth cracks and rough, knobby tubers (Nichols & Ruf, 1967; Thornton & Sieczka, 1980). Dry soil conditions during growth stage II can delay tuber initiation (Bradley & Pratt, 1955) and reduce the number of tubers per plant (Steckel & Gray, 1979).

During tuber bulking (growth stage III), water stress usually affects total yield more than tuber quality (Larsen, 1984; Miller & Martin, 1987a). A large, photosynthetically active leaf surface is necessary to maintain high tuber bulking rates for extended periods (Moorby & Milthorpe, 1975). Water stress can reduce tuber bulking rates by reducing leaf area and photosynthetic rates (Van Loon, 1981). Maintaining high tuber bulking rates also requires the regular development of new leaves to replace older, less efficient ones. Water stress hastens senescence of potato plants and inhibits the new leaf formation (Munns & Pearson, 1974; vander Zaag & Burton, 1978), thereby reducing leaf area duration and yield. Hang and Miller (1986) observed that drought stress early in the season sufficient to reduce plant size caused a tuber yield loss that could not be recovered.

Water stress also affects internal tuber quality, bruising, and tuber diseases. Stress during early tuber development can cause "translucent-end" or "sugar-end" tubers (Iritani & Weller, 1980). These have low starch and high sugar content in the basal (stem) end compared with the apical end. The sugar end fries darker than the other end, so the tubers are undesirable for processing. In contrast, the incidence of brown center and hollow heart can be increased by excessive early season irrigation (Hiller & Koller, 1984), particularly when soil temperatures are cool.

Water stress often reduces dry matter content or specific gravity as well as chipping and processing quality of tubers (Motes & Greig, 1970; Miller & Martin, 1987a). There is a direct relationship between the length of the tuber bulking period and high yields of high specific gravity potato (Thornton & Sieczka, 1980). Consequently, any stress-induced reduction in the length of the tuber bulking period or bulking rate can reduce specific gravity.

The drought hardening of potato is economically less relevant than for many other commercial field crops because of the strong dependence of tuber quality, and thus economic return, on water availability. The drought tolerance of potato varies among the many cultivars (Steckel & Gray, 1979) with the widely grown Russet Burbank being one of the least drought tolerant (Miller & Martin, 1987a). If drought conditions are anticipated, a drought-tolerant cultivar should be selected.

V. GROWTH AND SOIL WATER

Research efforts for many years have centered on determining the optimal soil water content for potato production. The literature indicates some uncertainty concerning minimum water levels, but there is general agreement that for maximum yields of high quality tubers, ASW should never drop below 50% in the zone of maximum root activity on any soil (Singh, 1969) and should preferably be kept well above 50%. Indeed, studies with the drought-sensitive cultivar Russet Burbank indicate that ASW should be maintained above 65% to avoid yield and quality losses (Larsen, 1984).

A. Soil Texture and Available Soil Water

Potato is frequently grown on coarse-textured soils with sprinkler irrigation. However, these soils not only have low water-holding capacities but also often restrict rooting to the plow layer (Kirkham et al., 1974; Lesczynski & Tanner, 1976; Miller & Martin, 1987b) and thus require careful irrigation management. Fulton (1970) established that the minimum soil matric potential permitting maximum yield was -50 kPa at a depth of 15 cm. Van Loon (1981) concluded from a review of available literature that the optimal soil matric potential for potato was between -20 and -60 kPa.

Examples of the estimated available soil water and allowable depletion for potato, for soils of different texture and with typical extension of rooting depths, are listed in Table 29-3. This table summarizes the relationship of available water to soil texture and rooting depth such as given by James et al. (1982, p. 44) and Ratliffe et al. (1983). Allowable depletion was as-

Table 29-3. Estimated allowable soil water depletion, as depth equivalent, for various soils according to estimated effective rooting depth of potato.

Soil texture class[†]	Available soil water	Allowable depletion	Days after planting			
			30	60	90	120
			Estimated rooting depth, cm			
			20	35	50	50
			Allowable root zone depletion			
	mm m^{-1}‡		mm			
s	80	28	6	10	14	14
ls	110	39	8	14	20	20
sl	140	49	10	17	25	25
sil	200	70	14	25	35	35
sicl	180	63	13	22	32	32

† s = sand, ls = loamy sand; sl = sandy loam; sil = silt loam; sicl = silty clay loam.
‡ Equivalent water depth, millimeters of water per meter depth of soil.

sumed to be 35%. The estimated rooting depth assumes the absence of a plow pan, and a rooting depth of 20 cm at planting and of 50 cm by growth stage III.

B. Soil Limitations

Tillage pans in soils that have been farmed for several years can severely restrict rooting depth (Timm & Flocker, 1966). Disrupting the pan on older soils may only increase effective rooting slightly (Van Loon et al., 1985), whereas roots seem to penetrate the subsoil on similar newly farmed land quite readily (Boone et al., 1985). DeRoo and Waggoner (1961) found that soil compaction due to tractor traffic restricted root growth more in a sandy loam than in a loamy sand soil. Miller and Martin (1983, 1985) found that a sandy soil having single grain structure restricted rooting to the upper 30 cm of soil because of hard subsoil. Subsoiling reduced soil strength below the 30-cm depth and promoted deeper rooting and greater withdrawal of water below the hard layer (Miller & Martin, 1987b). Effective rooting was greater in a loam soil than in sand, even without subsoiling. However, deep tillage may not be necessary for maximum yields of irrigated potato with proper land preparation (Buxton & Zalewski, 1983), and benefits from subsoiling may be economically inadequate to justify the practice if a reliable high frequency irrigation system (such as a center pivot or solid set sprinkler) is available (Miller & Martin, 1987b; Ross, 1986). DeRoo and Waggoner (1961) also found that a tillage pan formed in a coarse-textured soil reduced rooting. A complete shattering of the pan promoted deeper and more concentrated rooting below the pan. Timm and Flocker (1966) found that the effects of a compacted soil on water uptake were especially noticeable at soil water potentials < -70 kPa.

Excessive irrigation or poor soil drainage can interfere with tuber development because of reduced aeration. Potato tubers may be more sensitive to O_2 stress than the roots (Holder & Cary, 1984). The tubers create a relatively large O_2 sink and most of the O_2 must enter through the tuber lenticels. As a tuber grows, it may compact the soil around it, restricting aeration. Cary (1985) suggested that coarse-textured soils produce higher quality tubers than fine-textured soils because of better soil aeration. He studied the relative importance of lenticel spacing and soil compaction around tubers and concluded that reduced aeration was probably not a problem in well-drained loamy sand or porous silt loams, provided the bulk densities did not exceed 1.6 or 1.7 g cm^{-3}, but may be a problem in a silty clay.

VI. PLANT NUTRITION

Potato requires an ample nutrient supply to ensure rapid, steady growth and normal tuber development. Maximum tuber yields occur when nutrient and water levels are sufficient to maintain an active plant canopy throughout the season until normal maturation (Westermann & Kleinkopf, 1985b).

The maintenance of an active plant canopy and high tuber yields are strongly correlated with plant nutrient concentrations, uptake rates, and soil water availability. Thus, nutrient requirements are closely related to irrigation levels in arid climates (Reddy & Sastry, 1982). Monitoring plant growth and fertilizer uptake rates with fertilization and irrigation according to potato crop needs is essential to obtaining high fertilizer and water use efficiencies.

Nitrogen fertilizer is required on most soils to produce a profitable potato yield. The amount needed depends on the soil characteristics, crop residues, residual N levels in the soil, and specific crop management practices (Westermann & Kleinkopf, 1985a). Some N is necessary for important early tuber growth, but too much available soil N at planting can delay tuber set and increase the number of small tubers (Kleinkopf et al., 1981). In the past, fertilizer N was broadcast and incorporated into the soil before planting or banded at planting. Recent research results and grower experience have shown that in sprinkler-irrigated regions only part of the total N requirement need be applied at planting, while the remainder can be applied through the irrigation system during the season.

Scheduling N fertilizer applications with irrigation throughout the season allows growers to adjust their fertilization according to crop growth rates, local soil conditions, and the length of the growing season. Westermann and Kleinkopf (1985a) found with Russet Burbank potato grown on a silt loam soil in southern Idaho, that maximum early tuber growth occurred when the plant tops and roots contained between 80 and 140 kg of N ha^{-1}. Sufficient N was available to sustain this level when the soil NO_3-N concentration was >7.5 mg kg^{-1} (0.46-m soil depth), corresponding to 15 000 mg kg^{-1} of NO_3-N in the fourth petiole from the tip. Westermann et al. (1988) found further with field-grown Russet Burbank potatoes that N fertilizer efficiency increased when N fertilizer applications were split between preplant application and application during tuber growth, according to crop needs, as opposed to a single preplant application. Saffigna and Keeney (1977b) and Saffigna et al. (1977b) found in Wisconsin with Russet Burbank potato on a loamy sand soil, that N fertilizer efficiency was markedly increased with frequent N applications through a sprinkler system compared with conventional fertilizer practices. Lauer (1985, 1986) obtained fertilizer N use efficiencies of 90% with high-frequency sprinkler-applied N fertilizer on a sandy soil in Washington.

Phosphorus and K fertilizers are frequently needed on many soils. Soil analysis can be used to determine P and K requirements for potato (Painter et al., 1977), although subsequent monitoring of plant tissue concentrations is also useful (Lorenz et al., 1964; Roberts & Dow, 1982). As with N, P relationships in potato plants can be used to predict the petiole-soluble P concentration for the remainder of the growing season, following sampling, permitting prediction of when supplemental P fertilizer needs to be applied (Westermann & Kleinkopf, 1985b). Micronutrients such as Zn, Fe, B, and Mn may also be required, as indicated by soil analysis and plant tissue testing.

VII. IRRIGATION MANAGEMENT

In the past, growers have principally irrigated from experience, using either a schedule based on the calendar or visual observations of crop and soil water status. While this approach has served skilled irrigators well for years, it is prone to problems, especially with potato. Growers frequently irrigate excessively to prevent the detrimental effects of deficit irrigation. Excess irrigation usually results from applying too much water at a given irrigation rather than from irrigating too frequently. As indicated by the aforementioned yield vs. irrigation relationships, both under- and overirrigation reduce marketable yields. Overirrigation increases fertilizer requirements to compensate for N leached from the root zone and also increases pumping energy costs, wastes water, and thus, generally reduces the profit potential. Excess irrigation increases the movement of nitrates into groundwater or waste ditches (Saffigna & Keeney, 1977a).

The grower receives the best price for top-grade tubers and a much lower price for low grade ones. Thus, the irrigation management scheme needs to be oriented towards maximizing the percentage of top-grade tubers. This requires irrigation management to maintain optimum soil water contents meeting the crop water requirements. Such a management program includes (i) regular quantitative monitoring of soil water contents, (ii) scheduling irrigations according to crop water use and soil water holding capacity, and (iii) a water supply and irrigation system capable of providing the needed irrigation on schedule.

Furrow irrigation of potato is being largely replaced by sprinklers. Furrow irrigation can be used successfully under certain conditions, but sprinkler methods allow greater control of the amount of water applied, can apply small amounts frequently, and give uniform wetting of the hills early in the season while the root system is small. Sprinkler irrigation provides some cooling of the crop and soil, and can be used to apply fertilizer and other chemicals. Solid set, wheel line, center pivot, or linear move systems permit timely irrigation scheduling and minimize labor requirements. Drip irrigation is used to a limited extent, particularly in areas like Israel (Shalhevet et al., 1983), where water is in short supply. However, this technology is relatively new and quite expensive. Where drip and subsurface irrigation have been used with potato, results have been satisfactory (Dale, 1986; Malamud & Or, 1986; Sammis, 1980).

A. Determining Irrigation Needs

Only some of the available methods for determining irrigation needs are suitable for potato because of the critical threshold levels of available soil water and the limited root zone of the crop. Suitable methods of quantitatively monitoring soil water status include gravimetric sampling, tensiometers, and the neutron meter. These methods can be used successfully, but are labor intensive and require training and experience that are most often found in a consulting firm, a service company, or a large commercial farm operation to interpret the results and schedule irrigations. Campbell and Campbell (1982) concluded, from their experience, that the neutron probe offers the best combination of features for monitoring the soil water content.

Tensiometers have been used successfully to monitor soil water in potato fields (Flocker & Timm, 1966; Dubetz & Krogman, 1973) and are more applicable to potato production than to some other crops because of the need to keep the soil relatively wet and, thus, within the tensiometer range of from 0 to about -80 kPa. However, tensiometer placement is important, and poor results can occur with sandy soils because of poor conductance between the soil and the tensiometer cup. A useful practice is to install tensiometers in the hill at various depths, such as 20 and 50 cm below mean soil level, and to use them in sequence as the root system develops. The soil water potential corresponding to the allowable depletion of available soil water is approximately -30 kPa for sandy soils, -35 to -50 kPa for sandy loam soils, -55 kPa for silt loams, and -65 kPa for silty clay loams.

Visual appearance of the potato crop is usually not satisfactory for irrigation scheduling (Dubetz & Krogman, 1973), in part because the potato plant does not exhibit wilt symptoms as readily as do some other crops (Nelson & Hwang, 1975). Jackson (1982) suggested that foliage temperature may not be an adequately sensitive indicator of potato irrigation requirements because of the need to maintain relatively high soil moisture levels throughout the growing season. The results of Stark and Wright (1985) confirmed this. They found that, while infrared measurements of potato foliage temperature can be used to detect and quantify moderate to severe water deficits, the soil water potential must decrease below -70 kPa before any appreciable change occurs in the relationship between leaf temperature and relative humidity. Differences in the water potential of the leaf between nonstressed and moderately stressed potato also appeared to be too small to provide useful information for irrigators.

The use of climatic-based methods to determine irrigation needs involves determining the soil water balance from ET estimates obtained from meteorological data. The reference ET-crop coefficient concept was previously discussed (see section III.B of this chapter). Another method involves estimating water evaporation from a standard-size open pan with a pan factor used to adjust the pan evaporation to a reference ET. The data from evaporation pans are highly dependent on the surroundings of the pan and evaporation must be adjusted by means of a pan coefficient to estimate the reference ET (Doorenbos & Pruitt, 1977). The basic ET method of Jensen et al. (1971) uses a water budget computed from estimates of ET, drainage, and applied water to determine when irrigation is needed. Occasional measurements of soil water help assure that cumulative errors in the ET-climatological method do not lead to either crop stress or over-irrigation (Jensen & Wright, 1978).

B. Irrigation Scheduling

Profitable management of irrigated potato requires skill and the best-known management practices. Because of the tendency to over-irrigate potato, correct irrigation scheduling has the potential of increasing marketable yield while reducing irrigation costs. Other benefits include the conservation of water, energy, and N fertilizer, and reduction of groundwater contamination. Factors to consider in selecting an irrigation scheduling program for potato are water application method and irrigation system flexibility with respect to timing and amount.

Specific irrigation scheduling guidelines depend on soil types, general climatic conditions, cultivars commonly grown, and source of irrigation water. However, some guidelines apply to general potato production. Evidence strongly supports the need to maintain uniform and relatively high soil water levels. The ideal is to avoid large fluctuations in available soil water and to prevent excessive drainage. The soil profile should be near field capacity at planting and additional water should be applied in frequent light amounts during the growing season (Thornton & Sieczka, 1980). It is recommended to reduce soil water to 50% ASW by vine kill to promote maturing of the skin and closure of the lenticels.

An illustration of soil water depletion throughout the growing season in response to an irrigation scheme is presented in Fig. 29–6. The indicated allowable depletion was based on data of Table 29–3 for a silt loam soil. The actual potato ET and the irrigation schedule were as given in Fig. 29–2. At each irrigation, the soil water content was assumed to return to zero depletion.

Since the introduction of meteorologically based computerized irrigation scheduling procedures (Jensen et al., 1971), several practical, scientifically based irrigation scheduling programs and techniques have been developed

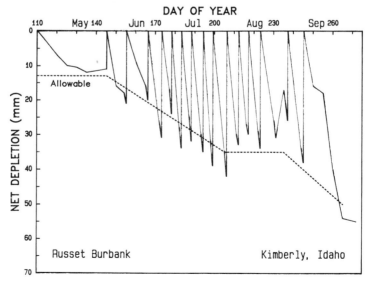

Fig. 29–6. Seasonal progression of net soil water depletion (ET − rainfall) for sprinkler-irrigated Russet Burbank potato in relation to estimated allowable depletion for a silt loam soil. The ET was measured with a weighing lysimeter.

and tested for potato (Curwen & Massie, 1984; Larsen, 1984; Vitosh, 1984). The programs range in complexity from the simple, hand-calculated water balance sheet to the more sophisticated mainframe computer and microcomputer programs. The program of choice depends on access to computers or scheduling services. Scheduling programs often use the potato crop coefficient data of Table 29-1 or those of Doorenbos and Pruitt (1977) or Jensen (1974). Several research trials have shown that scheduling irrigations on the basis of ET estimates, as compared with typical grower scheduling, results in a 40% decrease in irrigation amount while maintaining potato yield and quality (Curwen & Massie, 1984).

C. Scheduling for Limited Irrigation

Since irrigation scheduling for potato is aimed at maintaining optimum soil moisture regimes throughout the period of crop growth, the irrigation scheme should usually be directed to maximizing yield rather than spreading a limited water supply over a large area. Most irrigated potato is grown where water is adequate. If water resources are limited, it is advisable to schedule irrigations to cover the most critical early season periods and to select cultivars that are least sensitive to water stress (Hukkeri & Sharma, 1979; Martin & Miller, 1987). In scheduling for limited irrigation, some stress can be tolerated during early vegetative growth and late tuber bulking. Stress effects on yield depend on ET rate, soil type, crop growth stage, and cultivar. Larsen (1984) defined mild stress for a medium-textured soil as not irrigating for 10 d after 35% available water has been depleted, while severe stress was defined as a 15-d delay. A few observations indicated mild stress in the early, middle, and late growth stages reduced yield of U.S. No. 1's by 25, 20, and 0%, respectively, for the sensitive Russet Burbank variety. Severe stress reduced yields about 60, 35, and 30%, respectively. With a sandy soil, equivalent yield reductions would result from delays of only a few days.

D. Integrating with Other Management Practices

Because of the sensitivity of potato to high temperature and water stress, production in regions with hot climates presents special problems. In such climates, potato is usually planted to grow during the coolest months. Frequent, light sprinkler irrigations can provide some crop cooling. Burgers and Nel (1984) considered the benefits of irrigation cooling in a hot climate and concluded that these may explain part of the need to maintain relatively high water levels. They found that the cooling benefits could be obtained much more economically with a straw mulch applied after planting and hilling.

Knowledge of the nutritional needs of the potato crop and the development of monitoring techniques to determine the plant nutritional status now permits the scheduling of N and other fertilizer applications on almost the same basis as irrigation (see section V of this chapter). Irrigation systems are also sometimes used to apply pesticides to potato. Wyman et al. (1986)

concluded that the use of a center pivot irrigation system as a total pesticide delivery system on potato is an attractive alternative to ground rig or aerial applications.

On many soils, the instantaneous application rate by sprinkler irrigation with a center pivot or linear move system exceeds the infiltration rate, and runoff occurs from the application area. The shedding of water by the potato plants prior to full canopy cover accentuates this problem (Curwen & Massie, 1984; Saffigna et al., 1977b). In recent years, implements that produce small dikes or form small reservoirs between the rows have been used to reduce runoff. This practice, called *basin tillage*, is especially useful for irrigated potato on sloping fields (Aarstad & Miller, 1973; Garvin et al., 1986).

Since the planting and hilling operations require some tillage, potato is almost never grown on a no-till basis. Minimum tillage is sometimes practiced especially where sandy soils are subject to severe wind erosion. In such cases, grain stubble from a previous crop is left on the surface and potato is planted directly into disked or standing stubble.

VIII. OTHER FACTORS

Efforts to reduce potato water requirements through reduced transpiration presently seem impractical. This is largely because of the close relationship between the exchange of CO_2 between the atmosphere and plant leaves, in consequence of photosynthesis, and the associated loss of water through open stomata. Burton (1981) concluded from available literature that tuber yields of 90 Mg ha^{-1} would require the maximum possible gross production of dry matter on a clear day of about 480 kg ha^{-1}, which would in turn require a net photosynthetic assimilation of about 780 kg ha^{-1} of CO_2, all of which must diffuse into the leaves through open stomata. Burton concluded that this CO_2 influx would be accompanied by an inevitable water vapor loss in a dry climate, such as Idaho, of about 100 000 kg ha^{-1} per d, equivalent to 10 mm of rain or irrigation.

Potato is one of the most sensitive crops to soil salinity. Consequently, potato is usually not grown on soils or with irrigation water of high salinity. The threshold salinity level for potato, above which relative productivity is decreased, is an electrical conductivity of the saturated soil extract of 1.7 dS m^{-1} with 12% productivity decrease per dS m^{-1} increase in salinity (Carter, 1982).

The relatively high fertility levels required to produce near-maximum potato yields means that the soil solution is going to have relatively high levels of soluble fertilizer elements, with a corresponding potential for leaching. Kirkham et al. (1974) noted that a soil solution concentration of >50 g Mg^{-1} NO$_3$-N was required to maintain sufficient N for optimum potato growth. Consequently, the soil percolate during the growing season, in such production regions as central Wisconsin, may never contain less than the 10 g Mg^{-1} NO$_3$-N health limit. Therefore, avoiding excessive irrigation is particularly important with potato to prevent groundwater contamination. In humid and subhumid areas, it is important during irrigation to leave some soil water storage available for rain, since rain soon after an irrigation may otherwise cause leaching. This is less of a concern in arid, low-rainfall areas.

IX. SUMMARY

Proper irrigation management to provide a uniform and adequate supply of available soil water within the crop root zone is perhaps the most important factor in obtaining high yields of high-quality tubers in irrigated potato production. Since potato is more sensitive to water stress than most other crops, adequate soil water is necessary to ensure production of well-shaped tubers and avoid tuber disorders that are directly related to water stress.

While potato requires relatively high levels of available soil water, the actual ET of potato is less than that for many field crops. The seasonal ET of sprinkler-irrigated Russet Burbank potato in southern Idaho was 600 mm, as measured with a weighing lysimeter. Daily ET during the peak water use period averaged about 7.6 mm d^{-1}, or about 80% of daily alfalfa reference ET. Total seasonal ET was 57% of alfalfa reference ET and 52% of Class A pan evaporation. Seasonal water use and yield results indicate a greater variation in yield than in ET among locations, leading to a wide range in water use efficiencies.

Water stress effects on potato production are generally more complicated than for other field crops. Water stress during tuber initiation and early development has the greatest detrimental effect on tuber quality. Water stress during tuber bulking affects total yield more than tuber quality. The dry matter percentage, or specific gravity, of tubers is also dependent on soil water availability. Marketable tuber yield in the Pacific Northwest is essentially linearly related to seasonal ET from a minimum ET of about 300 mm, corresponding to zero yield, to a maximum yield at about 650 mm ET. Irrigation levels in excess of maximum ET usually depress the yield of marketable tubers, especially with commercially important cultivars. The relationship between seasonal water use and the irrigation system primarily concerns the effects of a wet soil surface on ET before development of a full crop canopy.

Sprinkler irrigation is presently the most widely used method of irrigating potato, permitting more frequent light irrigations and providing more uniform water distribution in the potato root zone than surface irrigation methods. Irrigations should be scheduled to maintain available soil water at 65% or above, particulary in arid regions with high ET demand. With sandy soils, effective rooting is often restricted by tillage pans to 30 cm. With silt loams, effective rooting may extend to 60 cm or deeper. Inefficiencies in potato irrigation usually result from applying excess water at a given irrigation rather than from irrigating more frequently than necessary.

Fertilizer requirements and practices are closely related to irrigation management in arid climates where irrigation supplies most of the water requirement. Monitoring plant growth and fertilizer uptake is essential to obtaining high water and fertilizer use efficiencies. Fertilizer N can be successfully applied through the sprinkler or drip irrigation systems to meet plant needs.

Potato production is particularly well suited to using an irrigation scheduling program that predicts when to irrigate and the amount to apply. Adopting such programs has reduced irrigation amounts without reducing yields, and has also decreased NO_3 leaching to the ground-water. Most scheduling programs are based on ET estimation through various procedures and the simulation of available soil water within the root zone. Since irrigtion scheduling is aimed at maintaining optimum soil water throughout the period of crop growth, the irrigation scheme should usually be designed to maximize yield per unit land area rather than spreading a limited water supply over a large area.

REFERENCES

Aarstad, J.S., and D.E. Miller. 1973. Soil management to reduce runoff under center-pivot sprinkler systems. J. Soil Water Conserv. 28(4):171–173.

Ackerson, R.C., D.R. Krieg, T.D. Miller, and R.G. Stevens. 1977a. Water relations and physiological activity of potatoes. J. Am. Soc. Hortic. Sci. 102:572–575.

Ackerson, R.C., D.R. Krieg, T.D. Miller, and R.E. Zartman. 1977b. Water relations of field grown cotton and sorghum: Temporal and diurnal changes in leaf water, osmotic, and turgor potentials. Crop Sci. 17:76–80.

Benoit, G.R., and W.J. Grant. 1980. Plant water deficit effects on Aroostock County potato yields over 30 years. Am. Potato J. 57:585–594.

Bishop, J.C., and D.W. Grimes. 1978. Precision tillage effects on potato root and tuber production. Am. Potato J. 55:65–71.

Boone, F.R., L.A.H. DeSmet, and D.D. Van Loon. 1985. The effect of ploughpan in marine loam soils on potato growth. I. Physical properties and rooting patterns. Potato Res. 28:295–314.

Bourgoin, T.L. 1984. Developing a potato irrigation scheduling program—The Maine situation. Am. Potato J. 61:195–203.

Boyer, J.S. 1970. Differing sensitivity of photosynthesis to low leaf water potentials in corn and soybean. Plant Physiol. 46:236–239.

Bradley, G.A., and A.J. Pratt. 1955. The effect of different combinations of soil moisture and nitrogen levels on early plant development and tuber set of the potato. Am. Potato J. 32:254–258.

Burgers, M.S., and P.C. Nel. 1984. Potato irrigation scheduling and straw mulching. South Afr. Tijdskr. Plant Grond. 1:111–116.

Burman, R.D., P.R. Nixon, J.L. Wright, and W.O. Pruitt. 1980. Water requirements. p. 187–232. In M.E. Jensen (ed.) Design and operation of irrigation systems. Monogr. No. 3. ASAE, St. Joseph, MI.

Burton, N.G. 1981. Challenges for stress physiology in potato. Am. Potato J. 58:3–14.

Buxton, D.R., and J.C. Zalewski. 1983. Tillage and cultural management of irrigated potatoes. Agron. J. 75:219–225.

Campbell, G.S., and M.D. Campbell. 1982. Irrigation scheduling using soil moisture measurements: Theory and practice. p. 25–42. In D. Hillel (ed.) Advances in irrigation. Vol. 1. Academic Press, New York.

Campbell, M.D., G.S. Campbell, R. Kunkel, and R.J. Papendick. 1976. A model describing soil-plant-water relations for potatoes. Am. Potato J. 53:431–441.

Carlson, R.E., N.N. Momen, O. Arjmand, and R.H. Shaw. 1979. Leaf conductance and leaf-water potential relationships for two soybean cultivars grown under controlled irrigation. Agron. J. 71:321–325.

Carter, D.L. 1982. Salinity and plant productivity. p. 117–133. *In* M. Rechcigl, Jr. (ed.) Plant productivity. CRC handbook of agricultural productivity. Vol. 1. CRC Press, Boca Raton, FL.

Cary, J.W. 1985. Potato tubers and soil aeration. Agron. J. 77:379–383.

Cho, J.L., and W.M. Iritani. 1983. Comparison of growth and yield parameters of Russet Burbank for a two-year period. Am. Potato J. 60:569–576.

Coleman, W.K. 1986. Water relations of the potato (*Solanum tuberosum* L.) cultivars Raritan and Shepody. Am. Potato J. 63:263–276.

Corey, A.T., and G.R. Blake. 1953. Moisture available to various crops in some New Jersey soils. Soil Sci. Soc. Am. Proc. 17:314–317.

Corey, G.L., and V.I. Myers. 1955. Irrigation of Russet Burbank potatoes in Idaho. Idaho Agric. Exp.Stn. Bull. 246.

Curwen, D., and L.R. Massie. 1984. Potato irrigation scheduling in Wisconsin. Am. Potato J. 61:235–241.

Dale, D. 1986. Drip irrigated spuds. Irrig. Age 21(3):4–5.

DeRoo, H.C., and P.E. Waggoner. 1961. Root development of potatoes. Agron. J. 532:15–17.

de Lis, B.R., I. Ponce, and R. Tizio. 1964. Studies on water requirement of horticultural crops. I. Influence of drought at different growth stages on the tuber's yield. Agron. J. 56:377–381.

Doorenbos, J., and W.O. Pruitt. 1977. Guidelines for predicting crop water requirements. FAO Irrig. Drain. Pap. 24. FAO, Rome.

Dubetz, S., and K.K. Krogman. 1973. Comparison of methods of scheduling irrigations of potatoes. Am. Potato J. 50:408–414.

Durrant, M.J., B.J.G. Love, A.B. Messeen, and A.P. Draycot. 1973. Growth of crop roots in relation to soil moisture extraction. Ann. Appl. Biol. 74:387–394.

Dwelle, R.B., P.J. Hurley, and J.J. Pavek. 1983. Photosynthesis and stomatal conductance of potato clones (*Solanum tuberosum* L.). Plant Physiol. 72:172–176.

Endrödi, G., and P.E. Rijtema. 1969. Calculation of evapotranspiration from potatoes. Neth. J. Agric. Sci. 17:283–299.

Epstein, E., and W.J. Grant. 1973. Water stress relations of the potato plant under field conditions. Agron. J. 65:400–404.

Erie, L.J., O.F. French, and K. Harris. 1965. Consumptive use of water by crops in Arizona. Arizona Agric. Exp. Stn. Bull. 169.

Flocker, W.J., and H. Timm. 1966. Effect of soil moisture tension and physical condition of soil on utilization of water and nutrients by potatoes. Agron. J. 58:290–293.

Frank, A.B., J.F. Power, and W.O. Willis. 1973. Effect of temperature and plant water stress on photosynthesis, diffusion resistance and leaf water potential in spring wheat. Agron. J. 65:777–780.

Fulton, J.M. 1970. Relationship of root extension to the soil moisture level required for maximum yield of potatoes, tomatoes and corn. Can. J. Soil Sci. 50:92–94.

Gandar, P.W., and C.B. Tanner. 1976. Leaf growth, tuber growth and water potential in potatoes. Crop Sci. 16:534–538.

Garvin, P.C., J.R. Busch, and D.C. Kincaid. 1986. Reservoir tillage for reducing runoff and increasing production under sprinkler irrigation. ASAE Pap. 86-2093. ASAE, St. Joseph, MI.

Hane, D.C., and F.V. Pumphrey. 1984. Yield-evapotranspiration relationships and seasonal crop coefficients for frequently irrigated potatoes. Am. Potato J. 61:661–668.

Hang, A.N., and D.E. Miller. 1986. Yield and physiological responses of potatoes to deficit, high frequency sprinkler irrigation. Agron. J. 78:436–440.

Hanks, R.J. 1983. Yield and water-use relationships: An overview. p. 393–411. *In* Limitations to efficient water use in crop production. ASA, CSSA, and SSSA, Madison, WI.

Harris, P.M. (ed.). 1978. The potato crop. The scientific basis for improvement. Chapman and Hall, London.

Hill, R.W., R.J. Hanks, and J.L. Wright. 1985. Crop yield models adapted to irrigation scheduling programs. Utah Agric. Exp. Stn. Res. Rep. 99.

Hiller, L.K., and D.C. Koller. 1984. Effect of early season soil moisture levels and growth regulator applications on internal quality of Russet Burbank potato tubers. Proc. Wash. State Potato Conf. 23:67–73.

Holder, C.B., and J.W. Cary. 1984. Soil oxygen and moisture in relation to Russet Burbank potato yield and quality. Am. Potato J. 61:67–75.

Horton, D., and R.L. Sawyer. 1985. The potato as a world food crop, with special reference to developing areas. p. 1–34. *In* P.H. Li (ed.) Potato physiology. Academic Press, New York.

Hsiao, T.C. 1973. Plant responses to water stress. Annu. Rev. Plant Physiol. 24:519–570.

Hukkeri, S.B., and A.K. Sharma. 1979. Tailoring the irrigation schedule for higher water-use efficiency in potato production. Indian J. Agric. Sci. 49:336–339.

Irrigation Survey Staff. 1986. 1985 irrigation survey. Irrig. J. 36(2):21–28.

Iritani, W.M., and L.D. Weller. 1980. Sugar development in potatoes. Washington State Univ. Ext. Bull. 717.

Jackson, R.D. 1982. Canopy temperature and crop water stress. p. 43–85. *In* D.E. Hillel (ed.) Advances in irrigation. Academic Press, New York.

James, D.W., R.J. Hanks, and J.J. Jurinak. 1982. Modern irrigated soils. John Wiley and Sons, New York.

Jensen, M.E. (ed.). 1974. Consumptive use of water and irrigation water requirements. Report of ASCE committee on irrigation water requirements. Am. Soc. Civ. Eng., New York.

Jensen, M.E., and J.L. Wright. 1978. The role of evapotranspiration models in irrigation scheduling. Trans. ASAE 21(1):82–87.

Jensen, M.E., J.L. Wright, and B.J. Pratt. 1971. Estimating soil moisture depletion from climate, crop and soil data. Trans. ASAE 14(5):954–959.

Kirkham, M.B., D.R. Keeney, and W.R. Gardner. 1974. Uptake of water and labeled nitrate at different depths in the root zone of potato plants grown on a sandy soil. Agro-Ecosystems 1:31–44.

Kleinkopf, G.E. 1982. Potato. p. 287–305. *In* I.D. Teare and M.M. Peet (ed.) Crop water relations. John Wiley and Sons, New York.

Kleinkopf, G.E., D.T. Westermann, and R.B. Dwelle. 1981. Dry matter production and nitrogen utilization by six potato cultivars. Agron. J. 73:799–802.

Kleinkopf, G.E., D.T. Westermann, M.J. Wille, and G.D. Kleinschmidt. 1987. Specific gravity of Russet Burbank potatoes. Am. Potato J. 64:579–587.

Larsen, D.C. 1984. Simplifying potato irrigation scheduling—the Idaho program. Am. Potato J. 61:215–227.

Lauer, D.A. 1985. Nitrogen uptake patterns of potatoes with high-frequency sprinkler-applied N fertilizer. Agron. J. 77:193–197.

Lauer, D.A. 1986. Russet Burbank yield response to sprinkler-applied nitrogen fertilizer. Am. Potato J. 63:61–69.

Lesczynski, D.B., and C.B. Tanner. 1976. Seasonal variation of root distribution of irrigated, field grown Russet Burbank potato. Am. Potato J. 53:69–78.

Li, P.H. (ed.). 1985. Potato physiology. Academic Press, New York.

Lorenz, O.A., K.B. Tyler, and F.S. Fullmer. 1964. Plant analysis for determining the nutritional status of potatoes. p. 226–240. *In* C. Bould et al. (ed.) Plant analysis and fertilizer problems. Am. Soc. Hortic. Sci., Alexandria, VA.

MacKerron, D.K.L., and T.D. Heilbroun. 1985. A method for estimating harvest indices for use in surveys of potato crops. Potato Res. 28:279–282.

Malamud, O.S., and U. Or. 1986. Micro-irrigation potential in potato cultivation. Am. Potato J. 63:442–443.

Martin, M.W., and D.E. Miller. 1987. Advantages of water-stress resistant genotypes in the Northwest. Am. Potato J. 64:448–449.

Miller, D.E., and M.W. Martin. 1983. Effect of daily irrigation rate and soil texture on yield and quality of Russet Burbank potatoes. Am. Potato J. 60:745–757.

Miller, D.E., and M.W. Martin. 1985. Effect of water stress during tuber formation on subsequent growth and internal defects in Russet Burbank potatoes. Am. Potato J. 62:83–89.

Miller, D.E., and M.W. Martin. 1987a. Effect of declining or interrupted irrigation on yield and quality of three potato cultivars grown on sandy soil. Am. Potato J. 64:109–117.

Miller, D.E., and M.W. Martin. 1987b. The effect of irrigation regime and subsoiling on yield and quality of three potato cultivars. Am. Potato J. 64:17–25.

Moorby, J., and F.L. Milthorpe. 1975. Potato. p. 225–257. *In* L.J. Evans (ed.) Crop physiology. Cambridge Univ. Press, London.

Moorby, J., R. Munns, and J. Walcott. 1975. Effect of water deficit on photosynthesis and tuber metabolism in potatoes. Aust. J. Plant Physiol. 2:323–333.

Motes, J.E., and J.K. Greig. 1970. Specific gravity, potato chip color and tuber mineral content as affected by soil moisture and harvest dates. Am. Potato J. 47:413–418.

Munns, R., and C.J. Pearson. 1974. Effect of water deficit on translocation of carbohydrate in *Solanum tuberosum*. Aust. J. Plant Physiol. 1:529–537.

Nelson, S.H., and K.E. Hwang. 1975. Water usage by potato plants at different stages of growth. Am. Potato J. 52:331–339.

Nichols, D.F., and R.H. Ruf. 1967. Relation between moisture stress and potato tuber development. Proc. Am. Soc. Hortic. Sci. 91:443–447.

Nkemdirim, L.C. 1976. Crop development and water loss—A case study over a potato crop. Agric. Meteorol. 16:371–388.

Painter, C.G., J.P. Jones, R.C. McDole, R.D. Johnson, and R.E. Ohms. 1977. Idaho fertilizer guide for potatoes. Univ. Idaho Coop. Ext. Serv. Current Info. Ser. 261.

Plaisted, P.H. 1957. Growth of the potato tuber. Plant Physiol. 32:445–453.

Ratliffe, L.F., J.T. Ritchie, and D.K. Cassel. 1983. Field-measured limits of soil water availability as related to laboratory-measured properties. Soil Sci. Soc. Am. J. 47:770–775.

Reddy, M.S., and V.V.K. Sastry. 1982. Uptake of nitrogen, phosphorus and potassium by potato as influenced by different levels of irrigation. Indian J. Agric. Chem. 15:165–169.

Ritter, W.F., T.H. Williams, and R.W. Scarborough. 1985. Water requirements for corn, soybeans, and vegetables in Delaware. Univ. Delaware Agric. Exp. Stn. Bull. 463.

Roberts, S., and A.I. Dow. 1982. Critical nutrient ranges for petiole phosphorus levels of sprinkler-irrigated Russet Burbank potatoes. Agron. J. 74:583–585.

Robins, J.S., and C.E. Domingo. 1956. Potato yield and tuber shape as affected by severe soil moisture deficits and plant spacing. Agron. J. 48:488–492.

Robins, J.S., J.T. Musick, D.C. Finfrock, and H.F. Rhoades. 1967. Grain and field crops. *In* R.M. Hagan et al. (ed.) Irrigation of agricultural lands. Agronomy 11:262–639.

Ross, C.W. 1986. The effects of subsoiling and irrigation on potato production. Soil Tillage Res. 7:315–325.

Ruf, R.H. 1964. The influence of temperature and moisture stress on tuber malformation and respiration. Am. Potato J. 41:377–381.

Saffigna, P.G., and D.R. Keeney. 1977a. Nitrate and chloride in groundwater under irrigated agriculture in central Wisconsin. Ground Water 15:170–177.

Saffigna, P.G., and D.R. Keeney. 1977b. Nitrogen and chloride uptake by irrigated Russet Burbank potatoes. Agron. J. 69:258–264.

Saffigna, P.G., D.R. Keeney, and C.B. Tanner. 1977a. Nitrogen, chloride, and water balance with irrigated Russet Burbank potatoes in a sandy soil. Agron. J. 69:251–257.

Saffigna, P.G., C.B. Tanner, and D.R. Keeney. 1977b. Non-uniform infiltration under potato canopies caused by interception, stemflow and hilling. Agron. J. 68:337–342.

Sammis, T.W. 1980. Comparison of sprinkler, trickle, subsurface, and furrow irrigation methods for row crops. Agron. J. 72:701–704.

Shalhevet, J., D. Shimshi, and T. Meir. 1983. Potato irrigation requirements in a hot climate using sprinkler and drip methods. Agron. J. 75:13–16.

Shimshi, D., J. Shalhevet, and T. Meir. 1983. Irrigation regime effects on some physiological responses of potato. Agron. J. 75:262–267.

Shimshi, D., and M. Susnoschi. 1985. Growth and yield studies of potato development in a semi-arid region. 3. Effect of water stress and amounts of nitrogen top dressing on physiological indices and on tuber yield and quality of several cultivars. Potato Res. 28:177–191.

Singh, G. 1969. A review of the soil-moisture relationship in potatoes. Am. Potato J. 46:398–403.

Smith, O. (ed.). 1968. Potatoes: Production storing, processing. AUI Publ. Co., Westport, CT.

Stark, J.C. 1987. Stomatal behavior of potatoes under nonlimiting soil water conditions. Am. Potato J. 64:301–309.

Stark, J.C., and J.L. Wright. 1985. Relationship between foliage temperature and water stress in potatoes. Am. Potato J. 62:57–68.

Steckel, J.R.A., and D. Gray. 1979. Drought tolerance in potatoes. J. Agric. Sci. 92:375–381.

Tanner, C.B. 1981. Transpiration efficiency of potato. Agron. J. 73:59–64.

Thomas, J.C., K.W. Brown, and W.R. Jordan. 1976. Stomatal response to leaf water potential as affected by preconditioning water stress in the field. Agron. J. 68:706–708.

Thornton, R.E., and J.B. Sieczka. 1980. Commercial potato production in North America. Am. Potato J. 57:1–36.

Timm, H., and W.J. Flocker. 1966. Responses of potato plants to fertilization and soil moisture tension under induced soil compaction. Agron. J. 58:153–157.

U.S. Department of Agriculture. 1985. Agricultural statistics. U.S. Gov. Print. Office, Washington, DC.

vander Zaag, P.E., and W.G. Burton. 1978. Potential yield of the potato crop and its limitations. p. 7–22. *In* Survey Papers, 7th Triennial Conf. European Assoc. Potato Res., Warsaw, Poland. 26 June–1 July. Instytut Ziemniaka, Bonin, Poland.

Van Loon, C.D. 1981. The effect of water stress on potato growth, development, and yield. Am. Potato J. 58:51–69.

Van Loon, C.D., L.A.H. Smet, and F.R. Boone. 1985. The effect of a ploughpan in marine loam soils on potato growth. 2. Potato plant responses. Potato Res. 28:315–330.

Vitosh, M.L. 1984. Irrigation scheduling for potatoes in Michigan. Am. Potato J. 61:205–213.

Ware, G.W., and J.P. McCollum. 1968. Producing vegetable crops. Interstate Printers and Publishers, Danville, IL.

Weaver, J.E. 1926. Root development of field crops. McGraw-Hill, New York.

Westermann, D.T., and G.E. Kleinkopf. 1985a. Nitrogen requirements of potatoes. Agron. J. 77:616–621.

Westermann, D.T., and G.E. Kleinkopf. 1985b. Phosphorus relationships in potato plants. Agron. J. 77:490–494.

Westermann, D.T., G.E. Kleinkopf, and J.K. Porter. 1988. Nitrogen fertilizer efficiencies on potatoes. Am. Potato J. 65:377–386.

Wolfe, D.W., E. Fereres, and R.E. Voss. 1983. Growth and yield response of two potato cultivars to various levels of applied water. Irrig. Sci. 3:211–222.

Wright, J.L. 1981. Crop coefficients for estimates of daily crop evapotranspiration. p. 18–26. *In* Irrigation scheduling for water and energy conservation in the 80's. ASAE, St. Joseph, MI.

Wright, J.L. 1982. New evapotranspiration crop coefficients. J. Irrig. Drain. Div. Am. Soc. Civ. Eng. 108(IR1):57–74.

Wright, J.L. 1985. Evapotranspiration and irrigation water requirements. p. 105–113. *In* Advances in evapotranspiration. Proc. Natl. Conf. on advances in evapotranspiration, Chicago, IL. 16–17 Dec. ASAE, St. Joseph, MI.

Wright, J.L., and M.E. Jensen. 1972. Peak water requirements of crops in southern Idaho. J. Irrig. Drain. Div. Am. Soc. Civ. Eng. 98(IR2):193–201.

Wright, J.L., and M.E. Jensen. 1978. Development and evaluation of evapotranspiration models for irrigation scheduling. Trans. ASAE 21(1):88–96.

Wyman, J.A., J.F. Walgenbach, W.R. Stevenson, and L.K. Binning. 1986. Comparison of aircraft, ground-rig and center pivot irrigation systems for application of pesticides to potatoes. Am. Potato J. 63:297–314.

30 Turfgrass

ROBERT N. CARROW

University of Georgia
Griffin, Georgia

ROBERT C. SHEARMAN

University of Nebraska
Lincoln, Nebraska

JAMES R. WATSON

The Toro Company
Minneapolis, Minnesota

Turfgrass irrigation with belowground systems is relatively new. In the 1950s, the turfgrass irrigation industry started to expand—first with quick coupler systems and eventually with automatic hydraulic or electrical systems. Growth was most pronounced on golf courses, but by the 1960s, many recreational sites, homelawns, business grounds, and park areas were irrigated (Fig. 30–1).

While turfgrass irrigation is most prevalent in semiarid or arid regions, irrigation is common in the humid climates, especially on recreational sites or areas where a season-long green color is desired. Even in climates with 100 to 150 cm of annual rainfall, uneven seasonal distribution necessitates some irrigation for adequate growth and color.

Irrigated turfgrass sites often present a number of problems not encountered for irrigation of agronomic or horticultural crops. Since turfgrasses are perennials, except for the case of sod production, permanent belowground systems are preferred (Fig. 30–2). This requires careful design to account for slopes, different soil types on a site, different turfgrasses, and the presence of trees and shrubs that may compete for water, as well as interfere with distribution patterns. Unusual landscape shapes have resulted in the use of spray heads in addition to rotary heads. With a diversity of areas that may vary in water needs (e.g., because of slope, shape, tree/shrub root competition, grasses, soil types) and the use of different head types (with differing outputs), turfgrass irrigation systems normally involve elaborate zoning of heads, numerous zones, and sophisticated control systems. Turfgrasses differ

Fig. 30–1. Golf course irrigation installation (courtesy The Toro Company).

from other crops in that they constitute total ground cover and are used daily. These eliminate irrigation methods common on other crops, such as flood and furrow procedures.

In this chapter, those aspects of irrigation that are unique to turfgrasses are emphasized: (i) water use rates of different turfgrass species and culti-

Fig. 30–2. Residential irrigation installation (courtesy The Toro Company).

Table 30-1. A classification of evapotranspiration rates for turfgrass. From J.B. Beard, 1985.

Relative ranking	Evapotranspiration rate	
	mm/d	mm/wk
Very low	<4.0	<25
Low	4.0–4.9	26–34
Medium-low	5.0–5.9	35–41
Medium	6.0–6.9	42–48
Medium-high	7.0–7.9	49–55
High	8.0–8.9	56–62
Very high	>9.0	>63

vars, (ii) the influence of cultural practices on water use, (iii) soil problems on turf sites that limit irrigation efficiency and correction of these problems, and (iv) a review of advances in turfgrass irrigation equipment technology.

I. TURFGRASS WATER USE

There is considerable interest in water conservation among turfgrass managers. Watson (1985) presented good reason for this concern in his review of water resources in the USA and future implications on turfgrass management. Turfgrass species and cultivars with reduced evapotranspiration (ET) rates could play an important part in turfgrass water conservation (Carrow, 1988).

Beard (1985) presented an interesting and complete review of turfgrass ET research. His summary of this review discussed concern among scientists that interspecific differences would not exist. Researchers have reported interspecific differences in ET rates (Biran et al., 1981; Danielson et al., 1981b; Fairborne, 1982; Kim & Beard, 1988; Kneebone & Pepper, 1982; Rogowsky & Jacoby, 1977; Tovey et al., 1969; Youngner et al., 1981). Beard's (1985) review indicated that data existed to support interspecific ET differences, but a number of species had no ET data substantiated in the turfgrass literature. This paucity of data was particularly true for cool-season turfgrass species. Research in Nebraska by Peterson (1985) and Aronson et al. (1985) in Rhode Island has expanded this cool-season turfgrass species ET data base, but more research is needed.

Based on his review of the literature, Beard (1985) developed a classification system for turfgrass ET rates (Table 30–1). This system offers a means to discuss turfgrass ET on a quantitiative rather than qualitative basis when comparing turfgrass species. Recent trends have developed which attempt to classify turfgrass ET on the basis of warm (C-4)- and cool (C-3)-season turfgrass species (Biran et al., 1981; Danielson et al., 1981b; Gibeault et al., 1985; Kneebone & Pepper, 1982; Youngner et al., 1981). These studies were conducted under conditions most suited to warm-season turfgrass species adaptation, making direct comparisons of species difficult, if not biased. Beard's (1985) summary demonstrated that cool-season turfgrass species did not have higher ET rates than warm-season species. St. Augustine grass [*Stenotaphrum secundatum* (Walt.) K.], a C-4 plant, had a high ET rate, ranging from 9.6 to 12.2 mm/d. Tall fescue (*Festuca arundinacea* Schreb.), a C-3 plant, was reported to have the highest ET rages (i.e., 10.6–12.6 mm/d) for the cool-season species. The tall fescue data were obtained from studies conducted in Arizona (10.6 mm/d) and Israel (12.6 mm/d), while St. Augustine grass data came from Arizona (9.6 mm/d) and Texas (12.2 mm/d). Kopec et al. (1988) reported a mean ET rate of 7.1 mm/d for 'Kenhy' tall fescue grown in Nebraska. Kentucky bluegrass (*Poa pratensis* L.), zoysiagrass (*Zoysia japonica* L.), and bermudagrass (*Cynodon* spp.) showed promise for water conservation through reduced ET rates (Beard, 1985). The C-3 vs. C-4 turfgrass species comparisons may be warranted, but broad statements comparing potential differences do not appear justified. More research is needed before conclusions can be made. For example, research conducted within species adaptability zones is not available for all cool-season turfgrass species.

Recent research has supported intraspecific differences in ET rates (Beard, 1985; Biran et al., 1981; Kopec et al., 1988; Schmidt & Everton, 1985; Shearman, 1986). These data support a strong potential to conserve water through the selection of low ET cultivars and to develop cultivars with reduced ET through breeding and selection programs. Shearman (1986) compared ET rates of 20 Kentucky bluegrass cultivars under controlled environment conditions. ET rates ranged from 3.9 mm/d for 'Enoble' to 6.3 mm/d for 'Merion', a difference of 64% from high to low ET rate. In a similar study conducted under field conditions, Kentucky bluegrass ET rates ranged from 7.67 to 9.99 mm/d with a difference of 30% in ET rates between high and low cultivars (Table 30–2). In these studies, verdure and shoot density were negatively correlated ($r = -0.83$ and -0.87, respectively) to cultivar ET rates, while vertical elongation rate was positively correlated ($r = 0.96$) to ET rate (Shearman, 1986). Kentucky bluegrasses with erect, open growth habits and rapid vertical elongation required more water than their counterparts.

Johns et al. (1983) introduced the term canopy or atmospheric resistance to the turfgrass literature. They demonstrated the major portion of resistance to turfgrass ET under well-watered conditions was contributed by shoot density, leaf angle, and leaf area which they termed *canopy resistance*. Kopec et al. (1988) reported that forage-type tall fescues had higher ET rates than turf types. These differences were attributed to canopy resistance based on differences in verdure density, verdure leaf area and vertical elongation rates (Kopec, 1985). Since most turfs receive irrigation to maintain desired quality and function (i.e., are maintained under well-watered or nonlimiting water conditions), canopy resistance mechanisms should play an important role in selection and maintenance of turfs for water conservation.

Turfgrass ET rates can be altered by changes in environment, soil water content, cultural practices, and disease and insect damage. These factors can easily influence vertical elongation rate, leaf area, leaf orientation, shoot density, and verdure. Peak ET rates can be strongly affected by turfgrass ability to adapt to evaporative demand, environmental stress, pest stresses, mow-

Table 30–2. Evapotranspiration (ET) rates and crop coefficients (K_c) of 14 Kentucky bluegrass cultivars under field conditions at the University of Nebraska John Seaton Anderson Turfgrass Research Facility.

Cultivar	ET rate†		K_c‡
	mm/d	Percentage of Baron	
Baron	7.67	0	0.80
Challenger	7.69	0	0.80
Touchdown	7.93	3	0.82
Sydsport	8.02	5	0.83
America	8.15	6	0.85
Ram I	8.15	6	0.85
Eclipse	8.97	17	0.93
Glade	9.00	17	0.93
Mystic	9.17	20	0.95
Nassau	9.23	20	0.96
Birka	9.34	22	0.97
Dormie	9.46	23	0.98
Park	9.91	29	1.03
Kenblue	9.99	30	1.04

LSD (0.05) = 0.67
ETp§, mm/d = 9.64

† ET based on lysimeter measurements made in the field, during 15 to 18 June 1987. Values are means of six replications per treatment and 4 d of measurement. Percentage was based on ET rate of cultivar minus ET rate of Baron/ET rate of Baron.
‡ Crop coefficients based on ET (actual)/ET (potential).
§ ETp based on modified Penman equation.

ing height and frequency, nutritional programs, and irrigation regimes. Substantial shifts in ET rate can result from these factors on an individual or interactive basis. Further research is needed to support our understanding of turfgrass ET rates and factors influencing the intensity of ET. Beard (1985) expressed concern over the variability that exists in turfgrass ET data and the need to conduct comparative ET rate studies under rigorously controlled and stress-free conditions. Under field conditions, this may not be reasonable, but careful description of environmental conditions and cultural practices is feasible and would be beneficial in attempts to compare ET rates of species. In addition, field data on water use under normal irrigation regimes, may vary substantially from controlled climate conditions because of rooting and soil water potential effects. Carrow and Johnson (1988) determined the water use of 'Tifway' bermudagrass, 'Meyer' zoysiagrass, and common centipedegrass [*Eremochloa ophiuroides* (Munro.) Hack.] in the Southeast under field conditions. When irrigated at a soil water potential of − 0.40 MPa based on a 15-cm-deep probe, water use in midsummer was 2.9, 3.3, and 4.1 mm/d, respectively, for bermudagrass, centipedegrass, and zoysiagrass. Under well-watered conditions, (i.e., soil water potential of −0.10 MPa) in midsummer, water use rates increased by 13, 39, and 17% for bermudagrass, centipedegrass, and zoysiagrass, respectively. The authors monitored root growth and found that increasing moisture stress reduced zoysiagrass rooting. Also, zoysiagrass exhibited little rooting into the acid, high bulk-density B horizon, which caused it to be shallow rooted. These differences in water use due to periods of limited soil water or regional soil factors that could influence root growth of a species, illustrate that data are needed from both controlled and field situations.

II. CULTURAL PRACTICE EFFECTS ON TURFGRASS WATER USE

Mowing, fertilizing, and irrigating are primary cultural practices that influence turfgrass water use because these practices have direct effects on the turfgrass growth rate, leaf surface area, canopy resistance, and depth and extent of rooting. Primary cultural practices alone or in combination can be manipulated by the turfgrass manager to minimize water loss and enhance water conservation. When applying cultural practices to reduce water use, they often influence drought resistance mechanisms. *Drought resistance* is defined as the means by which a plant is able to withstand periods of drought. Two major types of drought resistance are (i) *drought avoidance*— ability of a plant to avoid tissue damage in a drought period by postponement of dehydration by means of extracting more soil moisture and/or reducing water loss from the plant, and (ii) *drought tolerance*—ability of a plant to tolerate drought periods by escape (where a plant has a life cycle such that it lives through a drought in a dormant state of as a seed), greater genetic tissue/membrane tolerance to desiccation, and hardiness. *Hardiness* is where plant tissues develop a greater tolerance (hardiness) to tissue water deficits.

Sometimes a particular cultural practice will reduce turfgrass ET but adversely influence drought resistance. For example, close mowing can reduce ET but may also reduce rooting depth (decreased drought avoidance) and be less hardy (decreased drought tolerance). Thus, growers must be aware of the full impact a management regime may have on turfgrass water relations.

A. Mowing

Turfgrass mowing height, frequency, patterns, and equipment directly and indirectly influence turfgrass growth, development, and water use. Beard (1973) and Madison (1971) summarized many physiological and morphological turfgrass responses to variations in mowing height and frequency.

Increased water use has been associated with higher mowing heights in both warm- and cool-season turfgrasses (Doss et al., 1962; Feldhake et al., 1981; Madison & Hagan, 1962; Mitchell & Kerr, 1966; Shearman & Beard, 1973). Shearman and Beard (1973) found the water use rate of Penncross creeping bentgrass (*Agrostis stolonifera* var. *palustris*) increased by nearly twofold as mowing height was increased from 6 to 125 mm. Turfs mowed at 25 mm used 56% more water than those mowed at 6 mm. Well-watered bermudagrass [*Cynodon dactylon* (L.) Pers.] transpired 4.8 mm of water at a plant height of 150 mm and 3.0 mm daily at a plant height of 25 mm (Doss et al., 1962). Mitchell and Kerr (1966) reported similar results, a 37% decline in ET between perennial ryegrass (*Lolium perenne* L.) mowed at 50 and 25 mm. Increased water loss measured as soil moisture extraction beneath a Merion Kentucky bluegrass turf was related to increased mowing height and an associated increase in depth and extent of rooting (Madison, 1962a). Increased water use associated with higher mowing is likely related to added plant surface area exposed to desiccating conditions (Shearman & Beard, 1973). Studies with Seaside creeping bentgrass and Highland colonial bentgrass (*A. tenuis* Sibth.) indicated increased plant density but reduced verdure and rooting with lower mowing heights compared to higher height (Madison, 1962b). Similar effects of mowing treatments on leaf, shoot, and root development were reported by Evans (1949), Everson (1966), Oswalt et al. (1959), Robertson (1933), and Wood and Burke (1961).

Shearman (1986) found ET rates of Kentucky bluegrass grown under nonlimiting soil moisture were positively correlated with vertical elongation rate ($r = 0.96$) and were negatively correlated to shoot density ($r = -0.87$) and verdure ($r = -0.83$). Johns et al. (1983) had similar findings for St. Augustine grass grown in nonlimiting soil moisture conditions. These researchers attributed this response to canopy resistance. Low-cut turfs have a high shoot density and a dense, tight canopy. High-cut turfs have an open canopy with reduced shoot density. High-cut turfs have a greater leaf area and a lower canopy resistance than low-cut turfs. The reduced canopy resistance and increased leaf area contribute to high potential water loss through ET. Lower mowing heights could be used for water conservation when soil moisture is not limiting, within the realm of species and cultivar adaptation. Under limited soil moisture, higher mowing and the associated increase in ET would be offset by greater depth and extent of rooting. It is likely that these turfs would have higher drought avoidance, if not water conservation.

Studies have indicated that leaf area, shoot size, and extent of rooting decreased, but shoot density and tissue succulence increased with mowing frequency (Crider, 1955; Hart & Burton, 1966; Madison, 1960; Madison, 1962a, b; Madison & Hagan, 1962; Prine & Burton, 1956). Johns (1980) reported that turfgrass water use increased with days following mowing, indicating a greater leaf area attributed to regrowth between mowings. Shearman and Beard (1973) reported 41% increased water use for turfs mowed biweekly compared with those mowed six times weekly. These results agreed with those found by Shearman (1986) where ET increased with vertical elongation rate of Kentucky bluegrass cultivars. Johns et al. (1983), Shearman and Beard (1973), and Shearman (1986) indicated that the more frequently the turf is mowed, the less water it would use. The reduction in water use would be influenced by decreased vertical elongation rate and increased canopy resistance associated with frequent mowing.

It is apparent that within turfgrass species and cultivar tolerance ranges, there is room to manipulate mowing height and frequency for enhanced depth and extent of rooting and to reduce ET rates. The overall benefit of this manipulation would be increased drought avoidance and improved water conservation. This was demonstrated by Johns' research (1980), when he reported excessive leaf area was eliminated and transpiration was reduced by manipulating mowing frequency. Biran et al. (1981) demonstrated differences in water consumption between C-3 and C-4 pathway turfgrasses, with C-3 plants using more water, even though both groups had similar growth rates. Kneebone and Pepper (1982) and Youngner et al. (1981) reported similar results for warm- and cool-season species compared under conditions most suitable for C-4 plant (warm season) growth. More research relating mowing height and frequency responses to ET rates and canopy resistance to C-3 (cool season) and C-4 (warm season) turfgrasses is needed.

Beard (1973) speculated that water loss would be greater from turfs mowed with a dull or improperly adjusted mower than those mowed with a sharp mower blade. The potential increase in water loss was thought to be associated with increased tissue mutilation resulting from dull mower blade injury. Steinegger et al. (1983) compared dull and sharp mower-blade injury on turfgrass water use in field-grown 'Park' Kentucky bluegrass. Turfgrass water use was reduced by the dull mower treatment, which was accompanied by a decline in turfgrass quality and verdure. Water use was examined over a period of 1 to 96 h after treatment. Water loss increased initially with dull mower treatment, but this trend was overcome by the sharp mower treatment within 24 h. No comparisons were available in the turfgrass literature for rotary vs. reel-type mowers and their potential effect on turfgrass water use. It is quite likely that turfgrass water use would be influenced by mowing practices that change canopy resistance and vertical elongation rate.

B. Nutrition

Turfgrass nutrition influences growth rate, leaf surface area, depth and extent of rooting, and turfgrass water use. Researchers have reported increased water use with increased N nutrition (Carroll, 1943; Dexter, 1937; Feldhake et al., 1981; Krogman, 1967; Mantell, 1966; Shearman & Beard, 1973; Sprague & Graber, 1938). Water use of 'Penncross' creeping bentgrass increased from 0 to 10 g N/m^2, leveled off at 10 to 20 g N/m^2, and declined at rates above 20 g N/m^2. Shoot growth, shoot density, and root organic matter production followed identical trends. Leaf width ($r = 0.85$), shoot density ($r = 0.91$), and shoot growth ($r = 0.80$) were positively correlated to water use. These results demonstrated the importance of N nutrition on increased plant surface area and increased ET rates (Shearman and Beard, 1973). Root organic matter production was not associated with water use. This nonsignificant relationship would not be surprising, since there are confounding relationships of root and shoot growth with N nutrition. Beard (1973) reviewed the turfgrass literature and found that several researchers reported increased root and shoot growth as N nutrition was raised from a deficiency level and that this trend continued only until a critical growth stimulation point was reached. At this critical point and higher N levels, turfgrass root growth may decline or be significantly supressed. Turfs grown in this manner and exposed to high atmospheric evaporative demand conditions would likely wilt more rapidly than those grown under adequate N nutrition regimes. Common Kentucky bluegrass turfs fertilized with 10 g N/m^2 per growing season had a greater wilting tendency and were more susceptible to drought injury than unfertilized turfs (Juska & Hanson, 1967). In Arizona, Kneebone and Pepper (1982) reported that turfs receiving 5 g N/m^2 bimonthly used 340 mm less water per year than turfs receiving 5 g N/m^2 monthly. Water use reduction is available through proper manipulation of N nutrition within the range of species and cultivar adaptation, and the desired turfgrass quality which must be maintained. Krogman (1967) studied various combinations of N and P applied to mixed stands of grass. His work demonstrated that increased N rates increased water use. Phosphorus applications showed a slight increase, but the trend was not significant and there was no significant effect of the N-P interaction. Krogman (1967) suggested that year-to-year variation and weather conditions affect water consumption through ET at least as much as the increased plant surface area resulting from fertilization. Water use efficiency increased with N input. The increased yield from N fertilization offset the rise in ET. In most turfgrass situations, water use efficiency relative to yield is not a critical issue. Water conservation relative to verdure production is of primary concern.

Water use efficiency, as it related to N nutrition in turfs, is an interesting concept. Turfgrass managers are interested in nutritional programs that meet the needs of the turfgrass plant but avoid excessive growth. Water use efficiency relative to turfgrass nutrition should center around nutritional levels that maintain desired turfgrass quality, recuperative rate, and reduced water consumption. In this regard, turfgrass water use efficiency would be a relative term related to turfgrass function and quality, and would differ in comparison among levels of cultural practice intensity. Judicious and appropriately timed N fertilization oriented toward meeting turfgrass nutritional needs should be the turfgrass manager's goal for efficient use of available water.

Research relating N carrier, rate, and application timing relative to drought resistance is needed. Schmidt and Breuninger (1981) studied the influence of N, P, and K nutrition on Kentucky bluegrass drought recovery. They found that recovery was greater for turfs treated with fall-applied N than for those receiving spring treatments. Phosphorus increased recovery but was interactive with N rate and application timing. Potassium benefitted recovery regardless of N and P treatments. Powell et al. (1967) found bentgrass produced more root weight with fall N fertilization than spring. Snyder and Schmidt (1974) confirmed these findings.

Water use as it related to K nutrition has not been clearly delineated. Studies conducted at the Univ. of Nebraska have demonstrated decreased water use and wilting tendency with increased K nutrition in Kentucky bluegrass turfs (Fig. 30–3). No data are available in the turfgrass literature regarding the direct influence of K on turfgrass water use. Researchers have reported increased extent and depth of rooting, with increased K nutrition (Marklund & Roberts, 1967; Monroe et al., 1969). Increased rooting associated with K treatments could contribute to observed drought resistance and improved drought injury recovery in turfs, particularly those that are

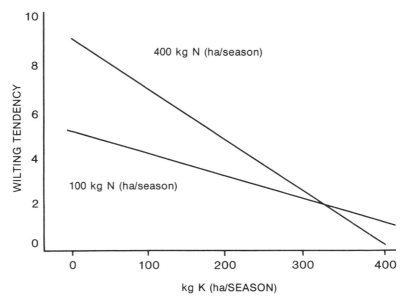

Fig. 30–3. Wilting tendency of a Fylking Kentucky bluegrass turf treated at two N regimes and five K levels. Wilting tendency was based on a 0 to 10 scale with 0 = no wilt and 10 = 100% of turf wilted.

deficient in K (Schmidt & Breuninger, 1981; Shearman, 1982a). Increased turgor and decreased tissue moisture content were reported with increased K fertilization (Shearman, 1975).

Studies on the influence of Fe on turfgrass water relations are limited. Snyder and Schmidt (1974) demonstrated improved root development with foliar-applied Fe made with late-season N applications. Iron treatments also reduced desiccation injury associated with high N nutrition levels. However, Glinski et al. (1988) demonstrated that foliar Fe applied each month on bentgrass decreased root length density in the summer by 17 and 36% at 0 to 10-cm and 10 to 20-cm depths, respectively, when averaged over three Fe carriers relative to the control. Water use in the summer varied with Fe carrier from 100 to 116% of the untreated control. Thus, root responses and water use were not well correlated and they noted that plant responses could be due to other substances in the three Fe carriers than the Fe component.

Nutrition plays a complicated and confounded role in turfgrass water use management. Its relationship with other cultural practices and interrelationship with nutrient interactions make it a difficult and complex research area. Turfgrass managers should adjust their nutritional programs to produce the least amount of excess top growth and the greatest amount of rooting possible if they wish to maximize their water conservation potential.

C. Irrigation

In many areas, natural precipitation must be supplemented with irrigation to provide desired turfgrass quality and function. As a cultural practice, irrigation interacts with other cultural practices to influence turfgrass water use (Danielson et al., 1981b). A commonly accepted turfgrass management recommendation is to irrigate deeply and infrequently (Hagan, 1955). This recommendation is difficult to practice since it does not provide an actual amount or frequency of application. Shearman and Beard (1973) reported a 33% reduction in water use for turfs receiving irrigation when visual wilt symptoms were evident compared to those receiving water three and seven times weekly. The decline in water use was correlated ($r = 0.98$) with decreased vegetative cover and turfgrass quality. In a Nevada study, Tovey et al. (1969) applied water every 3, 7, or 10 d to turfs growing on loam or sandy loam soils. They found Kentucky bluegrass/fine fescue turfs could be maintained with high quality by using irrigation twice weekly, regardless of soil type. Seven-day intervals were adequate for the loam soil, but 10-d intervals produced sparse, low-quality turfs, regardless of soil type. Marsh et al. (1980) and Youngner et al. (1981) used tensiometers and pan evaporation for irrigation scheduling on warm-season and cool-season turfgrass species. Tensiometer and evaporation pan scheduling were compared to a control irrigation treatment. The control treatment exemplified typical practices used by local turfgrass managers. Turfs receiving irrigations scheduled by tensiometer or evaporation pan received less water than the control, and they followed weather pattern changes more closely than the control. These researchers reported as much as 55% lower water use from tensiometer and evaporation pan treatments compared to the controlled treatment for the warm-season species. Youngner et al. (1981) found that tall fescue and Kentucky bluegrass turfgrass quality was influenced by irrigation scheduling. Kentucky bluegrass turfs scheduled to receive irrigation when tensiometer readings reached -0.055 MPa maintained poor quality throughout the study, having low shoot density and increased disease incidence. Tall fescue had reduced turfgrass quality in the first year of testing, but in subsequent years after the root system had developed, turfgrass quality did not decline with the -0.055-MPa tensiometer treatment. Research in Florida using microswitching tensiometers and electronic soil-moisture sensing devices for irrigation scheduling reduced water applications by as much as 89% over conventional approaches without reducing turfgrass quality (Augustine et al., 1981). Danielson et al. (1981b) studied Kentucky bluegrass responses to limited water under N deficiency. Turfgrass quality declined linearly with decreased water over a range of 100 to 40% maximum evaporation. With adequate N nutrition, quality was maintained until 70% of maximum of ET was obtained. These results indicated that there was a nitrogen-irrigation interaction. Kneebone and Pepper (1982) reported that water use rates of subirrigated turfgrasses followed Class A pan evaporation closely during periods of high water demand and active growth. Water use expressed as a percentage of Class A pan evaporation ranged from 42 to 80%, depending on the turfgrass species and intensity of management. They concluded that watering in excess of 50 to 80% of Class A pan evaporative losses for warm-season and 60 to 85% for cool-season turfgrass species was wasteful and resulted in unnecessary water use. Other investigations of sprinkler irrigation scheduling procedures have found similar plant responses (Busch & Kneebone, 1966; O'Neil & Carrow, 1982). O'Neil and Carrow (1982) compared tensiometer-controlled irrigation on Kentucky bluegrass to a set schedule. The tensiometer-controlled irrigation reduced water use by 28 to 48% over 16 wk without reducing turfgrass quality. The greatest savings (66%) occurred in the fall when evapotranspirational demands were reduced, and the least (11%) occurred in August when evapotranspirational demand was high. They concluded that the set irrigation schedule was inefficient and higher evapotranspirational water losses occurred with turfs receiving more water.

Adjusting irrigation frequency, using electronic moisture-sensing devices, tensiometers, Class A pan ET, or combinations of these devices would likely reduce unnecessary water use. The previously discussed research verifies this approach over a wide range of turfgrass species and growing conditions. Turfgrass managers interested in using these devices should approach them with caution, using local research data and information and on-site experimentation before embarking on irrigation programming that relies solely on such devices.

D. Soil Cultivation

Turfgrass is often exposed to traffic and subsequent wear and soil compaction stress. Soil compaction affects soil bulk density, aeration, and water retention. These soil effects in turn influence turfgrass growth and development. Researchers have observed reduced root growth and shoot growth as a result of soil compaction stress (Gore et al., 1979; O'Neil & Carrow, 1982; Rimmer, 1979; Sills & Carrow, 1983; Thurman & Pokorny, 1969). Research at Kansas State University with Kentucky bluegrass and perennial ryegrass demonstrated that compaction altered distribution and extent of turfgrass rooting (O'Neil & Carrow, 1982, 1983). O'Neil and Carrow (1982, 1983) reported that root viability decreased as a result of low oxygen diffusion rates (ODR) and that critical ODR levels were lower for shoot growth response than for root growth. Thus, shoot growth declined immediately after compaction, but root growth declined and root distribution was altered for a longer time. Turfgrass water use would decline with reduced shoot and root growth. It was reported that turfgrass water use declined with increasing levels of soil compaction (O'Neil & Carrow, 1982). Sills and Carrow (1983) found a decline in turfgrass quality, clipping yield, N use efficiency, ET rate, and root growth with increasing compaction treatment. Evapotranspiration rates declined by as much as 28% with compaction treatment. These stress studies demonstrated that soil compaction influenced turfgrass shoot and root growth, turfgrass quality, soil-moisture retention, and turfgrass water use.

Research on the effects of different turfgrass cultivation methods on water use has been limited. Recently, Wiecko et al. (1988) found that cultivation increased water use of 'Tifway' bermudagrass grown on a compacted soil by 3 to 22% (late summer) and 8 to 28% (early summer). Those cultivation methods that penetrated deepest in the soil resulted in deeper rooting and thereby higher water use under compacted soil conditions.

E. Chemicals

Various chemicals are used in turfgrass cultural programs. These chemicals may directly or indirectly influence turfgrass growth, leaf area, root development, and water use. Among these chemicals would be antitranspirants, pesticides, wetting agents, and plant growth regulators.

Antitranspirants have been used to control transpiration at the leaf-air interface. These materials may induce stomatal closure, cover the mesophyll surface with a thin, monomoleular film, or cover the leaf surface with a water-impervious film. Antitranspirants have potential detrimental effect on turfgrass photosynthesis and evapotranspirational cooling. Phenylmercuric acetate (PMA) has been used to induce stomatal closure in nongrass species (Shimishi, 1963; Slatyer & Bierhuizen, 1964; Waggoner et al., 1964; Zelitch & Waggoner, 1962). Stahnke (1981) conducted research on Penncross creeping bentgrass and 'Tifway' bermudagrass by using: abscisic acid (ABA), B-napthozyacetic acid, Stoma-seal[1] (a mixture of PMA and Aqua-Gro), Aqua-Gro (a soil wetting agent), and Experimental No. 913 (a monoglycerol ester of decinyl succinic acid in Aqua-Gro). Stahnke (1981) assessed their effectiveness by measuring transpiration rates differences of treated compared to untreated plants and whether plants were injured as a result of antitranspirant treatments. Transpiration rate was calculated based on difference between ET rate from live turf and evaporation from dead turf. It was found that ABA and Experimental No. 913 reduced transpiration on Penncross by 59 and 29%, respectively, without causing visual injury symptoms or increasing leaf temperatures. In Tifway, ABA at 1×10^{-4} M reduced transpiration by 12% and shoot growth rate by 23%. Abscisic acid was an effective antitranspirant, but its use on a large-scale basis was questionable because of economics.

Monomolecular films have been used on water reservoirs to minimize water loss from evaporation, and have been used with varying degrees of success to reduce plant transpiration (Aubertine & Grosline, 1964; Oerthi, 1963). When used on plants, these materials reduce photosynthesis to a greater extent than transpiration. Coatings that cover leaf surfaces and stomatal pores have been used as antitranspirants and winter antidesiccants (Comar & Barr, 1944; Marshall & Maki, 1946; Wooley, 1967). Use of these materials has been primarily restricted to ornamentals and there is a dearth of information for turf regarding their use. Shearman and Beard (1970, unpublished data) examined Wilt-pruf and Stoma-seal applied at manufacturer's rates to Penncross creeping bentgrass turfs. Water loss from treated turfs was not significantly different from that of the untreated control. It was concluded that the materials did not effectively coat the turfgrass leaf blades. Film-coating materials would be of limited benefit under turfgrass growing conditions, since it would be difficult to obtain uniform leaf-blade coating, and mowing would remove coated leaves and reduce effectiveness of treatments.

---

[1] The use of trade names in this chapter does not imply endorsement by the authors of the products named, nor criticism of similar ones not mentioned.

Johns (1980) expressed skepticism on the potential benefits of antitranspirants usage in turf based on research conducted with St. Augustine grass. He reported water use to be influenced to a greater extent by environmental factors external to the plant than by the stomatal aperture, and that canopy resistance manipulation offered more potential than antitranspirants for reducing ET rates.

Wetting agents have received limited testing in turf and little data are available in the turfgrass literature concerning their effects on water use. Stahnke (1981) observed the effects of Aqua-Gro on transpiration in studies conducted on creeping bentgrass and bermudagrass. Peterson et al. (1984) reported reduced ET rates from surfactant treatments of 18 to 29% 7 d after application, and 10 to 30% 21 d after treatment on Kentucky bluegrass. Treatment effects were reported to be transitory and were essentially gone by 4 wk after application. Letey et al. (1963) reported improved water infiltration rates on hydrophobic soils treated with nonionic wetting agents. Similar findings have been reported for hydrophobic sand, soils, and thatch (Paul & Henry, 1973; Pelishek et al., 1962; Wilkinson & Miller, 1978). Reduced evaporation and improved soil-water retention have also been found as results of wetting agent treatments (Engel & Alderfer, 1967; Law, 1964; Madison, 1966). Wetting agents may play a limited and indirect role in turfgrass water use through enhancement of soil wetting, water retention, evaporation, and increased availability of soil moisture for turfgrass use.

Pesticides may affect turfgrass water use. For example, PMA has been discussed as an antitranspirant, but also is used as an eradicative fungicide. The PMA fungicides applications could reduce transpiration by influencing stomatal aperture. A fungicide application might indirectly influence transpiration by increasing or decreasing the turfgrass canopy, root system, or both. The same case could be made for other pesticides, such as herbicides and insecticides. A decline of water use rate and water use efficiency was reported with Siduron [1-(2-methylcyclohexyl)-3-phenylurea] treatments on a Kentucky-31 tall fescue turf (Shearman et al., 1980). Siduron treatments reduced shoot and root dry matter production and increased wilting and drought injury at application rates of 13.6 kg/ha or greater. In a study conducted to determine the influence of successive preemergence herbicide applications on Kentucky bluegrass turfs, a reduction in water use associated with Prosulfalin and Benefin (N-butyl-N-ethyl-α,α,α,-trifluoro-2,6-dinitro-p-toluidine) treatments was noted (Reierson, 1979). The reduced water use rate for Kentucky bluegrass turfs was closely associated to reduced turfgrass density and quality. This study and another (Shearman et al., 1980) demonstrated the potential detrimental effects of pesticide applications on turfgrass water use and water use efficiency by influencing the turfgrass canopy and root systems. Detrimental effects of pesticide applications on water use apparently decrease water use efficiency and potential drought avoidance. In most cases, judicious pesticide applications enhance turfgrass quality and performance; their potential influence on turfgrass water-use efficiency would likely be beneficial.

Field experiments at Texas A&M University were conducted on Texas common St. Augustine grass and Tifway bermudagrass to test experimental plant growth regulators (PGR) in varying treatment combinations with and without mowing (Johns & Beard, 1982). Transpiration rate was assessed at 15-d intervals over 100 d. Leaf-area measurements were made on the 60th day and leaf area indices were calculated from these data. Significant reductions in ET rates of 11 to 29% were measured 14 wk after treatment on mowed and unmowed turfs treated with flurprimidol. Shearman (1982b) assessed ET rates in Kentucky bluegrass turfs treated with Paclobutrazol, Flurprimidol and Mefluidide compared with an untreated control. Turfs were mowed at 5 cm whenever they reached 7.5 cm in height. Water use was assessed with field lysimeters. Plant growth regulator treatments reduced water use by 23 to 44% through 28 d after treatment when compared to the untreated control. There was no significant difference among treatments after 42 d. Flurprimidol and Paclobutrazol [2RS,3RD(-1-(4-chlorophenyl)-4,4-dimethyl-2-(1H-1,2,4-triazol-1-yl)pentan-3-ol) were more effective in reducing ET rates than Mefluidide {N-[2,4-dimethyl-5-[[(trifluoromethyl)-sulfonyl]amino]-phenyl] acetamide}. Research has demonstrated that PGR's can be used to reduce ET rates effectively in warm- and cool-season turfgrass (Johns & Beard, 1982; Shearman, 1982b). These results offer a promising area of research investigation regarding plant growth regulator materials, application rates, application timing, and turfgrass species and cultivar responses. Similarly, combinations of PGR's and antitranspirants may offer additional water conservation potential.

Specific amounts of water conservation resulting from manipulation of various cultural practices on turfgrass species and cultivars cannot be stated precisely, since detailed research is not yet available. However, the relative magnitude of response should be comparable. Turfgrass managers should realize that no one cultural practice will reduce water consumption, since neglect of other critical practices can result in a decline of turfgrass quality and reduced water use efficiency. Therefore, it is important for the turfgrass manager to realize a water conservation program involving a systematic approach to the use of cultural practice manipulation.

III. TURFGRASS SOILS AND IRRIGATION

Turfgrass water use depends on dynamics in the soil-plant-atmospheric continuum (SPAC) (Phillip, 1966) (Fig. 30–4). Within this system, plants must obtain water from the soil for growth. Soil properties influence water use by two major means:

1. Turfgrass root growth, viability, and longevity are dramatically affected by soil physical, chemical, and biological characteristics. Examples of factors that restrict rooting are:

- Physical—high soil strength, low soil O_2, lack of soil moisture, compaction, layering, high soil temperatures.

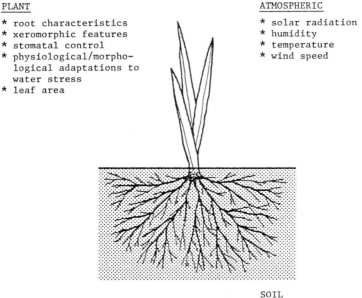

PLANT
* root characteristics
* xeromorphic features
* stomatal control
* physiological/morpho-
  logical adaptations to
  water stress
* leaf area

ATMOSPHERIC
* solar radiation
* humidity
* temperature
* wind speed

SOIL
* water holding capacity
* saturated/unsaturated
  water flow characteristics
* soil water potential
* infiltration
* soil properties influencing
  plant growth

Fig. 30-4. Soil-plant-atmospheric continum (SPAC). These soil, plant, and atmospheric factors form a dynamic and inter-related system that influences water movement, plant uptake,
and water potential (Carrow, 1985).

- Chemical—very acidic soil pH, Al or Mn toxicities, heavy metals, high
  salt levels, unfavorable nutrient levels or balances with other nutrients,
  alleopathy.
- Biological—disease organisms, root-feeding insects, nematodes, weeds
  competing for soil moisture and nutrients.

Restricted rooting due to adverse soil properties can greatly limit the volume
of soil reached by turfgrass roots for water uptake. In this section, these
aspects will not be discussed but they are in other sections of this chapter
and book. Through good management practices, many of these factors can
be alleviated or corrected as limiting for turfgrass root growth.

2. Soil water retention, activity, and movement are directly influenced
by the physical properties and sometines chemical properties of the soil (Carrow, 1985). It is these aspects that will be reviewed in this section. Of particular interest is how soil properties can be altered to positively influence
irrigation. For example, in the SPAC system, the irrigator desires soil properties such as rapid infiltration and drainage, good water-holding capacity, and

availability of the water in the soil for plant use. A soil with these characteristics will allow the grower to irrigate without excessive runoff, leaching,
or evaporation losses.

A. Infiltration

Before water from rainfall or irrigation can be used by a plant, it must
infiltrate into the soil. In turfgrass culture there are many factors that cause
adverse soil surface conditions, thereby resulting in low infiltration and water
loss from the SPAC system by runoff or evaporation. Also, proper irrigation scheduling is difficult on soils with low infiltration.

Soil texture is of prime importance in controlling infiltration. Sands have
infiltration rates of 2.50 to 20.00 cm/h, while clays are in the range of 0.02
to 0.25 cm/h. Soil structure can alter infiltration, especially if the structure
at the surface has been destroyed. A well-structured soil will have large pores
with good continuity between pores to allow for water movement.

Even on tight clay soils, infiltration can be quite high if cracks develop
because of the expanding-contracting nature of certain clays upon wetting
and drying. This illustrates the well-known fact that infiltration rates are not
a constant value but vary with surface and subsurface water contents. A dry
soil surface will result in an initially high infiltration that usually declines
to a steady-state value termed the *saturated hydraulic conductivity.*

In addition to crack formation, other factors that may alter infiltration
as soil moisture content changes are air entrapment ahead of the wetting front,
degradation of the surface structure, a reduced matrix potential gradient,
and perching of water above a subsurface layer of low permeability.

Sloping surfaces can greatly increase runoff, especially on rolling ter-
rains like a golf course. Irrigation design, zoning, and scheduling for optimum infiltration on such sites are complex. Another factor that reduces
infiltration is a hydrophobic surface. Thatch, often partially hydrophobic
when dry, will for the first few minutes after receiving irrigation repel water
until the thatch wets. Hydrophobic sands, caused by organic coatings on sand
particles, present an extreme case of water repellency (Wilkinson & Miller,
1978).

Surface layers that are finer textured than the underlying soil often contribute to poor water intake. These layers are usually relatively thin and may
arise from several sources: (i) wind or water deposition over an established
turfgrass, (ii) sodding a sandy soil with a sod containing a finer-textured soil,
(iii) organic sod over a mineral soil sometimes creates a layer effect, (iv) poor
choice of topdressed mixtures, (v) thatch, and (vi) soil compaction at the
surface.

In order to select appropriate management strategies to improve infiltration, the turfgrass grower must identify the specific causes for poor infiltration on his site. Beard (1973, 1982) provides detailed discussions on the theory
and applications of the various management practices recommended in this
chapter.

Cultivation improves infiltration on soils that are fine textured, compacted, layered, or sloped. Core aeration and slicing are most effective, but slicing may provide only short-term benefits. Sometimes sand topdressing is worked into the coring holes to provide long-term channels for water penetration. New cultivation procedures have been developed for turfgrasses in recent years. The grower needs to match the equipment capabilities with the specific problems on his site. For example, some cultivation techniques, such as deep drill aerifiers and deep slicing units, not only penetrate the soil surface but also are able to break up deeper layers that impede drainage. A good cultivation plan includes correct equipment, proper frequency of cultivation, cultivating at the right times of year, cultivating at the proper soil moisture content, and using supplemental N for recovery if turf injury occurs.

Another management strategy to obtain better infiltration is to allow the soil surface to dry between irrigations, especially if soil cracking occurs. Irrigators can program their systems to apply a larger quantity of water initially, followed by lower quantities with time between each application. The alternative would be irrigation at very low rates over long periods of time to match saturated hydraulic conductivity. Irrigation designers should consider designing systems to allow pulsed irrigation, which allows irrigation to be divided in increments with time in between for percolation. When the soil has a low infiltration rate, pulse irrigation could be spread over two consecutive evenings to provide sufficient time to apply the necessary quantity of water.

Addition of organic matter to a poorly structured soil can improve infiltration. Ideally, the organic matter should be incorporated prior to turfgrass establishment, but topdressing with 10% or less by volume organic matter in conjunction with core aeration (to prevent a layering problem) can be used on mature turfgrasses. Also, the routine return of clippings will improve soil surface conditions over a period of years.

If thatch limits water penetration, a wetting agent can resolve the problem on a short-term basis. Also, turfgrass managers can use a short prewetting cycle followed by a few minutes to allow thatch wetting before irrigation. A more permanent solution sould be to control the thatch by verticutting, topdressing, or other means (Carrow et al., 1987). The original reason for excessive thatch buildup should be corrected.

Wetting agents plus core aeration are beneficial on hydrophobic sands. Retreatment of several applications and careful irrigation to prevent the soil from drying may be needed for full recovery and prevention.

While there is much a grower can do to improve infiltration, the task is made easier with a well-designed irrigation system. Key features would be matching water application rates to the soil infiltration rate, designing for pulse irrigation, zoning similar areas together, a control system with suitable programing options, and application uniformity.

B. Percolation

Percolation is the downward movement of water in the rootzone once water infiltrates into the soil, while *drainage* is water movement beyond the root system. Many of the same factors causing low infiltration also affect percolation, but they can be separate problems. For example, a sandy soil may have a low infiltration due to a fine-textured layer at the soil surface, but once water penetrates, rapid percolation may occur. A sandy soil may have a high infiltration, but a fine-textured layer several inches deep would limit percolation and infiltration during heavy water application periods. Thus, for a grower to irrigate efficiently, consideration of the effects of both infiltration and percolation of the soil are necessary.

Fine-textured soil is a major cause of low percolation rates, unless the soil has a good, aggregated structure. Another important reason for poor drainage is a layer in the profile that impedes water movement. Even thin layers of only a few millimeters can restrict water movement. The formation of subsurface layers may result in several ways: (i) by natural soil formation processes, (ii) compaction during construction followed by incomplete tillage, (iii) deposition of topsoil over the site without cultivation, and (iv) buried thatch or topdressing layers.

In semiarid and arid regions, soil structure may deteriorate in response to Na accumulation (Levy, 1984). Sodium causes deflocculation and dispersion of the clay particles. These sodic soils exhibit low percolation and infiltration rates.

Turfgrass growers encounter many problems caused by poor internal drainage. Among them are: limited rooting because of poor soil aeration, soggy soils, scald, and intracellular freezing. Good subsurface drainage will aid irrigation scheduling and prevent numerous other management problems.

Once the cause of limited percolation has been determined, corrective measures may be formulated. Drainage tubes (tile) can remove subsurface moisture and increase percolation if the soil reaches saturation. A moist but unsaturated soil will exhibit little water drainage into a tile. If the water table is at or near the soil surface, tile drainage can be useful as long as an adequate outlet exists. To obtain maximum drainage from a tile system, it must be installed to the proper depth, spacing of lines, tile diameter, and outlet. Normally, tile lines are installed 1 to 1.25 m deep in humid regions and 1.6 to 1.9 m in arid climates. Deeper lines in arid regions are to avoid surface salination. On sports fields, tube drainage may be relatively close to the surface at about 0.30 m.

When a subsurface layer is the cause of poor drainage, tiling may not be beneficial since the layer would limit water movement to the tile line. However, once the layer is physically broken, water will seep into the tile line. Before tiles or tubes are installed, the grower should carefully evaluate the soil profile for layers. Disruption of a layer may provide adequate drainage without installing tile.

Deep cultivation into heavy soils or into semi-impervious layers improves percolation. Several types of cultivation techniques capable of deep penetration have evolved for turfgrass situations that can be used without destroying the sod. One device drills holes through the soil to a depth of 25 cm, while other cultivation equipment may use solid tines or blades to penetrate into the soil from 15 to 40 cm.

On some turfgrass sites, especially golf greens, the layers that impede internal drainage may be thin and near the surface. These often arise from past topdressing practices or by the burial of a thatch layer. If these layers are recognized before they are more than 7 to 8 cm deep, cultivation can be performed with several types of commonly available pieces of equipment, especially hollow or solid-tine core cultivators and slicing units. These machines can be used on heavy or compacted soils to improve percolation of the surface few centimeters. Topdressing with a sand to fill the core aeration holes can provide long-lasting channels for water infiltration and percolation.

Soil modification to improve internal drainage is another option. Addition of 5 to 10% by volume of well-decomposed organic matter to fine-textured soils is beneficial. Drainage will be improved with good mixing to the appropriate depth. Application of excessive quantities of organic matter should be avoided, since organic matter retains considerable moisture and can produce an excessive wet soil. Even after a fine-textured soil has been amended with organic matter, the surface may be susceptible to surface compaction. A good on-going cultivation program will alleviate this problem, although not correct it permanently.

Addition of sand to a fine-textured soil is sometimes attempted to improve infiltration and percolation. Unfortunately, noticeable improvement in drainage does not occur until the sand content reaches 80 to 85% by volume. Sufficient sand must be present for bridging between particles to occur, which then provides macropore space for percolation. Until bridging occurs, the solid sand particles essentially displace the volume of space occupied by clay or silt. The sand has no internal porosity, while the clay and silt have internal porosity composed of primarily small pores. Thus, the sand reduces what pore space was present. The quantity of sand needed to amend 3-cm depth of a clay to the point where significant sand particle bridging would occur would be 10 to 20 cm sand uniformly mixed with the 3 cm of clay. Madison (1971) provides a detailed discussion of soil amendments, especially sand, to improve turfgrass drainage.

Complete soil modification is often used on intensive traffic sites to develop a growing media with good internal drainge and infiltration. In complete soil modification, a growing media is manufactured that may not use existing soil. The most common systems are the U.S. Golf Association (USGA) golf green specifications and the PURR—WICK/PAT approaches (Beard, 1982). These mixes contain at least 85% sand, 5 to 15% well-decomposed organic matter, and 0 to 5% soil. Specifications indicate at least 10-cm/h infiltration and percolation rates. Obviously, irrigation programing will be much easier on such sites than a native soil with 0.25-cm/h percolation.

Table 30-3. Total, plant available, and unavailable water per 30 cm of soil for different soil texture classes as well as relative irrigation frequencies.

| Soil texture | Water-holding capacity (cm water per 30 cm of soil) | | | Days to use available water† |
	Total	Plant available	Unavailable	
Sand	1.5–4.6	1.0–2.5	0.5–2.0	2–5
Sandy loam	4.6–6.9	2.3–3.3	2.3–3.6	4–6
Loam	6.9–10.0	3.3–5.1	3.6–5.1	6–10
Silt loam	10.0–12.0	5.1–5.8	5.1–6.1	10–11
Clay loam	11.0–12.4	4.6–5.3	6.1–6.9	9–10
Clay	11.4–12.4	4.6–4.8	6.9–7.6	9–10

† Assuming a constant evapotranspiration rate of 0.50 cm/d and no other water losses. If the turfgrass root system is < 30 cm, less available water would result in more frequent irrigation.

Improvement of percolation on sodic soils requires several management practices. Gypsum can be incorporated to replace the Na with Ca. Also, severe cultivation is necessary to provide a temporary improvement in drainage followed by heavy leaching to remove Na.

C. Water-Holding Capacity

Infiltration and percolation characteristics of a soil directly influence irrigation by determining irrigation rate. Low application rates over long periods of time are often necessary on fine-textured soils to adequately recharge soil moisture. There is a high potential for runoff and evaporation losses on soils with low infiltration rates.

In contrast, soil water-holding capacity primarily influences the amount and frequency of water that can be applied before the soil becomes saturated and runoff or leaching losses occur. With a greater water-holding capacity, the grower would not need to irrigate as frequently. This is beneficial, especially on sandy soils. Attempting to enhance moisture retention on a fine-textured soil may not be beneficial, since these soils often have inadequate aeration but good natural water-holding properties.

The primary factor influencing water-holding capacity is soil texture (Table 30-3). These data reveal that loam soils have the most plant-available moisture. Clays have a high percentage of small pores, but the moisture in many of these is held too tightly by adhesive and cohesive forces to be available for plant uptake. In sands, few small pores are present for moisture retention.

Soil structure also strongly influences water retention, especially at soil water potentials above −0.10 MPa. *Structure* (arrangement of individual soil particles) greatly affects capillary forces, which result from surface tension of water and its contact angle with soil particles. As the soil dries to below −0.10 MPa, adhesive and cohesive forces associated with individual soil particles affect water availability to a greater extent than water held by surface tension between particles or structural units.

Organic matter content will improve water-holding capacity because of high water retention properties. Also, organic matter often improves soil structure and, thereby, affects moisture retention.

Sometimes the soil water content is higher than expected because of a water table within 0.6 to 1.2 m of the soil surface. Depending upon the soil texture, water can move a considerable distance by capillary rise above the free water table. The moisture zone above the water table is called the capillary fringe.

The term *perched* water table refers to a water table above the normal water level due to impeded drainage. A drainage barrier, such as a clay lens or fine-textured B horizon, will perch water above it and cause unusually high water content. A perched water table develops and water can move by capillary rise above the water level.

Turfgrass managers have a number of options for improving water retention of a soil. On a sandy site, the addition of 5 to 15% well-decomposed organic matter can significantly enhance water-holding capacity. Typically, the high-sand mixes used for golf green and athletic fields contain organic matter for better water relations (Beard, 1982). Too much organic matter (>20%) should be avoided because low aeration occurs under excessive moisture.

In some cases, such as a level organic sod farm, the grower can control the drainage system to maintain the water table within 0.6 to 1.2 m of the soil surface. The capillary fringe would then cause a higher water content than with free drainage. Care must be exercised with this form of subirrigation to avoid low aeration in the rootzone and salt accumulation at the soil surface. The water table may require adjustment over a season to fit the turfgrass root growth pattern.

Another form of subirrigation is through point or line sources instead of a free water table. Porous tubing or point emitters can be buried on a uniform grid under the turfgrass. As low pressures are applied to the irrigation lines, water is emitted. The same problems as from subirrigation via a uniform water table can occur for this type of irrigation, as well as several unique problems. While capillary water movement from a free water table is upward, capillary movement from a buried source will be uniformily away from the source as long as unsaturated flow conditions prevail. Thus, careful spacings of the buried lines or emitters are necessary to achieve relatively uniform soil wetting and avoid a striped appearance. If water is emitted too rapidly, saturated flow will occur and cause more water to move downward in response to gravitational forces, resulting in loss of water by leaching. Also, if water is applied too rapidly, a waterlogged situation can occur near the roots.

Water content in the soil can be increased by purposely constructing a drainage barrier. In the USGA golf green specifications, a sharp textural difference between the pea gravel blanket and the overlying coarse sand layer causes a perched water table (Beard, 1982). As water reaches this layer, drainage is impeded until several centimeters of the media above the interface becomes saturated. When sufficient pressure potential builds, the surface tension breaks and drainage starts. The PURR—WICK and PAT systems use a plastic barrier to retain more water than would normally be retained (Carrow, 1985).

D. Water Availability to Plants

Turfgrass plants must compete for soil moisture with the forces of adhesion and cohesion of the soil matrix and solutes. Thus, soil water content is not as good an indicator of plant-available moisture as is soil water potential (i.e., energy status). The relationship between soil water content and water potential is described by a water characteristic curve for a drying soil.

In a salt-affected soil, the soil water potential can be reduced by the solutes with the effect of decreasing plant-available water. The osmotic potential, because of solutes in soils, is normally minor, while the matrix potential dominates water relations. However, in semiarid or arid regions, salts may accumulate to the point where osmotic potential influences water availability. In order to prevent *physiological drought stress* (water stress caused by salts), turfgrass will require more frequent irrigation to maintain a higher soil water content. Evaporation losses are enhanced and irrigation scheduling complicated. Salts can reduce the effectiveness of turfgrass roots to extract water. If Na is a predominate ion, poor soil physical conditions are likely and restricted root growth will occur.

To produce a more favorable soil water potential vs. soil water content relationship, the excess salts must be leached from the turfgrass rootzone. One approach would be to apply irrigation amounts 10 to 20% greater than ET losses with each irrigation. Another option would be to apply high quantities of water periodically to leach salts. The former approach is often the most effective, especially on finer-textured soils.

If leaching of salts is to be successful, good drainage and a water table at least 1.2 o 1.8 m deep are essential. Without these conditions, salts will eventually accumulate in the subsoil and infuse into the rootzone. The grower should also evaluate the original cause for salt accumulation and take appropriate precautions. For example, irrigation water high in total salts can increase the osmotic potential rather quickly. If another water source is not available, the irrigator may need to apply significantly more of this water than if using a high-quality source.

In summation, management practices that focus on improving infiltration, percolation, water retention, and water availability characteristics will greatly enhance irrigation efficiency. To a significant degree, a successful irrigation program depends on the soil properties that determine soil-water retention, activity, and movement within the SPAC (Fig. 30–2) system.

IV. TURFGRASS IRRIGATION EQUIPMENT

Properly designed and correctly programmed, today's automatic irrigation systems apply water accurately and in precise amounts to landscaped areas. Also, today's systems conserve water and apply it in accordance with the grass plants' actual needs. Further, systems—especially those with single head control and a recycling capability—apply water commensurate with infiltration and percolation capacity of the soil (Anon. 1987; Madison, 1971; Watson, 1974, 1975).

A. System Components

Sprinkler systems are composed of sprinkler heads, valves, controllers and the needed pipe and fittings to connect them.

1. Sprinkler Heads

Sprinkler heads direct water to a specified area by forcing a stream or spray from the head. Sprays cover an area continuously, whereas, a stream is rotated over the area covered (Fig. 30–5). Streams will cover a greater distance from the head but with a reduced application rate.

Sprinkler heads are rotated by impulse, cams, and gear (Fig. 30–6). Gear drive heads operate smoothly and are usually more dependable because of the constant force applied to them. They, therefore, provide more uniform coverage.

Performance of sprinkler heads is often based on distance of throw (diameter or radius). However, these numbers are somewhat misleading and should not be used as an exact base for spacing. In order to obtain uniform coverage, heads must be spaced so that the streams overlap or be spaced no more than 55% of their diameter. Spacing must take into account wind speed and direction and landscape design.

2. Valves

Valves used in sprinkler systems are mainly of the globe type. Valves release water into the line or lines or, in the case of continually pressurized line, directly to the sprinklerhead (Beard, 1973; Watson, 1974). Valves are hydraulic or electrically actuated to apply water to a diaphragm or piston.

Valves may be used to control one or more sprinkler heads. In golf course-type sprinklers, valves often are incorporated into the head and may be serviced from the top of the head (i.e., valve-in-head). Valve-under-heads may control one or several heads; however, the fewer number of heads controlled, the more accurate the application of water. Single-head control is preferred.

Gate valves are used in the system to shut off downstream water flow in order to isolate a section or part of the system for repair without disrupting operation of the remainder of the system (Beard, 1982).

3. Controllers

Controllers are electrically actuated timing devices. They send an electric or hydraulic signal to the valve to either open or close to permit water to flow to the sprinkler heads (Madison, 1971; Watson, 1975). Controllers may be a simple mechanical clock-controlled device or an automatic device actuated by electromechanical or solid state timing units. Electromechanical controllers lack the accuracy of the solid-state controllers because their motor and gear wear, and cam and level wear affect the timing mechanism (Meyer & Camenga, 1985). Electronic controllers are usually accurate within 1 min, an obvious saving of water over the 1- to 4-min accuracy range of electromechanical ones.

Fig. 30–5. Examples of different spray heads (courtesy The Toro Company).

Fig. 30–6. Cut-away of a large-geared rotary head (courtesy The Toro Company).

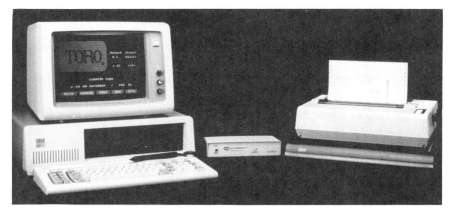

Fig. 30–7. A Modern central computer controller with graphics for operating an automatic irrigation system (courtesy The Toro Company).

Computer-controlled systems using advanced controllers may be operated from ET rates calculated from on-site weather stations thus providing a truly automatic system (Fig. 30–7 and 30–8). Other programs based on data from the California Irrigation and Management Information System, generalized average solar radiation, and historical data also may be used to operate the system (Anon. 1987; Sasso, 1988). Annual savings of up to 40% water use and 40% energy cost have been speculated when using low-pressure heads combined with a computerized controller (Anon., 1987).

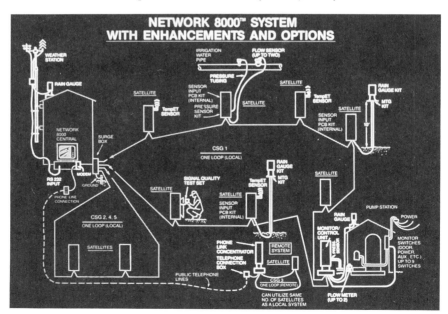

Fig. 30–8. Enhancements and options for a computer controller to allow calculation of ET rates from weather data (courtesy The Toro Company).

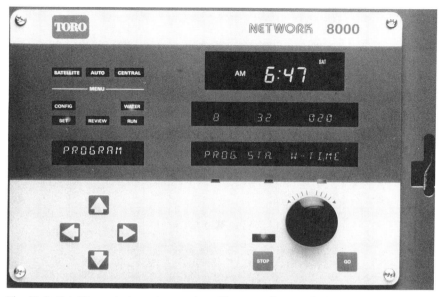

Fig. 30-9. Satellite control panel; several satellite controllers can be joined to a central controller (courtesy The Toro Company).

Golf course irrigation systems involve the use of many zones that may be out of sight of the central controller. A common practice is to use a central controller (Fig. 30-7) and several satellite controllers (Fig. 30-9) that are located at strategic locations on the course.

B. Soil Moisture Sensors

Moisture sensing as a means of determining frequency of operation for controllers is an emerging area that shows substantial promise of widespread and effective conservation of turfgrass irrigation water. Especially effective are on-site miniature weather stations that collect and store data at 1-min or less intervals, then download at a predetermined time for computer calculations of individual station running times, and then activate the irrigation system. Soil moisture sensors also may be used to activate controllers. These devices when placed in shallow and deep rootzones cancel predetermined start and run times when moisture levels reach a preselected level (Meyer & Camenga, 1985). They are subject to soil variability, but with proper programming of controllers have been shown to conserve 25 to 40% of the water normally recommended for home lawns (R.N. Carrow, 1988, personal communication).

C. Design

System components are brought together in the system design. They are connected with the appropriate size pipe and fittings. System layout is a major consideration in turfgrass irrigation. Design engineers should be consulted or, in the case of large turf areas, asked to prepare the design. Basically, a system design is always a compromise between cost and performance. The engineer in consultation with the turfgrass manager must determine the best performance by considering size of the area to be covered, hours available for watering, and amount of water to be applied. He or she also must consider precipitation rate of the sprinklers, wind speed and direction, soil type, infiltration, percolation, and service life of the equipment selected. A properly specified system with capacity to provide adequate water distribution and uniform coverage during stress periods ensures green turfgrass throughout the growing season.

REFERENCES

Anonymous. 1987. Irrigation revelations. Landscape Manage. 26(7)34–36.

Aronson, L.J., A.J. Gold, J.L. Cisar, and R.J. Hull. 1985. Water use and drought response of cool-season turfgrasses. p. 113. *In* Agronomy abstracts. ASA, Madison, WI.

Aubertine, G.M., and G.W. Gorsline. 1964. Effect of fatty alcohol on evaporation and transpiration. Agron. J. 56:50–52.

Augustine, B.J., G.H. Snyder, and E.O. Burt. 1981. Turfgrass irrigation water conservation using soil moisture sensing devices. p. 123. *In* Agronomy abstracts. ASA, Madison, WI.

Beard, J.B. 1973. Turfgrass: Science and culture. Prentice-Hall, Englewood Cliffs, NJ.

Beard, J.B. 1982. Turfgrass management for golf courses. Burgess Publ. Co., Minneapolis.

Beard, J.B. 1985. An assessment of water use by turfgrasses. p. 45–60. *In* V.B. Gibeault and S.T. Cockerham (ed.) Turfgrass water conservation. Univ. of Calif. Riverside Div. Agric. Nat. Resour. Publ. 21405.

Biran, I., B. Bravdo, I. Bushkin-Harav, and E. Rawitz. 1981. Water consumption and growth rate of 11 turfgrasses as affected by mowing height, irrigation frequency, and soil moisture. Agron. J. 73:85–90.

Busch, C.D. and W.R. Kneebone. 1966. Subsurface irrigation with perforated plastic pipe. Trans. ASAE 9:100–101.

Carroll, J.C. 1943. Effects of drought, temperature, and nitrogen on turfgrasses. Plant Physiol. 18:19–36.

Carrow, R.N. 1985. Soil/water relationships in turfgrass. p. 85–102. *In* V.B. Gibeault and S.T. Cockerham (ed.) Turfgrass water conservation. Univ. Calif. Riverside Div. Agric. Nat. Resour. Publ. 21405.

Carrow, R.N. 1988. A look at turfgrass water conservation. USGA Green Section Red. 26(4)7–9.

Carrow, R.N., and B.J. Johnson. 1988. Water use and growth of warm-season turfgrasses under different irrigation regimes. p. 234. *In* Agronomy abstract. ASA, Madison, WI.

Carrow, R.N., B.J. Johnson, and R.E. Burns. 1987. Thatch and quality of Tifway bermudagrass turf in relation to fertility and cultivation. Agron. J. 79:524–530.

Comar, C.L., and C.G. Barr. 1944. Evaluation of foliage injury and water loss in connection with use of was and oil emulsions. Plant Physiol. 19:90–104.

Crider, F.J. 1955. Root growth stoppage resulting from defoliation of grass. U.S. Tech. Bull. 1102. U.S. Gov. Print. Office, Washington, DC.

Danielson, R.E., C.M. Feldhake, and J.D. Butler. 1981a. Limited evapotranspiration by turfgrass management under water deficiencies. p. 124. *In* Agronomy abstracts. ASA, Madison, WI.

Danielson, R.E., C.M. Feldhake, and W.E. Hart. 1981b. Urban lawn irrigation and management practice for water saving with minimum effect on lawn quality. Compl. Office Water Res. Tilth Project no. A-043-Colo.

Dexter, S.T. 1937. The drought resistance of quackgrass under various degrees of fertilization with nitrogen. J. Am. Soc. Agron. 29:568–576.

Doss, B.D., O.L. Bennette, D.A. Ashley, and H.A. Weaver. 1962. Soil moisture regime effect on yield and evapotranspiration from warm-season perennial forage species. Agron. J. 54:239–242.

Engel, R.E., and R.B. Alderfer. 1967. The effect of cultivation, topdressing, lime, nitrogen and wetting agent on thatch development on 0.25 inch bentgrass turf over a ten-year period. 1967 rep. turfgrass research at Rutgers Univ. New Jersey Agric. Exp. Stn. Bull. 818.

Evans, M.W. 1949. Kentucky bluegrass. Ohio Agric. Exp. Stn. Bull. 681.

Everson, A.C. 1966. Effects of frequent clipping at different stubble heights on western wheatgrass (*Agropyron smithii* Rybd.) Agron. J. 58:33–35.

Fairborne, M.L. 1982. Water use by six grass species. Agron. J. 74:62–66.

Feldhake, C.M., R.E. Danielson, and J.D. Butler. 1981. Maximum evapotranspiration by turfgrass-environmental and management factors. p. 125. *In* Agronomy abstracts. ASA, Madison, WI.

Gibeault, V.A., J.L. Meyer, V.B. Youngner, and S.T. Cockeram. 1985. Irrigation of turfgrass below replacement of evapotranspiration as a means of water conservation: Performance of commonly used turfgrasses. p. 347–356. *In* F. Lemuir (ed.) Proc. 5th Int. Turfgrass Res. Conf., Avignon, France. July. INRA, Paris, and Int. Turfgrass Soc., Blacksburg, VA.

Glinski, D., K. Karnok, and R.N. Carrow. 1988. Influence of iron on root/shoot growth and water use of bentgrass. p. 151. *In* Agronomy abstracts. ASA, Madison, WI.

Gore, A.J.P., R. Cox, and T.M. Davies. 1979. Wear tolerance of turfgrass mixtures. J. Sports Turf Res. Inst. 55:45–68.

Hagan, R. 1955. Watering lawns and turf and otherwise caring for them. p. 462–467. *In* A. Stefferud (ed.) Water. USDA Yearb. Agric. U.S. Gov. Print. Office, Washington, DC.

Hart, R.H., and G.W. Burton. 1966. Prostrate vs. common dallesgrass under different clipping frequencies and fertility levels. Agron. J. 58:521–522.

Johns, D. 1980. Resistance to evapotranspiration from St. Augustinegrass [*Stenotaphrum secundatum* (Walt.) Kuntze] turf. Ph.D. thesis. Texas A&M Univ., College Station.

Johns, D., and J.B. Beard. 1982. Water conservation—a potentially new dimension in the use of growth regulators. Texas turfgrass research—1982. Texas A&M Univ., College Station.

Johns, D., J.B. Beard, and C.H.M. van Bavel. 1983. Resistance to evapotranspiration from a St. Augustinegrass turf canopy. Agron. J. 75:419–422.

Juska, F.V., and A.A. Hanson. 1967. Effect of nitrogen sources, rates and time of application on the performance of Kentucky bluegrass turf. Proc. Am. Soc. Hortic. Sci. 90:413–419.

Kim, K.S., and J.B. Beard. 1988. Comparative turfgrass evapotranspiration rates and associated morphological characteristics. Crop Sci. 28:328–332.

Kopec, D.M. 1985. Tall fescue soil moisture depletion, evapotranspiration and growth parameters. Ph.D. thesis. Univ. of Nebraska, Lincoln (Diss. Abstr. 85-26595).

Kopec, D.M., R.C. Shearman, and T.P. Riordan. 1988. Evapotranspiration of tall fescue turf. HortScience 23:300–301.

Kneebone, W.R. and I.L. Pepper. 1982. Consumptive water use by subirrigated turfgrasses under desert conditions. Agron. J. 74:419–423.

Krogman, K.K. 1967. Evapotranspiration by irrigated grass as related to fertilizer. Can. J. Plant Sci. 47:281–287.

Law, J.P. 1964. The effect of fatty alcohol and nonionic surfactant on soil moisture evaporation in controlled environment. Soil Sci. Soc. Am. Proc. 28:695–699.

Levy, R. (ed.). 1984. Chemistry of irrigated soils. Van Nostrand Reinhold Co., New York.

Letey, J., N. Welch, R.E. Pelishek, and J. Osborn. 1963. Effect of wetting agents on irrigation of water repellent soils. Calif. Turfgrass Culture 13(1):1–2.

Madison, J.H. 1960. Mowing of turfgrass. I. The effect of season, interval and height of mowing on the growth of Seaside bentgrass turf. Agron. J. 52:449–452.

Madison, J.H. 1962a. Mowing of turfgrass II. Responses of three species of grass. Agron. J. 54:250–253.

Madison, J.H. 1962b. Turfgrass ecology. Effects of mowing, irrigation, and nitrogen treatments of *Agrostis palustris* Huds.; 'Seaside' and *Agrostis tenuis* Sibth., 'Highland' on population, yield, rooting and cover. Agron. J. 54:407–412.

Madison, J.H. 1966. Effects of wetting agents on water movement in the soil. p. 35. *In* Agronomy abstracts. ASA, Madison, WI.

Madison, J.H. 1971. Principles of turfgrass culture. Van Nostrand Reinhold Co., New York.

Madison, J.H., and R.M. Hagan. 1962. Extraction of soil moisture by Merion bluegrass (*Poa pratensis* L. 'Merion') turf, as affected by irrigation frequency, mowing height and other cultural operations. Agron. J. 54:157–160.

Mantell, A. 1966. Effect of irrigation frequency and nitrogen fertilization on growth and water use of kikuyugrass lawn (*Pennisetum clandestinum* Hochst.). Agron. J. 58:559–561.

Marklund, F.E., and E.C. Roberts. 1967. Influence of varying nitrogen and potassium levels on growth and mineral composition of *Agrostis palustris* Huds. p. 53. *In* Agronomy abstracts. ASA, Madison, WI.

Marsh, A.W., R.A. Strohman, S. Spaulding, V. Youngner, and V. Gibeault. 1980. Turfgrass irrigation research at the University of California. Irrig. J. (July/Aug.) 1980:20–21, 32–33.

Marshall, H., and T.E. Maki. 1946. Transpiration of pine seedlings as influenced by foliage coatings. Plant Physiol. 21:95–101.

Meyer, J.L., and B.C. Camenga. 1985. Irrigation systems for water conservation. p. 103–112. *In* V.A. Gibeault and S.T. Cockerham (ed.) Turfgrass water conservation. Univ. of Calif. Riverside Div. Agric. Nat. Resour. Publ. 21405.

Mitchell, K.J., and J.R. Kerr. 1966. Differences in rate and use of soil moisture by stands of perennial ryegrass and white clover. Agron. J. 58:5–8.

Monroe, C.A., G.D.Coorts, and C.R. Skogley. 1969. Effects of nitrogen-potassium levels on the growth and chemical composition of Kentucky bluegrass. Agron. J. 61:294–296.

Oerthi, J.J. 1963. Effects of fatty alcohols and acids ontranspiration of plants. Agron. J. 55(2):137–138.

O'Neil, K.J., and R.N. Carrow. 1982. Kentucky bluegrass growth and water use under different soil compaction and irrigation regimes. Agron. J. 74:933–936.

O'Neil, K.J., and R.N. Carrow. 1983. Perennial ryegrass growth, water use and soil aeration status under soil compaction. Agron. J. 75:177–180.

Oswalt, D.T., A.R. Bertrand, and M.R. Teel. 1959. Influence of nitrogen fertilization and clipping on grass roots. Soil Sci. Soc. Am. Proc. 23:228–230.

Paul, J.L., and J.M. Henry. 1973. Nonwettable spots on greens. p. 65. *In* Proc. Calif. Golf Course Supt. Inst. Univ. Calif., Davis.

Pelishek, R.E., J. Osborn, and J. Letey. 1962. The effect of wetting agents on infiltration. Soil Sci. Soc. Am. Proc. 26:595–598.

Peterson, M.P. 1985. Evapotranspiration rates of cool-season turfgrass species. M.S. thesis. Univ. Nebraska, Lincoln.

Peterson, M.P., R.C. Shearman, and E.J. Kinbacher. 1984. Surfactant effects on Kentucky bluegrass evapotranspiration rates. p. 153. *In* Agronomy abstracts. ASA, Madison, WI.

Phillip, J.R. 1966. Plant water relations: Some physical aspects. Annu. Rev. Plant. Physiol. 17:245–268.

Powell, A.J., R.E. Blaser, and R.E. Schmidt. 1967. Effect of nitrogen on winter root growth of bentgrass. Agron. J. 59:529–530.

Prine, G.M., and G.W. Burton. 1956. The effect of nitrogen rate and clipping frequency upon yield, protein content, and certain morphological characteristics of coastal bermudagrass [*Cynodon dactylon* (L.) Pers.] Agron. J. 48:296–301.

Reierson, K.A. 1979. The effects of successive applications of Prosulfalin on Kentucky bluegrass turfs. M.S. thesis. Univ. Nebraska, Lincoln.

Rimmer, D.L. 1979. Effects of increasing compaction on grass growth in colliery spoil. J. Sports Turf. Res. Inst. 55:153–162.

Robertson, J.H. 1933. Effect of frequent clipping on the development of certain grass seedlings. Plant Physiol. 8:425–447.

Rogowsky, A.S., and E.L. Jacoby, Jr. 1977. Assessment of water loss patterns with microlysimeters. Agron. J. 69:419–424.

Sasso, C.M. 1988. Centralized irrigation control. Southwest Lawn Landscape 3(8):4–15.

Schmidt, R.D., and J.M. Breuninger. 1981. The effects of fertilization on recovery of Kentucky bluegrass turf from summer drought. p. 333–341. *In* R.W. Sheard (ed.) Proc. 4th Int. Turfgrass Res. Conf., Univ. of Guelph, Ontario. 19–23 July 1981.

Schmidt, R.E., and L.A. Everton. 1985. Moisture consumption of Kentucky bluegrass (*Poa pratensis* L.) cultivars. p. 373–379 *In* F. LeMaire (ed.) Prod. 5th Int. Turfgrass Res. Conf. Avignon, France. 1–5 July Inst. Nat. de la Res. Agron., Paris.

Shearman, R.C. 1975. Influence of nitrogen and potassium on turfgrass wear tolerance. p. 101. *In* Agronomy abstracts. ASA, Madison, WI.

Shearman, R.C. 1982a. Nitrogen and potassium nutrition influence on Kentucky bluegrass turfs. Proc. 7th Nebraska turfgrass field day and equipment show. Univ. of Nebraska Lincoln. Dep. Hortic. Publ. 82-2:67–70.

Shearman, R.C. 1982b. Turfgrass responses to plant growth regulators. Univ. Nebraska Lincoln. Dep. Hortic. Publ. 81-1:34–44.

Shearman, R.C. 1986. Kentucky bluegrass cultivar evapotranspiration rates. HortScience 21(3):455–457.

Shearman, R.C., and J.B. Beard. 1973. Environmental and cultural preconditioning effects on the water use rate of *Agrostis palustris* Huds., cultivar Penncross. Crop Sci. 13:424–427.

Shearman, R.C., E.J. Kinbacher, and K.A. Reierson. 1980. Siduron effects on tallfescue (*Festuca arundinacea*) emergence, growth and high temperature injury. Weed Sci. 28:194–196.

Shimishi, D. 1963. Effect of chemical closure of stomata on transpiration in varied soil and atmospheric environments. Plant Physiol. 38:709–712.

Sills, M.J., and R.N. Carrow. 1983. Turfgrass growth, N-use, and water use under soil compaction and N-fertilization. Agron. J. 75:488–492.

Slatyer, R.O., and J.F. Bierhuizen. 1964. The effect of several foliar sprays on transpiration and water use efficiency of cotton plants. Agric. Meterol. 1:42–53.

Snyder, V., and R.E. Schmidt. 1974. Nitrogen and iron fertilization of bentgrass. p. 176–185. *In* E.C. Roberts (ed.) Proc. 2nd Int. Turfgrass Res. Conf., Blacksburg, VA. 15–20 June 1973. ASA and CSSA, Madison, WI.

Sprague, V.G., and L.F. Graber. 1938. The utilization of water by alfalfa (*Medicago sativa*) and by bluegrass (*Poa pratensis*) in relation to managerial treatments. J. Am. Soc. Agron. 30:986–997.

Stahnke, G.K. 1981. Evaluation of antitranspirants of creeping bentgrass (*Agrostis palustris* Huds., cv. 'Penncross') and bermudagrass [*Cynodon dactylon* (L.) Pers. × *Cynodon transvaalensis* Burtt-Davy, cv. 'Tifway']. M.S. thesis. Texas A&M Univ., College Station.

Steinegger, D.H., R.C. Shearman, T.P. Riordan, and E.J. Kinbacher. 1983. Mower blade sharpness effects on turf. Agron. J. 75:479–480.

Thruman, P.C., and F.A. Pokorny. 1969. The relationship of several amended soils and compactionrates on vegetative growth, root development, and cold resistance of Tifgreen bermudagrass. J. Am. Soc. Hortic. Sci. 94:463–465.

Tovey, R., J.S. Spencer, and D.C. Muckel. 1969. Turfgrass evapotranspiration. Agron. J. 61:863–867.

Waggoner, P.E., J.L. Monteith, and G. Szeicz. 1964. Decreasing transpiration of field plants by chemical closure of stomata. Nature (London) 201:97–98.

Watson, J.R. 1974. Automatic irrigation systems for turfgrass. p. 241–245. *In* E.C. Roberts (ed.) Proc. 2nd Int. Turfgrass Res. Conf., Blacksburg, VA. 15–20 June 1973. ASA and CSSA, Madison, WI.

Watson, J.R. 1975. Hydraulic controls for turfgrass irrigation. p. 132–134. *In* Proc. Annu. Tech. Conf. Sprinkler Irrig. Assoc. Silver Spring, MD. 15–20 June. The Irrigation Assoc., Arlington, VA.

Watson, J.R. 1985. Water resources in the United States. p. 19–36. *In* V.A. Gibeault and S.T. Cockerham (ed.) Turfgrass water conservation. Univ. California Riverside, Div. Agric. Nat. Resour. Publ. 21405.

Wiecko, G., R.N. Carrow, K. Karnok, and B.J. Johnson. 1988. Turfgrass cultivation effects on growth and water/oxygen relations of bermudagrass. p. 81. *In* Agronomy abstracts. ASA, Madison, WI.

Wilkinson, J.F., and R.H. Miller. 1978. Investigation and treatment of localized dry spots on sand golf greens. Agron. J. 70:299–304.

Wood, G.M., and J.A. Burke. 1961. Effect of cutting height on turf density of Merion, Park, Delta, Newport, and common Kentucky bluegrass. Crop Sci. 1:317–318.

Wooley, J.T. 1967. Relative permeabilities of plastic films to water and carbon dioxide. Plant Physiol. 42:641–643.

Youngner, V.B., A.W. Marsh, R.A. Strohman, V.A. Gibeault, and S. Spaulding. 1981. Water use and turf quality of warm-season amd cool season turfgrasses. p. 251–257. *In* R.W. Sheard (ed.) Proc. 4th Int. Turfgrass Res. Conf., Univ. Guelph, Ontario, Canada.

Zelitch, I., and P.E. Waggoner. 1962. Effect of chemical control of stomata on transpiration and photosynthesis. Proc. Natl. Acad. Sci. 48:1101–1108.

31   Vegetables[1]

C. D. STANLEY AND D. N. MAYNARD

University of Florida
Gulf Coast Research and Education Center
Bradenton, Florida

Vegetable crops are a diverse grouping of herbaceous, mostly annual plants whose products are used for culinary purposes. Specific plant parts that are consumed include root and hypocotyl, stem and tuber, leaf, flower, seeds, and immature and mature fruit. Vegetables are a major component of the main and salad courses of the meal and may be featured in the appetizer and dessert portions as well. At least 30 different species in a dozen plant families are considered to be common vegetables. The number of vegetables increases several fold when specialty, ethnic, and exotic crops are included.

Vegetables are an important dietary component providing rich sources of essential elements and vitamins. In addition, most vegetables are low in caloric value. Recent emphasis has been placed on increased consumption of vegetables high in fiber because of the association between certain types of cancer and low-fiber diets. Finally, vegetables are important because of the variety of tastes, colors, and textures they provide in the meal.

Vegetables are grown commercially in every state of the USA and in virtually every country of the world. The industry is quite localized, however, into distinct growing regions dictated mostly by climate. Secondary considerations include availability of productive soils, labor, and resources such as water.

For the most part, vegetables are high-value crops that are grown intensively. Management, labor, and capital investments are high; accordingly the capability to irrigate vegetable crops is necessary and commonplace. In arid and semiarid areas, irrigation may supply all or most of the crop's water needs. In more humid production areas, irrigation is used primarily to supplement infrequent or irregular precipitation during short-term droughts. In any case, the ability to properly irrigate vegetable crops is mandatory for successful commercial production.

---

[1] Florida Agric. Exp. Stn. Journal Series No. 8044. The literature review for this chapter was completed on 28 Feb. 1987.

Table 31-1. Economically important plant parts for various vegetables.

Harvested plant part		
Leaves, flower parts, or stems	Underground parts	Fruits or seeds
Artichoke	Beet	Cucumber
Asparagus	Carrot	Eggplant
Beet	Garlic	Lima bean
Broccoli	Horseradish	Muskmelon
Brussels sprouts	Jerusalem artichoke	Okra
Cabbage	Leek	Pea
Cauliflower	Onion	Pepper
Celery	Potato	Pumpkin
Chive	Radish	Snap bean
Collard	Rutabaga	Southern pea
Endive	Sweet potato	Squash
Lettuce	Turnip	Sweet corn
Mustard	Yam	Tomato
Parsley		Watermelon
Rhubarb		
Spinach		
Turnip		
Watercress		

I. CROP GROWTH AND DEVELOPMENT

Since vegetable crops are grouped into a single crop category consisting of numerous diverse plant species, a general discussion of crop growth and development is difficult. No single description can cover the range of growth and development patterns found. Examples of the diversity of growth habits found within the vegetable group include the bulb-crop growth habit of onion (*Allium cepa* L.), the unique formation of miniature heads at the leaf axils of brussels sprouts [*Brassica oleracea* L. (Gemmifera Group)], the vining characteristic of vegetables within the Cucurbitaceae family, and the bush-canopy structure of vegetables within the Leguminosae family.

The relationship between irrigation and vegetable crop growth and development can be affected by several factors. These include the economically important portion of the vegetable crop (Table 31-1) and the stage of growth at which it is harvested. The harvested plant part can include immature flowers, stems, leaves, tubers, roots, seeds, or fruits. With most annual vegetable crops, irrigation is used only until the condition of market maturity is reached. With few exceptions, vegetable crops do not normally complete their natural growth cycle to the stage of leaf senescence or physiological maturity, as is the case with most agronomic crops. It is the goal of irrigation to avoid water stress, especially during the formation of the harvest plant part. Table 31-2 (Lorenz & Maynard, 1980) contains estimates of the time required for selected vegetable crops to reach market maturity from seeding or transplanting. This length of season can be extended somewhat during favorable market conditions for some hand-harvested, fresh market crops.

Table 31-2. Approximate time from planting to market maturity under optimum conditions for selected vegetables (Lorenz & Maynard, 1980).

| Vegetable | Time to market maturity, d | | |
	Early variety	Late variety	Common variety
Bean, snap	48	60	--
Bean, lima, bush	65	78	--
Beet	56	70	--
Broccoli†	55	78	--
Brussels sprouts†	90	100	--
Cabbage†	62	120	--
Carrot	50	95	--
Cauliflower†	50	125	--
Celery†	90	125	--
Collard	70	85	--
Corn, sweet	64	95	--
Cucumber, pickling	48	58	--
Cucumber, slicing	62	72	--
Eggplant†	50	80	--
Endive	85	100	--
Kale	--	--	55
Leek	--	--	150
Lettuce, head	70	85	--
Lettuce, leaf	40	50	--
Melon, mixed	--	--	110
Muskmelon	85	95	--
Mustard	35	55	--
Okra	50	60	--
Onion, dry	90	150	--
Onion, green	45	60	--
Parsley	70	80	--
Parsnip	--	--	120
Pea	56	75	--
Pepper, sweet†	65	80	--
Pumpkin	100	120	--
Radish	22	30	--
Spinach	37	45	--
Squash, summer	40	50	--
Squash, winter	85	110	--
Sweet potato	120	150	--
Tomato†	60	90	--
Turnip	40	75	--
Watermelon	75	95	--

† From transplanting.

The rooting characteristic of a vegetable crop is another growth factor which can affect irrigation practices. Table 31-3 (Lorenz & Maynard, 1980) contains information describing the characteristic rooting depths for specific vegetables grown on uniform, well-drained soil. Rooting depth information for crops grown on specific soils is important for irrigation scheduling decisions. For example, a shallow-rooted crop would normally be irrigated more frequently with lesser amounts of water than a deep-rooted crop. This irrigation schedule is practiced because less soil volume for water extraction is available, and avoidance of overirrigation is desired. Vegetable rooting

Table 31-3. Characteristic rooting depths of various vegetables (Lorenz & Maynard, 1980).

Shallow (46–61 cm)	Moderately deep (91–122 cm)	Deep (>121 cm)
Broccoli	Bean, bush	Artichoke
Brussels sprouts	Bean, pole	Asparagus
Cabbage	Beet	Bean, Lima
Cauliflower	Carrot	Parsnip
Celery	Chard	Pumpkin
Chinese cabbage	Cucumber	Squash, winter
Corn	Eggplant	Sweet potato
Endive	Muskmelon	Tomato
Garlic	Pea	Watermelon
Leek	Pepper	
Lettuce	Rutabaga	
Onion	Squash, summer	
Parsley	Turnip	
Potato		
Radish		
Spinach		

depths can be dramatically affected by irrigation practices such as method (i.e., subsurface vs. overhead sprinkler), amount of water applied, and schedule of applications.

II. EVAPOTRANSPIRATION

Many factors can affect the amount of evapotranspiration (ET) occurring for any particular vegetable crop. These include plant, soil, cultural, and environmental factors (Jones et al., 1984). Under nonlimiting irrigated conditions, daily ET rates for individual vegetable crops are directly related to the meteorological processes affecting evaporative demand, and to the existing stage of growth development or percent crop coverage. Many direct and indirect methods have been developed and evaluated (Jones et al., 1984; Blaney & Criddle, 1962) to estimate the potential or maximum ET for agricultural crops (see chapter 15 in this book). Estimates of actual ET for a particular crop are generally determined by a correction factor (specifically determined for that crop and its growing conditions) applied to the estimated potential ET.

Since widespread diversity of plant types, and environmental and cultural conditions under which they are grown, exists within the vegetable crop group, a general discussion of ET requirements for vegetable crops is as difficult as describing crop growth and development. Even within a plant species, water use can vary dramatically because of location and cultural method. For example, the major tomato (*Lycopersicon esculentum* Mill.) production areas in the USA range from the humid subtropical climate of southern Florida, to the humid temperate climate of the Midwest, to the arid subtropical climate of some parts of California. Estimates of ET requirements for vegetable crops range from 250 to 750 mm per crop based on location alone (Doorenbos & Pruitt, 1977). In addition, tomato can be grown

Table 31-4. Estimated crop coefficients (k_c) of selected vegetable crops at different crop development stages (Doorenbos & Kassam, 1979).

Crop	Initial	Crop establish-ment	Mid-season	Late season	At harvest	Total growing period
Bean, green	0.3-0.4	0.65-0.75	0.95-1.05	0.90-0.95	0.85-0.95	0.85-0.90
Cabbage	0.4-0.5	0.70-0.80	0.95-1.10	0.90-0.95	0.80-0.95	0.70-0.80
Corn, sweet	0.3-0.5	0.70-0.90	1.05-1.20	1.00-1.15	0.95-1.10	0.80-0.95
Onion, dry	0.4-0.6	0.70-0.80	0.95-1.10	0.85-0.90	0.75-0.95	0.80-0.90
Onion, green	0.4-0.6	0.60-0.75	0.95-1.05	0.95-1.05	0.95-1.05	0.65-0.80
Pea	0.4-0.5	0.70-0.85	1.05-1.20	1.00-1.15	0.95-1.10	0.80-0.95
Pepper	0.4-0.5	0.60-0.75	0.95-1.20	1.00-1.15	0.95-1.10	0.70-0.80
Tomato	0.4-0.5	0.70-0.80	1.05-1.25	0.80-0.95	0.60-0.65	0.75-0.90
Watermelon	0.4-0.5	0.70-0.80	0.95-1.05	0.80-0.95	0.65-0.75	0.75-0.85

for fresh market or processing, with or without polyethylene mulch, as staked or ground tomatoes, and with varying plant densities which result in 33 to near 100% maximum crop coverage.

The irrigation system that is used can also affect the actual ET of a vegetable crop under specific conditions. For example, water use from a widely spaced, polyethylene-mulched vegetable crop grown with microirrigation on a sandy soil with nominal rainfall would primarily consist of transpiration, with limited evaporation from the soil once a dry surface boundary layer formed. If the same crop were grown under the same conditions, except with a seepage subirrigation system (Geraldson, 1980), there would be a substantial amount of evaporation from the soil due to the continually moist soil surface condition, resulting in an increase in total crop ET.

Research studies performed to determine ET rates for selected vegetable crops grown under defined growing conditions basically have involved use of direct or indirect methods for estimating potential ET, and use of developed crop coefficients (k_c) particular to the desired crop in order to determine estimated crop ET. These k_c values can be derived for any specific period of crop development. Wright (1982), using the Penman method for determining potential ET along with weighing lysimeters, determined crop coefficients for a number of agronomic and vegetable crops at specific crop development stages. Erie et al. (1965) used the Blaney-Criddle method for estimating potential ET, with derived k_c values presented for selected vegetable crops grown in the semiarid climate of Arizona. Seasonal k_c values ranged from 0.63 for carrot (*Daucus carota* L.) to 0.98 for sweet corn (*Zea mays* L.) and were determined specifically for each crop and stage of development. Similar values derived for use with the pan evaporation method of potential ET estimation are presented in Table 31-4 for selected vegetable crops (Doorenbos & Kassam, 1979). These values can be useful for quantifying differences in water use among vegetable crops grown under similar environmental conditions at different stages of development.

As stated earlier, there can be considerable differences between estimations of ET requirements for crops from one location to another. These are generally the result of differences in local environmental conditions. Most such differences can be accounted for by the use of potential ET estimation methods. More difficult to standardize are the cultural conditions under which ET estimations are made. Any estimation of ET requirements for vegetable crops must be accompanied by a description of the associated conditions.

III. WATER-STRESS RESPONSE

Plant water stress caused by soil moisture deficits can affect plant physiological processes such as ion uptake and translocation, transpiration, photosynthesis, respiration, and overall plant growth and development (Crafts, 1968). As with any economically important agricultural or horticultural crop, avoidance of water stress which affects yield production is desired. This is particularly true for most vegetables because of the high investment costs associated with producing these crops.

Detection of plant water stress in vegetable crops can be accomplished by measurement of plant water content, leaf water potential, osmotic potential, turgor pressure, or stomatal aperature and conductance (Barrs, 1968). Recently, thermal-infrared techniques relating crop canopy temperatures to stress conditions (Jackson, 1982; Jackson et al., 1986) and other remotesensing techniques have been developed.

The use of such techniques allows critical plant water stress levels to be identified, along with specific growth development stages where stress particularly should be avoided with respect to yield of the economically important vegetable parts. Rudich et al. (1981) used leaf water potential measurements to describe the effect that soil water deficits have on tomato production. One study (Manning et al., 1977), where soil moisture deficits were maintained throughout the growing season for pea (*Pisum sativum* L.) production, showed that plant growth characteristics (stomatal density, leaf thickness, plant height, and seed yield) were affected by water stress conditions. Water stress imposed at prebloom, bloom, and postbloom growth stages of snap bean (*Phaseolus vulgaris* L.) caused overall yield reductions compared to nonstress treatments (Stansell & Smittle, 1980). Certain vegetable crops respond differently to water stress conditions, particularly at specific growth-development stages. Table 31–5 contains a listing of critical growth stages for selected vegetable crops where water stress can significantly affect yield.

Crop water stress can occur when soil moisture is excessive, as well as when it is deficient. When O_2 concentration levels in the soil atmosphere are lowered because of displacement by water for an extended period of time, severe damage to the root system can occur (Stolzy, 1972). This damage, in turn, can lower plant water uptake and cause stress to the plant as a whole.

Prevention of water stress for vegetable crops is primarily accomplished by proper use of irrigation. Vegetable crops can be especially susceptible to water stress because of the shallow rooting characteristics which many of them exhibit (Table 31–3). Efficient use of water to avoid stress thus requires

Table 31-5. Critical growth periods when moisture stress adversely affects vegetable yield and quality (Robson & Johnson, 1985).

Crop	Critical period
Asparagus	Brush
Bean, lima	Pollination and pod development
Bean, snap	Pod enlargement
Broccoli	Head development
Cabbage	Head development
Carrot	Root enlargement
Cauliflower	Head development
Corn	Silking and tasseling, ear development
Cucumber	Flowering and fruit development
Eggplant	Flowering and fruit development
Lettuce	Head development
Melon	Flowering and fruit development
Onion, dry	Bulb enlargement
Pea	Flowering and pod fill
Pepper	Flowering and fruit development
Potato, white	Tuber set and tuber enlargement
Potato, sweet	Root enlargement
Radish	Root enlargement
Squash, summer	Bud development and flowering
Tomato	Early flowering, fruit set and enlargment
Turnip	Root enlargement

irrigation scheduling to take into account crop water needs, critical growth stages, rooting characteristics, soil water holding and transmitting characteristics, and proper selection of irrigation system (Hiler & Howell, 1983).

IV. GROWTH AND SOIL WATER

The growth response of vegetable crops to varying soil moisture conditions is highly dependent on plant species (Williamson & Kriz, 1970). The goal of irrigation management is to maintain soil water potential levels in the -10 to -30 kPa range for vegetables grown under irrigated conditions in order to minimize water stress and maximize production for these high-value crops.

Excessive soil water conditions can have a profound effect on crop growth for many vegetable crops. Tomato has been often studied for growth response to waterlogged soil conditions. A common growth response is leaf epinasty, which is associated with the accumulation of high C_2H_4 concentrations under the anaerobic soil conditions (Bradford & Yang, 1981). Other growth responses to waterlogged conditions (Bradford & Yang, 1981) include: (i) reduction in stomatal conductance, (ii) increase in plant concentrations of indoleacetic acid and abscisic acid, (iii) reduction of concentrations of cytokinins and gibberellins, (iv) wilted appearance resulting in reduced leaf area for light interception, (v) unchanged leaf water potential levels, and (vi) the death of small roots. The degree to which crops express these growth characteristics depends on the longevity of the waterlogged conditions to which they are exposed (Poysa et al., 1987). Kahn et al. (1985) found that black bean (*P. vulgaris* L.) plants flooded for 4 to 7 d and then drained for 7 d showed a reduction in total root mass and a higher proportion of adventitious roots compared to basal roots. They concluded that young black bean plants recover from flooding injury by partitioning dry matter to their roots, resulting in a reduction of the shoot/root ratio compared to normal, unstressed plants. Kawase (1981) comprehensively discussed how plants can adapt to waterlogged soil conditions.

V. PLANT NUTRITION

In most cases, no other aspect of the management of vegetable crop production is affected as much by irrigation practices as is nutrient management. The interdependent nature of the availability of both water and essential nutrients to the crop is so strong that limited availability of one limits the vegetable crop response to the other. Irrigation management is important not only to provide for water stress avoidance, but also to provide nonlimited or more controlled uptake of these nutrients required for optimum production.

Because of the many cultural practices used in combination with various irrigation systems for vegetable production, application and placement of fertilizers is extremely important. For example, in Florida, fresh market tomato grown on sandy soils with a polyethylene-mulched bed using seepage irrigation generally requires banded application of N and K in the bed (Geraldson, 1980). The nutrients become available in the soil solution because of the presence of the capillary fringe above the underlying water table. Net movement of water and nutrients tends to be static or slightly upward unless heavy rainfall is experienced, when drainage can cause substantial amounts of fertilizer to be leached downward. Since this system is vulnerable to fertilizer leaching following excessive precipitation, the rates used often exceed plant requirements, but are necessary to provide nonlimiting nutrition to the crop throughout the growing season.

The same crop grown on the same soil by using microirrigation would have definite differences in nutrient management requirements. Generally, 40% of the N and K are applied as a dry fertilizer and 60% are applied as liquid fertilizer through the irrigation tubing (Kovach, 1984). Rate and timing of irrigation is important with microirrigation. Underirrigation can result in decreased water availability, and overirrigation can result in the movement of nutrients beyond the crop rooting zone, thereby decreasing nutrient availability to the crop. A distinct advantage of this system is its ability to provide the desired amount of nutrients at the desired stage of crop growth. This becomes increasingly important when markets are such that the extension of a particular vegetable crop's production season is desired.

Fertilizer, especially N, also can be applied with the irrigation water (fertigation) in surface and sprinkler systems (Gardner & Roth, 1984). With surface systems, the distribution of the fertilizer is directly related to the distribution of irrigation water. Thus, efficiency of nutrient application is

Tabel 31-6. Relative salt tolerance of vegetable crops (Maas, 1986).

| Vegetable | Electrical conductivity of saturated soil extract | | Rating† |
	Threshold	Slope	
	dS m^{-1}	% per dS m^{-1}	
Artichoke	--	--	MT
Asparagus	4.1	2.0	T
Bean	1.0	19.0	S
Beet	4.0	9.0	MT
Broccoli	2.8	9.2	MS
Brussels sprouts	--	--	MS
Cabbage	1.8	9.7	MS
Carrot	1.0	14.0	S
Cauliflower	--	--	MS
Celery	1.8	6.2	MS
Corn, sweet	1.7	12.0	MS
Cucumber	2.5	13.0	MS
Eggplant	--	--	MS
Kale	--	--	MS
Kohlrabi	--	--	MS
Lettuce	1.3	13.0	MS
Muskmelon	--	--	MS
Okra	--	--	S
Onion	1.2	16.0	S
Parsnip	--	--	S
Pea	--	--	S
Pepper	1.5	14.0	MS
Potato	1.7	12.0	MS
Pumpkin	--	--	MS
Radish	1.2	13.0	MS
Spinach	2.0	7.6	MS
Squash, scallop	3.2	16.0	MS
Squash, zucchini	4.7	9.4	MT
Strawberry	1.0	33.0	S
Sweet potato	1.5	11.0	MS
Tomato	2.5	9.9	MS
Turnip	0.9	9.0	MS
Watermelon	--	--	MS

† T = tolerant, MT = mostly tolerant, MS = moderately sensitive, S = sensitive.

directly related to irrigation water application efficiency. Erie and Dedrick (1979) found that, when properly managed, level basin flood irrigation systems can approach 92 to 96% distribution efficiencies. However, not all surface systems approach this level of efficiency. In general, fertigation is not recommended with surface irrigation systems.

Since the quality and appearance of the harvested vegetable plant part is extremely important to its market value, the effect of nutritional deficiencies (especially N) on quality attributes is also important (Locascio et al., 1984). Maynard (1979) provided a review of the literature concerning deficiencies of several nutrients and subsequent effects on vegetable crop growth.

Saline growing conditions—whether caused by soil conditions, the salinity of the irrigation water, or excessive fertilization—can also significantly affect vegetable production (Maas, 1985). The relative tolerance of various vegetable crops to saline conditions is given in Table 31-6 (Maas, 1986). Most of the crops listed fall into the sensitive to moderately sensitive classification groups. This indicates that vegetables, in general, are somewhat intolerant of saline growing conditions. Studies on sweet corn, a moderately sensitive crop, have shown that saline conditions can be managed by shortening irrigation intervals using microsprinklers (Shavhevet et al., 1986) and by proper placement of emitters when microirrigation is used (Goldberg & Shmueli, 1971b). The influence of saline irrigation water on foliage and root tolerance has also been studied for some vegetable crops. Maas (1986) concluded that, for sprinkler-irrigated crops such as cucumber (*Cucumis sativus* L.), tomato, and green pepper (*Capsicum annuum* L.), the threshold for tolerance to saline irrigation water was higher for the root systems than for the foliage. The relationship was reversed with strawberry (*Fragaria* × *ananassa* Duch.), in which the roots were much more susceptible to saline conditions than the foliage.

VI. CULTURAL PRACTICES

Vegetable production often utilizes many unique cultural practices which are not common to other agricultural crops. Many of these practices may affect crop water requirements and water use efficiency. These practices include the use of transplants as opposed to direct seeding, and use of raised production beds, bed mulches, varying planting densities, multiple cropping, and staking or trellising.

Transplants often are used for vegetable crops that have characteristically small seeds or have difficulty in stand or for earlier crop establishment. This cultural practice also decreases plant-stand variability and increases mature crop uniformity, thus facilitating harvest. Containerized transplants are commonly grown in flats which allow independent development of each plant's root system separate from that of each other plant, making the transplanting operation more efficient. Since this operation commonly occurs under intensively controlled growing conditions (such as in a greenhouse or hothouse), control of irrigation application can be easily managed to provide a high level of water use efficiency (WUE).

The use of raised beds for vegetable production is a common cultural practice which usually accompanies furrow irrigation systems to prevent excessive innundation in the primary rooting zone in the bed. It also facilitates soil aeration which can prevent problems with many soil-borne root diseases. Raised bed culture also can help prevent crop losses due to excessive rainfall. In areas of the country where vegetables are grown on shallow soils, raised beds increase the available rooting volume by concentrating much of the soil formerly located between the rows in order to build the bed.

Polyethylene mulch (especially with raised beds) is used extensively in many parts of the USA and elsewhere. This practice is used to conserve water, control weeds, increase crop quality, manipulate soil temperatures, and facilitate soil fumigation for nematode and disease control. Water conservation is achieved by decreasing evaporation from the mulched soil surface. More efficient management of applied nutrient loss can be accomplished through the use of mulches (Sweeney et al., 1987). Weed control is accomplished by providing a physical barrier to weed growth by eliminating sunlight and by allowing the use of fumigants. The mulch contains the fumigant for a period of time sufficient for the fumigant to be effective. This procedure is also used with polyethylene mulches for nematode and disease control.

Mulches can help increase crop quality by preventing maturing fruit of certain vegetable crops from coming in contact with the exposed soil surface. Furthermore, dark-colored mulches enhance warming of the soil which can promote earlier vegetable maturity. However, light-colored mulches may be necessary for production during warmer periods of the year to prevent excessive bed temperatures.

VII. IRRIGATION MANAGEMENT

A. Vegetable Irrigation Systems

Irrigation systems used for vegetable production range from surface systems such as overhead sprinkler, furrow, micro, and level basin irrigation, to subsurface systems such as seepage (water table control) and some micro-irrigation systems. The choice of system depends on soil type, crop, climate, water supply, economics, and personal preference (Finkle & Nir, 1983).

Overhead sprinkler irrigation is often used where other systems are not feasible on soils where the slope or infiltration rate is too great and water supply is too limited for use of furrow irrigation. Sprinkler irrigation is also of benefit for vegetable crops that may be intolerant of excessively wet soil conditions, and it can be used for transplant establishment, fertilizer application, crop cooling, and frost or freeze control. However, overhead sprinkler systems can also increase incidence of foliar diseases, which are enhanced by the presence of moisture on the leaf surfaces.

Other surface irrigation methods such as furrow irrigation and level basin irrigation require soils that are level or have gently sloping topography, and infiltration rates and subsequent water-holding capacities that are sufficient to support the crop for several days. Cropped areas are irrigated at frequencies corresponding to crop needs and soil water availability, and soil water-holding characteristics. Crops are commonly grown on raised beds when using such systems. The irrigation efficiency for these systems depends on design and management. For example, level basin irrigation eliminates runoff, thereby increasing the application efficiency.

One subirrigation system that can be used for vegetable production is the seepage irrigation system (Geraldson, 1980) in which water can be delivered either through surface irrigation ditches or through subsurface drains. The latter delivery system can also provide subsurface drainage when excessive rainfall is received. Seepage irrigation requires sandy or muck soils with either an impermeable layer located between 45 and 60 cm below the soil surface on which a perched water table can be maintained, or with a naturally shallow water table which can be artificially raised to a desired level. Water moves upward from the water table to the crop root system by capillary action. A careful balance between irrigation and drainage must be maintained to prevent excessive fluctuation of the water table and subsequent leaching of applied fertilizer. This is a common irrigation system for substantial areas of vegetable production in Florida on both sandy and organic muck soils.

Microirrigation is used primarily where water resources are restricted and high-value crops are grown or where steep slopes preclude the use of other irrigation methods (Finkle & Nir, 1983). Water is applied through polyethylene tubing with various emission-port configurations. These systems are operated under low pressure such that water is only delivered to soil areas where the crop roots are concentrated (Clark et al., 1987). Tubing may either be buried or located on the soil surface. With careful management, this system can reduce irrigation requirements substantially over other systems. Fertilizers and even some pesticides can be applied with the water. Another advantage of this system is that it can be easily automated to control irrigation amounts and frequency, and timing of application. However, initial costs are high (Prevatt et al., 1981, 1984) and the system requires careful maintenance to prevent emitter clogging and to insure uniform operation.

B. Irrigation Scheduling Considerations

Because of the high investment costs and economic value of most vegetable crops, scheduling of irrigation becomes important to prevent crop water stress which may cause economic production losses. Consideration must be given to the crop water requirements, soil conditions, crop rooting characteristics, water availability and irrigation system limitations when determining the most beneficial irrigation schedule.

Irrigation for vegetable crops can be scheduled in response to actual climatic demand conditions by using some method of estimating plant water requirements. Such scheduling requires flexibility of the irrigation system in order to adjust frequency and amount of water applied to correspond to changing plant growth stages and climatic conditions.

Scheduling according to soil moisture depletion requires monitoring of soil moisture levels as they change during the season. Allowable depletion in the rooting zone and amount of irrigation water needed to replenish the depleted moisture are factors that affect how frequently irrigation would be scheduled. For example, many vegetable crops are shallow-rooted—such as lettuce (*Lactuca sativa* L.), cabbage [*B. oleracea* L. (Capitata Group)], and radish (*Raphanus sativas* L.)—and thus require frequent, light irrigations. This type of schedule is especially important where fertilizer or other chemicals are applied through the irrigation water to avoid leaching.

C. Irrigation Management for Specific Vegetable Crops

As presented in earlier parts of this chapter, irrigation management of vegetable crops can vary dramatically with respect to plant species, cultural method, location, and climate. This section deals more specifically with irrigation management of individual vegetable crops grown under specific conditions.

1. Perennial Crops

Although most vegetable crops are grown as annuals, three exceptions are asparagus (*Asparagus officinalis* L.), globe artichoke (*Cynara scolymus* L.), and rhubarb (*Rheum rhabarbarum* L.). These crops are deep rooted and, consequently, response to irrigation is generally less than with most shallow-rooted annuals.

In the eastern USA, supplemental overhead irrigation is recommended (Robson & Johnson, 1985) to maintain vigorous asparagus fern growth during summer. This promotes high yields in the next and subsequent spring cutting seasons. Irrigation is usually not necessary during the cutting season because spring rains supply sufficient moisture. Furrow irrigation may be used for asparagus growing on mineral soils, while subirrigation is used on organic soils in California. Thorough wetting by irrigation every 10 to 14 d is generally practiced. Irrigation is stopped in October to allow the plants to become dormant (Takatori et al., 1977).

The potential use of geothermal water for asparagus irrigation was investigated by Robinson et al. (1984) on the Imperial East Mesa Desert, California. Average yields were severely restricted to 0.03 t ha^{-1} because of excessive total dissolved solids (2240 $\pm$ 1290 mg L^{-1}) in the geothermal water, in comparison to 3.09 t ha^{-1} obtained from asparagus irrigated with local groundwater (1430 $\pm$ 460 mg L^{-1}).

Commercial globe artichoke production in the USA occurs only in California. Topography dictates the method of water application; furrow irrigation is used on level land whereas sprinkler irrigation is used on rolling land. The timing of production may be controlled by the initiation of irrigation. May or June irrigation results in early winter production. Later initial irrigations result in later production. Usually five to eight thorough irrigations to the full rooting depth will carry this crop through to harvest. Moisture stress during the harvest period causes loose buds of inferior quality (Sims et al., 1977b).

Field rhubarb is grown mostly in the humid eastern areas of the northern USA and southern Canada. Supplemental irrigation is rarely used on this deep-rooted perennial.

2. Greens

Several vegetables are classed as greens including spinach (*Spinacia oleracea* L.), kale and collard [*B. oleracea* L. (Acephala Group)], mustard (*B. juncea* L.), and others of less economic importance. Although diverse botanically, they are all short-season annuals with shallow root systems which are adapted to cool weather when evapotranspiration is low. Little is known about the water requirements of these crops. Where irrigation is used, overhead sprinkler is usually the method of choice.

3. Salad Crops

Celery (*Apium graveolens* L.) and lettuce are the principal salad crops. Numerous other salad vegetables are of less economic importance. All are shallow-rooted and most, with the exception of celery, require a relatively short time to reach marketable size.

California and Florida are the leading celery-producing areas of the USA. Loam and clay loam soils in cool coastal California counties, and organic soils in central and south Florida are used for celery production. Irrigation practices vary with the soils. In California, celery is sprinkler-irrigated for transplant establishment or prevention of soil crusting for direct-seeded crops, whereas furrow irrigation is used for crop production (Sims et al., 1977c). On the other hand, subsurface irrigation is used for celery production in Florida following sprinkler irrigation for transplant establishment (Guzman et al., 1973).

Celery yields were higher with 500 mm microirrigation than with 400 mm in California (Feigin et al., 1982). Preplant-applied NH_4SO_4 or resin-covered, controlled-release fertilizer further resulted in higher yields at the lower water application rate whereas higher yields were obtained at the higher water application rate when urea-ammonium nitrate $[CO(NH_2)_2 \cdot NH_4NO_3]$ was applied through the microsystem. Excessive leaching of the preplant fertilizer was believed responsible for these divergent results (Feigin et al., 1982).

California produces the bulk of the lettuce in the USA. Furrow irrigation has been used for irrigation of lettuce there since the early 1900s. Because of higher water costs, increased salinity, soil-crusting problems, and excessive leaching of N fertilizer, alternatives to furrow irrigation have been explored for at least part of the crop's water needs.

The inefficiencies of furrow irrigation for lettuce were demonstrated by Moore (1970). About 1500 mm of water per season was applied to grow the lettuce. Surface drainage losses averaged 20% and percolation below the root zone accounted for 50% of the water applied. Leaching of NO_3-N averaged 100 kg ha^{-1}. Two-thirds of the water loss and three-fourths of the N loss occurred prior to thinning.

Robinson and McCoy (1965) compared furrow and sprinkler irrigation for lettuce grown in the Imperial Valley of California. Sprinkler irrigation reduced water use by 50% up to thinning. Furthermore, the soluble salt concentration in the seedling root zone was lowered slightly from preirrigation concentrations, and was much less than that in corresponding furrow-irrigated plots. More uniform seedling growth resulted and this uniformity continued to harvest so that the number of harvests was less for sprinkler-irrigated plots.

Although furrow irrigation is still used for California lettuce production, the crop is generally germinated with sprinkler irrigation. This practice alleviates the problems of excess water usage, N leaching, salinity, and soil crusting which were mentioned earlier. Expansion of sprinkler use to entire season production of lettuce is limited by capital and operating costs and by the likelihood of increased foliar-disease problems with frequent leaf wetting.

Yields and WUE were determined for sprinkler-, furrow-, surface micro-, and subsurface micro-irrigated lettuce (Sammis, 1980). Similar yields and WUEs were obtained with soil water potentials maintained at −20 kPa (sprinkler), and −20 kPa and −60 kPa (surface or subsurface micro irrigation). Comparable yields and WUEs were obtained from furrow irrigation using short runs which may not be practical in the field. Lower yields were consistently obtained from the drier −60 kPa soil water potential treatment.

4. Crucifers

Cabbage, broccoli [*B. oleracea* L. (Italica Group)] and cauliflower [*B. oleracea* L. (Botrytis Group)] are the economically most important vegetables in the Cruciferae family. Perhaps 20 additional species in this family are used as vegetables. Evapotranspiration is lower than for many other vegetables because of the thick, waxy leaf covering common to *Brassica* and the cooler weather in which most of these crops are grown. In the USA, winter production occurs in Florida and Texas, spring and fall production takes place in more northerly states, and production in various areas of California provides a year-round supply. These vegetables are shallow to moderately deep-rooted.

Vittum and Flocker (1967) stressed the importance of maintaining adequate, uniform soil moisture throughout the crop cycle. Water deficits, particularly in the 3 to 4 wk prior to harvest, lower crop yields and quality. On the other hand, excess water during this period may contribute to cabbage head bursting.

Cabbage water requirements vary from 380 to 500 mm per season depending on climate, cultivar, and growing season. Crop demand during the initial 20 to 30 d averages 0.4 to 0.5 times the potential evapotranspiration (ET_p), 0.7 to 0.8 × ET_p during the next 30 to 35 d, 0.95 to 1.1 × ET_p in the next 20 to 30 d, 0.9 to 1.0 × ET_p for the next 10 to 20 d, and 0.8 to 0.95 times ET_p at harvest. Hence, water deficits late in the crop season are more critical than those early in the crop season (Doorenbos & Kassam, 1979).

Water and N management are often inseparable and, together, exert a critical influence on crop performance. Studies were conducted in the lower Rio Grande Valley of Texas—an important cabbage-growing district—with respect to these variables (Thomas et al., 1970). Highest yields were obtained when cabbage was irrigated as the average soil water potential of the surface 61 cm approached −80 to −160 kPa. Yields were restricted when irrigation was delayed until soil water potentials reached −360 kPa and did not increase at this soil moisture level when N fertilization was increased from 0 to 358 kg ha^{-1}. At the higher irrigation rates, yield was maximized at 269 kg ha^{-1} N application.

Nitrogen and irrigation management of broccoli have been investigated in California (Letey et al., 1983). Furrow irrigation to replace evapotranspiration resulted in higher yields than irrigation levels 1.3 times ET_{crop}. At the lower irrigation rate, there was no difference between soil or irrigation-applied N. At the higher irrigation rate, irrigation-applied N gave higher yields.

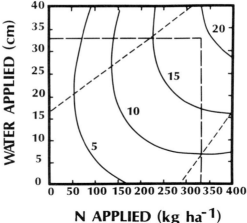

Fig. 31-1. Yield isoquants (Mg ha^{-1}) for water and N application treatments. Dashed lines denote treatment boundaries and dotted lines are ridgelines.

Regardless of irrigation rate or N application method, highest yields were obtained with 270 kg ha^{-1} N.

In more recent studies with broccoli (Beverly et al., 1986), a response surface (Fig. 31-1) was developed from the equation

$$Y = -1.03 + 0.0498(N) - 8.73 \times 10^{-5}(N^2)$$
$$+ 1.23 \times 10^{-3}(NW) + 0.272(W) - 8.20 \times 10^{-3}(W^2)$$

where Y = yield (Mg ha^{-1}), N = nitrogen (kg ha^{-1}), and W = water (cm). Use of the response surface permits varying of N and irrigation based on economic considerations to attain a particular yield.

5. Root Crops

Carrot is the most important root crop economically. Of significant, but lesser, value are radish and beet (*Beta vulgaris* L.). Several other minor root crops are also grown commercially.

Carrot is grown in deep sandy or sandy loam mineral soils or in organic soils because impediments to storage root elongation cause forked or misshapen roots. In commercial practice, carrots are irrigated at a rate equivalent to 25 mm wk^{-1} in California, which amounts to a seasonal total of up to 360 mm of water (Vittum & Flocker, 1967).

The effect of supplemental irrigation on carrot yield and quality was studied in Arkansas by Bradley et al. (1967). Application of 38 mm of water at 7-, 10-, or 14-d intervals was compared with no irrigation. Irrigation, regardless of frequency, increased carrot yields. A 7-d irrigation interval was superior to the longer intervals when harvest was delayed. Irrigation did not affect carotene content of the carrots, but solids content decreased with irrigation.

Supplemental irrigation was compared to no irrigation for carrots grown in Delaware (Orzolek & Carroll, 1978) for both tilled and no-till production systems. Irrigated and cultivated plots produced higher yields than did nonirrigated, noncultivated plots. Secondary root growth was lower in irrigated, conventional-tilled and rye-mulched plots, but not in soybean stubble.

In other experiments in Delaware (Orzolek & Carroll, 1976) fewer secondary roots were produced by irrigated carrots than by nonirrigated ones. In a comparison of cultivars, there was a differential yield response to irrigation. 'Spartan Fancy' yields increased only 1.1%, whereas 'Danvers 126' yields increased 144.2% with irrigation.

6. Bulb Crops

Onion and garlic (*A. sativum* L.) are the principal bulb crops. Onion has a shallow root system that is concentrated in the surface 0.3 m. Frequent irrigation is practiced to prevent soil moisture from being depleted below 25% of available water. Onion is most sensitive to water deficit during the bulb-enlargement period, which occurs 50 to 80 d after transplanting (Doorenhos & Kassam, 1979).

In California, all commercial onion production is irrigated. Sprinkler irrigation is commonly used for crop establishment and furrow irrigation is then used generally for crop production. However, some of the crop is sprinkler-irrigated throughout production. Subsurface irrigation is used for onion grown on organic soils. Total water application for an onion crop varies from 450 to 1800 mm of water, depending upon method of application, soil type, rainfall, and growing season temperatures. Irrigation ordinarily is terminated as the onion begins to mature to allow drying to proceed. Water deficits may result in increased pungency of onion (Voss, 1979).

A comparison of five irrigation methods for onion production in Colorado was made by Ells et al. (1986). Irrigation schedules were based on soil-water depletion estimates and were used for five application methods. Four of these received weekly irrigation (graded furrow, level furrow, overhead sprinkler, or microirrigation), and an additional treatment was microirrigated three times per week. Total yields were not affected by irrigation method, but minor grade-out differences did occur. Higher water use was measured with the graded furrow method compared to the other methods in 1983 because of a lower WUE. In 1984, however, total water use and WUEs did not vary greatly among irrigation methods. Based on these results, the authors concluded that, with good quality irrigation water, conversion from furrow irrigation systems to pressurized systems in order to achieve water savings was not justified for the uniform, heavy-textured soils on which these studies were conducted.

Riekels (1977) studied the effects of supplemental irrigation and N fertilizer topdressing on onion grown on organic soils in Ontario, Canada. Yield increased and maturity was advanced with increasing N and irrigation, whereas no N effect on yield was found for nonirrigated onion. Storage life was shortened with high N for irrigated onions, however. Maturity was earlier for nonirrigated plots, regardless of N treatment.

Antitranspirants reduced water use by greenhouse-grown onion and increased yields and bulb size of field-grown onion in Texas (Lipe et al., 1982). The favorable effect of antitranspirants was more pronounced for moisture-stressed onion than for unstressed ones.

Virtually all of the commercial garlic produced in the USA is grown in California. Garlic, like onion, is shallow rooted and responds to frequent sprinkler or furrow irrigations to insure adequate available moisture in the surface 0.5 m of soil. Irrigation is terminated as the tops begin to "break over" for fresh-market garlic and after most of the tops have fallen over for processing garlic (Sims et al., 1977a).

7. Sweet Potato

Sweet potato [*Ipomoea batatas* (L.) Lam.], once established in plantings, is relatively drought tolerant because of its extensive and moderately deep root system. Sensitivity to drought is most pronounced during the root-enlargement growth phase, which begins about 60 d after transplanting. Water requirements for sweet potato are approximately 2.5 mm d^{-1} early in the crop season, increasing to 6.5 mm d^{-1} during the period of most active growth. Seasonal water requirements are in the range of 450 to 600 mm of water. In irrigation studies in Alabama and Arkansas, yields were not restricted when soil water was depleted to 20% of available water before irrigation, as compared to more frequent irrigations (Steinbauer & Kushman, 1971).

Irrigation timing studies in Taiwan (Sajjapongse & Roan, 1982) for fall and spring crops suggested that a single irrigation, 60 d after planting, produces yields equivalent to all other single or multiple irrigation application times. The primary effect of irrigation was to increase root weight rather than the number of roots. Such results are in agreement with those of Steinbauer and Kushman (1971).

Soil moisture also affected sweet potato quality. Total solids, total carbohydrates, and starch content were all lower when moisture stress occurred, and the moisture content was higher than for roots grown without moisture stress (Steinbauer & Kushman, 1971). Somewhat conflicting results were obtained in Louisiana (Constantin et al., 1974). Dry matter, flesh color, protein, and firmness of sweet potatoes grown without irrigation in this case were higher than for sweet potatoes irrigated to maintain available soil water above 25% or 50%. No mention was made of the effects of irrigation on sweet potato yields or plant growth without irrigation.

8. Legume Crops

The principal vegetable legumes are snap bean, lima bean (*P. lunatus* L.) and English or garden pea. Bean is moderately deep-rooted, with a strong tap root and an extensive lateral root system. Although the tap root may extend to a depth of 1.5 m, the main zone of water extraction is to a depth of 0.5 to 0.7 m. The most critical plant-growth stages with respect to water deficits are the flowering and pod-production periods. General seasonal water requirements for maximum production are in the range of 300 to 500 mm (Doorenbos & Kassam, 1979).

An integrated irrigation/N-fertilization scheme for the production of snap beans was developed by Smittle (1976) in Georgia. Yields were maximized when 8 mm d^{-1} of irrigation water was applied when daily pan evaporation was 23 mm until one-third foliage cover, then irrigation water increased to 15 mm d^{-1} until two-thirds cover, and decreased again to 8 mm d^{-1} during flowering and fruiting. Petiole NO_3-N concentrations above 1500 mg kg^{-1} in the preblossom stage and above 1000 mg kg^{-1} at the fruit development stage were necessary for maximum yield at the irrigation levels indicated.

In other Georgia research, Stansell and Smittle (1980) reported that snap bean yields were restricted when irrigation was scheduled at soil water potentials of -50 or -75 kPa as compared to more frequent irrigation at a soil water potential of -25 kPa. The WUE was highest at the high soil water potential and decreased as soil water potential decreased. The WUE of 'Eagle' snap bean was more than twice that of 'Galagreen' snap bean, an indication of cultivar differences that exist as to water use and yield.

A strong correlation was found between soil water potential and snap bean yields by Bonanno and Mack (1983) in Oregon. Maximum yields were obtained when the soil water potential was not allowed to drop below 200 kPa. Yields of 'Oregon 1604' and 'Galamor' were similar under adequate soil moisture conditions, but Galamor yields were restricted to a greater extent than those of Oregon 1604 under conditions of water stress.

Water stress may also adversely affect snap bean pod quality. Low soil moisture restricted pod length and number of seeds per pod, whereas fiber content and seed weight were greater under water stress conditions (Bonanno & Mack, 1983). The incidence of a pod disorder, interlocular cavitation, was found to increase in proportion to the amount of supplemental irrigation applied between flowering and harvest (Lee et al., 1977). Accordingly, it is apparent that snap bean quality may be impaired by either limited or excessive soil moisture.

Irrigation method may also influence snap bean quality (Drake & Silbernagel, 1982). Sprinkler-irrigated bean in Washington was higher in $C_6H_8O_6$ content than rill-irrigated bean. On the other hand, rill-irrigated bean had better color and was more tender than those that were sprinkler-irrigated.

The impacts of plant population and irrigation as yield-controlling factors were investigated in Oregon (Mack & Varseveld, 1982; Mauk et al., 1983). Highest yields were obtained by irrigation at -61 kPa soil water potential, which corresponded to removal of 40 to 45% of the available water to a depth of 30 cm. This treatment required 60% more water than irrigation at -250 kPa soil water potential, and resulted in a 54% yield increase. Yields were also enhanced at high plant populations, but there were no irrigation $\times$ population interactions. However, more rapid soil water depletion was noted for high plant density than for low density. Oregon 1604 snap bean is highly productive at high plant populations, but the response is limited to only one of the two major pod-bearing nodes. On the other hand, both nodes responded to increased irrigation. The use of high plant populations with frequent irrigation may require further refinement because of the likelihood of increased plant disease under such cultural conditions.

Smittle (1979) developed an irrigation scheme for lima bean similar to the one described earlier for snap bean. Yield of shelled lima bean was highest when 8 mm d^{-1} of irrigation water was applied for each 23 mm d^{-1} of pan evaporation during the first one-third foliage cover period, for each 18 mm d^{-1} of pan evaporation in the second one-third foliage cover period, and for each 8 mm d^{-1} of pan evaporation in the third one-third foliage cover. Unlike snap bean, there was no effect of N fertilization on lima bean yield. Seasonal water requirements for lima bean production in Delaware are estimated to be 480 mm (Ritter et al., 1985).

Pea is moderately deep-rooted, with a tap root and many thin laterals. Although the tap root may extend to a depth of 1.5 m, the main area of water extraction extends only to 0.6 to 1.0 m. Water deficits during flowering and pod development most seriously restrict yields. General water requirements for maximum yields are in the range of 350 to 500 mm (Doorenbos & Kassam, 1979). Seasonal water requirements for pea production in Delaware are estimated at 381 mm (Ritter et al., 1985). In Oregon, irrigation at pod filling increased yield more than irrigation at any other growth stage, and produced the greatest annual yield increases: 470, 930, and 510 kg ha^{-1} per cm of water applied over a 3-yr study (Pumphrey & Schwanke, 1974).

Yield and quality of pea were higher for irrigated Arkansas plots than for nonirrigated plots (Smittle & Bradley, 1966). Furrow irrigation to field capacity whenever available soil moisture had been depleted by 50% resulted in delayed maturity, more pods and peas, better color of the frozen product, but poorer color of the raw product. These somewhat conflicting results of the effects of irrigation on pea color were later clarified by Ells and McSay (1977) and Pumphrey and Schwanke (1974). They found that irrigated pea had lower color scores and more "blond" peas than nonirrigated ones, and that the effect was intensified for later plantings.

9. Solanaceous Crops

Tomato, economically the most important crop discussed in this chapter, and pepper—both sweet or pungent (*Capsicum annuum* L.) and pungent (*C. frutescens* L.)—are the principal vegetables in this group.

Tomato is a deep-rooted plant wherever soil physical and moisture conditions permit full root extension. Most of the water extracted by the plant is from the surface 0.5 to 0.7 m, and growth is restricted when available water falls below 60% in this zone. Water stress most seriously affects yields during the plant establishment, flowering, and fruit enlargement periods. Total water requirements after transplanting are in the range of 400 to 600 mm (Doorenbos & Kassam, 1979). In Delaware, seasonal water requirements have been estimated at 540 mm (Ritter et al., 1985). For direct-seeded crops, the requirement would be substantially greater. Furrow irrigation is used almost exclusively in California for tomato production, but sprinkler irrigation may be used for stand establishment. In Florida, seepage irrigation predominates, though sprinkler irrigation is used in some portions of south Florida. In both states, microirrigation is becoming increasingly important where water is scarce or expensive, or where there is concern about groundwater quality.

To determine water usage for tomato production in west-central Florida, Marlowe et al. (1982) conducted a 3-yr grower survey. The tomato crop was seepage-irrigated to maintain a constant water table about 45 cm below the surface, and a full-bed, polyethylene-mulch cultural system was used. During the fall production season, growers used from 670 to 1720 mm of irrigation water with an average of 1180 mm. During the spring production season, 1140 to 1400 mm of water was used with an average of 1230 mm. The range reflects wide differences among farm sites with respect to soil type, drainage, efficiency of use, and water management. It is expected that the magnitude of variation in water usage among growers would be as large for other crops and with other irrigation application methods.

Furrow-, sprinkler-, or micro-applied irrigation was used to maintain available soil water at 50% or higher for the surface 60 cm of soil in Alabama experiments (Doss et al., 1980). Tomato yields from irrigated plots were higher than those from nonirrigated plots but there were no significant yield differences among application methods in this 3-yr study. The total amount of water applied was about the same for both furrow and sprinkler application, but water application with microirrigation was less than half of that applied with the other methods. However, since there was only one irrigation level and no yield difference, it is quite possible that the furrow and sprinkler treatments were overirrigated.

Furrow-irrigated and nonirrigated treatments were compared with microirrigation application when available soil water was at 80, 65, 50 or 25% in Taiwan (Lin et al., 1983). Tomato yields were similar for all microirrigation treatments and higher than those obtained from furrow-irrigated plots. All of the irrigation treatments resulted in higher yields than for the nonirrigated plots. Claims of higher quality fruit being produced with microirrigation were not substantiated since fruit pH, titratable acidity, total solids, total sugars, $C_6H_8O_6$, viscosity, beta-carotene, soluble solids, and color did not vary significantly among treatments.

Water (Locascio et al., 1981) and fertilizer (Locascio et al., 1985) application variables have been investigated in Florida. Depending upon the experiment and location, maximum yields were obtained with water application equivalent to 0.5 to 1.0 × pan evaporation. Preplant fertilization produced higher yields than treatments where 60% of the N and K was applied with the microirrigation at two locations, but no significant treatment effect was measured at a third location. Some Florida tomato growers now supply all crop fertilization through the microirrigation system to more precisely relate crop fertilization to crop requirements.

In California experiments, Stark et al. (1983) applied N fertilizer to microirrigated tomatoes at 120 to 585 kg N ha^{-1} when the soil water potential at 25 cm deep reached either -10 to -30 kPa. Total N uptake was less than the amount applied at rates exceeding 300 kg ha^{-1}. Yields were restricted by 10 to 15% at N appliction rates below this amount. Losses of N via denitrification were low when N application was below crop utilization, but increased when application exceeded crop use. These results suggest that the total N requirements of the tomato crop can be supplied with high-frequency N applications through the micro system, without large accompanying denitrification losses.

It appears that furrow, sprinkler, and seepage irrigation will all continue to be used for tomato production, but it also is quite certain that the proportion of the crop that is microirrigated will continue to increase. The rate of increase will be influenced by water costs and availability, technological advances and costs of equipment and tubing, and horticultural verification of its efficacy under various conditions.

Pepper has a taproot that may extend to 1.5 m when the crop is direct-seeded if soil physical and moisture conditions permit. The crop is frequently transplanted, which can lead to injury to the taproot and a predominance of lateral roots. Water uptake is from the surface 1.5 m in the former situation, but only to the 0.3 to 0.5 m depth in the latter case. Pepper is sensitive to water stress throughout the crop season, but particularly during flowering and fruiting. Total crop water requirements range from 600 to 1250 mm, depending on the number of harvests (Doorenbos & Kassam, 1979). Commercial pepper crops in the USA are currently irrigated by furrow, sprinkler, seepage or micro systems, depending on existent production systems.

Highest pepper yields were obtained in Georgia with frequent irrigation and N topdressings at plant populations above 27 000 ha^{-1} (Batal & Smittle, 1981). Soil NO$_3$-N concentrations of at least 20 mg L^{-1} for the spring crop and 30 mg L^{-1} for the fall crop were achieved by frequent topdressing of N, and were required to obtain a response from the irrigation treatments. Under these conditions, sprinkler irrigation applications of 6.5 mm at a soil water potential of 25 kPa produced higher yields than less frequent irrigations of 13 mm at a soil water potential of 50 kPa.

Goldberg and Shmueli (1971a) compared sprinkler and microirrigation at 1- and 5-d intervals in amounts sufficient to compensate for pan evaporation, for pepper production in the arid climate of Israel. Frequency of irrigation did not affect yields for either application method, but microirrigated pepper plants produced higher yields than those that were sprinkler irrigated. In a later report (Shmueli & Goldberg, 1972) where water application varied from 0.82 to 1.75 times pan evaporation, highest yields were obtained when water was applied at 1.33 times pan evaporation.

Microirrigation was used as a vehicle for N and K fertilization of hot peppers in Georgia (Jaworski et al., 1978). Yields were highest for the highest daily fertilization rate of 4.48 kg N ha^{-1} and 3.72 kg K ha^{-1}, and were significantly greater than at lower fertilization rates. Yield differences were slight early in the harvest season and great in the latter part of the harvest season.

10. Cucurbits

Cucumber, muskmelon (*Cucumis melo* L.), summer squash (*Cucurbita pepo* var. *melopepo* L.), and watermelon [*Citrullus lanatus* (Thunb) Matsum. & Nakai], the economically most important vegetables in this group, are discussed in this section.

Pickling cucumber yields were restricted slightly when the interval between irrigations was extended to more than 50% depletion of available water in the effective rooting zone, which was determined to be 90 cm (Loomis & Crandall, 1977). More frequent irrigation did not increase yields. Consumptive water use was 1.5 times pan evaporation during the late vegetative and early reproductive plant-growth stages. Total water use during the 8-wk growth period ranged from 30 to 400 mm. In Delaware, seasonal water requirement for slicing cucumber production was estimated to be 440 mm (Ritter et al., 1985).

Pickling cucumber yields were increased in Alabama by irrigation when 70% of the available soil water was depleted or when the soil water potential was -200 kPa, as compared to no irrigation. Additional yield increases did not occur with more frequent irrigation (Doss et al., 1977).

Most of the commercial muskmelon in the USA is produced in California, and furrow irrigation is the traditional method of water application. Flocker et al. (1965) irrigated muskmelon when soil moisture potentials were equivalent to -700, -300, or -50 kPa at 40 cm after three plant establishment irrigations. Yields were not affected by irrigation frequency, but fruit quality was affected somewhat since the number of small and sunburned fruit was larger with infrequent irrigation, and growth cracks were more common with frequent irrigation. The general lack of response to irrigation frequency was attributed to the deep-rooted nature of muskmelon, whereby water withdrawal at 1.2 m was recorded, and to the high water-holding capacity of the soil at this experimental site.

In Arizona (Pew & Gardner, 1983), higher muskmelon yields, larger fruit size, and earlier yields were obtained when furrow irrigation occurred at soil water potentials of -50 or -75 kPa at the 25-cm depth as compared to more frequent irrigation at -25 kPa. Fruit quality, as manifested by soluble solids and fewer decayed fruit, was higher with less frequent irrigation. On the other hand, the incidence of growth cracks was higher. A prethinning irrigation to simulate local grower practice was evaluated and found to be detrimental to root development, vine growth, fruit size, and yield. In general, it appears that three or four irrigations after germination are sufficient for muskmelon production on soils of high water-holding capacity in Arizona.

Sprinkler, furrow, and microirrigation for muskmelon production were compared in Israel by Shmueli and Goldberg (1971). Irrigations were applied every 2 or 3 d in amounts required to compensate for pan evaporation. Yields were higher for the micro-irrigated plots than for those furrow- or sprinkler-irrigated, which did not differ from each other. In Texas, Bogle and Hartz (1986) achieved equivalent yields with microirrigation and furrow irrigation, but water usage was lowered 58 to 75% with microirrigation, depending upon irrigation frequency.

Microirrigation, as compared to no irrigation, resulted in higher muskmelon yields in Pennsylvania (DeTar et al., 1983) during a dry year and also in Indiana (Bhella, 1985), where early yields and fruit size were also enhanced by microirrigation. Conversely, soluble solids were lower for melons from irrigated plots.

Summer squash yields were higher in 2 of 3 yr when the crop was sprinkler-irrigated at -30 kPa rather than -60 kPa soil water potential (Smittle & Threadgill, 1982). When combined with other cultural variables, frequent irrigation with 22.5 kg N ha^{-1} applied through the irrigation system at 2, 3, 4, 5, and 6 wk after planting, and moldboard-plow tillage, resulted in the greatest marketable fruit yields.

Yields of zucchini squash were increased by microirrigation during a dry year but not during a wet one in Indiana (Bhella & Kwolek, 1984) whereas black polyethylene mulch was effective in increasing yields over those for unmulched zucchini in both years. Overall yield and fruit quality were best with polyethylene mulch and microirrigation.

When direct-seeded, watermelon has an extensive and deep root system extending to a depth of 2 m. The surface 1.5 m may be depleted to 50% available water before signs of water deficit occur. Water deficit during establishment results in later and more variable production. Less leaf area, which causes yield reduction, results when water deficit occurs in the early vegetative stage. Water deficits in the reproductive period cause the most serious yield reductions. Watermelon water requirements vary from 40% of potential ET early in the season to more than 100% of potential ET during fruit development. Seasonal water requirements are in the range of 400 to 600 mm water (Doorenbos & Kassam, 1979, Ritter et al., 1985).

In Florida, the leading watermelon-producing state, seepage irrigation is used in the acid flatwoods growing areas; sprinkler irrigation is used on the sands and loams of central and north Florida. Sprinkler irrigation is used in other production areas of the humid East, and furrow irrigation is used in the western USA.

Microirrigation and sprinkler irrigation for watermelon production were compared at two locations in Florida (Elmstrom et al., 1981). No yield response to irrigation occurred at Gainesville on a soil with adequate water-holding capacity. On the other hand, microirrigation and sprinkler irrigation increased watermelon yields on a deep sand near Leesburg. About 40% less water was used for the microirrigation treatment than for the sprinkler irrigation treatment.

11. Sweet Corn

Seepage furrow, and sprinkler irrigation are used in commercial sweet corn production. Seasonal water use requirements in Delaware were estimated to be 280 mm (Ritter et al., 1985).

In Florida, most of the commercial sweet corn is grown on organic soils by using seepage irrigation, although sprinkler irrigation is sometimes used for stand establishment. The water table, in growing areas where seepage irrigation is used for establishment, is maintained at about 35 to 40 cm. Following emergence, the water table is lowered to 45 to 50 cm in sandy soils and to about 75 cm in organic soils (Guzman et al., 1967).

With sufficient preirrigation, the first sweet corn irrigation in California occurs when the plants are 8 to 15 cm tall. Subsequent irrigations of 508 to 1016 m^3 ha^{-1} are scheduled at 1- to 3-wk intervals, depending upon the season in which the sweet corn is being grown. Total crop requirements are estimated to be 3048 to 6350 m^3 ha^{-1} (Sims et al., 1978).

In Oregon, 380 to 460 mm of sprinkler irrigation water is applied to sweet corn grown for processing. This seasonal requirement is applied in four to six irrigations beginning before the corn is 30 cm tall (Mansour, 1975).

Sprinkler irrigation of sweet corn in Oregon was scheduled by available water depletion as measured by neutron probe, soil water depletion as estimated by the FAO Penman equation, and growth stages related to silking (Braunworth & Mack, 1987a). Water application in both years was greater when scheduled by the neutron probe, and yield in 1984 was significantly greater in the neutron probe treatment. In 1985, yields and quality were similar regardless of scheduling method. The authors concluded that any of the scheduling methods could be used satisfactorily, despite the apparent overestimation of water requirements by the neutron probe.

In complementary experiments, Braunworth and Mack (1987b), using sprinkler irrigation, compared sweet corn yield and quality reduction caused by irrigation treatments designed to maintain 70, 50, 24, and 0% of the root zone at field capacity. Yield and quality were maintained with reduced water use even when irrigation was only to 70 or 50% of the root-zone field capacity.

12. Strawberry

Although sometimes considered a fruit, strawberry is included in this chapter as a vegetable because its culture is more akin to that of vegetables than fruit. It is grown as an annual in the major producing states, and it is herbaceous rather than woody.

Strawberries are relatively shallow-rooted, with most of the roots located in the surface 20 to 25 cm of soil, though some roots may penetrate to 60 cm. Because of the shallow root system, strawberry is intolerant of drought and frequent irrigation is necessary (Albregts & Howard, 1984).

Furrow irrigation has been the traditional technique for strawberry production in California, the principal producing state, but a rapid transition to microirrigation has occurred recently. This change is particularly advantageous in areas with a marginal water supply or expensive water, as well as on soils having a high salt concentration, low infiltration rates, and poor drainage (Welch et al., 1982).

In Florida, sprinkler irrigation is used for strawberry production even though water conservation could be achieved and fruit quality would be improved with microirrigation. Sprinkler irrigation persists because it is essential for soil preparation, crop establishment, and frost protection. Until effective alternatives are available for crop establishment and frost protection, sprinkler systems will continue to be used; however, microsystems are being installed in conjunction with overhead sprinkler systems (Albregts & Howard, 1984).

The potential usefulness of microirrigation in Florida has been demonstrated (Locascio et al., 1977). They compared microirrigation, sprinkler irrigation, and no irrigation and reported higher early and total yields with microirrigation. Furthermore, water usage as compared to sprinkler irrigation was reduced by 35 to 66%, depending upon the year.

VIII. FUTURE RESEARCH NEEDS ON IRRIGATION OF VEGETABLE CROPS

The research information presented in the preceding sections pertains to the present status of vegetable irrigation management, primarily, conserving water while minimizing loss of production and quality. The extent to which irrigation is used and the level of management implemented depend greatly on the potential increased profit that results over a nonirrigated production system. In most cases, vegetable production costs per unit area are so high that use of irrigation to protect that investment is economically justified. Likewise, the degree to which present water resources for irrigation are used in a conservative manner has been primarily controlled by economics and the availability of the resource. The costs for obtaining water (energy for pumping or the cost for purchasing water), availability of water for irrigation, water quality, and governmental policies that affect the availability and distribution of water, all are likely to continue to affect the extent of vegetable crop irrigation. As urban population centers continue to grow into present vegetable production areas, increased conflict and competition between water users for this valuable resource will result. If allocation of water becomes severely reduced, constraints will be placed on growers to use water more efficiently with irrigation systems (such as microsystems) that make better use of the available water.

Although future irrigation research will continue to be aimed at developing methods to irrigate crops more efficiently, an increased effort into investigating the effects of water quality on production and, conversely, the effects of production on water quality will be likely. The public concern over environmental issues pretaining to the potential contamination of groundwater by chemicals used in agricultural production will likely cause an increased demand for accountability on the part of agricultural producers for protection of water resources. Understanding the fate and transport of applied chemicals and how they are affected by irrigation management practices will be important research areas in the future.

IX. SUMMARY

Vegetable crops comprise a wide range of plant species grown under a wide range of environmental conditions. In many cases, water can become limiting to optimum vegetable crop production for a number of reasons, including the climate, soil, or cultural practices used. Because of the large economic investment usually associated with the production of vegetable crops, the use of irrigation as a means of supplying or supplementing natural precipitation in order to meet the crop water requirements is a common management practice. An effort has been made in this chapter to discuss how vegetable crop water use can vary substantially because of the diverse plant species included in this crop group. Also discussed have been factors that affect irrigation management, including: crop vegetative and root growth characteristics, production location, cultural practices, crop reaction to deficit and excessive soil water conditions, nutrition, and salinity. Irrigation requirements for specific crops grown under specified conditions have also been presented.

REFERENCES

Albregts, E.E., and C.M. Howard. 1984. Strawberry production in Florida. Florida Agric. Exp. Stn. Bull. 841.

Barrs, H.D. 1968. Determination of water deficits in plant tissue. p. 236–368. *In* T.T. Kozlowski (ed.) Water deficits and plant growth. Vol. I. Development, control, and measurement. Academic Press, New York.

Batal, K.M., and D.A. Smittle. 1981. Response of bell pepper to irrigation, nitrogen and plant population. J. Am. Soc. Hortic. Sci. 106:259–262.

Beverly, R.B., W.M. Jarrell, and J. Letey, Jr. 1986. A nitrogen and water response surface for sprinkler-irrigated broccoli. Agron. J. 78:91–94.

Bhella, H.S. 1985. Muskmelon growth, yield and nutrition as influenced by planting method and trickle irrigation. J. Am. Soc. Hortic. Sci. 110:793–796.

Bhella, H.S., and W.F. Kwolek. 1984. The effects of trickle irrigation and plastic mulch on zucchini. HortScience 19:410–411.

Blaney, H.F., and W.D. Criddle. 1962. Determining consumptive use and irrigation water requirements. U.S. Agric. Res. Serv. Tech. Bull. 1275.

Bogle, C.R., and T.K. Hartz. 1986. Comparison of drip and furrow irrigation for muskmelon production. HortScience 21:242–244.

Bonanno, R.A., and H.J. Mack. 1983. Yield components and pod quality of snap beans grown under differential irrigation. J. Am. Soc. Hortic. Sci. 108:832–836.

Bradford, K.J., and S.F. Yang. 1981. Physiological responses of plants to waterlogging. HortScience 16:25–30.

Bradley, G.A., D.A. Smittle, A.A. Kattan, and W.A. Sistrunk. 1967. Planting date, irrigation, harvest sequence and varietal effects on carrot yields and quality. Proc. Am. Soc. Hortic. Sci. 90:223–234.

Braunworth, W.S., Jr., and H.J. Mack. 1987a. Evaluation of irrigation scheduling methods for sweet corn. J. Am. Soc. Hortic. Sci. 112:29–32.

Braunworth, W.S., Jr., and H.J. Mack. 1987b. Effect of deficit irrigation on yield and quality of sweet corn. J. Am. Soc. Hortic. Sci. 112:32–35.

Clark, G.A., C.D. Stanley, and A.G. Smajstrla. 1987. Microirrigation on mulched bed systems: Components, system capacities, and management. Univ. Florida Exp. Stn. Agric. Eng. Ext. Rep. 87-8.

Constantin, R.J., T.P. Hernandez, and L.G. Jones. 1974. Effects of irrigation and nitrogen fertilization on quality of sweet potatoes. J. Am. Soc. Hortic. Sci. 99:308–310.

Crafts, A.S. 1968. Water deficits and physiological processes. p. 85–134. *In* T.T. Kozlowski (ed.) Water deficits and plant growth. Vol. II. Plant water consumption and response. Academic Press, New York.

DeTar, W.R., D.F. Kibler, D.W. Grenoble, R. Daniels, R.H. Cole, L.D. Tukey, S.H. Hampson, and S.C. Geller. 1983. Trickle irrigation vs. no irrigation of five horticultural crops in Pennsylvania. Trans. ASAE 26:82–86.

Doorenbos, J., and A.H. Kassam. 1979. Yield response to water. FAO Irrig. Drain. Pap. 33. FAO, Rome.

Doorenbos, J., and W.O. Pruitt. 1977. Guidelines for predicting crop water requirements. FAO Irrig. Drain. Paper 24 (revised). FAO, Rome.

Doss, B.D., C.E. Evans, and J.L. Turner. 1977. Irrigation and applied nitrogen effects on snap beans and pickling cucumbers. J. Am. Soc. Hortic. Sci. 102:654–657.

Doss, B.D., J.L. Turner, and C.E. Evans. 1980. Irrigation methods and in-row chiseling for tomato production. J. Am. Soc. Hortic. Sci. 105:611–614.

Drake, S.R., and M.J. Silbernagel. 1982. The influence of irrigation and row spacing on the quality of processed snap beans. J. Am. Soc. Hortic. Sci. 107:239–242.

Ells, J.E., E.G. Kriuse, A.E. McSay, C.M.V. Neale, and R.A. Horn. 1986. A comparison of five irrigation methods on onions. Hort Science 21:1349–1351.

Ells, J.E., and A.E. McSay. 1977. A field study of color intensity in freezing peas. HortScience 12:558–560.

Elmstrom, G.W., S.J. Locascio, and J.M. Myers. 1981. Watermelon response to drip and sprinkler irrigation. Proc. Florida State Hortic. Soc. 94:161–163.

Erie, L.J., and A.R. Dedrick. 1979. Level-basin irrigation: A method for conserving water and labor. U.S. Dep. Agric. Sci. Educ. Adm. Farmers Bull. 2261. U.S. Gov. Print. Office, Washington, DC.

Erie, L.J., O.F. French, and K. Harris. 1965. Consumptive use of water by crops in Arizona. Arizona Agric. Exp. Stn. Tech. Bull. 169.

Feigin, A., J. Letey, and W.M. Jarrell. 1982. Celery response to type, amount, and method of N-fertilizer application under drip irrigation. Agron. J. 74:971–977.

Finkle, H.J., and D. Nir. 1983. Criteria for choice of irrigation method. p. 35–45. In H.J. Finkle (ed.) CRC handbook of irrigation technology. CRC Press, Boca Raton, FL.

Flocker, W.J., J.C. Lingle, R.M. Davis, and R.J. Miller. 1965. Influence of irrigation and nitrogen fertilization on yield, quality, and size of cantaloupes. Proc. Am. Soc. Hortic. Sci. 86:424–432.

Gardner, B.R., and R.L. Roth. 1984. Applying nitrogen in irrigation waters. p. 493–506. In R.D. Hauck (ed.) Nitrogen in crop production. ASA, CSSA, and SSSA, Madison, WI.

Geraldson, C.M. 1980. Importance of water control for tomato production using the gradient mulch system. Proc. Fla. State Hortic. Soc. 93:278–279.

Goldberg, D., and M. Shmueli. 1971a. Sprinkle and trickle irrigation of green pepper in an arid zone. HortScience 6:559–562.

Goldberg, D., and M. Shmueli. 1971b. The effect of distance from the tricklers on soil salinity and growth and yield of sweet corn in an arid zone. HortScience 6:565–567.

Guzman, V.L., H.W. Burdine, W.R. Forsee, Jr., E.D. Harris, Jr., J.R. Orsenigo, R.K. Showalter, C. Wehlburg, J.A. Winchester, and E.A. Wolf. 1967. Sweet corn production on the organic and sandy soils of south Florida. Florida Agric. Exp. Stn. Bull. 714.

Guzman, V.L., H.W. Burdine, E.D. Harris, Jr., J.R. Orsenigo, R.K. Showalter, P.L. Thayer, J.A. Winchester, E.A. Wolf, R.D. Berger, W.G. Genung, and T.A. Zitter. 1973. Celery production on organic soils of south Florida. Florida Agric. Exp. Stn. Bull. 757.

Hiler, E.A., and T.A. Howell. 1983. Irrigation option to avoid critical stress: An overview. p. 479–498. In H.M. Taylor et al. (ed.) Limitations to efficient water use in crop production. ASA, CSSA, and SSSA, Madison, WI.

Jackson, R.D. 1982. Canopy temperature and crop water stress. Adv. Irrig. 1:43–86.

Jackson, R.D., P.J. Printer, Jr., R.J. Reginato, S.B. Idso. 1986. Detection and evaluation of plant stresses for crop management decision. IEEE Trans. Geosci. Remote Sensing GE-24(1):99–106.

Jaworski, C.A., S.J. Kays, and D.A. Smittle. 1978. Effects of nitrogen and potassium fertilization in trickle irrigation on yield of pepper and pole bean. HortScience 13:477–478.

Jones, J.W., L.H. Allen, S.F. Shih, J.S. Rogers, L.C. Hammond, A.G. Smajstrala, and J.D. Martsolf. 1984. Estimated and measured evapotranspiration for Florida climate, crops and soils. Florida Agric. Exp. Stn. Tech. Bull. 840.

Kahn, B.A., P.J. Stoffela, R.F. Sandsted, and R.W. Zobel. 1985. Influence of flooding on root morphological components of young black beans. J. Am. Soc. Hortic. Sci. 110:623–627.

Kawase, M. 1981. Anatomical and morphological adaptation of plants to waterlogging. Hort-Science 16:30–34.

Kovach, S.P. 1984. Injection of fertilizers into drip irrigation systems for vegetables. Florida Coop. Ext. Serv. Circ. 606.

Lee, J.M., P.E. Read, and D.W. Davis. 1977. Effect of irrigation on interlocular cavitation and yield in snap bean. J. Am. Soc. Hortic. Sci. 102:276–278.

Letey, J., W.M. Jarrell, N. Valoras, and R. Beverly. 1983. Fertilizer application and irrigation management of broccoli production and fertilizer use efficiency. Agron. J. 75:502–507.

Lin, S.M., J.N. Hubbell, Samson C.S. Tsou, and W.E. Splittstoesser. 1983. Drip irrigation and tomato yield under tropical conditions. HortScience 18:460–461.

Lipe, W.N., K. Hodnett, M. Gerst, and C.W. Wendt. 1982. Effects of antitranspirants on water use and yield of greenhouse and field-grown onions. HortScience 17:242–244.

Locascio, S.J., J.M. Myers, and S.R. Kostowicz. 1981. Quantity and rate of water application for drip irrigated tomatoes. Proc. Florida State Hortic. Soc. 94:163–166.

Locascio, S.J., J.M. Myers, and F.G. Martin. 1977. Frequency and rate of fertilization with trickle irrigation for strawberries. J. Am. Soc. Hortic. Sci. 102:456–458.

Locascio, S.J., S.M. Olson, F.M. Rhoads, C.D. Stanley, and A.A. Csizinszky. 1985. Water and fertilizer timing for trickle-irrigated tomatoes. Proc. Fla. State Hortic. Soc. 98:237–239.

Locascio, S.J., W.J. Wiltbank, D.D. Gull, and D.N. Maynard. 1984. Fruit and vegetable quality as affected by nitrogen nutrition. p. 617–626. In R.D. Hauch (ed.) Nitrogen in crop production. ASA, CSSA, and SSSA, Madison, WI.

Loomis, E.L., and P.C. Crandall. 1977. Water consumption of cucumbers during vegetative and reproductive stages of growth. J. Am. Soc. Hortic. Sci. 102:124–127.

Lorenz, O.A., and D.N. Maynard. 1980. Knott's handbook for vegetable growers. 2nd ed. John Wiley and Sons, New York.

Maas, E.V. 1985. Crop tolerance to saline sprinkling water. Plant Soil 89:273–284.

Maas, E.V. 1986. Salt tolerance of plants. Appl. Agric. Res. 1:12–26.

Mack, H.J., and G.W. Varseveld. 1982. Response of bush snap beans (Phaseolus vulgaris L.) to irrigation and plant density. J. Am. Soc. Hortic. Sci. 107:286–290.

Manning, C.E., D.G. Miller, and I.D. Teare. 1977. Effect of moisture stress on leaf anatomy and water-use efficiency. J. Am. Soc. Hortic. Sci. 102:756–760.

Mansour, N.S. 1975. Commercial production of sweet corn for processing in Oregon. Oregon Ext. Circ. 863.

Marlowe, G.A., A.J. Overman, and S.G. Feinburg. 1982. Results of an irrigation survey for tomatoes and strawberries in Hillsborough and Manatee counties, 1979–82. Proc. Fla. State Hortic. Soc. 95:301–304.

Mauk, C.S., P.J. Breen, and H.J. Mack. 1983. Yield response of major pod-bearing nodes in bush snap beans to irrigation and plant population. J. Am. Soc. Hortic. Sci. 108:935–939.

Maynard, D.N. 1979. Nutritional disorders of vegetable crops: A review. J. Plant Nutr. 1:1–23.

Moore, F.D. III. 1970. Furrow irrigation of lettuce resulting in water and nitrogen loss. J. Am. Soc. Hortic. Sci. 95:471–474.

Orzolek, M.D., and R.B. Carroll. 1976. Method for evaluating excessive secondary root development in carrots. HortScience 11:479–480.

Orzolek, M.D., and R.B. Carroll. 1978. Yield and secondary root growth of carrots as influenced by tillage system, cultivation, and irrigation. J. Am. Soc. Hortic. Sci. 103:236–239.

Pew, W.D., and B.R. Gardner. 1983. Effects of irrigation practices on vine growth, yield, and quality of muskmelons. J. Am. Soc. Hortic. Sci. 108:134–137.

Poysa, J.W., C.S. Tan, and J.A. Stone. 1987. Flooding stress and the root development of several tomato genotypes. HortScience 22:24–26.

Prevatt, J.W., C.D. Stanley, and A.A. Csizinszky. 1981. An economic comparison of three irrigation systems for tomato production. Proc. Fla. State Hortic. Soc. 94:166–169.

Prevatt, J.W., C.D. Stanley, and S.P. Kovach. 1984. An economic comparison of vegetable irrigation systems. Proc. Fla. State Hortic. Soc. 97:213–215.

Pumphrey, F.V., and R.K. Schwanke. 1974. Effects of irrigation on growth, yield, and quality of peas for processing. J. Am. Soc. Hortic. Sci. 99:104–106.

Riekels, J.W. 1977. Nitrogen-water relationships of onions grown on organic soil. J. Am. Soc. Hortic. Sci. 102:139–142.

Ritter, W.F., T.H. Williams, and R.W. Scarborough. 1985. Water requirements for corn, soybeans, and vegetables in Delaware. Delaware Agric. Exp. Stn. Bull. 463.

Robinson, F.E., W.L. Berry, D.J. Scherer, and T.R. Thomas. 1984. Yield potential of asparagus irrigated with geothermal and ground water on Imperial East Mesa Desert, California. HortScience 19:407–408.

Robinson, F.E., and O.D. McCoy. 1965. The effect of sprinkler irrigation with saline water and rates of seeding on germination and growth of lettuce. Proc. Am. Soc. Hortic. Sci. 87:318–323.

Robson, M.G., and W.B. Johnson. 1985. Commercial vegetable production recommendations. New Jersey Coop. Ext. Serv. Circ. EOOIA.

Rudich, J., E. Rendon-Pablete, M.A. Stevens, and Abdel-Ilah Ambri. 1981. Use of leaf water potential to determine water stress in field-grown tomato plants. J. Am. Soc. Hortic. Sci. 106:732–736.

Sajjapongse, A., and Y.C. Roan. 1982. Physical factors affecting root yield of sweet potato (*Ipomoea batatas* (L.) Lam.). p. 203–208. *In* R.L. Villareal and T.D. Griggs (ed.) Sweet potato. Proc. First Int. Symp. Asian Vegetable Res. and Develop. Ctr., Taiwan.

Sammis, T.W. 1980. Comparison of sprinkler, trickle, subsurface, and furrow irrigation methods for row crops. Agron. J. 72:701–704.

Shalhevet, J., A. Vinten, and A. Meiri. 1986. Irrigation as a factor in sweet corn response to salinity. Agron. J. 78:539–545.

Shmueli, M., and D. Goldberg. 1971. Sprinkler, furrow, and trickle irrigation of muskmelon in an arid zone. HortScience 6:557–559.

Shmueli, M., and D. Goldberg. 1972. Response to trickle irrigated pepper in an arid zone to various water regimes. HortScience 7:241–243.

Sims, W.L., R.F. Kasmire, and O.A. Lorenz. 1978. Quality sweet corn production in California. Univ. California Div. Agric. Sci. Leafl. 2818.

Sims, W.L., T.M. Little, and R.E. Voss. 1977a. Growing garlic in California. Univ. California Div. Agric. Sci. Leafl. 2948.

Sims, W.L., V.R. Rubatzky, R.H. Sciaroni, and W.H. Lange. 1977b. Growing globe artichokes in California. Univ. California Div. Agric. Sci. Leafl. 2675.

Sims, W.L., J.E. Welch, and V.E. Rubatzky. 1977c. Celery production in California. Univ. California Div. Agric. Sci. Leafl. 2673.

Smittle, D.A. 1976. Response of snap bean to irrigation, nitrogen fertilization, and plant population. J. Am. Soc. Hortic. Sci. 101:37–39.

Smittle, D.A. 1979. Response of lima bean (*Phaseolus lunatus* L.) to irrigation, nitrogen fertilization, and seed grading. J. Am. Soc. Hortic. Sci. 104:176–178.

Smittle, D., and G. Bradley. 1966. The effects of irrigation, planting and harvest dates on yield and quality of peas. Proc. Am. Soc. Hortic. Sci. 88:441–446.

Smittle, D.A., and E.D. Threadgill. 1982. Response of squash to irrigation, nitrogen fertilization, and tillage systems. J. Am. Soc. Hortic. Sci. 107:437–440.

Stansell, J.R., and D.A. Smittle. 1980. Effects of irrigation regimes on yield and water use of snap bean (*Phaseolus vulgaris* L.). J. Am. Soc. Hortic. Sci. 105:869–873.

Stark, J.C., W.M. Jarrell, J. Letey, and N. Valoras. 1983. Nitrogen use efficiency of trickle-irrigated tomatoes receiving continuous injection of N. Agron. J. 75:672–676.

Steinbauer, C.E., and L.J. Kushman. 1971. Sweet potato culture and diseases. USDA Agric. Handb. 388. U.S. Gov. Print. Office, Washington, DC.

Stolzy, L.H. 1972. Soil atmosphere. p. 335–362. *In* E.W. Carson (ed.) The plant root and its environment. Univ. Press Virginia, Charlottesville.

Sweeney, D.W., D.A. Graetz, A.B. Bottcher, S.J. Locascio, and K.L. Campbell. 1987. Tomato yield and nitrogen recovery as influenced by irrigation method, nitrogen source, and mulch. HortScience 22:27–29.

Takatori, F.H., F.D. Souther, J.I. Stillman, and B. Benson. 1977. Asparagus culture in California. Univ. California Div. Agric. Sci. Bull. 1882.

Thomas, J.R., L.N. Namken, and R.G. Brown. 1970. Yield of cabbage in relation to nitrogen and water supply. J. Am. Soc. Hortic. Sci. 95:732–735.

Vittum, M.T., and W.J. Flocker. 1967. Vegetable crops. *In* R.M. Hagen et al. (ed.) Irrigation of agricultural lands. Agronomy 11:674–685.

Voss, R.E. (ed.). 1979. Onion production in California. Univ. California Div. Agric. Sci. Publ. 4097.

Welch, N.C., R. Bringhurst, A.S. Greathead, V. Voth, W.S. Seyman, N.F. McCalley, and H.W. Otto. 1982. Strawberry production in California. Univ. California Div. Agric. Sci. Leafl. 2959.

Williamson, R.E., and G.J. Kriz. 1970. Response of agricultural crops to flooding, depth of water table, and soil gaseous composition. Trans. ASAE 13:216–220.

Wright, J.L. 1982. New evapotranspiration crop coefficients. J. Irrig. Drain. Div. Am. Soc. Civ. Eng. 108:57–74.

32 Citrus Trees

JOSEPH SHALHEVET

Volcani Center
Bet Dagan, Israel

YOSEPH LEVY

Gilat Experiment Station
Mobile Post Negev, Israel

Citrus species belong to the subfamily Aurantioideae in the family Rutaceae. Most of the cultivated species within the genus *Citrus* appear to be indigenous to humid, tropical regions of southeast Asia, southern China, and the Philippines. The cultivated range is between 40 °N and 40 °S lat, in climates from desert to tropical. Most of the commercial citrus is grown under Mediterranean or summer-rain climates and requires irrigation for maximum production.

Citrus is one of the most important horticultural crops, with a mean annual world production for 1983 and 1984 of about 56.1 Tg (million t) (Anon., 1985a). About 50% of this production came from irrigated orchards. These include 14.2 Tg (25% of total world production) produced in the Mediterranean countries and 11.0 Tg (20%) produced in the USA. The U.S. production was sharply reduced by severe freezes which affected Florida and Texas during 1981 to 1985. The largest producer during the last 10 yr has been Brazil, with a mean annual production of 12.1 Tg (21.5%). Other important producers are Japan, which produced 2.9 Tg (5.2%); China, 1.9 Tg (3.5%); India, 1.9 Tg (3.0%); and Mexico, 1.3 Tg (2.4%).

I. CITRUS DEVELOPMENT AND WATER RELATIONS

A. Stomatal Aperture and Leaf Water Potential

In this chapter we will not deal with the basic aspects of citrus water relations, except as they pertain directly to the crop behavior under field conditions. Two excellent reviews were recently published by Kriedemann and Barrs (1981) and by Jones et al. (1985), to whom the reader is referred.

Citrus trees are evergreen but, unlike other evergreen fruit trees, such as avocado (*Persea americana* Mill.) or mango (*Mangifera indica* L.), continuous leaf replacement occurs as the tree grows. Leaves can remain on the tree up to 2 yr (Kriedemann & Barrs, 1981). Kalma and Fuchs (1976) considered citrus as a typical mesophyte, but the leaves have many xeromorphic characteristics. The adaxial epidermis is covered by a thick cuticle, without active stomata (Baker et al., 1980; Hirano, 1931; Levy & Horesh, 1984; Schneider, 1973). Few stomata can be found along the main vein on the adaxial side (Kriedemann & Barrs, 1981) and water vapor diffusion could not be detected when a porometer was placed on the adaxial side of healthy citrus leaves (Levy, 1980, 1983). The leaf is rigid, and wilting is usually seen in mature leaves only at low leaf water potentials (Ψ_{leaf}). Plants with xeromorphic leaves are capable of high transpiration rates under favorable conditions, but limit the transpiration under adverse edaphic and atmospheric moisture conditions, in comparison with plants with mesomorphic leaves that die or are abscised under such conditions.

Citrus stomata, like those of many other plants, close under conditions of high evaporation demand (Camacho-B et al., 1974a, b; Elfving & Kaufmann, 1972; Elfving et al., 1972; Hall, 1976; Hall et al., 1975; Kaufmann & Hall, 1974; Kaufmann & Levy, 1976; Kriedemann, 1971; Levy, 1980, 1983; Levy & Syvertsen, 1981).

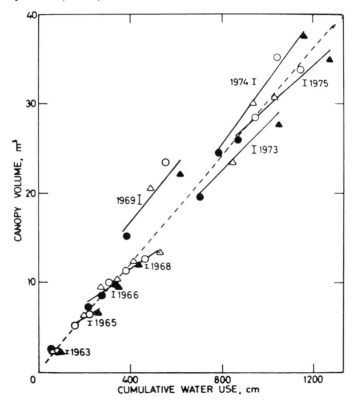

Fig. 32–1. Canopy volume of Marsh grapefruit as a function of ET. Dashed line indicates overall linear regression, V = 0.681 + 0.296ET; r = 0.964 (Levy et al., 1978b). Irrigation intervals: ▲ 14 to 18 d, ○ 21 to 24 d, △ 30 d, ● 40 d.

The effect of ambient humidity was demonstrated in observations of Hilgeman (1966) and Hilgeman et al. (1969), who found that the apparent transpirations and water stresses of orange [*C. sinensis* (L.) Osb.] trees were similar on summer days in Arizona and Florida, although the evaporative demands were much greater under Arizona desert conditions (Kaufmann, 1977). In Israel, consumptive use of citrus is similar in the desert and the humid coastal plain, although the evaporative demand is different, as will be discussed in the following sections.

The critical Ψ_{leaf} for stomata closure depends on leaf age, acclimation, and ambient humidity. Elfving et al. (1972) studied 'Valencia' orange trees under nonlimiting soil water conditions in California and reported stomatal closure at Ψ_{leaf} below -0.7 MPa for low leaf-to-air vapor pressure difference (VPD) and -1.2 MPa for high VPD. Syvertsen et al. (1981) studied mature 'Marsh' grapefruit (*C. paradisi* Macfad.) trees in Florida, and reported critical Ψ_{leaf} levels of -2.0 to -2.3 MPa for old leaves and -1.7 MPa for young leaves. Similar results were reported for lemon trees [*C. limon* (L.) Burm. f.] under desert conditions in the Negev region of Israel (Levy, 1980).

B. Vegetative Development

The vegetative development of young trees is closely dependent on the irrigation regime imposed on these trees. In a long-range irrigation experiment on 'Marsh' grapefruit trees in the Negev area, water stress was induced by irrigation at long intervals. The main effect of stress was to limit canopy development (Fig. 32-1). During the first 20 yr of orchard development, a good relationship was found between canopy volume and yield (Fig. 32-2; Levy et al., 1978b). However, as trees reach full size, excessive growth—

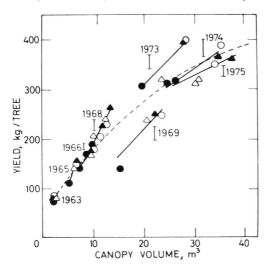

Fig. 32-2. Marsh grapefruit production in relation to canopy volumes (V) during a 9-yr experiment. Dashed line indicates overall regression; Yield = 1.449 + 0.397V + 0.004187V^2; r^2 = 0.869, irrigation treatments described in Fig. 32-1 (Levy et al., 1978b).

induced by intensive irrigation and fertilization—can lead to decreased yields, mainly because of shading and the need for severe hedging. Controlled water stress can be used to limit canopy development (Wiegand & Swanson, 1982c) or to induce vegetative growth flushes out-of-season (Goell et al., 1981).

Measurement of growth of wood, either the trunk or main branches, may be used to compare the response of the trees to different irrigation treatments in the same orchard (Cooper et al., 1964; Dasberg et al., 1981; Hilgeman, 1963, 1977; Levy et al., 1978b; Rodney et al., 1977; Wiegand & Swanson, 1982c). However, trunk sizes differ at different locations, and different rootstocks develop different trunk sizes. There may be poor correlation between trunk and canopy development. For example, rough lemon (*C. jambhiri* Lush) develops a large canopy and small trunk as compared with Cleopatra mandarin (*C. reshnii* Hort × Tan.), which has a large trunk but develops a small canopy, especially during the first years of orchard development.

C. Root Distribution and Water Uptake

Citrus develops a relatively shallow root system compared with some deciduous fruit trees, such as walnuts (*Juglans regia* L.). The maximum depth of rooting in deep, well-drained soils is 1.2 to 1.5 m, with the main root system spreading to a depth of 0.6 to 0.9 m. However, the roots can be much deeper in well-drained sandy soils. Castle and Krezdorn (1977) studied the root systems of different rootstocks in deep, sandy soils of central Florida and followed rough lemon roots to a depth of 3.6 m and sour orange (*C. aurantium* L.) to 2.7 m; water depletion studies confirmed that the roots were active at this depth. On the other hand, citrus trees can survive in shallow soils, but may develop smaller trees (Ford, 1954).

1. Water Stress

Hilgeman (1977) reported reduced root development in 'Valencia' orange trees that were irrigated at 95 ha-cm yr^{-1} in 15 irrigations per summer, compared to irrigation of 175 ha-cm yr^{-1} at the same interval or 135 ha-cm yr^{-1} in 10 irrigations. This is especially true when the upper soil layers dry out between irrigations at long intervals. Bielorai and Levy (1971) found higher water uptake from deep soil layers when irrigation was applied at 40-d interval, compared to shorter interval down to 14 d.

When < 40% of the soil volume was wet by irrigation, root density in the unirrigated soil, which was wet only by winter rains, was reduced by 54%, compared with fully irrigated plots (Moreshet et al., 1983).

2. Soil Salinity

The distribution of sour orange roots is affected by soil salinity. Increased salinity was reported to cause a 10% reduction in root development in the layer below 0.3 m and increased roots in the upper layers. Relative water uptake in the 0.6 to 0.9 m layer was reduced 20% as salinity increased from EC$_e$ 2.0 to 5.6 dS m^{-1} (Bielorai et al., 1978).

3. Soil Temperature

Root growth ceases at soil temperatures below 13° or above 35°, with an optimum near 26°C (Castle, 1978). Cary and Weerts (1978) reported doubling of the root growth-rate when soil temperature increased from 19° to 25°C. Root hydraulic conductance of mature Valencia orange trees in southern California was reduced at soil temperatures below 15°C when the soil water supply was adequate (soil matric potential [Ψ_m] of -0.03 MPa or higher) (Elfving et al., 1972). Ramos and Kaufmann (1979) described a continuous reduction in root conductance of rough lemon as soil temperature was reduced from 35 to 5°C. Water stress further decreased the hydraulic conductance of roots. The reduction was attributed to differences in the surface area of unsuberized roots. Moreshet and Green (1984) calculated the hydraulic conductance of a mature Valencia from lysimeter data, leaf conductance and Ψ_{leaf}. Their data confirm that citrus is sensitive to root environment temperature.

4. Vesicular-Arbuscular Mycorrhizae

Citrus is highly dependent on vesticular arbuscular mycorrhizae (VAM), which occur naturally in most citrus soils, though little is known about their efficiency in different locations. Inoculated seedlings grow much faster than seedlings in sterilized soil, mainly because of increased uptake of P from the soil (Menge et al., 1978). Levy and Krikun (1980) induced similar levels of P in VAM and non-VAM rough lemon seedlings and reported that when plant P levels were similar, VAM increased stomatal conductance and photosynthesis during the recovery from water stress without affecting the Ψ_{leaf}, suggesting an effect on stomatal regulation rather than on root conductance. This was confirmed by Levy et al. (1983b), who reported that VAM did not increase root conductance, which was actually reduced in some inoculated plants, with a concomitant increase in the transpiration. Increased transpiration of VAM-inoculated 'Troyer' citrange (*C. sinensis* L. × *Poncirus trifoliata* L.) seedlings was also reported by Johnson and Hummel (1985). Graham and Syvertsen (1984) confirmed the high transpiration of VAM plants, but challenged the assumption that VAM infection had a direct effect on transpiration, suggesting that this was an indirect effect of better P nutrition. The fact remains that VAM can increase transpiration significantly.

The irrigation regime can affect the amount of root colonization by VAM in the field. Increasing the interval between sprinkler irrigations from 18 to 30 d increased VAM infection of sour orange roots in the upper 0.3 m of the soil of a mature orchard in loess soil in the Negev, but decreased VAM in the deep layers. A further increase of the interval between irrigations to 40 d decreased VAM in the upper 0.3-m soil layer, which dried to the wilting point between irrigations. The VAM decreases with depth and soil salinity increases; however, the direct effect of salinity is not clear—only a slight inhibition of VAM was reported in the 0.9 to 1.2-m layer when EC_e was 5.7 dS m^{-1} (Levy et al., 1983a).

D. Fruit Quality

Fruit quality is an important factor in determining the value and marketability of citrus, either for fresh consumption or for processing as juice.

Fruit quality parameters can be divided into two main groups: (i) physical parameters, which include fruit size, peel thickness, color, and juice content and (ii) chemical parameters of the juice, which include acid, sugar, their ratio, and the amount of minor constituents, including bitter or aromatic compounds which may influence fruit palatability.

The effect of changes in irrigation regime on yield and tree development is cumulative and slow to become evident. In contrast, the changes in fruit quality can be detected within a season (A. Cohen et al., 1968; Levy et al., 1978a, 1979; Young & Garnsey, 1977). This makes the comparison of fruit quality parameters a useful diagnostic tool for rapid detection of differences between the effects of different irrigation regimes (Levy et al., 1979), or year-to-year weather patterns.

1. Physical Properties of Fruit

a. Fruit Size and Juice Contents. Fruit size is considered the major fruit characteristic influenced by irrigation (Fishler, 1976; Legaz et al., 1981; Marsh, 1973). The enlargement of fruit by increased irrigation quantities is not always linear (Levy et al., 1979).

Valencia orange that developed under luxurious moisture conditions (microjets every 3 d) had tender rag (8.4-kg force) compared with trees that were drip irrigated or otherwise mildly water stressed (9.9 kg) (Bredell & Barnard, 1977). Hilgeman (1966) concluded that, in Florida, fruit had thinner peels because it was exposed to less internal water stress. Drought increased the peel/pulp ratio in Marsh grapefruit (A. Cohen et al., 1968) and Valencia orange (Hilgeman, 1977). Irrigation increased juice percent and decreased peel thickness in navel orange (Mougheith et al., 1977). Increased salinity, which reduced water uptake by the trees and increased plant water stress during the summer did not affect the physical characteristics of Marsh grapefruit, although it affected the internal quality (Levy et al., 1979). Extending the interval between irrigations decreased juice percent and increased peel thickness of Marsh grapefruit, but increased juice percent and rind thickness in Valencia orange (Bredell & Barnard, 1977). Shortening the interval between irrigations from 40 to 21 d increased Marsh grapefruit size, but further shortening of the interval to 14 d decreased fruit size (Levy et al., 1979). Drip, when compared with sprinkler irrigation, reduced the size of grapefruit in the inland valleys of Israel. Also, in most years drip irrigation of trees resulted in smaller fruit than did flood irrigation (Fishler, 1976).

b. Fruit Color. The rate of rind color development decreases when the irrigation is increased. Bielorai et al. (1981) studied 'Shamouti' orange in the coastal plain of Israel under different soil wetting regimes. Irrigation with stationary microsprinklers, which wet 35% of the soil surface area, caused 72% of the fruits to reach full orange color, compared to 64.5% full color in sprinkler irrigation which wet 90% of the soil area. Huff et al. (1981) showed that drip irrigation of 'Ruby Red' grapefruit increased the chlorophyll and lycopene content of the rind, caused more regreening, but did not increase β carotene. It is not clear, however, if this was the effect of the drip irrigation water regime or an indirect effect of drip irrigation on improved N uptake, leading to increase in leaf N. Such effect of N on fruit color was reported by Bielorai et al. (1981). Kuriyama et al. (1981) enhanced color development in 'Satsuma' mandarin (*C. unshiu* Marc) by inducing water stress, using plastic sheets to drain rain water away from trees during the rainy months of October and November in Japan.

2. Internal Quality

a. Sugar. The amount of sugar in the juice is often expressed as total soluble solids (TSS, usually measured with a refractometer) or as brix (based on juice specific gravity, usually measured with a hydrometer), and is an important parameter in determining the price of processed fruit. Acid concentration and the ratio between TSS and acid are important parameters in defining fruit quality, picking time, and value. In general, water shortage causes increased concentration of TSS in the juice (Cruse et al., 1982; Koo & Smajstrla, 1985; Kuriyama et al., 1981; Levy et al., 1978b, 1979; Mougheith et al., 1977; Young & Garnsey, 1977).

b. Acid. Drought increases citrus juice acidity (Cruse et al., 1982; Kuriyama et al., 1981; Levy et al., 1978a, 1979; Maotani & Machida, 1980; Mougheith et al., 1977; Sakomoto & Okuchi, 1968; Young & Garnsey, 1977). Acidity may increase more than the TSS, lowering the TSS/acid ratio and thereby diminishing fruit quality (Levy et al., 1978a, 1979; Maotani & Machida, 1977; Mougheith et al., 1977; Scuderi & Raciti, 1977; Young & Garnsey, 1977). Grapefruit may be more sensitive than orange (Wiegand & swanson, 1982d). Levy et al. (1978a) reported that the effect of summer water stress on acid accumulation in Marsh grapefruit lasted long after the stress was relieved and was more pronounced than the effect of stress on TSS. Koo (1981b) observed that fertigation, compared with dry fertilization, reduced the acid concentration in orange juice. The effect was more apparent in low-volume fertigation.

c. Naringin. This is one of the principal bitter compounds in grapefruit. Reducing the amount of water application from 12 to 6 ha-cm per irrigation significantly reduced the concentration of naringin in Ruby Red grapefruit (Cruse et al., 1982).

3. Peel Disorders

a. Oleocellosis. This is caused by the rupture of rind oil glands and is a major problem when citrus fruits are picked green. Susceptibility diminishes with the disappearance of the green peel color which, in turn, is induced by cold weather. Susceptibility to *oleocellosis*, defined as the pressure needed to rupture the oil glands (rind oil rupture pressure) (Cahoon et al.,

1964), is closely related to the Ψ_{fruit}. Conditions of decreased water stress (high Ψ_{leaf} and high Ψ_{fruit}) greatly increase susceptibility (Oberbacher, 1966). The usual practice is to avoid picking fruit when environmental conditions favor relatively high Ψ_{leaf}, e.g., shortly after irrigation or rainfall, early in the morning, or under weather conditions that reduce water loss: fog, dew, high humidity, or clouds (Cahoon et al., 1964; Grierson et al., 1965; Levy, 1967; Wardowski et al., 1976). Ethylene-releasing compounds can reduce the susceptibility of the fruit by accelerating chlorophyll breakdown in the peel. However, it was observed (Levy et al., 1979) that the same compounds reduced transpiration, increased Ψ_{leaf} and turgor, and caused higher suscep-tibility to oleocellosis prior to complete color break.

b. Other Peel Blemishes. Water stress decreased puffiness of Satsuma mandarins in rainy Japan (Kuriyama et al., 1981). Autumn irrigation decreased russeting and creasing but increased blemishes of 'Temple' (*C. temple* Hort. ex Tan.) fruits in Florida (Koo & Reese, 1977).

4. Effect of Irrigation Technique

Changing the irrigation technique can affect fruit quality. As a rule, ir-rigation methods that wet the soil partially induce a higher concentration of juice. Drip irrigation increased acid and TSS in Valencia orange when compared with jet irrigation in Florida (Koo & Smajstrla, 1985) and reduced the TSS/acid ratio in South Africa (Bredell & Barnard, 1977). Irrigation methods that wet only a portion of the soil surface tend to increase the TSS/acid ratio, as reported for furrow-irrigated Valencia orange in Califor-nia (Jordan, 1981) and for drip-irrigated early Marsh grapefruit in the Jor-dan Valley of Israel, in comparison with sprinkler or flood irrigation (Fishler, 1976). Daily sprinkler irrigation at frequent pulses (every 90 min) has been found to reduce the TSS/acid ratio (Fishler, 1976; Zur et al., 1974).

II. IRRIGATION AND ORCHARD MANAGEMENT

A. Diseases

1. Gummosis

Gummosis, or citrus foot-rot, a fungal disease caused by *Phytophthora parasitica* Dastur, is closely related to conditions of high soil water content, since the zoospores of *Phytophthora* spp. need free water to infect healthy bark. Citrus rootstocks differ in their susceptibility to this disease, with sweet orange, rough lemon, and Palestine sweet lime (*C. limettioides* Tan.) among the most susceptible and sour orange, alemow (*C. macrophylla* Westr.), and trifoliage orange [*Poncirus trifoliata* (L.) Raf] among the tolerant rootstocks. Planting in heavy soils and in old orchard soils greatly aggravates the problem. The best preventive measure is soil drainage, keeping the soil around the trunk dry and avoiding irrigation regimes that wet the trunk for long periods. Use of flood irrigation or frequent irrigation with microsprinklers should be avoided, especially in heavy soils. If this is not done, even comparatively tolerant rootstocks, such as sour orange, may be affected, as experienced in the Rio Grande Valley of Texas (Timmer & Leyden, 1976). A preventive measure is to plant on ridges and avoid deep planting. Klotz et al. (1967) observed that trees in one-furrow systems had only 25% of the rot found in two-furrow systems.

Drip-irrigated trees were found in Israel to suffer less from foot-rot than trees irrigated with microsprinklers, but drippers should not be placed in contact with the trunk. Timmer and Leyden (1976) studied foot-rot on young 'Star Ruby' and 'Redblush' grapefruit and on sour orange planted on a fine-textured, poorly drained soil considered unsuitable for citrus. Drip-irrigated trees planted on level soil suffered from increased foot-rot, apparently because freeze-protection tree-wraps became saturated by irrigation water. Planting on ridges greatly reduced infection.

The use of the systemic fungicides Efosite-Al (Al tris-O-ethyl phosphonate) and metalaxyl [N-(2,6-dimethylphenyl)-N-(methoxyacetyl)-alanine methyl ester] to control gummosis was considered a solution to *Phytophthora* (Davies, 1981; Lavilee & Chalandon, 1981). However, Menge (1986) warns that the buildup of a microbial population, capable of breaking down these chemicals in the soil, may reduce their effectiveness.

Surface irrigation water may be contaminated by *Phytophthora*, a common problem where drainage water is collected at the end of the furrows ("tail water") and returned to the water distribution system to be reused.

2. Flooding

A different problem is the death of trees from what was termed by M. Cohen et al. (1981) as "water damage" (asphyxiation of the roots as a result of excessive water and inadequate oxygenation after heavy rains in heavy soils). Sour orange and 'Carrizo' citrange [$C. sinensis$ (L.) × *Poncirus trifoliate* (L.)] rootstocks may be sensitive to this problem while rough lemon and sweet lime are comparatively tolerant (Cohen & Mendel, 1965; M. Cohen et al., 1981). Citrus can tolerate flooding better when the soil pH is 7.0 as compared with 4.5 (Phung & Knipling, 1976) and during the winter rather than during periods of high soil temperatures (M. Cohen et al., 1981; Ford, 1964).

3. Lime-Induced Chlorosis

This disorder is closely related to irrigation and soil aeration, and can be aggravated by high soil-water content. Hilgeman and Sharples (1957) found that frequent irrigation in Arizona caused chlorosis and Chapman (1968) regarded "over moist" soil conditions as one of the most common causes of chlorosis in California. Levy (1984) reported that chlorosis was increased by shortening the interval between sprinkler irrigations, but was reduced by drip irrigation.

Gummosis of trunks and branches of grapefruit trees, similar to "Rio Grande gummosis" (Childs, 1978), is common in Marsh grapefruit trees suffering from lime-induced chlorosis in the Negev region of Israel. This disorder often disappears when the irrigation method is changed from sprinkler to drip irrigation (Y. Levy, 1980, unpublished data).

4. Wilt Diseases

Citrus blight, or young tree decline, is a wilt-like disease of unknown cause, which is characterized by restricted water movement. It affects mainly trees on rough lemon, apparently only in tropical climates (Florida, South America). The numerous publications describing this disease failed to report a link between the disease and cultural practices, including irrigation. The disease causes water-stress symptoms in the affected tree (Allen & Cohen, 1974) and one of the methods for diagnosing the disease is to measure the rate of uptake of water injected into the xylem (M. Cohen, 1974; Young & Garnsey, 1977).

B. Pests

Overhead irrigation, when compared with undercanopy irrigation, was found to affect the mite populations in citrus orchards; overhead irrigation suppressed citrus rust mite (*Phyllocoptruta oleivora* Ashmead) and Texas citrus mite (*Eutetranychus citri* McGregor) but enhanced the population of Citrus red mite (*Panonychus citri* McGregor) (Knapp et al., 1982).

In the greenhouse, red mite (*Tetranychus telarius* Lin.) often attack water-stressed plants more than well-watered plants. Similarly, the oriental red mite (*E. orientalis* Klein) causes more leaf drop to water-stressed trees (Y. Levy, 1979, unpublished data).

C. Chemigation

1. Fertigation

The most common chemicals applied through the irrigation system are fertilizers (the procedure termed *fertigation*). Koo (1980) applied up to 20 kg of fertilizer per m^3 of water through overhead sprinklers without leaf damage, provided that the irrigation system was operated for 4 to 6 h after the fertilizer was applied. Koo (1981b) worked with Valencia orange on rough lemon in central Florida, in deep sandy soils wet by summer rains, with a fully developed root system. Under this condition, trees that received low-volume fertigation had higher leaf P and lower leaf Mg, compared with dry fertilization. This also had an effect on fruit quality as discussed earlier. The success of fertigation in small-volume irrigation may depend on the size of the active root system which, in turn, depends on the amount of summer rain in the area. Koo (1984), working in humid Florida and sandy soils, suggested coverage of 40% of the ground or 80% of the tree canopy is necessary in order to obtain results comparable to broadcasting dry fertilizer. Drip irrigation, which often covers much less of the ground, cannot be practical without fertigation. In summer-rain areas it may be important to apply fertilizer frequently through the drip system, even during the rainy season.

2. Herbigation and Other Chemicals

Uda and Morimoto (1977) used overhead impact sprinklers to apply fertilizers, fungicides, herbicides (herbigation), and pesticides with excellent results. They stressed that some of the chemicals should be applied without the surfactants or emulsifiers commonly used in spray treatment. Good results were obtained when NAA was applied by sprinklers for fruit thinning of 'Satsuma' mandarin. The herbicides Diuron [3-(3,4-dichlorophenyl)-1,1 dimethylurea], DCPA or chlorthal dimethyl (dimethyl tetrachloroteraphthalate), Terbacil (3-tert-Butyl-5-chloro-6-methyluracil), and Bromacil (5-bromo-3-sec-butyl-6-methyluracil) were used safely for herbigation via overhead sprinklers. Water should be supplied after application of the chemical to wash the herbicide off the citrus leaves.

Under-canopy irrigation is particularly suited for applying herbicides to orchards in arid areas, where weeds do not develop outside the irrigated area. Oren and Israeli (1977) reported that herbigation, using microsprinklers, with Bromacil plus Diuron, gave better control of bermudagrass [*Cynodon dactylon* (L.) Pers.] and other weeds with lower quantities of herbicides, compared with conventional application methods. Good results were reported for drip herbigation. Del Amor et al. (1981) stressed the need to apply herbicides in small concentrations continuously. However, a problem with drip herbigation in some areas in Israel is the development of weeds at the boundary of the wet area.

Under-canopy sprinklers were used successfully for applying Cu-based fungicides for the control of brown rot on fruit. The best treatment was of Bordeaux mixture, applied at a concentration of 1 to 2 kg m^{-3} of water for 9 min, followed by 1 min of water to clean the system of Cu. It is not clear whether the suppression of brown rot was effected by direct protection of the fruit by the Cu, similar to the effect of sprayed Cu, or because of Cu that dripped to the soil suppressed the soil-borne fungus (Oren et al., 1977).

D. Climate Control

1. Frost

a. Cold Hardening. Citrus is vulnerable to freeze damage below $-2.2\,°C$ (Yelenosky, 1985). Unfortunately, many of the major citrus-growing areas around the world [Japan, the Yangtse Valley in China (Wen-Cai, 1981), the Rio Grande Valley, Texas, California, Arizona, Florida, Spain, and Italy] suffer periodically from severe freezes.

The susceptibility of citrus to freeze damage is dependent on the dormancy of the trees when the subfreezing temperatures occur. Dormant or inactively growing (resting) trees are much less sensitive than others to frost. Water stress can enhance cold hardiness when comparatively high temperatures (above 12 °C) would otherwise encourage continued growth of buds, a typical situation during many winter months in Florida. Yelenosky (1979b) observed that water stress induced cold hardening of young citrus trees, and found free proline accumulation in leaves of cold-hardened trees (Yelenosky, 1979a). Proline also accumulates during water stress (Levy, 1980). Cold-hardened trees accumulate carbohydrates rapidly (Yelenosky, 1985), a process which is also enhanced by water stress. Trees that were not irrigated in the fall, but were irrigated during the freeze were less damaged than trees that were irrigated all the time or not at all (Davies et al., 1981). Koo (1981a) observed that trees that were irrigated late enough in the fall to avoid growth flushes were in better condition to withstand freezes than nonirrigated trees.

b. Frost Protection. Irrigation was regarded as the best method for frost protection in 1901 (Turrell, 1973). The rising prices of fuel made direct heating of citrus orchards uneconomical (Parsons et al., 1985, 1986; Yelenosky, 1985). Wet soil has a greater heat capacity and heat conductance than dry soil. Thus, surface irrigation can increase the ability of the soil to accumulate heat during the day and radiate it back to the canopy by night; trees in dry top soil will be more damaged than those where top soil is moist. In surface irrigation, the water heats the orchard mainly by the release of sensible heat, which depends on the water's initial temperature. However, water will cool and usually cannot be left standing around the trees for long periods, because aeration problems might damage the trees. This is less of problem if day temperatures remain low, since O_2 solubility is increased and root respiration is decreased by low temperatures. In many locations, the water supply system cannot deliver the large amounts of water needed for good protection of large areas. Sprinklers cause a temperature drop, due to evaporative cooling, which is dependent on the ambient humidity and wind velocity (Turrell, 1973; Parsons et al., 1985). Rapid cooling, induced by strong winds at low humidities, is greater than the sensible heat released by the water. Thus, the temperature of sprinkled leaves will be near wet bulb temperature and below dry bulb temperature (Turrell, 1973). In contrast to the situation with deciduous trees, when citrus is subjected to overhead irrigation on a freeze night, more than 500 kg of ice may accumulate on the dense foliage of a single tree, causing severe breakage (Parsons et al., 1986). For these reasons, overhead sprinkling is not recommended.

Much data on the use of microsprinklers for reducing leaf damage accumulated during the severe freezes of the 1980s in Florida. With radiation freeze during a calm night, typically with a strong temperature inversion, the effect of evaporative cooling is small and microsprinklers can protect the lower limbs of a tree. Temperature increases of 6 to 8 °C have been reported (Parsons et al., 1981, 1982). Irrigation water warms the trees through the release of sensible heat and the heat of fusion. The heat is released as irrigation water freezes on the trunk and lower limbs. The dense canopy of the evergreen citrus tree helps contain the heat released by the water so that microsprinklers might be more effective for cold protection of citrus than of deciduous trees. However, during an advective freeze with high winds and low dew point, evaporative cooling caused more damage to the wetted area than to the dry area. Prevention of damage to young trees may still be possible under such freeze conditions if water is applied directly to the trunks and main scaffold branches at high enough volumes, up to 2.5 ha-cm^{-1} (Parsons & Tucker, 1985). The microsprinklers should be located on the upwind side of the tree at a distance not >0.7 m (Parsons et al., 1986). Insulated wraps, combined with microsprinklers, can increase the protection of trunks and save young trees from total destruction (Jackson & Davies, 1985).

2. Cooling

Heat spells can cause damage to flowers and young fruitlets and increase "June drop." 'Washington' navel orange is especially prone to such damage. Reduction in net photosynthesis by temperatures above 30 °C may cause low-carbohydrate induced fruit abscission (Palmer et al., 1977). Overhead irrigation can lower the air and canopy temperature considerably. Koo and Reese (1975) reported that overhead irrigation lowered midday air temperatures from 34 to 28 °C and increased relative humidity from 40 to 60%, relative to nonirrigated control. Brewer et al. (1977) reported that in the San Joaquin Valley, intermittent application of overhead sprinklers reduced air temperature by 5 to 6 °C, and leaf temperature was reduced more than air temperature. On a hot afternoon, on the southwest side of the tree, leaf temperature was reduced from 51.7 to 29.4 °C when the ambient temperature was 37.8 °C. Sprinkler irrigating large orchard areas increased the cooling efficiency and resulted in a 40% increase in yields of Washington navel in a 3-yr experiment. A 4-min low-volume irrigation pulse, applied every 16 min, increased fruit set and size (Palmer et al., 1977).

Overhead sprinkling may increase the hazard of Na and Cl damage to leaves (as discussed in chapter 32, section VII) and should be practiced only with water of low salinity.

E. Using Water Stress to Induce Flowering

Rainfall, after a period of drought, can induce flowering in many trees in tropical climates, while in subtropical climates cold weather usually induces dormancy, followed by flowering in spring. Citrus trees growing in tropical climates can bloom continuously all year round, and rainfall or irrigation, after a period of drought, can trigger a flush of flowering. In subtropical climates, some citrus species, notably lemon, lime [*C. aurantifolia* (Christm.) Swingle], citron (*C. medica* L.), and pumello [*C. maxima* (Burm.) Mer.] are capable of flowering all year round. The capability of orange trees to reflower after a severe drought is important in Brazil, where citrus is usually nonirrigated and may suffer from a spring drought, which harms the normal spring flowering flush (Y. Levy, 1984, personal observation). Irregular irrigation can cause out-of-season flowering in orange and mandarin. The fruit that may set is usually undesirable, and heavy out-of-season flowering can reduce the amount of main-season fruit.

A traditional orchard management practice in lemon culture in Sicily is *forzatura* (forcing), i.e., withholding irrigation for periods of 30 d or longer in order to cause wilting and induce summer bloom. The fruit that will set from such flowers will be picked next summer as green (*verdelli*) fruits (Barbera et al., 1985). Lemon cultivars differ in their response to this treatment. Good responders are the cultivars Eureka, Femminello, Villafranca, Verna, and Fino, as compared with cultivars that tend to flower only in the spring—Lisbon and Interdonato.

Nir et al. (1972) detected flower bud differentiation in Eureka lemon during the drought treatment before irrigation was resumed and suggested that the effect of water stress may inhibit production of GA_3 gibberellin in the water-stressed root system. The amount of stress necessary to induce flowering is important, since excessive stress can be harmful. Torrisi (1952) [cited by Barbera et al. (1981)], noticed that 64% of the flowers aborted after excessive water stress, compared with only 20% after a moderate stress. Calabrese and Di Marco (1981) noticed that excessive stress also harmed the development of fruits already on the trees. In their work, they treated Femminello lemon in western Sicily and reported that soil water content in the 0 to 0.6-m depth should be decreased from 22 to 15–16%. Barbera et al. (1981) also studied the cultivar Femminello growing in western Sicily, on a typical reddish-brown sandy loam and reported that the leaves should reach a maximum (predawn) Ψ_{leaf} of -1.3 MPa. Goodall and Silviera (1981) were unable to adapt the Italian verdelli process to 'Bears' lime production in the upper foothills of Santa Barbara County, California (far away from coastal fogs). Flowering was induced in October, after the trees had wilted for 17 d. However, the verdelli crop amounted to only 13 kg per tree, or <10% of the total annual production.

III. IRRIGATION METHOD

Technically, irrigation methods differ in the way water is distributed over the field. Gravity irrigation, where water is conveyed to the point of consumption directly on the land surface, is the traditional method of irrigation of citrus, as of many other crops. This method is gradually being replaced by closed conduit, pressurized distribution systems, the most important of which is sprinkler irrigation. Although sprinkler irrigation is still the most prevalent method used in citrus-growing regions, there has been a rapid shift, recently, to microirrigation methods—drip, minisprinklers, and sprayers.

A. Gravity Irrigation

Water may be distributed in furrows or border strips along the tree rows or in flood basins around each tree. Water distribution by this method is inherently less uniform than by alternative methods, and usually larger quantities of water need to be applied to obtain reasonable coverage. Hoffman et al. (1984) used one-third less water in a drip-irrigated orange plot in Yuma, AZ, at 20% leaching, than in flood-irrigated plots. The yield obtained was similar for the two methods. The water use after 5 yr of experimentation in a Valencia orange orchard near Yuma, AZ, was greater under flood irrigation (170 ha-cm yr^{-1}) than with 39, 36, and 43 ha-cm yr^{-1} for sprinkler, basin, and drip irrigation, respectively (Rodney et al., 1977). The yield on *C. macrophylla* rootstock was 112, 146, 232, and 194 kg tree^{-1} for the four methods, respectively. The yield on Troyer rootstock showed the same trend but was 40% lower.

Bingham et al. (1984) analyzed outflow from a 273-ha irrigated citrus watershed over two time periods—1967 to 1970 and 1980 to 1981. During the first period, half of the watershed was irrigated by sprinklers and half by furrows. During the second period, drip irrigation occupied 60% of the area. The overall leaching percentage changed only slightly, from 45 to 41% between the periods. The drip-irrigated area, however, received one-third less water than areas irrigated by the two other methods.

Rawlins (1977) proposed a novel water distribution system—bubbler irrigation. The method combines the advantages of a closed conduit system of low labor requirements and good water distribution, with the low energy requirements of flood irrigation. The method uses a large diameter (7.5–10 cm) corrugated, buried main delivery pipe connected to 9.5-mm diam. hoses which are attached to each tree trunk. The head requirements are usually satisfied by the normal slope of the land and uniformity of discharge is achieved by moving the hose outlets along the trunks.

B. Sprinkler Irrigation

Sprinkler irrigation may be applied by overhead solid-set sprinklers, traveling gun, under-canopy low-angle solid-set, or portable systems. Overhead sprinkler irrigation is still a common method in Florida orchards (Harrison, 1978). The advantages of the system are its low cost compared with an under-tree solid-set system, and its use for frost protection. When the irrigation water contains even a low concentration of salts (see chapter 32, section VII), foliar damage may occur. For this reason overhead sprinkling has been abandoned in Australia (Cole & Till, 1977) and in Israel. Traveling gun puts more water in a small area, then moves on. Consequently, evaporation is greater from solid-set above-the-canopy system, which keeps the whole canopy area wet for a longer period. Koo and Reese (1975) compared the water distribution uniformity under traveling gun (a common irrigation method in Florida) with solid-set overhead and found the latter to give a 5% lower Christiansen coefficient. Day evaporation was 10% greater with the solid-set overhead system.

Under-the-canopy sprinklers do not provide the distribution uniformity that is obtained with overhead systems. It seems, however, that distribution uniformity in orchards is of little significance. Since the soil volume occupied by the root system of each tree is large, it is unlikely that a tree will not receive water even if the distribution is poor. Heller et al. (1973)

used under-the-canopy sprinkler irrigation, which, because of canopy interference, wetted only the space between the rows and the sprinkler line, while the other side of the row remained dry. Alternate irrigation of the tree row of Shamouti orange reduced water use by 18% with no reduction in yield. The water uptake pattern depends on the availability of water in any portion of the root zone. When water is deficient or depleted in one portion of the soil, the roots take up water from other, still-wet portions (Dirksen et al., 1979; Myers et al., 1976) and thus the trees do not suffer from water deficiency.

C. Microirrigation

Microirrigation is a relative newcomer to orchards. The method includes microjets or minisprinklers, sprayers, and drip/trickle irrigation. The advantages of microirrigation are high application efficiency, low pressure requirements, ease of operation, and good control of soil aeration. Minisprinklers and sprayers are used in many new plantations because they can be tailor-designed to the size of the tree; the discharge and the wetted diameter can be increased as the tree grows. The diameter of wetted soil surface under many minisprinklers ranges from 4 to 8 m, a distance similar to normal tree spacing. Likewise, the number of drippers per tree can be adjusted to cover any desired land area. The conversion of existing mature orchards from flood and sprinkler to minisprinkler and drip irrigation must be done with due consideration to changes in the wetted surface area. Sprinkler and gravity methods result in the wetting of a greater surface area than drip and minisprinklers. The conversion of Florida orchards from flood to microjet resulted in a 37% increase in yield. A 10% yield increase was realized when a Marsh grapefruit orchard was converted from a pull-over sprinkler system to microjet (Anon., 1976). Trials in Nelspruit, South Africa (Bredell & Barnard, 1977), showed the microjet system to be superior in both yield and fruit quality to drip irrigation. The yield of a Washington navel orange on trifoliate rootstock orchard in California remained the same when the orchard was converted from furrow to drip irrigation (Aljibury et al., 1977). The amount of water used under drip was 25% less than under furrow or sprinkler irrigation, and the direct energy use was about 915 Kwh ha^{-1} for drip and furrow compared with 2270 Kwh ha^{-1} for sprinkler irrigation. Total energy use was least for drip irrigation (Aljibury, 1981).

A few experiments were conducted in Israel to determine the minimum surface area that can be wetted without causing yield reduction. Bielorai (1977) showed that the minimum area depended on the irrigation interval. The yield of Marsh grapefruit trees on sour orange rootstock was the same for drip irrigation, which wetted 30% of the soil surface area, as for sprinkler irrigation, which wetted 70% of the area, when the irrigation interval was 3 d for drip and 7 d for sprinkler irrigation. For the 7-d interval, yield under drip was 7% less than under sprinkler irrigation. Increasing the sprinkler irrigation interval to 21 d had no effect on yield (Bielorai, 1982).

In another experiment with Shamouti on sweet lime rootstock, Bielorai (1982) found no yield differences among three irrigation methods—microsprayers (30% wetted area), minisprinklers (67% wetted area) and under-tree sprinklers (90% wetted area). The mean yield was 72 Mg ha^{-1}. Moreshet et al. (1983), on the other hand, obtained a 21% yield reduction of Shamouti on sour orange when the wetted area was reduced to 40% by the use of sprayers instead of sprinklers. The total soil-to-atmosphere conductance was reduced under partial wetting (Y. Cohen et al., 1987), probably because of a reduction in root concentration.

Koo and Smajstrla (1985) compared no irrigation with supplemental irrigation of a Valencia on rough lemon rootstock orchard in Florida using two or four drippers per tree (5 and 10% wetted area, respectively) and one or two microsprayers per tree (28 and 50% wetted area, respectively). Yield increase over no irrigation was 43% for the two drip irrigation treatments and 65% for the spray irrigation treatments. Rainfall during the year was 125 ha-cm yr^{-1}. Yagev (1977) and Ben-Meir and Israeli (1985) found no substantial yield differences in Marsh grapefruit irrigated by sprinkler or by one or two lateral drippers. This was true as long as sufficient water was applied to satisfy the evapotranspiration needs of the tree.

IV. IRRIGATION WATER REQUIREMENTS

Crop evapotranspiration is controlled by climatic factors. This is true for citrus as well as for other crop plants. However, there are some specific crop characteristics which may alter the level of water loss from crop canopies. Under similar climatic conditions, citrus trees are known to have lower transpiration rates than other crop plants. For example, the midsummer evapotranspiration in Israel was found to be 0.7 to 0.8 ha-cm d^{-1} for many field crops, 0.85 ha-cm d^{-1} for apple (*Malus sylvestris* Mill.) orchards, but only 0.45 ha-cm d^{-1} for citrus orchards (Shalhevet et al., 1981). The low transpiration is due to low citrus canopy conductance. Van Bavel et al. (1967) calcualted canopy conductance from evapotranspiration E_t estimated from soil water depletion data, and potential evapotranspiration (E_o) estimated from a modified Penman (1948) equation. The values thus obtained were 0.22 to 0.29 cm s^{-1} compared with 2.00 to 3.30 cm s^{-1} for a number of field crops. Direct measurement of leaf diffusion conductance was in agreement with the calculated values. Stanhill (1972) reported annual E_t values of 84 ha-cm for a Shamouti orchard in Israel when E_o was 157 ha-cm yr^{-1}. The difference was due to canopy conductance of 0.62 cm s^{-1}.

A. Irrigation During Crop Establishment

As with annual crops, the degree of ground cover (leaf area index) has a decisive influence on E_t. However, it takes a number of years (5–6) to achieve full ground cover and maximum E_t for a newly planted orchard. The amount of water applied during the period of establishment should take into account the canopy size. The recommendations of the Irrigation Extension Service in the central citrus region of Israel are given in terms of the water required per tree (Anon., 1985b). With modern microirrigation methods, unlike the situation with the standard under-tree sprinklers, it is possible to irrigate individual trees and increase the size of the wetted area or the number of emitters as the tree develops. The recommendation for mid-summer weekly applications is 10, 15, 25, 45, and 65 L d^{-1} per tree from the first to the sixth year, respectively. From the sixth year on, the full amount of 0.40 to 0.45 ha-cm d^{-1} is applied (100 L d^{-1} per tree).

Leyden (1975) compared the water needs during establishment of Star Ruby grapefruit in the lower Rio Grande Valley, using three water application techniques. Strip irrigation required 10 m^3 per tree, while ring irrigation (1.5 m diam.) and drip irrigation required 37 and 23% of this amount, respectively. Trunk diameter 2 yr after planting was largest for ring irrigation (6.0 cm) and smallest for drip irrigation (5.3 cm).

Toledo et al. (1981) found mean summer water requirements for young Valencia orange trees in Cuba to be 0.22 for seedlings, 0.33 for 4-yr-old trees and 0.40 ha-cm d^{-1} for fully grown trees.

B. Irrigation Scheduling

The total seasonal amount of irrigation water needed by a fully grown orchard for optimum yield depends on the daily course of evapotranspiration, the rainfall distribution, and the citrus variety grown.

In a Mediterranean-type climate, rainfall is concentrated in the winter months, November to April, with little or no summer precipitation. Normally, during the winter period there is no need for irrigation except under extreme drought conditions. Therefore, there is little reference in the technical literature of such climates regarding irrigation needs during the winter. Van Bavel et al. (1967) estimated E_t for a Valencia orchard during the winter months in Tempe, AZ. Using such estimates and maintaining a water balance sheet of rainfall and E_t, one can compute the soil water deficit and use this to decide on winter irrigation scheduling. Information on the relationship of yield to changes in winter irrigation scheduling is, however, not available. The E_t values given by Van Bavel et al. (1967) for Tempe, AZ, range from 1.0 during January to 5.2 mm d^{-1} during June. These values corresponded to the ratio of evapotranspiration to Class A pan evaporation (E E_p^{-1}) of 0.40 and 0.62, respectively. In October, the value was 0.71. Thus, the maximum ratio is not obtained at the time of maximum transpiration, unlike the situation with most field crops. Under the conditions of Phoenix, AZ, the consumptive use of Marsh grapefruit ranged from 0.15 in January to 0.56 ha-cm d^{-1} in July, and from 0.10 to 0.44 ha-cm d^{-1} for navel oranges. The total annual E_t's for the two crops were 121.7 and 99 ha-cm yr^{-1}, respectively (Erie et al., 1965). For Yuma, AZ, Hoffman et al. (1984) reported E_t values for Navel oranges of 0.08 ha-cm d^{-1} in December and 0.75 ha-cm d^{-1} in August. The high summer E_t values reflect the dry and hot climate of southern Arizona.

Stylianou (1974) compared irrigation intervals ranging between 25 and 50 d for Valencia on sour orange rootstocks grown on a sandy loam soil in Cyprus. Only the 50-d interval resulted in a significant 12% yield reduction over the 32-d interval. The common irrigation interval in Israel for sprinkler irrigation of orange on similar soils is 21 to 30 d (Shalhevet et al., 1981).

In Florida and Texas there may be a considerable amount of rainfall during the summer growing season. Yet, in most years, rainfall is insufficient to satisfy crop water requirements and supplemental irrigation is beneficial. Koo (1979) found a response of Valencia on rough lemon rootstock to irrigation in 8 out of 9 yr of experiments, with an average yield increase of 22% when 27.6 ha-cm yr^{-1} of irrigation water was added to the 110 ha-cm yr^{-1} of rain. Most of the rain in Florida falls from June to September, while during the critical period of January to June, rainfall is low. For 50% rainfall probability, the net irrigation requirements were calculated to be 28 and 53 ha-cm yr^{-1} when the usable soil water-holding capacity was 10 and 2.5 ha-cm, respectively. The total E_t for a 10-yr-old orchard under these conditions was 107 to 122 ha-cm yr^{-1} (Anon., 1977). The recommendation for allowable available water depletion was 50 to 75% of soil water-holding capacity.

In a 7-yr irrigation experiment using three citrus varieties, 'Marrs' and Valencia oranges, and Ruby Red grapefruit, Wiegand and Swanson (1982a, b, c) supplemented the 73.2 cm yr^{-1} of rainfall with 21.6 to 67.6 ha-cm yr^{-1} of irrigation water. Water was applied at an available water depletion of 30 to 40%, respectively. An optimum irrigation interval of 28 d was recommended for grapefruit and 35 d for orange for the rainless summer season. The daily water use for summer and winter, respectively, was 0.40 and 0.13 ha-cm. There was no yield response to irrigation when rainfall exceeded 60 cm yr^{-1}, if it was well distributed during the season. Lyons (1977) recommended under Rio Grande Valley conditions, an application of 10 to 15 ha-cm five to seven times during the year, with a summer irrigation interval of 21 d and a winter interval of 60 d.

C. Water Production Function

The construction of reliable water production functions require many data points of yield and water use under controlled experimental conditions. The size and the number of years required for such experiments with citrus orchards are large and therefore little information is available. The best data available are those of Shalhevet et al. (1981), who summarized results from several experiments conducted in Israel over a number of years. The data for some locations are presented in Fig. 32–3 for Shamouti orange on sour orange and sweet lime rootstock, and for Marsh grapefruit on sour orange rootstock. There is considerable scatter in the data because of year-to-year variability and because of soil water content variability in orchards irrigated by under-tree sprinklers. Nevertheless, the general response curve is quite apparent, showing a water requirement for optimum yield in the coastal plain of Israel of 60 to 66 ha-cm d^{-1} and 50 to 57 ha-cm d^{-1} for Shamouti on sour orange and sweet lime, respectively.

Marsh grapefruit is grown in the more inland areas of Israel. Therefore, the water requirement of 65 to 70 ha-cm yr^{-1} for the western Yizre'el Valley and 80 to 86 ha-cm yr^{-1} for the northern Negev estimated from Fig. 32–3 cannot be compared directly with the other varieties. The evapotranspiration (E) data for grapefruit normalized on the basis of Class A pan evaporation (E_p) (Shalhevet & Bielorai, 1978) can be described by the equation:

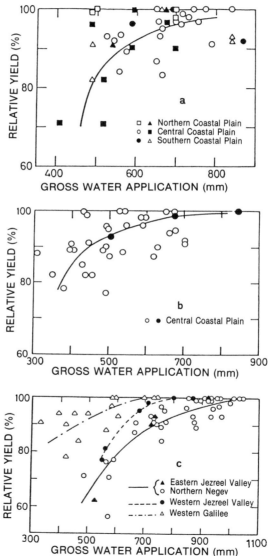

Fig. 32–3. Relation between relative yield of citrus and gross water application in several regions of Israel (Shalhevet et al., 1981). (a) Shamouti orange on sour orange; (b) Shamouti orange on sweet lime; and (c) Marsh grapefruit on sour orange.

$$Y = 37.7 + 102.2 \, (E \, E_p^{-1}) \qquad r = 0.73$$

where Y is the relative yield in percent.

The positive x-axis intercept indicates that not all the water used up by the trees for yield production was taken into account. Indeed, the relationship presented includes only summer irrigation and not winter precipitation.

The slope of the production function shows a unit increase in relative yield for each unit increase in relative evapotranspiration. For sorghum [*Sorghum bicolor* (L.) Moench] and cotton (*Gossypium hirsutum* L.) there was a two-unit increase in yield per unit increase in relative evapotranspiration (Shalhevet & Bielorai, 1978).

V. INDICATORS FOR IRRIGATION TIMING

A. Irrigation According to Plant Physiological Indicators

Kriedemann and Barrs (1981) suggested that trees can be used as biological indicators, since they are continuously solving their own water balance equations and, in principle, an appropriate physiological parameter is all that should be required. Many such indicators have been suggested and used in citriculture.

1. Fruit Growth

The use of fruit growth to time irrigation was suggested by Taylor and Furr (1937) and by Furr and Taylor (1939), who calculated fruit volume from fruit circumference in order to obtain reliable data for a long period. Fruit volume increase was correlated with stomatal aperture (Oppenheimer & Elze, 1941) and with leaf water deficit and soil suction (Lombard et al., 1965). Ashizawa et al. (1979) reported that fruit growth of Satsuma mandarin stopped when maximum Ψ_{leaf} dropped to -0.8 MPa. The main obstacle to the use of this indicator is that fruit growth is not uniform throughout the irrigation season, or that no fruit is available. Good agreement was found between fruit growth and shrinkage and Ψ_{leaf} (Ashizawa et al., 1981).

2. Stomatal Opening

This parameter was used by Oppenheimer and Elze (1941). The rate of kerosene infiltration into the leaf of Shamouti orange and Marsh and 'Duncan' grapefruit was used as a measure of stomatal aperture. This was correlated with soil water content and fruit growth. Stomatal aperture measurement proved particularly useful when fruit was small and its growth could not be measured as an indicator (Mendel, 1951). Van Bavel et al. (1967) confirmed that actual water use agrees with stomatal opening measured with a diffusion porometer.

Table 32-1. Critical Ψ_{leaf} (MPa) value suggested by different authors for initiating irrigation.

Cultivar	Ψ_{max}	Ψ_{min}	Author
	———	MPa ———	
Satsuma mandarin	−0.7		Maotani & Machida, 1980
Satsuma mandarin	−0.8	−1.2 to −2.0	Ashizawa et al., 1979, 1981
Valencia orange	−1.1	−1.9	Green & Meyer, 1980
Toroco orange	−0.3	−0.6	Sardo & Germana, 1985
Marsh grapefruit			Syvertsen et al., 1981
Young leaves		−1.7	
Old leaves		−2.5	
Marsh grapefruit		−3.5	Syvertsen, 1982
Eureka lemon		−2.8	Levy, 1980

3. Trunk Growth

Monitoring of trunk growth with a dendrometer is a good means of following the tree water relations (Cooper et al., 1964; Hilgeman, 1963) and can even detect oscillations in the tree water-status (Levy & Kaufmann, 1976). The diurnal changes in trunk diameter are related to the tree Ψ_{xylem} (Levy & Kaufmann, 1976). Periods in which the tree shrinks during the day and does not recover at night indicate severe water stress.

4. Leaf Water Potential Ψ_{leaf}

Kaufmann (1968) found good agreement between pressure chamber readings and Ψ_{leaf}, estimated with a thermocouple psychrometer. Thus, the pressure chamber can be used in the field for rapid estimation of leaf water potential. The Ψ_{leaf} depends on leaf exposure and location in the canopy. Ideally, Ψ_{leaf} should be determined before sunrise (maximum Ψ_{leaf}), to prevent the effect of differing exposure and differing transpiration in different leaves. Some critical Ψ_{leaf} values are presented in Table 32-1.

5. Fruit Turgor

The measurement of the width of the gap that opens after the peel of citrus fruit is cut was proved by Kaufmann (1969) to be related to the fruit turgor and was suggested as an indicator for irrigation. However, the changes in slice gap are usually small under normal irrigation practices.

6. Measurement of Intact Leaf Moisture Level

Measurement of the attenuation of β rays by intact leaf was suggested by Peynado and Young (1968) and used by Bielorai (1968). Kadoya et al. (1975) demonstrated that leaf thickness is related to the plant water balance and changes with stem diameter. Syvertsen and Levy (1982) correlated leaf thickness to Ψ_{leaf}.

7. Sap Velocity

The measurement of sap velocity in the trunk by the heat pulse method was perfected by Y. Cohen et al.(1981, 1983). The use of this or other noninvasive methods, such as magneto hydrodynamic and doppler, should be ideal for estimating the tree water use, and such a procedure could be connected with an automated irrigation system, providing an automatic feedback system.

8. Canopy-to-Air Temperature Difference

The rate of evaporative cooling because of transpiration and stomatal opening is indicated by ΔT, and is sensitive enough to detect cycling in stomatal opening (Levy & Kaufmann, 1976). The ΔT was measured with an infrared thermometer on 'Toroco' orange in Sicily by Sardo and Germana (1985), and ΔT and Ψ_{leaf} were found to be correlated. However, ΔT measurement is detectable only at high levels of stress. The authors suggest the critical values of ΔT should be 1 to 2 °C, corresponding to Ψ_{leaf} of -0.3 to -0.6 MPa. The lower values of stress should not be exceeded during fruit set.

B. Irrigation According to Soil Instrumentation

1. Gravimetric Soil Water Measurement

Soil water content can be determined directly by taking core samples from the soil and determining weight differences before and after the soil has been dried in an oven. This is the simplest and most effective method for monitoring soil moisture. However, it is labor intensive and time consuming and cannot be used routinely to determine irrigation requirements.

2. Tensiometers

Kaufmann and Elfving (1972) found a good correlation between tensiometer readings and Ψ_{leaf}. Gerard and Sleeth (1960) recommended installation of tensiometers at a depth of 0.6 m and irrigation when the matric potential was reduced to -0.05 to -0.07 MPa. Hilgeman and Howland (1955) suggested that the critical value of Ψ_{leaf} was -0.05 MPa (-50 cb) at a depth of 0.75 m. Marsh (1968) advocated the use of electrotensiometers for automatic scheduling of irrigation. Electrotensiometers, coupled to automatic irrigation systems, were used in an attempt to create a restricted root system and cause dwarfing (A. Golomb, 1986, personal communication).

3. Neutron Scattering

The use of this method to measure soil water content has been limited mainly to research purposes and has not been used in commercial orchards. Lower prices and simpler operation and calibration procedures may change this.

VI. IRRIGATION AND SALINITY

Citrus is a salt-sensitive crop, as are most of the woody perennial fruit trees. Its high sensitivity is assumed to be related to a specific toxic effect of the accumulation of two ionic species, Cl and Na, in the leaves. The considerable variability in sensitivity among various citrus rootstocks, contrary to the experience with most field crops, is cited in support of this assumption (Bernstein, 1980).

A. Rootstocks and Salinity

The pioneering work of Cooper and Gorton (1952) at Weslaco, TX, showed that there are distinct differences in the rate of Cl and Na uptake among 13 rootstocks tested. Rangpur lime (*C. limonia* Osb.) and Cleopatra mandarin accumulated Cl at a slow rate (71–124 mg kg^{-1} d^{-1}), rough lemon and sour orange at a medium rate of (248–298 mg kg^{-1} d^{-1}), and calamondin (*C. madurensis* Lour) and 'Etrog' citron at the fastest (348–515 mg kg^{-1} d^{-1}). In the above experiment, 1-yr-old 'Shary Red' grapefruit trees grafted on the various rootstocks were used. The results of this and other experiments by Cooper et al. (1951) and Cooper and Peynado (1959) were the basis for the salt-tolerance ratings of citrus rootstock given by Bernstein (1980). These results were later verified by many other investigators (e.g., Cerda et al., 1976; Furr & Ream, 1969; Ream & Furr, 1976; Rokba et al., 1979; Spiegel-Roy et al., 1986).

According to Bernstein (1980), the following values of the Cl concentration in the soil saturation extract are levels which will result in not more than 10% yield reduction: Cleopatra mandarin and Rangpur lime, 25 mol m^{-3}; rough lemon, sour orange, and tangelo (*C. paradisi* Macfad × *C. reticulata* Blanco), 15 mol m^{-3}; and sweet orange and citrange, 10 mol m^{-3}. Measurements of vegetative development of seedlings of the various rootstock correlated with the rate of Cl accumultion. For example, the dry weight of Cleopatra mandarin was reduced by 45% as the Cl concentration of the irrigation water increased to 62 mol m^{-3}, while growth of sour orange was reduced by 62% and that of rough lemon by 74% (Rokba et al., 1979).

The comparative tolerance of the young seedlings of the various rootstocks has not been verified quantitatively for mature orchards. The only comparable work known to us is that of Levy and Shalhevet (1985) on 25-yr-old Marsh grapefruit and Washington navel orange trees grafted on Cleopatra mandarin, sour orange, and rough lemon rootstocks. Results obtained after 3 yr indicated that Cleopatra indeed gave higher tolerance to salinity than the two cultivars, as attested by water use and Cl accumulation in the leaves.

B. Salinity Response Function

Seed germination under conditions of salinity was found to be completely unrelated to the sensitivity of the seedlings grown (Mobayen & Milthorpe, 1977). On the contrary, the most sensitive rootstock (*Poncirus trifoliata*) was least affected by NaCl concentration of 100 mol m^{-3} (no reduction in final germination), while final germination of Cleopatra mandarin seeds was reduced by 20% and that of 'Bakraie' mandarin (*C. reticulata* Blanco) seeds by 53%.

Maas and Hoffman (1977) reassessed the previous salinity ratings of Bernstein (1980) and reported the tolerance of two citrus varieties, orange and grapefruit, in terms of the total salt concentration [electrical conductivity (EC_e) of the soil saturation extract]. They ignored the differences in sensitivity among the various rootstocks. Their analysis resulted in a threshold salinity of 1.7 dS m^{-1} (1.8 dS m^{-1} for grapefruit) and a yield reduction of 16% per 1 dS m^{-1} increase in EC_e.

Figure 32-4 presents data of experiments and surveys in California, Israel, and Australia on the response of various citrus varieties to the increase in EC_e. The uniformity in response of the various varieties and rootstocks is striking and supports the approach taken by Maas and Hoffman (1977).

The regression analysis of these data yielded a relationship of relative yield (Y/Y_m) to EC_e

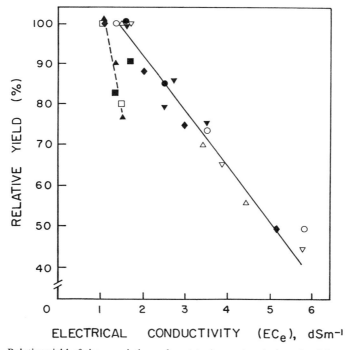

Fig. 32-4. Relative yield of citrus varieties and rootstocks vs. electrical conductivity of the soil saturation extract (EC_e). Regression equation $Y/Y_m = 1.0 - 0.129 (EC_e - 1.28)$. ▽—Valencia × Rough lemon (Francois & Clark, 1980); ▼—Valencia × Troyer (Bingham et al., 1974); △—Valencia (Chapman et al., 1969); ▲—Valencia × Rough lemon (Cole, 1985); ◆—Washington (Harding et al., 1958); ○—Grapefruit × Cleopatra (Pearson et al., 1957); ●—Grapefruit × Sour orange (Bielorai et al., 1978); □—Shamouti × Sweet lime (Hausenberg et al., 1974); and ■—Shamouti × Sour orange (Heller et al., 1973).

$$Y/Y_m = 1.0 - 0.129 \, (EC_e - 1.28)$$

giving a threshold salinity of 1.3 dS m^{-1} and a slope of 13% per 1 dS m^{-1} increase in EC_e. Exceptions to this relationship are the data obtained by Cole (1985) in a 5-yr field experiment in Australia and by Hausenberg et al. (1974) in a 10-yr salinity survey in Israel. The results of Francois and Clark (1980), using the same rootstock and variety as Cole, fit well into the general relationship. The difference between the results of Cole (1985) and Hausenberg et al. (1974) and those of Francois and Clark (1980) is in the predominance of Cl in the former experiments and of SO$_4$ in the latter.

Bingham et al. (1974) found a greater yield reduction of Valencia oranges due to Cl than to SO$_4$ salinity. The specific toxicity of Cl may have played a part in these differences.

C. Specific Toxicity and Total Salt Effect

There is still some ambiguity relating to the dominant effect of salinity on the response of citrus trees. The accumulation of Cl in the leaves of varieties grafted on certain rootstocks is well established.

The symptoms of Cl injury to citrus leaves are not very marked. Marginal leaf burn and necrotic spots are usually absent. Leaves normally become bronze in color, especially on the sunny side of the trees, and in severe cases, leaves dry out and drop. Sodium toxicity is manifested in necrotic spots and marginal burn, but Na toxicity is uncommon except following direct contact with the spray solution.

The significance of Cl accumulation and the level at which it becomes toxic are not always clear. Chapman et al. (1969) indicated 7.5 g kg^{-1} Cl on a dry-weight basis as the maximum permissible level in the leaves. Cole (1985), on the other hand, reported that concentrations above 2.5 g kg^{-1} in Valencia on rough lemon cause a reduction in yield. Furr and Ream (1969) found no leaf burn below a concentration of 20 g kg^{-1}. Obviously there is no clear cut level of toxicity in the leaves: it depends on time, climatic conditions, size of the root system (Altman, 1969), and possibly other factors. In the Israel Salinity Survey (Hausenberg et al., 1974) there was a low correlation between leaf Cl concentration and tree response to salinity.

Goell (1967) could not find a correlation between the vegetative development of 'Eureka' lemon grafted on various rootstocks and leaf drop due to high Cl concentrations. Lemon on rough lemon rootstock showed the strongest growth as well as the highest Cl uptake and leaf drop. Lemon on Cleopatra mandarin, on the other hand, developed strongly with little Cl accumulation and leaf drop. Eremocitrus (*Eremocitrus glauca* × *Citrus* sp.) showed strong Cl uptake and weak growth, but little leaf drop.

The Israel Salinity Survey (Hausenberg et al., 1974; Shalhevet et al., 1974) revealed a similar response to soil salinity of Shamouti orange grafted on either sour orange or Palestine sweet lime rootstocks. At moderate Cl concentration in the irrigation water, sweet lime is a Cl transmitter while sour orange is a Cl excluder. The response functions were, respectively, $Y = 61.4 - 2.6C$ and $Y = 61.0 - 3.0C$, where Y is yield in Mg ha^{-1} and C is Cl concentration in the irrigation water in mol m^{-3}.

The response of Marsh grapefruit on sour orange was shown by Bielorai et al. (1978) to be dominated by the total salt (osmotic) effect. Water uptake and yield were reduced proportionally with low leaf accumulation of Cl. Chloride was the dominant anion in the irrigation water. They found, furthermore, that the EC$_e$ at the bottom of the root zone reached similar levels to the extrapolated salinity intercept at the zero-yield level (8 dS m^{-1}). At this EC$_e$, water uptake ceased, as is the case with other crops (Bernstein & Francois, 1973). The data reported by du Plessis (1985) from a lysimeter experiment with Valencia orange support the above findings.

After the experiment with Cl, the grapefruit orchard was irrigated for 4 yr with water of variable Sodium Absorption Ratios (SAR) (Bielorai et al., 1983). No specific toxicity due to Na was observed up to 16 M Na m^{-3} in the saturation extract [SAR of 10 $(M$ m$^{-3})^{0.5}$].

A 3-yr experiment in a 20-yr-old Shamouti on Palestine sweet lime rootstock with sour orange inarch revealed no differences in yield in response to up to 13 M m^{-3} Cl in the irrigation water (EC$_i$ = 1.8 dS m^{-1}) (Bielorai et al., 1986; Dasberg et al., 1986b). However, there was some reduction in water use in the high Cl treatment, which correlated with a decrease in the extent of the root system. There was also a small increase in the Cl content of the leaves, from 1.6 to 3.1 g kg^{-1}.

It seems that when the dominant anion in the soil solution is Cl and the rootstock readily absorbs this anion, its toxicity overshadows the total salt effect. Rootstocks that are "tolerant" to high Cl cannot, however, be expected to do better than the relationship given in Fig. 32–4.

D. Direct Foliar Damage Through Sprinkler Spray

Sprinkler irrigation wets the foliage, either partially or fully. Harding et al. (1956) were the first to describe severe damage to the leaves in the skirt of the trees of under-tree sprinkler-irrigated orchards. They reported results from three orchards—grapefruit, Valencia, and Washington navel—where Cl and Na concentrations of the lower leaves were about four times greater than those of the upper leaves. The water used in these orchards was of good quality (2–3.3 M m^{-3} Cl). The lowest concentration of either Na or Cl generally associated with leaf burn was 2.5 g kg^{-1}. Shortly after the report by Harding et al. (1956), Eaton and Harding (1959) and Ehlig and Bernstein (1959) showed through controlled experiments that citrus leaves easily accumulate Cl and Na from direct contact with water drops. Eaton and Harding further demonstrated that the accumulation was greater from intermittent than continuous wetting (Cl: 8.9 vs. 0.5, Na: 10.2 vs. 2.9 g kg^{-1}) and from daytime than nighttime irrigation (Cl: 3.4 vs. 1.7, Na: 4.7 vs. 2.7 g kg^{-1}). The obvious conclusion is that accumulation was a function of the rate of evaporation, which resulted in increased salt concentration of the water film on the leaves. The range of toxic concentration was 7.0 to 10.5 g kg^{-1} dry weight (Ehlig & Bernstein, 1959). Washington navel orange leaves accumulated salts slower than plum (*Prunus domestica* L.) or apricot (*P. armeniaca* L.), but considerably faster than avocado. Thus, the sensitivity of a crop to injury through direct foliar contact bears no relationship to its general tolerance to soil salinity.

Reports from South Australia (Cole & Till, 1977) are in complete agreement with the above findings. When overhead sprinkler irrigation was changed from daytime to nighttime, or to under tree, Cl foliar concentration was reduced from 4.2 to 2.8 g kg^{-1}.

VII. SUMMARY AND CONCLUSION

Citrus is one of the major fruit crops, with a 1983 to 1984 mean world production of 56.1 Tg. It is a subtropical mesophytic tree with many xerophytic characteristics.

The leaves are rigid, with a thick cuticle and no effective stomata on the adaxial side, capable of reaching low Ψ_{leaf} (> -3.5 MPa) with no permanent damage. Stomatal conductance decreases under low humidity.

Irrigation water, at least as a supplemental source, is essential for full production in most of the main citrus-growing regions of the world. The water requirements of citrus are lower than those of most other major crops. Typical midsummer values range from 0.4 to 0.5 ha-cm d^{-1}, compared with 0.7 to 0.8 ha-cm d^{-1} for other crops. The low transpiration is due to low stomatal and canopy conductance. The production function showed a unit increase in relative yield per unit increase in relative evapotranspiration.

Root development is suppressed by water deficiency, salinity, low temperature (below 13 °C), and high temperature (above 35 °C). Vegetative development is closely related to the water use and to fruit yield. Nevertheless, while the suppression of vegetative growth under water deficiency can be detected within one season, it may take two seasons or more before yield reduction is observed.

Fruit quality is clearly affected by water stress. Water-stressed fruit is normally smaller with a thicker peel, greater peel/pulp ratio, lower juice content, greater total soluble solids (TSS) content, higher juice acidity, and lower TSS/acid ratio. Salinity, within the range tolerated by the trees, has negligible effects on quality.

Irrigation may be used to control the microclimate within the orchard (i.e., cooling it during heat spells and protecting it against freeze damage). It may also be used to manipulate flowering and fruit set.

Some plant growth characteristics may be used as a guide in determining irrigation timing. They are: fruit growth rate, changes in trunk diameter (dendrometry), fruit turgor, stomatal opening, leaf water content, Ψ_{leaf}, and stem hydraulic conductance (heat pulse measurements). None of the methods is used commercially.

All three major irrigation methods—gravity, sprinkler, and microirrigation—are used for citrus irrigation. Microirrigation techniques (drip, minisprinklers, sprayers) are gaining in popularity because they provide good control over water application. Water is applied only to a portion of the orchard surface area along the tree rows. Under full irrigation, at least 30% of the surface area should be covered; under supplemental irrigation the minimum area may be less.

Fertigation is commonly practiced in citrus orchards. This seems to be more important under microirrigation, especially in arid climates. Chemigation of herbicides, and to a lesser extent pesticides and growth regulators, is becoming feasible as the delivery and control systems improve.

The sensitivity of citrus to *Phytophthora* root rot is affected by the irrigation practices. Any irrigation method that wets the trunk for a long period may cause foot rot to sensitive rootstocks; this is especially true for flood irrigation in heavy soils. The concern about this disease may change with the development of new fungicides.

Citrus is sensitive to salinity, and especially to high concentration of Cl salts. There are substantial differences among citrus rootstocks in Cl accumulation. Some rootstocks (Rangpur lime, Cleopatra mandarin, sour orange) limit Cl transport to the scion, while others (sweet lime, trifoliate, Troyer) do not. The relationship of yield to Cl accumulation is not always clear. There is a 13% reduction in yield for each 1-dS m^{-1} increase in the electrical conductivity of the soil saturation extract beyond a threshold value of 1.3 dS m^{-1}.

REFERENCES

Aljibury, F.K. 1981. Water and energy conservation in drip irrigation. Drip/Trickle Irrig. 18:26–28.

Aljibury, F.K., F. Arabzadah, and T. Hawkins. 1977. Conversion of citrus from furrow to drip. Citrograph 62:275–276.

Allen, L.H., and M. Cohen. 1974. Water stress and stomatal diffusion resistance in citrus affected with blight and young tree decline. Proc. Fla. State Hortic. Soc. 87:96–101.

Altman, A. 1969. Chloride uptake by the roots of citrus rootstocks. (In Hebrew, English summary.) Ph.D. thesis. Hebrew Univ. Jerusalem, Israel.

Anonymous. 1976. A new look in irrigation systems. Citrus Veg. Mag. 40:20, 28.

Anonymous. 1977. Water requirements for citrus. Citrus Veg. Mag. 41:6–7, 32.

Anonymous. 1985a. FAO 1984 production yearbook. Vol. 38. FAO Stat. Ser. 61.

Anonymous. 1985b. Irrigation and fertilizer recommendations for citrus (In Hebrew.) Isr. Minist. Agric. Ext. Serv.

Ashizawa, M., T. Goto, and K. Manabe. 1979. Studies on leaf water stress in citrus trees. I. Effects of sunlight, temperature and drought on leaf water potential of Satsuma mandarin trees. Kagawa Univ. Faculty Agric. Tech. Bull. 30(2):133–141.

Ashizawa, M., G. Kondo, and T. Chujo. 1981. Effect of soil moisture on the daily change of fruit size in summer season of Satsuma mandarin. Kagawa Univ. Faculty Agric. Tech.Bull. 32(2):87–94.

Baker, E.A., J. Procopiou, and G.M. Hunt. 1980. The cuticles of some "citrus" species. Composition of leaf and fruit waxes. J. Sci. Food Agric. 55:85–87.

Barbera, G., G. Fatta del Bosco, and B. Lo Cascio. 1985. Effects of water stress on lemon summer bloom: The "forzatura" technique in the Sicilian citrus industry. Acta Hortic. 171:391–397.

Barbera, G., G. Fatta del Bosco, B. Lo Cascio, and G. Occorosco. 1981. Some aspects on water stress physiology of forced lemon (*Citrus limon*) trees. Proc. Int. Soc. Citric. (Tokyo) 2:522–523.

Ben-Meir, Y., and E. Israeli. 1985. Irrigating grapefruit—drip or sprinkler. Water Irrig. Rev. (January) p. 8–9.

Bernstein, L. 1980. Salt tolerance of fruit crops. USDA Info. Bull. 292 (revised).

Bernstein, L., and L.E. Francois. 1973. Leaching requirement studies: Sensitivity of alfalfa to salinity of irrigation and drainage waters. Soil Sci. Soc. Am. Proc. 37:931–943.

Bielorai, H. 1968. Beta-ray gauging technique for measuring leaf water content changes of citrus seedlings as affected by the moisture status of the soil. J. Exp. Bot. 19:489–495.

Bielorai, H. 1977. The effect of drip and sprinkler irrigation on grapefruit yield, water use and soil salinity. Proc. Int. Soc. Citric. (Orlando) 1:99–103.

Bielorai, H. 1982. The effect of partial wetting of the root zone on yield and water use efficiency in a drip and sprinkler-irrigated mature grapefruit grove. Irrig. Sci. 3:89–100.

Bielorai, H., S. Dasberg, Y. Erner, and M. Brum. 1981. The effect of various soil moisture regimes and fertilizer levels on citrus yield under partial wetting of the root zone. Proc. Int. Soc. Citric. (Tokyo) 1:585–589.

Bielorai, H., S. Dasberg, Y. Erner, and M. Brum. 1986. Effect of various soil moisture regimes and fertilizer levels on citrus yield response under partial wetting of the root zone. Nordia—1977–1984. (In Hebrew.) Alon haNotea 9:1–27.

Bielorai, H., and Y. Levy. 1971. Irrigation regimes in a semiarid area and their effects on grapefruit yield, water use and soil salinity. Isr. J. Agric. Res. 21:3–12.

Bielorai, H., J. Shalhevet, and Y. Levy. 1978. Grapefruit response to variable salinity in irrigation water and soil. Irrig. Sci. 1:61–70.

Bielorai, H., J. Shalhevet, and Y. Levy. 1983. The effect of high sodium irrigation water on soil salinity and yield of a mature grapefruit orchard. Irrig. Sci. 4:255–266.

Bingham, F.T., B.A. Glaubig, and E. Shade. 1984. Water salinity—nitrate relations of a citrus watershed under drip, furrow and sprinkler irrigation. Soil Sci. 138:306–313.

Bingham, E.T., R.J. Mahler, J. Parra, and L.H. Stolzy. 1974. Long-term effects of irrigation salinity management on 'Valencia' orange orchard. Soil Sci. 117:369–377.

Bredell, G.S., and C.Y. Barnard. 1977. Microjets for macro efficiency. Proc. Int. Soc. Citric. (Orlando) 1:87–92.

Brewer, R.F., K. Optiz, F. Aljibury, and K. Hench. 1977. The effects of cooling by overhead sprinkling on "June drop" of navel oranges in California. Proc. Int. Soc. Citric. (Orlando) 3:1045–1048.

Cahoon, G.A., B.L. Grover, and I.L. Eaks. 1964. Cause and control of oleocellosis in lemons. Proc. Am. Soc. Hortic. Sci. 84:188–198.

Calabrese, F., and L. Di Marco. 1981. Researches on the "Forzatura" of lemon trees. Proc. Int. Soc. Citric. (Tokyo) 2:520–521.

Camacho-B, S.E., A.E. Hall, and M.R. Kaufmann. 1974a. Efficiency and regulation of water transport in some woody and herbaceous species. Plant Physiol. 54:169–172.

Camacho-B, S.E., M.R. Kaufmann, and A.E. Hall. 1974b. Leaf water potential response to transpiration in citrus. Physiol. Plant. 31:101–105.

Cary, P.R., and P.G.J. Weerts. 1978. Factors affecting growth, yield and fruit composition of Washington navel and Late Velencia orange trees. Proc. Int. Soc. Citric. (Sydney) 1:72–86.

Castle, W.S. 1978. Citrus root systems: Their structure, function, growth and relationship to tree performance. Proc. Int. Soc. Citric. (Sydney) 1:62–69.

Castle, W.S., and AH. Krezdorn. 1977. Soil water use and apparent root efficiencies of citrus trees on four rootstocks. J. Am. Soc. Hortic. Sci. 102:403–406.

Cerda, A., M. Caro, F.G. Fernandez, and M.G. Guillen. 1976. Foliar content of sodium and chloride in citrus rootstocks irrigated with saline water. p. 155–163. In H.E. Dregne (ed.) Managing saline water for irrigation. Proc. Int. Salinity Conf., Lubbock, TX. 16–20 Aug. 1976. Texas Tech. Univ., Lubbock, TX.

Chapman, H.D. 1968. The mineral nutrition of citrus. p. 161–171. In W. Reuther (ed.) The citrus industry. Vol. 2. Univ. California, Berkeley.

Chapman, H.D., H. Joseph, and D.S. Reyner. 1969. Effect of variable maintained chloride levels on orange growth, yield, and leaf composition. Proc. First Int. Citrus Symp. (Riverside) 3:1811–1817.

Childs, J.F.L. 1978. Rio Grande gummosis disease of citrus trees. I. A brief review of the history and occurrence of Rio Grande gummosis. Plant Dis. Rep. 62:390–399.

Cohen, A., A. Goell, A. Rassis, and H. Gokkes. 1968. Effects of irrigation regimes on grapefruit peel and pulp relations. Isr. J. Agric. Res. 18:155–160.

Cohen, A., and K. Mendel. 1965. Methods for the rapid evaluation of rootstocks for citrus. Volcani Inst. Agric. Res. Pam. 87.

Cohen, M. 1974. Diagnosis of young tree decline, blight and sand-hill decline of citrus by measurements of water uptake using gravity injection. Plant Dis. Rep. 58:801–805.

Cohen, M., D.V. Calvert, and R.R. Pelosi. 1981. Disease occurrence in citrus rootstock and drainage experiment (SWAP) in Florida. Proc. Int. Soc. Citric. (Tokyo) 1:366–368.

Cohen, Y., M. Fuchs, and S. Cohen. 1983. Resistance to water uptake in a mature citrus tree. J. Exp. Bot. 34:451–460.

Cohen, Y., M. Fuchs, and G.L. Green. 1981. Improvement of the heat pulse method for determining sap flow in trees. Plant Cell Environ. 4:391–397.

Cohen, Y., S. Moreshet, and M. Fuchs. 1987. Changes in hydraulic conductance of citrus trees following a reduction in a wetted soil volume. Plant Cell Environ. 10:53–57.

Cole, P.J. 1985. Chloride toxicity in citrus. Irrig. Sci. 6:63–71.

Cole, P.J., and M.R. Till. 1977. Evaluation of alternatives to overhead sprinklers for citrus irrigation. Proc. Int. Soc. Citric. (Orlando) 1:103–106.

Cooper, W.C., and B.S. Gorton. 1952. Toxicity and accumulation of chloride salts in citrus on various rootstocks. Proc. Am. Soc. Hortic. Sci. 59:143–146.

Cooper, W.C., B.S. Gorton, and C. Edwards. 1951. Salt tolerance of various rootstocks. Proc. Rio Grande Valley Hortic. Soc. 5:46–52.

Cooper, W.C., R.H. Hilgeman, and G.K. Rasmussen. 1964. Diurnal and seasonal fluctuations of trunk growth of 'Valencia' orange as related to climate. Proc. Fla. State Hortic. Soc. 77:101–106.

Cooper, W.C., and A. Peynado. 1959. Chloride and boron tolerance of young-line citrus trees on various rootstocks. J. Rio Grande Valley Hortic. Soc. 13:89–96.

Cruse, R.R., C.L. Wiegand, and W.A. Swanson. 1982. The effect of rainfall and irrigation management on citrus juice quality in Texas. J. Am. Soc. Hortic. Sci. 107:767–770.

Dasberg, S., H. Bielorai, and Y. Erner. 1981. Partial wetting of the root zone and nitrogen effect on growth and quality of Shamouti orange. Acta Hortic. 119:103–105.

Dasberg, S., H. Bielorai, Y. Erner, and M. Brum. 1986a. Irrigation of citrus (Shamouti on sweet lime) with saline water in the coastal region. (In Hebrew.) Res. Rep. Agric. Res. Org. Isr.

Dasberg, S., H. Bielorai, Y. Erner, and M. Brum. 1986b. The effect of saline irrigation water on Shamouti oranges. (In Hebrew.) Volcani Cent. Agric. Res. Org. Rep. 304-0064.

Davies, F.S., D.W. Buchanan, and J.A. Anderson. 1981. Water stress and cold hardiness in field-grown citrus. J. Am. Soc. Hortic. Sci. 106:197–200.

Davies, R.H. 1981. Phytophthora foot rot control with the systemic fungicides Metalaxyl and Fosetyl aluminum. Proc. Int. Soc. Citric. (Tokyo) 2:349–351.

Del Amor, F., A. Leon, A. Torerecillas, and A. Ortuno. 1981. Herbicide application through drip irrigation systems. Proc. Int. Soc. Citric. (Tokyo) 2:493–496.

Dirksen, C., J.D. Oster, and P.A.C. Raats. 1979. Water and salt transport, water uptake and leaf water potentials during regular and suspended high frequency irrigation of citrus. Agric. Water Manage. 2:241–256.

du Plessis, H.M. 1985. Evapotranspiration of citrus as affected by soil water deficit and soil salinity. Irrig. Sci. 6:51–61.

Eaton, F.M., and R.B. Harding. 1959. Foliar uptake of salt constituents of water by citrus plants during intermittent sprinkling and immersion. Plant Physiol. 33:22–26.

Ehlig, C.F., and L. Bernstein. 1959. Foliar absorption of sodium and chloride as a factor in sprinkler irrigation. Proc. Am. Soc. Hortic. Sci. 74:661–670.

Elving, D.C., and M.R. Kaufmann. 1972. Diurnal and seasonal effects of environment on plant water relations and fruit diameter of citrus. J. Am. Soc. Hortic. Sci. 97:566–570.

Elfving, D.C., M.R. Kaufmann, and A.E. Hall. 1972. Interpreting leaf water potential measurements with a model of the soil-plant-atmosphere continuum. Physiol. Plant. 27:161–168.

Erie, L.J., O.F. French, and K. Harris. 1965 (reprinted 1976). Consumptive use of water by crops in Arizona. Univ. Arizona Agric. Exp. Stn. Tech. Bull. 169.

Fishler, M. 1976. Research and field experiments on grapefruit irrigation in the inland valleys. (In Hebrew.) Minis. Agric., Bet Shean.

Ford, H.W. 1954. Root distribution in relation to the water table. Proc. Fla. State Hortic. Soc. 67:30–33.

Ford, H.W. 1964. The effect of rootstock, soil type, and soil pH on citrus growth in soils subjected to flooding. Proc. Fla. State Hortic. Soc. 86:41–45.

Francois, L.E., and R.A. Clark. 1980. Salinity effect on yield and fruit quality of 'Valencia' orange. J. Am. Soc. Hortic. Sci. 105:190–202.

Furr, J.R., and C.L. Ream. 1969. Breeding citrus rootstocks for salt tolerance. Proc. First Int. Citrus Symp. 1:373–379.

Furr, J.R., and C.A. Taylor. 1939. Growth of lemon fruits in relation to moisture content of the soil. USDA Tech. Bull. 640.

Gerard, C.I., and B. Sleeth. 1960. The use of tensiometers in the irrigation of citrus groves. J.Rio Grande Valley Hortic. Soc. 14:99–103.

Goell, A. 1967. Salinity effects on citrus trees. Proc. First Int. Citrus Symp. 3:1819–1823.

Goell, A., A. Golomb, D. Kalmar, A. Mantell, and S. Sharon. 1981. Moisture stress—a potent factor for affecting vegetative growth and tree size in citrus. Proc. Int. Soc. Citric. (Tokyo) 2:503–506.

Goodall, G.E., and K.G. Silviera. 1981. Adapting the Italian verdelli process to Persian lime production in California. Proc. Int. Soc. Citric. (Tokyo) 2:518–520.

Graham, J.M., and J.P. Syvertsen. 1984. Influence of V.A. mycorrhiza on hydraulic conductivity of roots of two citrus rootstocks. New Phytol. 97:277–284.

Green, G.C., and W.S. Meyer. 1980. The use of leaf water potential measurements to estimate water loss from orange trees. Gewasproduksie 9:93–96.

Grierson, W., A.A. McCornack, and F.W. Hayward. 1965. Tangerine handling. Univ. Florida Agric. Res. Exp. Serv. Circ. 285.

Hall, A.E. 1976. Temperature and humidity effects on net photosynthesis and transpiration of citrus. Physiol. Plant. 36:29–34.

Hall, A.E., S.E. Camacho-B, and M.R. Kaufmann. 1975. Regulation of water loss by citrus leaves. Physiol. Plant. 33:62–65.

Harding, R.B., M.P. Miller, and M. Fireman. 1956. Sodium and chloride absorption by citrus leaves from sprinkler-applied water. Citrus Leaves 36:6–8.

Harding, R.B., P.F. Pratt, and W.W. Jones. 1958. Changes in salinity, nitrogen and soil reaction in differentially fertilized irigated soil. Soil Sci. 88:177–184.

Harrison, D.S. 1978. Irrigation methods and equipment for production of citrus in Florida. Citrus Ind. 59:10–12, 14.

Hausenberg, I., Y. Pozin, and M. Boaz. 1974. Salinity survey, final report. Spring 1964–Spring 1973. (In Hebrew.) Isr. Minis. Agric. Ext. Serv., Tel Aviv.

Heller, J., J. Shalhevet, and A. Goell. 1973. Response of a citrus orchard to soil moisture and soil salinity. p. 409–419. In A. Hadas et al. (ed.) Physical aspects of soil water and salts in ecosystems. Ecol. Stud. 4. Springer-Verlag, Berlin.

Hilgeman, R.H. 1963. Trunk growth of 'Valencia' orange in relation to soil moisture and climate. Proc. Am. Soc. Hortic. Sci. 82:193–198.

Hilgeman, R.H. 1966. Effect of climate of Florida and Arizona on grapefruit fruit enlargement and quality, apparent transpiration and internal water stress. Proc. Fla. State Hortic. Soc. 79:99–106.

Hilgeman, R.H. 1977. Response of citrus to water stress in Arizona. Proc. Int. Soc. Citric. (Orlando) 1:70–74.

Hilgeman, R.H., C.E. Ehrler, C.E. Everling, and F.O. Sharp. 1969. Apparent transpiration and internal water stress in 'Valencia' oranges as affected by soil water, season and climate. Proc. First Int. Citrus Symp. (Riverside) 3:1713–1723.

Hilgeman, R.H., and L.H. Howland. 1955. Fruit measurements and tensiometers can tell you when to irrigate citrus trees. Progr. Agric. Arizona 7(1):10–11 (cited by Marsh, 1973).

Hilgeman, R.H., and G.C. Sharples. 1957. Irrigation trials with Valencia in Arizona. Calif. Citrograph 42:404–407.

Hirano, E. 1931. Relative abundance of citrus stomata and some related genera. Bot. Gaz. 92:296–310.

Hoffman, G.J., J.D. Oster, E.V. Maas, J.D. Rhoades, and J. van Schilfgaarde. 1984. Minimizing salt in drain water by irrigation management—Arizona field studies with citrus. Agric. Water Manage. 9:61–78.

Huff, A., M.Z. Abdel-Bar, D.R. Rodney, and R.L. Roth. 1981. Enhancement of citrus regreening and peel lycopene by trickle irrigation. HortScience 16:301–302.

Jackson, L.K., and F.S. Davies. 1985. Maximize cold protection of young citrus trees with microsprinkler/wrap combinations. Citrus Veg. Mag. 48:28, 90.

Johnson, C.R., and R.L. Hummel. 1985. Influence of mycorrhizae and drought stress on growth of Poncirus × Citrus seedlings. HortScience 20:754–755.

Jones, H.G., A.N. Lasko, and J.P. Syvertsen. 1985. Physiological control of water status in temperate and subtropical fruit trees. Hortic. Rev. 7:301–343.

Jordan, L.S. 1981. Weeds affect citrus growth, physiology, yield and fruit quality. Proc. Int. Soc. Citric. (Tokyo) 2:481–483.

Kadoya, K., K. Kameda, S. Chikaigumi, and K. Matsumoto. 1975. Studies on the hydrophysiological rhythms of citrus. I. Cyclic fluctuations of leaf thickness and stem diameter of Natsudaidai seedlings. J. Jpn. Soc. Hortic. Sci. 44:260–264.

Kalma, J.D., and M. Fuchs. 1976. Citrus orchards. p. 309–328. In J.L. Monteith (ed.) Vegetation and the atmosphere. Academic Press, New York.

Kaufmann, M.R. 1968. Evaluation of the pressure chamber method for the measurement of water stress of citrus. Proc. Am. Soc. Hortic. Sci. 93:186–190.

Kaufmann, M.R. 1969. Relation of slice gap width in oranges and plant water stress. J. Am. Soc. Hortic. Sci. 94:161–163.

Kaufmann, M.R. 1977. Citrus—a case study of environmental effects on plant water relations. Proc. Int. Soc. Citric. (Orlando) 1:57–62.

Kaufmann, M.R., and D.C. Elfving. 1972. Evaluation of tensiometers for estimating plant water stress in citrus. HortScience 7:513–514.

Kaufmann, M.R., and A.E. Hall. 1974. Plant water balance—its relationship to atmospheric and edaphic conditions. Agric. Meteorol. 14:85–98.

Kaufmann, M.R., and Y. Levy. 1976. Stomatal response of Citrus jambhiri to water stress and humidity. Physiol. Plant 38:105–108.

Klotz, L.J., S.J. Richards, and T.A. DeWolfe. 1967. Irrigation effects on root rot of young citrus trees. Calif. Citrograph 52:91.

Knapp, J.L., T.R. Fasulo, D.P.H. Tucker, and L.R. Parsons. 1982. The effects of different irrigation and weed management practices on mite populations in a citrus grove. Proc. Fla. State Hortic. Soc. 95:47–50.

Koo, R.C.J. 1979. The influence of N, K and irrigation on tree size and fruit production of 'Valencia' orange. Proc. Fla. State Hortic. Soc. 92:10–13.

Koo, R.C.J. 1980. Results of citrus fertigation studies. Proc. Fla. State Hortic. Soc. 93:33–36.

Koo, R.C.J. 1981a. The effect of fall irrigation on freeze damage to citrus. Proc. Fla. State Hortic. Soc. 94:37–39.

Koo, R.C.J. 1981b. Results of citrus fertigation studies. Citrus Ind. 62(5):6–7, 13–15.

Koo, R.C.J. 1984. The importance of ground coverage by fertigation for citrus on sandy soils. Fertil. Issues 1:75–78.

Koo, R.C.J., and R.L. Reese. 1975. Water distribution and evaporation loss from sprinkler irrigation in citrus. Proc. Fla. State Hortic. Soc. 88:5–9.

Koo, R.C.J., and R.L. Reese. 1977. Fertility and irrigation effects on 'Temple' orange. II. Fruit quality. J. Am. Soc. Hortic. Sci. 102:152–155.

Koo, R.C.J., and A.G. Smajstrla. 1985. Effects of trickle irrigation and fertigation on fruit production and juice production and juice quality of 'Valencia' orange. Proc. Fla. State Hortic. Soc. 97:8–10.

Kriedemann, P.E. 1971. Photosynthesis and transpiration as a function of gaseous diffusive resistances in orange leaves. Physiol. Plant. 24:218–225.

Kriedemann, P.E., and M.D. Barrs. 1981. Citrus orchards. p. 325–417. In T.T. Kozlowski (ed.) Water deficit and plant growth. Vol. 6. Academic Press, New York.

Kuriyama, T., M. Shimoosako, M. Yoshida, and S. Shiraishi. 1981. The effect of soil moisture on the fruit quality of Satsuma mandarin (Citrus unshiu Marc.). Proc. Int. Soc. Citric. (Tokyo) 2:524–527.

Lavilee, E.Y., and A.J. Chalandon. 1981. Control of Phytophthora gummosis in citrus with foliar sprays of Fosetyl Al, a new systemic fungicide. Proc. Int. Soc. Citric. (Tokyo) 1:346–349.

Legaz, F., D.G. Ibanez, D. deBarreda, and E. Primo milo. 1981. Influence of irrigation and fertilization on productivity of the 'Navelate' sweet orange. Proc. Int. Soc. Citric. (Tokyo) 2:591–595.

Levy, Y. 1967. Factors influencing the occurrence of oleocellosis in Washington navel oranges in the Negev area. (In Hebrew, English abstract.) Natl. Univ. Inst. Agric. Prelim. Rep. 593.

Levy, Y. 1980. Effect of evaporative demand on water relations of Citrus limon. Ann. Bot. 46:695–700.

Levy, Y. 1983. Acclimation of citrus to water stress. Sci. Hortic. 20:267–273.

Levy, Y. 1984. The effect of sprinkler and drip irrigation on lime-induced chlorosis of citrus. Sci. Hortic. 22:249–255.

Levy, Y., A. Bar-Akiva, and Y. Vaadia. 1978a. Influence of irrigation and environmental factors on grapefruit acidity. J. Am. Soc. Hortic. Sci. 103:73–76.

Levy, Y., H. Bielorai, and J. Shalhevet. 1978b. Long-term effects of different irrigation regimes on grapefruit tree development and yield. J. Am. Soc. Hortic. Sci. 103:680–683.

Levy, Y., J. Dodd, and J. Krikun. 1983a. Effect of irrigation, water salinity and rootstock on the vertical distribution of vesicular-arbuscular mycorrhiza in citrus roots. New Phytol. 95:397–403.

Levy, Y., and I. Horesh. 1984. Importance of penetration through stomata in the correlation of chlorosis with iron salts and low-surface-tension surfactants. J. Plant Nutr. 7:279–281.

Levy, Y., and M.R. Kaufmann. 1976. Cycling of leaf conductance in citrus exposed to natural and controlled environments. Can. J. Bot. 54:2215–2218.

Levy, Y., and J. Krikun. 1980. The effect of vesicular-arbuscular mycorrhiza on *Citrus jambhiri* water relations. New Phytol. 85:25–31.

Levy, Y., and J. Shalhevet. 1985. Irrigation of a mature orchard with high salinity water—comparison of two cultivars and three rootstocks. (In Hebrew.) Report for years 1982–1984. Isr. Agric. Res. Org., Inst. Hortic. Soil Water Gilat Exp. Stn.

Levy, Y., J. Shalhevet, and H. Bielorai. 1979. Effect of irrigation regime and water salinity on grapefruit quality. J. Am. Soc. Hortic. Sci. 104:356–359.

Levy, Y., and J.P. Syvertsen. 1981. Water relations of citrus in climates with different evaporative demands. Proc. Int. Soc. Citric. (Tokyo) 2:501–503.

Levy, Y., J.P. Syvertsen, and S. Nemec. 1983b. Effect of drought stress and vesicular-arbuscular mycorrhiza on citrus transpiration and hydraulic conductivity of roots. New Phytol. 93:61–66.

Leyden, R.F. 1975. Comparison of three irrigation systems for young citrus trees. J. Rio Grande Valley Hortic. Soc. 29:31–36.

Lombard, P.B., L.H. Stolzy, M.J. Garber, and T. Szuszkiewicz. 1965. Effect of climatic factors on fruit volume increase and leaf water deficits of citrus in relation to soil suction. Soil Sci. Soc. Am. proc. 29:205–208.

Lyons Jr., C.J. 1977. Water management in Texas citrus. Proc. Int. Soc. Citric. (Orlando) 1:117.

Maas, E.V., and J.G. Hoffman. 1977. Crop salt tolerance—current assessment. J. Irrig. Drain. Div. Am. Soc. Civ. Eng. 103(IR2):115–134.

Maotani, T., and Y. Machida. 1977. Studies on leaf water stress in fruit trees, VII. Effects of summer water potential on Satsuma mandarin trees on fruit characteristics at harvest time. J. Jpn. Soc. Hortic. Sci. 46(2):145–152.

Maotani, T., and Y. Machida. 1980. Leaf water potential as an indicator of irrigation timing for Satsuma trees in summer. J. Jpn. Soc. Hortic. Sci. 49:41–48.

Marsh, A.W. 1968. Automatic tensiometer signalled irrigation systems for orchards. Calif. Citrograph 54:2, 12.

Marsh, A.W. 1973. Irrigation. p. 230–279. *In* W. Reuther (ed.) The citrus industry. Vol. 3. Univ. California, Berkeley.

Mendel, K. 1951. Orange leaf transpiration under orchard conditions. III. Prolonged drought and the influence of stocks. Palest. J. Bot. Rehovot Ser. 8:45–53.

Menge, J.A. 1986. The use of systemic fungicides metalaxyl and efosite aluminum on citrus. Proc. 2nd World Congr. Int. Soc. Citrus Nurserymen 1:112–127.

Menge, J.A., E.L. Johnson, and R.G. Platt. 1978. Mycorrhizal dependency of several citrus cultivars under three nutrient regimes. New Phytol. 85:553–559.

Mobayen, R.G., and F.L. Milthorpe. 1977. Citrus and germination as influenced by water potential and salinity. Proc. Int. Soc. Citric. (Orlando) 1:247–249.

Moreshet, S., Y. Cohen, and M. Fuchs. 1983. Response of mature Shamouti orange trees to irrigation of different soil volumes at similar levels of available water. Irrig. Sci. 3:223–236.

Moreshet, S., and G.C. Green. 1984. Seasonal trends in hydraulic conductance of field-grown 'Valencia' orange trees. Sci. Hortic. 23:169–180.

Mougheith, M.G., M. El-Ashram, G.W. Amerhom, and M. Madbouly. 1977. Effect of different rates of irrigation on navel orange trees. II. Yield and fruit quality. Ann. Agric. Sci. Moshtohor 8:119–127.

Myers, M.J., D.S. Harrison, and W.J. Phillips, Jr. 1976. Soil moisture distribution in a sprinkler irrigated orange grove. Proc. Fla. State Hortic. Soc. 89:23–26.

Nir, I., R.Goren, and B. Leshem. 1972. Effects of water stress, gibberellic acid and 2-chloroethyltrimethylammoniumchloride (CCC) on flower differentiation in 'Eureka' lemon trees. J. Am. Soc. Hortic. Sci. 97:774–778.

Oberbacher, M.F. 1966. A method to predict the post-harvest incidence of oleocellosis in lemons. Proc. Fla. State Hortic. Soc. 78:237–240.

Oppenheimer, H.R., and D.A. Elze. 1941. Irrigation of citrus according to physiological indicators. Agric. Res. Stn. Rehovot Bull. 31.

Oren, Y., and E. Israeli. 1977. Herbicide application through irrigation systems (herbigation) in citrus. Proc. Int. Soc. Citric. (Orlando) 1:152–154.

Oren, Y., Z. Solel, and A. Ben-Gal. 1977. Administration of fungicides through the irrigation system—efficient for controlling brown rot in citrus (In Hebrew.) Hassadeh 56:2173–2178.

Palmer, R.L., Z. Hanscom, III, and W.M. Dugger. 1977. Studies of high temperature effects on fruit drop from Washington navel orange: I. Interaction of temperature and leaf water potential. Proc. Int. Soc. Citric. (Orlando) 3:1048–1052.

Parsons, L.R., B.S. Combs, and D.P.H. Tucker. 1985. Citrus freeze protection with microsprinkler irrigation during an advective freeze. HortScience 20:1078–1080.

Parsons, L.R., and D.P.H. Tucker. 1985. Sprinkler irrigation for cold protection in citrus groves and nurseries during an advective freeze. Proc. Fla. State Hortic. Soc. 97:28–30.

Parsons, L.R., T.A. Wheaton, and D.P.H. Tucker. 1986. Florida freezes and the role of water in citrus cold protection. HortScience 21(1):cover page.

Parsons, L.R., T.A. Wheaton, and J.D. Whitney. 1981. Low volume microsprinkler undertree irrigation for frost protection of young citrus trees. Proc. Fla. State Hortic. Soc. 91:55–59.

Parsons, L.R., T.A. Wheaton, and J.D. Whitney. 1982. Undertree irrigation for cold protection with low-volume microsprinklers. HortScience 17:799–801.

Pearson, G.A., J.A. Gross, and H.E. Hayward. 1957. The influence of salinity and water on growth and mineral composition of young grapefruit trees. J. Am. Soc. Hortic. Sci. 69:197–203.

Penman, H.L. 1948. Natural evaporation from open water, bare soil, and grass. Proc. R. Soc. London Ser. A 193:120–146.

Peynado, A., and R.H. Young. 1968. Moisture changes in intact citrus leaves monitored by a beta gauge technique. Proc. Am. Soc. Hortic. Sci. 92:211–220.

Phung, H.T., and F.B. Knipling. 1976. Photosynthesis and transpiration of citrus seedlings under flooded conditions. HortScience 11:131–133.

Ramos, C., and M.R. Kaufmann. 1979. Hydraulic resistance of rough lemon roots. Physiol. Plant 45:311–314.

Rawlins, S.L. 1977. Uniform irrigation with a low-head bubbler system. Agric. Water Manage. 1:167–178.

Ream, C.L., and J.R. Furr. 1976. Salt tolerance of some citrus species, relatives and hybrids tested as rootstocks. J. Am. Soc. Hortic. Sci. 101:265–267.

Rodney, D.R., R.L. Roth, and B.R. Gardener. 1977. Citrus response to irrigation methods. Proc. Int. Soc. Citric. (Orlando) 1:106–110.

Rokba, A.M., M.N. Abdel-Messih, and M.A. Mohammed. 1979. Breeding and screening some citrus rootstocks for salt tolerance in Egypt. Egypt. J. Hortic. 6:69–79.

Sakomoto, T., and S. Okuchi. 1968. Effects of rainfall on soluble solids and acids in Satsuma oranges. J. Jpn. Soc. Hortic. Sci. 37:212–220.

Sardo, V.B., and C. Germana. 1985. Environmental and physiological parameters in scheduling irrigations of orange trees. Acta Hortic. 171:405–416.

Schneider, H. 1973. The anatomy of citrus. p. 1–85. In W. Reuther et al. (ed.) The citrus industry. Vol. 2. Univ. California, Berkeley.

Scuderi, A., and G. Raciti. 1977. Citrus trickle irrigation trials. Proc. Int. Soc. Citric. (Orlando) 1:244–247.

Shalhevet, J., and H. Bielorai. 1978. Crop water requirement in relation to climate and soil. Soil Sci. 25:240–247.

Shalhevet, J., A. Mantell, H. Bielorai, and D. Shimshi. 1981. Irrigation of field and orchard crops under semi-arid conditions (revised). Int. Irrig. Inf. Cent. Bet Dagan Publ. 1.

Shalhevet, J., D. Yaron, and J. Horowitz. 1974. Salinity and citrus yield—an analysis of results from a salinity survey. J. Hortic. Sci. 49:15–27.

Spiegel-Roy, P., A. Vardi, G. Ben-Hayyim, and J. Shalhevet. 1986. Breeding of citrus rootstocks for salinity tolerance. (In Hebrew.) Volcani Cent. Inst. Hortic. Rep. 204-231.

Stanhill, G. 1972. Recent developments in water relation studies: Some examples from Israel citriculture. Proc. 18th Int. Hortic. Congr. (Tel Aviv) 4:367–379.

Stylianou, Y. 1974. Irrigation requirements of Valencia oranges as affected by the frequency of water application. Agric. Res. Inst. Tech. Bull. 16. Nicosia, Cyprus.

Syvertsen, J.P. 1982. Minimum leaf water potential and stomatal closure in citrus leaves of different ages. Ann. Bot. 49:827–834.

Syvertsen, J.P., and Y. Levy. 1982. Diurnal changes in citrus leaf thickness, leaf water potential and leaf to air temperature difference. J. Exp. Bot. 33:783–789.

Syvertsen, J.P., M.L. Smith, and J.C. Allen. 1981. Growth rate relations of citrus leaf flushes. Ann. Bot. 47:97–105.

Taylor, C.A., and J.R. Furr. 1937. Use of soil moisture and fruit growth records for checking irrigation practices of citrus orchards. USDA Circ. 426. U.S. Gov. Print. Office, Washington, DC.

Timmer, L.W., and R.F. Leyden. 1976. Effect of irrigation and soil management practices on the incidence of Phytophthora foot rot of citrus. J. Rio Grande Valley Hortic. Soc. 30:19–25.

Toledo, E., R. Cardenas, R. Ray, and Y. Delibaltor.1981. Evapotranspiracion de la meranja 'Valencia late' en las condiciones de Jagiiey Grande (In Spanish.) Agrotec. Cuba 13(1):131–139.

Torrisi, M. 1952. Indagini fisioloiche sugli agrumi: la differenziazione della gemme a fiore del limone e dati preleminari sulla forzatura. (In Italian.) Ann. Sper. Agrar. 881–894.

Turrell, F.M. 1973. The science and technology of frost protection. p. 338–446. In W. Reuther (ed.) The citrus industry. Vol. 3. Univ. California, Berkeley.

Uda, H., and J. Morimoto. 1977. Studies on the multi-purpose utilization of sprinkler irrigation systems in citrus groves of Wakayama prefecture. Proc. Int. Soc. Citric. (Orlando) 1:93–103.

Van Bavel, C.H.M., J.E. Newman, and R.H. Hilgeman. 1967. Climate and the estimated water use by an orange orchard. Agric. Meteorol. 4:27–37.

Wardowski, W.F., A.A. McCornack, and W. Grierson. 1976. Oil spotting (oleocellosis) of citrus fruits. Univ. Florida Circ. 410.

Wen-Cai, Z. 1981. Development and outlook of the citrus industry in China. Proc. Int. Soc. Citric. (Tokyo) 2:987–990.

Wiegand, C.L., and W.A. Swanson. 1982a. Citrus response to irrigation. I. Irrigation requirements; daily, monthly and annual evapotranspiration amounts; and water management recommendations. J. Rio Grande Valley Hortic. Soc. 35:72–85.

Wiegand, C.L., and W.A. Swanson. 1982b. Citrus response to irrigation. II. Fruit yield, size and number. J. Rio Grande Valley Hortic. Soc. 35:87–95.

Wiegand, C.L., and W.A. Swanson. 1982c. Citrus response to irrigation. III. Tree trunk and canopy growth. J. Rio Grande Valley Hortic. Soc. 35:97–107.

Wiegand, C.L., and W.A. Swanson. 1982d. Marrs, Valencia and Ruby Red juice quality as affected by irrigation plus rainfall. J. Rio Grande Valley Hortic. Soc. 35:109–120.

Yagev, E. 1977. Drip irrigation in citrus orchards. Proc. Int. Soc. Citric. (Orlando) 1:110–113.

Yelenosky, G. 1979a. Accumulation of free proline in citrus leaves during cold hardiness of young trees in controlled temperature regimes. Plant Physiol. 64:425–427.

Yelenosky, G. 1979b. Water-stress-induced cold hardening of young citrus trees. J. Am. Soc. Hortic. Sci. 104:270–273.

Yelenosky, G. 1985. Cold hardening in citrus. Hortic. Rev. 7:201–238.

Young, R.H., and S.M. Garnsey. 1977. Water uptake patterns in blighted citrus trees. J. Am. Soc. Hortic. Sci. 102:751–756.

Zur, B., S. Sharon, D. Kalmar, A. Golomb, and A. Goell. 1974. Observations in daily pulse irrigation of grapefruit orchards in the western Galilee. Technion-Isr. Inst. Tech., Faculty Agric., Haifa. Publ. 214.

33   Deciduous Fruit and Nut Trees

ELIAS FERERES

University of Cordoba
Cordoba, Spain

DAVID A. GOLDHAMER

University of California
Parlier, California

Most deciduous fruit and nut orchards are located in temperate zones where rainfall is insufficient to meet tree water demand throughout the season. Irrigation practices that supply orchards with adequate water must be based on quantitative knowledge of tree water requirements and of tree responses to soil water supply. Such knowledge must be acquired through experimentation which, in the case of trees, should span a substantial fraction of the active life of the orchard. Long-term field experiments are the exception rather than the norm in agricultural research activities of recent decades. In addition, fruit and nut trees do not have the economic and social importance of the major herbaceous crops, and research efforts on deciduous fruit and nut trees have been restricted to specific world areas.

Irrigation studies with deciduous fruit and nut trees started long ago (Veihmeyer, 1927) and information on physiological responses to water deficits has substantially increased in recent years (Jones et al., 1985). The amount of available information varies with species, with apple (*Malus sylvestris* Mill.) receiving the most attention. It is evident from the information summarized below that much more research is needed in all other species to achieve a level of understanding of irrigation responses of tree crops similar to that existing for the major irrigated herbaceous crops.

I. GROWTH AND DEVELOPMENT OF DECIDUOUS FRUIT AND NUT TREES IN RELATION TO IRRIGATION

The typical patterns of growth and development of deciduous fruit and nut trees over an annual cycle are shown in Fig. 33–1. Bud growth resumes in the spring following a cold period usually necessary for bud break. Root growth usually accelerates at the same time, with maximum growth occurring in soil areas of low mechanical resistance and of high water content and oxygen supply. Flowering and pollination take place during low evaporative demand periods and when tree transpiration is small; thus irrigation requirements at full bloom are normally minimal. As spring advances, faster shoot and fruit growth rates lead to full canopy development and increasing irrigation needs in the absence of rainfall. Fruit bud development and morphogenesis are sensitive to water deficits although there is substantial variation among species in their sensitivity. Irrigation not only affects fruit growth and current year harvest but influences following years' crops through its effects on bud development, morphogenesis, and shoot growth. After fruit harvest, vegetative growth may proceed well into the fall depending on the availability of soil water. Postharvest irrigation management is a critical issue as additional shoot growth and carbohydrate storage may or may not be needed to obtain maximum yields in the following year.

Figure 33–1 does not fully illustrate the dynamics of growth and development as they change with environmental conditions and internal plant factors. The competition between fruit growth and other growth forms depends on tree species and the stage of development. Little is known about the dynamics of root growth in relation to shoot development in fruit trees. This is particularly important as the water supply and demand of a fruit tree is determined by the relative size of the shoot and root system. Chalmers and van den Ende (1975) showed in peach that a strong relationship exists between top and root dry weight. The relationship is linear on a logarithmic basis in peach trees varying in age from 3 to 15 yr.

Adequate soil water supply via irrigation should maximize growth and radiation interception in developing tree canopies, shortening the planting to maturation period. Assimilate partitioning is as important as full radiation interception. As in other perennials (Schulze, 1982), there are ontogenetic changes in the distribution of assimilates in fruit trees. In general, the proportion to total dry weight that goes into vegetative growth decreases with age in fruit trees, from 70% for a young peach tree to <30% for a mature tree (Chalmers & van den Ende, 1975). Chalmers (1987) speculated that as the

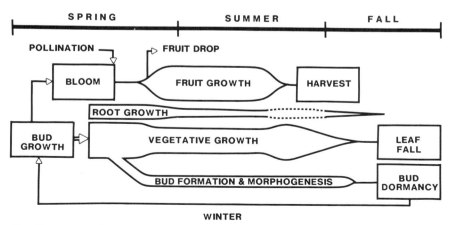

Fig. 33–1. Generalized diagram of the growth and development processes in deciduous fruit and nut trees.

top/root ratio decreases with age, a reduced root system may limit vegetative shoot growth, leading to greater allocation of assimilates to fruits. It follows that harvest indices in fruit crops would increase as trees age or when the growth of the root system is limited by unfavorable soil conditions (Chalmers, 1987).

While the growth and development of all major deciduous tree crops follow more or less the simplified diagram of Fig. 33–1, there are substantial differences among species in the chronology of events. Among nut trees, almond [*Prunus dulcis* (Mill.) D.A. Webb] trees flower more than 4 wk earlier than pistachio (*Pistacia vera* L.) or walnut (*Juglans regia* L.) and bud development if also completed in a shorter time period. Similar differences may be found among deciduous fruit tree species and cultivars (e.g., peach [*Prunus persica* (L.) Batsch] vs. apple).

II. ASPECTS OF WATER RELATIONS OF DECIDUOUS FRUIT AND NUT TREES

The water relations of fruit trees may be studied from many viewpoints; aspects that are particularly relevant to irrigation include the evaluation of various measures of tree water status, the role of stomata in controlling transpiration, the interactions between roots and soil water, and adaptation to drought. Two recent articles on the water relations of fruit trees (Jones et al., 1985; Syvertsen, 1985) provide comprehensive reviews on the physiological aspects of tree water relations.

A. Indicators of Tree Water Status

Leaf water potential (ψ) is commonly accepted as an indicator of fruit tree water status (Syversten, 1985) despite its questionable relevance in the control of plant physiological processes as affected by water status (Sinclair & Ludlow, 1985; Jones et al., 1985). A major reason for the acceptance of ψ is related to the speed of ψ measurement using a pressure chamber (Ritchie & Hinkley, 1975). Leaf ψ in fruit trees varies on a daily basis (Klepper, 1968; Smart & Barrs, 1973) and on a seasonal basis (Goode & Higgs, 1973), regardless of soil water status. Factors affecting leaf ψ values, other than soil water supply, include evaporative demand, stomatal conductance, hydraulic resistance to water flow and have been reviewed by Jones et al. (1985). Because leaf ψ is influenced by so many variables and can directly be used only to indicate the direction of water flow in the tree, its use as a tree water status indicator is often questioned (Garnier & Berger, 1985). Evidence that similar values could represent quite different plant stress levels has been provided by Bates and Hall (1981), and by Bethell et al. (1979) and Jones et al. (1983) in apple and Garnier and Berger (1985) in peach. In contrast, leaf ψ has proven to be a reliable indicator of tree stress in almond (Fereres et al., 1981), pistachio (Goldhamer et al., 1985b), apple (Peretz et al., 1984), and in studies of pear (*Pyrus communis* L.) and peach on relatively shallow soils (E. Fereres, 1979, unpublished data).

Sampling for leaf ψ in trees is more difficult than in herbaceous crops as leaf exposure (Jones & Cumming, 1984), height and position within the canopy (Olsson & Milthorpe, 1983), leaf age (Lakso, 1979), presence of fruit (Chalmers et al., 1983; DeJong, 1986), and leaf type (Jones & Cumming, 1984) all affect leaf ψ values. Nevertheless, leaf ψ is often the only practical way of assessing tree water status and irrigation needs under situations of variable soil water content, high water tables, saline soils, or in the case of localized irrigation.

In addition to other leaf water parameters frequently quoted as alternatives to leaf ψ (i.e., leaf relative water content and turgor pressure) (Jones et al., 1985), there is the possibility of monitoring the water content of the stems as a water status indicator in fruit trees—either directly (Brough et al., 1986) or indirectly as trunk shrinkage (Garnier & Berger, 1986). The latter authors found that treatments differing in irrigation levels showed differences in maximum stem shrinkage well before the growth patterns of peach trees were affected by water deficits. Thus, stem shrinkage could be a sensitive indicator if a standard measurement (i.e., a tree with unlimited water supply) were available. Fluctuations in stem diameter may be easily correlated with xylem water potential (Klepper et al., 1971) although there is a lag time in trees due to the capacitance of the stem tissue (Brough et al., 1986). Infrared thermometry (Idso et al., 1981) could be used in fruit trees along the same lines as in herbaceous crops—the assumption being that the partial stomatal closure required to decrease transpiration and thus, elevate canopy temperature, does not decrease marketable yield. Relative differences in tree canopy temperature are related to ψ differences in olive trees (Fereres, 1986, unpublished data). This technique as well as stem diameter monitoring shows promise in improving the assessment of irrigation needs directly from tree or orchard measurements.

B. Stomatal Conductance, Leaf Water Potential, and Transpiration

Stomatal behavior in the field reflects the interactions between the C assimilation and transpiration processes (Jarvis & Morison, 1981). Studies in fruit trees demonstrate that radiation, temperature, air humidity, and tree water status all affect stomatal opening (Jones et al., 1985). Of particular interest is the response to air humidity, first described for apricot (*Prunus armeniaca* L.) trees under desert conditions (Schulze et al., 1972) which appear to be responsible for the continuous decline in leaf conductance during the course of the day, irrespective of tree water status (e.g., Castel & Fereres, 1982). As evaporative demand increases during the day, diurnal leaf ψ values reflect the balance between stomatal conductance and transpiration rates.

Long-term tree water deficits apparently exert control of leaf ψ via adjustments in stomatal conductance. Figure 33–2a presents data for two treatments in peach that differed in soil water supply (33% of ET supplied via a drip system vs. dryland) that had essentially the same leaf ψ throughout most of the day (E. Fereres, 1980, unpublished data). Leaf conductance was markedly different, suggesting that stomatal conductance controlled leaf ψ in these trees.

It is likely that when long-term water deficits affect stomatal control of tree water status, leaf ψ will not reflect the level of water stress being experienced by the tree. Recently, evidence of the direct influence of soil drought on stomatal conductance has been reported (Schulze, 1986). Additional evidence of the long-term effects of drought on leaf conductance not mediated by tree water status is presented in Fig. 33–2b (E. Fereres, 1981, unpublished data). After 3 yr of drought, peach trees exhibited reduced stomatal conductance in the following spring, even though the soil profile was replenished by winter rains (at least partially). The presence of fruit is known to influence stomatal conductance and water potential in peach (Chalmers et al., 1983; DeJong, 1986). Lower stomatal conductance and higher leaf ψ values have been observed following fruit removal in peach, irrespective of soil and tree water status (E. Fereres, 1979, unpublished data). Additional work is needed to identify the basis for responses such as those described above.

Increases in the rate of transpiration produce a greater drop in the leaf ψ of fruit trees than in that of herbaceous crop plants (Schulze & Hall, 1982). Thus, it appears that hydraulic resistance is of greater magnitude in fruit trees than in herbaceous plants (Jones et al., 1985). Higher resistances in fruit trees may be due to sparser root systems (Atkinson, 1980). Evidence of increased hydraulic resistance caused by localized irrigation in apple trees may be inferred from the data of Peretz et al. (1984).

The impact of stomatal control on transpiration rates depends on the degree of coupling between the crop canopy and the atmosphere (Monteith, 1981). Jarvis (1985) indicated that tree crops have aerodynamically rough canopies, are subjected to higher windspeeds because of their height, have

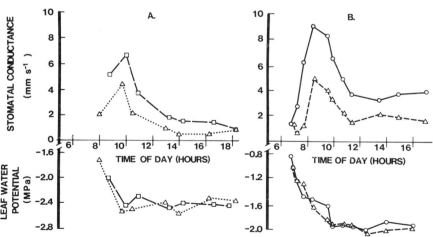

Fig. 33–2. A. Daily course of stomatal conductance (g_l) and leaf water potential (ψ) in peach trees for deficit drip irrigation (33% of ET, squares) and dryland (triangles) treatments on 31 Aug. 1979 in Winters, CA. B. Daily course of g_l and ψ for the fully irrigated (100% ET, circles) and dryland (triangles) treatments on 31 May 1980 after three experimental years in Winters, CA.

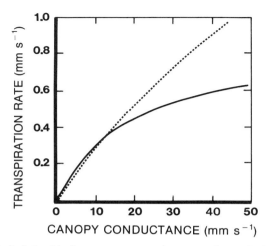

Fig. 33–3. Calculated relationships between canopy conductance and transpiration rate for boundary layer conductances of 50 mm s^{-1}, typical of field crops (solid line), and of 300 mm s^{-1}, typical of trees (dashed line). After Jarvis (1981).

high boundary layer conductances, and therefore, are well coupled to the atmosphere. In contrast, herbaceous crop canopies are smooth, have low boundary layer conductances, and are not coupled to the environment. Figure 33–3 shows data from Jarvis (1981) that demonstrate a change in stomatal conductance will influence the transpiration rate of trees more than that of field crops. Thus, stomatal conductance is effective in controlling transpiration in tree canopies (Jarvis, 1985) and reductions in stomatal conductance should be paralleled by decreases in transpiration and CO_2 assimilation.

C. Root Growth and Water Uptake

It is generally accepted that root density values encountered in fruit trees are less than those found in annual crops (Atkinson, 1980). Root density values per unit ground surface (LA) of 10 to 20 cm cm^{-2} are common in fruit trees while reports of LA values exceeding 300 cm cm^{-2} have been published for crops such as corn (*Zea mays* L.) (e.g., Hsiao & Acevedo, 1974). The sparse top growth in trees could be responsible for the low LA, although Chalmers et al. (1983) reported LA values of only 100 cm cm^{-2} for peach with a Leaf Area Index (LAI) of 10. The perennial growth habit of deciduous trees may be associated with the low root density values. Chalmers and van den Ende (1975) indicated that mature peach trees allocate only 1% of the annual growth increment to root growth. Low rooting densities may limit water uptake in deciduous trees and be partly responsible for relatively large decreases in leaf ψ as transpiration rates increase. However, as will be discussed in section III of this chapter, measured peak transpiration rates from deciduous irrigated orchards commonly exceed those observed in herbaceous crops. Thus, transpiration of fruit trees may not always be limited by low LA values, at least under irrigated conditions.

Irrigation enhances root growth probably because of the concomitant increase in top growth (Atkinson, 1980). Root growth takes place preferentially in soil volumes that have favorable conditions for root proliferation (Taylor, 1983). As soil water influences not only the availability of water and nutrients but also soil strength, variations in the soil water regime largely determine the distribution of roots in the soil.

Root proliferation in the soil surface layers is common in herbaceous crops (Taylor, 1983) as well as in deciduous trees (Chalmers et al., 1983) when such layers are wetted often by irrigation or rainfall. Water extraction in both crop types occurs preferentially from such layers, presumably because of the higher root densities and low axial resistances to water flow in the root system. As the surface soil dries, water uptake patterns change, with the maximum rates of extraction displaced down to the deeper layers (Levin et al., 1972). Drought also reduces the hydraulic conductivity of root systems in trees (Wiersum & Harmany, 1983). Water uptake rates from trees are similar to those described for herbaceous crops (Taylor, 1983) under full coverage irrigation. However, under localized irrigation, rates of water absorption are quite high as roots concentrate around the wetted zones (Nakayama & Bucks, 1986).

Three factors related to irrigation affect root distribution and water uptake in orchards: (i) frequency of irrigation, (ii) depth of applied water, and (iii) pattern of water application (localized or full coverage). The success of localized irrigation, even though water is applied to only a fraction of the potential root zone, has evidenced the plasticity of fruit tree root systems and their capability to adjust to limited wetting patterns. The effects of irrigation method on root systems are further discussed in section IV of this chapter.

D. Adaptation to Drought

In recent years, substantial interest has developed in elucidating the mechanisms of plant adaptation to drought (Fereres, 1984). However, most of the research has been carried out with annual crops (Turner, 1986; Fereres, 1987). At the leaf level, deciduous fruit and nut trees exhibit adaptive mechanisms similar to those described in annual plants (Fereres, 1984; Jones et al., 1985). Water extraction deep in the profile in the absence of irrigation has been described in deciduous trees (Syverstein, 1985) and must have adaptive value. Osmoregulation occurs in apple (Goode & Higgs, 1973), almond (Castel & Fereres, 1982), pistachio (Behboudian et al., 1986), peach and pear (Fereres, 1979, unpublished data), and probably, to some degree, in most other species. Lakso (1979) associated a large seasonal change in solute potential (around 2 MPa) to the maintenance of stomatal opening in apple trees subjected to water deficits. Pistachio potted plants under water stress continued to photosynthesize at leaf values down to -5 MPa and maintained positive turgor with leaf values as low as -6 MPa (Behboudian et al., 1986). This represents an extreme observation on the drought resistance of deciduous fruit and nut tree species. The long-term effects of drought on stomatal conductance (Fig. 33–2b), probably mediated by the effects of dry soil on the root system (Schulze, 1986), also must have adaptive significance. When stomatal control of transpiration is insufficient to prevent the development of potantially damaging leaf ψ values, leaf shedding is another mechanism frequently observed in deciduous trees that is effective in controlling transpiration. Castel and Fereres (1982) showed that radiation intercepted by almond canopies decreased 30% because of leaf shedding in the presence of increasing drought while midday leaf ψ stayed more or less constant (around -3.5 MPa). Obviously, recovery is not possible after leaf shedding and productivity must be sacrificed in exchange for tree survival. It appears that it takes more than 1 yr for deciduous trees to fully recover from severe water stress (Proebsting et al., 1981). Prune (*Prunus domestica* L.) trees subjected to 1 yr of deficit irrigation (15% of maximum demand) required 2 yr of recovery before attaining normal growth and yield. In the same experiment, growth and production of cherry (*Prunus avium* L.) trees did not recover after 2 yr (Proebsting et al., 1981). Functional recovery of citrus trees following severe water stress also required substantial time (Fereres et al., 1979). While pistachio trees showed almost complete vegetative growth recovery following 1 yr of severe water stress, the yield of marketable product was only half that of unstressed trees (Goldhamer et al., 1985b). This was attributed primarily to increased production of blank and aborted nuts.

III. IRRIGATION MANAGEMENT

A. Evapotranspiration and Water Requirements

Evapotranspiration (ET) losses from orchards are caused by the combination of direct evaporation from the soil surface (E) and transpiration (T) losses from the trees and cover crops if that technique of vegetation management is used. Transpiration losses in orchards without a cover crop are directly related to the amount of radiation intercepted by tree foliage and thus, to canopy size and architecture. Direct E losses are primarily a function of the frequency of surface soil wetting and of the irrigation method (localized or full coverage).

The economics of irrigation developments usually mandate that the maximum possible yields are obtained from irrigated lands. Since productivity and water use are usually considered to be directly related, crop water deficits have traditionally been viewed as detrimental to yield. In herbaceous crops, the full ET requirements are estimated relative to a maximum or reference ET (ET_0) as defined by Doorenbos and Pruitt (1977). ET_0 represents the ET losses from a grass crop where the upper limit is set solely by the evaporative demand of the environment. However, orchard ET is controlled not only by evaporative demand but also by stomatal conductance (see section II.B in this chapter) even if the soil water supply is not limiting. Thus, the use of ET_0 for orchards may be inappropriate. The ET rate of a well-irrigated orchard will be determined by the interactions between evaporative demand

Table 33-1. Midseason crop coefficients (K_c) for mature, clean cultivated deciduous fruit and nut orchards. Based on Doorenbos and Pruitt (1977), Fereres and Puech (1981), and Goldhamer and Snyder (1989).

Species	K_c
Almond, apricot, peach, pear, and plum	0.9 –0.95
Apple, cherry	0.95–1.0
Walnut, pecan	1.05–1.15
Pistachio	1.15–1.20

(radiation balance, windspeed, and vapor pressure gradient) and the canopy conductance. Orchard canopy conductance reflects direct stomatal responses to air humidity (Schulze et al., 1972), to the presence of fruits and their stage of development (Chalmers et al., 1983; DeJong, 1986), and possibly to long-term drought (Fig. 33-2b) among other factors.

Notwithstanding the above statement, most researchers have found it convenient to relate measured ET either to ET_0 or to pan evaporation (E_0). The use of E_0 as a reference adds uncertainty, as the environment around the pan is seldom standardized in orchards, and bare soil, tree shade, etc., all influence the E_0 rate for a given evaporative demand (Doorenbos & Pruitt, 1977).

Information on estimated orchard ET is scant. Most ET data from orchards have been obtained through a water balance technique that is often too crude to provide accurate estimates for short time periods. There are few cases where the ET of deciduous fruit or nut trees has been measured in lysimeters (e.g., Worthington et al., 1984). An added complication is that the spatial variation in root distribution in orchards increases the number of measurements required by a water balance technique. Summary information reviewed by Doorenbos and Pruitt (1977) and by Fereres and Puech (1981) permits the estimation of empirical crop coefficients relative to ET_0 (K_c) which can be used to obtain a first approximation of the typical, maximum ET requirements of mature orchards. Table 33-1 presents the midseason crop coefficients for different tree species. If a K_c curve needs to be developed, the procedure proposed by Doorenbos and Pruitt (1977) may be used, assuming that it takes from 8 to 12 wk from leafout to reach the K_c values of Table 33-1. The initial K_c would depend on the degree of soil surface wetness and soil management practices (Doorenbos & Pruitt, 1977). Sometime before leaf fall, K_c values tend to drop, reflecting leaf aging effects on T. The time at which K_c starts to drop is affected by irrigation management and by variation among tree species. Snyder et al. (1987) developed a leaflet for California conditions based on the data and procedures discussed above.

The higher K_c values reported in Table 33-1 for some nut tree species (walnut, pecan, and pistachio) deserve attention. Tree height is greater in mature walnut and pecan trees than in other species and that may increase windspeed and boundary layer conductance (Jarvis, 1985). The high midseason K_c of pistachio is probably related to the relatively high stomatal conductance values of this species (up to 16 mm s^{-1}) (Goldhamer et al., 1985b) and to the presence of stomata on the abaxial and adaxial sides of the leaf. Usual stomatal conductance values during peak evaporative demand periods for other deciduous trees normally do not exceed 10 mm s^{-1} (Jones et al., 1986).

The fact that peak ET rates from most deciduous fruit trees are similar to ET_o (Table 33-1) despite the higher windspeeds and boundary layer conductances of orchards may be explained by the declining diurnal trends in stomatal conductance discussed above. Since trees are well coupled to the atmosphere (Jarvis, 1985), the decrease in stomatal conductance throughout the day may compensate the increases in radiation load and vapor pressure deficits, resulting in T reduction.

There are occasional reports of tree ET rates exceeding the typical values of Table 33-1. High density peach trees reportedly had an estimated K_c of 1.3 (Natali et al., 1985). Crop coefficients approaching 1.40 have been reported for a high-density, tall pecan orchard (Miyamoto, 1983) and values varying from 1.3 to 1.6 have been estimated in an apple experiment (Beukes & Weber, 1982). Certainly, orchards with cover crops use 25 to 30% more water than clean cultivated orchards (Doorenbos & Pruitt, 1977; Bethell et al., 1979). In most orchards, some space between the trees must be cleared to allow for orchard operations (spraying, harvesting, etc.). Thus, trees do not intercept all incoming radiation but only a fraction, ranging from 60 to 80% in mature orchards. Thus, 20 to 40% of the incoming radiation is available for additional E if the soil surface is wet or if there is a cover crop. However, if the ground surface is clean cultivated and dry, the incoming radiant energy heats up the soil and the air around it, contributing energy to the transpirational process of the tree leaves. This microadvection (Fereres

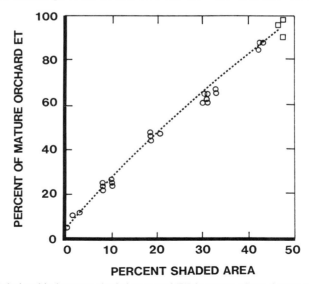

Fig. 33-4. Relationship between shaded area and ET in young almond trees relative to that of a mature orchard. Data obtained from drip-irrigated trees from 1 to 4 yr old (circles) and in a 6-yr-old orchard (squares) by using soil water balance techniques. After Fereres et al. (1989).

et al., 1981) is probably responsible for the reported lack of differences in the ET of mature orchards differing in canopy cover once they exceed 50 to 60% cover (Black et al., 1977; Doorenbos & Pruitt, 1977).

In the case of newly planted orchards, the percent cover by the trees is minimal and increases with age. Thus, the ET of young orchards is highly dependent on the irrigation frequency and the soil wetting patterns. The ET of young almond trees, experimentally determined under drip irrigation (Fereres et al., 1982), was related to the surface area shaded by the trees at 1200 h as shown in Fig. 33–4. The data indicate a high degree of microadvection (an orchard with 20% cover had an ET 50% that of a mature orchard) and provide estimates for the irrigation requirements of young almond trees (Fereres et al., 1982). The factors developed in northern California (Fereres et al., 1982) were found to be accurate in the San Joaquin Valley (Sharples et al., 1985). This suggests that the data of Fig. 33–4 may be used in similar climates with drip irrigation and even perhaps with other tree species (Renquist, 1987). However, values reported for young apple trees under drip irrigation (Middleton et al., 1979) do not agree with the data of Fig. 33–4. There is an urgent need to increase the number and accuracy of ET determinations in deciduous fruit and nut trees of various ages to resolve some of the issues stated above.

B. Irrigation Scheduling

Developing irrigation regimes requires knowledge of both the timing and amount of water to apply in order to replenish ET losses. The procedures used to schedule irrigations in orchards may be classified into those using soil or plant measurements to determine irrigation timing and those based on a water budget to estimate both depth of application and timing (Goldhamer & Snyder, 1989).

Measurement of soil water status may be done by evaluating the soil matric potential at a given depth with tensiometers, gypsum blocks, or other types of soil matric potential sensors. Tensiometers and gypsum blocks have been used quite often by farmers as indicators of irrigation timing. The threshold reading varies with soil type, crop, and stage of development and with the extent of soil wetting. The trends of a series of readings from two sensors at two different depths usually provide better information as they indicate the direction of water movement, as well as the rate of depletion (Fereres et al., 1981). Deep soils of medium to fine texture under conventional irrigation methods are better suited for the use of gypsum blocks while tensiometers are more practical under high frequency irrigation and/or in shallow soils of coarse texture (Smith & Fereres, 1988). The major limitation for the use of soil sensors is the spatial variation in soil water properties and the uncertainty of the root distribution in the orchard. Once a representative location has been identified through trial and error, soil matric potential sensors provide an effective means for irrigation scheduling. Klein (1983) showed that the use of tensiometers for peach irrigation reduced irrigation needs from 12 to 23% relative to irrigation scheduling based on E_o readings. The cause of the savings was that daily orchard ET variations were not detected when weekly coefficients were used with the latter method. When localized high-frequency irrigation is used in orchards, the use of soil water potential or water content sensors is critical (Fereres et al., 1981). Use of tree measurements such as leaf ψ for irrigation scheduling is another alternative as discussed in section I.A of this chapter.

The water budget method requires, in addition to orchard ET estimates (section III.A), the determination of the allowable soil water depletion (Smith & Fereres, 1988), also referred to as *yield threshold depletion* (Goldhamer & Snyder, 1989). The principles of orchard irrigation were first developed in the deep soils of California's Central Valley (Veihmeyer, 1927; Veihmeyer & Hendrickson, 1949). There, tree roots extend to considerable depth, extracting water from the subsoil as the surface layers are depleted. Under those conditions, a large portion of the root zone may be fully depleted before an irrigation is needed. Thus, the allowable depletion of orchard soils has traditionally been considered quite high (70–80% of the available water capacity of the soil). A consequence of the high depletion values is the large interval between irrigations, typical of surface irrigation practices in valley soils. Evaluation of the allowable depletion requires knowledge of the extent and density of the tree root system, as well as of the soil water characteristics. Such information is not normally available and must be obtained if an accurate water budget is sought. On the other hand, if the irrigation interval is long, it may not be possible to fully replenish the ET losses because of irrigation system or infiltration rate limitations. Thus, it is not uncommon for orchards on deep soils to start the season with a wet profile and to end it with the profile almost totally depleted (Viehmeyer, 1972).

Irrigation scheduling models may be used in orchards (Miyamoto, 1984). However, orchard-to-orchard variation precludes this approach without site-specific calibration (Bethell et al., 1979). It is encouraging, though, that an irrigation scheduling program was successfully operated for over 100 orchardists during several years in an area of diverse topography and microclimate (Bethell et al., 1980). The use of the neutron probe to evaluate soil water depletion (Campbell & Campbell, 1982) is another alternative to the water budget method that has similar limitations to those discussed for other soil water status sensors. It can be used in combination with the water budget method, allowing a specific calibration of every orchard site (Bethell et al., 1980). Under high-frequency methods, declining trends in subsoil water content as measured with the neutron probe suggest insufficient water application.

The only information required to schedule irrigations with high-frequency systems that do not refill the soil water reservoir is orchard ET and irrigation system efficiency (section IV of this chapter). To avoid obtaining such estimates, many high-frequency systems are scheduled on the basis of soil water instruments which may be connected to the system for automatic irrigation when a predetermined soil water level is reached (e.g., Feyen et al., 1985).

C. Effects of Reduced Irrigation on Growth, Development, Yield, and Fruit Quality

Appropriate irrigation scheduling is required to minimize plant water deficits throughout the crop life cycle. In orchards, there are many situations where irrigation applications are insufficient to meet the tree demand, either because the supply of water is limited or because of management-imposed water deficits at certain growth stages. Detecting tree water deficits is now possible by a combination of tree and soil water measurements although limited irrigation does not always induce water stress, particularly when soil water storage is substantial. Tree responses to reduced irrigation are reviewed below with emphasis on yield-determining processes. Salter and Goode (1967) and Uriu and Magness (1967) summarized the older literature, and Landsberg and Jones (1981) thoroughly reviewed the responses of apple trees to water deficits.

1. Growth

As in herbaceous crop plants (Hsiao et al., 1976), the process of expansive growth in fruit trees is very sensitive to water deficits (apple, Landsberg & Jones, 1981; pear, Brun et al., 1985a; almond, Fereres et al., 1981; peach, Chalmers et al., 1983; and pistachio, Goldhamer et al., 1986). The reviews by Salter and Goode (1967) and Uriu and Magness (1967) concluded that shoot growth was the first process to be affected by water stress. Mild water deficits early in the season reduce shoot elongation which takes place during the first few months of the season in deciduous fruit and nut trees. Lötter et al. (1985) found significant differences in apple shoot extension growth between two treatments that had average soil water ψ values of -0.02 and -0.06 MPa, respectively, during the shoot growth phase. Unfortunately, they did not measure leaf ψ. Shoot extension in the spring has been related to water supply conditions late in the previous season. Conditions favoring high water uptake rates until late in the season also promote shoot growth in the following season (Salter & Goode, 1967).

Shoot and trunk thickening may proceed throughout the growing season and are also sensitive to water deficits (see review by Salter & Goode, 1967). There are suggestions that trunk growth is less sensitive to water deficits than shoot extension (e.g., Maggs, 1961 cited by Landsberg & Jones, 1981) although the recovery of trunk growth upon relief of water stress that may occur at any time is probably the cause of such observations. Leaf size is normally less affected by water stress in fruit trees than are both shoot extension and thickening (Landsberg & Jones, 1981; Goldhamer et al., 1986). The cumulative effect of water stress on vegetative growth in deciduous fruit and nut trees is a reduction in tree size (Salter & Goode, 1967).

In addition to shoot elongation, another critical aspect of vegetative growth on tree productivity is shoot initiation. Goldhamer et al. (1987) found that the rate of shoot initiation in pistachio was directly related to the degree of the water deficits imposed; even mild water deprivation resulted in reduced initiation. In species where the number of fruiting positions per shoot is relatively insensitive to water stress, the number of shoots initiated may be one of the primary yield components affected by deficit irrigation.

Reproductive development interacts with vegetative growth in fruit trees (Chalmers, 1987). Thus, the final effects of water deficits on growth are variable, depending on crop size. In addition to having a relatively large C demand, trees with large fruit numbers may transpire at high rates (Chalmers et al., 1983) and develop low leaf ψ values. In those trees, vegetative growth suffers more when water supply is limited, particularly during the fruit growth periods.

2. Development

Aspects of development that are relevant to fruit and nut production include flower bud formation, floral development, and fruit set. As shown in Fig. 33–1, the formation of flower buds and differentiation of floral organs in deciduous trees generally occurs during the previous season. Water stress during summer and fall affects the initiation and differentiation of flower buds in most species (Uriu & Magness, 1967). In apple, severe drought prevents flower bud formation but moderate water deficits do not appear to have a significant effect (Landsberg & Jones, 1981). Flower bud formation in apricot trees appears to be sensitive to water deficits resulting from the lack of postharvest irrigation (Uriu & Magness, 1967). In some nut tree species—such as almond, walnut, and pistachio—summer water stress that reduces leaf ψ to values between -2.5 and -3.0 MPa does not appear to have a negative effect on flower bud formation per unit shoot length in the following year (Fereres & Goldhamer, 1979, 1987, unpublished observations). On the contrary, there are reports suggesting that return bloom of some species increases after trees are subjected to water deficits the previous summer. Chalmers et al. (1985) observed such response in peach under deficit irrigation. Similar effects of increased flowering induced by water stress in the previous season have been observed in pear by Mitchell et al. (1986) and in peach by Larson et al. (1988). Water stress during flower development in deciduous trees is an unlikely event due to minimal tree water requirements at that stage and to the low evaporative demand of the spring. If water deficits are severe enough during the summer to cause premature leaf fall, irrigation after that time is known to promote flowering during the fall, thereby reducing flowering the next spring. Fruit set as a percentage of flower numbers increases as tree water supply improves in apple (Landsberg & Jones, 1981). If water deficits develop immediately after petal fall, fruit set will be substantially reduced (Salter & Goode, 1967). Fruit shedding caused by water stress has been described in pear and apple trees (Uriu & Magness, 1967) but it is seldom observed under field conditions, unless the stress is severe.

3. Fruit Quality

Fruit size is one aspect of fruit quality negatively affected by water stress. The fruit enlargement process is sensitive even to mild tree water deficits

(Landsberg & Jones, 1981; Daniell, 1982). Fruit numbers interact with fruit size, and water deficits reduce fruit size more when fruit numbers are high (Landsberg & Jones, 1981). Brun et al. (1985a) suggested that slower fruit growth early in the season in pear is caused by water deficits during the previous summer.

Eating quality of many pome and stone fruits is enhanced by mild water deficits (e.g., Proebsting et al., 1984; Chalmers et al., 1983). Soluble solids increase and water content decreases in fruits from trees subjected to water deficits (Uriu & Magness, 1967). Mild water deficits also enhance apple color and reduce titratable acidity (Proebsting et al., 1984). Lötter et al. (1985) found that irrigating apple trees when only 15 to 35% of available water was depleted increased the incidence of a number of physiological disorders such as bitter pit, superficial scald, and water core. Landsberg and Jones (1981) discussed other aspects of apple quality as affected by water supply. Fruit disorders in pear, such as cork spot, have been correlated with increased irrigation (Brun et al., 1985b). Sunburn in prune increases under water deficits (Uriu & Magness, 1967).

Nut quality is generally reduced by water deficits. Sunburn in heat-sensitive walnut cultivars increases under water stress (Ramos et al., 1978). Goldhamer et al. (1986) showed a general decrease in nut quality parameters of pistachio subjected to water deficits, including greater unsplit nut production. Interestingly, even severe deficit irrigation did not materially affect kernel filling in pistachio. On the other hand, water deficits are known to affect the kernel-filling process in almond, reducing nut quality.

4. Fruit Yield

Obtaining the highest marketable yield at the lowest possible cost is the primary objective of the orchardist. This objective must be achieved throughout the life of the orchard (at least 15–20 yr). Thus, experiments designed to fully evaluate irrigation strategies should be conducted over a number of years to provide definitive answers. Unfortunately, few long-term experiments on irrigation of fruit trees have occurred in recent decades and we still must rely on earlier work (e.g., Viehmeyer, 1972, 1975). His early experiments with peach, apricot, and apple on deep, fertile soils can now be interpreted to indicate that if trees have access to subsoil moisture, it is possible to achieve maximum yields without irrigating to meet the maximum T demand (Viehmeyer, 1975). In other words, deficit irrigation may not result in reduced T if deep reserves of soil moisture are available for plant uptake. On the other hand, tree water stress can occur if a dry soil profile is irrigated precisely to meet the T demand because of competition between soil absorptive forces and roots for soil moisture.

Debate continues regarding the effects of deficit irrigation on yields. Recent results of Chalmers et al. (1985) suggest that limited irrigation in shallow soils can be a valuable strategy to obtain maximum yields in some peach and pear cultivars. Notwithstanding this regulated deficit irrigation approach, the application of full or nearly full irrigation requirements has resulted in maximum yields in a number of experiments also conducted recently (Lötter et al., 1985; Brun et al., 1985a; Bethell et al., 1979; Assaf et al., 1984; Layne & Tan, 1984; Goode et al., 1978; Daniell, 1982). There is little doubt that water deficits severe enough to seriously affect any of the yield-determining processes discussed in this section will result in economic losses.

The evaluation of marketable yields requires an assessment of all parameters that have an influence on tree productivity and crop quality. The sensitivity of such parameters to water deficits may be variable and a comprehensive study is needed for full evaluation of the impact of water deficits on tree yield and performance. An example of one such study was conducted in pistachio by Goldhamer et al. (1987) involving 3 yr of deficit irrigation, the summary of which is illustrated in Fig. 33–5. The excellent reputation of pistachio as a drought-resistant tree (Behboudian et al., 1986) contrasts with the sensitivity of yields to moderate water deficits. Studies such as that shown in Fig. 33–5 are needed for other deciduous fruit and nut trees to elucidate their responses to reduced irrigation applications.

5. Irrigation Management Under Limited Water Supply

In situations where water is limited or its supply is purposely regulated (Chalmers et al., 1985), knowledge of tree responses to water deficits may be applied to develop irrigation strategies directed at obtaining maximum marketable yields of orchard crops.

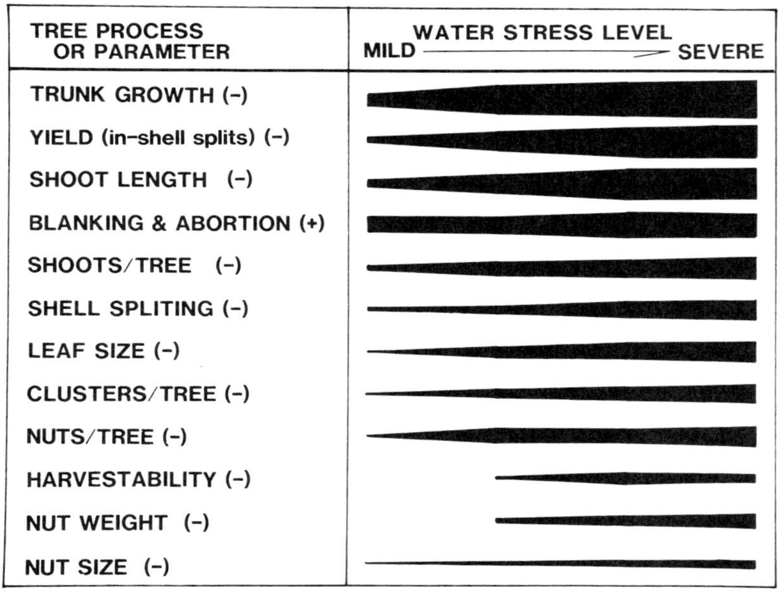

Fig. 33–5. Generalized diagram of the sensitivity to water stress of selected pistachio plant processes and parameters. Widths of the horizontal bars at given stress levels represent the relative magnitudes of the response and were developed by evaluating effects under dryland, 25, 50, and 75% ET relative to 100% ET (dryland illustrated as "severe" stress and 75% ET shown as "mild"). Signs indicate whether water deficits resulted in a decrease (−) or increase (+) in the magnitude of the process or parameter.

As a general principle, water deficits should be avoided at all times during the first few years of the orchard. Canopy development should not be limited by water stress effects on expansive growth as they delay the achievement of maximum orchard yields. Fereres et al. (1981) showed that young almond trees irrigated throughout the season outyielded those where irrigation was cut off 6 wk before the beginning of leaf fall. Minimum pruning and high tree densities (Mitchell & Chalmers, 1982) are additional practices that may be used to shorten the time for establishing a mature orchard.

The water storage capacity of the orchard root zone plays an important role in defining the limited irrigation program. In deep soils, it is desirable to start the season with a fully charged profile and to apply water at a rate below the consumptive use until about harvest. Postharvest irrigation practices are discussed separately below. In shallow soils, it is possible to control growth and tree water status by using high-frequency deficit irrigation. Decreasing the application rate to a small fraction of the tree requirements (about 10–20% in some stages) markedly reduces shoot growth without seriously affecting fruit growth, at least with certain tree species or cultivars. Mitchell and Chalmers (1982) and Mitchell et al. (1984) showed that yields of late-season maturing peach and pear are not affected by deficit irrigation during the shoot growth stage, while pruning and irrigation costs are reduced as compared to fully irrigated trees. This is possible because by the time the shoot growth period in these trees is nearly complete, about 75% of the fruit growth period remains (Chalmers et al., 1985). Thus, full irrigation after the deficit induces compensatory fruit growth, leading to about the same fruit size. Chalmers et al. (1985) indicated that it is possible for the extreme treatments to decrease vegetative growth by three-fourths and irrigation applications by one-third in peach and pear trees without reducing yields or fruit size. The approach outlined above is difficult to use in deep soils as limited irrigation applications do not prevent tree uptake if there is deep moisture available. Precise, controlled water applications are required to induce water stress at the desired stage. Nevertheless, as a general philosophy, the suggestions of Chalmers et al. (1985) deserve attention and more experiments should be conducted with different tree species, cultivars, soils, and irrigation systems to evaluate the feasibility of regulated deficit irrigation.

In deep soils, the use of continuous high-frequency deficit irrigation has proven useful to reduce irrigation requirements. In a 5-yr experiment with peach trees in Winters, CA, applications of 25 to 33% of the ET requirements resulted in the maintenance of leaf photosynthetic rates close to those exhibited by trees that received 100% of the ET requirement (Fig. 33–6). In contrast, the nonirrigated treatment had low photosynthesis throughout the season, even though its leaf ψ had similar values to those of the deficit treatment (Fig. 33–2a; Fereres, Beutel, and Uriu, 1979, unpublished data). The maintenance of higher stomatal conductance and of photosynthesis in the deficit irrigation treatment was associated with substantial water extraction deep in the profile as shown in Fig. 33–7. Subsoil water extraction in the deficit treatment exceeded that of the dryland treatment by more than 25% (Table 33–2). Deficit-drip irrigation, combined with high soil-water storage,

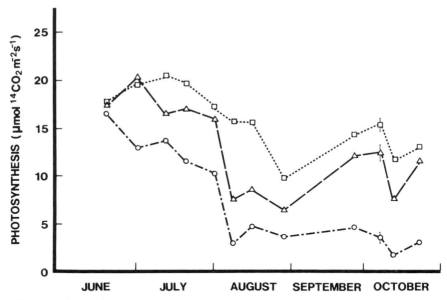

Fig. 33–6. Seasonal course of midday leaf photosynthesis rates for fully irrigated (100% ET, squares), deficit irrigated (33% ET, triangles) and dryland (circles) treatments in peach trees (cv. Suncrest) measured with a C-14 technique in Winters, CA during 1979.

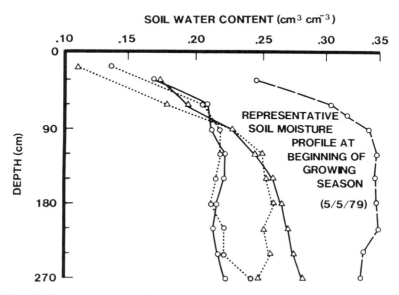

Fig. 33–7. Soil water content as a function of depth at the beginning and end of the 1978 and 1979 growing seasons for dryland (triangles) and deficit-irrigated (25 or 33% ET, circles) peach trees in Winters, CA.

Table 33-2. Evapotranspiration and soil water extraction of peach trees under various irrigation treatments in Winters, CA, for a period between 1 May and 31 Oct. 1979 (Fereres et al., 1979, unpublished data).

Treatment	Applied water	Soil water Extraction	Evapotranspiration
		mm	
Flood	690	166	856
Drip 100%	501	202	703
Drip 33%	146	321	467
Dryland	--	243	243

kept peach trees functional throughout several seasons and allowed for the complete depletion of soil water reserves. Even though applied water was only 25 to 33% of ET requirements, the seasonal ET of such treatments ended up being 65% of maximum ET because of the substantial contribution of stored soil water.

A general recommendation, already practiced by many orchardists around the world, is to reduce irrigation after harvest. Postharvest irrigation of mature, early season peach trees grown in California was reduced to a single application after 15 June for 3 yr without negative effects on fruit load. However, marketable yields of the second and third years were reduced because of fruit doubling, i.e., two joined fruit per stem (Handley & Johnson, 1986, unpublished data). Phene et al. (1987) showed that in over a 3-yr period, it is possible to end irrigation on pistachio at about 75% of the seasonal water requirement without marketable yield losses. Common irrigation practices in almond and walnut orchards in California's Central Valley result in the development of substantial water deficits during and after harvest. It is not known if, in such cases, additional irrigation would improve yields. Reducing postharvest irrigation may have little or no impact on yields of deciduous fruit and nut trees. However, there is variation among species in their sensitivity to postharvest water stress (Uriu & Magness, 1967) and more research is needed to quantify the stress levels that are acceptable for the various species and situations.

If the reduction in water supply is drastic, severe pruning in peach and pear trees has proven instrumental in tree survival as it prevents the development of severe water deficits that cause tree death (Proebsting & Middleton, 1980).

D. Frost and Climate Control

Overhead sprinkler irrigation may be used for frost control by spraying fruit trees during freezing temperatures (Barfield & Gerber, 1979; Jarrett & Morrow, 1984). Hamer (1986) calculated the water requirements for frost protection by using a heat balance equation that permits the accurate estimation of water needs. Cooling the buds during winter delays bud development until frost probabilities are negligible (Anderson et al., 1973). Such a technique delays apple bloom by 5 to 15 d in southern England, although the sprinkled buds are more frost sensitive at every stage, probably due to their higher water content (Hamer, 1983).

Climate control by evaporative cooling has been investigated in warm climates in many crops such as pear (Lombard et al., 1966), apple (Unrath and Sneed, 1974), grape (Gilbert et al., 1970), and prune (Henderson, Aldrich & Uriu, 1978, unpublished data). Goode et al. (1979) have shown that overhead irrigation by misting reduces apple tree environmental stress even in temperate regions. Misting in combination with conventional irrigation increased fruit bud production, fruit number, and yield in years of high evaporative demand. Nevertheless, the strict requirements of good quality water, the limited benefits on fruit quality, and the questionable economic feasibility have all prevented widespread application of this technique. Chilling requirements in warm climates may be fully achieved by sprinkling trees during the dormant period (Gilreath & Buchanan, 1981). Erez and Couvillon (1983) reduced bud temperatures by about 3 °C during the day, leading to enhanced bud break and to an early and more uniform bloom in nectarine.

E. Irrigation-Disease Relationships

Water management influences the predisposition of deciduous fruit and nut trees to disease infection. Notably, excess soil moisture and poor aeration enhance the probability of root or crown infection by *Phytophthora* spp. in walnut, cherry, avocado (*Persea americana* Mill.), and citrus (MacDonald & Duniway, 1978; Wilcox & Mircetich, 1985; Sterne et al., 1977; Stolzy et al., 1965). While the extent of disease damage depends on interactions between the *Phytophthora* spp., the rootstock, and the behavior of the pathogen propagules, irrigation management strongly influences the latter factor and thus, plays a key role in the longevity and productivity of infected orchards.

Soil water status largely affects the development of *Phytophthora* by enhancing the release and dispersal of zoospores and by creating root conditions favorable for host infection by zoospores (Matheron & Mircetich, 1985). Working with cherry, Wilcox and Mircetich (1985) found that soil water levels near saturation increased the rate of sporangia initiation and zoospore discharge which resulted in severe infection rates by numerous *Phytophthora* spp. Lowering the soil matric ψ to about -2 kPa almost completely eliminated pathogen infection, presumably because of low sporangium formation or failure of the sporangia to discharge zoospores. By flooding the soil for short periods of time to create conditions favorable for zoospore creation and discharging, Wilcox and Mircetich (1985) failed to observe high disease levels, presumably because zoospore mobility was limited by the short flooding duration. Zoospores move most readily in large water-filled soil pores. Excessive moisture may also lead to root changes due to prolonged anoxia that may increase root susceptibility to infection. Indeed, the buildup of ethanol in oak roots because of anaerobic soil conditions resulted in increased infection by *Armillaria mellea* (Vahl.:Fr.) Kumme *s. str.* fungus (Wargo & Montgomery, 1983).

The extent of orchard infection due to *Phytophthora* is highly influenced by the species and the rootstock, as well as water managment, especially for walnut. With the *Juglans hindsii* (Jeps.) Jeps. rootstock, a susceptible cultivar, infection by *P. cryptogea* Pethybr. & Laff. and *P. citrophthora* (R.E. Sm. & E.H. Sm.) Leonian can be minimized by avoiding prolonged or repeated periods of soil saturation around the trunks (Matheron & Mircetich, 1985). This may require physical or chemical modification of soils to improve infiltration or internal drainage. Avoidance of saturated conditions for prolonged periods can also be accomplished by changing from flood to high-frequency, localized irrigation methods, such as low-volume sprinkler. On the other hand, massive and widespread infection can occur in orchards on *J. hindsii* rootstock infected with *P. citricola* Sawada and *P. cinnamomi* Rands., even if careful water management is practiced. If *Phytophthora* problems are anticipated when planning an orchard, a more resistant rootstock, such as Paradox (*J. hindsii × J. regia*), which has proven to be more resistant to several *Phytophthora* species, should be planted.

While *Phytophthora* is commonly enhanced by high soil water levels, deep bark canker is associated with both excessive and inadequate moisture in the root zone of deciduous fruit and nut trees. Irrigating during the winter to supplement chronic deficit irrigation during the summer reduces the extent of active deep bark canker symptoms in mature walnut trees (Teviotdale & Sibbett, 1982). This response is attributed to improved tree vigor.

Another soil-borne fungus that is influenced by water management is *Verticillium dahliae* Kleb. In this case, it's temperature, rather than soil oxygen status, that appears to be the primary factor controlling infection by the pathogen. The infection rate of guayule (*Parthenium argentatum* A. Gray) by *V. Dahliae* is directly related to irrigation frequency; maximum damage occurs with weekly irrigation and least occurs under dryland conditions (Schneider, 1948). This is attributed to the fact that each irrigation reduces soil temperature which enhances disease infection. The temperature influence on plant expression of *V. dahliae* infection was also observed by Wilhelm and Taylor (1965) in olive (*Olea europaea* L.) trees. High midsummer temperatures resulted in recovery from *Verticillium* infection symptoms in the canopy, followed by renewed infection with cooler fall temperatures. They concluded that temperature is a major factor in determining the severity of this wilt due to its effect on the growth rate of the pathogen.

Orchardists with *Verticillium*-infested soils and sensitive tree species can use the experience of cotton (*Gossypium hirsutum* L.) growers afflicted with this problem as a guide to develop optimal irrigation strategies. Preplant cotton irrigations are recommended to fill the profile, followed by the use of as few heavy irrigations as possible during the season (Raney, 1973). Frequent, light irrigations that reduce soil temperature should be avoided. A practice emerging among California pistachio growers with high levels of *V. dahliae* in the soil is to apply a heavy winter irrigation and delay the first postleafout irrigation as long as possible. This practice presumably allows soil temperatures to remain relatively high during the spring—a time when pistachio seems particularly susceptible to infection. It should be noted that withholding water for too long can create tree water deficits that may ultimately affect orchard yield more than the disease.

F. Irrigation-Insect Relationships

Both increases and decreases in insect populations attributed to water management have been observed over a wide range of crops. With herbaceous crops, plant water stress-induced increases in spider mite (*Tetranychus telarius* Koch) populations were found by Hollingsworth and Berry (1982) on peppermint (*Mentha* × *piperita* L.), Chandler et al. (1979) on corn, and Kattes and Teetes (1978) on sorghum [*Sorghum bicolor* (L.) Moench]. Tree crops under water stress that manifested increased mite pressures include almond (Youngman & Barnes, 1986; Youngman et al., 1988; Feese & Wilde, 1977), walnut (D.A. Goldhamer, 1986, unpublished data), and peach (D. Handley, 1987, unpublished data). Two hypotheses are usually offered to account for increased mite pressures from deficit irrigation:

1. Since stressed leaves are usually warmer than unstressed leaves due to stomatal control of transpiration, insect development rates are enhanced (van de Urie et al., 1972).
2. Stressed leaves may provide a more nutritious meal that favors insect growth and reproductive success (McNeil & Southwood, 1978). Increased protein synthesis in plants at the onset of leaf wilting may temporarily increase available N to feeding insects (Henckel, 1964).

While irrigation-related plant water stress is usually associated with increased insect problems, numerous cases have been observed where water stress reduced insect pressures on both herbaceous and tree crops, including European red mite on apple (Specht, 1965) and two-spotted spider mite (*Tetranychus urficae* Koch) on soybean (*Glycine max* L.) (Mellors et al., 1984). Scribner (1978) found that larval growth of a variety of leaf-chewing *Lepidoptera* slowed as leaf water content decreased because of water stress.

The predisposition of tree crops to increased insect damage depends largely on the feeding nature of the insects involved, whether they are larvae or adults. Goldhamer et al. (1986) found that epicarp lesion damage on pistachio nuts, believed to be primarily the result of feeding by adult leaf-footed plant bugs (*Leptoglossus clypealis* Heldemann), was least in trees under severe water stress and greatest in fully irrigated trees. They suggested two possible explanations. One, microclimatic differences in the tree canopies made the stressed tree environment undesirable for the feeding insects. For example, peak canopy temperatures on a midsummer day were 38.2 and 28.9 °C, for dryland and fully irrigated trees, respectively. Thus, when given the choice, flying adult nut-feeding insects preferred the cooler canopy environment of nonstressed trees. The second hypothesis involves the fact that insect wounds apparently trigger the production of substances that destroy nut tissue, and transport of these substances within the nuts depends on concentration or hydraulic gradients. Plant water stress may affect the latter, resulting in less movement of wound-related substances and thus, less epicarp lesion damage.

IV. METHODS OF ORCHARD IRRIGATION

Knowledge of tree irrigation requirements and optimal irrigation timing can only be of benefit if water is delivered uniformly through an efficient irrigation system. Tree crops respond more to soil water levels and irrigation scheduling regimes than to the irrigation method. However, there are some effects unique to certain methods, thus the need to briefly review the basic features of orchard irrigation systems. Extensive reviews are available on irrigation system design (ASAE, 1981) and, specifically, on trickle irrigation systems (Nakayama & Bucks, 1986).

A. Surface Irrigation Systems

Traditionally, fruit trees have been irrigated by a surface method (basin, border, or furrow irrigation). The fact that orchards in the past have been established preferentially on relatively good soils with flat topography facilitated the use of surface methods. Most surface methods require high labor inputs. As the number of irrigations determines the seasonal labor needs, irrigating as infrequently as possible is a widespread attitude in orchard surface irrigation.

Basin irrigation is commonly practiced around the world, with basins varying in size from a single tree to over 0.4 ha. Obviously, filling individual tree basins requires substantial labor which can be minimized if an underground pipe network is installed to deliver equal amounts of water to individual basins. Rawlins (1981) designed a low-head bubbler system that performs such a function with high uniformity and minimum labor and energy costs.

Border irrigation requires high instantaneous flow rates for efficient orchard irrigation. Water is normally delivered to border checks through a network of low-head underground pipes and valves. Furrow irrigation can be used under steeper slopes and requires less land grading than the two other surface methods; it also may be used with smaller flow rates and/or low intake rate soils. A number of configurations may be used to increase intake opportunity time (Booher, 1974). Water is delivered to the furrows through siphons over the ditch or through plastic or aluminum gated pipe. Furrow irrigation of orchards can be considered as one form of localized irrigation as it is possible to wet only a fraction (20–40%) of the soil surface.

Surface irrigation systems in orchards already established cannot be upgraded without tree disturbance and thus, are difficult to improve. Recent developments in surface irrigation technology such as precise laser-beam land leveling (Erie & Dedrick, 1979), surge irrigation (Bishop et al., 1981), and fully automated surge irrigation valves may be used to design highly efficient surface irrigation systems for newly developed orchards located on relatively uniform soils of flat topography. The evaluation and improvement of existing systems (Merriam & Keller, 1978; Liss et al., 1984) is a prerequisite for more efficient use of water in existing surface-irrigated orchards.

B. Pressurized Irrigation Systems

Pressurized irrigation systems for orchards include a variety of portable and permanent sprinkler systems—both under-tree and overhead—and microsprinkler systems, as well as drip or trickle systems. Portable aluminum pipes with under-tree sprinklers are seldom used nowadays because of their extremely high labor requirements. Permanent or solid set systems with overhead sprinklers are versatile as they can also be used for frost protection and environmental control. However, for the full coverage of the orchard needed for such applications, the instantaneous flow rate required is several times that needed for irrigation alone. Investments for permanent overhead sprinkler systems are quite high but may be warranted in some areas and with certain crops because they provide the highest possible degree of frost protection (Barfield & Gerber, 1979) and cooling, as well as the possibility of applying agrochemicals directly to the tree foliage.

Permanent, under-tree sprinkler systems should avoid foliage wetting, reducing the risk of diseases in susceptible species (Michailides & Ogawa, 1986). The systems provide some degree of frost protection (1 to 2 °C; R. Snyder, 1988, personal communication) but cannot be used for closely spaced orchards or hedgerow plantings.

The maximum possible spacing between sprinklers and laterals in orchard systems is required to reduce investment costs. If frost protection is desired, overhead systems must have high uniformity for full protection. If this form of frost protection is not important, full-coverage uniformity may be partially sacrificed in view of the success of high-frequency localized irrigation systems (see below). In any case, uniform distribution above the trees does not ensure uniform infiltration as the trees act as water collectors and the water is conveyed through the limbs to specific points at the ground surface.

A large proportion of agricultural lands irrigated by localized irrigation methods is devoted to deciduous fruit and nut trees. The cost of a localized system (drip and microsprinklers) is directly proportional to the number of emission points per unit ground surface, which is a minimum under the usual spacings of fruit trees (Fereres et al., 1978). A major advantage of the localized methods in orchard irrigation is related to the water savings potential during the first few years of the orchard (Middleton et al., 1979; Fereres et al., 1982). The savings have been estimated to range between 3300 mm for an almond orchard (Fereres et al., 1982) to over 6000 mm for an apple orchard compared to a full-coverage sprinkler system that maintained a cover crop (Middleton et al., 1979). Localized irrigation systems include drippers as well as a variety of microsprinklers which range from 10 to 200 L/h in flow rates and from 1 to 8 m diam. of the wetted area (Vermeiren & Jobling, 1983).

The fact that crop responses to drip irrigation were initially positive, despite partial wetting of the root system, generated numerous research efforts aimed at evaluating the minimum fraction of the root system that could be watered for optimum crop performance under localized irrigation (e.g., Black & West, 1974). Such experiments conducted under controlled conditions have shown qualitatively that T rates may not be limited by partial wetting of the root zone in apple (Black & West, 1974) and peach (Tan & Buttery, 1982). The issue has not been resolved for all species and situations although field experience has shown that wetting 20 to 30% of the potential root zone of mature deciduous trees appears sufficient for maximum production, provided enough water is applied to meet the ET demand (Henderson & Fereres, 1981).

Localized irrigation systems allow for the precise control of tree water status in areas of low rainfall and in shallow, marginal soils. As an extreme case, leaf ψ of almond trees dropped from -1.5 to -3.0 MPa in 4 d when the root system was restricted to the area wetted by two emitters (Fereres et al., 1981). A number of experiments with apple trees on relatively shallow soils (Evans & Proebsting, 1985) have shown that even though water is applied at rates sufficient to meet the tree demand, leaf ψ is lower than that observed on trees under full coverage irrigation (sprinkler and furrow). The mild water stress, presumably attributed to higher root zone resistance due to the limited wetting, induces higher soluble solids in the fruits, early flowering in young trees, and sometimes, higher cumulative apple yields in the first few years of the orchard (Evans & Proebsting, 1985). Such differential responses due to the irrigation method have not been documented for other deciduous fruit or nut tree species, although they are likely to occur when drip irrigation is used on marginal soils that restrict root development beyond the zone wetted by the emitters (Chalmers et al., 1985). Other methods could be used to impose similar mild water stress patterns although their management would be difficult.

Use of microsprinklers has become widespread in recent years for irrigation of deciduous fruit and nut orchards. Major advantages of microsprinklers over conventional drip systems include less clogging, lower maintenance requirements, larger wetted surface areas from one emission point, better weed control due to less frequent applications, and the fact that it is easier to identify the initial stages of emitter clogging. Direct E losses from microsprinklers amount to only a few percent of applied water (Goldhamer & Cantrell, 1982) and are not a significant disadvantage of this method. Poor surface distribution uniformity within the wetted pattern (Post et al., 1986) of most microsprinklers does not result in equally poor subsurface soil water distribution; lateral movement and redistribution of water substantially improve root zone moisture uniformity (Goldhamer, 1985a).

In recent years, it has been common to change irrigation systems from full coverage surface or sprinkler to localized irrigation. Concern is usually expressed about the possibility that stress may develop before the root system adapts to the new, localized wetting pattern. Evidence gathered so far (Bielorai et al., 1985) suggests that, in most cases, the conversion may be made without development of a significant stress in the converted orchard, even in high evaporative demand periods.

The method of irrigation interacts with the vegetation management program of the orchard. Fischer et al. (1986) showed that the effectiveness of preemergence herbicides is directly related to irrigation frequency, and thus, to the irrigation method. Although high-frequency drip systems result in weed proliferation that is generally limited to the wetted surface areas, the weeds are more difficult to control. Strip weed control is needed for drip irrigation, in part, to facilitate system surveillance. Permanent levees are desirable in surface irrigation but can interfere with tillage operations. A cover crop may be used for full coverage sprinkler and border irrigation but may not be suited for furrow or localized irrigation systems.

REFERENCES

American Society of Agricultural Engineers. 1981. Design and operation of farm irrigation systems. Monogr. 3. ASAE, St. Joseph, MI.

Anderson, J.L., E.A. Richardson, G.L. Ashcroft, J. Keller, J. Alfar, G. Hanson, and R.E. Griffin. 1973. Reducing freeze damage to fruit by overhead sprinkling. Utah Sci. (December) 1973:108–109.

Assaf, R., E. Levin, and B. Bravdo. 1984. Effect of drip irrigation on the yield and quality of "Golden Delicious and Jonathan apples." J. Hortic. Sci. 59:493–499.

Atkinson, D. 1980. The distribution and effectiveness of the roots of tree crops. Hortic. Rev. 2:424–490.

Barfield, B.J., and J.F. Gerber. 1979. Systems for reducing environmental stress. p. 13–21. In B.J. Barfield et al. (ed.) Modifications of the aerial environment of crops. ASAE, No. 2.

Bates, L.M., and A.E. Hall. 1981. Stomatal closure with soil water depletion not associated with changes in bulk leaf water status. Oceologia 50:62–65.

Behboudian, M.H., R.R. Walker, and E. Törökfalvy. 1986. Effects of water stress and salinity on photosynthesis of pistachio. Sci. Hortic. 29:251–261.

Bethel, R., E. Fereres, R. Buchner, and B. Fitzpatrick. 1979. Water conservation and management for foothill orchards. Calif. Agric. 33:7–10.

Bethel, R., E. Fereres, R. Buchner, and R. Mansfield. 1980. Irrigation management for the Sierra Nevada foothills of California. Final rep. to U.S. Bur. Reclam. Univ. California Coop. Ext., Placerville.

Beukes, D.J., and H.W. Weber. 1982. The effects of irrigation at different soil water levels on the water use characteristics of apple trees. J. Hortic. Sci. 57:383–391.

Bielorai, H., S. Dasberg, and Y. Erner. 1985. Long-term effects of partial wetting in a citrus orchard. p. 568–573. In Proc. 3rd Int. Drip/Trickle Irrig. Congr., Vol. 2, Fresno, CA. November.

Bishop, A.A., W.R. Walker, N.L. Allen, and G.J. Poole. 1981. Furrow advance rates under surge flow systems. J. Irrig. Drain Div. Am. Soc. Civ. Eng., 107:257–264.

Black, J., P.O. Mitchel, and P. Newgreen. 1977. Optimum irrigation for young trickle irrigated peach trees. Austr. J. Agric. Anim. Husb. 17:342–345.

Black, J., and D. West. 1974. Water uptake by an apple tree with various proportion of the root systems supply with water. p. 432–436. In Proc. 2nd Int. Drip Irrig. Congr., San Diego, CA. July.

Booher, L.J. 1974. Surface irrigation. FAO, Rome.

Brough, D.W., H.G. Jones, and J. Grace. 1986. Diurnal changes in water content of the apple trees as influenced by irrigation. Plant Cell Environ. 9:1–7.

Brun, C.A., J.T. Raese, and E.A. Stahly. 1985a. Seasonal response of "Anjou" pear trees to different irrigation regimes. I. Soil moisture, water relations, tree and fruit growth. J. Am. Soc. Hortic. Sci. 110:830–834.

Brun, C.A., J.T. Raese, and E.A. Stahly. 1985b. Seasonal response of "Anjou" pear trees to different irrigation regimes. II. Mineral composition of fruit and leaves, fruit disorders, and fruit set. J. Am. Soc. Hortic. Sci. 110:835–840.

Campbell, G.S., and M.D. Campbell. 1982. Irrigation scheduling using soil moisture measurements: Theory and practice. p. 25–42. In Daniel Hillel (ed.) Advances in irrigation. Vol. 1. Academic Press, New York.

Castel, J.R., and E. Fereres. 1982. Responses of young almond trees to two drought periods in the field. J. Hortic. Sci. 57:175–187.

Chalmers, D.J. 1987. Opportunities for improving crop yields through research—A physiological perspective. p. 1–8. In 4th Austr. Agron. Conf. Proc. La Trobe Univ., Melbourne.

Chalmers, D.J., P.D. Mitchell, and P.H. Jerie. 1985. The relation between irrigation, growth, and productivity of peach trees. Acta Hortic. 173:283–288.

Chalmers, D.J., K.A. Olsson, and T.R. Jones. 1983. Water relations of peach trees and orchards. p. 197–232. In T.T. Koxlowski (ed.) Water deficits and plant growth. Vol. 3. Academic Press, New York.

Chalmers, D.J., and B. van den Ende. 1975. Productivity of peach trees: Factors affecting dry-weight distribution during tree growth. Ann. Bot. 39:423–432.

Chandler, L.D., T.L. Archer, C.R. Ward, and W.M. Lyler. 1979. Influences of irrigation practices on spider mite densities on field corn. Environ. Entomol. 8:196–201.

Daniell, J.W. 1982. Effect of trickle irrigation on the growth and yield of "Loring" peach trees. J. Hortic. Sci. 57:393–399.

DeJong, T.M. 1986. Effects of reproductive and vegetative sink activity on leaf conductance and water potential in "Prunus persica" L. Batsch. Sci. Hortic. (Amsterdam) 29:131–137.

Doorenbos, J., and W.O. Pruitt. 1977. Crop water requirements. FAO Irrig. Drain. Pap. 24 (revised). FAO, Rome.

Erez, A., and G.A. Couvillon. 1983. Evaporative cooling to improve rest breaking of nectarine buds by counteracting high daytime temperatures. HortScience 18:480–481.

Erie, L.J., and A.R. Dedrick. 1979. Level-basin irrigation: A method for conserving water and labor. Farmers' Bull. 2261:24.

Evans, R.G., and E.L. Proebsting. 1985. Response of red delicious apples to trickle irrigation. p. 231–239. In Proc. 3rd Int. Drip/Trickle Irrig. Congr. Vol. 1, Fresno, CA. November.

Feese, H., and G. Wilde. 1977. Factors affecting survival and reproduction of the Banks grass mite, Oligonychus practensis. Environ. Entomol. 6:53–56.

Fereres, E. 1984. Variability in adaptative mechanisms to water deficits in annual and perennial crop plants. Bull. Soc. Bot. Fr. Actual Bot. 131:17–32.

Fereres, E. 1987. Responses to water deficits in relation to breeding for drought resistance. p. 253–273. In J.P. Srivastava (ed.) Drought tolerance in winter cereals. John Wiley and Sons, New York.

Fereres, E., T.M. Aldrich, H. Schulbach, and D.A. Martinich. 1981. Responses of young almond trees to late-season drought. Calif. Agric. 35(7, 8):11–12.

Fereres, E., G. Cruz-Romero, G.J. Hoffman, and S.L. Rawlins. 1979. Recovery of orange trees following severe water stress. J. Appl. Ecol. 16:833–842.

Fereres, E., D.A. Martinich, T.M. Aldrich, J.R. Castel, E. Holzapfel, and H. Schulbach. 1982. Drip irrigation saves money in young almond orchards. Calif. Agric. 36:12–13.

Fereres, E., and I. Puech. 1981. Irrigation management program. Univ. Calif. Coop. Ext. Serv. and Calif. Dep. Water Resour. 1981. Irrigation scheduling guide. California Dep. Water Resour., Sacramento, CA.

Fereres, E., A.D. Reed, J.L. Meyer, F.K. Aljibury, and A.W. Marsh. 1978. Irrigation costs. Univ. Calif. Div. Agric. Sci. Leaflet 2875.

Feyen, J., D. Crabbé, N. Kihupi, and P. Michels. 1985. Irrigation timing through micro-computer controlled tensiometers. p. 773–781. In Proc. 3rd Int. Drip/Trickle Irrig. Congr. Vol. 2, Fresno, CA. November.

Fischer, B.B., D.A. Goldhamer, T. Babb, and R. Kjelgren. 1986. Weed control under drip and low-volume sprinkler irrigation. Calif. Agric. 39(11, 12):24–25.

Garnier, E., and A. Berger. 1985. Testing water potential in peach trees as an indicator of water stress. J. Hortic. Sci. 60:47–56.

Garnier, E., and A. Berger. 1986. Effect of water stress on stem diameter changes of peach trees growing in the field. J. Appl. Ecol. 23:193–209.

Gilbert, D.E., J.L. Meyer, J.J. Kissler, P.D. Latine, and C.V. Carlson. 1970. Evaporative cooling of vineyards. Calif. Agric. 24:12–14.

Gilreath, P.R., and D.W. Buchanan. 1981. Floral and vegetative bud development of "Sungold" and "Sunlite" nectarine as influenced by evaporative cooling by overhead sprinkling during rest. J. Am. Soc. Hortic. Sci. 106:321–324.

Goldhamer, D.A., and R. Cantrell. 1982. Evaporative spray losses from low volume sprinklers. Univ. Calif. Davis. Soil Water Newsl. 50:3–5.

Goldhamer, D.A., R.K. Kjelgren, R. Beede, L. Williams, J.M. Moore, J. Lane, G. Weinberger, and J. Menezes. 1985b. Water use requirements of pistachio trees and response to water stress. Calif. Pistachio Ind. Annu. Rep. 1984–85:85–92.

Goldhamer, D.A., R.K. Kjelgren, J.M. Moore, and J. Lane. 1985a. Low volume sprnkler surface and subsurface distribution uniformity. p. 851–858. *In* Proc. 3rd Int. Drip/Trickle Irrig. Congr. Vol. 2, Fresno, CA. November.

Goldhamer, D.A., B.C. Phene, R. Beede, L. Scherlin, J. Brazil, R.K. Kjelgren, and D. Rose. 1986. Water management studies of pistachio: I. Tree performance after two years of sustained deficit irrigation. Calif. Pistachio Ind. Annu. Rep. 1985–86:104–112.

Goldhamer, D.A., B.C. Phene, R. Beede, L. Scherlin, S. Mahan, and D. Rose. 1987. Effects of sustained deficit irrigation on pistachio tree performance. Calif. Pistachio Ind. Annu. Rep. 1986–87:61–66.

Goldhamer, D.A., R.S. Snyder (ed.). 1989. Irrigation scheduling. Univ. California, Berkeley. Leafl. 21454.

Goode, J.E., and K.H. Higgs. 1973. Water, osmotic and pressure potential relationships in apple leaves. J. Hortic. Sci. 48:202–215.

Goode, J.E., K.H. Higgs, and K.J. Hyrycz. 1978. Nitrogen and water effects on the nutrition, growth, crop yield and fruit quality of orchard-grown Cox's Orane Pippin apple trees. J. Hortic. Sci. 53:295–306.

Goode, J.E., K.H. Higgs, and K.J. Hyrycz. 1979. Effects of water stress control in apple trees by misting. J. Hortic. Sci. 54:1–11.

Hamer, P.J.C. 1983. Evaporative cooling of apple buds: The effect of timing of water application on bud development and frost resistance of the cv. Cox's Orange Pippin. J. Hortic. Sci. 58:153–159.

Hamer, P.J.C. 1986. The heat balance of apple buds and blossoms. Agric. For. Meteorol. 37:159–174.

Henckel, P.A. 1964. Physiology of plants under drought. Annu. Res. Plant Physiol. 15:363–386.

Henderson, D., and E. Fereres. 1981. Crop responses to drip irrigation. p. 6–9. *In* Drip irrigation management. Univ. California Div. Agric. Sci. Leafl. 21259.

Hollingsworth, C.S., and R.E. Berry. 1982. Twospotted spider mite (Acari: Tetranyohidae) in peppermint: Population dynamics and influence of cultural practices. Environ. Entomol. 11:1280–1284.

Hsiao, T.C., and E. Acevedo. 1974. Plant responses to water deficits, water use efficiency and drought resistance. Agric. Meteorol. 14:59–84.

Hsiao, T.C., E. Acevedo, E. Fereres, and D.W. Henderson. 1976. Water stress, growth and osmotic adjustment. Phil. Trans. R. Soc. Lond. B. 273:479–500.

Idso, S.B., R.D. Jackson, P.J. Pinter, R.J. Reginato, and J.L. Hatfield. 1981. Normalizing the stress-degree-day parameter for environmental variability. Agric. Meteorol. 24:45.

Jarrett, A.R., and C.T. Morrow. 1984. Distribution of frost protection water applied to apple trees during bud and leaf development. Trans. ASAE 27:89–92, 98.

Jarvis, P.G. 1981. Stomatal conductance, gaseous exchange and transpiration. p. 175–204. *In* J. Grace et al. (ed.) Plants and their atmospheric environment. Blackwells, Oxford.

Jarvis, P.G. 1985. Coupling of transpiration to the atmosphere in horticultural crops: The omega factor. Acta Hortic. 171:187–205.

Jarvis, P.G., and J.I.L. Morison. 1981. The control of transpiration and photosynthesis by the stomata. p. 247–279. *In* P.G. Jarvis and T.A. Mansfield (ed.) Stomatal physiology, Cambridge Univ. Press, England.

Jones, H.G., and I.G. Cumming. 1984. Variation of leaf conductance and leaf water potential in apple orchards. J. Hortic Sci. 59:329–336.

Jones, H.G., A.N. Lakso, and J.P. Syvertsen. 1985. Physiological control of water status in temperate and subtropical fruit trees. Hortic. Rev. 7:301–344.

Jones, H.G., M.T. Luton, K.H. Higgs, and P.J.C. Hamer. 1983. Experimental control of water status in an apple orchard. J. Hortic. Sci. 58:301–316.

Kattes, D.H., and G.L. Teetes. 1978. Selected factors influencing the abundance of Banks grass mite in sorghum. Texas Agric. Exp. Stn. Bull. 1186.

Klein, I. 1983. Drip irrigation based on soil matric potential conserves water in peach and grape. HortScience 18:942–944.

Klepper, B. 1968. Diurnal pattern of water potential in woody plants. Plant Physiol. 43:1931–1934.

Klepper, B., V.D. Browing, H.M. Taylor. 1971. Stem diameter in relation to plant water status. Plant Physiol. 48:683–685.

Lakso, A.N. 1979. Seasonal changes in stomata response to leaf water potential in apple. J. Am. Soc. Hortic. Sci. 104:58–60.

Landsberg, J.J., and H.G. Jones. 1981. Apple orchards. p. 419–469. *In* T.T. Kozlowski (ed.) Water deficits and plant growth. Vol. VI. Academic Press, New York.

Larson, K.D., T.M. DeJong, and R.S. Johnson. 1988. Physiological and growth responses of mature peach trees to postharvest water stress. J. Am. Soc. Hortic. Sci. 113:296–300.

Layne, R.E.C., and C.S. Tan. 1984. Long-term influence of irrigation and tree density on growth, survival and production of peach. J. Am. Soc. Hortic. Sci. 109:795–799.

Levin, I., R. Assaf, and B. Bravdo. 1972. Effect of irrigation treatment for apple trees on water uptake from different soil layers. J. Am. Hortic. Sci. 97:521–526.

Liss, H.J., B.A. Faber, and E. Fereres. 1984. A survey of irrigation scheduling services in five California countries. Calif. Dep. Water Resour. Final Rep. B54104. Sacramento.

Lombard, P.B., P.H. Westigard, and D. Carpenter. 1966. Overhead sprinkler system for environmental control and pesticide application in pear orchards. HortScience 1:95–96.

Lötter, J.V., D.J. Beukes, and H.W. Weber. 1985. Growth and quality of apples as affected by different irrigation treatments. J. Hortic. Sci. 60:181–192.

MacDonald, J.D., and J.M. Duniway. 1978. Temperature and water stress effects on sporangium viability and zoospore discharge in *Phytophthora cryptogea* and *P. megasperma*. Phytopathology 68:1449–1455.

Matheron, M.E., and S.M. Mircetich. 1985. Influence of flooding duration on development of Phytophthora root and crown rot of *Juglans hindsii* and Paradox walnut rootstocks. Phytopathology 75:973–976.

McNeil, S., and T.R.E. Southwood. 1978. The role of nitrogen in the development of insect/plant relationships. p. 77–79. *In* J. Harbone (ed.) Biochemical aspects of plant and animal co-evolution. Academic Press, London.

Mellors, W.K., A. Allegro, and A.N. Hsu. 1984. Effects of carbofuran and water stress on growth of soybean plants and twospotted mite (Acari: Tetrahychidae) populations under greenhouse conditions. Environ. Entomol. 13:561–567.

Merriam, J.L., and J. Keller. 1978. Farm irrigation system evaluation: A guide for management. Agric. Irrig. Eng. Dept. Utah State Univ., Logan.

Michailides, T.J., and J.M. Ogawa. 1986. Sources of inoculum, epidemiology, and control of Botryosphaeria shoot and panicle blight of pistachio. Calif. Pistachio Ind. Ann. Rep. 1985–86:87–91.

Middleton, J.E., E.L. Proebsting, and S. Roberts. 1979. Apple orchard irrigation by trickle and sprinkler. Trans. ASAE 22:582.

Mitchell, P.D., and D.J. Chalmers. 1982. The effect of reduced water supply on peach tree growth and yields. J. Am. Soc. Hortic. Sci. 107:853–856.

Mitchell, P.D., D.J. Chalmers, P.H. Jerie, and G. Burge. 1986. The use of initial withholding of irrigation and tree spacing to enhance the effect of regulated deficit irrigation on pear trees. J. Am. Soc. Hortic. Sci. 111:858–861.

Mitchell, P.D., P.H. Jerie, and D.J. Chalmers. 1984. The effect of regulated water deficits on pear tree growth, flowering, fruit growth and yield. J. Am. Soc. Hortic. Sci. 109:604–606.

Miyamoto, S. 1983. Consumptive water use of irrigated pecans. J. Am. Soc. Hortic. Sci. 108:676–681.

Miyamoto, S. 1984. A model for scheduling pecan irrigation with microcomputers. Trans. ASAE 27:456–463.

Monteith, J.L. 1981. Coupling of plants to the atmosphere. p. 1–29. *In* J. Grace et al. (ed.) Plants and their atmospheric environment. Blackwell Scientific Publ., Oxford.

Nakayama, F.S., and D.A. Bucks (ed.) 1986. Trickle irrigation for crop production. Design, operation and management. Development in agricultural engineering. No. 9. Elsevier, New York.

Natali, S., C. Xiloyannis, and M. Mugano. 1985. Water consumption in high density peach trees. Acta Hortic. 173:413–420.

Olsson, K.A., and F.L. Milthorpe. 1983. Diurnal and spatial variation in leaf water potential and leaf conductance of irrigated peach trees. Aust. J. Plant. Physiol. 10:291–298.

Peretz, J., R.G. Evans, and E.L. Proebsting. 1984. Leaf water potentials for management of high frequency irrigation on apples. Trans. ASAE 27:437–442.

Phene, B.C., D.A. Goldhamer, J. Menezes, R. Beede, G. Weinberger, and Z. Cervantes. 1987. Response of pistachio trees to three consecutive years of irrigation cut off. p. 67–70. Annu. Rep. Calif. Pistachio Ind.

Post, S.E.C., D.E. Peck, R.A. Brendler, N.J. Sakovich, and L. Waddle. 1986. Evaluation of low flow sprinklers. Calif. Agric. 40:27–29.

Proebsting, E.L., S.R. Drake, and R.G. Evans. 1984. Irrigation management, fruit quality and storage life of apples. J. Am. Soc. Hortic. Sci. 109:229–232.

Proebsting, E.L., and J.E. Middleton. 1980. The behavior of peach and pear trees under extreme drought stress. J. Am. Soc. Hortic. Sci. 105:380–385.

Proebsting, E.L., J.E. Middleton, and M.O. Mahan. 1981. Performance of bearing cherry and prune trees under very low irrigation rates. J. Am. Soc. Hortic. Sci. 106:243–246.

Ramos, D.E., L.C. Brown, K. Uriu, and B. Marangoni. 1978. Water stress affects size and quality of walnuts. Calif. Agric. 32:5–6.

Raney, C.D. 1973. Cultural control. p. 98–104. *In* C.D. Raney (ed.) Verticillium wilt of cotton. USDA Publ. ARS-S-14.

Rawlins, S.L. 1981. Other methods of localized irrigation. p. 33–35. *In* E. Fereres (ed.) Drip irrigation management. Univ. Calif., Leafl. 21259.

Renquist, R. 1987. Evapotranspiration calcultions for young peach trees and growth responses to irrigation amount and frequency. HortScience 22:221–223.

Ritchie, G.A., and T.M. Hinkley. 1975. The pressure chamber as an instrument for ecological research. p. 165–254. *In* J.B. Cragg (ed.) Advances in ecological research. Vol. 12. Academic Press, London.

Salter, P.J., and J.E. Goode. 1967. Crop responses to water at different stages of growth. Commonw. Agric. Bur. Res. Rev. No. 2. East Malling, England.

Schneider, H. 1948. Susceptibility of quayle to Verticillium wilt and influence of soil temperature and moisture on development on infection. J. Agric. Res. 76:129–143.

Schulze, E.D. 1982. Plant life forms and their carbon, water and nutrient relations. p. 615–676. *In* O.L. Lange et al. (ed.) Encyclopedia of plant physiology. New series, Vol. 12B.

Schulze, E.D. 1986. Carbon dioxide and water vapor exchange in response to drought in the atmosphere and in the soil. Annu. Rev. Plant Physiol. 37:247–274.

Schulze, E.D., and A.E. Hall. 1982. Stomatal responses, water loss and CO_2 assimilation rates of plants in contrasting environments. p. 181–230. *In* O.L. Lange et al. (ed.) Encyclopedia of plant physiology. New series, Vol. 12B.

Schulze, E.D., O.L. Lange, U. Buschbom, L. Kappen, and M. Evenari. 1972. Stomatal responses to changes in humidity in plants growing in the desert. Planta 108:259–270.

Scribner, J.M. 1978. The effects of larval feeding specialization and plant growth form upon the consumption and utilization of plant biomass and nitrogen: An ecological consideration. Entomol. Exp. Appl. 24:694–710.

Sharples, R.A., D.E. Rolston, J.W. Biggar, and H.I. Nightengale. 1985. Evapotranspiration and soil water balances of young trickle-irrigated almond trees. p. 792–797. *In* Proc. 3rd Int. Drip/Trickle Irrig. Congr., Vol. 2, Fresno, CA. November.

Sinclair, T.R., and M.M. Ludlow. 1985. Who taught plants thermodynamics? The unfulfilled potential of plant water potential. Aust. J. Plant Physiol. 12:213–217.

Smart, R.E., and H.D. Barrs. 1973. The effect of environmental and irrigation interval on leaf water potential of four horticultural species. Agric. Meteorol. 12:332–346.

Smith, M., and E. Fereres. 1988. Irrigation programming. FAO Irrig. Drain. Pap. FAO, Rome. (In press.)

Snyder, R.L., W.O. Pruitt, and D.A. Shaw. 1987. Determining daily reference evapotranspiration (ET_0). Univ. California Coop. Ext. Div. Agric. Nat. Resour. Leafl. 21426.

Specht, H.B. 1965. Effect of water-stress on the reproduction of European red mite *Panonychus ulmi* (Koch) on young apple trees. Can. Entomol. 97:82–85.

Sterne, R.E., G.A. Zentmyer, and M.R. Kaufman. 1977. The influence of matric potential, soil texture, and soil amendment on root disease caused by *Phytophthora cinnamomi*. Phytopathology 67:1495–1500.

Stolzy, L.H., J. Letey, L.J. Klotz, and C.K. Labanouskas. 1965. Water and aeration as factors in root decay of *Citrus sinensis*. Phytopathology 55:270–275.

Syvertsen, J.P. 1985. Integration of water stress in fruit trees. HortScience 20:1039–1043.

Tan, C.S., and B.R. Buttery. 1982. The effect of soil moisture stress to various fractions of the root system on transpiration, photosynthesis, and internal water relations of peach seedlings. J. Am. Soc. Hortic. Sci. 107:845–849.

Taylor, H.M. 1983. Managing root systems for efficient water use: An overview. p. 87–113. *In* H.M. Taylor et al. (ed.) Limitations to efficient water use in crop production. ASA, CSSA, and SSSA, Madison, WI.

Teviotdale, B.L., and G.S. Sibbett. 1982. Midwinter irrigation can reduce deep bark canker of walnuts. Calif. Agric. 36(5, 6):6–7.

Turner, N.C. 1986. Crop water deficits: A decade of progress. Adv. Agron. 59:1–51.

Unrath, C.R., and R.E. Sneed. 1974. Evaporative cooling on "Delicious" apples—the economic feasibility of reducing environmental heat stress. J. Am. Soc. Hortic. Sci. 99:372–375.

Uriu, K., and J.R. Magness. 1967. Deciduous tree fruit and nuts. *In* R.M. Hagan et al. (ed.) Irrigation of agricultural lands. Agronomy 11:686–703.

van de Urie, M., J.A. McMutry, and C.B. Huffaker. 1972. Ecology of tetranychid mites and their natural enemies: A review. III. Biology, ecology, and pest status, and host-plant relations of tetranychids. Hilgardia 41:343–432.

Veihmeyer, F.J. 1927. Some factors affecting the irrigation requirements of deciduous orchards. Hilgardia 2:125–284.

Veihmeyer, F.J. 1972. The availability of soil moisture to plants: Results of empirical experiments with fruit trees. Soil Sci. 114:268–294.

Veihmeyer, F.J. 1975. The growth of fruit trees in response to different soil moisture conditions measured by widths of annual rings, and other means. Soil Sci. 119:448–460.

Veihmeyer, F.J., and A.H. Hendrickson. 1949. Methods of measuring field capacity and permanent wilting percentage of soil. Soil Sci. 68:75–94.

Vermeiren, L., and G.A. Jobling. 1983. Localized irrigation. FAO Irrig. Drain. Pap. 36. FAO, Rome.

Wargo, P.M., and M.E. Montgomery. 1983. Colonization by Armillarea mellea and Agrilus bilineatus of oaks injected with ethanol. For. Sci. 29:848–857.

Wiersum, L.K., and K. Harmany. 1983. Changes in the water permeability of roots of some trees during drought stress and recovery, as related to problems of growth in an urban environment. Plant Soil 75:443–448.

Wilcox, W.F., and S.M. Mircetich. 1985. Effects of flooding duration on the development of Phytophthora root and crown rots of cherry. Phytopathology 75:1451–1455.

Wilhelm, S., and J.B. Taylor. 1965. Control of verticillium wilt of olive through natural recovery and resistance. Phytopathology 55:310–316.

Worthington, J.W., M.J. McFarland, and P. Radrigue. 1984. Water requirements of peach as recorded by weighing lysimeters. HortScience 19:90–91.

Youngman, R.R., and M.M. Barnes. 1986. Interaction of spider mites (Acari: Tetranychidae) and water stress on gas-exchange rates and water potential of almond leaves. Environ. Entomol. 15:594–600.

Youngman, R.R., J.P. Sanderson, and M.M. Barnes. 1988. Life history parameters of *Tetranychus pacificus* McGregor (Acari: Tetranychidae) on almonds under differential water stress. Environ. Entomol. 17:488–495.

34 Grapevine

L. E. WILLIAMS AND M. A. MATTHEWS

University of California
Davis, California

More than 10 million ha of grapevines are grown worldwide, ranging from 50 °N lat, through the tropics, to 43 °S lat and on all continents except Antarctica. Accordingly, the climates in which grapes are grown are quite varied, with freezing temperatures limiting the northernmost range for different *Vitis* spp. The production of grapes in the Southern Hemisphere is determined by where fruit will reach maturity. The wide distribution of grape production may be possible due to the large genetic diversity of available species and cultivars and to a low, if any, chilling requirement for release from dormancy of the vine's buds. In general, the grapevine needs a long growing season to mature its fruit, although the actual length for a given cultivar varies. The importance of climate to fruit quality also dictates that cultivars be chosen with care for a given site.

Grapes primarily are used for the fresh (table) grape market, raisins (dried grapes), fruit juice, and wine production. The species *V. vinifera* L. produces over 90% of the world's grapes (Weaver, 1976) and 99.9% of the world's wine (Robinson, 1986). Approximately 60% of the annual world production of >63 Tg of grapes is produced in Europe (FAO, 1985). Sixty four percent of the world's wine (342 million hL) is produced by Italy, France, the USSR, and Spain (FAO, 1986). Spain leads the world in acreage devoted to vineyards. However, because of low yields it ranks fourth behind France, Italy, and the USSR in wine production. Spain, the USSR, Italy, and France all have >1 million ha planted to grapevines. The top 20 wine grape cultivars planted worldwide on an area basis are dominated by several cultivars planted in Spain (Robinson, 1986). Airen and Granacha Tinta (Grenache) are the number one and two cultivars, respectively, in terms of acreage, with most of this found in Spain. The third most widely planted variety is Rkatsiteli, which is found mainly in the USSR. The wine grape cultivar more familiar to North Americans, Cabernet Sauvignon, is seventh on the list of wine grape varieties with 135 000 ha distributed evenly throughout most of the wine-producing countries. The list does not include cultivars used primarily for table grape or raisin production. Thus, the cultivar Thompson Seedless (Sultanina), which leads in acreage (112 830 ha) for grape production in California (Tippett et al., 1986), is not included in the top 20 wine grape variety list (Robinson, 1986).

The grapevine is a woody perennial generally propagated via vegetative means, i.e., cuttings, rootings, and grafting. Once a vineyard is planted the vines are considered mature after 4 to 5 yr of growth. Commercial production of grapes in the USA and in many areas of the world commonly uses some form of permanent support (trellis) although older vineyards may have free-standing vines. The kind of trellis used is dependent upon the variety, type of pruning, final use of the grape, method of harvesting, and other cultural and environmental factors. Trellis systems may range from a single wooden stake at each vine, with a single wire atop or along the stake running the length of the row, to an overhead arbor in which the shoots of the vine are trained to form a horizontal layer of foliage.

The cycle of vine growth starts with budbreak in the spring and is followed by rapid shoot elongation from a previously dormant compound bud. The time between budbreak and bloom generally is about 8 wk (Winkler et al., 1974), differing somewhat from many temperate deciduous fruit trees, which flower early in the spring before much vegetative growth has occurred (Kramer & Kozloski, 1979; Westwood, 1978). After harvest the vine retains its leaves and may continue its vegetative growth if conditions are favorable. In temperate climates extensive periderm develops, the leaves abscise, and the vine becomes dormant. During the dormant period the vine will be pruned to regulate next year's crop load. Last season's shoots, now leafless and woody and called *canes*, are the structures that bear the buds which give rise to next year's crop. The basal buds on some cultivars are fruitful so that the fruiting canes can be pruned to two-node spurs. Other cultivars will be pruned to canes that are 12 to 15 nodes in length because of low basal bud fruitfulness, as in Thompson Seedless, or to the small size of the clusters that ultimately form, as in Cabernet Sauvignon. During the next growing season the whole process is repeated.

This chapter will review grape production and how irrigation management may affect vegetative growth of the vine and, more importantly, fruit quality. A review of the water relations of grapevines recently has been published (Smart & Coombe, 1983). Therefore, this review will concentrate on information that has been published since that time and on water management practices as used in the irrigated grape-growing regions of the western and southwestern USA.

I. VINE GROWTH AND DEVELOPMENT

A. Phenology

The development of the grapevine may be influenced by many factors to include environmental variability, management practices, grafting differences, and pest problems. There have been numerous studies, though, to quantify the effect of climatic factors on vine development. McIntyre et al. (1982) demonstrated that individual grape cultivars tended to develop at consistent rates relative to other cultivars, regardless of seasonal conditions. Each stage of the vine's development was located in regular positions within the whole population of each cultivar examined in that study. A low amount of chilling may be needed in breaking organic dormancy for many European wine grape cultivars (Antcliff & May, 1961). In temperate climates, budbreak generally occurs when the daily mean maximum temperature exceeds 10 °C. Models to predict this event have used temperature summation following the period of vine dormancy to estimate the time of budbreak (Baldwin, 1966; Williams et al., 1985b). Degree day summations also have been shown to accurately predict the time between budbreak and bloom for grapevines (Christensen, 1969; Williams et al., 1985b; Williams, 1987c). However, McIntyre et al. (1982) found that cumulative maximum temperature between budbreak and bloom was the best measure of determining the duration between these two phenological stages. Fruit maturation of grapevines, while closely correlated with degree day summations in several studies, may be more dependent upon cultural practices than date of either budbreak or bloom. Such factors as crop load and water availability may be the more important determinants of fruit development once fertilization takes place (Morris & Cawthon, 1982; Williams & Grimes, 1987; Winkler et al., 1974).

B. Aerial Growth

1. Vegetative

The growth of the vegetative parts of the vine generally has been described as being sigmodial when time was measured as calendar days (Alexander, 1958; Coombe, 1960; De La Harpe & Visser, 1985). The increment in stem length throughout the season has been shown to be curvilinear when plotted as a function of calendar days (De La Harpe & Visser, 1985) or growing degree days (GDD's) (Williams, 1987a). The rate of shoot elongation is greatest early in the growing season and then steadily decreases (Fig. 34-1).

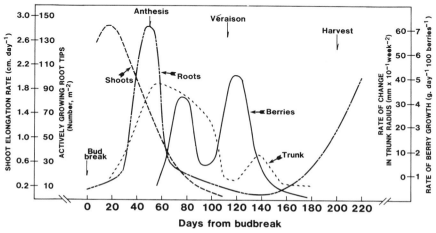

Fig. 34-1. Growth rate of various organs of Colombar grapevines grown in South Africa throughout the season. After Van Zyl (1984).

The accumulation of both leaf and stem dry matter is linear from shortly after budbreak until the start of fruit growth when expressed as a function of GDD's under California growing conditions (Gutierrez et al., 1985; Williams et al., 1985a; Williams, 1987a). In all three studies, the partitioning of dry matter to the vegetative structures of the vine was linear until approximately 1000 GDD's ($> 10\,°C$) after budbreak (Fig. 34-2). It was at this time that the rate of fruit growth was rapidly increasing. While fruit command most of the photosynthate from berry set to harvest, significant amounts of C are still partitioned to the leaves and stems at this time. The increase in the vine's leaf weight is due to a linear increase in weight per unit leaf area throughout the season for the cultivar Thompson Seedless (Williams, 1987a). Similar results were found for Chenin blanc, a seeded wine grape cultivar, whether using mature field-grown vines (L.E. Williams, 1986, unpublished data) or young potted vines (Conradie, 1980). The increase in stem dry weight per unit length is curvilinear, with the greatest increase in dry weight occurring early in the season, followed by a smaller but steady increase thereafter.

The development of the vine's canopy with regards to leaf area is similar to the accumulation of leaf dry weight throughout the season for mature field-grown vines (Fig. 34-2). Leaf area development was shown to be linearly related to GDD's from budbreak to the period when fruit growth rapidly increased (Williams, 1987a). Once full canopy is obtained, 33 to 83% of the total leaf area has been shown to be on the outside of the canopy depending upon the configuration of the trellis system (Smart et al., 1985a; Williams,

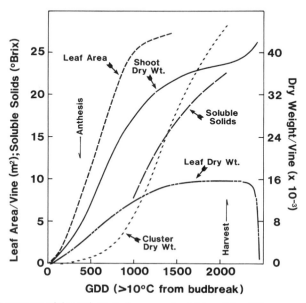

Fig. 34-2. Increment of dry weight, leaf area, and soluble solids of Thompson Seedless grapevines grown in the San Joaquin Valley of California as a function of growing degree days (GDD) $> 10\,°C$. Average date of budbreak was 9 March. After Williams (1987a).

1987a). While the total number of leaves on lateral stems was shown to be approximately 80% that of leaves on the primary stem, area of lateral leaves accounted for only 25% of the total leaf area of the vine at maximum canopy for Thompson Seedless vines (Williams, 1987a). With a different variety and using no irrigation, Smart et al. (1985a) found lateral leaf area to account for approximately 6 to 9% of the total vine area. In both studies, individual lateral leaves were smaller than individual leaves on the primary stem.

The growth of the vine's trunk in a field situation has received little attention. van Zyl (1984) found that the rate of increase in the trunk's diameter increases to a maximum at anthesis and then declines throughout the remainder of the season (Fig. 34–1). Studies of potted vines indicate that the permanent aerial structures of the vine increase in dry matter throughout the season. The trunk dry weight of Chenin blanc vines more than doubled from budbreak to just after harvest (Conradie, 1980). A study in the field on 2-yr-old vines showed that the trunk tripled in weight during the same period as reported by Conradie (Araujo & Williams, 1988). Between 10 and 30% of the total ^{14}C assimilated by young vines is translocated to the trunk of the vine depending upon the time of the year the labeled CO_2 is fed to the vines (Yang & Hori, 1979).

2. Reproductive

As with most perennial crops, floral differentiation for the current year's crop takes place the preceding year. The differentiation of fruit buds begins early in the growing season the year preceeding the development of flowers (Winkler et al., 1974). The differentiation of floral primordia begins first at the base of the shoot in buds that are farthest along in their development, and proceeds on up the shoot as the season progresses. The rudimentary clusters that are formed develop until late summer, when they enter into a dormant period with no further development taking place. Numerous factors contribute to the extent of fruit bud differentiation during any one season. The formation of fruit buds on vines in growth chambers has been shown to be dependent upon temperature, irradiance level, and the duration of high-intensity radiation during the time that differentiation takes place (Buttrose, 1969a, b, c). Field studies indicate good correlations between the number of fruitful buds and the number of hours of bright solar radiation and mean daily temperatures during the time fruit primordia are initiated (Baldwin, 1964). It appears that photoperiod does not play an important role in regulating floral differentiation (Buttrose, 1970). The effect of light on fruit bud differentiation is one of a direct response rather than an indirect one mediated through photosynthesis and carbohydrate accumulation (May, 1965; Sartorius, 1968). The mode of action of these environmental factors with respect to this differentiation has yet to be determined.

The development of the cluster subsequent to budbreak has been shown to be closely linked to temperature summations (Buttrose & Hale, 1973; McIntyre et al., 1982; Williams et al., 1985a; Williams, 1987a). This seems reasonable as cluster development and shoot growth are closely linked (Pratt &

Coombe, 1978) and shoot growth and dry matter accumulation are linearly related to GDD's early in the growing season (Gutierrez et al., 1985; Williams et al., 1985a; Williams, 1987a).

The flowers of most *V. vinifera* vines are perfect. There is a question as to whether grapevines are self- or cross-pollinated but most of the evidence indicates that self-pollination predominates with *V. vinifera* grapes (Winkler et al., 1974). The growth of berries takes place after fertilization, in conjunction with the development of the seed. Berries of several grape cultivars set via *stenospermocarpy*, that is, pollination and fertilization followed by embryo abortion. Many raisin and table grape cultivars, such as Thompson Seedless, Perlette, and Flame Seedless, develop in this manner. The cultivar Black Corinth develops by stimulative *parthenocarpy* in which there is no ovule development. It appears that the stimulus of pollination is satisfactory to set the fruit (Olmo, 1936). High temperatures during this period of fruit development may reduce set in grapes (Kobayashi et al., 1960; Buttrose & Hale, 1973). Winkler (1958) also demonstrated that the amount of leaf area present at time of fertilization can significantly influence the final number of berries per vine.

Once set has occurred, the grape berry rapidly increases in size and weight. The growth curve characteristic of grape berries is a double sigmoid type in which growth occurs in three stages (Fig. 34-3). This type of growth also is characteristic of stone fruits such as cherry (*Prunus ovium* L.), peach

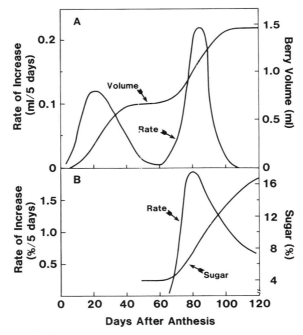

Fig. 34-3. (A) Double sigmoid growth curve and rate of berry increase for Seedless Emperor grapevine throughout reproductive development. (B) Accumulation of sugar and its rate of increase during fruit growth. After Coombe (1960).

[*Prunus persica* (L.) Batsch], and plum (*Prunus domestica* L.) (Kramer & Kozoloski, 1979). Stage I is characterized by the enlargement of the pericarp because of both cell division and enlargement (Harris et al., 1968; Coombe, 1960). The seed also increases in size at this time. During Stage II the rate of pericarp growth slows but the embryo continues to develop rapidly and reaches its maximum size (Winkler et al., 1974). Expansive growth of the pericarp resumes in Stage III. Growth in Stage III may be equal to or exceed that ocurring during Stage I. The relative growth which occurs in Stages I and III and the duration of Stage II is dependent upon cultivar. The double sigmoid pattern in fresh weight or volume occurs in both seeded and seedless berries (Coombe, 1960). However, the increase in berry dry weight has been reported as a double (Cawthon & Morris, 1982; Considine & Knox, 1979; Matthews et al., 1987b) and a single (De La Harpe & Visser, 1985) sigmoid. The increase in cluster and berry dry weight also has been shown to be a linear function of both calendar and GDD's (Gutierrez et al., 1985; Williams et al., 1985a; Alexander, 1958; Williams, 1987a).

There is a dramatic change in the development of the grape berry at the end of Stage II which is termed *veraison*. Veraison usually has been associated with the change in color of the berry at this time; however, it has been convenient to use it to represent the time when several other physiological events occur simultaneously. As mentioned earlier, growth of the berry accelerates rapidly during Stage III, with the majority of the increase in weight due to water accumulation (Coombe, 1960). The berry softens at this time, in addition to an accumulation of fructose and glucose in all cultivars (Fig. 34–3) and anthocyanins in red and black cultivars. There is a decrease in the acid concentration of the berry and a concommitant increase in pH of the expressed sap. The control of ripening in grape is not fully understood but there is preliminary evidence of differential gene expression at veraison (M.A. Matthews, 1988, unpublished data). The in vitro activity of several enzymes is altered, especially those associated with sugar metabolism.

The physiological changes that take place starting with veraison continue as the berry matures until such time that the fruit is judged to be harvestable. As the berry matures it becomes sweet, due to the large increase in the amount of fructose relative to glucose and the fact that acid levels decline. The criteria used to determine the acceptability of fruit for harvest is dependent upon the ultimate use of the crop, e.g., fresh or dried fruit or for juice to be fermented to make wine. For dried grapes, sugar is the main criteria used to determine when to harvest. However, other factors such as the method of drying or availability of labor may dictate harvest date. The production of natural raisins in California requires that the grapes be sun dried, thus the probability of avoiding fall rains decreases the later the grapes are harvested. The amount of sugar and/or the sugar/acid ratio is used to determine when to harvest grapes used for the fresh fruit market. Color development also is used as a criteria for red- and black-fruited table grapes. Lastly, there are legal requirements regulating the proper picking conditions for table grapes, such as minimum sugar levels or sugar/acid ratio. While sugar or the sugar/acid ratio may be used to determine time of harvest for wine grapes, this becomes inadequate when other factors contribute to the quality of the wine. Acid composition and pH of the juice have been shown to play important roles in the sensory properties of wine, microbial stability, extent of malo-lactic fermentation, and several other chemical properties associated with the wine's composition (Boulton, 1980a).

The potential yield of grapevines is determined in large part by the pruning pattern established during the dormant period of the vine's annual life cycle. However, the capacity of the vine to mature a given crop is dependent upon the vine's canopy and the environment. For example, May et al. (1973) demonstrated that doubling the number of nodes per vine, by altering pruning pattern, resulted in an increase of about 30% in cluster number and only 20% in fruit yield per vine. Williams (1987a) found that increasing the shoot number 70% from one year to another resulted in a 70% increase in cluster number but no significant increase in fruit weight per vine. Thus, while the number of floral primordia already are determined at the beginning of the season, conditions that affect the number of flowers that set—and later berry size—also have a significant effect on final yield.

C. Root Growth

Most field studies on vine root growth have been limited to observations on the periodicity of new root initiation and turnover (Hiroyasu, 1961; Freeman & Smart, 1976; McKenry, 1984; van Zyl, 1984). Data from those studies indicate that the peak of root initiation (root flushes) occurs after shoot growth begins to slow down and before rapid berry growth commences (Fig. 34–1). A second flush occurs after fruit harvest, but it is of a smaller magnitude than that occurring earlier in the season. From these data it would appear that aerial and root growth alternates such that root growth only occurs when other sinks on the vine are not using large amounts of photosynthate. However, the amount of weight accumulated in the roots involved in the flushes is only a minor fraction of the total weight of the entire root system (McKenry, 1984). When the total root dry weight was measured throughout the season on 2-yr-old vines, the accumulation of dry weight commenced shortly after budbreak and continued at the same rate until fruit harvest, when the experiment was terminated (Araujo & Williams, 1988). Thus, the importance of the root flushes to the overall growth of the vine and their contribution to the uptake of water and mineral nutrients remains to be determined.

The distribution of grapevine roots is dependent upon physical and chemical soil characteristics including soil strength, compaction, and impervious layers. The rooting habit also is influenced by the type of irrigation system used, planting density, and the rooting characteristics of the individual cultivar. Most studies have demonstrated that the majority of the vine's roots are found in the top meter of soil (Barnard, 1932; McKenry, 1984; Perry et al., 1983; Saayman & van Huyssteen, 1980; van Zyl & van Huyssteen, 1980; van Zyl & Weber, 1981). However, grape roots also have been found at depths of 2 to 6 m (Seguin, 1972; Richards, 1983). It was demonstrated that the type of soil preparation had little effect on total vine root mass for Chenin blanc vines grafted on '101-14 Mgt.' rootstock grown on a Hutton/Sterkspruit soil (fine, montmorillonitic, frigid Cumulic Haplaquolls/unknown) (Saayman & van Huyssteen, 1980). McKenry (1984) found that the distribution of roots between rows was affected by compaction because of the passage of machinery down the row with little root mass found in the areas corresponding to the positions of the wheels. However, Nagarajah (1987) did not find a marked reduction of root growth in the traffic row. Planting density was shown to affect both root size and distribution within the soil profile (Archer & Strauss, 1985). As the number of vines increased per unit land area, the root system became smaller; however, root density increased. In addition, the angle of root penetration became greater as the planting density increased.

II. VINEYARD EVAPOTRANSPIRATION

The use of water in a vineyard, as with other crops, is dependent upon the interaction of numerous physical atmospheric and edaphic factors with the physiological processes of the plant. The loss of water from a single leaf is dependent upon temperature, leaf/air vapor pressure difference, age, soil water status, position within the canopy, and other factors too numerous to mention. Thus the transpiration rate of an individual leaf may differ hourly, daily, or seasonally. The extrapolation of individual leaf water loss to the entire vine only would be possible if one knew the position of each leaf within the canopy.

The use of water by vines throughout the growing season is characterized by lowered use in the early (spring) and late (fall) parts of the season when compared to that during the middle (summer) of the season (Hardie & Considine, 1976; Freeman & Smart, 1976; Prior & Grieve, 1987; Van Rooyen et al., 1980; Williams & Grimes, 1987). The development of the vine's canopy during the early part of the growing season and leaf fall during the late part of the season (Williams, 1987a; Williams et al., 1987a) and decreased evaporative demand early and late in the season may account for the lowered water use characteristics of vines at those times. A slow increase in vine water use compared to Class A pan evaporation until full canopy development was noted by Stanhill (1962). The vine's trellis system may affect water use of the vineyard. van Zyl and van Huyssteen (1980) found that vineyard evapotranspiration (ET) of bush-trained vines was greater than that in which the trellis system spread the canopy. The higher ET of the bush vines was attributed to greater ambient air temperatures, more air movement, and less shading of the soil surface when compared to the other treatments.

Vineyard ET in the semiarid southwestern USA is approximately 500 mm for table grape cultivars during the period budbreak to harvest (Table 34–1). Full-season vineyard ET in Arizona and central California ranges from 660 to 800 mm. It also has been shown that ET of Sultana (Thompson Seedless) vines in Australia grown under climatic conditions similar to those in

Table 34-1. Vineyard evapotranspiration (ET) in California and Arizona for several table and raisin grape cultivars.

Cultivar	Location	Year data collected	Evapo-ration	ET_0	ET	Data source
				mm		
Thompson Seedless & Cardinal	Mesa, AZ	1961–64	--	--	498†	Erie et al., 1981
Perlette	Phoenix, AZ	1973–77	--	--	712	Bucks et al., 1985
Thompson Seedless	Arvin, CA	1967–69	1443‡	1179§	660	State of California, 1975
Thompson Seedless	Delano, CA	1984	1446	1128	685	MacGillivray et al., 1985
Thompson Seedless	Parlier, CA	1985	1565‡	1231¶	799	Williams & Grimes, 1987

† Vineyard ET for this study represents water use from budbreak to fruit harvest; the remaining values in this column are for the entire growing season.
‡ These values were obtained from averages of Class A pan evaporation measured over a period of 3 and 11 yr in Bakersfield (for Arvin) and Fresno (for Parlier), respectively.
§ The average ET of a grass pasture near Arvin (State of California, 1975) was used as the value for ET_0 at this location.
¶ Potential ET was obtained from the California Irrigation Management Information System (CIMIS). ET_0 was calculated with information from a weather station at the Parlier site.

the San Joaquin Valley of California is between 700 and 800 mm when ET_0 is approximately 1250 mm (Prior & Grieve, 1987). The annual amounts of precipitation for the Fresno and Bakersfield areas in California are 250 and 125 mm, respectively. This would indicate that seasonal vineyard irrigation requirements in this area to be approximately 550 mm. Irrigation deliveries measured by the California Department of Water Resources during the period from 1973 to 1983, in the San Joaquin and Tulare Lake Hydrologic Regions of California, indicated that an average of 853 mm of water was used to irrigate vineyards in those regions (State of California, 1986). This average was obtained from the irrigation of approximately 23 000 ha of vineyards and 227 individual measurements. The least and greatest amounts of delivered irrigation water for a particular vineyard were 650 and 1650 mm/season, respectively. This information indicates that vineyards in the San Joaquin Valley of California may be receiving 1.6 times, on the average, to greater than 3.0 times the amount of irrigation water actually required by the vines.

III. VINE GROWTH AND WATER STATUS

A. Diurnal and Seasonal Patterns of Vine Water Status

Many viticultural regions have Mediterranean climates in which rainfall occurs primarily during the dormant season, with subsequent vine growth being dependent upon stored soil water (including parts of Portugal, Spain,

France, South Africa, Australia, and California). Therefore, in most of the above regions, soil water content is high at budbreak from winter rains. The perennial nature of most grape cultivation systems provides a significant root mass available for a relatively small transpirational surface area for water uptake early in the season. Thus, early in the growing season, vine water status generally does not restrict growth.

High predawn water potentials also may be due partially to the significant root pressures created in grapevines. Scholander et al. (1965) reported root pressures of 0.4 MPa in native *Vitis*, and these values may be greater for cultivated vines. Root pressure develops before budbreak and often results in sap exudate from cut surfaces and later from hydathodes at leaf margins. Although root pressure may contribute to the capacity for water transport to transpiring shoots, the extent to which significant root pressures persist diurnally and seasonally has not been established. It is interesting that these root pressures are created with suberized roots and occur despite the lack of active root growth (Scholander et al., 1965). Vine root pressure also may be important for the recovery of shoot water status at night and in refilling cavitated vessels.

Grapevines undergo relatively large diurnal fluctuations in leaf water potential (ψ_l). Under well-watered conditions, leaf water potential declines during the day to values of approximately -0.5 MPa as evaporative demand increases, and recovers rapidly in late afternoon to near predawn values. Midday leaf water potential becomes more negative as the season progresses, even under well-watered conditions. As soil water is depleted, the absolute change in diurnal leaf water potential may approach 1.2 MPa. The rate of change of leaf water potential has been correlated with solar radiation (Smart & Barrs, 1973). By taking into account diurnal variation in radiation, temperature, and leaf-to-air vapor difference, 75 to 96% of the variation in leaf water potential could be explained in a linear model. If only the morning decline in water potential was considered, radiation alone accounted for greater than 75% of the decline in water potential. It was suggested that a series of these relationships, developed at various soil water contents, could be used to schedule irrigation (Smart & Barrs, 1973).

Water deficits can develop early in the season for nonirrigated vines despite dormant season rainfall which may fully recharge the soil profile. It was shown that all of the plant-available soil moisture was depleted by veraison in a South African vineyard that initially contained approximately 85 mm of available soil water (van Zyl & Weber, 1981). Midday leaf water potential declines early in the season regardless of whether vines are maintained well watered by daily or weekly irrigation (Matthews et al., 1987a; Smart & Coombe, 1983). Midday leaf water potential declines from initial measurements of approximately -0.4 MPa before anthesis to -1.0 to -1.2 at veraison. Leaf water potential is relatively stable after veraison (Williams and Grimes, 1987).

The early season decline in vine water status from initial measurements may indicate osmotic adjustment (During, 1984) or that transpiration exceeded the capacity of the root system to supply water to the transpiring leaves despite a high soil water content. The latter possibility is supported by the large diurnal fluctuations of ψ_l in vineyards in New York (Liu et al., 1978), South Africa (van Zyl & Weber, 1981), and California (Matthews et al., 1987a; Grimes & Williams, 1990). In addition, the inhibition of shoot elongation by soil water deficits was reversed by spraying the leaves with water (During & Broquedis, 1980). These observations indicate a significant plant resistance to water transport. It is reasonable, therefore, to assume that water deficits in shoots are a frequent component of grape production in many areas.

The supply of irrigation water is insufficient to meet seasonal vineyard requirements in some wine grape production areas. It is clear that adequate water is essential in the establishment of young vines with limited root systems. Water deficits during vineyard establishment can delay vine maturation by one or more seasons (Kasimatis, 1967). For mature vines, water deficits affect both present and subsequent season's growth. The question of when to apply limited water has been addressed by withholding water at different developmental stages. Matthews et al. (1987a) reported that when water was withheld until veraison, midday ψ_l of vines was more negative than the irrigated vines by 0.3 MPa but recovered to a similar or higher ψ_l than the irrigated vines after water was resupplied. However, the recovery required in excess of 2 wk. The cause of the slow recovery is not clear but may have been due to either inactive roots in the upper (previously dry) soil layers, or diminished function of root and stem xylem created by cavitation (Byrne et al., 1977). When water was withheld after veraison, midday ψ_l declined rapidly and was approximately -1.6 MPa at harvest (Matthews et al., 1987a).

B. Symptoms of Water Deficits

Symptoms of water deficits in vineyards are dependent upon the stage of vine growth and rate at which the deficits develop (Kasimatis, 1967). Early symptoms include a decrease in the angle formed by the axis of the leaf petiole and the plane of the lamina (Smart, 1974). When deficits persist, internode growth is inhibited, the relative effect being greater for the older internodes that are still elongating. As water deficits progress, necrosis of the shoot apex and of tendrils is commonly observed. These losses of tissue can occur at water potentials of -1.0 to -1.2 MPa (Smart, 1974). Nevertheless, tissue necrosis only occurs after the soil water content is well below that required for adequate water uptake and cannot be relied upon to indicate soil water availability. Internode length is shortened by water deficits, as is the duration of shoot growth when these deficits occur early in the growing season (Matthews et al., 1987a). In extreme situations and during late-season stress, leaf abscission is induced, originating with the most mature leaves and progressing apically. Susceptibility to stress-induced abscission may increase after veraison. Water potentials more negative than -1.3 MPa generally are required before leaf drop occurs.

Grape leaves are hypostomatous. Cuticular conductance of leaves is about 0.01 cm/s (Kriedemann & Smart, 1971). Under nonlimiting conditions leaf conductance to water vapor in full sunlight is approximately 1.0 cm/s. Conductance declines with developing water deficits in two respects. Absolute conductance in full sunlight decreases to a minimum of approximately 0.2 cm/s at a ψ_l of -1.6 MPa. In addition, stomata of water-stressed vines remain closed throughout more of a 24-h period than for well-watered vines. Consequently, midday leaf water potentials have seldom been observed more negative than approximately -1.6 MPa. Grape berries have few stomata or lenticels, giving rise to a low conductance to water vapor—approximately 0.02 cm/s (Considine & Kriedemann, 1972; Nobel, 1975). Therefore, the majority of the water lost by vines occurs through the leaf surfaces with the fruit playing a minor role in vine water use (Blanke & Leyhe, 1987).

C. Effects of Water Deficits on Expansive Growth

Moderate differences in midday vine water status throughout the season have little effect on vine phenology (Matthews et al., 1987a; Williams & Grimes, 1987) but the degree of vegetative and reproductive growth is quite dependent upon vine water status. Early shoot growth may not be water limited since the rate of shoot elongation increased rapidly after budbreak despite declining vine water status (Matthews et al., 1987a). However, differences in the rate of shoot elongation were observed when differences in midday ψ_l developed due to irrigation treatments (Matthews et al., 1987a; van Zyl, 1984; Williams & Grimes, 1987). Since grapevines do not form terminal buds, axial shoot growth may continue as long as environmental conditions permit (e.g., until first frost). As a result of differences in vine water status, elongation of shoots continued approximately 45 d longer in drip irrigated vines than in nonirrigated vines in a study in Napa Valley, California (Matthews et al., 1987a). Similarly, water deficits inhibited shoot elongation rate and duration when drip irrigation was scheduled to maintain various fractions of plant-available moisture in the soil (van Zyl, 1984) or vineyard ET (Williams & Grimes, 1987). The differences in shoot length due to irrigation treatments are reflected in total vine leaf area (Table 34–2). It is interesting to note that water deficits, imposed as described in Table 34–2, reduced leaf area on lateral shoots to a greater extent than leaf area on the primary shoots.

In addition to canopy size and structure, periderm development or "wood ripening" is an important aspect of shoot growth and vine development. Periderm development and cane diameter—important factors for cold tolerance of dormant canes (Perold, 1927; Wolpert & Howell, 1985a, b)—are frequently used as criteria for pruning decisions and selection of grafting or planting material (Winkler, 1927). The rate of radial expansion of the shoot reaches a maximum at approximately the same time as maximum shoot elongation rate and decreases rapidly thereafter. Radial growth ceases well before axial growth in well-watered vines but ceases at approximately the same time as axial growth when early-season water deficits occur (Matthews

Table 34-2. Leaf area of Thompson Seedless grapevines irrigated at various fractions of vineyard ET (L.E. Williams & D.W. Grimes, 1986, unpublished data).

Location†	ET treatment‡	Primary shoot leaf area	Lateral shoot leaf area	Total vine leaf area
			m²/vine	
Parlier, CA	1.0	18.3§	6.9	25.2
	0.4	13.7	2.3	16.0
Five Points, CA	1.2	16.4	5.7	22.1
	0.8	11.9	2.4	14.3
	0.4	10.1	1.4	11.5

† Vineyards were located on two Univ. of California field stations in the San Joaquin Valley. Soil types were a Hanford sandy loam and Panoche clay loam (fine-loamy, mixed [calcareous], thermic Typic Torriorthents) at the Parlier and Five Points field stations, respectively.
‡ Treatments represent water applications throughout the season based upon fractions of vineyard ET, calculated and measured as described by Williams and Grimes, 1987.
§ Entire vines were harvested 6, 7, and 8 Aug. 1986, and leaves were separated into those on shoots derived from compound or latent buds (primary shoots) and those on shoots derived from lateral buds (lateral shoots). Values represent the mean of five individual vine replicates.

et al., 1987a). In that study, cane diameter was only 7% less in nonirrigated than in well-watered vines. This indicates that radial growth is less sensitive to water deficits than axial growth.

It is commonly observed that vines with excessive vegetative growth fail to ripen canes adequately for overwintering or for propagation. Visual scor-

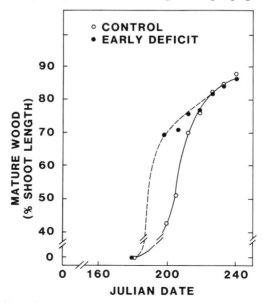

Fig. 34-4. The effect of an early water deficit on wood maturation of Cabernet franc grapevines grown in Napa Valley, California. Control vines were irrigated with water at 45 L/wk throughout the growing season, while the early deficit treated vines did not receive any water until veraison, Julian Day 200. After Matthews et al. (1987).

ing of internodes for the presence of bark indicated that early-season water deficits induced early periderm development (Fig. 34–4). The onset of periderm development may be related to the cessation of axial growth since bark initially was observed at approximately the same time that shoot elongation and new node production ceased, regardless of vine water status. The difference in periderm development was not an artifact of expressing the data as a percent of shoot length and was unlikely to be due to increased incident irradiance upon the shoot internodes. Similar results were obtained in controlled experiments with single, vertical shoots (cv. Thompson Seedless) in which there was no difference in incident irradiance (M.A. Matthews, 1987, unpublished data). Thus, moderate water deficits may result in improved propagation material and overwintering viability of canes since radial growth is relatively insensitive and periderm development is stimulated.

D. Reproductive Growth and Yield Responses to Water Deficits

Expansion of the cluster rachis is an important factor in determining cluster compactness (Weaver & McCune, 1959; Weaver et al., 1962). Less compact clusters are desirable because improved aeration and accessibility of pesticides aid in the control of bunch rot complexes—a major production concern in viticulture. However, linear expansion of the rachis is rapid during anthesis and is complete by 30 d after anthesis. Therefore, unusually early water deficits would be required to observe an effect upon timing or degree of expansion of the rachis. Similarly, significant water deficits during anthesis are uncommon.

Water deficits during berry development affect growth and composition of the fruit, but not the characteristic double-sigmoid growth curve (Fig. 34–5). Water deficits during Stage I of berry growth generally have a much greater effect on final berry size than water deficits during Stage II or III (Hardie & Considine, 1976; Matthews et al., 1987a; Smart et al., 1974; van Zyl, 1984). Water deficits during Stage I probably diminish the potential for berry growth due to reduced cell number per berry, whereas deficits that occur after this should only affect cell size (Coombe, 1976). This may be seen in the data of Matthews et al. (1987a) which demonstrate that at harvest, berries on vines that were exposed to a water deficit during Stages I and II approached their maximum size, whereas berries on vines exposed to a deficit during Stage III or irrigated weekly still were expanding at significant rates. However, when plant-available soil moisture was maintained at 25 and 90% of well-watered conditions (Fig. 34–5), or when vines were irrigated daily at a fraction of full ET (Williams & Grimes, 1987), there was no clear difference in the timing of maximum berry size. the onset of the lag phase (Stage II) does not appear to be sensitive to moderate water deficits since it will begin at the same time, regardless of either vine (Matthews et al., 1987a) or soil (van Zyl, 1984) water status. The continued growth of berries on stressed vines during Stage III indicates that fruit growth can be maintained despite low and decreasing vine water potentials (Matthews et al., 1987a; Vaadia & Kasimatis, 1961; Williams & Grimes, 1987).

The effects of seasonal water deficits upon berry size are reflected in vine yields (Freeman et al., 1979; Smart et al., 1974; Vaadia & Kasimatis, 1961; Williams & Grimes, 1987). Vines irrigated weekly produced greater yields than vines exposed to post veraison deficits, and the post-veraison treatments had greater yields than vines exposed to preveraison deficits (Hardie & Considine, 1976; Matthews et al., 1987a). Hardie and Considine (1976) imposed deficits at several different stages of development and observed that yield loss was greater the earlier water deficits occurred in potted vines. A water production function developed for Thompson Seedless grapevines in the San Joaquin Valley (Fig. 34–6) indicates that vineyard yield without supplemental irrigation is 60% that of full ET (800 mm). The application of approximately 250 mm additional water (total available water = 630 mm) results in vineyard yields that are >85% that of full ET. Prior and Grieve (1987) found that good yields were obtained with 150 mm effective rainfall and 300 mm irrigation water, which agrees with the ET/yield data obtained in California (Fig. 34–6). Their results were obtained in Dareton, N.S.W., Australia, where vineyard irrigation requirements are approximately 700 to 800 mm, minus rainfall.

An important consideration in the irrigation of many perennial fruit crops, including grapes, is that the differentiation of reproductive structures is initiated in the season prior to the season in which those structures mature fruit. Hence, water deficits of the previous season may affect productivity in the current season, but there are reports of positive, negative, and nil

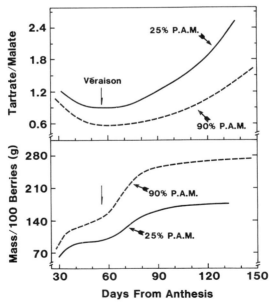

Fig. 34–5. The effect of irrigating Colombar grapevines to predetermined soil moisture levels on berry growth and the tratrate/malate ratio. A soil moisture regime of 25% meant that 75% of the Plant Available Moisture (P.A.M.) contained to a depth of 1 m was depleted by evapotranspiration. After Van Zyl (1984).

responses. This may be largely due to the failure to quantitate vine water status. Barnard and Thomas (1933) suggested that a depression of vegetative growth due to water stress could benefit flower bud initiation. Carbonneau and Casteran (1979) found a depressed potential fertility (number of reproductive bud primordia) due to irrigation, even though the general vigor of the plant was increased. Freeman et al. (1979) suggested that fruit bud differentiation is not adversely affected by the levels of water deficit normally encountered in the field. Fruitfulness increased or decreased for irrigated vines compared to nonirrigated controls in different years with no consistent trend. Significant water deficits were probable for the nonirrigated vines since the data show a large effect of irrigation on berry weight. In a comparison of irrigation delivery systems there was no significant difference in fruitfulness between the most- and least-irrigated vines in the first of the 2-yr study (Smart et al., 1974). In the second year, there was significantly greater fruitfulness in the last-irrigated treatment. In contrast, Huglin (1960) showed an inverse relationship for several years between fruitfulness and rainfall during the previous season.

The above cited discrepancies may be due to the timing of imposed water deficits since the initiation of reproductive primordia is believed to occur near anthesis. The sensitivity of reproductive effort to water deficits recently was investigated by withholding water before or after veraison (Table 34–3). In the first season, clusters/vine were similar in all treatments. Under normal production practices for this vineyard (weekly drip irrigation), clusters/vine in the second season were 30% greater than in the initial season (Table 34–3). When water was withheld before veraison, clusters/vine in the second season were only 15% greater than in the first season. Fruitfulness also was

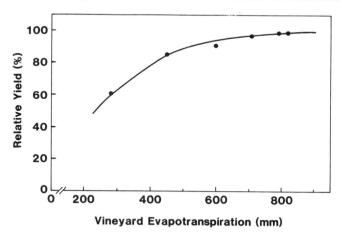

Fig. 34–6. Water production function for Thompson Seedless grapevines grown in the San Joaquin Valley of California on a Hanford sandy loam soil (coarse-loamy, mixed, nonacid, thermic Typic Xerorthents). Data were obtained by irrigating vines at fractions of estimated vineyard ET (from 0–1.2) and each point represents the mean of three replicates per yr for 2 yr. Maximum yield, averaged over the 2-yr period, was approximately 30 t/ha. Soil moisture and effective rainfall averaged 280 mm for the nonirrigated vines during the 2 yr the study was conducted. After Grimes & Williams (1990).

Table 34-3. Relative yield and yield components of Cabernet franc vines in 1985 which received different irrigation treatments in 1984 (Matthews & Anderson, 1989).†

Irrigation treatment‡	Clusters/vine	Berries/cluster	Vine yield
		% of 1984 results	
Control	129	127	133
Two X	144	141	162
Early deficit	115	93	96
Late deficit	123	106	130
Full deficit	108	85	85

† The vineyard was located in Napa Valley, California. There were no significant differences in yield components among treatments in 1984. Yields in 1984 ranged from 9.7 to 11.5 t/ha. Vines were pruned to the same number of buds both years.

‡ Treatments were as follows: Control, weekly water application of 45 L/vine throughout the season; Two X, two times the weekly application of water to the control treatment (90 L/vine); Early deficit, water withheld until veraison at which time weekly irrigations (45 L/vine) were initiated; Late deficit, weekly irrigation (45 L/vine) until veraison after which water was withheld; and Full deficit, no irrigation throughout the season.

decreased by postveraison water deficit, but to a lesser extent. When irrigation was supplied at two times the normal practice, number of clusters increased 44% from the initial season to the following season (Table 34-3). The relative changes in berries/cluster among treatments were of a similar magnitude as that for clusters/vine. These results clearly show a dependence of fruitfulness on vine water status. The extent to which these effects are due to altered initiation and to abortion of initiated reproductive structures needs to be pursued.

E. Effects of Water Deficits on Fruit Composition and Quality

As a horticultural crop, grape quality is of more concern than, for example, grain crops, since consumption is greatly influenced by perceived quality. Quality criteria depends upon the intended grape product (as discussed previously) and, therefore, influences irrigation strategies. Freedom from mold, mildew, and other defects is important for all grapes.

Grape quality is determined largely by the composition of the fruit. In contrast to table and raisin grapes, the quality of wine grapes is not influenced by blemishes produced by weather or handling. While berry size is an important quality factor for table and raisin grapes, size also may be a quality factor in red wine grapes since fermentation is conducted with the skins (Singleton, 1972). Small berry size (large surface/volume ratio) is likely to be preferred since the dermal tissue (skin) contains most of the color- and flavor-producing compounds. Clearly, smaller berries are produced on vines that experience water deficits than on vines that are continually well watered. Fruit composition and maturation of all grape types may be altered by vegetative growth (Freeman & Kliewer, 1983; Neja et al., 1977) and C partitioning (Hofaker, 1977) responses to irrigation treatments. Crop load also has been shown to be negatively correlated with rate of crop maturation (Weaver & McCune, 1960; Winkler, 1954), fruit quality (Kasimatis, 1977), and wine sensory characteristics (Sinton et al., 1978).

Wine grapes typically are purchased on a fresh weight basis after meeting minimal standards for concentration of soluble solids and defects such as rot, with some wineries paying bonuses for grapes of higher soluble solids concentrations. As the raw material for the fermentation process, the composition of the resultant wine is clearly dependent upon the composition of the fruit, since much of the *must* (crushed fruit) is carried over to the end product. Some flavor characteristics, though, may be altered by choice of yeasts, additives, and other fermentation conditions. The solute composition of wine grapes at harvest is sensitive to the vine water status throughout fruit development. Hence, in wine grape production water deficits may be beneficial to the quality of the final product. Nevertheless, it is clear that irrigaion is required to obtain a harvestable crop in many grape-producing areas of the USA.

The concentration of sugars in wine grapes determines the final alcohol content and its influence on wine flavor. Wine grapes typically contain 21 to 24% (w/v) fermentable sugars at harvest. Although fruit growth appears to be less sensitive to water deficits than shoot growth, the accumulation of sugar is much less sensitive to water deficits than fruit growth. Hence, supplemental irrigation generally increases yield while frequently having little effect on sugar concentration. When sugar accumulation is affected, it is delayed by increased water status. However, severe water deficits also may delay maturity (Williams & Grimes, 1987).

Organic acid levels normally range from 6 to 12 g/L (expressed as tartaric acid equivalents) in wine grapes at harvest. The acid level in grapes is important to the balance or acidity of wine. A moderate decrease in titratable acidity was always observed in studies where measurements of soil (Van Rooyen et al., 1980; van Zyl, 1984) or plant (Bravdo et al., 1985) water status indicated significant water deficits. In those studies, water deficits had the greatest effect on malate concentration, with early season (before veraison) deficits decreasing malate concentration more than deficits after veraison for 'Colombard' (van Zyl, 1984) and Cabernet Sauvignon (Bravdo et al., 1985).

Slight increases in the concentration of tartaric acid at harvest have been observed when the volume of water applied to Colombard (van Zyl, 1984) and Cabernet Savignon vines (Bravdo et al., 1985) was < 1.0 ET. The result of these relative changes in the two acids has been large increases in the ratio of tartrate/malate (Fig. 34–5). These two acids account largely for the free organic acid pool in wine grapes. The losses of malate may be balanced by similar decreases in counter-balancing cations or by accumulation of other acidic moieties such as amino acids. However, while the concentration of proline, a major storage form of amino N in grapes, increases throughout berry growth, water deficits had no effect on its concentration in 'Riesling' berries (Coombe & Monk, 1979). Freeman and Kliewer (1983) also reported that the concentration of proline in the fruit of irrigated 'Carignane' vines was higher than in the fruit of nonirrigated vines. Data in Table 34–4, though, demonstrates that proline concentration can be increased by water deficits and that this increase is not due to a decrease in berry volume.

Table 34-4. Berry composition of Cabernet franc vines subjected to various irrigation treatments during the 1984 growing season (Matthews & Anderson, 1988).

Irrigation treatment[†]	Juice			Skin	
	pH	Phenolics	Proline	Phenolics	Anthocyanins
		mg/L‡	mM	mg/cm^2‡	mg/cm^2
Control	3.48	149	3.95	0.46	0.51
Early deficit	3.49	194	4.66	0.56	0.61
Late deficit	3.47	202	5.76	0.52	0.59
Full deficit	343	198	4.79	0.57	0.65

† Treatments are the same as those described in Table 34-3.
‡ Phenolics are expressed as mg gallic acid equivalents.

The pH of juice and wine is an important stability parameter. Wines of high pH also are more susceptible to growth of spoilage organisms. Supplemental irrigation may have an effect on juice pH. Berry pH was increased by irrigation for 'Shiraz' (Freeman, 1983), Carignane (Freeman & Kliewer, 1983), Chenin blanc (Vaadia & Kasimatis, 1961) and Cabernet Sauvignon (Neja et al., 1977), but not for 'Cabernet franc' (Matthews & Anderson, 1988), Cabernet Sauvignon (Bravdo et al., 1985) or Thompson Seedless (Williams & Grimes, 1987).

A recurring problem for many wine and table grape producing areas and for some cultivars such as Pinot noir, is the development of adequate color. A generalization made possible by numerous field studies is that irrigation scheduling may be used to improve grape color since water deficits increase the production of pigments (anthocyanins) in red wine cultivars (Bravdo et al., 1985; Freeman, 1983; Freeman & Kliewer, 1983; Hardie & Considine, 1976). This effect is not due simply to a decrease in berry size since the effect is observed when anthocyanin concentration is expressed on a berry surface area basis (Table 34-3; Freeman & Kliewer, 1983). Phenolics, of which one-half the total found in black or red grape berries are anthocyanins, contribute to the astringency of wines and to wine aging characteristics (Singleton & Esau, 1969). The concentration of total phenolics in grape juice and skin extracts also responds to water deficits (Table 34-4). Although early and late-season deficits were similarly effective in increasing total phenolic content (approximately 33%). Irrigation also has been shown to increase the total terpene (free + bound) content in Riesling berries (McCarthy & Coombe, 1985). Irrigation did, however, result in a reduction in the concentration of bound terpenes on most sampling dates in that study.

The effects of water deficits in the vineyard on fruit composition are reflected in the composition of the subsequent wine. Studies that demonstrated a lack of irrigation treatment effect on juice titratable acidity and pH also showed little effect in the wine. Wines had greater color and total phenolic concentrations when made from vines that experienced early- or late-season water deficits than from vines irrigated weekly (Table 34-4). Similar observations were made by Hepner et al. (1985). Unfortunately, current methodologies for assay of soluble phenolics do not discriminate between flavor-producing and nonflavor-producing phenolics.

IV. VINE NUTRITION

A. Mineral Nutrient Requirements

The requirement of mineral nutrients, especially N, by grapevines is considerably less than that required by many major agricultural crops (Olson, 1978; Olson & Kurtz, 1982). The amount of N required for the growth of shoots and fruit of Thompson Seedless (Williams, 1987b) and Sultana (Alexander, 1958) grapevines was 84 kg/ha, with 30 kg/ha of that needed for fruit production. Lafon et al. (1965) determined that 68 kg N/ha was required for a 'St-Emilion' vineyard with 28 kg/ha needed for fruit growth. Conradie (1981b) and Marocke et al. (1976) reported that 1.43 and 1.70 kg N/t of fruit were needed for the production of wine grapes in South Africa and the Alsace region of France, respectively.

The amount of K and P found in the vegetative and reproductive structures of the vine also are less than that required for several annual row crops (Olson, 1978). Williams et al. (1987a) determined that 56 kg K/ha was found in the fruit of Thompson Seedless grapevines at harvest, which was about 60% of the total K found in the current season's growth at that time. This percent of the total vine K found in clusters at harvest was similar to that reported by Lafon et al. (1965) and Smart et al. (1985a). The amount of P found in the structures of field-grown grapevines is approximately 10 kg/ha (Lafon et al., 1965) or 0.25 kg P/t fresh fruit (Conradie, 1981b). The approximate level of other mineral elements required by the grapevine including Ca, Mg, S, and several other micronutrients may be found in papers by Lafon et al. (1965) and Conradie (1981a).

The requirement of mineral nutrients for shoot growth early in the growing season by the grapevine before sufficient soil warming or significant root absorption takes place indicates that these elements, especially N, may be remobilized from the permanent structures at that time (Conradie, 1980, 1986). It was estimated that 20% of the N required for 2-yr-old potted Chenin blanc vines was remobilized from the roots or other permanent structures (Conradie, 1986). Approximately 10% of the seasonal N accumulated in the vegetative and reproductive structures by 10-yr-old, field-grown 'Barbera' grapevines was due to remobilization from the root system (Williams et al., 1987b). All of the remobilization took place prior to bloom.

The accumulation of N in the vegetative parts of the vine has been shown to be linearly related to GDD's from budbreak until 1000 GDD's later (Williams, 1987b). A linear increase in stem and leaf dry weight also occurred during this time (Williams, 1987a). After 1000 GDD's the accumulation of N by the vine is due to that accumulating in the clusters. While there is an increase in vine N content throughout the growing season (Alexander, 1958; Williams, 1987b), N concentration decreases in the leaves, stems and clusters during the same time (Alexander, 1958; Conradie, 1980; Williams, 1987b). It has been shown that the decrease in N concentration is due to a dilution effect as the weight per unit leaf area and stem length increases throughout the season (Williams, 1987a). Thus little, if any, N may be redistributed from the leaves or shoots to the clusters during fruit growth as previously had been suggested (Alexander, 1958; Conradie, 1980). The amount of N lost from the vine because of leaf fall and winter pruning and incorporated back into the soil is equivalent to 35 to 40 kg/ha (Alexander, 1958; Williams, 1987b).

The accumulation of K in the fruit of Thompson Seedless grapevines was almost linear from the time of bloom until harvest (Williams et al., 1987a). Conradie (1981a) found a similar increase in the K content of clusters from vines grown in pots. Lafon et al. (1965) and Conradie (1981a) observed appreciable translocation of K from the shoots and leaves to the other structures of the vine, while Levy et al. (1972) and Williams et al. (1987a) found that little K was redistributed within the vine prior to harvest. The redistribution of K from the vegetative structures of the vine to the fruit during ripening has been implicated in affecting wine quality (Smart et al., 1985b). It has been suggested that monovalent cations, especially K, are exchanged for H ions in the berry with a resultant increase in juice pH (Boulton, 1980b, c; Somers, 1977) which may decrease the quality of the fruit and ultimately the wine.

B. Fertilization Practices

The macronutrients and micronutrients that are most commonly limiting to vine growth and productivity under California conditions are N, K, and Zn (Christensen et al., 1978; Winkler et al., 1974). The application of these nutrients to vineyards usually is dependent upon visual deficiency symptoms, vine vigor, reductions in yield, or vine tissue analysis. While vine nutrient applications based upon the approximate vine nutrient requirement and time of utilization, as noted in the preceding section, are helpful, one also needs to consider the time in which loss of these fertilizers (particularly N) from the soil is minimized. Past recommendations for N fertilizer application were during the vine's dormant period so that the N would be available to support the vine during most of the vegetative growth phase (Christensen et al., 1978; Winkler et al., 1974). Recent studies have indicated that this may not be the most efficient time to apply fertilizer. The use of ^{15}N-labeled fertilizer on both almond [*Prunus dulcis* (Mill.) D.A. Webb] trees (Weinbaum et al., 1984) and grapevines (Conradie, 1986) indicates that significant amounts of fertilizer applied one year to a perennial crop may be utilized the next year because of remobilization from the permanent structures of the plant to support both vegetative and reproductive growth. In addition, the fruit of the vine constitute a large sink for mineral nutrients (Alexander, 1958; Conradie, 1981b; Lafon et al., 1965; Williams, 1987b; Williams et al., 1987a), indicating a possible large requirement of soil-derived nutrients throughout the growing season. Future recommendations for N fertilization may call for frequent, small applications during the season with the goal of controlling vegetative growth to optimize fruit production and quality.

Fertilization of grapevines, as with other horticultural crops (Claypool, 1975; Embleton et al., 1975), may have complex effects on the quality of the fruit as well as increasing the quantity. An increase in vine growth and yield in response to N fertilization was shown to occur only under circumstances where a deficiency existed (Alexander & Woodham, 1970; Kliewer & Cook, 1971). Pouget (1984) found that the annual growth of shoots and vine yield were unaffected by doubling the concentration of a nutrient solution the vines were watered with. Cook and Kishaba (1956) reported that N fertilization of vineyards with high soil NO_3 depressed yields below those of the control plot. Thompson Seedless grapevines supplied with various amounts of NO_3 responded with increased concentrations of arginine in the berry juice up to about 4 mM NO_3 in the nutrient solution, at which time it leveled off with further increases in NO_3 concentration (Kliewer, 1971). Total soluble solids, acidity, and tartrate in the berries were not significantly affected by the N treatment. Pouget (1984) and Delas and Pouget (1984) demonstrated that while overall vine growth and yield were unaffected by doubling the concentration of a nutient solution, the wine made from vines grown with the more concentrated solution had inferior sensory characters. A fertilization and rootstock trial showed that the application of 337 kg N/ha had no measurable effect on wine quality when compared with vines that had not been fertilized (Ough et al., 1968b). Juice N concentration has been shown to be related to the rate of fermentation (Cantarelli, 1957). The fermentation rates were greater for wines made from vines fertilized with N (337 kg/ha) than those not receiving N (Ough et al., 1968a).

Potassium deficiency in California vineyards is confined to areas low in available soil K, such as sandy soils, or where root growth is restricted (Christensen, 1975a). This deficiency may be alleviated by a single large application of K fertilizer, approximately 1000 kg K/ha (Christensen, 1975a; Cook, 1960; Kasimatis & Christensen, 1976) with subsequent increased yields (Cook, 1960, 1966). Potassium fertilization will increase K in the vine's petioles even when adequate K is available in the soil (Mattick et al., 1972; Morris et al., 1980). More recently, though, Kliewer et al. (1983) found that the application of 1090 kg K/ha per yr over a 3-yr period had relatively little effect on petiole K of Carignane vines grown on a deep fertile loam soil. They also found that K fertilization of those vines reduced yields slightly when compared to the control treatment. Potassium fertilization may (Morris et al., 1980) or may not (Kliewer et al., 1983) increase fruit K concentration. Recent evidence indicates that wine quality is reduced by high pH level (Boulton, 1980b; Somers, 1975). The juice pH of fruit from vines may be related to the fruit's concentration of K (Boulton, 1980b). It has been shown that cultural practices may help in regulating the partitioning of K within the vine and its distribution to the fruit (Smart et al., 1985a, b).

Other mineral nutrients often limiting growth of vines in California are P and Zn. Phosphorus deficiency occurs in certain acid soils found in vineyards planted on the hillsides of the coastal mountain range and foothills of the Sierra Nevada (Skinner et al., 1987). Application of P fertilizer has been shown effective either during the dormant or growing season (Skinner et al., 1987). Zinc deficiency is most common in San Joaquin Valley vineyards grown on sandy soils and on vines in which rootstocks impart high vegetative growth (Christensen et al., 1978). Methods used to correct a Zn deficiency include painting all fresh pruning cuts with a solution of zinc sulfate ($ZnSO_4$) for spur-pruned varieties, a foliar spray of Zn shortly before bloom (Christensen, 1980), or a soil application of the Zn fertilizer.

C. Vine Response to Saline/Toxic Soils

Grapes are grown in several areas of the arid southwestern USA which are affected by salinity problems. In California alone, approximately 1.2 million of the 4.1 million irrigated hectares experience yield losses due to saline soil or water, and it is estimated that this will increase to 1.5 million hectares by the year 2000 if significant corrective measures are not taken (Blacklund & Hoppes, 1984). Vineyards in the saline affected areas are primarily table grapes in the Coachella Valley and raisin and wine grapes in the San Joaquin Valley of California (Christensen et al., 1974; Halsey et al., 1963).

Unlike most herbaceous crop species that are affected primarily by the lowered osmotic potential of saline soil solutions, most woody crops are subject to specific ion toxicities, primarily that of Cl (Maas, 1987). Grapes have been classified as moderately sensitive to salinity based upon yield and foliar symptom responses (Maas & Hoffman, 1977). Yield decreases of approximately 10% per dS m^{-1} have been shown (Mass, 1987).

Toxicity symptoms appear first as marginal chlorosis on leaves, followed by progressive necrosis towards the midrib and petiole until only the veins remain green. There is a sharp demarcation between necrotic and green areas. Toxicity symptoms are probably due to Cl since grapes grown on sodic soils rarely exhibit symptoms. Petiole content of Cl, which is higher than lamina Cl, is generally used to indicate leaf Cl status (Woodham, 1956; Christensen et al., 1978).

The effects of salinity on fruit composition and quality are not clearly established. Salinity treatments increase the salt accumulated in fruit and this may be perceived in the subsequent wine (Downton, 1977b). Severe saline conditions inhibit growth and photosynthesis and delay ripening. However, there are two reports of moderate salt treatments (50 mM NaCl or less) causing an acceleration of ripening (Downton & Loveys, 1978; Hawker & Walker, 1978).

All grape cultivars are susceptible to salt injury but fruiting stock and rootstock vary significantly in uptake and sensitivity. In sand culture, Cardinal and Black Rose accumulated Cl two to three times faster than Thompson Seedless and Perlette (Ehlig, 1961). In another study, shoot growths of Cabernet Sauvignon and Doradillo were more sensitive to a saline nutrient culture than those of Sultana, Shiraz, and Palomino (West & Taylor, 1984). In solution culture, Shiraz and Cabernet Sauvignon were less sensitive (shoot elongation) and accumulated less Cl than the less vigorous Doradillo and Palmonio varieties (Obbink & Alexander, 1973).

Rootstocks generally are considered to exert a major effect on ion uptake in fruit trees and the same is likely true of grapevines. In sand culture, the Cl content of Cardinal and Thompson Seedless leaves varied from 27.2

to 1.7 cmol kg^{-1} dry wt. when grown on their own roots or on 'Dog Ridge', '1613-3', or 'Salt Creek' rootstocks (Bernstein et al., 1969). Cardinal on its own roots accumulated Cl to a concentration three times that present in the soil solution, whereas on Salt Creek rootstock, Cl was excluded to 25% of that present in the soil solution (Bernstein et al., 1969). Sauer (1968) made similar observations in the field, where salt-tolerant rootstocks (Dog Ridge and Salt Creek) decreased Cl uptake three to six times compared to own-rooted Sultana vines. In a survey of potential rootstock material, Downton (1977a) ranked *Vitis* spp. for Cl accumulation (low accumulation to high accumulation) as follows: *V. rupestris* < *V. champini* < *V. longii* < *V. cineria, V. cordifolia* < *V. vinifera*.

V. IRRIGATION AND CULTURAL MANAGEMENT PRACTICES

A. Wine Grape Vineyards

Wine grapes commonly are irrigated in regions with dry summers. Nevertheless, many vineyards are not irrigated in those areas that produce premium wines. Indeed, irrigation generally is illegal in most of France and Germany. The effects of irrigation on various aspects of the subsequent wine have been reported (and reviewed earlier in this chapter) as either positive (Hofaker, 1977; Kliewer et al., 1983; Neja et al., 1977) or negative to nil (Kliewer et al., 1983; Morris & Cawthon, 1982; Spayd & Morris, 1978). It is clear that irrigation is required in some climates and soils to produce economic yields. However, the question remains as to what vine or soil water status is required during the crop's development to produce the optimum structure and composition for wine grapes. This is due in part to the failure to adequately quantify plant and soil water status in irrigation studies and the nebulosity of perceived wine quality.

Wine grapes are irrigated by flood, furrow, overhead sprinkler, and drip delivery systems in the western USA. Irrigation practices are based upon water cost and availability, soil type, tillage practices, and irrigation district scheduling to mention a few considerations. Flood- and furrow-irrigated vineyards are often irrigated on a calendar date schedule with some seasonal modifications. In some cases, soil water content is monitored and irrigation scheduled at approximately 50% depletion of available soil water, ascertained most often by tensiometers or estimated ET. In some wine grape areas, including the coastal valleys of California, shallow soil and limited supply of irrigation water have increased interest in prudent use of water (Wildman et al., 1976).

In a direct comparison of furrow, sprinkler, and drip irrigation of wine grapes in the San Joaquin Valley, Peacock et al. (1977) found that drip increased irrigation efficiency while maintaining yield and fruit quality parameters similar to that observed under furrow and sprinkler irrigation. Soil salt levels were similar under furrow and sprinkler irrigation but slightly higher under drip. Under sprinkler irrigation, light and frequent applications, based upon soil water content at various depths, produced greater yields and higher sugar content than heavy and infrequent applications based upon calendar date (Wildman et al., 1976). These differences consistently were observed in vineyards at four soil depths and over several seasons. Stevensen (1981) recently demonstrated that the delivery system could be changed from sprinkler to drip without significant effects on vine growth and yield. Drip irrigation at 0.4 times the rate of Class A pan evaporation produced similar yields to furrow irrigation with a crop factor of 0.5 (Smart et al., 1974). In that study, there was no difference in yield when water was applied through the drip system daily or every other day; however, water applications using a crop factor of 0.2 decreased yield 16%.

Regardless of the irrigation delivery system, soil water content generally is allowed to diminish during fruit ripening (Kasimatis, 1967). The extent to which vine water status typically declines has not been defined, but is clearly an important factor in determining fruit composition at harvest. In some cases, moderate water deficits can accelerate ripening and enhance color development (Hepner et al., 1985). In addition, water deficits are used to inhibit shoot growth. Excessive vegetative growth is a primary concern in many U.S. viticultural areas. Irrigation late in the season can maintain shoot growth and promote growth of lateral shoots (Smart & Coombe, 1983). Late-season vegetative growth is undesirable since it shades developing clusters and the shoots often fail to ripen properly for overwintering (Spiegel-Roy & Bravdo, 1964).

B. Raisin Grape Vineyards

The California raisin industry produces approximately 209 000 t of dried fruit annually with the variety Thompson Seedless accounting for 90% of that total. The production of sun-dried (natural) raisins is limited to the central San Joaquin Valley with 80% of the vineyards used for that production located within 50 km of Fresno, CA. This area is typified by an environment that can mature the crop and allow adequate time before the chance of significant precipitation can damage the grapes once they are harvested to be sun-dried.

Environmental conditions in the Fresno area at harvest, which generally occurs the last week in August to first week in September, are characterized by high temperatures and solar radiation—ideal to dry the grapes once they are laid on paper trays between rows. Most raisin vineyards are planted in east-to-west rows to optimize the use of solar radiation to dry the fruit. Trellis systems used in vineyards to produce raisins are such that the height or width of the vine's canopy does not cause shading on the grapes between the rows. Prior to harvest, the portion of the shoots nearest the soil surface are trimmed in order to facilitate ground preparation for making soil terraces. This procedure can remove approximately 6 m^2 of the vine's leaf area (Williams et al., 1987a). The terraces are sloped to the north so that the grape trays will be exposed to maximum solar radiation during the grape-drying process.

Irrigation management of raisin vineyards is similar to that of wine vineyards planted in the San Joaquin Valley. Water applications are a necessity in this area of California where summer rainfall is negligible. At present, most raisin vineyards are furrow- or flood-irrigated with only a small portion under drip irrigation. The approximate water requirements of raisin vineyards were covered previously in this chapter (Table 34–1). During the grape-ripening period it has been recommended that the soil moisture supply gradually should be lessened (Kasimatis & Lynn, 1975). The timing of the last irrigation prior to harvest, for vineyards that are not under drip, depends upon soil type, probable harvest date, and opinion. Christensen (1975b) found that an early (early July) irrigation cutoff had no advantage over a later one (early August) on deep fine sandy loam to loamy sand Thompson Seedless vineyards. He concluded that irrigation cutoff need only be early enough to slow vine growth and to provide for adequate surface soil drying to prepare terraces for fruit drying.

C. Table Grape Vineyards

The table (fresh) grape production areas of California include the San Joaquin Valley and the Coachella Valley in the desert of southern California. Production practices for table grapes differ from those of wine and raisin grapes. This is due in part to the fact that the physical appearance of the fruit is a major determinant of the quality of the crop. Thus the control of diseases and pests is of utmost importance to the table grape vineyard manager. The suppression of dust in many vineyards may call for the use of a grass cover crop between rows. Other quality factors also include berry size, color, sugar, and sugar/acid ratio of the fruit. The use of *girdling* (the removal of a narrow strip of phloem/bark entirely around either the trunk or a fruiting cane) long has been used to increase the size of seedless grape varieties (Jacob, 1931). Girdling is most effective in increasing berry size if applied to the vine at fruit set, approximately 2 wk after bloom in the San Joaquin Valley (Winkler, 1953). Girdling also has been used to help improve color and advance fruit maturation of colored table grape varieties when applied once the first traces of color appear on the fruit at early veraison (Weaver, 1955). The use of gibberellic acid (GA_3) also is used to increase the size of seedless table grape varieties. Two applications of GA are applied, one at bloom, which thins the cluster by reducing berry set and increases berry size, and the other at fruit set, which further increases berry size (Christodoulou et al., 1968). Vines also are girdled at fruit set in conjunction with a GA spray, with a resultant increase in berry size that is additive (Harrell & Williams, 1987b).

Due to the location in California and Arizona where most of the table grape vineyards are located and to the soil types and depth on which they are planted, frequent irrigations are needed for maximum production and quality (Winkler et al., 1974). The irrigation requirement for the production of Perlette table grapes in Arizona on a sandy loam soil was approximately 380 mm from budbreak to harvest and 330 mm from harvest to leaf fall

(Bucks et al., 1985). It is interesting to note that the use of water by individual Perlette vines decreased from 35 L/d just prior to girdling to 25 L/d 2 wk after girdling (Fig. 34–7). The decrease in water use by the girdled vines may be due to decreased stomatal conductance brought about by an increase in abscisic acid (During, 1978). It also has been shown that the rate of individual leaf photosynthesis of field-grown vines was significantly decreased by trunk girdling 2 wk after the treatment was imposed (Harrell & Williams, 1987a). Once the girdle wound starts healing, the daily water use may (Bucks et al., 1985) or may not (Erie et al., 1982) increase to levels greater than that before the vines are girdled. Even though water demand is reduced by girdling, it has been recommended that the frequency of irrigation be increased during the time the girdle remains open to reduce the weakening effect the girdle has on the vine (Winkler et al., 1974).

In the southern desert table grape-growing regions of California and Arizona, vine development proceeds rapidly. In addition, most table grapes are picked when the soluble solids levels are 15 to 17% w/v, which is considerably less than sugar levels required for wine or raisin grapes. Thus the harvest of table grapes generally can be completed by June in these areas. This leaves an additional 16 to 20 wk in which the vineyard must be managed without a harvestable crop. It is important to note that next year's crop has already been differentiated, thus management practices have to take this into account when irrigation and pest control decisions are made. The extreme high temperatures and low humidity indicate that evaporative demand is great and that irrigation must proceed throughout the remainder of the season. Initial results from a study in Arizona indicate that partial removal of the vine's leaves 4 to 8 wk after harvest may be beneficial in promoting bud fruitfulness (Durate, 1983). It was suggested that the carbohydrate status of the vine is greater for defoliated vines than for vines in which the leaves are allowed to remain until normal leaf fall, which helps maintain next year's cluster

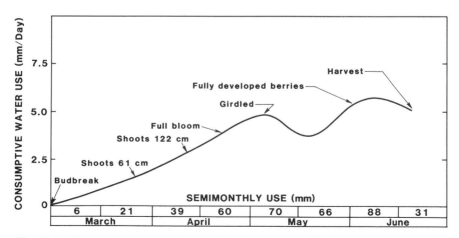

Fig. 34–7. Consumptive water use from budbreak to harvest of Perlette grapevines grown in Arizona. After Bucks et al. (1985).

primordia. The use of irrigation may play an important role in regulating the vine's growth during this period in this grape production area.

D. Rootstocks

Phylloxera [*Daktylosphaira vitifoliae* (Fitch)], a root-feeding louse, decimated French vineyards during the late 1800s. The insect had been indigenous to the eastern USA and was probably imported to Europe during the 1850s. The insect spread throughout Europe and quickly destroyed most vineyards as the species *V. vinifera* had little or no resistance to this pest. The only effective means of controlling phylloxera, to this day, is to graft *V. vinifera* to resistant native North American *Vitis* rootstocks (Pongracz, 1983).

While the use of rootstocks is common today, viticulturists in France initially found that not all of the North American rootstocks responded favorably to the growth conditions there. It was found that close attention had to be paid to different species and varieties obtained from the USA. The original North American species and hybrids were adapted to various physical and chemical properties of the soil, soil water status, and compatibility with the scion. Thus the use of rootstocks, in addition to providing protection from the soil pests, phylloxera, and nematodes, may provide a means to control vegetative growth of the scion and ultimately affect the vine's fruit production and quality. This includes the selection of rootstocks for drought tolerance (Carbonneau, 1985).

VI. SUMMARY

Although supplemental irrigation is recognized as a factor in determining grape composition, the solutes that impart varietal character and high quality fruit are not well defined or understood (Webb, 1981). Research towards improved cultural practices, including irrigation scheduling, and improved cultivars is hampered by the inability to define goals, i.e., the identity and optimum concentration of chemical components which impart quality attributes of grapes. In addition, objective measurements that correlate with sensory measures of wine quality would greatly facilitate comparisons among various vineyard management practices. It appears now that such measurements are available for red wine grape cultivars (Somers, 1975; Somers & Evans, 1977) and cultivars in which terpenes are a significant part of the berry's composition (McCarthy & Coombe, 1985).

Numerous studies cited in this chapter have described the effects of irrigation on fruit quality; however, few of them have quantified vine water status in enough detail so as to be useful for irrigation scheduling at different locations. Recently, the measurements of both soil and vine water status in irrigation studies have received more attention and attempts to establish vine "stress indices" (Stevens & Cole, 1987; van Zyl, 1986; Grimes & Williams, 1990) may provide a means to transfer irrigation results from one vineyard to another. The use of remote sensing equipment may allow vineyard managers in the future to assess the water status of their vineyards rapidly and schedule irrigations according to desired results.

van Zyl and coworkers (van Zyl, 1984, 1986; van Zyl & van Huyssteen, 1980; van Zyl & Weber, 1981) have conducted the most extensive investigations on vineyard irrigation. They conclude that irrigation management can be a powerful tool to control vegetative growth and improve the growth and quality of the fruit. However, in order for this to happen, management systems that concentrate and restrict the root system are required before such regulation by irrigation can be effective. One means of accomplishing this is by the use of drip irrigation. Within the last 10 to 15 yr, drip irrigation has been adopted to a limited degree by all segments of the grape-growing industry in California. In some cases this was done out of necessity because of the planting of vineyards in locations where surface irrigation was not feasible. However, many growers are using drip to enable them to more accurately irrigate based upon best estimates of vineyard ET (Peacock et al., 1977, 1987; Synder et al., 1987) and to control water applications at various stages of vine development.

ACKNOWLEDGMENT

We appreciate the helpful comments on the manuscript from L.P. Christensen, W.M. Kliewer, and D.W. Grimes. We also thank M. Benham for typing the manuscript.

REFERENCES

Alexander, D.M. 1958. Seasonal fluctuations in the nitrogen content of the Sultana vine. Aust. J. Agric. Res. 8:162–178.

Alexander, D.M., and R.C. Woodham. 1970. Chemical composition of leaf tissues of Sultana vines grown in nutrient solutions deficient in macro-elements. Vitis 9:207–217.

Antcliff, A.J., and P. May. 1961. Dormancy and bud burst in Sultana vines. Vitis 3:1–14.

Araujo, F.J., and L.E. Williams. 1988. Dry matter and nitrogen partitioning and root growth of young "Thompson Seedless" grapevines grown in the field. Vitis 27:21–32.

Archer, E., and H.C. Strauss. 1985. Effect of plant density on root distribution of three-year-old grafted 99 Richter grapevines. South Afr. J. Enol. Vitic. 6:25–30.

Baldwin, J.G. 1964. The relation between weather and fruitfulness of the Sultana vine. Aust. J. Agric. Res. 15:920–928.

Baldwin, J.G. 1966. The effect of some cultural practices on nitrogen and fruitfulness in the Sultana vine. Am. J. Enol. Vitic. 17:58–62.

Barnard, C. 1932. The root system of the Sultana. J. Counc. Sci. Ind. Res. Aust. 5:88–93.

Barnard, C., and J.C. Thomas. 1933. Fruit bud studies. II. The sultana: Differentiation and development of the fruit buds. J. Counc. Sci. Ind. Res. Aust. 6:285–294.

Bernstein, L., C. Ehlig, and R.A. Clark. 1969. Effect of grape rootstocks on chloride accumulation in leaves. J. Am. Soc. Hortic. Sci. 94:584–590.

Blacklund, V.A., and R.R. Hoppes. 1984. Status of soil salinity in California. Calif. Agric. 38:8–9.

Blanke, M.M., and A. Leyhe. 1987. Stomatal activity of the grape berry cv. Riesling, Muller-Thurgau and Ehrenfelser. J. Plant Physiol. 127:451–460.

Boulton, R. 1980a. The relationships between total acidity, titratable acidity and pH in wine. Am. J. Enol. Vitic. 31:76–80.

Boulton, R. 1980b. The general relationship between potassium, sodium and pH in grape juice and wine. Am. J. Enol. Vitic. 31:182–186.

Boulton, R. 1980c. A hypothesis for the presence, activity, and role of potassium/hydrogen, adenosine triphosphatases in grapevines. Am. J. Enol. Vitic. 31:283–287.

Bravdo, B., Y. Hepner, C. Loinger, S. Cohen, and H. Tabacman. 1985. Effect of irrigation and crop level on growth, yield and wine quality of Cabernet Sauvignon. Am. J. Enol. Vitic. 36:132–139.

Bucks, D.A., O.F. French, F.S. Nakayama, and D.D. Frangmeier. 1985. Trickle irrigation management for grape production. p. 204–211. *In* Proc. 3rd Int. Drip/Trickle Irrig. Congr., Fresno, CA. 18–21 November. ASAE Publ. 10-85 Vol. I. ASAE, St. Joseph, MI.

Buttrose, M.S. 1969a. Fruitfulness in grapevines: Effects of light intensity and temperature. Bot. Gaz. 130:166–173.

Buttrose, M.S. 1969b. Fruitfulness in grapevines: Effects of changes in temperature and light regimes. Bot. Gaz. 130:173–179.

Buttrose, M.S. 1969c. Fruitfulness in grapevines: Effects of daylength. Vitis 8:188–190.

Buttrose, M.S. 1970. Fruitfulness in grape-vines: The response of different cultivars to light, temperature and daylength. Vitis 9:121–125.

Buttrose, M.S., and C.R. Hale. 1973. Effect of temperature on development of the grapevine inflorescence after bud burst. Am. J. Enol. Vitic. 24:14–16.

Byrne, G.F., J.E. Begg, and G.K. Hansen. 1977. Cavitation and resistance to water flow in plant roots. Agric. Meteorol. 18:21–25.

Cantarelli, C. 1957. On the activation of alcoholic fermentation in wine making. Am. J. Enol. Vitic. 8:167–175.

Carbonneau, A. 1985. The early selection of grapevines rootstocks for resistance to drought conditions. Am. J. Enol. Vitic. 36:195–198.

Carbonneau, A., and P. Casteran. 1979. Irrigation-depressing effect on floral initiation of Cabernet Sauvignon grapevines in Bordeaux area. Am. J. Enol. Vitic. 30:3–7.

Cawthon, D.L., and J.R. Morris. 1982. Relationship of seed number and maturity to berry development, fruit maturation, hormonal changes, and uneven ripening of 'Concord' (*Vitis labrusca* L.) grapes. J. Am. Hortic. Sci. 107:1097–1104.

Christensen, P. 1969. Seasonal changes and distribution of nutritional elements in Thompson Seedless grapevines. Am. J. Enol. Vitic. 20:176–196.

Christensen, P. 1975a. Long-term responses of 'Thompson Seedless' vines to potassium fertilizer treatment. Am. J. Enol. Vitic. 26:179–183.

Christensen, P. 1975b. Response of 'Thompson Seedless' grapevines to the timing of preharvest irrigation cut-off. Am. J. Enol. Vitic. 26:188–194.

Christensen, P. 1980. Timing of zinc foliar sprays. I. Effects of application intervals preceding and during the bloom and fruit-set stages. II. Effects of day vs. night application. Am. J. Enol. Vitic. 31:53–59.

Christensen, P., R.S. Ayers, and A.N. Kasimatis. 1974. Boron and salinity in the vineyards of the west side, Fresno county. Calif. Agric. 28:10–11.

Christensen, P., A.N. Kasimatis, and F.L. Jensen. 1978. Grapevine nutrition and fertilization in the San Joaquin Valley. Agric. Sci. Univ. California Berkeley Div. Agric. Sci. Publ. 4087.

Christodoulou, A.J., R.J. Weaver, and R.M. Pool. 1968. Relation of gibberellin treatment to fruit-set, berry development, and cluster compactness in *Vitis vinifera* grapes. Proc. Am. Soc. Hortic. Sci. 92:301–310.

Claypool, L.L. 1975. Plant nutrition and deciduous fruit crop quality. Hortic. Sci. 10:45–47.

Conradie, W.J. 1980. Seasonal uptake of nutrients by Chenin blanc in sand culture: I. Nitrogen. South Afr. J. Enol. Vitic. 1:59–65.

Conradie, W.J. 1981a. Seasonal uptake of nutrients by Chenin blanc in sand culture: II. Phosphorus, potassium, calcium and magnesium. South Afr. J. Enol. Vitic. 2:7–13.

Conradie, W.J. 1981b. Nutrient consumption by Chenin blanc grown in sand culture and seasonal changes in the chemical composition of leaf blades and petioles. South Afr. J. Enol. Vitic. 2:15–18.

Conradie, W.J. 1986. Utilisation of nitrogen by the grape-vine as affected by time of application and soil type. South Afr. J. Enol. Vitic. 7:76–83.

Considine, J.A., and R.B. Knox. 1979. Development and histochemistry of the cells, cell walls and cuticle of the dermal system of the fruit of the grape, *Vitis vinifera* L. Protoplasma 99:347–365.

Considine, J.A., and P.E. Kriedemann. 1972. Fruit splitting in grapes: Determination of the critical tugor pressure. Aust. J. Agric. Res. 23:17–24.

Cook, J.A. 1960. Vineyard fertilizers and covercrops. Univ. California Berkeley Agric. Exp. Stn. Leafl. 128.

Cook, J.A. 1966. Grape nutrition. p. 777-812. *In* N.F. Childers (ed.) Nutrition of fruit crops. Horticultural Publ., Rutgers Univ., NB.

Cook, J.A., and T. Kishaba. 1956. Petiole nitrate analysis as a criterion of nitrogen needs in California vineyards. Proc. Am. Soc. Hortic. Sci. 68:131-140.

Coombe, B.G. 1960. Relationship of growth and development to changes in sugars, auxins, and gibberellins in fruit of seeded and seedless varieties of *Vitis vinifera*. Plant Physiol. 35:241-250.

Coombe, B.G. 1976. The development of fleshy fruits. Annu. Rev. Plant Physiol. 27:207-228.

Coombe, B.G., and P.R. Monk. 1979. Proline and abscisic acid content of the juice of ripe Riesling grape berries: Effect of irrigation during harvest. Am. J. Enol. Vitic. 30:64-67.

De La Harpe, A.C., and J.H. Visser. 1985. Growth characteristics of *Vitis vinifera* L. cv. Cape Riesling. South Afr. J. Enol. Vitic. 6:1-6.

Delas, J., and R. Pouget. 1984. Action de la concentration de la solution nutritive sur quelques caracteristiques physiologiques et technologiques chez *Vitis vinifera* L. cv. Cabernet-Sauvignon II. Composition minerale des organes vegetatifs, du mout et du vin. Agronomie 4:443-450.

Downton, W.J.S. 1977a. Chloride accumulation in different spp. of grapevine. Sci. Hortic. 7:249-253.

Downton, W.J.S. 1977b. Salinity effects on the ion composition of fruiting 'Cabernet Sauvignon' vines. Am. J. Enol. Vitic. 28;210-219.

Downton, W.J.S., and B.R. Loveys. 1978. Compositional changes during grape berry development in relation to abscisic acid and salinity. Aust. J. Plant Physiol. 5:415-423.

Duarte, M. 1983. Budbreak and fruitfulness of desert grapes (*Vitis vinifera* L.). Ph.D. diss. Univ. Arizona, Tempe (Diss. Abstr. 84-04662).

During, H. 1978. Studies on the environmentally controlled stomatal transpiration in grape vines. II. Effects of girdling and temperatures. Vitis 17:1-9.

During, H. 1984. Evidence for osmotic adjustment to drought in grapevines (*Vitis vinifera* L.). Vitis 23:1-10.

During, H., and M. Broquedis. 1980. Effects of abscissic acid and benzyladenine on irrigated and non-irrigated grapevines. Sci. Hortic. 12:253-260.

During, H., and B.R. Loveys. 1982. Diurnal changes in water relations and abscisic acid in field grown *Vitis* vinifera cvs. I. Leaf water potential components and leaf conductance under humid temperate and semiarid conditions. Vitis 21:223-232.

Ehlig, C.F. 1961. Effects of salinity on four varieties of table grapes grown in sand culture. Proc. Am. Soc. Hortic. Sci. 76:323-331.

Embleton, T.W., W.W. Jones, and R.G. Platt. 1975. Plant nutrition and citrus fruit crop quality and yield. Hortic. Sci. 10:48-50.

Erie, L.J., O.F. French, D.A. Bucks, and K. Harris. 1982. Consumptive use of water by major crops in the southwestern United States. U.S. Dept. Agric. Conserv. Res. Rep. 29. Washington, DC.

Food and Agriculture Organization of the United Nations. 1985. Commodity review and outlook 1984-85. FAO Econ. Social Develop. Ser. 36. FAO, Rome.

Food and Agriculture Organization of the United Nations. 1986. FAO production yearbook. FAO Stat. Ser. 70. FAO, Rome.

Freeman, B.M. 1983. Effects of irrigation and pruning of Shiraz grapevines on subsequent red wine pigments. Am. J. Enol. Vitic. 34:23-26.

Freeman, B.M., T.H. Lee, and C. Turkington. 1979. Interaction of irrigation and pruning level on growth and yield of shiraz vines. Am. J. Enol. Vitic. 30:218-223.

Freeman, B.M., and W. Kliewer. 1983. Effect of irrigation, crop level and potassium fertilization on Carignane vines. II. Grape and wine quality. Am. J. Enol. Vitic. 34:197-207.

Freeman, B.M., and R.E. Smart. 1976. Research note: A root observation laboratory for studies with grapevines. Am. J. Enol. Vitic. 27:36-39.

Grimes, D.W., and L.E. Williams. 1990. Irrigation effects on plant water relations and productivity of Thompson Seedless grapevines. Crop Sci. 30:(in press).

Gutierrez, A.P., D.W. Williams, and H. Kido. 1985. A model of grape growth and development: The mathematical structure and biological considerations. Crop Sci. 25:721-728.

Halsey, D., J.R. Spencer, R.L. Branson, and A.W. Marsh. 1963. Vineyard salinity problems corrected with special leaching in Coachella Valley. Calif. Agric. 17:2-3.

Hardie, W.J., and J.A. Considine. 1976. Response of grapes to water-deficit stress in particular stages of development. Am. J. Enol. Vitic. 27:55–61.

Harrell, D.C., and L.E. Williams. 1987a. Net CO_2 assimilation rate of grapevine leaves in response to trunk girdling and gibberellic acid application. Plant Physiol. 83:457–459.

Harrell, D.C., and L.E. Williams. 1987b. The influence of girdling and gibberellic acid application at fruitset on Ruby seedless and Thompson seedless grapes. Am. J. Enol. Vitic. 38:83–88.

Harris, J.M., P.E. Kriedemann, and J.V. Possingham. 1968. Anatomical aspects of grape berry development. Vitis 7:106–119.

Hawker, J.S., and R.R. Walker. 1978. The effect of sodium chloride on the growth and fruiting of 'Cabernet Sauvignon' vines. Am. J. Enol. Vitic. 29:172–176.

Hepner, Y., B. Bravdo, C. Loinger, S. Cohen, and H. Tabacman. 1985. Effect of drip irrigation schedules on growth, yield, must composition and wine quality of Cabernet Sauvignon. Am. J. Enol. Vitic. 36:77–85.

Hiroyasu, T. 1961. Nutritional and physiological studies on the grapevine. J. Jpn. Soc. Hortic. Sci. 30:111–116.

Hofaker, W. 1977. Investigation on the substance production of vines under the influence of changing soil water supply. Vitis 16:163–172.

Huglin, P. 1960. Causes determinant les alterations de la floraison de la vigne. Ann. Amelior. Plant. 10:351–358.

Jacob, H.E. 1931. Girdling grapevines. California Agric. Ext. Serv. Circ. 56.

Kasimatis, A.N. 1967. Grapes and berries. Part I—Grapes. In R.M. Hagon et al. (ed.) Irrigation of agricultural lands. Agronomy 11:719–739.

Kasimatis, A.N. 1977. Differential cropping levels of Zinfandel vines. A progress report on some effects on vine growth, fruit composition and wine quality. p. 189–196. In Proc. Int. Symp. Quality Vintage, Cape Town, South Afr. 14–21 Feb.

Kasimatis, A.N., and L.P. Christensen. 1976. Response of Thompson seedless grapevines to potassium application from three fertilizer sources. Am. J. Enol. Vitic. 27:145–149.

Kasimatis, A.N., and C.L. Lynn. 1975. How to produce quality raisins. Univ. California Berkeley Div. Agric. Sci. Leafl. 2277.

Kliewer, W.M. 1971. Effect of nitrogen on growth and composition of fruits from 'Thompson Seedless' grapevines. J. Am. Soc. Hortic. Sci. 96:816–819.

Kliewer, W.M., and J.A. Cook. 1971. Arginine and total free amino acids as indicators of the nitrogen status of grapevines. J. Am. Soc. Hortic. Sci. 96:581–587.

Kliewer, W.M., B.M. Freeman, and C. Hossom. 1983. Effect of irrigation, crop level and potassium fertilization on Carignane vines. I. Degree of water stress and effect on growth and yield. Am. J. Enol. Vitic. 34:186–196.

Kobayashi, A., H. Yukenoya, T. Fukushima, and H. Wada. 1960. Studies on the thermal conditions of grapes. II. Effects of night temperatures on the growth, yield and quality of Delaware grapes. Bull. Res. Inst. Food Sci. Kyoto Univ. 24:29–42.

Kramer, P.J., and T.T. Kozlowski. 1979. Physiology of woody plants. Academic Press, Orlando, FL.

Kriedemann, P.E., and R.E. Smart. 1971. Effects of irradiance, temperature and leaf water potential on photosynthesis of vine leaves. Photosynthetica 5:6–15.

Lafon, J., P. Couillaud, F. Gay-Bellile, and J.F. Levy. 1965. Rythme de l'absorption minerale de la vigne au cours d'un cycle vegetatif. Vignes Vins 140:17–21.

Levy, J.F., G. Chaler, E. Camhaji, and C. Hego. 1972. Neue statistische Untersuchungen uber die Zusammenhange zwischen dem Mineralsoffgehalt der Blatter und den Ernahrungsbedingungen er Rebe. Vignes Vins 212:21–25.

Liu, W.T., R. Pool, W. Wenkert, and P.E. Kriedeman. 1978. Changes in photosynthesis, stomatal resistance and abscisic acid of Vitis labruscana through drought and irrigation cycles. Am. J. Enol. Vitic. 29:239–246.

Maas, E.V. 1987. Salt tolerance of plants. p. 57–75. In B.R. Christie (ed.) Handbook of plant science in agriculture. Vol. II. CRC Press, Boca Raton, FL.

Maas, E.V., and G.J. Hoffmann. 1977. Crop salt tolerance—current assessment. J. Irrig. Drain Div. Am. Soc. Civ. Eng. 103:115–134.

MacGillivray, N.A., J.D. Gonzales, and D.L. Scruggs. 1985. Irrigation scheduling from evaporation data—method development and field performance. p. 786–791. In Proc. 3rd Int. Drip/Trickle Irrig. Congr. Fresno, CA. 18–21 November. ASAE Publ. 10-85, Vol. II. ASAE, St. Joseph, MI.

Marocke, R., J. Balthazand, and G. Correge. 1976. Exportations en elements fertilisants des principaux cepages cultives en Alsace. C.R. Agric. 420–429.

Matthews, M.A., and M.M. Anderson. 1988. Fruit ripening in grape (*Vitis vinifera* L.): Responses to seasonal water deficits. Am. J. Enol. Vitic. 39:313–320.

Matthews, M.A., and M.M. Anderson. 1989. Reproductive development in grape (*Vitis vinifera* L.): Responses to seasonal water deficits. Am. J. Enol. Vitic. 40:52–60.

Matthews, M.A., M.M. Anderson, and H.R. Schultz. 1987a. Phenologic and growth responses to early and late season water deficits in Cabernet Franc. Vitis 26:147–160.

Matthews, M.A., G. Cheng, and S.A. Weinbaum. 1987b. Changes in water potential and dermal extensibility during grape berry development. J. Am. Soc. Hortic. Sci. 112:314–319.

Mattick, L.R., N.J. Shaulis, and J.C. Moyer. 1972. The effect of potassium fertilizer on the acid content of Concord grape juice. Am. J. Enol. Vitic. 23:26–30.

May, P. 1965. Reducing inflorescence formation by shading individual sultana buds. Aust. J. Biol. Sci. 18:463–473.

May, P., M.R. Sauer, and P.B. Scholefield. 1973. Effect of various combinations of trellis, pruning, and rootstock on vigorous Sultana vines. Vitis 12:192–206.

McCarthy, M.G., and B.G. Coombe. 1985. Water status and winegrape quality. Acta Hortic. 171:447–456.

McIntyre, G.N., L.A. Lider, and N.L. Ferrari. 1982. The chronological classification of grapevine phenology. Am. J. Enol. Vitic. 33:80–85.

McKenry, M.V. 1984. Grape root phenology relative to control of parasitic nematodes. Am. J. Enol. Vitic. 35:206–211.

Morris, J.R., and D.L. Cawthon. 1982. Effects of irrigation, fruit load and potassium fertilization on yield, quality and petiole analysis on Concord grapes. Am. J. Enol. Vitic. 33:145–148.

Morris, J.R., D.L. Cawthon, and J.W. Fleming. 1980. Effects of high rates of potassium fertilization on yield, quality, and petiole analysis of Concord (*Vitis labrusca* L.) grapes. Am. J. Enol. Vitic. 33:145–148.

Nagarajah, S. 1987. Effects of soil texture on the rooting patterns of Thompson Seedless vines on own roots and on Ramsey rootstock in irrigated vineyards. Am. J. Enol. Vitic. 38:54–60.

Neja, R.A., W.E. Wildman, R.S. Ayers, and A.N. Kasimatis. 1977. Grapevine response to irrigation and trellis treatments in the Salinas Valley. Am. J. Enol. Vitic. 28:16–26.

Nobel, P.S. 1975. Effective thickness and resistance of the air boundary layer adjacent to spherical plant parts. J. Exp. Bot. 26:120–130.

Obbink, J., and D.M. Alexander. 1973. Response of grapevine cultivars to a range of chloride concentrations. Am. J. Enol. Vitic. 24:65–68.

Olmo, H.P. 1936. Pollination and the setting of fruit in the black Corinth grape. Proc. Am. Soc. Hortic. Sci. 34:402–404.

Olson, R.A. 1978. The indispensable role of nitrogen in agricultural production. p. 1–31. *In* P.F. Pratt (ed.) Natl. conf. manage. nitrogen irrig. agric. U.S. Natl. Sci. Found., USEPA, and Univ. California, Riverside.

Olson, R.A., and L.T. Kurtz. 1982. Crop nitrogen requirements, utilization, and fertilization. *In* F.J. Stevensen (ed.) Nitrogen in agricultural soils. Agronomy 22:566–605.

Ough, C.S., J.A. Cook, and L.A. Lider. 1968b. Rootstock-scion interactions concerning wine making. II. Wine compositional and sensory changes attributed to rootstock and fertilizer level differences. Am. J. Enol. Vitic. 19:254–265.

Ough, C.S., L.A. Lider, and J.A. Cook. 1968a. Rootstock-scion interactions concerning wine making. I. Juice composition changes and effects on fermentation rate with St. George and 99-R rootstocks at two nitrogen fertilizer levels. Am. J. Enol. Vitic. 19:213–227.

Peacock, W.L., L.P. Christensen, and H.L. Andris. 1987. Development of a drip irrigation schedule for average canopy vineyards in the San Joaquin Valley. Am. J. Enol. Vitic. 38:113–119.

Peacock, W.L., D.E. Rolston, F.K. Aljibury, and R.S. Rauschkolb. 1977. Evaluating drip, flood, and sprinkler irrigation of wine grapes. Am. J. Enol. Vitic. 28:193–195.

Perold, A.I. 1927. A treatise on viticulture. Macmillian, London.

Perry, R.L., S.D. Lyda, and H.H. Bowen. 1983. Root distribution of four *Vitis* cultivars. Plant Soil 71:63–74.

Pongracz, D.P. 1983. Rootstocks for grapevines. David Philip Publ., Cape Town, Rep. of South Africa.

Pouget, R. 1963. Physiology of vegetative dormancy in the vine. Bud dormancy and the mechanism of its breaking. Ann. Amelior. Plant. 13:247.

Pouget, R. 1972. Considerations generales sur le rythme vegetatif et la dormace de bourgeions de la vigne. Vitis 11:198–217.

Pouget, R. 1984. Action de la concentration de la solution nutritive sur quelques caracteristiques physiologiques et technologiques chez *Vitis vinifera* L. cv. Cabernet-Sauvignon. I. Vigueur, rendement, qualite du mout et du vin. Agronomie 4:437–442.

Pratt, C., and B.G. Coombe. 1978. Shoot growth and anthesis in *Vitis*. Vitis 17:125–133.

Prior, L.D., and A.M. Grieve. 1987. Water use and irrigation requirements of grapevines. p. 165–168. *In* T. Lee (ed.) Proc. 6th Aust. Wine Ind. Tech. Conf. Adelaide, Aust. 14–17 July 1986. Australian Industrial Publ., Adelaide.

Richards, D. 1983. The grape root system. Hortic. Rev. 5:127–168.

Robinson, J. 1986. Vines, grapes and wines. Alfred A. Knopf, New York.

Saayman, D., and L. Van Huyssteen. 1980. Soil preparation studies: I. The effect of depth and method of soil preparation and of organic material on the performance of *Vitis vinifera* (var. Chenin Blanc) on Hutton/Sterkspruit Soil. South Afr. J. Enol. Vitic. 1:107–121.

Sauer, M.R. 1968. Effects of vine rootstocks on chloride concentration in 'Sultana' scions. Vitis 7:223–226.

Scholander, P.F., H.T. Hammel, E.D. Bradstreet, and A.E. Hemingsen. 1965. Sap pressure in vascular plants. Science 148:339–346.

Seguin, M.G. 1972. Repartition dans lespace du systeme rediculaire de la vigne. Comp. Rendus Acad. Sci. Paris D:2178–2180.

Singleton, V.L. 1972. Effects on red wine quality of removing juice before fermentation to simulate variation in berry size. Am. J. Enol. Vitic. 23:106–113.

Singleton, V.L., and P. Esau. 1969. Phenolic substances in grapes and wine, and their significance. Advances in food research, Suppl. 1. Academic Press, New York.

Sinton, T.H., C.S. Ough, J.J. Kissler, and A.N. Kasimatis. 1978. Grape juice indicators for prediction of potential wine quality. I. Relationship between crop level, juice and wine composition and wine sensory ratings and scores. Am. J. Enol. Vitic. 29:267–271.

Skinner, P.W., M.A. Matthews, and R.M. Carlson. 1987. Phosphorus requirements of wine grapes: Extractable phosphate of leaves indicates phosphorus status. J. Am. Soc. Hortic. Sci. 112:449–454.

Smart, R.E. 1974. Aspects of water relations of the grapevines (*Vitis vinifera*). Am. J. Enol. Vitic. 25:84–91.

Smart, R.E., and H.D. Barrs. 1973. The effect of environment and irrigation interval on leaf water potential of four horticultural species. Agric. Meteorol. 12:337–346.

Smart, R.E., and B.G. Coombe. 1983. Water relations of grapevines. p. 137–196. *In* T.T. Kozlowski (ed.) Water deficit and plant growth. Vol. 7. Academic Press, New York.

Smart, R.E., J.B. Robinson, G.R. Due, and C.J. Brien. 1985a. Canopy microclimate modification for the cultivar Shiraz I. Definition of canopy microclimate. Vitis 24:17–31.

Smart, R.E., J.B. Robinson, G.R. Due, and C.J. Brien. 1985b. Canopy microclimate modification for the cultivar Shiraz II. Effects on must and wine composition. Vitis 24:119–128.

Smart, R.E., C.R. Turkington, and J.C. Evans. 1974. Grapevine response to furrow and trickle irrigation. Am. J. Enol. Vitic. 25:62–66.

Snyder, R.L., B.J. Lanini, D.A. Shaw, and W.O. Pruitt. 1987. Using reference evapotranspiration (ET_o) and crop coefficients to estimate crop evapotranspiration (ET_c) for trees andvines Univ. California Coop. Ext. Leafl. 21428.

Somers, T.C. 1975. In search of quality for red vines. Food Technol. Aust. 27:49–56.

Somers, T.C. 1977. A connection between potassium in the harvest and relative quality in Australian wines. Aust. Wine Brew. Spirit Rev. 7:32–34.

Somers, T.C., and M.E. Evans. 1977. Spectral evaluation of young red vines, anthocyan equilibria, total pherolics, free and molecular SO_2, chemical age. J. Sci. Food Agric. 28:279–287.

Spayd, S.E., and J.R. Morris. 1978. Influence of irrigation, pruning severity and nitrogen on yield and quality of 'Concord' grapes in Arkansas. J. Am. Soc. Hortic. Sci. 103:211–216.

Spiegel-Roy, P., and B. Bravdo. 1964. Le regime hydrique de la vigne. Bull. Office Int. Vigne 37:232–248.

Stanhill, G. 1962. The control of field irrigation practice from measurements of evaporation. Isr. J. Agric. Res. 12(2):51–62.

State of California. 1975. Vegetative water use in California, 1974. Calif. Dep. Water Resour. Bull. 113-3.

State of California. 1986. Crop water use in California. California Dep. Water Resour. Bull. 113-4.

Stevens, R., and P. Cole. 1987. Grape must composition depends on irrigation management. p. 159-164. *In* T. Lee (ed.) Proc. 6th Aust. Wine Ind. Tech. Conf. Adelaide, Australia. 14-17 July 1986. Australian Industrial Publ., Adelaide.

Stevenson, D.S. 1981. Responses of 6-year-old Diamond grapevines to the change from sprinkler to trickle irrigation and to the time and method of applying nitrogen. Can. J. Soil Sci. 61:571-575.

Tippett, J., R. Radenz, D. Kleweno, and R. Kurosaka. 1987. California grape acreage 1986. California Agric. Stat. Serv., Sacramento.

Vaadia, Y., and A.N. Kasimatis. 1961. Vineyard irrigation trials. Am. J. Enol. Vitic. 12:88-98.

Van Rooyen, F.C., H.W. Weber, and I. Levin. 1980. The response of grapes to a manipulation of the soil-plant-atmosphere continuum. II. Plant-water relationships. Agrochemophysica 12:69-74.

van Zyl, J.L. 1984. Response of colombar grapevines to irrigation as regards quality aspects and growth. South Afr. J. Enol. Vitic. 5:19-28.

van Zyl, J.L. 1986. Canopy temperature as a water stress indicator in vines. South Afr. J. Enol. Vitic. 7:53-60.

van Zyl, J.L.,and L. van Huyssteen. 1980. Comparative studies on wine grapes on different trellising systems: I. Consumptive water use. South Afr. J. Enol. Vitic. 1:7-14.

van Zyl, J.L., and H.W. Weber. 1981. The effect of various supplementary irrigation treatments on plant and soil moisture relationships in a vineyard (*Vitis vinifera* var. Chenin blanc). South Afr. J. Enol. Vitic. 2:83-99.

Weaver, R.J. 1955. Relation of time of girdling to ripening of fruit of Red Malaga and Ribier grapes. Proc. Am. Soc. Hortic. Sci. 65:183-186.

Weaver, R.J. 1976. Grape growing. John Wiley and Sons, New York.

Weaver, R.J., A.N. Kasimatis, and S.B. McCune. 1962. Studies with gibberellin on wine grapes to decrease bunch rot. Am. J. Enol. Vitic. 13:78-82.

Weaver, R.J., and S.B. McCune. 1959. Effect of gibberellin on seeded *Vitis vinifera* and its translocation within the vine. Hilgardia 28:625-645.

Weaver, R.J., and S.B. McCune. 1960. Effects of overcropping Alicante Bouschet grapevines in relation to carbohydrate nutrition and development of the vine. Proc. Am. Soc. Hortic. Sci. 75:341-353.

Webb, A.D. 1981. Quality factors in California grapes. p. 1-9. *In* R. Teranishi and H. Barrera-Benitez (ed.) Quality of selected fruits and vegetables of North America. ACS Symp. Ser. 170. Am. Chem. Soc., Washington, DC.

Weinbaum, S.A., I. Klein, F.E. Broadbent, W.C. Micke, and T.T. Muroka. 1984. Effects of time of nitrogen application and soil texture on the availability of isotopically labeled fertilizer nitrogen to reproductive and vegetative tissue of mature almond trees. J. Am. Soc. Hortic. Sci. 109:339-343.

West, D.W., and J.A. Taylor. 1984. Response of six grape cultivars to the combined effects of high salinity and rootzone waterlogging. J. Am. Soc. Hortic. Sci. 109:844-851.

Westwood, M.N. 1978. Temperate-zone pomology. W.H. Freeman and Co., San Francisco.

Wildman, W.E., R.A. Neja, and A.N. Kasimatis. 1976. Improving grape yield and quality with depth-controlled irrigation. Am. J. Enol. Vitic. 27:168-175.

Williams, D.W., L.E. Williams, W.W. Barnett, K.M. Kelley, and M.V. McKenry. 1985a. Validation of a model for the growth and development of the Thompson seedless grapevine. I. Vegetative growth and fruit yield. Am. J. Enol. Vitic. 36:275-282.

Williams, D.W., H.L. Andris, R.H. Beede, D.A. Luvisi, M.V.K. Norton, and L.E. Williams. 1985b. Validation of a model for the growth and development of the Thompson seedless grapevine. II. Phenology. Am. J. Enol. Vitic. 36:283-289.

Williams, L.E. 1987a. Growth of 'Thompson seedless' grapevines: I. Leaf area development and dry weight distribution. J. Am. Soc. Hortic. Sci. 112:325-330.

Williams, L.E. 1987b. Growth of 'Thompson seedless' grapevines: II. Nitrogen distribution. J. Am. Soc. Hortic. Sci. 112:330-333.

Williams, L.E. 1987c. The effect of cyanamide on budbreak and vine development of Thompson seedless grapevines in the San Joaquin Valley of California. Vitis 26:107-113.

Williams, L.E., P.J. Biscay, and R.J. Smith. 1987a. Effect of interior canopy defoliation on berry composition and potassium distribution of Thompson seedless grapevines. Am. J. Enol. Vitic. 38:287-292.

Williams, L.E., P.J. Biscay, and G.Y. Yokota. 1987b. Growth of Barbera and Chenin blanc grapevines as influenced by row spacing and trellis type. p. 25. *In* Tech. Abstr. Am. Soc. Enol. Vitic. 38th Annu. Meet.

Williams, L.E., and D.W. Grimes. 1987. Modelling vine growth—development of a data set for a water balance subroutine. p. 169-174. *In* T. Lee (ed.) Proc. 6th Aust. Wine Ind. Tech. Conf. Adelaide, Aust. 14-17 July 1986. Australian Industrial Publ., Adelaide.

Winkler, A.J. 1927. Some factors influencing the rooting of cuttings. Hilgardia 2:329-349.

Winkler, A.J. 1953. Producing table grapes of better quality. Blue Anchor 30:28-31.

Winkler, A.J. 1954. Effects of overcropping. Am. J. Enol. 5:4-12.

Winkler, A.J. 1958. The relation of leaf area and climate to vine performance and grape quality. Am. J. Enol. Vitic. 9:10-23.

Winkler, A.J., J.A. Cook, W.M. Kliewer, and L.A. Lider. 1974. General viticulture. 2nd ed. Univ. California Press, Berkeley.

Wolpert, J.A., and G.S. Howell. 1985a. Cold acclimation of Concord grapevines. I. Variation in cold hardiness within the canopy. Am. J. Enol. Vitic. 36:185-188.

Wolpert, J.A., and G.S. Howell. 1985b. Cold acclimation of Concord grapevines. II. Natural acclimation pattern and tissue moisture decline in canes and primary buds of bearing vines. Am. J. Enol. Vitic. 36:189-194.

Woodham, R.C. 1956. The chloride status of the irrigated 'Sultana' vine and its relation to vine health. Aust. J. Agric. Res. 7:414-427.

Yang, Y., and Y. Hori. 1979. Studies on retranslocation of accumulated assimilates in 'Delaware' grapevines. Tohoku Agric. Res. 30:43-56.

35 Drainage and Return Flows in Relation to Irrigation Management[1]

K. K. TANJI

University of California
Davis, California

B. R. HANSON

University of California
Davis, California

Surface and subsurface irrigation return flows are considered as significant components of water resources. In the past, irrigated agriculture had the opportunity to discharge its return flows to the hydrologic system with little or no constraints. Now, however, increasing constraints are being placed on the discharge of return flows to the environment because of two major factors. First, there is increasing awareness of water quality and pollution. Water quality constituents of concern in irrigation return flows include salinity, sediments, nutrients (N and P), pesticide residues, and toxic or harmful trace elements. Second, there is an increasing perception that irrigated agriculture is not using water efficiently. Competing needs for water by various sectors of society, and movement to enhance environmental quality, will require irrigated agriculture to take appropriate actions to use water conservatively.

This chapter focuses on the type and quantity of irrigation return flows as affected by irrigation application methods and their management. Effects of irrigation on the environment due to salinity are covered in chapters 36 (Rhoades & Loveday) and 38 (Hornsby), of this book, erosion and sediment production in chapter 37 (Carter), and pesticide residues and other toxic or harmful constituents in chapter 38 (Hornsby).

This chapter first describes the components and flow pathways of surface and subsurface drainage from irrigated agriculture, the associated problems of unmeasured flows, and differentiation between storm runoffs and irrigation return flows. Next, it identifies factors influencing the type and quantity of return flows relative to various irrigation application methods and their management. Beneficial and deleterious effects of irrigation return flows are briefly enumerated. Finally, a case study on agricultural drainage,

[1] Contribution from Division of Agriculture and Natural Resources, Univ. of California, Davis, CA 95616.

salinity, and Se problems in the San Joaquin Valley of California is presented and various drainage water management options under investigation are described.

I. COMPONENTS AND FLOW PATHWAYS OF RETURN FLOWS

A. The Practice of Irrigation

The practice of irrigation is to provide supplemental water to effective precipitation to satisfy the evapotranspirational needs of crops. In arid to semiarid areas, irrigation is the primary source of water for crop plants. In some semiarid areas, however, crops are grown without the benefit of irrigation but typically require water conservation measures such as summer fallow. In subhumid areas, irrigation may be practiced for a brief period,

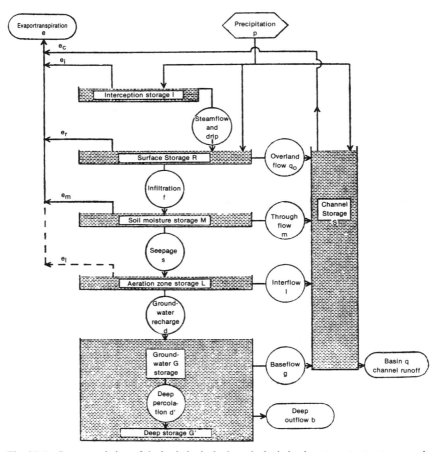

Fig. 35-1. Conceptual view of the basin hydrologic cycle depicting inputs, outputs, storage units, and transfer processes (More, 1969).

depending on effective precipitation and available water-holding capacity of soils, while in humid areas, irrigation may be needed during periods of low precipitation. Thus, return flows from irrigated agriculture may exist in arid to subhumid areas and may become commingled to varying degrees with runoffs from natural precipitation.

B. Basin Hydrologic Cycle

The gross components of the natural basin hydrologic cycle may be defined as Outflow = Inflow ± Change in Storage. Figure 35-1 from More (1969) gives one perspective of the components, flow pathways, and processes involved in the basin hydrologic cycle. Inflow is defined by precipitation and outflow by evapotranspiration (ET), stream runoff, and deep groundwater outflow. The shaded portions represent various system storage units. The transfer processes and flow pathways of water are depicted respectively by the circles and arrows. Computer simulation models for these pools and fluxes of water are available, e.g., Stanford Watershed Model IV (1966).

C. Regional Irrigation System

In a manner similar to the basin hydrologic cycle, the components and pathways for the irrigated agriculture system may be conceptualized. For instance, Fig. 35-2 (Utah State Univ. Foundation, 1969) depicts a reach (length) of a river from which water is diverted for irrigation, and ET losses and a number of return flow components to the river are identified. Figure 35-3 (Hornsby, 1973) shows another conceptual version including on-farm elements. These two diagrams illustrate that a regional irrigated system consists of various subsystems: surface water hydrology, water conveyance and distribution system, soil water, and groundwater hydrology. Not shown in these diagrams are outflows to surface water impoundments and lakes and to the ocean.

D. Farm Irrigation System

Figure 35-3 shows components of farm irrigation systems beginning with lateral diversions, delivery of water to the farm gate, and tracing its path over and through the root zone. Figure 35-4 (Tanji, 1976) gives a more detailed version focusing on the crop root zone and is applicable to smaller field scale spatial entities.

The symbol Q refers to volume of water per unit area and time. The inputs to the crop root zone are infiltrated (effective) applied irrigation water (Q_{eaiw}), infiltrated (effective) precipitation (Q_{ep}), and stored soil water (Q_{ssw}). The outputs include evaporation of precipitation (Q_{etp}), ET (Q_{et}), and surface runoff from precipitation (Q_{pro}), as well as from irrigation (Q_{iwro}), deep percolation (Q_{dp}), and collected subsurface drainage water (Q_{sdw}). In this scheme, surface runoff and collected subsurface drainage water constitute surface irrigation return flow (Q_{sirf}). The principal storage component in the crop root zone is the soil water (Q_{sw}).

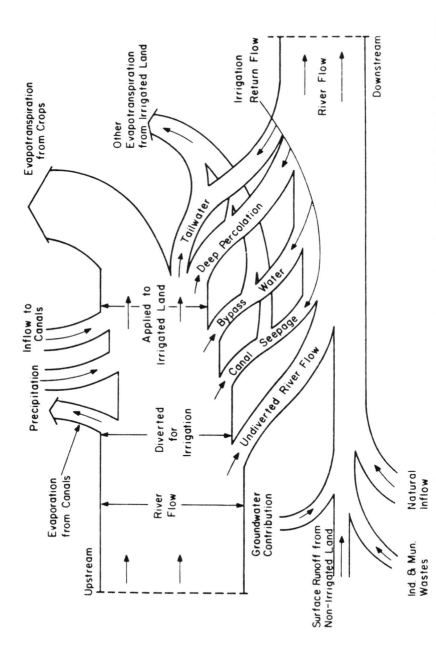

Fig. 35–2. Conceptual diagram of the irrigation return flow system for a given reach of a river system (Utah State Univ. Foundation, 1969).

E. Problems of Unmeasured Flows

Figures 35–1 to 4 identify a number of components and flow pathways for natural precipitation (rainfall and snowmelt), diverted irrigation water, and irrigation return flows. The components most frequently gaged in a watershed are rainfall and stream flow. For irrigated agriculture, rainfall, water deliveries, operational spills, and surface return flows from irriga-

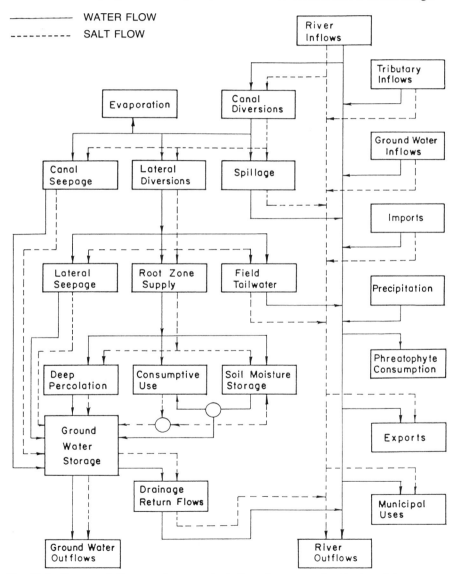

Fig. 35–3. Conceptual diagram of water and salt flows for a generalized hydrosalinity model (Hornsby, 1973).

tion/drainage districts are most frequently measured. All other components and transfer fluxes are typically estimated—some, such as crop ET, with more confidence than others, such as deep percolation. Except for experimental watersheds and extensively monitored irrigation projects, most of the components and fluxes of water are unmeasured because of costs and technical problems. However, water managers and agencies report "ball park" estimates on water flows based on long-run experience.

F. Flood Vs. Irrigation Return Flows

Having been appraised that many water components and transfer fluxes are unmeasured, the next leading question is whether or not we can differentiate between storm and irrigation return flows. In subhumid areas in which irrigation is practiced once or only for a short duration, assessing storm vs. irrigation return flows is difficult. In arid to semiarid areas where little summer precipitation occurs and irrigation is practiced in the spring and summer, a clearer distinction exists between storm runoffs and irrigation return flows. However, the separation between nonirrigation and irrigation return flows may be diffuse due to lag time for flows to reach points at which measurements are made.

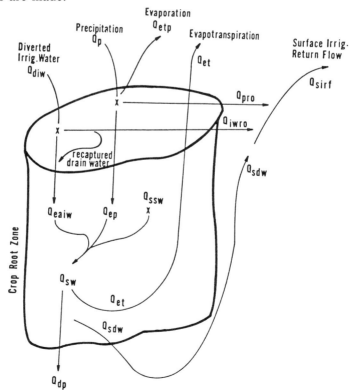

Fig. 35–4. Conceptual view of water flow components in the crop root zone of irrigated field (Tanji, 1976).

For illustrative purposes, the hydrology and irrigation are presented for the 662-km^2 Glenn-Colusa Irrigation District (GCID) in the Sacramento Valley, California (Tanji et al., 1983). Irrigation season in GCID begins in March and ends in October of each year. Figure 35–5 contains plots of monthly precipitation, irrigation deliveries, and surface drainage outflows from GCID for the period November 1977 through October 1980. Much of the precipitation (rainfall) occurs from about November through April. The annual precipitation was 639, 427, and 589 mm for the 1978, 1979, and 1980 hydrologic years (October through September), respectively. Water deliveries were 987 300, 851 700, and 1 031 450 dam^3 (1 acre-ft = 0.8107 dam^3) for

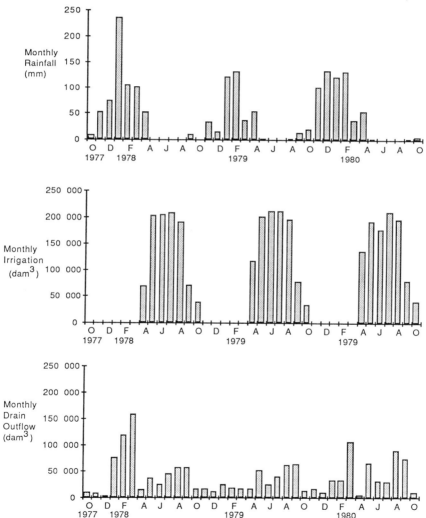

Fig. 35–5. Monthly distribution of rainfall and irrigation and surface drainage outflow in Glenn-Colusa Irrigation District, California (Tanji et al., 1983).

the 1978, 1979, and 1980 irrigation seasons, respectively, applied to 52 080, 55 400, and 57 430 ha of irrigated lands. Paddy rice (*Oryza sativa* L.), the major crop, consisted of 71% of the cropland in each of these 3 yr. The drainage outflows from GCID for the nonirrigation seasons of 1978, 1979, and 1980 hydrologic years were, respectively, 374 040, 87 630, and 210 700 dam^3 and for irrigation seasons, respectively, 251 800, 261 000, and 300 300 dam^3. Drainage during the nonirrigation season consists of runoffs from rainfall within the district as well as runoffs from an adjacent watershed.

As noted in Fig. 35-5, precipitation during the nonirrigation seasons and water deliveries during irrigation seasons are clearly discernible. The drainage outflows (surface runoffs and natural subsurface drainage) for winter flood runoffs are also clearly distinguished for January through March of 1978 and 1980, but not for 1979 because of lower rainfall. Irrigation return flows generally are smaller in April when the rice fields are flooded for aerial seeding and there is little or no rainfall, and larger in August and September when the rice fields are drained before harvest. The fractions contributed to runoff by rainfall and irrigation would be more difficult to assess for areas where significant precipitation occurs during the irrigation season. This is generally the case in subhumid and other areas where the precipitation season overlaps the irrigation season.

II. QUANTITY AND MANAGEMENT OF RETURN FLOWS

A. Factors Influencing Quantity of Irrigation Return Flows

Irrigation return flows consist of surface as well as subsurface components. Surface irrigation runoff, sometimes referred to as *tailwater*, is applied irrigation water which does not infiltrate into the soil and runs off the lowest portions of the irrigated field. *Subsurface irrigation return flow* is irrigation water draining through the crop root zone. Typically, a portion of the subsurface return flow percolates vertically into the underlying groundwater storage zone and the remainder moves laterally along the hydraulic gradient. The latter portion of the subsurface return flow in shallow groundwater systems may be intercepted and collected by underdrainage systems or open channels, and frequently commingles with surface runoffs. The quantity and type of irrigation return flow depend primarily on the irrigation application method and its management.

In both arid and humid areas, surface irrigation (furrow, border, and basin), sprinkler irrigation, and drip/trickle irrigation are used. These methods each have advantages and disadvantages, but the potential attainable application efficiency for each system is nearly the same for properly designed and managed systems. In addition, subirrigation is practiced in humid areas where a high water table exists. Subirrigation involves manipulating the depth of the water table using a subsurface drainage system to supply water to the crop and to remove excess water from the soil. Irrigation water is added to the groundwater through the drainage system during periods of high crop water use. This method is practical where the salinity of the shallow groundwater is low, and leaching from rainfall is sufficient.

A major difference between arid and humid areas is irrigation scheduling. In arid areas, the crop water needs are met by irrigation with little contribution from rainfall. Irrigation scheduling in humid areas, however, is complicated by uncertainties in rainfall occurring during the growing season. Since the rainfall can only be expressed by probabilities, a timing uncertainty exists in applying irrigation water. Even with this uncertainty, however, Dickey (1986, personal communication) recommended irrigating when the allowable depletion occurs. Adequate subsurface and surface drainage should be provided to remove surface and subsurface return flows if substantial rainfall occurs just after an irrigation. Fuoss and Carter (1987) have demonstrated the use of computer simulation using climate and soil conditions, including rainfall forecast, for the dual control of subsurface drainage and subirrigation.

B. Irrigation Systems and Return Flows

Performance characteristics of irrigation systems are (Walker & Skogerboe, 1986):

1. Application efficiency = ratio of the average amount of water stored in the root zone to the average amount of water applied. Good efficiency depends on good uniformity and proper management of irrigation water.
2. Uniformity = a measure of how uniformly water is applied throughout a field. Uniformity is normally expressed as Christiansen's coefficient of uniformity (CU) or distribution uniformity (DU). Drip irrigation also uses emissions uniformity (EU). Good uniformity depends on proper selection, design, operation and maintenance of the irrigation system.
3. Deep percolation ratio = ratio of the average amount of deep percolation or subsurface drainage to the average amount of water applied. Subsurface drainage is water that percolates below the root zone.
4. Tailwater ratio = ratio of the average surface runoff to the average amount of water applied. Surface runoff is excess irrigation water that runs off at the lower end of the field.

Some potential attainable uniformities and application efficiencies are listed in Table 35-1.

Adequate irrigation at the potential attainable application efficiency is the criterion normally used for irrigation system design and management. Typically, irrigation systems are designed to apply at least the desired amount of water (usually the soil moisture depletion) to 80% of the field, which is defined as an adequate irrigation. However, for high-value crops, the percent adequately irrigated may exceed 90%. The actual amount applied depends on the system management. However, where irrigation return flows

Table 35-1. Potential attainable irrigation uniformities and application efficiencies (assumes adequate irrigation as previously defined). Sources: Jensen (1980), Burt (1987, personal communication), and Hanson (1987).

System	Uniformity (DU)	Application efficiency
	%	
Sprinklers		
Periodic move	70–80	65–80
Continuous move	70–90†	75–85
Solid set	90–95	85–90
Drip/trickle	80–90	75–90
Surface		
Furrow	80–90‡	60–90§
Border	70–85‡	65–80§
Basin	90–95‡	75–90

† Higher value for systems using spray nozzles on booms or using impact sprinklers.
‡ Uniformity values do not include nonuniform water infiltration because of soil variability. Some evidence (Wallender, 1986) suggests that DU's based on infiltration opportunity times only should be reduced five to ten percentage points to account for the effect of soil variability.
§ Higher values for systems with tailwater recovery systems or using cutback flow.

cause adverse environmental concerns and impacts, a design and management criterion should be reduction of these impacts. The most desirable design and management strategy may not be to achieve the application efficiency, but instead may be to reduce adverse environmental impacts from irrigation return flows. For example, where municipal wastewater is used for irrigation, a system might be designed for minimum surface return flows and maximum subsurface return flows. However, where disposal of subsurface drainage water is a problem, design criteria might be to minimize subsurface drainage, which for surface irrigation may increase surface runoff. For both cases, the most desirable design and management strategy may not be at the potential application efficiency.

The application method and the management of the irrigation system determines the type of irrigation return flow and its quantity and quality. Surface return flows occur when the application rate of the system exceeds the infiltration rate of the soil. Subsurface return flows are caused by over-irrigation (least-watered areas receive excess water) and by nonuniform water applications.

Some properties of irrigation systems affecting return flows (Hanson, 1987) are explained below.

1. Drip/Trickle Irrigation

This irrigation method applies water precisely, both in volume and location. No surface runoff should occur for a properly designed system, however, nonuniform water application throughout the system may cause subsurface return flow. Nonuniformity is caused by hydraulic losses due to elevation differences and friction losses along laterals and mainlines, plugging of emitters, and by variability of emitter discharge from wear and manufacturing variability. Nonuniformity is minimized by maintaining pressure losses <20% of the average pressure, using adequately filtered water and using emitters with a manufacturing coefficient of variation of <5% (Von Bernuth & Solomon, 1986). An advantage of drip/trickle irrigation is the precise application of fertilizers, nutrients, and herbicides, which may reduce the mass loading of these chemicals in subsurface return flows, provided good system uniformity is maintained.

2. Sprinkler Irrigation

Little or no surface runoff should occur for a properly designed sprinkler system. In reality, however, matching the application rate to the changing infiltration rate may be difficult, and thus on sloping ground, surface runoff can occur.

Nonuniformity is caused by hydraulic losses along laterals and mainlines from friction losses and elevation differences, by poor system maintenance which results in leakage, sprinkler head and nozzle wear, and mixing of nozzle sizes; and by the areal distribution of the applied water. The areal distribution depends on sprinkler spacing, wind speed and direction, pressure, and sprinkler head and nozzle types. Nonuniformity can be minimized by limiting pressure losses to 20% of the design pressure, by timely system maintenance, and by a proper combination of pressure, sprinkler head, and nozzle type; and by spacing to provide good areal uniformity.

Continuous-move sprinkler machines such as center-pivot and linear-move systems are assumed to have higher uniformities than periodic-move systems, particularly under high-wind and low-pressure conditions. However, Hanson and Wallender (1986) found moderate uniformities, reasons for which include inadequate overlap of the spray pattern of spray nozzles and the start-stop sequence of the machines (although these machines are called continuous-move, movement occurs in a series of starts and stops controlled by a guide tower).

3. Surface Irrigation

Furrow and border irrigation cause both surface and subsurface return flows, while basin irrigation causes subsurface flows only. The performance of these systems is affected by soil infiltration rate and its variability, surface slope, length of run of field, surface roughness, and unit or furrow inflow rate. A primary factor is the infiltration rate, which varies spatially and temporally, and is difficult to estimate reliably in some soils using conventional methods such as ring and blocked furrow infiltrometers.

Uniformity of surface irrigation systems depends on the time required for water to advance across the field (advance time), the time required for water to recede across the field (recession time), and on the variability of the soil infiltration rate. The advance time causes variation in depths of infiltered water along the field length. These differences can be reduced by decreasing the length of run, increasing the unit inflow rate, modifying the slope, or reducing the infiltration rate through furrow compaction. The maximum uniformity, however, is limited by the spatial variability of the infiltration rate, which had a CU nearly equal to 80% for a field described as uniform in soil texture (Wallender, 1986).

Application efficiency of surface irrigation does not differentiate between surface and subsurface return flows. Surface return flows, however, can be recovered using tailwater recovery systems, which collect the surface runoff at the lower end of the field and return the water at the upper end for subsequent reuse or for use elsewhere. *Cutback irrigation*, which involves decreasing the unit or furrow inflow rate after the water advance is complete, also reduces this return flow but can create management difficulties.

Reducing irrigation return flows from surface irrigation systems involves trade-offs, since surface and subsurface flows are competitive. Subsurface return flows can be reduced by decreasing the length of run, but this will increase the surface return flow (unless cutback irrigation or tailwater recovery systems are used), and vice versa (Fig. 35–6).

4. Subirrigation

Surface return flow can occur in a subirrigation system if sufficient rainfall occurs after the water table has been raised to its maximum height (Smith et al., 1985). For high water-table conditions, little storage capacity exists in the soil for the rainfall and, as a result, the potential for surface runoff is increased. The possibility also exists for subsurface lateral flow from the field under irrigation to the surrounding fields, caused by a regional flow if the water table of the irrigated field is higher than that of the surrounding fields.

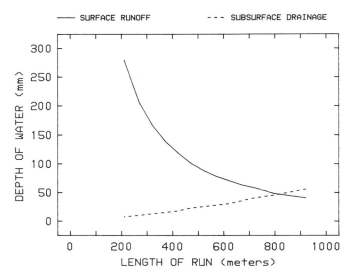

Fig. 35–6. Amounts of surface runoff and subsurface drainage vs. length of run. No overirrigation is assumed to occur.

C. Control and Management of Surface Return Flows

In general, both surface runoff and drainage from the crop root zone are more prominent in gravity flow systems (e.g., furrow and border) as compared to pressurized systems (e.g., sprinkler and drip irrigation). The reasons for this observation are pressurized systems allow for better control of amounts of water applied and are not as significantly affected by variabilities in soil infiltration rates.

Chapter 37 (Carter) of this book presents considerable data on tail water and sediment loading from furrow-irrigated fields. A 2-yr study on four commercial rice fields in the Sacramento Valley showed that seasonal runoffs from flooded rice fields averaged 0.52 m³ m⁻² from 2.72 m³ m⁻² of water inflow or 19% surface outflow of the total surface inflow (applied + rainfall) (Tanji et al., 1977). In another California study in the San Joaquin Valley, tailwater tests made on a 631-ha furrow-irrigated tomato (*Lycopersicon esculentum* Mill.) field equipped with a tailwater recovery system produced 10 000 m³ of tailwater from an application of 35 000 m³ or 29% runoff for one irrigation event (Tanji et al., 1980).

Conclusions from the above and other studies (Tanji et al., 1980) are summarized as follows. In most instances it is difficult to accurately measure tailwater production from a given field because tailwater is produced intermittently. Unless tailwater is collected in a sump or drain it is usually discharged diffusively. Tailwater is not usually collected in a drain separate from other collected return flows such as tile drainage effluents. This tailwater may not always reach a receiving water body because of evaporation losses, infiltration into the land surface, ET by phreatophytes, and recovery and reuse by downstream irrigators and other water uses (e.g., wildlife refuges). In areas where water is scarce or expansive, tailwater is seldom discharged or, if produced, is generally reused at or close to the site of production. In contrast, in areas where water is plentiful or inexpensive, there is a tendency for larger water applications and, consequently, larger production of tailwater. This observation does not necessarily mean inefficient water management if downstream reuse is commonly practiced.

In furrow- and border-irrigated fields, tailwater and sediment loading are affected by soil texture and infiltration rates, length of run, slope of field, and stream size. Based on studies in southern Idaho, Carter and Bondurant (1977) recommended reducing sediment loads to receiving waters by: (i) reducing the stream size; (ii) using cutback irrigation, (iii) changing direction of irrigation to one of lower slope, (iv) changing some cultivation practices, and (v) using tailwater recovery ponds, vegetative buffer strips, and/or sediment retention ponds.

Cablegation is an automated surface irrigation system with a potential for reducing surface runoff. This system consists of a pipeline with orifices at intervals equal to the furrow spacing installed at the head of the field. A plug inside the pipeline is allowed to slowly move downstream, which causes a gradual reduction in the furrow inflow rate. This reduction, in turn, decreases the surface runoff rate compared to a system with a constant inflow rate. Goel et al. (1982) found that for a cablegation system, both the volume of surface runoff and the runoff rate (which was fairly constant) were < 50% of the volume of runoff and the maximum runoff rate, respectively, of a conventional system, thus reducing the potential for soil erosion during an irrigation.

Surface return flow under sprinkler irrigation occurs when the application rate of the sprinkler system exceeds the infiltration rate of the soil. Factors affecting runoff and erosion under sprinkler irrigation include application rate and uniformity, amount applied, tillage system, crop type and rotation, soil compaction, soil type, and field topography (Crowley & Loudon, 1986). Reducing runoff involves changing system characteristics to reduce the application rate and to improve uniformity or modifying the soil characteristics to reduce erosion. Studies such as those by Laflen et al. (1978); Laflen and Tabatabai (1984); Barias et al. (1978); Haan (1971); and McDowell and McGregor (1980) have shown that sediment losses caused by rainfall or sprinkler irrigation can be greatly reduced using methods of conservation tillage.

D. Control and Management of Subsurface Return Flows

The primary hazard of subsurface return flows is the accumulation of soluble salts and other constituents of concern as the drainage water from irrigation percolates through the soil. The problems can be compounded if this return flow causes a high water table, thus requiring subsurface drainage systems to control soil salinity and to prevent waterlogging. Managing subsurface return flows consists of several possible aspects—one, reducing the amount of return flow through better irrigation management and two, reduc-

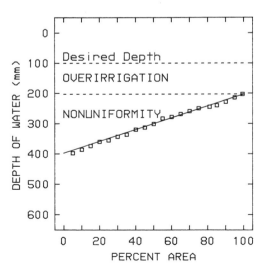

Fig. 35–7. Distribution of applied water showing amount of subsurface drainage because of overirrigation and amount due to irrigation system nonuniformity.

ing return flows in high water table areas by using subsurface drainage water for crop ET.

Sources of subsurface drainage from irrigation are illustrated in Fig. 35–7, the cumulative distribution of applied water. The cumulative distribution shows the area of the field that receives at least a given amount of water. For example, Fig. 35–7 shows that 100% of the field received at least 200 mm of water, which is in excess of the desired depth. The area labeled "over-irrigation," between the dashed lines, is the amount of subsurface drainage from the excess application. Figure 35–7 also shows that 10% of the field received at least 380 mm of water, the result of nonuniform water application. The area labeled "nonuniformity," between the solid slanted line and the bottom dashed line, is the subsurface drainage from the nonuniform application. Proper irrigation management can reduce overirrigation, while proper irrigation system selection, design, operation, and maintenance can control the uniformity.

Keys to reducing subsurface return flows through irrigation water management are improving the uniformity of the applied water and decreasing the average depth of water applied (Hanson, 1987). Potential attainable uniformities of the various irrigation systems were presented earlier in Table 35–1. The higher the uniformity, the higher the potential of the irrigation system for subsurface drainage reduction. Figure 35–8 shows the approximate cumulative water distributions for CU's of 60 and 80%. The area between the dashed line labeled "desired depth" and the solid line representing the depth of applied water is the amount of subsurface drainage. For an adequately irrigated field, the amount of subsurface drainage is about 42% (53 mm) of the average depth applied (127 mm) for a CU of 60%. If the CU

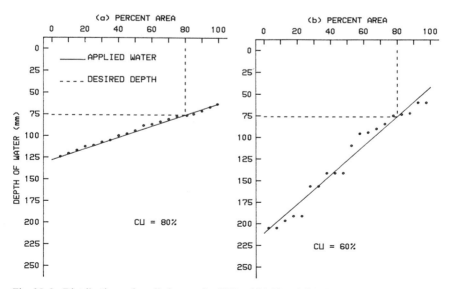

Fig. 35–8. Distributions of applied water for CU's of (a) 80 and (b) 60%. Area between dashed and solid line is the subsurface drainage.

is increased to 80%, then the subsurface drainage is reduced to 22% (21 mm) of the average depth (96 mm). If the uniformity is not improved and if no overirrigation occurs, subsurface drainage can only be reduced by deficit irrigating a considerable part of the field. For the CU of 60% (Fig. 35-8), deficit irrigating 60% of the field will reduce the subsurface drainage to about 7 mm.

If overirrigation occurs, then the cumulative distribution in Fig. 35-7 applies. Substantial drainage reductions may occur simply by reducing the average depth applied by decreasing the irrigation time.

Several studies have evaluated various irrigation methods for reducing subsurface return flows. Wierenga (1977) compared the effect of furrow and trickle irrigation on the reduction and salinity of subsurface return flows. His experiment consisted of applying water at three application efficiencies and at three irrigation schedules. Crop type was cotton (*Gossypium hirsutum* L.) and the soil type was a silty clay loam with sand below 80 cm. He found that trickle irrigation used about 20 to 28% less water than the furrow and increased yield by about 8%, but the impact of the irrigation method on soil salinity was inconclusive. Irrigation treatments showed no significant effect on cotton quality parameters.

Bliesner et al. (1977) evaluated the effect of different leaching fractions on subsurface return flows by using a sprinkler system. Alfalfa (*Medicago sativa* L.), grown on a sandy clay loam, was irrigated with three salinities of water under three depths to a water table. Results showed that the desired leaching fractions were attained, but the measured changes in soil salinity were less than expected; little seasonal change occurred, which was believed to occur because of salt precipitation.

In conjunction with this study, Willardson et al. (1977), discussed the ability of commercial sprinkler systems to attain low leaching fractions (2-3%). The dominant factor in controlling subsurface return flows is the areal uniformity of application. They pointed out that the minimum realistically attainable leaching fraction would be about 10%, and for an overall field leaching fraction of 15%, it would vary from zero to 34% throughout the field because of the uniformity of applied water.

Wendt et al. (1977) compared furrow, sprinkler, and subsurface (manual and automatic) drip irrigation systems to determine the effect of the irrigation system on subsurface return flow quality. Plots about 61 m long were planted with corn (*Zea mays* L.) in a loamy fine sand. They found that the water use was the lowest for the automatic drip and was the highest for the furrow system. They also found that the quality of the subsurface return flows under the subsurface system was better than those of the furrow and sprinkler systems. Quality was measured in terms of soluble salts and N.

Dawood and Hamad (1985) measured the irrigation efficiency and uniformity of a trickle system, a solid-set sprinkler system, and a furrow system. The trickle and sprinkler laterals were 45 and 81 m long, respectively, while the furrow length was 70 m. Similar efficiencies were found for the trickle and sprinkler systems (about 95%), while that of the furrow system was about 50%. They indicated that surface runoff, not measured, contributed substantially to the low efficiency of the furrow system. Distribution uniformity was about 90% for the trickle system and about 80% for the sprinkler and furrow systems (a uniform soil was assumed).

French et al. (1985) compared trickle irrigation with level-basin irrigation. Plot length for both systems was 17 m. Little difference between irrigation systems was found in crop yield of cotton and in the water applied. For the level-basin system, no surface return flows occurred.

Surge irrigation is a recent development with potential for reducing subsurface return flows. This irrigation method applies the water in pulses in contrast to the conventional continuous flow method, thereby apparently decreasing the soil infiltration rate. The end result is that less water is required for complete advance. Research has shown that 50 to 70% of the water needed by continuous-flow systems is required by surge irrigations (Coolidge et al., 1982; Goldhamer et al., 1987). These differences reflect the drainage reduction potential of surge irrigation, although its impact on the uniformity is uncertain.

One should be cautious in extrapolating these results to field conditions. First, such comparisons are valid only if all systems are optimally operated. Naturally, one would expect a poorly managed furrow system to use more water than a highly managed drip system. Second, an experimental design that reflects a field-wide design should be used. The effect of an irrigation system on return flows may be considerably different for small plots than for a field-scale system. While little differences may be found for small plots, significant differences may exist on a field-wide basis.

Walker et al. (1978) discussed methods for controlling subsurface return flows. Their suggestions include:

1. Improving water delivery systems through canal and ditch lining.
2. Improving the management of water delivery system through irrigation scheduling by demand.
3. Improving the uniformity of surface irrigation system by using larger stream flows such that the advance time equals 25% of the minimum intake opportunity time.
4. Using tailwater return systems and cutback irrigation.
5. Improving land grading.
6. Changing to sprinkler and trickle irrigation.
7. Improving irrigation scheduling.

They concluded, however, that while the most effective means of controlling subsurface return flows is by source management, no economic incentive for better water management exists because of cheap and plentiful water.

Irrigation scheduling has also been suggested as a means for controlling subsurface return flows (Jensen, 1975). Irrigation scheduling will reduce subsurface drainage if better timing of irrigations decreases the number of irrigations and if better estimates of soil moisture depletion between irrigations decrease the average depth applied. However, Walker et al. (1978) stated that improved irrigation scheduling by itself is insufficient for substantially reducing subsurface return flows.

Subsurface return flows may also be reduced using subsurface drainage systems to recover the flows for crop production. Rhoades et al. (1987a) reviewed a number of studies investigating crop production using saline water and found upper-limit recommendations of electrical conductivity of 7 to 8 dS m^{-1} for successful production of salt-tolerant crops. Rhoades et al. (1987a, b) also advocated a strategy for successful use of saline water which involves using low-salinity water for stand establishment and then changing to higher salinity water during a salt-tolerant growth stage of the plant. A rotation of crops could be used where relatively salt-tolerant crops would be grown using the high-salinity water and then the soil would be reclaimed by irrigating a relatively sensitive crop only with low-salinity water. A long-term project in the Imperial Valley of California showed no yield or quality declines from this strategy, where about 25 to 50% of the water applied was saline (3000 mg L^{-1}).

On-farm strategies for irrigating with saline water include: (i) irrigating with undiluted saline water after crop establishment, as suggested by Rhoades et al. (1988a, b); and (ii) irrigating with diluted saline water, where low-salinity water is used for crop establishment, and then blending the saline water with the low-salinity water for the remaining irrigations. While Rhoades et al. (1988a, b) felt that the first strategy is the best, the most feasible one may be determined by on-farm constraints. Irrigating with undiluted subsurface drainage water may require a storage facility to accumulate sufficient water for irrigation, whereas irrigating with diluted drainage water may require a distribution system to continuously convey the water to the fields being irrigated.

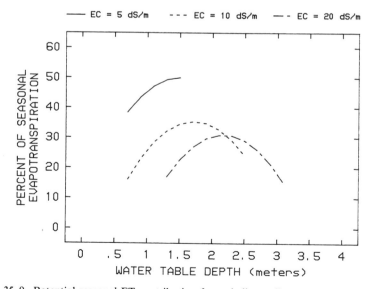

Fig. 35–9. Potential seasonal ET contribution from shallow saline groundwater.

Where high water tables exist, subsurface return flows can be reduced through water table management, which involves controlling the depth of the water table to maximize the crop's use of the shallow groundwater. This approach has been advocated by Benz et al. (1981) for areas where leaching from rainfall is sufficient. However, Hanson (1984) advocated using this method in arid areas such as the San Joaquin Valley to reduce the volume of saline subsurface drainage water for disposal. Grimes et al. (1984), found that up to 50% of the crop's seasonal ET can come from saline shallow groundwater, and that the depth at which maximum use occurs depends on the salinity of the shallow groundwater (Fig. 35-9). Thus, through water table management, the irrigation efficiency can be increased by recovering subsurface return flows. Water table management also may improve the uniformity of the applied irrigation water by redistribution of deep percolation through localized lateral flow of the shallow groundwater. Water table management, however, requires that subsurface drainage systems be designed and managed to control the water table depth and that irrigation systems be managed such that seasonal leaching is sufficient.

III. EFFECTS OF IRRIGATION RETURN FLOWS ON THE ENVIRONMENT

A. Beneficial Uses and Impacts

Irrigation return flows are oftentimes beneficially used—either by plan or incidentally (Tanji et al., 1977). This reuse may occur at several spatial levels, e.g., on farm, regional, river basin, and up to interbasin levels. The types of beneficial uses of irrigation return flows include irrigation of crop and pasture lands, livestock water supply, wetland/wildlife habitats, groundwater recharge, maintenance of summer (low) flows in streambeds that otherwise would become dry, repulsion of salinity in tidal habitats and coastal groundwater bodies, warm water fish habitat, anadromous fish migration and spawning, navigation, recreation and aesthetics, and municipal and industrial water supply. Some of these beneficial uses are partly to wholly dependent on irrigation return flow as the principal source of supply water.

B. Deleterious Impacts

Irrigation return flows may have detrimental impact(s) on the environment, some of which are because of quality characteristics while others may be due to quantity. The latter may be illustrated by promotion of mosquito (*Aedes aegypti*) and other pests from flooding, as well as contribution to waterlogging of downslope lands. The former include the presence of excessive levels of dissolved mineral salts, sediments, nutrients, pesticide residues, and toxic trace elements such as B, Se, As, Cd, and other heavy metals (Tanji et al., 1986).

As depicted in Fig. 35–3 and 35–4, irrigation return flows consist of surface runoffs and subsurface waters either percolating past the root zone, intercepted by natural or human-made open drains, and collected from underdrainage systems such as tile drainage. The quality characteristics of these components of return flows vary (Tanji et al., 1977). For instance, operational spills have qualities similar to the delivered irrigation water; surface runoffs are only slightly degraded with respect to salinity and nutrients but may be markedly higher in sediments and certain pesticide residues. Subsurface waters generally contain greater concentrations of salts and nitrates than applied irrigation waters, and are lower in sediments and certain pesticide residues than surface runoffs. Sediments are addressed in chapter 37 (Carter) of this book, salinity in chapter 36 (Rhoades & Loveday), and nutrients, pesticide residues and some trace elements in chapter 38 (Hornsby).

IV. CASE STUDY ON DRAINAGE WATER MANAGEMENT

A. Background

A case study on drainage, salinity, Se, and other toxic elements is presented to illustrate the complex water quantity-water quality relationships in salt-affected lands, and the various options being explored to reduce drainage water production as well as treatment and disposal of drain waters (Tanji et al., 1986; Letey et al., 1987).

Figure 35–10 (San Joaquin Valley Drainage Program, 1987) shows the extent of the drainage problem area in the west side of the San Joaquin Valley where irrigation has been practiced for about 15 to as long as 125 yr. Professor E.W. Hilgard of the University of California initiated alkali soil investigations around 1877 and concluded in 1889 that subsurface drainage was needed in the San Joaquin Valley for soils that were waterlogged and salt affected. But it was not until the early 1950s when the first subsurface drainage systems were installed in this area (Tanji et al., 1986).

The drainage problem was escalated by irrigation water imported from the Sacramento River system to the north and delivered by the federal Central Valley Project (CVP). The northern portion (Fig. 35–10) of the CVP, known as the Delta-Mendota Service Area, was completed in 1950 to serve over 100 000 ha of croplands with about 105 000 ha-m of water delivered annually by the Delta-Mendota Canal. Most of the surface and subsurface irrigation return flows are either used in wetland habitats or discharged into the middle reaches of the San Joaquin River. The southern portion of the CVP, the San Luis Service Area, was completed in 1969 to serve the 243 000-ha Westlands Water District (WWD) with about 148 000 ha-m of irrigation water delivered annually from the California Aqueduct. The San Luis Service Area currently has no means of disposing drainage waters. Surface return flows are contained within the service area.

In 1957, the California Department of Water Resources (DWR) recommended that a joint state-federal Valley-wide drain was needed. Plans were developed in the mid-1960s to construct a 450-km $86 million master drain from near Bakersfield to an outfall in the Sacramento-San Joaquin Delta. However, state funding for the master drain did not materialize and the upper 137-km reach of the drain, now known as the San Luis Drain (SLD), was built by the U.S. Bureau of Reclamation (USBR) between 1968 and 1975

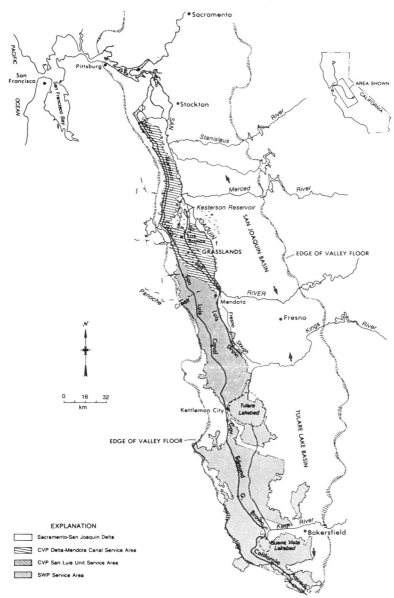

Fig. 35–10. Drainage-impacted area in the west side of San Joaquin Valley showing locations of Kesterson Reservoir and the San Luis Drain (San Joaquin Valley Drainage Program, 1987).

to provide drain water disposal for the WWD (Fig. 35–11; Letey et al., 1987). The SLD terminates at Kesterson Reservoir, which was originally planned as a flow-regulating reservoir for the master drain.

In 1975 the USBR, DWR, and California State Water Resources Control Board (SWRCB) formed the San Joaquin Valley Interagency Drainage Program (IDP) to develop alternatives for in-valley and off-valley disposal of saline agricultural drain waters. After evaluating 18 alternative options, the IDP recommended the completion of the SLD with an outfall in the western portion of the delta (San Joaquin Valley Interagency Drainage Program, 1979).

In 1980, USBR requested a waste discharge permit for the SLD from the SWRCB, which in turn, requested an environmental impact report from USBR by 1984. Concerns over agricultural drain water disposal options centered on salinity and B in the 1950s, then nitrates in the 1960s, pesticide residues in the 1970s, and now Se in the 1980s (Tanji et al., 1986).

B. Problem Area

From 1976 to 1980, a regional drainage collector system was constructed to service 17 000 ha in the northeast area of the WWD (Fig. 35–11), west and south of Mendota. The collector system was to convey subsurface

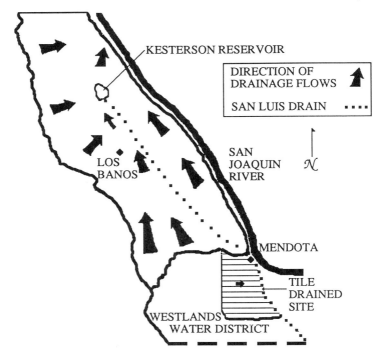

Fig. 35–11. Drainage flow pattern in the west side of San Joaquin Valley and the 17 000-ha tile drained area in Westlands Water District that contributed drainage waters to the Kesterson Reservoir, Westlands Water District.

drainage water from farm drainage systems to the San Luis Canal. This water was then eventually discharged into Kesterson Reservoir. The reservoir, consisting of 12 ponds with a surface area of 490 ha and an average depth of 1.2 m, was to serve as an interim facility for evaporating the drainage waters. At this time, the ponds were managed by the U.S. Fish and Wildlife Service (FWS) as a wetland/wildlife refuge, attracting both resident and migratory waterfowl.

The SLD conveyed only surface runoffs and fresh irrigation water between 1971 and 1977, and then increasing flows of subsurface drainage occurred from 1978 as on-farm tile drainage systems were installed and connected to WWD's drainage collector system. The SLD from 1981 conveyed exclusively about 900 ha-m of tile drainage effluents annually from about 3250 ha in the 17 000-ha site. The Se concentration was about 300 μg L^{-1}. The salinity of the water generally ranged between 10 and 15 dS m^{-1}. However, in 1981 the USBR imposed a moratorium on additional farm tile connections to WWD's drainage collector system because of limiting capacity to dispose of drainage in the Kesterson ponds.

C. Selenium Toxicity

In 1981 the USBR detected high concentrations of Se in SLD and Kesterson ponds. In the following year, 1982, FWS found Se concentrations in mosquito fish (*Gambusia affinis*) in Kesterson ponds 100 times greater than those from nearby Volta Wildlife Refuge, which receives mainly freshwater inflows. In May 1983, FWS's investigation of 350 waterfowl nests at Kesterson showed a high incidence of deformed and dead embryos as well as hatchlings—mainly coot (*Fulica americana*), pied-billed grebe (*Podilymbus podiceps*), and black-necked stilt (*Himantopus mexicanus*)—attributed to Se toxicosis. The FWS estimated that about 1000 migratory birds probably died in 1983 and 1984 from Se contamination of the food chain at the Kesterson National Wildlife Refuge.

These findings are the first documented effects of Se poisoning from irrigation return flows. A critical environmental and agricultural problem now exists in the San Joaquin Valley. Similar problems appear to exist throughout the western USA where soils derived from Se-containing rocks and geologic formations are irrigated.

D. Drainwater Disposal Constaints

In May 1984, James Claus, a duck club owner adjacent to the Kesterson ponds, filed a petition with the SWRCB alleging that the regional board had failed to take action to regulate drain water flows into Kesterson ponds, and requested enforcement of water quality regulations against USBR. After board hearings were conducted (ending in February 1985) the SWRCB concluded that wastewater in Kesterson ponds posed a hazard to the environment because the ponds had been improperly located, constructed, and managed, and contained hazardous constituents. The SWRCB also concluded the following: (i) wastewater in Kesterson ponds had caused death and deformities of waterfowl because of high Se concentrations; (ii) location of the ponds posed a hazard to resident and migratory waterfowl; (iii) Kesterson Reservoir posed a threat to human health from potential consumption of contaminated waterfowl; (iv) USBR was discharging wastewater that is causing and threatening to pollute waters of the state; and (v) Kesterson Reservoir must be upgraded to meet the requirements for a Class I surface impoundment or other acceptable waste disposal methods must be found.

The SWRCB on 5 Feb. 1985, then issued Cleanup and Abatement Order No. WQ-85-1 to the USBR to clean up and abate the effects of the Kesterson ponds, develop and submit a monitoring program within 12 wk, either file a revised report of waste discharge or a closure maintenance plan within 24 wk, and complete compliance within 3 wk. This order also directed a technical committee to develop proposed water quality objectives for the San Joaquin River Basin and effluent limitations for agricultural drainage discharges in the river basin, and submit a plan to regulate the discharge of drain waters in the basin.

On 15 Mar. 1986 at a congressional hearing held by the House Subcommittee on Water and Power Resources, the Office of the Secretary of the U.S. Department of Interior (USDI) made a surprise announcement ordering immediate closure of Kesterson Reservoir and SLD, and termination of irrigation water deliveries to the 17 000 ha in WWD overlying the drainage collector system. Potential criminal violation by USDI personnel of the Migrating Bird Treaty Act of 1916 was cited as the basis for this monumental announcement.

On 3 Apr. 1985, after intensive negotiations, an agreement was reached between WWD and USDI for continued delivery of irrigation supply water to the 17 000-ha drainage-impacted lands and with a phased reduction of drainage flows into SLD starting September 1985 and ceasing by June 1986.

In October 1985, WWD met the first scheduled 20% reduction in drainage flows into SLD. They also adopted a water conservation and management program for the 17 000-ha drainage-impacted land that included a reimbursement of $20 ha^{-1} to the water user who retains a qualified irrigation consultant during 1986. In November 1985, WWD released a draft environmental impact report proposing to eliminate drainage into SLD by intensive farm water management, farm recycling of tile effluents for crop irrigation, and plugging of the drainage collector system. The WWD began installing the first of 115 plugs in the drainage collector system and disconnecting pumps in the collector system in March 1986. This was completed in May 1986 to completely off drainage flows into SLD to meet the 30 June 1986 deadline. Impacts of plugging thus far are isolated cases of a sudden rise in water table.

E. San Joaquin Valley Drainage Program

In August 1984 the Governor of the State of California and Secretary of USDI agreed to form a joint-federal-state task force to produce alterna-

Table 35–2. Drainage management options.

- Discharge untreated drain waters to the Sacramento-San Joaquin Delta or to the Pacific Ocean.
- Take croplands out of production in drainage-impacted areas containing toxic constituents.
- Provide fresh water instead of saline drain water to marshland habitats.
- Reduce drain-water production by more efficient irrigation-water management and irrigation-application systems.
- Reuse of drain waters if of suitable quality for crop irrigation or other beneficial uses.
- Use on-farm and regional evaporation ponds to dispose of drain waters.
- Dispose of drain water by deep-well injection.
- Remove Se and other constituents biologically with bacteria and microalgae.
- Desalt drain waters through reverse osmosis, ion exchange, electrodialysis, or freeze crystallization.
- Remove toxic elements by chemical coagulation with alum, lime soda, sulfides, or FeOOH.
- Remove toxic elements by adsorption with activated C, activated Al, or cellulose xanthate.
- Apply electrochemical treatments such as chemical reduction, metallic reduction, or electrolytic or electron excitation.

---

tive solutions to the San Joaquin Valley drainage, salinity, and Se problems. The San Joaquin Valley Drainage Program (SJVDP) was formed in April 1985 to coordinate the activities of USBR, FWS, U.S. Geological Survey, DWR, California Dep. of Fish and Game, and several other governmental agencies and universities.

A \$23 million investigation for 1985 to 1986 was launched in July 1985 by the SJVDP for the cleanup of Kesterson ponds as ordered by SWRCB's Order WQ-85-1, alternative treatment and disposal of drain waters, and the protection of human health and natural resources. Table 35–2 contains a list of potential drainage management options.

F. Ongoing Activities and Studies

The current thrust for drainage management options in the San Luis Service Area may be broadly categorized into two activities: drainage water reduction, and treatment and disposal. Initially not considered by federal, state and local agencies addressing the disposal problem, reduction in subsurface drainage water production is now being promoted as a source control that would reduce the costs of collection, treatment, and disposal. Also, because this area has no surface drainage outlet and the problem of ultimate disposal remains, several options are being examined: desalination and reuse of desalted water, and removal of Se, B, and other constituents of concern by biological and chemical processes, evaporation ponds, and deep-well injection.

The Univ. of California Committee of Consultants on Drainage Reduction (1987), as well as the Agricultural Water Management Technical Subcommittee for the SJVDP (1987), studied drainage water management and came up with several findings. Drainage water production is highest during preplant irrigation (preirrigation) in the winter months, when measured application efficiencies have been as low as 30%, as well as during the initial postplant irrigation—both of which have high infiltration rates. Irrigators along the west side of the San Joaquin Valley commonly preirrigate cotton, cantaloupe (*Cucumis melo* L.), corn, bean (*Phaseolus vulgaris* L.), and small grains to fill the root zone prior to planting to help meet ET demand during the summer when infiltration rates typically become low. Preirrigations are mainly done by furrow irrigation along with hand-move sprinkler systems. The root zone for most crops in this area has a water-holding capacity of about 10 to 18 cm. Taking 1983 to 1984 as the typical irrigation season, annual rainfall, which occurs mainly between November through March, was 15 cm. Assuming effective precipitation to be about 65%, about 9.8 cm was infiltrated. The average preirrigation for this same period was 33 cm. Assuming a uniformity coefficient of 60%, preirrigation contributed nearly 20 cm of infiltrated water. The sum of effective precipitation and preirrigation exceeds the water-holding capacity by 12 to 20 cm, of which excess is subsurface drainage water.

The annual subsurface drainage past the root zone over the whole region has been estimated to range from 21 to 36 cm. The two study groups mentioned previously are of the opinion that with improved irrigation management, subsurface drainage could be reduced by about 50%. Drainage reduction with furrow irrigation, the most prevalent irrigation system currently used, may be accomplished by shortening the length of runs and decreasing set times, increasing the furrow inflow rate during the advance, establishing a uniform grade in the field, using scientific irrigation scheduling, and taking into account root-water extraction from shallow groundwaters to satisfy a portion of the ET needs of the crop. Surge, sprinkler, and drip/trickle irrigation methods are also being considered for drainage reduction.

The other major effort on drainage water management is treatment and disposal. In 1971 the DWR began testing various desalting techniques from bench to pilot scales, and from 1982 to 1986 operated a demonstration desalting facility at Los Baños, south of Kesterson Reservoir. The heart of this desalting facility is a reverse osmosis process (CH$_2$M Hill, 1986). Because the feed water (tile effluents in SLD) contains constituents that clog the cellulose membrane (e.g., scaling by $CaSO_4$, SiO_2, and sediments, as well as bacteria), it was necessary to expand the desalination process to include filtration by marshes and ponds—as well as sand filtration, ion exchange to soften the water, and chlorination before desalination. Despite these pretreatment processes, the economic and technical feasibility of desalinating agricultural drainage waters of this region are quite uncertain. One cost estimate based on off-the-shelf equipment and technologies is $150 ha-m^{-1} ($1220 acre-ft^{-1}) for capitol, operation, and maintenance. The removal of excessive levels of B, Se, and other toxic elements results in additional costs.

A second option under investigation is the use of on-farm and regional evaporation ponds (Tanji et al., 1985). Other than drain water reuse, this option is currently the only short-term option available to growers where discharge to surface water is not permitted. Under the current drainage production rates, a pond surface area from 10 to 15% of the total area drained is needed to evaporate the collected drainage. These ponds are constructed with shallow inside side slopes and are kept devoid of vegetation to discourage attraction to waterfowl and wildlife. The Regional Water Quality Control Board requires that pond seepage be controlled to no more than 10^{-6} cm s^{-1} for ponds overlying usable groundwaters. There is much concern that drainage water under evaporative salinization will accumulate toxic trace elements (e.g., B, Se, As, Pb, and Mo) to concentration levels exceeding water quality and public health regulations. Moreover, there is a long-run question of how to dispose of the accumulated deposits in ponds containing salts and toxic ions.

A third option under investigation is deep-well injection. This option is faced with many of the problems associated with deep-well injection of oilfield brines and industrial waters. A number of possible injection sites have been identified. Potential problems include the need for pretreatment to prevent plugging of pores in the deep geologic formations, the large volume of drainage water that requires disposal, the life span of injection wells, and potential enhancement of seismic activities. The estimated cost of deep-well injections is comparable to desalination. Public and institutional acceptance for such an alternative disposal option suggest that it should be considered as a last resort. Some feel that deep-well injection should be reserved for wastes that are more toxic than agricultural drains (URS Corp., 1986).

A fourth option is the chemical and biological removal of Se. A number of proposals have been made to remove Se by ion exchange, adsorption by activated C or Al as well as Fe_2O_3 surfaces, electrochemical processes involving metallic reduction, and coagulation with alum, lime soda, and other chemicals. The first two methods mentioned appear to have the most promise and have been demonstrated on a bench scale. Removal by Se by microalgae, bacteria, and fungi are extensively being studied. Laboratory and pilot studies indicate high-rate algal growth ponds may be feasible. A number of green and blue-green algae have the capability to assimilate Se, heavy metals, and N equivalent to physicochemical methods. Oswald (1985) recommended a pilot integrated ponding system that includes a facilitative pond, a paddle wheel-mixed high-rate pond, algal separation, algae fermentation, and sludge drying beds. He projected a cost of about $28 ha-m^{-1} ($228 acre-ft^{-1}).

Binnie California, Incorporated (Squires & Johnston, 1987) successfully tested a Se-removal system using another biological treatment and removal process. The pilot plant accepts drainage water into a feed tank dosed with a soluble C source which is then passed into a fluidized bed unit where Se is immobilized by bacteria similar to those present in conventional activated sludge treatment systems. The treated water then enters a flocculation tank and is pumped through a cross-flow microfilter. Boron is removed by an ion exchange column. This test facility is said to produce treated water containing 10 to 66 ppb Se. The cost for Se removal is reported to be about $19 ha-m^{-1} ($152 acre-ft^{-1}).

Most recently, Frankenberger and Karlson (1987, personal communication) discovered that indigenous fungi present in soils can volatilize Se.

Maximum volatilization was obtained with addition of C as an energy source for fungi, and favorable environmental conditions (moisture, temperature, and pH). Several fungi are capable of methylating selenate (Se +6) and selenite (Se +4) and then volatilizing $(CH_3)_2Se$ and $(CH_3)_2Se_2$. This laboratory finding is now being field tested at the Kesterson pond and at an on-farm evaporation pond.

A preliminary assessment on treatment and disposal options by the SJVDP reveals that a number of technologies are available, but the expected costs are too high for irrigated agriculture to bear the total costs. However, WWD in collaboration with Binnie California, Incorporated, is launching an independent investigation on drainage water collection, treatment, and disposal. This pilot-scale study includes a holding reservoir for collection of drainage waters, removal of Se and B, evaporation of saline waters, deep-well injection, and byproduct recovery.

The case study described herein is only a small portion of the regional drainage and Se problem. The 17 000-ha drainage-impacted portion (Fig. 35-11) of the Panoche Creek Fan lies in the northeastern portion of the 243 000-ha WWD (Fig. 35-10). Other areas in WWD on the Cantua Creek fan, south of Panoche Creek, also contain elevated concentrations of Se. The Panoche Creek fan extends into a 38 260-ha area of the Delta-Mendota Canal Service Area, directly north of the San Luis Service Area (Fig. 35-10). This second drainage-impacted area has traditionally discharged its return flows into the middle reaches of the San Joaquin River. Most recently, the SWRCB (1987) adopted revised water quality objectives on salinity, B, Se, and Mo. The interim Se water quality objective has been set at 5 μg L^{-1} as the maximum mean monthly level and 26 μg L^{-1} as the maximum instantaneous level, effective October 1991. To meet this Se standard, the irrigation return flows from the 38 260-ha drainage area would need to be reduced by 33% according to some sources. A concerted effort is being made to assess how irrigated agriculture could meet these more stringent water quality objectives placed on the middle reach of the San Joaquin River.

Since Se is naturally present in the western USA and other locations in the world, this case study is of significance in dealing with a critical environmental and agricultural problem. As more constraints are placed on discharge of irrigation return flows containing constituents of concern (e.g., salts, nitrates, trace elements, and pesticide residues) there will be an increasing challenge to researchers and growers to meet these constraints.

V. SUMMARY

A case study on drainage, salinity, Se, and other toxic trace elements in the west side of the San Joaquin Valley of California illustrated the complex problem of the environmental impact of agricultural drainage, and the management and control expected by regulatory agencies and the public. This case study demonstrated that a new perspective on irrigation and drainage management is required as discharge of nonpoint sources of pollutants is increasingly regulated.

This chapter provided an overview on the principal components and flow pathways of surface and subsurface irrigation return flows, most of which are unmeasured or at best only estimated. The quantity and type of irrigation return flow is primarily dependent on the irrigation application method and its management, as well as soil hydraulic properties.

Surface irrigation such as furrow, border and basin, sprinkler irrigation, drip/trickle irrigation, and subirrigation were evaluated for their potential performance in reducing drainage water production. Each of these methods of application systems has their advantages and limitations, but the practical attainable efficiencies are nearly the same for properly designed and managed systems. Central to the performance evaluation is uniformity of the application system and infiltration rate as well as average depth of water application relative to the water-holding capacity in the crop root zone.

Specific methods for controlling subsurface drainage for the various irrigation application systems are delineated with an emphasis on source control. These control measures also include irrigation and drainage management under high water table conditions as well as management strategies on reuse of degraded drainage waters.

REFERENCES

Agricultural Water Management Technical Subcommittee. 1987. Draft report on agricultural drainage on-farm water management alternatives. Calif. Dep. Water Resour., Sacramento.

Barias, D.G., J.L. Baker, H.P. Johnson, and J.M. Laflen. 1978. Effect of tillage systems on runoff losses of nutrients, a rainfall simulation study. Trans. ASAE 21:893–897.

Benz, L.C., E.J. Doering, and G.A. Reichman. 1981. Watertable management saves water and energy. Trans. ASAE 24(4):995–1001.

Bliesner, R.D., R.J. Hanks, L.G. King, and L.W. Willardson. 1977. Effects of irrigation management on the quality of irrigation return flow in Ashley Valley. Soil Sci. Soc. Am. Proc. 42(2):424–428.

Carter, D.L., and J.A. Bondurant. 1977. Management guidelines for controlling sediments, nutrients, and adsorbed biocides in irrigation return flows. Proc. Natl. Conf. on Irrig. Return Flow Quality Manage. Fort Collins, CO. 16–19 May.

CH$_2$M Hill. 1986. Review osmosis desalting of the San Luis Drain, conceptual study. Final rep. to U.S. Bur. Reclam., Sacramento, CA.

Coolidge, P.S., W.R. Walker, and A.A. Bishop. 1982. Advance and runoff-surge flow furrow irrigation. J. Irrig. Drain. Div. Am. Soc. Civ. Eng. 108, IR1:35–41.

Crowley, P.A., and T.L. Loudon. 1986. Tillage and infiltration under moving sprinkler systems. ASAE Pap. 86-2566. Presented at 1986 winter meeting. Chicago. 16–19 December.

Dawood, S.A., and S.N. Hamad. 1985. A comparison of on-farm irrigation systems performance. p. 540–545. *In* Drip/trickle irrigation in action. Proc. 3rd Int. Drip/Trickle Irrig. Cong. Fresno, CA. 18–21 November.

French, O.F., D.A. Bucks, R.L. Roth, and B.R. Gardner. 1985. Trickle and level-basin irrigation management for cotton production. p. 555–561. *In* Drip/trickle irrigation in action. Proc. 3rd Int. Drip/Trickle Irrig. Congr. Fresno, CA. 18–21 November.

Fuoss, J.L., and C.E.Carter. 1987. Controlled drainage supplements subirrigation in humid regions. p. 1–8. *In* Proc. irrigation systems for the 21st century. Specialty conf. sponsored by Irrig. Drain. Div., Portland, OR 28–30 July. Am. Soc. Civ. Eng., New York.

Goel, M.C., W.D. Kemper, R. Worstell, and J. Bondurant. 1982. Cablegation: III. Field assessment of performance. Trans. ASAE 25(5):1304–1309.

Goldhamer, D.A., M.H. Alemi, and R.C. Phene. 1987. Surge vs. continuous flow irrigation. Calif. Agric. 41(9, 10):29–32.

Grimes, D.W., R.L. Sharma, and D.W. Henderson. 1984. Developing the resource potential of a shallow water table. Univ. Calif. Water Resour. Cent. Contrib. 188.

Haan, C.T. 1971. Movement of pesticides by runoff and erosion. Trans. ASAE 14(2):445–447.

Hanson, B.R. 1984. A systems approach to managing irrigation and drainage water. ASAE Pap. 84-2571.

Hanson, B.R. 1987. A systems approach to drainage reduction. Calif. Agric. 41(9, 10):19–24.

Hanson, B.R., and W.W. Wallender. 1986. Bidirectional uniformity of water applied by continuous-move sprinkler machines. Trans. ASAE 29(4):1047–1053.

Hornsby, A.G. 1973. Prediction modeling for salinity control in irrigation return flows. USEPA-R2-73-168. U.S. Gov. Print. Office, Washington, DC.

Jensen, M.E. 1975. Scientific irrigation scheduling for salinity control of irrigation return flows. EPA-600/2-75-064. U.S. Gov. Print. Office, Washington, DC.

Jensen, M.E. 1980. Irrigation methods and efficiencies. Pap. presented World Bank Sem. Washington, DC. 8 January.

Laflen, J.M., J.L.Baker, R.O. Hartwig, W.F. Buchele, and H.P. Johnson. 1978. Soil and water losses from conservation tillage systems. Trans. ASAE 21:881–885.

Laflen, J.M., and M.A. Tabatabai. 1984. Nitrogen and phosphorus losses from corn-soybean rotation as affected by tillage practices. Trans. ASAE 27(1):58–63.

Letey, J., C. Roberts, M. Penberth, and C. Vasek. 1987. An agricultural dilemma: Drainage water and toxics disposal in the San Joaquin Valley. Univ. California Agric. Exp. Stn. Spec. Publ. 3319.

McDowell, L.L., and K.C. McGregor. 1980. Nitrogen and phosphorus losses in runoff from no-till soybeans. Trans. ASAE 23(3):643–648.

More, R.J. 1969. The basin hydrological cycle. p. 67–75. In R.J. Chorley (ed.) Water, earth and man. Methuen & Co., London.

Oswald, W.J. 1985. Potential for treatment of agricultural drain water with microalgal—bacteria systems. Rep. to U.S. Bur. Reclam., Sacramento, CA.

Rhoades, J.D., F.T. Bingham, J. Letey, A.R. Dedrick, M. Bean, G.J. Hoffman, W. Alves, R.V. Swain, P.G. Pecheco, and R.D. Lemert. 1988a. Reuse of drainage water for irrigation: Results of Imperial Valley study. I. Hypothesis, experimental procedures and cropping results. Hilgardia 56:1–16.

Rhoades, J.D., F.T. Bingham, J. Letey, G.J. Hoffman, W. Alves, R.V. Swain, P.G. Pecheco, R.D. Lemert, P.J. Pinter, and J.A. Replogle. 1988b. Reuse of drainage water for irrigation: Results of Imperial Valley study. II. Soil salinity and water balance. Hilgardia 56:17–44.

San Joaquin Valley Drainage Program. 1987. Prospectus, U.S. Bur. Reclam., U.S. Fish Wildl. Serv., U.S. Geolog. Surv., Calif. Dep. Fish Game, and Calif. Dep. Water Resour., Sacramento.

San Joaquin Valley Interagency Drainage Program. 1979. Agricultural drainage and salt management in the San Joaquin Valley. Final Rep. U.S. Bur. Reclam., Calif. Dep. Water Resour., and Calif. State Water Resour. Control Board, Sacramento.

Smith, M.C., R.W. Skaggs, and J.E. Parson. 1985. Subirrigation system control for water use efficiency. Trans. ASAE 28:489–493.

Squires, R.C., and W.R. Johnson. 1987. Selenium removal—can we afford it? p. 455–466. In L.G. Jones and M.J. English (ed.) Proc. of irrigation systems for the 21st century. Spec. conf. sponsored by Irrig. Drain. Div. Am. Soc. Civ. Eng. Portland, OR. 28–30 July.

State Water Resources Control Board. 1987. Regulation of agricultural drainage to the San Joaquin River, executive summary. Sacramento, CA.

Tanji, K.K. 1976. A conceptual hydrosalinity model for predicting salt load in irrigation return flows. p. 49–65. In H.E. Dregne (ed.) Managing saline water for irrigation. Texas Tech. Univ., Lubbock.

Tanji, K.K., J.W. Biggar, R.J. Miller, W.O. Pruitt, and G.L. Horner. 1980. Irrigation tailwater managment. Final rep. to USEPA. Robert S. Kerr Environ. Res. Lab. Ada, OK. NTIS PB 81-196925.

Tanji, K.K., M.E. Grismer, and B.R. Hanson. 1985. Subsurface drainage evaporation ponds. Calif. Agric. 39(9, 11):10–12.

Tanji, K.K., M.M. Iqbal, A.F. Quek, R.M. Van De Pol, L.P. Wagenet, R. Fujii, R.J. Schnagl, and D.A. Prewitt. 1977. Surface irrigation return flows vary. Calif. Agric. 31(5):30–31.

Tanji, K.K., A. Läuchli, and J. Meyer. 1986. Selenium in the San Joaquin Valley. Environment 28(6):6–11.

Tanji, K.K., M.J. Singer, L.D. Whittig, J.W. Biggar, D.W. Henderson, S.A. Mirbagheri, A.F. Quek, J. Blackard, and R. Higashi. 1983. Nonpoint sediment production in the Colusa Basin drainage area, California. Final rep. to USEPA. Robert S. Kerr Environ. Res. Lab. Ada, OK. NTIS PB 83-193920.

University of California Committee of Consultants on Drainage Water Reduction. 1987. Draft report on opportunities for drainage water reduction. Univ. Calif. Div. Agric. Nat. Resour.

URS Corporation. 1986. Deep well injection of agricultural drain waters. Rep. to U.S. Bur. Reclam. Sacramento, CA.

Utah State University Foundation. 1969. Characteristics and pollution problems of irrigation return flow. Rep. to Federal Water Pollution Control Admin. Ada, OK.

Van Bernuth, R.D., and K.H. Solomon. 1986. Design principles: Emitter construction. p. 27–52. *In* F.S. Nakayama and D.A. Bucks (ed.) Trickle irrigation for crop production: Design, operation, and management. Elsevier, Amsterdam.

Walker, W.R., and G.V. Skogerboe. 1986. The theory and practice of surface irrigation. Utah State Univ., Logan.

Walker, W.R., G.V. Skogerboe, and R.G. Evans. 1978. "Best management practices" for salinity control in Grand Valley. EPA-600/2-78-162. U.S. Gov. Print. Office, Washington, DC.

Wallender, W.W. 1986. Furrow model with spatially varying infiltration. Trans. ASAE 29(4):1012–1016.

Wendt, C.W., A.B. Onken, O.C. Wilke, R. Hargrove, W. Bausch, and L. Barnes. 1977. Effect of irrigation systems on water use efficiency and soil-water solute concentrations. Proc. Natl. Conf. Irrig. Return Flow Quality Manage. Fort Collins, CO. 16–19 May.

Wierenga, P.J. 1977. Influence of trickle and surface irrigation on return flow quality. EPA-600/2-77-093. U.S. Gov. Print. Office, Washington, DC.

Willardson, L.S., R.J. Hanks, and R.D. Bliesner. 1977. Field evaluation of sprinkler irrigation for management of irrigation return flow. Proc. Natl. Conf. Irrig. Return Flow Quality Manage. Fort Collins, CO. 16–19 May.

36

Salinity in Irrigated Agriculture[1]

J. D. RHOADES AND J. LOVEDAY

USDA-ARS, U.S. Salinity Laboratory
Riverside, California

Saline soils and waters contain excessive amounts of soluble salts for the practical and normal production of most agricultural crops. Most major irrigation schemes throughout the world suffer to some degree from the effects of salinity and, historically, many once-productive areas have become saline wastelands. Although accurate data are not available, FAO and UNESCO (reported by Szabolcs, 1985) have estimated that about one-half of all existing irrigation systems (totalling about 250 million ha) are seriously affected by salinity and waterlogging and that 10 million ha of irrigated land are abandoned annually. The spread of salinity affects not only older irrigation schemes but also more recently developed areas.

The effects of salinity are manifested in loss of stand, reduced rates of plant growth, reduced yields, and in severe cases, total crop failure. Salinity limits water uptake by plants by reducing the osmotic potential and thus the total potential of the soil water; additionally, certain salts may be specifically toxic to plants or may upset nutritional balance if they are present in excessive amounts or proportions. The salt composition of the soil water influences the composition of cations on the exchange complex of the soil colloids, and jointly, salinity level and exchangeable cation composition influence soil permeability and tilth. When Na on the exchange complex becomes excessive, permeability and tilth are deleteriously affected if the concentration of salts in the infiltrating water is below some threshold level.

The immediate source of the salt found in saline soils can be the parent material, the irrigation water, the shallow groundwater, or the fertilizers or amendments applied to the soil. All irrigation waters contain some salt which concentrates in the root zone as the water, but little of the salt, is extracted by the crop. For example, each application of a 100-mm depth of water containing as little as 500 mg salt/L adds 500 kg of salt to each hectare of irrigated land; this salt will progressively increase in the root zone over time with each irrigation, unless it is removed through leaching (overirrigation) and drainage. The processes of leaching and drainage are "necessary evils," so-to-speak, since they are the causes of the salt loading (pollution) of the waters receiving the drainage. Salts within the root zone may be redistributed towards the soil surface through the upward flux of water driven by evapo-

---

[1]Contribution from U.S. Salinity Lab., USDA, Riverside, CA 92501 and CSIRO Division of Soils, G.P.O.Box 639, Canberra, A.C.T., 2601.

ration, in the absence of a net downward flux through leaching and drainage. This process can create saline soils in the presence of shallow, saline groundwaters. Shallow groundwaters typically develop in irrigated lands (usually in downslope positions) when the portion of applied water that passes beyond the root zone (including channel seepage, excessive rainfall, and deep percolation) is not dissipated through drainage. Where irrigation water is derived from wells within the area being irrigated, development of shallow groundwater is much less likely, except where impermeable strata occur above the aquifer from which the irrigation water is drawn.

Thus, a prime requirement for salinity control in irrigation projects is that leaching and natural or artificial drainage be adequate to ensure that the net flux of water (and salt) is downward (with respect to the rootzone); additionally, the water table should be deep enough to provide adequate root development, aeration, and trafficability. Drainage often brings with it the necessity to dispose of the drainage water. The volume needing disposal can be reduced through improved irrigation efficiency and reuse for irrigation, but ultimately some disposal is still required. Ideally, the manner of disposal should neither damage other water supplies nor reduce the amenity and productivity of other lands. Typically, drainage water is simply discharged to surface waters or groundwaters of good quality, causing a reduction in their suitability for use in proportion to the increase in their salt concentration. Often these consequences of pollution are borne by other, off-site users of the receiving waters.

The best means of controlling soil and water salinity is the provision of efficient irrigation with adequate but minimum leaching and drainage management that maintains over time the net flux of water in the soil downward and which minimizes off-site pollution through interception, isolation, reuse, and finally ultimate disposal to appropriate treatment or waste depository facilities.

The reclamation of saline soils is accomplished through leaching with water of lower salinity and the provision of drainage if necessary. Resalination is prevented by eliminating the cause of salination. Selection of the appropriate management is site specific and requires an understanding of the sources of salt, the processes of salt mobilization and redistribution operative within the crop root zone, the farm field, the irrigation project, and the overall river basin. For sodic soils, application of appropriate amendments may also be required, in addition to leaching, to increase soil permeability and to reduce exchangeable Na levels.

In this chapter we discuss various concepts involved in defining and diagnosing soil and water salinity and the associated problem of sodicity, and we outline the effects of salts on soils and plants, and the processes influencing salinity and sodicity within the soil-plant-water system. This material provides the background needed for the discussions that follow of practices to control salinity in irrigated lands and to reclaim salt-affected soils. The all-important associated activity of drainage, without which reclamation cannot proceed, is discussed in the preceding chapter, and the pollution of receiving waters by drainage from irrigated areas in the following one.

I. EFFECTS OF SALTS ON PLANTS AND SOILS, AND CRITERIA AND GUIDELINES FOR ASSESSING SOIL SALINITY AND SODICITY

Methods for measuring soil salinity and sodicity are not dealt with in this chapter. They are described in detail in Loveday (1974), Rhoades (1982a, 1984a), Rhoades and Oster (1986), and Rhoades and Miyamoto (1989).

A. Salinity Hazard

Excess salinity in the root zone adversely affects the growth of established plants by a general reduction in growth rate. Salt stress essentially increases the energy that must be expended by the plant to extract water from the soil and to make the biochemical adjustments necessary to grow relative to the nonsaline condition (Maas, 1984a, 1985). This energy is wasted in the sense that it is diverted from the processes that lead to normal growth and yield. The potential adverse effects of salinity on crop production should be assessed using information of salinity levels in the regions of the soil where roots actively extract water. A problem is likely if the level of salinity in the vicinity of the active rootlets exceeds the tolerances of the crops for adequate growth and yield.

Thus, to assess the likelihood of a salinity problem for conventionally irrigated and established crops, one should first estimate the mean salinity of the major root zone (i.e., where the majority of water extraction occurs). When high frequency or drip irrigation is used, the water-uptake-weighted salinity (i.e., the salinity within each soil depth should be weighted in proportion to the fraction of root zone water extracted from it) should be estimated (Rhoades, 1982b). Secondly, the mean and the variability of salinity within the area of concern should be determined. If there are a sufficient number of measurements (at least 10), the standard deviation should be computed. To avoid a salt problem anywhere in the cropped area, the maximum salinity observed in the area should be taken as the salt index for the field. If salt problems are permitted to occur in about 15% of the area, the sum of mean and standard deviation should be used as the salt index. If the standard deviation is small (e.g., $<15\%$ in coefficient of variability), the sample mean may be used as the salt index. Thirdly, the value of the salt index obtained should be compared with that of the salt tolerance of the crop(s) to be grown. The salt tolerances of established crops based on growth and yield are given in Tables 36–1 and 36–2, after Maas (1986), in terms of their threshold values and percentage decreases in yield per unit increase of soil salinity in excess of the threshold. Soil salinity is expressed herein in terms of the electrical conductivity of the extract of a saturated soil paste (EC_e in dS/m). Salt tolerances of ornamental shrubs, trees, and ground cover are not given here; they may be obtained from Maas (1986). Figure 36–1 gives the divisions used in these tables for classifying crop tolerance to salinity (Maas, 1986). It should be recognized that such salt tolerance data cannot provide fully accurate, quantitative crop yield losses to be expected from salinity for every situa-

Table 36-1. Salt tolerance of herbaceous crops.†

Crop		Electrical conductivity of saturated soil extract		
Common name	Botanical name‡	Threshold§	Slope	Rating¶
		dS/m	% per dS/m	
	Fiber, grain, and special crops			
Barley#	Hordeum vulgare L.	8.0	5.0	T
Bean	Phaseolus vulgaris L.	1.0	19.	S
Broadbean	Vicia faba L.	1.6	9.6	MS
Corn	Zea mays L.	1.7	12.	MS
Cotton	Gossypium hirsutum L.	7.7	5.2	T
Cowpea	Vigna unguiculata (L.) Walp	4.9	12.	MT
Flax	Linum usitatissimum L.	1.7	12.	MS
Guar	Cyamopsis tetragonoloba (L.) Taub			MT
Millet, foxtail	Setaria italica (L.) Beauvois	--	--	MS
Oat	Avena sativa L.			MT*
Peanut	Arachis hypogaea L.	3.2	29.	MS
Rice, paddy††		3.0‡‡	12.‡‡	S
Rye	Secale cereale L.	--	--	MT*
Safflower	Carthamus tinctorius L.	--	--	MT
Sesame	Sesamum indicum L.	--	--	S
Sorghum	Sorghum bicolor (L.) Moench	6.8	16.	MT
Soybean	Glycine max (L.) Merrill	5.0	20.	MT
Sugarbeet§§	Beta vulgaris L.	7.0	5.9	T
Sugarcane	Saccharum officinarum L.	1.7	5.9	MS
Sunflower	Helianthus annuus L.	--	--	MS*
Triticale	X Triticosecale			T
Wheat#	Triticum aestivum L.	6.0	7.1	MT
Wheat (semidwarf)¶¶	T. aestivum L.	8.6	3.0	T
Wheat, durum	T. durum Desf.	5.9	3.8	T
	Grasses and forage crops			
Alfalfa		2.0	7.3	MS
Alkaligrass, Nuttall	Puccinellia airoides (Nutt.)	--	--	T*
Alkali sacaton	Sporobolus airoides	--	--	T*
Barley (forage)#		6.0	7.1	MT
Bentgrass	Agrostis stolonifera L., palustris	--	--	MS
Bermudagrass##	Cynodon dactylon L.	6.9	6.4	T
Bluestem, Angleton	Dichanthium aristatum (Poir) C.E. Hubb.	--	--	MS*
Brome, mountain	Bromus marginatus Nees ex Steud.	--	--	MT*
Brome, smooth	B. inermis Leyss	--	--	MS
Buffelgrass	Cenchrus ciliaris L.	--	--	MS*
Burnet	Poterium sanguisorba	--	--	MS*
Canarygrass, reed	Phalaris arundinacea L.	--	--	MT
Clover, alsike	Trifolium hybridum L.	1.5	12.	MS
Clover, berseem	T. alexandrinum L.	1.5	5.7	MS
Clover, hubam	Melilotus alba	--	--	MT*
Clover, ladino	Trifolium repens L.	1.5	12.	MS
Clover, red	T. pratense L.	1.5	12.	MS
Clover, strawberry	T. fragiferum L.	1.5	12.	MS
Clover, sweet	Melilotus Mill.	--	--	MT*
Clover, white Dutch	Trifolium repens L.	--	--	MS*

(continued on next page)

Table 36–1. Continued.

Crop		Electrical conductivity of saturated soil extract		
Common name	Botanical name‡	Threshold§	Slope	Rating¶
		dS/m	% per dS/m	
Corn (forage)		1.8	7.4	MS
Cowpea (forage)		2.5	11.	MS
Dallisgrass	*Paspalum dilatatum* Poir.	--	--	MS*
Fescue, meadow	*F. pratensis* Huds.	--	--	MT*
Fescue, tall	*Festuca arundinacea* Schreb.	3.9	5.3	MT
Foxtail, meadow	*Alopecurus pratensis* L.	1.5	9.6	MS
Grama, blue	*Bouteloua gracilis* (HBK) Lag.	--	--	MS*
Harding grass	*Phalaris stenoptera* L.	4.6	7.6	MT
Kallargrass	*Diplachne fusca*	--	--	T*
Lovegrass†††	*Eragrostis* sp. n.m. Wolf	2.0	8.4	MS
Milkvetch, cicer	*Astragalus cicer* L.	--	--	MS*
Oatgrass, tall	*Arrhenatherum elatius* Beauvois, Danthonia	--	--	MS*
Oats (forage)		--	--	MS*
Orchardgrass	*Dactylis glomerata* L.	1.5	6.2	MS
Panicgrass, blue	*Panicum antidotale* Retz.	--	--	MT*
Rape	*Brassica napus* L.	--	--	MT*
Rescuegrass	*Bromus unioloides* HBK	--	--	MT*
Rhodesgrass	*Chloris gayana* Kunth	--	--	MT
Rye (forage)	*Secale cereale* L.	--	--	MS*
Ryegrass, Italian	*Lolium italicum* L. *multiflorum*	--	--	MT*
Ryegrass, perennial	*L. perenne* L.	5.6	7.6	MT
Saltgrass, desert	*Distichlis stricta*	--	--	T*
Sesbania††	*Sesbania exaltata* (Raf.) V.L.Cory	2.3	7.0	MS
Siratro	*Macroptilium atropurpureum* (DC.)	--	--	MS
Sphaerophysa	*Sphaerophysa salsula*	2.2	7.0	MS
Sudangrass	*Sorghum sudanense* (Pipee) Stapf	2.8	4.3	MT
Timothy	*Phleum pratense* L.	--	--	MS*
Trefoil, big	*Lotus uliginosus* L.	2.3	19.	MS
Trefoil, narrowleaf birdsfoot	*L. corniculatus* L., *tenuifolium*	5.0	10.	MT
Trefoil, broadleaf birdsfoot‡‡‡	*L. corniculatus* L., *arvenis*	--	--	MT
Vetch, common	*Vicia sativa* L.	3.0	11.	MS
Wheat (forage)¶¶		4.5	2.6	MT
Wheat, durum (forage)	*T. durum*	2.1	2.5	MT
Wheatgrass, standard crested	*Agropyron desertorum* A.	3.5	4.0	MT
Wheatgrass, fairway crested	*A. cristatum* (L.) Gaertn.	7.5	6.9	T
Wheatgrass, intermediate	*A. intermedium* (Host) Beauv.	--	--	MT*
Wheatgrass, slender	*A. trachycaulum* (Link) Malte	--	--	MT
Wheatgrass, tall	*A. elongatum* (Hort) Beauv.	7.5	4.2	T
Wheatgrass, western	*A. smithii* Rydb.	--	--	MT*
Wildrye, Altai	*Elymus angustus* Trin.	--	--	T
Wildrye, beardless	*E. triticoides* Buckl.	2.7	6.0	MT
Wildrye, Canadian	*E. canadensis* L.	--	--	MT*
Wildrye, Russian	*Psathyrostachys juncea* (Fisch.)	--	--	T

(continued on next page)

Table 36-1. Continued.

	Crop	Electrical conductivity of saturated soil extract		
Common name	Botanical name‡	Threshold§	Slope	Rating¶
		dS/m	% per dS/m	
	Vegetable and fruit crops			
Artichoke, Jerusalem	Helianthus tuberosus L.	--	--	MT*
Asparagus	Asparagus officinalis L.	--	--	T
Bean	Phaseolus vulgaris L.	1.0	19.	S
Beet, red§§	Beta vulgaris L.	4.0	9.0	MT
Broccoli	Brassica oleracea B., Botrytis	2.8	9.2	MS
Brussels sprouts	B. oleracea gemmifera B.	--	--	MS*
Cabbage	B. oleracea capitata B.	1.8	9.7	MS
Carrot		1.0	14.	S
Cauliflower	Brassica oleracea B., botrytis	--	--	MS*
Celery	Apium graveolens L.	1.8	6.2	MS
Corn, sweet		1.7	12.	MS
Cucumber	Cucumis sativus L.	2.5	13.	MS
Eggplant	Solanum melongena L.	--	--	MS*
Kale	Brassica oleracea B., acephala	--	--	MS*
Kohlrabi	B. oleracea gongylode	--	--	MS*
Lettuce	Lactuca sativa L.	1.3	13.	MS
Muskmelon	Cucumis melo L.	--	--	MS
Okra	Abelmoschus esculentus (L.) Moench	--	--	S
Onion	Allium cepa L.	1.2	16.	S
Parsnip	Pastinaca sativa L.	--	--	S*
Pea	Pisum sativum L.	--	--	S*
Pepper	Capsicum annuum L.	1.5	14.	MS
Potato	Solanum tuberosum L.	1.7	12.	MS
Pumpkin	Cucurbita pepo pepo L.	--	--	MS*
Radish	Raphanus sativus L.	1.2	13.	MS
Spinach	Spinacia oleracea L.	2.0	7.6	MS
Squash, scallop	Cucurbita pepo melopepo	3.2	16.	MS
Squash, zucchini		4.7	9.4	MT
Strawberry	Fragaria sp. L.	1.0	33.	S
Sweet potato	Ipomoea batatas (L.) Lom.	1.5	11.	MS
Tomato		2.5	9.9	MS
Turnip	Brassica rapa L.	0.9	9.0	MS
Watermelon	Citrullus lanatus (Thunb.)	--	--	MS*

† These data serve only as a guideline to relative tolerances among crops. Absolute tolerances vary depending upon climate, soil conditions, and cultural practices.

‡ Botanical and common names follow the convention of *Hortus Third* where possible.

§ In gypsiferous soils, plants will tolerate EC_e's about 2 dS/m higher than indicated.

¶ Ratings are defined by the boundaries in Fig. 36-1. Ratings with an * are estimates. For references consult the indexed bibliography by Francois and Maas (1985).

\# Less tolerant during emergence and seedling stage. EC_e at this stage should not exceed 4 or 5 dS/m.

†† Less tolerant during emergence and seedling stage.

‡‡ Because paddy rice is grown under flooded conditions, values refer to the electrical conductivity of the soil water while the plants are submerged.

§§ Sensitive during germination. EC_e should not exceed 3 dS/m.

¶¶ Data from cultivar Probred.

\#\# Average of several varieties. Suwannee and Coastal are about 20% more tolerant and Common and Greenfield are about 20% less tolerant than the average.

††† Average for Boer, Wilman, Sand, and Weeping cultivars. Lehmann seems about 50% more tolerant.

‡‡‡ Broadleaf birdsfoot trefoil seems less tolerant than narrowleaf.

Table 36-2. Salt tolerance of woody crops.†

Crop		Electrical conductivity of saturated soil extract		
Common name	Botanical name‡	Threshold§	Slope	Rating¶
		dS/m	% per dS/m	
Almond#	*Prunus dulcis* (Mill.)	1.5	19	S
Apple	*Malus sylvestris* Mill.	--	--	S
Apricot#	*Prunus armeniaca* L.	1.6	24	S
Avocado#		--	--	S
Blackberry	*Rubus* spp.	1.5	22	S
Boysenberry	*Rubus ursinus* Cham. and Schlechtend	1.5	22	S
Castorbean	*Ricinus communis* L.	--	--	MS*
Cherimoya	*Annona cherimola* Mill.	--	--	S*
Cherry, sweet	*Prunus avium* L.	--	--	S*
Cherry, sand	*P. besseyi* L.	--	--	S*
Currant	*Ribes* sp.	--	--	S*
Date palm	*Phoenix dactylifera* L.	4.0	3.6	T
Fig	*Ficus carica* L.	--	--	MT*
Gooseberry	*Ribes* sp.	--	--	S*
Grape#	*Vitis* sp.	1.5	9.6	MS
Grapefruit	*Citrus paradisi* Macfad.	1.8	16	S
Guayule		--	--	T
Jojoba#	*Simmondsia chinensis* (Link) C. Schneid.	--	--	T
Jujube	*Ziziphus jujuba* Mill.	--	--	MT*
Lemon#	*Citrus limon* (L.) Burm. f.	--	--	S
Lime	*C. aurantiifolia* (Christm.) Swingle	--	--	S*
Loquat	*Eriobotrya japonica*	--	--	S*
Mango	*Mangifera indica* L.	--	--	S*
Olive	*Olea europaea* L.	--	--	MT
Orange	*Citrus sinensis* (L.) Osb.	1.7	16	S
Papaya#	*Carica papaya* L.	--	--	MT
Passion-fruit	*Passiflora mollissima* (HBK) L.H. Bailey	--	--	S*
Peach	*Prunus persica* (L.) Batsch	1.7	21	S
Pear	*Pyrus communis* L.	--	--	S*
Persimmon	*Diospyros virginiana* L.	--	--	S*
Pineapple	*Ananas comosus* (L.) Merr.	--	--	MT*
Plum; Prune#	*Prunus domestica* L.	1.5	18	S
Pomegranate	*Punica granatum* L.	--	--	MT*
Pummelo	*Citrus grandis* C.	--	--	S*
Raspberry	*Rubus idaeus* L.	--	--	S
Rose apple	*Syzygium jambos* (L.) Alston	--	--	S*
Sapote, white	*Casimiroa edulis* Llave	--	--	S*
Tangerine	*Citrus reticulata* Blanco	--	--	S*

† These data are applicable when rootstocks are used that do not accumulate Na or Cl rapidly or when these ions do not predominate in the soil.
‡ Botanical and common names follow the convention of *Hortus Third* where possible.
§ In gypsiferous soils plants will tolerate EC_e's about 2 dS/m higher than indicated.
¶ Ratings are defined by the boundaries in Fig. 36-1. Ratings with an * are estimates. For references consult the indexed bibliography by Francois and Maas (1985).
# Tolerance is based on growth rather than yield.

tion, since actual response to salinity varies with growing conditions, e.g., climate, irrigation management, agronomic management, and crop variety. Such salt tolerance data are useful, however, for semiquantitative estimates of yield loss and especially to diagnose the likelihood of salinity problems and to predict how one crop might fare relative to another under similar conditions of salinity. For more information on the tolerance of specific crops to salinity, see Francois and Maas (1978) or the references given in Maas (1986).

Improvement in diagnosis can be made by using salinity of the soil solution per se (EC_{sw}), rather than that of the saturation extract, since salinity of the saturation extract does not account for the increase in salinity that the soil water undergoes between irrigations due to soil water depletion (Rhoades et al., 1981). With use of high-frequency irrigation methods, the use of the saturation extract is even less appropriate as a measure of soil water salinity. Soil solution-based salinity appraisal is also preferred for sandy soils, especially those irrigated with saline waters. The EC_e values of such soils are considerably less than those of medium-textured soils containing soil waters of similar salinities (e.g., Longenecker, 1973). For this reason, it is common to underrate the salt hazard of sandy soils irrigated with saline waters using criteria based on EC_e. However, the use of soil solution-based salinities necessitates the conversion of crop salt tolerance data from EC_e to EC_{sw} and practical measurement techniques for measuring EC_{sw}. Such methods are lacking for soil samples, but instrumental techniques can be used to measure EC_{sw} directly in the field (Rhoades & Oster, 1986; Rhoades et al., 1989). For more information on salinity diagnosis procedures, see Rhoades and Miyamoto (1989).

The salt tolerance data of Tables 36–1 and 36–2 apply to surface-irrigated crops and conventional irrigation management. Sprinkler-irrigated crops may suffer additional damage from foliar salt uptake and "burn" caused by direct

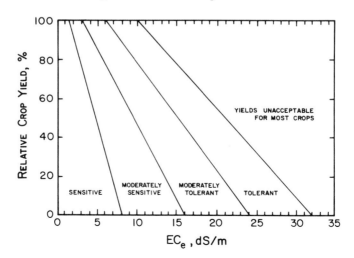

Fig. 36–1. Divisions for classifying crop tolerance to salinity. After Maas (1986).

Table 36–3. Relative susceptibility of crops to foliar injury from saline sprinkling waters.†

Na or Cl concentrations causing foliar injury, mmol$_c$/L‡			
<5	5–10	10–20	>20
Almond	Grape	Alfalfa	Cauliflower (*Gossypium hirsutum* L.)
Apricot	Pepper	Barley	Cotton (*Beta vulgaris* L.)
Citrus	Potato	Corn	Sugarbeet (*Helianthus annuus* L.)
Plum	Tomato	Cucumber	Sunflower
		Safflower (*Carthamus tinctorius* L.)	
		Sesame (*Sesamum indicum* L.)	
		Sorghum [*Sorghum bicolor* (L.) Moench]	

† Susceptibility based on direct accumulation of salts through the leaves. Data compiled from Maas (1986).
‡ Foliar injury is influenced by cultural and environmental conditions. These data are presented only as general guidelines for daytime sprinkling.

contact of leaves with the spray. The available data base for predicting yield losses from foliar spray effects is limited. Most of the available data are given in Maas (1985, 1986). These data are summarized in Table 36–3. The degree of foliar injury depends not only on the salinity of the irrigation water but also upon atmospheric conditions, the size of sprinkler droplets, crop type, and growth stage. The tolerance of crops to foliar-induced salt damage does not generally coincide with that of root-induced damage. For this reason, it is necessary to evaluate irrigation water salinity in addition to soil salinity for the case of sprinkler irrigation. The data of Table 36–3 should be consulted in this regard.

Besides the above-mentioned effects on established crops, salinity also adversely influences crop establishment, especially under conditions of furrow irrigation. In fact, obtaining a good stand of plants is often the factor

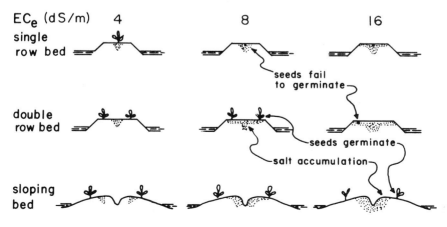

Fig. 36–2. Pattern of salt buildup as a function of irrigation management, bedshape, and level of soil salinity. After Bernstein et al. (1955).

that most limits crop production in saline areas. Once the stand is established, management risks are generally substantially reduced. The problem of reduced seedling establishment is due in part to the generally lower salt tolerance of seedlings compared to established plants. Additionally, the problem is enhanced because the seeds or small seedlings are exposed to locally excessive salinity in the seed bed that often occurs there as a result of evaporation (Miyamoto et al., 1985a, b). Salt concentrations in crop beds vary markedly with depth and time (Bernstein et al., 1955; Bernstein & Francois, 1973). As shown in Fig. 36-2, the tendency of salt to accumulate near the seed during irrigation is greatest in single-row, flat-topped beds. Sufficient salt may move laterally and concentrate in the seedling zone to prevent seedling establishment, even if the average salt content of the soil is relatively low. With double-

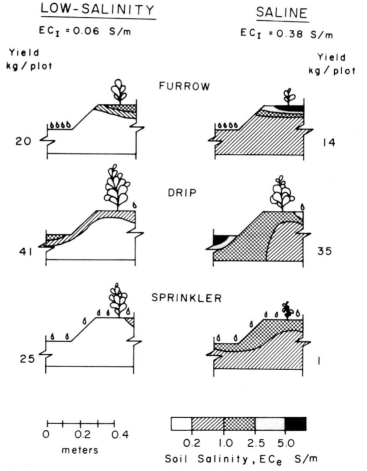

Fig. 36-3. Influence of the irrigation system on the soil salinity pattern and yield of bell pepper (*Capsicum anuum* L.) at two levels of irrigation water quality. After Bernstein and Francois (1973).

row beds, under moderately saline conditions, most of the salt is also carried into the center of the bed, leaving the shoulders relatively free of salt for seedling establishment. Sloping beds are best for saline soils because the seedling can be safely established on the slope below the zone of salt accumulation. The salt is moved away from around the seedling instead of accumulating near it. Planting in furrows or basins is satisfactory from the stand point of salinity control but is often unfavorable for the emergence of many row crops because of crusting or poor aeration. As shown in Fig. 36–3, preemergence irrigation by sprinklers or drip lines placed close to the seed may be used to keep the soluble salt concentration low in the seed bed during germination and seedling establishment. Special temporary furrows may also be used in place of drip lines during the seedling establishment period. After the seedlings are established, the special furrows may be abandoned and conventional furrows made between the rows, likewise, sprinkling may be replaced by furrow irrigation. For the above reasons, it is essential to have a clear specification of the volume of active rooting at the time of sampling when interpreting salinity data based on soil samples taken from irrigated beds. Much misinterpretation occurs because this is not taken into account.

For many crops, stand problems begin after the seed has germinated. Hypocotyls emerging from the seed may have trouble passing through the overlying soil which has had its salt content increased through evaporation. During the emergence process, hypocotyl mortality often occurs with crops sensitive to foliar salt damage. The levels of salinity that cause hypocotyl mortality vary widely, e.g., EC_e as low as 5 dS/m in the top 5 mm of soil

Table 36–4. Relative salt tolerance of various crops at emergence and during growth to maturity. After Maas (1986).

Common name[†]	Electrical conductivity of saturated soil extract	
	50% Yield[‡]	50% Emergence[‡]
	dS/m	
Barley	18	16–24
Cotton	17	15
Sugarbeet	15	6–12
Sorghum	15	13
Safflower	14	12
Wheat	13	14–16
Beet, red	9.6	13.8
Cowpea	9.1	16
Alfalfa	8.9	8–13
Tomato	7.6	7.6
Cabbage	7.0	13
Corn	5.9	21–24
Lettuce	5.2	11
Onion	4.3	5.6–7.5
Rice	3.6	18
Bean	3.6	8.0

† Common names follow the convention of *Hortus Third* where possible.
‡ Emergence percentage of saline treatments determined when nonsaline control treatments attained maximum emergence.

for guayule (*Parthenium argentatum* A. Gray), 10 dS/m for carrot (*Daucus carota* L.), and over 40 dS/m for tomato (*Lycopersicon esculentum* Mill.) (Miyamoto et al., 1986). One approach to eliminate this problem is to remove the surface crust of salts by mechanical means prior to seedling emergence (Miyamoto et al., 1986). More typically, this salt buildup is avoided through irrigation planting and bed shaping techniques (Bernstein et al., 1955; Rhoades, 1982b).

Emerged seedlings can also undergo mortality if seedling roots are exposed to excessive salinity in the ridge region of furrow-irrigated beds, or if rain or sprinkler irrigation leaches the surface-accumulated salts back into the seedling root zone (Bernstein, 1974). Mortality of emerged seedlings also often occurs when seedling leaves are exposed to saline splatters resulting from the light application of water by sprinklers on salted soil surfaces (Miyamoto et al., 1986).

Firm criteria to diagnose the extent of seedling mortality from all of the above-mentioned processes have not been developed. The data given in Table 36–4 may be helpful in this regard.

B. Specific Ion and Toxicity Hazards

Certain salt constituents are specifically toxic to some crops. Boron is toxic to certain crops when present in the soil solution at concentrations of only a few milligrams per liter (Maas, 1986; Keren et al., 1985; Bingham et al., 1985). In some woody crops, Cl may accumulate in the tissue to toxic levels (Bernstein, 1974; Maas, 1984b, 1985, 1986). These toxicity problems are not generally major ones but may be important in certain cases. Guidelines for judging the problems of B and Cl toxicities are given in Tables 36–5 and 36–6, respectively.

Direct toxicity effects related to Na are generally limited to perennial woody species. Sodium toxicity injury is common in avocado (*Persea americana* Mill.), citrus (*Citrus* spp.), and stone-fruit (*Prunus* spp.) trees and can occur at Na concentrations as low as 5 $mmol_c/L$ in soil water. Sodium is often retained in the roots and lower bases of trees, but over time enough may accumulate and be transported to the leaves to cause leaf burn (Maas, 1986). In sodic, nonsaline soils, total salt concentrations are low and consequently, Ca and or Mg concentrations may be nutritionally inadequate for certain crops. Such deficiencies, rather than sodium toxicity per se, are usually the primary cause of poor plant growth among nonwoody species. Sodicity may induce Ca and various micronutrient deficiencies because the associated high pH and HCO_3 levels repress their solubilities and concentrations. As a general rule, Ca and Mg concentrations in the soil solution are nutritionally adequate in saline, sodic soils. As the total salt concentration reaches the saline range, Ca concentrations become adequate for most plants and osmotic effects dominate (Rhoades, 1982b). The effects of salinity and toxic solutes on the physiology and biochemistry of plants are discussed in more detail by Maas and Nieman (1978) and Maas (1984a).

Table 36–5. Boron tolerance limits for agricultural crops. After Maas (1986).

Common name (Botanical name not included in text)	Threshold†
	mg/L
Very sensitive	
Lemon‡	<0.5
Blackberry‡	<0.5
Sensitive	
Avocado‡	0.5–0.75
Grapefruit‡	0.5–0.75
Orange‡	0.5–0.75
Apricot‡	0.5–0.75
Peach‡	0.5–0.75
Cherry‡	0.5–0.75
Plum‡	0.5–0.75
Persimmon, Japanese (*Diospyros kaki* L.f.)	0.5–0.75
Fig, kadota‡	0.5–0.75
Grape‡ (*Vitis vinifera*)	0.5–0.75
Walnut‡ (*Juglans regia* L.)	0.5–0.75
Pecan‡ [*Carya illinoensis* (Wangenh.) K. Koch]	0.5–0.75
Cowpea‡	0.5–0.75
Onion	0.5–0.75
Garlic (*Allium sativum* L.)	0.75–1.0
Sweet potato	0.75–1.0
Wheat	0.75–1.0
Sunflower	0.75–1.0
Bean, mung‡ [*Vigna radiata* (L.) R. Wilcz.]	0.75–1.0
Sesame‡	0.75–1.0
Lupine‡ (*Lupinus hartwegii* Lindl.)	0.75–1.0
Strawberry‡ (*Fragaria* sp. L.)	0.75–1.0
Artichoke, Jerusalem‡	0.75–1.0
Bean, kidney‡ (*Phaseolus vulgaris*)	0.75–1.0
Bean, lima‡ (*Phaseolus lunatus*)	0.75–1.0
Peanut (*Arachis hypogaea*)	0.75–1.0
Moderately sensitive	
Broccoli	1.0–2.0
Pepper, red	1.0–2.0
Pea‡	1.0–2.0
Carrot	1.0–2.0
Radish	1.0–2.0
Potato	1.0–2.0
Cucumber	1.0–2.0
Moderately tolerant	
Lettuce‡	2.0–4.0
Cabbage‡	2.0–4.0
Celery‡	2.0–4.0
Turnip	2.0–4.0
Bluegrass, Kentucky‡ (*Poa pratensis* L.)	2.0–4.0
Barley	2.0–4.0
Oat	2.0–4.0
Corn	2.0–4.0
Artichoke, globe‡ (*Cynara scolymus* L.)	2.0–4.0
Tobacco‡ (*Nicotiana tabacum* L.)	2.0–4.0
Mustard, Indian‡ [*Brassica juncea* (L.) Czerni.]	2.0–4.0
Clover, sweet‡	2.0–4.0
Squash	2.0–4.0
Muskmelon‡	2.0–4.0

(continued on next page)

Table 36-5. Continued.

Common name (Botanical name not included in text)	Threshold†
	mg/L
Cauliflower	2.0-4.0
Tolerant	
Tomato	4.0-6.0
Alfalfa‡	4.0-6.0
Vetch, purple‡ (*Vicia benghalensis*)	4.0-6.0
Parsley‡ (*Petroselinum crispum*)	4.0-6.0
Beet, red	4.0-6.0
Sugarbeet	4.0-6.0
Very tolerant	
Sorghum	6.0-10.0
Cotton	6.0-10.0
Asparagus‡	10.0-15.0

† Maximum permissible concentration in soil water without yield reduction. Boron tolerances may vary depending upon climate, soil conditions, and crop varieties.
‡ Tolerance based on reductions in vegetative growth.

C. Permeability and Tilth Hazards

For successful irrigated cropping, soil permeability and tilth must be adequate for good water and air movement, for water storage, and for seedling emergence and root extension. The processes of irrigation and cultivation affect soil permeability and tilth. Aggregates, especially when wetted rapidly during irrigation, may break up into smaller units of structure (a process called *slaking*) when forces associated with osmotic swelling and air entrapment exceed binding forces (for more detail about these forces see reviews by Quirk, 1986; Shainberg & Letey, 1984; and Shainberg, 1984). Pore dimensions may be further reduced as a result of clay swelling and, clay particles may be dispersed into suspension, move and subsequently lodge in pore interstices. There is further opportunity for clay migration with the soil water and for general rearrangement of subaggregates and soil particles during successive periods of slumping on wetting and consolidation on drying. The physical condition resulting from irrigation is one characterized by a general loss of porosity (and hence permeability) and increase in dry strength (*hard setting* or *crusting*). Additional mechanical stresses to which aggregates are subjected include those due to waterdrop impact and to cultivation and traffic (McIntyre, 1958). The mechanical breaking of bonds between soil particles enables disaggregation and clay dispersion to proceed more readily (Emerson, 1983; Shainberg, 1984; Oster & Schroer, 1979). Soil crusts, typically form when waterdrop impact combines with clay dispersion and deposition (Shainberg, 1984). In contrast to saline and normal soils, sodic soils typically have reduced permeabilities and poorer tilth and hence are less suitable for irrigation and as a medium for crop growth.

All of the deleterious processes described above are influenced significantly by the salt (electrolyte) concentration of the soil solution, the level

Table 36-6. Chloride tolerance limits of some fruit-crop cultivars and rootstocks. After Maas (1986).

Crop	Rootstock or cultivar	Maximum permissible Cl⁻ in soil water without leaf injury
	Rootstocks	
Avocado	West Indian	15
	Guatemalan	12
	Mexican	10
Citrus (*Citrus* spp.)	Sunki mandarin, grapefruit, Cleopatra mandarin, Rangpur lime	50
	Sampson tangelo, rough lemon,‡ sour orange, Ponkan mandarin	30
	Citrumelo 4475, trifolate orange, Cuban shaddock, Calamondin, sweet orange, Savage citrange, Rusk citrange, Troyer citrange	20
Grape (*Vitis* spp.)	Salt Creek, 1613-3	80
	Dog ridge	60
Stone fruit (*Prunus* spp.)	Marianna	50
	Lovell, Shalil	20
	Yunnan	15
	Cultivars	
Berries§ (*Rubus* spp.)	Boysenberry	20
	Olallie blackberry	20
	Indian summer raspberry	10
Grape (*Vitis* spp.)	Thompson seedless, Perlette	40
	Cardinal, Black rose	20
Strawberry (*Fragaria* spp.)	Lassen	15
	Shasta	10

† For some crops these concentrations may exceed the osmotic threshold and cause some yield reductions.
‡ Data from Australia indicate that rough lemon is more sensitive to Cl⁻ than sweet orange.
§ Data available for one variety of each species only.

of soil sodicity and their interactions with the clay (colloidal) fraction of the soil (Emerson, 1984; Yousaf et al., 1987a, b).

The capacity of a soil to adsorb exchangeable cations is referred to as the cation exchange capacity and the major cations involved in irrigated soils are Ca, Mg, and Na. The relative proportions of cations on the exchange complex result from equilibria established by exchange with cations in the soil solution. In the case of Na, this proportion is referred to as the exchangeable Na percentage, P_{Na}. It is approximately numerically equal to the Na adsorption ratio, R_{Na}, in the soil solution as follows:

$$R_{Na} = Na/\sqrt{(Ca + Mg)/2},$$ [1]

where the concentrations are expressed in $mmol_c/L$. The R_{Na} can be used approximately interchangeably with P_{Na} over the normal range of P_{Na} found in irrigated soils.

The classic criterion used for defining a sodic soil has been a P_{Na} value of ≥ 15 (U.S. Salinity Laboratory Staff, 1954). Others have used lower values, for example a P_{Na} value of 6 has been advocated to separate nonsodic and sodic soil classes for Australian conditions (Northcote & Skene, 1972; Loveday & Pyle, 1973; McIntyre (1979). However, no single value of P_{Na} should be used in this regard since the presence of sufficient electrolyte can counteract the adverse effects of Na on permeability and tilth. The soil physical problems associated with exchangeable sodium (P_{Na} or R_{Na}) should be assessed only in conjunction with a consideration of the accompanying level of salinity. The term *sodicity*, as used herein, refers to combinations of exchangeable Na and electrolyte concentration which result in degradation of soil permeability and tilth substantially in excess of those of normal and saline soils. Setting standards for the assessment of sodicity is complicated by the interactive effects of certain other soil properties, especially pH, and of management and climatic conditions.

Most of the definitive data concerning the effects of sodicity on soil permeability have come from laboratory studies, usually employing disturbed soil samples. Such studies have shown, for example, the following. For soils of mixed mineralogy, when the electrical conductivity of the percolating solution is 10 $mmol_c/L$, a 15 to 25% reduction in hydraulic conductivity is caused by P_{Na}'s ranging from 5 to 25 (McNeal & Coleman, 1966a, b; Quirk & Schofield, 1955; Oster & Schroer, 1979). Reductions in the hydraulic conductivity of soils often occur even with P_{Na}'s as low as 5, when salinity is lower than about 3 to 5 $mmol_c/L$ (McIntyre, 1979; Shainberg, 1984). Soil compaction and shearing generally accentuate the adverse effects of Na (e.g., Frenkel et al., 1978; Emerson, 1983). Under unsaturated flow conditions, sodicity exerts less influence on hydraulic conductivity than for saturated conditions (e.g., Russo & Bresler, 1977). The field behavior of many undisturbed soils, with their typically complex and variable networks of macropores (earthworm and root channels, cracks) which provide opportunity for considerable bypass flow (Bouma, 1984), is unlikely to be well related to laboratory results. Such data cannot be directly extrapolated to the field, but they should be valid for predicting the relative behavior of cultivated surface soils to sodicity (Reeve & Tamaddoni, 1965).

The sodicity hazard is best assessed when certain processes which occur during irrigation are taken into account. When water infiltrates into the soil surface, the soil solution of the immediate topsoil is essentially that of the infiltrating water, while the exchangeable Na percentage is essentially that pre-existent in the soil (since R_{Na} is buffered against rapid change by the relatively large reservoir of exchangeable Na). Except with the case of cracked soils, all irrigation water and rain entering the soil must pass through the soil surface; hence the stability of the aggregates of the topsoil greatly influences the water entry rate of the soil. After irrigation water enters the soil to become soil water, its composition (hence the value of R_{Na}) changes in response to several factors, including evapotranspiration, the dissolution and/or precipitation of salts, mineral weathering, and organic matter decomposition. There may be loss of Ca and Mg from solution due to the precipitation of alkaline earth carbonates or a gain of Ca, Mg, and HCO_3 ions in solution from the dissolution of these minerals and/or by the weathering of soil minerals and decomposition of organic matter. For most irrigation waters with electrical conductivities (EC_{iw}) < 1.0 dS/m, the effects of salt precipitation are generally insignificant in influencing soil water composition at leaching fractions of 0.2 or greater (Rhoades et al., 1974). Under such conditions the Na adsorption ratio of the irrigation water itself may be a suitable index of R_{Na} in the resultant upper root zone soil water (Bingham et al., 1979; Suarez, 1981). However, at leaching fractions of 0.1 or less, precipitation may be significant, even more so in the lower root zone, depending on the composition of the irrigation water. Thus for sodic waters, an adjusted R_{Na} is recommended as a generally better index; it may be calculated by either of two methods which give essentially equivalent results (Rhoades, 1982b; Suarez, 1981, 1982; Oster & Rhoades, 1977). For near-surface soil conditions, the adjusted R_{Na} may be calculated as:

$$\text{adjusted } R_{Na} = Na/\sqrt{(Ca_X + Mg)/2}, \qquad [2]$$

where Na and Mg are the sodium and magnesium concentrations in the irrigation water in $mmol_c/L$ and Ca_x is an estimate of the calcium concentration that will result in the soil solution upon equilibration of the soil and irrigation water. The value of Ca_x to use in Eq. [2] is obtained in Table 36-7 in terms of the HCO_3/Ca ratio ($mmol_c/L$) and the electrical conductivity (dS/m) of the irrigation water. To calculate the adjusted R_{Na} for lower depths see Rhoades (1982b) or Suarez (1981, 1982). For waters having EC_{iw}'s of ≤ 0.4 dS/m, dissolution of minerals may exert more influence on the composition of the soil water than that of the irrigation water. Lack of appropriate descriptions of mineral weathering currently prevents inclusion of the effects of weathering in the predictions of soil solution composition (Jurinak, 1984).

Obviously soil permeability and tilth responses to irrigation should be assessed in terms of the salinity of the infiltrating water and the exchangeable Na percentage of the topsoil known to exist, or predicted to result (using the adjusted R_{Na}) after irrigation. A number of other soil properties have been identified as affecting the acceptable levels of these parameters. Clay mineralogy and content have both been implicated, as have certain "cementing" materials (various poorly defined calcareous, siliceous, and oxidic compounds) and organic matter which tend to stabilize aggregates. Nevertheless, a useful general guideline relationship for assessing sodicity hazard is given in Fig. 36-4, after Rhoades (1982b). The likelihood of permeability or tilth problems occurring with irrigation, from sodicity causes, is expressed in terms of the R_{Na} (or P_{Na}) of the topsoil and the electrical conductivity of the infiltrating water (EC_{iw}). The "threshold" relation curves downward below R_{Na} values of 10 and intersects the EC_{iw} axis at a value of about 0.3 dS/m,

Table 36-7. Calcium concentration (Ca_x) expected to remain in near-surface soil water following irrigation with water of given HCO_3/Ca ratio and EC_w. Adapted from Suarez (1981).[†]

	Salinity of applied water (EC_w), dS/m											
	0.1	0.2	0.3	0.5	0.7	1.0	1.5	2.0	3.0	4.0	6.0	8.0
0.05	13.20	13.61	13.92	14.40	14.79	15.26	15.91	16.43	17.28	17.97	19.07	19.94
0.10	8.31	8.57	8.77	9.07	9.31	9.62	10.02	10.35	10.89	11.32	12.01	12.56
0.15	6.34	6.54	6.69	6.92	7.11	7.34	7.65	7.90	8.31	8.64	9.17	9.58
0.20	5.24	5.40	5.52	5.71	5.87	6.06	6.31	6.52	6.86	7.13	7.57	7.91
0.25	4.51	4.65	4.76	4.92	5.06	5.22	5.44	5.62	5.91	6.15	6.52	6.82
0.30	4.00	4.12	4.21	4.36	4.48	4.62	4.82	4.98	5.24	5.44	5.77	6.04
0.35	3.61	3.72	3.80	3.94	4.04	4.17	4.35	4.49	4.72	4.91	5.21	5.45
0.40	3.30	3.40	3.48	3.60	3.70	3.82	3.98	4.11	4.32	4.49	4.77	4.98
0.45	3.05	3.14	3.22	3.33	3.42	3.53	3.68	3.80	4.00	4.15	4.41	4.61
0.50	2.84	2.93	3.00	3.10	3.19	3.29	3.43	3.54	3.72	3.87	4.11	4.30
0.75	2.17	2.24	2.29	2.37	2.43	2.51	2.62	2.70	2.84	2.95	3.14	3.28
1.00	1.79	1.85	1.89	1.96	2.01	2.09	2.16	2.23	2.35	2.44	2.59	2.71
1.25	1.54	1.59	1.63	1.68	1.73	1.78	1.86	1.92	2.02	2.10	2.23	2.33
1.50	1.37	1.41	1.44	1.49	1.53	1.58	1.65	1.70	1.79	1.86	1.97	2.07
1.75	1.23	1.27	1.30	1.35	1.38	1.43	1.49	1.54	1.62	1.68	1.78	1.86
2.00	1.13	1.16	1.19	1.23	1.26	1.31	1.36	1.40	1.48	1.54	1.63	1.70
2.25	1.04	1.08	1.10	1.14	1.17	1.21	1.26	1.30	1.37	1.42	1.51	1.58
2.50	0.97	1.00	1.02	1.06	1.09	1.12	1.17	1.21	1.27	1.32	1.40	1.47
2.00	0.85	0.89	0.91	0.94	0.96	1.00	1.04	1.07	1.13	1.17	1.24	1.30
3.50	0.78	0.80	0.82	0.85	0.87	0.90	0.94	0.97	1.02	1.06	1.12	1.17
4.00	0.71	0.73	0.75	0.78	0.80	0.82	0.86	0.88	0.93	0.97	1.03	1.07
4.50	0.66	0.68	0.69	0.72	0.74	0.76	0.79	0.82	0.86	0.90	0.95	0.99
5.00	0.61	0.63	0.65	0.67	0.69	0.71	0.74	0.76	0.80	0.83	0.88	0.93
7.00	0.49	0.50	0.52	0.53	0.55	0.57	0.59	0.61	0.64	0.67	0.71	0.74
10.00	0.39	0.40	0.41	0.42	0.43	0.45	0.47	0.48	0.51	0.53	0.56	0.58
20.00	0.24	0.25	0.26	0.26	0.27	0.28	0.29	0.30	0.32	0.33	0.35	0.37
30.00	0.18	0.19	0.20	0.20	0.21	0.21	0.22	0.23	0.24	0.25	0.27	0.28

[†] Assumes a soil source of Ca from lime ($CaCO_3$) or silicates; no precipitation of Mg, and partial pressure of CO_2 near the soil surface (P_{CO_2}) is 0.0007 kPa Ca_x, HCO_3, Ca are reported in $mmol_c/L$; EC_w is in dS/m.

because of the predominant effect of electrolyte concentration on soil aggregate stability and clay dispersion at such low salinities. Because significant differences exist in the sodicity response among soils, this relation should only be used as an approximate guideline in the absence of more site-specific data.

In so far as water impact enhances the loss of permeability and tilth, the guideline relationship given in Fig. 36-4 for assessing the sodicity hazard may be inappropriate. In this regard, a diagram produced for an Australian group of red-brown earths (Rengasamy et al., 1984) may be of value. In this diagram in which R_{Na} is plotted against total cation concentration (both being determined in a 1:5 soil water extract), guidelines are given to separate dispersive, potentially dispersive, and nondispersive soils.

In many irrigation areas the irrigation season is followed by a rainy season, so that the ability of the soil to resist aggregate collapse and crusting

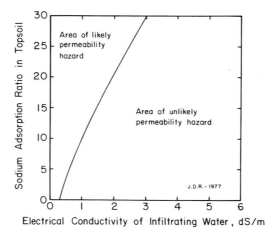

Fig. 36–4. Threshold values of Na adsorption ratio of topsoil and electrical conductivity of infiltrating water associated with the likelihood of substantial losses in permeability. After Rhoades (1982b).

needs to be predicted for both the irrigation water and the rain water. Insufficient research has been directed to predicting sodicity standards for conditions of alternating rain and irrigation, which often differ greatly in quality. Because of these limitations, any of the above mentioned guidelines must be used with caution. It is recommended that, whenever possible, analogous relations be established for the specific conditions. Methods are available to test the susceptibility of soils to sodicity-induced disaggregation, clay dispersion, and losses of permeability (Yousaf et al., 1987a, b; Abu-Sharar et al., 1987a, b; Emerson, 1984; Shainberg, 1984; Shainberg & Letey, 1984).

II. SALINITY-RELATED PROCESSES OPERATING IN IRRIGATED SOIL-PLANT-WATER SYSTEMS

The effective management of salinity under irrigated conditions requires an understanding not only of the more or less direct effects of salt on plants and soils but also of the effects and interactions of irrigation on soil and water salinity and, hence, on plant response.

A. Leaching Requirement and Salt Balance

Salts accumulate in the irrigated root zone when they are left behind as the soil water is used by the plant in transpiration or lost by evaporation. By drying the upper soil relative to deeper layers, evapotranspiration also creates the potential for an upward flow of water into the root zone. The root zone and the soil surface may become salinized by this process, especially where shallow saline water tables are present. The removal of salt from the root zone to maintain the soil solution at a salinity level compatible with the cropping system is referred to as maintaining *salt balance*. The fraction of infiltrated water that passes through the root zone is called the *leaching fraction* (L), i.e., D_{dw}/D_{iw} where D_{dw} and D_{iw} are, respectively, the depths of drainage water and of irrigation water. The L can be estimated for steady-state conditions from EC_{iw}/EC_{dw}, where EC_{iw} and EC_{dw} are the electrical conductivities of the irrigation water and the drainage water, respectively, based on a salt balance model and the following assumptions (Rhoades, 1974). A salt balance relation (Eq. [3]) may be obtained by algebraically summing the various inputs and outputs of salt to the soil water salinity (S_{sw}) of the root zone:

$$V_{iw} C_{iw} + V_{gw} C_{gw} + S_m + S_f - V_{dw} C_{dw} - S_p - S_c = \Delta S_{sw} \qquad [3]$$

where V_{iw}, V_{gw}, V_{dw}, and C_{iw}, C_{gw}, C_{dw} are volume and total salt concentration of irrigation, ground, and drainage water, respectively. The V_{gw} refers to that water which moves up into the root zone from the water table. S_m is the amount of salt brought into solution from weathering soil minerals or dissolving salt deposits, S_f is the quantity of soluble salt added in agricultural chemicals (fertilizers and amendments) and animal manures, S_p is the quantity of applied soluble salt (in the irrigation water) that precipitates in the soil after application, and S_c is the quantity of salt removed from the soil water in the harvested portion of the crop. The net difference between these inputs and outputs gives the resultant change in soil-water salinity (ΔS_{sw}). Under steady-state conditions ($\Delta S_{sw} = 0$), assuming: (i) no appreciable contribution of salts from the dissolution of soil minerals or salts, or loss of soluble salts by precipitation processes and crop removal (or alternatively that the net effect of these opposing reactions is approximately compensating), (ii) uniform areal application of water in the field, and (iii) where the water table depth is sufficient to prevent the introduction of salts into the root zone from capillary rise processes, Eq. [3] reduces to

$$D_{dw}/D_{iw} = EC_{iw}/EC_{dw} \qquad [4]$$

where the equivalent depth of water D is substituted for volume, and concentration is replaced by electrical conductivity (EC), since the EC of a water is a reliable index of its total solute concentration within practical limits (U.S. Salinity Laboratory Staff, 1954). This equation shows the approximate equality, under the given conditions, between the leaching fraction ($L = D_{dw}/D_{iw}$) and the ratio of salinity in the irrigation and drainage waters (EC_{iw}/EC_{dw}). Thus, by varying the fraction of applied water that is percolated through the root zone, it is possible to control the concentration of salts in the drainage water within certain limits, and hence, to control either the average or the maximum salinity of the soil water in the profile at some desired level (intermediate between the EC_{iw} and EC_{dw} levels).

Evidence for the validity of this latter statement is given in Fig. 36–5 and 36–6. These data were obtained in controlled lysimeter experiments by Bower et al. (1969). Figure 36–5 shows how the irrigation water salinity and

Key to treatment symbols: $\dfrac{\text{EC of irr. water, dS/m}}{\text{Leaching fraction}}$

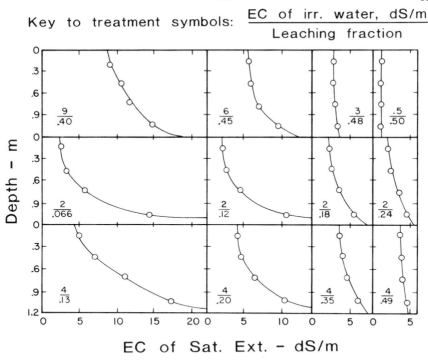

Fig. 36–5. Steady-state salt profiles expressed as electrical conductivity (EC) of the soil-saturation extract, as influenced by EC of irrigation water and leaching fraction. Key to treatment symbols: EC of irrigation water, dS/m per leaching fraction (Bower et al., 1969a).

leaching fraction affect the distribution and accumulation of soluble salts in soil profile (as expressed by EC of saturation extracts) irrigated to steady-state conditions. Figure 36–6 presents the data of Fig. 36–5 in terms of average root zone salinity. Figure 36–7 relates alfalfa (*Medicago sativa* L.) yield data obtained in this experiment to average root zone salinity. Data like these and field experience with conventional irrigation management show that for fields irrigated to steady-state conditions:

1. The salt content of the soil water increases with depth in the root zone region of soil profiles, except when irrigating with low-salt waters ($EC_{iw} < 0.2$ dS/m) and high leaching fractions ($L > 0.5$).
2. Soil water salinity for irrigation waters of a given salt content is essentially uniform near the soil surface regardless of leaching fraction, but increases with depth as L is decreased.
3. At approximately equal EC_{iw}/L ratios, soil-water salinity is proportional to EC_{iw} near the soil surface, but is nearly independent of EC_{iw} at the bottom of the root zone.
4. Average root zone soil-water salinity increases and crop yield decreases as EC_{iw} increases and as L decreases.
5. The first increments of leaching are the most effective in preventing salt accumulation in the soil water of the root zone (see Fig. 36–6).

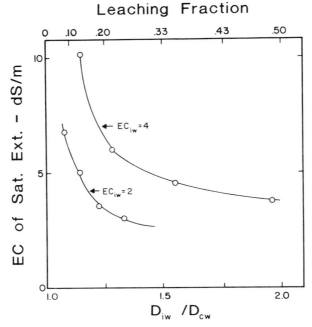

Fig. 36–6. Relation between average root zone salinity, expressed as electrical conductivity (EC) of the soil-saturation extract, and leaching fraction for two irrigation water concentrations (Bower et al., 1969).

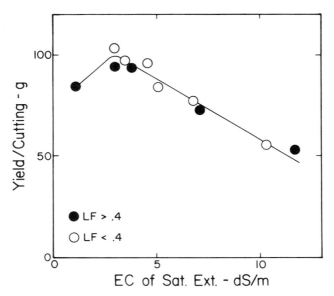

Fig. 36–7. Relation between alfalfa (*Medicago sativa* L.) yield and average salinity of the root zone expressed as electrical conductivity of the soil-saturation extract (Bower et al., 1969).

The average root zone salinity is also affected by the degree to which the soil water is depleted between irrigations. The kinds of changing conditions of salinity that occur with moisture depletions between irrigations are illustrated in Fig. 36–8 after Rhoades (1972). The time-averaged soil water salinity is greater in soils that are irrigated less frequently than in soils irrigated more frequently, other things being equal (Rhoades & Merrill, 1976). In any system of irrigated agriculture, the soil water should be maintained in a range that will give the greatest net return for the crop being grown. But, because the effects of matric and osmotic potentials on crop response are approximately additive (Wadleigh & Ayers, 1945; Bresler, 1987), the soil water content under saline conditions should be kept somewhat higher by irrigating more frequently than would be required under nonsaline conditions.

From the preceding salt balance considerations it is obvious that sufficient irrigation water (or rainfall) must be applied over and above the evapotranspiration needs of the particular crop, so that there is excess water to pass through and beyond the root zone and to carry away salts with it. This leaching prevents excessive salt accumulation in the root zone. This excess water is referred to as the *leaching requirement*, Lr (U.S. Salinity Laboratory Staff, 1954). The Lr for salinity control may be determined using the relations given in Fig. 36–9, the tolerance of the crop (the threshold levels given in Tables 36–1 and 36–2), and the salinity of the irrigation water and type of irrigation management (after Rhoades, 1977, 1982b). The curve labelled "conventional irrigation" is used where the soil is allowed to dry out between irrigation, i.e., where significant matric stress occurs along with the salinity induced osmotic stress; the curve labelled "high frequency irrigation" is used where the soil does not dry out significantly between irriga-

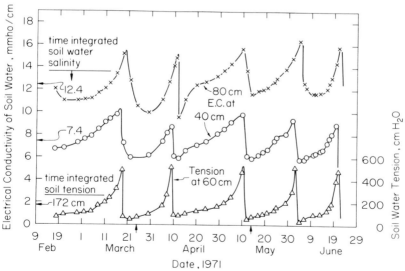

Fig. 36–8. Variations in in situ soil water electrical conductivity (EC) and tension (centimeter of water) in the root zone of an alfalfa crop during the spring of the year and their respective integrated values (Rhoades, 1972).

tions. Inherent within these two curves is a change in the index of salinity used to relate crop response to salinity. The linear average salinity within the root zone is used for conventional irrigation; water uptake-weighted salinity within the root zone is used for high-frequency irrigation. Smith and Hancock (1986) recently presented the same rationale and approach as Rhoades (1982b) to calculate Lr; they suggested that the water-uptake-weighted relation is the only acceptable means of calculating Lr. Rationale and evidence for the need to change the parameter of salinity depending on irrigation management have been given by Ingvalson et al. (1976), Rhoades and Merrill (1976), and Rhoades (1977, 1982b, 1984d). The results of the recent evaluation by Bresler (1987) of crop response to salinity and irrigation management may be interpreted as supportive of the "dual Lr" concept of Rhoades (1982b). The dual Lr concept is an indirect way of adjusting Lr for matric stress effects on crop response to salinity and is recommended herein, until a more convincing method is developed. There remains a need to develop a simple model that adequately integrates the response of plants to time- and space-varying salinity and matric stresses, and to irrigation

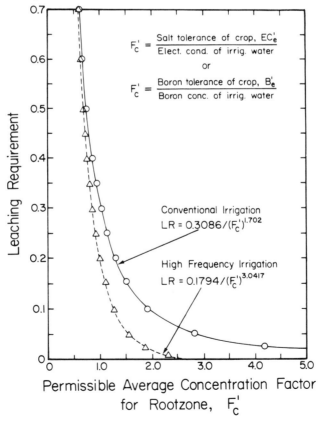

Fig. 36–9. Relation between leaching fraction and permissible root zone concentration factor for use in determining leaching requirement for conventional irrigation. After Rhoades (1982b).

management, as well as more appropriate crop tolerance to salinity data than that now present (Bresler, 1987; Bresler & Hoffman, 1986).

Leaching requirements for Cl and B can be determined analogously to salinity also using Fig. 36-9, when the tolerances of the crops to Cl or B are used to calculate F_c'. These tolerances may be obtained from data given in Table 36-5 and 36-6 after converting the threshold values given there in terms of soil water content to a saturation extract water content basis (by dividing by 2 for Cl and by 1.4 for B). The required leaching may be achieved either by applying sufficient water at each irrigation to meet the Lr or by applying, less frequently, a "leaching irrigation" sufficient to remove the salt accumulated from previous irrigations (Nightingale et al., 1986). For a more detailed treatment of leaching requirement see Rhoades (1982b) and Smith and Hancock (1986).

In order to achieve the desired degree of leaching, either natural or artificial drainage must be adequate to convey the drainage water (leachate) away from the root zone. Moreover, watertables, if present, must be controlled at an appropriate depth both to enable leaching and to prevent any appreciable upward flow of water (with its salt) into the root zone. While the appropriate water table depth is a function of the hydraulic properties of the soil substrate and climate, it is also dependent on irrigation management (van Schilfgaarde, 1976). For an exhaustive treatment of the practice of drainage and its application to irrigated lands, see van Schilfgaarde (1974) and chapter 35 in this book.

B. Root Zone Salinity, Crop Water Availability, and Evapotranspiration

Under irrigated field conditions, *soil water salinity* (i.e., the osmotic component of total soil water potential) is seldom uniform with depth throughout the root zone; typically it is low in the upper layers and increases, often considerably, towards the lower root zone, depending on the leaching fraction as previously discussed (see Fig. 36-5). Furthermore, in the time interval following an irrigation, as water is used by the crop and lost by evaporation, the total soil water potential changes at any depth. The potential decreases because of reductions in both the matric potential (as the soil dries) and the osmotic potential (as salt is concentrated in the reduced volume of soil water). Thus the salinity level varies both in time and depth, as shown in Fig. 36-8, depending on the degree to which water is depleted between irrigations and the leaching fraction, as discussed in Rhoades (1972) and Rhoades and Merrill (1976).

Crop yields have been shown to be closely correlated with the time- and water uptake-averaged soil water potential of the root zone (Ingvalson et al., 1976; Bresler, 1987). This finding is in keeping with the observation that following irrigation, plant roots preferentially absorb water from regions of high total potential, i.e., of low matric plus osmotic stress (Shalhevet & Bernstein, 1968). Thus water is used from the upper, less saline root zone, until the total water stress becomes greater in the upper than in the lower root zone; at such time water is used from the lower root zone (Wadleich & Ayers, 1945). Once this situation is reached, salinity effects upon crop growth become increasingly apparent, unless irrigation intervenes for the reasons given elsewhere in more detail (Rhoades & Merrill, 1976). Because salinity effects are predominantly water stress effects, there are obvious implications for irrigation management. For example, plants can tolerate higher levels of salinity under conditions of low matric stress resulting from high-frequency irrigation (such as drip and trickle). Again, deleterious effects of high soil water salinities in the lower root zone can be minimized if sufficient, low-salinity water is added to the upper root zone fast enough to satisfy the crop's evapotranspiration needs so that it does not need to extract water from the deeper, more saline part of the root zone. Thus, irrigation management affects permissible levels of salinity in soils and irrigation waters (Rhoades, 1972, 1976, 1982b). Furthermore, crop salinity tolerances determined under flood or furrow irrigation may not be directly applicable to trickle and sprinkler irrigation because of the typically higher water potentials achieved with these latter forms of irrigation.

The relation between crop yield and seasonal amount of applied water (crop-water production function) is required to design irrigation systems and to plan irrigation management. This relation is complicated by the presence of salinity. Sufficient water must be applied to provide for leaching in addition to evapotranspiration (ET). The leaching requirement is somewhat uncertain because matric stress effects on crop response to salinity are not explicitly included, though they can be with the use of somewhat-complicated models (Rhoades & Merrill, 1976; Bresler, 1987). Recently, several investigators have attempted to develop crop-water production functions for irrigation under saline conditions (Letey et al., 1985; Solomon, 1985; Bresler, 1987). These models combine relations between yield and ET, between yield and root zone salinity and between root zone salinity and leaching fraction. Some models assume crop response is related to average root zone salinity, and some assume it is related to water uptake-weighted salinity. All models assume the relation between yield and ET is independent of whether the yield is primarily affected by matric stress (soil water content) or osmotic stress (salinity). All models ignore the chemical reactions occurring during irrigation and leaching, i.e., they assume mass balance of total salts. [If the model of Rhoades & Merrill (1976) were combined with ET-yield relations, this deficiency could be eliminated.] These models have been tested using results of various lysimeter and field experiments. The correspondence is reasonable, suggesting their utility for planning irrigation management. The details of the calculation procedures are not given herein. The simplest to use would seem to be that of Letey et al. (1985). All of these models show that for saline conditions, more water must be applied, extra water is needed for leaching, and the frequency of irrigation must be increased relative to nonsaline conditions.

C. Assessing Water Suitability for Irrigation

The suitability of an irrigation water should be evaluated on the basis of the specific conditions of use, including the crops grown, soil properties, irrigation management, cultural practices, and climatic factors. The "ultimate" method for assessing the suitability of a water for irrigation consists of: (i) predicting the composition and matric potential of the soil water both in time and space resulting from irrigation and cropping and (ii) interpreting such information in terms of how soil conditions are affected and how the crop would respond to such conditions under any set of climatic variables (Rhoades, 1972). A method using these criteria has been developed (Rhoades & Merrill, 1976). The basic approach is to: (i) predict the salinity, sodicity, and toxic-solute concentration of the soil water within a simulated crop root zone resulting from use of the particular irrigation water of given composition at a specified leaching fraction and (ii) evaluate the effect of this salinity level (or solute concentration) on crop yield and of the sodicity level on soil permeability. This scheme has been simplified for steady-state conditions, both for computer and non-computer users (Rhoades, 1984d), by eliminating the matric stress factor and varying the index of salinity depending on whether matric stress is significant or not. The rationale for this approximation is given in Rhoades and Merrill &1976) and Rhoades (1982b). The computer program "watsuit" calculates soil water compositions [concentrations of Ca, Mg, Na, CO_3, HCO_3, Cl, SO_4 and sum of cations and anions; the Ca/Mg ratio; pH; EC; R_{Na}; and osmotic potential (π)] for the soil surface and the four quarter-depth intervals of the root zone, for each of five (0.05, 0.1, 0.2, 0.3, and 0.4) leaching fractions. These calculations are made assuming a water uptake pattern by the crop of 40, 30, 20, and 10% of total water used by quarter fractions of its root zone and P_{CO2} levels for the surface and four lower depths of 0.0007, 0.005, 0.015, 0.023, and 0.030 KPa, in order of increasing depth, respectively. The equilibrium chemistry program described in Oster and Rhoades (1975) was used for these calculations. The calculations are repeated with the water of soil being hypothetically amended with various chemicals. These latter calculations are made to determine the potential benefits of various amendments to improve the utilization of sodic waters where soil permeability or Na "toxicity" are limiting factors.

In addition, watsuit calculates average profile EC, upper profile EC, average profile R_{Na}, upper profile R_{Na}, and water uptake-weighted salinity, in both $mmol_c/L$ (C) and osmotic potential (π), corrected for salt precipitation and mineral dissolution.

Prognosis of problem likelihood is made after the soil water compositions are predicted. A soil salinity problem is deemed likely if the predicted root zone salinity exceeds the tolerance level of the crop to be grown. Use of the water will result in a yield reduction unless there is a change in crop and/or L. If yield reduction can be tolerated, then a higher salinity tolerance level can be used in place of the no yield-loss threshold value. The salt tolerances of crops have been conveniently summarized by Maas (1986). The

Table 36–8. Relative concentration or electrical conductivity of soil water (saturation paste extract basis) at steady-state compared to that of irrigation water (F_c').

Root zone interval	F_c'					
	Leaching fraction					
	0.05	0.10	0.20	0.30	0.40	0.50
	Linear average†					
Upper quarter	0.65	0.64	0.62	0.60	0.58	0.56
Whole root zone	2.79	1.88	1.29	1.03	0.87	0.77
	Water uptake weighted‡					
Whole root zone	1.79	1.35	1.03	0.87	0.77	0.70

† Use for conventional irrigation management.
‡ Use for high-frequency irrigation management or where matric potential development between irrigations is insignificant.

effect of salinity under conditions of frequent irrigation management (i.e., when little matric stress exists) is evaluated from water uptake-weighted salinity, EC, C, or π. For infrequent irrigation (i.e., conventional management where significant matric stress occurs over the irrigation interval), one uses average profile EC. To assess toxicity, specific solute concentrations are used in place of EC or π. To evaluate potential sodicity (permeability/crusting) problems, use is made of soil-surface R_{Na} and EC of the infiltrating water and appropriate R_{Na} − EC threshold relations for the soils of concern. In the absence of such specific data in this regard, use Fig. 36–4. The permeability hazard is determined by observing whether the adjusted R_{Na} − EC_{iw} combination lies to the left (problem likely) or right (no problem likely) of the threshold line in Fig. 36–4. The benefits of adding amendments are evaluated from the predicted soil solution compositions after "treatment."

The noncomputer version of watsuit is described in more detail elsewhere (Rhoades, 1982b, 1984d). With this procedure, steady-state salinity (or solute concentration) is estimated by multiplying the EC of the irrigation water (or solute concentration) by the relative concentration factor (F_c') appropriate to the leaching fraction given in Table 36–8 or from Fig. 36–10 or 36–11 (the graphical equivalents of Table 36–8). These estimates are less accurate than watsuit because they do not take account of precipitation − dissolution reactions or ion-pair effects.

Use of Table 36–8 is illustrated with the following example. Given an irrigation water with an EC_{iw} of 2 dS/m and a leaching fraction of 0.1 with conventional irrigation frequency, the appropriate concentration factor from Table 36–8 is 1.88. The maximum predicted EC_e for the average root zone salinity at steady state then becomes (1.88)(2) = 3.8 dS/m. Alternatively, the value of 3.8 dS/m could be obtained from Fig. 36–10. An analogous approach is used to predict specific solute concentrations (such as Cl or B) when they are of concern. Evaluation of suitability for cropping then proceeds as previously explained for the case of watsuit. The likelihood of a permeability hazard is also assessed as previously discussed but the adjusted R_{Na} value (see page 1105) is used in place of the R_{Na} value provided by watsuit.

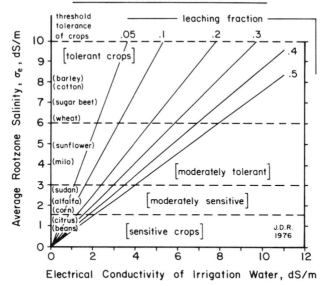

Fig. 36–10. Relations between average root zone salinity (saturation extract basis), electrical conductivity of irrigation water, and leaching fraction to use for conditions of conventional irrigation management (Rhoades, 1982b).

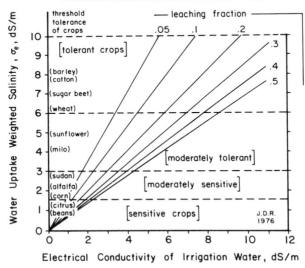

Fig. 36–11. Relations between water uptake-weighted salinity (saturation extract basis), electrical conductivity of irrigation water, and leaching fraction to use for conditions of high-frequency irrigation (Rhoades, 1982b).

D. Salination of Drainage and Receiving Waters

As discussed above, an appropriate amount of leaching, the Lr, is needed to maintain salt balance in the profile of irrigated crops. Unfortunately leaching water that drains from the soil root zone frequently conveys salt to groundwater or surface water systems. Often this process degrades the quality of the receiving waters and with repeated cycles of water diversion, irrigation use, drainage, percolation and return flow of such waters, the degradation progresses. This degradation is sufficient to create actual or potential salinity problems to other water users and conflicts may develop between the various user groups—urban, industrial, and agricultural. The topic of drainage and return flows and their effects upon water quality are dealt with in detail in chapter 38 (Hornsby) of this book. Hence we discuss briefly how the undesirable consequences of pollution from irrigated agriculture may be reduced by minimizing the degree of leaching and deep drainage.

Typically L exceeds Lr in most irrigation projects, giving the opportunity to improve irrigation management so as to reduce L without harming crop production (van Schilfgaarde et al., 1974). Commensurate with a reduction in L, the volume of applied irrigation water is reduced, and hence the added salt required to be leached is also reduced. Minimizing L results in the smallest possible total salt load in the drainage flow from the root zone because it maximizes the precipitation of Ca, Mg, HCO_3, and SO_4 as carbonates and gypsum in the soil, and it minimizes mineral weathering and dissolution of previously deposited salts (Rhoades & Suarez, 1977). To illustrate, it was demonstrated in a lysimeter study (Rhoades et al., 1974) that reducing L from 0.3 to 0.1 reduced the salt load leaving the root zone by amounts varying between 2 and 12 t (ha/yr) for a range of irrigation water types.

The fact that the amount of salt discharged from irrigated lands will always be less with reduced leaching means that minimizing leaching is clearly beneficial in those situations where the salt load of the drainage waters must be directly addressed. Obviously there will be reduced volumes of saline water for disposal or desalting. However, where drainage waters do return to groundwaters or streams, reduced leaching may or may not cause a reduction in the salinity degradation of the receiving waters. A range of typical situations have been examined by Rhoades and Suarez (1977) and the outcome for any particular case is shown to depend on the geohydrologic situation at hand, the chemistries of the drainage and receiving waters, and especially on their degree of saturation with $CaCO_3$ and/or gypsum. Other factors involved include the quality and volume of other sources of recharge for the groundwater and surface water bodies. But in general, increased irrigation efficiency and reduced leaching will be beneficial with respect to water conservation, pollution, and drainage disposal problems (Rhoades, 1984e; Rhoades & Suarez, 1977).

III. MANAGEMENT PRACTICES FOR RECLAIMING SALT-AFFECTED SOILS

Sometimes excessive levels of salt in soils cannot be reduced over time by irigation and crop management practices. For such situations, it is more usual to set aside cropping temporarily and to speed the removal process by use of so-called reclamation practices. Selection of appropriate reclamation practices requires knowledge of the cause of the salt-related problem. Procedures for diagnosing the cause of a given problem will not be discussed herein, but reclamation should not be undertaken until the cause has been identified and measures implemented to preclude its recurrence. Methods used to test and diagnose salt-affected soils and waters are described elsewhere (Rhoades & Miyamoto, 1987).

The first requisite for the reclamation of any salt-affected soil is adequate drainage. After drainage is provided, salinity can be reduced to an acceptable level by leaching; for sodic soils, application of appropriate amendments may be required, in addition to leaching, to reduce the exchangeable Na content. Principles of reclamation will be described separately for saline soils and sodic soils. Techniques and guidelines for dealing with drainage problems per se are described elsewhere (van Schilfgaarde, 1974; chapter 35 [Tanji & Hanson] of this book).

A. Reclamation of Saline Soils

The only practical way to reduce the content of salts in soils is to leach them out by the passage of water (usually of a lower salinity than the soil solution) through the root zone. The amount of leaching required in this regard is a function of the initial level of soil salinity, of the final level desired and the depth of soil to be reclaimed (which are largely determined by the crops to be grown), of certain soil and field properties, and of the method of water application and its level of salinity. The overall salt transport process is made up of *convection* (transport with the bulk solution) and *diffusion* (movement under a concentration gradient) processes. Convection is usually the predominant process; it results in solute dispersion because differential water flow velocities occur in the soil pores as a consequence of nonuniform pore size distributions. Diffusion acts to equalize the salt distribution between large and small pores, but it is sufficiently slow that it limits salt removal under high-velocity flow conditions. During leaching, cation adsorption by the negatively charged colloid surfaces retards salt transport on the one hand, while anion exclusion hastens it on the other. Boron also undergoes adsorption reactions which retard its movement (Keren et al., 1985). Wagenet (1984) reviewed the theory of these processes and the relations used to describe them mathematically.

Soils with well-developed structure are especially variable in their water and solute transport properties, because the structural pore space (interpedal voids, earthworm channels, root holes, planar voids, or cracks) provides preferred pathways for flow, especially under flooded conditions. Hence much of the water and salt in interaggregate pores can be essentially bypassed during saturated leaching, as occurs with continuous ponding. With intermittent flooding, there is more time for diffusion processes to transport salt from immobile to mobile regions. For soils irrigated by sprinklers, flow velocity and water content are typically lower, so that less salt is bypassed and the efficiency of leaching is increased.

1. Direct Determination of Leaching Required to Reclaim Saline Soils

Leaching requirements for the reclamation of saline soils can be ascertained directly by measurement in field trials. The obvious, straight-forward way to determine the actual amount of leaching required for the reclamation of a particular combination of field situation and water application method is to leach a representative test site(s) by the intended method and to follow the change in salinity through the soil profile with amount of water infiltration. A convenient approach to follow the salinity changes is to measure bulk soil electrical conductivity (EC_a) (Rhoades, 1979). Under conditions of reference soil water content, EC_a is a direct measure of soil salinity and is easily and instantaneously measured in the field using either in situ, four-electrode probes of various types (burial or protable) or remote-sensing electromagnetic induction devices held by hand aboveground (Rhoades, 1984a; Rhoades & Oster, 1986). The progress of salt removal with leaching amount is immediately evident from the EC_a readings; this clearly and simply establishes the rate of reduction in salinity with water application and time, and hence, describes the leaching requirements of the particular field and water application condition.

2. Estimation of Leaching Required to Reclaim Saline Soils

In an ideal porous matrix system, i.e., without pore bypass, dissolution of precipitated salts, salt diffusion constraints, or hydrodynamic dispersion, the salt concentration of soil water passing a given depth in the soil profile should decrease to the concentration of the applied water when the volume of applied water equals the pore space of the soil volume to be leached [1 pore volume (PV)]. However, as previously discussed, soils seldom behave in such an ideal way. Considerable bypass may occur during leaching, especially in structured soils, because of water moving preferentially through interpedal macropores relative to micropores within the peds (McIntyre et al., 1982; van Hoorn, 1981). Generally the extent of bypass is greater the higher the clay content, the greater the rate of leaching, the higher the content of soil water maintained during leaching, and the greater the content of precipitated salts in the profile. Various estimates of appropriate factors have been made to account for such inefficiencies under field conditions (van der Molen, 1956; Gardner & Brooks, 1957; van Hoorn, 1981). Complex dynamic models have been developed for simulating the movement and reactions of salts in soils during leaching (Bresler et al., 1982; Jury et al., 1979; Oster & Halvorson, 1978; Oster & Frenkel, 1980; Dutt et al., 1972; Tanji et al., 1972). Reviews of these modelling processes are described in detail by Bresler et al. (1982) and Wagenet (1984). The prevalent models of salt reactions and transport in soils suffer the deficiency of not appropriately representing the large variations in the processes described above that often occur under actual, natural field conditions. Only recently has this problem been approached directly by measuring, on a large scale, solute distributions in undisturbed soil profiles. The results to date indicate that we do not yet have a suitable method to quantitatively integrate and describe mathematically all of the processes operating during salt transport through the soil on a field basis (Jury, 1984).

For the reasons given above, predictions based on transport theory are not widely used to estimate the amounts of leaching needed to reclaim saline soils. Estimates are usually based on guidelines established from empirical relationships derived from field reclamation experiments and experience. The following relations are useful in this regard.

Jury et al. (1979), in large outdoor lysimeter column studies, found that 1.5 PV of applied water or 1 PV of drainage water, removed essentially all Cl salt from both sandy loam and clay loam soils by either ponded or unsaturated leaching. (The difference is due to the water required to wet the soil from its initial condition). However, total salt removal, including dissolving soil gypsum, required considerable additional leaching. Total salt removal was well described by a single curve,

$$(C/C_o)(d_l/D_s\theta) = 0.8,$$ [5]

for all treatments, where C is the salt concentration of the effluent, C_o is the initial salt concentration of the soil water, d_l and d_s are depth of water applied and depth of soil respectively, and θ is the soil volumetric water content. The term $(d_l/d_s\theta)$ is equal to pore volumes of leaching water applied. In this case about 60 and 80% total salt removal occurred with applications of 2 and 4 PV equivalents of leaching water, respectively.

Alternative relations have been developed by Hoffman (1980) who summarized leaching results of a number of field reclamation trials. These data fit the following relation:

$$(C/C_o)(d_l/d_s) = k,$$ [6]

where C/C_o is the fraction of the initial salt concentration remaining in the profile after application of the amount of water per unit depth of soil, d_l/d_s. Equation [6] (also Eq. [5]) can be refined by taking the salt concentration of the applied water (C_i) into account by substituting $(C - C_i)/(C_o - C_i)$ for C/C_o. This improves the assessment of d_l as C_i increases or as complete reclamation (i.e., $C = C_i$) is approached. The constant k varies with soil type and method of water application. Representative values of k for continuous ponding by soil type are: peat (0.45), clay loam (0.3), and sandy loam (0.1), as shown in Fig. 36–12. In general, k is assumed to be equal to 0.3 with continuous ponded leaching. Compared to continuous flooding, intermittent flooding may reduce by about one-third the volume of water required to achieve the same degree of salt removal (Miller et al., 1965). Even more efficient leaching can be achieved by sprinkler irrigation (Nielsen et al., 1966). For leaching by intermittent ponding or sprinkler application, k is not highly soil type-dependent, as is shown in Fig. 36–13 and 36–14. For such leaching conditions, k is assumed to be equal to 0.1.

Leaching efficiency may often be increased by reducing the soil water content maintained during leaching, because the more unsaturated flow conditions reduce large pore bypass. Thus, cropping the soil during or between leachings is expected to enhance efficiency of salt removal because the soil water content is reduced by evapotranspiration. Removal of salts by crop harvesting, per se, is not an effective method for removing salt from soil Crops generally remove < 5% of the amount of soluble salts present in the root zone and less than the amount which normally would be applied in the irrigation water to meet their evapotranspiration needs (Hoffman, 1980).

Boron is more difficult to leach from soils than are Cl and SO_4 salts because it is adsorbed by the soil matrix. Boron is also toxic at low concentrations (1–10 mg/L), so much more leaching is required for B reclamation. Based on limited field studies (Reeve et al., 1955; Bingham et al., 1972), Boron removal by leaching can be described by Eq. [6], where k is equal to about 0.6 and is not highly dependent on method of water application. Thus,

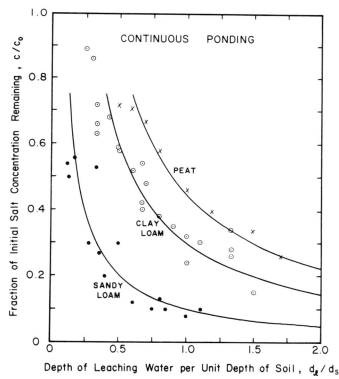

Fig. 36–12. Depth of leaching water per unit depth of soil required to reclaim a saline soil by continuous ponding. After Hoffman (1980).

for soils high in B, the amount of water required to remove a given fraction of B is about twice that required to remove soluble salts by continuous ponding. In addition, subsequent or periodic leachings may be required to remove additional B released with time from soil constituents by mineral weathering (Rhoades et al., 1970). Boron removal has been demonstrated in laboratory studies to be enhanced appreciably by use of H_2SO_4 (Prather, 1977), but this has not yet been tested in the field.

The above relations for the reclamation of saline soils should be used only as guidelines because, as previously discussed, leaching efficiency is influenced by many variables including spatial variation in water application and intake rates across the field, which were not addressed in the above studies. But as a guide, it may be assumed that about 70% of the soluble salt initially present in the soil will be removed by the continuous ponded leaching of medium-textured soils with a depth of applied water equivalent to the depth of soil to be reclaimed. In terms of PV, between 1.5 and 2.0 PV of water must pass through the soil to lower the salt concentration by this amount. About one-third this amount of leaching is needed if the water is applied by intermittent ponding or by sprinklers; however, these methods usually extend the time of reclamation.

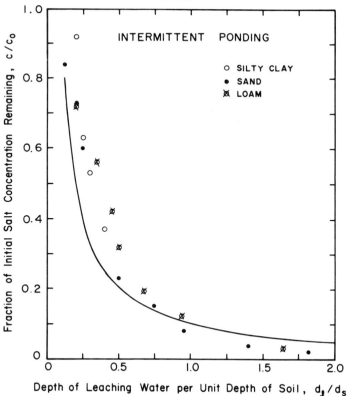

Fig. 36-13. Depth of leaching water per unit depth of soil required to reclaim a saline soil by ponding water intermittently. After Hoffman (1980).

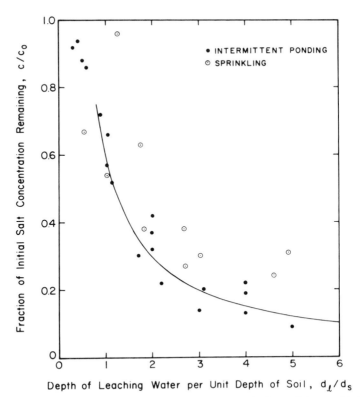

Fig. 36–14. Depth of leaching water per unit depth of soil required to reclaim a soil inherently high in B. After Hoffman (1980).

B. Reclamation of Sodic Soils

The interference to crop production caused by soil sodicity is generally due to adverse soil physical conditions, such as surface crusting, reduced permeability to water and air, and increased mechanical resistance to root penetration. As discussed earlier, a few crops, i.e., deciduous fruits and citrus, are extremely sensitive and show symptoms of toxicity in the range of exchangeable Na percentage (P_{Na}) of 2 to 10 (Bernstein, 1974; Maas, 1986). Most crops, however, are much less sensitive and, if soil structure is artificially maintained, only small to moderate yield losses occur at high P_{Na}'s of 30 to 60. At these P_{Na} levels the structure of field soils would normally have been severely affected so that, as Bernstein (1974) commented, "poor structure becomes limiting regardless of the varying ability of plants to function nutritionally with increasing exchangeable sodium." Paddy rice (*Oryza sativa* L.) is an important exception, in that it is not adversely affected by poor structure which limits the production of other crops.

1. The Reclamation Process

The purpose of reclaiming a sodic soil is to produce a stable structure which provides adequate porosity for water penetration and storage and which permits the drainage of any excess water to ensure aeration throughout the root zone. This usually requires increasing the Ca level on the cation exchange complex at the expense of Na, the replaced Na being removed either to lower levels or out of the profile by leaching. Thus, there are two prime requirements—a source of Ca and adequate leaching. Pertinent considerations in meeting these requirements are the physicochemical properties of the soil, the desired extent and rate of reclamation, the amount and quality of water available for leaching, and the cost involved. Several of these points are dealt with generally, before discussing in some detail the possible sources of Ca.

To achieve leaching, the soil's permeability and drainage capacities must be adequate (frequently they are not). Often permeability must be increased and drainage provided. The soil's porosity, and hence permeability, can be increased by tillage, but it subsequently declines rapidly in the presence of exchangeable Na unless a sufficiently high electrolyte concentration is maintained in the leaching solution (the required level can be estimated from Fig. 36–4). Thus in selecting an amendment, besides giving attention to the source of Ca, the choice must be such as to provide the appropriate electrolyte level to achieve and maintain adequate permeability. Gypsum is often inadequate in this regard, so reclamation is often not accomplished with its use in a reasonable time. Therefore, before undertaking reclamation, the permeability of the soil to the anticipated leaching water should be determined first, with and without the selected amendment, either in a small field trial or in a laboratory column experiment.

The economic practicality of reclaiming sodic soils is influenced by the level to which P_{Na} needs to be reduced and the depth to which the reduction is required, because these determine the time and amount of amendment needed. There is no single level of P_{Na} that can be specified for all situations, but generally the goal should be to reduce P_{Na} to as low a level as practical. The root zone sodicity profile that provides adequate conditions varies with crop species and with associated soil properties, such as clay and organic matter content and type. With regard to the depth of soil to be reclaimed, the upper root zone depth is usually of greatest importance. Miyamoto et al. (1975a) suggested that reclaiming a depth of 40 cm, encompassing the main root zone, is adequate for most practical purposes, but these authors did not clearly specify the actual P_{Na} profile to be achieved. Shainberg and Oster (1978) suggested P_{Na}'s of 5 and 15 at depths of 0.2 and 1 m, respectively, as being adequate for most purposes. It seems doubtful, however, whether any appropriate experimental evidence is available upon which such specification for different crops, soils, and irrigation waters can be based.

In many situations where an amendment is necessary for reclamation, the required amount (for even shallow depths of soil) will be beyond the eco-

Table 36-9. Reclamation methods appropriate to different sodic soil conditions.

Soil condition	Method
Saline-sodic, with soluble Ca in the A horizon	Leaching
Sodic B horizon, $CaCO_3$ or $CaSO_4$ in B/C horizons	Profile mixing followed by leaching
Calcareous sodic soils	Acids or acid-formers (e.g., S)
Moderately sodic soils ($P_{Na} < 20-25$)	Gypsum
Strongly sodic soils ($P_{Na} > 20-25$)	$CaCl_2$, H_2SO_4, or high-salt water dilution method

nomic capacity of the farmer in any one year. Smaller applications can often provide immediate permeability benefits (from the electrolyte effect) and some initial exchange of Na which can be built upon by later applications over a period of years (Loveday, 1976). This strategy of gradual, incremental reclamation—not requiring the total amount of amendment to be applied in one application—is commonly practiced in the reclamation of sodic soils (Rhoades, 1974).

2. The Source of Calcium

The source of Ca for replacing Na may be the soil itself, but many soils contain insufficient amounts of Ca-rich minerals, so an external source is required. The first three soil conditions listed in Table 36-9 are situations in which the soil generally provides the required Ca, whereas a Ca amendment is needed for the other two.

a. The Soil as a Source of Calcium. Leaching alone can often be successful in reclaiming saline-sodic soils (e.g., Overstreet et al., 1951; Jury et al., 1979), providing soil permeability is adequate and there is good drainage through and beyond the profile, an adequate supply of leaching water, and a soil source of Ca that can be mobilized. The Ca-supplying minerals include $CaSO_4 \cdot 2H_2O$, $CaCO_3$ and, in some instances, silicates (Rhoades et al., 1968). During the leaching of saline-sodic soils, some Ca-mineral solubilities may be enhanced as much as fivefold or more (Oster, 1982), because the exchange complex acts as a sink for the dissolving Ca. When such minerals are absent from the surface horizon but present in the B and C horizons, the profile may be deep plowed to invert and mix the minerals into the surface horizon before leaching (Rasmussen & McNeal, 1973; Toogood & Cairns, 1978). Rice is relatively tolerant to saline and sodic conditions and is frequently grown during the process of reclamation to provide some cash return. Although it has been suggested that the crop itself aids reclamation (Chabra & Abrol, 1977), this does not seem to have been clearly established under field conditions.

While leaching alone may be sufficient for the reclamation of saline-sodic soils, it is usually found to be too slow to be economic for the reclamation of sodic soils (Abrol & Bhumbla, 1973) and amendments are generally required to supply the Ca needed for Na exchange and the electrolyte level needed for permeability enhancement. Acid or acid-forming amendments can be used to produce Ca when $CaCO_3$ is present in the soil. A wide range of such amendments has been tested, often with positive results, though their commercial use does not appear to be widespread. Sulfuric acid, which suffers from handling difficulties because of corrosivity, may be more effective than gypsum in increasing water penetration into calcareous sodic soils (Prather et al., 1978), presumably because of the higher electrolyte levels achieved with the acid. In some areas acidic waste products from mining and industrial activities (including pollution abatement) are available in quantity and their disposal by use as soil amendments could be safe and economical (Miyamoto et al., 1975b).

b. Gypsum. When an external source of Ca is required, the possibilities include gypsum, $CaCl_2$, $CaCO_3$, or a water supply containing Ca. Of these, gypsum is by far the most widely used and will be discussed in more detail than the others.

Gypsum usually is advantageous in the reclamation of moderately sodic soils because of its relatively low cost, general availability, and ease of handling. However, for highly sodic soils (P_{Na}'s > 20–25) the dissolution of gypsum often does not provide a sufficiently high electrolyte concentration to create and maintain adequate permeability for the required leaching. Potentially achievable electrolyte concentrations in the soil solution saturated by gypsum range from 30 to 260 $mmol_c/L$ as the P_{Na} varies from 0 to 40 (Oster, 1982). Actual concentrations achieved may range from near saturation, as was apparently the case in the field experiment of Dutt et al. (1972), to much less—in the order of one-third to one-half of saturation. At the lower end of this range, the electrolyte concentration may be only about 12 to 16 $mmol_c/L$ and is influenced by factors affecting the rate of gypsum dissolution (Quirk & Schofield, 1955; Chaudry & Warkentin, 1968). Gypsum mined from natural deposits is distinguished from phosphogypsum, a byproduct of the manufacture of superphosphate. Because of particle size differences affecting surface area, the dissolution rate of the latter is up to ten times the former (Oster, 1982), yielding solution concentrations two to three times those of mined gypsum (Keren & Shainberg, 1981). Other factors that influence the dissolution rate of gypsum are the water flow velocity, the depth of mixing in the soil, and the presence and composition of other salts.

The effects of differences in particle size (really surface area) have been investigated by Hira and Singh (1980) and Chawla and Abrol (1982), among others. With fine particles, an initial high electrolyte level and hydraulic conductivity may be followed by sharp decreases, particularly in soils containing Na_2CO_3. This phenomenon has been attributed to the precipitation of dissolved Ca as $CaCO_3$. An additional cause may be $CaCO_3$ coating and inhibiting further dissolution of the gypsum particles. Coarser gypsum particles, on the other hand, result in initially lower levels of electrolyte, which are better maintained over time. For these reasons, Chawla and Abrol (1982) recommended that a mix of gypsum particle sizes be used, with an upper limit of 2 mm, to obtain the dual benefit of an initial rapid dissolution of some fine gypsum followed by the longer, more sustained release of Ca from the coarser particles.

When control is possible (i.e., when sprinkler irrigating), the water flow velocity should be reduced to increase the time of contact between water and gypsum fragments. This increases the rate of gypsum dissolution and the efficiency of cation exchange (Keren & O'Connor, 1982).

Depth of mixing influences the ability of applied gypsum to provide sufficient electrolyte to maintain surface infiltration rates. Khosla et al. (1973) found that applied gypsum was less effective the greater the depth of mixing, presumably both because of dilution of the added gypsum and repression of its dissolution by the presence of $CaCO_3$. Sulfates, if present in the soil solution, will also limit the dissolution of gypsum (Redley et al., 1980). On the other hand, the effective solubility of gypsum is increased when mixed with sodic soil, because the exchange phase acts as a sink for the dissolving Ca until a new equilibrium is reached (Oster & Frenkel, 1980). For improving infiltration and controlling runoff by reducing crusting, Agassi et al. (1982) found surface applications of gypsum more beneficial than applications mixed into the surface soil. Surface applications have also been reported to be effective, even for the reclamation of sodic subsoils (Bridge & Kleinig, 1968; Carter et al., 1978). However, Sharma et al. (1974) concluded that the Huey silt loam (fine-silty, mixed, mesic typic Natraqualfs) of Illinois (containing a natric B horizon) could only be effectively reclaimed by mixing gypsum to 90 cm and by providing tiles for drainage.

Generally, the evidence seems fairly clear that for sodic soil problems involving surface crusting and poor infiltration, a surface application, i.e., placement directly on the surface or mixed shallowly, is best. For impervious sodic B horizons, the combination of mechanical disruption and gypsum admixture is perhaps best; surface application may also be effective, but it requires that the dissolved gypsum be first leached down into the subsoil.

Gypsum may also be dissolved in the irrigation water. In some instances this is beneficial because it reduces the high R_{Na} value of the water itself (Gobran et al., 1982); in other instances, the benefit is due to the increase in electrolyte concentration of the infiltrating water that, in turn, promotes clay flocculation and increases permeability (Davidson & Quirk, 1961). Dissolving gypsum in the irrigation water may reduce the cost of reclamation by reducing both the amount of gypsum to be applied in any one year and the need for crushing and spreading of the mined gypsum. Aspects of the dissolution of large gypsum fragments in flowing water have been examined (e.g., Kemper et al., 1975; Keisling et al., 1978) with a view to providing criteria for designing beds of appropriate length, cross section, and gypsum fragment size for enhancing and sustaining dissolution.

c. Calcium Chloride and Calcium Carbonate. Calcium chloride is expensive, and is seldom economical unless available as an industrial waste product. Yet reclamation can be achieved rapidly with this amendment, even of sodic soils found to be essentially unreclaimable with the use of gypsum (Reeve & Doering, 1966a; Alperovich & Shainberg, 1973; Magdoff & Bresler, 1973). High solubility provides high initial electrolyte levels, which makes $CaCl_2$ a more effective amendment than gypsum for high P_{Na} soils. There may be situations where use of the two amendments in combination will provide nearly as effective reclamation but at a lesser cost (Mozheiko, 1969; Prather et al., 1978; Rhoades, 1982b).

Calcium carbonate has generally been considered of doubtful value for reclaiming sodic soils because its solubility is too low to provide sufficient electrolyte concentration for increasing permeability or enough Ca for exchange of Na, unless an acid or acid former is applied concurrently. However, Shainberg and Gal (1982) conducted laboratory experiments with finely (< 44 diam.) powdered $CaCO_3$. They found a mix of about 0.5% in the soil increased the electrolyte concentration of the soil solution sufficiently to facilitate leaching with low electrolyte waters. The effect was most pronounced for soils with P_{Na}'s near 20.

d. High Salt Water Dilution Method. This method is essentially reclamation by means of leaching with a Ca-containing water of an initially high enough electrolyte concentration to increase the permeability of sodic soils to a level needed to achieve leaching, followed by progressive dilution to enhance Na exchange (Reeve & Bower, 1960). The R value (ratio of divalent to total cations on an equivalent basis) of the water used in this method should be at least 0.3 (Doering & Reeve, 1965), and the greater R is, the less water required to effect reclamation (Reeve & Doering, 1966b). If the water available does not meet the R value criterion, additional Ca will be needed. Sea water saturated with gypsum can be effective for reclaiming sodic soils by this method (Muhammed et al., 1969). By maintaining or increasing soil permeability, the high salt water dilution method greatly reduces the time required for reclamation compared with more conventional amendments (e.g., van der Merwe, 1969; Reeve & Doering, 1966b). However, the total depth of water required for leaching is substantially greater than with other methods. The depth of leaching water required can be reduced by using less than the full equilibrium volume (e.g., two-thirds) at each dilution step and by minimizing the number of dilutions (Roos et al., 1976).

3. Amendment Rates and Leaching Required to Reclaim Sodic Soils

The conventional approach for calculating the application rate of a particular amendment has been to determine the so-called "gypsum requirement" (U.S. Salinity Laboratory Staff, 1954) and, for any amendment other than gypsum, to adjust it using a conversion factor (see Table 36–10). Even though Na replacement is not such a stoichiometric process, the following relation is used in the calculation:

$$\text{kg gypsum/ha} = (8.5)d \cdot \rho_B \cdot E_c \, (R_{Nai} - R_{Naf}), \quad [7]$$

where d is depth of soil to be claimed in meters, ρ_B is soil bulk density in megagrams per cubic meter, E_c is cation exchange capacity in millimoles of charge per kilogram, R_{Nai} and R_{Naf} are initial and final Na adsorption ratios, respectively. In theory, the amount calculated according to Eq. [7] needs to be multiplied by an appropriate factor to compensate for inefficiencies

Table 36-10. Equivalent amounts of common amendments for reclaiming sodic soils.

Amendment	Amount equivalent to 1 kg gypsum
	kg
Gypsum	1.00
S	0.19
H_2SO_4	0.57
CaS_5 (24% S)	0.77
$CaCO_3$	0.58
Calcium chloride dihydrate ($CaCl_2 \cdot 2H_2O$)	0.85
Ferrous sulfate ($FeSO_4$)	1.61
Aluminum sulfate [$Al_2(SO_4)_3$]	1.29

in the cation exchange process. A figure of 1.25 is often used when gypsum is the amendment (U.S. Salinity Laboratory Staff, 1954; Chaudry & Warkentin, 1968). The efficiency of exchange is different for the other amendments, and varies with the P_{Na} of the soil to be reclaimed and the electrolyte concentration of the leaching water. More appropriate ways to estimate the value of this efficiency factor are given in Rhoades (1982b) and Oster (1982). Additionally, Ca displaces exchangeable Mg so that the efficiency of exchange of Na is even further reduced (Loveday, 1976; Manin et al., 1982). Thus, the actual amount of amendment needed to effect reclamation may differ considerably from the calculated amount.

Dutt et al. (1972) produced, and tested in a field experiment, a computer model for calculating the gypsum requirement. This model accounts for some of the various chemical reactions occurring during reclamation and salt leaching, including $CaCO_3$ dissolution and water quality and gypsum effects on cation distribution within the profile. For a situation where the efficiency of exchange was high, they found that predicted and measured P_{Na} values to a depth of 30 cm and water quantities required for leaching were in good agreement. More recently, Hira et al. (1981) developed an equation for predicting the water requirement for dissolution of gypsum of a nominated particle diameter and density, taking account of the increase in gypsum solubility that results when it is mixed with a sodic soil. Other models were described in a thorough treatise by Bresler et al. (1982). None of these models are applicable to amendments other than gypsum nor do they account for the salt transport processes that occur under field conditions, as previously discussed.

Since there is an inadequate quantification of the processes of leaching and cation exchange under field situations and insufficient knowledge of the desired nature of the reclaimed P_{Na} profile for different crops and soils, application rates, in practice, are most often determined by local experience and financial considerations. Especially when gypsum is being used, application and reclamation are often spread over a number of years. Besides spreading the costs associated with the amendment, this approach reduces the amount of leaching required in any one year (and hence the drainage requirement) and it takes account of the limited solubility of gypsum. About 1 m of water is needed to dissolve 7.3 t/ha of gypsum having a fineness such that 85% passes a 100-mesh sieve. Thus, common practice for the reclamation of sodic soils is to apply about 10 t/ha of gypsum in the first year and use about 1.5 m of leaching water. In the subsequent 2 to 3 yr, additional annual applications of about 4 t/ha should be applied, with some leaching, until the root zone is eventually reclaimed.

IV. PRACTICES TO CONTROL SALINITY IN IRRIGATED LANDS

While there are areas of saline land (both natural and irrigation induced) for which the previously described reclamation procedures are needed, there is a greater need to undertake appropriate practices to prevent the development of excessive salination in irrigated lands. Management need not necessarily aim to control salinity at the lowest possible level, but rather to keep it within limits commensurate with sustained productivity. Crop, soil, and irrigation practices can be modified to achieve these limits. To maintain a check on the efficacy of the control practices, some system of monitoring soil salinity and drainage adequacy is advisable. This section deals with management and monitoring for salinity control for this purpose.

A. Crop Management

Because crops and different cutlivars of the same crop vary considerably in their tolerance to salinity (e.g., Maas & Hoffman, 1977; Maas, 1986), crops can be selected that produce satisfactorily for the particular conditions of salinity in the root zone. As discussed earlier, it is important to consider the crop's salt tolerance during seedling development. This is often the most sensitive growth stage (Shannon, 1982), and optimum yields are impossible without satisfactory establishment. Salt present in the seed bed reduces the rate of germination (Shannon & Francois, 1977) and thus increases the time to emergence. The stand may suffer as a consequence because of the occurrence of crusting resulting from surface drying, as well as from the increased opportunity for disease problems to develop. When a crust is likely to develop, sowing rate may be increased to facilitate seedling emergence. Other techniques to combat crusts include various forms of mulching and, in sodic soils, application of certain amendments, such as gypsum. Plant density may be increased to compensate for smaller plant size that exists under saline conditions. This increases the interception of the incoming energy of the sun and therefore, crop yield relative to normal densities.

B. Soil and Land Management

Where irrigation is by flood or furrow methods, careful land grading, such as is possible using laser-controlled earth-moving equipment, is desirable to achieve more uniform water application and consequently better salinity control. Barren or poor areas, in otherwise productive fields, are often high or low spots that receive insufficient or excessive water for good plant growth. Where perennial crops are planned, planting should be delayed after land grading for 1 or 2 yr during which time annual crops are grown and the fill areas allowed to settle prior to regrading for the permanent planting.

Salt accumulation can be especially damaging to germination and seedling establishment when raised beds or ridges are used and "wet-up" by furrow irrigation, even when the average salt levels in the soil and irrigation water are moderately low. The salt accumulates progressively towards the center of the surface of the bed or ridge and is most damaging when a single row of seeds is planted in the central position. With double-row beds, the seeds are planted on the shoulders to germinate in a region of relatively lower salinity. Sloping beds are perhaps best for saline soils (Bernstein et al., 1955) because the seed can be safely planted on the slope below the zone of salt accumulation. Salt accumulation in the seed bed may be prevented with pre-emergence irrigations using sprinklers or special furrows placed close to the seed. Once the seedlings are established, the sprinklers or special furrows may be replaced by conventional furrows.

During irrigation, sodic soils are especially prone to clay dispersion and, on drying and consolidation, to surface crusting. Frequently the surface sets into a massive layer, or the aggregates fuse together to form a coarse cloddy tilth. Application of various chemical amendments, as discussed earlier, can be used to alleviate such conditions, thus enabling better seedling emergence, improved water entry and storage, increased leaching of soluble salts, reductions in tillage costs, and greater flexibility of other operations. Practices that maintain high organic matter levels, e.g., green manuring and incorporation of crop residues, also help in the maintenance of good tilth. Where structural conditions are likely to hinder crop emergence and establishment, more frequent light irrigations may be applied to soften crusts.

For sodic soils that are especially liable to structural damage, but for other soils too, it is important to avoid tillage at high water contents. The most suitable water content for tillage is usually described as "moist," and is defined by the plastic and shrinkage limits (Archer, 1975). To reduce compaction, heavy machinery traffic should also be avoided. For row crops, Hillel (1980) suggested recognizing three distinct zones—a planting zone with conditions optimal for germination and seedling establishment, a management zone between rows where structure should be coarse and open for maximum water and air intake, and a traffic zone to which all compaction is confined. For more on the management of sodic soils, see the review by Loveday (1984).

C. Irrigation and Drainage Management

Although the prime requirements of irrigation management for salinity control are frequent irrigations, adequate leaching, drainage, and water table depth control, there are other significant contributing and interacting factors that should be considered. These include the delivery system and the method and manner of the irrigation.

1. The Delivery System

In many irrigation projects, the water supply canals are constructed in soil materials of sufficient permeability that considerable seepage losses occur. Such seepage is often a major cause of the development of high water tables and excessive soil salinity in irrigated lands. Seepage losses may be reduced by compacting the canal floor and walls or by lining them with less permeable materials. In some instances, concrete linings or pipeline construction may be the preferred solution.

For efficient control of a supply system, the water volume passing critical points, including the outlets to individual farms, needs to be known. This demands installation of effective flow-measuring devices, without which seepage losses are difficult to identify, and oversupply to farms is likely to occur. Additionally, many delivery systems encourage overirrigation because the water is supplied for fixed periods, or in fixed amounts, irrespective of seasonal variations in on-farm needs. Ideally, water delivery should be on demand, and to accomplish this there needs to be, in addition to delivery facilities, close coordination between the water distributing agency and the users.

2. On-farm Irrigation Practices

Earlier chapters have been devoted to this topic; various systems for irrigation have been described and the concepts and techniques for scheduling irrigation have been discussed. Still, some special aspects related to salinity merit further discussion.

In general, improvements in salinity control come from improvements made in on-farm irrigation efficiency, i.e., by providing the appropriate amount of water at the appropriate time with high uniformity of application (van Schilfgaarde, 1976). The ideal irrigation scheme would provide water more or less continuously to the plant to match evapotranspiration losses and to keep the water content in the root zone within narrow limits commensurate with adequate aeration and minimum loss in deep percolation for leaching. By this means the salinity of the soil water is prevented from increasing significantly between irrigation events as evapotranspiration proceeds. The availability of the water to the crop is maximized since the matric and osmotic potentials are maximized. To achieve such an ideal system requires delivery of the water to the field on demand at appropriate flow rates and volumes. To know what volume of replenishment water is needed for irrigation, evapotranspiration rates need to be accurately known or else "feedback" devices are needed to measure water and salt content in the soil (Phene et al., 1981; Rhoades et al., 1981).

Well-designed trickle or drip irrigation systems come close to being the ideal application system, as they maintain high water content, move accumulating salts out to the periphery of the wetted zone, and allow for high uniformity of application. Higher levels of salinity in the irrigation water can be tolerated with these systems than with other methods of irrigation. Good volume control and uniform distribution are also possible with certain types of fixed and moving sprinkler systems, but they suffer the disadvantage of tending to produce drop impact-induced soil crusts, especially in sodic soils, thereby restricting seedling emergence and water entry. Gravity flood systems can also achieve good water and salinity control, if designed and operated properly, though variations in soil properties within irrigated field units can mitigate against uniform intake. Laser-controlled precision leveling and level-basin methods of irrigation help to achieve high distribution efficiency for such food systems of irrigation (Dedrick et al., 1978). In furrow irrigation systems, use of closed conduits instead of open waterways for laterals gives the advantage of effective "off-on" control. Reducing furrow lengths, as in Worstell's (1979) multiset system, improves intake distribution and minimizes tail water losses. Surge irrigation can also improve uniformity of intake in furrow-irrigated fields (Bishop et al., 1981). With furrows, as indicated previously, salts tend to accumulate in the beds, and periodic flooding, along with crop rotation, is recommended for salinity control. Subirrigation, in which the water table is maintained high enough so that the capillary fringe and the root zone coincide, is generally not suitable over the long term when salts are high in the water supply. If subirrigation is to be used, the water table should be lowered periodically to allow leaching of accumulated salts by rainfall or surface water applications. For more discussion on irrigation management for salinity control, see the reviews of van Schilfgaarde (1976), van Schilfgaarde and Rawlins (1980), and Rhoades (1985a, b).

3. Drainage and Its Reuse for Irrigation

For any irrigation area to remain viable in the long term, drainage (either natural or artificial) must be able to cope with waters percolating beneath the irrigated land. Without such drainage, groundwaters eventually rise to levels that allow salts to accumulate in the soil and the root zone to become waterlogged. Contributions to deep percolation come from leaching water, canal and lateral waterway seepage, waters invading the irrigated area from elsewhere, and excessive rainfall in the area itself. Management practices that reduce these contributions also reduce the volume of drainage water and the degradation of the waters that receive it, as previously discussed. Such practices include increasing irrigation efficiency, adoption of the concept of "minimized leaching," recovery and reuse of tail water for irrigation, and interception and reuse of subsurface drainage flows for irrigation or diversion to appropriate waste sites.

With the minimized leaching approach, the aim is to make the maximum use of each volume of the applied irrigation water in evapotranspiration, thus producing minimum drainage and salt return (van Schilfgaarde et al., 1974). Where the drainage water can be intercepted, such as by groundwater pumping or tile drainage, it is often of a quality which permits reuse on irrigated crops of higher salt tolerance. Rhoades (1977, 1984b, c, 1986a) and Rhoades et al. (1989) demonstrated the feasibility of substituting high-saline drainage water for some of the conventional irrigation water in a cyclic reuse strategy which also involves the rotation of salt-tolerant crops and salt-sensitive crops. The strategy succeeds because: (i) preplant and initial irrigations of the tolerant crops are made with the lower salinity water, thereby leaching salts out of the soil in the vicinity of the emerging seedling (the drainage water being substituted after seedling establishment), (ii) the maximum salinity in the root zone possible with long-continued use of the more saline drainage water does not result since it is used for only part of the rotation, and (iii) the salt accumulated in the soil profile from irrigation with the drainage water is leached out during the subsequent period of irrigation of a sensitive crop with the lower-salinity water. In situations where the normal water is of particularly low salinity, crusting, and permeability problems may develop if its electrolyte concentration is too low for the level of sodicity developed during the period of irrigations with drainage water.

An alternate strategy often advocated is to dilute the drainage waters with better-quality waters and to use the blend for irrigation. In fact this is the natural process operative when drainage waters move by diffuse flow back to river or groundwater systems. Whether by intentional blending or by diffuse flow, this process of blending generally reduces the supply of water suitable for irrigation, especially when the drainage water salinity is high. In attempting to meet transpiration, a plant can only extract soil water up to its tolerance limit of salt concentration (this is the usable portion of the blend); the remaining water is unusable and must pass once again out of and beyond the root zone, often displacing or dissolving more salt in the process. Thus, any addition of salt to a water supply reduces the degree to which it can be consumed in crop growth. In terms of crop production, greater flexibility and use of the total water supply can be achieved by intercepting and keeping drainage waters isolated from surface water or groundwater supplies of better quality (Rhoades, 1989).

Reuse of a drainage water for irrigation eventually increases its salinity to the point that further reuse is no longer possible and it must be disposed of by some other means. Desalination of agricultural drainage water is not generally economically feasible and normally is only undertaken for political reasons. Discharge to evaporation ponds, outfall to the ocean, or placement in deep aquifers are more generally suitable as the means of disposal.

D. Monitoring for Salinity Control

The proper operation of a viable, permanent irrigated agriculture requires periodic information on the levels and distributions of soil salinity within the root zones and fields of the irrigation project. The salt level within the root zone must be kept below harmful levels. Gross salt balance evaluations on a project scale (i.e., measurement of salt load out vs. in) generally do not provide information on salinity changes occurring within root zone. They provide no information on the absolute level of salinity within the root zone and, therefore, are inadequate for assessing the adequacy of irrigation, leaching, and drainage practices and facilities for salinity control (Rhoades, 1974; Kaddah & Rhoades, 1976). Direct monitoring of root zone salinity is recommended to evaluate the effectiveness of various management programs (Rhoades et al., 1981). The shape of the salinity-depth relation of the soil profile and information on water table depth provide direct information of the direction of net water flux, and hence, of the adequacy of the irrigation/drainage system (Rhoades, 1976).

Changes in soil salinity can be determined from periodic measurements made: (i) on extracts of soil samples; (ii) on soil water samples collected in situ, usually with vacuum extractors; (iii) in soil, using buried porous salinity sensors which imbibe and equilibrate with the soil water; (iv) in soil, using four-electrode probes, or (v) remotely by electromagnetic induction techniques. These methods of monitoring soil salinity were recently critically reviewed, and appropriate methods were recommended for different situations (Rhoades, 1984a, b, 1979; Rhoades & Oster, 1986).

Especially useful is the measurement of soil electrical conductivity, EC_a, since EC_a is a measure of both soil water content and soil water salinity (Rhoades, 1980; Rhoades et al., 1981). Soil salinity in irrigated agriculture is normally low at shallow soil depths and increases through the root zone. Thus, measurements of EC_a in shallow depths of the soil profile made over an irrigation cycle are relatively more indicative of changing soil water content there, while measurements of EC_a deeper in the profile, where little water uptake occurs, are more indicative of salinity. Depletion of soil water to a set-point level, depth of water penetration from an irrigation or rainfall, and leaching fraction can all be determined from EC_a measurements made within the root zone over time (Rhoades et al., 1981). However, measurements of both volumetric soil water content and soil water salinity, from which the total water potential can be estimated (matric plus osmotic), are more ideally suited for these needs. The combined use of time domain reflectometric and four-electrode sensors offer good potential in this regard.

REFERENCES

Abrol, I.P., and D.R. Bhumbla. 1973. Leaching alone does not pay: Apply gypsum in alkali soils. Indian Farming 23:13–14.

Abu-Sharar, T.M., F.T. Bingham, and J.D. Rhoades. 1987a. Reduction in hydraulic conductivity in relation to clay dispersion and disaggregation. Soil Sci. Soc. Am. J. 51:342–346.

Abu-Sharar, T.M., F.T. Bingham, and J.D. Rhoades. 1987b. Stability of soil aggregates as affected by electrolyte concentration. Soil Sci. Soc. Am. J. 51:309–314.

Agassi, M., J. Morin, and I. Shainberg. 1982. Laboratory studies of infiltration and runoff control in semi-arid soils in Israel. Geoderma 28:345–356.

Alperovich, N., and I. Shainberg. 1973. Reclamation of alkali soils with $CaCl_2$ solutions. p. 441–452. In A. Hadas et al. (ed.) Physical aspects of soil water and salts in ecosystems. Vol. 4. Springer-Verlag, Berlin.

Archer, J.R. 1975. Soil consistency. p. 284–297. In Soil physical conditions and crop production. Great Britain Minist. Agric. Fish. Food Tech. Bull. 29.

Bernstein, L. 1974. Crop growth and salinity. In J. van Schilfgaarde (ed.) Drainage for agriculture. Agronomy 17:39–54.

Bernstein, L., M. Fireman, and R.C. Reeve. 1955. Control of salinity in the Imperial Valley, California. ARS-41-4, USDA-ARS, Washington, DC.

Bernstein, L., and L.E. Francois. 1973. Comparisons of drip, furrow, and sprinkler irrigation. Soil Sci. 115:73–86.

Bingham, F.T., R.J. Mahler, and G. Sposito. 1979. Effects of irrigation water composition on exchangeable sodium status of a field soil. Soil Sci. 127:248–252.

Bingham, F.T., A.W. Marsh, R. Bronson, R. Mahler, and G. Ferry. 1972. Reclamation of salt-affected high boron soils in western Kern County. Hilgardia 41:195–211.

Bingham, F.T., J.D. Rhoades, and R. Keren. 1985. An application of the Maas-Hoffman salinity response model for boron toxicity. Soil Sci. Soc. Am. J. 49:672–674.

Bishop, A.A., W.R. Walker, N.L. Allen, and G.J. Poole. 1981. Furrow advance rates under surge flow systems. J. Irrig. Drain. Div. Am. Soc. Civ. Eng. 107:(IR3)257–264.

Bouma, J. 1984. Using soil morphology to develop measurement methods and simulation techniques for water movement in heavy clay soils. *In* J. Bouma and P.A.C. Raats (ed.) Water and solute movement in heavy clay soils. Proc. ISSS symp. ILRI Publ. 37. Wageningen, Netherlands.

Bower, C.A., G. Ogata, and J.M. Tucker. 1969. Rootzone salt profiles and alfalfa growth as influenced by irrigation water salinity and leaching fraction. Agron. J. 61:783–785.

Bresler, E. 1987. Application of a conceptual model to irrigation water requirement and salt tolerance of crops. Soil Sci. Soc. Am. J. 51:788–793.

Bresler, E., and G.J. Hoffman. 1986. Irrigation management for soil salinity control: Theories and tests. Soil Sci. Soc. Am. J. 50:1552–1560.

Bresler, E., B.L. McNeal, and D.L. Carter. 1982. Saline and sodic soils, principles-dynamics-modeling. Springer-Verlag New York, New York.

Bridge, B.J., and C.R. Kleinig. 1968. The effect of gypsum on the water storage in a sandy loam soil under an irrigated perennial pasture. Int. Congr. Soil Sci. Trans. 9th (Adelaide) 1:312–313.

Carter, M.R., R.R. Cairns, and G.R. Webster. 1978. Surface application of gypsum and ammonium nitrate for amelioration of black solonetz soil. Can. J. Soil Sci. 58:279–282.

Chabra, R., and I.P. Abrol. 1977. Reclaiming effect of rice grown in sodic soils. Soil Sci. 124:49–55.

Chaudhry, G.H., and B.P. Warkentin. 1968. Studies on exchange of sodium from soils by leaching with calcium sulphate. Soil Sci. 105:190–197.

Chawla, K.L., and I.P. Abrol. 1982. Effect of gypsum fineness on the reclamation of sodic soils. Agric. Water Manage. 5:41–50.

Davidson, J.L., and J.P. Quirk. 1961. The influence of dissolved gypsum on pasture establishment on irrigated sodic clays. Aust. J. Agric. Res. 12:100–110.

Dedrick, A.R., J.A. Replogle, and L.J. Erie. 1978. On-farm level basin irrigation—save water and energy. Civ. Eng. 48(1):60–65.

Doering, E.J., and R.C. Reeve. 1965. Engineering aspects of the reclamation of sodic soils with high-salt waters. J. Irrig. Drain. Div. Am. Soc. Civ. Eng. 9:57–72.

Dutt, G.R., R.W. Terkeltoub, and R.S. Rauschkolb. 1972. Prediction of gypsum and leaching requirements for sodium-affected soils. Soil Sci. 114:93–103.

Emerson, W.W. 1983. Interparticle bonding. p. 477–498. *In* Soils: An Australian viewpoint. CSIRO Div. Soil. Melbourne/Academic Press, London.

Emerson, W.W. 1984. Soil structure in saline and sodic soils. p. 65–76. *In* I. Shainberg and J. Shalhevet (ed.) Soil salinity under irrigation. Ecol. Stud. Vol. 51. Springer-Verlag New York, New York.

Francois, L.E., and E.V. Maas. 1985. Plant responses to salinity: A supplement to an indexed bibliography. ARS-24. USDA-ARS, Washington, DC.

Frenkel, J., J.O. Goertzen, and J.D. Rhoades. 1978. Effect of clay type and content, ESP and electrolyte concentration on clay dispersion and soil hydraulic conductivity. Soil Sci. Soc. Am. J. 42:32–39.

Gardner, W.R., and R.H. Brooks. 1957. A descriptive theory of leaching. Soil Sci. 83:295–304.

Gobran, G.R., J.E. Dufey, and H. Laudelot. 1982. The use of gypsum for preventing soil sodification: Effect of gypsum particle size and location in the profile. J. Soil Sci. 33:309–316.

Hillel, D. 1980. Application of soil physics. Academic Press, New York.

Hira, G.S., M.S. Bajwa, and N.T. Singh. 1981. Prediction of water requirements for gypsum dissolution in sodic soils. Soil Sci. 131:353–358.

Hira, G.S., and N.T. Singh. 1980. Irrigation water requirement for dissolution of gypsum in sodic soil. Soil Sci. Soc. Am. J. 44:930–933.

Hoffman, G.J. 1980. Guidelines for reclamation of salt-affected soils. p. 49–64. *In* Proc. Inter-American Salinity Water Manage. Tech. Conf. Juarez, Mexico. 11–12 December.

Ingvalson, R.D., J.D. Rhoades, and A.L. Page. 1976. Correlation of alfalfa yield with various indices of salinity. Soil Sci. 122:145–153.

Jurinak, J.J. 1984. Thermodynamic aspects of the soil solution. p. 15–31. *In* I. Shainberg and J. Shalhevet (ed.) Soil salinity under irrigation. Ecol. Stud. Vol. 51. Springer-Verlag New York, New York.

Jury, W.A. 1984. Field scale water and solute transport through unsaturated soils. p. 115–125. *In* I. Shainberg and J. Shalhevet (ed.) Soil salinity under irrigation. Ecol. Stud. Vol. 51. Springer-Verlag New York, New York.

Jury, W.A., W.M. Jarrell, and D. Devitt. 1979. Reclamation of saline sodic soils by leaching. Soil Sci. Soc. Am. J. 43:1100–1106.

Kaddah, M.T., and J.D. Rhoades. 1976. Salt and water balance in Imperial Valley, California. Soil Sci. Soc. Am. J. 40:93–100.

Keisling, T.C., P.S.C. Rao, and R.E. Jessup. 1978. Pertinent criteria for describing the dissolution of gypsum beds in flowing water. Soil Sci. Soc. Am. J. 42:234–236.

Kemper, W.D., J. Olsen, and C.J. Demooy. 1975. Dissolution rate of gypsum in flowing water. Soil Sci. Soc. Am. Proc. 39:458–464.

Keren, R., F.T. Bingham, and J.D. Rhoades. 1985. Plant uptake of boron as affected by boron distribution between the liquid and the solid phases in soil. Soil Sci. Soc. Am. J. 49(4):297–302.

Keren, R., and G.A. O'Connor. 1982. Gypsum dissolution and sodic soil reclamation as affected by water flow velocity. Soil Sci. Soc. Am. J. 46:726–732.

Keren, R., and I. Shainberg. 1981. Effect of dissolution rate on the efficiency of industrial and mined gypsum in improving infiltration of a sodic soil. Soil Sci. Soc. Am. J. 45:103–107.

Khosla, B.K., K.S. Dargan, I.P. Abrol, and D.R. Bhumbla. 1973. Effect of depth of mixing gypsum on soil properties and yield of barley, rice and wheat grown on a saline-sodic soil. Indian J. Agric. Sci. 43:1024–1031.

Letey, J., A. Dinar, and K.C. Knapp. 1985. Crop-water production function model for saline irrigation waters. Soil Sci. Soc. Am. J. 49:1005–1009.

Longenecker, D.E. 1973. The influence of soil salinity upon fruiting and shedding, boll characteristics, fiber quality and yields of two cotton species. Soil Sci. 115:94–302.

Loveday, J. (ed.). 1974. Methods for analysis of irrigated soils. Commonw. Bur. Soils Tech. Commun. 54.

Loveday, J. 1976. Relative significance of electrolyte and cation exchange effects when gypsum is applied to a sodic clay soil. Aust. J. Soil Res. 14:361–371.

Loveday, J. 1984. Amendments for reclaiming sodic soils. p. 220–237. *In* I. Shainberg and J. Shalhevet (ed.) Soil salinity under irrigation. Ecol. Stud. Vol. 51. Springer-Verlag New York, New York.

Loveday, J., and J. Pyle. 1973. The Emerson dispersion test and its relation to hydraulic conductivity. CSIRO Aust. Div. Soils Tech. Pap. 15.

Maas, E.V. 1984a. Salt tolerance of plants. p. 57–75. *In* B.R. Christie (ed.) Handbook of plant science in agriculture. CRC Press, Boca Raton, FL.

Maas, E.V. 1984b. Crop tolerance. Calif. Agric. 38(10):20–21.

Maas, E.V. 1985. Crop tolerance to saline sprinkling waters. Plant Soil 89:273–284. [Also *In* D. Pasternak and A. San Pietro (ed.) Biosalinity in action: bioproduction with saline water. Martinus Nijhoff, Dordrecht, Holland 17:273–284].

Maas, E.V. 1986. Salt tolerance in plants. Appl. agricultural research. Appl. Agric. Res. 1:12–26.

Maas, E.V., and G.J. Hoffman. 1977. Crop salt tolerance—current assessment. J. Irrig. Drain. Div. Am. Soc. Civ. Eng. 103(IR2):115–134.

Maas, E.V., and R.H. Nieman. 1978. Physiology of plant tolerance to salinity. p. 277–299. *In* G.A. Jung (ed.) Crop tolerance to suboptimal land conditions. ASA Spec. Publ. 32. ASA, CSSA, and SSSA, Madison, WI.

Magdoff, F., and E. Bresler. 1973. Evaluation of methods for reclaiming sodic soils with $CaCl_2$. p. 441–452. *In* A. Hadas et al. (ed.) Physical aspects of soil water and salts in ecosystems. Vol. 4. Springer-Verlag, Berlin.

Manin, M., A. Pissarra, and J.W. van Hoorn. 1982. Drainage and desalinization of heavy clay soil in Portugal. Agric. Water Manage. 5:227–240.

McIntyre, D.S. 1958. Permeability measurements of soil crusts formed by raindrop impact. Soil Sci. 84:185–189.

McIntyre, D.S. 1979. Exchangeable sodium, subplasticity and hydraulic conductivity of some Australian soils. Aust. J. Soil Res. 17:115–120.

McIntyre, D.S., J. Loveday, and C.L. Watson. 1982. Field studies of water and salt movement in an irrigated clay soil. III. Salt movement during ponding. Aust. J. Soil Res. 20:101–105.

McNeal, B.L., and N.T. Coleman. 1966a. Effect of solution composition on soil hydraulic conductivity. Soil Sci. Soc. Am. Proc. 30:308–312.

McNeal, B.L., and N.T. Coleman. 1966b. Effect of solution composition on the swelling of extracted soil clays. Soil Sci. Soc. Am. Proc. 30:313–317.

Miller, R.J., D.R. Nielsen, and J.W. Biggar. 1965. Chloride displacement in Panoche clay loam in relation to water movement and distribution. J. Water Resour. Res. 1:63–73.

Miyamoto, S., G.R. Gobran, and K. Piela. 1985a. Salt effects on seedling growth and in uptake of three bean rootstock cultivars. Agron. J. 7:383–388.

Miyamoto, S., K. Piela, and J. Petticrew. 1985b. Salt effects on germination and seedling emergence of several vegetable crops and guayule. Irrig. Sci. 6:159–170.

Miyamoto, S., K. Piela, and J. Petticrew. 1986. Seedling mortality of several crops induced by root, stem or leaf exposure to salts. Irrig. Sci. 7:97–106.

Miyamoto, S., R.J. Prather, and J.L. Stroehlein. 1975a. Sulfuric acid and leaching requirements for reclaiming sodium-affected calcareous soils. Plant Soil 43:573–585.

Miyamoto, S., J. Ryan, and J.L. Stroehlein. 1975b. Potentially beneficial uses of sulfuric acid in south-western agriculture. J. Environ. Qual. 4:431–437.

Mozheiko, A.M. 1969. Chemical reclamation of sodic solonetzes in the southern part of the Middle Dneiper region by the application of gypsum and calcium chloride. Agrokem. Talajtan 18:(Suppl.) 310–314.

Muhammed, S., B.L. McNeal, C.A. Bower, and P.F. Pratt. 1969. Modification of the high-salt water method for reclaiming sodic soils. Soil Sci. 108:249–256.

Nielsen, D.R., J.W. Biggar, and J.N. Luthin. 1966. Desalination of soils under controlled unsaturated flow conditions. 6th Congr. Int. Comm. Irrig. Drain. (New Delhi). Question 19, p. 15–24.

Nightingale, H.I., K.R. Davis, and C.J. Phene. 1986. Trickle irrigation of cotton: Effect on soil chemical properties. Agric. Water Manage. 11:159–168.

Northcote, K.H., and J.K.M. Skene. 1972. Australian soils with saline and sodic properties. CSIRO Aust. Div. Soils Soil Publ. 27.

Oster, J.D. 1982. Gypsum usage in irrigated agriculture: A review. Fert. Res. 3:73–89.

Oster, J.D., and H. Frenkel. 1980. The chemistry of the reclamation of sodic soils with gypsum and lime. Soil Sci. Soc. Am. J. 44:41–45.

Oster, J.D., and A.D. Halvorson. 1978. Saline seep chemistry. p. 2–7 to 2–9. In H.S.A van der Pluym (ed.) Proc. Dryland-Saline-Seep Control. Edmonton, Canada. June.

Oster, J.D., and J.D. Rhoades. 1975. Calculated drainage water compositions and salt burdens resulting from irrigation with river waters in the western United States. J. Environ. Qual. 4:73–79.

Oster, J.D., and J.D. Rhoades. 1977. Various indices for evaluating the effective salinity and sodicity of irrigation waters. p. 1–14. In H.E. Dregne (ed.) Proc. Inst. Salinity Conf. Texas Tech. Univ., Lubbock. 16–20 Aug. 1976.

Oster, J.D., and F.W. Schroer. 1979. Infiltration as influenced by irrigation water quality. Soil Sci. Soc. Am. J. 43:444–447.

Overstreet, R., J.C. Martin, and H.M. King. 1951. Gypsum, sulfur and sulfuric acid for reclaiming an alkali soil of the Fresno series. Hilgardia 21:113–127.

Phene, C.J., J.L. Fouss, T.A. Howell, S.H. Patton, M.W. Fisher, J.O. Bratcher, and J.L. Rose. 1981. Scheduling and monitoring irrigation with the new soil matric potential sensor. Trans. ASAE 23-81:91–105.

Prather, R.J. 1977. Sulfuric acid as an amendment for reclaiming soils high in boron. Soil Sci. Soc. Am. J. 41:1098–1101.

Prather, R.J., J.O. Goertzen, J.D. Rhoades, and H. Frenkel. 1978. Efficient amendment use in sodic soil reclamation. Soil Sci. Soc. Am. J. 42:782–786.

Quirk, J.P. 1986. Soil permeability in relation to socidity and salinity. Phil. Trans. R. Soc. London A326:297–317.

Quirk, J.P., and R.K. Schofield. 1955. The effect of electrolyte concentration on soil permeability. J. Soil Sci. 6:163–178.

Rasmussen, W.W., and B.L. McNeal. 1973. Predicting optimum depth of profile modification by deep plowing for improving saline-sodic soils. Soil Sci. Soc. Am. Proc. 37:432–437.

Redly, M., K. Darab, and J. Csillag. 1980. Solubility and ameliorative effect of gypsum in alkali soils. p. 313–321. In D.R. Bhrumbla and J.S.T. Yadav (ed.) Proc. Int. Soil Sci. Soc. Salinity Conf. Karnal, India. 18–21 February.

Reeve, R.C., and C.A. Bower. 1960. Use of high-salt waters as a flocculant and source of divalent cations for reclaiming sodic soils. Soil Sci. 90:139–144.

Reeve, R.C., and E.J. Doering. 1966a. Field comparison of the high-salt-water dilution method and conventional methods for reclaiming sodic soils. p. 19.1–19.14. *In* 6th Congr. Int. Comm. Irrig. Drain. New Delhi. Question 19R1.

Reeve, R.C., and E.J. Doering. 1966b. The high-salt water dilution method for reclaiming sodic soils. Soil Sci. Soc. Am. Proc. 30:498–504.

Reeve, R.C., A.F. Pillsbury, and L.V. Wilcox. 1955. Reclamation of a saline and high boron soil in the Coachella Valley of California. Hilgardia 24:69–91.

Reeve, R.C., and G. Tamaddoni. 1965. Effect of electrolyte concentration on laboratory permeability and field intake rate of a sodic soil. Soil Sci. 99:261–266.

Rengasamy, P., R.S.B. Greene, G.W. Ford, and A.H. Mehanni. 1984. Identification of dispersive behaviours and the management of red-brown earths. Aust. J. Soil Res. 22:413–431.

Rhoades, J.D. 1972. Quality of water for irrigation. Soil Sci. 113:277–284.

Rhoades, J.D. 1974. Drainage for salinity control. *In* J. van Schilfgaarde (ed.) Drainage for agriculture. Agronomy 17:433–461.

Rhoades, J.D. 1976. Measuring, mapping and monitoring field salinity and water table depths with soil resistance measurements. FAO Soils Bull. 31:159–186.

Rhoades, J.D. 1977. Potential for using saline agricultural drainage waters for irrigation. p. 85–116. *In* Proc. Water Manage. Irrig. Drain. Reno, NV. July. Am. Soc. Civ. Eng., New York.

Rhoades, J.D. 1979. Inexpensive four-electrode probe for monitoring soil salinity. Soil Sci. Soc. Am. J. 43:817–818.

Rhoades, J.D. 1980. Determining leaching fraction from field measurements of soil electrical conductivity. Agric. Water Manage. 3:205–215.

Rhoades, J.D. 1982a. Soluble salts. *In* A.L. Page et al. (ed.) Methods of soil analysis. Part 2. Agronomy 9:167–178.

Rhoades, J.D. 1982b. Reclamation and management of salt-affected soils after drainage. p. 123–197. *In* Proc. 1st Annu. Western Provincial Conf. Rationalization Water Soil Res. Manage. Lethbridge, Alberta, Canada. November.

Rhoades, J.D. 1984a. Principles and methods of monitoring soil salinity. p. 130–142. *In* I. Shainberg and J. Shalhevet (ed.) Soil salinity and irrigation—processes and management. Vol. 5. Springer-Verlag New York, New York.

Rhoades, J.D. 1984b. New strategy for using saline waters for irrigation. p. 231–236. *In* J.A. Replogle and K.G. Renard (ed.) Water today and tomorrow. Proc. Am. Soc. Civ. Eng. Irrig. Drain. Spec. Conf., Flagstaff, AZ. 24–26 July.

Rhoades, J.D. 1984c. Reusing saline drainage waters for irrigation: A strategy to reduce salt loading or rivers. p. 455–464. *In* R.H. French (ed.) Salinity in watercourses and reservoirs. Buttersworth Publ., Boston.

Rhoades, J.D. 1984d. Using saline waters for irrigation. p. 22–52. *In* Proc. Int. Workshop on Salt Affected Soils of Latin America. Maracay, Venezuela. 23–30 Oct. 1983. (Also in Sci. Rev. Arid Zone Res. 2:233–264.)

Rhoades, J.D. 1984e. Salt problems from increased irrigation efficiency. J. Irrig. Drain. Div. Am. Soc. Civ. Eng. 111(3):218–229.

Rhoades, J.D. 1985a. Principles of salinity control on food production in North America. *In* W.R. Jordan (ed.) Water and water policy in world food supplies. Proc. Conf., College Station, TX. 26–30 May. Texas A&M Univ. Press, College Station.

Rhoades, J.D. 1985b. Management practices for the control of soil and water salinity. p. 449–478. *In* Int. Symp. Reclamation of Salt-affected Soils. Jinan, China. 8–26 May.

Rhoades, J.D. 1986a. Use of saline water for irrigation. Am. Soc. Agric. Eng. Summer. Meet.

Rhoades, J.D. 1986b. Use of saline water for irrigation. Water Qual. Bull. 12:14–20.

Rhoades, J.D. 1989. Intercepting, isolating and reusing drainage waters for irrigation to conserve water and protect water quality. Agric. Water Manage. 16:37–52.

Rhoades, J.D., F.T. Bingham, J. Letey, G.J. Hoffman, A.R. Dedrick, P.J. Printer, and J.A. Replogle. 1989a. Use of saline drainage water for irrigation: Imperial Valley study. Agric. Water Manage. 16:25–36.

Rhoades, J.D., D.L. Corwin, and G.J. Hoffman. 1981. Scheduling and controlling irrigations from measurements of soil electrical conductivity. p. 106–115. *In* Proc. ASAE Irrig. Scheduling Conf. Chicago. 14–15 December. ASAE, St. Joseph, MI.

Rhoades, J.D., R.D. Ingvalson, and J.T. Hatcher. 1970. Laboratory determination of leachable soil boron. Soil Sci. Soc. Am. Proc. 34:871–875.

Rhoades, J.D., D.B. Kreuger, and M.J. Reed. 1968. The effect of soil mineral weathering on the sodium hazard of irrigation waters. Soil Sci. Soc. Am. Proc. 32:643–674.

Rhoades, J.D., N.A. Manteghi, P.J. Shouse, and W.J. Alves. 1989b. Soil electrical conductivity and soil salinity: New formulations and calibrations. Soil Sci. Soc. Am. J. 53:433–439.

Rhoades, J.D., and S.D. Merrill. 1976. Assessing the suitability of water for irrigation: Theoretical and empirical approaches. FAO Soils Bull. 31:69–109.

Rhoades, J.D. and S. Miyamoto. 1990. Testing soils for salinity and sodicity. In R.L. Westerman (ed.) Soil testing and plant analysis. SSSA, Madison, WI. (In press.)

Rhoades, J.D., and J.D. Oster. 1986. Solute content. In A. Klute (ed.) Methods of soil analysis. Part I. 2nd ed. Agronomy 9:985–1006.

Rhoades, J.D., J.D. Oster, R.D. Ingvalson, J.M. Tucker, and M. Clark. 1974. Minimizing the salt burdens of irrigation drainage waters. J. Environ. Qual. 3:311–316.

Rhoades, J.D., and D.L. Suarez. 1977. Reducing water quality degradation through minimized leaching management. Agric. Water Manage. 1:127–142.

Roos, S.A.A., E.A. Awadalla, and M.A. Khalaf. 1976. Reclamation of a sodic soil by the high-salt water dilution method. Z. Pflanzenernaehr. Bodenkd. 6:731–737.

Russo, D., and E. Bresler. 1977. Effect of mixed Na/Ca solutions on the hydraulic properties of unsaturated soils. Soil Sci. Soc. Am. J. 41:713–717.

Shainberg, I. 1984. The effect of electrolyte concentration on the hydraulic properties of sodic soils. p. 49–64. In I. Shainberg and J. Shalhevet (ed.) Soil salinity under irrigation. Ecol. Stud. Vol. 51. Springer-Verlag New York, New York.

Shainberg, I., and M. Gal. 1982. The effect of lime on the response of soils to sodic conditions. J. Soil Sci. 33:489–498.

Shainberg, I., and J. Letey. 1984. Response of soils to sodic and saline conditions. Hilgardia 52(2):1–57.

Shainberg, I., and J.D. Oster. 1978. Quality of irrigation water. Volcani Cent. Int. Irrig. Inf. Cent. Publ. 2. Bet Dagan, Israel.

Shalhevet, J., and L. Bernstein. 1968. Effects of vertically heterogeneous soil salinity on plant growth and water. Soil Sci. 106:85–93.

Shannon, M.C. 1982. Genetics of salt tolerance: New challenges. p. 271–282. In A. San Pietro (ed.) Biosaline research: A look into the future. Plenum Publ. Corp., New York.

Shannon, M.C., and L.E. Francois. 1977. Influence of seed pretreatments on salt tolerance of cotton during germination. Agron. J. 69:619–622.

Sharma, A.K., J.B. Fehrenbacker, and B.A. Jones. 1974. Effect of gypsum, soil disturbance and tile spacing on the amelioration of Huey silt loam, a matric soil in Illinois. Soil Sci. Soc. Am. Proc. 38:628–632.

Smith, R.J., and N.H. Hancock. 1986. Leaching requirement of irrigated soils. Agric. Water Manage. 11:13–22.

Solomon, K.H. 1985. Water-salinity-production functions. Trans. ASAE 28(6):1975–1980.

Suarez, D.L. 1981. Relationship between pH_c and SAR and an alternative method of estimating SAR of soil or drainage water. Soil Sci. Soc. Am. J. 45:469–475.

Suarez, D.L. 1982. Graphical calculation of ion concentration in $CaCO_3$ and/or gypsum solutions. J. Environ. Qual. 11:302–308.

Szabolcs, I. 1985. Salt affected soils, as world problem. p. 30–47. In The reclamation of salt-affected soils. Proc. Int. Symp. Jinan, China. 13–21 May. Beijing Agric. Univ., Beijing, China.

Tanji, K.K., L.D. Doneen, G.V. Ferry, and R.S. Ayers. 1972. Computer simulation analysis on reclamation of salt-affected soil in San Joaquin Valley, California. Soil Sci. Soc. Am. proc. 36:127–133.

Toogood, J.A., and R.R. Cairns (ed.). 1978. Solonetzic soils technology and management. Univ. Alberta Bull. B-78-1. Edmonton, Canada.

U.S. Salinity Laboratory Staff. 1954. Diagnosis and improvement of saline and alkali soils. Handb. 60. U.S. Gov. Print. Office, Washington, DC.

van der Merwe, A.J. 1969. Reclamation of the black alluvial soils of Riet River using the threshold concentration concept. Agrochemophysica 1:67–72.

van der Molen, W.H. 1956. Desalination of saline soils as a column process. Soil Sci. 81:19–27.

van Hoorn, J.W. 1981. Salt movement, leaching efficiency, and leaching requirement. Agric. Water Manage. 4:409–428.

van Schilfgaarde, J. (ed.). 1974. Drainage for agriculture. Agronomy 17.

van Schilfgaarde, J. 1976. Water management and salinity. FAO Soils Bull. 31:53–67, 69–109.

van Schilfgaarde, J., L. Bernstein, J.D. Rhoades, and S.L. Rawlins. 1974. Irrigation management for salt control. J. Irrig. Drain. Div. Am. Soc. Civ. Eng. 100(IR3):321–338.

van Schilfgaarde, J., and S.L. Rawlins. 1983. Water resources management in a growing society. p. 517–530. *In* H.M. Taylor et al. (ed.) Limitations to efficient water use in crop production. ASA, CSSA, and SSSA, Madison, WI.

Wadleigh, C.H., and A.D. Ayers. 1945. Growth and biochemical composition of bean plants as conditioned by soil moisture tension and salt concentration. Plant Physiol. 20:106–132.

Wagenet, R.J. 1984. Salt and water movement in the soil profile. p. 100–114. *In* I. Shainberg and J. Shalhevet (ed.) Soil salinity under irrigation. Ecol. Stud.Vol. 51. Springer-Verlag New York, New York.

Worstell, R.V. 1979. Selecting a buried gravity irrigation system. Trans. ASAE 22(1):110–114.

Yousaf, M., O.M. Ali, and J.D. Rhoades. 1987a. Clay dispersion and hydraulic conductivity of some salt-affected arid land soils. Soil Sci. Soc. Am. J. 51(4):905–907.

Yousaf, M., O.M. Ali, and J.D. Rhoades. 1987b. Dispersion of clay from some salt-affected, arid land soil aggregates. Soil Sci. Soc. Am. J. 51(4):920–924.

37   Soil Erosion on Irrigated Lands

D. L. CARTER

USDA-ARS
Kimberly, Idaho

Soil erosion is a serious agricultural problem. Most of the available literature on the subject concerns nonirrigated cropland where natural precipitation and snowmelt water produce the forces needed to erode soil and transport sediment. Most of the water providing those forces for erosion on irrigated land is that applied by humans to supply water to growing crops. The purpose of this chapter is to provide a review of irrigation-induced soil erosion and to supply some insight into its hazards and control.

I. GENERAL OBSERVATIONS

A. Irrigation-Induced Erosion Began With Irrigation

Irrigation-induced erosion began when water was first applied to the soil surface where the land slope was sufficient that the moving water had enough shear force energy to detach soil particles from the soil mass and transport them as suspended sediment or bedload. Some of the first irrigating was by wild flooding (see chapter 16 in this book). Using this method to irrigate grass meadows seldom caused much erosion, but when farmers began to till the soil to produce crops of greater value, serious erosion resulted from wild flooding.

Most high-value crops are planted in rows, and to surface irrigate them, small ditches are made parallel to these crop rows; thus, furrow irrigation has become a common method. The concept that water could be controlled by these parallel furrows has extended the use of furrow irrigation to non-row or solid stand crops, such as alfalfa (*Medicago sativa* L.) and cereals. Today, furrow irrigation is widely practiced and in most areas where it is practiced, furrow erosion is a problem.

Sprinkler irrigation has been developed to irrigate areas that cannot be irrigated by other methods and to improve irrigation efficiency. Soil erosion can also be serious under sprinkler irrigation if the water application rate exceeds the soil infiltration capacity.

B. Recognizing Erosion as a Problem on Irrigated Land

Furrow erosion was recognized as a serious problem more than 50 yr ago in Utah. Isrealson et al. (1946) reported that furrows near the head ditches eroded 2.5 to 10 cm in one irrigation season. These researchers, along with Gardner et al. (1946) and Gardner and Lauritzen (1946), developed graphic relationships and equations relating erosion to stream size and furrow slope. They discouraged furrow irrigating slopes that were too steep and encouraged the use of smaller stream sizes to reduce erosion. Even earlier, Taylor (1940) published information on the mean furrow stream velocities at which different sizes of soil aggregates began to be transported.

Following the early work in Utah, investigations were conducted in other western states (Mech, 1949; Evans & Jensen, 1952). The USDA Soil Conservation Service, Division of Irrigation conducted tests throughout the western USA from 1948 to 1952 to determine the maximum nonerosive stream size as a function of slope.

Mech (1949) conducted several studies of the effect of stream size and slope on furrow erosion in Washington, and his results were similar to those reported earlier in Utah. All of these researchers indicated that erosion was not a serious problem on slopes < 1%, and that by carefully controlling stream size, furrow erosion can be controlled reasonably well on slopes up to 2%. Irrigation erosion research conducted before 1967 was been reviewed by Mech and Smith (1967), and their paper should be consulted for more detailed information before that date.

Renewed interest in irrigation-induced erosion arose from water quality legislation in the early 1970s, directed at cleaning up rivers and streams. Sediment was recognized as the most serious pollutant in most rivers and streams (Robinson, 1971; Wadleigh, 1968), and some of that sediment was traced to irrigation erosion. Brown et al. (1974) investigated sediment inflows and outflows for two large irrigation tracts and found a net sediment loss of 1.42 t/ha for an 82 000-ha tract and a net sediment inflow for a 65 350-ha tract. This study stimulated further investigations on individual fields and small watersheds. Results from these studies supported the need for developing and implementing erosion control practices.

C. Differences in Erosion on Irrigated and Nonirrigated Cropland

Erosion on nonirrigated cropland occurs when the rainfall or snowmelt rate, or both, exceed the soil infiltration capacity. Serious erosion occurs when the soil is frozen and has an infiltration rate near zero, and rain falls on snow, causing melting, or when unseasonably high temperatures occur suddenly and snow melts rapidly on frozen soil. Under these situations, water moves downslope, and the streams become larger from the addition of more water from the source. Usually rainfall and snowmelt involve the entire soil surface. An exception is where water from melting snowdrifts runs downslope on to non-frozen soil. The crop canopy also plays an important role in reducing erosion on nonirrigated cropland. As the canopy cover increases, it has a continuously greater reducing effect on erosion because it intercepts more and more of the raindrops, thereby reducing the raindrop impact effects on the soil surface.

Erosion on irrigated land can occur by the same processes as described for nonirrigated land, but because irrigated land generally receives less precipitation, this type of erosion is infrequent. On these lands, irrigation-induced erosion is the primary erosion problem.

Erosion on furrow-irrigated land is different from that on nonirrigated land because when water is placed in furrows, only the furrows erode. The water source is at one point, and the streams become smaller as a result of infiltration as they move downslope. Sprinkler irrigation is similar to rainfall except that only a portion of a field receives water at any given time. If water is applied by the sprinklers at a rate greater than the soil infiltration capacity, concentrated flow may begin downslope. If these concentrated flow streams run out of the area being irrigated, they begin to decrease in size as a result of infiltration.

II. FURROW EROSION

Furrow erosion is a dynamic process influenced by many factors and having multiple impacts. The following sections will discuss the furrow erosion process, factors influencing it, the impact of furrow erosion on crop yield, practices to control furrow erosion, and measures to lessen the effect of furrow erosion. The approach will be to report what is known today and project what can be expected in the future.

A. The Furrow Erosion Process

Furrow erosion begins when water enters the furrow, creating forces caused by soil wetting and water flow that exceed cohesive forces holding soil particles together and in place. The condition of the soil in the furrow when it is contacted by water largely determines if erosion will occur. When soils are dry, O_2 and N_2 are adsorbed on the internal surfaces of aggregates. When dry soils are wetted suddenly, water molecules rapidly displace the adsorbed O_2 and N_2 molecules. These gases join entrapped air in the gaseous phase, causing pressure forces sufficient to break soil aggregates apart along planes which constitute their internal surface area. The bursting of small clods resembles tiny explosions. Many times, I have observed dust in the air from this process along the flowing front of furrow streams. When water is applied in the early morning after a cool night, much less erosion results than when the water is applied during or following hot, dry afternoons (Kemper et al., 1985a). As the relative humidity rises above 50% during cool nights, more strongly adsorbed water molecules displace O_2 and N_2 molecules gradually.

Other forces involved are shear forces caused by the flowing water and transported materials and the opposing cohesive factors that bind soil aggregates and their combinations together. When shear forces exceed cohesive forces, erosion occurs. In contrast, soils are stable when cohesive forces exceed shear forces. These two counteracting forces are discussed below.

1. Cohesion factors

Bonds between primary soil particles hold soil aggregates together. The strength of these bonds represents the soil cohesion or stability which varies with clay content and type, organic matter content, compaction, adsorbed ions, time and water content since the last disruption, the wetting rate, and the chemical composition of the water wetting the soil. Soil bond formation and disruption processes are not yet well understood, but recent research has provided considerable enlightenment about factors controlling bond strength or cohesive forces. Kemper and Rosenau (1984) demonstrated that soil cohesion increases with time since the last disruption. Soil disintegrated less in water if a few days were allowed between cultivation and irrigation than when irrigation immediately followed cultivation. Dry soils, high in clay concentration, had greater cohesion than silty soils. Gums and resins from decomposing plant residues also tend to form organic bonds that increase soil stability.

Aggregate stability increases in the spring and summer months in areas where soils freeze (Bullock et al., 1988). In contrast, when minimum daily temperatures fall below 0 °C during winter and early spring months, soil cohesion—as measured by aggregate stability—decreases because of compacting and shearing forces caused by freezing and thawing at high water content. As a result, furrow erosion is higher in the spring when the first and second irrigations are applied and then decreases as the season progresses into summer and fall (Berg & Carter, 1980; Brown et al., 1974). Spring tillage also breaks soil bonds and contributes to greater early-season erosion, particularly when excessive tillage is done.

Conclusions about the erosivity of specific soils at specific sites from single time measurements can be misleading. For example, a measurement made in the spring may give much higher erosion and soil loss values than a similar measurement at the same site in the fall. Similarly, soil may be more erosive immediately following tillage than a week or so later.

2. Shear Forces

The furrow stream size and the slope are factors affecting water velocity that causes the shear forces on the furrow perimeter. Under slow velocities there may be practically no particle detachment. Erosion begins when flow velocities increase, causing shear forces to increase and eventually exceed the critical shear stress (Foster & Lane, 1983; Kemper et al., 1985b; Trout & Neibling, 1989) required to overcome cohesion of soil aggregates and particles. Because there is a wide range of soil cohesion, there is also a wide range of critical shear stresses to be overcome before erosion begins. Because of these variations, analytical predictions of this effect of shear forces on erosion have not been accomplished (Kemper et al., 1985b; Trout & Neibling, 1989).

B. Factors Affecting Furrow Erosion

There are several factors that affect furrow erosion. Some of these have been recognized for more than 50 yr while others have been discovered recently. These factors are difficult to quantify because most of them are complexly interrelated, and the literature contains mostly qualitative statements about them. Recent attempts to quantify some of these factors and their relationships to others have greatly expanded our understanding of irrigation furrow erosion. Following are discussions of some factors known to have significant impacts on furrow erosion.

1. Slope Along the Furrow

The slope along the irrigation furrow was recognized as one of the most important factors in furrow irrigation erosion in the 1930s, and the first attempts to develop relationships between slope and erosion were initiated (Gardner & Lauritzen, 1946; Isrealson et al., 1946). These relationships usually also included the furrow stream size.

A misconception about the relationship between slope and furrow erosion resulted from the fact that measured papameters were generally field slopes taken as an average from the head to the lower end of the furrow, and the sediment loss from the lower end of the furrow was called erosion. This simplification has resulted in the misconception for several reasons. First, few fields have uniform slope along the full furrow length, and erosion along a furrow is dynamic. Where the slope is steeper than the average, erosion will be greater than average, and where the slope is less than average, erosion will be less than average. Another factor is that the furrow stream size decreases from the point of entry to the furrow outlet. The furrow stream size, combined with the slope, determines if the stream will generate sufficient velocity to produce the shear force required to cause erosion along any furrow segment and maintain enough energy to transport the sediment load it has accumulated.

A second reason for this misconception is that convex end patterns have developed on most furrow-irrigated fields (Carter & Berg, 1983). A convex end is an increasing slope beginning approximately 5 to 15 m upslope from the lower end of the furrow. This pattern developed because farmers generally maintained drainage ditches deeper than the furrow ends along the lower ends of fields to allow for free flow of drainage water. As a result, eroding head-cuts begin at the drainage ditch and move along the furrow in response to the increased energy of water because of the increased slope. This is a self-perpetuating process that becomes more severe with time. On many fields, the majority of the sediment loss from the field is from this portion.

A third reason involves factors not fully understood. Brown (1985b) found that erosion may occur along a specific furrow segment during one irrigation and deposition may occur along that same segment during a subsequent irrigation, with the same stream size. These single irrigation effects

Table 37-1. Estimated sediment losses from fields of different crops irrigated from concrete-lined ditches with siphon tubes. Run length was 200 m.

Crop	Average field slope, %							
	0.5–1		1–2		2–3		>3	
	N†	S	N	S	N	S	N	S
	t/ha							
Alfalfa	0.0	0.0	1.6	2.7	5.2	9.2	12.6	22.0
Cereal grain or pea	2.5	4.0	7.2	12.6	14.3	25.1	23.3	40.8
Dry bean or corn	5.6	9.9	19.5	34.3	41.2	72.2	62.8	109.8
Sugarbeet	7.2	12.6	27.1	47.5	59.2	103.6	98.6	172.6

† N = No convex end; S = severe convex end.

are often hidden when only seasonal totals are reported, but they do indicate that there are factors involved that we do not yet understand.

Kemper et al. (1985b). concluded that erosion is approximately a two- to three-power function of furrow slope. They reached this conclusion from studies where sediment losses at the end of the furrows were measured for different slopes. More appropriately, the conclusion should be that sediment loss is about a two- to three-power function of the average furrow slope, and a better definition of the length of the slope concerned should be given. Carter and Berg (1983) reported considerably higher sediment losses for fields with convex ends than from fields without convex ends with the same average slope (Table 37-1).

2. Stream Size

The furrow stream size or flow rate is an important erosion factor. As the stream size diminishes along the furrow length, its energy to erode and capacity to transport sediment decreases. Hence, erosion is greatest on the upper one-third of the furrow and sedimentation generally results on the lower half (Carter et al., 1985; Carter, 1989; Farnstrom et al., 1985).

The furrow stream must be large enough to supply adequate water to irrigate the entire furrow length. Furthermore, the infiltration time should be, as nearly as possible, the same over the entire furrow length to provide uniform amounts of water to the crop. A stream large enough to flow to the lower end of the furrow in a few hours is almost always erosive over the upper one-third of the furrow. Once water has reached the lower end of the furrow, the stream can be cut back to reduce future erosion. Some automated stream size cutback systems have been developed (Humpherys, 1971).

Kemper et al. (1985b) concluded that erosion is commonly about a 1.5-power function of stream size. Again, this relationship should more appropriately be stated as one between furrow stream size and sediment loss.

The variability in stream sizes among furrows averages about 25% and the infiltration rate in a wheel track furrow may be only half that in a non-travelled furrow (Trout & Mackey, 1988; Trout & Kemper, 1983). These two factors add to the difficulty of controlling erosion on any particular field.

The normal response of the irrigator is to apply large enough streams to assure that the water reaches the lower end in all furrows within 2 to 4 h. This practice usually assures erosive stream sizes along the upslope reaches on most fields. As a result, erosion, sediment loss, and runoff are greater than necessary.

Changes in infiltration rate during an irrigation can cause problems. If the furrow stream is reduced after water has reached the lower end of the furrow to reduce runoff, and then the infiltration rate increases (Trout & Mackey, 1988), the stream may no longer reach the lower end of the furrow. Thus, farmers use streams large enough to protect against such an event they believe may occur because of past experience.

3. Residue

Small quantities of straw or other crop residue in irrigation furrows reduce soil erosion and increase infiltration. Aarstad and Miller (1981) showed that as little as 60 kg straw/ha placed in clumps along the furrow greatly reduced sediment loss from irrigation furrows along a 3% slope. Berg (1984) applied small amounts of straw uniformly along 4% slope sections of furrows in a corn (*Zea mays* L.) field to reduce erosion on that portion of the field and to reduce sedimentation downstream where the slope decreased to about 1.5%. This practice not only decreased erosion and sediment loss but also increased corn silage yields. The more uniform infiltration along the furrows improved the irrigation effectiveness in supplying water needed by the corn crop. Brown (1985a) placed 1.5 kg straw/100 m of furrow and measured both infiltration and sediment loss for six irrigations with two stream sizes. The straw increased 10-h infiltration by 50% and decreased sediment loss by 52% at a furrow inflow rate of 13.2 L min^{-1} as compared to furrows without straw and with an inflow rate of 10.3 L min^{-1}. Differences were even greater at higher inflow rates. Brown and Kemper (1987) later demonstrated that the increased infiltration resulting from placing straw in furrows significantly increased dry bean (*Phaseolus vulgaris* L.) yields. They concluded that cereal straw in furrows conserves soil, water, and plant nutrients, reduces labor, and increases crop yields.

4. Surface Roughness

When furrow surfaces are rough, water flow is retarded, and generally the water depth in the furrow is increased. The slow flow velocity decreases erosion. The deeper furrow stream increases the infiltrating area, thereby increasing infiltration. This reduces the furrow stream size and velocity, further decreasing erosion. Furrow roughness can be caused by purposely tilling soils and forming furrows at selected water contents wet enough to form clods (Kemper et al., 1982).

The effective roughness or the roughness coefficient is increased by crop residues and by crop plants or weeds growing in furrows. Plant foliage hanging into the furrow stream, as often occurs with sugarbeet (*Beta vulgaris* L.), dry bean, potato (*Solanum tuberosum* L.), and alfalfa late in their growing seasons, increases the roughness coefficient and reduces erosion. Plant roots can also increase furrow roughness, as well as physically hold the soil against shear forces.

Where soils have high infiltration rates and farmers have difficulty forcing the water to the lower ends of the furrows, implements are pulled along the furrows to reduce surface roughness. Such a practice may increase furrow erosion because furrow stream flow velocities are higher in smooth than in rough furrows when the same stream size is applied. The higher kinetic energy associated with higher flow velocities results in greater shear forces.

5. Tillage

The kind and amount of tillage determine the fineness of the soil aggregates in the furrow, or the roughness of the furrow surface as previously discussed. Extensive tillage physically breaks bonds holding soil particles together, decreases aggregate size, and increases soil erodibility. One of the first cautions issued to irrigators to reduce erosion was to avoid pulverizing the soil by excessive tillage operations (Taylor, 1940). We might consider that warning as the first recommendation for use of reduced tillage on furrow-irrigated land, and it is still applicable today. Tillage also affects furrow roughness through its effect on crop residues. Where moldboard plowing buries all crop residues, erosion is greater than where tillage practices leave residues in the irrigation furrows to increase the resistanct to water flow and reduce shear forces.

The kinds of tillage implements used today are diverse compared to those used a few decades ago. Chisel plows, roller harrows, sweeps, rototillers, slot planters, various disks, no-till planters, provide a wide range of soil and residue conditions. Some tillage operations reduce crop residues to small pieces, while others leave rather large pieces. Soils may be left either highly pulverized or cloddy, depending upon the implements used and the soil water content at the time of tillage. Other operations disrupt soils to depths of 30 cm or more, whereas others may reach to only 10 cm. Furrow erosion depends upon the extent to which soil particle-to-particle bonds have been broken and the condition of the remaining crop residue.

6. Cropping Sequence

The erosion and sediment loss from any given field of any particular crop depends upon previous cropping, particularly the most recent crop. Where significant quantities of crop residue remain in the furrows, erosion will be less than where none remains. During studies to determine sediment loss rates for various crops at different slopes and stream sizes, sediment loss was always greater where dry beans followed dry beans than where dry beans followed cereals (Berg & Carter, 1980). The impact of the previous crop on erosion is always greater for row crops than for cereals. Cereal plants slow the furrow stream flow velocity and reduce erosion sufficiently such that previous crop effects are masked and of little significance.

The kind of tillage used influences the previous crop effect on erosion and sediment loss. Generally, the less tillage used, the greater the effect. Hence, the greatest influence will be manifested with no-tillage cropping. Reduced tillage and no-tillage practices have only recently been introduced to furrow-irrigated land (Aarstad & Miller, 1979; Allen et al., 1976; Musick et al., 1977; Carter et al., 1989b).

C. The Impact of Erosion on Crop Yield

Reduced crop yields generally accompany decreases in topsoil depth resulting from soil erosion. There have been recent reports of 40% productivity loss from erosion of some USSR soils, 30% less production on eroded than noneroded Haiti soils, and a 50% yield decline from erosion of 5 cm of surface soil from some Nigeria soils (Wolman, 1985).

Most reports of the detrimental impact of soil erosion on crop production in the USA have been made in the 1980s, and they represent all regions of the country. White et al. (1985) reported that severely eroded southern Piedmont soils produced crop yields only 50% as high as on adjacent, noneroded areas of the same soils. McDaniel and Hajek (1985) found that crop yields were reduced on moderately eroded sites in 65% of the fields studied in Alabama, with an average yield decrease of 22%. Erosion reduced corn yields 12% on Maury soil (fine, mixed, mesic Typic Paleudalf) and 21% on Crider soil (fine-silty, mixed, mesic Typic Paleudalf) in Kentucky, and the yield decreases for winter crops on the eroded Maury soil ranged from 17 to 36% (Frye et al., 1985). Krauss and Allmaras (1982) reported that a loss of 13 cm of topsoil over 90-yr at a site in Whitman County, Washington, decreased wheat (*Triticum aestivum* L.) yields 50%. Papendick et al. (1985) reviewed research results for the northwestern USA and reported both linear and curvilinear relationships between wheat yield and topsoil thickness. White et al. (1985), Bruce et al. (1987), and Daniels et al. (1987) reminded us that the effect of erosion upon the crop productivity of the landscape is difficult to accurately estimate, particularly with depth of topsoil criteria. In some situations, soil water is the controlling factor, and soybean [*Glycine max* (L.) Merr.] and gain sorghum [*Sorghum bicolor* (L.) Moench] yields are about equal on severely and slightly eroded sites. In other cases, correction of nutrient deficiencies can restore yield potential. Therefore, caution should be exercised to avoid drawing incorrect conclusions in these kinds of studies.

The foregoing reports were for nonirrigated lands. The impact of irrigation erosion on crop yield has been reported by Carter (1985b, 1988) and Carter et al. (1985). After 80 yr of furrow irrigation, 75% of the fields in the study area exhibited exposure of subsoil evidenced by a whitish color from the head ditch or pipe downslope for approximately one-third of the field length (Fig. 37–1). The actual distance varied, but commonly ranged from 50 to 100 m. The whitish color came from the topsoil being eroded away and moldboard plowing mixing the white subsoil with topsoil. Crop yields were severely reduced on these areas where topsoil and subsoil were

Fig. 37–1. Erosion along the (A) upper ends of furrows where reduced topsoil thickness allows plowing to mix topsoil and subsoil. Where subsoils are almost white, the effects of these processes becomes visible. (B) Ground level. (C) Aerial of one field. (D) Aerial of several fields.

mixed as well as on areas where the topsoil depth was reduced—but not sufficiently for plowing to mix subsoil with topsoil. Crop yield potential has been
reduced 25% by 80 seasons of irrigation furrow erosion on approximately
1 million ha of furrow-irrigated land (Carter et al., 1985).

1. Topsoil Redistribution

Erosion on the upper portion and sedimentation on the lower portion
of fields redistributes topsoil. The results of these processes become visible
when the color of the subsoil differs from that of the topsoil (Fig. 37-1).
The visual evidence of topsoil redistribution would be lacking where subsoil
and topsoil are nearly the same color. Furrow erosion can cause a major
topsoil redistribution on any field and have a simultaneous, severe, negative
impact on crop production.

Typical fields that have been irrigated for about 80 yr are illustrated
in Fig. 37-1, showing the color change as whitish subsoil is mixed with darker
topsoil. The topsoil distribution varies depending upon the field length and
irrigation practice used over the 80 yr. The deepest topsoil areas, resulting
from deposition, vary from field to field from about the midpoint to the
extreme lower end. Also, there has been a net topsoil loss from most fields,
thereby negatively impacting crop yield.

Topsoil depth originally averaged 38 cm in the study area when irrigation began in 1905. The gray topsoil, underlain by a nearly white, caliche
and silica-cemented hardpan, varies in thickness from 0 to 30 cm. The hardpan may contain as much as 30% $CaCO_3$. Root growth is limited by this
hardpan layer over much of the area. Below the hardpan layer is nearly white
subsoil with little structure. Before irrigation was introduced, soil below the
hardpan was seldom wetted with water from precipitation and was powdery.
The natural fertility of the hardpan and the subsoil beneath is low, but can
be corrected. Phosphorus requirements to raise available P to adequate levels
are high. Zinc is also needed for dry bean production, and N has to be added
according to the crops grown. Other nutrients are adequate according to soil
test values, and deficiencies have not been noted in growing crops.

The first fields exhibiting the exposure of subsoil were generally those
of slopes exceeding 3%, which were among the steepest irrigated. Gradually, fields having lower slopes began to exhibit the color change until today
almost all fields with slopes along the furrows >1% exhibit the phenomenon, as well as some fields with slopes <1%. Studies (Carter et al., 1985)
have shown that some fields have lost as much as 90 to 100 cm depth of
soil near the head ditch and have deposition as much as 180 cm deep.
Commonly, 30 to 40 cm have been lost from the upper ends and 20 to 40
cm have been deposited at some point downslope.

2. Effects on Yields of Major Crops

Investigations indicated that a topsoil depth of about 38 to 40 cm is the
minimum depth for maximum yields of all crops in one large study area (Carter et al., 1985). Where topsoils are deeper because of deposition, no signifi-

cant yield increases are found for any crop. In contrast, significant crop production decreases are evident for all crops where topsoil depth is <38 cm. The relationship between topsoil depth and crop yield is approximated by the Mitscherlich-Spillman type equation, $y = a + b (1 - e^{cx})$, where y is yield, x is topsoil depth, and a, b and c, are contants (Carter, 1988). Dry bean and corn are the crops most detrimentally affected by reduced topsoil depth, followed by wheat. Barley (*Hordeum vulgare* L.) and alfalfa are less severely affected and sugarbeet is least affected by decreased topsoil depth (Fig. 37–2).

The factors responsible for the yield reduction are not known. In efforts to restore productivity, adequate plant nutrients were applied on the many fields studied. None of the crops exhibited nutrient deficiencies. Soil water was monitored in some of these studies and adequately supplied by irrigation toa void moisture stress. Several organic matter amendments were tried without significant response. The only effective treatment was to replace 30 to 40 cm of topsoil. This restored yields to levels where topsoil had not been removed by erosion.

The soil erosion topsoil redistribution process is progressing in all areas where irrigation furrow erosion is occurring. Impacts on crop yields will not be as pronounced where the crop yield potential of subsoil is nearly that of topsoil. However, the overwhelming evidence indicates that topsoil losses will ultimately lead to serious crop yield losses, because subsoils are generally less productive than topsoils. There are many areas in the western USA

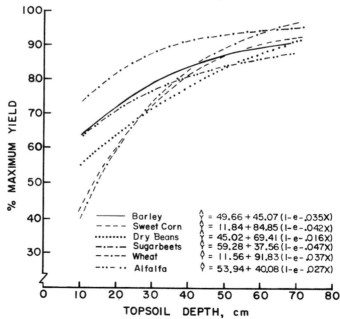

Fig. 37–2. Effect of topsoil depth on relative crop yield for six crops and associated Mitscherlich—Spellman type equations.

where soils have been furrow irrigated for < 80 yr. We hope that our findings will stimulate both interest and action towards applying conservation practices to prevent the potential crop yield loss already experienced.

D. Controlling Furrow Erosion and Soil Loss

The concern for improved water quality during the late 1960s, and since that time, stimulated legislation aimed at reducing water pollution and improving water quality. Sediment was recognized as the most important nonpoint source water pollutant from the standpoint of quantity (Robinson, 1971; Wadleigh, 1968). Irrigation return flows were identified as serious nonpoint pollution sources and attempts were made during the 1970s to require permits based upon quality standards before irrigation return flows could be discharged to navigable streams.

Brown et al. (1974) reported sediment balances from two large furrow irrigated tracts. Subsequently, Carter (1976) reviewed the available literature and published some guidelines for controlling sediments in irrigation return flows. Continued interest has resulted in many studies aimed at developing practices to control irrigation erosion and sediment loss. The earlier efforts were on controlling the quality of drainage water after leaving a field. More recently, efforts have been aimed at controlling furrow erosion on the field.

1. Sediment Retention Basins

Basins or ponds constructed in drainage ways to temporarily pond irrigation runoff water can effectively trap sediment and prevent sediment loss into streams and rivers. These basins range in size from about 1.0 ha on main drains to minibasins receiving runoff from only a few furrows. They vary in size and shape, and have been given different names. All are effective, and each type has its best application. Large basins on main drains are usually formed by constructing an earthen dam across a drainage at a suitable site and installing a proper outlet. These large basins trap or remove 65 to 95% of the incoming sediment (Brown et al., 1981; Carter, 1985a). This sediment removal efficiency depends upon the sediment concentration, the particle size of the sediment, and the time required for water to pass through the basin (Brown et al., 1981; Carter et al., 1989a).

Medium-sized sediment retention basins are often excavations in drain ditches receiving runoff water from one or more fields. Their sediment removal efficiencies range from 75 to 95%. Often they are placed at the lowest corner of a field. Unfortunately many of them are undersized and fill with sediment as a result of one or two irrigations. As a basin fills with sediment the water retention time decreases, resulting in a decrease in sediment removal efficiency. The capacity of these basins to remove sediment can change rapidly during a single irrigation as they fill with sediment.

Minibasins are formed by excavating a sequence of small basins along the lower end of a field or by placing earthen checks across the tailwater drainage ditch. If each basin has an outlet into a separate drainage ditch, sediment removal efficiencies range from 85 to 95%. If the water is allowed to pass from one basin to the next, each successive basin becomes less efficient, and the overall sediment removal efficiency of a series of basins is only 40 to 70%. Often the accumulated flow volume washes out earthen checks and basins (Brown et al., 1981; Carter & Berg, 1983).

A common disadvantage of all sediment retention basins is that costly cleaning is required for them to remain effective. In some instances basins are constructed in low areas and fill with sediment. Fields can then be combined or expanded by rerouting the water after a basin is filled and farming the sediment accumulated as part of the field.

Where farmers own equipment, the sediment may be economically hauled back to the upper ends of fields, or onto a rocky, nonfarm area to expand cropping area.

Sediment retention basins have been an effective educational tool for encouraging farmers to implement erosion and sediment control practices. Few farmers are aware of the quantity of sediment they are losing from their fields until they construct a sediment retention basin and watch it fill with sediment. As they learn how much soil they are losing, they become more interested in implementing practices to reduce soil loss.

2. Buried Pipe Erosion and Sediment Loss Control System

A system comprised of a buried pipe with vertical inlets at intervals to correct the convex field end erosion problem has been developed (Carter & Berg, 1983). The buried pipe replaces the tailwater drainage ditch, and the vertical inlets serve as indivdual outlets for minibasins formed by placing earthen dams across the convex portion of the field, as illustrated in Fig. 37–3.

The minibasins of this system initially function the same as other minibasins with individual outlets, but with a sediment removal efficiency of 80 to 95%. As these minibasins fill with sediment, their efficiency decreases. At the same time, the convex end of the field is being corrected by filling with sediment. This decreases the erosion rate on the convex end. The sediment deposition depth is controlled by the elevation of the top of the vertical inlet into the buried pipe. The convex end was entirely eliminated by the end of the first irrigation season in all but two of 40 systems studied.

After the minibasins are filled with sediment and the convex end corrected, the sediment removal efficiency of this system decreases to about 70%. However, the sediment involved is from further upslope instead of that generated from the convex end. The sediment load in the water is usually much lower than before. The end of the field becomes flat, and that flat area gradually extends further upslope. Drainage water is carried away through the inlets and the buried pipe, preventing water ponding in these flat areas. Several systems have been in operation for 10 yr and continue to function effectively.

The buried pipe erosion and sediment loss control system has been shown to be a cost-effective practice. Initially, installation costs are higher than for some other sediment loss control practices. But, in contrast to some other

Fig. 37-3. Convex field end showing (A) waste water ditch, (B) operating buried pipe erosion and sediment loss control system during the first irrigation, and (C) convex end corrected after four irrigations.

systems, this alternative increases the productive area of fields by eliminating the tailwater ditch. This also facilitates ingress and egress of farming equipment. With the open drainage ditch, equipment could not enter or leave the field except at constructed crossings,and also had to turn around inside the field perimeter (Fig. 37–3). Eliminating the ditch also improves convenience for cultivating part of the field while another part is being irrigated, and reduces weed problems associated with wet drainage ditches (Carter & Berg, 1983).

Correcting convex ends improves crop production near the lower ends of fields. Many fields with convex ends often erode to the plow depth, resulting in furrow streams 20 to 30 cm below the soil surface, where lateral water movement doesn't reach the roots of small plants. These plants die from drought, and commonly the lower few meters of these areas produce little or no crops. Correcting the convex end eliminates this problem.

Increased crop yields resulting from increasing the harvested area and reducing drought losses increased income sufficiently to pay the costs of installing a buried pipe system in 4 to 8 yr. After that, the increased returns will add to farm profits (Carter & Berg, 1983).

3. Vegetative Filters

A simple, inexpensive erosion and sediment loss control practice is planting a strip of cereal, grass, or alfalfa along the lower end of a field in row crops. These densely planted crops at the lower ends, and in a few cases the upper ends of fields, are called *vegetative filters*. When properly placed and managed, these vegetative filters remove 40 to 60% of the sediment from furrow runoff water when at the lower end of a field and can reduce erosion when at the upper end, but no quantitative data are available for the later

Fig. 37–4. Vegetative filter strip of wheat along the downslope end of a dry edible bean field.

situation. Proper placement and management include planting the vegetative filter close to the drainage ditch and forming the irrigation furrows about one-half the way through the filter strip. Leaving the last 1 to 2 m between the furrow end and the drainage ditch allows the water to spread out through the densely planted crop before entering the drainage ditch. If furrows are made all the way through the vegetative filter, effectiveness is lost. If the furrows are not pulled far enough into the densely planted filter strip, sediment soon accumulates in the upslope side of the strip and water accumulates just ahead of the filter strip. This generally results in eroding a new channel immediately upslope from the strip, parallel to the drainage ditch.

The advantages of vegetative filter strips are simplicity, low cost, and the filter crop can be harvested. An example of a wheat filter strip at the lower end of a dry bean field is illustrated in Fig. 37-4.

4. Placing Straw in Furrows

The effectiveness of straw placed in irrigation furrows for reducing erosion and increasing infiltration was discussed earlier in this chapter. The most effective application of this practice is to apply straw to the steeply sloping segments of the furrows. Berg (1984) and Brown and Kemper (1987) reported significant crop yield increases, infiltration increases, and sediment loss decreases using this approach. Based upon this research, a commercial machine is now available to spread straw in furrows.

The application of straw to furrows should be viewed as an alternative when residues from the previous crop are not available on the field. Where previous crop residue is present, it is better to alter tillage operations to keep some of that residue in the soil surface than to expend the cost and time to bring straw from a source outside the field and spread it in the furrows.

5. Irrigation Management

The relationships among the furrow stream size, furrow slope, and sediment loss were discussed earlier. These relationships illustrate that the larger the furrow stream size, the greater the amount of erosion. One irrigation management tool is to apply the smallest possible stream to accomplish the irrigation. The required stream size is determined by the infiltration rate, slope along the furrow, and the run length. In some instances, reducing the run length by adding a midfield gated pipe may be the best alternative. Other situations may be better controlled by compacting furrows to reduce the infiltration rate. This compacting can be accomplished by traversing the furrow with the tractor wheel or with furrow packers on a tool bar (Kemper et al., 1982; Trout & Mackey, 1988). Another approach is to use surge flow (Kemper et al., 1988), surge flow with crop residues (Evans et al., 1987), or to use a manual or automated stream-size cut-back system (Humpherys, 1971). Automated cut-back systems generally apply furrow streams to one set of furrow until the water reaches the lower furrow ends, then to another adjacent equal number of furrows for the same amount of time. After that the water is applied to all these wetted furrows simulatneously, resulting in a stream size one-half the original, until sufficient water is infiltrated for crop needs.

Cablegation systems (Kemper et al., 1987) provide for a gradual stream-size reduction. Sediment loss is significantly reduced by cablegation as compared with irrigating with the same stream size for a given set time.

Carefully controlling the stream size in each furrow and selecting either wheel track or non-wheel track furrows are important parameters. The minimum required furrow stream sizes to irrigate adjacent wheel track and non-wheel track furrows are different because of differing infiltration rates (Kemper et al., 1982; Trout & Mackey, 1988). Also furrow-to-furrow variability is 25% greater using gated pipe than using siphon tubes from a cement-lined ditch. Knowing the furrow condition relative to the wheel tracks and knowing the best system for controlling stream size help to make decisions about the stream size to use.

Unfortunately, many farmers operate on a highly regimented time schedule and are limited by the particular irrigation system they have on each field. The general tendency is to apply streams that are erosive to assure that the water transverses the entire furrow length in 2 to 4 h so that adequate infiltrating time to add the appropriate amount of water to the soil reservoir will be certain. When this approach is used, 40 to 50% of the applied water runs off the field as surface drainage, and furrow erosion is often severe (Berg & Carter, 1980).

Changing the direction of irrigating a field to one of less slope can reduce erosion and sediment loss. Also, where slopes exceed 3%, consideration should be given to converting to sprinkler irrigation.

6. Conservation Tillage

Conservation tillage, including no-tillage and reduced or minimum tillage systems, has been applied successfully to rainfed cropland. Until recently, there has been little interest in trying these systems on furrow-irrigated lands. Farmers have long practiced clean tillage to provide clean furrows for irrigation, and it has been unthinkable to consider a tillage system that leaves residue on the soil surface.

Conservation tillage was first introduced to furrow-irrigated land in Washington (Aarstad & Miller, 1979; Miller & Aarstad, 1983). These authors found that conservation tillage significantly reduced sediment losses from furrow-irrigated land. Carter et al. (1989b) introduced no-tillage practices to the Northwest where irrigation furrows are so small that some farmers call them "marks in the soil." These furrows, commonly called *corrugates,* are generally 5 to 8 cm deep and 6 to 8 cm wide at the top. The initial study area in southern Idaho produces garden and commercial bean seed, sugarbeet, and corn as row crops, and alfalfa and cereal as dense-stand crops.

Alfalfa is generally grown in rotation with other crops in this area, and preparing the land for a row crop following alfalfa usually involves 8 to 12 tillage operations, including moldboard plowing when using conventional methods. The first study (Carter et al., 1988b) demonstrated that wheat and corn could be successfully produced without tillage after killing alfalfa with herbicide (Fig. 37–5). Both winter and spring wheat and silage corn produced the same yields when grown without tillage as when grown under conventional tillage. It was necessary to clean the small furrows to remove rodent mounds and clumps of grass that had invaded the alfalfa and had been killed by herbicide. Wheat was seeded with a conventional, irrigated land-type double-disk drill. Corn was seeded with a double tool bar arrangement with small, chisel-type bull tongues on the leading bar to make a groove in the soil. Flex-type corn planters were attached to the second tool bar, so that they followed directly behind the bull tongue chisels. These no-tillage crops irrigated with better uniformity and required only about one-half as much water as the adjacent, conventionally tilled plots for the first irrigation (Carter et al., 1989b).

Subsequent no-tillage studies have included no-tillage corn following cereal, cereal following corn, and a variety of investigations involving dry bean, corn, cereal, and sugarbeet. The general conclusions from 3 yr of study are that furrow erosion and sediment loss can be reduced 80 to 100% by no-tillage systems and 50 to 80% by reduced tillage systems. Direct crop production costs can be reduced 20 to 30% by using no-tillage practices and 10 to 20% by using reduced tillage practices.

Wide application of conservation tillage on furrow-irrigated land has the potential to reduce erosion and sediment loss 80 to 90%. Such widespread acceptance could almost eliminate the need for the erosion and sedi-

Fig. 37–5. No-till winter wheat growing in a herbicide-killed alfalfa field.

ment loss control practices discussed earlier in this chapter. However, such wide acceptance will require many years of educating farmers, if it is to ever be achieved (Carter et al., 1989b).

III. EROSION UNDER SPRINKLER IRRIGATION

Soil erodes under sprinkler irrigation by processes similar to those reported under rainfall. These are soil particle detachment caused by falling water drops and flowing water, and transport by water drop splash and flowing water (Meyer & Wischmeier, 1969; Trout & Neibling, 1989). However, conditions are often quite different under sprinkler irrigation than under rainfall because: (i) only a small part of a field is receiving water at any given time, (ii) water drops from sprinklers vary considerably depending upon the type of system used, and (iii) sprinkler irrigation is generally applied only when the soil water reservoir needs replenishing for a growing crop or in preparation for tillage. These systems can be properly designed for any particular soil conditions to minimize runoff and erosion.

The most serious erosion under sprinkler irrigation usually occurs with center pivot systems where the application rate at the outer end may exceed the soil infiltration capacity, creating runoff and the potential for erosion. Recently developed low-pressure sprinklers and spray heads also increase the potential for runoff and erosion because the application rate per unit area on the smaller wetting areas must be greater to achieve the same total application. In any case, sprinkler irrigation systems should be designed and operated according to soil characteristics of the field to be irrigated.

A. The Erosion Process

When water drops strike the soil surface, erosion may result. Impacting water drops may detach soil particles from the soil mass. Detached particles are splashed in all directions from the impact point, with a net movement downslope.

Soil particle detachment by water drop impact is proportional to the intensity squared (Meyer & Wischmeier, 1969), or to a product of the momentum and number of drops, both raised to a power (Park et al., 1983). Splash erosion measured by simulated rainfall is proportional to rainfall, or sprinkler intensity to a power that varies with soil type from 1.6 to 2.1 (Meyer, 1981; Park et al., 1983).

An alternative method of evaluating erosion from raindrop impact is to relate it to the kinetic energy of the rainfall. Simulated rainfall with drop diameters of 2.2, 3.2, and 4.9 mm from several heights has been used to study soil detachment from a silty clay, a loamy sand, and two silt loam soils. The regression equation relating soil splash (SS) to kinetic energy (KE), rainfall intensity (I), and percent clay (PC) was

$$SS = 7.50 \; (I)^{0.41} \; (KE)^{1.14} \; (PC)^{-0.52}$$

with a correlation coefficient of 0.93. Kinetic energy was by far the most significant of these three parameters. Adding other soil parameters did not significantly improve the correlation coefficient (Bubenzer & Jones, 1971).

The general erosion potential from various sprinklers can be evaluated by converting the mean drop diameter to kinetic energy using the procedure of Stillmunkes and James (1982). The kinetic energy values can then be used in the above equation to estimate soil detachment by drop splash, and the relative erosivity of any particular sprinkler can be estimated by this method. Recent research has provided limited information on drop size distribution from various sprinkler nozzle designs and the effects of nozzle size or pressure on drop size distribution (Dadio & Wallender, 1985; Kohl & DeBoer, 1984; Kohl et al., 1985).

The preceding discussion concerned the processes governing the sediment produced at a particular site. Sediment transport processes generally determine how much of that sediment is moved from the site. The sediment transported by overland flow depends upon the water application rate in excess of infiltration. The infiltration rate can be reduced by water drop impact and increase the amount of runoff.

When the water application rate exceeds infiltration, water ponds in small surface depressions until they become full. Then water begins to flow downslope as shallow overland flow. This flow seldom produces sufficient shear forces to detach particles, but it does transport some sediment detached by water impact. Usually considerably more soil particles are detached by water drop impact than are transported by this shallow overland flow. Many transport equations have been applied in attempts to describe this part of the overall erosion process (Foster, 1982; Neibling, 1984).

As overland flow moves downslope it concentrates in tillage marks, previous erosion channels, or, as a result of the natural microtopography, it forms new rills. The detachment and transport processes in these rills are similar to those in irrigation furrows. One difference is that the flow rate in rills increases downslope as a result of increasing collection areas, whereas the flow rate in irrigation furrows decreases downslope. Thus, the transport capacity in rills increases until the water flows out of the area receiving water.

Water may flow downslope in rills to an area just previously irrigated, into a dry area not yet irrigated, or along the operating sprinkler line where water is being applied. These three situations can all produce different erosion and sediment transport results. Flows from rills tend to concentrate into fewer, larger channels in natural drainage ways called *ephemeral channels or gullies*. Sediment detachment and transport processes in ephemeral gullies are similar to those for rills. Typically, an ephemeral gully will erode downward to a tillage pan or a less erodible layer and then enlarge to an equilibrium width during the first significant erosion event following tillage. Unless tillage occurs, additional erosion will be minimal for subsequent events smaller or equal in size to the event that formed the channel.

Usually the amount of erosion during each center pivot sprinkler irrigation is relatively small because only 30 to 40 mm of water is applied. This amount of water normally will not cause extensive erosion. Most of that water will infiltrate and not run off the field. The amount of runoff depends on how much the application rate exceeds the infiltration rate. Large amounts of water are applied with wheel line and hand-moved sprinklers, and erosion can be severe.

1. Cohesion Factors

The relationships between soil cohesion factors and erosion are the same under both furrow and sprinkler irrigation. The erosion difference between the two situations involves the forces acting against the soil-bonding forces. The bombardment of water drops on the soil under sprinkler irrigation is a different type of force than the shear force of a flowing stream in an irrigation furrow.

2. Tillage

Extensive tillage that breaks soils into small aggregates also breaks many particle-to-particle bonds and makes the soil more susceptible to erosion under sprinkler irrigation, just as it does under furrow irrigation. Fewer tillage operations generally result in less erosion under sprinkler irrigation. The direction of the final tillage or planting marks can have an important impact on rill and subsequent gully formation under sprinkler irrigation. Such marks up and down the slope should be avoided. This, of course, is not always possible, particularly on rolling topography where much of the sprinkler irrigation is practiced. Tillage and planting marks should follow level contours to the extent possible. No-tillage and reduced-tillage practices can be applied more easily to sprinkler-irrigated land than to furrow-irrigated land because rougher surfaces can be tolerated better under sprinkler irrigation.

3. Surface Condition Effects

The condition of the soil surface can have a major effect on erosion under sprinkler irrigation. Rough, cloddy surfaces have higher infiltration rates. As a result, runoff is decreased or eliminated and erosion is decreased. Overtilled, smooth surfaces are more erodible and generally have lower infiltration rates, greater runoff, and more erosion than rougher surfaces. To be effective as an erosion control measure, soil clods must be large and stable enough to keep infiltration at a high level until the crop canopy covers the soil surface. Such cloddy surfaces can be attained by tilling at selected soil water contents. Also, tilling compacted soils generally results in greater cloddiness than does tilling noncompacted soils (Johnson et al., 1979; Römkens & Wang, 1986).

Residue on the soil surface decreases the amount of water drop impact erosion, increases infiltration, and decreases runoff. As a result, overland flow erosion is also decreased by residue on and in the soil surface. Conservation tillage practices increase quantities of surface residues and decrease erosion under sprinkler irrigation.

B. Controlling Sprinkler Irrigation Erosion

Any practice that will reduce the impact energy of water drops striking the soil surface, maintain infiltration, reduce overland flow, and protect against rill and gully formation will decrease soil erosion under sprinkler irrigation. There are several approaches that can be used towards accomplishing these goals. Usually a combination of practices leads to the best results.

1. Irrigation Management

The most important aspect of sprinkler irrigation management is the proper design of the system. Infiltration characteristics of the soil should be evaluated and the results used to select a sprinkler system that will apply water at a rate less than the infiltration capacity of the soil (Bruce et al., 1980; 1985). This is usually easier with wheel line, lateral move systems than with center pivot systems. If the application rate is less than the infiltration capacity and adequate to supply sufficient water to meet crop needs, the only erosion that will occur is that from water drop impact.

The area covered per segment of line increases with distance from the pivot point of a center pivot system. Therefore, to apply the water needed by the crop, the application rate increases with distance from the pivot point. The most serious erosion usually occurs at the outer end of a center pivot system, because the application rate often exceeds that of infiltration.

Once a properly designed sprinkler irrigation system has been installed, it is important to operate it correctly. Operating at pressures different from those designed, improper set times, or operating center pivots at improper travel speeds can also lead to erosion problems.

Another important factor in the design and operation of sprinkler systems is that nozzles or heads should be designed to distribute water drops of the lowest possible kinetic energy to the soil. Water drops with the lowest kinetic energy will cause the least water drop splash erosion and soil surface compaction.

2. Conservation Tillage

Conservation tillage has been practiced for erosion control on rainfed soils for over 20 yr, but only recently have conservation tillage systems been developed for sprinkler-irrigated lands. The same basic systems used for erosion control on rainfed soils will also control erosion on sprinkler-irrigated soils. Such systems are easier to apply under sprinkler irrigation because water can be applied when needed instead of depending upon nature to provide rainwater. For example, deep-furrow drills used to place seed into moist soil on rainfed lands are not required on sprinkler-irrigated land where water can be applied to wet the soils to germinate the seed if necessary.

Conservation tillage systems for sprinkler-irrigated land should leave crop residues on and in the soil surface, provide a rough cloddy surface, and leave drill or tillage marks on level contours to the extent possible. Crop residues are the most important consideration and tend to mask the effects of the other two parameters.

Crop rotations impact the application of conservation tillage on irrigated land. Usually more crop rotating is required to minimize crop disease on irrigated land than on rainfed land. The cropping sequence should be adjusted if necessary to assure the production of crop residues throughout the rotation. Conservation tillage is then required to maintain these residues on or near the soil surface.

3. Reservoir Tillage

Aarstad and Miller (1973) first demonstrated that making small water storage basins in the soil surface to catch and temporarily store water until it infiltrates was a useful technique to prevent runoff and increase irrigation uniformity. This also almost eliminates erosion under sprinkler irrigation (Longley, 1984). In recent years, tillage equipment has been developed and used effectively for that purpose, and the process of forming the basins has become known as *reservoir tillage*. These small reservoirs function best when they are depressions formed by scooping or pressing rather than being formed by earthen dams in furrows. The latter are not as stable when nearly filled as the former.

Reservoir tillage is generally done after planting the crop but can be done in the same operation for cereals. The tiny reservoirs are placed between rows of row crops, but can be randomly placed in solid cover crops, such as cereals. In either case, once installed, 1 h of land will have thousands of these small reservoirs on the surface (Fig. 37-6). When water is applied by a sprinkler

Fig. 37-6. Reservoir tillage on a potato field.

system, water not immediately infiltrated accumulates in the tiny reservoirs where it gradually infiltrates. Runoff can be prevented or reduced even when the water application rate far exceeds the infiltration rate. Since runoff is eliminated, so are erosion and sediment transport that would have occurred with overland flow. Therefore, erosion is confined to that caused by water drop impact. The use of reservoir tillage has had a major impact on both irrigation uniformity and erosion control under sprinkler irrigation. It compensates for design and operation errors and is of particular importance in areas covered by the outer sections of center pivot systems and on steep slopes. Crop yields have been dramatically increased and soil erosion almost eliminated by reservoir tillage of sprinkler-irrigated land.

IV. SUMMARY AND CONCLUSION

Irrigation-induced erosion began with irrigation and has continued largely unabated until the past 10 yr. The problem was recognized as serious by scientists in the late 1930s and 1940s, but work done then was given little attention by irrigated land farmers. During the late 1960s and early 1970s, sufficient attention was given to water quality that legislation was set forth to control irrigation return flow quality. This stimulated research on polluting sediment sources because sediment was defined as the most serious water pollutant from the standpoint of quantity. This continuing research has provided much-needed information about erosion on irrigated lands. It has now progressed to the point that effective erosion control practices have been developed for irrigated lands.

Although significant erosion can occur under improperly designed and operated sprinkler irrigation systems, the most serious erosion occurs under furrow irrigation.

Soil erosion results when shear forces are sufficient to overcome cohesive bonds between soil particles, allowing soil particles to be broken off the soil mass and transported by flowing water. Both the erosive shear and sediment transporting forces increase exponentially with stream size and flow velocity. Therefore, the furrow stream size, furrow roughness, and the slope in the direction of irrigation are important factors affecting the energy of the stream to exert shear forces. Controlling these factors, is of primary importance in furrow irrigation erosion. Of these factors, humans can control the furrow stream size to a limited extent. However, the furrow stream size must be large enough to provide water to infiltrate along the entire furrow length in a reasonable time to accomplish the purposes of irrigation. Usually, best results can be attained by starting the irrigation with a furrow stream that will reach the lower end of the furrow in 2 to 4 h, and then decreasing it to about one-half the original.

Crop residues in irrigation furrows and rough furrows both decrease erosion because they reduce the kinetic energy of the stream. In contrast, excessive tillage, leaving a fine soil and resulting in smooth furrows without residue, increases furrow erosion.

Cropping sequences affect irrigation furrow erosion by influencing the amount of residue remaining in the soil while producing the subsequent crop. Tillage plays the most important role in the presence or absence of residue. Moldboard plowing, which buries crop residues, is the worst tillage practice commonly used on irrigated land from the erosion standpoint.

Furrow erosion redistributes topsoil by removing soil from the upper reaches of furrows and depositing it downslope. This reduces topsoil depth on the upper 25 to 40% of each furrow-irrigated field, causing serious decreases in crop production.

During the past 15 yr, erosion and sediment loss control technology has been developed and evaluated for furrow-irrigated land. The first practices developed and evaluated were aimed primarily at sediment loss control. These included sediment retention basins ranging in size from 1.0 ha to minibasins receiving runoff water from only a few irrigation furrows. These sediment retention basins remove 65 to 95% of the inflowing sediment from the water. Vegetative filters comprised of cereal, grass, or alfalfa crops planted along the lower ends of fields can filter out about 40 to 60% of the sediment if properly managed.

One important discovery made in the mid-1970s was that large amounts of erosion and sediment loss from furrow-irrigated fields were resulting from mismanagement of the tailwater ditch, thereby creating convex field ends. A buried pipe erosion and sediment loss control system was developed to completely eliminate this problem, as well as remove the tailwater ditch and field access problems associated with it. This system increases the productive area of the field where installed. Increased crop production from that area will generally pay installation costs in 4 to 8 yr.

Placing crop residues in irrigation furrows increases infiltration and reduces furrow erosion. Equipment has been developed to accomplish this straw placement. However, a far more logical approach is to use tillage practices that will leave crop residues on and in the soil surface. Moldboard plowing must be eliminated to avoid burying crop residue.

Reduced-tillage and no-tillage systems introduced to furrow-irrigated land for erosion control in 1978 and 1984, respectively, have great potential for erosion control on irrigated land. A series of studies, beginning in the fall of 1984, have demonstrated that no-tillage and reduced tillage can be used effectively on furrow-irrigated land without causing irrigation problems. These conservation tillage systems reduce soil erosion and sediment loss from 60 to 100%, with the best results coming from no-tillage systems. Direct cropping input costs are decreased 10 to 30% when compared to conventional tillage systems. These savings translate into net profits because crop yields are generally the same as for conventional tillage. Widespread application of conservation tillage on furrow-irrigated lands has the potential to reduce erosion and sediment loss about 85 to 90%. This would also eliminate the need for sediment retention basins, vegetative filters, and placing straw in furrows. The buried pipe erosion and sediment loss control system may still be used in combination with conservation tillage, but the need for such a system would be decreased.

Soil erosion processes under sprinkler irrigation are similar to those under rainfall, with some differences. The amount of water applied by a single irrigation is controlled and it is applied only when needed. Generally, only a small part of the field is receiving water at any given time. Therefore, water flow resulting from runoff is confined. Rill and gully erosion are therefore limited when compared to erosion under rainfall.

The most important erosion control practice for sprinkler-irrigated land is the proper design and operation of sprinkler irrigation systems, which means using reliable soil water transmission and retention data. The use of conservation tillage and the application of recently developed reservoir tillage will also greatly reduce the erosion potential.

REFERENCES

Aarstad, J.S., and D.E. Miller. 1973. Soil management to reduce runoff under center-pivot sprinkler systems. J. Soil Water Conserv. 28:171–173.

Aarstad, J.S., and D.E. Miller. 1979. Corn residue management to reduce erosion in irrigation furrows. J. Soil Water Conserv. 33:289–391.

Aarstad, J.S., and D.E. Miller. 1981. Effects of small amounts of residue on furrow erosion. Soil Sci. Soc. Am. J. 45:116–118.

Allen, R.R., J.T. Musick, and A.F. Wiese. 1976. Limited tillage of furrow irrigated winter wheat. Trans. ASAE 19:234–236, 241.

Berg, R.D. 1984. Straw residue to control furrow erosion on sloping irrigated cropland. J. Soil Water Conserv. 39:58–60.

Berg, R.D., and D.L. Carter. 1980. Furrow erosion and sediment losses on irrigated cropland. J. Soil Water Conserv. 35:267–270.

Brown, M.J. 1985a. Effect of grain straw and furrow irrigation stream size on soil erosion and infiltration. J. Soil Water Conserv. 40:389–391.

Brown, M.J. 1985b. Within furrow erosion and deposition of sediment and phosphorous. p. 113–118. In S.A. El-Swaity et al. (ed.) Soil erosion and conservation. Soil Conserv. Soc. Am., Ankeny, IA.

Brown, M.J., J.A. Bondurant, and C.E. Brockway. 1981. Ponding surface drainage water for sediment and phosphorus removal. Trans. ASAE 24:1479–1481.

Brown, M.J., D.L. Carter, and J.A. Bondurant. 1974. Sediment in irrigation and drainage waters and sediment inputs and outputs for two large tracts in southern Idaho. J. Environ. Qual. 3:347–351.

Brown, M.J., and W.D. Kemper. 1987. Using straw in steep furrows to reduce soil erosion and increase dry bean yields. J. Soil Water Conserv. 42:187–191.

Bruce, R.R., J.L. Chesness, T.C. Keisling, J.E. Pallas, Jr., D.A. Smittle, J.R. Stansell, and A.W. Thomas. 1980. Irrigation of crops in the southeastern United States. Principles and practice. Agric. Rev. Man. (USDA-SEA) ARM-S-9.

Bruce, R.R., A.W. Thomas, V.L. Quisenberry, H.D. Scott, and W.M. Snyder. 1985. Irrigation practice for crop culture in the southeastern United States. p. 51–106. In D. Hillel (ed.) Advances in irrigation. Academic Press, Orlando, FL.

Bruce, R.R., S.R. Wilkinson, and G.W. Langdale. 1987. Legume effects on soil erosion and productivity. p. 127–138. In The role of legumes in conservation tillage systems. Soil Conserv. Soc. Am., Ankeny, IA.

Bubenzern, G.D., and B.A. Jones, Jr. 1971. Drop size and impact velocity effects on the detachment of soils under simulated rainfall. Trans. ASAE 14:625–628.

Bullock, M.S., W.D. Kemper, and S.D. Nelson. 1988. Soil cohesion as affected by freezing, water content, time and tillage. Soil Sci. Soc. Am. J. 52:770–776.

Carter, D.L. 1976. Guidelines for sediment control in irrigationreturn flows. J. Environ. Qual. 5:119–124.

Carter, D.L. 1985a. Controlling erosion and sediment loss on furrow-irrigated land. p. 355–364. *In* S.A. El-Swaity et al. (ed.) Soil erosion and conservation. Soil Conserv. Soc. Am., Ankeny, IA.

Carter, D.L. 1985b. Furrow irrigation erosion effects on crop production. p. 39–47. *In* Proc. natl. symp. on erosion and soil productivity, New Orleans. 10–11 Dec. 1984. ASAE, St. Joseph, MI.

Carter, D.L. 1989. Furrow irrigation erosion lowers soil productivity. J. Irrig. Drain. Div. Am. Soc. Civ. Eng. (In press.)

Carter, D.L., and R.D. Berg. 1983. A buried pipe system for controlling erosion and sediment loss on irrigated land. Soil Sci. Soc. Am. J. 47:749–752.

Carter, D.L., R.D. Berg, and B.J. Sanders. 1985. The effect of furrow irrigation erosion on crop productivity. Soil Sci. Soc. Am. J. 49:207–211.

Carter, D.L., R.D. Berg, and B.J. Sanders. 1989a. Crop sequences and conservation tillage to control irrigation furrow erosion and increase farmer income. J. Soil Water Conserv. (In press.)

Carter, D.L., C.E. Brockway, and K.K. Tanji. 1989b. Controlling erosion and sediment loss in irrigated agriculture. J. Irrig. Drain. Div. Am. Soc. Civ. Eng. (In press.)

Dadio, C., and W.W. Wallender. 1985. Droplet size distribution and water application with low pressure sprinklers. Trans. ASAE 28:511–516.

Daniels, R.B., J.W. Gilliam, D.K. Cassel, and L.A. Wilson. 1987. Quantifying the effects of past soil erosion on present soil productivity. J. Soil Water Conserv. 42:183–187.

Evans, R.G., J.S. Aarstad, D.E. Miller, and M.W. Kroeger. 1987. Crop residue effects on surge furrow irrigation hydraulics. Trans. ASAE 30:424–429.

Evans, N.A., and M.E. Jensen. 1952. Erosion under furrow irrigation. North Dakota Agric. Exp. Stn. Bimon. Bull. 15:7–13.

Farnstorm, K.J., J. Borrelli, and J. Long. 1985. Sediment losses from furrow irrigated croplands in Wyoming. Univ. Wyoming, Laramie.

Foster, G.R. 1982. Modeling the erosion process. p. 297–380. *In* C.T. Haan et al. (ed.) Hydrologic modeling of small watersheds. ASAE Monogr. 5. ASAE, St. Joseph, MI.

Foster, G.R., and L.J. Lane. 1983. Erosion by concentrated flow in farm fields. p. 9.65–9.82. *In* R.-M. Li and P. LaGasse (ed.) Proc. D.B. Simons symp. erosion and sedimentation, Simons and Li Assoc. Inc., Fort Collins, CO.

Frye, W.W., O.L. Bennett, and G.J. Buntley. 1985. Restoration of crop productivity on eroded or degraded soils. p. 335–356. *In* R.F. Follett and B.A. Stewart (ed.) Soil erosion and crop productivity. ASA, CSSA, and SSSA, Madison, WI.

Gardner, W., J.H. Gardner, and C.W. Lauritzen. 1946. Rainfall and irrigation in relation to soil erosion. Utah Agric. Exp. Stn. Bull. 326.

Gardner, W., and C.W. Lauritzen. 1946. Erosion as a function of the size of the irrigating stream and the slope of the eroding surface. Soil Sci. 62:233–242.

Humpherys, A.S. 1971. Automatic furrow irrigation systems. Trans. ASAE 14:466–481.

Isrealson, O.W., G.D. Clyde, and C.W. Lauritzen. 1946. Soil erosion in small furrows. Utah Agric. Exp. Stn. Bull. 320.

Johnson, C.B., J.V. Mannering, and W.C. Moldenhauer. 1979. Influence of surface roughness and clod size and stability on soil and water losses. Soil Sci. Soc. Am. J. 43:772–777.

Kemper, W.D., and R.C. Rosenau. 1984. Soil cohesion as affected by time and water content. Soil Sci. Soc. Am. J. 48:1001–1006.

Kemper, W.D., R.C. Rosenau, and S. Nelson. 1985a. Gas displacement and aggregate stability of soils. Soil Sci. Soc. Am. J. 49:25–28.

Kemper, W.D., B.J. Ruffing, and J.A. Bondurant. 1982. Furrow intake rates and water management. Trans. ASAE 25:333–339, 343.

Kemper, W.D., T.J. Trout, M.J. Brown, and R.C. Rosenau. 1985b. Furrow erosion and water and soil management. Trans. ASAE 28:1564–1572.

Kemper, W.D., T.J. Trout, A.S. Humpherys, and M.S. Bullock. 1988. Mechanisms by which surge irrigation reduces furrow infiltration rates in a silty loam soil. Trans. ASAE 31:821–829.

Kemper, W.D., T.J. Trout, and D.C. Kincaid. 1987. Cablegation: Automated supply for surface irrigation. Adv. Irrig. 4:1–66.

Kohl, R.A., and D.W. DeBoer. 1984. Drop size distributions for a low pressure spray type agricultural sprinkler. Trans. ASAE 27:1836–1840.

Kohl, R.A., D.W. DeBoer, and P.D. Evanson. 1985. Kinetic energy of low pressure spray sprinklers. Trans. ASAE 18:1526-1529.

Krauss, H.A., and R.R. Allmaras. 1982. Technology masks the effects of soil erosion on wheat yields—a case study in Whitman County, Washington. p. 75-86. *In* B.L. Schmidt et al. (ed.) Determinates of soil loss tolerance. ASA Spec. Publ. 45. ASA and SSSA, Madison, WI.

Longley, T.S. 1984. Basin tillage for irrigated small grain production p. 338-345. *In* Water today and tomorrow. Proc. Irrig. Drain. Div. Am. Soc. Civ. Eng. Flagstaff, AZ. 24-26 July. Am. Soc. Civ. Eng., New York.

McDaniel, T.A., and B.F. Hajek. 1985. Soil erosion effects on crop productivity and soil properties in Alabama. p. 48-57. *In* Proc. nat. symp. erosion and soil productivity. New Orleans. 10-11 Dec. 1984. ASAE, St. Joseph, MI.

Mech, S.J. 1949. Effect of slope and length of run on erosion under irrigation. Agric. Eng. 30:379-383, 389.

Meyer, L.D. 1981. How rainfall affects interrill erosion. Trans. ASAE 24:1472-1475.

Meyer, L.D. and W.H. Wischmeier. 1969. Mathematical simulation of the process of soil erosion by water. Trans. ASAE 12:754-762.

Miller, D.E., and J.S. Aarstad. 1983. Residue management to reduce furrow erosion. J. Soil Water Conserv. 38:366-370.

Musick, J.T., A.F. Wiese, and R.R. Allen. 1977. Management of bed-furrow irrigated soil with limited- and no-tillage systems. Trans. ASAE 20:666-672.

Neibling, W.H. 1984. Transport and deposition of soil particles by shallow flow on concave slopes. Ph.D. diss. Purdue Univ., West Lafayette, IN.

Papendick, R.I., D.L. Young, D.K. McCool, and H.A. Krauss. 1985. Regional effects of soil erosion on crop productivity—the Palouse area of the Pacific Northwest. p. 305-320. *In* R.F. Follett and B.A. Stewart (ed.) Soil erosion and crop productivity. ASA, CSSA, and SSSA, Madison, WI.

Park, S.W., J.K. Mitchell, and G.D. Bubenzer. 1983. Rainfall characteristics and their relation to splash erosion. Trans ASAE 26:795-804.

Robinson, A.R. 1971. Sediment. J. Soil Water Conserv. 26:61-62.

Römkens, M.J.M, and J.Y. Wang. 1986. Effect of tillage on surface roughness. Trans. ASAE 29:429-433.

Stillmunkes, R.T., and L.G. James. 1982. Impact energy of water droplets from irrigation sprinklers. Trans. ASAE 25:130-133.

Taylor, C.A. 1940. Transportation of soil in irrigation furrows. Agric. Eng. 21:307-309.

Trout, T.J., and W.D. Kemper. 1983. Factors which influence furrow intake rates p. 302-312. *In* Advances in infiltration. Proc. nat. conf. advances in infiltration, Chicago, IL. 12-13 Dec. ASAE, St. Joseph, MI.

Trout, T.J., and B.E. Mackey. 1988. Furrow inflow and infiltration variability. Trans. ASAE. 31:531-537.

Trout, T.J., and W.H. Neibling. 1989. Erosion and sedimentation processes in irrigation. J. Irrig. Drain. Div. Am. Soc. Civ. Eng. (In press.)

Wadleigh, C.H. 1968. Wastes in relation to agriculture and forestry. U.S. Dep. Agric. Misc. Publ. 1065.

White, A.W., Jr., R.R. Bruce, A.W. Thomas, G.W. Langdale, and H.F. Perkins. 1985. Characterizing productivity of eroded soils in the southern Piedmont. p. 83-95. *In* Proc. nat. symp. erosion and crop productivity. New Orleans. 10-11 Dec. 1984. ASAE, St. Joseph, MI.

Wolman, M.G. 1985. Soil erosion and crop productivity: A worldwide perspective. p. 9-21. *In* R.F. Follett and B.A. Stewart (ed.) Soil erosion and crop productivity. ASA, CSSA, SSSA, Madison, WI.

38 Pollution and Public Health Problems Related to Irrigation[1]

ARTHUR G. HORNSBY

University of Florida
Gainesville, Florida

Irrigation of agricultural lands may result in return flows which have deteriorated with respect to water quality because of increased concentration of dissolved or suspended constituents. These constituents may include sediments, nutrients, dissolved salts, pesticides and other toxic organic chemicals, pathogens, or other toxic leachates (e.g., Se, B). The presence of these constituents in the return flow in excessive amounts may impair the use of water in receiving streams for some activities. The impairment may manifest itself in the area being irrigated or in a location removed from that immediate area. Such off-site impairment is the focus of state and federal environmental regulations.

Although the Rivers and Harbors Act of 1899 (U.S. Congress, 1899) provided authority to control sedimentation from agricultural lands that interfered with transportation in waterways, the Federal Water Pollution Control Act (U.S. Congress, 1948), with amendments, marked the entry of the federal government into active participation in water quality control. It has as its principal goal "to restore and maintain the chemical, physical, and biological integrity of the Nation's waters." The Water Quality Act of 1965 (U.S. Congress, 1965) required each state to establish water quality standards for interstate streams and coastal waters within its boundaries, designate allowable uses for specific waters, and develop implementation schedules and enforcement programs necessary to achieve these goals. Thus irrigation return flows are specifically included. In several western river systems the flows are composed of irrigation return flows from multiple diversions of water for irrigation, with seepage and operational spills making up the flow in the streams.

The passage of Public Law 92-500 (U.S. Congress, 1972) entitled "Federal Water Pollution Control Act Amendments" in 1972 was directed primarily at obvious point source discharges of municipal and industrial wastes into interstate streams. Point source agricultural discharges were identified and National Pollutant Discharge Elimination System (NPDES) per-

---

[1] Contribution from the Institute of Food and Agricultural Sciences, University of Florida, Gainesville, FL 32611.

mits were required to discharge into waterways. Irrigation return flow discharge permits were included. Subsequently, the Clean Water Act of 1977 (PL 95-217, U.S. Congress, 1977) amended Section 208 of PL 92-500 by adding return flows from irrigated agriculture, and their cumulative effects. General definitions in Section 502(14) of PL 92-500 were amended at the end to state, "This term does not include return flows from irrigated agriculture." Finally, Section 402 of PL 92-500 (NPDES) was amended by adding at the end a new subsection as follows: "(1) The Administrator shall not require a permit under this section for discharges composed entirely of return flows from irrigated agriculture, nor shall the Administrator directly or indirectly, require any State to require such a permit." Thus irrigation return flows are subject to Section 208 "best management practices" (BMP) as were other nonpoint sources of discharge.

With the designation of irrigation return flows as "nonpoint sources" of pollutants and the public desire to support the national goal of restoring and maintaining the chemical, physical, and biological integrity of the nation's waters, irrigation of agricultural lands must be carefully managed to avoid adverse environmental effects on water quality. This presents an interesting conundrum in irrigated agriculture. Consumptive use of irrigation water results in increased total dissolved solids in the drainage water. Fertilizers and pesticides used to assure economical yields likewise are not totally removed by crop harvest. Agrichemical residues are an unavoidable consequence of agricultural production. At the same time, the public demands quality foodstuffs and safe drinking water. The individual consumer of agricultural products and the agricultural producer are co-conspirators in this dilemma. One does not exist without the other.

Obviously, there is both the need and opportunity to manage irrigation to minimize negative water quality impacts. Pollution of streams and groundwater supplies from sediment production, excess leaching of total dissolved solids (salinity), nutrients, pesticides and other toxic elements, and biological agents must be addressed by appropriate management practices and regulatory direction. These aspects will be discussed in subsequent sections.

I. PROCESSES LEADING TO POLLUTANT LOADING

A. Consumptive Use of Irrigation Water

Deterioration of the quality of water used in irrigation is a direct result of physical and chemical processes occurring in the soil-water-plant-atmosphere system. Plants consumptively use water by evapotranspiration, leaving behind residual salts originally contained in the applied irrigation water (concentrating effect). With approximately 60% of applied water transpired, the opportunity for accumulation of residual salts in the soil is significant. Natural rainfall and irrigation that exceed the soil-water deficit leaches these residues deeper into the soil profile, resulting in increased salinity in the groundwater and, subsequently, in the base flow of streams in arid and semiarid regions. While an increase in salinity of receiving streams is not desirable, it is an inevitable consequence of irrigating arid lands.

B. Reclamation of Saline Soils Formed Under Deficit Rainfall

Soils that have developed under arid and semiarid conditions contain considerable residual salts. Reclamation of the lands is required for economic crop production. When these lands are placed in production, the precipitated salts are purposefully solubilized and leached from the root zone by reclamation practices (leaching effect). Such soils would require an appropriate leaching fraction to maintain acceptable salinity levels in the root zone. With continued irrigation, these salts may find their way to groundwater or surface water bodies.

C. Natural Deposits and Mineral Weathering

In some geological settings, such as the Coast Ranges in California, alluvial deposits contain significant quantities of soluble mineral salts and trace elements, such as As, B, Cd, Cr, Cu, Pb, Se, and Zn. On the western slope of the Rocky Mountains in Utah, Colorado, and New Mexico, marine sediments underlying alluvial soils contain massive amounts of saline salts. These salts may be released as a result of mineral weathering of parent material or simple solubilization as water percolates through the stratum containing these soluble salts (salt pickup effect). Irrigation of agricultural lands in these settings may cause these elements to accumulate in the drainage water or in groundwater. In some cases the leaching process is enhanced by the oxidation-reduction state of the soil caused by irrigation applications.

D. Excessive Leaching

Overirrigation further exacerbates the leaching of soluble salts in drainage water. This is particularly counterproductive in areas underlaid with saline marine shales, such as is the case in the Grand Valley in Colorado. In this region, extensive shale beds contain crystalline NaCl that is readily solubilized by excess irrigation drainage. This region of the Colorado River basin contributes a significant portion of the total salinity that flows over Hoover Dam.

II. DETRIMENTAL EFFECTS OF SPECIFIC POLLUTANTS

For the most part, constituents that become water pollutants are displaced from a setting where they were considered a resource and their presence deemed desirable. Problems emerge in environmental compartments where their presence or concentrations are detrimental to public health or the environment in general. The nature of these problems may be acute but transitory or may be chronic and long lasting.

A. Sediments

The off-site impacts of soil erosion on water quality include: reduction of water storage capacity of lakes and reservoirs, clogging of streams and drainage channels, deterioration of aquatic habitats, impairment of recreational waters, increased water treatment costs, and transport of agricultural chemicals associated with sediments into water systems (CAST, 1982). Although evidence does not show that suspended sediments have any direct adverse human health effects, organic matter and fine sediments (clays) may carry adsorbed nutrients, pesticides, viruses, bacteria, and other toxic compounds (OTA, 1983). These constituents could present serious problems where surface water bodies are used for drinking water supplies.

Suspended sediments in surface water can impair aquatic organism productivity by influencing light penetration, temperature, solubility products, and aquatic life. Sedimentation during the spawning season severely reduces fry hatch and survival. Sedimentation of high organic matter content can create O_2 depletion in the water column and can serve as a reservoir of sorbed toxic compounds that may be slowly released into the water. More complete details of the effects of suspended sediments in surface waters are discussed in *Water Quality Criteria,* 1972 (USEPA, 1973).

B. Nutrients

Excess nutrients, especially N and P, may lead to accelerated eutrophication of surface water bodies. At sufficient concentrations, algal blooms may occur and subsequent O_2 depletion in streams and reservoirs may result in fish kills. Esthetic and recreational value will be diminished.

Nitrates may accumulate in groundwater and present a public health problem, especially to infants younger than 12 wk of age. Because NO_3 is converted to the toxic form, NO_2, in the digestive systems of human infants and some livestock, NO_3-contaminated drinking water supplies are a serious problem. The NO_2 reacts with blood hemoglobin (which carries O_2 to all parts of the body) to form methemoglobin, which does not carry O_2. Around the age of 12 wk, an increase in the HCl in the infant's stomach kills most of the bacteria that convert NO_3 to NO_2, thereby reducing the risk of NO_2 toxicity. The drinking water standard for humans is 10 mg/L of NO_3 and 1 mg/L for NO_2 (USEPA, 1973).

Ruminant animals and infant monogastrics also have NO_3-converting bacteria in their digestive systems, thus NO_3 poisoning affects them in the same manner as it affects human infants. Research results suggest a limit of 100 mg/L for livestock (USEPA, 1973).

C. Salinity

While increased salinity is a consequence of the concentrating effect of evaporation and transpiration of applied irrigation water, it is the downstream user that has the burden of managing increasingly greater yield reductions, limited range of crops that can be grown, and more rapid salinization of soil resources. Yield decreases have been estimated as a function of increased electrical conductivity of soil-water extracts and are presented in Table 36–1 of chapter 36 in this book (Rhoades and Loveday). While no specific limit is placed on total dissolved solids (TDS) for public water supplies, the constituents Cl and SO_4 have recommended limits of 250 mg/L. For livestock, 3000 mg of soluble salts per liter or less should be satisfactory under most circumstances.

D. Pesticides

Pesticide residues in irrigation return flows become both management and health problems. In tailwater pumpback systems, residual pesticides may impair the value of this water when used on different crops with less tolerance than the original target crop. Residues discharged into surface streams and impoundments may impact aquatic species directly or accumulate in the food chain.

Pesticides in drinking water supplies present a health risk to the public. Health advisory levels (HAL) are being proposed by the U.S. Environmental Protection Agency (USEPA) for certain pesticides. These HAL values are relative measures of the acute and chronic toxicities and the carcinogenic nature of these compounds. Table 38–1 gives the currently proposed values for 64 pesticides plus NO_3 and NO_2.

E. Toxic Trace Elements

Trace elements have been found in irrigation return flows at concentrations that severely impair subsequent uses of the water. The Kesterson Reservoir in California exemplifies the water quality impact of Se leached from agricultural soils. Other trace elements such as As, B, Cd, Cr, Cu, Pb, Mn, Hg, Mo, Ni, and Zn are present in some soils and occur naturally in spring water in geothermal areas that contribute to the total water supply available for irrigation. These elements are essential to human and animal health at appropriate concentrations. However, at elevated levels they cause toxicosis and damage to vital organs. These trace elements produce symptoms of toxicity in humans that include reproductive anomalies, respiratory difficulties, and kidney, renal, cardiovascular, and gastrointestinal damage. Toxicity problems associated with Cd, Pb, Hg, and Se are more serious because these metals bioaccumulate in the food chain. Drinking water standards and recommended irrigation water concentration limits are given in Table 38–2.

F. Pathogens

Many microorganisms, pathogenic to animals, humans, or both, may be carried in irrigation water, particularly if the source is from surface supplies. A large variety of bacteria, spirochetes, protozoa, helminths, and viruses find their way into irrigation water from municipal wastes, feedlots, farm-

Table 38-1. Lifetime Health Advisory Levels (HAL) for pesticides in drinking water (USEPA, 1989a, b).

Common and chemical name	Concentration, μg/L
Aciﬂuorfen	1
Sodium 5-[2-chloro-4-(triﬂuoromethyl) phenoxy]-2-nitrobenzoate	
Alachlor†	0.4
2-Chloro-2'-6'-diethyl-N-(methoxy methyl) acetanilide	
Aldicarb	10
2-Methyl-2(methylthio) propionaldehyde O(methylcarbamoyl) oxime	
Ametryn	60
2-(Ethylamino)-4-(isopropylamino)-6-(methylthio)-1,3,5-triazine	
Ammonium Sulfamate	1500
Ammonium sulfamate	
Atrazine	3
2-Chloro-4-ethylamino-6-isopropylamino-1,3,5 triazine	
Baygon (Propoxur)	3
2-(1-Methylethoxy)phenyl methylcarbamate	
Bentazon	20
3(1-Methylethyl)-1H-2,1,3-benzothiodiazin-4(3H)-one-2,2-dioxide	
Bromacil	90
5-Bromo-3-sec-butyl-6-methyluracil	
Butylate	700
S-Ethyl diisobutylthiocarbamate	
Carbaryl	700
1-Napthyl methylcarbamate	
Carbofuran	40
2,3-Dihydro-2,2-dimethyl-7-benzofuranyl methylcarbamate	
Carboxin	700
5,6-Dihydro-2-methyl-N-phenyl-1,4-oxathiin-3-carboxamide	
Chloramben	100
3-Amino-2,5 dichlorobenzoic acid	
Chlordane†	0.03
1,2,3,4,5,6,7,8,8-octachloro-2,3,3a,4,7,7a-hexahydro-4,7-methanoindene	
Chlorothalanil†	2
Tetrachloroisophthalonitrile	
Cyanazine	10
2-{[4-Chloro-6-(ethylamino)-S-triazin-2-yl]amino}-2-methylproprionitrile	
Dacthal (DCPA)	3500
Dimethyl tetrachloroterephthalate	
Dalapon	200
2,2-Dichloropropionic acid	
2,4-D	70
(2,4-Dichlorophenoxy) acetic acid	
DBCP†	0.03
1,2-Dibromo-3-chloropropane	
Diazinon	0.6
O,O-Diethyl O-2-isopropyl-6-methylpyrimidin-4-yl phosphorothioate	
Dicamba	200
3,6-Dichloro-o-anisic acid	
1,2-Dichloropropane	0.6
1,2-Dichloropropane	
1,3-Dichloropropene (Telone)	0.2
1,3-Dichloropropene	
Dieldrin†	0.002
1,2,3,4,10,10-Hexachloro-6,7-epoxy-1,4,4a,5,6,7,8,8a-octahydro-$endo$-1,4,-exo-5,8-dimethanonaphthalene	

(continued on next page)

Table 38–1. Continued.

Common and chemical name	Concentration, μg/L
Dimethrin 2,4-Dimethylbenzyl-2,2-dimethyl-3(2-methyl propenyl) cyclopropanecarboxylate	2100
Dinoseb 2-(sec-Butyl)-4,6-dinitrophenol (alkanolamino salts)	7
Diphenamid N,N-Dimethyl-2,2-diphenylacetamide	200
Disulfoton O,O-Diethyl S-[2-(ethylthio)ethyl] phosphorodithioate	0.3
Diuron 3-(3,4-Dichlorophenyl)-1,1-dimethylurea	10
Endothall 7-Oxabicyclo-(2,2,1)heptane-2,3-dicarboxylic acid	140
Endrin 1,2,3,4,10,10-Hexachloro-6,7-epoxy-1,4,4a,5,6,7,8,8a-octahydro-exo- 1,4,-exo-5,8-dimethanonaphthalene	0.3
Ethylene Dibromide (EDB)† 1,2-Dibromoethane	0.0004
Ethylene Thiourea† 2-Imidazolidinethione	0.2
Fenamiphos Ethyl 3-methyl-4-(methylthio) phenyl (1-methylethyl) phosphoramidate	2
Fluometuron 1,1-Dimethyl-3-(α,α,α-trifluoro-m-tolyl) urea	90
Fonofos O-Ethyl-S-phenylethylphosphonodithioate	14
Glyphosate N-(Phosphono-methyl) glycine	700
Heptachlor† 1,4,5,6,7,8,8-Heptachloro-3a,4,7,7a-tetrahydro-4,7-methanoindene	0.008
Hexachlorobenzene Hexachlorobenzene	0.02
Hexazinone 3-Cyclohexyl-6-(dimethylamino)-1-methyl-1,3,5-triazine- 2,4(1H, 3H)-dione	200
Maleic Hydrazide 1,2-Dihydropyridazine-3,6-dione	3500
Methomyl S-Methyl-N-[(methylcarbamoyl)oxy] thioacetimidate	200
Methoxychlor 1,1,1-Trichloro-2,2-bis(4-methoxyphenyl)ethane	400
Methyl Parathion O,O-Dimethyl-O-4-nitrophenyl phosphorothioate	2
Metolachlor 2-Chloro-N-(2-ethyl-6-methylphenyl)-N-(2-methoxy-1-methylethyl) acetamide	100
Metribuzin 4-Amino-6-(1,1-dimethylethyl)-3-methylthio-1,2,4-triazin-5(4H)-one	200
Oxamyl S-Methyl N',N'-dimethyl-N-(methylcarbamoyloxyl)-1-thiooxamimdate	200
Paraquat 1,1'-Dimethyl-4,4'-bipyridinium-dichloride	30
Pentachlorophenol Pentachlorophenol	200

(continued on next page)

Table 38-1. Continued.

Common and chemical name	Concentration, μg/L
Picloram	500
4-Amino-3,5,6-trichloropicolinic acid	
Prometon	100
2,4-Bis(isopropylamino)-6-methoxy-s-triazine	
Pronamide	50
3,5-Dichloro-N(1,1-dimethyl-2-propynyl) benzamide	
Propachlor	90
2-Chloro-N-isopropylacetanilide	
Propazine	10
2-Chloro-4,6-bis(isopropylamino)-S-triazine	
Propham	100
Isopropyl carbanilate	
Simazine	4
2-Chloro-4,6-bis(ethylamino)-s-triazine	
2,4,5-T	70
2,4,5-Trichlorophenoxy-acetic acid	
Tebuthiuron	500
N-[5-(1,1-dimethylethyl)-1,3,4-thiadiazol-2-yl]-N,N'-dimethylurea	
Terbacil	90
3-tert-Butyl-5-chloro-6-methyluracil	
Terbufos	1
S-tert-butylthiomethyl o,o-diethyl phosphorodithioate	
2,4,5-TP (Silvex)	50
2-(2,4,5-Trichlorophenoxy) propionic acid	
Trifluralin	2
α,α,α-Trifluoro-2,6-dinitro-N,N-dipropyl-p-toluidine	

† No HAL established: Lifetime exposure at this level represents an excess cancer risk of one in one million.

Table 38-2. Drinking water standards and maximum recommended irrigation water concentration of various toxic trace elements.

Trace element	Drinking water standard†	Irrigation recommendation‡
	mg/L	
As	0.10	0.10
Cd	0.010	0.010
Cr	0.050	0.10
Pb	0.050	--
Mn	0.050	0.200
Hg	0.002	--
Mo	--	0.010
Se	0.010	0.020
V	--	0.100

† USEPA, (1973).   ‡ Letey et al. (1986).

steads, and industrial wastes such as food-processing plants, slaughterhouses, and poultry-processing plants.

Diseases associated with these organisms include bacillary and amoebic dysentery, Salmonella gastroenteritis, typhoid and paratyphoid fevers, leptospirosis, cholera, vibriosis, and infectious hepatitis. Other less common infections are tuberculosis, brucellosis, listeriosis, cocidiosis, swine erysipelas, ascariasis, cysticerciosis and tapeworm disease, facioliasis, and schistosomiasis (USEPA, 1973).

III. METHODS OF ASSESSING OFF-SITE EFFECTS

The adverse on-site environmental effects of irrigation are more easily recognized and more often addressed than are off-site effects, since the farmer is most directly affected by the on-site impacts. Such problems as rising water tables, increased salinity, eroded fields, and excessive irrigation cause decreased yields and/or increased production costs. Farmers are more likely to seek solutions to problems that directly impact farm productivity and profitability than to problems that occur as a result of their irrigation activities but are manifest off the production unit. Such off-site environmental impacts result in impaired use of groundwater and surface waters.

Federal requirements to conduct environmental impact assessments of water projects and to reduce nonpoint source pollution from irrigated agricultural activities have led to the development of technology to assess the technical and economic aspects of these activities on the environment. These technologies include: resource inventory procedures, baseline water quality assessment procedures, simulation models, and integrated physical and economic resource analysis tools. These tools have been used to assess the impact of irrigation of new lands on water quality and other environmental compartments, such as riparian habitat.

A. Conceptual Framework for Making Assessments

Assessing the off-site effects of nonpoint source pollution emanating from irrigation of agricultural lands requires a broad framework that encompasses the major processes that lead to impaired uses of receiving streams, lakes, and aquifers. The effects associated with irrigation activities must be put into perspective with other pollution sources within the watershed or study area in order to determine the potential impact of implementation of improved management practices in the irrigated sector. Through interagency efforts, the agencies that are involved in Rural Clean Water Projects (RCWP) have developed a conceptual framework for assessing agricultural nonpoint source water quality impacts (NCSU, 1981). The conceptul framework provides a guide to defining water quality impairments and developing an experimental approach to determining the effects of BMP implementation on water quality.

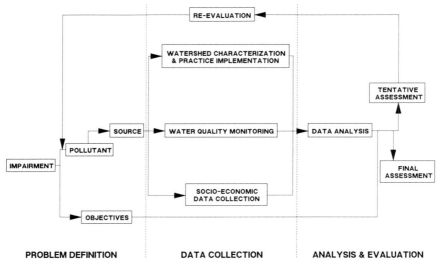

Fig. 38-1. Outline of conceptual framework for evaluating nonpoint source pollution (NCSU, 1981).

This framework identifies the process by which water quality problems or water use impairments are defined, describes how these may be traced to specific sources and pollutants, and suggests methods of analysis and evaluation that can be used to arrive at an acceptable water quality control plan. An outline of the procedure is given in Fig. 38-1.

Sometimes institutional constraints are greater than technical or economic constraints to implementing BMP's. The impact of water law upon allocation and use of waters within the western USA is currently recognized as one of the major constraints to adaptation by irrigated agriculture of more efficient water use practices. Radosevich (1977) and Radosevich and Skogerboe (1978) evaluated the relationships between western water laws and irrigation return flows and indicated that innovative approaches would be required to resolve these conflicts. Procedures have been developed to evaluate these conflicts (Vlachos et al., 1977). These procedures include: (i) definition of the problem in terms of its legal, physical, and economic and social parameters, (ii) identification of potential solutions in relation to the parameters of the problem, (iii) testing the implementability of these potential solutions for diverse situations, and (iv) specifying those solutions or groups of solutions that are the most effective in reducing pollution and that are implementable.

The preceding studies provide guidance and methodology needed to assess the off-site impacts of irrigation under a wide variety of environmental, economic, and institutional constraint settings. In the next section, specific assessment tools are described that can be used to evaluate physical, economic, and institutional aspects of water quality impacts resulting from irrigation of agricultural lands.

B. Examples of Assessment Tools

Development of new federal water projects for irrigation requires an assessment of the impact on water quality in the region being developed. The U.S. Department of Interior, Bureau of Reclamation (USBR) developed a comprehensive irrigation return flow quality simulation model as an assessment tool to aid in evaluating the potential effects of irrigation of large land areas on the water quality (Shaffer et al., 1977). This model simulates the plant-soil-aquifer system from the soil surface to a tile or open drain. The process-oriented model includes evapotranspiration, unsaturated and saturated water flow, solution and precipitation of slightly soluble salts, ion exchange, ion pairing, N transformations, crop uptake of N, and the movement and redistribution of salts and nutrients.

This dynamic non-steady-state model predicts the concentrations of Ca, Mg, Na, NH_4, HCO_3, CO_3, Cl, SO_4, NO_3-N, and urea contained in soil-, aquifer-, and drain-waters. Concentrations of organic N; exchangeable Ca, Mg, Na, and NH_4; and gypsum are predicted within the soil and aquifer materials. This model has been used to assess the water quality aspects of irrigation in the San Luis area of southwestern Colorado (Shaffer et al., 1977), the Souris Loop area (Ribbens & Shaffer, 1976) and the Garrison Diversion Project (USBR, 1976) in North Dakota, the Oahe Project in South Dakota (USBR, 1975), the Grand Valley in Colorado (Ayers et al., 1977), and in the middle Rio Grande Valley of New Mexico (Gelhar et al., 1983).

Tanji and co-workers developed an assessment approach and sediment/nutrient transport model which has been validated in the Colusa Basin drainage area of northern California (Mirbagheri & Tanji, 1981; Tanji et al., 1983). The production of sediments from both natural and human-induced activities, as well as the transport of these sediments and associated nutrients (PO_4, NO_3, and NH_3) in an integrated 405 000 ha agricultural and watershed system were assessed. Natural drainage, irrigation return flows, and operational waste from surface irrigation supplies are commingled in the 113-km-long Colusa Basin Drain (CBD) which is discharged into the Sacramento River. The results of this assessment procedure was used to develop recommendations for BMP's to reduce sediment, salinity, nutrient, and pesticide loading to the Sacramento River.

One of the more important, but often slighted, aspects of off-site water quality impact assessments is that of quantification of damages and benefits. In some cases, neither the nature of the damages nor the beneficiaries of potential abatement are well understood. However, in most cases the off-site effects do not directly affect the economics of the production unit that contributes to those effects (e.g., yield reductions experienced by downstream users of water of increased salinity). Gutema and Whittlesey (1982) presented an analytical model consisting of a water quality submodel and an economic submodel to assess the economic benefits of controlling water pollution in an irrigated river system. The water quality submodel consists of three elements: parameters, water quality index, and an aggregation rule. The parameters define water quality as a multidimensional vector, with each component representing some aspect of the physical, chemical, biological, and esthetic characteristics of water affecting water uses. The water quality index functions translate the measured levels of parameters into numerical values of quality levels which water users can interpret. The aggregation rule provides a way for combining the numerical values of water quality and parameters into an overall water quality index for each use. The economic submodel views water as a multiple use resource, with each use having its own quality requirements. This submodel provides estimates of aggregate net social benefits to be derived from water quality changes.

The model has been tested in the USA in the Yakima River Basin of south-central Washington. The model was applied to three typical water quality improvement policies: stream flow augmentation, reduced sediment levels, and reduced NO_3 levels. The social benefits from flow augmentation exceeded the socical benefits derived from policies reducing sediments or NO_3 levels. For both streamflow augmentation and reduced NO_3 levels, the annual benefits fell short of annual costs at the highest level of water quality, implying that water quality standards may be too high and that achieving these standards may not be economically efficient. Not only does this model permit assessment of off-site water quality impacts, but it also permits assessing alternative control strategies. Figure 38–2 depicts the relationship between the environment and the economy. This model explores the social balance between residual wastes that flow from the economy to the natural environment and the flow of productive resources and environmental services to sectors of the economy.

The above studies represent a range of assessment tools available to address water quality impairment resulting from irrigation activities. Often these same tools can be used to develop alternative management practices or evaluate the effectiveness of proposed management changes.

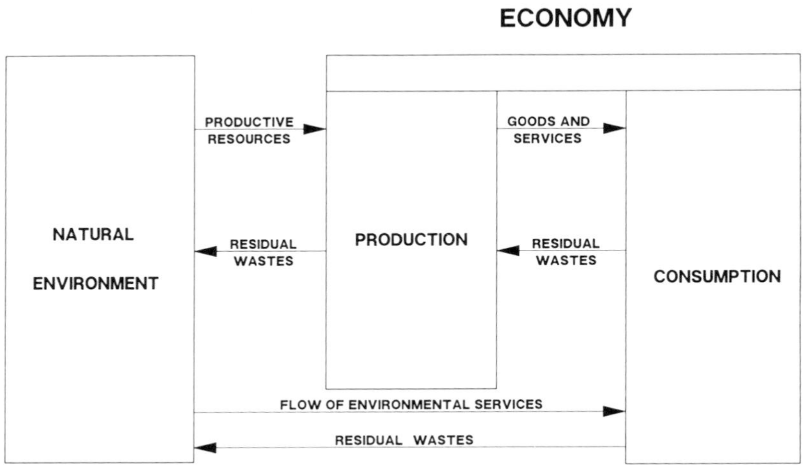

Fig. 38–2. Flow of materials between the natural environment and the economy (Gutema & Whittlesey, 1983).

IV. MANAGEMENT ALTERNATIVES AND CONTROL
STRATEGIES TO REDUCE OFF-SITE IMPACTS

Development of management alternatives and control strategies to reduce water quality impairment from irrigated agriculture has been a major thrust of federal and state water quality control agencies. Sound management practices developed from research studies are available; however, they are not universally applied when needed. The reasons for lack of implementation are usually economic or lack of perceived relevance. Unless an individual perceives that his or her irrigation activities are creating a water quality problem, he or she is unlikely to participate in resolution of the problem (Vlachos et al., 1977).

Reduction in salinity of irrigation return flows can be achieved by improved on-farm management of water. Reduction of excessive leaching by more uniform water application (Karmeli, 1977; Willardson et al., 1977), improving irrigation efficiency (van Schilfgaarde, 1977; Wendt et al., 1977), irrigation scheduling (Skogerboe et al., 1974; Jensen, 1975), and conveyance lining (Skogerboe & Walker, 1972) are proven technologies. Salt pickup from excess irrigation application and conveyance seepage below the root zone where saline marine shales underlie the soil mantle can be reduced by reduction of excess leaching, which is, in part, a result of overirrigation.

Sediment production in irrigated agriculture arises primarily from furrow irrigation. Sediment loss from furrow irrigation can be reduced or eliminated by controlling land slope, controlling furrow stream size, reducing the run length, controlling irrigation frequency and duration, and controlling tailwater (Carter & Bondurant, 1977). Sediment and associated nutrient and pesticide losses can be reduced by edge-of-field retention structures, tailwater reclaim pits, minibasins, and vegetated buffer strips (Fitzsimmons et al., 1977; NCSU, 1982b). When economically feasible, conversion to properly designed and operated sprinkler and drip irrigation systems greatly reduces sediment production and associated nutrient and pesticide losses.

Leaching of salts, nutrients, pesticides, and other toxic elements into subsurface drainage water or aquifers can be reduced by proper on-farm water management. Proper scheduling of the time and amount of water to replenish depleted soil moisture, and scheduling of amount and timing of fertilizer (NCSU, 1982a) and pesticide applications to avoid conflicts with irrigation applications can reduce mass emission of these constituents into groundwater.

Strategies have been developed to reduce water quality impairment by irrigation activities. Regional strategies have been suggested for irrigated lands in the Great Plains of the USA (Quinn, 1982). This strategy document assesses the problems, identifies the available management alternatives, examines the economic feasibility of the practices, and suggests control program strategies. Walker (1977, 1978) developed a strategy for integrating desalting technologies with agricultural management practices to develop an optimal salinity control program in the Grand Valley of Colorado.

An innovative "Influent Contol Approach" has been suggested as a strategy for addressing the complex integration of technical, economic, and institutional issues to improve water quality of return flows (Radosevich & Skogerboe, 1977). This strategy consists of the following elements to be carried by the states:

1. Designate areas for irrigation return flow quality management and the responsible entity.
2. Develop standards and criteria for beneficial use in designated areas.
3. Introduce incentives to use water more efficiently.
4. Include the element of water quality in new or transferred and changed water rights.
5. Adopt and enforce a reporting and recording system for water rights.
6. Recognize reasonable degradation from agricultural water use.
7. Adopt an agricultural practices act.
8. Promote the close cooperation or integration of state water agencies.

Horner and Dudek (1980) developed an integrated physical-economic resource analysis system to examine the tradeoffs between alternative management practices, water quality, and profitability in a regional setting. This analytical system allows for an assessment of potential control strategies to meet water quality goals in the context of optimization of physical, institutional, and economic resources.

Resolution of water quality impairment from irrigation of agricultural lands will require considerable cooperation on the part of many agencies and will cut across many disciplines. Technical solutions that are available must be linked with credible economic analyses and equitable implementation strategies developed in a political arena that weighs the relative merits in the context of present-day values.

V. SUMMARY

Irrigation of agricultural crops will require more attention to the on-site and off-site environmental consequences than has been the case in the past. Protection of groundwater from contamination by NO_3 fertilizer and pesticides is necessary due to health threats and the extreme cost of aquifer rehabilitation. Recent experiences with aldicarb on Long Island in New York and with ethylene dibromide (EDB) in Florida underscore this statement. Nevertheless, much has been learned in the past two decades that can be used to avoid excess leaching of agrichemicals from the crop root zone. To a large extent, the knowledge and technology needed is on the shelf. The challenge is to integrate these technologies into management schemes that take a holistic approach to crop production, resource use efficiency, and environmental quality, and that are accepted by the agricultural community.

REFERENCES

Ayers, J.E., D.B. McWorter, and G.V. Skogerboe. 1977. Modeling salt transport in the irrigated soils of Grand Valley. p. 369–374. *In* J.P. Law and G.V. Skogerboe (ed.) Irrigation return flow quality management. Proc. Natl. Conf., Ft. Collins, CO. 16–19 May. Colorado State Univ., Ft. Collins.

Carter, D.L., and J.A. Bondurant. 1977. Management guidelines for controlling sediments, nutrients, and adsorbed biocides in irrigation return flows. p. 143–152. *In* J.P. Law and G.V. Skogerboe (ed.) Irrigation return flow quality management. Proc. Natl. Conf., Ft. Collins, CO. 16–19 May. Colorado State Univ., Ft. Collins.

Council for Agricultural Science and Technology. 1982. Soil erosion: Its agricultural, environmental and socioeconomic implications. Council Agric. Sci. Tech., Ames, IA.

Fitzsimmons, D.W., C.E. Brockway, J.R. Busch, G.C. Lewis, G.M. McMaster, and C.W. Berg. 1977. On-farm methods for controlling sediment and nutrient losses. p. 183–191. *In* J.P. Law and G.V. Skogerboe (ed.) Irrigation return flow quality management. Proc. Natl. Conf., Ft. Collins, CO. 16–19 May. Colorado State Univ., Ft. Collins.

Gelhar, L.W., P.J. Wierenga, K.R. Rehfeldt, C.J. Duffy, M.J. Simonett, T.C. Yeh, and W.R. Strong. 1983. Irrigation return flow water quality monitoring, modelling, and variability in the middle Rio Grande Valley, New Mexico. Environ. Prot. Tech. Ser. EPA-600/2-83-072. USEPA, Washington, DC.

Gutema, Y., and N.K. Whittlesey. 1983. Economic benefit of controlling water pollution in an irrigated river basin: Methodology and application. Environ. Prot. Tech. Ser. EPA-600/2-83-008. USEPA, Washington, DC.

Horner, G.L., and D.J. Dudek. 1980. An analytical system for the evaluation of land use and water quality policy impacts upon irrigated agriculture. p.537–568. *In* D. Yaron and C. Tapiero (ed.) Operations research in agriculture and water resources. North-Holland Publ. Co. IFORS, Lyngby, Denmark.

Jensen, M.E. 1975. Scientific irrigation scheduling for salinity control of irrigation return flows. Environ. Prot. Tech. Ser. EPA-600/2-75-064. USEPA, Washington, DC.

Karmeli, D. 1977. Water distribution patterns for sprinkler and surface irrigation systems. p. 233–252. *In* J.P. Law and G.V. Skogerboe (ed.) Irrigation return flow quality management. Proc. Natl. Conf., Ft. Collins, CO. 16–19 May. Colorado State Univ., Ft. Collins.

Letey, J., C. Roberts, M. Penberth, and C. Vasek. 1986. An agricultural dilemma: Drainage water and toxics disposal in the San Joaquin Valley. Kearney Found. Soil Sci. Univ. Calif., Riverside.

Mirbagheri, S.A., and K.K. Tanji. 1981. Sediment characterization and transport modeling in Colusa Basin Drain. Univ. California, Davis Dep. Land Air Water Resour. Water Sci. Eng. Pap. 4021.

North Carolina State University 1981. Conceptual framework for assessing agricultural nonpoint source project. North Carolina Agric. Ext. Serv. WQ-04.

North Carolina State University. 1982a. Best management practices for agricultural nonpoint source control. II. Commercial fertilizer. North Carolina Agric. Ext. Serv. WQ-07.

North Carolina State University. 1982b. Best management practices for agricultural nonpoint source control. III. Sediment. North Carolina Agric. Ext. Serv. WQ-08.

Office of Technology Assessment (OTA). 1983. Water related technologies for sustainable agriculture in U.S. arid/semiarid lands. OTA-F-212. U.S. Congress, Washington, DC.

Quinn, M.L. (ed.). 1982. Strategies for reducing pollutants from irrigated lands in the Great Plains. EPA-600/2-81-108. Nebr. Water Resour. Cent., Lincoln.

Radosevich, G.E. 1977. Interface of water quantity and quality laws in the west. p. 405–422. *In* J.P. Law and G.V. Skogerboe (ed.) Irrigation return flow quality management. Proc. Natl. Conf., Ft. Collins, CO. 16–19 May. Colorado State Univ., Ft. Collins.

Radosevich, G.E. 1978. Western water law and irrigation return flow. Environ. Prot. Tech. Ser. EPA-600/2-78-180. USEPA, Washington, DC.

Radosevich, G.E., and G.V. Skogerboe. 1977. An influent control approach to irrigation return flow quality management. p. 423–434. *In* J.P. Law and G.V. Skogerboe (ed.) Irrigation return flow quality management. Proc. Natl. Conf., Ft. Collins, CO. 16–19 May. Colorado State Univ., Ft. Collins.

Ribbens, R.W., and M.J. Shaffer. 1976. Irrigation return flow modeling for the Souris Loop. p. 545–557. *In* Environmental aspects of irrigation and drainage. Proc. ASCE Spec. Conf., Ottawa, Canada. 21–23 July. ASCE, New York.

Shaffer, M.J., R.W. Ribbens, and C.W. Huntly. 1977. Prediction of mineral quality of irrigation return flow. Vol. 5. Detailed return flow salinity and nutrient simulation model. Environ. Prot. Tech. Ser. EPA-600/2-77-179. USEPA, Washington, DC.

Skogerboe, G.V., and W.R. Walker. 1972. Evaluation of canal lining for salinity control in Grand Valley. Office Res. Monitor. USEPA-R2-72-047. USEPA, Washington, DC.

Skogerboe, G.V., W.R. Walker, J.H. Taylor, and R.S. Bennett. 1974. Evaluation of irrigation scheduling for salinity control in Grand Valley. Environ. Prot. Tech. Ser. EPA-660/2-74-052. USEPA, Washington, DC.

Tanji, K.K., M.J. Singer, L.D. Whittig, J.W. Biggar, D.W. Henderson, S.A. Mirbagheri, A.F. Quek, J. Blackard, and R. Higashi. 1983. Nonpoint sediment production inthe Colusa Basin drainage area, California. Environ. Prot. Tech. Ser. EPA-600/2-83-025. USEPA, Washington, DC.

U.S. Bureau of Reclamation. 1975. Summary report. Initial stage Oahe unit. Water quality and return flow study. U.S. Bur. Reclam., Denver.

U.S. Bureau of Reclamation. 1976. Water quality study, Garrison diversion unit, North Dakota. U.S. Bur. Reclam., Denver.

U.S. Congress. 1899. Rivers and harbors act of 1899. 3 Mar. 1899, Chapter 425, 30 Stat. 1121.

U.S. Congress. 1948. Federal water pollution control act. 30 June 1948. Chapter 758, 62 Stat. 1155.

U.S. Congress. 1965. The water quality act. Pub. Law 86-234, 79 Stat. 903.

U.S. Congress. 1972. The federal water pollution control act amendments of 1972, Public Law 92-500, 86 Stat. 816.

U.S. Congress. 1977. The clean water act. Pub. Law 92-217. 91 Stat. 1566.

U.S. Environmental Protection Agency. 1973. Water quality criteria 1972. Ecolog. Res. Ser. EPA/R3/73/033. USEPA, Washington, DC.

U.S. Environmental Protection Agency. 1989a. Health advisory summaries. Office of Water. USEPA, Washington, DC.

U.S. Environmental Protection Agency. 1989b. Drinking Water Health Advisory: Pesticides. USEPA. Office of Drinking Water Health Advisories. Lewis Publ., Chelsea, MI.

van Schilfgaarde, J. 1977. Minimizing salt in return flow by improving irrigation efficiency. p. 81–98. In J.P. Law and G.V. Skogerboe (ed.) Irrigation return flow quality management. Proc. Natl. Conf., Ft. Collins, CO. 16–19 May. Colorado State Univ., Ft. Collins.

Vlachos, E., W.H. Barrett, P. Huzar, J.J. Lawton, G.E. Radosevich, M. Sabey, G.V. Skogerboe, and W. Trock. 1977. A process for identifying, evaluating and implementing solutions for irrigation return flow problems. p. 435–445. In J.P. Law and G.V. Skogerboe (ed.) Irrigation return flow quality management. Proc. Natl. Conf., Ft. Collins, CO. 16–19 May. Colorado State Univ., Ft. Collins.

Walker, W.R. 1977. Combining agricultural improvements and desalting of return flows to optimize local salinity control policies. p. 203–213. In J.P. Law and G.V. Skogerboe (ed.) Irrigation return flow quality management. Proc. Natl. Conf., Ft. Collins, CO. 16–19 May. Colorado State Univ., Ft. Collins.

Walker, W.R. 1978. Integrating desalination and agricultural salinity control alternatives. Environ. Prot. Tech. Ser. EPA-600/2-78-074. USEPA, Washington, DC.

Wendt, C.W., A.B. Onken, O.C. Wilke, R. Hargrove, W. Bausch, and L. Barnes. 1977. Effect of irrigation systems on water use efficiency and soil-water solute concentrations. p. 123–131. In J.P. Law and G.V. Skogerboe (ed.) Irrigation return flow quality management. Proc. Natl. Conf., Ft. Collins, CO. 16–19 May. Colorado State Univ., Ft. Collins.

Willardson, L.S., R.J. Hanks, and R.D. Bliesner. 1977. Field evaluation of sprinkler irrigation for management of irrigation return flow. p. 109–114. In J.P. Law and G.V. Skogerboe (ed.) Irrigation return flow quality management. Proc. Natl. Conf., Ft. Collins, CO. 16–19 May. Colorado State Univ., Ft. Collins.

SUBJECT INDEX